LIPOPROTEINS – ROLE IN HEALTH AND DISEASES

Edited by **Saša Frank** and **Gerhard Kostner**

Lipoproteins – Role in Health and Diseases
http://dx.doi.org/10.5772/2931
Edited by Saša Frank and Gerhard Kostner

Contributors

Ruth Prassl, Peter Laggner, Benjamin Dieplinger, Hans Dieplinger, D.S. Mshelia, A.A. Kullima, Stanislav Oravec, Johannes Mikl, Kristina Gruber, Elisabeth Dostal, Göran Walldius, Assia Rharbi, Khadija Amine, Zohra Bakkoury, Afaf Mikou, Anass Kettani, Abdelkader Betari, Jelena Umbrasiene, Ruta-Marija Babarskiene, Jone Vencloviene, Adebowale Saba, Olayinka Oridupa, Xiaoyan Zhang, Hainsworth Y. Shin, Summer F. Acevedo, Karl-Erik Eilertsen, Rune Larsen, Hanne K. Mæhre, Ida-Johanne Jensen, Edel O. Elvevoll, Eduardo Guimarães Hourneaux de Moura, Ivan Roberto Bonotto Orso, Bruno da Costa Martins, Guilherme Sauniti Lopes, Somayeh Hosseinpour-Niazi, Parvin Mirmiran, Fereidoun Azizi, Maaike Kockx, Leonard Kritharides, Edward Loane, Vikram Jairam, Koji Uchida, Vasanthy Narayanaswami, Mohammad Z. Ashraf, Swati Srivastava, Ivana Pejin-Grubiša, Andriy L. Zagayko, Anna B. Kravchenko, Mykhaylo V. Voloshchenko, Oxana A. Krasilnikova, Amany M. M. Basuny, Isaac Karimi, Masashi Shiomi, Tomonari Koike, Tatsuro Ishida, Armando Sena, Carlos Capela, Camila Nóbrega, Véronique Férret-Sena, Elisa Campos, Rui Pedrosa, Sanja Stankovic, Milika Asanin, Nada Majkic-Singh, Etsuro Matsubara, Caryl J. Antalis, Kimberly K. Buhman, Adebowale Bernard Saba, Temitayo Ajibade, Vladana Vukojević, Ludmilla A. Morozova-Roche, Dalibor Novotný, Helena Vaverková, David Karásek, Yoshio Aizawa, Hiroshi Abe, Kai Yoshizawa, Haruya Ishiguro, Yuta Aida, Noritomo Shimada, Akihito Tsubota, Haruhiko Sakamoto, Masaki Ueno, Wu Bin, Yumiko Nagai, Kouichi Matsumoto, Takao Yamanaka, Sumiko Tanaka, Anna Gries

Published by InTech

Janeza Trdine 9, 51000 Rijeka, Croatia

Copyright © 2012 InTech

Notice

Publishing Process Manager Vedran Greblo
Technical Editor InTech Prepress, Novi Sad
Cover InTech Design team

Additional hard copies can be obtained from orders@intechopen.com

Lipoproteins – Role in Health and Diseases, Edited by Saša Frank and Gerhard Kostner
p. cm.
ISBN 978-953-51-0773-6

For EU product safety concerns:
IN TECH d.o.o., Prolaz Marije Krucifikse Kozulić 3, 51000 Rijeka, Croatia,
info@intechopen.com or visit our website at intechopen.com.

Meet the editors

Saša Frank is Associate Professor at the Institute of Molecular Biology and Biochemistry, Medical University of Graz, Austria. He was born in Zagreb, Croatia where he studied Molecular Biology at the Faculty of Natural Sciences. Dr. Frank got his PhD in Biochemistry from the Medical University in Graz.

Gerhard Kostner is Emeritus Full Professor of Biochemistry and Molecular Biology at the Medical University of Graz. After his studies of Chemistry and Physics he got a PhD in Biochemistry and spent two years at OMRF, USA with Pierre Alaupovic and half a year as guest professor at the University of Frankfurt, Germany.

Contents

Preface

Lipids are water insoluble in nature and need to be transported in body fluids in the form of lipid:protein complexes, i.e. lipoproteins. Lipoproteins consist of a lipid core of triglycerides and cholesteryl esters, and an amphiphilic surface of free cholesterol and phospholipids. Apolipoproteins (apo-Lp) are interchelated into surface lipids of the lipid droplet forming the mature plasma lipoproteins (Fig.1).

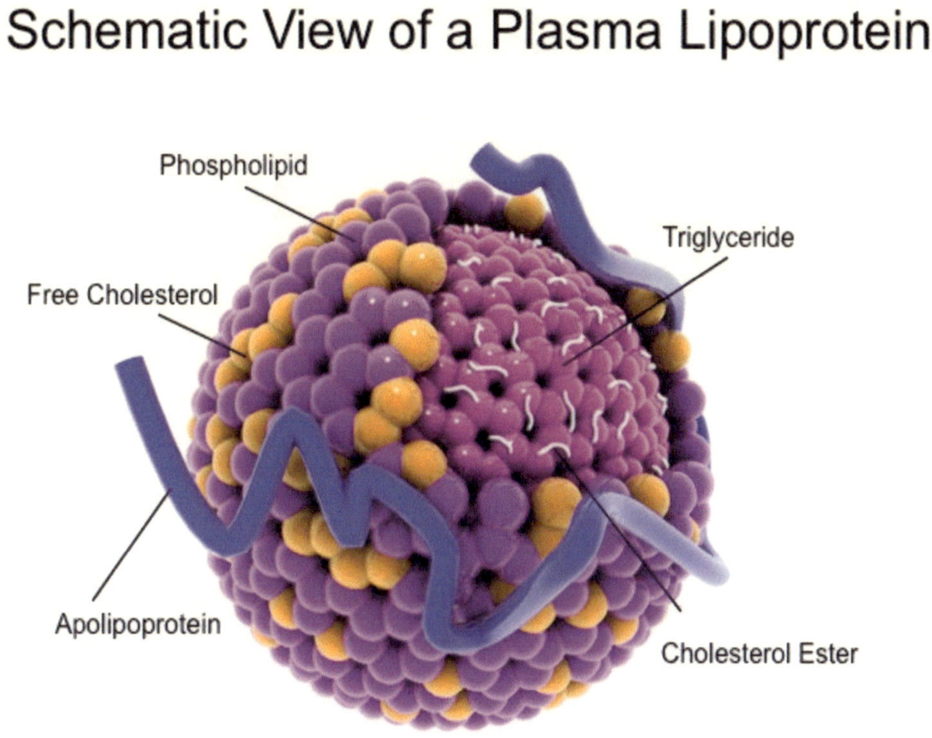

Figure 1.

Nascent lipoproteins on the other hand lack the lipid core and have a disc like structure. There are four main lipoprotein density classes - chylomicrons, very low density lipoproteins, low density lipoproteins (LDL) and high density lipoproteins (HDL) - and yet each of them may be divided into numerous subfractions. The major function of lipoproteins is the delivery of nutrient lipids, i.e. triglycerides (TG), phospholipids (PL) and cholesterol, to various organs and tissues. Whereas dietary TG and PL may be absorbed up to almost 100%, the absorption rate of cholesterol ranges from 30 – 60% only and is influenced by genes and other nutritional factors. There exist some 15 or more proteins associated with lipids in the form of apo-Lp that function as "structural proteins", co-factors and inhibitors of enzymes, ligands for specific cell surface receptors and possibly others. Although unsaturated and polyunsaturated fat is considered to be beneficial, these lipids are prone to oxidation and degradation. These products contribute significantly to common diseases found in civilized countries such as atherosclerosis, coronary heart disease, Type-2 diabetes mellitus, stroke, Alzheimer disease but also auto-immune diseases and cancer. A major trigger for these diseases is LDL whose mass consist to approx. 50% of cholesterol. Oxidized and modified LDL are taken up by scavenger receptors of macrophages and transform them into foam cells. Foam cells in turn synthesize inflammatory cytokines and enzymes hat lead to a vicious cycle of self-perpetuation that triggers smooth muscle cell proliferation, lipid deposition and plaque formation in the arterial intima (Fig.2).

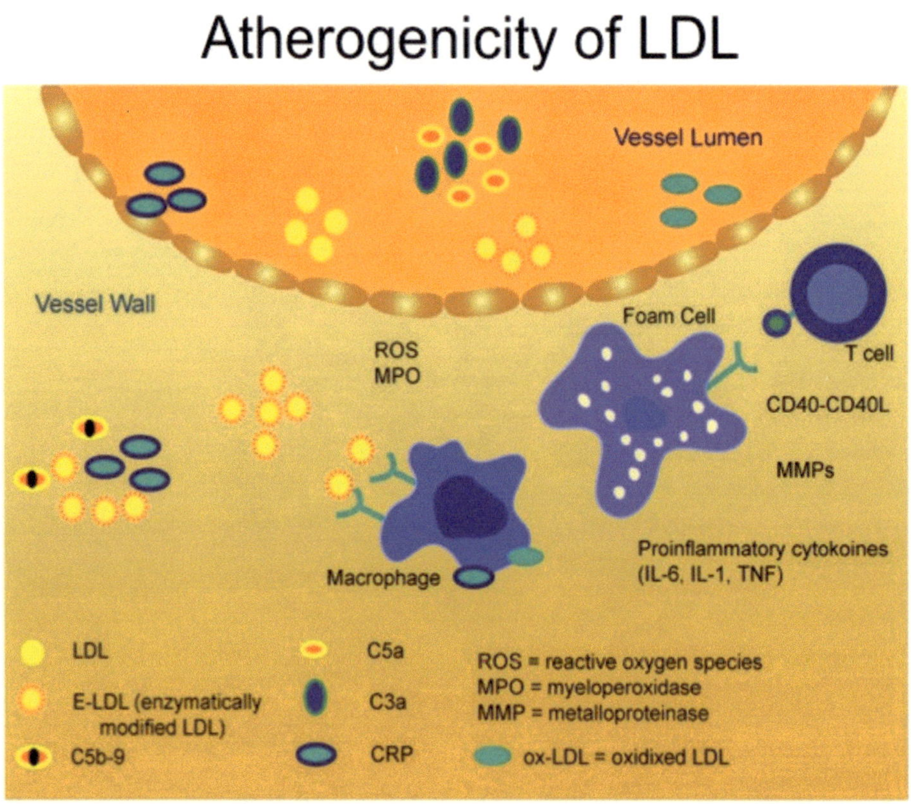

Figure 2.

Similar pathways appear to be also involved in Alzheimer disease and cancer. HDL on the other hand have anti-atherogenic, anti-thrombotic anti-oxidative effects and therefore are considered to be beneficial (Fig.3).

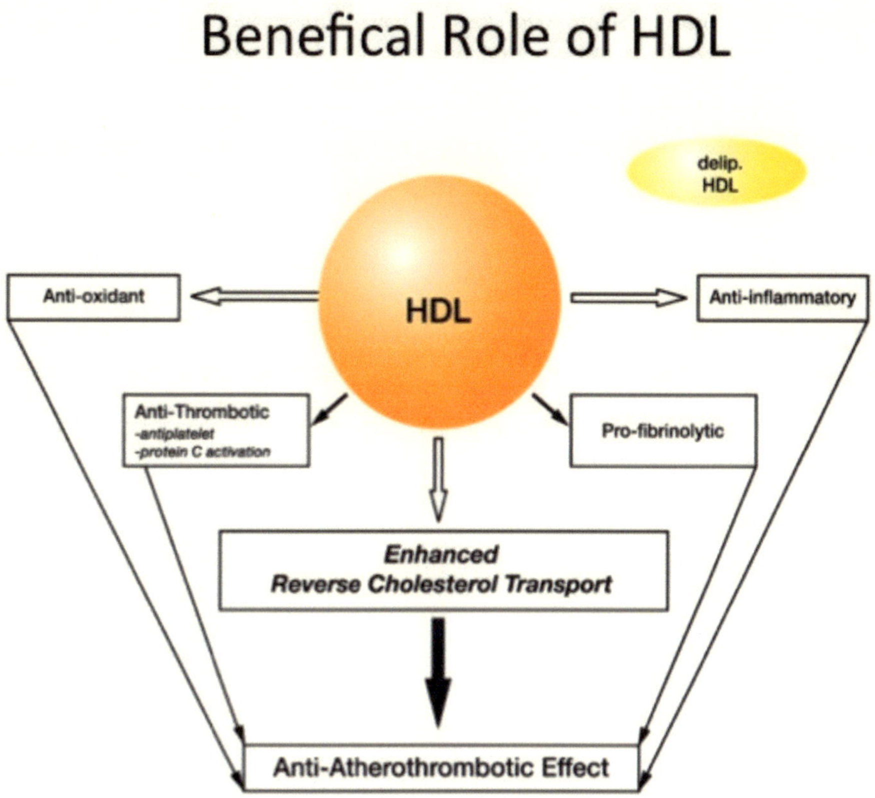

Figure 3.

As we know there exist so called "LDL" creatures such as primates or rabbits that easily develop atherosclerosis and myocardial infarction upon overfeeding with lipid rich diet. On the other hand, "HDL" creatures such as rats, mice and dogs hardly develop atherosclerosis by feeding them an atherogenic Western-type diet. Thus numerous attempts are made to interfere with elevated LDL and low HDL concentrations therapeutically in individuals at increased risk for myocardial infarction and stroke.

By typing into databases such as Medline or PubMed the word "lipoprotein" one gets more than 100.000 hits that highlights the common interest in this topic. It is actually impossible to cover all aspects of lipoprotein structure, function, metabolism and pathophysiology in one issue like the present volume, but attempts have been made to concentrate on topics that are in focus of current lipoprotein research. These topics have been divided into 10 sections.

Section 1 deals with important issues of lipoprotein structure, the assembly and kinetics of apoB containing lipoproteins, and the role of Lp(a) in kidney patients.

Section 2 reviews clinical chemical methods for diagnosing patients at increased risk for atherosclerotic diseases, myocardial infarction and stroke. In addition, hints are given how to approach biological databases and how to interprete complex data sets.

In **Section 3**, the characterization of dys- and hyperlipoproteinemias is described in detail and their impact on caedivascular disease and mortality is presented.

In **Section 4**, the impact of lipid lowering drugs, of long chain polyunsaturated fatty acids and of endoscopic treatments on lipoprotein metabolism and atherogenesis is described.

Section 5 highlights the important issue of lipid oxidation and prevention by anti-oxidants. There is no doubt that the oxidative stress per se and the supply of anti-oxidative vitamins and plant compound such as polyphenols play a eminent role in atherogenesis, yet also other factors mainly genes and environment influence the development of atherosclerosis , cardiovascular diseases and stroke.

Section 6: It is obvious that without animal models it would have been impossible to study the function and metabolism of lipoproteins in detail. Thus in this chapter, models for dys-and hyperlipoproteinemia are described in addition to the influence of diet on obesity and atherosclerosis. Finally, genetically modified animals and animals with inborn errors of lipoprotein metabolism such as the WHHL rabbit are described for studying the role of lipoproteins in the development of atherosclerotic diseases.

Neurodegenerative diseases are a burden for the civilized world and still in progress. **Section 7** highlights different forms of brain diseases with major emphasis to stroke and Alzheimer disease that without doubt are causally related to lipid rich diet, lipid oxidation and derangements in lipoprotein metabolism.

Section 8 summarized the most important theories of the involvement of lipids and lipoproteins in the development of cancer.

Adipose tissue has been recognized as an important organ for hormone and cytokine production. One of the best studied adipokinine, adiponectin is characterized in **Section 9**. In addition hyperlipoproteinemias associated with chronic virus hepatitis is outlined in that chapter.

Last but not least, lipids and lipoproteins play an eminent role in platelet and leucocyte function, and hemostasis. This is the topic of **Section 10** that also focuses on one of the most atherogenic lipoprotein, Lp(a). Lp(a) has been studied for more than 50 years and we still do not know the physiological function of that particle that is found only in primates, and in a somewhat different structure in hedgehogs.

Intense investigations in all these areas are still going one and we wait for exciting new developments mediated by "omics" methods and translational research. This volume will help new investigators in the field to get acquainted with the general topic of lipoprotein research and guides scientists interested in this area to emerging new fields.

Saša Frank and Gerhard Kostner
Institute of Molecular Biology and Biochemistry,
Medical University of Graz, Graz,
Austria

Section 1

Lipoprotein Structure and Assembly

Chapter 1

Lipoprotein Structure and Dynamics: Low Density Lipoprotein Viewed as a Highly Dynamic and Flexible Nanoparticle

Ruth Prassl and Peter Laggner

Additional information is available at the end of the chapter

http://dx.doi.org/10.5772/48145

1. Introduction

Low density lipoproteins (LDLs) are the principal transporter of cholesterol and fat in human blood. Circulating LDLs guarantee a constant supply of cholesterol for tissues and cells, whereas cholesterol is required for membrane synthesis, modulation of membrane fluidity and the regulation of cell signaling pathways. The function of LDL in metabolism is mediating by cellular uptake via receptor-mediated endocytosis followed by lysosomal degradation [1,2], and is strongly dependent on the lipid distribution, the structure of LDL particles and on the proper conformational orientation of apolipoprotein B100 (apo-B100). Apo-B100 is the sole protein component of LDL being mainly located on the surface of LDL. Apart from their well established role as lipid transporter, LDL particles are intimately involved in the progression of cardiovascular diseases such as atherosclerosis or stroke, which are among the most prevalent causes of death in developed countries [3]. In particular, raised plasma levels of LDL are linked to an increased risk for disease. Moreover, dysregulations of LDL due to abnormalities in LDL structure have been identified as independent predictors of risk for coronary heart diseases [4,5]. LDL particles by themselves are highly heterogeneous in nature, varying in their buoyant densities, size, surface charge and chemical composition [6,7], and these biochemical characteristics determine the fate of LDL in the subendothelial space [8,9]. For example, small, dense LDL subclasses are more atherogenic than their light counterparts, which are more susceptible to modifications [5,10]. Modifications of LDL, primarily through oxidation, enzymatic degradation or lipolysis are the initiating factors in early atherosclerosis. In that case, LDL particles accumulate in the intima of the arterial wall where apo-B100 binds to proteoglycans of the extracellular matrix through ionic interactions. As a consequence, LDL becomes trapped in the subendothelium, where it is prone to oxidation processes, aggregation and fusion. Bioactive lipids, such as oxidized phospholipids, lysolipids or oxidized cholesterylester, are released from LDL particles, which are simultaneously non-specifically altered. A broad spectrum of diverse LDL particles with non-defined physicochemical properties is generated that, in turn, promotes a rapid uptake of these particles by macrophages to form foam cells [11]. This is one of the key steps in the progression of atherosclerosis. Today, atherosclerosis is known to be a chronic inflammatory disorder of the blood vessels and recognized as a prevailing cause of cardiovascular disorders, the leading causes of morbidity and mortality worldwide [12]. Since the early initiation of atherosclerosis strongly depends on the metabolism of LDL, which is predominantly triggered by molecular characteristics of LDL, it is of paramount biomedical importance to explore structural features of LDL particles in great detail. However, mostly due to the complex nature of LDL particles many questions concerning molecular details are still unanswered.

This article will review our current knowledge on the structure and dynamics of LDL particles. In fact, several recent studies revealed that the molecular organisation and dynamics of LDL core lipids, in close relationship to the intrinsic dynamics of LDL surface components, control not only the metabolism of lipids in humans, but determine the role of LDL in the pathogenesis of cardiovascular diseases. In this article, we will give a short historical review on LDL structure and then present prevailing concepts on the self-organisation of LDL. Special emphasis will be paid to dynamic features of LDL particles. In particular, we will discuss the interplay between structure and dynamics in more detail. Finally, we will give an outlook to promising future strategies to clarify the molecular structural details of LDL and how to exploit LDL nanoparticles for medical needs.

2. Molecular architecture of LDL

LDLs are composed of lipids and protein, which assemble to form a supramolecular complex with a molecular mass exceeding 2.5 - 3.0 million Da and involving 2000 to 3000 lipid molecules. Thus, LDL particles are commonly described as micellar complexes, macromolecular assemblies, self-organized nanoparticles or microemulsions. Regardless of diverse definitions, it is generally accepted that assembled LDL particles are organized into two major compartments, namely an apolar core, comprised primarily of cholesteryl esters (CE), minor amounts of triglycerides (TG) and some free unesterified cholesterol (FC). The core is surrounded by an amphipathic outer shell. This shell is composed of a phospholipid (PL) monolayer containing the larger part (>2/3) of the FC molecules and one single copy of apo-B100, which is one of the largest known monomeric glycoproteins [13]. Figure 1 provides an overview on characteristic properties of LDL together with a schematic presentation of an LDL particle. Since molecular interactions between different kinds of lipids have turned out to be highly complex, it is almost impossible to separate the surface and core regions exactly from each other. Accordingly, in some recent reports an additional hydrophobic interfacial layer composed of phospholipid acyl chains, FC, some CE molecules and hydrophobic protein domains is defined. This description takes account for the interplay between neutral core lipids and the surface layer [14].

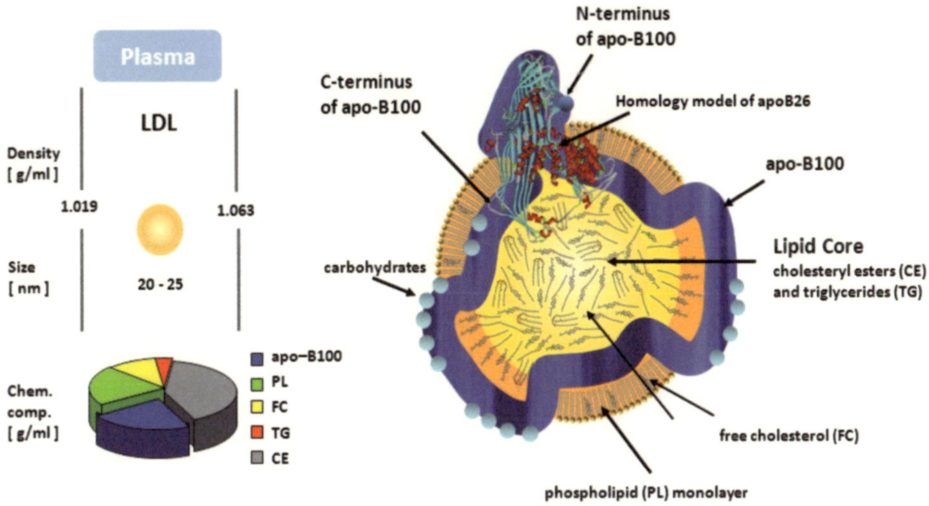

Figure 1. Molecular organisation of LDL. LDL particles are isolated from human plasma within a defined density range. Their particle size varies between 20 to 25 nm. LDLs are built up by a hydrophobic lipid core of cholesterylester (CE) and triglyceride (TG) molecules, which make up more than 40% of particle mass surrounded by a phospholipid (PL) monolayer corresponding to about 20% of particle mass. Varying amounts of free cholesterol (FC) are incorporated in the shell and the core regions. One single copy of apo-B100 (550 kDa) is embedded in the surface monolayer, partially penetrating the core and covering about 40 to 60% of the surface area. The carbohydrate moieties are distributed along the protein chain and are surface exposed. The N-terminal end of apo-B100 (about 26% of total) is hydrophilic and shows a high homology to lamprey lipovitellin. The C-terminal end was shown to be located close to the N-terminus.

Since LDL particles are highly heterogeneous, especially with respect to the chemical composition of the core lipids, the actual size of LDL particles varies between 20 to 25 nm, with an average particle diameter of about 22 nm. This intrinsic heterogeneity allows a subdivision of LDLs into distinct highly homogeneous LDL subspecies, which are identified on the basis of their hydrated densities, which normally lies between the extremes of d, 1.019 and 1.063 g ml^{-1} [15]. These subspecies also differ in their physico-chemical characteristics, receptor binding affinity [16], susceptibility to oxidative modifications [17,18], and in their atherogenic behaviour. Following these lines, it is important to consider LDL as a flexible construct, which needs to respond to changing environmental conditions during lipid exchange. Hence, during particle remodelling, apo-B100 and the surface PL molecules have to rearrange to compensate for changes in the surface area and surface pressure [6]. It is known, that apo-B100 predominantly resides on the surface of LDL and displaces PL molecules, concomitantly changing the diffusion and order parameter of lipids as shown in a recent near atomistic simulation study [19]. Based on simple geometrical considerations taking into account the surface PL monolayer (about 700 PL molecules) with an average area per lipid of 0.65 nm^2 and an LDL particle diameter of 22 nm, large parts of the surface layer must be covered by the protein to avoid unfavourable hydrophobic contacts. In support of these considerations, a loose surface packing of PL molecules was derived from molecular dynamics simulations [19]. This low surface pressure enables hydrophobic amino acid regions of apo-B100 to penetrate into the interfacial regions, predominantly formed by the acyl chains of PLs. Consequently, apo-B100 might interact more readily with the neutral core lipids, and indeed it was shown that some of the CE molecules align along the β-sheet structures of apo-B100 [20], thereby driving CE molecules to the surface, where they become part of the interfacial layer. Particularly noteworthy is the fact that the lipids within the interfacial layer are not homogeneously distributed but form local microenvironments [14]. More precisely, two nanodomains were identified, one rich on sphingomyelin and FC, the other one rich in phosphatidylcholine and poor in FC. The latter was shown to be associated more closely with apo-B100 [21]. Even though, one has to keep in mind that these domains are not static or confined in size and number and co-determine the intrinsic dynamics of LDL. Based on these types of findings, it seems reasonable to suggest that variations in the molecular organisation of lipid/apo-B100 impact the structure of LDL, and have to be considered to act as physiological determinants of LDL function.

3. Structural models of LDL

Our present understanding of the structure of LDL particles has emerged from the concerted application of different physico-chemical techniques with early ground-breaking findings derived from neutron- or X-ray small angle scattering data [22-25] complemented by results from negative staining electron micoscopic (e.m.) [26,27] and spectroscopic techniques [28,29]. For comprehensive reviews on different biophysical studies applied on LDL species see refs. [30,31]. In recent years structural investigations using cryo-e.m. reconstruction techniques have become prevalent and with time 3-dimensional models with improved resolution were presented [32-37]. While in earlier studies LDLs are described as quasi-spherical particles, later studies presented a new view of the overall particle structure displaying an oblate elliptical particle shape. Moreover, recent 3D-images show convincing data that LDL can be considered as discoidal-shaped particle with two flat surfaces on opposite sides. In this model, apo B100 encircles LDL at the edge of the particle, while the PL monolayer is rather located at the flat surfaces which are parallel to the CE layers in the core [36,37]. To get a better impression of what LDL looks like in a structure map obtained by 3D-reconstruction from cryo-e.m, we show some images in Figure 2 revealing the surface density distribution on LDL. It has to be stated that this model strictly holds true for LDL particles with the core lipids being in a frozen liquid-crystalline state.

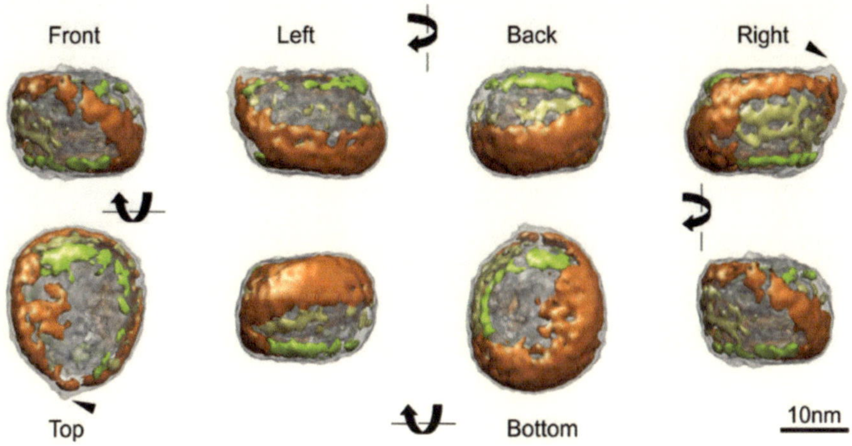

Figure 2. Density distribution at the surface of LDL. The 3D-density map derived from cryo e.m. images by reconstitution reveals the oblate overall particle shape of LDL shown in gray. The overlaid high density regions represent the backbone of apo-B100, colored in orange. The belt surrounds the particle to form an enclosed circle. The second group of high density regions (green) contours the rims and complements the backbone enclosing lower-density regions. The high density regions on the sidewall (yellow) are structures extending from the backbone. A knob-like protrusion is visible at the pointed end (indicated by triangles in the right and top views). The 3D-map is turned 90° in each frame. Reprinted with permission from ref. [37].

4. Core lipid packing and lipid phase transition

Despite of compositional heterogeneity, LDL particles share one common feature: the CE molecules in the core undergo a structural transition from an ordered liquid-crystalline phase to a fluid oil-like state as function of temperature and chemical composition [38]. More precisely, the actual transition temperature, which is close to body temperature, is inversely correlated to the content of triglycerides within the lipid core [22,39]. Based on these characteristics, several models for CE packing have been suggested including a spherical concentric layer model derived primarily from X-ray and neutron scattering data [40,41]. More recently, the concept of a flat lamellar structure came up. This model is derived from single-particle reconstructions from cryo-e.m. images of LDL in vitreous ice [32,34]. An ordered three-layer internal lamellar structure with a distance of about 3.6. nm between the single lamellae was reported [32], in agreement with repeat distances derived from X-ray scattering patterns for LDL below the transition temperature. While these images were observed for LDL particles being in the liquid crystalline phase before snap-freezing, diverse results were reported for LDL particles frozen from a state above the phase transition temperature [42,43]. One plausible explanation for these discrepancies might be that the melting rate of the core lipids proceeds extremely fast. It has been shown that the physical state of core lipids changes within milliseconds [44]. This fast kinetics has caused experimental difficulties for long time, however, a recent experimental approach by speeding up freezing allowed to trap the lipids in the molten state [45]. The authors report on a co-existing phase of layered and broken shells for LDL particles, which are shock-frozen in a state above the phase transition. This is the first time to visualize the nucleation process of CEs within LDL. Most interesting, the images indicate intermediate states between the order/disorder phase transition. Figure 3 shows the dynamic model of the core CE packing during the phase transition and gives a comparison of the internal features of reconstructed 3D-volumes of LDL.

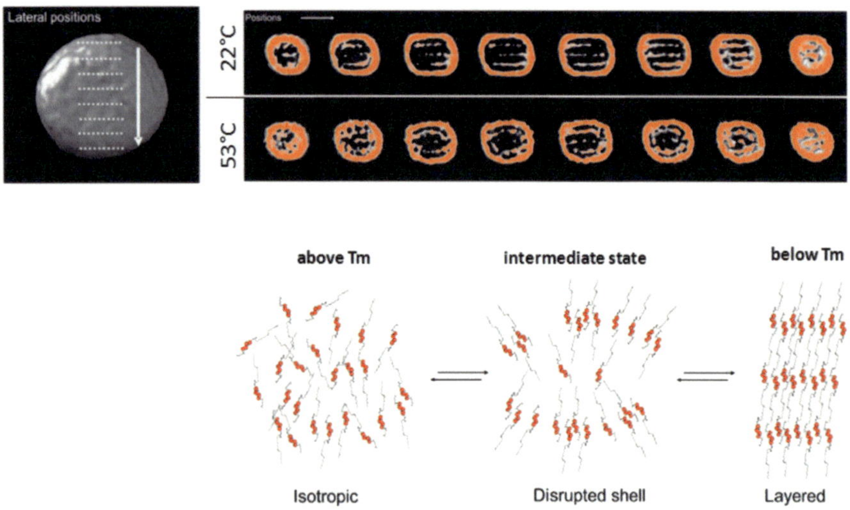

Figure 3. Schematic picture of the dynamic model of LDL core lipid packing during the phase transition. Comparision of the internal features of the reconstructed 3D-volume of LDL snap-frozen from below (22°C) and above (53°C) the phase transition temperature (Tm). Samples prepared from 22°C show a layered organisation while samples prepared from 53°C reveal a disorded shell like structure, which is concentric to the surface. Note, the overall shape of LDL has also changed slightly. The lower panel shows a hypothetical model for the core lipid packing depicting the dynamic process of the core lipid phase transition upon cooling from isotropic to layered passing through an intermediate state. Modified with permission from ref. [45].

In summary, it seems reasonable to argue that both the overall shape and core lipid packing of LDL particles are highly sensitive to changes in temperature and lipid composition. Indeed, this newly proposed patch nucleation behavior permits the temporary formation of local molecular microenvironments as suggested previously by our group in terms of trigylceride segregation [46]. In the next paragraph we will address some interesting questions in support of above hypotheses.

Does a lipid microphase separation occur in LDL particles as a function of the relative core content of CE and TG ?

As already mentioned, the transition temperature correlates with the lipid composition, however, a discontinuity in the concentration dependence was observed [46]. A break in the concentration dependence of a transition temperature in a mixed lipid system constitutes an index for the existence of a phase separation at the break point. In isolated triglyceride - cholesteryl ester systems no indication of a phase separation at similar compositions was found [39,47]. It appears therefore, that structural constraints within the LDL particle determine this effect. Experimental data provide evidence that at low TG content (below 12%) the TG molecules separate into distinct hydrophobic nanoenvironments while the CEs form a smectic liquid crystalline layer. With increasing TG content the thermal stability of the CE layer is decreased by intermixing with TG [46]. This hypothesis implies that the TG-rich fluid nanodomains can serve as a reservoir for lipophilic minor constituents, such as vitamins (tocopherol, carotenoids etc.) below the phase transition. The local concentration of these antioxidants and hence their efficiency in scavenging lipophilic free radicals is higher than if they were dissolved in the bulk volume of total apolar lipids. At the same conditions the CE molecules are strongly immobilized and the intracellular degradation of LDL is decelerated [48], equally the activity of lipid transfer proteins is diminished [49,50]. Based on these considerations it is tempting to speculate that circulating LDL, as a consequence of the variation in blood temperature, periodically undergoes a thermal transition resulting in a transient increase in the local core concentration of minor constituents [46]. Here, it should be emphasized that a periodic redistribution of lipophilic solutes, and also for example of drugs, into the confined LDL core volume could represent an attractive approach to the modulation of biochemical reactions, which would not occur at sufficient rates under the normal conditions of relative concentration. Studies along these lines could indeed verify the long missing physiological role of the thermal LDL transition.

Can LDL structure follow quasi-isothermal changes in blood temperature during its circulation, or does it remain adiabatically metastable in the molten-lipid state?

In order to provide evidence to answer this question we have applied time resolved X-ray scattering experiments using a high flux synchrotron generated X-ray beam. Thus, we have been able to trigger the thermal transition in either direction (heating and cooling) simultaneously monitoring associated structural changes in sub-second time intervals. With our special instrumental setup we managed to evaluate the kinetics of core-transition by T-jump and T-drop experiments [44]. We found that the melting transition proceeds faster than 10 milliseconds indicating that thermal-induced lipid reorganisation takes place at the time scale of blood circulation. As the velocity of blood-flow can be as low as 0.3 mm/s in peripheral blood capillaries the residence time for LDL particles in cooler regions of the body can be several seconds. Consequently, LDL can easily follow periodic temperature changes during blood circulation and assist the redistribution of lipophilic constituents within its core nanodomains forming fluid defect zones. For biomedicine, this strengthens the hypothesis that the core lipids of LDL not only act as passive chemical substrates in metabolism, but that their physical state within the LDL nanoparticles has the potential to control their metabolic fate in normal and atherosclerotic cholesterol transport.

Does the core lipid transition have a physiological meaning ?

Despite its occurrence conspicuously close to blood temperature and the variation of the transition temperature of LDL among different subjects, no clear evidence for a physiological or patho-biochemical role of this transition has so far been found. It is now generally accepted in literature that the rearrangement of the core lipids also affects the overall structure and shape of the LDL particle. Morphological changes in turn can impact receptor-binding activity as well as the action of lipid hydrolyzing enzymes. Equally, the susceptibility of LDL particles to oxidative modifications and lipid peroxidation might be correlated to temperature [18]. As oxidized LDL play a crucial role in the pathogenesis of atherosclerosis, any contribution to the comprehension of antioxidant efficiency may be of therapeutic potential [2,51], further pointing to the physiological relevance of the lipid core organisation. However, this vital question still remains unanswered.

5. Apo-B100 is a flexible string wrapped around the surface of LDL

As already indicated above, the physicochemical state of the core lipids is intimately related to the structure and dynamics of the particle surface, which consists of about 700 phospholipid molecules and one single copy of apo-B100. Apo-B100 is a huge glycoprotein and its polypeptide chain consists of 4536 amino acid residues with an estimated molecular mass of about 550 kDa for the glycosilated form [52,53]. Apo-B100 is a single chain protein with a total contour length of about 70 nm [54] and can be viewed as a highly flexible molecular string composed of single domains [20]. Five consecutive domains were identified based on secondary structure elements representing the main conformational motifs of apo-B100. The single domains are connected by flexible interdomain linker regions, which allow relative movements of domains to each other. The feasibility of such motions was shown in a low resolution model of detergent solubilized apo-B100, which was derived from small angle neutron scattering data [55]. In this model, compact rigid domains are visible being connected by flexible interdomain linkages, which possess a substantial degree of freedom in their spatial orientation. A hypothetical spatial arrangement of the apo-B100 molecule on a spherical LDL particle was created after assigning the secondary structure elements, which were deduced from a secondary structure prediction, to the surface of apo-B100 (Figure 4). Likewise, the averaged surface shape of the 3D-model would allow for variations in the thickness of the apo-B100 molecule by about 1 nm. Such variations are most likely required to compensate for changes in the surface area upon lipid exchange and particle shrinking during endogeneous lipoprotein conversion from very low density lipoprotein (VLDL) to LDL.

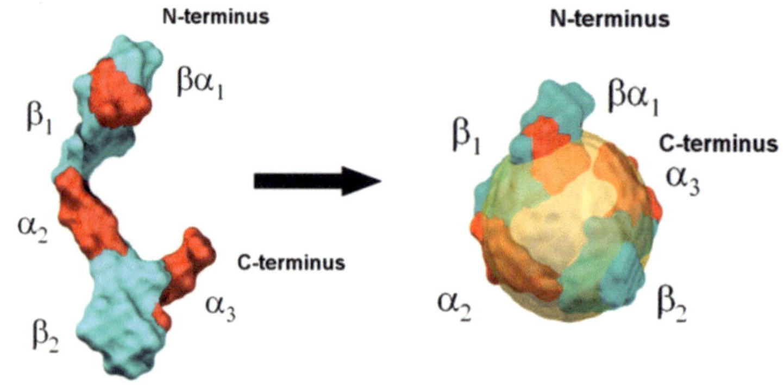

Figure 4. Reconstituted low resolution model of lipid-free apo-B100 derived from small angle neutron scattering data. Apo-B100 shows an elongated arch-like morphology indicating single domains and mobile less defined linker regions. A hypothetical model of a spherical LDL particle after superposition of the structural model of apo-B100 is shown (adapted from ref. [55]). Secondary structure modules are assigned to the surface after a secondary structure prediction was performed. The results nicely correspond to the pentapartite model suggested by [20].

Concerning the topology of apo-B100 on the surface of LDL the most detailed information is obtained from cryo-e.m. images (see also Figure 2). Chatterton et al. were among the first to visualize apo-B100 as string circumventing LDL, and to report on mapped epitopes of apo-B100 distributed over one hemisphere of the LDL particle [56,57]. Recent single particle 3D-reconstructions from immuno cryo e.m. images delineated a more accurate picture of apo-B100 revealing a looped topology of the protein backbone with distinct epitopes identified along the protein chain. According to this model, epitopes in the LDL receptor binding domain are located on one side of LDL, whereas epitopes located in the N-terminal and C-terminal domains are in close vicinity to each other on the opposite side of LDL [36]. In addition, a prominent protrusion is visible in the images at the pointed end of the particle. A similar knob-like region was apparent in the low resolution model of lipid-free apo-B100 shown in Figure 4. This protrusion most probable represents the non-lipid associated globular N-terminal domain of apo-B100, which shows a high homology to lamprey lipovitellin [58]. Except for the N-terminal domain, little is known about the molecular organisation of the structural motifs, whose amphipathic nature determine lipid association. However, to evaluate lipid-protein interactions physical parameter like interfacial elasticity or molecular dynamics have to be considered. In this context, it was suggested that the hydrophobic β-sheet domains of apo-B100 act as elastic lipid anchors, whereas the amphipathic α-helical domains respond rapidly to changes in surface pressure [59,60]. In any case, it can be assumed that alterations in the adsorption and penetration depth of apo-B100 in the phospholipid monolayer and in the lipid core are accompanied by structural rearrangements of the domains and changes in the orientation of the domains relative to each other. In the course of such elastic motions, intramolecular rearrangements are likely to alter the overall hydrophobicity and surface activity of single protein domains. These modifications not only affect lipid-protein interaction, but are equally important for molecular and cellular recognition of apo-B100.

6. Apo-B100 containing lipoproteins are very soft and flexible

LDL particles are formed in the circulation by lipolytic conversion of TG-rich VLDL particles. This enzyme mediated endogenous transformation is accompanied by an extensive shrinking in particle size from about 50-80 nm for VLDL to ~20 nm for LDL. In the course of remodelling, apo-B100 remains bound to its nanocarrier stabilizing the lipid assembly by maintaining structural integrity. To accomplish this, apoB100 has to become more condensed or relaxed depending on the lipid packing density. Likewise, this dynamic process is modulated by the molecular mobility of the surrounding microenvironment. To test for this hypothesis we have recorded temperature dependent molecular motions in VLDL and LDL particles using elastic incoherent neutron scattering [61]. With this technique, motions in the nano- to picosecond time scale can be recorded. The calculated dynamic force constants are a direct measure for the resilience of the particles. The results show that at physiological temperatures VLDL particles are very soft, elastic and mobile as compared to LDL, which is more rigid (see Figure 5). This observation supports the notion that apo-B100 in VLDL is loosely packed at the interface covering a large surface area with low interfacial surface tension [59]. During particle conversion from VLDL to LDL, however, the relative number of surface molecules increases and a higher molecular packing density leads to a compression of the lipid anchored protein regions and an overall stiffening of the LDL particle [60].

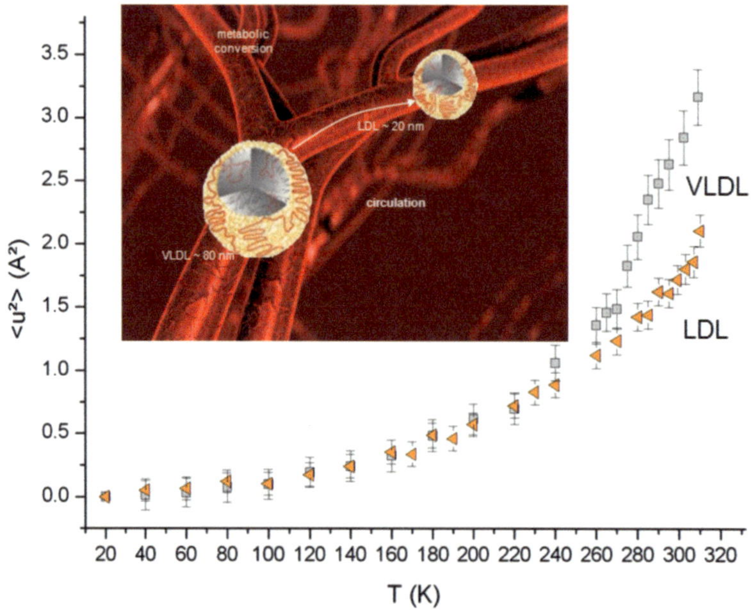

Figure 5. Molecular motions in LDL and VLDL. Elastic temperature scans are recorded with elastic incoherent neutron scattering. The mean square thermal fluctuations ($\langle u^2 \rangle$) are shown as function of temperature. The molecular resiliences are derived from the slopes in the curves. It is seen that VLDL has an increased motion at elevated temperatures compared to LDL. Parts of this figure are reproduced, with permission, from ref. [60].

To conclude, the intrinsic conformational flexibility and elasticity of apo-B100 containing lipoprotein particles is most likely critical for specific affinities of lipoproteins to receptors, antibodies or enzymes. Moreover, it would seem that the susceptibility of lipoproteins to oxidative modifications and hence their atherogenicity is influenced by their dynamic nature.

7. LDLs are flexible nanotransporters circulating in blood

In the search for new and improved therapeutics, the field of nanomedicine dealing with functionalized nanoparticles for molecular imaging and therapy is rapidly emerging. Nanoparticles offer new opportunities to transfer active substances directly to the diseased site in the body. By additional surface coatings or functionalizations, the properties of nanoparticles can be tuned to specific needs. Within the last two decades, a variety of artificial nanoparticles have been designed for targeted delivery of drugs or contrast agents. Many of these nanoconstructs are developed for cancer therapy taking advantage of the leaky vasculature of tumours. Apart from tumour targeting, increasing efforts are devoted to the treatment and imaging of atherosclerotic plaques (for a recent review see ref. [62]). Over time, a broad and versatile nanoparticle platform was created in which liposomes and biodegradable polymers have turned out to be the most promising candidates. It is important to mention that several nanomedicine products have already been established on the market and numerous products are successfully applied in clinical trials [63]. However, inherent problems of nanoparticles are biocompatibility and low stability in vivo, since most nanoparticles become rapidly cleared by the reticuloendothelial system. In contrast to artificial systems, lipoproteins are naturally occurring nanoparticles evading recognition by the body´s immune system. Hence, lipoproteins are excellent candidates with attractive properties to be considered as molecular transporters. A great advantage of LDL over other nanoparticles is the fact that LDL particles stay in circulation for several days, and are not cleared immediately by the mononuclear phagocyte system of the liver and spleen. The average lifetime of an LDL particle is 2-3 days and this time span is about three times longer as reported for long-circulating liposomes, currently applied in chemotherapy [64]. It was recognized that certain tumor cells overexpress LDL receptor, however, the targeting specificity is limited as the LDL receptor is ubiquitously expressed throughout the body, most prominent in the liver. However, using apo-B100 as inherent targeting sequence the enhanced circulation times in blood enable drug-loaded LDL particles to bind to specific receptors exposed on the surface of e.g. tumor or atherosclerotic plaque. Once recognized by the receptor, the functionalized LDL particles become internalized, accumulate in the tissue and exert an enhanced effect (reviewed by [65]). The intrinsic targeting properties of LDL to atherosclerotic plaques are already utilized for early diagnosis and detection of atherosclerotic lesions by different imaging modalities (for reviews see refs. [66,67]). However, to modify lipoprotein particles for medical purposes, care has to be taken not to compromise essential biophysical and structural features of LDL with the goal to preserve the biological activity. In general, there are several possibilities to create multi-functionalized lipoprotein particles. Some representative examples are shown in the scheme in Figure 7. One possibility is to load hydrophobic drugs (e.g. chemotherapeutics, antibiotics, vitamins, signal emitting molecules or small nanocrystals) in the lipophilic inner core of LDL. This can be accomplished by different techniques including lyophilisation, solvent evaporation and reconstitution procedures [68,69]. However, LDL particles can not be reconstituted so easily and remote drug/contrast agent loading into native lipoprotein particles is still a tedious approach currently not being standardized. Amphiphilic substances (drugs or marker molecules) or fatty acid modified chelator complexes can be incorporated in the PL monolayer [70,71]. This has successfully been done in numerous biophysical studies and for diagnostic purposes. Finally, the surface of LDL can be modified by protein labeling. This is done by covalent attachment of substances to the lysine and cysteine amino acid residues of apo-B100. Such substances include fluorophores, radionuclides or metal ions for molecular imaging [65]. Alternatively, targeting sequences (e.g. folic acid) can be coupled to apo-B100 with the purpose to reroute LDL to alternate receptors, which, in case of folate, are more specifically expressed in tumor cells [72].

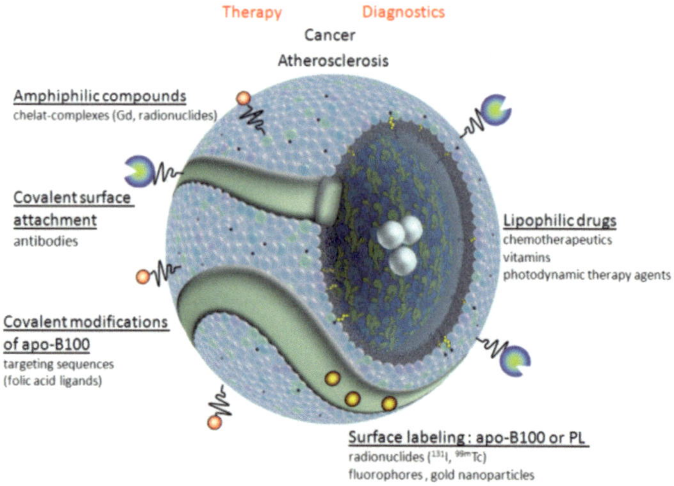

Figure 6. Scheme giving some examples of how LDL particles can be modified to act as natural endogenous nanoparticles for targeted drug delivery or multifunctional molecular imaging.

To construct lipoprotein mimetic particles, also referred to as lipoprotein related particles, artificial lipoprotein particles have to be reassembled from individual lipid and protein entities. This approach was highly successful for high density lipoproteins using apo-AI mimetic peptide sequences [73]. For LDL, this approach was not pursued yet and will be much more complicated considering the complex dynamic nature of apo-B100.

Over the last few years, a promising nanoparticle platform was established, which exploits the endogenous properties of natural lipoproteins being non-toxic, non-immunogenic and biodegradable. Although this platform still offers vast potential for improvements, first promising results in enhanced multimodal imaging of tumors and atherosclerotic plaques are achieved giving hope that further endeavors to combine diagnostics and personalized therapeutics will also be successful.

8. Conclusions and future directions

The intrinsic flexibility and dynamics of LDL lipids and protein in conjunction with the inherent compositional heterogeneity of LDL particles has hitherto hampered successful structure determinations at atomic level. Recent technological developments, however, allowed to restore characteristic structural features of individual LDL particles at low resolution. In particular, using cryo e.m. 3D-reconstruction techniques several groups have succeeded in imaging morphological and topological details of LDL to a resolution limit of approximately 2 nm [34-36]. Now, new concepts will be needed to make further progress in the development of high resolution models of LDL. One promising way is to put stronger emphasis on protein crystallography in combination with computational modelling and molecular dynamics simulations. X-ray crystallography apprears to be a hopeless pursuit with heterogeneous and flexible particles like LDL. Nevertheless, our earlier attempts of crystallisation have been partially successful [74]. Additional efforts, however, have to be focussed on the stabilization of apo-B100 in a more rigid state, perhaps by co-crystallisation with monoclonal antibodies. An alternative way ahead would be to work with lipid-free apo-B100 stabilized by detergent-mimetic systems, e.g. amphipathic designer peptides, or to proceed with truncated fragments of apo-B100.

At present there is still a deficit in our knowledge concerning the molecular lipid trafficking mechanisms of LDL. To know the atomic structure of LDL, in particular of apo-B100, may well contribute to a better understanding of biologic aspects of cardiovascular diseases, especially with respect to future strategies towards rational pharmaceutical interventions.

Author details

Ruth Prassl and Peter Laggner
Institute of Biophysics and Nanosystems Research,
Austrian Academy of Sciences, Graz, Austria

Acknowledgement

This manuscript is based in part upon work supported by the Austrian Science Fund under grant number P-20455.

9. References

[1] Brown MS and Goldstein JL (1976) Receptor-mediated control of cholesterol metabolism. Science 191: 150-154.
[2] Steinberg D, Parthasarathy S, Carew S, Khoo JC, Witztum JL (1989) Beyond cholesterol. Modifications of low-density lipoprotein that increase its atherogenicity. New Engl.J.Med. 320: 915-924.
[3] Lusis AJ (2000) Atherosclerosis. Nature 407: 233-241.
[4] Packard C, Caslake M, Shepherd J (2000) The role of small, dense low density lipoprotein (LDL): a new look. Int.J.Cardiol. 74 Suppl 1: S17-S22.
[5] Packard CJ (2006) Small dense low-density lipoprotein and its role as an independent predictor of cardiovascular disease. Curr.Opin.Lipidol. 17: 412-417.
[6] McNamara JR, Small DM, Li ZL, Schaefer EJ (1996) Differences in LDL subspecies involve alterations in lipid composition and conformational changes in apolipoprotein B. J.Lipid Res. 37: 1924-1935.
[7] Chapman MJ, Guerin M, Bruckert E (1998) Atherogenic, dense low-density lipoproteins. Pathophysiology and new therapeutic approaches. Eur.Heart J. 19 Suppl A: A24-A30.
[8] Pentikainen MO, Oksjoki R, Oorni K, Kovanen PT (2002) Lipoprotein lipase in the arterial wall: linking LDL to the arterial extracellular matrix and much more. Arterioscler.Thromb.Vasc.Biol. 22: 211-217.

[9] Skalen K, Gustafsson M, Rydberg EK, Hulten LM, Wiklund O, Innerarity TL, Boren J (2002) Subendothelial retention of atherogenic lipoproteins in early atherosclerosis. Nature 417: 750-754.
[10] Hurt-Camejo E, Camejo G, Sartipy P (2000) Phospholipase A2 and small, dense low-density lipoprotein. Curr.Opin.Lipidol. 11: 465-471.
[11] Williams KJ and Tabas I (2005) Lipoprotein retention--and clues for atheroma regression. Arterioscler.Thromb.Vasc.Biol. 25: 1536-1540.
[12] Hansson GK and Hermansson A (2011) The immune system in atherosclerosis. Nature Immunology 12: 204-212.
[13] Kostner, G. M. and Laggner, P. (1989) in Human Plasma Lipoproteins - Clinical Biochemistry, Principles, Methods, Applications 3 (Fruchart, J. C. and Shepherd, J., eds.), pp. 23-54, Walter de Gruyter, Berlin - New York.
[14] Hevonoja T, Pentikainen MO, Hyvonen MT, Kovanen PT, Ala-Korpela M (2000) Structure of low density lipoprotein (LDL) particles: basis for understanding molecular changes in modified LDL [In Process Citation]. Biochim.Biophys.Acta 1488: 189-210.
[15] Chapman MJ, Laplaud PM, Luc G, Forgez P, Bruckert E, Goulinet S, Lagrange D (1988) Further resolution of the low density lipoprotein spectrum in normal human plasma: physicochemical characteristics of discrete subspecies separated by density gradient ultracentrifugation. J.Lipid Res. 29: 442-458.
[16] Nigon F, Lesnik P, Rouis M, Chapman MJ (1991) Discrete subspecies of human low density lipoproteins are heterogeneous in their interaction with the cellular LDL receptor. J.Lipid Res. 32, 1741-1753.
[17] Dejager S, Bruckert E, Chapman MJ (1993) Dense low lipoprotein subspecies with diminished oxidative resistance predominate in combined hyperlipidemia. J.Lipid Res. 34, 295-308.
[18] Schuster B, Prassl R, Nigon F, Chapman MJ, Laggner P (1995) Core lipid structure is a major determinant of the oxidative resistance of low density lipoprotein. Proc.Natl.Acad.Sci.USA 92: 2509-2513.
[19] Murtola T, Vuorela TA, Hyvonen MT, Marrink SJ, Karttunen M, Vattulainen I (2011) Low density lipoprotein: structure, dynamics, and interactions of apoB-100 with lipids. Soft Matter 7: 8135-8141.
[20] Segrest JP, Jones MK, De Loof H, Dashti N (2001) Structure of apolipoprotein B-100 in low density lipoproteins. J.Lipid Res. 42: 1346-1367.
[21] Sommer A, Prenner E, Gorges R, St³tz H, Grillhofer H, Kostner GM, Paltauf F, Hermetter A (1992) Organization of phosphatidylcholine and sphingomyelin in the surface monolayer of low density lipoprotein and lipoprotein(a) as determined by time-resolved fluorometry. J.Biol.Chem. 267: 24217-24222.
[22] Atkinson D, Deckelbaum RJ, Small DM, Shipley GG (1977) Structure of human plasma low-density lipoproteins: Molecular organization of the central core. Proc.Natl.Acad.Sci.USA 74: 1042-1046.
[23] Laggner P, Degovics G, Müller KW, Glatter O, Kostner GM, Holasek A (1977) Molecular packing and fluidity of lipids in human serum low density lipoproteins. Hoppe-Seyler's Z.Physiol.Chem. 358: 771-778.

[24] Laggner P and Kostner GM (1978) Thermotropic changes in the surface structure of lipoprotein B from human-plasma low-density lipoproteins. A spin-label study. Eur.J.Biochem. 84: 227-232.

[25] Laggner P, Kostner GM, Rakusch U, Worcester DL (1981) Neutron small-angle scattering on selectively deuterated human plasma low density lipoproteins. The location of polar phospholipid headgroups. J.Biol.Chem. 256, 11832-11839.

[26] Gulik-Krzywicki T, Yates M, Aggerbeck LP (1979) Structure of serum low-density lipoprotein. J.Mol.Biol. 131: 475-484.

[27] Spin JM and Atkinson D (1995) Cryoelectron microscopy of low density lipoprotein in vitreous ice. Biophys.J. 68: 2115-2123.

[28] Laggner P, Chapman MJ, Goldstein S (1978) An X-Ray Small-Angle Scattering Study of Trypsin Treated Low Density Lipoprotein from Human Serum. Biochem.Biophys.Res.Commun. 82: 1332-1339.

[29] Lund-Katz S, Ibdah JA, Letizia JY, Thomas MT, Phillips MC (1988) A ^{13}C NMR characterization of lysine residues in apolipoprotein B and their role in binding to the low density lipoprotein receptor. J.Biol.Chem. 263: 13831-13838.

[30] Prassl, R., Schuster, B., and Laggner, P. (1997) in Supramolecular Structure and Function 5 (Pifat, G., ed.), pp. 47-73, Balaban Publishers.

[31] Prassl R and Laggner P (2009) Molecular structure of low density lipoprotein: current status and future challenges. Eur.Biophys.J.Biophys.Lett. 38: 145-158.

[32] Orlova EV, Sherman MB, Chiu W, Mowri H, Smith LC, Gotto AM (1999) Three-dimensional structure of low density lipoproteins by electron cryomicroscopy. Proc.Natl.Acad.Sci.U.S.A 96: 8420-8425.

[33] Van Antwerpen R (2004) Preferred orientations of LDL in vitreous ice indicate a discoid shape of the lipoprotein particle. Arch.Biochem.Biophys. 432: 122-127.

[34] Ren G, Rudenko G, Ludtke SJ, Deisenhofer J, Chiu W, Pownall HJ (2010) Model of human low-density lipoprotein and bound receptor based on cryoEM. Proc Natl Acad Sci U S A 107: 1059-1064.

[35] Kumar V, Butcher SJ, Oorni K, Engelhardt P, Heikkonen J, Kaski K, Ala-Korpela M, Kovanen PT (2011) Three-Dimensional cryoEM Reconstruction of Native LDL Particles to 16 angstrom Resolution at Physiological Body Temperature. PLoS ONE 6.

[36] Liu YH and Atkinson D (2011) Enhancing the Contrast of ApoB to Locate the Surface Components in the 3D Density Map of Human LDL. Journal of Molecular Biology 405: 274-283.

[37] Liu YH and Atkinson D (2011) Immuno-electron cryo-microscopy imaging reveals a looped topology of apoB at the surface of human LDL. J.Lipid Res. 52: 1111-1116.

[38] Deckelbaum RJ, Shipley GG, Small DM, Lees RS, George PK (1975) Thermal transitions in human plasma low density lipoproteins. Science 190, 392-394.

[39] Deckelbaum RJ, Shipley GG, Small DM (1977) Structure and interactions of lipids in human plasma low density lipoproteins. J.Biol.Chem. 252: 744-754.

[40] Laggner P and Müller K (1978) The structure of serum lipoproteins as analysed by X-ray small-angle scattering. Q.Rev.Biophys. 11: 371-425.

[41] Laggner P, Kostner GM, Degovics G, Worcester DL (1984) Structure of the cholesteryl ester core of human plasma low density lipoproteins: Selective deuteration and neutron small- angle scattering. Proc.Natl.Acad.Sci.USA 81: 4389-4393.
[42] Sherman MB, Orlova EV, Decker GL, Chiu W, Pownall HJ (2003) Structure of triglyceride-rich human low-density lipoproteins according to cryoelectron microscopy. Biochemistry 42: 14988-14993.
[43] Coronado-Gray A and Van Antwerpen R (2005) Lipid composition influences the shape of human low density lipoprotein in vitreous ice. Lipids 40: 495-500.
[44] Prassl R, Pregetter M, Amenitsch H, Kriechbaum M, Schwarzenbacher R, Chapman JM, Laggner P (2008) Low density lipoproteins as circulating fast temperature sensors. PLoS ONE 3: e4079 .
[45] Liu Y, luo D, Atkinson D (2010) Human LDL core cholesterol ester packing: 3D image reconstruction and SAXS simulation studies. J Lipid Res 51.
[46] Pregetter M, Prassl R, Schuster B, Kriechbaum M, Nigon F, Chapman J, Laggner P (1999) Microphase separation in low density lipoproteins. Evidence for a fluid triglyceride core below the lipid melting transition. J.Biol.Chem. 274: 1334-1341.
[47] Small, D. M. (1986) in The Physical Chemistry of Lipids - From Alkanes to Phospholipids pp. 395-473, Plenum Press, New York and London.
[48] Lusa S and Somerharju P (1998) Degradation of low-density-lipoprotein cholesterol esters by lysosomal lipase in-vitro - effect of core physical state and basis of species selectivity. Bba-Lipid Lipid Metab 1389: 112-122.
[49] Morton RE and Parks JS (1996) Plasma cholesteryl ester transfer activity is modulated by the phase transition of the lipoprotein core. J.Lipid Res. 37: 1915-1923.
[50] Zechner R, Kostner GM, Dieplinger H, Degovics G, Laggner P (1984) In vitro modification of the chemical composition of human plasma low-density lipoproteins: Effects on morphology and thermal properties. Chem.Phys.Lipids 36: 111-119.
[51] Esterbauer H, Dieber-Rotheneder M, Waeg G, Striegl G, Jürgens G (1990) Biochemical, Structural, and Functional Properties of Oxidized Low-Density Lipoprotein. Chem.Res.Toxicol. 3: 77-92.
[52] Chen S-H, Yang C-Y, Chen PF, Setzer D, Tanimura M, Li W-H, Gotto AM, Jr., Chan L (1986) The complete cDNA and amino acid sequence of human apolipoprotein B-100. J.Biol.Chem. 261: 2918-2921.
[53] Knott TJ, Pease RJ, Powell LM, Wallis SC, Rall SC, Innerarity TL, Blackhart B, Taylor WH, Marcel Y, Milne R, Johnson D, Fuller M, Lusis AJ, McCarthy BJ, Mahley RW, Levy-Wilson B, Scott J (1986) Complete protein sequence and identification of structural domains of human apolipoprotein B. Nature 323: 734-738.
[54] Phillips ML and Schumaker VN (1989) Conformation of apolipoprotein B after lipid extraction of low-density lipoproteins attached to an electron microscope grid. J.Lipid Res. 30: 415-422.
[55] Johs A, Hammel M, Waldner I, May RP, Laggner P, Prassl R (2006) Modular structure of solubilized human apolipoprotein B-100. Low resolution model revealed by small angle neutron scattering. J.Biol.Chem. 281: 19732-19739.

[56] Chatterton JE, Phillips ML, Curtiss LK, Milne RW, Marcel YL, Schumaker VN (1991) Mapping apolipoprotein B on the low density lipoprotein surface by immunoelectron microscopy. J.Biol.Chem. 266: 5955-5962.
[57] Chatterton JE, Schlapfer P, Bütler E, Gutierrez MM, Puppione DL, Pullinger CR, Kane JP, Curtiss LK, Schumaker VN (1995) Identification of apolipoprotein B 100 Polymorphisms that affect low-density lipoprotein metabolism: Description of a new approach involving monoclonal antibodies and dynamic light scattering. Biochemistry 34: 9571-9580.
[58] Mann CJ, Anderson TA, Read J, Chester SA, Harrison GB, Kochl S, Ritchie PJ, Bradbury P, Hussain FS, Amey J, Vanloo B, Rosseneu M, Infante R, Hancock JM, Levitt DG, Banaszak LJ, Scott J, Shoulders CC (1999) The structure of vitellogenin provides a molecular model for the assembly and secretion of atherogenic lipoproteins. J.Mol.Biol 285: 391-408.
[59] Wang L, Walsh MT, Small DM (2006) Apolipoprotein B is conformationally flexible but anchored at a triolein/water interface: a possible model for lipoprotein surfaces. Proc.Natl.Acad.Sci.U.S.A 103: 6871-6876.
[60] Wang L, Martin DD, Genter E, Wang J, McLeod RS, Small DM (2009) Surface study of apoB1694-1880, a sequence that can anchor apoB to lipoproteins and make it nonexchangeable. J Lipid Res 50: 1340-1352.
[61] Mikl C, Peters J, Trapp M, Kornmueller K, Schneider WJ, Prassl R (2011) Softness of atherogenic lipoproteins: a comparison of very low density lipoprotein (VLDL) and low density lipoprotein (LDL) using elastic incoherent neutron scattering (EINS). J Am Chem Soc 133: 13213-13215.
[62] Lobatto ME, Fuster V, Fayad ZA, Mulder WJM (2011) Perspectives and opportunities for nanomedicine in the management of atherosclerosis. Nature Reviews Drug Discovery 10: 835-852.
[63] Duncan R and Gaspar R (2011) Nanomedicine(s) under the Microscope. Molecular Pharmaceutics 8: 2101-2141.
[64] Allen TM and Cullis PR (2004) Drug delivery systems: entering the mainstream. Science 303: 1818-1822.
[65] Ng KK, Lovell JF, Zheng G (2011) Lipoprotein-Inspired Nanoparticles for Cancer Theranostics. Accounts of chemical research 44: 1105-1113.
[66] Frias JC, Lipinski MJ, Lipinski SE, Albelda MT (2007) Modified lipoproteins as contrast agents for imaging of atherosclerosis. Contrast.Media Mol.Imaging 2: 16-23.
[67] Cormode DP, Skajaa T, Fayad ZA, Mulder WJ (2009) Nanotechnology in medical imaging: probe design and applications. Arterioscler Thromb Vasc Biol 29: 992-1000.
[68] Hammel M, Laggner P, Prassl R (2003) Structural characterisation of nucleoside loaded low density lipoprotein as a main criterion for the applicability as drug delivery system. Chem.Phys.Lipids 123: 193-207.
[69] Song LP, Li H, Sunar U, Chen J, Corbin I, Yodh AG, Zheng G (2007) Naphthalocyanine-reconstituted LDL nanoparticles for in vivo cancer imaging and treatment. International Journal of Nanomedicine 2: 767-774.

[70] Corbin IR, Li H, Chen J, Lund-Katz S, Zhou R, Glickson JD, Zheng G (2006) Low-density lipoprotein nanoparticles as magnetic resonance imaging contrast agents. Neoplasia 8: 488-498.
[71] Chen LC, Chang CH, Yu CY, Chang YJ, Hsu WC, Ho CL, Yeh CH, Luo TY, Lee TW, Ting G (2007) Biodistribution, pharmacokinetics and imaging of Re-188-BMEDA-labeled pegylated liposomes after intraperitoneal injection in a C26 colon carcinoma ascites mouse model. Nuclear Medicine and Biology 34: 415-423.
[72] Zheng G, Chen J, Li H, Glickson JD (2005) Rerouting lipoprotein nanoparticles to selected alternate receptors for the targeted delivery of cancer diagnostic and therapeutic agents. Proc.Natl.Acad.Sci.U.S.A 102: 17757-17762.
[73] Zhang ZH, Chen J, Ding LL, Jin HL, Lovell JF, Corbin IR, Cao WG, Lo PC, Yang M, Tsao MS, Luo QM, Zheng G (2010) HDL-Mimicking Peptide-Lipid Nanoparticles with Improved Tumor Targeting. Small 6: 430-437.
[74] Prassl R, Chapman JM, Nigon F, Sara M, Eschenburg S, Betzel C, Saxena A, Laggner P (1996) Crystallization and preliminary X-ray analysis of a low density lipoprotein from human plasma. J.Biol.Chem. 271: 28731-28733.

Chapter 2

New Insights into the Assembly and Metabolism of ApoB-Containing Lipoproteins from *in vivo* Kinetic Studies: Results on Healthy Subjects and Patients with Chronic Kidney Disease

Benjamin Dieplinger and Hans Dieplinger

Additional information is available at the end of the chapter

http://dx.doi.org/10.5772/51865

1. Introduction

Lipoproteins are complexes consisting of a lipid core of mainly triglycerides and cholesterol esters surrounded by a surface monolayer of phospholipids, free cholesterol and specific protein components named apolipoproteins [1]. Most apolipoproteins undergo complex exchange reactions and serve many metabolic functions including transport, enzyme cofactors and receptor ligands. Except for the covalently linked apolipoprotein(a)-apolipoproteinB-100 (apo(a)-apoB) complex in Lipoprotein(a) [Lp(a)], apolipoproteins are non-covalently associated with each other and the lipid core.

Lipoprotein disorders are often associated with cardiovascular disease (CVD), atherosclerosis and other organ dysfunctions [2, 3]. To prevent and treat these diseases and to fully understand their cause, it is necessary to characterise the underlying metabolic disorders [1]. The conventional initial approach to do this is by measuring concentrations of plasma lipids or apolipoproteins. However, abnormal concentrations of lipids and apolipoproteins can result from changes in the production, conversion or catabolism of lipoprotein particles. Therefore, although static measurements and functional assays are important techniques to gain first in vivo functional insights, it is necessary to study their metabolic pathway to understand the complexity of lipoprotein function and pathophysiology [4, 5].

Animal models cannot sufficiently replace human studies to explore lipoprotein metabolism due to substantial species specificity. This holds particularly true for conventional laboratory animals such as mice and rats which – unless genetically modified or induced by special diet - do not develop atherosclerosis (see review [6]). The same argument is valid for investigations using cellular model systems. Since the liver is the central organ responsible for lipoprotein metabolism and primary human hepatocytes are of only limited use in research, most cellular studies in lipoprotein metabolism have been conducted in human hepatoma cells lines. These lines express, secrete and assemble a lipoprotein pattern which is substantially different from the respective human counterpart [7].

For all these reasons, the in vivo investigation of metabolic pathways in human subjects is the ultimate approach to elucidate physiological or pathological functions of metabolites in the human body. Historically, such human kinetic studies were performed using radioactive tracers; this methodology is, however, nowadays of only restricted use. Therefore, stable-isotope tracer kinetic studies in human subjects with clear advantages regarding safety and technical issues have replaced the radiotracer methods to become an important research tool for achieving a quantitative understanding of the dynamics of metabolic processes in vivo.

The aim of this review is to shortly describe the methodology and illustrate how the approach has expanded our understanding of physiological mechanisms as well as the pathogenesis of disorders of human lipoprotein metabolism. We will then specifically address the assembly mechanism of the atherogenic Lp(a) complex and focus on the kinetics of apoB-containing lipoproteins in patients with chronic kidney disease. This patient group is well-known for its high risk for atherosclerotic complications and a 10- to 20-fold increased cardiovascular mortality compared to the general population [8].

2. Principles of tracer technology

Exogenous and endogenous labelling techniques have been used to study the in vivo metabolism of an endogenous molecule, the tracee (see review [4]). In the exogenous method, the same molecule, in form of a usually radioactively labelled tracer, is introduced into the bloodstream [9]. In lipoprotein studies, this methodology first requires purification of the target molecule or particle and ex-vivo radiolabelling followed by reinfusion into the circulation. The physological integrity of the target molecule might, however, suffer from such procedure. Furthermore, in case of multiprotein complexes (which most lipoproteins are), the kinetics of individual protein components cannot be investigated by this approach. As an example, the investigation of in vivo kinetics of both protein components of Lp(a), as described in this article, to study its assembly mechanism would not be possible with the exogenous labelling approach.

In contrast, in endogenous labelling, a labelled precursor of the molecule of interest, in case of proteins usually a labelled amino acid, is used to label the target molecule by infusion into the circulation of a suitable proband. Ideally, the tracer can easily be detected and quantified, has the same kinetic behaviour as the tracee, and does not perturb the system. Usually, kinetic studies are performed in steady state, where the rates of input and output for a given unlabelled tracee substance are equal and time invariant. Thus, the information provided by the tracer reflects the behaviour of the tracee [10, 11]. At various times, the target protein or particle has to be purified from the blood of human probands and the amount of tracer is quantified to provide a kinetic curve. A mathematical model is then constructed to extract all the information contained in the kinetic curve. By fitting a model to the data, it is possible to calculate the parameters of the model that characterize the flux of molecules between kinetically homogeneous pools. For example, it is thus possible to investigate the whole pathway including production, conversion or catabolism of lipoprotein particles, information that cannot be obtained by static measurements alone.

The term stable isotope refers to a non-radioactive isotope of a given atom that is less abundant in a molecule within a biological system than the lightest naturally occurring isotope. The most common stable isotope used as metabolic tracer for apolipoprotein kinetic studies is [2H3]-leucine. Stable isotope tracers are much safer than radioactive tracers for both the study subject and the investigator. Furthermore, the duration of stable isotope experiments is normally less than 24 hours which is much shorter compared to radiotracer techniques which may need up to 14 days of examination [9].

2.1. Tracer administration

A tracer can be administered intravenously as either a single bolus injection, a primed constant infusion (i.e., a constant infusion given immediately after a priming bolus), or as a combination of both. The tracer bolus administration offers superior dynamics compared with the primed constant infusion, because the enrichment curves (the tracer/tracee ratios) after a bolus injection correspond to the impulse response of the system. It is therefore suitable to study components of lipoprotein metabolism with a slow rate of turnover. Another advantage of bolus administration is that it facilitates the determination of newly synthesized particles, as the intracellular precursor enrichment is greater at the start of the study. This argument therefore counts particularly when investigating kinetics of particle assembly, as described in 3.1.1. Practically, the bolus infusion is also most convenient for both subjects and investigators.

2.2. Multicompartment models for data analysis

Multicompartment modelling is a superior method to dissect the complexities of lipoprotein metabolism, and has been widely applied to systems in which material is transferred over time between compartments connected in a specific structure to permit the movement of material amongst the compartments [12].

Each compartment is assumed to be a homogenous entity within which the entities being modelled are equivalent. For instance, the compartments may represent different types of lipoprotein particles that are kinetically homogeneous and distinct from other material in the system. Very often, the data can be described by more than one model. To ensure that the best model is selected, it is necessary to carefully examine the fitting of the kinetic curve, to determine the precision of the parameter estimates, and to perform statistical tests to compare results obtained with different models. However, the complexity of a multicompartment model is usually a compromise for what is practically possible. A very simple model may not adequately describe the kinetic heterogeneity present within the system. A model that is too complex, on the other hand, will not be supported by experimental data and, hence, will have little predictive value. Furthermore, even if the development of models is based on experimental data, several assumptions are required in order to derive the model that is to be used. Thus, mathematical models do not determine the kinetics of lipids directly; rather, they derive an indirect approximation.

The software SAAM (Epsilon Group, Charlottesville, VA, USA) has become the first choice for modelling lipoprotein kinetic studies. The SAAM II program was recently developed by SAAM Inst., Inc., Seattle, WA, USA, and is frequently used to analyse lipoprotein tracer data using compartmental models [13, 14]. The primary kinetic parameter resulting after modelling with SAAM II is the fractional synthesis rate (FSR) which, under steady state conditions, is identical to the fractional catabolic rate (FCR) and has the dimension of pools/day. The reciprocal value of FSR/FCR is called retention time (RT, given in days) and indicates the residence time of the investigated tracee (the target apolipoprotein in our cases) in the circulation. The product of FSR multiplied by the concentration of tracee is called production rate (PR) and is usually expressed as mg/kg body weight/day.

3. Metabolism of apoB-containing lipoproteins

Dietary lipids are absorbed in the intestine and packaged into large, triglyceride-rich chylomicrons which undergo lipolysis to form chylomicron remnants. In the last step of the so-called exogenous lipoprotein pathway, these particles are finally taken up by the liver. The liver then secretes triglyceride-rich lipoproteins known as very low-density lipoproteins (VLDLs) representing the first step oft the endogenous lipoprotein pathway (Figure 1). Lipoprotein kinetic studies have shown that VLDLs are metabolically heterogeneous. Following lipolysis by endothelium-bound lipoprotein lipase (LPL) and hepatic lipase (HL), these particles are converted via intermediate-density lipoproteins (IDL, also called VLDL remnants) to low-density lipoprotein (LDL) or taken up by the liver. LDL is catabolized mainly by the liver or peripheral tissues via the LDL receptor. Increased plasma concentrations of LDL are a major risk factor for CVD. ApoB-100 is the major apolipoprotein of chylomicrons, VLDL, IDL and LDL.

Lipoprotein(a) [Lp(a)] consists of an LDL-like particle which is covalently bound to the glycoprotein apolipoprotein(a) [apo(a)] by disulfide linkage and derives from the liver [15] (Figure 2). Among individuals, Lp(a) plasma concentrations vary more than 1000-fold, ranging from less than 0.1 mg/dl to more than 300 mg/dl. Depending on the investigated population and the used genetic approach, it has been shown that between 30% and 90% of this variation in plasma concentrations of Lp(a) is determined by the apo(a) gene locus, encoding proteins from <300 to >800 kDa [16-18]. Apo(a) size is negatively correlated with Lp(a) concentrations, such that low-molecular-weight (LMW) apo(a) isoforms express on average high Lp(a) plasma concentrations, while high-molecular-weight (HMW) isoforms are usually associated with lower concentrations (reviewed in reference [15]). Elevated plasma concentrations of Lp(a) have been found associated with an increased risk of developing CVD in many studies which was confirmed by recent large meta-analyses [19, 20]. In vivo kinetic studies using radio-labeled Lp(a) indicated that the large differences in Lp(a) concentrations seen among individuals are determined by synthesis and not degradation [9, 21].

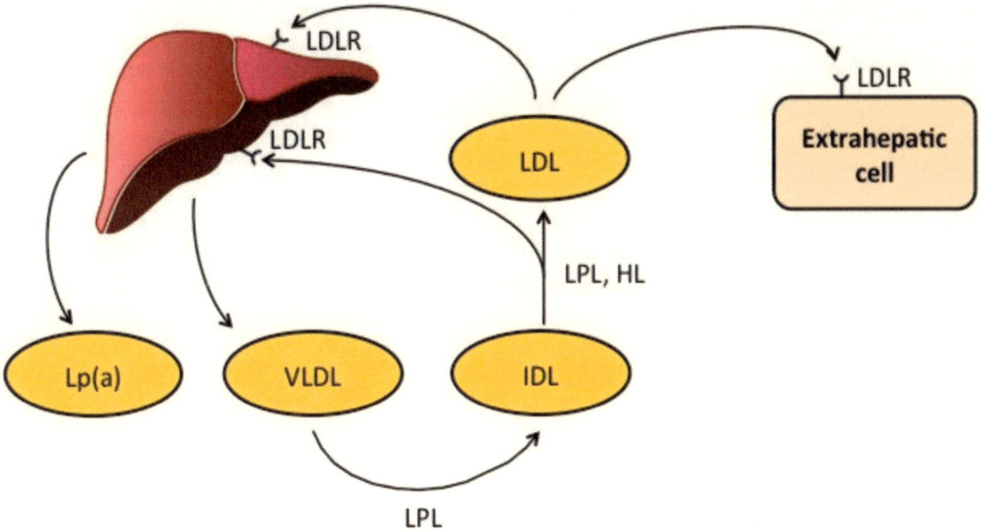

Figure 1. Endogenous metabolic pathway of apolipoprotein B (apoB)-containing lipoproteins. Triglyceride-rich very-low-density lipoproteins (VLDL) are synthesized and secreted by the liver into the blood stream and their triglycerides catabolized by the endothel-bound enzyme lipoprotein lipase (LPL) resulting in intermediate-densitly lipoproteins (IDL). LPL and hepatic lipase (HL) further convert IDL to low-density lipoproteins (LDL) which are removed from the circulation by the liver and extrahepatic tissue cells via LDL-receptor (LDLR)-mediated endocytosis. Lipoprotein(a) [Lp(a)] is synthesized and secreted by the human liver into circulation.

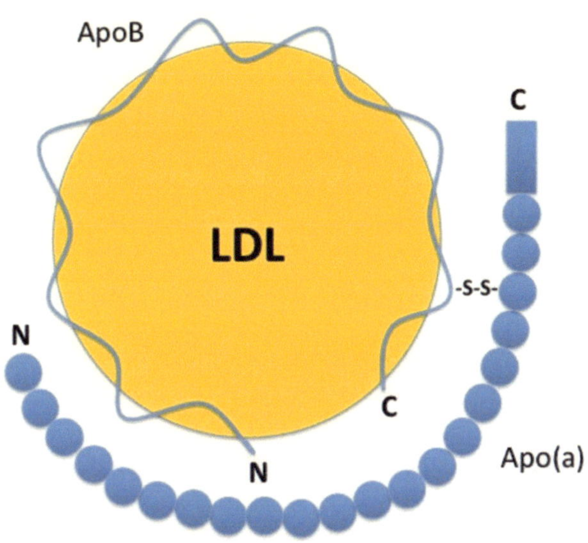

Figure 2. Structure of lipoprotein(a) [Lp(a)]. Lp(a) consists of an LDL-like particle and the disulfide-bridge-linked glycoprotein apolipoprotein(a) [apo(a)] which exerts high sequence homology to plasminogen. Apo(a) consists of an inactive protease domain (blue rectangle) and identical as well as non-identical repeats of kringle domains (blue circles). The number of identical kringles vary among individuals and gives rise to a genetically determined molecular size polymorphism of apo(a).

3.1. Biosynthesis of Lp(a)

Lp(a) has been the target of extensive and successful research particularly with respect to the unusually high degree of genetic control of its expression. In contrast, metabolism and physiological roles as well as pathogenicities of Lp(a) are still poorly understood, as recently reviewed by Dubé et al. [22]. The mechanisms that control Lp(a) secretion and assembly were investigated mostly by means of cellular hepatocyte model systems, yielded contrasting results and thus remain highly controversially discussed (see review [23]). Assembly of apo(a) and apoB to Lp(a) is generally viewed as a two-step procedure [24, 25]. In a first step, distinct domains within the apoB molecule initially associate with apo(a) in a non-covalent interaction to bring the two molecules into close proximity. In a second step, a disulfide bond is formed between apo(a) cysteine 4057 and apoB cysteine 4326 residues [24, 26]. Whether this disulfide bond is formed through a spontaneous oxidation reaction or through a specific enzymatic reaction is unclear [27, 28].

The location of this assembly process is the subject of controversial discussion as well. Intracellular, extracellular and/or plasma membrane-associated assembly procedures have been reported to occur in various cell systems [23]. Lp(a), like many other oligomeric protein complexes, may assemble in the endoplasmic reticulum of the hepatocyte and be secreted as a whole particle [29, 30]. Alternatively, newly synthesized apo(a) could bind extracellularly to preexisting LDL or VLDL circulating in the plasma. Most authors postulate an extracellular assembly of Lp(a) based on studies conducted in various cellular model systems. White *et al.* could not detect an intracellular apo(a)-apoB complex by adding anti-apo(a) antiserum to the culture medium of primary baboon hepatocytes, but found such complexes attached to the plasma membrane. The authors therefore concluded that, in that cellular system, Lp(a) is primarily assembled after secretion and to some extent also on the plasma membrane [31]. This conclusion has to be, however, critically evaluated since baboon hepatocytes secrete most of their apoB as VLDL, which does not associate with apo(a) [32]. Similar studies in apo(a)-transfected HepG2 cells could not demonstrate an intracellular apo(a)-apoB assembly for this human hepatocyte model and thus confirmed the results from the baboon studies [24, 33, 34]. Nevertheless, there is also evidence for intracellular assembly of Lp(a) in cell culture systems. Bonen *et al.* were able to detect an intracellular apo(a)-apoB complex in HepG2 cells transfected with an apo(a) minigene [35]. HepG2 cells have been reported to secrete a triglyceride-rich lipoprotein particle with an LDL density that does not exist at all in human plasma [36]. Taken together, the extracellular Lp(a) assembly proposed by numerous *in vitro* studies needs to be reviewed with caution, because these studies used cellular models that do not reflect the physiological lipoprotein metabolism.

3.1.1. *In vivo metabolism of Lp(a) and LDL in healthy subjects*

Kinetic *in vivo* studies in humans have unfortunately also produced controversial results. Krempler *et al.* injected radiolabeled VLDL in Lp(a)-positive healthy probands and found no metabolic relationship between apoB in VLDL or LDL and apoB in Lp(a). The authors therefore concluded that Lp(a) seems to be synthesized as a separate lipoprotein independently of other apoB-containing lipoproteins [37, 38]. Two *in vivo* turnover studies using stable-isotope labeling techniques came to the same conclusion: Morrisett *et al.* and Su *et al.* observed similar synthesis rates of Lp(a)-apo(a) and Lp(a)-apoB [39, 40]. While these findings are compatible with an intracellular assembly of nascent apo(a) and apoB to Lp(a), two other kinetic studies concluded that Lp(a) originates from *de novo* hepatic LDL as well as from plasma LDL [41, 42].

We investigated by stable-isotope technology the metabolism of apo(a) and apoB-100, the two major Lp(a) protein components, in comparison to apoB of LDL in nine healthy probands. The metabolic data accumulating in this study after appropriate modeling present a scenario of virtually complete intracellular assembly of Lp(a) [43].

Mean FSR, RT and PR values of apo(a) from Lp(a) were similar to those of apoB from Lp(a) but significantly different from the kinetic parameters of LDL-apoB. The differences were particularly large between the PR values of LDL and Lp(a) since this parameter takes into account plasma concentrations that are much higher for LDL than for Lp(a).

Tracer/tracee data from Lp(a)-apo(a), Lp(a)-apoB, LDL-apoB and VLDL-apoB were analyzed based on the multicompartment model shown in Figure 3 in order to investigate whether Lp(a) assembles from circulating LDL or from *de novo* produced "hepatic" LDL. 92% of leucine in Lp(a)-apoB originated from the hepatic apoB pool. The remaining 8% derived from plasma LDL-apoB. LDL-apoB stemmed from two sources, namely from VLDL-apoB (54%) and from de novo synthesis (46%).

The kinetic parameters obtained from this *in vivo* turnover study of Lp(a) metabolism in healthy men allow three major conclusions: i) Since FSRs of both protein components of Lp(a) is very similar and different from those of LDL, an almost exclusive intracellular hepatic Lp(a) assembly can be assumed. This analysis, however, does not allow any conclusions to be drawn on where (inside the hepatocyte, at its plasma membrane or, eventually, in the space of Dissé) this assembly takes place. ii) Apo(a) FSR/FCR is positively related to the number of apo(a) kringle 4 repeats (e.g. apo(a) molecular size), suggesting that plasma Lp(a) concentrations are controlled not only by synthesis but also to some smaller extent by degradation. iii) Longer plasma RT of apo(a) from probands with LMW apo(a) isoforms compared to those with HMW apo(a) isoforms help to explain the potential atherogenicity of higher concentrations in carriers with LMW apo(a) isoforms.

The *de novo* synthesis of LDL is an absolute prerequisite for the postulated (intra)cellular hepatic assembly of Lp(a). Such a "direct" LDL production has been questioned by some investigators who presume that it may instead be the consequence of a very fast lipolytic pathway [45]. However, metabolic studies of apoB metabolism using stable-isotope technology fitted by multicompartmental modeling support a significant "direct" LDL production by the liver [46, 47]. A substantial amount of nascent LDL production was also detected in cultured primary human hepatocytes [48] but not in HepG2 cells [36]. Lp(a) secretion was previously demonstrated in such cells, thus additionally supporting the view of "direct" LDL synthesis by human hepatocytes [49].

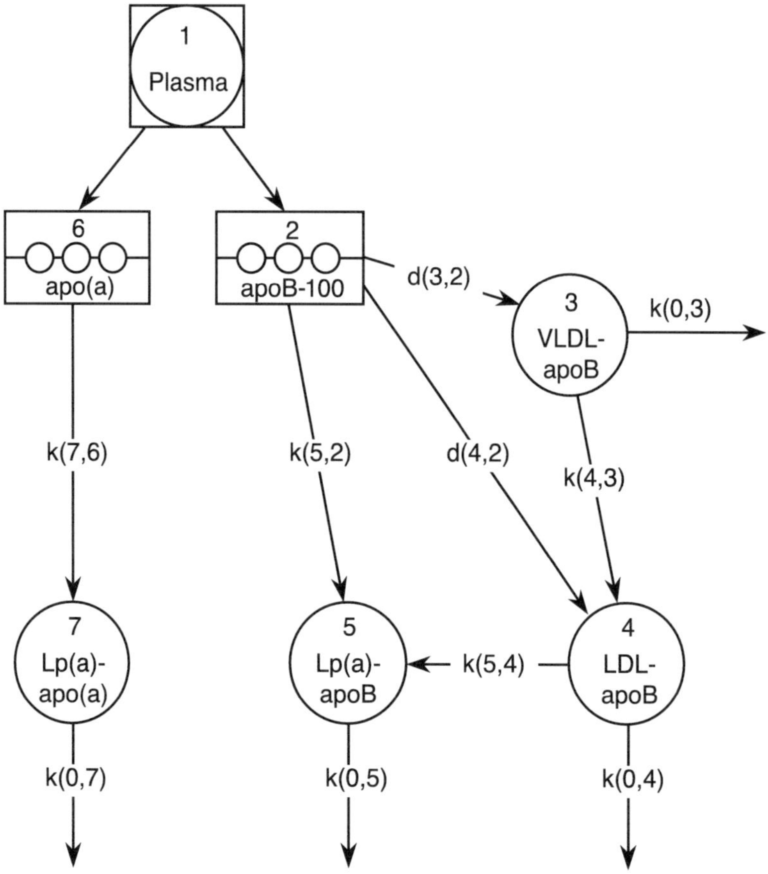

Figure 3. Multicompartmental model for apoB-100 and apo(a) metabolism. A plasma leucine pool (compartment 1) was used as a forcing function, and delay compartments that account for assembly and subsequent secretion of apoB-100 (compartment 2) and apo(a) (compartment 6), respectively. ApoB in VLDL, LDL, Lp(a), and apo(a) in Lp(a) consist all of single compartments. The input of apoB in Lp(a) is twofold: one via *de novo* synthesis from the liver and one from LDL-apoB. d(i,j) denotes the distribution of transfer from the delay compartment j to compartment i and k(i,j) represents the rate constant from compartment j to compartment i. In this model, tracer/tracee data for VLDL, LDL, Lp(a) apoB, and Lp(a) apo(a) as well as leucine masses (nmol/L) in these compartments were fitted simultaneously. The fractional catabolic rate (FCR) of apo(a)-Lp(a) and apoB-Lp(a) was equal to k(0,7) and k(0,5), respectively (taken from Frischmann et al. [44], with permission).

3.2. Lipoprotein metabolism in HD patients suffering from chronic kidney disease (CKD)

Dyslipidemia in patients with CKD and hemodialysis (HD) patients is distinct from other organ-specific diseases with far-reaching therapeutic consequences (see review [2]). It involves all lipoprotein classes, shows considerable variations depending on the stage of CKD [50, 51] and is further modified by concurrent diseases such as diabetes [52] and nephrotic syndrome [53]. In addition, major qualitative compositional changes in lipoprotein particles, such as oxidation, glycation, carbamylation and formation of small dense LDL (sdLDL – see below) which render the particles more atherogenic, have been observed [54]. Reduced activities of plasma cholesterol esterification and cholesterol ester transfer between lipoproteins – key factors for the so-called „reversed cholesterol transport" – result in substantially abnormal lipid composition of virtually all lipoprotein classes in HD patients [55].

Plasma triglycerides start to increase in early stages of CKD and show the highest concentrations in nephrotic syndrome and in patients treated with peritoneal dialysis (PD). In pre-dialysis CKD patients, the accumulation of triglycerides is the consequence of both an increased PR and a decreased FCR of triglyceride-rich lipoproteins [56]. The increased production of triglyceride-rich lipoproteins is possibly a consequence of impaired carbohydrate tolerance and enhanced hepatic VLDL synthesis [57]. The reduced catabolism is likely due to decreased activities of LPL and HL [58, 59], two endothelium-associated lipases that cleave triglycerides into free fatty acids for energy production or storage.

Diminished catabolism results in the accumulation of IDL particles contributing to compositional and size heterogeneity of triglyceride-rich lipoproteins in plasma of CKD patients. IDL are rich in apoE, a ligand that is important for removal from the circulation by binding to the LDL receptor [60]. The arterial wall therefore is exposed to high plasma concentrations of IDL which may predispose to atherosclerosis [54].

Elevated plasma concentrations of LDL cholesterol and –apoB are common in nephrotic syndrome and PD but do not occur in patients with advanced CKD, treated with HD. There are, however, qualitative changes in LDL in patients with CKD and dialysis patients. The fraction of sdLDL, which is considered to be highly atherogenic, is increased in HD patients. sdLDL is a subtype of LDL which penetrates the vessel wall more efficiently than normal LDL, becomes oxidized, and triggers atherosclerotic processes. In addition, sdLDL exert a high affinity for macrophages promoting their entry into the vascular wall to participate in the formation of foam cells and atherosclerotic plaques [61].

In kidney disease, elevated plasma Lp(a) concentrations are not only genetically determined but also a consequence of kidney failure [62]. In predialysis CKD patients, Lp(a) concentrations are influenced by the glomerular filtration rate (GFR). In patients with HMW apo(a) isoforms but not in those with LMW apo(a) isoforms, plasma Lp(a) concentrations begin to increase in stage 1 CKD before GFR starts to decrease [50]. This isoform-specific increase in plasma Lp(a) concentrations was observed in several but not all studies in CKD and HD patients [50, 62-66]. In contrast, in patients with nephrotic syndrome [67, 68] and in PD patients [63], increases in plasma Lp(a) concentrations occur in all apo(a) isoform groups, probably as a consequence of the pronounced protein loss and a subsequently increased production in the liver [69]. After successful kidney transplantation, a decrease in plasma Lp(a) can be observed in HD patients with HMW apo(a) isoforms [70, 71] and in PD patients with all apo(a) isoform groups [72]. Thus, the elevation of Lp(a) in CKD is due to non-genetic causes, mostly influenced by the degree of proteinuria [50, 67] and less by the cause of kidney disease [63].

In summary, the hallmarks of uremic dyslipidemia include hypertriglyceridemia and increased circulating concentrations of IDL, sdLDL and Lp(a). HD patients are characterised by normal LDL concentrations, whereas patients with nephrotic syndrome and CKD patients treated by PD are diagnosed with elevated LDL concentrations.

3.2.1. *Dyslipidemia and CVD in CKD*

Forty years ago, Lindner and colleagues described in their seminal report the excessive risk of CVD in HD patients for the first time [73]. Later, Foley et al. extended these observations by reporting a 10 to 20 times higher mortality rate in HD patients compared to the general population [8]. While in the general population high plasma concentrations of apoB-containing lipoproteins, low concentrations of HDL cholesterol and high total triglyceride concentrations are associated with an increased atherosclerotic cardiovascular risk [74], most investigations, including cross-sectional [75-78] and longitudinal [66, 79-87] studies, do not support the association between dyslipidemia and CVD in hemodialysed CKD populations or even observe opposite associations. Indeed, a worse survival among HD patients has been observed with low rather than high BMI [88], blood pressure [89] and serum/plasma concentrations of cholesterol [90]. This seemingly paradoxical phenomenon is often called „reverse epidemiology" [91] and exemplified in crossing curves when relating BMI with mortality in HD patients and the general population [92].

While the BMI-associated death risk shows an almost linear negative gradient in HD patients [92], the relationship between plasma total cholesterol and mortality has been found to be U-shaped [93]. The group with total cholesterol between 200 and 250 mg/dl had the lowest risk for death, whereas those with levels >350 mg/dl had a relative risk of 1.3-fold and those with levels <100 mg/dl had a relative risk of 4.2-fold. The association between low total cholesterol and increased mortality, however, was reduced after statistical adjustment for plasma albumin levels. This dichotomous relationship was confirmed in the Choices for Healthy Outcomes in Caring for ESRD (CHOICE) study [94], which showed a nonsignificant negative association of cardiovascular mortality with plasma total as well as non–HDL cholesterol levels in the presence of inflammation and/or malnutrition; in contrast, there was a positive association between total and non–HDL cholesterol and mortality in the absence of inflammation or malnutrition. These observations are compatible with the hypothesis that the inverse association of total cholesterol levels with mortality in dialysis patients is mediated by the cholesterol-lowering effect of malnutrition and/or systemic inflammation and not due to a protective effect of high cholesterol concentrations.

The association of Lp(a) with atherosclerotic complications and CVD has been investigated in numerous studies in dialysis patients. Like other atherogenic lipoproteins, Lp(a) has been found to contribute to the high cardiovascular burden [66, 79, 84, 95-97]. When apo(a) phenotyping was performed along with plasma Lp(a) concentrations, an association between the apo(a) K-IV repeat polymorphism and CV complications was consistently observed.

Two final considerations regarding the impact of classical risk factors for the development of CVD in CKD patients are, however, worth mentioning: the cardiovascular risk for an individual CKD patient at a given time point is the sum (or combination) of risk exposure before and after developing CKD. When taking Lp(a) concentrations and apo(a) isoforms as an example, a previously healthy subject with low Lp(a) concentrations and a HMW apo(a) isoform develops CKD with subsequently rising Lp(a) concentrations covering a relatively short period of his lifespan. A subject with LMW apo(a) isoform, on the other hand, has genetically caused elevated Lp(a) concentrations for his whole life which do not substantially increase after developing CKD. Since the HMW apo(a) carrier is exposed to elevated atherogenic Lp(a) for a much shorter period of his life, this condition has to be considered less CVD-prone than having LMW apo(a). This example demonstrates the importance of the „longitudinal" factor when considering risk factors for CVD in CKD patients.

Finally, as already discussed in the introduction, the quantification of a target parameter deemed to be associated with or predictive for a disease can only provide a static picture and hardly reflects the true in vivo metabolism. Seemingly normal blood concentrations of suspected marker candidates can only be validated by kinetic studies in humans and have been therefore performed also in CKD patients. They have provided novel and unexpected information regarding the physiology and pathology of atherogenic apoB-containing lipoproteins (see review [98]).

3.2.2. Delayed in vivo catabolism of LDL and IDL in HD patients as potential cause of premature atherosclerosis

For better understanding the atherogeneity of apoB-containing lipoproteins in HD patients and to resolve the apparent discrepancy between their obviously impaired lipoprotein metabolism and e.g. normal LDL plasma concentrations, we studied the in vivo kinetics of VLDL, IDL and LDL by stable isotope technology in HD patients and compared them to those of healthy controls [12].

This study demonstrated for the first time severely decreased FCRs of IDL- and LDL-apoB in HD patients as compared to controls (Figure 4), whereas the in vivo kinetics of VLDL did not change significantly. A decreased FCR of IDL- and LDL-apoB is identical to a prolonged RT of these highly atherogenic particles. The longer RT of these lipoproteins results in an extended exposure to oxidation for IDL and LDL in a highly oxidative environment. This is in line with experimental data showing a highly significant correlation of 5-hydroxy-2-aminovaleric acid (HAVA) in LDL, an oxidation product of apoB, with LDL RT in normolipidemic controls [99]. In accordance with these results, two previously conducted randomized placebo-controlled studies revealed a significant reduction in composite cardiovascular disease endpoints when HD patients were treated for two years with supplementation of antioxidants such as vitamin E [100] or acetylcysteine [101].

Most remarkably, the observed impaired metabolism of apoB-containing lipoproteins is accompanied by normal concentrations of LDL-apoB and elevated levels of IDL-apoB (Figure 4), in line with previous reports which found increased concentrations of IDL as an independent risk factor for atherosclerosis in HD patients [102]. A closer look at the kinetic data reveals that the normal concentrations of LDL are the result of a combination of decreased FCR and PR. This pattern therefore demonstrates convincingly the strength of kinetic studies in contrast to simply quantifying blood concentrations of a target marker such as LDL concentrations. Its normal concentrations are masked by two metabolic disorders which neutralise each other and result in normal values such as observed in the general population. The altered lipoprotein metabolism therefore puts HD patients at high risk for developing atherosclerotic disease despite their normal total and LDL cholesterol concentrations. Since most lipid-lowering drugs act by "normalising" the RT of the major atherogenic lipoproteins IDL and LDL [103], these drugs are expected to correct some of the basic defects of the severely disturbed lipoprotein metabolism in HD patients. Therefore, kinetic studies on the impact of lipid-lowering medication on the lipoprotein metabolism in CKD patients were a logic consequence of the observed, above-described findings (see chapter 3.2.5.).

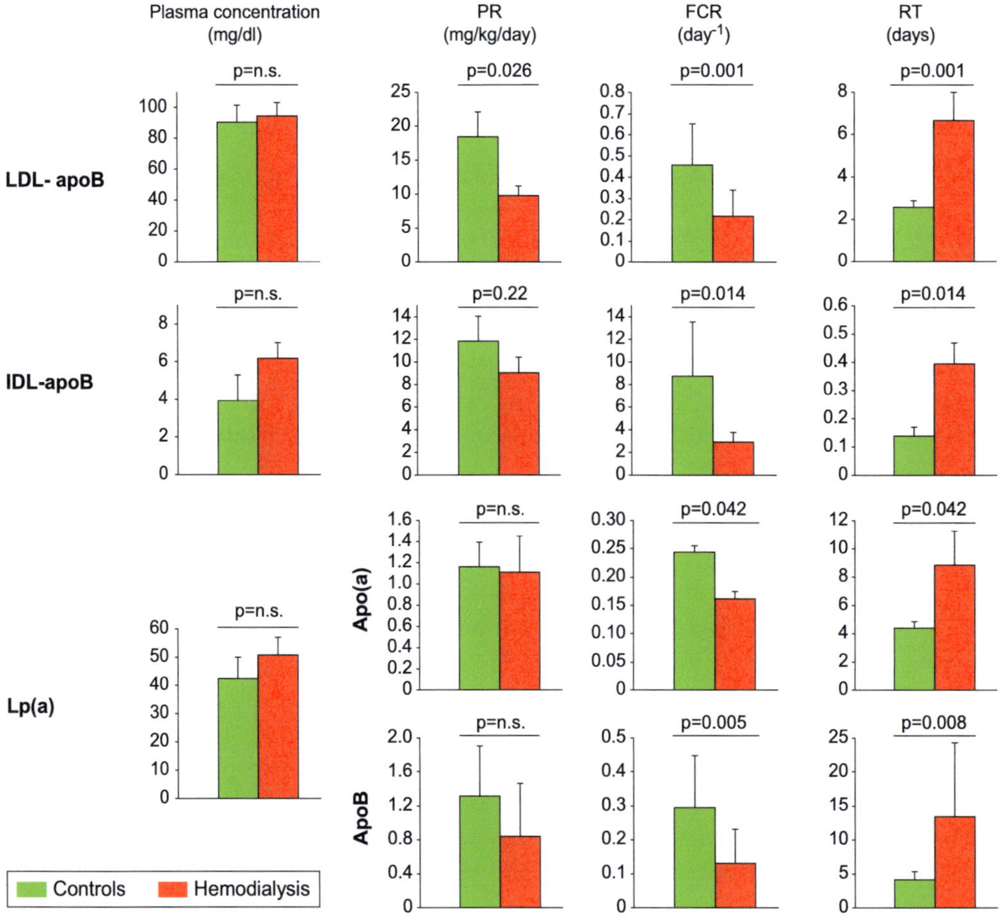

Figure 4. Kinetic parameters of apoB in LDL, IDL and Lp(a) and apo(a) in Lp(a). Plasma concentrations, production rates (PR), fractional catabolic rates (FCR) and residence times (RT) are given for healthy controls (green columns) and HD patients (red columns). Columns represent mean values ± SD. Results for LDL and IDL are taken from Ikewaki et al. [12], those for Lp(a) from Frischmann et al. [44].

Due to the laborious nature of these studies and the complexity of the metabolic modeling, only few studies have been performed so far in CKD patients either by radiotracer or stable isotope technology. Our kinetic data seem to contrast with a previously published turnover study in Finnish HD patients performed with conventional radiotracer techniques. While the authors found decreased LDL clearance rates in predialysis CKD patients [104] they could not find a significant difference in LDL-apoB FCR between HD patients and controls [105]. More recently, Prinsen et al., by using stable isotopes, found unchanged FCRs for LDL-apoB in CKD patients treated with peritoneal dialysis [106]. Chan et al. injected radio-labeled VLDL into HD patients with or without hyperlipidemia and found decreased FCRs of VLDL-apoB and IDL-apoB (the latter only in hyperlipidemic patients) [107]. LDL kinetics were not investigated in this study. The reason for these discrepancies is not clear. There might be ethnic differences in the lipoprotein metabolism between the investigated patient populations of different ethnic origin. One major difference between our and the Finnish study is an age difference between patients and controls in our but not in the Finnish study. Our control subjects were considerably younger than the HD patients (35 vs. 51 years). At first glance, this age difference might explain to some extent the dramatic differences found in our study, since LDL clearance rates have been repeatedly described to decrease with age presumably due to down-regulated hepatic LDL receptor expression in the elderly [108, 109]. Based on the results of these studies, an age difference of 15 years (as observed in our work) would result in an approximately 10% change in FCR values and could therefore not explain the more than two-fold difference in our study. The observed differences in kinetic parameters can therefore not be explained by age differences between study groups.

Several mechanisms may contribute to our observations. First, the diminished LDL catabolism in HD patients might be explained by a possible contribution of LDL uptake by the healthy human kidney which does not function appropriately (or at all) in chronic kidney failure. In fact, glomerular cells like mesangial or epithelial cells have been shown *in vitro* to express lipoprotein receptors and to take up LDL comparably to fibroblasts and hepatocytes [110]. It is, however, completely unclear whether the kidney plays a significant role in LDL catabolism *in vivo*. Perfusion studies in rat kidneys indicate that virtually no intact LDL is cleared from the circulation by the kidney [111]. Second, an impaired lipolytic cascade in HD patients most likely also contributes to our results. The relatively normal VLDL concentrations and kinetic parameters and the correspondingly impaired IDL parameters are in good accordance with previous findings of normal lipoprotein lipase (LPL) but significantly decreased activities of hepatic triglyceride lipase (HL) in HD patients [59]. Since HL promotes the conversion of IDL to LDL, a decrease in HL activity might contribute to the accumulation of IDL and reduced production rates of LDL (without accumulating small, dense LDL) in HD patients.

3.2.3. *Kinetics of Lp(a) in hemodialysis patients*

We previously performed *in vivo* kinetic studies using stable-isotope techniques to elucidate the mechanism for increased plasma Lp(a) concentrations in HD patients [44]. PRs of apo(a) and apoB, the two apolipoproteins contained in Lp(a), were normal, when compared to control subjects with similar plasma Lp(a) concentrations (Figure 4). The FCR of these apolipoproteins was, however, significantly reduced compared to controls resulting in a much longer plasma RT for apo(a) of almost 9 days, compared to only 4.4 days in controls. Since the PR of Lp(a) did not differ between HD patients and controls, its decreased clearance in HD patients leads to increased Lp(a) plasma concentrations and is likely the result of loss in kidney function [44]. A role of the kidney in the catabolism has been previously supported by the observation of renovascular arteriovenous differences in Lp(a) concentrations [112] as well as apo(a) fragments in urine [113, 114].

Comparing kinetic data in HD patients [44] with those in patients with nephrotic syndrome [69] points to fundamental differences in the metabolism of Lp(a) and other proteins between these two patient groups. Patients with nephrotic syndrome do not differ with respect to the FCR of Lp(a) compared to controls but have increased Lp(a) PRs [69]. It is well known that nephrotic patients show a generally increased lipoprotein synthesis of lipoproteins [115]. Since kidney function is relatively well preserved in nephrotic syndrome, a decreased clearance of Lp(a) in these patients is not likely to be expected. Metabolic differences between nephrotic and dialysis patients are not only evident for Lp(a) but also for albumin. Whereas the FCR of albumin in HD patients is similar or even reduced compared to controls, the FCR in patients with nephrotic syndrome is increased [116, 117].

3.2.4. Consequences of the impaired metabolism of atherogenic lipoproteins in HD patients

The observation of markedly decreased FCRs of apoB of LDL and IDL as well as apo(a) and apoB in Lp(a) causes a prolonged RT of these highly atherogenic lipoproteins. Due to the long retention period, "aged" lipoprotein complexes are thus more susceptible for alterations such as oxidation damage, which was shown to be associated with accelerated atherogenesis in HD patients [118]. Previous kinetic studies investigated the metabolism of the two LDL subclasses, "buoyant" LDL1 and the smaller cholesterol-poor "dense" LDL2, in subjects with familial defective apoB-100 (FDB). The authors found a more than four-fold longer RT for small dense LDL2 in those patients as compared to normolipidemic controls [99]. It was therefore suggested that oxidative damage of an "aged" LDL2, which is present in large concentrations in both blood and the subendothelial space, may be an important mechanism for the development of premature atherosclerosis in patients with familial defective apoB-100. Since the LDL-like particle of Lp(a) is compositionally similar to LDL2 [41], it is tempting to speculate that the increased RT of circulating Lp(a) might pose an additional risk factor for the increased incidence of cardiovascular disease in HD patients.

3.2.5. Influence of statin treatment on kinetic parameters in hemodialysed patients

In the general population, therapy with HMG-CoA-reductase inhibitors (statins) which inhibit endogenous cholesterol biosynthesis has shown to improve outcome in several atherosclerotic diseases [119, 120]. The inhibition of cholesterol biosynthesis subsequently leads to up-regulation of LDL receptors and therefore increased clearance and thus reduced RT of circulating LDL [103]. Statins also have a beneficial role as anti-inflammatory agents, which is independent of their lipid-lowering effect. Inflammation is highly prevalent in patients with CKD and is consistently associated with cardiovascular morbidity and mortality. In line with this metabolic background, the first studies in HD patients demonstrated a substantial normalisation of the dyslipidemic plasma profile and reduced progression of renal disease [121, 122] and in one study also reduced mortality [123] in these patients.

In contrast and quite surprisingly, three previously conducted large, randomized, placebo-controlled trials on statin treatment in CKD patients had not led to significant benefits regarding their primary cardiovascular outcome. Two of those studies, the German Diabetes Dialysis (4D) Study and the Study to Evaluate the Use of Rusovastatin in Subjects on Regular Hemodialysis (AURORA) were performed on HD patients, one study, the Assessment of Lescol in Renal Transplantation (ALERT) Study on patients who had undergone kidney transplantation. Their primary endpoints (death of cardiovascular cause, nonfatal myocardial infarction or nonfatal stroke) were virtually unchanged [124-126]. However, there has been a promising risk reduction in the secondary endpoint 'all cardiac events combined' in one study [126]. A simulated study of exactly the same trial using a large historical database with more than 10.000 patients also demonstrated that statin use was associated with some benefit [127]. A comprehensive review of outcome data from the 4D and AURORA trials found no benefit of statin therapy in either the whole study group of HD patients or after stratification for inflammatory marker levels [128]. More recently, another, much larger trial including 9270 patients with chronic kidney disease, the Study of Heart and Renal Protection (SHARP) could show a significant risk reduction in cardiovascular events in a mixed population of patients with kidney disease including 2/3 predialysis and 1/3 HD patients treated with a combination of a statin and ezitimibe. This effect did not differ between HD and predialysis patients [129].

Based on our previous studies on lipoprotein kinetics in HD patients and the above-described conflicting results regarding their cardiovascular risk profile after statin treatment, we examined by stable-isotope technology the in vivo kinetics of apoB-containing particles in HD patients before and after treatment with atorvastatin (Schwaiger et al., unpublished).

In this study we described for the first time effects of HMG-CoA reductase inhibition on apoB metabolism in CKD patients treated with HD. Low-dose atorvastatin, given for three months to six male patients, lowered, as expected, concentrations of VLDL- and LDL-apoB, both accompanied by a significant increase of their FCR, while hepatic production of both apolipoproteins was not altered. This led, as expected, to a lower RT of these atherogenic apoB-containing particles comparable to RT values of healthy subjects with normal kidney function. The observed findings therefore argue for a beneficial effect of statin therapy regarding cardiovascular events in HD patients similar to described for the general population.

To understand why statins have surprisingly failed to reduce cardiovascular events in HD patients, the basic mechanisms underlying the pathophysiology of CVD in CKD must be critically considered. In contrast to the general population, CKD patients suffer, in addition to dyslipidemia, from several further complex comorbid conditions including diabetes mellitus, hypertension, oxidative stress, inflammation, insulin resistance, anemia and disturbances in mineral metabolism. Lipid lowering therapy by statins have the potential to ameliorate only some but no all of those conditions (see review [130]). Taken together, statin therapy in CKD maybe recommended based on our kinetic studies on apoB-containing lipoproteins, optimally combined with medication to treat atherogenic non-lipid factors in HD patients.

4. Conclusion

Kinetic in vivo studies in human subjects are superior to many methodological approaches including animal and cell culture models and thus represent the ultimate approach to understand basic metabolic pathways in humans. They have clearly revolutionized human lipoprotein research and have particularly resulted in novel insights into the metabolism of atherogenic apoB-containing lipoproteins some of which have been the subject of our previous investigations and object of this review.

Author details

Benjamin Dieplinger*
Department of Laboratory Medicine, Konventhospital Barmherzige Brüder Linz, Austria

Hans Dieplinger**
Division of Genetic Epidemiology, Department of Human Genetics and Molecular Pharmacology, Medical University of Innsbruck, Austria

Acknowledgement

We would like to thank Ramona Berberich, Linda Fineder, Michael E. Frischmann, Tatsuo Hosoya, Katsunori Ikewaki, Paul König, Florian Kronenberg, Seibu Mochizuki, Yoshinobu Nakada, Ulrich Neyer, Keio Okubo, Hermann Salmhofer Jürgen R. Schäfer, Johannes P. Schwaiger, Horst Schweer, Alex Starke, Evi Trenkwalder and Emanuel Zitt for invaluable contributions over many years within the various projects covered in this chapter. Funding support from the Austrian Science Fund (P10090-MED, P12358-MED), the Austrian National Bank (6721/4, 9331) and Pfizer-Austria (ATV-A-02-007G) is greatly appreciated.

5. References

[1] Mahley RW, Innerarity TL, Rall SCJ, Weisgraber KH. Plasma lipoproteins: Apolipoprotein structure and function. J Lipid Res 1984;25:1277-1294.
[2] Kwan BC, Kronenberg F, Beddhu S, Cheung AK. Lipoprotein metabolism and lipid management in chronic kidney disease. J Am Soc Nephrol 2007;18:1246-1261.
[3] Williams KJ, Tabas I. The response-to-retention hypothesis of early atherogenesis. Arterioscler Thromb Vasc Biol 1995;15:551-561.

* benjamin.dieplinger@bs-lab.at
** Corresponding Author, hans.dieplinger@i-med.ac.at

[4] Boren J, Taskinen MR, Adiels M. Kinetic studies to investigate lipoprotein metabolism. J Intern Med 2012;271:166-173.

[5] Adiels M, Olofsson SO, Taskinen MR, Boren J. Overproduction of very low-density lipoproteins is the hallmark of the dyslipidemia in the metabolic syndrome. Arterioscler Thromb Vasc Biol 2008;28:1225-1236.

[6] Jawien J, Nastalek P, Korbut R. Mouse models of experimental atherosclerosis. J Physiol Pharmacol 2004;55:503-517.

[7] Javitt NB. Hep G2 cells as a resource for metabolic studies: Lipoprotein, cholesterol, and bile acids. FASEB J 1990;4:161-168.

[8] Foley RN, Parfrey PS, Sarnak MJ. Clinical epidemiology of cardiovascular disease in chronic renal disease. Am J Kidney Dis 1998;32:S112-119.

[9] Krempler F, Kostner GM, Bolzano K, Sandhofer F. Turnover of lipoprotein(a) in man. J Clin Invest 1980;65:1483-1490.

[10] Barrett PH, Chan DC, Watts GF. Thematic review series: patient-oriented research. Design and analysis of lipoprotein tracer kinetics studies in humans. J Lipid Res 2006;47:1607-1619.

[11] Ikewaki K, Rader DJ, Sakamoto T, et al. Delayed catabolism of high density lipoprotein apolipoproteins A-I and A-II in human cholesteryl ester transfer protein deficiency. J Clin Invest 1993;92:1650-1658.

[12] Ikewaki K, Schaefer JR, Frischmann ME, et al. Delayed in vivo catabolism of intermediate-density lipoprotein and low-density lipoprotein in hemodialysis patients as potential cause of premature atherosclerosis. Arterioscler Thromb Vasc Biol 2005;25:2615-2622.

[13] Barrett PH, Bell BM, Cobelli C, et al. SAAM II: Simulation, analysis, and modeling software for tracer and pharmacokinetic studies. Metabolism 1998;47:484-492.

[14] Cobelli C, Foster DM. Compartmental models: theory and practice using the SAAM II software system. Adv Exp Med Biol 1998;445:79-101.

[15] Utermann G. Lipoprotein(a). In: Scriver C. R., Beaudet A. L., Sly W. S.,Valle D. (eds), The metabolic bases of inherited disease, New York, McGraw Hill Inc., 2001:2753-2787.

[16] Clarke R, Peden JF, Hopewell JC, et al. Genetic variants associated with Lp(a) lipoprotein level and coronary disease. N Engl J Med 2009;361:2518-2528.

[17] Kraft HG, Köchl S, Menzel HJ, Sandholzer C, Utermann G. The apolipoprotein(a) gene: a transcribed hypervariable locus controlling plasma lipoprotein(a) concentration. Hum Genet 1992;90:220-230.

[18] Schmidt K, Kraft HG, Parson W, Utermann G. Genetics of the Lp(a)/apo(a) system in an autochthonous Black African population from the Gabon. Eur J Hum Genet 2006;14:190-201.

[19] Bostom AG, Gagnon DR, Cupples LA, et al. A prospective investigation of elevated lipoprotein (a) detected by electrophoresis and cardiovascular disease in women: The Framingham Heart Study. Circulation 1994;90:1688-1695.

[20] Erqou S, Kaptoge S, Perry PL, et al. Lipoprotein(a) concentration and the risk of coronary heart disease, stroke, and nonvascular mortality. JAMA 2009;302:412-423.

[21] Rader DJ, Cain W, Ikewaki K, et al. The inverse association of plasma lipoprotein(a) concentrations with apolipoprotein(a) isoform size is not due to differences in Lp(a) catabolism but to differences in production rate. J Clin Invest 1994;93:2758-2763.

[22] Dube JB, Boffa MB, Hegele RA, Koschinsky ML. Lipoprotein(a): more interesting than ever after 50 years. Curr Opin Lipidol 2012;23:133-140.

[23] Dieplinger H, Utermann G. The seventh myth of lipoprotein(a): where and how is it assembled? Curr Opin Lipidol 1999;10:275-283.

[24] Brunner C, Kraft HG, Utermann G, Müller HJ. Cys4057 of apolipoprotein(a) is essential for lipoprotein(a) assembly. Proc Natl Acad Sci U.S.A. 1993;90:11643-11647.

[25] Trieu VN, McConathy WJ. A two-step model for lipoprotein(a) formation. J Biol Chem 1995;270:15471-15474.

[26] McCormick SP, Ng JK, Taylor S, Flynn LM, Hammer RE, Young SG. Mutagenesis of the human apolipoprotein B gene in a yeast artificial chromosome reveals the site of attachment for apolipoprotein(a). Proc Natl Acad Sci U.S.A. 1995;92:10147-10151.

[27] Becker L, Cook PM, Koschinsky ML. Identification of sequences in apolipoprotein(a) that maintain its closed conformation: a novel role for apo(a) isoform size in determining the efficiency of covalent Lp(a) formation. Biochemistry 2004;43:9978-9988.

[28] Becker L, Nesheim ME, Koschinsky ML. Catalysis of covalent Lp(a) assembly: evidence for an extracellular enzyme activity that enhances disulfide bond formation. Biochemistry 2006;45:9919-9928.

[29] Gething MJ, Sambrook J. Protein folding in the cell. Nature 1992;355:33-45.

[30] Hurtley SM, Helenius A. Protein oligomerization in the endoplasmic reticulum. Annu Rev Cell Biol 1989;5:277-307.

[31] White AL, Lanford RE. Biosynthesis and metabolism of lipoprotein(a). Curr Opin Lipidol 1995;6:75-80.

[32] White AL, Rainwater DL, Lanford RE. Intracellular maturation of apolipoprotein[a] and assembly of lipoprotein[a] in primary baboon hepatocytes. J Lipid Res 1993;34:509-517.

[33] Lobentanz EM, Krasznai K, Gruber A, et al. Intracellular metabolism of human apolipoprotein(a) in stably transfected Hep G2 cells. Biochemistry 1998;37:5417-5425.

[34] Koschinsky ML, Côté GP, Gabel B, Van der Hoek YY. Identification of the cysteine residue in apolipoprotein(a) that mediates extracellular coupling with apolipoprotein B-100. J Biol Chem 1993;268:19819-19825.

[35] Bonen DK, Hausman AML, Hadjiagapiou C, Skarosi SF, Davidson NO. Expression of a recombinant apolipoprotein(a) in HepG2 cells. Evidence for intracellular assembly of lipoprotein(a). J Biol Chem 1997;272:5659-5667.

[36] Dashti N, Alaupovic P, Knight-Gibson C, Koren E. Identification and partial characterization of discrete Apolipoprotein B containing lipoprotein particles produced by human hepatoma cell line HepG2. Biochemistry 1987;26:4837-4846.

[37] Krempler F, Kostner GM, Bolzano K, Sandhofer F. Lipoprotein(a) is not a metabolic product of other lipoproteins containing apolipoprotein B. Biochim Biophys Acta 1979;575:63-70.

[38] Krempler F, Kostner G, Bolzano K, Sandhofer F. Studies on the metabolism of the lipoprotein Lp(a) in man. Atherosclerosis 1978;30:57-65.

[39] Morrisett J, Gaubatz J, Nava L, et al. Metabolism of apo(a) and apoB-100 in human lipoprotein(a). In: Catapano A., Gotto Jr. A. M., Smith L. C.,Paoletti R. (eds), Drugs affecting lipid metabolism, Dordrecht, The Netherlands, Kluwer Academic Publishers and Fondazione Giovanni Lorenzini, 1993:161-167.

[40] Su W, Campos H, Judge H, Walsh BW, Sacks FM. Metabolism of Apo(a) and ApoB100 of lipoprotein(a) in women: effect of postmenopausal estrogen replacement. J Clin Endocrinol Metab 1998;83:3267-3276.

[41] Demant T, Seeberg K, Bedynek A, Seidel D. The metabolism of lipoprotein(a) and other apolipoprotein B-containing lipoproteins: a kinetic study in humans. Atherosclerosis 2001;157:325-339.

[42] Jenner JL, Seman LJ, Millar JS, et al. The metabolism of apolipoproteins (a) and B-100 within plasma lipoprotein (a) in human beings. Metabolism 2005;54:361-369.

[43] Frischmann ME, Ikewaki K, Trenkwalder E, et al. In vivo stable-isotope kinetic study suggests intracellular assembly of lipoprotein(a). Atherosclerosis 2012;(in press).

[44] Frischmann ME, Kronenberg F, Trenkwalder E, et al. In vivo turnover study demonstrates diminished clearance of lipoprotein(a) in hemodialysis patients. Kidney Int 2007;71:1036-1043.

[45] Shames DM, Havel RJ. De novo production of low density lipoproteins: Fact or fancy. J Lipid Res 1991;32:1099-1112.

[46] Pietzsch J, Wiedemann B, Julius U, et al. Increased clearance of low density lipoprotein precursors in patients with heterozygous familial defective apolipoprotein B-100: a stable isotope approach. J Lipid Res 1996;37:2074-2087.

[47] Packard CJ, Demant T, Stewart JP, et al. Apolipoprotein B metabolism and the distribution of VLDL and LDL subfractions. J Lipid Res 2000;41:305-318.

[48] Bouma ME, Pessah M, Renaud G, Amit N, Catala D, Infante R. Synthesis and secretion of lipoproteins by human hepatocytes in culture. In Vitro Cell Dev Biol 1988;24:85-90.

[49] Kagawa A, Azuma H, Akaike M, Kanagawa Y, Matsumoto T. Aspirin reduces apolipoprotein(a) (apo(a)) production in human hepatocytes by suppression of apo(a) gene transcription. J Biol Chem 1999;274:34111-34115.

[50] Kronenberg F, Kuen E, Ritz E, et al. Lipoprotein(a) serum concentrations and apolipoprotein(a) phenotypes in mild and moderate renal failure. J Am Soc Nephrol 2000;11:105-115.

[51] Kronenberg F, Kuen E, Ritz E, et al. Apolipoprotein A-IV serum concentrations are elevated in patients with mild and moderate renal failure. J Am Soc Nephrol 2002;13:461-469.

[52] Krentz AJ. Lipoprotein abnormalities and their consequences for patients with type 2 diabetes. Diabetes Obes Metab 2003;5 Suppl 1:S19-27.

[53] Kronenberg F. Dyslipidemia and nephrotic syndrome: recent advances. J Ren Nutr 2005;15:195-203.

[54] Shoji T, Ishimura E, Inaba M, Tabata T, Nishizawa Y. Atherogenic lipoproteins in end-stage renal disease. Am J Kidney Dis 2001;38:S30-33.

[55] Dieplinger H, Schoenfeld PY, Fielding CJ. Plasma cholesterol metabolism in end-stage renal disease. Difference between treatment by hemodialysis or peritoneal dialysis. J Clin Invest 1986;77:1071-1083.
[56] Batista MC, Welty FK, Diffenderfer MR, et al. Apolipoprotein A-I, B-100, and B-48 metabolism in subjects with chronic kidney disease, obesity, and the metabolic syndrome. Metabolism 2004;53:1255-1261.
[57] Appel G. Lipid abnormalities in renal disease. Kidney Int 1991;39:169-183.
[58] Chan MK, Persaud J, Varghese Z, Moorhead JF. Pathogenic roles of post-heparin lipases in lipid abnormalities in hemodialysis patients. Kidney Int 1984;25:812-818.
[59] Oi K, Hirano T, Sakai S, Kawaguchi Y, Hosoya T. Role of hepatic lipase in intermediate-density lipoprotein and small, dense low-density lipoprotein formation in hemodialysis patients. Kidney Int Suppl 1999;71:S227-228.
[60] Brown MS, Goldstein JL. A receptor-mediated pathway for cholesterol homeostasis. Science 1986;232:34-47.
[61] Littlewood TD, Bennett MR. Apoptotic cell death in atherosclerosis. Curr Opin Lipidol 2003;14:469-475.
[62] Dieplinger H, Lackner C, Kronenberg F, et al. Elevated plasma concentrations of lipoprotein(a) in patients with end-stage renal disease are not related to the size polymorphism of apolipoprotein(a). J Clin Invest 1993;91:397-401.
[63] Kronenberg F, König P, Neyer U, et al. Multicenter study of lipoprotein(a) and apolipoprotein(a) phenotypes in patients with end-stage renal disease treated by hemodialysis or continous ambulatory peritoneal dialysis. J Am Soc Nephrol 1995;6:110-120.
[64] Milionis HJ, Elisaf MS, Tselepis A, Bairaktari E, Karabina SA, Siamopoulos KC. Apolipoprotein(a) phenotypes and lipoprotein(a) concentrations in patients with renal failure. Am J Kidney Dis 1999;33:1100-1106.
[65] Stenvinkel P, Heimburger O, Tuck CH, Berglund L. Apo(a)-isoform size, nutritional status and inflammatory markers in chronic renal failure. Kidney Int 1998;53:1336-1342.
[66] Zimmermann J, Herrlinger S, Pruy A, Metzger T, Wanner C. Inflammation enhances cardiovascular risk and mortality in hemodialysis patients. Kidney Int 1999;55:648-658.
[67] Kronenberg F, Lingenhel A, Lhotta K, et al. The apolipoprotein(a) size polymorphism is associated with nephrotic syndrome. Kidney Int 2004;65:606-612.
[68] Wanner C, Rader D, Bartens W, et al. Elevated plasma lipoprotein(a) in patients with the nephrotic syndrome. Ann Intern Med 1993;119:263-269.
[69] de Sain-van der Velden MG, Reijngoud DJ, Kaysen GA, et al. Evidence for increased synthesis of lipoprotein(a) in the nephrotic syndrome. J Am Soc Nephrol 1998;9:1474-1481.
[70] Kronenberg F, König P, Lhotta K, et al. Apolipoprotein(a) phenotype-associated decrease in lipoprotein(a) plasma concentrations after renal transplantation. Arterioscler Thromb 1994;14:1399-1404.
[71] Kronenberg F, Lhotta K, Konig P, Margreiter R, Dieplinger H, Utermann G. Apolipoprotein(a) isoform-specific changes of lipoprotein(a) after kidney transplantation. Eur J Hum Genet 2003;11:693-699.

[72] Kerschdorfer L, Konig P, Neyer U, et al. Lipoprotein(a) plasma concentrations after renal transplantation: a prospective evaluation after 4 years of follow-up. Atherosclerosis 1999;144:381-391.

[73] Lindner A, Charra B, Sherrard DJ, Scribner BH. Accelerated atherosclerosis in prolonged maintenance hemodialysis. N Engl J Med 1974;290:697-701.

[74] Wilson PW, D'Agostino RB, Levy D, Belanger AM, Silbershatz H, Kannel WB. Prediction of coronary heart disease using risk factor categories. Circulation 1998;97:1837-1847.

[75] Cheung AK, Sarnak MJ, Yan G, et al. Atherosclerotic cardiovascular disease risks in chronic hemodialysis patients. Kidney Int 2000;58:353-362.

[76] Guz G, Nurhan Ozdemir F, Sezer S, et al. Effect of apolipoprotein E polymorphism on serum lipid, lipoproteins, and atherosclerosis in hemodialysis patients. Am J Kidney Dis 2000;36:826-836.

[77] Koch M, Kutkuhn B, Trenkwalder E, et al. Apolipoprotein B, fibrinogen, HDL cholesterol, and apolipoprotein(a) phenotypes predict coronary artery disease in hemodialysis patients. J Am Soc Nephrol 1997;8:1889-1898.

[78] Stack AG, Bloembergen WE. Prevalence and clinical correlates of coronary artery disease among new dialysis patients in the United States: a cross-sectional study. J Am Soc Nephrol 2001;12:1516-1523.

[79] Cressman MD, Heyka RJ, Paganini EP, O'Neil J, Skibinski CI, Hoff HF. Lipoprotein(a) is an independent risk factor for cardiovascular disease in hemodialysis patients. Circulation 1992;86:475-482.

[80] Degoulet P, Legrain M, Reach I, et al. Mortality risk factors in patients treated by chronic hemodialysis. Report of the Diaphane collaborative study. Nephron 1982;31:103-110.

[81] Hocher B, Ziebig R, Altermann C, et al. Different impact of biomarkers as mortality predictors among diabetic and nondiabetic patients undergoing hemodialysis. J Am Soc Nephrol 2003;14:2329-2337.

[82] Iseki K, Fukiyama K. Predictors of stroke in patients receiving chronic hemodialysis. Kidney Int 1996;50:1672-1675.

[83] Koda Y, Nishi S, Suzuki M, Hirasawa Y. Lipoprotein(a) is a predictor for cardiovascular mortality of hemodialysis patients. Kidney Int Suppl 1999;71:S251-S253.

[84] Kronenberg F, Neyer U, Lhotta K, et al. The low molecular weight apo(a) phenotype is an independent predictor for coronary artery disease in hemodialysis patients: a prospective follow-up. J Am Soc Nephrol 1999;10:1027-1036.

[85] Ohashi H, Oda H, Ohno M, Watanabe S, Sakata S. Lipoprotein(a) as a risk factor for coronary artery disease in hemodialysis patients. Kidney Int Suppl 1999;71:S242-S244.

[86] Schwaiger JP, Lamina C, Neyer U, et al. Carotid plaques and their predictive value for cardiovascular disease and all-cause mortality in hemodialysis patients considering renal transplantation: a decade follow-up. Am J Kidney Dis 2006;47:888-897.

[87] Shoji T, Emoto M, Shinohara K, et al. Diabetes mellitus, aortic stiffness, and cardiovascular mortality in end-stage renal disease. J Am Soc Nephrol 2001;12:2117-2124.

[88] Kalantar-Zadeh K, Kopple JD, Kilpatrick RD, et al. Association of morbid obesity and weight change over time with cardiovascular survival in hemodialysis population. Am J Kidney Dis 2005;46:489-500.

[89] Kalantar-Zadeh K, Kilpatrick RD, McAllister CJ, Greenland S, Kopple JD. Reverse epidemiology of hypertension and cardiovascular death in the hemodialysis population: the 58th annual fall conference and scientific sessions. Hypertension 2005;45:811-817.

[90] Nishizawa Y, Shoji T, Ishimura E, Inaba M, Morii H. Paradox of risk factors for cardiovascular mortality in uremia: is a higher cholesterol level better for atherosclerosis in uremia? Am J Kidney Dis 2001;38:S4-7.

[91] Kalantar-Zadeh K. What is so bad about reverse epidemiology anyway? Semin Dial 2007;20:593-601.

[92] Kalantar-Zadeh K, Block G, Humphreys MH, Kopple JD. Reverse epidemiology of cardiovascular risk factors in maintenance dialysis patients. Kidney Int 2003;63:793-808.

[93] Lowrie EG, Lew NL. Death risk in hemodialysis patients: the predictive value of commonly measured variables and an evaluation of death rate differences between facilities. Am J Kidney Dis 1990;15:458-482.

[94] Longenecker JC, Coresh J, Powe NR, et al. Traditional cardiovascular disease risk factors in dialysis patients compared with the general population: the CHOICE Study. J Am Soc Nephrol 2002;13:1918-1927.

[95] Longenecker JC, Klag MJ, Marcovina SM, et al. Small apolipoprotein(a) size predicts mortality in end-stage renal disease: The CHOICE study. Circulation 2002;106:2812-2818.

[96] Kronenberg F, Kathrein H, König P, et al. Apolipoprotein(a) phenotypes predict the risk for carotid atherosclerosis in patients with end-stage renal disease. Arterioscler Thromb 1994;14:1405-1411.

[97] Longenecker JC, Klag MJ, Marcovina SM, et al. High lipoprotein(a) levels and small apolipoprotein(a) size prospectively predict cardiovascular events in dialysis patients. J Am Soc Nephrol 2005;16:1794-1802.

[98] Kronenberg F, Ikewaki K, Schaefer JR, Konig P, Dieplinger H. Kinetic studies of atherogenic lipoproteins in hemodialysis patients: do they tell us more about their pathology? Semin Dial 2007;20:554-560.

[99] Pietzsch J, Lattke P, Julius U. Oxidation of apolipoprotein B-100 in circulating LDL is related to LDL residence time. In vivo insights from stable-isotope studies. Arterioscler Thromb Vasc Biol 2000;20:E63-67.

[100] Boaz M, Smetana S, Weinstein T, et al. Secondary prevention with antioxidants of cardiovascular disease in endstage renal disease (SPACE): randomised placebo-controlled trial. Lancet 2000;356:1213-1218.

[101] Tepel M, van der Giet M, Statz M, Jankowski J, Zidek W. The antioxidant acetylcysteine reduces cardiovascular events in patients with end-stage renal failure: a randomized, controlled trial. Circulation 2003;107:992-995.

[102] Shoji T, Nishizawa Y, Kawagishi T, et al. Intermediate-density lipoprotein as an independent risk factor for aortic atherosclerosis in hemodialysis patients. J Am Soc Nephrol 1998;9:1277-1284.

[103] Vega GL, Grundy SM. Influence of lovastatin therapy on metabolism of low density lipoproteins in mixed hyperlipidaemia. J Intern Med 1991;230:341-350.

[104] Hörkkö S, Huttunen K, Korhonen T, Kesäniemi YA. Decreased clearance of low-density lipoprotein in patients with chronic renal failure. Kidney Int 1994;45:561-570.

[105] Hörkkö S, Huttunen K, Kesäniemi YA. Decreased clearance of low-density lipoprotein in uremic patients under dialysis treatment. Kidney Int 1995;47:1732-1740.

[106] Prinsen BH, Rabelink TJ, Romijn JA, et al. A broad-based metabolic approach to study VLDL apoB100 metabolism in patients with ESRD and patients treated with peritoneal dialysis. Kidney Int 2004;65:1064-1075.

[107] Chan PC, Persaud J, Varghese Z, Kingstone D, Baillod RA, Moorhead JF. Apolipoprotein B turnover in dialysis patients: its relationship to pathogenesis of hyperlipidemia. Clin Nephrol 1989;31:88-95.

[108] Ericsson S, Eriksson M, Vitols S, Einarsson K, Berglund L, Angelin B. Influence of age on the metabolism of plasma low density lipoproteins in healthy males. J Clin Invest 1991;87:591-596.

[109] Millar JS, Lichtenstein AH, Cuchel M, et al. Impact of age on the metabolism of VLDL, IDL, and LDL apolipoprotein B-100 in men. J Lipid Res 1995;36:1155-1167.

[110] Quaschning T, Koniger M, Kramer-Guth A, et al. Receptor-mediated lipoprotein uptake by human glomerular cells: comparison with skin fibroblasts and HepG2 cells. Nephrol Dial Transplant 1997;12:2528-2536.

[111] Pegoraro AA, Gudehithlu KP, Cabrera E, et al. Handling of low-density lipoprotein by the renal tubule: release of fragments due to incomplete degradation. J Lab Clin Med 2002;139:372-378.

[112] Kronenberg F, Trenkwalder E, Lingenhel A, et al. Renovascular arteriovenous differences in Lp[a] plasma concentrations suggest removal of Lp[a] from the renal circulation. J Lipid Res 1997;38:1755-1763.

[113] Kostner KM, Maurer G, Huber K, et al. Urinary excretion of apo(a) fragments. Role in apo(a) catabolism. Arterioscler Thromb Vasc Biol 1996;16:905-911.

[114] Mooser V, Seabra MC, Abedin M, Landschulz KT, Marcovina S, Hobbs HH. Apolipoprotein(a) kringle 4-containing fragments in human urine. Relationship to plasma levels of lipoprotein(a). J Clin Invest 1996;97:858-864.

[115] Kaysen GA, de Sain-van der Velden MG. New insights into lipid metabolism in the nephrotic syndrome. Kidney Int Suppl 1999;71:S18-21.

[116] Giordano M, De Feo P, Lucidi P, et al. Increased albumin and fibrinogen synthesis in hemodialysis patients with normal nutritional status. J Am Soc Nephrol 2001;12:349-354.

[117] Kaysen GA. Albumin turnover in renal disease. Miner Electrolyte Metab 1998;24:55-63.

[118] Shoji T, Fukumoto M, Kimoto E, et al. Antibody to oxidized low-density lipoprotein and cardiovascular mortality in end-stage renal disease. Kidney Int 2002;62:2230-2237.

[119] Randomised trial of cholesterol lowering in 4444 patients with coronary heart disease: the Scandinavian Simvastatin Survival Study (4S). Lancet 1994;344:1383-1389.
[120] Cannon CP, Braunwald E, McCabe CH, et al. Intensive versus moderate lipid lowering with statins after acute coronary syndromes. N Engl J Med 2004;350:1495-1504.
[121] Fried LF, Orchard TJ, Kasiske BL. Effect of lipid reduction on the progression of renal disease: a meta-analysis. Kidney Int 2001;59:260-269.
[122] Nishizawa Y, Shoji T, Tabata T, Inoue T, Morii H. Effects of lipid-lowering drugs on intermediate-density lipoprotein in uremic patients. Kidney Int Suppl 1999;71:S134-136.
[123] Seliger SL, Weiss NS, Gillen DL, et al. HMG-CoA reductase inhibitors are associated with reduced mortality in ESRD patients. Kidney Int 2002;61:297-304.
[124] Fellstrom BC, Jardine AG, Schmieder RE, et al. Rosuvastatin and cardiovascular events in patients undergoing hemodialysis. N Engl J Med 2009;360:1395-1407.
[125] Holdaas H, Holme I, Schmieder RE, et al. Rosuvastatin in diabetic hemodialysis patients. J Am Soc Nephrol 2011;22:1335-1341.
[126] Wanner C, Krane V, Marz W, et al. Atorvastatin in patients with type 2 diabetes mellitus undergoing hemodialysis. N Engl J Med 2005;353:238-248.
[127] Chan KE, Thadhani R, Lazarus JM, Hakim RM. Modeling the 4D Study: statins and cardiovascular outcomes in long-term hemodialysis patients with diabetes. Clin J Am Soc Nephrol 2010;5:856-866.
[128] Krane V, Wanner C. Statins, inflammation and kidney disease. Nat Rev Nephrol 2011;7:385-397.
[129] Baigent C, Landray MJ, Reith C, et al. The effects of lowering LDL cholesterol with simvastatin plus ezetimibe in patients with chronic kidney disease (Study of Heart and Renal Protection): a randomised placebo-controlled trial. Lancet 2011;377:2181-2192.
[130] Epstein M, Vaziri ND. Statins in the management of dyslipidemia associated with chronic kidney disease. Nat Rev Nephrol 2012;8:214-223.

Section 2

Diagnosis of Lipoprotein Disorders

The Importance of Lipid and Lipoprotein Ratios in Interpretetions of Hyperlipidaemia of Pregnancy

D.S. Mshelia and A.A. Kullima

Additional information is available at the end of the chapter

http://dx.doi.org/10.5772/46064

1. Introduction

In spite of the fact that the hyperlipidaemia of pregnancy is usually considered physiological [1-12], all pregnant women develop hypertriglyceridaemia with subsequent formation of small, dense low-density lipoprotein(LDL) particles, both of which are an independent risk factor of coronary heart disease(CHD) [13]. By 3rd trimester most women have a lipid profile which could be considered highly atherogenic in the nonpregnant state [14]. Similarly, animal model studies showed that maternal hypercholesterolaemia during pregnancy even when temporary and limited to pregnancy triggers pathogenic events in the fetal aorta, greatly enhanced atherogenesis later in life[14, 15]. On the other hand, intrauterine growth retardation (IUGR) has been associated with pre-eclampsia [16], as a result of decreased maternal lipid transfer to the fetus secondary to placental abnormalities. IUGR has also been associated with failure of development of hyperlipidaemia during pregnancy with subsequent reduction in maternal lipid reaching the fetus in a normal placenta [17, 18]. Generally, serum lipid and lipoprotein levels in pregnancy are modulated by complex interactions between genetics, medical complications of pregnancy, co-existing medical conditions, and other maternal factors [9, 19]. This underscores the need to take a meticulous and decisive approach in interpreting hyperlipidaemia of pregnancy. In searching for an emergent or new cardiovascular risk factor, concerning lipid and lipoprotein in adult males and nonpregnant women, several lipoprotein ratios or atherogenic indices have been defined[20]. These ratios were found to provide information on risk factors difficult to quantify by routine analyses and could be a better mirror of the metabolic and clinical interactions between lipid fractions[21]. Despite findings of [22] in a registry study of heterozygous familial hypercholesterolaemia(FH) mothers, who observed no significant untoward effect of lipid-lowering drugs during pregnancy, the current trend is that Statins, classified by FDA as category X, should be avoided in pregnancy[23, 24]. The use of lipid and lipoprotein ratios in interpreting pregnancy associated hyperlipidaemia may provide a balanced hyperlipidaemia not only in normal pregnancy but also in the other modulators of lipid metabolism in pregnancy.

2. Pathophysiology of hyperlipidaemia of pregnancy

Pregnancy is a dynamic state consequent of the fact that normal fetal development needs the availability of essential nutrients such as glucose, free fatty acids(FFAs), long-chain polyunsaturated fatty acids(LCPUFAs), amino acids, minerals, vitamins, to be continuously supplied to the growing fetus despite intermittent maternal food intake[10,25]. The dynamism of the gestational period support fetal growth and development while maintaining maternal homeostasis and preparation for lactation. This is achieved by complex and continuously evolving adjustments in maternal nutrient metabolism occurring throughout gestation.

Many of these maternal adjustments occur in the early stages of pregnancy when the fetus is too small to make considerable metabolic demands of the mother, resulting in the maternal metabolism working from a different baseline compared with the nonpregnant state. This period is called the anabolic phase. In late pregnancy, however, the maternal metabolic processes become more complicated because of the two-way interaction between the mother and the developing fetus. This is caked the catabolic phase.

The changes in nutrient metabolism can be described by several general concepts[8]: (a) adjustments in nutrient metabolism are driven by hormonal changes, fetal demands and maternal nutrient supply; (b) more than one potential adjustment exists for each nutrient; (c) maternal behavioural changes augment physiologic adjustment; and (d) a limit exists in the physiologic capacity to adjust nutrient metabolism to meet pregnancy needs, which when exceeded, fetal growth and development are impaired. Subsequently, metabolic adaptations, during pregnancy are essential [26]: 1, To ensure adequate growth and development of the fetus; 2, to provide the fetus with adequate energy stores and substrates that are needed following birth; 3, and, to provide the mother with sufficient energy stores and substrates to cope with the demands of pregnancy as well as those of labour and lactation. One of the maternal metabolic adjustments during pregnancy includes accumulation of fat depots in maternal tissues[26]. During this anabolic phase, the number of insulin receptors on the adipocytes increases, culminating into increased insulin sensitivity, increase lipoprotein lipase(LPL) activity which hydrolyses circulating triglycerides for tissue uptake, enhanced lipogenesis and marked maternal fat deposition(about 3.5 to 6.0kg) which is used as energy sources for the mother so that glucose is spared for the developing fetus in the catabolic part of the pregnancy[27, 28]. Lean women increase their fat stores more than obese women per kg body weight, likely due to higher insulin sensitivity in them, in early pregnancy, promoting lipid uptake and de novo synthesis.

The important attributes of fat deposits during the anabolic phase in pregnant women are :(1) Hyperphagia, present in pregnant women and increases as gestational time advances. This progressive increase in the availability of exogenous substrates actively contributes to maternal accumulation of fat depots [29]; (2) Promotion of lipogenesis and suppression of lipolysis mediated by progressive increase in insulin and its sensitivity and enhanced by progesterone and cortisol [30]; (3) The proportional increase in adipose tissue lipoprotein lipase (LPL) activity [1,12,31) which hydrolyzes triglycerides(TGs) in form of TG-rich lipoproteins, chylomicron and very-low density lipoprotein(VLDL), which are respectively converted into remnant particles and intermediate-density lipoprotein (IDL). The hydrolytic products, non-esterified fatty acids(NEFA) and glycerol, are partially taken up by subjacent tissues [11, 12, 32, 33); (4) the unique capacity of tissue to utilize intracellularly the glycerol released during lipolysis. Under normal circumstances, the negligible glycerol kinase activity in adipose tissues hampers the utilization of glycerol for glycerol-3-phosphate synthesis and its use for the synthesis of TGs [34,35]. However, an increase in glycerol kinase activity and its subsequent capacity to metabolize glycerol has been found in rodents under condition of hyperinsulinaemia and enhanced fat accumulation, such as occurs in obesity [35, 36]. The lower lipolytic activity together with the augmented capacity of the tissues for the synthesis of glycerol-3-phosphate for uses in TG synthesis from both glucose and intracellular released glycerol results in net intracellular accumulations of TGs. Since all these pathways are stimulated by insulin, it is proposed that the enhanced insulin responsiveness [37] in the presence of an augmented response of the pancreatic β-cells to the insulinotropic stimulus of glucose that has been found in early pregnant women [38] would be the principal driving forces for the net fat depot accumulation at this stage of pregnancy. These ultimately lead to maternal fat accumulations in the anabolic phase of gestation.

The anabolic condition of adipose tissue during early pregnancy switches to a net catabolic state during the last 1/3 of gestation. The signals responsible for this switch from lipid storage to lipid mobilization are not well understood; however, placental hormones that increase with advancing gestation, known to induce maternal insulin resistance, may play a major role. Placental growth hormone, human placental lactogen, leptin, and tumour necrosis factor-α(TNF-α) are placental hormones that induce insulin resistance. The presence of high plasma levels of placental hormones, known to have lipolytic effects, human placental lipase (HPL), an augmented production of catecholamine secondary to maternal hypoglycemia [38], and the insulin-resistant condition present at this stage [39, 40], appear to be responsible for the net breakdown of maternal fat depots, consistently causing increments of plasma nonesterified fatty acids(NEFA) and glycerol levels during the 3rd trimester of gestation. The main destination of these lipolytic products released from maternal adipose tissue is the maternal liver. They are converted in the liver into their respective active forms, acyl-CoA and glycerol-3-phosphate, to become partially re-esterified for the synthesis of triglycerides, which are transferred to native VLDL particles and released into the circulation. Acyl-CoA can also be converted throughout the β-oxidation pathway to acetyl-CoA for energy production and ketone body synthesis, whereas glycerol may also be used for glucose synthesis. Fetal-placental glucose and amino acids utilization rates are highest at 22 to 26weeks decreasing near term. In contrast lipid transport is maximal in the 3rd trimester coincident with rapid fetal fat accretion, this spares the mother to utilize glucose during this period. Humans are born with the highest percentage of fat (12 to 15%) compared to any species and 90% deposition occurs in the last 10weeks of gestation, exponentially increasing to 7g/day near term. The preferential use of glycerol released from maternal adipose tissue for gluconeogenesis acquire greater importance during maternal fasting period, when circulating glucose levels are lower than under nonpregnant conditions[34]. Under fed condition in early gestation, plasma ketone body values are even lower in pregnant than in nonpregnant condition [41], indicating an enhanced use of these fuels by maternal tissues as alternative substrate to glucose. However, during fasting period maternal ketogenesis become highly accelerated, as indicated by the exaggerated increase in plasma ketone bodies that occur [41]. This benefit the fetus in two ways: (1) ketone bodies are used by maternal tissues, thus, saving glucose for essential function and delivery to the fetus, (2) placental transfer of ketone bodies is very efficient, attaining the same concentration in fetal plasma as in maternal circulation[42]. In addition, ketone bodies may be used by the fetus as oxidative fuels as well as substrate for brain lipid synthesis [43].

Insulin is well known to inhibits adipose tissue lipolytic activity, hepatic gluconeogenesis and ketogenesis but to increases adipose tissue LPL activity. Thus, it is not surprising that all of these pathways change in the opposite direction which is consistent with insulin resistance occurring in later part of pregnancy. These pathways become even further modified under uncontrolled gestational diabetes mellitus(GDM), where insulin resistance is further enhanced [44].

3. Maternal hyperlipidaemia of pregnancy

The enhanced net breakdown of fat depots during late pregnancy is associated with hyperlipidaemia, which chiefly corresponds to plasma rises in TGs with smaller rise in phospholipids and cholesterol [44]. The greatest increased in plasma TGs corresponds to VLDL values but TGs also accumulates in other lipoprotein fractions, which do not normally transport them, such as LDL and HDL [45]. The high TGs concentration secondary to lipolysis in the presence of increased cholesteryl ester transfer protein(CETP) activity, occurring in midgestation[45], contributes to the accumulation of TGs in the lipoprotein fractions of higher densities, LDL and HDL[44, 45]. CETP facilitates the exchange of TGs by esterified cholesterol between VLDL and either LDL or HDL. Furthermore, during late gestation the activity of hepatic lipase (HL) greatly decreased [45]. HL converts the buoyant HDL-2-TG-rich particles into small HDL-3-TG-poor particle allowing a proportional accumulation of buoyant HDL-2-TG-rich particle.

Other hormonal dynamism occurring during pregnancy contributing to maternal hypertriglyceridaemia are, **table** 1, consistently increasing oestrogen concentration almost throughout the gestation period and oestrogen has been shown to (1) increase endogenous production of VLDL-TGs [46]; (2) reduce adipose tissue LPL activity [33, 45], and (3) inhibition of hepatic TG lipase activity [33, 44]. Thus, the oestrogenic influence over TG metabolism suggests an increased circulating VLDL-TG. Although the role of progesterone in TG metabolism is not certain, its administration in rats had a lipid neutral effect. Thus, the interaction between oestrogen and progesterone would favour hypertriglyceridaemia. Prolactin may inhibit adipose tissue LPL while stimulating breast LPL in late gestation [45]. Thus, the physiologic outcome of increasing concentration of Prolactin with advancing pregnancy would be a shift in storage from the adipocytes to the breast in preparation for lactation.

Enzymes	Activities
Adipose tissue lipoprotein lipase(LPL)	decrease
Diacylglycerol acetyltransferase	decrease
Cholesterol 7-alpha hydroxylase	decrease
Placental lipoprotein lipase (PLPL)	increase
Placental triglycerides hydroxylase	Increase
Phospholipase A2(PLA2)	Increase
Cholesterol ester transfer protein(CETP)	Increase
Hepatic lipase	Decrease
Hormones and cytokines	Concentrations
Estradiol	Increase
Insulin	Sensitivities increase during first trimester but subsequently decreases from second trimester to end of gestation
Human placental lactogen	Increase
Prolactin	Increase
Cortisol	Increase
Glucagon	Increase
Porgesterone	Increase
Leptin	Increase
Tumor Necrosis Factor-alpha(TNF-alpha)	Increase
Human chorionic somatomammotropin(HCS)	Increase

Table 1. Hormone and enzyme changes during the course of pregnancy.

The combined effects of enhanced liver production of VLDL [47, 48], decreased removal of these particles from the circulation due to low LPL activity [45,49], high CETP activity and low HL activity, would not only be responsible for the accumulation of TGs in LDL but also for the proportional accumulation of TG in buoyant TG-rich HDL-2b subfractions at the expense of the cholesterol-rich and TG-poor HDL-2a or HDL-3[45].

The increasing insulin-resistance in late gestation and continuously increasing plasma oestrogen levels occurring during pregnancy are the main hormonal factors responsible for these metabolic changes resulting into the development of maternal hypertriglyceridaemia, see table 2.

Lipid and lipoproteins (mg/dl)	First trimester	Second trimester	Third trimester	Nonpregnant controls
HDL-C	67±12	83±19	81±17	69±10
LDL-C	90±17	130±46	136±33	99±23
TGs	79±27	151±80	245±73	77±34
TC	173±18	243±53	267±30	183±23
ApoA-1	170±27	204±22	196±28	163±24
ApoA-2	49±7	52±6	49±5	47±6
ApoB	70±21	91±25	113±29	61±22
ApoC-11	265±13	299±18	314±21	237±11
ApoC-111	141±3	188±5	217±6	121±19
ApoE	41±12	42±9	49±19	42±20
Lp(a)	60(0-1440)	63(2-1210)	54(0-1230	86(11-473)
VLDL-1	19(12-55)	47(26-110)	109(38-170)	23(5-85)
VLDL-2	17(7-45)	36(20-77)	103946-168)	23(13-44)
IDL	26(13-54)	58(24-100)	124(79-157)	35(18-62)
Total LDL	200(135-323)	292(206-410)	353(244-534)	207(150-363)
LDL-1	33(16-52)	49(37-70)	67(27-96)	50(22-130)
LDL-11	143(95-231)	160(103-287)	201(59-316)	135(72-258
LDL-111	28(15-56)	32(24-165)	123(43-192)	31(5-68)

Table 2. The increasing lipid and lipoproteins during course of gestation(courtesy: Ahmet Basaran, MD)

4. Placental transfer of maternal lipid and lipoproteins and their metabolites to the fetus

The human placenta contains VLDL, LDL, HDL, and scavenger receptors as well as LDL receptor-related proteins. The placenta also has LPL, phospholipase-A2 (PLA-2) and intracellular lipase activities as well as plasma membrane fatty acid-binding protein (FABP/GOT2), fatty acid translocase (CD36), fatty acid transfer protein (FATP) and different cytoplasmic FABPs [29, 42,50, 51]. Thus, lipid and lipoproteins in maternal plasma can be taken up and handled by the placenta, allowing LCPUFAs associated with plasma lipoproteins to be transferred to the fetus. The human placenta is capable of transporting free fatty acids(FFAs) by diffusion and selectively increases the transport of essential fatty acids (EFAs) and their long-chain polyunsaturated fatty acids (LCPUFA) derivatives by fatty acid carrier proteins.

Although lipoprotein TGs does not directly cross the placental barrier, the placenta has mechanisms to release fatty acids(FAs) circulating in maternal plasma lipoproteins into the fetus. In addition, high levels of TGs in maternal circulation may create a steep concentration gradient across the placenta, which accelerates their transport and deposition in fetal tissues. In term human trophoblasts, insulin and fatty acids have been shown to enhance the expression of adipophylin, which is associated with cellular lipid droplets and implicated in cellular fatty acid uptake and storage of neutral lipids

4.1. Fatty acids transfer

The supply of LCPUFA is important for fetal growth and tissue development especially for the development of the nervous system and the considerable requirements of these LCPUFAs in the fetus must be provided by their placental transfer [52]. The plasma membrane fatty acid-binding proteins present in human placental membrane [51,52] are responsible for the preferential uptake of LCPUFAs. A selective cellular membrane of certain FAs may also contributes to the placental transfer process, as would the conversion of a certain proportion of arachidonic acid(AA) to prostaglandins(PGs)[52], the incorporation of some FAs into phospholipids[50-52], the oxidation of placental fatty acids[53] and the synthesis of FAs[52,53]. Even though essential fatty acids(EFA) as well as LCPUFAs are transferred across the placenta, the fetus needs to receive substantial amounts of preformed AA and docosahexaenoic acid(DHA) which can be synthesized to a limited extent from the EFA. The two dietary EFAs are linoeic acid(18:2ω-6) and α-linolenic acid(18:3ω-3), which are precursors of the ω-6 and ω-3 LCPUFA, respectively. The synthesis of AA and DHA do not take place in the fetus or the placenta in substantial amounts, owing to the low activities of the desaturating enzymes. Both AA and DHA are abundant in the brain and the retina and their appropriate supply during pregnancy and the neonatal period is critical for proper function [1,54]. Maternal plasma NEFA, though in smaller proportion than lipoprotein TGs, is an important source of polyunsaturated fatty acids(PUFAs) for the fetus [51,52]. Maternal plasma NEFAs correlates with those in the fetus and maternal adipocytokines have been associated with fetal growth[1]. The combination of these processes determines the actual rate of placental FAs transfer and its selectivity, consequent to the proportional enrichment of certain LCPUFAs, such as AA and DHA in fetal as compared with maternal compartments [52, 54].

4.2. Cholesterols

Cholesterol plays a key role in embryonic and fetal development hence the demands for cholesterol in the embryo and fetus is relatively high. Cholesterol is an essential component of cell membrane influencing the fluidity and passive permeability by interacting with phospholipids and sphingolipids [55]. It's the precursor of bile acids and steroid hormones. It is also required for cell proliferation and development of the growing body, cell differentiation, and cell-to-cell communications, and is the precursor of oxysterol, which regulates key metabolic processes. Available cholesterol in fetus is contributed by: (1) transfer from the mother especially during the first half of the gestation and too little cholesterol due to lack of maternal cholesterol or reduced expressions of placental lipoprotein receptors is correlated with small fetuses and a trends for microcephally; and (2) Fetal synthesis especially during the last half of gestation. Too little cholesterol due to lack of synthesis leads to a spectrum of congenital defects as seen in infants with Smith-Lomli-

Opitz Syndrome(SLOS) who are unable to synthesize cholesterol at normal rate due to null/null mutations in 3β-hydroxysteriod Δ7-reductase, the enzyme that converts 7-dehydrocholesterol to cholesterol. The placental endothelial cells are capable of transporting substantial amounts of cholesterol to the fetal circulation and this mechanism is further enhanced by liver-X receptors and induced up regulation of ATP-binding cassette transporter, ABCA1 and ABCG1[56].

4.3. Glycerol

Maternal Plasma glycerol levels are consistently elevated during late pregnancy, but crosses the placenta less than glucose or L-alanine [1,25, 57] though they all have similar molecular weights. Transfer of maternal glycerol via the placenta is by simple diffusion (2). However, its effective and rapid utilization through other pathways, such as gluconeogenesis and glyceride glycerol synthesis in the mother[10,25] results in its low plasma concentration and this very active kinetics impede the formation of the adequate gradient to create the appropriate driving forces for its placental transfer.

4.4. Ketone bodies

In the 3rd trimester of pregnancy, under fed conditions, plasma ketone body concentrations remain low although are greatly increase compared to nonpregnant condition under fasting [58] consequent to enhanced adipose tissue lipolysis. The lipolysis accelerates delivery of NEFA to the liver and enhanced ketogenesis. Ketone bodies can easily cross the placenta and be used as fuels and lipogenic substrates by the fetus. The transfer of ketone bodies across the placenta occurs either by simple diffusion or by a low-specificity carrier-mediated process [25]. The activities of ketone body metabolizing enzyme are present in fetal tissues (brain, liver and kidneys)[1,25] and can be increased by conditions of maternal ketonaemia such as occurs in starvation, during late pregnancy[39] or high-fat feeding[25]. Ketone bodies are used by the fetus as oxidative fuels as well as substrates for brain lipid synthesis [25]. However, in maternal hyperketonaemia as occurs in poorly controlled diabetes patients associated with transfer of excessive arrival of ketone bodies to the fetus seems to be responsible for the major damages [10], increasing stillbirth rate, incidence of malformations, and impaired neurophysiologic development [10]. Subsequently, it could be recommended that pregnant mothers, if possible, should avoid starvation and high fat diet especially in the 3rd trimester.

5. The importance of lipid and lipoprotein ratios in hyperlipidaemia in adult male and nonpregnant females

While cholesterol is a key component of the development of atherosclerosis, LDL-C concentration has been the prime index of cardiovascular disease(CVD) risk and the main target for therapy[21]. However, currently, there is almost unanimous agreement among epidemiologists and clinicians that coronary risk assessment based exclusively on LDL-C is not optimal[59]. Therefore in the recent past, efforts have been made in seeking emergent or new cardiovascular risk factors to improve cardiovascular disease prediction[20] and in an attempt to optimize the predictive capacity of lipid profile, several lipoprotein ratios or "atherogenic indices" have been defined. In the Framingham study, the TC:HDL-C ratio, a useful summary of the joint contribution of total cholesterol(TC) and HDL-C to coronary heart disease(CHD) risk[60], was also found to be an excellent predictor of CHD risk, with a hazard ratio of 1.21 for a 1.0 increment in ratio[60]. The value of this ratio should be emphasized when lipid profile is within desirable range. It was shown that patients with high-risk LDL-C levels >160mg/dl(4.2mmol/L) and low TC: HDL-C ratio (≤5.0) had an incidence of CHD of 4.9%. This was similar to those with low levels of both LDL-C(≤130mg/dl, 3.4mmol/L) and TC:HDL-C ratios, 4.6%[60]. By contrast, subjects with low-risk LDL-C levels(≤130mg/dl, 3.4mmol/L) and high TC:HDL-C ratio(>5.0) had a 2.5-fold higher incidence of CHD than those with similar LDL-C levels but low TC:HDL-C ratio[60]. For example, TC of 231mg/dl(5.89mmol/L) and HDL-C of 42mg/dl(1.09mmol/L) gives a TC:HDL-C ratio of 5.5, which indicate moderate atherogenic risk[61]. On the other hand, with the same level of TC, if HDL-C were 60mg/dl(1.55mmol/L), the ratio would be 3.8[61]. However, in the Helsinki Heart Study[62], it was demonstrated that the LDL-C:HDL-C ratio that paints the most relevant picture of a person's cardiovascular health risk especially when triglyceridaemia is taken into account and the risk is significantly higher in the presence of hypertriglyceridaemia. When there is no reliable calculation of LDL-C, especially when triglyceridaemia exceeds 300mg/dl(3.36mmol/L), it is preferable to use the TC:HDL-C ratio. Similarly individuals with high concentration of triglycerides, VLDL fraction shows cholesterol enrichment and thus the LDL-C:HDL-C ratio may underestimate the magnitude of the lipoprotein abnormalities in them[21]. Subsequently, both TC:HDL-C, known as the atherogenic or Castelli index, and LDL-C:HDL-C ratios are two important components and indicators of vascular risk, the predictive values of which is greater than isolated parameters used independently, particularly LDL-C. These ratios can provide information on risk factors difficult to quantify by routine analyses and could be a better mirror of the metabolic and clinical interactions between lipid fractions. Their applications therefore in interpreting hyperlipidaemia of pregnancy cannot be over emphasized.

5.1. ApoB:ApoA-1 ratio

Apolipoprotein-B(apoB) represents most of the protein contents in LDL and is also present in IDL and VLDL. ApoA-1 is the principal apolipoprotein in HDL and is believed to be a more reliable parameter for measuring HDL than cholesterol content since it is not subject to variation. Therefore, the apoB:apoA-1 ratio is also highly valuable for detecting atherogenic risk, and there is currently sufficient evidence to demonstrate that it is better for estimating vascular risk than the TC:HDL-C ratio[63-65]. The apoB:apoA-1 ratio was found to be stronger than the TC:HDL and LDL:HDL ratios in predicting risk[63]. ApoB:ApoA-1 ratio reflects the balance between two completely opposite processes. Transport of cholesterol to peripheral tissues, with its subsequent arterial internalization, and reverse transport to the liver[66]. Consequently, a larger ratio will implies higher amount of cholesterol from atherogenic lipoprotein circulating through the plasma compartment and likely to induce endothelial dysfunction and trigger the atherogenic process. On the other hand, a lower ratio will indicate less vascular aggression by plasma cholesterol and increased more effective reverse transport of cholesterol, as well as other beneficial effects, thereby reducing the risk of CVD. However, its use is limited by the fact that apolipoprotein measurement methods are not widely used as lipoprotein methods

5.2. TG:HDL ratio

Known as the atherogenic plasma index shows a positive correlation with HDL-C estimation rate(FER$_{HDL}$) and an inverse correlation with LDL size[67]. Therefore, the phenotype of LDL and HDL particles is clearly synchronized with the FER$_{HDL}$. The simultaneous use of TG and HDL in this ratio reflects the complex interaction of lipoprotein metabolism overall and can be useful for predicting plasma atherogenecity especially in pregnant women who manifesting with hypertriglyceridaemia of pregnancy. An atherogenic plasma index[Log(TGs:HDL)] over 0.5 has been proposed as the cutoff point indicating atherogenic risk[67].

5.3. LDL-C:apoB ratio

Although apoB is not an apolipoprotein exclusive to LDL, since it is present in other atherogenic lipoproteins such as IDL and VLDL, the LDL:apoB ratio provides approximate information on LDL particle size. A ratio of <1.3 indicate the predominance of LDL particle with low cholesterol content, consistent with small, dense LDL particle[68].

Variations in plasma lipid and lipoprotein ratios in adult men and nonpregnant women have been associated with more substantial alterations in metabolic indices predictive of future consequences of hyperlipidaemia than individual components of plasma lipid profile alone[69, 70] and as discussed above. Given the physiological role of gestational hyperlipidaemia in fetal development and the fact that the adaptations in maternal lipid metabolisms taking place throughout gestation is not without consequences, an urgent establishment of reference values for lipid and lipoprotein ratios in normal pregnancy is highly recommended.

5.4. The hyperlipidaemia of pregnancy, a dyslipidaemia? Find out!

5.4.1. The importance of lipid and lipoprotein ratios in interpreting the hyperlipidaemia of pregnancy

In normal nonpregnant adult population, higher concentrations of plasma triglycerides are associated with preferentially higher VLDL-1 concentration [71]. This particle is secreted by the liver to supply tissues with TGs fatty acids in the post absorptive state. The concentration of VLDL-2, the principal precursor in the circulation to IDL and LDL, does not change as dramatically. In addition, in normal nonpregnant adult population, a higher concentration of VLDL-1 is associated with a failure of insulin action and increased risk of CHD. In contrast, in pregnant women, as pregnancy progresses and high TG levels developed, VLDL-1 and VLDL-2 rose together so that the ratio, instead of increasing 2-fold, as would be predicted from population studies in the nonpregnant subjects (VLDL-1 o VLDL-2 ratio at a plasma TGs of 0.5mmol/L is 1.0 compared to 2.0 at plasma TGs of 2.5mmol/L)[71], remain constant. Sattar[33], *et al*, found a parallel increase in the small cholesterol-rich VLDL-2(17 to 103mg/dl) and the larger TG-rich VLDL-1(19 to 109mg/dl) at 35 weeks. Similarly, the relationships of VLDL constituents expressed as ratios were not significantly different comparing antepartum and postpartum observations, however, the TG/C ratio was higher at all of these times compared to controls, but the composition of these fractions was similar to that seen in a recent cross-sectional survey of healthy adults (19). The increase in VLDL-TG during gestation is likely due to an increase in VLDL synthesis rather to a compositional change in the VLDL particle, as a study showed no significant increase in VLDL TG/C ratio over time, and the ratios is similarly lower in all the trimesters compared to nonpregnant period(see table 3)

Ratios	First trimester	Second trimester	Third trimester	Nonpregnant control
TC:HDL-C	2.56	3.37	3.90	3.29
LDL-C:HDL-C	1.44	1.95	2.37	1.79
TGs:HDL-C	0.56	0.79	1.16	0.64
VLDL-TGs:CL	1.64±1.53	2.47±3.91	2.57±3.60	3.69±3.48
LDL-TG:CL	0.46±	1.24±2.68	0.56±0.29	0.14±0.08
HDL-TG:CL	0.58±0.21	0.60±0.19	0.69±0.32	0.21±0.09
HDL-TG:ApoA-1	4.09±1.55	5.24±1.43	6.13±1.28	2.73±0.71
LDL-CL:ApoB	2230±339	2222±228	2113±305	2506±167
LDL-TG:ApoB	217±59	256±41	332±60	157±32
LDL-PL:ApoB	748±123	753±66	727±109	824±64
IDL-TG:ApoB	2026±1085	1666±360	1550±202	1530±371
VLDL-TG:ApoB	6272±1924	6278±1629	5551±1416	7040±2778

Table 3. Lipid and lipoprotein ratios in the three trimesters of normal pregnancy.

Taken together, and as shown in **table 3**, although one of the consequences of pregnancy is that maternal lipid metabolism is specifically altered, using the lipid and lipoprotein ratios, the hyperlipidaemia occurring in the later part of pregnancy appears to be a balanced hyperlipidaemia. These are discussed below

During the course of normal pregnancy, plasma TGs and cholesterol rise by 200-400% and 25-50% respectively. The total LDL mass increased during gestation (median concentration increased by about 70%, 200-353mg/dl) between 10 to 35weeks, see table 4. The lipid become enriched with TGs and depleted in cholesterol. The larger, more buoyant subclasses of LDL (LDL-1 and LDL-2) predominant in healthy pregnant females and may in the reproductive age, whereas smaller, denser LDL-3 often occur after menopause [11, 72]. Several studies showed there to be an association between elevated plasma TG concentrations, small, dense LDL [11, 73] and decreased HDL cholesterol [74], in particular HDL-2 cholesterol [73]

VARIABLES	10 WEEKS OF GESTATION	35 WEEKS OF GESTATION
Triglyceride(mean)	69.65mg/dl	227.69mg/dl
Total cholesterol(mean)	172.57mg/dl	282.38mg/dl
HDL-C(mean)	64.73mg/dl	65.50mg/dl
VLDL-1(mean)	19mg/dl	109mg/dl
VLDL-2(mean)	17mg/dl	103mg/dl
IDL(mean)	26mg/dl	124mg/dl
LDL	200mg/dl	333mg/dl
LDL-1	33mg/dl	67mg/dl
LDL-11	143mg/dl	201mg/dl
LDL-111	28mg/dl	123mg/dl
LDL-1	17% of total LDL	20% of total LDL
LDL-11	69% of total LDL	49% of total LDL
LDL-111	14% of total LDL	32% of total LDL

Table 4. Magnitude of changes in lipid and liporpotein values from first to third trimester.

In men and nonpregnant females, plasma TG is the major determinant of small, dense LDL, occurring for 40-60% of the variability of this fraction in the plasma [71,75,76]. In addition, recent cross-sectional studies [70,74] have prompted the suggestion that, within the relationship between plasma TGs and LDL subfractions profile, there is a threshold effect. At low-normal plasma TG concentrations, there is a positive association between LDL-2(the major LDL species) concentration and plasma TGs. Above a certain plasma value, however(reportedly about 1.5mmol/L in men)[71,75], LDL-2 concentration correlates negatively with plasma TGs, and LDL-3 concentration which had been relatively constant below this TG concentrations, correlates positively with plasma TG. Generally, percent LDL-3(and LDL-3 mass) changed little in early gestation despite increasing TG concentrations. However, there appeared to be considerable variation between individuals in the gestational age and plasma TGs intervals at which change in the LDL profile first manifested—the elevated TG levels already present in the first trimester may be responsible for the increased in dense LDL

In line with the alarming observations in LDL subclasses and total LDL mass, LDL-1 mass increased around 2-fold, from 33 to 67mg/dl; LDL-2 mass increased least by around 40% from a median of 143 to 201mg/dl, reaching a maximum of 218mg/dl at 30weeks gestation, whereas in sharp contrast, LDL-3 mass increased by greater than 4-fold from 23 to 123mg/dl.

However, as concentration of LDL-2 is declining, that of LDL-3 is increasing and implying that the ratio may tend towards a unit.

Towards end of second trimester to end of gestation, the concentrations of VLDL, IDL, and LDL-1/LDL-2 further increased, producing a distribution of lipoproteins dominated by buoyant lipoprotein species, in particular LDL-1. In line with this, Winkler's[11], et al data do not support the idea that the same mechanisms as those described for the atherogenic lipoprotein phenotypes govern lipid metabolism in late pregnancy. Therefore, in uncomplicated pregnancy there appears to be a balance between potentially damaging factors such as altered lipid metabolism and as yet poorly understood protective mechanisms [11,33,75]. However, the clinical significance of gestational lipoprotein metabolisms may arise if this balance is compromised as in hypertensive disorders of pregnancy. It is in these circumstances then when the application of these ratios is very important, for example; Toescu, [77] et al while comparing lipid levels between pregnant diabetic women(types 1 and 2 and GDM) and pregnant nondiabetic counterparts did not demonstrated any significant differences among the groups according to trimesters, implying that the observed hyperlipoproteinaemia during pregnancy is independent of diabetes status[10]

Kilby,[78] *et al* although observed higher lipid levels and increased in TC, TGs, VLDL/LDL ratio, HDL-C with gestational age in type 1 DM, similarly found no significant difference from gestationally matched controls[78] in their study. Investigations are required to characterize lipid and lipoprotein profile using ratios in the other modulators, particular these will assists clinicians while dealing with hyperlipidaemia of pregnancy considering the limited quantification opportunities.

Currently the applications of lipid and lipoprotein ratios in interpreting the hyperlipidaemia of pregnancy are limited particularly in the poor developing nations. In spite of the fact that the hyperlipidaemia of pregnancy is usually considered physiological, serum lipid and lipoprotein levels in pregnancies are generally modulated by complex interactions between genetics, medical complications of pregnancy, co-existing medical conditions and other maternal factors[19], **table 5**.

Therefore the hyperlipidaemia during pregnancy could be classified according to clinical implications and future prospects as in **fig 1,** particularly where there is limited opportunity of investigations do to poverty.

In our laboratory [79] the ratios were found to be important particularly where measurement of lipid and lipoprotein is not routinely done due to poverty. In addition hyperlipidaemia in pregnancy is confounded by other conditions that may predispose to hyperlipidaemia, such as obesity, diabetes mellitus, chronic renal insufficiency, and pre-eclampsia. Similarly subfractions of lipoproteins are usually not done due to limited methodology. Without the use of lipid and lipoprotein ratios particularly considering these confounding conditions which are also likely to present with hyperlipidaemia, interpreting the hyperlipidaemia of pregnancy is encountered with difficulties.

Medical complications of pregnancy
1. Pre-eclampsia
2. Pregnancy-induced hypertension
3. Gestational diabetes mellitus
4. Intra-uterine growth restriction(retardation)
5. Prelipaemia
Co-existing medical conditions
1. Obesity
2. Types 1 and 2 diabetes mellitus
3. Hypothyroidism
4. Hypertension
5. Renal diseases, particularly nephritic syndrome
6. Alcoholism
7. Medications, eg LMWt-heparin and glucocorticoid
Other maternal factors
1. BMI(Obesity)
2. Maternal weight gain in the index pregnancy
3. Maternal nutrition
4. Pre-pregnancy lipid levels

Table 5. Factors that can also modulate lipid and lipoprotein concentrations in pregnancy (genetic factors not mentioned)

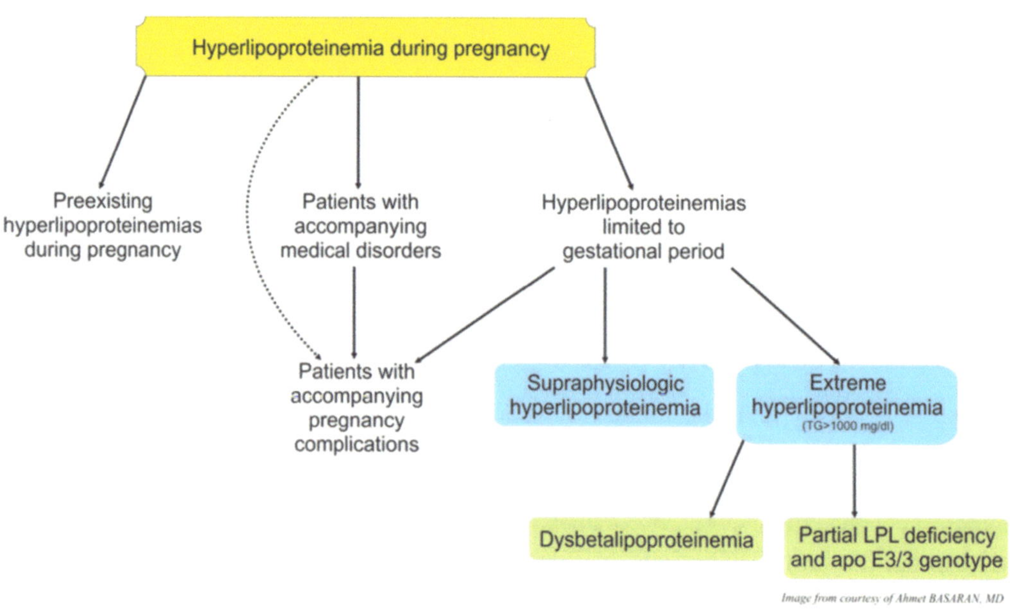

Figure 1. Classification of hyperlipidaemia of pregnancy

Whilst the hyperlipidaemia of pregnancy is considered physiological, studies have demonstrated that deviations present as a two-edged sword. On one hand, development of the physiological hyperlipidaemia out of proportion could be associated with many consequences and on the other hand failure to development the required proportion of physiological hyperlipidaemia of pregnancy could also be associated with some consequences, **Table 6** and these will be discussed subsequently

Consequences of hyperlipidaemia
Complications associated with hyperlipidaemia of pregnancy:
Cholesterol gallstones
Intrahepatic cholestasis
Acute pancreatitis
Endothelial dysfunction
Unanswered questions concerning hyperlipidaemia of pregnancy
Is hyperlipidaemia of pregnancy atherogenic?
Is hyperlipidaemia of pregnancy a dyslipidaemia?
Future effect of pregnancy-induced Supraphysiologic hyperlipidaemia
Pre-eclapmsia and hyperlipidaemia of pregnancy
Consequences of failure of development of hyperlipidaemia
IUGR—Intra-uterine growth retardation
Future development of metabolic syndrome in affected fetus

Table 6. Consequences of deviations of Hyperlipidaemia of pregnancy.

6. Is normal pregnancy atherogenic?[80]

The change in triglycerides in normal pregnancy is important in relation to lipoprotein subclasses, such as LDL. These lipoproteins contain subfractions of various sizes, densities and compositions, which differ in their ability to initiate atherogenesis [81]. One of the subfractions of LDL (LDL-3) is small, dense LDL particles which do not bind readily to the LDL receptors and therefore remain in the circulation for longer time, penetrate the arterial intima better than do larger ones[82] and are more readily oxidized, probably because they contain less vitamin E and other antioxidants[83]. Finally, their uptake into macrophages to create form cells, and initiate atherogenesis, is facilitated [84]. This may explain their identification as an independent risk factor for coronary heart disease [82-84].

Plasma triglycerides are the major determinant of small, dense LDLs, accounting for 40-60% of the variability of this fraction in the plasma [75]. VLDL represents the major precursor of LDL and reflects plasma TGs levels. Two subclasses of VLDL have been defined: a large and buoyant fraction enriched with TGs (VLDL-1) and a smaller, denser fraction(VLDL-2). It follows from the association between LDL subclasses and raised TGs that VLDL-1 may be important as a vehicle in the process of neutral lipid exchange and generation of small, dense LDLs. Cholesterol esters are transferred from LDL and HDL to VLDL-1 by cholesterol ester transfer protein(CETP) in exchange for TGs and the increased concentration of VLDL-1, due to hypersecretion by the liver promote TG transfer into LDL during pregnancy[33]. TG-enriched LDL particles subsequently undergo a size reduction through the action of hepatic lipase, resulting in the formation of small, dense LDL subfractions. In addition Lippi[2], et al demonstrated in their study that advanced pregnancy is associated with an increased prevalence of undesirable or abnormal values for total cholesterol, LDL-C and TGs in the second trimester, and total cholesterol, LDL-C, TGs TC:HDL ratio in third trimester demonstrating that physiological pregnancy is associated with a substantial modification of lipid and lipoproteins metabolism from the second trimester, providing reference ranges for traditional and emerging cardiovascular risk predictors throughout the gestational period. Therefore, is normal pregnancy atherogenic?

All pregnant women develop a transient hyperlipidaemia associated with hypertriglyceridaemia, and subsequent formation of small, dense LDL particles, both of which are an independent risk factor for CHD, and by 3rd trimester most women have a lipid profile which would be considered highly atherogenic in the non-pregnant state[13]. Increased prevalence of angina, cholesterol gallstone, and obesity in postmenopausal women who have had several pregnancies has been observed [85]. Yet the long-term consequences of multiple pregnancy, gestational diabetes or maternal obesity in LDL subfractions and lipid profile are unknown. Further studies are recommended to determine if certain women are at increased risk of CVD in later life because of effects on their lipid profile during pregnancy. In contrast, increasing suggestions are that maternal hypercholesterolaemia during pregnancy even when temporary and limited to pregnancy triggers pathogenic events in the fetal aorta, greatly enhanced fatty streak formation and that may influence atherogenesis later in life[14,15]. Fetal plasma cholesterol levels are high and are proportional to the maternal cholesterol levels [14] in second trimester, decline with increasing fetal age[14] and are even lower at term birth. This is supported by the fact that lipid levels observed in umbilical cord blood(UCB) from normal pregnancy were significantly lower than those found in maternal blood with exception of HDL-C, and that LDL:HDL ratio in neonate of normal pregnancy are much lower than the value in normal pregnant mothers[16]. The high HDL levels and a lower LDL:HDL ratio in UCB suggest that the fetus of a normal pregnancy is protected against atherogenic lipoprotein[16]. Despite these findings, studies at autopsy demonstrated that atherosclerosis progresses much faster in offsprings of hypercholesterolaemic mother than in offsprings of normocholesterolaemic mothers[86]. Same studies observed that at each time point, offsprings of hypercholesterolaemic mothers had 1.5 to 3-fold larger lesions than offsprings of normocholesterolaemic mothers, and they suggested that, pathogenic programming in utero increases the susceptibility to atherogenic risk factors later in life and maternal intervention with cholesterol-lowering agents reduce postnatal lipid peroxidation and atherosclerosis in their offsprings[87]. A registry study by Toleikyte,[22] *et al*, of heterozygous familial hypercholesterolaemic(FH) mothers observed that: the serum levels of cholesterol in the nonpregnant, nontreated women were 370mg/dl(9.59mmol/L); no maternal cardiovascular deaths were observed; the children of mothers with FH were no more likely than the general population to be born prematurely, have low birth weight, or have congenital malformations; and that no congenital malformations were observed in the 19 pregnancies associated with the use of lipid-lowering drugs during pregnancy. However, the current trend is that Statins, classified by FDA as category X in pregnancy, should be avoided in pregnancy [23,24]. Although there are observations for and against the maternal hyperlipidaemia being atherogenic to the fetus and increasing tendencies of future atherosclerosis, a long-term follow-up studies of offsprings of mothers with FH who did not inherited the disease is recommended. The result will demonstrate evidence of effects of maternal hyperlipidaemia on fetal atherosclerosis and or predisposition to future atherosclerosis in these offsprings.

6.1. Fetal lipoproteins in pre-eclampsia

Successful placental development is very important for normal fetal growth, and may condition health and well being during childhood and even adulthood [88], because it forms the interface for nutrients, fluids and gas exchange between mother and fetus. Pre-eclampsia (PE), a human pregnancy specific vascular disorder, defines a pathologic condition that affects the mother and can adversely influence the feto-placental unit. PE is associated with placental dysfunction, oxidative stress[1, 89], dyslipidaemia[16,90] and endothelial cell activation, and is a major cause of maternal and fetal morbidity and mortality[88] A pro-atherogenic lipid profile, characterized by increased TG levels with reduced HDL concentration[90, 91] and increased small, dense LDL particles[90] has been described. In contrast other studies demonstrated a dominance of buoyant LDL-1 and a significant decreased of small, dense LDL, namely LDL-5 and LDL-6[92]. Notwithstanding, it has been suggested that an abnormal lipid metabolism may not only be a manifestation of PE, but dyslipidaemia may play an essential role in its pathogenesis

Apart from genetic predisposition, the second group of disorders associated with an increased risk of PE includes a variety of chronic conditions such as dyslipidaemia, diabetes mellitus, hypertension, renal diseases, and various thrombophilias[93]. These disorders can be convincingly grouped together based on their common association with vascular endothelial dysfunction, especially those which have been included in the proposed metabolic syndrome [93]. Ironically also all are associated with dyslipidaemia. Although PE is a multifactorial disorder, it is therefore tempting to ask, could dyslipidaemia be the central focal point linking these disorders into pathogenesis of PE? One of the abnormalities found in the abnormal placental bed is presence of acute atherosis in desidual vessels, characterized by accumulations of foam cells and perivascular mononuclear cell infiltration. Reduced placental perfusion and placental/fetal hypoxia may develop.

Catarino[16], et al, while comparing lipid and lipoproteins in normal pregnant and PE pregnant women found an enhanced physiologic hyperlipidaemia. However, the most striking difference noted in PE women was the rise TGs that almost double the median value compared to normal pregnant women. Higher TGs value has been associated with endothelial dysfunction and may therefore play an important role in the pathogenesis of PE. Considering the placental dysfunction and lipid changes occurring in PE, fetal lipid metabolism can be affected due to an altered placental transfer of lipids. Maternal TGs does not cross the placenta. It has to be hydrolyzed by human placental LPL into FFAs which is then transported to the fetus. Fetal TG levels are therefore dependent on maternal TGs. Moreover, the placenta also contains receptors for VLDL, LDL and HDL which also transport TGs and other esterified lipid to the fetus (23)

Catarino[16],et al observed that lipid levels observed in umbilical cord blood (UCB) from normal pregnancy were significantly lower than those found in maternal blood except for HDL, which was similar. In addition, LDL:HDL ratio in neonates of normal pregnancies are much lower than the values in normal pregnant mothers. In contrast, lower values of HDL and ApoA-1 and higher TG levels were noted in neonates of PE mothers. In addition, higher LDL:HDL ratio, a decreased HDL which is more pronounced than ApoA-1, suggest that fetal loading of ApoA-1 with cholesterol is significantly less in PE. Hence fetal HDL composition is likely to be altered due possibly to enrichment with the enhanced hypertriglyceridaemia. Also observed in the PE is a significantly higher value of TGs in UCB which is parallel with significant increased TGs in maternal blood. Since hypertriglyceridaemia is considered a maternal adaptation in order to assure fetal growth in normal pregnancy, the exaggerated hypertriglycedaemia noticed in PE mothers could be a compensation pathway to face the uteroplacental hypoperfusion in order to enhance FAs transfer to the fetus. In line with this, it seems LPL expression is also enhanced in PE as was observed in IUGR [94]. Taken together, it appears lipid transfer from maternal side in PE mothers to their fetus are altered in both quantity and quality and does not seems to be protective as noticed in neonates of normal pregnancies. In support of this PE has been associated with reduced fetal birth weight [95, 96] and the expression of lipoprotein receptors are decreased in the placenta of women with PE.

PE pregnancies is associated with an enhanced hypertriglyceridaemia, which seems to have a negative impact on fetal lipid profile, as reflected by a higher atherogenic LDL:HDL ratio, decreased HDL, disproportionate decreased in HDL and ApoA-1 and enhanced hypertriglycedaemia, children born in pregnancies associated with PE deserve a closer clinical follow-up later in life.

6.2. Role of lipid metabolism in pathogenesis of intrauterine growth restriction (IUGR)[17]

It was proposed that the abnormal lipid metabolism noted in pre-eclampsia was in an attempt to compensate for the placental insufficiency [97], given the physiological role of gestational hyperlipidaemia in supplying both cholesterol and triglycerides to the rapidly developing fetus [98]. In contrast Dabi[17], et al demonstrated that concentration of total cholesterol (TC), TGs, LDL and VLDL observed to increase with increasing gestational age in normal pregnancies, these lipids and lipoproteins decreased with increasing gestational age in pregnancies with IUGR. HDL did not change significantly. These findings certainly indicated that pregnancies complicated by IUGR are associated with an abnormal lipid profile, particularly decreased levels of TC, TGs, LDL and HDL(Dabi [17],et al Sattar[18], et al), see table 7

	Group A			Group B		
Weeks	28-31	32-36	37-40	28-31	32-36	37-40
TC	216.3+29.2	202.5+26.0	191.5+34.5	238.0+17.0	250.7+27.6	260.3+23.6
TGs	173.83+78.18	168.23+51.73	137.83+18.25	148.25+15.31	160.36+24.43	171.14+41.56
HDL-C	42.0+5.3	43.0+4.4	46.4+3.2	44.4+4.5	43.0+6.0	41.6+8.2
LDL	139.4+23.9	126.3+21.8	117.0+31.8	165.9+26.5	174.8+26.5	184.5+23.4
VLDL	34.78+15.63	33.64+10.34	27.26+3.81	29.77+3.06	32.05+4.91	34.22+8.31

Table 7. Lipid and Lipoprotein changes with advancing gestational age(Group A=Pts with IUGR and Group B=Pt with normal pregnancy)

It is well known that normal fetal development needs the availability of both essential fatty acids and long chain polyunsaturated fatty acids (LCPUFAs), thus making a persuasive case indicating a relationship between nutritional status of mother during gestation reflecting her lipid profile and fetal growth. From observations in study by Dabi[17], *et al* and similar findings in other studies, it is possible that the decreased concentrations of TC, TGs, VLDL and LDL may have decreased availability of glycerol, LCPUFAs and essential fatty acids to the fetuses of mothers with otherwise normal pregnancy ultimately leading to IUGR. In addition to above findings, Sattar[18], *et al* observed a decreased in levels of VLDL-2 and IDL in IUGR pregnancies, which are precursors of LDL. Taken together, the decreased cholesterol levels (mainly reflected as decrease LDL) may be due to decreased synthesis of LDL in women with IUGR. The HDL: apoA and apoB:apoA ratios were found to be higher in the IUGR and was suggested that blood lipid modifications in the IUGR group are partly secondary to changes in HDL metabolism and the competitive inhibition of fibrinolysis by apoB which is increased in pregnancy with IUGR. This indicated that apoA: apoB ratio could be a good marker for the early detection of IUGR. Taken together also, these findings definitely generated considerable interest in certain aspects of fetal growth and its relationship to blood lipid levels during pregnancy. However, more study is recommended aiming at analyzing the otherwise normal pregnancies associated with IUGR, particularly to elucidate the hypothesis that the decrease in TGs(and particularly LDL and VLDL-2) compromises the supply of substrates for energy production to the growing fetus resulting in IUGR. The effect of the changes in lipid profile and its translation in changes in blood viscosity needs more extensive research including detailed analyses of apoA and apoB levels in these patients.

6.3. Pregnancy-induced Supraphysiologic hyperlipidaemia

It is well known in literature that hyperlipidaemia is a normal metabolic adjustment in pregnancy benefiting both mother and the fetus. However, some women may not be able to adapt to the hyperlipidaemic stress of pregnancy. In addition, in similarity with other gestational metabolic syndromes such as gestational diabetes mellitus (GDM), pregnancy-induced hypertension (PIH), pre-eclampsia, eclampsia, etc, some of them may develop a state of Supraphysiologic hyperlipidaemia, defined as lipid levels greater than 95th percentile for the corresponding gestational age, because of failed adaptation to requirement of pregnancy. Supraphysiologic hyperlipidaemia may serve as a marker for what is cited by Montes[99], et al, a 'prelipaemia' in the same way that GDM is a marker for pre-diabetes.

The characteristics of Supraphysiologic hyperlipidaemia, as observed by Montes[99], et al, are that, the antepartum hyperlipidaemia may return to normal levels postpartum more slowly than normal, the presence of HDL cholesterol concentrations that are persistently low antepartum and postpartum, and the patients do have hyperlipidaemic family members. In contrast, hypercholesterolemia is not greatly exaggerated in pregnancy among these women. Are there future consequences of the pregnancy-induced Supraphysiologic hyperlipidaemia? Long-term follow-up studies of women with genetically well-characterized disorders of lipoprotein metabolism are required to determine if an abnormal lipoprotein response in pregnancy can identify prelipaemic subjects and distinguish among the major disorders of lipoprotein metabolism. Identification of the prelipaemia will provide an opportunity to study prospectively the natural progression, potential for atherosclerosis, and possible treatment of hyperlipidaemia from early adulthood.

Author details

D.S. Mshelia and A.A. Kullima
University of Maiduguri/University of Maiduguri Teaching Hospital, Nigeria

Acknowledgement

I, D.S. Mshelia, greatly appreciate the support given to me by my lovely wife, Aisha A Buba and my children. Similarly, I, A.A. Kullim, sincerely appreciate the support given to me by my wife, Yagana Umara and my children (Inna and Yaanama).

7. References

[1] Herrera E and Ortega-Senovilla H(2010). Maternal lipid metabolism during normal pregnancy and its implications to fetal development. Clin Lipidol; 5(6):899-911.

[2] Lippi G, Albiero A, Montagnana M, et al (2007). Lipid and lipoprotein profile in physiological pregnancy; 53(3-4):173-177.

[3] Brizzi P, Tonolo G, Esposito F, Puddu L, Dessole S, Maioli M, et al, (1999). Lipoprotein metabolism during normal pregnancy; 181:430-434

[4] Hadden DR, and McLaughlin C (2009). Normal and abnormal maternal metabolism during pregnancy. Seminars in fetal and neonatal medicine; 4:66-71

[5] Hytten F, Chamberlain G (1980). Clinical physiology in obstetrics. Oxford, United Kingdom: Blackwell Scientific Publications;---

[6] Food and Nutrition Board, Committee on Nutrition of mother and preschool child. Laboratory indices of nutritional status in pregnancy. Washington, DC: National Academy of Sciences, 1978

[7] Denne SC, Patel D, Kalhan SC (1991). Leucine kinetics and fuel utilization during a brief fast in human pregnancy. Metabolism 40:1249-1256

[8] King JC (2000). Physiology of pregnancy and nutrient metabolism. Am J Clin Nutr. 719Supll0: 1218S-1225S.

[9] Basaran A(2009). Pregnancy-induced hyperlipoproteinaemia: Review of the literature. Reproductive Sciences. 1695):431-437.

[10] Butte NF (2000). Carbohydrate and lipid metabolism in pregnancy: normal compared with gestational diabetes mellitus. Am J Clin Nutr. 71(suppl): 1256S-1261S

[11] Winkler K, Wetzka B, Hoffmann MM, Fredrich I, Kinner M, Baumstark MW, et al (2000). Low density lipoprotein(LDL) subfractions during pregnancy: Accumulation of buoyant LDL with advancing gestation. Metab 85: 4543-4550

[12] Alvarez JJ, Montelongo A, Iglesias, A, Lasuncion MA, and Herrera E (1996). Longtudinal study on lipoprotein profile, high density lipoprotein subclass, and postheparin lipase during gestation in women. J Lipid Research; 37:299308

[13] Lyall F and Greer AI(1996). The vascular endothelium in normal pregnancy and pre-eclampsia. Reviews of reproduction 1:107-116.

[14] Napoli C, D'Armiento FP, Mancini FP, Witztum JL, Palumbo G, Palinski W(1997). Fatty streak formation occurs in human fetal aortas and is greatly enhanced by maternal hypercholesterolaemia. Intimal accumulation of low density lipoprotein and its oxidation precede monocyte recruitment into early atherosclerotic lesions. J. Clin. Invest. 100:2680-2690.

[15] Napoli C, Witztum JL, de Nigris F, Palumbo G, D'Armiento FP, Palinski W.(1999). Intracranial ateries of human fetuses are more resistant to hypercholesterolemia-induced fatty streak formation than extracranial arteries. Circulation 99:2003-2010.

[16] Catarino C, Rebelo I, Belo L, Rocha-Pereira P, Rocha S, Casto EB, Patricio B, et al (2008). Fetal lipoprotein changes in pre-eclampsia. Acta Obstetricia et Gynecologica 87:628-634.

[17] Dabi DR, Manish P and Vikas G (2004). A cross-sectional study of lipids and lipoproteins in pregnancies with intrauterine growth retardation. 64(5):467-472.

[18] Sattar N, Greer IA, Galloway PJ, Packard CJ, Stepherd J, Kelly T, et al (1999). Lipid and lipoprotein concentrations in pregnancies complicated by intrauterine growth restriction. J clin Endocrinol Metab. 84:128-130.

[19] Hegele RA (1991). Hyperlipidaemia in pregnancy. Can Med Assoc J. 145(12):1596.

[20] Yusuf S, Hawken S Qunpuu S, et al (2004). Effect of potentially modifiable risk factors associated with myocardial infarction in 52 countries(the INTERHEART study): case-control study. Lancet. 364:937-952.

[21] Millan J, Pinto X, Munoz A, Zuniga M, Rubies-Prat J, Pallardo LF, et al(2009). Lipoprotein ratios: Physiological significance and clinical usefulness in cardiovascular prevention. Vascular health and risk Management 5:757-765

[22] Toleikyte I, Retterstol K, Leren PT, Iversen PO (2011). Pregnancy outcome s in familial hypercholesterolaemia: a registry-based study. Circulation 124:1606-1614.

[23] Ito MK, McGowan MP, Moriarty PM (2011). Management of familial hypercholesterolaemia in adult patients: recommendations from the National Lipid

association Expert Panel on familial Hypercholesterolaemia. I Clin Lipidol. 5(3 suppl): S38-S45.

[24] Edison RJ, Muenke M (2004). Central nervous system and limb anomalies in case reports of first-trimester statin exposure. N. Engl. J Med. 350:1579-1582.

[25] Herrera E. Lipid metabolism in pregnancy and its consequences in the foetus and newborn. Endocrine 2000; 19:43-55.

[26] Blackburn ST, Loper DL (1992). Carbohydrate, fat, and protein metabolism. In: Blacburn ST, Loper DL, Editors, Maternal, fetal, and neonatal physiology: a clinical perspective, Philadelphia:W.B. Saunders: p, 583-613

[27] Clapp j, Seaward BL, Seamaker RH, and Hiser j (1982). Maternal physiologic adaptations to early human pregnancy. Lancet 1:588-592

[28] Hytten FE (1991). Weight gain in pregnancy. In: Hytten F. Chambelain G, editors. Clinical physiology in obstetric. 2nd ed. Oxyford Blacwell: p. 174

[29] Herrera E (2002). Lipid metabolism in pregnancy and its consequences in the fetus and newborn. 19(1): 43-55.

[30] Salameh W and Mastrogianis D (1994). Maternal hyperlipidaemia in pregnancy. Clin Obstet. Gynaecol 37:66

[31] Couch SC, Philipson EH, Bendel RB, Pujda LM, Milvae RA and Lammi-Keefe CJ (1998). Elevated lipoprotein lipids and gestational hormones in women with diet-treated gestational diabetses mellitus compared to healthy pregnant controls. J Diabetes and Its Complications 12(1): 1-9

[32] Mazur A, Ozgo M and Rayssiguier Y (2009). Altered plasma triglyceride-rich lipoproteins and triglyceride secretion in feed-restricted pregnant ewes. Veterinani Medicina, 54(9): 412-418

[33] Sattar N, Greer IA, Louden J, Lindsay G, Mcconnell M, Shepher J and Packard CJ (1997). Lipoprotein subfraction changes in normal pregnancy: threshold effect of plasma triglyceride on appearance of small, dense low density lipoprotein. J Clin Endocrinol Metab. 82(8):2483-2491

[34] Herrera E, Lasuncion MA, Zorzano A (1992): Changes with starvation in the rat of the lipoprotein lipase activity and hydrolysis of triacylglycerols from triacylglycerol-rich lipoproteins in adipose tissue preparations. Biochem J 210(3):639-643

[35] Koschinsky T, Gries FA, and Herberg L (1971): Regulation of glycerol kinase by insulin in isolated fat cells and liver of Bar Harbor obese mice. Diabetologia 7(5): 316-322

[36] Thenen SW and Mayer J (1975): adipose tissue glycerokinase activity in genetic and acquired obesity in rats and mice. Proc Soc. Exp. Boil Med. 148(4):953-957

[37] Ramos MP, Crespo-Solans MD, del Campo S, Cacho J and Herrera E (2003): fat accumulation in the rat during early pregnancy is modulated by enhanced insulin responsiveness. Am J Physiol Endocrinol Metba. 285(2):E318-E328

[38] Buch I, Hornnes PJ, Kuhl C (1986): Glucose tolerance in early pregnancy. Acta Endocrinol (Copenh.) 112(2):263-266

[39] Herrera EM, knopp RH and Freikel N (1969). Urinary excretion of epinephrine and norepinephrine during fasting in late pregnancy in the rat. Endocrinology 84(2): 447-450

[40] Sivan E and Boden g (20030. Free fatty acids, insulin resisitance, and pregnancy. Curr. Diab. Rep. 3(4): 319-322

[41] Lopez-Saldado I, Betancor-Fernandez A and Herrera E (2002). Differential metabolic response to 4hr food deprivation at different periods of pregnancy in the rat. J Physiol Biochem 58(2): 75-83.

[42] Herrera E and Lasuncion MA(2004). Maternal-fetal transfer of lipid metabolites. In: fetal and neonatal physiology (vol 1). Polin RA, fox WW, Abman SH(Eds). Saunders, Philadelhia, PA, USA

[43] Shambaugh GE, Metzger BE, Radosevich JA(1992). Nurient metabolism and fetal brain development. In: Perinatal biochemistry, Herrera E and Knopp RH(Eds). CRC Press, Posca Raton, FL, USA

[44] Monlelango A, Lasuncion MA, Pallardo LF and Herrera E(1992). Longitudinal study of plasma lipoproteins and hormones during pregnancy in normal and diabetic women. Diabetes 42(12): 1651-1659

[45] Alvarez JJ, Montelongo A, Iglesias A, Lasuncion MA and Herrera E(1996): Longitudinal study on lipoprotein profile, high density lipoprotein subclass, and postheparin lipases during gestation in women. J Lipid Res 37(2):299-308

[46] Desoye G, Schweditsch MO, Pfeiffer KP, Zechner R and Kostner GM(1987). Correlation of hormones with lipid and lipoprotein levels during normal pregnancy and postpartum. J Clin Endocrinol Metab. 64:704-712

[47] Wasfi I, Weinstein I and Heimberg M(1980). Increased formation of triglyceride from oleate in perfused livers from pregnant rats. Endocrinology 107(2): 584-590

[48] Wasfi I, Weinstein I and Heimberg M(1980). Hepatic metabolism of [1-14C]oleate in pregnancy. Biochim. Biophys. Acta 619(3):471-481

[49] Ramos P and Herrera E(1995). Reversion of insulin resistance in the rat during late pregnancy by 72h glucose infusion. Am J Physiol. 269(5 Pt 1), E858-E863)

[50] Jones HN, Powell TL, Jansson T(2007). Regulation of placental nutrient transport—a review. Placenta 28(8-9):763-774

[51] Koletzko B, Larque E and Demmelmair H(2007). Placental transfer of long-chain polyunsaturated fatty acids(LU-PUFA). J Perinat Med 35(suppl 1):S5-S11

[52] Dutta-Roy AK(2000). Transport mechanism for long-chain polyunsaturated fatty acids in the human placenta. Am J clin Nutr. 71(Suppl 1):315S-322S

[53] Herrera E, Ortega H, Alvino G, Giovannini N, Amusquivar E, and Cetin I(2004). Relationship between plasma fatty acids profile and antioxidant vitamins during normal pregnancy. Eur J Clin Nutr. 58(9): 1231-1238

[54] Koletzko B and Braun M(1991). Arachidonic acid and early human growth:Is there a relation?. Ann nutr Metab 35(3):128-131

[55] Woollett LA(2005). Maternal cholesterol in fetal development: transport of cholesterol from the maternal to fetal circulation. Am J clin nutr. 82:1155-1161

[56] Stefulj J, Panzenboeck U, Becker T, et al, (2009). Human endothelial cells of the placental barrier efficiently deliver cholesterol to the fetal circulation via ABCA1 and ABCG 1. Circ Res 104950:600-608

[57] Lasuncion MA, Lorenzo J, palacin M and Herrera E(1987). Maternal factors modulating nutrient transfer to fetus. Boil Neonate 51(2):86-93

[58] Martin-Hidalgo A, Holm C, Belfrage P, Schotz MC and Herrera E(1994). Lipoprotein lipase and hormone-sensitive lipase activity and mRNA in rat adipose tissue during pregnancy. Am J Physiol 266(6 Pt 1):E930-E935

[59] Superko HR and King S 111 (2008). Lipid management to reduce cardiovascular risk: a new strategy is required. Circulation 117:560-568

[60] Loshak D. Ratio of Total to LDL Cholesterol is Best Predictor of Coronary heart Disease: A DG Review of: "Efficacy of cholesterol levels and ratios in predicting future coronary heart disease in a Chinese population". Am. J. Cardiol. 2001; 88:737-743.

[61] Gotto AM, Assmann G, Carmena R, et al(2000). The ILIB lipid hand-book for clinical practice: blood lipids and coronary heart disease. 2nd ed New York, NY: International Lipid Information Bureau; p 52, 53, 201

[62] Manninen V, Tenkanen L, Koskinen P, et al(1992). Joint effects of serum triglyceride and LDL cholesterol and HDL cholesterol concentrations on coronary heart disease risk in the Helsinki Heart Study. Implications for treatment. Circulation 85:37-45

[63] Walldius G, Junter I, Aastveit A, Holme I, Furberg CD, Sniderman AD(2004). The ApoB-ApoA-1 ratio is better than the cholesterol ratios to elstimate the balance between the plasma proatherogenic and antiatherogenic lipoproteins and to predict coronary risk. Clin Chem Lab Med 42:1355-1363

[64] Sniderman AD, Junter I, Holme I, Aastveit A, Walldius G(2006). Error that result from using the ApoB/ApoA-1 ratio to identify the lipoprotein-related risk of vascular disease. J Intern Med 259:455-461

[65] Walldius G, Junter I(2006). The ApoB/apoA-1 ratio: a stronger, new risk factor for cardiovascular disease and a target for lipid-lowering therapy---a review of evidence. J Intern Med 259:493-519

[66] Thompson A and Danesh J(2006). Association between apolipoprotein B, apolipoprotein A1, the apolipoprotein B/A-1 ratio and coronary heart disease: a literature-based meta-analysis of prospective studies. J Intern Med 259:481-492

[67] Dobiasova M and Frohlich J(2001). The plasma parameter log(TG/HDL-C) as an atherogenic index: correlation with lipoprotein particle size and esterification rate in apoB-lipoprotein-depleted plasma(FERHDL). Clin Biochem 34:583-588

[68] Vega GL, Beltz WF, Grundy SM(1985) Low density lipoprotein metabolism in hypertriglyceridaemic and normolipidemic patients with coronary heart disease. J Lipid Res 26:115-126

[69] Linn S, Fulwood R, Carroll M, Brook JG, Johnson C, Kalsbeek WD, and Rifkind BM(1991). Serum total cholesterol:HDL cholesterol ratios in US white and Black adults by selected demographic and socioeconomic variables(HANES II). Am J Public Health; 81(8): 1038-1043

[70] American Heart Association. Cholesterol statistics from National Health and Nutrition Examination Survey(NHANES), 1999-2004, National Center for Health Statistics and the NHLB1. Available at:
http://www.americanheart.org/presenter.jhtml?identfier=4506. Accessed on 12-11-07.

[71] [71]Tan CE, Squires L, Caslake MJ, et al(1995). Relationship between very low and low density lipoprotein subfractions in normolipaemic men and women. Arterioscler Thomb Vasc Biol 13:1839-1848

[72] McNamara JR, Campos H, Ordovas JM, Peterson JH, Wilson PWF, Schaefer E(1987). Effect of gender,age, and lipid status on low density lipoprotein subfraction distribution. Arteriosclerosis 7:483-490

[73] Austin MA, King MC, Vraizian KM, Krauss RM(1990). The atherogenic lipoprotein phenotypes: a proposed gentic marker for coronary heart disease risk. Circulation 82:495-506

[74] Krauss RM(1991). The tangled web of coronary risk factors. Am J Med 90:36S-41S

[75] Griffin BA, freeman DJ, Tait GW, et al(1994). Role of plasma triglyceride in the regulation of plasma low density lipoprotein(LDL) subfractions: relative contribution of small, dense LDL to coronary heart disease risk. Atherosclerosis 106:241-253

[76] Austin MA and Edward KL(1996). Small, dense low density lipoprotein, and insulin resistance syndrome and noninsulin-dependent diabetes. Curr Opin Lipidol 7:167-171

[77] Toescu V, Nuttall SL, Martin U, et al(2004). Changes in plasma lipids and markers of oxidative stress in normal pregnancy and pregnancies complicated by diabetes. Clin Sci(Lond). 106:93-98

[78] Kilby MD, Neary RH, Mackness MI, Durrington PN(1998). Fetal and maternal lipoprotein metabolism in human pregnancy complicated by type 1 diabetes mellitus. J Clin Endocrinol Metab. 83:1736-1741

[79] Mshelia DS, Kulima AA, Gali RM, Kawuwa MB, Mamza YP, Habu SA, et al(2010).The use of plasma lipid and lipoprotein ratios in interpreting the hyperlipidaemia of pregnancy. 30(8):804-808

[80] Martin U, Davies C, Hayavi S, Hartland A and Dunne F(1999). Is normal pregnancy atherogenic?. Clin Sci 96: 421-425

[81] Rajman I, Maxwell S, Cramb R and Kendall MJ(1994). Particle size: the key to the atherogenic lipoprotein? Q J Med 87:709-720

[82] Dejager S, Brukert E and Chapman MJ (1993). Dense LDL subspecies with diminished oxidative resistance predominate in combined hyperlipidaemia. J Lipid Res 34:295-308

[83] Tribble DL, Van der Berg JIM, Motchnik PA et al (1994). Oxidative susceptibility of low density lipoprotein subfractions is related to their ubiquinol-10 and alpha-tocopherol content. Proc Natl. Acad. Sci USA 91: 1183-1187

[84] Campos H, Genest JJ, Blijlevens E et al (1992). Low density lipoprotein particles size and coronary artery disease. Arteriosclerosis Thomb 12:187-195

[85] Bengtsson C, Rybo G and Westerberg H (19730. Number of pregnancies, use of contraceptives and menopausal age in women with ischaemic heart disease, compared to a population sample of women. Acta Med Scand Suppl 549: 75-81

[86] Napoli C, Glass CK, Witztum JL, Deutsch R, D'Armiento FP Palinski W(1999). Influence of maternal hypercholesterolaemia during pregnancy on progression of early atherosclerotic lesions in childhood: Fate of early lesions in Children(FELIC) study. Lancet 354:1234-1241

[87] Palinski W, D'armiento FP, Witztum JL, de Nigris F, Casanada F, Condorelli M, et al.(2001). Maternal hypercholesterolaemia and treatment during pregnancy influence the long-term progression of atherosclerosis in offspring of rabbits 89:991-996
[88] Barker D(1998). In utero programming of chronic disease . Clin Sci. 95:115-128
[89] Serdar Z, Gur E, Colakodullary M, Develiodlu O, Saradol E (2003). Lipid and protein oxidation and antioxidant function in women with mild and severe pre-eclampsia. Arch Gynaecol Obstet 268:19-25
[90] Belo L, Caslake M, Gaffney D, Santos-Silva A, Pereira LL, Quintanilha A, et al(2002). Cahnges in LDL size and HDL concentration in normal and pre-eclamptic pregnancies. Atherosclerosis 162:425-432
[91] Williams M, Woelk G, King I, Jenkis I and Mahomed K(2003). Plasma carotenoids, retinol, tocopherols and lipoproteins in pre-eclamptic and normotensive pregnant Zimbabwean women. Am J Hypertens 16:665-672
[92] Winkler M, Wetzka B, Hoffmann M Friedrich I, Kinner M, Baumstark M et al(2003). Triglyceride-rich lipoproteins are associated with hypertension in pre-eclampsia. J Clin Endocrinol Metab 88:1162-1166
[93] Cudihy D and Lee RV(2009). The Pathophysiology of pre-eclampsia: current clinical concepts. 29970: 576-582
[94] Tabano S, Alvino G, Antonazzo P, Grati F, Miozzo M, Cetin I(2006). Placenta LPL gene expression is increased in severe intrauterine growth-restricted pregnancies. Pediatr Res. 59:250-253
[95] Lau T, Pang M, Sabota D, Leung T(2005). Impact of hypertensive disorders of pregnancy at term on infant birth weight. 84:875-877
[96] Sanchez S, Zhang C, Williams M(2003). The influence of maternal triglyceride lvels on infant birth weight in Peruvian women with pre-eclampsia. J Mater Fetal Noenatal Med. 13:328-333
[97] Sitadevi C, Patrudu MB, Kumar Ym et al(1981). Longitudinal study of serum lipids and lipoproteins in normal pregnant and puerperium. Trop Geogr Med 33:319-323
[98] Franz H and Wendler D(1992). A controlled study of maternal serum concentration of lipoproteins in pregnancy-induced hypertension. Arch gynecol Obstet 252:1-6
[99] Montes A, Walden CE, knopp RH, Cheung M, Chapman MB and Albers JJ(1984). Physiologic and Supraphysiologic increases in lipoprotein lipids and apolipoproteins in late pregnancy and postpartum: Possible markers for the diagnosis of 'prelipemia'. Arteriosclerosis 4:407-417.

Chapter 4

A Non-Atherogenic and Atherogenic Lipoprotein Profile in Individuals with Dyslipoproteinemia

Stanislav Oravec, Johannes Mikl, Kristina Gruber and Elisabeth Dostal

Additional information is available at the end of the chapter

http://dx.doi.org/10.5772/46067

1. Introduction

The 1985 Nobel Prize in Medicine was awarded to American lipidologists Goldstein and Brown for their work in identifying the role of the LDL receptor pathway in lipoprotein metabolism and in maintaining the homeostasis of blood cholesterol (Goldstein & Brown 1985).

The discovery of the LDL receptor and an understanding of its role in lipid metabolism in health and illness were a milestone in research into metabolic disorders in lipids. At the same time, some other successes in lipoprotein research were also reported: a new understanding of the role of oxidized LDL in atherosclerosis pathogenesis (Steinberg 1987, Witztum & Steinberg 1991); an update of the Ross theory on atherosclerosis genesis (Ross 1986); studies with hypolipidemics; a cholestyramine study, the Coronary Drug Project with niacin, and the Helsinki Heart Study with gemfibrosil. The next two decades was devoted to the effort to create sophisticated criteria for determining risk groups in populations, developing a consensus about cholesterol, and adopting pharmacological uniformity to achieve so-called target lipid values in at-risk individuals with dyslipidemia. The well-defined criteria as a result of these efforts gave hope to at-risk individuals for longer-term survival without ischemic vascular accidents (Canner *et al.* 1986, Frick *et al.* 1987, Expert panel 2001).

Generally, it was confirmed that hypercholesterolemia represents a risk factor for the development of cardiovascular diseases. In addition to arterial hypertension and nicotine abuse, hypercholesterolemia is considered one of three cardinal risk factors.

Cholesterol in plasma is transported by a sophisticated lipoprotein complex system and is also an active part of this lipoprotein system. Different parts of the lipoprotein system are called lipoprotein families. Every lipoprotein family transports different concentrations of cholesterol in blood plasma, but the major conveyor of cholesterol in plasma is the family of Low Density Lipoproteins, i.e., the LDL family. LDL is considered an atherogenic part of the lipoprotein system (Kwiterovich 2002a, 2002b).

LDL transports a major cholesterol load from the liver to the peripheral cells of the body. Under conditions of impaired LDL catabolism in the periphery, LDL particles persist in the circulation, their physical-chemical characteristics are modified, and the physiological pathway of LDL degradation - via LDL receptors - fails. The consequence of this sequence of events is the formation of an alternative metabolic pathway of LDL degradation through scavenger receptors and the formation of cholesterol deposits in the subendothelial space of the arterial wall. In this way, the process of atherogenesis and atherothrombosis begins; and LDL particles play a crucial role at the beginning and in the development of this injury process in the vessel walls (Berneis & Krauss 2002, Haffner 2006, Fruchart *et al.* 2008).

LDL-cholesterol became a criterion for the degree of atherogenic risk for the development of atherothrombosis. A high LDL-cholesterol concentration in plasma correlates positively with the premature onset of cardiovascular diseases, and is considered a strong cardiovascular risk factor. From this point of view, the aim of treatment of hypercholesterolemia, in secondary as well as in primary prevention, is the reduction of LDL concentration in plasma and a lowering of the cholesterol level to the ′′target reference values" (Expert panel 2001, Backers 2005).

However, in the last few decades, lipoprotein research has focused on the phenomenon of atherogenic and non-atherogenic lipoproteins, atherogenic and non-atherogenic lipoprotein profiles, and on the phenotype A vs. phenotype B characterization (Austin *et al.* 1990, Chait *et al.* 1993, Van *et al.* 2007). The traditional approach to hypercholesterolemia as an atherogenic risk factor for the development of degenerative diseases of the cardiovascular system became a target of criticism. Castelli published evidence that more than 75 percent of patients with an acute coronary syndrome or a myocardial infarction had normal plasma values of cholesterol, LDL cholesterol and/or HDL cholesterol (Castelli 1988, 1992, 1998). Thus, it was necessary to look for other risk factors in plasma, at levels that could cause an acute coronary event. An increased cholesterol level, as an universal explanation for the origin of atherogenesis, was no longer valid.

A reasonable explanation was found in atherogenic lipoprotein subpopulations, the presence of which in plasma, even in very low concentrations, could impair the integrity of the vessel wall and lead to endothelial dysfunction with its fatal consequences: formation of atherothrombotic plaques, acute myocardial infarction, stroke, and sudden death (Nichols & Lundmann 2004, Rizzo & Berneis 2006, Shoji *et al.* 2009, Zhao *et al.* 2009).

Those laboratory analysis methods became an essential contribution to the identification of atherogenic lipoprotein entities, which simplified the analysis and quantification of the atherogenic lipoprotein subfractions. Gradient gel electrophoretic separation of LDL and HDL subclasses or proton nuclear magnetic resonance spectroscopy were the methods of choice for the analysis of these entities (Rainwater *et al.* 1997, Alabakovska *et al.* 2002, Otvos *et al.* 2003).

Recently, electrophoresis of plasma lipoproteins on the polyacrylamide gel (PAG) Lipoprint LDL System is one of several diagnostic analytical methods for the identification and quantitative evaluation of lipoprotein subfractions, i.e., the atherogenic and non-atherogenic lipoproteins (Hoefner *et al.* 2001).

The LDL System has become a staple in routine laboratory analysis and in the diagnosis of lipoprotein metabolism disorders, and has also been recommended by the FDA for human medicine. Lipoprint LDL enables the analysis of 12 lipoprotein subfractions: VLDL; IDL1; IDL2; IDL3; LDL1; LDL2; LDL 3-7; HDL; and determines an atherogenic lipoprotein profile phenotype B versus a non-atherogenic lipoprotein profile phenotype A.

Atherogenic lipoprotein profiles are characterized by a predominance of atherogenic lipoproteins, namely very low density (VLDL), intermediate density IDL1, and IDL2, and particularly by the presence of small dense lipoproteins with low density (LDL). Profiles identify highly atherogenic LDL subfractions that form the LDL 3-7 fractions (Tab.1). These subfractions are smaller, with a diameter < 26.5 nm (265 Angström) and they float within a density range of 1.048 – 1.065 g/ml, i.e., a higher density than LDL1 and LDL2. On the PAG they are detected as subtle bands on the anodic end of the gel right behind HDL that migrate to the head of separated lipoproteins.

Small dense LDL are highly atherogenic for ((Berneis&Krauss 2002, Lamarche *et al.* 1999, Packard 2003, Carmena *et al.* 2004):	
*low recognition by LDL-receptors (configuration alterations Apo B) →	
*enhanced aptitude for oxidation and acetylation →	
*Oxid-LDL	→ release of pro-inflammatory cytokines → muscle cell apoptosis
*Oxid-LDL	→ release of metalloproteinase → collagen degradation
*Oxid-LDL	→ enhanced aptitude for trapping by macrophages (scavenger-receptors) → stimulation of foam cell formation
*easier penetration into the subendothelial space and formation of cholesterol deposits	

Table 1.

On the basis of lipoprotein separation by the Lipoprint LDL System, a non-atherogenic normolipidemia, an atherogenic normolipidemia, a non-atherogenic hyperbetalipoproteinemia and an atherogenic hyperlipoproteinemia can be characterized (Oravec 2006a, 2006b, 2007a, 2007b).

Two of these are identified as new lipoprotein profiles with high clinical significance: an **atherogenic normolipidemia** and a **non-atherogenic hyper-betalipoproteinemia LDL1,2.**

A non-atherogenic hyperbetalipoproteinemia LDL1,2 involves individuals with a high concentration of plasma cholesterol, predominantly transported by LDL1 and LDL2 subfractions. However, these individuals are at low risk for a cardiovascular event based on cardiologic and angiologic examimation results, and have familial history negative for cardiovascular diseases.

Conversely, an atherogenic normolipidemia was identified in a group of individuals with normal cholesterol and triglyceride concentrations in plasma, who had a high concentration of strongly atherogenic small dense LDL in the lipoprotein profile. These individuals could be at higher risk for a cardiovascular event despite normolipidemia.

In our clinical study, we characterized hypercholesterolemic individuals with untreated hypercholesterolemia, who had a non-atherogenic hyperbetalipoproteinemia, as well as normolipemic individuals who were currently without clinical or laboratory signs of damage to the cardiovascular system, but who, nevertheless, had an atherogenic lipoprotein profile. All these subjects underwent a medical examination to identify the extent of the arterial vessel damages caused by hypercholesterolemia, or dyslipidemia.

2. Patients and methods

The hypercholesterolemic individuals with untreated hypercholesterolemia were tested by Lipoprint LDL analysis. In this group of hypercholesterolemic subjects, 145 individuals with a non-atherogenic lipoprotein profile were identified.

Of the total number, 15 individuals were under 40 years of age without clinically apparent impairment and no laboratory signs of cardiovascular disease. These subjects formed one subgroup of younger people (34 years +- 5 years). The subgroup of younger subjects was separated from the older individuals with hypercholesterolemia because a separate analysis of the older subjects with hypercholesterolemia was performed to confirm that undamaged vessels in older individuals persist even into old age, and that diagnosed hypercholesterolemia does not cause an atherogenic impairment in the vessels. The subgroup of older subjects consisted of 130 individuals (32 males, 57 +-11 years of age; and 98 females, 62 +- 9 years of age).

The medical examination, which included a physical examination, blood pressure, and ECG examination, bicycle stress test, echocardiography, and duplex ultrasound examination of the carotid arteries, confirmed that there was no impairment of the cardiovascular system. Only mild signs of clinically irrelevant aortic valve sclerosis were found in the subgroup of older subjects.

Individuals with hyperglycemia, diabetics, and those individuals who were being treated with lipid-lowering drugs were excluded from the study.

The control group consisted of 165 normolipidemic volunteers, all nonsmokers, who had no clinically apparent impairment, or laboratory signs of cardiovascular disease. Volunteers were recruited from medical students at the medical facility. The average age of the subjects was 21.5 ± 2.5 years, and the group involved 65 males and 100 females. All subjects gave written, informed consent, and the study was approved by the local ethics committee.

A blood sample from the antecubital vein was collected in the morning after a 12-hour fasting period. EDTA-K_2 plasma was obtained and the concentration of total cholesterol and triglycerides in plasma was analyzed, using an enzymatic CHOD PAP method (Roche Diagnostics, Germany).

The quantitative analysis of lipoprotein families and lipoprotein subfractions included: VLDL; IDL1; IDL2; IDL3; LDL1; LDL2; LDL3-7; and HDL. A non-atherogenic lipoprotein profile, phenotype A, versus an atherogenic lipoprotein profile, phenotype B, was determined using the Lipoprint LDL System (Quantimetrix Corp., USA; (Hoefner *et al.* 2001). The analysis of HDL subclasses, with their subpopulations, including large HDL-, intermediate HDL-, and small HDL- subclasses in plasma, was also performed using the Lipoprint HDL System (Morais *et al.* 2003).

The Score of the Anti-Atherogenic Risk (SAAR) was calculated as the ratio between non-atherogenic and atherogenic lipoproteins in plasma (Oravec 2007a). SAAR values over 10.8 represented a non-atherogenic lipoprotein profile, whereas values under 9.8 represented an atherogenic lipoprotein profile. The cut-off values for a non-atherogenic lipoprotein profile and an atherogenic lipoprotein profile were calculated from the results of 940 Lipoprint LDL analyses. Using the Quantimetrix Lipoprint LDL system interpretation, all 940 individuals were examined (general group of subjects) and tested for the occurrence of atherogenic and non-atherogenic lipoprotein profiles, and were then divided into the two subgroups of subjects with an LDL profile:

- Indicative of Type A, i.e., a non-atherogenic lipoprotein profile phenotype A
- Not indicative of Type A, i.e., an atherogenic lipoprotein profile, phenotype B (Hoefner *et al.* 2001).

For practical interpretation of the analysed lipoprotein profiles using the Lipoprint LDL System, for the non-atherogenic lipoprotein phenotype A, the subtypes 1a, 1b, 2a, 2b, 3, and 4 were introduced, because of the large profile heterogeneity in the non-atherogenic lipoprotein profile phenotype A. With regard to the atherogenic lipoprotein phenotype B, only subtype 5 and subtype 6 were introduced. (Oravec 2007b). (Tab.2)

Statistical evaluation of obtained values was performed by an unpaired student´s t-test. The level of significance was set at $p < 0.05$.

3. Results

The subjects with a non-atherogenic hypercholesterolemia had a significantly increased concentration of total cholesterol and lipoprotein parameters ($p<0.0001$), except for LDL 3-7 subfractions (small dense LDL), which were significantly lower ($p<0.0001$), compared to the control group (Tab.3). The highest increase of concentrations was found for total cholesterol, LDL cholesterol, HDL cholesterol, IDL3, and LDL1 subfractions. The concentration of LDL1 exceeded the LDL1 concentration in the control group by more than 88 percent. The LDL1 concentration in the younger hypercholesterolemic subjects reached 1.84 mmol/l, i.e., more than twice, comparing to 0.89 mmol/l in the control group. (Tab.3, Tab.5). The rise of LDL2 concentration (32 percent in younger hypercholesterolemic subjects), did not match the increase in LDL1 concentrations (Tab.3 - Tab.6).

A. Non-atherogenic lipoprotein profile, phenotype A ……	**59 %**
1a. Subtype: Non-atherogenic lipoprotein profile phenotype A…..	11 %
Atherogenic lipoproteins absent	
LDL cholesterol normal	
1b. Subtype: Non-atherogenic lipoprotein profile phenotype A…..	10 %
Atherogenic lipoproteins absent	
LDL cholesterol elevated	
2a. Subtype: Non-atherogenic lipoprotein profile phenotype A……	12%
Atherogenic lipoproteins present in traces	
LDL cholesterol normal	
2b. Subtype: Non-atherogenic lipoprotein profile phenotype A…	11%
Atherogenic lipoproteins present in traces	
LDL cholesterol elevated	
3. Subtype: Non-atherogenic lipoprotein profile phenotype A……	3%
Atherogenic lipoproteins present	
LDL cholesterol normal	
4. Subtype: Non-atherogenic lipoprotein profile phenotype A …….	12%
Atherogenic lipoproteins present	
LDL cholesterol elevated	
B. Atherogenic lipoprotein profile phenotype B …………	**41 %**
5. Subtype: Atherogenic lipoprotein profile phenotype B……………	12%
Atherogenic lipoproteins present	
LDL cholesterol normal	
6. Subtype: Atherogenic lipoprotein profile phenotype B…..............	29%
Atherogenic lipoproteins present	
LDL cholesterol elevated	

An atherogenic lipoprotein profile was identified in 41% of examined individuals in a general group of subjects (n = 940), (Oravec 2007b).

Table 2. Incidence rate of non-atherogenic vs. atherogenic lipoprotein subtypes in a general group of subjects (n = 940)

	T-Chol (mmol/l ±SD)	TAG	VLDL	IDL1	IDL2	IDL3	LDL1	LDL2	LDL3-7	T-LDL	HDL	SAAR
Control	4.31	1.16	0.62	0.39	0.28	0.33	0.89	0.41	0.04	2.34	1.33	36.1
n = 165	±0.62	±0.39	±0.16	±0.16	±0.09	±0.12	±0.28	±0.21	± 0.04	±0.54	±0.32	±20.6
H-βLP	6.71	1.29	0.74	0.55	0.51	0.82	1.68	0.52	0.01	4.09	1.88	76.0
n= 145	±0.90	±0.49	±0.21	±0.16	±0.12	±0.23	±0.36	±0.21	±0.01	±0.69	±0.46	±17.0
Control vs. HLP	p< 0.0001	n.s.	<........................p< 0.0001........................>									

Legend: T-cholesterol: total cholesterol, T-LDL: total LDL-cholesterol, H-βLP: hyperbetalipoproteinemia

Table 3. Plasma concentration of lipids, lipoproteins, and SAAR score in the group of hypercholesterolemic subjects vs. control normolipidemic subjects

	T-Chol (mmol/l ±SD)	TAG	VLDL	IDL1	IDL2	IDL3	LDL1	LDL2	LDL3-7	T-LDL	HDL	SAAR
Control	4.31	1.16	0.62	0.39	0.28	0.33	0.89	0.41	0.04	2.34	1.33	36.1
n = 165	36.1	±0.39	±0.16	±0.16	±0.09	±0.12	±0.28	±0.21	± 0.04	±0.54	±0.32	±20.6
H-βLPs	6.73	1.30	0.73	0.55	0.52	0.80	1.67	0.52	0.01	4.08	1.93	76.5
n= 130	±0.91	±0.48	±0.19	±0.16	±0.13	±0.23	±0.35	±0.22	±0.01	±0.69	±0.45	±18.1
Control vs. HLP	p< 0.0001	n.s.	<........................p< 0.0001........................>									

Legend: H-βLPs : hyperbetalipoproteinemia subgroup of seniors

Table 4. Plasma concentration of lipids, lipoproteins, and SAAR score in the subgroup of older hypercholesterolemic subjects and controls

	T-Chol (mmol/l ±SD)	TAG	VLDL	IDL1	IDL2	IDL3	LDL1	LDL2	LDL3-7	T-LDL	HDL	SAAR
Control	4.31	1.16	0.62	0.39	0.28	0.33	0.89	0.41	0.04	2.34	1.33	36.1
n = 165	±0.62	±0.39	±0.16	±0.16	±0.09	±0.12	±0.28	±0.21	± 0.04	±0.54	±0.32	±20.6
H-βLPjr	6.62	1.20	0.84	0.58	0.44	0.80	1.84	0.54	0.01	4.20	1.46	71.1
n= 15	±0.80	±0.59	±0.28	±0.18	±0.01	±0.25	±0.42	±0.18	±0.01	±0.64	±0.23	±13.2
Control vs. HLP	p< 0.0001	n.s.	<..........p< 0.0001..........>					n.s.	n.s.	p< 0.0001	n.s.	p< 0.0001

Legend: H-βLP jr : hyperbetalipoproteinemia subgroup of younger hypercholesterolemic subjects

Table 5. Plasma concentration of lipids, lipoproteins, and SAAR-score in the subgroup of younger hypercholesterolemic subjects and controls

	T-Chol (mmol/l ±SD)	TAG	VLDL	IDL1	IDL2	IDL3	LDL1	LDL2	LDL3-7	T-LDL	HDL	SAAR
H-βLP jr	6.62	1.20	1.20	0.58	0.44	0.80	1.84	0.54	0.01	4.20	1.46	71.1
n= 15	±0.80	±0.59	±0.28	±0.18	±0.01	±0.25	±0.42	±0.18	±0.01	±0.64	±0.23	±13.2
H-βLPs	6.73	1.30	0.73	0.55	0.52	0.80	1.67	0.52	0.01	4.08	1.93	76.5
n= 130	±0.91	±0.48	±0.19	±0.16	±0.13	±0.23	±0.35	±0.22	±0.01	±0.69	±0.45	±18.1
	<.. n.s..>										p< 0.001	n.s.

juniors v.s. seniors

Legend: H-βLP jr.: Hyperlipoproteinemia subgroup of younger subjects
H-βLP s.: Hyperlipoproteinemia subgroup of older subjects

Table 6. Plasma concentration of lipids, lipoproteins, and SAAR-score in the subgroup of younger (n=15) versus older (n=130) hypercholesterolemic subjects

The lipid and lipoprotein parameters in younger and older hypercholesterolemic subjects were very similar, and the results were not statistically significantly different between the groups, except that HDL cholesterol in the older hypercholesterolemic individuals was statistically significant higher (p<0.001) compared to the control group (Tab.6). Results similar to those in older hypercholesterolemic subjects were obtained when the group of younger hypercholesterolemic subjects was compared to the control group (Tab.5), except for LDL2, LDL 3-7, and HDL lipoproteins, where the changes in the cholesterol concentrations - increased in LDL2- and decreased in LDL3-7 subfractions were not significant.

	T-HDL mmol/l ± SD	HDL large	HDL intermediate	HDL small
Control	1.31	0.59	0.56	0.15
(n=103)	± 0.29	± 0.23	± 0.10	± 0.09
H-βLP	1.51	0.70	0.65	0.15
(n=110)	± 0.34	± 0.46	± 0.42	± 0.12
	p< 0.0001	p< 0.005	p< 0.005	n.s.

Legend: T-HDL: total HDL

Table 7. Plasma concentration of HDL lipoprotein subclasses

Tab.7 shows the HDL-cholesterol concentration and HDL subclasses, analysed by the Lipoprint HDL System. The concentration of total HDL cholesterol (T-HDL) in the group of hypercholesterolemic subjects was significantly higher (p<0.0001), compared to the control group. There was an increased concentration of both HDL subclasses, i.e. the HDL large subclass (p<0.005) and the HDL intermediate subclass (p<0.005) in the hypercholesterolemia subjects. The difference in the concentration of the HDL small subclass between hypercholesterolemic subjects and the control group was not confirmed.

	T-Chol (mmol/l ±SD)	TAG	VLDL	IDL1	IDL2	IDL3	LDL1	LDL2	LDL3-7	T-LDL	HDL	SAAR
Subjects with a non atherogenic profile, n = 155	4.31 ±0.62	1.12 ±0.38	0.62 ±0.16	0.39 ±0.17	0.28 ±0.09	0.33 ±0.12	0.91 ±0.27	0.40 ±0.21	0.03 ±0.03	2.33 ±0.54	1.33 ±0.32	38.1 ±19.6
Subjects with an atherogenic profile, n = 10	4.37 ±0.50	1.63 ±0.30	0.72 ±0.14	0.36 ±0.08	0.28 ±0.06	0.27 ±0.08	0.67 ±0.17	0.55 ±0.14	0.25 ±0.06	2.37 ±0.34	1.27 ±0.36	5.3 ±2.0
All subjects n = 165	**4.31** ±0.62	**1.16** ±0.39	**0.62** ±0.16	**0.39** ±0.16	**0.28** ±0.09	**0.33** ±0.12	**0.89** ±0.28	**0.41** ±0.21	**0.04** ± 0.04	**2.34** ±0.54	**1.33** ±0.32	**36.1** ±20.6
nonath.vs.athero		p<0.001					p< 0.01	p< 0.02	p< 0.0001			p<0.0001

Table 8. Plasma concentration of lipids, lipoproteins, and the SAAR score in the subgroups of normolipemic control volunteers

Tab.8 shows the lipid and lipoprotein values obtained and the Score for Anti-Atherogenic Risk (SAAR) in the examined group of 165 control subjects.

In a subgroup of 155 subjects, a non-atherogenic lipoprotein profile phenotype A was identified. In a subgroup of 10 subjects, an atherogenic lipoprotein profile phenotype B was identified. Both lipoprotein phenotypes were confirmed by the Lipoprint LDL method. All examined subjects had normal values of cholesterol and triglycerides. The highest significant difference ($p<0.0001$) between the subgroup with an atherogenic lipoprotein profile phenotype B and a non-atherogenic lipoprotein profile phenotype A was found in the subfractions LDL 3-7, i.e., small dense LDL ($p< 0.0001$), which represent strongly atherogenic lipoproteins. The SAAR score also showed highly significant differences in the values between the atherogenic and the non-atherogenic subgroup ($p<0.0001$). There was a higher concentration of triglycerides ($p<0.001$) in the atherogenic subgroup. LDL1 was higher in the non-atherogenic subgroup ($p<0.01$) and LDL2 was higher in the atherogenic subgroup.

4. Discussion

The identification of atherogenic and non-atherogenic lipoproteins in the plasma lipoprotein spectrum represents a deeper analysis of lipoprotein parameters than a routine analysis of plasma cholesterol, triglycerides, or lipoproteins like LDL, HDL, and VLDL. These lipid parameters only provide limited information about the percentage of subjects in the general population (general group of subjects) who are at-risk for a sudden attack for cardiovascular or cerebral-vascular event. The 41 percent of the subjects from our large population of 950 individuals, who were identified by this analytical method, would not otherwise have been identified, confirming the value of this information for physicians (Tab.2) know that, based on mortality statistics, approximately 50 percent of deaths are caused by cardiovascular events. It may be that this 41 percent represents a major part of the 50 percent of deaths attributable to a cardiovascular cause, and the individuals with atherogenic dyslipidemia are surely at risk for a sudden cardiovascular event. Thus these individuals could be target for close monitoring, have a follow-up examination, and the optimal treatment could be recommended.

In addition, the identification of six percent of normolipidemic young healthy individuals with an atherogenic lipoprotein profile among clinically healthy volunteers questions our knowledge and generally accepted belief that normolipidemia, 'per se', represents an optimal health lipid constellation (Tab.8). An atherogenic normolipidemia in the lipoprotein profile of our clinically healthy subjects represents a new phenomenon. These individuals are also at risk for the development of premature cardiovascular ischemic disease and should undergo close medical follow-up. If these individuals receive no preventive anti-atherothrombotic measures, the manifestation of cardiovascular ischemic diseases is certain later in life.

The findings of hypercholesterolemia in clinically healthy subjects, without clinically apparent signs of cardiovascular disease or laboratory confirmation of cardiovascular disease, and with a negative history for the occurence of cardiovascular events, stimulated an active search for hypercholesterolemic indviduals and the initiation of a medical examination of these subjects.

For the identification of the hypercholesterolemic individuals with a non-atherogenic lipoprotein profile, a new innovative electrophoretic method for the analysis of plasma lipoproteins on polyacrylamide gel (PAG) was used (Hoefner *et.al* 2001). The method can analyze the total lipoprotein spectrum of examined subjects, identify an atherogenic/non-atherogenic lipoprotein profile, and quantify the atherogenic lipoprotein subpopulations in plasma, including strongly atherogenic LDL subpopulations, i.e., the small dense LDL, which form the subfractions LDL 3-7. In the absence of atherogenic lipoproteins, or when the atherogenic lipoproteins form a minor part of the whole lipoprotein spectrum, a non-atherogenic lipoprotein profile exists.

The identification of a non-atherogenic hypercholesterolemia offers new information, which suggests a re-evaluation of the belief that the whole LDL family is an atherogenic lipoprotein part of the plasma lipoprotein spectrum. Our results confirme the results of several previous research studies. They show that only a part of the LDL is atherogenic. Atherogenic are small dense LDL, subfraction of LDL, which are associated with the premature development of ischemic cardiovascular diseases. In contrast, LDL1 and LDL2, even in higher concentrations in plasma, do not represent a high cardiovascular risk. Also negative cardiological examination with normal results: only milde signs of clinically irrelevant aortic valve sclerosis, support and confirm the non-atherogenicity of large 'buoyant' LDL subfractions in the individuals with hyperbetalipoproteinemia LDL1,2. Fig.1 - 4. Based on these laboratory results and medical findings, the medical approach to these hypercholesterolemic individuals needs to be revised. The intensive hypolipidemic treatment should not be recommended, and the question also remains, whether any treatment at all, in cases of non-atherogenic hypercholesterolemia, in general, is a reasonable clinical decision. The reduction of total LDL-cholesterol as a target for hypolipidemic treatment for prevention of atherogenesis and atherothrombosis seems to be no longer necessary.

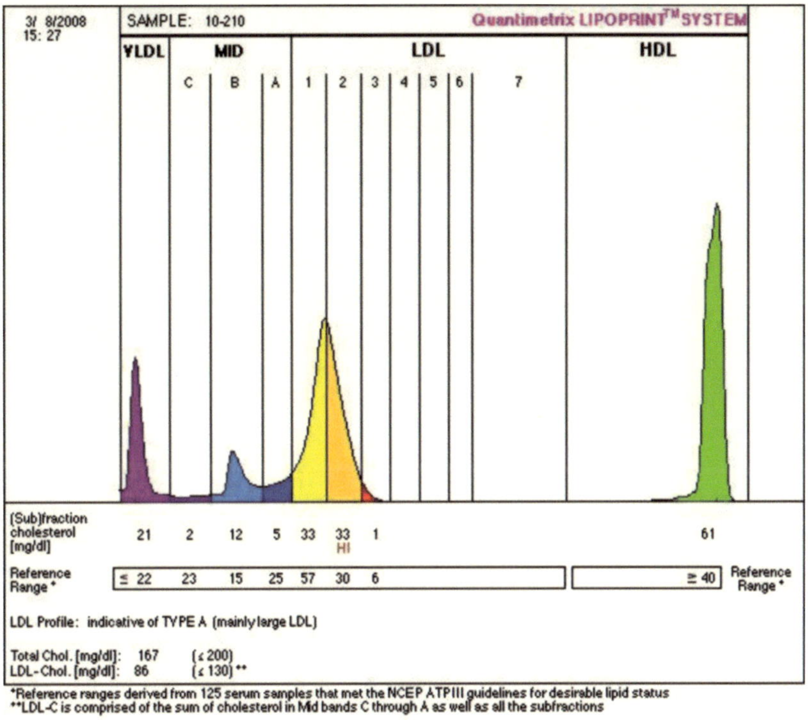

Figure 1. Non-atherogenic normolipidemia

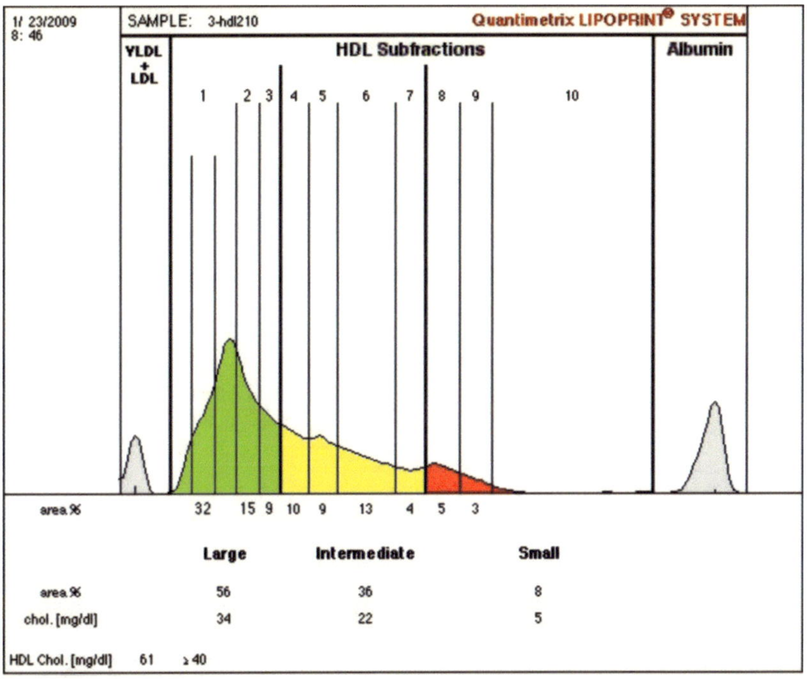

Figure 2. Non-atherogenic normolipidemia HDL subfractions

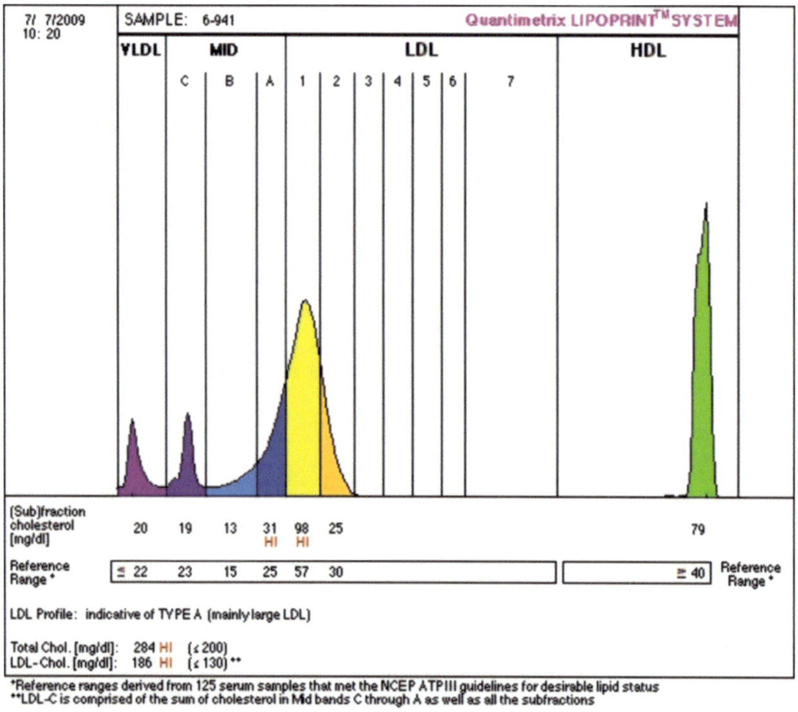

Figure 3. Non-atherogenic hyperbetalipoproteinemia LDL1,2

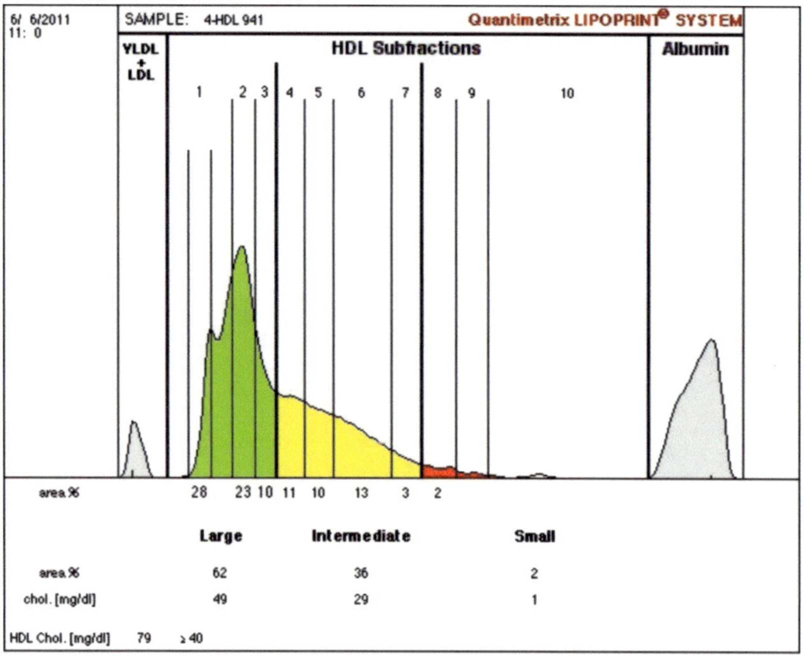

Figure 4. Non-atherogenic hyperbetalipoproteinemiaLDL1,2 HDL subfractions

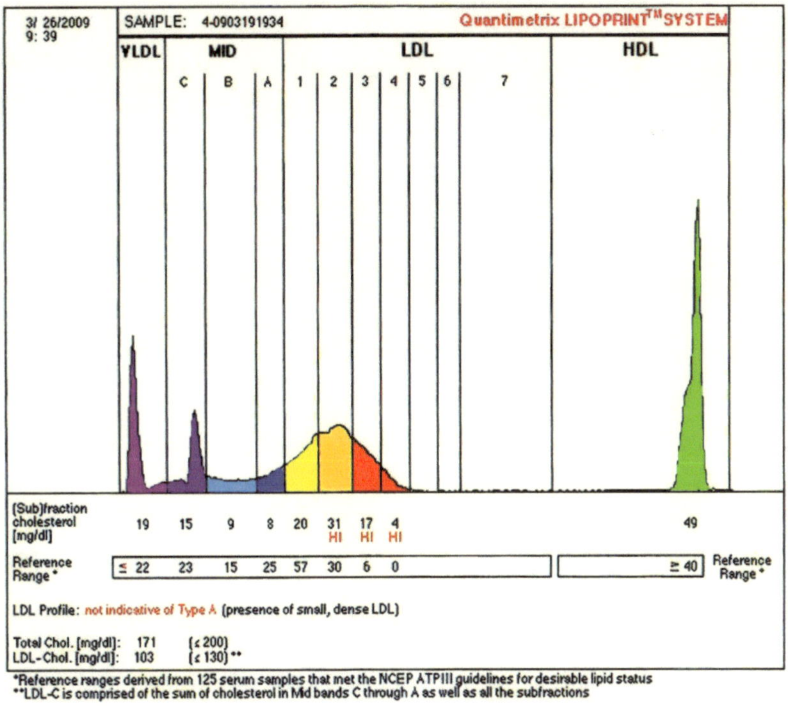

Figure 5. Atherogenic normolipidemia

LDL represent a lipoprotein family created by several LDL subfractions with different characteristics and different role in the intermediary metabolism and in the atherothrombogenesis. LDL1 and LDL2 subfractions are important physiological major conveyors of cholesterol in plasma. These subfractions are an important source for the biosynthesis of highly physiologically effective drugs and structures in the body (steroid hormones, bile acids, vitamin D3, membranes of cells and of subcellular structures). Lowering of the concentration of LDL1 and LDL2 by using a non-specific hypolipidemic treatment has a negative effect on several physiological processes, which create the optimal maintenance of healthy equilibrium in the body. LDL1 and LDL2 seem to be a not atherogenic part of LDL. The non-specific lowering of total cholesterol reduces in the first step the concentration of cholesterol in LDL1, LDL2 subfractions. A protective part of LDL (LDL1, LDL2) is removed and the strong atherogenic small dense LDL persist.

The non-specific hypolipidemic treatment does not form a non-atherogenic lipoprotein constellation. On the contrary, along with the impairment of endocrine steroid synthesis in the body, with an unjustified hypolipidemic treatment approach, the atherogenicity of the plasma will be increased. Figure 6 - 8 shows a Lipoprint LDL picture of atherogenic normolipidemia obtained frequently after hypolipidemic treatment of atherogenic hypercholesterolemia.

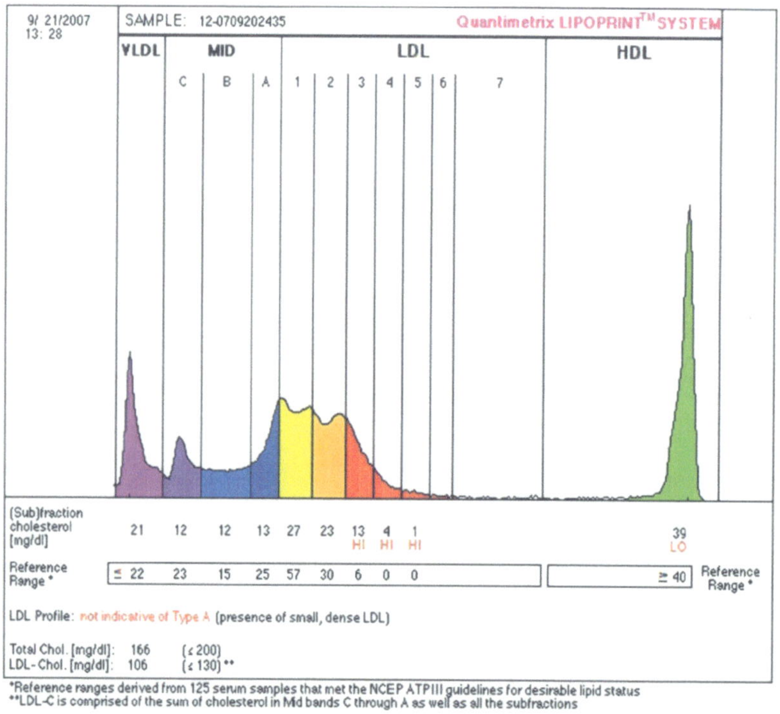

Figure 6. Atherogenic normolipidemia obtained frequently after hypolipidemic treatment of atherogenic hypercholesterolemia

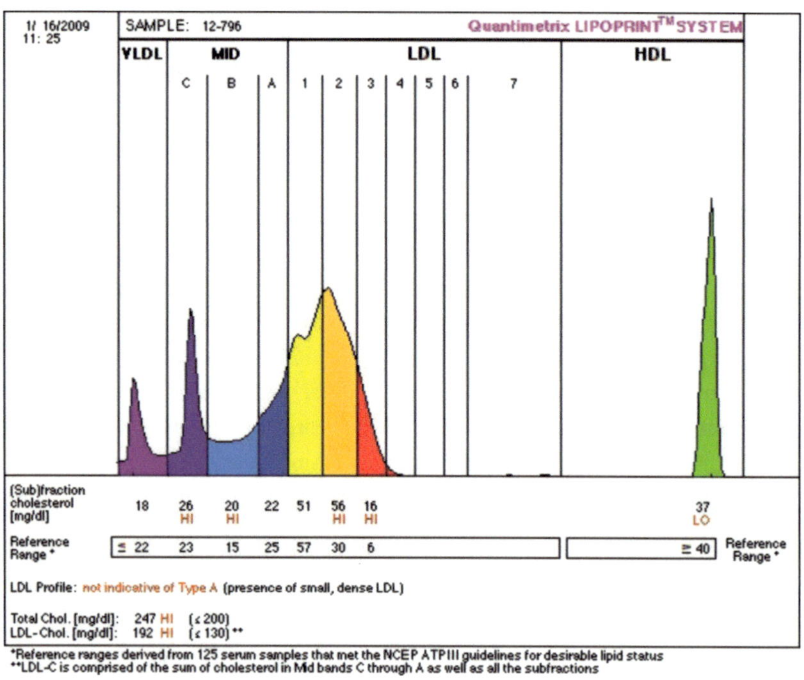

Figure 7. Atherogenic hypercholesterolemia

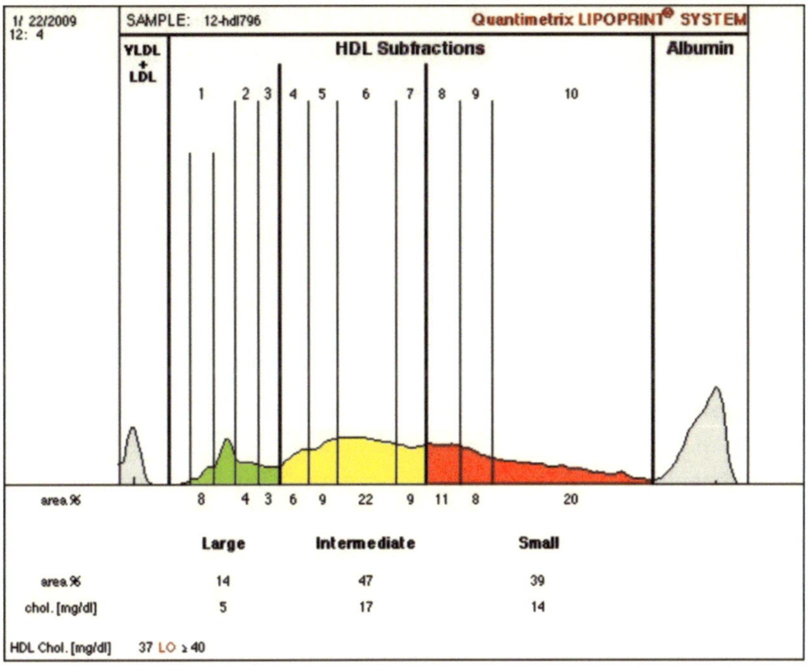

Figure 8. Atherogenic hypercholesterolemia HDL subfractions

In our study a group of individuals with hypercholesterolemia was divided into two subgoups: younger and older subjects (Tab.3-6). The reason was to differentiate the influence of the age factor on the lipoprotein constellation and on the quality of the vascular wall, especially in the group with older subjects. The quality of the arteries was evaluated by medical examination. Tested individuals were examined, including physical examination, blood pressure, and ECG examination, a bicycle stress test, echocardiography, and duplex ultrasound examination of the carotid arteries. The medical results confirmed that the vessel wall was not seriously impaired, not even in older subjects with hypercholesterolemia, which is why a hyper-betalipoproteinemiaLDL1,2 does not represent a serious cardiovascular risk for individuals with this type of hypercholesterolemia.

The results of HDL subclass analysis (Lipoprint HDL System (Morais *et al.* 2003) in individuals with a non-atherogenic hyperbetalipoproteinemia LDL1 confirm a supposition of low atherogenicity in hyperberalipoproteinemia LDL1,2 (Tab.7), The lipoprotein profile of HDL typically contains a predominance of HDL large and HDL intermediate subclasses, which confer a protective, anti-atherogenic effect on the vessel wall (Morais 2005, Muniz & Morais 2005 Oravec *et al.* 2011c). The small HDL subclass with atherogenic characteristics was present in the lipoprotein profile in low concentrations only, compared to the control group of healthy volunteers. Fig.3 – 4.

The major findings can be summarized as follows:

1. In examined subjects with hypercholesterolemia, a non-atherogenic lipoprotein profile, phenotype A was confirmed with a high concentration of LDL1 and LDL2 subfractions. In particular, the LDL1 subfraction was nearly double that of the LDL1 of the control group, and, in some individual cases, three times that of the control group average (Oravec et al. 2011b).
2. The lipoprotein electrophoresis confirmed only a trace concentration of LDL3-7 subpopulations (1mg LDL 3-7 cholesterol/dl, i.e., 0.0256 mmol/l). In the overwhelming majority of subjects (60%) indeed, there was an absence of the atherogenic LDL 3-7 in the lipoprotein profile of these subjects. (Plasma lipoprotein profiles for patients with confirmed cardiovascular disease are generally characterized by a high concentration of small dense LDL) (Kwiterovich 2000, Maslowska 2005, Oravec 2010, Oravec *et al.* 2010a, 2010b, Oravec *et al.* 2011a).
3. The concentration of HDL was significantly increased ($p<0.0001$) compared to the control group, with an overwhelming majority of the non-atherogenic HDL subpopulations, HDL large and HDL intermediate. The concentration of small dense HDL was not increased (Tab.7), Fig 1-4. Small dense HDL form an atherogenic part of the HDL lipoprotein spectrum, and their higher plasma concentration corelates with the development of cardiovascular diseases (Luc *et al.* 2002, St Pierre *et al.* 2005, Morais 2005, Muniz & Morais 2005, Oravec *et al.* 2011d), Fig.7,8. The structural representation of HDL subpopulations confirmed a non-atherogenic type of lipoprotein profiles in our examined group of hypercholesterolemic subjects.
4. The examined individuals, despite increased total cholesterol and LDL cholesterol values, were healthy, without apparent clinical signs of cardiovascular disease (angina pectoris, cardiac insufficiency, myocardial infarction, or other survived cardiovascular events). There is evidence that an optimal anti-atherogenic LDL profile (see the lipoprotein results) could actually have a vasoprotective effect in tested hypercholesterolemic individuals. Based on the present results, a further, more extensive study will continue to evaluate the Lipoprint electrophoretic method as a standard method for the diagnosis of cardiovascular risk, along with the standard tests now used (ECG examination, bicycle stress test, echocardiography, and duplex ultrasound examination of the carotid arteries).
5. The newly introduced SAAR, a ratio of non-atherogenic/atherogenic lipoproteins, also confirmed a non-atherogenic lipoprotein constellation in the plasma of hypercholesterolemic individuals (Oravec 2007a).

Based on the results of examined individuals with hypercholesterolemia, these conclusions can be drawn:

1. LDL1 and LDL2 do not fulfill the criteria of atherogenicity for lipoprotein entities that is usually ascribed to LDL lipoproteins.
2. LDL1 and LDL2 subfractons in hypercholesterolemic indidviduals, in our study group, created a non-atherogenic hypercholesterolemia - a non-atherogenic hyperbetalipo-

proteinemia LDL1,2 without the presence of atherogenic small dense LDL (or with traces only) that are typically associated with a high concentration of cardiovascular protective HDL subfractions in the plasma lipoprotein spectrum.

We report the existence of a newly described type of hypercholesterolemia, **a non-atherogenic hyperbetalipoproteinemia LDL 1,2**, characterized by a minimal onset of cardiovascular complications, even in those individuals who are not treated with hypolipidemic therapy.

The hypercholesterolemic subjects of the study group are still undergoing follow-up examinations.

4.1. Atherogenic normolipidemia

Generally, a normolipidemia is interpreted as an equilibrated state of lipoprotein metabolism, characterized by total cholesterol and triglyceride values within reference ranges. We know from clinical experience that patients with normolipidemia are better protected from development of cardiovascular diseases and degenerative vessel changes, a source of cardio-vascular disease.

In normolipidemia, of the goal is to create a non-atherogenic lipoprotein profile and to lower or eliminate the risk of atherosclerosis development and prevent the rise of an acute cardiovascular event. However, the existence of an atherogenic normolipidemia disproves the theory that normolipidemia provides protection against the development of atherosclerotic vessel impairment. A premature atherosclerosis development can be found even in young people, adolescents with the high risk (Backers 2005; Rizzo & Berneis 2006).

An atherogenic lipoprotein profile is characterized by the rich presence of atherogenic lipoproteins, very low density lipoprotein (VLDL), intermediate density lipoproteins (IDL1, IDL2), and especially, by the presence of small dense low-density lipoproteins (sdLDL), which form LDL 3-7 subfractions, and which are strongly atherogenic (Lamarche *et al.* 1997; Gardner et al. 1996; Rajman *et al.* 1996; Halle *et al.* 1998, Austin *et al.* 1994).

An analysis of the lipoprotein profile by the Lipoprint LDL system reveals a new lipoprotein composition in lipoprotein profile and focuses authors on a new clinical-diagnostic phenomenon: an **atherogenic normolipidemia**. Compared to the well known atherogenic dyslipidemia, or atherogenic hyperlipoproteinemia, this new **atherogenic normolipidemia** (Oravec *et al.* 2010; Oravec *et al.* 2011d) is not identifiable by common biochemical diagnostic analysis.

This phenomenon represents a serious cardiovascular risk for individuals with this profile, and these individuals at high cardiovascular risk are not currently identified, diagnosed, medically registered, or treated. The presence of an **atherogenic normolipidemia** enlarges the portion of the population at increased risk for a cardiovascular event, however these individuals at risk do not participate on the protective measures of primary cardiovascular prevention. Fig.5. Medical community does not know till now, that the individuals with an atherogenic normolipimia are at-risk individuals for the development of premature ischemic cardiovascular diseases. Identification of the type of lipoprotein profile (atherogenic vs. non-atherogenic) by this innovative electrophoretic method for lipoprotein analysis in plasma represents a beneficial contribution to actual lipid diagnostics. This system provides the analysis of lipoprotein parameters but also offers new interpretation for lipoprotein profiles, including an actual framework of the practising scheme for diagnostics and treatment of dyslipidemias.

The Score of Anti-Atherogenic risk SAAR, newly introduced parameter, a ratio of non-atherogenic/atherogenic lipoproteins, also confirmes atherogenic lipoprotein constellation and determines the degree of the atherogenic risk of subjects with atherogenic normolipidemia (Oravec 2007a; Oravec 2007b; Oravec 2010).

5. Summary

A new method of electrophoretic lipoprotein separation on polyacrylamide gel (PAG)using the Lipoprint LDL System can quantify non-atherogenic and atherogenic plasma lipoproteins, including small dense LDL, i.e. strong atherogenic lipoprotein subpopulations.

With respect to the predominance of a non-atherogenic or atherogenic lipoproteins in thewhole lipoprotein profile, this method distinguishes a non-atherogenic lipoprotein profilephenotype A from an atherogenic lipoprotein profile phenotype B.

The contribution of this method is to confirm the existence of a non-atherogenic type of hyper-betalipoproteinaemia and the existence of normolipidemia with atherogenic lipoprotein profile, along with the common and well-known atherogenic hyperlipoproteinemia and non-atherogenic normolipidemia.

According to our preliminary analysis of a normolipidemic population, an atherogenic lipoprotein profile was revealed in 6% of normolipidemic young healthy individuals.

More than 40% of the examined individuals in the general group of subjects had an atherogenic lipoprotein profile phenotype B. These people represent an at-risk population.

However, the tools by which is possible to identify these individuals at risk for a cardiovascular event are limited.

A non-atherogenic hyperbetalipoproteinemiaLDL1,2 can be identified, which represents approxmately 20% of examined individuals with hypercholesterolemia and 10% of individuals in a general group of subjects. HyperbetalipoproteinemiaLDL1,2 is not associated with the premature development of arterial vascular impairment.

Author details

Stanislav Oravec
2nd Department of Internal Medicine,
Faculty of Medicine, Comenius University, Bratislava, Slovak Republic

Johannes Mikl
Department of Cardiology, Hietzing Hospital, Austria

Kristina Gruber
Department of Internal Medicine, Landesklinikum Thermenregion Baden, Austria

Elisabeth Dostal
Krankenanstalten Dr. Dostal, Vienna, Austria

Acknowledgement

This study was supported by an EU structural research fund Interreg III AT-SR, project code: 1414-02-000-28 in years 2006-2008.

We would like to acknowledge the excellent technical assistance of MTA Barbara Reif, MTA Judith Trettler and MTA Karin Waitz, Krankenanstalten Dr.Dostal, Vienna, Austria and also to acknowledge the excellent technical assistance of MTA Olga Reinoldova, 2nd Department of Internal Medicine, Faculty of Medicine, Comenius University, Bratislava, Slovakia

6. References

[1] Goldstein JL, Brown MS. Receptor mediated endocytosis:concepts emerging from the LDL-receptor system. Ann Rew Cell Biol 1985; 1: 1-39

[2] Steinberg D. Lipoproteins and the pathogenesis of atherosclerosis. Circulation 1987; 76: 504 - 7

[3] Witztum JL, Steinberg D. Role of oxidized low density lipoprotein in atherosclerosis. J Clin Invest 1991; 84: 1086 - 95

[4] Ross R. The pathogenesis of atherosclerosis – an update. N Engl J Med 1986; 314: 365 - 374

[5] Canner PL, Berge KG, Wenger NK, Stamler J, Friedman L, Prineas RJ et al. Fifteen year mortality in Coronary Drug Project patients, long term benefit with niacin. J Amer Coll Cardiol 1986; 8: 1245 - 55

[6] Frick MH, Elo O, Haapa K, Heinonen OP, Heinsalmi P, Helo P, Huttunen JK, Kaitaniemi P, Koskinen P, Manninen V et al. Helsinki Heart Study: primary prevention trial with gemfibrozil in middle aged men with dyslipidemia. N Engl J Med 1987; 317: 1237 - 45

[7] Kwiterovich PO. Clinical Relevance of the Biochemical, Metabolic and Genetic Factors that influence Low density Lipoprotein Heterogeneity. Am J Card 2002; 90 (Suppl 8A): 30i-48i

[8] Kwiterovich PO. Lipoprotein Heterogeneity: Diagnostic and Therapeutic Implications. Am J Card 2002; 90 (Suppl 8A): 1i-10i

[9] Berneis KK, Krauss RM. Metabolic origins and clinical significance of LDL heterogeneity. J Lipid Res. 2002; 43: 1363-79.

[10] Expert Panel on Detection Evaluation and Treatment of High Blood Cholesterol in Adults. Executive summary of the third report of the National Cholesterol Education Program (NCEP) expert panel of detection, evaluation and treatment of high blood cholesterol in adults (Adult Treatment Panel III). JAMA 2001; 285: 2488 - 97
[11] Backers JM. Effect of Lipid-Lowering Drug Therapy on Small-dense Low-Dense Lipoprotein. Ann Pharmacol 2005; 39: 523 - 26.
[12] Austin MA, King MC, Vranizan KM, Krauss RM. Atherogenic lipoprotein phenotype. A proposed genetic marker for coronary heart disease risk. Circulation 1990; 82: 495-506
[13] Chait A, Brazo RL, Tribble DL, Krauss RM. Susceptibility of small, low- density lipoproteins to oxidative modification in subjects with the atherogenic lipoprotein phenotype, pattern B. Amer J Med 1993; 94: 350-6
[14] Van J, Pan J, Charles MA, Krauss R, Wong N, Wu X. Atherogenic lipid phenotype in a general group of subjects. Arch Pathol Lab Medicine 2007; 131: 1679 – 85
[15] Castelli WP. Cholesterol and lipids in the risk of coronary artery disease – The Framingham Heart Study. Can J Cardiol 1988; (Suppl A): 5A-10A.
[16] Castelli WP. Epidemiology of triglycerides; a view from Framingham. Am J Cardiol 1992; 70: 43-49
[17] Castelli WP. The new pathophysiology of coronary artery disease. Am J Cardiol 1998; 82: (Suppl 2): 60-85
[18] Nicholls S, Lundmann P (2004). The emerging role of lipoproteins in atherogenesis: beyond LDL cholesterol. Semin Vasc Med 2004; 4: 187-195
[19] Rizzo M, Berneis K. Low density lipoprotein size and cardiovascular prevention. Europ J Int Med 2006; 17: 77 - 80.
[20] Shoji T, Hatsuda S, Tsuchikura S, Shinohara K, Komoto E, Kovama H, Emoto M, Nishizhawa Y. Small dense low-density lipoprotein cholesterol concentration and carotid atherosclerosis. Atherosclerosis 2009; 202: 582 - 588.
[21] Zhao ChX, Cui YH, Fan Q, Wang PH, Hui R, Cianflone K, Wang DW. Small Dense Low-Density Lipoproteins and Associated Risk Factors in Patients with Stroke. Cerebrovasc Dis 2009; 27: 99-104.
[22] Rainwater DL, Moore PH jr, Shelledy WR, Dyer TD, Slifer SH. Characterization of a composite gradient gel for the electrophoretic separation of lipoproteins. J Lipid Res 1997; 38: 1261-1266
[23] Alabakovska SB, Todorova BB, Labudovic DD, Tosheska KN. Gradient gel electrophoretic separation of LDL and HDL subclasses on BioRad Mini Protean II and size phenotyping in healthy Macedonians. Clin Chim Acta 2002; 317: 119-123.
[24] Otvos JD, Jeyarajah EJ, Bennet SW, Krauss RM.Development of a proton nuclear magnetic resonance spectroscopic method for determining plasma protein concentrations and subspecies distribution from a single, rapid measurement. Clin Chem 1992; 38: 1632- 38
[25] Hoefner DM, Hodel SD, O'Brien JF, Branum EL, Sun D, Meissner I, McConnell JP. Development of a rapid quantitative method for LDL subfraction with use of the Quantimetrix Lipoprint LDL system. Clin Chem 2001; 472: 266-274.

[26] Lamarche B, Lemieux I, Despres JP. The small, dense LDL phenotype and the risk of coronary heart disease : epidemiology, patho-physiology and therapeutic aspects. Diabetes Metab 1999; 25: 199-211

[27] Packard CJ. Triacylglycerol-rich lipoproteins and the generation of small dense low-density lipoprotein. Biochem Soc Transactions 2003; 31: 1066 - 69

[28] Carmena R, Duriez P, Fruchart JC. Atherogenic lipoprotein particles in atherosclerosis. Circulation 2004; 109: III2-III7

[29] Oravec S. Nová laboratórno-medicínska pomoc v diagnostike dyslipoproteinemií a kardiovaskulárnych ochorení: Identifikácia LDL podskupín. Med Milit Slov 2006a; 8: 28-32.

[30] Oravec S. Identifikácia subpopulácií LDL triedy –Aktuálny prínos v diagnostike porúch metabolizmu lipoproteínov a ochorení kardiovaskulárneho systému. Med Milit Slov 2006b; 8: 32-34.

[31] Oravec S. Nové perspektívy v diagnostike porúch metabolizmu lipoproteínov - prínos v interpretácii výsledkov. Med Milit Slov 2007a; 9: 42-45

[32] Oravec S. Nové možnosti posúdenia kardiovaskulárneho rizika u pacientov s obezitou a metabolickými ochoreniami. Med Milit Slov 2007b; 9: 46-49.

[33] Morais J, Neyer G, Muniz N. Measurement and Distribution of HDL subclasses with the new Lipoprint® HDL Method (pdf format). Presented at AACC, Philadelphia, PA , June 2003

[34] Morais J. Quantimetrix shows that all HDL subfractions may not protect against heart disease. AACC international congress of Clinical Chemistry, Orlando, FL, June 2005

[35] Muniz N, Morais J. Coronary heart disease. High density lipoprotein subclasses associated with heart disease. Medical Letter on the CDL and FDA, July 31st, 2005

[36] Maslowska M, Wang HW, Cianflone K. Novel roles for acylation stimulatory protein/C3ades Atg: a review of recent in vitro and in vivo evidence. Vitam Horm 2005; 70: 309-32

[37] Kwiterovich PO jr. The metabolic pathways of HDL,LDL and triglycerides. A current review. Am J Card 2000; 86 (Suppl 1): 5-10

[38] Oravec S. Den drohenden Herztod erkennen- und vermeiden. Der Mediziner 2010; 4: 6-7

[39] Oravec S, Dukát A, Gavorník P, Caprnda M, Kucera M. Lipoproteínový profil séra pri novozistenej arteriálnej hypertenzii. Úloha aterogénnych lipoproteínov v patogenéze ochorenia. Vnitr Lek 2010a; 56: 967-971.

[40] Oravec S., Dukát A., Gavorník P., Čaprnda M, Reinoldová O.Zmeny v lipoproteínovom spektre pri končatinovo-cievnej ischemickej chorobe.Vnitř. Lék 2010b; 56(6): 620-623

[41] Oravec S, Dukat A, Gavornok P, Caprnda M, Kucera M, Ocadlik I. Contribution of the atherogenic lipoprotein profile to the development of arterial hypertension. Brat Lek Listy 2011a; 112: 4-7

[42] 42) Luc G, Bard J-M, Ferriéres J, Evans A, Amouyel P, Arveiler D, Fruchart J-Ch, Ducimetière P, Prime Study Group. Value of HDL-cholesterol, apolipoprotein A-I, Lipoprotein A-I, Lipoprotein A-I/A-II in prediction of coronary heart disease . The Prime Study. Arterioscler Thromb Vasc Biol. 2002; 22: 1155- 61

[43] St-Pierre AC, Cantin B, Daganais GR, Mauriege P, Bernard PM, Despres JP, Lamarche B. Low density lipoprotein subfractions and the long-term risk of ischemic heart disease in

men : 13-year follow-up data from the Quebec Cardiovascular Study. Arterioscler Thromb Vasc Biol. 2005; 25: 553-559

[44] Fruchart JC, Sacks FM, Hermans MP et al. The residual risk reduction initiative: a call to action to reduce residual vascular risk in dyslipidaemic patients. Diabetes Vasc Res 2008; 5: 319-335

[45] Chun Xia Zhao, Ying Hua Cui, Qiao Fan, Pei Hua Wang, Ruitai Hui, Cianflone K, Dao Wen Wang. Small Dense Low-Density Lipoproteins and Associated Risk Factors in Patients with Stroke. Cerebrovasc Dis 2009; 27: 99-104

[46] Haffner SM. The metabolic syndrome: inflammation, diabetes mellitus and cardiovascular disease. Am J Cardiol 2006; 97: 3A-11A

[47] Lamarche B, Tchernof A, Moorjani S, Cantin B, Dagenais GR, Lupien PJ, Despres JP. Small dense LDL lipoprotein particles as a predictor of the risk of ischemic heart disease in men. Prospective results from the Quebec Cardiovascular Study. Circulation 1997; 95: 69-75

[48] Gardner CD, Fortman SP, Krauss RM. Association of small low-density lipoprotein particles with the incidence of coronary artery disease in men and women. JAMA 1996; 276: 875-881

[49] Rajman I, Kendall MJ, Cramb R, Holder RL, Salih M, Gammage MD. Investigation of low density lipoprotein subfractions as a coronary risk factor in normotriglyceridaemic men. Atherosclerosis 1996; 125: 231-42

[50] Halle M, Berg A, Baumstark MW, Keul L. LDL-Subfraktionen und koronare Herzerkrankung – Eine Übersicht. Zeitschrift Kardiol 1998; 87: 317-30

[51] Austin MA, Hokanson JE, Brunzell JD. Characterization of low-density lipoprotein subclasses: methodologic approaches and clinical relevance. Curr.Opinion Lipidol 1994; 5: 395-403

[52] Oravec S, Gruber K, Dostal E, Mikl J. Hyper-betalipoproteinenmia LDL1,2: a newly identified non-atherogenic hypercholesterolemia in a group of hypercholesterolemic subjects. Neureoendocrinol Lett 2011b; 32: 322-327

[53] Oravec S, Dostal E, Dukat A, Gavorník P, Kucera M, Gruber K. HDL subfractions analysis: A new laboratory diagnostic assay for patients with cardiovascular diseases and dyslipoproteinemia. Neuroendocrinol Lett 2011c; 32: 502-509

[54] Oravec S, Dukat A, Gavorník P, Lovásová Z, Gruber K. Atherogenic normolipidemia – a new phenomenon in the lipoprotein of clinically healthy subjects. Neuroendocrinol Lett 2011d; 32:317-321

Chapter 5

The apoB/apoA-I Ratio is a Strong Predictor of Cardiovascular Risk

Göran Walldius

Additional information is available at the end of the chapter

http://dx.doi.org/10.5772/47869

1. Introduction

In the present paper the rationale for including apolipoprotein (apo)B and apoA-I into clinical practice is reviewed. ApoB and apoA-I are the two major apolipoproteins involved in lipid transport and in the processes causing atherosclerosis and its complications. ApoB is the major protein in Very Low Density (VLDL), Intermediate Density (IDL) and Low Density Lipoproteins (LDL), one protein per particle (1). ApoA-I is the major protein in High Density Lipoprotein (HDL) particles **(Figure 1)**. The apoB number indicates the total number of atherogenic particles, the higher the number the higher is the cardiovascular (CV) risk.

ApoA-I reflects the anti-atherogenic potential in HDL particles, the higher the value the better protection of CV risk. The apoB/apoA-I ratio (apo-ratio) indicates the balance between atherogenic and anti-atherogenic particles, the higher the value, the higher is the CV risk. In previous papers we (2-6) and others (7-12) have reviewed the importance of apolipoproteins, mainly apoB and apoA-I, but also other apolipoproteins like apoC-II and apoCIII, apoE, and Lp(a) as markers of atherogenic risk. In this review many new data on apoB, apoA-I and the apo-ratio and their relations to cardiovascular (CV) risk are presented. The majority of these studies were published in the last 6 year period.

The debate today (mid 2012) is about whether LDL-C should remain as the primary variable for CV risk evaluation and target for lipid-lowering therapy. During the last few years non-HDL-C has been found and proposed to be the next primary target for CV risk evaluation and target for treatment (9-11,13,14). Notably, although LDL-C and non-HDL-C are considered the best CV risk markers most large studies of CV risk have shown that the lipid ratios, i.e. the TC/HDL-C, the LDL-C/HDL-C and the non-HDL-C/HDL-C ratios, are stronger than any specific single lipid fraction (2,3,4,6,15). The major aim of this paper is therefore to review papers on apoB, apoA-I and the apo-ratio related to risk of atherosclerosis and various clinical complications like myocardial infarction (MI), stroke and other severe events to find out if there is evidence for using apoB and apoA-I, and especially the apoB/apoA-I ratio (apo-ratio) motivating clinical use of these risk markers/predictors. Both similarities, but mainly differences between apos and conventional lipids to predict CV risk, will be highlighted. Methodological aspects and the role of apoB and apoA-I, the two determinants of the apo-ratio, will first be commented. The major part of the paper describes the role of the apo-ratio as a CV risk marker/predictor. The overall conclusion from this paper will be that apoB, apoA-I and the apo-ratio merit to be included in future guidelines in order to be recognized and used in clinical practice.

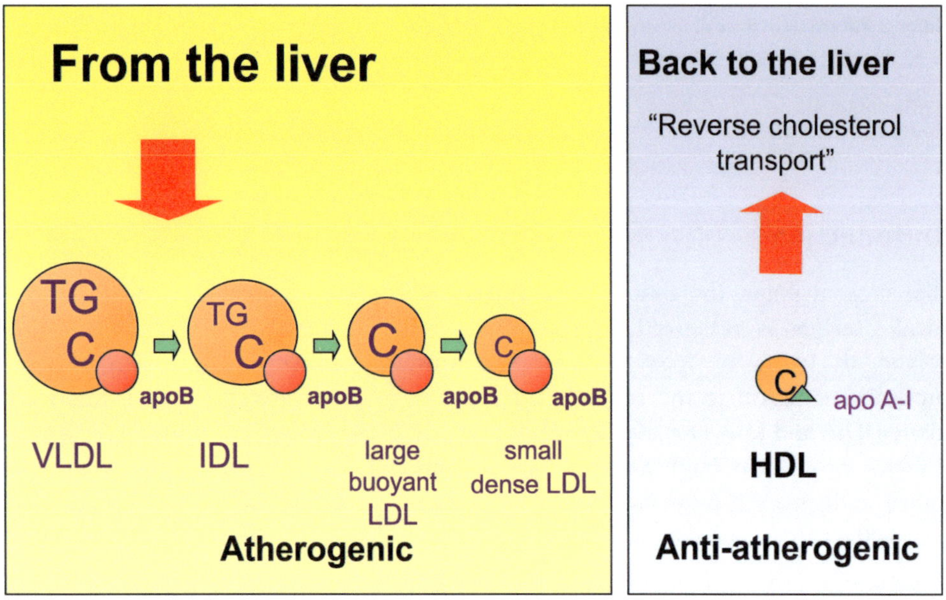

Figure 1. The figure shows the atherogenic particles containing one apoB protein per VLDL, IDL, large buoyant LDL, small dense LDL particles and the anti-atherogenic lipoproteins containing apoA-I. The balance between apoB and apoA-I, i.e., the apoB/apoA-I ratio, reflects the balance between the "bad cholesterol particles and the good cholesterol particles". This apo-ratio is strongly related to cardiovascular risk, the higher the ratio, the higher is the risk. (From reference 3).

2. Methodological pros and cons for using apoB, apoA-I and the apo-ratio versus conventional lipids

2.1. Methodological problems for various lipids

The most commonly used method world-wide to measure LDL-C is based on the Friedewald formula (16). However, errors are common and the methodological problems and shortcomings are not commonly recognized but have been discussed in many papers (17-25). Thus, the formula (LDL-C = TC – HDL-C – TG/5) is not valid for blood samples having triglycerides (TG) above 3.5-4 mmol/L, for patients with type III hyperlipoproteinemia or chylomicronemia or non-fasting specimens (17-19). The errors for

LDL-C can be false positive in the range of 2-17% or false negative between 12-15% if TG levels are very low or closer to 4 mmol/L. This may create large problems for both clinicians and patients since patients may be misclassified as being at risk or not at risk according to guidelines. Similarly, it may be difficult for the clinician to evaluate if a patient has been adequately treated to the target of LDL-C. Newer so called "direct LDL-C methods" have been developed and they are homogeneous methods, that is, assays that do not require a preliminary separation step, such as ultracentrifugation, or manual manipulation of the sample for determining LDL-C (9,18-20). However, these methods, although standardized at a given laboratory, do not always correlate well over the whole range of lipid values, and they are not even internationally standardized like those for apolipoproteins.

The practical problems of measuring HDL-C are also of concern and correlation between various methods are sometimes even worse than those for LDL-C (18,19,26). Consequently, the values for non-HDL-C (TC minus HDL-C) may also be subject to large variations due to the errors mainly for measuring HDL-C. However, there is an advantage for non-HDL-C over LDL-C determined by the Friedewald formula since non-HDL-C is not subject to influence of non-fasting that may distort the TG levels and make it difficult to obtain a correct value for LDL-C (27). Furthermore, non-HDL-C contains C from all atherogenic fractions i.e. VLDL, IDL and various forms of LDL. Thus, non-HDL-C which indicates the total mass of C is more likely to reflect the variation of atherogenic particle set up for many patients with various genotypes and phenotypes. Such patients may have a greater chance to be correctly identified as risk individuals based on non-HDL-C, rather than to an imprecise measure of only LDL-C. For the interested reader of methods and concerns of validity, see further excellent reviews (18,19,26,27).

2.2. Methodological advantages for apolipoproteins

There are methodological advantages of using apoB and apoA-I compared to LDL-C and HDL-C since the apo-methods have been internationally standardized according to WHO-IFCC already in 1990-ies (26,28,29). The standardization initiatives for apo B have proceeded more quickly and more successfully than for LDL-C. The WHO-IFCC collaboration has resulted in the development of secondary reference material to ensure traceability of manufacturer calibrators to an approved standard. The bias and imprecision for 22 immunonephelometric and immunoturbidimetric assays ranged were usually below 5%. These errors are commonly smaller than that for calculated LDL-C and lipid ratios. Costs for measuring apos can be much reduced if apos are introduced as routine methods. However, pedagogical aspects (education of physicians, patients and laymen), and the well documented and cemented LDL-paradigm will make it difficult to convince guideline committees to introduce apoB and apoA-I as CV risk predictors. Importantly, this should not invalidate that apos are accepted as strong risk markers especially since so many other methods determining LDL-C and HDL-C are accepted in guidelines despite rather weak correlations between various methods due to incomplete standardizations.

3. Physiological and pathophysiological aspects of apoB

3.1. ApoB production, circulation and distribution

ApoB-100 is produced in the liver and apoB-48 is synthesized in the gut (3,12). ApoB-100 is the dominating protein in plasma compared with minute amounts of apoB-48 even in the postprandial state. In most conditions, more than 90% of all apoB in blood is found in LDL. There are excellent reviews of how apoB-100 assembles VLDL in the liver, more details on VLDL composition (12), and some comments on the genetics of apoB (30-33). ApoB is present in VLDL, IDL large buoyant LDL, and small dense LDL (sdLDL), with one molecule of apoB in each of these atherogenic particles (1). Importantly, apoB does not occur on HDL particles. Thus, total apoB reflects the total number of potentially atherogenic particles **(Figure 1).** This is principally different from non-HDL-C which indicates the total mass of C. ApoB produced in the liver stabilizes and allows the transport of C and TG in plasma VLDL, IDL, large buoyant LDL and sdLDL. ApoB also serves as the ligand for the apoB and apoB,E receptors thereby facilitating uptake of C in peripheral tissues and in the liver as reviewed (2,3,12). ApoB may provoke atherogenesis since it can be entrapped in the arterial wall of the coronary arteries and also as exemplified by findings in femoral plaques (12,34,35) where it may be modified, oxidized and glucosylated and therefore also contribute in the process of plaque formation. In this process LDL-C with apoB infiltrates the arterial wall and many factors like adhesion molecules, cytokines, growth factors are involved in oxidation processes leading to inflammation and growth of plaques unless HDL bound apoA-I can neutralize these processes (see elsewhere in this paper). Interestingly, already in 1976 Hoff presented data showing that apoB and apoA-I were found in the arterial wall of the coronary and carotid arteries as well as in the aorta (35). Olofsson et al (12) discuss the intra-arterial metabolism of apoB and apoA-I and also Fogelstrand and Borén (36).

3.2. Plasma levels of apoB and target values for therapy

The levels of apoB in plasma may vary from 0.2 to above 3 g/L, with highest values for those with hereditary hypercholesterolemias. In the "normal case" the values for males and females do not differ much. Reference values have been published by Cantois et al. already in 1996 (37). The values slowly increase from childhood to adult life (2,3). Those who live to ages above 75 years commonly have relatively low apoB values since those with higher values may have died due to various CV events. During lipid-lowering therapy apoB targets have been recommended to be < 0.90 g/L for those at moderate risk and < 0.80 for those at a high risk, see further below (3,9,11,38). Values should be given in two decimals.

3.3. ApoB versus LDL-C and risk for CV events

One of the first publications on clinical risks during the course of myocardial infarction (MI) related to apoB and also to apoA-I was presented by Avogaro already in 1978 (39,40). In 1980 Sniderman et al. presented data indicating that hyperapoB with normal C levels was related to coronary atherosclerosis (41). Since then many reports have been published indicating that apoB is involved in atherogenesis and its complications like MI. In 1996 Lamarche et al. (42) showed that apoB was strongly associated with onset of coronary heart disease in 2,155 men aged 45–76 years followed for 5 years (Quebec Cardiovascular Study). The predictive effect of apoB remained after adjustment for TG, HDL-C and TC/HDL-C. ApoA-I was protective, but not as strong as the harmful apoB in multivariate analysis. In the 10-year follow up of the Atherosclerosis Risk in Communities (ARIC) study, apoB was measured in 12,339 middle-aged participants (43) and had predictive power above that of LDL-C, TG and HDL-C. However, despite strong univariate associations for apoB and LDL-C, apoB did not contribute to risk prediction in subgroups with elevated TG, with lower LDL-C, or with high apoB relative to LDL-C. This may be due to the error for apoB determination which was estimated at 17% which is considerably higher than the approximate 5% that is common in most recent trials.

Importantly, apoB has been found to have a stronger relation with CV risk than LDL-C in several other studies as reported in coming sections. These include the AMORIS study (44), especially at low values of LDL-C (see below), the Thrombo Study (45), the Thrombo Metabolic Syndrome Study (46), the Northwick Park Heart Study (47), the Nurses' Health Study (48) and amongst patients with type 2 diabetes in the Health Professionals Follow-up Study (49).

In the Copenhagen City Heart Study Benn et al. (50) studied 9,231 asymptomatic women and men from the Danish general population followed prospectively for 8 years and observed the following incident events: ischemic heart disease 591, MI 278, ischemic cerebrovascular disease 313, ischemic stroke 229, and any ischemic CV event 807. ApoB, adjusted for multiple common confounding risk factors, had a higher predictive ability than LDL-C in all these various ischemic events ($p < 0.03$ to < 0.001). They suggested that prediction of future ischemic cardiovascular events could be improved by measuring apoB.

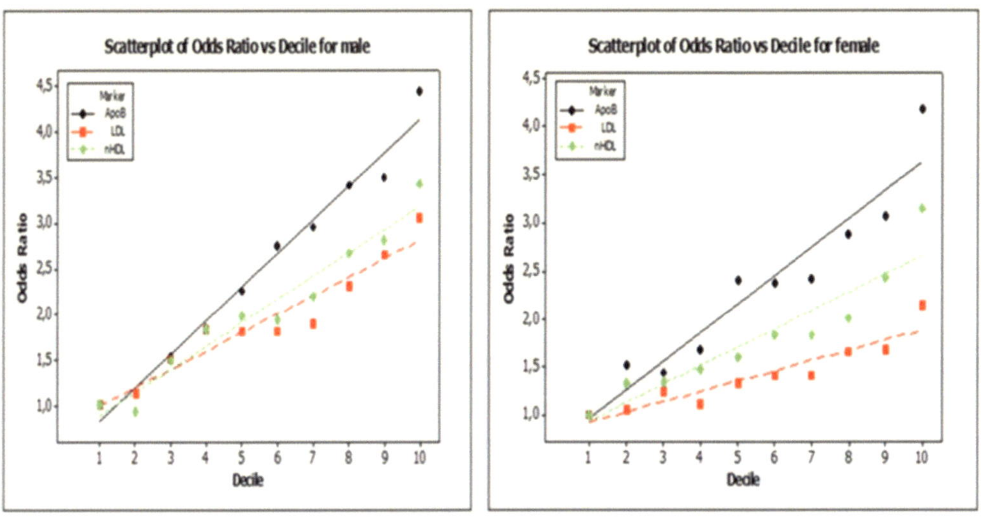

Figure 2. The AMORIS study; apoB, non-HDL-C and LDL-C (x-axis, deciles) versus risk of myocardial infarction (Odds Ratio) (y-axis) in males (left) and females (right).

In addition, in the placebo groups of several major statin clinical trials such as 4S (51), AFCAPS/TexCAPS (52,53) and LIPID (54) apoB was more informative than LDL-C as an index of the risk of CV events. Taken together, this strongly indicates that apo B is superior to

LDL-C in recognizing the risk of CV disease and effects of statin therapy. Additional results (55) also favor apoB over LDL-C, and others are also reported in the section on the apo-ratio below. Such major studies are the AMORIS (3,44,56,57). In our study we found the steepest risk-relationship for MI with increasing values of apoB followed by non-HDL-C and the lowest increase in relation to LDL-C values with similar risk progressions for men and women **(Figure 2 and Figure 3 left).** Also in the INTERHEART (58,59) and ISIS-studies (60) as well as those summarized in the ERFC-meta-analyses (8,10) apoB was strongly related to risk of MI. In meta-analyses similar strong findings for apoB versus LDL-C are summarized by a large number of international scientists and clinicians in more detail (4,13,61,62).

4. ApoB versus non-HDL-C

ApoB indicates the number of atherogenic particles whereas non-HDL-C indicates the C mass from all atherogenic fractions like VLDL, IDL and the large buoyant LDL and the most atherogenic sdLDL fractions. But is apoB similar to or better than non-HDL-C in predicting risk? Although there is a similarity between apoB and non-HDL-C, they may have different metabolic fate and thus impact on risk. The rationale for using non-HDL-C is based on the fact that there is a close relationship between non-HDL-C and apoB values. Usually the correlation is about 0.80–0.85. However, correlation is not the same as concordance. In fact, two variables can be highly correlated but also be highly discordant, i.e. they do not correspond well. Either they are too high or too low compared with the other variable. Importantly, discordance produces major errors in the middle of the population distribution. Sniderman has frequently presented data with explanations of the advantages of apoB over LDL-C and non-HDL-C (4,13,14,21,62,63). Commonly, the sdLDL-particles contribute much to the large numbers of atherogenic particles, i.e. the apoB number is high and these small particles can easily penetrate into the arterial wall. However, in conditions with high non-HDL-C due to high VLDL-C and high large buoyant LDL-C the sdLDL-particles may be rather low in numbers indicating comparatively low numbers of apoB particles. These larger cholesterol-containing VLDL and IDL particles, although rich in C, do not easily penetrate into the arterial wall.

Of interest, the number of apoB is more closely associated with insulin resistance or markers of the metabolic syndrome than either LDL-C or non-HDL-C (3,62,63). Thus, in patients with hypertriglyceridaemia with normal, or even low LDL-C values, i.e. patients with the metabolic syndrome (MetS), and in patients with overt diabetes, apoB has been shown to be superior to non-HDL-C in predicting vascular risk (3,62,63). Again, even if non-HDL-C and apoB correlate, they are not the same biologically or clinically. In most cases apoB is associated with a higher CV risk than non-HDL-C as well as LDL-C. Furthermore, non-HDL-C may not be that easy to understand or explain for the clinician or the patient once they have learnt that the bad C is LDL-C.

Many studies and clinical trials have been published showing that apoB has a stronger capacity to identify all different phenotypes and to better predict CV risk than both LDL-C and non-HDL-C (3,11,61-63). Sniderman et al. (62) have published a convincing meta-analysis of results from published epidemiological studies that contains estimates of the relative risks of LDL-C, non-HDL-C and apoB of fatal or non-fatal ischemic CV events. Twelve reports including 23,455 subjects and 22,950 events, were analyzed. Whether analyzed individually or in head-to-head comparisons, apoB was the most potent marker of CV risk RR = 1.43; (95% CI, 1.35-1.51), LDL-C was the least RR = 1.25; (1.18-1.33), and non-HDL-C was intermediate RR = 1.34; (1.24-1.44). Only HDL-C accounted for any substantial portion of the variance of the results among the studies. They commented that in patients in whom LDL composition is normal, the cholesterol markers and apoB are equivalent markers of risk, i.e. correlation between the three markers is high. However, when the markers are discordant, that is, when LDL-C is normal but LDL-particles (P)(= apoB) is high or, alternatively, when LDL-C is high but LDL-P are normal, then apoB and non-HDL-C are better markers of risk than LDL-C. They calculated the number of clinical events prevented by a high-risk treatment regimen of all those >70th percentile of the US adult population using each of the 3 markers. Over a 10-year period, a non-HDL-C strategy would prevent 300,000 more events than an LDL-C strategy, whereas an apoB strategy would prevent 500,000 more events than a non-HDL-C strategy. These examples emphasize the greater potential for using apoB rather than non-HDL-C and LDL-C.

However, in another major meta-analysis by Boekholdt et al. (64) they studied 62,154 patients enrolled in 8 statin trials published between 1994 and 2008. Among 38,153 statin treated patients 158 developed fatal MI, 1,678 non-fatal MI, 615 fatal events from other coronary artery disease, 2,806 hospitalizations for unstable angina, and 1,029 fatal or nonfatal strokes occurred during follow-up. The adjusted HRs for major CV events per 1-SD increase were 1.13 (95% CI, 1.10-1.17) for LDL-C, 1.16 (1.12- 1.19) for non-HDL-C, and 1.14 (1.11-1.18) for apoB. These HRs were significantly higher for non-HDL-C than LDL-C ($p = 0.002$) and apoB ($p = 0.02$). Thus, from both these meta-analyses non-HDL-C stands out as a stronger predictor of CV diseases than LDL-C. The explanation for the different findings of apoB in these two meta-analyses is unclear but may be explained by the fact that the first study is based on data from a prospective risk studies, whereas the second study reflects effects of statins on lipid and lipoprotein metabolism. Further comments are given in the discussion.

5. Physiological and pathophysiological aspects of apoA-I

There are many subgroups of particles of HDL with different lipid and apo compositions (3,12,29). Beyond apoA-I there are other apos such as apoA-II, apoA-III, apoC-III, apoD and apoM. ApoA-I is the major protein in HDL and this protein is taken to represent HDL metabolism since it occurs almost exclusively in HDL particles. By measuring HDL-C the amount of C transported in blood is indicated to represent the reverse cholesterol transport (RCT), a major protective aspect of HDL metabolism – by laymen named "the good cholesterol". ApoA-I initiates the RCT process in peripheral tissues. ApoA-I has also many other functions beyond RCT since apoA-I is involved in anti-inflammation, anti-oxidation, anti-infectious activity, anti-proteas activity, anti-apoptotic, and anti-thrombotic functions (3). Furthermore, apoA-I can initiate the endothelial production of nitric oxide that is of vital help in producing vasodilation (3). Furthermore apoA-I may help to regulate glucose-insulin homeostasis. Thus, by measuring apoA-I you may get additional "protective" effects above those given only by the HDL-C number. For methodological reasons Warnick and others prefer to use apoA-I rather than HDL-C methods (19). HDL and apoA-I metabolism are reviewed in more detail, see ref. (3,12,29,65-67).

5.1. Plasma levels of apoA-I and target values for therapy

The plasma concentration of apoA-I can vary from 0.1 to over 3 g/L. Reference values have been published by Cantois et al. already in 1996 (37). There is little variation with fasting-non-fasting (68). Normally women have 0.1-0.3 g/L higher apoA-I values than men, similar to the higher HDL-C values for women. After menopause apoA-I values commonly decrease in parallel with HDL-C. However, there have been few published recommendations regarding what should be a "normal apoA-I value". A normal value for any adult should be at least close to 1 g/L or above. So far, there have been few recommendations on valid cut values indicating increased CV risk and target values. Values should be given in two decimals. For further comments, see the section on the apo-ratio.

5.2. Biological variation of apoA-I

ApoA-I and ApoA-II may also enter the cerebrospinal flow via the choroid plexus (69). Reduction in the HDL apoA-I/apoC-III ratio, changes in the HDL subpopulation distribution and an increase in HDL oxidation potential correlated with the development of MI in young patients (70). In a Korean study of 15,154 healthy subjects higher CRP levels were associated with significantly lower HDL-C and apoA-I levels, and also higher apoB values (71). In a US population of 8,708 apparently healthy population apoA-I was strongly positively associated, whereas apoB was significantly reduced with alcohol intake. Similarly the transaminases AST, ALT and gamma-GT increased with higher alcohol consumption (72).

5.3. ApoA-I and risk for CV events

Already in 1978,1979, Avogaro et al. (39,40) showed that apoA-I was as good as lipids in predicting myocardial infarction (MI) in those under 50 years of age but apoA-I was a better predictor in those over 60 years of age. In the Swedish APSIS study (73) Held et al. in 1994 studied patients with angina pectoris. During a median follow-up time of 3.3 years (2,663 patient years), 37 patients suffered a CV death, 30 suffered a non-fatal MI and 100 underwent a revascularization. Apo-I and TG were predictors of CV death or non-fatal MI in univariate analyses, but only apoA-I remained as an independent predictor in multivariate analyses. All lipid variables except LDL-C were related to the risk of revascularization in univariate analyses, but only apoA-I and apoB were independent predictors of such events. They concluded that apolipoprotein levels were better predictors of CV events than other lipid parameters in patients with stable angina pectoris.

Many studies have shown an inverse relationship between apoA-I and MI (3,44,74,52,53,59,60). In a study of Japanese Americans apoA-I predicted coronary heart disease only at low concentrations of HDL (75). High apoA-I values have been found to correlate with low risk for MI in AMORIS as indicated **(Figure 3, right).** Luc et al. also found that apoA-I is the best prospective risk marker of several other apoproteins in HDL (76). In the large INTERHEART case-control study apoA-I had a greater protective effect of MI at a wider range of apoA-I values than HDL-C (58,59). Patel et al. (77) found that ApoA-I levels are a consistent discriminator of atherosclerotic burden among patients with stable CAD.

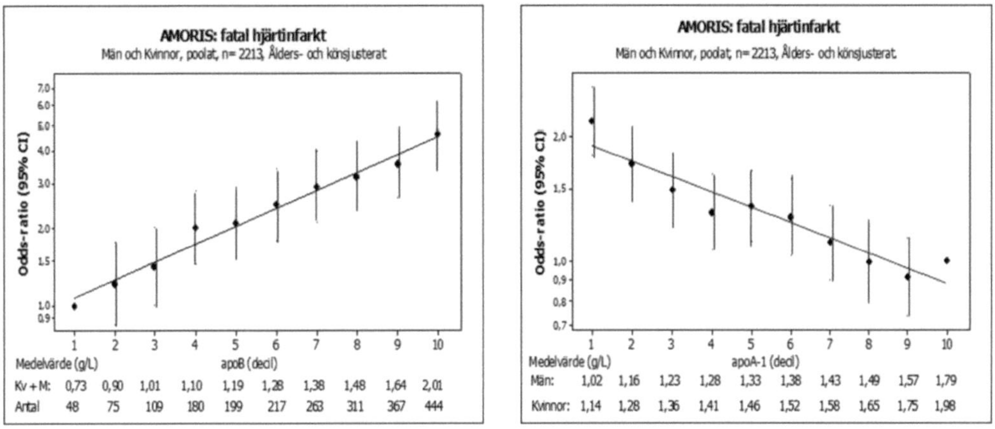

Figure 3. Left; The AMORIS study: Fatal myocardial infarction (Odds Ratio) is related to increasing values of apoB. The values are adjusted for age, TC and TG. Right; The AMORIS study: Fatal myocardial infarction is related to decreasing values of the apoA-I. The values are adjusted for age, gender, TC and TG. Similar pattern for men (män) and women (kvinnor).

In the CORONA study performed in patients with severe heart failure (placebo versus rosuvastatin) apoA-I, in univariate analysis, was the second best (after apoB plus apoA-I) of all different lipid fractions in predicting total death and MACE (MAjor Coronary Events). Furthermore, in a multivariate stepwise analysis apoA-I ranked fifth, better than high sensitivity CRP (hsCRP), of all 14 predictors of outcome where no conventional lipid fraction was significant. The best predictor was pro-BNP (78).

In a study of risk of stroke in Taiwan it was shown that apoA-I but not apoB levels may serve as an effect modifier of hypertension for the risk of stroke events (79).

In the combined analysis of data from the IDEAL statin trial and the Epic-Norfolk case-control study (80) very high HDL-C due to enlarged HDL-particles values were associated with increased rather than decreased CV risk. However, in contrast, apoA-I appears not to turn into a significant risk factor at high plasma concentrations. They conclude that apoA-I is associated with CHD risk independently from HDL size suggesting that the cardioprotective role of large HDL might be more closely related to its apoA-I content than to HDL size per se. These observations may have important consequences for future CAD risk assessment and novel treatment strategies. Indeed, several experimental studies have pointed to a crucial role for apoA-I in protection against atherosclerosis (3,12,65,81).

In the AFCAPS /TexCAPS statin trial (placebo versus lovastatin) multivariate analysis showed that apoA-I was better than HDL to predict outcome (52,53). In addition, the apo-ratio was the best of all lipids and apo-fractions to explain CV risk reduction, see further below.

6. General comments on the validity of using a ratio as a primary marker of risk

Lipid and lipoprotein ratios like TC/HDL-C and LDL-C/HDL-C have been used in various international guidelines for decades to define CV risk. However, LDL-C has in the vast number of guidelines dominated as the primary risk marker why ratios rarely are used today in clinical practice. One major reason why the lipid ratios are questioned as relevant risk markers is due to the fact that HDL-C is included in the value for TC, so HDL-C occurs both in the nominator and denominator of the ratio. Similarly, since LDL-C most commonly is derived by the Friedewald formula, HDL-C is involved as a factor for calculating LDL-C and therefore also indirectly in the nominator and denominator of that ratio. Therefore physicians are hesitant to the mathematical way of dividing various lipid numbers to obtain a mathematical, but, in their mind, not a biologically relevant ratio. When so called direct methods are used for measuring LDL-C this problem is less. In recent years non-HDL-C has been recommended as the next primary risk variable and the new non-HDL-C/HDL-C ratio has been defined. Interestingly, this ratio gives the same final number of risk as that of the TC/HDL-C ratio.

Most researchers and guidelines recommend the use the TC/HDL-C ratio since calculation of this ratio is not dependent on that blood sampling has been performed in the fasted state. This is the same argument as for using non-HDL-C rather than calculated LDL-C. The challenge now is can the apo-ratio, which summarizes the CV risk related to all atherogenic and all anti-atherogenic variables into one number, be the next rational choice as a primary risk variable? Does the apo-ratio add to information already obtained by lipids and lipid ratios? And are the values for apoB, apoA-I and especially the apo-ratio much influenced by other confounding risk factors? These and many other questions are addressed in the sections below based on a vast number of publications.

7. Prospective cardiovascular risk studies – Relations to apoB, apoA-I and the apoB/apoA-I-ratio

7.1. The AMORIS prospective study and risk of myocardial infarction (MI)

The Swedish AMORIS (Apolipoprotein-related MOrtality RISk) study is the largest of all studies in which apoB and apoA-I have been measured in more than 175.000 individuals followed prospectively for up to 25 years. The participating subjects were recruited from health check-ups during 1985-1996. Their age ranged from below 10 years to above 90 years. In these years health screening was very common in Sweden. Subjects included in the database called AMORIS were mainly healthy, not acutely ill or hospitalized and no subject participated in clinical trials. They were all treated by their general practitioners in the greater Stockholm area and they constitute a valid socio-economical cohort of the greater Stockholm population as indicated in several of our papers presented below. Large blood screening programs were used including some 8,000 determinations of LDL-C according to Friedewald. Simultaneously apoB and apoA-I were analyzed by automated immunoturbidimetric methods in all 175.000 subjects according to the WHO-IFCC protocol and in collaboration with their representatives (82). LDL-C was calculated according to the

Jungner formula (44). The Jungner formula yields the same LDL-C values as those obtained by using the Friedewald formula (16) as confirmed by Talmud et al. (47). Also HDL-C values were determined by the Friedewald formula once LDL-C was calculated by the Jungner formula. All analyses in the AMORIS laboratory database (confounding clinical risk factors like hypertension, diabetes and obesity were available in cohorts) were performed by automated methods at the same CALAB laboratory headed by Ingmar Jungner. Several papers were published describing the apoB, apoA-I and the apo-ratio characteristics of the population and the methods (3,21,44,82-85). In an early AMORIS study we have previously noted that patients with type IIB dyslipidemias, i.e. combined hypercholesterolemia and hypertriglyceridemia, had the highest apo-ratio (86). The subsequent CV manifestations were related to the laboratory variables obtained at the first visit to the physician.

In 2001 we presented the first endpoint paper based on 98,722 men and 76,831 women in the Lancet (44). We found a strong direct relationship between apoB and an indirect inverse relationship between apoA-I and risk of MI (men = 864, women = 359) **(Figure 3)**. Furthermore, apoB was a stronger risk factor than LDL-C especially at low values of LDL-C. The apo-ratio was the strongest lipid-related factor **(Figure 4).** In the left part of the figure the values for apoB and apoA-I divided into quartiles are displayed in a three dimensional way. Thus, in those with highest values of apoB and in those with lowest apoA-I values the risk increased about 6-fold in a stepwise fashion compared to those with lowest apoB and highest apoA-I values. The highly significant results were similar for men and women and remained after adjusting for age, TC and TG. The figure clearly illustrate that the risk is about the same for those with an increased apoB at highest apoA-I levels, as the risk for those with lowest apoA-I levels but with low apoB values. Thus the figure illustrates the importance of measuring both apoB and apoA-I to get correct information on MI risk level. In the right part of the figure the same results can also be depicted as a straight line (semi-log scale) showing the impact of higher apo-ratio versus increased risk of MI. We also found that apoB was significantly better to predict risk than LDL-C especially for those with low values for LDL-C.

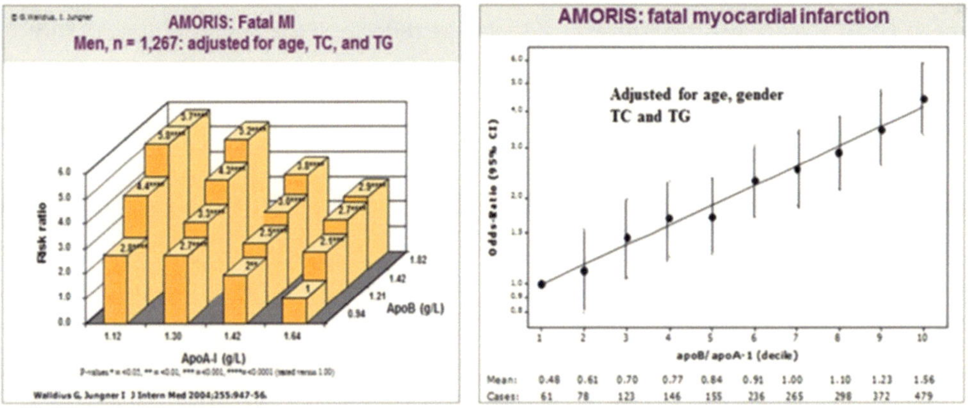

Figure 4. Left; The AMORIS study: Fatal myocardial infarction (Risk ratio) is related to increasing values of apoB and decreasing values of apoA-I. The values are adjusted for age, TC and TG. Similar pattern is seen for men and women (reference 3). Right; The AMORIS study: Fatal myocardial infarction is related to increasing values of the apoB/apoA-I ratio. The values are adjusted for age, gender, TC and TG. (Both figures from reference 3).

With increasing values of the apo-ratio there was a parallel increase in apoB, LDL-C, non-HDL-C and TG **(Figure 5, left)** and a decrease in apoA-I and HDL-C values **(Figure 5, right).** This figure illustrates that an increasing apo-ratio indirectly also indicate the contribution of the other lipids as risk factors. In multivariate analyses the apo-ratio is the strongest of all lipid-related variables and is thus the best summarizing risk variable.

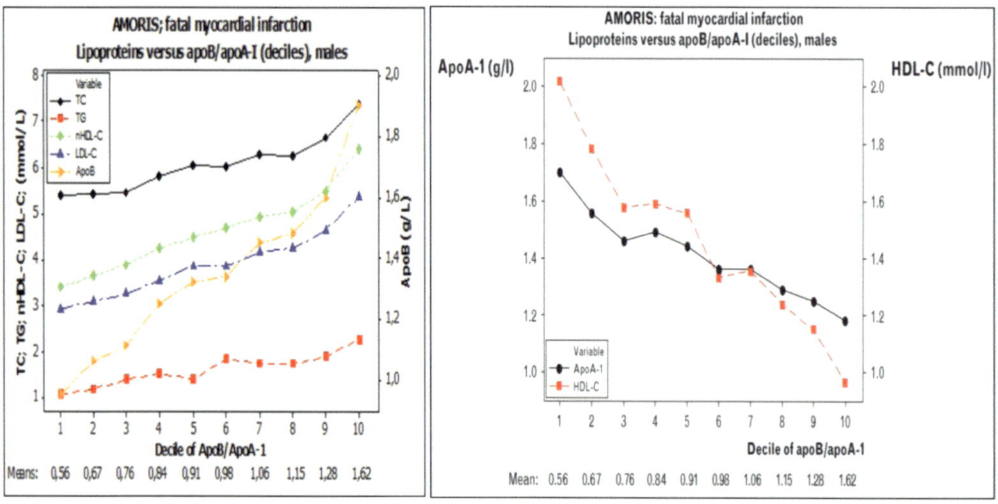

Figure 5. Left; the apo-ratio in deciles (x-axis) versus different atherogenic lipid fractions (y-axis). Right; the apo-ratio in deciles (x-axis) versus values for HDL-C and apoA-I (y-axis). Both figures from the AMORIS study (in reference 3).

In collaboration with Sniderman we have also published data from AMORIS showing that the apo-ratio has a significantly stronger relation with MI than any other lipid-based ratio (3,4,21).

In another AMORIS cohort including 69,029 men and 57,167 women who were followed for a mean of 10.3 years we determined LDL size as reflected by the LDL-C/apoB ratio (87). Because LDL size did not add predictive information to the apo-ratio, it appears that this apo-ratio also captures the risk related to LDL size. These findings add to our previously published results from AMORIS that indicates that the apo-ratio is the best single lipid-related summary index of risk and that TC, TG, non-HDL-C, and LDL-C do not add significant predictive power to the apo-ratio.

7.2. The apo-ratio and inflammatory risk factors – relations to MI, stroke and heart failure in the AMORIS study

The risk relation to age during prolonged follow-up was also studied in an AMORIS population (n = 149,121) free of previous MI at blood sampling. They were followed from 1985 to 2002 with respect to n = 6,794 first cases of MI. The mean value of the apo-ratio for men was 1.0 and for women 0.85 at baseline. In collaboration with Holme we found that the apo-ratio was somewhat stronger for those developing non-fatal than fatal MI (88). The risk was also stronger associated with the apo-ratio in those < 65 years of age than above, but risk remained significantly related to the apo-ratio also in the older population. In multivariate analyses the apo-ratio was a better predictor than TC/HDL-C. Furthermore, the apo-ratio added clinically significant information to TC/HDL-C in men as reflected by a net reclassification improvement (NRI) of 9.4% (P < 0.0001). Furthermore, also in patients developing heart failure, a common complication after MI, the apo-ratio is the best lipid-related variable to classify risk especially in men (89).

Subsequently we have shown that for the inflammation marker haptoglobin (Hp) has strong relations with MI, stroke and heart failure in the AMORIS cohort (90). There were 11,216 men and 4,291 women who had a first MI, 8,463 men and 6,072 women who had a first stroke, and 4,670 and 3,634 who had a first heart failure, respectively. Based on 4,254 MI cases the risk of MI was about 4.5 times higher in the upper joint quartile of the apo-ratio as compared to the lower, whereas this relative risk for Hp was about 4.1. However, the attributable risk for the apo-ratio is higher since more subjects were classified into the top joint quartile of TC and the apo-ratio (12.8%) than that of TC and Hp (8.8%) and into the lower joint quartiles (12.1%) and (6.4%), respectively.

In another AMORIS-based cohort of 65,050 subjects Holme et al. (91) developed an inflammatory score comprising white blood cell count, haptoglobin and in a subgroup also CRP. After 11.8 years follow-up 3,649 MI, 2,663 stroke, 2,690 heart failure, in total 7, 456 MACE, occurred. In multivariate Cox proportional hazards analysis the inflammatory scores added predictive information over and above classical lipids such as TC and TG. Based on the apo-ratio, which was a stronger marker of CVD risk than conventional lipids, the inflammatory score added significant information value measured by net reclassification improvement, especially for those with the higher values for these variables. However, there was no statistically significant biological interaction between lipoproteins and the inflammatory markers. These data indicate that routinely used markers of inflammation in combination with the apo-ratio could be used in daily medical practice to assess CV risk.

We have also published data of lipid- and the apo-ratio from three cultures (Sweden, Iran, US) showing that the apo-ratio is highest in the Swedes (the AMORIS cohort) but similar in the Americans (NHANES) and Iranians (92). By contrast, the TC/HDL-C ratio is highest in the Iranians, intermediate in the Americans and lowest in the Swedes. There were similar associations of the pro-atherogenic and anti-atherogenic lipoproteins between the genders and variation with age in these three different cultures. These data indicate that complete characterization of lipoproteins requires measurement of apoB and apoA-I as well as lipoprotein lipids.

7.3. The apo-ratio in relation to chronic kidney disease and MI risk in the AMORIS study

Some previous studies have shown apoB to be increased and apoA-I to be decreased in patients with renal insufficiency. In the much larger AMORIS study Holzmann et al. (93) performed in 142,394 middle-aged mainly healthy men and women it was shown that the apo-ratio, the TC/HDL-C ratio, and non-HDL-C all are strong predictors of first MI, among both men and women, with or without chronic kidney disease (CKD). Those with the lowest glomerular filtration rate (estimated GFR mL/min/1.73 m^2, n = 5,838) had the highest apo-ratio. In Receiver Operator Characteristics (ROC) analysis the area under the curve (AUC) for the apo-ratio was 0.77 for men and 0.83 for women without CKD, and 0.65 and 0.74 among men and women with CKD, respectively analyses. These and other data reflect a certain advantage in the prediction of MI for the apo-ratio as compared to conventional lipids. Furthermore, the findings also indicate the presence of severe atherosclerosis both in the kidney and in the coronary arteries.

7.4. The apo-ratio and risk of stroke in the AMORIS study

High LDL-C is a major risk factor for MI. However, LDL-C is rarely increased in those who suffer any type of stroke. A low HDL-C and some abnormalities in either apoB and/or apoA-I have previously been found in patients with ischaemic stroke (94-99). In 2006 Walldius et al. published the first report on risk of stroke based on the AMORIS-population (100). The relationships between different types of fatal stroke and the lipid fractions, apoB, apoA-I and the apo-ratio were examined in 98,722 men and 76,831 women followed for a mean of 10.3 years. High apoB and low apoA-I values were significantly related to risk of stroke. The odds ratio comparing the upper 10th vs. the 1st decile of the apo-ratio for all strokes adjusted for age, gender, TC and TG was 2.07 (95% CI: 1.49–2.88, $p < 0.0001$). The apo-ratio was linearly related to the risk of stroke although the slope was less than observed for the risk of fatal MI (**Figure 6, left**). Low apoA-I was a common abnormality in all stroke subtypes including subarachnoidal and haemorrhagic strokes. In multivariate analyses the apo-ratio was a significantly stronger risk predictor than TC/HDL-C and LDL-C/HDL-C ratios.

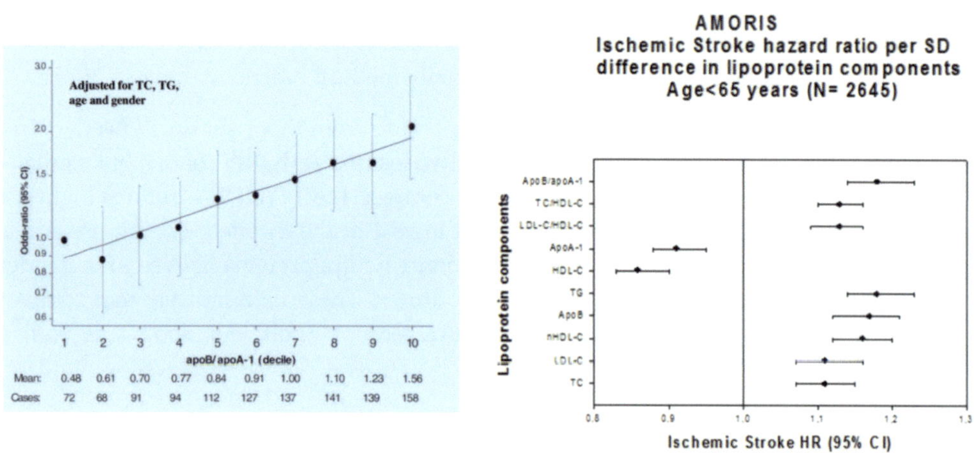

Figure 6. Risk of total stroke (left) (reference 3 and 100) and ischemic stroke (right) (reference 101). Both figures from the AMORIS study.

In a prospective follow-up study (mean observation age 11.8, range 7–17 years) based on the AMORIS population (n = 148,600). Holme et al. focused on risk of fatal and non-fatal ischaemic and haemorrhagic stroke in relation to all lipids and apos (101). Hazard ratio of non-fatal and fatal ischaemic and haemorrhagic stroke for 1 SD difference in lipoprotein components was calculated by gender, adjusted for age, MI, diabetes and hypertension. Ischaemic stroke was more common than haemorrhagic stroke (5:1), but case fatality was higher in haemorrhagic stroke. The apo-ratio, non-HDL-C and TG as well as low HDL-C and a high TC/HDL-C ratio were all predictors of ischemic stroke **(Figure 6, right)** and all cerebrovascular events (n=7,480) with somewhat stronger relations for non-fatal than fatal events. The apo-ratio was significantly stronger than the TC/HDL-C ratio in the patients with ischaemic stroke as reflected by chi-squared information value, adjusted for hypertension, diabetes, AMI, age and gender. The strongest association was for ischaemic stroke in those < 65 years of age and also for those with LDL-C < 3.0 mmol/L. There were no lipid relations to risk of haemorrhagic stroke other than a high apo-ratio related to risk in women.

7.5. Other findings from AMORIS indicating that the apo-ratio predicts CV risk

In addition, the risk of death from aortic aneurysms (n = 241) was significantly related to the apo-ratio ($p < 0.0039$) (3) adding to the importance of the apo-ratio as a predictor of severe ischaemic complications related to atherosclerosis. In that paper we also noted that, there was no relationship between the apo-ratio and risk of cancer (n = 4,423), motor vehicle accidents (n =100) or dementia (n = 255).

8. The apo-ratio in case-control CV risk studies

8.1. The INTERHEART study and risk of MI

The largest case-control study which has been performed is the INTERHEART study (58) comprising 15,152 patients with a first MI compared to 14,820 subjects from 52 countries world-wide matched for age, gender, ethnicity and continent. The aim of the study was to investigate which of the nine most common risk factors had the strongest relation to risk of MI and also which of the factors was most prevalent (highest Population Attributable Risk). These factors were: lipids primarily measured as the apoB/apoA-I ratio, smoking, diabetes, hypertension, abdominal obesity, psychosocial, fruits and vegetables, exercise, and alcohol. They found that all these risk factors were statistically related to risk.

The strongest **(Figure 7, left, (Table 1, left)** and also the most prevalent risk factor **(Table 1, right)**, was the apo-ratio both in men and women in each of the 52 countries worldwide. The apo-ratio plus smoking variables explained 70% of the entire risk which amounted to 90% for all nine risk factors taken together.

In a subsequent paper (59) they also showed that the apo-ratio had the strongest relation to MI-risk of all other measured lipids **(Figure 7, right, top panel).** They also showed a significantly stronger relationship to MI risk for the apo-ratio than the TC/HDL-C ratio **(Figure 7, right; bottom panel).** It was also shown that apoA-I had better diagnostic power than HDL-C over a wider range of low to high values.

Based on the findings and impact of these risk factors on risk of MI the INTERHEART Modifiable Risk Score (IHMRS) was developed based on age, the apo-ratio, smoking –

present, smoking – second hand, diabetes and hypertension with a range of points from 0-32 (102).

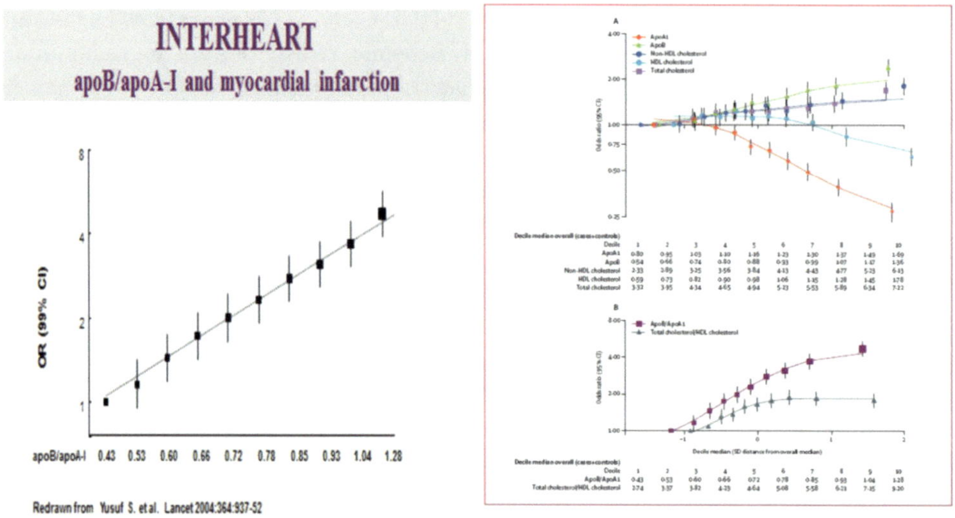

Figure 7. The INTERHEART study. Risk (Odds ratio, y-axis) versus the apoB/apoA-I ratio (left) (reference 58), and single lipids, apolipoproteins and their ratios (right) (reference 59).

INTERHEART
Risk of AMI in relation to common risk factors (Yusuf S et al. Lancet 2004;364:937)

Risk factor adj. for all	OR (99 % CI)
ApoB/apoA-I	3.25 (2.81-3.76)
Smoking	2.87 (2.58-3.19)
Diabetes	2.37 (2.07-2.71)
Hypertension	1.91 (1.74-2.10)
Abd.obesity	1.62 (1.45-1.80)
Psychosocial	2.67 (2.21-3.22)
Veg.&Fruits	0.70 (0.62-0.79)
Exercise	0.86 (0.76-0.97)
Alcohol	0.91 (0.82-1.02)
Combined	129 (90-185)

INTERHEART
Risk of AMI in relation to common risk factors (Yusuf S et al. Lancet 2004;364:937)

Risk factor adj. for all	PAR (99 % CI)
ApoB/apoA-I	49.2 (43.8-54.5)
Smoking	35.7 (32.5-39.1)
Diabetes	9.9 (8.5-11.5)
Hypertension	17.9 (15.7-20.4)
Abd.obesity	20.1 (15.3-26.0)
Psychosocial	32.5 (25.1-40.8)
Fruits & Veg.	13.7 (9.9-18.6)
Exercise	12.2 (5.5-25.1)
Alcohol	6.7 (2.0-20.2)
Combined	90 (88-92)

Table 1. The INTERHEART study. Risk of myocardial infarction (AMI); Odds ratios for nine conventional risk factors (left) and Population Attributable Risk for nine conventional risk factors (right) (booth tables reprinted from reference 58).

The IHMRS was positively associated with incident MI in a large cohort of people at low risk for CV disease (12% increase in MI risk with a 1-point increase in score). The data were internally validated and the discrimination was tested (ROC c-statistic 0.69, 95% CI: 0.64-0.74) or even higher values up to 0.79 in certain global areas. Results were consistent across ethnic groups and geographic regions. A non-laboratory-based score has also been supplied. The IHMRS demonstrated clinical credibility, evidence of accuracy, and evidence of generality.

In an analysis of 15,780 patients from the INTERHEART study (103) it was shown that HbA1c was a useful diagnostic tool of risk and the levels increased with increasing apo-ratio from 0.75-0.84 for each quintile increase of HbA1c from <5.4 – >6.12% ($p < 0.0001$). Most of the MI patients had values in the highest HbA1c quintile. The advantage of using the apo-ratio in India (104), Latin America (105), Puerto Rico (106), and Africa (107) based on the INTERHEART study designs has been useful for evaluating CV risk and should be valuable in treating risk in these countries but also elsewhere in the world.

8.2. The INTERSTROKE study

The standardized INTERSTROKE case-control study was performed in 22 countries worldwide (108). Cases were patients with acute first stroke (within 5 days of symptoms onset and 72 hours of hospital admission). Controls had no history of stroke, and were matched with cases for age and sex. In 3,000 cases (n = 2,337, 78%, with ischaemic stroke; n = 663, 22% with intracerebral haemorrhagic stroke) and 3,000 controls, significant risk factors for all stroke were: history of hypertension, current smoking, waist-to-hip ratio, diet risk score, regular physical activity, diabetes mellitus, alcohol intake, psychosocial stress and depression, cardiac causes, and the apo-ratio in falling order. Together, these risk factors accounted for 88.1% of the population attributable risk for all stroke. Increased concentration of non-HDL-C was not associated with risk of ischaemic stroke, but was associated with reduced risk of intracerebral haemorrhagic stroke, whereas increased concentration of apoB was associated with increased risk of ischaemic stroke, but was not associated with risk of intracerebral haemorrhagic stroke. The apo-ratio was a stronger predictor of ischaemic stroke than was ratio of non-HDL-C/HDL-C.

8.3. Other studies on stroke and atherosclerosis in the carotid arteries

Kostapanos et al. (109) studied 163 patients aged 70 years (88 men) with a first-ever acute ischemic/non-embolic stroke and 166 volunteers (87 men) with no history of CV disease. Compared with subjects with an apo-ratio in the lowest quartile, those within the highest quartile had a 6.3-fold increase in the odds of suffering an ischemic stroke ($p<0.001$). This association remained significant after controlling for sex, age, smoking status, body mass index, waist circumference, glucose and insulin levels, the presence of hypertension and diabetes mellitus, and lipid profile parameters (adjusted OR = 3.02; 95% CI 5.16-7.83; $p = 0.02$). The findings support elevated apo-ratio as an independent predictor of ischemic stroke in individuals over age 70.

Park et al. (110) studied 464 statin or fibrate naïve Korean patients with acute ischemic stroke: intracranial (ICAS, n = 236), extracranial (n = 44), and no cerebral atherosclerotic stenosis (n = 184). The ICAS group showed a significantly higher apo-ratio than the other two groups. The apo-ratio of 0.93 was substantially increased in patients with advanced ICAS (3 or more intracranial stenoses), the highest quartile of the apo-ratio was an independent predictor of ICAS (OR, 2.13; 95% CI, 1.05 - 4.33). A dose–response relationship (multivariate analysis) was observed between the presence of advanced ICAS and the apo-ratio quartiles (ORs, 4.03, 4.88, and 7.79, for the fourth quartile versus the first quartile). Patients having more metabolic syndrome components indicating MetS were more likely to have ICAS, advanced ICAS, and a higher apo-ratio ($p < 0.001$ for all). Thus, a higher apo-ratio is a predictor of ICAS rather than of extracranial atherosclerotic stenosis or no cerebral atherosclerotic stenosis. The apo-ratio might be a biomarker for ICAS in Asian patients with stroke.

8.4. The ISIS-study relating the apo-ratio to risk of MI

This ISIS case–control study was conducted among 3,510 acute MI patients (without prior vascular disease, diabetes, or statin use) in UK hospitals and 9,805 controls (60). Relative risks (age, sex, smoking, and obesity-adjusted) were more strongly related to apoB than to LDL-C and, given apoB, more strongly negatively related to apoA-I than to HDL-C. The apo-ratio was substantially more informative about risk than LDL-C/HDL-C, TC/HDL-C, non-HDL-C, and TC. Relative risks within several subgroups of patients showed no clear heterogeneity of effect with respect to sex, smoking, or BMI. The strongest effects were seen in those aged 30-49 years but even at ages 70-79, a 2SD higher apo-ratio was associated with a highly significant ($P < 0.00001$) relative risk. Furthermore, the apo-ratio, if untreated, is stable over time. Given the usual value of apoB, the usual value of LDL-C (indicating sdLDL particles) the risk was significantly higher. They concluded that single measurements of apoB and apoA-I are more predictive than single measurements of LDL-C and HDL-C and that the apo-ratio is the single best predictor of all lipid fractions is consistent with previously reviewed results including the AMORIS study (3,44).

9. Other studies showing strong prediction of CV risk by the apo-ratio

In our previous review from 2006 (3) we commented results from several prospective risk studies all showing an important diagnostic improvement of CV risk using apos and the apo-ratio over conventional lipids most commonly also adjusted for other confounders. The Dutch EPIC-Norfolk study (111) published in 2007 was performed in 1,511 apparently healthy controls and in 869 cases who had developed a non-fatal or fatal MI. They showed that in a head to head analysis of TC/HDL-C ratio versus the apo-ratio the Odds ratio for linear trend for quartiles was non-significant for the lipid-ratio but strongly significant for the apo-ratio, $p < 0.006$. These analyses were adjusted for sex, age, and time of enrollment and was adjusted for diabetes (yes or no), body mass index, smoking status (yes or no), systolic blood pressure, C-reactive protein level, and log-transformed triglyceride level. The apo-ratio added significant predictive value above that of the Framingham risk score since the area under the receiver-operating characteristic curve was 0.594 for Framingham risk score alone vs. 0.613 for Framingham risk score plus the apo-ratio, $p < 0.001$. Despite the fact that the difference was strongly significantly in favor of the apo-ratio the authors concluded that this was only a small increase. However, the authors pointed out that the apo-ratio is also useful since it can be applied in non-fasting samples.

The German MONICA/Kora Augsburg study (112) showed that in 1,414 men and 1,436 women without prior MI and a median follow up of 13 years the TC/HDL-C ratio predicted

MI risk. In addition, the apo-ratio was significantly related to increased risk of MI adjusted for age, smoking, alcohol, BMI, diabetes and hypertension.

In the American Thrombo study and its follow-up (113) both high apoB and low apoA-I predicted risk of re-infarction. In a follow-up they found that apoB was the strongest risk factor in those who manifested the MetS (114). However, in the German GRIPS (115), the results were negative in that LDL-C in multivariate analysis was found to be a stronger determinant of risk than apoB and the apo-ratio. This is, in fact, one of the very few studies to be found that shows LDL-C to be significantly better than apos in predicting risk. In the South Wales Cearphilly studies (116), although significant prediction was seen for apoB and apoA-I, the addition of apos did not improve prediction MI. In both of these two studies the number of events was below 300.

In the Swedish ULSAM studies (117,118) they showed that the risk of MI increased in parallel with increasing values of the apo-ratio. In those who had values for the ratio of <0.67 the incidence of MI was 9.5%, those who had ratios of 0.67–0.86 had an incidence of 17.7%, those with ratios of 0.87–1.23 had an incidence of 30.7%, and those with apo-ratio values >1.24 had an incidence of 44.8%. These risk values correspond well with those found in the AMORIS study (3,44). A risk prediction score was derived from one half of the population sample from the ULSAM cohort including systolic blood pressure, smoking, family history of MI, serum pro-insulin, and the apo-ratio. The score was highly predictive for future MI in the other half of the population that was not used for generating the score. The ULSAM score performed slightly better than the Framingham and PROCAM scores (evaluated as areas under the receiver operating curves; Framingham, 61%; PROCAM, 63%; ULSAM, 66%; $p < 0.08$). The authors also reported from the 30-year follow up of patients in the ULSAM study that ECG abnormalities were risk markers after the first 20 years of follow up but also that the apo-ratio and blood pressure remained significant risk predictors over three decades (118).

Ingelsson et al. (119) in the US Framingham study found that after a median follow-up of 15.0 years, 291 participants, 198 of whom were men, developed various manifestations of CHD. In multivariate models adjusting for non-lipid risk factors, the apo-ratio predicted CHD (HR per SD increment, 1.39; 95% CI 1.23-1.58 in men and HR, 1.40; 1.16-1.67 in women), but risk ratios were similar for the TC/HDL-C ratio (HR, 1.39; 1.22- 1.58 in men and HR, 1.39; 1.17-1.66 in women) and for LDL-C/HDL-C (HR, 1.35; 1.18-1.54 in men and HR, 1.36; 1.14-1.63 in women). In both genders, models using the apo-ratio were comparable with but not better than that for other lipid ratios. The apo-ratio did not predict CHD risk in a model containing all components of the Framingham risk score including the TC/HDL-C ratio. They concluded that the apo-ratio adds no incremental utility over this lipid ratio. Notably, there were few hard events in this small study, a fact that may restrict the interpretation of the results.

In India Goswami et al. (120) studied 100 patients with MI who were age-matched with 100 healthy control subjects. The exponential value of the regression coefficient beta for the apo-ratio was 11.9, as compared to 4.4 for the LDL-C/HDL-C ratio, 3.5 for the TC/HDL-C ratio and 2.2 for the TC/HDL-C ratio. The findings suggested that the apo-ratio is a better discriminator of CAD risk in the atherosclerosis-prone Indian population, than any of the conventional lipid ratios. They suggested that the apo-ratio should be an alternative to other lipid ratios in the risk assessment in patients with CAD.

In a comparative observational study by Agoston-Coldea et al. (121) on 289 subjects were divided into two groups: 144 subjects with old MI, and 145 subjects without CHD, but with CV risk factors. The multivariate analysis indicated that apoB over 1.7 g/L are closely correlated with MI ($p = 0.001$) independent of age, smoking, diabetes, hypertension, lipid TC/HDL-C and the LDL-C/HDL-C ratio. The protective effect of apoA-I was also significant ($p = 0.004$) in multivariate analysis. They concluded that the predictive value of the apo-ratio is superior to that of serum lipid fractions and that the apo-ratio therefore should be introduced in current clinical practice.

In the prospective case-cohort study (PREVEND cohort) (122) 6,948 subjects without previous CHD they studied the risk factors predicting major coronary events. The age- and sex-adjusted HR was 1.37 (95% CI, 1.26-1.48) for the apo-ratio and 1.24 (1.18-1.29) for the TC/HDL-C ratio (both $p < 0.001$). The risks of the two ratios were only marginally attenuated by additional controlling for traditional risk factors TG, hypertension, diabetes, obesity and smoking), hs-CRP and albuminuria.

In a Korean study by Kim et al (123) they studied the association between plasma lipids, and apolipoproteins and coronary artery disease: a cross-sectional study in a low-risk Korean population in 544 subjects. In the lowest quartile of TC, TG and LDL-C, and the highest quartile of HDL-C, only the apo-ratio was associated with CAD in both men and women. They concluded that the apo-ratio is the only variable that differentiates the patients with CAD from those without and, furthermore, gives additional information to that supplied by traditional lipid risk factors in a low-risk Korean population.

Agoston-Coldea et al. (124) studied 208 patients (100 men and 108 women), with and without previous MI by coronary angiography. They showed that the apo-ratio had a stronger correlation with MI than the TC/HDL-C ratio. Multivariate analysis performed with adjustments for conventional risk factors, showed that the levels of apoB, the apo-ratio and Lp(a), are significant independent CV risk factors. Therefore they recommend that the apo-ratio and Lp(a) should be included in clinical practice.

10. Meta-analysis of studies on CV risk

In 2006 Thomson and Danesh published a meta-analysis based on data from 23 relevant prospective studies in which apoB, apoA-I and the apo-ratio were associated with risk of MI (8). They compared risk in the top versus the bottom tertile of baseline values. The relative risks were; apoB 1.86 (95% CI 1.55-2.22, cases n = 6,320), apoA-I 1.62 (1.43-1.83, cases n = 6,333), and the apo-ratio 1.86 (1.55-2.22, cases n = 3,730). ApoB and the apo-ratio were directly related to risk, whereas apoA-I was protective. In that study no results were given for any lipids.

In 2009 the Emerging Risk Factor Collaboration (ERFC) published an extended meta-analysis in which they included 302,430 men and women without previous vascular disease from 68 long-term prospective studies, mostly in Europe and North America (10). During 2.79 million person-years of follow-up, there were 8,857 nonfatal MI, 3,928 coronary heart disease deaths, 2,534 ischemic strokes, 513 hemorrhagic strokes, and 2,536 unclassified strokes. Half of the studies included less than 100 events, and the largest study (ARIC) included 871 cases. In 22 studies on risk of MI and in 8 studies on risk of ischemic stroke they had also measured apoB, apoA-I and the apo-ratio. In 91,307 individuals with 4,499 MI and in 8 studies with 60,571 individuals and 1,192 cases they could compare how well the TC/HDL-C ratio and apoB, apoA-I and the apo-ratio were related to these CV events. In all of these comparisons non-HDL-C, HDL-C, the non-HDL-C/HDL-C ratio, apoB, apoA-I and the apo-ratio, adjusted for age and sex, were significantly related to risk of both MI and ischemic stroke. When additionally adjusted also for blood pressure, smoking, BMI, hypertension, and other lipid markers the HR was 1.50 (95%CI, 1.38-1.62) for the non-HDL-C/HDL-C ratio and 1.49 (1.39-1.60) for the apo-ratio. Interestingly, adjusting for these confounders changed the HR only marginally. These data show that the lipid- and apo-ratios give similar and significant prediction of risk. Furthermore they also found that apoB had similar risk as non-HDL-C, and apoA-I had similar risk as HDL-C. The ERFC authors concluded that both lipid- and apo-ratios can be used even in the non-fasted state since the apo-ratio and the lipid ratio give similar information. Furthermore, they also discuss that there may be important advantages for using apolipoproteins.

Importantly, the ERFC did not include the three largest studies on risk of MI and stroke related to lipids, apos and the apo-ratio. These are the studies; AMORIS, n = 6,794 first cases of AMI (88), and n = 4,470 first ischemic stroke (101), INTERHEART, cases n = 15, 152 for first MI (59), n = 2,337 for first stroke (108), and the ISIS study, n = 3,510 for first MI (60). These studies were excluded because a complete set of confounding variables were not available (AMORIS), or that two studies were case-control studies (INTERHEART and ISIS). The findings in ERFC are therefore restricted to the results based on only prospective studies with many fewer number of events (total n = 5,691) compared to these much larger studies also covering a world-wide population (AMORIS, INTERHEART, INTERSTROKE and ISIS) (total n = 32,263). So adding all these results to those obtained in the ERFC studies the advantages of the apo-ratio as risk predictor may be even more compelling. Such advantages for clinical use are commented in several sections below and are summarized in the discussion. Results from a recent ERFC publication are also included and discussed in page 39.

11. Relations between the apo-ratio and the metabolic syndrome, glucose - insulin metabolism and diabetes - Risk predictors for CV manifestations

11.1. Metabolic syndrome (MetS) and diabetes

In subjects with the MetS and in patients with diabetes several studies have been performed indicating advantages of using apos, especially apoB, over conventional lipids. In our previous review (3) we summarized these results from Stewart et al. (125), Korean studies (126-128) and studies from India (129) and Canada (130). In these papers the highest values for the apo-ratio were found in those who had most manifestations of the MetS. The apo-ratio was also related to atherosclerosis verified by angiography even if LDL-C values were low. In the Swedish ULSAM study (131) at the 26.8 year follow-up 462 patients had developed MI. The apo-ratio was highest in those who developed a MetS, and their apo-ratio was inversely related to glucose disposal. These findings were independent of LDL-C and smoking. Both the apo-ratio and MetS independently predicted MI.

Sierra-Johnson et al. (132) studied 2,955 adults (mean age 47 years; 1,457 women) without diabetes from the US NHANS III population. The apo-ratio was an independent predictor of insulin resistance after adjustment for age and race, and remained significant after further adjustment for MetS components including TG, HDL-C, traditional and inflammatory risk factors. They recommended that the apo-ratio should be recognized and implemented in future clinical guidelines. In the follow-up paper (133) of a multi-ethnic representative subset of 7,594 US adults (mean age 45 years; 3,881 men, 3,713 women) there were 673 CV deaths of which 432 were from CHD. Both the apo-ratio (HR 2.14, 95% CI, 1.11 – 4.10) and the TC/HDL-C ratio (HR 1.10, 1.04 – 1.16) were related to CHD death. Only apoB (HR 2.01, 1.05 – 3.86) and the apo-ratio (HR 2.09, 1.04 – 4.19) remained significantly associated with CHD death after adjusting for CV risk factors **(Figure 8 left)**. This suggested that the measurement of apolipoproteins has superior clinical utility over traditional risk markers such as the TC/HDL-C ratio in identifying subjects at risk for fatal CV disease. In addition, the combined elevation of glucose and a high apo-ratio increases the risk of MI as documented in the AMORIS study (3) **(Figure 8, right)**.

Zhong et al. (134) found also in China that the apo-ratio increased significantly with number of MetS components. Belfki et al (135) have shown in a Tunisian population that the apo-ratio increased significantly with each of the components as well as with increasing numbers of components of the MetS after adjusting for age and gender. Similarly, the apo-ratio was associated with insulin resistance.

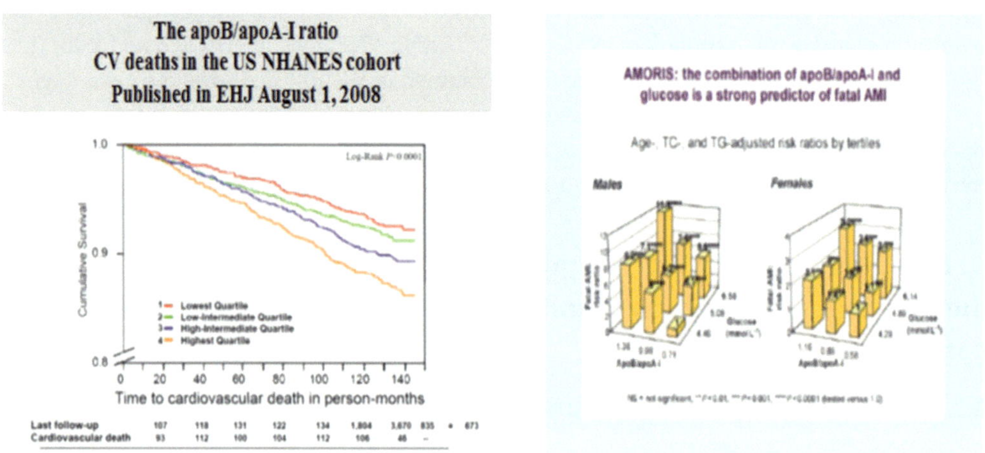

Figure 8. Cumulative survival (y-axis) in relation to quartiles of the apoB/apoA-I ratio in patients with the metabolic syndrome (left) (from NHANES cohort, reference 133). Risk of myocardial infarction in relation to glucose and the apoB/apoA-I ratio (right) (AMORIS study, reference 3).

Based on the findings in subjects with MetS Sniderman and Faraj (136) have argued for including both apoB and apoA-I as stronger risk markers especially compared to LDL-C (often low in MetS), TG and HDL-C. These apos also have strong relations to glucose and insulin homeostasis. Therefore the apo-ratio should be a valid component of the MetS especially since the apo-ratio has so strong predictive value of CV risk. The apo-ratio also summarizes the risk for individuals with MetS into one simple and predictive risk number. In another paper Sniderman et al. (137) have also analyzed pros and cons for using the apo-ratio.

Bruno et al. (138) studied diabetic subjects and they found that apoB and the apo-ratio were associated with CV mortality independently of non-HDL-C. They recommended that apoB and apoA-I should be measured routinely in all people with diabetes, particularly in the elderly.

Bayu et al. (139) studied 224 diabetic patients (85 type 1 and 139 type 2). After adjusting for age, sex, diabetes duration, systolic blood pressure and diabetes medications they found that the apo-ratio was the best predictor of diabetic retinopathy. Traditional lipids improved the ROC area by only 1.8 % whereas the apo-ratio improved the area by 8.2 %.

Enkhma et al. (140) have studied several ethnic groups of European and African Americans and developed a CV risk score which was found to be significantly increased across tertiles of the apo-ratio. They concluded that the apo-ratio differed across ethnicities and was associated with presence of the MetS in both groups. Among African Americans, an elevated apo-ratio independently predicted a greater risk of CAD.

Ounis et al. (141) studied thirty-two obese 13 years old children with 16 subjects who participated in a 8-week training period and 16 subjects serving as a control group. The apo-ratio was positively correlated with TG ($r = 0.46$, $p < 0.01$), blood glucose ($r = 0.48$, $p < 0.01$), waist circumference ($r = 0.34$, $p < 0.01$), systolic ($r = 0.31$, $p < 0.01$) or diastolic ($r = 0.29$, $p < 0.05$) blood pressure and was negatively correlated with HDL-C ($r = 0.51$, $p < 0.01$), Fat max ($r = 0.45$, $p < 0.01$) and VO^2 peak ($r = 0.39$, $p < 0.01$). When adjusted for pubertal stage, the relationships between the apo-ratio and other variables were not significantly altered. The multiple regression analysis showed that the change in total HDL-C is the most significant predictor of the change of the apo-ratio explaining 82% of the variance of its change over the training program.

Gatz et al. (142) studied thirty same-sex twin pairs in which both members were assessed at baseline and one twin subsequently developed dementia, at least 3 years subsequent to the baseline measurement, while the partner remained cognitively intact for at least three additional years. Eighteen of the 30 cases were diagnosed with Alzheimer's disease. Baseline assessments were conducted when twins' average age was 70.6 (SD = 6.8) years. Which twin would develop dementia was predicted by less favorable lipid values defined by higher apoB and higher apo-ratio, poorer grip strength, and — to a lesser extent — higher emotionality on the EAS Temperament Scale. Given the long preclinical period that characterizes Alzheimer's disease, these findings may suggest late life risk factors for dementia. Alternatively, there may be early development of atherosclerosis in critical cerebral arteries based on an elevated apo-ratio over time.

Carnevale-Schianca et al. (143) enrolled 616 patients with normal glucose tolerance (NGT) (273 men and 343 women), and measured insulin resistance, lipid profile, the apo-ratio and the factors compounding the MetS. An unfavorable apo-ratio (> 0.90 for males and > 0.80 for females) was present in 13.9 % of 108 patients with LDL-C < 100 mg/dL. Compared to subjects with lower apo-ratio, they had more elements of MetS and their lipid profile strongly correlated with high CV risk. In NGT individuals with LDL-C < 100 mg/dL, a higher apo-ratio indicated an atherogenic lipid profile, suggesting that LDL-C alone is insufficient to define CV risk. This study demonstrates that the apo-ratio is at least complementary to LDL-C in identifying a more correct CV risk profile of asymptomatic NGT subjects.

Wen et al. (144) measured high sensitive hsCRP, apoB, apoA-I, and the profiles of coronary angiograms, echocardiography and oral glucose tolerance tests (OGTT)s as well as traditional risk factors in 1,757 cardiology patients. The hsCRP or the apo-ratio were significantly correlated with the presence and severity of angiographic profiles, the levels of left ventricular (LV) ejection fraction, LV mass and LV mass index, and the presence of abnormal OGTT. The combination of the apo-ratio and hsCRP had greater correlation with abnormal glucose metabolism than its individual components in patients with normal fasting glucose, and was an independent predictor for coronary artery disease.

12. The apo-ratio and relations to atherosclerosis, vascular functions and inflammation

In many clinical conditions coronary arteriography, carotid ultrasound (CIMT), endothelial function, calcium scoring (CAC) and even more recently Intra Vascular Ultrasound (IVUS) studies of the coronary arteries has been related to lipid- and apo-abnormalities. Coronary and femoral plaques also contain apos (34-36). Many of these studies indicate that apos are more closely related to the amount of atherosclerosis than conventional lipids. Relevant studies are commented below.

In the Uppsala PIVUS study by Andersson et al. (145) the prevalence of carotid plaque was investigated. In 942 free living 70 year old men (n = 469) and women (n = 473) an ultrasound was performed. A plaque was defined by at least 50% increase of the intima-media thickness (IMT). Plaques were slightly more prevalent in men (n = 322) than in women (n = 293). Individuals with plaques had significantly higher the apo-ratio (p = 0.013), LDL-C/HDL-C-ratio (p = 0.04), LDL-C (p = 0.02), higher levels of fasting blood glucose (p = 0.02), Framingham risk score (p < 0.0001), higher levels of systolic blood pressure, (p < 0.0001), and also a higher average of pack-years of cigarette smoking (p = 0.008) after adjustment for gender and statin use. No significant differences were seen for HDL-C, diastolic blood pressure or BMI. The inflammatory markers oxidized LDL, TNF alpha, and leucocyte count as well as insulin resistance (HOMA) were increased.

In another subsample of 70 years old men (n = 124) and women (n = 123) who did not use lipid-lowering drugs from the PIVUS study (146) were investigated whether the amount of visceral (VAT) or subcutaneous adipose tissue (SAT) independently of the other can determine the apo-ratio. VAT and SAT areas were assessed using magnetic resonance imaging. Their adipose tissue areas were related to their levels of apoB, apoA-I and the apo-ratio. ApoA-I levels were independently related to the VAT area ($r = -0.33$, $p < 0.0001$) whereas the apoB levels were not ($r = 0.102$, $p = 0.07$). The VAT area was independently significantly ($r = 0.25$, $p = 0.001$) related to the apo-ratio in the multiple regression analysis whereas the SAT area was not. This observation may indicate that VAT is metabolically active possibly through decreased adiponectin levels. The VAT metabolism seems more related to abnormalities in the apo-ratio which also may be a consequence of abnormal glucose-insulin metabolism as discussed above in other studies on the MetS.

Schmidt and Wikstrand (147) reported that in a multi-variable analysis including all baseline variables only the apo-ratio ($p = 0.003$) and serum insulin ($p = 0.026$) were significantly related to IMT composite progression rate indicating that the apo-ratio is an important risk factor for predicting atherosclerotic progression rate during very long-term follow-up in clinically healthy middle-aged men.

Reis et al (148) have studied factors that may influence MetS and development of obesity. They performed weighted Pearson partial correlation coefficients for waist circumference, log-transformed leptin, and insulin vs. metabolic, inflammatory, and thrombogenic CV risk factors among men and women aged 40 years and older, NHANES III. They found that apoB was positively correlated with waist, leptin and insulin both in men and women, whereas apoA-I was significantly and negatively related to these risk markers. These findings may indicate that the apo-ratio can summarize the lipid abnormalities into one number. The results were adjusted for age, ethnicity, smoking, physical activity, alcohol intake and time of fasting.

Junyent et al. (149) assessed carotid intima-media thickness (CIMT) and plaque in relation to classical risk factors and apoA-I and apoB levels in 131 unrelated patients with familial hypercholesterolemia (FCHL), 27 with prior CVD and 190 age- and sex-matched control subjects. By multivariate analysis in a model with all risk factors, inclusive of the MetS, independent associations of CIMT were age, the apo-ratio, systolic blood pressure, fasting glucose, family history of CVD and TC/HDL-C ratio ($r^2 = 0.475$, $p < 0.001$). The strongest determinant of IMT was the apo-ratio ($\beta = 0.422$, $p < 0.001$). The findings support the atherogenicity of the lipid phenotype in FCHL beyond associated risk factors. They also have implications for diagnosis and management of CVD risk in this condition.

Vladimirova-Kitova et al. (150) have found that carriers of a LDL-receptor defective gene have a higher carotid IMT and apo-ratio than non-carriers, whereas no difference between the groups was found with respect to the level of other lipid parameters, ADMA, total homocysteine, cell adhesion molecules, and % flow mediated dilation. Thus the apo-ratio is a predictor of IMT in carriers of this LDL-receptor gene.

Dahlen et al. (151) performed the CARDIPP-1 primary care study a study in 247 patients with type 2 diabetes, aged 55-66 years. They found that there was a significant association between the apo-ratio and CIMT in middle-aged patients with in type 2 diabetes. The association was independent of conventional lipids, hsCRP, glycaemic control and use of statins.

In the study by Rasouli et al. (152) 138 men and 126 women aged 40-70 years, were classified as CAD cases or controls, according to the results of coronary angiography. The severity of CAD was scored on the basis of the number and extent of lesions in coronary arteries. The results indicate that the apo-ratio, apoB and Lp(a) are independent risk factors for CAD and are superior to any of the cholesterol ratios. They suggested using the apo-ratio as the best marker of CAD in clinical practice.

Smith et al. (153) compared the body composition and the apo-ratio in migrant Asian Indians white Caucasians in Canada. Indian men and women had a higher apo-ratio than Caucasians ($p = 0.0003$)]. Of interest, there were also significant correlations between the apo-ratio and WHR in all groups, except the Indian women.

Both in children and adults obesity either defined by BMI or waist/hip ratio has been found to be directly related to apoB and the apo-ratio, and indirectly to apoA-I levels (154-156).

In the Cardiovascular Risk in Young Finns Study (157) they measured CIMT and brachial endothelial function in 879 subjects. They determined whether apoB and apoA-I measured in childhood and adolescence could predict atherosclerosis in adulthood. In subjects aged 12 to 18 years at baseline, apoB and the apo-ratio were directly ($p < 0.001$) related and apoA-I was inversely ($p = 0.01$) related with adulthood IMT. In subjects aged 3 to 18 years at baseline, apoB ($p = 0.02$) and the apo-ratio ($p < 0.001$) were inversely related, and apoA-I ($p = 0.003$) was directly related to adulthood flow mediated dilatation. Adjustment for age, gender, blood pressure, BMI, TG, insulin, CRP and brachial diameter at baseline did not change these relations. The apo-ratio measured in adolescence was stronger than the LDL-C/HDL-C or non-HDL-C/HDL-C ratios (c-values, 0.623 vs. 0.569, $p = 0.03$) in predicting increased CIMT in adulthood. The authors concluded that apoB and apoA-I measured in children and adolescents reflect an abnormal lipoprotein profile that may predispose to the development of subclinical atherosclerosis later in life. These markers are therefore useful in pediatric lipid risk assessment.

In a cross-sectional and 6-year prospective data from the cardiovascular risk in young Finns study (aged 24 to 39 years) (158) they studied metabolic risk variable MetS and their associations with CIMT. ApoB, CRP, and type II secretory phospholipase A2 enzyme activity were significantly higher and apoA-I lower in subjects with MetS ($n = 325$) than in subjects without MetS ($n=858$) indicating that the apo-ratio may summarize the risk into one number. In prospective analysis both MetS and high apoB predicted ($p < 0.0001$) incident high CIMT. The association between MetS and incident high CIMT was attenuated by about 40% after adjustment with apoB. Adjustments with apoA-I, CRP, or type II secretory phospholipase A2 did not diminish the association. Thus, the atherogenicity of MetS in this population assessed by incident high CIMT is mainly mediated by elevated apoB, but not inflammatory markers.

In the Swedish study Wallenfeldt et al. studied the relationships between abnormalities in lipoprotein concentrations in 338 apparently healthy 58-year-old men with manifestations of the MetS (159). Those who had an apo-ratio > 0.74, irrespective of blood pressure and smoking, had a significant progression (untreated) of the IMT values of the carotids over a

3-year period. Thus CIMT is a non-invasive simple, sensible and useful method to follow dynamic progression of atherosclerosis. Furthermore, the level of the apo-ratio is a strong predictor of these atherosclerotic changes in the arterial wall. Thus, values of the apo-ratio > 0.74 may alert the treating doctor to the need of adequate lipid-lowering therapy.

In a Japanese study (160) sixty-six type 2 diabetic patients with carotid atherosclerosis and 66 age- and sex-matched patients without carotid atherosclerosis were compared. They concluded that the combination of apoB and HOMA-R is a superior marker of carotid atherosclerosis compared with LDL-C alone in patients with type 2 diabetes.

Kim et al. (161) have studied 757 stroke patients undergoing coronary artery bypass grafting. They found that prevalence of asymptomatic carotid stenosis > 50% and > 70% was 26.4 % and 8.6%, respectively. In multivariate analysis, plasma levels of the apo-ratio and homocysteine were independently associated with carotid stenosis. Receiver operating characteristic curve (ROC) analysis indicated area under the curve values of 0.708 (the apo-ratio), 0.678 (Lp(a)), and 0.689 (homocysteine).

Ajeganova et al. (162) have studied patients with rheumatoid arthritis (RA) that commonly are affected by premature atherosclerosis including development of xanthomas. They studied 114 patients, age 50.6 years, 68.4% women, with recent RA (< 12 months after symptoms onset) and they were assessed at 0, 3, 12, 24 and 60 months after RA diagnosis. Plaque detection was positively associated with age and smoking (ever). After adjustment, a longitudinal approach demonstrated an independent positive prediction of CIMT by the apo-ratio ($p = 0.030$), but negative prediction by apoA-I ($p = 0.047$). Higher levels of the pro-atherogenic apo-ratio and apoB and low anti-PC (IgM antibodies against phosphorylcholine) were independently associated with bilateral carotid plaque $p = 0.002$, 0.026 and 0.000, respectively). Both baseline and longitudinal levels of other inflammatory/disease-related factors failed to show significant associations with the study outcomes.

13. Effects of lipid-lowering therapy on change of apoB, apoA-I and the apo-ratio

The mode of actions of statins and their effects on lipids and apos is reviewed in more detail elsewhere (163-165). The most commonly used drugs today are the statins that can reduce apoB synthesis and increase apoA-I synthesis and turnover. In clinical practice simvastatin and pravastatin are the most commonly used statins since they are now available as generics. They can reduce apoB up to about 20% and increase apoA-I by about 2-5% and a bit more for simvastatin. The most effective apoB-reducing statins are atorvastatin and rosuvastatin which lower the apoB-values by about 40-45% and 45-50%, respectively. Best increase in apoA-I concentrations is obtained by rosuvastatin which can increase the value by about 10-15% depending on baseline values, the lower the higher is the increase (163-165). Commonly for all statins there is a strong dose-response relationship, except for atorvastatin where higher doses commonly result in lowering of HDL-C and apoA-I values. The strongest lowering effects of the apo-ratio is obtained by rosuvastatin which lowers this ratio by about 50 %, followed by atorvastatin about 40-45 %, and simvastatin and pravastatin up to 30 %.

14. Prediction of outcome in statin trials using LDL-C or the apo-ratio

LDL-C has been the primary focus in lipid-lowering trials for more than two decades. A vast number of studies, both in primary and secondary prevention, have shown that there is a close relationship between LDL-C and CV event rates, the lower the LDL-C, the lower is the risk (163-165). In several of these trials also apoB, apoA-I and the apo-ratio have been measured. When explaining the relationship of each lipid fraction and each apo-fraction to CV event reduction virtually all lipids as well as apoB and apoA-I and the apo-ratio are significantly related to outcome. However, LDL-C is much weaker predictor than apoB and any lipid ratio. The best relationship with CV risk reduction is the apo-ratio. Examples from several trials are presented below.

In the AFCAPS/TexCAPS study (52,53), lovastatin 20-40 mg/d or placebo were given to 3,304 patients with rather normal LDL-C but low HDL-C values. ApoB decreased by 18.9 % and apoA-I increased by 7.2 %. At 5 years, there was a 37 % decrease in the relative risk for having a first acute coronary event in the lovastatin versus placebo group. In a head to head analysis it was found that apoB was better than LDL-C, $p < 0.01$, apoA-I was better than HDL-C, $p < 0.01$, and the apo-ratio was better than the TC/HDL-C ratio, $p < 0.01$ in explaining the event reduction **(Figure 9, left).** In this study it made no difference to which treatment group the patients were assigned, conventional diet – placebo or the lovastatin group. The apo-ratio value on treatment was the only lipid-related marker that was significantly related to outcome **(Figure 9, right).**

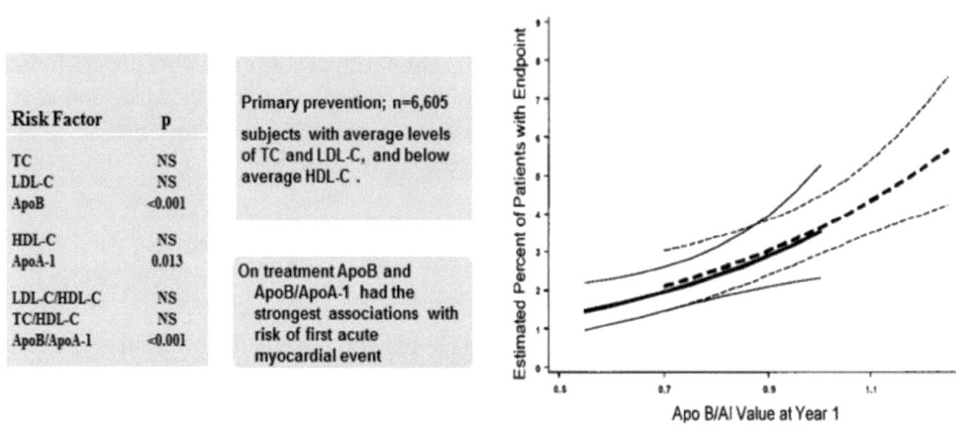

Figure 9. The AFCAPS/TexCAPS study. The apoB/apoA-I ratio was the best predictor of outcome expressed in head to head analyses versus lipids (left) (reference 52). Risk remaining during treatment (lovastatin versus placebo) in relation to obtained values for the apoB/apoA-I ratio (right) (reference 53).

In the LIPID trial pravastatin reduced CHD mortality by 24 % and total mortality by 22 % (3,4,54). The TC/HDL-C and the apo-ratios on treatment were considerably better in explaining outcome than either LDL-C or HDL-C. The values of the apo-ratio had strongest relations to event reduction.

To which target LDL-C values should lipid-lowering aim? In the ACCESS study (166) therapy reduced LDL-C levels to 'normal – target levels'. However, such therapy only reached apoB levels to about the 50th percentile of a population. This means that the patients were not optimally treated by using LDL-C at recommended guideline levels. These results may illustrate that apoB would be a better target if proper target levels have been proposed.

To reach targets in guidelines has been further investigated in a more recent paper by Vodnala et al. (167). They applied the ATP III guidelines, including Framingham Risk Scores to determine whether patients met non-HDL-C goals upon referral. In order to reach targets for non-HDL-C among patients (n = 5,692) most high- and many intermediate-risk patients goals would require more aggressive treatment to reach either the TC/HDL-C = 3.5 or the apo-ratio = 0.50 goals. Thus, a more intense therapy using better target goals, i.e. apoB or the apo-ratio, than the conventional LDL-C or non-HDL-C would most likely add clinical value and better treatment effects.

Van den Bogaard et al. (168) studied 9,247 patients (mean age 61 years, 81% males), participating in the Treatment to New Targets (TNT) trial in which the effects of 80 mg versus 10 mg atorvastatin was compared. The association between lipoprotein components and the risk of cerebrovascular events after the first year into the trial was investigated. All lipoprotein components, except LDL-C, showed a significant gradient for incidence of cerebrovascular events with increasing quartiles of the lipoprotein component. If the lipoprotein components were treated as continuous variables, the adjusted HR for cerebrovascular events for 1 SD difference in 1-year lipoprotein components were for LDL-C 1.13 (95%CI, 1.02–1.25), for HDL-C 0.86 (0.76–0.97), for apoB 1.17 (1.04–1.28), for apoA-I 0.83 (0.74–0.94), for TC / HDL-C 1.22 (1.10–1.34) and for the apo-ratio 1.24 (1.12–1.37). The apo-ratio was superior to TC/ HDL-C, because adding the apo-ratio to TC/HDL-C improved prediction, whereas adding TC/HDL-C to the apo-ratio did not. These findings are consistent with the AMORIS study linking the apo-ratio to risk of stroke (99,100), and are also similar to results from the combined data of TNT and IDEAL showing that TC/HDL-C and the apo-ratios are more closely associated with CVD than any of the individual lipoprotein parameters. They concluded that in coronary heart disease patients receiving intensive lipid-lowering treatment, the on-treatment apo-ratio provides the strongest association with incidence of cerebrovascular events followed by TC / HDL-C. They also stated that as current European and US guidelines only acknowledge LDL-C as a therapeutic target and HDL-C and triglycerides as risk markers it will be up to future guideline committees to implement these new parameters as risk predictors and to define new treatment targets based on these apolipoproteins.

Kastelein et al. (169) showed in a post hoc analysis that combined data from 2 prospective, randomized clinical trials in which 10 001 TNT and 8,888 ("Incremental Decrease in End Points through Aggressive Lipid Lowering"- IDEAL) patients with established coronary heart disease were assigned to atorvastatin 10 mg/d or atorvastatin 80 mg/d. In models with LDL-C, non-HDL-C and apoB were positively associated with cardiovascular outcome, whereas a positive relationship with LDL-C was lost. In a model that contained non-HDL-C and apoB, neither was significant owing to collinearity. Inclusion of measurements of apoA-I further strengthened the relationships. The TC/HDL-C and the apo-ratio in particular were each more closely associated with outcome than any of the individual pro-atherogenic lipoprotein parameters **(Table 2).** In a pair-wise COX model comparison of the two ratios the TC/HDL-C was non-significant but the apo-ratio was significant, p<0.001. However, the authors mainly conclude that these data support the use of non-HDL-C or apoB as novel treatment targets for statin therapy, but do not believe that the apo-ratio is yet a valid risk variable because of uncertainness of the impact of risk of HDL-C and apoA-I. Furthermore, they state that in the absence of interventions that have been proven to consistently reduce CVD risk through raising plasma levels of HDL-C or apoA-I, it seems premature to consider the ratio variables as clinically useful. These conclusions merit further comments in the discussion. However, clearly the apo-ratio comes out as the best CVD predictor as manifested by their data when all head-to-head comparisons are performed between various lipids and apos.

LDL cholesterol	0.95	0.87–1.05	0.33	Apolipoprotein B	0.94	0.85–1.04	0.26
Apolipoprotein B	1.24	1.13–1.36	<0.001	Apolipoprotein B/A-I	1.30	1.20–1.39	0.001
Non-HDL cholesterol†	1.00	0.93–1.07	0.94	Total/HDL cholesterol	1.00	0.92–1.10	0.91
Apolipoprotein B/A-I	1.24	1.17–1.32	0.001	Apolipoprotein B/A-I	1.24	1.13–1.36	0.001

Table 2. TNT-IDEAL pooled data. Head to head comparisons between various lipids, apolipoproteins and ratios (redrawn from reference 169).

Holme et al. (170) studied the ability of apolipoproteins to predict new-onset of congestive heart failure (HF) in statin-treated patients with coronary heart disease (CHD) in the IDEAL study based on 8,326 patients of whom 185 subjects had a HF event. Variables related to LDL-C carried less predictive information than those related to HDL-C, and apoA-I which was the single variable most strongly associated with HF. LDL-C was less predictive than both non-HDL-C and apoB. The apo-ratio was most strongly related to HF after adjustment for potential confounders, among which diabetes had a stronger correlation with HF than did hypertension. The apo-ratio was 2.2 times stronger associated than that of diabetes. Calculation of the net reclassification improvement (NRI) index revealed that about 3.7 % of the patients had to be reclassified into more correct categories of risk once the apo-ratio was added to the adjustment factors. The reduction in risk by intensive lipid-lowering treatment as compared to usual-dose simvastatin was well predicted by the difference in apo-ratio on-treatment levels mostly through the reductions in apoB. Thus, both apoB, apoA-I and the apo-ratio had additional clinical value above lipids in predicting risk of HF.

Holme et al. (171) also looked into the ability of apoB, apoA-I or the apo-ratio to predict new coronary heart disease (CHD) events in patients with CHD on statin treatment in the IDEAL trial comparing the effects of atorvastatin 80 mg/d to that of simvastatin 20-40 mg/d to prevent CHD subsequent major coronary events (MACE). Variables related to LDL-C carried more predictive information than those related to HDL-C, but LDL-C was less predictive than both non-HDL-C and apoB. Of all lipoprotein variables, the apo-ratio was the best predictor of MACE during statin treatment. The apo-ratio carried as much information as apoB, apoA-I, LDL-C, and HDL-C together. However, for estimating differences in relative risk reduction between the treatment groups, apoB and non-HDL-C were the strongest predictors. They recommended that measurements of apoB and apoA-I should be more widely available in clinical praxis.

Results from the recently published ASTEROID Trial (172) showed that in patients with acute coronary syndromes treated with rosuvastatin 40 mg daily for 2 years a significant ($p < 0.001$) regression was found of the atherosclerotic burden in the coronary arteries (intravascular ultrasound). In these patients LDL-C was reduced from 3.35 mmol/L (130 mg/dl) to 1.55 mmol/L (60 mg/dl), $p < 0.001$ and the apo-ratio was reduced from high 0.95 to low 0.49, $p < 0.001$. These results indicate that the risk related to the apo-ratio risk was reduced from the eighth risk decile to the first decile, i.e. to normality.

Nicholls et al. (173) presented data based on 4 studies in which IVUS was used in 1,455 coronary patients. They were given lipid-lowering with either atorvastatin, simvastatin, pravastatin and rosuvastatin (strongest lipid-lowering). A highly significant regression of coronary atheroma volume over a two year period was recorded. They stated that "Reducing the ratio of apoB to apoA-I was the strongest lipid predictor of changes in atheroma burden in patients treated with a statin". Thus, even small, but clinically important changes in atheroma volume, can be identified by IVUS techniques and also by closely related changes in the strongest marker of lipoprotein metabolism, i.e. the apo-ratio.

Tani et al. in Japan performed a 6-month prospective study of 64 patients with coronary artery disease treated with pravastatin (174). The plaque volume, assessed by IVUS, decreased by 12.6% ($p < 0.0001$). A significant decrease of 6.4 % and 14.6 % was found in the serum level of apoB and the apo-ratio ($p < 0.0001$ and $p <0.0001$, respectively, vs baseline), and apoA-I increased by 14.0 % ($p < 0.0001$). A stepwise regression analysis revealed that the change in the apo-ratio was an independent predictor of the change in coronary plaque volume ($p < 0.0023$). They concluded that a decrease in the apo-ratio is a simple predictor for coronary atherosclerotic regression: the lower the apo-ratio, the lower the risk of coronary atherosclerosis.

Taskinen et al. studied diabetic patients treated with fenofibrate (the FIELD study,175). Lipid ratios and the apo-ratio performed significantly better than any single lipid or apolipoprotein in predicting CVD risk during treatment. In the placebo group, the variables best predicting CVD events were non-HDL-C/HDL-C, TC/HDL-C (HR 1.21, $p < 0.001$ for both), the apo-ratio (HR 1.20, $p < 0.001$), LDL-C/HDL-C (HR 1.17, $p < 0.001$), HDL-C (HR 0.84, $p < 0.001$) and apoA-I (HR 0.85, $p < 0.001$). In the fenofibrate group, the first four predictors were very similar (the apo-ratio was fourth), followed by non-HDL-C and apoB.

In the JUPITER primary prevention trial (176) rosuvastatin 20 mg versus placebo was given to patients with initial LDL-C levels < 3.4 mmol/L and hsCRP > 2 mg/dL. Already after a medium time of treatment of 1.9 years, the trial was stopped for safety reasons since the actively treated patients benefitted by a highly significant risk reduction in MACE by 50 %. It should be pointed out that several thousand patients, those first recruited into the trial, participated for more than three to four years in the trial. LDL-C was reduced to 1.4 mmol/L and the apo-ratio was reduced from 0.95 to 0.49, $p < 0.001$. This indicates that "normal values" for the apo-ratio should be in the order of < 0.50 in order to obtain as low future risk as possible.

In a recent publication from JUPITER the authors reported that LDL-C, non-HDL-C, apoB and lipid-ratios as well as the apo-ratio had about similar predictive value of remaining risk during treatment with rosuvastatin (177). However, in subgroup analyses they reported that apoA-I had a greater capacity to define remaining risk than HDL-C. Furthermore, they also found that any lipid-related ratios had a greater predictive value than single values of LDL-C, non-HDL-C or apoB. In addition, if LDL-C values reached < 100 mg/dL or < 70 mg/dL, or if non-HDL-C targets were reached < 130 mg/dL or < 100 mg/dL, the only lipid-related variable or ratio that still was associated with remaining significant risk was the apo-ratio. These data, although the number of events is small in the sub-cohorts, indicate that the apo-ratio is a realistic and a valid predictor of risk and may be better than conventional lipids. However, the authors indicated that differences were small and that LDL-C and non-HDL-C were still sufficiently good as targets for treatment despite the fact that the results were in favor of the apo-ratio.

15. Treating CV risk patients to new targets using apolipoproteins

The apo-ratio, as shown in this paper, has commonly been shown to predict CV risk equally well or, in fact, more commonly even significantly better than conventional lipids in both prospective and treatment studies. So, which cut levels and targets of the apo-ratio should be recommended in the clinic to indicate CV risk before and after treatment? Since there is an almost linear increase (semi-log scale) in risk with increasing values of the apo-ratio from both AMORIS and the INTERHEART studies **(Figure 10)** it is clear that at values of the apo-ratio > 0.90 (values should be given in two decimals in order not to lose important information) there is a considerable increase in risk, whereas values from 0.70 to about 0.90 are indicative of a moderate risk. Values for men < 0.70 and for fertile females < 0.60 can be more normal especially if no other risk factors are present. The "ideal-biologically normal values" are rather < 0.50 as also documented in lipid-lowering trials in which CV events have been successfully reduced (176,177). So the target values during therapy must focus on these levels, the lower the apo-ratio the better is the therapy.

Lipids and apos are commonly correlated as also manifested in the AMORIS study (3 and others). In order to simplify for the physicians to learn what a value of LDL-C corresponds to regarding apoB **(Figure 11, left)**, a table has been compiled based on data from AMORIS also for the relationship between LDL-C and the apo-ratio **(Figure 11, right)**. A value of the apo-ratio of 0.80 roughly corresponds to a value for LDL-C of 3.0 mmol/L, and an apo-value of about 0.50 corresponds to LDL-C value 1.6 mmol/L for men and about 0.1 units lower for females. Notably, there is a large deviation from this correlation line. Those having a higher apoB or a higher apo-ratio at all levels of LDL-C (above the line) in general have a much higher CV risk than those below the line. Further details and relations between apolipoproteins, lipids and their relations to CV risk, and cut- and target levels of apoB and apoA-I have been reviewed (3). Since the target level for LDL-C according to many guidelines is set at LDL-C < 1.6 mmol/L, a target and normal value of the apo-ratio < 0.50 seems to be a realistic number.

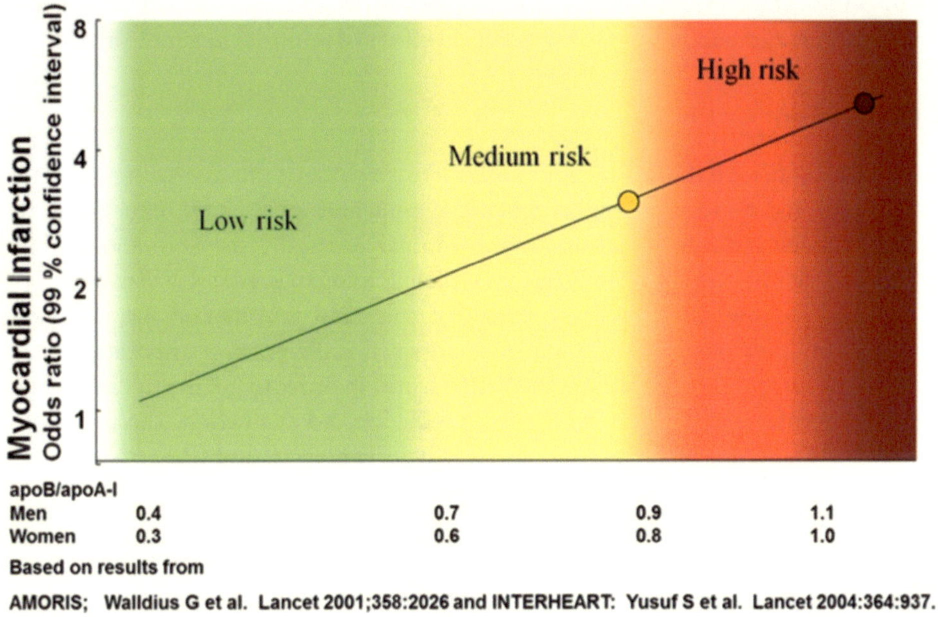

Figure 10. This line of risk of myocardial infarction is based on the findings in the AMORIS (reference 3) and the INTERHEART (reference 58) studies. Tentative cut-values are indicated in green (low risk), yellow (medium risk), and red (high risk). Values for a particular patient can be indicated by the dots on the line. During lipid-lowering treatment it is easy to monitor how a patient moves upwards or downwards in the risk line for the apo-ratio.

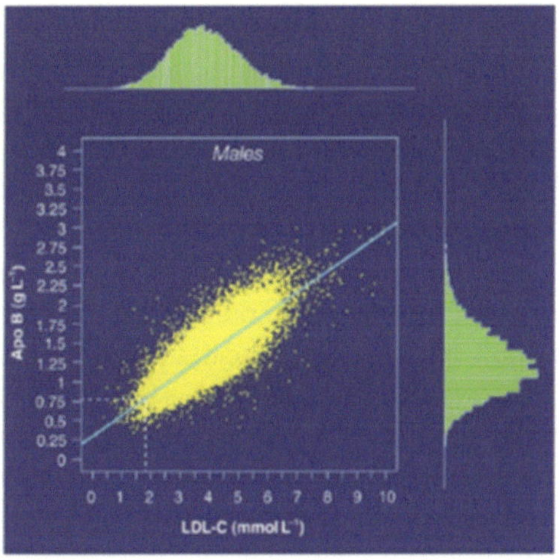

AMORIS; LDL-C correlates to the apoB/apoA-I ratio

LDL-C (mmol/L)	apoB/apoA-I men	apoB/apoA-I women
1.6	0.510	0.414
2.0	0.595	0.491
2.6	0.722	0.606
3.0	0.806	0.684
3.6	0.933	0.799
4.0	1.017	0.877

Figure 11. Data from the AMORIS study. Relations between LDL-C and apoB (left), and LDL-C and the apoB/apoA-I levels (right). Various cut-levels of LDL-C correspond to apoB and apoB/apoA-I values (both figures from reference 3).

How much can effective lipid-lowering therapy reduce apoB and the apo-ratio and how much can apoA-I be increased? Physical exercise and diet, if effective and longstanding, can reduce apoB by 5-10 % at the most and the apo-ratio by about 5% and increase apoA-I by about 5%. For more information see reference 3 and data on effects of statins in section 13 above.

16. Discussion

Today, LDL-C, non-HDL-C and lipid ratios are prioritized in international guidelines although apoB has also been mentioned in a few guidelines (9,18,38,178-180). In this review evidence is given indicating that apoB, apoA-I, and especially the apo-ratio, are at least equally good, or often even better than conventional lipids to predict CV risk prospectively and during lipid-lowering treatment. Much of this new information has not yet been included in any previously published meta-analyses. It is therefore of importance to review these data obtained from countries in the whole world in order to get the full information of what apolipoproteins can deliver for CV risk prediction and evaluation. These findings and advantages are summarized below and also briefly in **Table 3 (I) and (II).**

The biological relevance of using apoB and apoA-I as markers of CV risk is convincing since these proteins are carriers of lipids in the circulation and deliver C to peripheral tissues including the arterial wall (mediated by apoB). ApoA-I can remove C from the subendothelial space for breakdown and removal through the bile and further GI excretion. ApoA-I has also my other protective actions as summarized above and can thereby modify or inhibit inflammation and atherogenesis provoked by oxidation and modification of LDL-C with apoB (3,64). Thus, both apoB and apoA-I are biologically and patho-physiologically strongly active in normal biology and in plaque formation.

There are also methodological advantages of using these apos (3,23,28,65). Direct measurements of apoB and apoA-I by internationally standardized methods are available, analysis can be performed even if taken from patients in the non-fasted state, apos can be trustfully analyzed on frozen samples, the errors of the methods are not dependent on TG levels, and the methodological errors are usually low. Furthermore, costs for the direct analysis can be low as is the case in many countries. Importantly, the apo-ratio reflects the whole lipoprotein spectrum of virtually all phenotypes (type III patients need additional definition of risk) into one number. No sort like mg/dL or mmol/L has to be given for the ratio, which may otherwise be difficult to convert to understandable numbers in different countries. In summary, just one number of the apoB/apoA-I ratio indicates the risk level, similar numbers in all parts of the world. The higher the number, the higher is the risk (3) as also supported by a vast number of studies summarized in this paper. The cardiovascular risk line related to increasing value of the apo-ratio seems to be very similar world-wide **(Figure 10).**

ApoB, which indicates the number of potentially atherogenic particles, mainly sdLDL particles, has in a majority of publications been shown to be a better predictor of CV risk than LDL-C (9,11,13,14,44,59-62,88,89 and others) but in several instances apoB and non-HDL-C seem to indicate similar CV risk (10). One explanation why apoB may be better than non-HDL-C in risk prediction may be due to the fact that larger VLDL- and IDL-C-containing particles may have less potential to penetrate into the arterial walls than

Biological relevance
ApoB and apoA-I are carriers of lipids into and out of the arterial wall. Major pathophysiological mechanims are dependent on these proteins and how they can be modified (apoB) and be protective due to defensive actions.

Methodological advantages
Methods for apoB and apoA-I internationally standardized. Methodological errors of apoB and apoA-I are generally <5%. Fasting is not needed. High TG does not interfere. Frozen samples can be analyzed. The apo-ratio; no sort like mmol/L or mg/dL is needed. One number indicates the "cholesterol balance", easy to remember and act upon. Identifies sdLDL particle numbers. The apo-ratio reflects the risk associated with an imbalance between atherogenic and anti-atherogenic lipoproteins.

Relations to CV diseases
Strong predictors of myocardial infarction, stroke, heart failure, and also related to risk of renal failure and aortic aneurysms

Table 3 (I)

Relations to CV risk factors
Strong associations with abdominal obesity, metabolic syndrome and both diabetes

Relations to CV risk in univariate and multivariate analyses
Risk relationships for individual apos ant the apo-ratio commonly remained after adjustment for multiple conventional risk factors.

Relations to lipids and lipoproteins as predictors of CV risk
Lipids and lipid-based ratios are rarely significantly better than apos or the apo-ratio. However, apos and the apo-ratio are at least as good as lipids and lipid ratios, but commonly significantly better than lipids to predict CV risk

Relations to atherosclerosis
Strong associations with atherosclerosis in carotid arteries (IMT), coronary atherosclerosis (angiography and IVUS), femoral plaques and impaired endothelial function. Predicts progression and regression of carotid and coronary atherosclerosis

Relations to lipid-lowering treatment
Predicts outcome in statin trials equally well or commonly better than conventional lipids and lipoproteins

Table 3 (II)

Table 3. (I) and (II). Summary of findings supporting the use of apoB, apoA-I and the apo-ratio.

sdLDL particles. In fact, in a number of large publications including meta-analyses apoB has been shown to be a stronger predictor than the next best predictor non-HDL-C (9,11,13,14,61,62). LDL-C is only the third best predictor of future CV risk according to major analyses (9,62) and so also during statin treatment (169-171). However, in another meta-analysis they found non-HDL-C to be better than LDL-C and apoB during statin treatment (64). In a majority of these studies data have been adjusted for age and gender as well as other confounding risk factors like blood pressure, smoking, obesity and commonly also diabetes and other lipids.

Direct comparative data of HDL-C versus apoA-I is more sparse and is still much debated due to the complexity of HDL metabolism. ApoA-I has often similar predictive value as HDL-C as presented in the ERFC meta-analysis (10). However, especially in the large INTERHEART study, apoA-I over the whole range of HDL-values was a better predictor of risk than HDL-C (59). Similarly, in the AFCAPS/TexCAPS statin study apoA-I was a stronger determinant of risk than HDL-C (52,53).

What about the lipid-ratios versus the apo-ratio, which has strongest relations to CV risk? Importantly, all ratios and especially the apo-ratio predict prospective risk better than any single lipid variable (3,4,7,22,44,52,53,56,59,60,80,88,89,100,101,169-171 and others). Similarly, during statin treatment ratios also beat single lipoproteins in predicting risk (168-171,174). That should be obvious since ratios has a greater potential to find subjects at risk in whom the anti-atherogenic capacity of HDL-C or apoA-I are deranged. These data are also obtained when controlling for confounders i.e. conventional risk factors. Thus, the apo-ratio may have a better potential to identify subjects with different phenotypes than a single lipoprotein fraction. These strong findings in favor of any ratio, especially the apo-ratio, are strangely enough, not considered important in any international guideline despite the fact that ratios virtually in all studies in which ratios have been used outperform the results obtained by single lipoprotein fractions. Why this unscientific approach by guidelines committees?

Results from meta-analyses are generally well trusted but can also be questioned regarding selection criteria for including studies, acceptance of analytical and diagnostic methods used in each of the studies, primary and secondary variables used as major endpoints as well as the general conclusions drawn from the analyses. The results from the first ERFC meta-analysis have been taken to indicate that the apo-ratio and the TC/HDL-C ratio are equally good predictors of risk (10) and that apoB is equally good as risk predictor as non-HDL-C and apoA-I is equally predictive as HDL-C. The authors also open for future use of apolipoproteins especially in evaluation of risk of MI. In these risk conditions they found that apoB and apoA-I could be more useful in men than women, and in subjects with high TC, in those with low HDL-C, in individuals with hypertension, and in those with intermediate CV risk (Framingham risk score) apos can also be useful. They also found that the apo-ratio was a better predictor of CHD than stroke. However, in a recent publications in JAMA they conclude that the TC/HDL-C ratio had stronger predictive power than the apo-ratio when these ratios were added to conventional risk factors.

Two major critical views against these JAMA (see footnotes a/ and b/ below) papers may be raised that unfavorably affect the trust of using apos as risk predictors. Many early studies on apos included in these meta-analyses had large methodological errors (not internationally standardized) which may affect the conclusions on the credibility to use apos as risk predictors. This is unfair to the modern apo-technology which has much lower methodological errors.

Furthermore, in the ERFC studies they pooled non-fatal MI, all CHD fatalities, peripheral vascular diseases, and even all strokes, especially haemorrhagic stroke and unidentified stroke in very old people, into the primary variable "cardiovascular events". Such pooling of events considerably dilute the potential of adequate information yielded by an appropriately measured apos and the apo-ratio. This is especially the case for patients with risk of MI and in those suffering ischemic strokes in which positive diagnostic values have been obtained for apos as summarized in previously commented studies.

The authors also discuss some potential problems with introduction of apos such as need for education, lack of availability of apo-methods in the most laboratories, standardization problems as well as additional costs for such methods. All these aspects and possible problems must obviously be considered when new diagnostic tests shall be introduced for clinical use in risk evaluation. Yes, education is mandatory and may take time, but such problems must not over shade the importance of innovation of analytical tools. Costs can be significantly reduced if apo-tests become standard analyses. In fact, many biochemists already now favor these analyses over conventional lipids as documented previously in this paper.

Another criticism of the ERFC-studies is related to which studies were excluded (lack of confounding variables or case-control studies) from the meta-analyses in ERFC. Thus, major studies like AMORIS (44), INTERHEART (58) and ISIS (60) were not included in the ERFC meta-analysis. Neither were their positive results commented in the discussion on risk of MI despite the fact that these three studies have six times as many well defined events than those in the ERFC studies. In all these large studies apoB, apoA-I and especially the apo-ratio, due to their large number of events, were each significantly stronger predictors than conventional lipids. In ERFC there were many studies, but few of these studies showed significant differences between lipids and apos due to few well defined hard events. Neither did ERFC point out that the apo-ratio also seems to be the best variable to describe the remaining CV risk after statin treatment. This has been shown especially in the statin trials like AFCAPS/TexCAPS (52,53), IDEAL (169-171), TNT (168-170), CARDS (181), and JUPITER (176) as well as in studies on regression of atherosclerosis during lipid-lowering therapy (172-174) as pointed out above. In most of these studies the data were also adjusted for age, gender, conventional lipids and lipoproteins as well as other major risk factors like blood pressure, smoking, obesity and diabetes. Grundy simply concludes in the JAMA editorial

---

a/ The Emerging Risk Factors Collaboration. Lipid-Related Markers and Cardiovascular Disease Prediction. JAMA. 2012;307(23):2499-2506.

b/ Scott M. Grundy. Editorial: Use of Emerging Lipoprotein Risk Factors in Assessment of Cardiovascular Risk. 2012;307(23):2540-2541.

that conventional risk factors plus LDL-C, and possibly one more risk factor, is enough as tools for prediction of risk – a very conservative approach which is so (too!) common in US!

How much do confounders/risk factors impact on the results from all these CV risk studies? Importantly, in ERFC (10) and also in the majority of studies cited above, the impact of adjusting for major confounders was very small and only changed the risk (HR, RR or OR) to a minor degree. In fact, the apolipoproteins added value, measured as net reclassification index (NRI) in several large studies (88,91,170). This indicates that apolipoproteins, especially the apo-ratio, could change the numbers of individuals either to a higher or a lower CV risk compared to conventional lipids. Newly developed risk algorithms based on the apo-ratio have also been developed showing at least equal predictive or even better values than conventional risk algorithms (102,117,118). Thus, apoB, apoA-I and the apo-ratio can already now be used in clinical settings.

In the present review the apo-ratio has been shown to be closely related to many different types of CV events in prospective studies. These common diseases are myocardial infarction, stroke, especially ischemic stroke, heart failure, renal failure, aortic aneurysms, development of diabetes, including retinopathy **(Table 3 I and II)**. However, in the meta-analyses published so far only CV events have been chosen as endpoints and other manifestation of CV risk related to atherosclerosis have been excluded.

Is the apo-ratio useful in predicting various metabolic and inflammatory conditions commonly underlying atherosclerosis and its future consequences? In fact, the apo-ratio has also been found to be a valuable summarizing index of lipid-abnormalities and their complications in a large number of studies of the MetS and/or diabetes (125-139,143,144). In addition, the apo-ratio values are also increased in patients with hypertension, obesity, in pubertal children and in those with heredity for CV diseases (130,141). The apo-ratio is also more closely than lipids related to atherosclerosis in a large number of studies in which different techniques like coronary angiography, arterial wall thickness obtained by ultrasound techniques in the carotid arteries (CIMT values) or even in the coronary arteries by intravascular ultrasound (IVUS) and arterial abnormalities such as the endothelial dysfunction have been used (145-162). Thus, in all these disease or risk situations the apo-ratio may identify those at an increased risk even better than what is currently performed by using LDL-C or the recently recommended non-HDL-C.

The newest research data on the apo-ratio have not yet been reviewed by international guideline committees. Thus, so far, in the newest guidelines developed over the last few years non-HDL-C and apoB are mentioned, and accepted for clinical use, whereas the apo-ratio is still waiting for acceptance (9,35,178,179).

In conclusion; with all the new knowledge presented in this paper about the strong relations between apoB, apoA-I, and the apo-ratio, and CV risk as well as other disease manifestations, it is proposed, as many researchers have already done, that these strong risk predictors/factors/markers are included in new guidelines. In many disease conditions and manifestations of atherosclerosis apolipoproteins are at least equally informative, and often better than LDL-C, non-HDL-C and lipid ratios in predicting risk. It is realized that there will be pedagogical hurdles, but it should be possible to educate physicians, patients and health providers to understand that these apolipoproteins are markers of normal and abnormal cholesterol metabolism. The apo-ratio simply reflects the "balance between the bad cholesterols and the good cholesterols" technically measured by apolipoproteins. The apo-ratio is a valid cardiovascular risk index (CRI) that reflects the level of CV risk for virtually all patients with different lipid phenotypes, the higher the value of the apo-ratio, the higher is the risk. Finally, targeting lower values (about 0.50) of the apo-ratio during therapy may more correctly identify who is at risk or not at risk, and how high is the risk? Does the risk depend on the atherogenic apoB, or the anti-atherogenic apoA-I or rather on the most informative value i.e. the apo-ratio which summarizes the level of risk in a simple way? Since physicians usually only manage to effectively evaluate and trust one laboratory marker, the apo-ratio is such a valid marker. By simply plotting the value for a given patient on the risk line you can easily follow improvement during therapy and also motivate the patient to improve values to normal levels **(Figure 10)**. New guidelines should at least contain equally objective information (cut-values and target values) on how to use apoB, apoA-I, and the apo-ratio as on lipids so that physicians can choose whichever diagnostic marker of risk they prefer. Gradually this new apolipoprotein-based risk classification with a focus on the apoB/apoA-I ratio may, or rather should be, introduced in clinical practice.

Author details

Göran Walldius
Department of Epidemiology, Institute of Environmental Medicine (IMM),
Karolinska Institutet, Stockholm, Sweden

Acknowledgement

The close collaboration since 1985 with Ingmar Jungner, the former head of CALAB laboratories in Stockholm, Sweden and the founder of the AMORIS database, is gratefully acknowledged. Many of the publications from the AMORIS study presented in this paper were also developed together with Ingar Holme, Are Aastveit, both from Oslo, Norway, Werner Kolar and Eugen Steiner, CALAB, Stockholm, Sweden, Allan Sniderman, Montreal, Canada and Niklas Hammar, Stockholm, Sweden. Their scientific knowledge and friendship is much appreciated. Much of the AMORIS work has been supported by grants from Gunnar and Ingmar Jungner Foundation for Laboratory Medicine, Stockholm, Sweden, Bure Hälsa & Sjukvård Ltd, Stockholm; and AstraZeneca, Mölndal, Sweden.

Abbreviations

apos	apolipoproteins
apoB	apolipoprotein B
apoA-I	apolipoprotein A-I
apo-ratio	apoB/apoA-I ratio
C	cholesterol

CV	cardiovascular
VLDL	Very Low Density Lipoprotein
IDL	Intermediate Density Lipoprotein
LDL	Low Density Lipoprotein
sdLDL	small dense LDL
TC	total cholesterol
HDL	high density lipoprotein
hsCRP	high sensitivity CRP (C-reactive protein)
IVUS	intravascular ultrasound
CIMT	carotid intima media thickness
CRI	Cardiovascular Risk Index
HR	Hazards Ratio
NRI	Net Reclassification Index
MACE	Major Coronary Events
RR	Relative Risk

17. References

[1] Elovson J, Chatterton JE, Bell GT, Schumaker VN, Reuben MA, Puppione D L, Reeve Jr JR, Young NL. Plasma very low density lipoproteins contain a single molecule of apolipoprotein B. J Lipid Res 1988; 29:1461-1473.

[2] Walldius G, Jungner I. Apolipoprotein B and apolipoprotein A-I: risk indicators of coronary heart disease and targets for lipid-modifying therapy. J Intern Med 2004; 255(2): 188-205.

[3] Walldius G, Jungner I. The apoB/apoA-I ratio: a strong, new risk factor for cardiovascular disease and a target for lipid-lowering therapy – a review of the evidence. J Intern Med 2006; 259: 493-519.

[4] Sniderman AD, Furberg CD, Keech A, Roeters van Lennep JE, Frohlich J, Jungner I, Walldius G. Apolipoproteins versus lipids as indices of coronary risk and as targets for statin therapy treatment. Lancet 2003; 361: 777-780.

[5] Walldius G, Jungner I. Is there a better marker of cardiovascular risk than LDL cholesterol? Apolipoprotein B and A-I – new risk factors and targets for therapy. Nutr Metab Cardiovasc Dis 2007; 17: 565-571.

[6] Walldius G, Jungner I. ApoB, apoA-I, and the apoB/apoA-I ratio as predictors, markers, and factors for cardiovascular risk. The Fats of Life, LVDD/AACC. www.aacc.org. 2007; XXI(1): 11-21.

[7] Rader DJ, Hoeg JM, Brewer HB Jr. Quantitation of plasma apolipoproteins in the primary and secondary prevention of coronary artery disease. Ann Intern Med 1994; 120: 1012-1025.

[8] Thompson A, Danesh J. Associations between apolipoprotein B, apolipoprotein AI, the apolipoprotein B/AI ratio and coronary heart disease: a literature-based meta-analysis of prospective studies. J Intern Med. 2006; 259: 481-492.

[9] Contois JH, McConnell JP, Sethi AA, Csako G, Devaraj S, et al. Apolipoprotein B and cardiovascular disease risk: position statement from the AACC Lipoproteins and Vascular Diseases Division Working Group on Best Practices. Clin Chem 2009; 55(3): 407-419.
[10] Di Angelantonio E, Sarwar N, Perry P, Kaptoge S, Ray KK, et al. The Emerging Risk Factors Collaboration. Major lipids, apolipoproteins, and risk of vascular disease. JAMA 2009; 302(18): 1993-2000. Note also the added supplement.
[11] Contois JH, Warnick GR, Sniderman AD. Review Article; Reliability of low-density lipoprotein cholesterol, non-high-density lipoprotein cholesterol, and apolipoprotein B measurement. J Clin Lipidol 2011; 5: 264-272.
[12] Olofsson SO, Wiklund O, Borén B. Apolipoproteins A-I and B: biosynthesis, role in the development of atherosclerosis and targets for intervention against cardiovascular disease. Vascular Health and Risk Management 2007; 3(4): 491-502.
[13] Sniderman A, McQueen M, Contois J, Williams K,. Furberg CD. Why is non-high-density lipoprotein cholesterol a better marker of the risk of vascular disease than low-density lipoprotein cholesterol? Journal of Clinical Lipidology 2010; 4: 152-155.
[14] Sniderman A, Williams K, de Graaf J. Non-HDL C equals apolipoprotein B: except when it does not! Curr Opin Lipidol 2010; 21: 518-524.
[15] Rader DJ, Davidson M, Caplan RJ, Pears JS. Lipid and apolipoprotein ratios: association with coronary artery disease and effects of rosuvastatin compared with atorvastatin, pravastatin, and simvastatin. Am J Cardiol 2003; 91(Suppl.): 20C-24C.
[16] Friedewald WT, Levy RI, Fredrickson DS. Estimation of the concentration of low-density lipoprotein cholesterol in plasma, without use of the preparative ultracentrifuge. Clin Chem 1972; 18: 499-502.
[17] Stein EA, Sniderman AD, Laskarzewski P. Assessment of Reaching Goal in Patients with Combined Hyperlipidemia: Low-Density Lipoprotein Cholesterol, Non-High-Density Lipoprotein Cholesterol, or Apolipoprotein B. Am J Cardiol 2005; 96(suppl): 36K–43K.
[18] Myers GL, Christenson RHM, Cushman M, Ballantyne CM, Cooper GR, et al. National Academy of Clinical Biochemistry Laboratory Medicine Practice Guidelines: Emerging Biomarkers for Primary Prevention of Cardiovascular Disease. Clin Chem 2009; 55: 378-384.
[19] Warnick GR. Editors´Notes. Fats of Life, LVDD/ AACC, www.aacc.org. 2010; XXIV(3):2-5, 33-37.
[20] Sniderman AD, Walldius G, Jungner I. The four horsemen of cholesterol versus ApoB: Good care versus better care. Fats of Life. LVDD/ AACC, www.aacc.org. 2003; XVII(4): 9-24.
[21] Sniderman AD, Jungner I, Holme I, Aastveit A, Walldius G. Errors that result from using the TC/HDL-C ratio rather than the apoB/apoA-I ratio to identify the lipoprotein-related risk of vascular disease. J Intern Med 2006; 259: 455-461.
[22] Walldius G, Jungner I, Aastveit AH, Holme I, Furberg CD, Sniderman AD. The apoB/apoA-I ratio is better than the cholesterol ratios to estimate the balance between plasma pro-atherogenic and anti-atherogenic lipoproteins and to predict coronary risk. Clin Chem Lab Med 2004; 42: 1355-1363.

[23] Scharnagl H, Nauck M, Wieland H, März W. The Friedewald formula underestimates LDL cholesterol at low concentrations. Clin Chem Lab Med 2001; 39: 426-431.
[24] Otvos JD. Why cholesterol measurements may be misleading about lipoprotein levels and cardiovascular disease risk – clinical implications of lipoprotein quantification using NMR spectroscopy. J Lab Med 2002; 26: 544-550.
[25] van Deventer HE, Miller WG, Myers GL, Sakurabayashi I, Bachmann LM, Caudill SP, Dziekonski A, Edwards S, Kimberly MM, Korzun WJ, Leary ET, Nakajima K, Nakamura M, Shamburek RD, Vetrovec GW, Warnick GR, Remaley AT. Non–HDL Cholesterol Shows Improved Accuracy for Cardiovascular Risk Score Classification Compared to Direct or Calculated LDL Cholesterol in a Dyslipidemic Population. Clinical Chemistry 2011; 57(3): 490-501.
[26] Marcovina SM, Albers JJ, Henderson LO, Hannon WH. International Federation of Clinical Chemistry standardization project for measurements of apolipoproteins A-I and B: III. Comparability of apolipoprotein A-I values by use of International Reference Material. Clin Chem 1993; 39: 773-781.
[27] Liu J, Sempos CT, Donahue RP, Dorn J, Trevisan M, Grundy SM. Non–High-Density Lipoprotein and Very-Low-Density Lipoprotein Cholesterol and Their Risk Predictive Values in Coronary Heart Disease. Am J Cardiol 2006; 98: 1363-1368.
[28] Marcovina SM, Albers JJ, Kennedy H, Mei JV, Henderson LO, Hannon WH. International Federation of Clinical Chemistry standardization project for measurements of apolipoproteins A-I and B: IV. Comparability of apolipoprotein B values by use of International Reference Material. Clin Chem 1994; 40: 586-592.
[29] Marcovina S, Packard CJ. Measurement and meaning of apolipoprotein AI and apolipoprotein B plasma levels. J Intern Med 2006; 259: 437-446.
[30] Zambon A, Brown BG, Deeb SS, Brunzell JD. Genetics of apolipoprotein B and apolipoprotein AI and premature coronary artery disease. J Intern Med 2006; 259: 473-480.
[31] Gatz M, Reynolds CA, Finkel D, Pedersen NL, Walters E. Dementia in Swedish Twins: Predicting Incident Cases. Behav Genet 2010; 40:768-75, DOI 10.1007/s10519-010-9407-4.
[32] Vladimirova-Kitova LG, Deneva-Koicheva TI. Increased Intima-Media Thickness in Carriers of the LDL-Receptor Defective Gene versus Noncarriers with Newly Detected Asymptomatic Severe Hypercholesterolemia. Echocardiography 2011; 28: 223-234.
[33] Middelberg RPS, Spector TD, Swaminathan R, Snieder H. Genetic and Environmental Influences on Lipids, Lipoproteins, and Apolipoproteins Effects of Menopause. Arterioscler Thromb Vasc Biol. 2002; 22: 1142-1147.
[34] Schmidt C, Fagerberg, B Wikstrand J, Hulthc J. ApoB/apoA–I ratio is related to femoral artery plaques and is predictive for future cardiovascular events in healthy men. Atherosclerosis 2006; 189: 178-185.
[35] Hoff HF. Apolipoprotein Localization in Human Cranial Arteries, Coronary Arteries, and the Aorta. Stroke 1976; 7(4): 390-393.
[36] Fogelstrand P, Borén J. Retention of atherogenic lipoproteins in the artery wall and its role in atherogenesis; Review. Nutrition, Metabolism & Cardiovascular Diseases 2012; 22: 1-7.

[37] Contois JH, McNamara JR, Lammi-Keefe CJ, Wilson PWF, Massov T, Schaeffer EJ. Reference intervals for plasma apolipoprotein B determined with a standardized commercial immunoturbidimetric assay: results from the Framingham Offspring Study. Clinical Chemistry 1996; 42: 515-523.
[38] Grundy SM, Cleeman JI, Merz CNB et al. Implications of recent clinical trials for the National Cholesterol Education Program Adult Treatment Panel III Guidelines. J Am Coll Cardiol 2004; 44: 720-732.
[39] Avogaro P, Bon GB, Cazzolato G, Quinci GB, Sanson A, et al. Variations in apolipoproteins B and AI during the course of myocardial infarction. Eur J Clin Invest 1978; 8: 121-129.
[40] Avogaro P, Cazzolato G, Bon GB, Berlussi F, Quinci GB. Values of apoA-I and apoB in humans according to age and sex. Clin Chim Acta 1979; 95: 511-515.
[41] Sniderman A, Shapiro S, Marpole S, Skinner B, Teng B, et al. Association of coronary atherosclerosis with hyperapobetalipoproteinemia (increased protein but normal cholesterol levels in human plasma low density (beta) lipoproteins. Proc Natl Acad Sci, USA 1980; 77: 604-608.
[42] Lamarche B, Moorjani S, Lupien PJ et al. Apolipoprotein A-1 and B levels and the risk of ischemic heart disease during a five- year follow-up of men in the Que´bec Cardiovascular Study. Circulation 1996; 94: 273-278.
[43] Sharrett AR, Ballantyne CM, Coady SA et al. Coronary heart disease prediction from lipoprotein cholesterol levels, triglycerides, lipoprotein(a), apolipoproteins A-I and B, and HDL density subfractions: The Atherosclerosis Risk in Communities (ARIC) Study. Circulation 2001; 104: 1108-1113.
[44] Walldius G, Jungner I, Holme I, Aastveit AH, Kolar W, Steiner E. High apolipoprotein B, low apolipoprotein A-I, and improvement in the prediction of fatal myocardial infarction (AMORIS study): a prospective study. Lancet 2001; 358: 2026-2033.
[45] Moss AJ, Goldstein RE, Marder VJ, Sparks CE, Oakes D, Greenberg H, et al. Thrombogenic factors and recurrent coronary events. Circulation 1999; 99: 2517-2522.
[46] Corsetti JP, Zareba W, Moss AJ, Sparks CE. Apolipoprotein B determines risk for recurrent coronary events in postinfarction patients with metabolic syndrome. Atherosclerosis 2004; 177: 367-373.
[47] Talmud PJ, Hawe E, Miller GJ, Humphries SE. Nonfasting apolipoprotein B and triglyceride levels as a useful predictor of coronary heart disease risk in middle-aged UK men. Arterioscler Thromb Vasc Biol 2002; 22: 1918-1923.
[48] Shai I, Rimm EB, Hankinson SE et al. Multivariate assessment of lipid parameters as predictors of coronary heart disease among postmenopausal women. Potential implications for clinical guidelines. Circulation 2004; 110: 2824-2830.
[49] Pischon T, Girman CJ, Sacks FM, Rifai N, Stampfer MJ, Rimm EB. Non-high-density lipoprotein cholesterol and apolipoprotein B in the prediction of coronary heart disease in men. Circulation 2005; 112: 3375-3383.
[50] Benn M, Nordestgaard BG, Boje Jensen G, Tybjærg-Hansen A. Improving Prediction of Ischemic Cardiovascular Disease in the General Population Using Apolipoprotein B. Arterioscler Thromb Vasc Biol. 2007; 27: 661-670.

[51] Pedersen TR, Olsson AG, Faergeman O, et al. Lipoprotein changes and reduction in the incidence of major coronary heart disease events in the Scandinavian Simvastatin Survival Study (4S). Circulation 1998; 97: 1453-1460.

[52] Gotto AM, Whitney E, Stein EA et al. Relation between baseline and on-treatment lipid parameters and first acute major coronary events in the Air Force/Texas Coronary Atherosclerosis Prevention Study (AFCAPS/TexCAPS). Circulation 2000; 101: 477-484.

[53] Gotto, AM Jr. Establishing the benefit of statins in low-to-moderate—risk primary prevention: The Air Force/Texas Coronary. Atherosclerosis Prevention Study (AFCAPS/TexCAPS). Atherosclerosis Supplements 2007; 8: 3-8.

[54] Simes RJ, Marschner IC, Hunt D et al. Relationship between lipid levels and clinical outcomes in the long-term intervention with pravastatin in ischemic disease (LIPID) trial. To what extent is the reduction in coronary events with pravastatin explained by on-study lipid levels? Circulation 2002; 105: 1162-1169.

[55] van Lennep JE, Westerveld HT, van Lennep HWO, Zwinderman AH, Erkelens DW, van der Wall EE. Apolipoprotein concentrations during treatment and recurrent coronary artery disease events. Arterioscler Thromb Vasc Biol 2000; 20: 2408-2413.

[56] Holme I, Aastveit AH, Hammar N, Jungner I, Walldius G. Relationships between lipoprotein components and risk of ischaemic and haemorrhagic stroke in the Apolipoprotein MOrtality RISk study (AMORIS). J Intern Med 2009; 265: 275-287.

[57] Sniderman AD, Jungner I, Holme I, Aastveit A, Walldius G. Errors that result from using the TC/HDL-C ratio rather than the apoB/apoA-I ratio to identify the lipoprotein-related risk of vascular disease. J Intern Med 2006; 259: 455-461.

[58] Yusuf S, Hawken S, Ôunpuu S, Dans T, Avezum A, et al., on behalf of the INTERHEART Study Investigators. Effect of potentially modifiable risk factors associated with myocardial infarction in 52 countries (the INTERHEART study): case-control study. Lancet 2004; 364: 937-952.

[59] McQueen MJ, Hawken S, Wang X, Ounpuu S, Sniderman A, et al. Lipids, lipoproteins, and apolipoproteins as risk markers of myocardial infarction in 52 countries (the INTERHEART study): a case-control study. Lancet 2008; 372: 224-233.

[60] Parish S, Peto R, Palmer A, Clarke R, Lewington S, et al. The joint effects of apolipoprotein B, apolipoprotein A1, LDL cholesterol, and HDL cholesterol on risk: 3510 cases of acute myocardial infarction and 9805 controls. Eur Heart J 2009; 30(17): 2137-2146. First published online June 11, 2009,doi:10.1093/eurheartj/ ehp221.

[61] Barter PJ, Ballantyne CM, Carmena R, Castro Cabezas M, Chapman MJ, Couture P, De Graaf J, Durrington PN, Faergeman O, Frohlich J, Furberg CD, Gagne C, Haffner SM, Humphries SE, Jungner I, Krauss RM, Kwiterovich P. Marcovina S, Packard CJ, Pearson TA, Reddy KS, Rosenson R, Sarrafzadegan N, Sniderman AD, Stalenhoef AF, Stein E, Talmud PJ, Tonkin AM, Walldius G, Williams KMS. ApoB versus cholesterol in estimating cardiovascular risk and in guiding therapy: report of the thirty-person/ten-country panel. J Intern Med 2006; 259: 247-258.

[62] Sniderman AD, Williams K, Contois JH, Monroe HM, McQueen MJ, de Graaf J, Furberg CD. A Meta-analysis of LDL-C, non-HDL-C, and apoB as markers of cardiovascular risk. Circ Cardiovasc Qual Outcomes. 2011; 4: 337-345.

[63] Sniderman AD, St-Pierre A, Cantin B, Dagenais GR, Depre´s J-P, Lamarche B. Concordance/discordance between plasma apolipoprotein B levels and the cholesterol indexes of atherosclerotic risk. Am J Cardiol 2003; 91: 1173-1177.

[64] Boekholdt SM, Arsenault BJ, Mora S, Pedersen TR, LaRosa JC, Nestel PJ, Simes RJ, Durrington P, Hitman GA, Welch KMA, DeMicco DA, Zwinderman AH, Clearfield MB, Downs JR, Tonkin AM, Colhoun HM, Gotto Jr AM, Ridker PM, Kastelein JJP. Association of LDL Cholesterol, Non–HDL Cholesterol, and Apolipoprotein B Levels With Risk of Cardiovascular Events Among Patients Treated With Statins. A Meta-analysis. JAMA 2012; 307(12): 130213-130209.

[65] Barter PJ, Rye K-A. The rationale for using apoA-I as a clinical marker of cardiovascular risk. J Intern Med 2006; 259: 447-454.

[66] Walldius G, Jungner I. Apolipoprotein A-I versus HDL cholesterol in the prediction of risk for myocardial infarction and stroke Curr Opin Cardiol 2007; 22: 359-367.

[67] Asztalos BF, Tani M, Schaefer EJ. Metabolic and functional relevance of HDL subspecies. Curr Opin Lipidol 2011; 22: 176-185.

[68] Larsson A, Carlsson L , Axelsson J. Low diurnal variability of apolipoprotein A1, apolipoprotein B and apolipoprotein B/apolipoprotein A1 ratio during normal sleep and after an acute shift of sleep. Clinical Biochemistry 2008; 41: 859-862.

[69] Elliott DA, Shannon Weickert C, Garner B. Apolipoproteins in the brain: implications for neurological and psychiatric disorders. Clin Lipidol. 2010; 51(4): 555-573. doi:10.2217/CLP.10.37

[70] Kavoa AE et al. Qualitative characteristics of HDL in young patients of an acute myocardial infarction. Atherosclerosis 2012; 220: 257-264.

[71] Sung K-C, Rhee E-J , Kim H, Park JB, Kim Y-K, Rosenson RS. Prevalence of low LDL-cholesterol levels and elevated high-sensitivity C-reactive protein levels in apparently healthy Korean adults. Nutrition, Metabolism & Cardiovascular Diseases. Available on line July 23, 2011. 2011; doi:10.1016/ j.numecd.2011.03.006.

[72] Liangpunsakul S, Rong QI, M.S, Crabb DW, Witzmann F. Relationship Between Alcohol Drinking and Aspartate Aminotransferase: Alanine Aminotransferase (AST:ALT) Ratio, Mean Corpuscular Volume (MCV), Gamma-Glutamyl Transpeptidase (GGT), and Apolipoprotein A1and B in the U.S. Population. J. Stud. Alcohol Drugs, 2010; 71: 249-252.

[73] Schlitt A, Blankenberg S, Bickel C et al. Prognostic value of lipoproteins and their relation to inflammatory markers among patients with coronary artery disease. Int J Cardiol 2005; 102: 477-485.

[74] Held C, Hjemdahl P, Rehnqvist N , Björkander I, Forslund L, Brodin U , Berglund L, Angelin B. Cardiovascular prognosis in relation to apolipoproteins and other lipid parameters in patients with stable angina pectoris treated with verapamil or metoprolol. Results from the Angina Prognosis Study in Stockholm (APSIS) Atherosclerosis 1979; 135: 109-118.

[75] Sharp DS, Burchfield CM, Rodriguez BL, Sharrett AR, Sorlie PD, Marcovina SM. Apolipoprotein A-I predicts coronary heart disease only at low concentrations of high-density lipoprotein cholesterol: an epidemiological study of Japanese-Americans. Int J Clin Lab Res 2000; 30: 39-48.

[76] Luc G, Bard J-M, Ferrie`res J et al. Value of HDL cholesterol, apolipoprotein A-I, lipoprotein A-I, and lipoprotein A-I/A-II in prediction of coronary heart disease: the PRIME Study.Prospective Epidemiological Study of Myocardial Infarction. Arterioscler Thromb Vasc Biol 2002; 22: 1155-1161.
[77] Patel JV, Abraheem A, Creamer J, Gunning M, Hughes EA, Lip GYH. Apolipoproteins in the discrimination of atherosclerotic burden and cardiac function in patients with stable coronary artery disease. European Journal of Heart Failure 2010; 12: 254-259. doi:10.1093/eurjhf/hfp202.
[78] Wedel H, McMurray JJV, Lindberg M, Wikstrand J, Cleland JGF, Cornel JH, Dunselman P, Hjalmarson Å, Kjekshus J, Komajda M, Kuusi T, Vanhaecke J, Waagstein F on behalf of the CORONA Study Group. Predictors of fatal and non-fatal outcomes in the Controlled Rosuvastatin Multinational Trial in Heart Failure (CORONA): incremental value of apolipoprotein A-1, high-sensitivity C-reactive peptide and N-terminal pro B-type natriuretic peptide. European J Heart Fail 2009; 11: 281-291 doi:10.1093/eurjhf/hfn046.
[79] Chien K-L, Sung F-C, Hsu H-C, Su T-C, Lin R-S, Lee Y-T. Apolipoprotein A-I and B and Stroke Events in a Community-Based Cohort in Taiwan. Report of the Chin-Shan Community Cardiovascular Study. Stroke. 2002; 33: 39-44.
[80] Van der Steeg WA, Boekholdt SM, Stein EA, El-Harchaoui K, Stroes ESG, Sandhu MS, Wareham NJ, J. Jukema JW, Luben R, Zwinderman AH, Kastelein JJP, Khaw K-T. Role of the Apolipoprotein B–Apolipoprotein A-I Ratio in Cardiovascular Risk Assessment: A Case–Control Analysis in EPIC-Norfolk. Ann Intern Med. 2007; 146: 640-648.
[81] Walldius G, Jungner I. Apolipoprotein A-I versus HDL cholesterol in the prediction of risk for myocardial infarction and stroke. Curr Opin Cardiol 2007; 22: 359-367.
[82] Jungner I, Marcovina SM, Walldius G, Holme I, Kolar W, Steiner E. Apolipoprotein B and A-I values in 147 576 Swedish males and females, standardized according to the World Health Organization – International Federation of Clinical Chemistry First International Reference Materials. Clin Chem 1998; 44: 1641-1649.
[83] Walldius G, Jungner I, Kolar W, Steiner E. Apolipoprotein B and total serum cholesterol levels in 41 000 males and females. Clin Chem 1990; 36(6): 952-956.
[84] Jungner I, Walldius G, Holme I, Kolar W, Steiner E. Apolipoprotein B and A-1 in relation to serum cholesterol and triglycerides in 43,000 Swedish males and females. Int J Clin Lab Res 1992; 21: 247-255.
[85] Walldius G, Jungner I, Kolar W, Holme I, Steiner E. High cholesterol and triglyceride values in Swedish males and females: increased risk of fatal myocardial infarction: First report from the AMORIS (Apolipoprotein related MOrtality RISk) Study. Blood Press 1992; I (suppl 4): 35-42.
[86] Walldius G, Aastveit A, Jungner I. Hypercholesterolemia and hypertriglyceridemia—greatest cardiac risk in subjects with high apoB/apoA-I levels. Original Research Article. International Congress Series, 2004; 1262: 203-206.
[87] Jungner I, Sniderman AD, Furberg C, Aastveit AH, Holme I, Walldius G. Does Low-Density Lipoprotein Size Add to Atherogenic Particle Number in Predicting the Risk of Fatal Myocardial Infarction? Am J Cardiol 2006; 97: 943-946.

[88] Holme I, Aastveit AH, Jungner I, Walldius G. Relationships between lipoprotein components and risk of myocardial infarction: age, gender and short versus longer follow-up periods in the Apolipoprotein MOrtality RISk study (AMORIS). J Intern Med 2008; 264: 30-38.

[89] Holme I, Aastveit AH, Hammar N, Jungner I, Walldius G. Lipoprotein components and risk of congestive heart failure in 84 470 men and women in the Apolipoprotein MOrtality RISk study (AMORIS). Eur J Heart Fail 2009; 11: 1036-1042. Published Oct 3, 2009. Doi:10.1093/eurjhf/hfp129

[90] Holme I, Aastveit AH, Hammar N, Jungner I, Walldius G. Haptoglobin and risk of myocardial infarctiion, stroke and congestive heart failure in 342,125 men and women in the Apolipoprotein MOrtality RISk study (AMORIS). Ann Med 2009; 41: 522-532.

[91] Holme I, Aastveit AH, Hammar N, Jungner I, Walldius G. Inflammatory markers, lipoprotein components and risk of major cardiovascular events in 65,005 men and women in the Apolipoprotein MOrtality RISk study (AMORIS). 2010: 213(1): 299-305. Atherosclerosis 2010; doi:10.1016/j.atherosclerosis.2010.08.049.

[92] Solhpour A, Parkhideh S, Sarrafzadegan N, Asgary S, Williams K, Jungner I, Aastveit A, Walldius G, Sniderman A. Levels of lipids and apolipoproteins in three cultures. Atherosclerosis 2009; 207: 200-207.

[93] Holtzmann MJ, Jungner I, Walldius G, Ivert T, Nordqvist T, Östergren J, Hammar N. Dyslipidemia is a Strong Predictor of Myocardial infarction in Subjects with Chronic Kidney Disease. Ann Med 2010; Early Online, 1–8. Published in Annals of Medicine, 2012; 44: 262–270.

[94] Bhatia M, Howard SC, Clark TG et al. Apolipoproteins as predictors of ischaemic stroke in patients with a previous transient ischaemic attack. Cerebrovasc Dis 2006; 21: 323-328.

[95] Qureshi AI, Giles WH, Croft JB, Guterman LR, Hopkins LN. Apolipoproteins A-1 and apoB and the likelihood of non-fatal stroke and myocardial infarction – data from The Third National Health and Nutrition Examination Survey. Med Sci Monit 2002; 8: CR311-316.

[96] Chien KL, Sung FC, Hsu HC, Su TC, Lin RS, Lee YT. Apolipoprotein A-I and B and stroke events in a community-based cohort in Taiwan: report of the Chin-Shan Community Cardiovascular Study. Stroke 2002; 33: 39-44.

[97] Sharobeem KM, Patel JV, Ritch AES, Lip GYH, Gill PS, Hughes EA. Elevated lipoprotein (a) and apolipoprotein B to AI ratio in South Asian patients with ischaemic stroke. Int J Clin Pract 2007; 61: 1824-1828.

[98] Koren-Morag N, Goldbourt U, Graff E, Tanne D. Apolipoproteins B and AI and the risk of ischemic cerebrovascular events in patients with pre-existing atherothrombotic disease. J Neurol Sci 2008; 270: 82-87.

[99] Hankey GJ. Potential new risk factors for ischemic stroke. What is their potential? Stroke 2006; 37: 2181-2188.

[100] Walldius G, Aastveit AH, Jungner I. Stroke mortality and the apoB/apoA-I ratio: results of the AMORIS prospective study. J Intern Med 2006; 259: 259-266.

[101] Holme I, Aastveit AH, Hammar N, Jungner I, Walldius G. Relationships between lipoprotein components and risk of ischaemic and haemorrhagic stroke in the

Apolipoprotein MOrtality RISk study (AMORIS). J Intern Med. 2009; 265: 275-287. doi: 10.1111/j.1365-2796.2008.02016.xJ.

[102] McGorrian C, Yusuf S, Islam S, Jung H, Rangarajan S, Avezum A, Prabhakaran D, Almahmeed W,Rumboldt Z, Budaj A, Dans AL, Gerstein HC, Teo K, Anand SS on behalf of the INTERHEART Investigators Estimating modifiable coronary heart disease risk in multiple regions of the world: the INTERHEART Modifiable Risk Score. European Heart Journal Advance Access published December 22, 2010. European Heart Journal, doi:10.1093/eurheartj/ehq448. Published in European Heart Journal 2011; 32: 581-590.

[103] Gerstein HC, Islam S, Anand S, Almahmeed A, Damasceno A, Dans A, Lang CC, Luna MA, McQueen M, Rangarajan S, Rosengren A, Wang X, Yusuf S. Dysglycaemia and the risk of acute myocardial infarction in multiple ethnic groups: an analysis of 15,780 patients from the INTERHEART study. Diabetologia 2010; 53: 2509-2517, DOI 10.1007/s00125-010-1871-0.

[104] Joshi P, Islam S , Pais P, Reddy S, Dorairaj P, Kazmi K, Pandey MR, Haque S, Mendis S, Rangarajan S, Yusuf S. Risk Factors for Early Myocardial Infarction in South Asians Compared With Individuals in Other Countries. JAMA 2007; 297: 286-294.

[105] Lanas F, Avezum A, Bautista LE, Diaz R, Luna M, Islam S, Yusuf S for the INTERHEART Investigators in Latin America. Risk Factors for Acute Myocardial Infarction in Latin America. The INTERHEART Latin American Study. Circulation 2007; 115: 1067-1074.

[106] Kabagambe EK, Baylin A, Campos H. Nonfatal Acute Myocardial Infarction in Costa Rica. Modifiable Risk Factors, Population-Attributable Risks, and Adherence to Dietary Guidelines. Circulation. 2007; 115: 1075-1081.

[107] Steyn K, Sliwa K, Hawken S, Commerford P, Onen C, Damasceno A, Ounpuu S, Yusuf S for the INTERHEART Investigators in Africa. Risk Factors Associated With Myocardial Infarction in Africa The INTERHEART Africa Study. Circulation 2005; 112: 3554-3561.

[108] O'Donnell MJ, Xavier D, Liu L, Zhang H, Chin SL, Rao-Melacini P, Rangarajan S, qul Islam S, Pais P, McQueen MJ, Mondo C, Damasceno A, Lopez-Jaramillo P, Hankey GJ, Dans AL, Yusoff K, Truelsen T, Diener HC, Sacco RL, Ryglewicz D, Czlonkowska A, Weimar C, Wang X, Yusuf S on behalf of the INTERSTROKE investigators. Risk factors for ischaemic and intracerebral haemorrhagic stroke in 22 countries (the INTERSTROKE study): a case-control study. Published Online June 18, 2010 DOI:10.1016/S0140- 6736(10)60834-3. Published in Lancet 2010; 376: 112–23.

[109] Kostapanos MS, Christogiannis LG, Bika E, Bairaktari ET, Goudevenos JA, Elisaf MS, Milionis HJ. Apolipoprotein B-to-A1 Ratio as a Predictor of Acute Ischemic Nonembolic Stroke in Elderly Subjects. J Stroke Cerebrovasc Dis. 2010; 19: 497-502.

[110] Park J-H, Hong K-S, Lee E-J, Lee J, Kim D-E. High Levels of Apolipoprotein B/AI Ratio Are Associated With Intracranial Atherosclerotic Stenosis. Stroke 2011; 42: 3040-3046.

[111] Boekholdt SM, van der Steeg WA, Stein EA et al. The ratio of apolipoproteins B to A-I and the risk of future coronary artery disease in apparently healthy men and women; the EPIC-Norfolk prospective population study. Ann Intern Med. 2007; 146: 640-648.

[112] Meisinger C, Loewel H, Mraz W, Koenig W. Prognostic value of apolipoprotein B and A-I in the prediction of myocardial infarction in middle-aged men and women: results from the MONICA/KORA Augsburg cohort study. Eur Heart J 2005; 26: 271-278.

[113] Moss AJ, Goldstein RE, Marder VJ et al. Thrombogenic factors and recurrent coronary events. Circulation 1999; 99: 2517-2522.

[114] Corsetti JP, Zareba W, Moss AJ, Sparks CE. Apolipoprotein B determines risk for recurrent coronary events in postinfarction patients with metabolic syndrome. Atherosclerosis 2004; 177: 367-373.

[115] Cremer P, Nagel D, Mann H et al. Ten-year follow-up results from the Goettingen Risk, Incidence and Prevalence Study (GRIPS): I. Risk factors for myocardial infarction in a cohort of 5790 men. Atherosclerosis 1997; 129: 221-230.

[116] Sweetnam PM, Bolton CH, Downs LG et al. Apolipoproteins A-I, A-II and B, lipoprotein(a) and the risk of ischaemic heart disease: the Caerphilly Study. Eur J Clin Invest 2000; 30: 947-956.

[117] Dunder K, Lind L, Zethelius B, Berglund L, Lithell H. Evaluation of a scoring scheme, including proinsulin and the apolipoprotein B/apolipo-protein AI ratio, for the risk of acute coronary events in middle-aged men: Uppsala Longitudinal Study of Adult Men (ULSAM). Am Heart J 2004; 148: 596-601.

[118] Ström-Möller C, Zethelius B, Sundström J, Lind L. Impact of follow-up time and re-measurement of the electrocardiogram and conventional cardiovascular risk factors on their predictive value for myocardial infarction. J Intern Med 2006; 260: 22-30.

[119] Ingelsson E, Schaefer EJ, Contois JH, McNamara JR, Sullivan L,Keyes MJ, Pencina MJ, Schoonmaker C, Wilson PWF, D'Agostino RB, Vasan RS. Clinical Utility of Different Lipid Measures for Prediction of Coronary Heart Disease in Men and Women. JAMA 2007; 298(7): 776-785.

[120] Goswami B, Rajappa M, Mallika V, Kumar S, Shukla DK. Apo-B/apo-AI ratio: a better discriminator of coronary artery disease risk than other conventional lipid ratios in Indian patients with acute myocardial infarction. Acta Cardiol 2008; 63(6): 749-755.

[121] Agoston-Coldea L. Apolipoproteins A-I and B-markers in coronary risk evaluation. Rom J Intern Med 2007; 45(3): 251-258.

[122] Kappelle P.J.W.H., Gansevoort R.T., Hillege J.L., Wolffenbuttel B.H.R., Dullaart R.P.F. on behalf of the PREVEND study group. Apolipoprotein B/A-I and total cholesterol/high-density lipoprotein cholesterol ratios both predict cardiovascular events in the general population independently of non- lipid risk factors, albuminuria and C-reactive protein. J Intern Med 2011; 269(2): 232-242.

[123] Kim H-K, Chang S-A, Choi E-K, Kim Y-K, Kim H-S, Sohn D-W, Oh B-H, Lee M-M, Park Y-B, Choi Y-S. Association between plasma lipids, and apolipoproteins and coronary artery disease: a cross-sectional study in a low-risk Korean population. International Journal of Cardiology 2005; 101: 435-440.

[124] Agoston-Coldea L, Mocan T, Gatfossé M, Dumitrascu DL. The correlation of apolipoprotein B, apolipoprotein B/apolipoprotein A-I ratio and lipoprotein(a) with myocardial infarction. Cent Eur J Med 2008; 3(4): 422-429. DOI: 10.2478/s11536-008-0057-3.

[125] Stewart MW, Humphriss DB, Mitcheson J, Webster J, Walker M, Laker MF. Lipoprotein composition and serum apolipoproteins in normoglycaemic first-degree relatives of noninsulin dependent diabetic patients. Atherosclerosis 1998; 139: 115-121.
[126] Kim BJ, Hwang ST, Sung KC et al. Comparison of the relationships between serum apolipoprotein B and serum lipid distributions. Clin Chem 2005; 51: 2257-2263.
[127] Sung K.-C., Rhee E.-J., Kim H., Park, J.-B., Kim, Y.-K., Woo, S. Wilson A.M. An elevated apolipoprotein B/AI ratio is independently associated with microalbuminuria in male subjects with impaired fasting glucose. 2011; 21(8): 610-616.
[128] Kim H-K, Chang S-A, Choi E-K et al. Association between plasma lipids, and apolipoproteins and coronary artery disease: a cross-sectional study in a low-risk Korean population. Int J Cardiol 2005; 101: 435-440.
[129] Snehalatha C, Ramachandran A, Sivasankari S et al. Is increased apolipoprotein B-A major factor enhancing the risk of coronary artery disease in type 2 diabetes? J Assoc Physicians India (JAPI) 2002; 50: 1036-1038.
[130] Solymoss BC, Bourassa MG, Campeau L et al. Effect of increasing metabolic syndrome score on atherosclerotic risk profile and coronary artery disease angiographic severity. Am J Cardiol 2004; 93: 159-164.
[131] Lind L, Vessby B, Sundström J. The apolipoprotein B/AI ratio and the metabolic syndrome independently predict risk for myocardial infarction in middle-aged men. ArterioscIerThromb Vasc Biol 2006; 26: 406-410.
[132] Sierra-Johnson J, Romero-Corral A, Somers VK., Lopez-Jimenez F, Wall¬dius G, et al. ApoB/apoA-I ratio: an independent predictor of insulin resistance in US non-diabetic subjects. Eur Heart J 2007; 28: 2637-2643. doi:10.1093/eurheartj/ehm360.
[133] Sierra-Johnson J, Fisher RM, Romero-Corral A, Somers VK, Lopez-Jimenez F, Öhrvik J, Walldius G, Hellenius ML, Hamsten A. Concentration of apolipoprotein B is comparable with the apolipoprotein B/apolipoprotein A-I ratio and better than routine clinical lipid measurements in predicting coronary heart disease mortality: findings from a multi-ethnic US population. Eur Heart J. 2009; 30: 710-717.
[134] Zhong L, Li Q, Jiang Y, Cheng D, Liu Z, Wang B, Luo R, Cheng Q, Qing H. The ApoB/ApoA1 Ratio is Associated with Metabolic Syndrome and its Components in a Chinese Population. Inflammation 2010; 33(6): 353-358. DOI:10.1007/s10753-010-9193-4.
[135] Belfki H , Ben Ali S , Bougatef S, Ben Ahmed D, Haddad N , Jmal A, Abdennebi M , Ben Romdhane H. The Apolipoprotein B/Apolipoprotein A 1 ratio in relation to metabolic syndrome and its components in a sample of the Tunisian population. Experimental and Molecular Pathology 2011; 91: 622-625.
[136] Sniderman AD, Faraj M. Apolipoprotein B, apolipoprotein A-I, insulin resistance and the metabolic syndrome. Curr Opin Lipidol 2007; 18: 633-637.
[137] Sniderman AD, Kiss R. The strength and limitations of the apoB/apoA-I to predict the risk of vascular disease: a Hegelian analysis. Curr atheroscl rep. 2007; 9(4): 261-265.
[138] Bruno G, Merletti F, Biggeri A, Bargero G, Prina-Cerai S, Pagano G, Cavallo-Perin P. Diabetologia 2006; 49: 937-944.
[139] Bayu M, Sasongo TY, Wong STY, Nguyen TT, Kawaskai R, Jenkins A, Shaw J, Wang JJ. Serum Apolipoprotein AI and B are stronger biomarkers of diabetic retinopathy than traditional lipid. Diabetes Care 2011; 34: 474-479.

[140] Enkhma B, Anuurad E, Zhanga Z, Pearson TA, Berglund L. Usefulness of Apolipoprotein B/Apolipoprotein A-I Ratio to Predict Coronary Artery Disease Independent of the Metabolic Syndrome in African Americans. Am J Cardiol 2010; 106: 1264-1269.

[141] Ounis OB, Elloumi M, Makni E, Zouhal H, Amri M, Tabka Z, Lac G. Exercise improves the ApoB/ApoA-I ratio, a marker of the metabolic syndrome in obese children. Acta Pædiatrica 2010; 99: 1679-1685.

[142] Gatz M, Reynolds CA, Finkel D, Pedersen NL, Walters E. Dementia in Swedish Twins: Predicting Incident Cases. Behav Genet 2010; 40: 768-775, DOI 10.1007/s10519-010-9407-4.

[143] Carnevale Schianca GP, Pedrazzoli R, Onolfo S, Colli E, Cornetti E, Bergamasco L, Fra GP, Bartoli E. ApoB/apoA-I ratio is better than LDL-C in detecting cardiovascular risk. Nut Metabol Cardiovasc Dis 2011; 21: 406-411.

[144] Wen ZZ, Geng DF, Luo JG, Wang JF. Combined use of high-sensitivity C-reactive protein and apolipoproteins B/apolipoprotein A-1 ratio prior to elective coronary angiography and oral glucose tolerance tests Clinical Biochemistry Clinical Biochemistry 2011; 44: 1284-1291.

[145] Andersson J, Sundström J, Kurland L, Gustavsson T, Hulthe J, et al. The Carotid Artery Plaque Size and Echogenicity are Related to Different Cardiovascular Risk Factors in the Elderly. Lipids 2009; 44: 397-403.

[146] Barbier CE, Lind L, Ahlström H, Larsson A, Johansson L. Apolipoprotein B/A-I ratio related to visceral but not to subcutaneous adipose tissue in elderly Swedes Atherosclerosis 2010; 211: 656-659.

[147] Schmidt C, Wikstrand J. High apoB/apoA-I ratio is associated with increased progression rate of carotid artery intima-media thickness in clinically healthy 58-year-old men: Experiences from very long-term follow-up in the AIR study. Atherosclerosis 2009; 205: 284-289.

[148] Reis JP, Macera CA, Wingard DL, Araneta MRG, Lindsay SP, Marshall SJ. The relation of leptin and insulin with obesity-related cardiovascular risk factors in US adults. Atherosclerosis 2008; 200: 150-160.

[149] Junyent M , Zamb´on D , Gilabert R , Cof´an M , N´u˜nez I , Rosa E. Carotid atherosclerosis in familial combined hyperlipidemia associated with the APOB/APOA-I ratio. Atherosclerosis 2008; 197: 740-746.

[150] Vladimirova-Kitova LG, Deneva-Koicheva TI. Increased Intima-Media Thickness in Carriers of the LDL-Receptor Defective Gene versus Noncarriers with Newly Detected Asymptomatic Severe Hypercholesterolemia. Echocardiography 2011; 28: 223-234.

[151] Dahlén E. M., Länne T, Engvall J, Lindström T, Grodzinsky E, Nyström FH, Östgren CJ. Carotid intima-media thickness and apolipoprotein B⁄apolipoprotein A-I ratio in middle-aged patients with Type 2 diabetes 247 patients with Type 2 diabetes, aged 55–66 years, in the Cardiovascular Riskfactors in Patients with Diabetes—a Prospective study in Primary care (CARDIPP-1) study. Diabet Med 2009; 26: 384-390.

[152] Rasouli M, Kiasari AM, Mokhberi V. The ratio of apoB/apoAI, apoB and lipoprotein(a) are the best predictors of stable coronary artery disease. Clin Chem Lab Med 2006; 44(8): 1015-1021.

[153] Smith J, Cianflone K, Al-Amri M, Sniderman A. Body composition and the apoB/apoA-I ratio in migrant Asian Indians and white Caucasians in Canada. Clin Sci 2006; 111: 201-207. doi:10.1042/CS20060045 201.

[154] Maffeis C, Pietrobelli A, Grezzani A, Provera S, Tato` L. Waist circumference and cardiovascular risk factors in prepubertal children. Obes Res 2001; 9: 179-87.

[155] Gardner CD, Tribble DL, Young DR, Ahn D, Fortmann SP. Associations of HDL, HDL2, and HDL3 cholesterol and apolipoprotein A-I and B with lifestyle factors in healthy men and women: The Stanford Five City Project. Prev Med 2000; 30: 346-356.

[156] Okosun IS, Prewitt TE, Liao Y, Cooper RS. Association of waist circumference with ApoB to ApoA-I ratio in black and white Americans. Int J Obes Relat Metab Disord 1999; 23: 498-504.

[157] Juonala M, Viikari JSA, Kähönen M, Solakivi T, Helenius H, Jula A, Marniemi J, Taittonen L, Laitinen T, Nikkari T, Raitakari OT. Childhood Levels of Serum Apolipoproteins B and A-I Predict Carotid Intima-Media Thickness and Brachial Endothelial Function in Adulthood The Cardiovascular Risk in Young Finns Study. J Am Coll Cardiol 2008; 52: 293-299.

[158] Mattsson N, Magnussen CG, Rönnemaa T, Mallat Z, Benessiano J, Jula A, Taittonen L, Kähönrn M, Juonala M, Viikari JSA,Raitakari OT. Metabolic Syndrome and Carotid Intima-Media Thickness in Young Adults: Roles of Apolipoprotein B, Apolipoprotein A-I, C-Reactive Protein, and Secretory Phospholipase A2: The Cardiovascular Risk in Young Finns Study. Arterioscler Thromb Vasc Biol. 2010; 30: 1861-1866.

[159] Wallenfeldt K, Bokemark L, Wikstrand J, Hulthe J, Fagerberg B. Apolipoprotein B/apolipoprotein A-I in relation to the metabolic syndrome and change in carotid artery intima-media thickness during 3 years in middle-aged men. Stroke 2004; 35: 2248-2252.

[160] Matsumoto K, Fujita N, Nakamura K, Senoo T, Tominaga T, Ueki Y. Apolipoprotein B and insulin resistance are good markers of carotid atherosclerosis in patients with type 2 diabetes mellitus. Diabet Res Clin Pract 2008; 82: 93-97.

[161] Kim SJ, Song P, Park JH, Lee YT, Kim WS, Park YG, Bang OY, Chung C-S, Lee KH, Kim G-M. Biomarkers of Asymptomatic Carotid Stenosis in Patients Undergoing Coronary Artery Bypass Grafting. Stroke 2011; 42: 734-739, DOI: 10.1161/STROKEAHA.110.595546.

[162] Ajeganova S, Ehrnfelt C, Alizadeh R, Rohani M, Jogestrand T, Hafström I, Frostegård J. Longitudinal levels of apolipoproteins and antibodies against phosphorylcholine are independently associated with carotid artery atherosclerosis 5 years after rheumatoid arthritis onset—a prospective cohort study. Rheumatology 2011; 50: 1785-1793. Advance Access published July 9, 2011.

[163] Koha KK, Sakumab I, Quonc MJ. Review; Differential metabolic effects of distinct statins. Atherosclerosis 2011; 215: 1-218.

[164] Nicholls SJ, Brandrup-Wognsen G, Palmer M, Barter PJ. Meta-analysis of Comparative Efficacy of Increasing Dose of Atorvastatin Versus Rosuvastatin Versus Simvastatin on Lowering Levels of Atherogenic Lipids (from VOYAGER). Am J Cardiol 2010; 105: 69-76.

[165] Faergeman O, Hill L, Windler E, Wiklund O, Asmar R, Duffield E, Sosef F on behalf of the ECLIPSE study investigators. Efficacy and Tolerability of Rosuvastatin and

Atorvastatin when Force-Titrated in Patients with Primary Hypercholesterolemia. Results from the ECLIPSE Study. Cardiology 2008; 111: 219-228. DOI: 10.1159/000127442.

[166] Ballantyne CM, Andrews TC, Hsia JA, Kramer JH, Shear C for the ACCESS Study Group. Correlation of non-high-density lipoprotein cholesterol with apolipoprotein B: effect of 5 hydroxymethylglutaryl coenzyme A reductase inhibitors on non-high-density lipoprotein cholesterol levels. Am J Cardiol 2001; 88: 265-269.

[167] Vodnala R, Bard RL, Krishnan SM, Jackson EA, Rubenfire M, Brook RD. Potential effects on clinical management of treatment algorithms on the basis of apolipoprotein-B/A-1 and total/high-density lipoprotein-cholesterol ratios. J Clin Lipidol 2011; 5: 159-165. DOI: 10.1111/j.1365-2362.2010.02387.x

[168] Van den Bogaard B, Van den Born B-JH, Fayyad R, Waters DD, DeMicco DA, et al., on behalf of the Treating to New Targets investigators. On-treatment lipoprotein components and risk of cerebrovascular events in the Treating to New Targets study. Eur J Clin Invest 2010 Sep 27, doi:10.1111/j.1366-2362.2010.02387.x On line.www TNT. Published Eur J Clin Invest 2011; 41(2): 134-142.

[169] Kastelein JJP, van der Steeg WA, Holme I, Gaffney M, Cater NB, Barter P, Deedwania P, Olsson AG, Boekholdt M, Demicco DA, Szarek M, LaRosa JC, Pedersen TR, Grundy SM, for the TNT and IDEAL Study Groups. Lipids, Apolipoproteins, and Their Ratios in Relation to Cardiovascular Events With Statin Treatment. Circulation 2008; 117: 3002-3009.

[170] Holme I, Strandberg TE, Faergeman O, Kastelein JJP, Olsson AG, Tikkanen MJ, Lytken Larsen M, Lindahl C, Pedersen TR, on behalf of the Incremental Decrease in End Points Through Aggressive Lipid Lowering Study Group. Congestive heart failure is associated with lipoprotein components in statin-treated patients with coronary heart disease. Insights from the Incremental Decrease in End points Through Aggressive Lipid Lowering Trial (IDEAL). Atherosclerosis 2009; 205: 522-527.

[171] Holme I, Cater NB, Faergeman O, Kastelein JJP, Olsson AG, et al., on behalf of the Incremental decrease in endpoints through aggressive lipid-lowering study group. Lipoprotein predictors of cardiovascular events in statin-treated patients with coronary heart disease. Insights from the Incremental Decrease in End-points through Aggressive Lipid-lowering Trial (IDEAL). Ann Med 2008; 40: 456-464.

[172] Nissen, SE, Nicholls SJ, Sipahi et al. Effect of very highintensity statin therapy on regression of coronary atherosclerosis. The ASTEROID trial. JAMA 2006; 295: 1556-1565. March 13, 2006; Epub ahead of print.

[173] Nicholls SJ, Tuzcu EM, Sipahi I, Grasso AW, Schoenhagen P, et al. Statins, High-Density Lipoprotein Cholesterol, and Regression of Coronary Atherosclerosis. JAMA 2007; 297: 499-508.

[174] Tani S, Nagao K, Anazawa T, Kawamata H, Furuya S, Takahashi H, Iidaa K, Matsumoto M, Washioa T, Kumabe N, Hirayama A. Relation of Change in Apolipoprotein B/Apolipoprotein A-I Ratio to Coronary Plaque Regression After Pravastatin Treatment in Patients With Coronary Artery Disease. Am J Cardiol 2010; 105: 144-148.

[175] Taskinen MR, Barter PJ, Ehnholm C, Sullivan DR, Mann K, Simes J, Best JD, Hamwood S, Keech AC on behalf of the FIELD study investigators. Ability of traditional lipid ratios and apolipoprotein ratios to predict cardiovascular risk in people with type 2 diabetes. Diabetologia 2010; 53: 1846-1855, DOI 10.1007/s00125-010-1806-9.

[176] Ridker PM, Danielson E, Fonseca FAH, et al, on behalf of the JUPITER Trial Study Group. Reduction in C-reactive protein and LDL cholesterol and cardiovascular event rates after initiation of rosuvastatin: a prospective study of the JUPITER trial. Lancet 2009; published online March 29, 2009. DOI:10.1016/S0140-6736(09)60447-5. Published in Lancet 2009; 373: 1175–82

[177] Mora S, Glynn RJ, Boekholdt M,. Nordestgaard BG, Kastelein JJP, Ridker PM.On-Treatment Non–High-Density Lipoprotein Cholesterol, Apolipoprotein B, Triglycerides, and Lipid Ratios in Relation to Residual Vascular Risk After Treatment With Potent Statin Therapy JUPITER (Justification for the Use of Statins in Prevention: An Intervention Trial Evaluating Rosuvastatin) J Am Coll Cardiol, 2012; 59: 1521-1528, doi:10.1016/j.jacc.2011.12.035

[178] Reiner Z, De Backer G, Graham I, Taskinen M-R, Wiklund O, Agewall S, Alegria E, Chapman MJ, Durrington P, Erdine S, Halcox J, Hobbs R, Kjekshus J, Perrone Filardi P, Riccardi G, Storey RF, Wood D.Review: ESC/EAS Guidelines for the management of dyslipidaemias The Task Force for the management of dyslipidaemias of the European Society of Cardiology (ESC) and the European Atherosclerosis Society (EAS). Atherosclerosis 2011; 217S: S1-S44.

[179] Genest J, Frohlich J, Fodor G, McPherson R for the Working Group on Hypercholesterolemia and other Dyslipidemias. Recommendations for the management of dyslipidemia and the prevention of cardiovascular disease: summary of the 2003 update. Can Med Assoc J 2003; 169: 921-924.

[180] Connelly PW, Poapst M, Davignon J et al. Reference values of plasma apolipoproteins A-I and B, and association with nonlipid risk factors in the populations of two Canadian provinces: Quebec and Saskatchewan. Can J Cardiol 1999; 15: 409-418.

[181] Charlton-Menys V, Betteridge DJ, Colhoun H, Fuller J, France M, Hitman GA, Livingstone SJ, Neil HAW, Newman CB, Szarek M, DeMicco DA, Durrington PN. Apolipoproteins, cardiovascular risk and statin response in type 2 diabetes: the Collaborative Atorvastatin Diabetes Study (CARDS). Diabetologia 2009; 52: 218-225. DOI 10.1007/s00125-008-1176-8.

Chapter 6

Approaches to Access Biological Data Sources

Assia Rharbi, Khadija Amine, Zohra Bakkoury,
Afaf Mikou, Anass Kettani and Abdelkader Betari

Additional information is available at the end of the chapter

http://dx.doi.org/10.5772/48740

1. Introduction

In recent years, technological revolutions in genomics and proteomics have revolutionized the work of researchers in molecular biology. Through various techniques of data generation, they have at their hand in the web a very large amount of information contained in public and heterogeneous data sources. Each source has content organized around a particular data type like sequences in Uniprot (for proteins) and Genbank (for gene and mRNA), protein structure in PDB (Protein Data Bank) and publications in biomedical Medline. Their content is heterogeneous in the sense that a similar data can be represented differently in two data sources (eg different names for the same gene). More data sources have a variety in terms of structure, and there are sources of structured data, such as relational databases or semi-structured sources like XML and unstructured sources such as databases composed of flat files. That is to say that a biologist who wishes to obtain information from these sources have to question these one by one, then copy and analyze the data collected, and manage redundancy, complementarities of the information and inconsistencies. Today, one of the greatest challenges of bioinformatics is to enable biologists to effectively access multiple data sources, each with a different pattern. Various approaches have been adopted to unify access to various data sources given a query. Several systems have been produced from data warehouses, a federation of databases or mediators.

In this work, we are interested in mediation systems. Such systems offer to the user a uniform and centralized view of distributed data, this view may also reflect a more abstract, condensed, qualitative data and therefore more meaningful to the user. These mediation systems are also very useful in the presence of heterogeneous data, because they seem to use a homogeneous system.

We aim to assist biologists in their research through the development of a generic tool for the integration of heterogeneous genomic data distributed over the web, and we are placed in a very particular context that is the study of cardiovascular disease and especially familial hypercholesterolemia. This is a disorder of high LDL ("bad") cholesterol that is passed down through families, which means it is inherited. This disease is caused by a genetic mutation of certain lipoproteins. Indeed, these lipoproteins (called LDL) carry the 2/3 of cholesterol circulating in the blood; they deliver cholesterol to tissues by a system of recognition between Apo lipoprotein Band a receiver: the LDL receptor (lock and key system) that allows the entry of LDL and their cholesterol content in cells. When the LDL receptor (LDL-R) is weak (about one mutation), LDL accumulates in the blood and artery walls causing familial hypercholesterolemia (HF). So knowing these different mutations by biologists, can greatly facilitate the molecular screening of the disease and therefore to find the proper treatment. However, to answer such a query: "What are the mutations that cause familial hypercholesterolemia (HF)?" The biologist hasto make a fastidious search in disparate and heterogeneous databases which requires a considerable investment time.

This chapter is structured as following:

- First we present the background of the project
- Second we focus on the problem of heterogeneity of data sources and biological characteristics of these sources.
- Third we present the state of the art of data integration, problems and constraints of this integration and the various existing approaches to solve this problem.
- And fourth we expose studied scenario, the realization and perspectives.

2. General context

Since the completion of the human genome sequencing in April 2003, we observe the accumulation of an outsize amounts of genomic and proteomic data on the web often syntactically and semantically heterogeneous and difficult to capitalize.

Information about genes provides access to their corresponding proteins. In addition, all diseases are associated with alterations in the structure or function of such proteins. A good knowledge of protein structure provides insight into their function.

Bioinformatics has become an important tool to explore genomic data by relying heavily on computer systems. It suggests methods and software's for biological data storage and processing. Actually, it is acquiring and organizing data, developing software for the analysis, comparison and modeling of these data and analysis results produced by bioinformatics software to infer new biological knowledge, in collaboration with biologists.

This work contributes to facilitate to biologists searching among heterogeneous and distributed data in public and / or private data sources on the web. In particular, it helps them to analyze proteins, by building a platform for integrating biological data. This will provide a tracking system to target special proteins involved in a disease known as familial hypercholesterolemia and thus, to better understand the biological activity of these macromolecules.

Familial hypercholesterolemia disease results from mutations in the LDLR gene. The LDLR gene provides instructions for making a protein called a low-density lipoprotein receptor.

This type of receptor binds to particles called low-density lipoproteins (LDLs), commonly known as bad cholesterol. By removing low-density lipoproteins from the bloodstream, these receptors play a critical role in regulating cholesterol levels. When the LDL receptor (LDL-R) is deficient, LDL accumulated in arteries induces the familial hypercholesterolemia (HF) pathology. So, in biology knowing these different mutations can greatly facilitate the molecular screening of the disease and thus find appropriate treatment.

3. Biological data sources

Number of data sources and tools available to biologists on the web has grown dramatically in recent years. This huge number of available data along with heterogeneous information generated wide variety of access interfaces, and also a profound heterogeneity.

3.1. Genomic databases

There are two types of databanks, those that correspond to a set of heterogeneous data so-called "databases" and those more homogeneous established around a specific theme.

Also, to avoid confusion we will distinguish between semantic databases, general [2] and specialized [3]databases.

For specific requirements related to the activity of a group, or to bibliographic compilations, many specific databases were created in laboratories. In some cases, these databases have been developed continuously; others have not been updated and disappeared as they represented a specific need. Still others are unknown or poorly known and are waiting for further investigation.

All these specialized databases of interest may vary considerably from one base to another according to their size. In most of the case, these bases correspond to a combination compared of generalist databases such as: Swiss-Prot, GenBank. DDBJ (DNA Data Bank of Japan), EMBL (European Molecular Biology Laboratory) which are used very often. It is important to know, that according to the field of activity or the genomics research, the surveyed banks are not necessarily the same. The genomic libraries contain various information that may include:

- Characteristics of proteins or genes such as localization of the gene in the cell: LocusLink5, the 3D structure of protein: Protein Data Bank (GDP) and Molecular Modeling Database (MMDB) or its biological function. More specifically, some databases contain information about a specific family of protein such as "Enzyme8" which include exclusively enzyme type proteins.
- Some phenotypes (specific genes, morphological feature, clinical syndrome ...) or more specifically some genetic diseases: Online Mendelian Inheritance in Man (OMIM);
- Specific species or families of species: FlyBase, Reptilia, Saccharomyces Genome Database (SGD), Mouse Genome Database (MGD);
- The medical literature (banks abstracts): Medline, PubMed.

Table1 and Table 2 give two examples of genomic databases along with protein database.

Designation	Location	Roles	Comments	Web sites and references
Nucleic bases				
EMBL	European Bioinformatics Institute (EBI) Europe	More than 1 million records (January 1998) for more than 15,500 species. The predominant species: Homo sapien,Caenorhabditis elegans, Saccharomyces cerevvisae ...	Information's search tools: SRS, System Retrieval System and via a web interface on EBI, through BLAST et FASTA software	Accessible via the web site: http://www.ebi.ac.uk/ebi_docs/embl_db/ebi/topembl.html
Specific genomic resources				
SGD Saccharomyces Genome Database	Works of Cherry and al, 1998.	Online resources on molecular biology and S.cerevisiae genetic	Numerous research help functions on line	Accessible via the web site: http://genome-www.stanford.edu/Saccharomyces/

Table 1. Genomic databases

Designation	Location	Roles	Comments	Web sites and references
Primary databases				
PIR Protein Information Resource	National Biomedical Research Foundation	Sequences collecting to detect evolutionary relationship between proteins	The current structure includes 4 compartments : PIR1, PIR2, PIR3 and PIR4	Accessible via the web site http://nbrfa.georgetown.edu/pir/
Composite databases				
NRDB Non-redundant Database	National Center Biotechnology Information USA	Composed of GenPept (derived from GenBank), PDB sequences, SWISS-PROT, SPupdate, PIR, and GenPeptupdate (update of GenPept). NRDB is the default database of BLAST and NCBI service	Accessible via the web site: http://www.ncbi.nlm.nih.gov/Web/NRDB/	NRDB Non-redundant Database

Table 2. Protein databases

3.2. Characteristics of biological data sources

The diversity of information distributed sources and their heterogeneity are the one of the main problem that the web users have to face. This heterogeneity may result from the size or structure of the sources (structured sources: relational databases, partially structured sources: XML documents, or unstructured: texts), the access mode and query, or semantic heterogeneity: between concept maps, and implicit or explicit underlying ontology's.

Biological sources have a large heterogeneity at different levels:

- Syntactic: because of the different formats for describing the content sources usually ASN.1 (formal notation for describing data transmitted via exchange protocols), (eg Enter), but also more standard formats such as XML (eg GenBank).
- Semantic which covers several aspects. First, it concerns the focus. However, each base focuses on a type of biological object (eg, the focus of Swiss-Prot is the protein, the focus of GenBank is gene, and the PDB is the 3D protein structure).
- Then, according to the base, the same information is not represented with the same level of detail: some bases are generalists (eg Swiss-Prot in general on proteins) while others are more specialized (eg SGD (Saccharomyces Genome Database) on yeast proteins).
- The final aspect of semantic heterogeneity is related to the diversity of nomenclature modes. Different vocabularies are used to annotate the sequences and the reliance on such annotations is seldom complete. Moreover, within a same database, there are a several names for each single entity (protein, gene). The name of an entity may depend on the disease to which it is linked or to its inventor.
- Source query language: another form of heterogeneity comes from query languages. Languages are often simple forms (combinations of words to search in a text), in the case of portals or simple databases. But one can also find structured languages such as SQL or OQL.
- Protocols for collecting data that are different such as CGI / HTTP or FTP. Access to web sources is limited to the entry forms and their underlying programs
- The tools offered by the Web: there are many tools for text searching and sequence comparison algorithms such as BLAST (Basic Local Alignment Search Tool), FASTA15 or LASSAP16.

4. State of the art of approaches to integration

A data integration system remedies to the problems associated with the expansion of public data sources by giving the possibility to have a unified view of them. Such a system is the interface between user and data sources simplifying requests to perform (a request to query all sources covered by the system). The user is not obliged to know where the data are and how they are structured.

4.1. Current integration approaches

There are two major approaches for integration of information: (1) the data warehouse (DW) or materialized approach and (2) virtual approach (mediator based).In DW approach, huge amount of historic data is stored in the DW. In the virtual approach, on the other hand, the data is not materialized, but rather is globally manipulated using views. Each of these approaches is suitable in some kinds of applications.

4.1.1. Data warehouse

DW is a powerful tool for decision support and querying the data because it explicitly stores information from heterogeneous sources locally. However, some external data, such as new product announcements from opponents and currency exchange rates, may be needed to support the accuracy of the business decisions. We should not neglect the importance of such data to avoid the problems of incompleteness, inexact, or sometimes wrong results. Warehousing huge and frequently changed information is a big challenge for the following reasons.

Firstly, since the data in the DW is loaded in snapshots and the DW is a huge information repository. Secondly, as the data sources change frequently, the maintenance becomes a complicated and costly issue

Here are two examples of using data warehouses:

- Genomics Unified Schema, GUS [4] is a system for creating a data warehouse focused on molecular biology;
- Gene Expression Data Warehouse, GEDAW [5] is a warehouse dedicated to the analysis of the transcriptome of human liver.

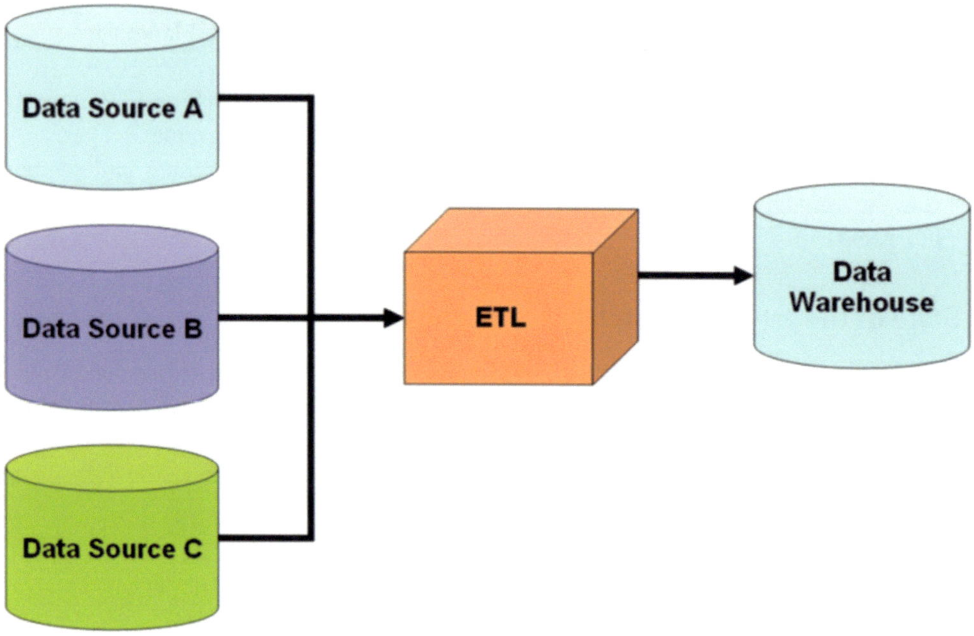

Figure 1. Simple schematic for a data warehouse

4.1.2. The virtual approach (mediator based)

In this approach, the actual data resides in the sources, and queries against the integrated 'virtual' view will be decomposed into sub queries and posed to the sources. This approach is preferred over the materialized approach DW when the information sources change very often. On the other hand, the DW approach may be desired when a quick query answer is required and the information sources change rarely.

The most important step in the construction of a mediator is the creation of the global schema. The mapping consists on the relations between the global schema and local sources. Specification of this mapping, depending on the method, determines the difficulty of query reformulation and the facility of adding or removing sources within the system. Two methods are commonly used to determine the global schema

- GAV (Global As View) approach: In this approach, each concept of the global schema is mapped to a query over data sources. In other words, when the user presents his/her query over the integrated schema, the data corresponds to a concept in the integrated schema, which can actually be answered from the data sources through a specific query. The query processing in GAV is easy, since it just unfold each concept in the integrated schema in the user query with the associated query over the sources, but this approach does not help much when the sources change or grow very often, since these factors affect the mappings and require restricting the integrated schema.
- LAV (Local As View) approach: LAV approach defines the mapping in the other way around; each concept in the data sources is defined in terms of a query over the integrated schema. This makes query processing more difficult, since in this case, the system does not know explicitly how to reformulate the concepts in the integrated view expressed in the user query in terms of the data sources. On the other hand, changes or incremental growth in the sources will not lead to reconstruction of the integrated schema, and need only to modify the mappings

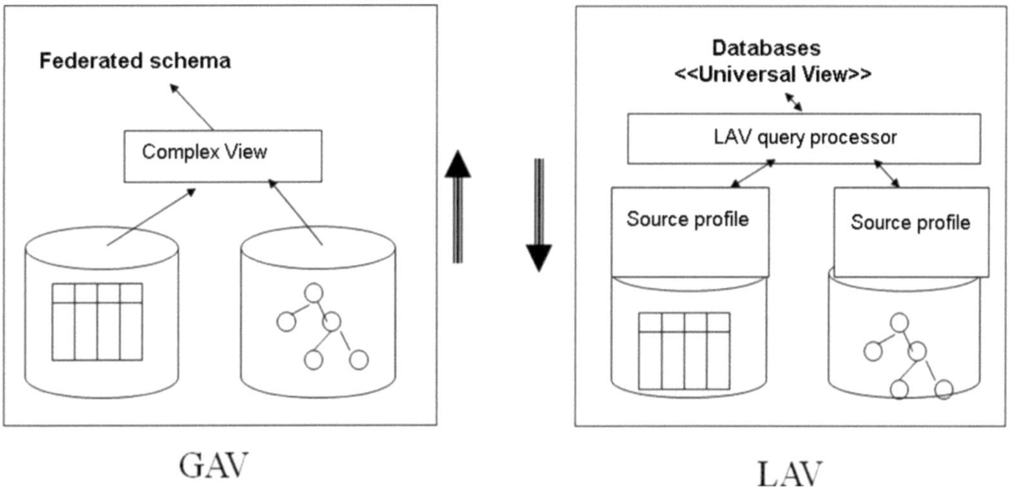

Figure 2. GAV vs. LAV

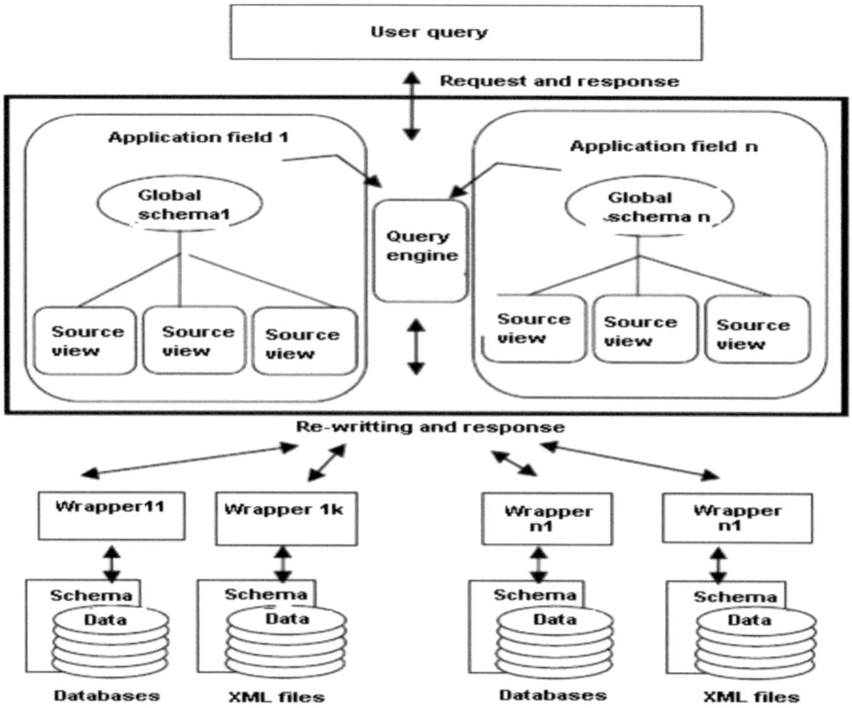

Figure 3. Architecture of a mediator

In fact, these two approaches are not opposite, but complementary; depending on the problem to be solved. To integrate a few sources, most of which are stable, better to use the GAV method. By cons, as part of a large-scale integration, the LAV method is preferable as a material change at a local source with little or no impact on the global schema.

Two examples of systems integration based mediator:

- *Tambis (Transparent Access To Multiple Bioinformatics Information Sources) [6]* is an integration system coupled to an ontology that allows for better interoperability between sources;

K2/BioKleisli [7] is a system based on CPL (Collection Programming Language) is a query language for high-level querying multiple sources.

4.1.3. The multi agents approach

This approach was used in GID-IGC *(Integrated Genomic Database - Genome Information System) project.* The proposed architecture uses a network of agents communicating each with other via CORBA and KQML. All have a specific function, such as *EIA* (External Interface Agent) that manages the user interface, or *SCA* (Dial Selector

Agent) witch decompose the global query into sub-queries for local data sources. This approach is very modular and easily extensible.

4.1.4. Navigating between sources

This approach is based on what users usually do when searching for information on the web, which involves a search page to page by clicking the mouse. In practice, queries generated for this type of tool are converted into path expressions. The data banks are then integrated based on their cross-references. These expressions can answer the query of the user according to different levels of satisfaction.

A reference is a link between two data sources (Figure 4), a bridge between the information relating on the same object or the same concept. It can be done through an identifier of an external source or a URL (Unified Resource Locator). If the link can be browsed in both directions it is a cross-reference ("cross-reference").

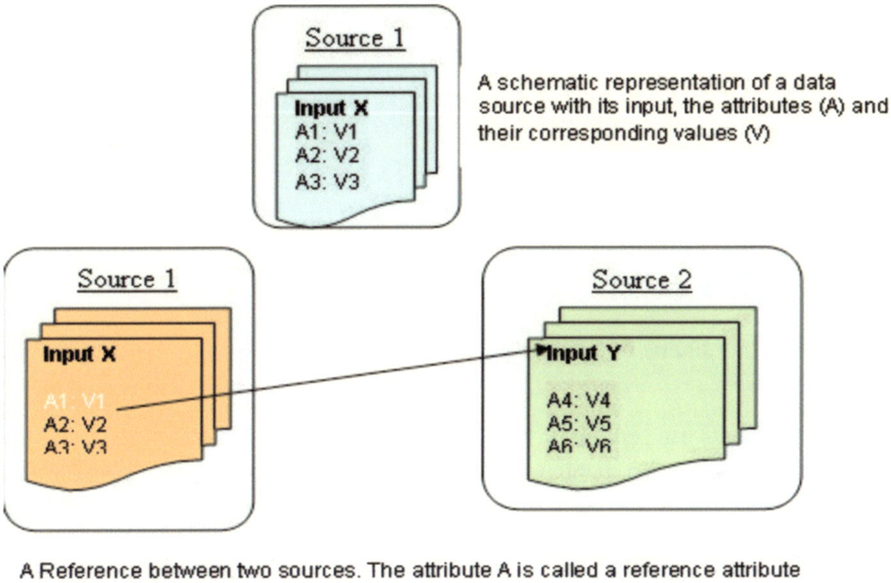

Figure 4. Navigating between sources

4.2. Adopted approach

In this work, we are interested in mediation systems. Such systems offer a uniform and centralized view of distributed data. This view may also reflect a more abstract, condensed, qualitative data and therefore more meaningful to the user. These mediation systems are also very useful in the presence of heterogeneous data, because they seem to use a homogeneous system.

In this architecture, each component provides a set of features, which, together will help to satisfy the user request at the end.

- The mediator is a software module that directly receives the user's request. It has to locate the necessary information to answer the query, resolve schematic and semantic conflicts, query different sources and integrate the partial results in a consistent and coherent response. This is the most complex component but only one instance of it is necessary (unlike multiple adapters). It provides access to multiple data sources as if it was a single one and offers this consultation through multiple languages and ontologies.

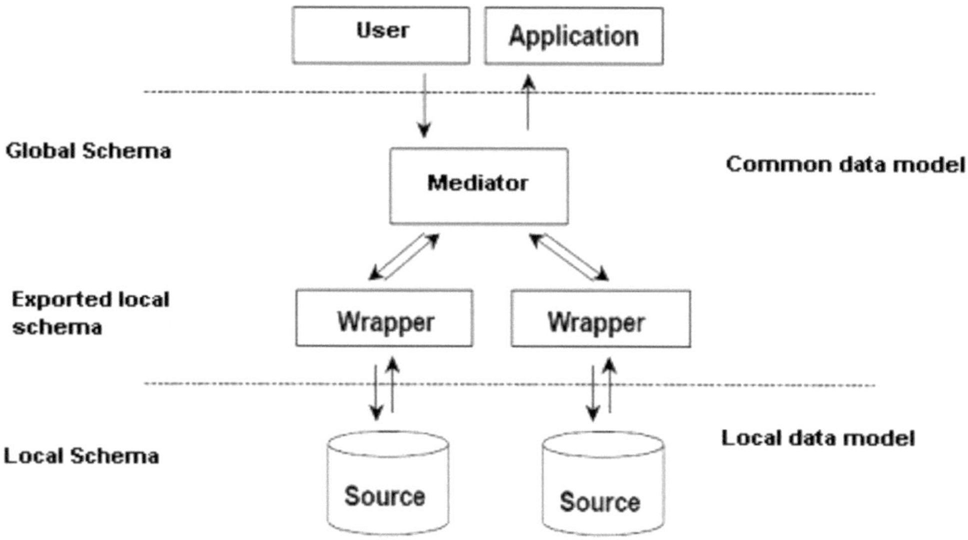

Figure 5. Adopted approach

This is a crucial component that allows a local system to distribute its information to a community of users.

- The adapter allows the presentation of data in the mediation's syntactic format. So it's an interface for querying a database using a standardized language (pivot language).
- Data source: Represent the sources and banks of biological information. A data source can be described by its:
  - Location: Reference, communication protocol, access technique (JDBC, ODBC API), support (DBMS, web pages)
  - Type of data it manages: structured (relational, object), semi-structured (XML, OEM), unstructured (image, multimedia)
  - Ability to query: SQL, OQL, search
  - Results Format: XML, HTML, relationships, texts

5. Studied scenario

After exploring the different integration systems that solve the problem of heterogeneous biological data sources, this section describes the scenario we have chosen to work on.

5.1. Biologist's need

The objective of our work is to develop an integration system for biological data with an application on familial hypercholesterolemia. Such system should facilitate access to multiple data sources available on the Web, in a transparent and uniform way by giving biologists a single virtual source that summarizes the sites of interest to the application.

In order to satisfy this biologist's need we studied their current way to work. We first focused on existing tools, data sources they use and their functional specifications.

Tools: they use, mainly:

- CHARMM (Chemistry at Harvard Macromolecular Mechanics) [8]: This program offers a wide choice for the production and analysis of molecular simulations. It simulates the standard energy minimization of a given structure and the production of a molecular dynamics trajectory.
- VMD (Visual Molecular Dynamic) [9]: This software is used to visualize the molecules available on the web.

Data sources: The focus was mainly on the following sources:

- PDB (Protein Data Bank) [10]: It's a worldwide collection of data on three-dimensional structure of biological macromolecules: Proteins and nucleic acids. The PDB is the primary source of structural biological data. It allows access to 3D structures of pharmaceutical interest proteins.
- PubMed [11]: it's a free search engine giving access to the MEDLINE bibliographic database, gathering citations and abstracts of biomedical research.

5.2. Adopted scenario

The adopted scenario consists on building a local database "homemade" that would include unorganized data already available in the biologist's laboratory and for our particular case data related to LDL receptor mutations. This goes along with the research we're going to do on the Web using the mediation system:

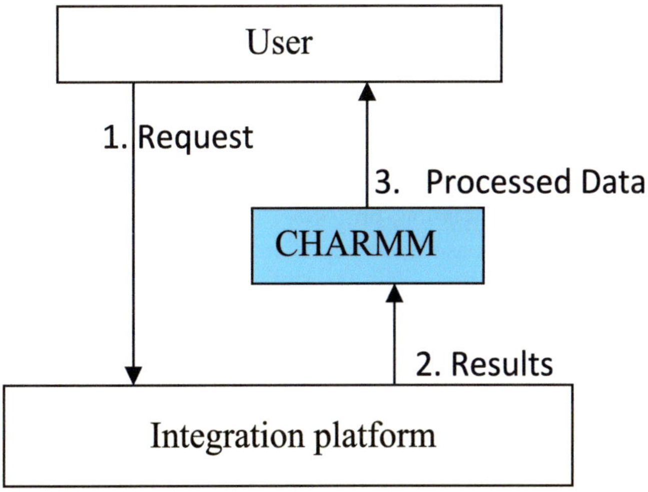

Figure 6. The adopted scenario

The integration platform is the mediation system. It will query data sources related to cardiovascular diseases especially familial hypercholesterolemia namely: PDB and PubMed. The results of the query will be processed by the tool CHARMM before being presented to the user [12, 13].

5.3. Selected data sources

PubMed [8]: Is the leading bibliographic data search engine of all fields of biology and especially medicine. It was developed by the National Center for Biotechnology Information (NCBI), and is hosted by the National Library of Medicine U.S. National Institutes of Health. PubMed is a free search engine giving access to the MEDLINE bibliographic database, gathering citations and abstracts of biomedical research.

The MEDLINE database in April 2007 had more than 15 million citations from 1950 published in 5000 biomedical journals (journals in biology and medicine) distinct. It is the database of reference for biomedical sciences. As with other indexes, including a citation in PubMed has no content. In addition to MEDLINE, PubMed also provides access to:

- OldMedline for articles before 1966
- Citations of all articles, even "irrelevant" (that is to say, covering topics such as plate tectonics or astrophysics) from certain MEDLINE journals, primarily those published in major newspapers of general science or biochemical (such as Science and Nature, for example).
- Citations being listed before indexing in MEDLINE or MeSH, or passage or status "off topic"
- Older citations selected for MEDLINE journal from which they arise (when they are supplied electronically by the publisher)
- Articles submitted to PubMed Central for free

Most citations include a link to the full article when it is available (eg PubMed Central). PubMed is a search engine that allows users to search in the MEDLINE database; this information is also available from private organizations such as Ovid and Silverplatter, among others. PubMed is free since the mid-1990s. For optimal use of PubMed, it is necessary to have an understanding of his core, MEDLINE, and especially the MeSH vocabulary used for indexing articles in MEDLINE.

We can also find in PubMed information about the log, which can search by title, subject, short title, NLM ID, ISO abbreviation, and ISSN (International Standard Serial Number) written and electronic. The database "newspaper" includes all newspapers Enter Base.

The major interest of these bibliographic databases is that:

- Their bodies are used to identify recent publications in scientific journals.
- They help to establish bibliographies (lists of relevant articles) on a subject or author.
- They are portals to access full text documents available on the Internet.
- The bibliographic databases used to find references to documents, select, print or export them to other software. They may also propose to order documents or provide access to full text.

PDB (Protein Data Bank): The databank on proteins of Research Collaborator for Structural Bioinformatics, more commonly known as Protein Data Bank or PDB is a worldwide collection of data on the three-dimensional (or 3D structure) of biological macromolecules : protein essentially, and nucleic.

Founded in 1971 by Brookhaven National Laboratory, the Protein Data Bank was transferred in 1998 to the Research Collaborator for Structural Bioinformatics (RCSB), which consists of Rutgers University, the University of Wisconsin at Madison, National Institute of Standards and Technology (NIST) and the "San Diego Supercomputer Centre." The PDB originally contained (in 1971) 7 structures. The number of structures deposited has grown since the 1980s. Indeed, at that time, the crystallographic techniques have improved, the structures determined by NMR have been added, and the scientific community has changed its view on data sharing.

The PDB containedon 28-04-2008, 50480 structures. The data are from the original pdb format, and in recent years are also mmCif format, specifically developed for structural data from the PDB. From 2000 to 3000 structures are added each year. The bank contains files for each molecular model. These files describe the exact location of each atom of the macromolecule studied, that is to say, the Cartesian coordinates of the atom in a three-dimensional coordinate.

Each model is referenced in the bank by a unique identifier to 4 characters, the first is always a numeric character, the next three being alphanumeric characters. This identifier is called "**pdb code**".

Several formats exist for PDB files:

The PDB format: it is the original format. The guide of this format has been revised several times; the current version is version 2.2[14], which has existed since 1996. Originally pdb format was dictated by the width and the use of punch cards for computers. Consequently, each line contains exactly 80 characters.

Pdb file format is a text file where each column has its meaning: Each parameter is positioned so immutable. Thus, the first 6 columns, that is to say the first 6 characters for a given line, determine the scope of the file. Found for example in the fields " TITLE_ "(That is to say, the title of the macromolecule of interest)," KEYWDS "(The keywords of the entry)," EXPDTA "Which provides information on the experimental method used," SEQRES "(The sequence of the protein under study)," ATOM_ "Or" HETATM "Fields containing all information related to a particular atom.

Pdb format limitations: Format in 80 columns pdb files is relatively restrictive. The maximum number of atoms in a pdb file is 99999, since there are only 5 columns allocated for the numbers of atoms. Similarly the number of residues per chain is at most 9999: There are only 4 columns allowed for this figure. The number of channels is limited to 62: A single column is available, and possible values are one of the 26 letters of the alphabet in upper or lower case, or one of the digits 0 through 9. As this format has been defined, these limitations did not seem restrictive, but they have been taken several times during the deposition of extremely large structures, such as viruses, ribosome, and multienzyme complexes.

MmCIF format:The growing interest in the development of database and electronic publications in the late 1980s has created the need for a more structured, standardized, open-ended and high quality data from the PDB. In 1990, the International Union of Crystallography IUCr extended to macromolecules data representation used to describe crystal structures of molecules of low molecular weight. This representation is called CIF, for Crystallographic Information File. The dictionary mmCIF (macromolecular Crystallographic Information File) published in 1996, was then developed.

In MmCIF format, each field of each section of a pdb file is represented by a description of a characteristic of an object, which includes both the name of the characteristic (eg _struct.entry_id), and the content of the description (pdb code: 1cbn). Which we can call "name-value". It is easy to convert, without loss of information, an mmCIF file format pdb, since all information is directly analyzed. It is not possible, however, to completely automate the conversion of a pdb file format mmCIF, since many mmCIF descriptors are either absent from the PDB file, either in this field " REMARK "Who can not always be analyzed. The contents of fields " REMARK " is indeed separated according to different mmCIF dictionary entries, in order to preserve the completeness of the information contained in such Materials and Methods section (crystal characteristics, refinement method ...) or in the description of the biologically active molecule or other molecules (substrate, inhibitor, ...)

The mmCIF dictionary contains over 1700 entries, which are much safer not all used in a single PDB file. All field names are preceded by the character "underscore"(_), In order to differentiate the values themselves. Each name corresponds to an mmCIF dictionary entry, where the characteristics of the object are exactly defined.

Pdbml format: This format is pdbml adaptation to XML data format bps and contains the entries described in the dictionary "PDB Exchange Dictionary". This dictionary contains the same entries as the mmCIF dictionary, in order to take into account all data managed and distributed by the PDB. This format can store much more information on models than pdb format.

Data retrieval: The files describing molecular models can be downloaded from the website of the PDB and visualized using various software such as Rasmol [15], Jmol [16], chime [17] or an extension VRML [18] (plugin) a browser. The website of the PDB also contains resources for teaching, on structural genomics and other useful software.

5.4. The global schema

By studying and exploring the previous sources and by combining data from genome sources, we have identified all data that define the dictionary related to familial hypercholesterolemia disease (Table 3). From this data dictionary and business rules (as defined and established by experts in the field of biology), we extracted the major biological entities useful for our study. These entities are not independent and form a semantic graph with nodes reflecting relationships between these entities.

Property	Description
code_biblio	Library code
auteur_biblio	Author of publication
date_biblio	Date of publication
volume_biblio	Volume of structure
langue_biblio	Language
contribution_biblio	Contribution
journal_biblio	Newspaper
revue_scint_biblio	Journal
livre_biblio	Book
Cd_proceeding_biblio	CD procedure
nom_recepteur	Name of receiver
nom_mutation	Name of mutation
classe_mutation	Class of mutation
nom_proteine	Name of the protein
longueur_proteine	Length of the protein
type_proteine	Protein type
structure_summary	Summary of the structure
structure_title	Under the structure
nom_molecule	Name of the molecule
author_name	Name of author
date_depot	Date Filed
date_release	Date of publication
derniere_release	Last update
Resolution	Resolution
Compound	Compound
Classification	Classification
molecule_chain_type	Channel Type
experimental_methode	Experimental method of resolution (RX, NMR)

Table 3. Data Dictionaries

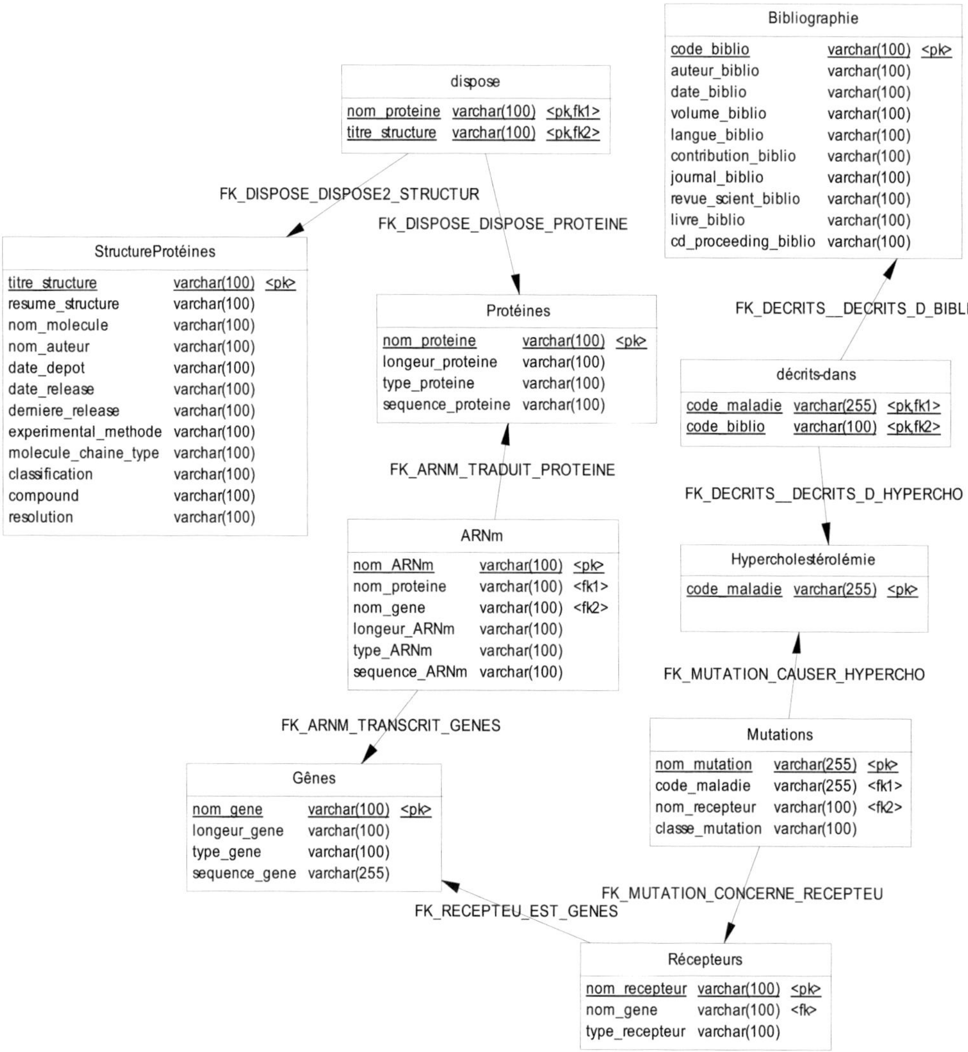

Figure 7. Global Schema

From the global schema, it is possible to make our request and submit it to SQL mediator for treatment. For example, the query that givesthe associated protein mutated gene and publications on familial hypercholesterolemia is expressed as follows in SQL:

```
Select nom_proteine, journal_biblio, auteur_biblio, date_biblio, langue_biblio
From mRNA, gene g, b bibliography
Where a.nom_gene = g.nom_gene
And g.nom_gene in (select nom_gene from recepteurs r, mutation m
                    Where r.nom_recepteur = m.nom_recepteur)
```

For its execution, the query is first submitted to the mediator which is responsible for locating sources and queries them through the wrapper or the associated adapter.It should be noted that the only access point to our sources for interrogation is a web form that, once processed through a wrapper, gives us the local sources that we describe below.

5.5. Analysis of the query

In the global query, 6 attributes are involved: Nom_proteine, nom_gene, journal_biblio, auteur_biblio, date_biblio, langue_biblio shown in the following table along with the sources (PubMed (S1), and PDB (S2).

Attributes	Lists of sources
journal_biblio, auteur_biblio, date_biblio,langue_biblio	S1
nom_proteine	S2
nom_gene	S2, S1

Table 4. Identification of sources

From these sources, we can extract a local schema, for example:

S1_L (journal_biblio, auteur_biblio, date_biblio, langue_biblio, nom_gene),

S2_L (nom_proteine, nom_gene)

From a programming point of view, S1_L and S2_L represent wrapper sources.

Next section describes the realization in which we develop wrappers, submit queries to the mediator, and combine the final result to be presented to the user.

6. Realization

We define four steps in the realization:

- The development of wrapper
- Definition of "global schema"
- Correspondence between local tables and global schema
- Results analysis

6.1. Step 1: Development of wrappers

A wrapper is a program that envelops the execution of another program in the way that the environment can be more suitable. The mediator requests the various databases via wrappers that will extract information from websites of interest. It is necessary to create a wrapper for each specific database.

The sources that we identified in section 4 will be integrated through wrappers. There are different types of wrappers depending on the type of pages they incorporate. These can be either text files or XML files (Extensible Mark-up Language). It is necessary to know the structure of these files and know where the information is located (after any tag, for example). Developing wrappers is linked to functional specifications of the sources presented earlier.

Wrappers allow therefore the extraction of data to be represented in tables. Indeed, we declare the objects and their attributes for each site based on data provided. From all this information, local schemas (relational) for each of these databases are established.

Various programs were written in java. Even if a wrapper has been created for each database, they all have the same main structure. To fill out the fields of tables, the wrapper accesses the Web site to integrate the page and look for keywords behind which is the value to extract. Wrappers are of two types depending on the format of the sources : Either text wrappers or XML wrappers (Figure 8).

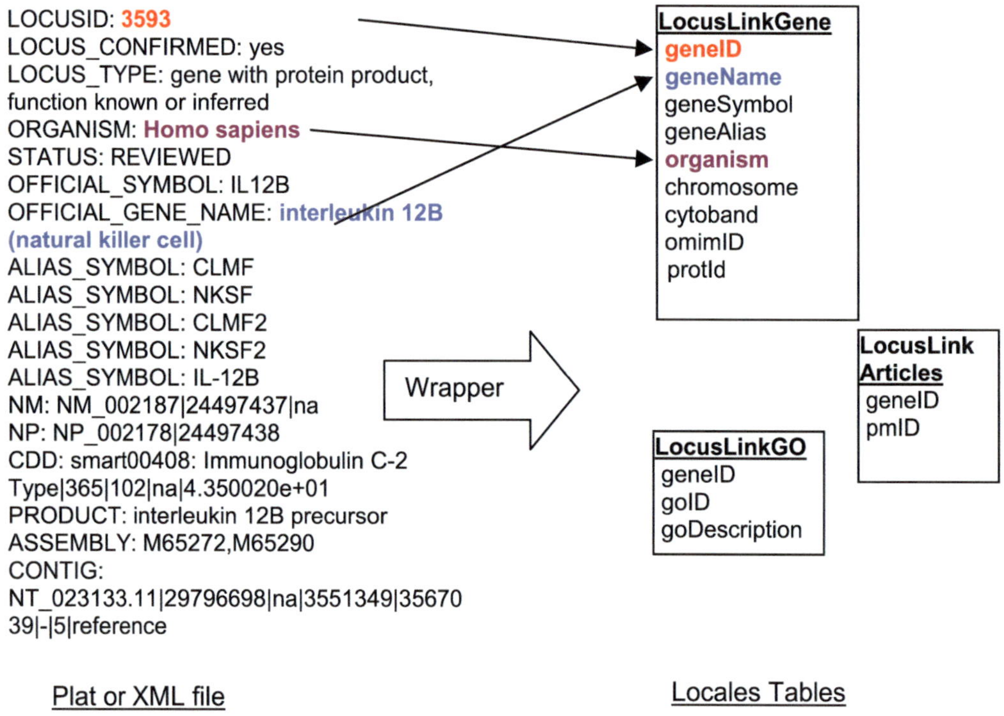

Figure 8. Presentation of a wrapper

Finally, a program that generates and initializes (gives the starting values for all wrappers) is created to coordinate everything. This program is also written in Java and integrates all the wrappers and their relationships. We thus obtain a set of local tables (Provisional) performed by the wrappers.

6.2. Step 2: Definition of "global schema"

It is therefore necessary to build the global schema that will be the only interface for user. Indeed, the user does not know absolutely how the data are integrated. The global schema is a set of relational tables that are defined using local tables (for information). This schema was introduced in the previous section.

6.3. Step 3: Matches between local tables and global schema

As the various local tables have been filled by the wrappers and the global schema has been established, we should now define the correspondence rules between them in order to implement global schema with the extracted information from local schemas. The problem is that several sources may correspond to a business table (we must then join conditions on these tables) or otherwise a source may have several tables trades. For this, we use the Medience server tool.

Medience Server (Figure 5.2) is a complete environment that treats all matching problems (different formats, different representation of business information, dispersal of information described in a single business table). It is a "virtual database", because it does not store information but analyse the user needs. This tool will serve as a mediator that is to say that it will be the unique interface for the user as it will both integrate databases, present data and also offers possibility to loop and see only some information tables of interest to the user. The use of this tool goes in three steps:

1. The first step consists on recording data sources and creating associated source tables using source files provided by wrappers. The global schema is also implemented by local tables. We define the attributes of all tables.
2. In the second part we define the correspondence rules from source tables to the global schema tables. The supply of each table in the global database is defined from the records of source tables. It is thus possible to standardize results coming from various sources by the definition of a standard type.
3. The final step is the verification of all components and installation of all matching rules to make it operational (Figure 9).

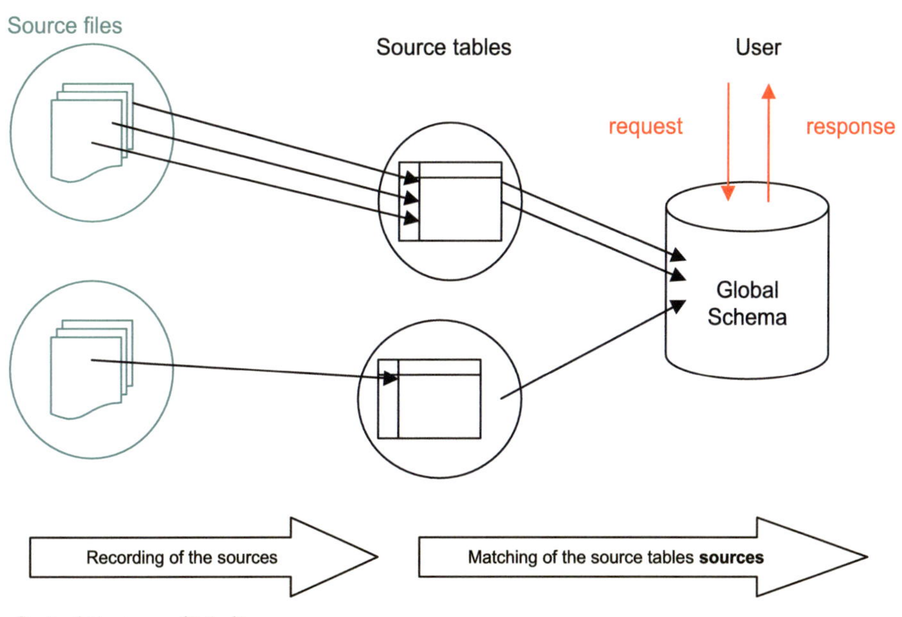

Figure 9. Architecture of Medience

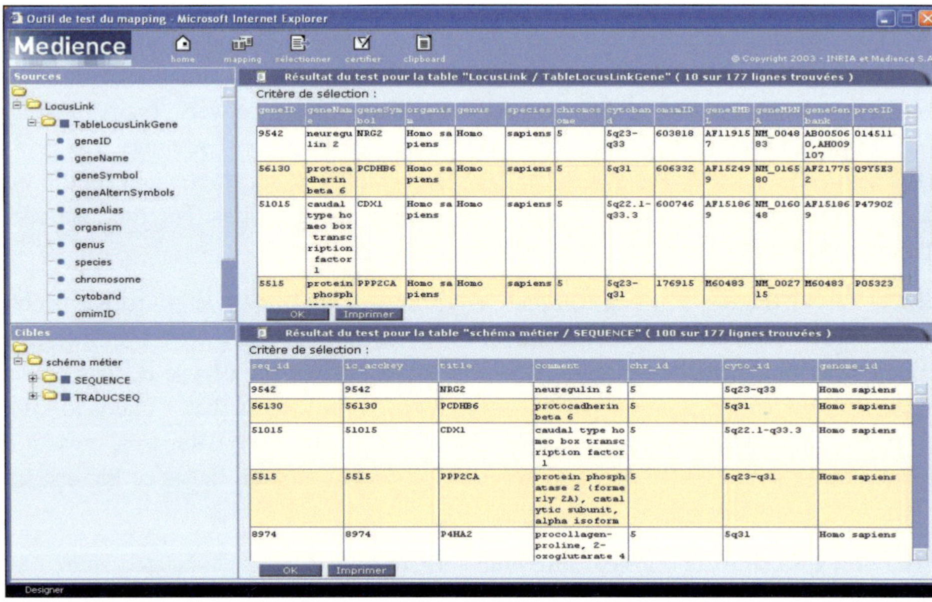

Figure 10. Anexample of Medience interface [19]

6.4. Step 4: Data analysis

It is therefore possible through a platform like Medience to integrate data sources (BD, Excel files, and text files) and view the results in a tabular form. Now, we can process to the analysis of the results. For this, the definition of demand in terms of mining must be decided: How can we use the data provided. Medience offers the possibility to ask tables on the global schema in SQL way. It also offers the ability to define views on these tables and keep a small part that is particularly interesting. It is possible to use our tool to answer the question like what is the protein associated with the mutated gene responsible for familial hypercholesterolemia and related publications.

7. Conclusion

The objective of this work was to develop a system for integrating biological data with an application on familial hypercholesterolemia disease. Such a system should facilitate access to multiple data sources available on the Web, in a transparent and uniform way, giving biologists a single virtual source that summarizes all relevant data sources for the application.

This chapter describes the solution adopted to achieve such a system, where the main elements have been identified, and a computer deployment scenario developed. Among different existing integration approaches, we adopted the mediator approach to integrate data sources. In this approach the most important step is the construction of the global schema as the mediator has to process queries at runtime in order to integrate data sources. We first studied the biologists' needs by exploring different scenarios and we identified with their help various data sources involved.

A study of these sources was necessary in order to build our global schema. From the diagram established, we formulated our SQL query as we built various adapters associated with different sources and at the end we have submitted this request to the mediator for treatment.

As prospects, we have to implement and test this solution and combine the final result of the mediator and that of the tool CHARMM before presenting to the user.

We are currently expanding the platform by integrating other proteins involved in cardiovascular diseases which are the main cause of mortality in the world. In particular, we are investigating a protein called paraoxonase-1 (PON1) which plays an important role in the cardiovascular diseases prevention.

PON1 is an HDL associated enzyme synthesized in the liver and distributed in the blood. It catalyzes the hydrolysis of modified lipids in both HDL (known as good cholesterol) and LDL (known as bad cholesterol) particles and protects them from oxidative modifications, and subsequently reducing the risk of atherosclerosis.

Further bioinformatics analysis including molecular simulations are performed on the PON1 enzyme to better understand the structure activity relationship and also to explore the mutated proteins (genetic polymorphism associated with heart disease) responsible for the weak activity revealed trough the clinical study in both diabetic and coronary patients from Morocco.

Author details

Assia Rharbi and Zohra Bakkoury
Equipe : AMIPS Ecole Mohammadia des Ingénieurs,
Université Mohammed V, Agdal, Rabat – Morocco

Afaf Mikou
Laboratoire GAIA, Spectroscopie Faculté des Sciences Ain Chock –
Université Hassan II, Casablanca – Morocco

Khadija Amine
Laboratoire GAIA, Spectroscopie Faculté des Sciences Ain Chock -
Université Hassan II, Casablanca – Morocco
Laboratoire de Recherche sur les Lipoprotéine et l'Athérosclérose, Unité de Recherche Associée au CNRS-URAC 34-, Faculté des Sciences Ben Msik-Casablanca, Université Hassan II Mohammedia, Morocco

Anass Kettani
Laboratoire de Recherche sur les Lipoprotéine et l'Athérosclérose, Unité de Recherche Associée au CNRS-URAC 34-, Faculté des Sciences Ben Msik-Casablanca, Université Hassan II Mohammedia, Morocco

Abdelkader Betari
ENSA Oujda, Université Mohammed Premier Oujda, Morocco

8. References

[1] M. El Messal, K. Aït Chihab, R. Chater, JC. Vallvé, F. Bennis, A. Hafidi, J. Ribalta, M. Varret, M. Loutfi, JP. Rabès, A. Kettani, C. Boileau, L. Masana, A. Adlouni. Familial Hypercholesterolemia in Morocco: first report of mutations in the LDL receptor gene. J Hum Genet.48 (4):199-203, 2003

[2] http://www.dsi.univ-paris5.fr/bio2/autof2/cha2_1.htm : Bases de données biologiques / Banques généralistes

[3] http://www.dsi.univ-paris5.fr/bio2/autof2/cha2_2.htm : Bases de données biologiques / Banques spécialisées

[4] www.gusdb.org : Genomics Unified Schema (GUS)

[5] Emilie Guérin, Gwenaëlle Marquet, Anita Burgun, Olivier Loréal et Fouzia Moussouni, GEDAW : un environnement intégré pour l'analyse du Transcriptome, JOBIM 2005

[6] http://www.cs.man.ac.uk/~stevensr/tambis/ : TAMBIS

[7] Thomas Hernandez, Subbarao Kambhampati, Integration of Biological Sources: Current Systems and Challenges Ahead, SIGMOD september 2004

[8] http://www.charmm.org/: CHARMM (Chemistry at HARvard Macromolecular Mechanics)

[9] http://www.ks.uiuc.edu/Research/vmd/: VMD (Visual Molecular Dyamic)

[10] www.rcsb.org/pdb/

[11] http://www.ncbi.nlm.nih.gov/sites/entrez?db=PubMed&itool=toolbar

[12] Assia Rharbi, Zohra Bakkoury, Afaf Mikou, Anass Kettani, Abdelkader Betari, and Omar Boucelma, Intégration des données génomiques pour la maladie d'hypercholestérolémie familiale, Journées Scientifiques en Bio-informatique (JSB'2007)

[13] Assia Rharbi, Zohra Bakkoury, Afaf Mikou, Anass Kettani, Abdelkader Betari, and Omar Boucelma, Intégration des données appliquée au domaine biologique, Cinquième Conférence sur les Systèmes Intelligents : Théories et Application (SITA'08)

[14] http://www.rcsb.org/pdb/file_formats/pdb/pdbguide2.2/guide2.2_frame.html

[15] http://rasmol.org/

[16] http://jmol.sourceforge.net/

[17] http://fr.wikipedia.org/w/index.php?title=Chime&action=edit&redlink=1

[18] http://www.w3.org/MarkUp/VRML/

[19] F.-M. Colonna, Thèse : "Intégration de données hétérogènes et distribuées sur le Web et applications à la biologie", Université Paul Cézanne (Aix-Marseille III), Décembre 2008

Section 3

Hyper- and Dyslipoproteinemias

Chapter 7

Lipoproteins Impact Increasing Cardiovascular Mortality

Jelena Umbrasiene, Ruta-Marija Babarskiene and Jone Vencloviene

Additional information is available at the end of the chapter

http://dx.doi.org/10.5772/47404

1. Introduction

Cardiovascular diseases – the main result of the generalized atherosclerosis are the leading cause of global mortality all over the world [1,2]. The number of atherosclerotic diseases - an ischemic stroke, coronary heart disease and peripheral artery disease increases every year [1]. Possibly, due to increase in the population age, better health care and improved survival the prevalence of heart diseases is still so high [3]. The cardiovascular mortality in the most developed countries also is very high [2,3]. About half of all deaths occures due to cardiovascular diseases, it's an over 4,35 million deaths each year in the 53 member states of the World Health Organization European Region and more than 1,9 million deaths each year in the European Union [2]. Moreover there is a 35 billion euros damage due to working people production loss regarding to cardiovascular morbidity and mortality [2]. The cardiovascular mortality is still a problem not only in the European Union, but in the other developed countries as well. Atherosclerotic coronary artery disease was the most common cause of death in the United States in 2004. Men were more often affected, than women by a ratio of 4:1 and after age of 70 by ratio 1:1 [4]. In 2000 about 37 % of death in Canada were due to cardiovascular diseases [3]. They are still the main cause of mortality in Lithuania, as in the older Western European countries as well [5,6]. At the last decade, cardiovascular morbidity and mortality in Lithuania has not declined (Figure 1) [7,8].

In 2008 in Lithuania standartized cardiovascular mortality rate was 520,1 per 100 000 population (Figure 2) [8]. Although in the last years cardiovascular mortality has a tendency to decrease, it's still very high [7]. Lithuanian mortality from coronary artery disease rate in 2008 was 321,29 per 100 000 population (Figure 3) [7,8]. By the statistic data from the Lithuanian Institute of Hygiene, in 2011 56,3% of the people have died from cardiovascular disease in Lithuania. In 2011, 20944 men and 20093 women have died, 47,7% and 62,7% due to coronary artery disease respectively [9].

The main cardiovascular disease - coronary heart disease - highly associated with an increased cardiovascular mortality, hospitalisation and patients disability, significantly raising the cost of medical care [6]. In 2009 it was 4283,39 per 100 000 population hospital discharge for cardiovascular diseases and 1311,8 for coronary artery disease in Lithuania (Figure 4,5) [8].

In 2000 in Canada 7,3 billion dollars (17%) of total direct health care costs and 12,3 billion (14,5%) dollars of total indirect health care costs for all disease categories were attributed to cardiovascular diseases [3]. In the European Union, the economic cost of cardiovascular diseases in direct and indirect healthcare goes to 192 billion euros annually [1]. A total annual cost for person is vary from 50 euros in Malta to 600 euros in Germany, and 372 euros in average [2].

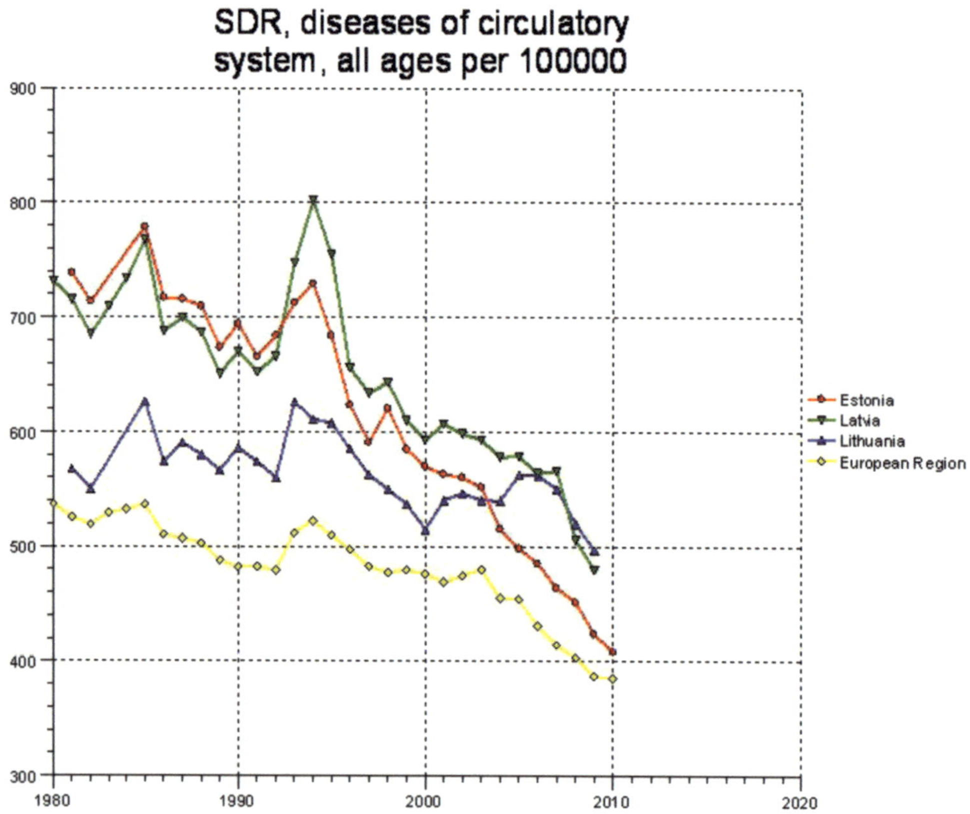

SDR – standartized death rate

Figure 1. Age standartized cardiovascular mortality rate for Baltic States and all European Region dynamic.

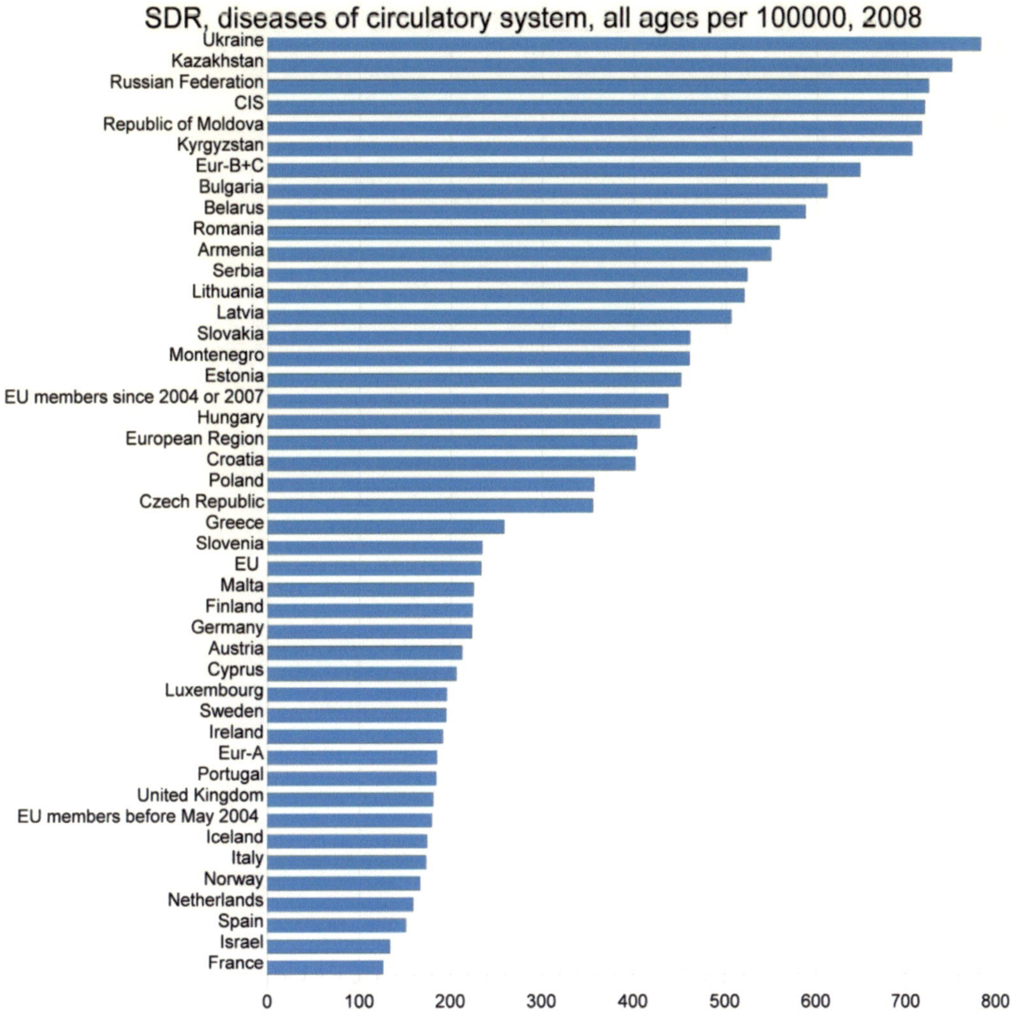

SDR – standartized death rate

Figure 2. Age standartized cardiovascular mortality rate per 100 000 population, 2008.

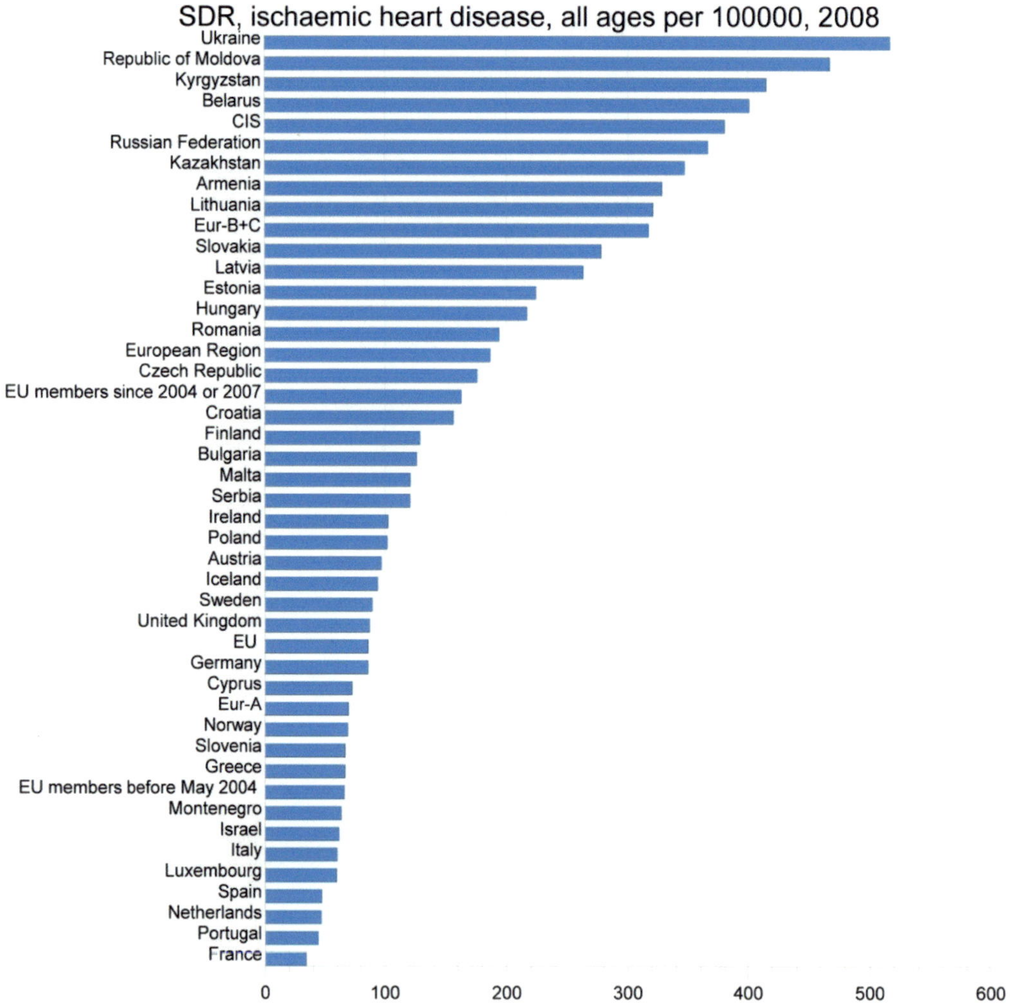

SDR – standartized death rate

Figure 3. Age standartized mortality rate for coronary artery disease per 100 000 population, 2008.

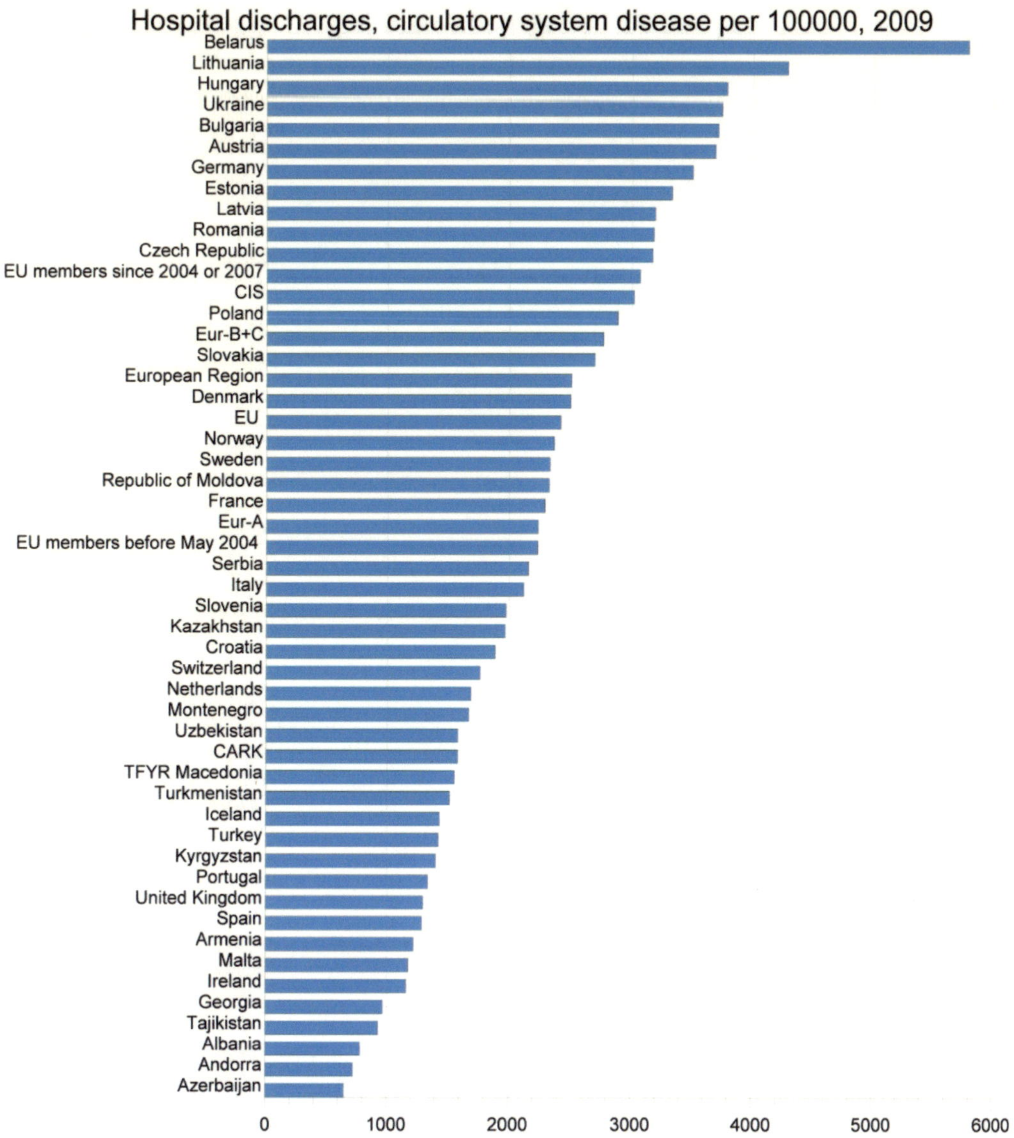

Figure 4. Hospital discharges for the patients with cardiovascular diseases in 2009, per 100 000 population.

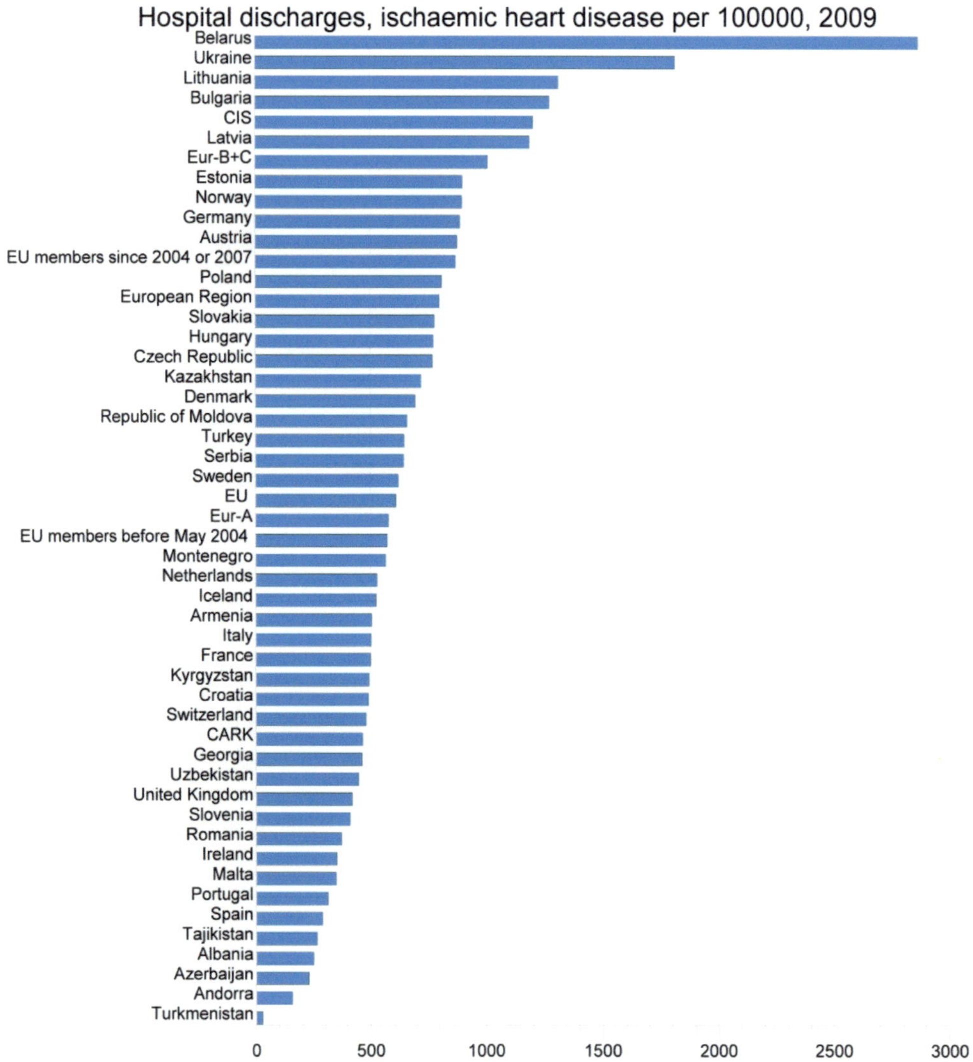

Figure 5. Hospital discharges for the patients with coronary artery disease in 2009, per 100 000 population.

Epidemiological studies have evaluated a number of important risk factors for coronary artery disease, such as positive family history, particulary in the age less than 40 for men, and 50 for women, age, male gender, blood lipids abnormalities, diabetes, hypertension, loss of physical activity, smoking and others, not so substantial (high sensitivity C-reactive protein, hyperfibrinogenemia etc.) [4,6,10-12]. Reducing one or more of these risk factors reduces the risk of major cardiac event accordingly [4]. There are a lot of evidence that lipoprotein disorder is the main pathogenesis of atherosclerosis. This relationship that was estimated century ago by Anitschkow is still important today [3,13]. Variuos epidemiological studies demonstrated a strong association between dyslipoproteinemia and coronary heart disease. There is a strong relation between serum cholesterol concentration level and the coronary heart disease risk [14]. The Multiple Risk Factor Intervention Trial (MRFIT) in USA with 356222 men with different cardiovascular risk factors and 6-year follow-up period have shown that elevated total cholesterol blood concentration significantly increases cardiovascular risk [14,15]. In 2008, the authors published the report about the continuous follow-up for 25 years. The main finding was that total cholesterol is continuous and strong independent predictor for cardiovascular mortality. Estimated increased cardiovascular mortality risk at every total cholesterol level from 160 mg/dl (about 4,14 mmol/l) and higher [15]. Abnormal lipids metabolism or excessive intake of cholesterol especially with a genetic predisposition, initiates the atherosclerosis. A lot of clinical studies established total cholesterol and low density lipoprotein cholesterol are associated with a great risk of coronary heart disease. The reduce of total cholesterol by 10% decreasing the risk of ischemic heart disease by 25% within 5 years [16]. Low density lipoprotein cholesterol reduction not only decreases cardiovascular events, but reduce total mortality as well [3,17]. Furthermore, large randomized controlled clinical trials established the low density lipoprotein cholesterol lowering benefits [10]. It is proved, the reduce of low density lipoprotein cholesterol by 1 mmol/l, decreasing the risk of acute cardiac events by 20%, cardiovascular mortality by 22% [1,11,16-18]. Treatment of lipoproteins disorder also decrease the development of new lesion, regenerates endothelial function and signally reduce cardiovascular events in treated patients [4]. However, the data based on the National Health And Nutrition Examination Survey (NHANES) study from 2005-2008 have estimated that 71 million adults (33,5%) in the USA had elevated low density lipoprotein cholesterol level, but only 34 million (48,1%) were treated and 23 million (33,2%) had reached target low density lipoprotein value. Though, comparing this data to the data from NHANES study in 1999-2002, the number of people with elevated low density lipoprotein level treated with lipids-lowering medications increased from 28,4% to 48,1% between 1999-2002 and 2005-2008 periods. The prevalence of controled low density lipoprotein increased from 14,6% to 33,2% [17]. Although, statins significantly reduce low density lipoprotein cholesterol and coronary heart disease risk, substantial residual cardiovascular risk remains, even with very aggressive low density lipoprotein cholesterol values reduction [11,19,20]. However, atherosclerosis pathogenesis is multiple. It depends not only on low density lipoprotein cholesterol level, but also on genetic, environmental factors, infections, lifestyle factors and other diseases or condition [10-12]. More than a hundred different risk factors for atherosclerosis are estimated today. Although it is known many risk factors for coronary heart disease, the most of them are modifiable. Such as smoking cessation, treatment of dyslipidaemia, lowering of blood pressure can prevent the progression of atherosclerosis and major cardiovascular events [4]. One of the most important mechanisms of the atherosclerosis pathogenesis is Endothelial dysfunction [21]. In the early stages of atherosclerosis endothelian-dependent vasorelaxation disturbes due to oxidative stress and reduced nitric oxide bioavailability. Monocytes and T-lymphocytes adhesion occures. These inflammatory cells penetrate the cell wall, as well as lipid accumulation in the walls of blood vessels takes place. The inflammation and lipids accumulation make a plaque unstable, so it may occlude the vessel. Endothelial dysfunction is observed not only in the initial stage, but also in all other stages of atherosclerosis as well [21-23]. However, the main risk factors still are male gender and older age (more common in women in menopause), heredity, hypertension, diabetes, smoking, stress, obesity, lack of physical activity, elevated low density lipoprotein cholesterol and total cholesterol and decrease high density lipoprotein cholesterol levels [6,10-12]. Numerous epidemiological studies have found reduced high density lipoprotein cholesterol as an independent risk factor for cardiovascular disease [24]. The Framingham study evaluated 43-44% increasing coronary events in patients with high density lipoprotein cholesterol < 40 mg/dL (1,03 mmol/l) [25]. Patients whose high density lipoprotein cholesterol less than 0,9 mmol/l (35 mg/dL) have 8 times higher risk of cardiovascular disease, versus those, whose high density lipoprotein cholesterol more than 1,68 mmol/l (65 mg/dL) [26]. Studies demonstrates that declined high density lipoprotein cholesterol levels are relatively common in general population. 16-18% of men and 3-6% of women have a high density lipoprotein cholesterol level less than 0,9 mmol/l (35 mg/dL) [20]. Moreover, the reduced high density lipoprotein cholesterol level is a component of the metabolic syndrome – the great predictor of high cardiovascular risk. Experimental studies have found high density lipoprotein cholesterol as a potential antiatherogenic by following characteristics. Estimated high density lipoprotein cholesterol facilitates reverse cholesterol transport and delivers cholesterol from the smooth muscles into hepatic cholesterol uptake. So, harmfull atherogenic cholesterol parts, such as low density lipoprotein cholesterol, are catabolized and neutralized [27-29]. High density lipoprotein cholesterol acts as an antioxidant, reducing vascular oxidative stress and has anti-inflammatory properties, reducing vascular inflammation due to atherosclerosis. There are evidence high density lipoprotein cholesterol has a vasoprotective effect, facilitates blood vessel relaxation, play an important role in the inhibition of white blood cells chemotaxis and adhesion. Also it is known about an anti-apoptotic effect of high density lipoprotein cholesterol on endothelial cells. High density lipoprotein cholesterol enhances the proliferation and migration of Endothelial cells and endothelial progenitor cells and thereby promotes the restoration of the endothelium's integrity. Finally, it has an antiplatelet/profibrinolitic effect, in this way reducing platelet aggregation and inactivating coagulation cascade [20,27-29]. Despite the evidence that reduced high density lipoprotein cholesterol is associated with an increased cardiovascular morbidity and mortality, the major guidelines in cardiology still do not recommend to initiate the treatment of dyslipidemia on high density lipoprotein cholesterol.

So, dyslipoproteinemia is a major risk factor for atherosclerosis and coronary artery disease. Its' proper recognition and management can significantly reduce cardiovascular and total mortality rates [12]. Follow the American Heart Association and the National Heart, Lung and Blood Institute and the Adult Treatment Panel III guidelines it is recommended to start treat from the low density lipoprotein cholesterol. Recent clinical studies provide supporting evidence for low density lipoprotein cholesterol target values of less than 2,5 mmol/l (< 100 mg/dl) for the prevention of coronary artery disease for the high cardiovascular risk patients and less thant 1,8 mmol/l (< 70 mg/dl) for the very high cardiovascular risk patients [1]. Studies demonstrate the significant decrease of atherosclerosis with aggressive reduction of low density lipoprotein cholesterol level in patients with coronary artery disease [3]. Only achieved target low density lipoprotein cholesterol value it is recommended to take care of high density lipoprotein cholesterol. Studies evaluated, that high density lipoprotein cholesterol level more than 60 mg/dl (about 1,5 mmol/l) significantly reduce cardiovascular risk and can be named as „inverse risk factor" [21]. The target high density lipoprotein cholesterol is over 1,03 mmol/l (40 mg/dL) for men and more than 1,29 mmol/l (50 mg/dL) for women [20].

2. Lipoproteins disorder as a risk factor for cardiovascular mortality

Today there are more than one lipoproteins disorder classification. The Frederickson, Lees and Levy's one was based on the lipoprotein fraction after separation by electrophoresis. This classification recognized chylomicrons, very low density cholesterol and low density cholesterol. However, the main limitation of this classification, that it does not include high density lipoprotein cholesterol. That's why the World Health Organisation, the European Atherosclerosis Society and the National Cholesterol Education Program have classified lipoproteins disorder on the basis of the absolute plasma level of lipids (total cholesterol and trygliceride) and lipoprotein cholesterol level (low density lipoprotein cholesterol and high density lipoprotein cholesterol) [12,30]. This classification sustained on biochemical characteristics of lipoproteins and lipids. The plasma lipids do not circulate freely in plasma. They are bound to proteins and transported as macromolecular complexes called lipoproteins [24]. In these complexes lipids are surrounded by a stabilizing coat of phospholipid. There are five principal types of lipoprotein particles in the blood: very low density lipoproteins, intermediate density, low density, high density lipoproteins and chylomicrons. They are structurally different by electrophoretic mobility and density after separation in the ultracentrifuge and by the function [14,24]. The lipoprotein density depends on amount of fats contained within it [31].

Chylomicrons are the largest lipoproteins and synthesized in the small intestine from dietary fat and cholesterol [14,24,31]. They contain triglyceride from the intestine and a small amount of cholesterol. The main task of chylomicrons to transport the digestion products of dietary fat to the liver and peripheral tissue, where they are needed as a source of energy. In the circulation triglycerides are removed from chylomicrons via the action of lipoprotein lipase. If present in large amounts, such as after a fatty meal, chylomicrons cause the plasma to appear milky. Very low density lipoproteins are synthesized in the liver continuously and consists of triglyceride and cholesterol. Like chylomicrons they function primarily to distribute triglycerides to target sites such as adipose tissue and skeletal muscle where they are used for storage and energy [31]. It is the main body source of energy in prolonged fasting [14]. Like chylomicrons, they are removed due to lipoprotein lipase action. With removal of triglycerides and protein, very low density lipoproteins are converted to low density lipoproteins. High plasma levels of very low density lipoprotein cholesterol are to be found in familial hypertriglyceridaemia, diabetes mellitus, in people with a depressed thyroid function and in people with a high alcohol intake [31]. Intermediate density lipoproteins – one of the source of low density lipoproteins production. Last-mentioned are the main particles of lipids. They can deposit lipids into the arterial wall and initiate atherosclerosis. Low density lipoprotein cholesterol are cholesterol-rich particles. About 70% of plasma cholesterol find in this form. Low density lipoprotein cholesterol have a main role in transporting the cholesterol manufactured in the liver to the tissues, where it is used. When low density lipoprotein cholesterol binds to low density lipoprotein cholesterol receptors on the cell surface, low density lipoprotein cholesterol is taken into the cell and broken down into free cholesterol and amino acids. Disorders involving a defect in or lack of low density lipoprotein cholesterol receptors are usually characterised by high plasma cholesterol levels. In the case of the inherited familial hypercholesterolemia the cholesterol excess cannot be cleared efficiently from the blood and therefore accumulates, caused coronary heart disease. And the last particles – high density lipoproteins are produced in liver and intestine. They are composed of 50% protein, with phospholipid and cholesterol as the remainder [31]. They transport lipids away from the periphery. The transfer of pro-atherogenic particles, such as very low density lipoproteins to the liver for the reverse cholesterol transport is one of the most important role of high density lipoprotein cholesterol. In this process harmful pro-atherogenic particles are transporting from the periphery to the liver for the reverse cholesterol transport and neutralizing [14,21]. It is well known low density lipoprotein cholesterol is one of the major factor for the development of atheroma. Atherosclerotic plaque consist of accumulated intracellular and extracellular lipids, smooth muscle cells, connective tissue, and glycosaminglycans.

There are two main hypotheses to explain the pathogenesis of atherosclerosis: the lipid hypothesis and the chronic endothelial injury hypothesis. Both of them are interrelated. The endothelial dysfunction is an initial stage of atherosclerosis, occures due to oxidative stress and sub-endothelial accumulation of lipids. Low density lipoprotein cholesterol undergo oxidation and become local cytotoxic. Macrophages migrate into the sub-endothelial space, take up lipids and become "foam" cells. The earliest detectable lesion of atherosclerosis is the fatty strip. This strip consists of foam cells full of lipids. As the process progress, the smooth muscle cells also migrate into the lesion. At this stage, the lesion may be hemodynamically insignificant. But endothelial dysfunction exists and it's ability to limit the entry of lipoproteins into the vessel is impaired [4]. So, the elevation of plasma low density lipoprotein cholesterol level results in penetration of low density lipoprotein cholesterol into the vessel wall, lipids accumulation in macrophages and smooth muscle cells. Endothelial injury produces loss of endothelium, adhesion of platelets to subendothelium, aggregation of platelets, chemotaxis of monocytes and T-cell lymphocytes, and release of growth factors that induce migration of smooth muscle cells from media to intima, where synthesize connective tissue and proteoglycans and forms a fibrous plaque. Low density lipoprotein cholesterol is cytotoxic and may cause endothelial injury and stimulate smooth muscle growth. Touched endothelial cell are functionally impaired and increase the uptake of low density lipoprotein cholesterol from plasma [24]. Growing atherosclerotic plaque may cause a severe stenosis that can progress to total arterial occlusion. Eventually the plaque may become calcified. Some plaques, reached in lipids and inflammatory cells, as macrophages, covered with a thin fibrous cap may undergo spontaneous rupture, resulting in cascade of events, stimulates thrombosis and ends in acute ischaemic event [4,24].

Hypercholesterolemia occures either from overproduction or defective clearance of very low density lipoprotein cholesterol or from increased conversion of very low density lipoprotein cholesterol to low density lipoprotein cholesterol. Overproduction of very low density lipoprotein cholesterol by liver may be caused by obesity, diabetes, alcohol consumption, nephrotic syndrome or genetic disorders. Each of this conditions can result in increased low density lipoprotein cholesterol and trygliceride levels. When dietary cholesterol reaches the liver, elevates intracellular cholesterol level. Due to this, low density lipoprotein cholesterol receptor synthesis is suppressed. This suppression occurs at the level of transcription of the low density lipoprotein cholesterol gene. A reduced number of receptors results in higher levels of plasma low density lipoprotein cholesterol and total cholesterol [24]. Today, it appears, that high density lipoprotein cholesterol assist in the mobilization of low density lipoprotein cholesterol [4]. Cardiovascular risk increases progressively with elevated of low density lipoprotein cholesterol and with a decrease in high density lipoprotein cholesterol level. Studies demonstrated that each decrease in high density lipoprotein cholesterol level by 1 mg/dl (0,0259 mmol/l) elevating cardiovascular risk by 2-3%. [20]. And contrarily, each increase in high density lipoprotein cholesterol level by 1 mg/dl (about 0,02 mol/l) lowering cardiovascular mortality by 6%, independently of low density lipoprotein cholesterol level [20,32]. Similarly, The Treating to New Target study evaluated high density lipoprotein cholesterol as a more significantly predictive for cardiovascular events comparing with low density lipoprotein cholesterol [21,33]. High density lipoprotein cholesterol is protective through multiple mechanisms. There are some new points in the high density lipoprotein cholesterol role and effects on atherosclerosis. Recently published studies showed the antooxidative role of high density lipoprotein cholesterol. Due to this effect the reduction of vascular oxidative stress is occured. It is thought, this can contribute to the atheroprotective effects. Supposedly, high density lipoprotein cholesterol decreases inflammatory process, stops the proliferation and migration of endothelial cells and has anti-apoptotic effects on them. All of this contributes to the anti-atherosclerotic effect [21]. High density lipoprotein cholesterol also affects the platelets function and haemostatic cascade [14].

The prevalence of hypercholesterolemia differ in the world. In 1996 in Taiwan 41,5% men and 19,6% of women had abnormal rates of plasma cholesterol [34]. In 1995 in Holand 19,2% of men and 12,4% of women have total cholesterol more than 6,5 mmol/l [35]. In 1999-2000 in Europe the European Action on Secondary Prevention through Intervention to Reduce Events (EUROASPIRE) study have been performed. The prevalence of high total cholesterol have been declined in Europe 1995-2000 from 86,2 till 58,8% In 2000 58,8% of the population have total cholesterol more than 5,0 mmol/l [36]. In 2006-2007 The European Society of Cardiology carried out the EUROASPIRE III survey in 76 medical centers in 22 European countries (Belgium, Bulgaria, Cyprus, Croatia, The Czech Republic, Finland, France, Germany, Greece, Hungary, Ireland, Italy, Latvia, Lithuania, Poland, the Netherlands, Romania, Russian Federation, Slovenia, Spain, Turkey and UK). The total 13935 participants with established coronary artery disease were reviewed and 8966 were interviewed 6 months after acute coronary event. 76,5% of all the patients had elevated total cholesterol and low density lipoprotein cholesterol. For 51% the total cholesterol more than 4,5 mmol/l was estimated and only for one half of them a total cholesterol goal (<4,5 mmol/l) was reached with lipid-lowering medications, despite of the rather high rate of the statins prescription (80,7%). So, this study showed that one of the most important risk factor for cardiovascular mortality - dyslipoproteinemia control was inadequate and most of the patients did not achieve the targets defined in the guidelines [37]. By the data from MONICA project with 39 population from the 21 countries, in 2003 the main total cholesterol in Kaunas, Lithuania was near 6 mmol/l for men and 6,5 mmol/l for women. Females from Lithuania were at the top of all countries, whereas the men were about an average. 15 countries evaluated higher levels of total cholesterol for men [38]. One of the meta-analysis showed that the highest protective effect can be get treating high risk patients with very high total cholesterol level [39]. Regarding to the existing evidence on lipoproteins disorder, treatment and also due to increased cardiovascular mortality in Lithuania we started the clinical data. At this study we have evaluated independent risk factors for one year cardiovascular mortality for the patients with acute and chronic coronary syndromes. Lipoprotein disorder was one of the most important risk factor for one year cardiovascular mortality.

3. Methods

A total of 3268 patients with coronary heart disease who were selected for this study. The data was collected by a standardized questionnaire. A total of 1865 (728 women and 1137 men) with acute and chronic coronary heart disease, male and female, aged from 20 years till more than 80 years were reexamined after one year. Risk factors for coronary heart disease were evaluated. Lipoprotein disorder was definable as low density lipoprotein cholesterol level in twelve-hour fasting venous blood samples more than 3 mmol/l, total cholesterol level – more than 5,2 mmol/l, high density lipoprotein cholesterol level less than 1,2 mmol/l for women and less than 1,0 mmol/l for men. Due to medical history and data on admission patients were attributed to chronic or acute coronary syndrome. The myocardial infarction and unstable angina were attributed to acute coronary syndromes. Stable angina – to chronic coronary syndrome. Myocardial infarction was diagnosed according to the World Health Organisation guidelines: angina pain and equivalent, ischemic signs on ECG (Q wave, ST and T changes) and an increase in troponin I more than 0,05 mg/l. Unstable angina diagnosis confirmed with the angine syndrome, ischemic changes on the ECG without increasing enzymes in the blood and with angiography assessment of the coronary artery. Stable angina determined using a standard clinic, ECG, exercise test and angiography.

4. Statistical analysis

The statistical analysis was performed using SPSS (Statistical Package for Social Science) version 13 and Microsoft Office Excel 2003 statistical programs. Descriptive statistics was used for the quantitative data analysis. Categorical data have been summarized as frequencies and percentages, and for comparisons, chi-square test have been used. Univariate and multivariate logistic regression analysis was used for the risk assessment. One year mortality risk was evaluated by isolated and standardized odds ratios with 95% confidence interval (CI).

5. Results

The data from 1865 patients with chronic and acute coronary syndromes was analysed. For more than a half of the patients an acute coronary syndrome was diagnosed. The participants were mostly men (61%). 54,7% of the patients had a reduced level of high density lipoprotein cholesterol (less than 1,0 mmol/l for male, and less than 1,2 mmol/l for female), for about 32% an increased total cholesterol and low density lipoprotein cholesterol levels for each were evaluated. About 20,5% of the patients with decreased high density lipoprotein cholesterol level have elevated low density lipoprotein cholesterol level together. Nearly 90% of women with diagnosed dyslipidemia had a reduced high density lipoprotein cholesterol, whereas total cholesterol and low density lipoprotein cholesterol levels were elevated in about one-third of the females. The proportion in these atherogenic lipids in men was about 30% for everyone. 7,6% of the patients had died within one year. The one year cardiovascular mortality was similar for men and women, also for the patients with acute or chronic coronary syndrome. Nearly 22% of died patients had an increased levels of total cholesterol and low density lipoprotein cholesterol. For more than 67% of them the decrease of high density lipoprotein cholesterol was evaluated. The majority of the patients (50,5%) with acute coronary syndrome and more than 80% with stable angina had a reduced high density lipoprotein cholesterol (Table 1).

	High LDL[1] n(%)	**Low HDL[2] n(%)**	**High TC[3], n(%)**	**Total, n(%)**	**one-year CV mortality, n(%)**
Total	588 (31,5)	1021 (54,7)	594 (31,8)	1865 (100)	
Medical history					
Acute coronary syndrome	371(35,3)	531 (50,5)	359 (34,2)	1050(56,3)	90(8,6)
Chronic coronary syndrome	217(36,5)	490(82,5)	235(39,6)	815(43,6)	52(8,7)
Gender					
Female	241(33,1)	654(89,8)	264(36,2)	728(39)	55(7,5)
Male	347(30,5)	367(32,2)	330(29)	1137(61)	87(7,6)
Age groups					
< 70 years	419(35)	585(48,8)	433(36,2)	1197(64,2)	70(49,3)
70-80 years	144(25,6)	368(65,6)	136(24,2)	561(30,1)	56(39,4)
> 80 years	25(24,7)	63(62,4)	24(23,8)	101(5,4)	15(10,6)
One year CV mortality	31(21,8)	96(67,6)	31(21,8)	142(7,6)	

LDL- low density lipoprotein cholesterol, HDL – high density lipoprotein cholesterol, CV – cardiovascular; [1]>3,0 mmol/l, [2]< 1,0 for males; < 1,2 mmol/l for females, [3]> 5,2 mmol/l

Table 1. Patients baseline characteristics.

For the 34-40% of the patients elevated low density lipoprotein cholesterol and total cholesterol were diagnosed (Figure 6,7). Patients distribution due to age shows Figure 8.

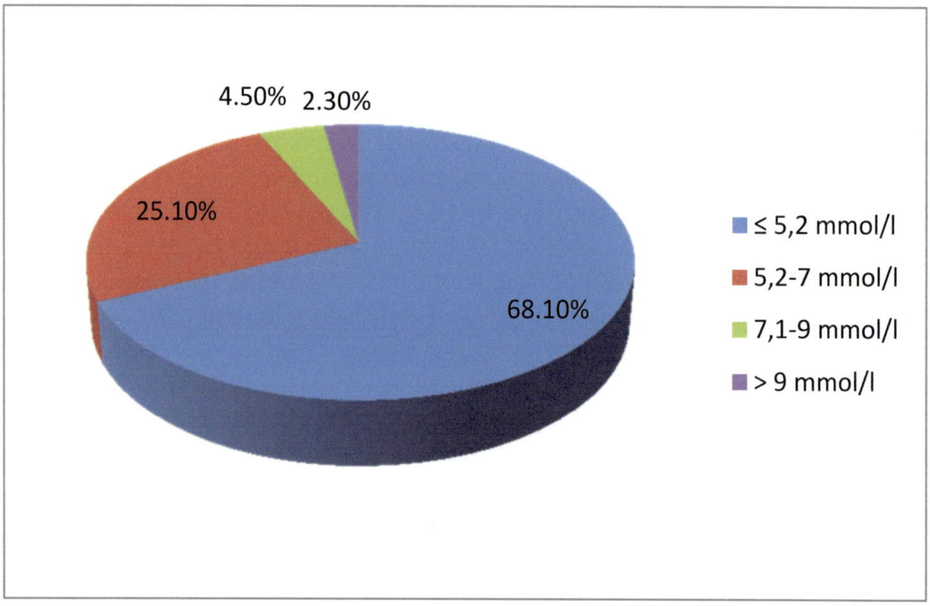

Figure 6. Part of the patients with different total cholesterol level.

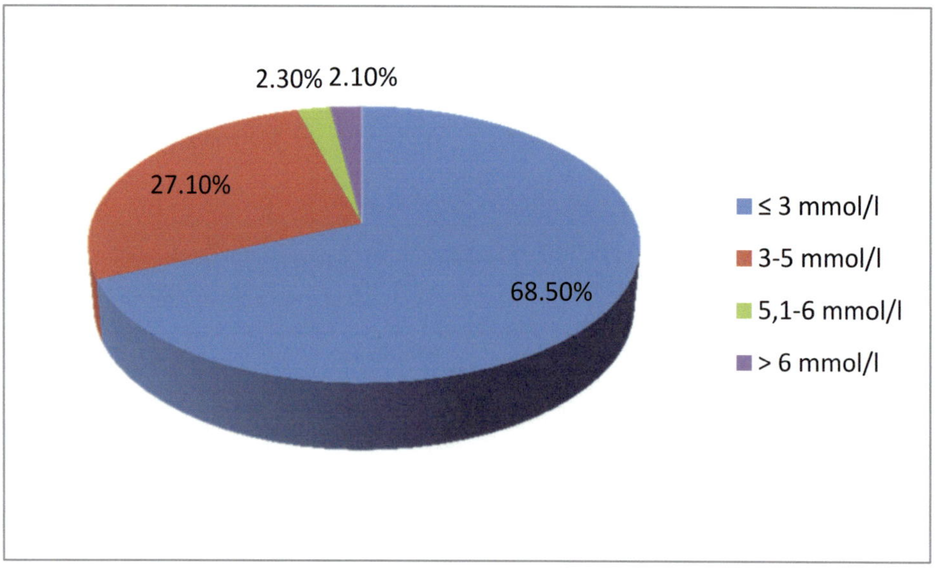

Figure 7. Part of the patients with different low density lipoprotein cholesterol level.

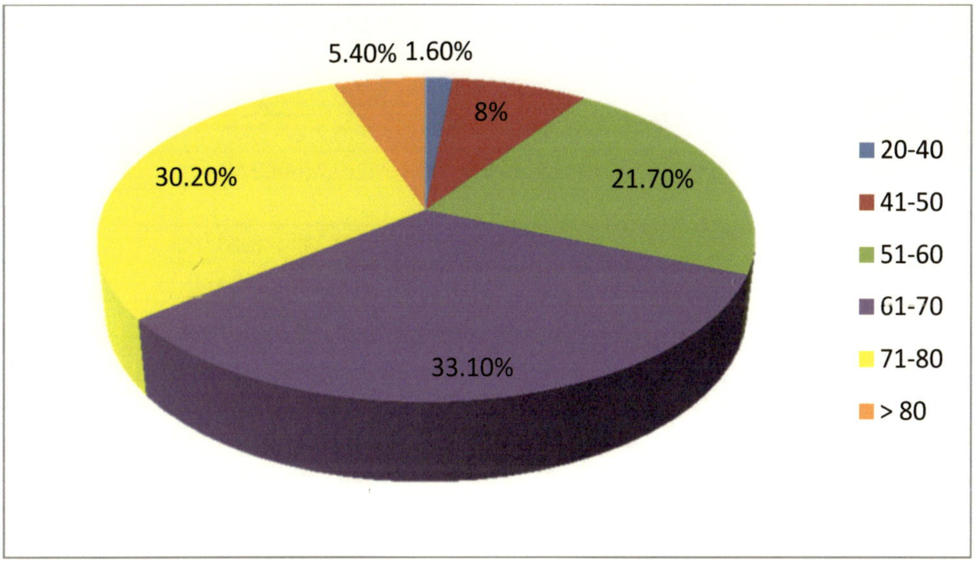

Figure 8. Patients distribution due to the age.

About 35-36% of the patients younger than 70 years, high total cholesterol and low density lipoprotein cholesterol levels were evaluated, the high density lipoprotein cholesterol have been decreased in nearly 49%. For the seniors (more than 70 years), the elevated total cholesterol and low density lipoprotein cholesterol were not so common (24-25% of the patients), but the reduced high density lipoprotein cholesterol was present more frequently (62-65%) (Table 1). Our data evaluated 1,8 times greater independent one year cardiovascular mortality risk for the patients with decreased high density lipoprotein cholesterol level (1,800, 95%CI 1,251-2,591, p=0,002). However, the assessment of the increased general values of the total cholesterol (more than 5,2 mmol/l) and low density lipoprotein cholesterol (more than 3,0 mmol/l) reduced mortality risk (0,575, 95%CI 0,382-0,868, p=0,008 and 0,585, 95%CI 0,388-0,882, p=0,01 respectively) (Table 2).

Risk factor	OR (95%CI)	p value
TC	0,575(0,382-0,868)	0,008
LDL	0,585 (0,388-0,882)	0,01
HDL	1,800 (1,251-2,591)	0,002

TC – total cholesterol, LDL- low density lipoprotein cholesterol, HDL – high density lipoprotein cholesterol

Table 2. Independent cardiovascular one year mortality rate.

Though, one year cardiovascular death risk elevates with the increase of these parameters (total cholesterol more than 9,0 mmol/l, low density lipoprotein cholesterol more than 6,0 mmol/l), although not significant (1,742, 95%CI 0,718-4,224, p=0,22 for total cholesterol more than 9,0 mmol/l and 1,167, 95%CI 0,408-3,339, p=0,773 for low density lipoprotein cholesterol more than 6,0 mmol/l). It is believed, the absence of the statistical significans due to small sampe size (Table 3).

Risk factor	OR (95%CI)	p value
TC		
5,2-7 mmol/l	0,465(0,286-0,759)	0,002
7,1-9 mmol/l	0,670(0,266-1,689)	0,396
> 9 mmol/l	1,742(0,718-4,224)	0,22
LDL		
3-5 mmol/l	0,524(0,333-0,825)	0,005
5,1-6 mmol/l	0,788(0,240-2,588)	0,694
> 6 mmol/l	1,167(0,408-3,339)	0,773
HDL		
< 1,0 for males; < 1,2 mmol/l for females	1,800(1,251-2,591)	0,002

TC – total cholesterol, LDL- low density lipoprotein cholesterol, HDL – high density lipoprotein cholesterol

Table 3. Independent cardiovascular one year mortality rate depending on lipoproteins level.

For the patients with acute coronary syndrome one year cardiovascular mortality rate insignificantly increases with the total cholesterol more than 9,0 mmol/l (2,578, 95%CI 0,931-7,136, p=0,068) and low density lipoprotein cholesterol more than 5 mmol/l (1,030, 95%CI 0,305-3,481, p=0,963 for the level of 5-6 mmol/l, and 2,023, 95%CI 0,668-6,130, p= 0,213 for the level more than 6,0 mmol/l). Simillary, high density lipoprotein cholesterol less than 1,0 mmol/l for men and less than 1,2 mmol/l for women increases one year cardiovascular mortality 1,4 times insignificantly (1,444, 95%CI 0,932-2,239, p=0,1) (Table 4).

Risk factor	OR (95%CI)	p value
TC		
>5,2 mmol/l	0,765(0,476-1,230)	0,269
5,2-7 mmol/l	0,633(0,364-1,102)	0,106
7,1-9 mmol/l	0,768(0,269-2,195)	0,623
> 9 mmol/l	2,578(0,931-7,136)	0,068
LDL		
> 3 mmol/l	0,724(0,451-1,164)	0,183
3-5 mmol/l	0,613(0,360-1,041)	0,07
5,1-6 mmol/l	1,030(0,305-3,481)	0,963
> 6 mmol/l	2,023(0,668-6,130)	0,213
HDL		
< 1,0 for males; < 1,2 mmol/l for females	1,444(0,932-2,239)	0,1

TC – total cholesterol, LDL- low density lipoprotein cholesterol, HDL – high density lipoprotein cholesterol

Table 4. Cardiovascular one year mortality rate for the patients with acute coronary syndrome depending on lipoproteins level.

For the patients with chronic coronary artery disease, only reduced high density lipoprotein cholesterol increased mortality risk, and this was great and significant (3,378, 95%CI 1,623-7,028, p= 0,001). It surprised the increase of the total cholesterol and low density lipoprotein cholesterol reduced one year cardiovascular mortality (almost in all groups significantly, p<0,05) (Table 5).

Risk factor	OR (95%CI)	p value
TC		
>5,2 mmol/l	0,247(0,097-0,628)	0,003
5,2-7 mmol/l	0,183(0,056-0,595)	0,005
7,1-9 mmol/l	0,420(0,056-3,160)	0,005
> 9 mmol/l	0,667(0,087-5,124)	0,697
LDL		
> 3 mmol/l	0,276(0,108-0,705)	0,007
3-5 mmol/l	0,320(0,125-0,818)	0,017
5,1-6 mmol/l	-	-
> 6 mmol/l	-	-
HDL		
< 1,0 for males; < 1,2 mmol/l for females	3,378(1,623-7,028)	0,001

TC – total cholesterol, LDL- low density lipoprotein cholesterol, HDL – high density lipoprotein cholesterol

Table 5. Cardiovascular one year mortality rate for the patients with chronic coronary syndrome depending on lipoproteins level.

Both in men and women with reduced high density lipoprotein cholesterol elevated mortality risk was evaluated (2,044, 95%CI 0,622-6,716, p= 0,239 for women, and 2,303, 95%CI 1,483-3,577, p< 0,001 for men respectively). For females high total cholesterol (more than 9,0 mmol/l) and low density lipoprotein cholesterol (more than 6,0 mmol/l) insignificantly increased one-year cardiovascular mortality risk. For men, relevant total cholesterol level was more than 7,0 mmol/l and low density lipoprotein cholesterol more than 5,0 mmol/l, insignificantly (Table 6,7).

Risk factor	OR (95%CI)	p value
TC		
>5,2 mmol/l	0,236(0,105-0,530)	0
5,2-7 mmol/l	0,123(0,038-0,4)	0
7,1-9 mmol/l	0,263(0,035-1,963)	0,193
> 9 mmol/l	2,0(0,550-7,239)	0,292
LDL		
> 3 mmol/l	0,372(0,179-0,773)	0,008
3-5 mmol/l	0,283(0,119-0,674)	0,004
5,1-6 mmol/l	0,533(0,07-4,082)	0,544
> 6 mmol/l	1,743(0,375-8,106)	0,479
HDL		
< 1,0 for males; < 1,2 mmol/l for females	2,044(0,622-6,716)	0,239

TC – total cholesterol, LDL- low density lipoprotein cholesterol, HDL – high density lipoprotein cholesterol

Table 6. Cardiovascular one year mortality rate for females depending on lipoproteins level.

Risk factor	OR (95%CI)	p value
TC		
>5,2 mmol/l	0,926(0,568-1,510)	0
5,2-7 mmol/l	0,844(0,484-1,470)	0,548
7,1-9 mmol/l	1,050(0,366-3,017)	0,928
> 9 mmol/l	1,540(0,450-5,272)	0,491
LDL		
> 3 mmol/l	0,755(0,458-1,246)	0,008
3-5 mmol/l	0,722(0,421-1,239)	0,238
5,1-6 mmol/l	1,014(0,233-1,239)	0,985
> 6 mmol/l	0,892(0,207-3,852)	0,879
HDL		
< 1,0 for males; < 1,2 mmol/l for females	2,303(1,483-3,577)	0

TC – total cholesterol, LDL- low density lipoprotein cholesterol, HDL – high density lipoprotein cholesterol

Table 7. Cardiovascular one year mortality rate for males depending on lipoproteins level.

It was noticed, that a decrease in high density lipoprotein cholesterol – is an important and reliable cardiovascular mortality risk factor in middle-aged patients (40-60 years). For the 41-50 years patients the mortality risk increases nearly 5 times when high density lipoprotein cholesterol level declines less than 1,0 mmol/l for men, and less than 1,2 mmol/l for women (4,985, 95%CI 1,230-20,196, p<0,05). In the 51-60 year group the risk of death increases 2,5 times with a similar levels of high density lipoprotein cholesterol significantly (2,572, 95%CI 1,094-6,106, p< 0,05) and with a total cholesterol more than 5,2 mmol/l insignificantly (1,073, 95%CI 0,462-2,495, p=0,87). A similar trend for the high density lipoprotein cholesterol was evaluated for the elderly patients, without significance (due to small sample size) (Table 8,9,10).

Years	OR (95%CI)	p value
20-40		
41-50	0,591(0,147-2,383)	0,46
51-60	0,610(0,238-1,564)	0,303
61-70	0,581(0,258-1,312)	0,191
71-80	0,771(0,395-1,504)	0,445
>80	0,421(0,088-2,012)	0,279

Table 8. Cardiovascular one year mortality rate depending on age groups for the patients with increased low density lipoprotein cholesterol level.

Years	OR (95%CI)	p value
20-40	1,545(0,087-27,358)	0,767
41-50	4,985(1,230-20,196)	0,024
51-60	2,572(1,094-6,106)	0,032
61-70	1,155(0,568-2,347)	0,691
71-80	1,227(0,675-2,232)	0,502
>80	2,745(0,721-10,445)	0,139

Table 9. Cardiovascular one year mortality rate depending on age groups for the patients with decreased high density lipoprotein cholesterol level.

Years	OR (95% CI)	p value
20-40	2,111(0,118-37,722)	0,611
41-50	0,265(0,054-1,293)	0,1
51-60	1,073(0,462-2,495)	0,87
61-70	0,485(0,207-1,137)	0,096
71-80	0,654(0,321-1,334)	0,243
>80	0,448(0,094-2,142)	0,314

Table 10. Cardiovascular one year mortality rate depending on age groups for the patients with increased total cholesterol level.

6. Discussion

In the last decade, lack of evidence on low density lipoprotein cholesterol and high density lipoprotein cholesterol in the pathogenesis of coronary heart disease have appeared. Mostly long-term outcomes were evaluated by the previous studies on lipoprotein disorder. We decided to estimate impact of the dyslipoproteinemia to the one year survival. It is proved by another studies, that patients with very low high density lipoprotein cholesterol have much higher risk of severe cardiovascular event or cardiovascular death comparing with patients with normal high density lipoprotein cholesterol level. Lower high density lipoprotein cholesterol values are associated with a higher great cardiovascular events risk and a greater burden of atherosclerosis, even among the patients with reduced low density lipoprotein cholesterol level [33,40,41]. In another side, very low low density lipoprotein cholesterol level is a significant prognostic factor, improved survival for the patient with acute coronary syndrome and may be a target for the treatment. In this study first and foremost we found that reduced high density lipoprotein cholesterol are highly prevalent in a large cohort of the patients with coronary artery disease and tend to be associated with a significantly higher cardiovascular mortality risk. More than a half of the patients in our study had decreased high density lipoprotein cholesterol, and therefore the higher cardiovascular events and mortality risk, especially for the patients with stable angina. These data are similar to another studies [40]. Results from another studies showed that the prevalence of the elevated low density lipoprotein cholesterol increases with age [17]. By data from our study it is not only the problem for the eldery patients. The prevalence of the impaired low density lipoprotein cholesterol by the gender was similar both for men and women and it was high for the patients with established coronary artery disease, taking notice that elevated low density lipoprotein cholesterol can be managed and controlled successfully with lifestyle changes, medications or a combination both of them. We have found that decreased high density lipoprotein cholesterol level is a significant independent risk factor for cardiovascular one year mortality. Interestingly, in another similar studies reduced high density lipoprotein cholesterol more often were found in young men. In our study 90% of females with coronary artery disease had a decreased high density lipoprotein cholesterol level. Also, insufficient high density lipoprotein cholesterol level more often have been found in eldery people. Although high density lipoprotein cholesterol less than 1,3 mmol/l for women has been widely considered as a cardiovascular risk factor, in the present study we selected a cutoff point of less than 1,2 mmol/l as a lowest high density lipoprotein cholesterol value that allowed us to identify those females at risk of cardiovascular one year mortality. It have been evaluated that about 20% of participants of our study had reduced high density lipoprotein cholesterol with elevated low density lipoprotein cholesterol level together. So, it is let to suspect, that one year cardiovascular mortality risk for them have to be much higher. There are a lot of evidence, that decreased high density lipoprotein cholesterol significantly increases cardiovascular mortality risk in stable patients. Also, there are some studies, showed that reduced high density lipoprotein cholesterol is associated with a higher risk of adverse outcomes [40]. Some reports on lipoproteins did not evaluated cardiovascular mortality due to acute or chronic ischaemic syndrome. Comparing acute coronary syndrome and chronic coronary artery disease patients we have been evaluated the more important role of total cholesterol and low density lipoprotein cholesterol on cardiovascular one year mortality for acute patients, though not significant. In contrast, high density lipoprotein cholesterol was strong independent risk factor both for acute (not significant) and chronic patients. Suprisingly, total cholesterol more than 5,2 mmol/l and low density lipoprotein cholesterol more than 3,0 mmol/l reduced one year mortality risk both for acute and chronic patients significantly. Additionally, the previous studies showed the increased mortality rate due to elevated low density lipoprotein cholsterol, have not comprehensively evaluated the impact of different low density lipoprotein cholesterol and high density lipoprotein cholesterol lipoproteins levels on cardiovascular mortality. Lehto and al. evaluated, that among 35-64 years females with acute myocardial infarction total cholesterol more that 8 mmol/l significantly increases reccurence cardiovascular disease risk [42]. It was a reason to search an impact of different levels of low density lipoprotein cholesterol and total cholesterol on cardiovascular mortality risk for men and women. Our hypothesis was confirmed, as it became clear, that one year cardiovascular mortality risk sharply rises when signally increased total cholesterol more than 9 mmol/l and low density lipoprotein cholesterol more than 6 mmol/l, especially in women. The future major research need to evaluate a different lipoprotein and total cholesterol levels impact in cardiovascular mortality, not only in short term, but in long-term outcomes as well. It seems, the highest levels of lipids, that could be attributed to hereditary dyslipoproteinemia, may be very important predicting cardiovascular mortality rates and reducing a cardiovascular death risk. As it was find earlier, high density lipoprotein cholesterol more predictive for middle-aged men. Similarly, our study evaluated the more important role of decreased high density lipoprotein cholesterol, especially for 51-60 years men with the chronic coronary artery disease for one year cardiovascular mortality.

7. Conclusion

Lipoproteins disorder is the main factor for development of the atherosclerosis and predicts cardiovascular mortality. The most important findings from our data concerns the inverse relationship between the high density lipoprotein cholesterol and cardiovascular mortality rates. This association is characterized by a high degree of generality and strength.

Author details

Jelena Umbrasiene*
Medical Academy, Lithuanian University of Health Sciences, Kaunas, Lithuania,
Kaunas Region Cardiology Society, Kaunas, Lithuania,

Ruta-Marija Babarskiene
Kaunas Region Cardiology Society, Kaunas, Lithuania
Department of Cardiology, Medical Academy, Lithuanian University of Health Sciences, Kaunas, Lithuania

Jone Vencloviene
Institute of Cardiology, Medical Academy, Lithuanian University of Health Sciences, Kaunas, Lithuania,
Vytautas Magnus University, Kaunas, Lithuania

Acknowledgement

We would like to show our gratitude to the company Berlin-Chemie Menarini Group, whose financial encouragement enabled us to publish this paper. The Company had no interest in this research.

8. References

[1] Reiner Z, Catapano AL, Backer GD, Graham I, Taskinen MR, Wiklund O et al. ESC/EAS Guidelines for the management of dyslipidaemias. The Task Force for the management of dyslipidaemias of the European Society of Cardiology (ESC) and the European Atherosclerosis Society (EAS). European Heart Journal 2011;32,1769–1818.
[2] European Heart Health Charter. Available: http://www.heartcharter.eu

* Corresponding Author

[3] Fodor JG, Frohlich JJ, Genest JJG, McPherson PR. Recommendations for the management and treatment of dyslipidemia. Report of the Working group on Hypercholesterolemia and Other Dyslipidemias. CMAJ 2000;162(10):1441-7.
[4] Tierney LM, McPhee SJ, Papdakis MA. Current medical diagnosis and treatment. 43 edition. Lange medical books/McGraw-Hill. 2004, p. 326-327.
[5] OECD iLibrary. Health at a glance: Europe 2010. Available: http://ec.europa.eu/health/reports/docs/health_glance_en.pdf
[6] Statins for the prevention of cardiovascular events. National Institute for Health and Clinical Excellence (NICE). January 2006. Available: http://www.nice.org.uk/nicemedia/pdf/TA094guidance.pdf
[7] National Health Council Annual Report 2010. Available: http://www3.lrs.lt/pls/inter/w5_show?p_r=697&p_k=1
[8] European health for all database (HFA-DB). Reviewed 2012-04-09. Available: http://data.euro.who.int/hfadb/
[9] Lithuanian Institute of Hygiene database. Reviewed 2012-04-09. Available: http://www.hi.lt/news/391.html
[10] Grundy SM, Cleeman JI, Merz CN, Brewer HB, Clark LT, Hunninghake DB et al. Implications of recent clinical trials for the National Cholesterol Education Program Adult Treatment Panel III guidelines. Circulation. 2004;110(2):227-239.
[11] Cannon CP, Braunwald E, McCabe CH, Rader DJ, Rouleau JL, Belder R et al. Intensive versus moderate lipid lowering with statins after acute coronary syndromes. N Engl J Med. 2004;350(15):1495-1504.
[12] Braunwald E, Douglas P. Braunwald Heart Disease. A textbook of cardiovascular medicine. The 6th edition. 2005 p.921-939.
[13] Anitschkow N. Über die Veranderungen der Kaninchenaorta bei experimenteller Cholesterinsteatose. Beitr Pathol Anat 1913;56:379-404.
[14] Kumar P, Clark M. Clinical medicine. Fifth edition. Saunders, 2002 p.1104-1107.
[15] Stamler J, Neaton JD. The Multiple Risk Factor Intervention Trial (MRFIT)—importance then and now (reprinted). JAMA 2008 (300);11:1343-1345.
[16] European guidelines on cardiovascular disease prevention in clinical practice: executive summary. Eur Heart J 2007;28:2375-2414.
[17] Kuklina EV, Shaw KM, Hong Y. Vital Signs: Prevalence, Treatment, and Control of High Levels of Low-Density Lipoprotein Cholesterol — United States, 1999–2002 and 2005–2008. Morbidity & Mortality Weekly Report. 2011;60(4):109-114.
[18] Cholesterol Treatment Trialists' (CTT) Collaboration. Efficacy and safety of more intensive lowereing of LDL-cholesterol: a meta-analysis of data from 170,000 participants in 26 randomised trials. Lancet 2010; 376:1670–81.
[19] Ballantyne C, ed. Clinical lipidology. A Companion to Braunwald's Heart Disease. Philadelphia,PA. Saunders 2009, p.56 - 129.
[20] LaRosa JC, Grundy SM, Waters DD, et al. Inten-sive lipid lowering with atorvastatin in patients with stable coronary disease. N Engl J Med. 2005;352 (14):1425-1435.
[21] Singh IM, Shishehbor MH, Ansell BJ. High-density lipoprotein as a therapeutic target. JAMA 2007;298(7):786-798.

[22] Werner N., Bohm M. Importance of high density lipoprotein cholesterol in atherosclerotic disease. www.escardio.org. E-journal. Vol.6, N.18. Available: http://www.escardio.org/communities/councils/ccp/e-journal/volume6/Pages/vol6n18.aspx
[23] Libby P. Inflammation in atherosclerosis. Nature. 2002;420:868-874.
[24] Schachinger V, Britten MB, Zeiher AM. Prognostic impact of coronary vasodilator dysfunction on adverse long-term outcome of coronary heart disease. Circulation. 2000;101:1899-1906.
[25] Beers MH, Berkow R. The Merck manual of diagnosis and therapy. The seventeenth edition. 1999, p. 201-203.
[26] Castelli WP, Garrison RJ, Wilson PW, Abbott RD,Kalousdian S, KannelWB. Incidence of coronary heart disease and lipoprotein cholesterol levels: the Framing-ham Study. JAMA. 1986;256(20):2835-2838.
[27] Sharrett AR, Ballantyne CM, Coady SA, et al. Coro-nary heart disease prediction from lipoprotein cho-lesterol levels, triglycerides, lipoprotein(a), apolipo-proteins A-I and B, and HDL density subfractions: the Atherosclerosis Risk in Communities (ARIC) Study. Circulation. 2001;104(10):1108-1113
[28] Thompson MM, Reed SC, Cockerill GW. Therapeutic approaches to raising plasma HDL-cholesterol levels. Nat Clin Pract Cardiovasc Med. 2004;1:84-89.
[29] Sumi M, Sata M, Miura S, Rye KA, Toya N, Kanaoka Y, Yanaga K, Ohki T, Saku K, Nagai R. Reconstituted high-density lipoprotein stimulates differentiation of endothelial progenitor cells and enhances ischemia-induced angiogenesis. Arterioscler Thromb Vasc Biol. 2007;27:813-818.
[30] Noor R, Shuaib U, Wang CX, Todd K, Ghani U, Schwindt B, Shuaib A. High-density lipoprotein cholesterol regulates endothelial progenitor cells by increasing eNOS and preventing apoptosis. Atherosclerosis. 2007;192:92-99.
[31] Joint British Recommendations on prevention of coronary heart disease in practice: summary. BMJ 2000;320:705-708.
[32] Rosengren A, Eriksson H, Larsson B, Svärdsudd K, Tibblin G, Welin L et al. Secular changes in cardiovascular risk factors over 30 years in Swedish men aged 50: the study of men born in 1913, 1923, 1933 and 1943. J Int Med 2000;247:111-118.
[33] Barter P, Gotto AM, LaRosa JC, Maroni J, Szarek M, Grundy SM, Kastelein JJ, Bittner V, Fruchart JC. HDL cholesterol, very low levels of LDL cholesterol, and cardiovascular events. N Engl J Med. 2007;357:1301-1310.
[34] Wu DM, Chu NF, Sung PK, Sung PK, Lee MS, Tsai JT, et al. Prevalence and clustering of cardiovascular risk factors among healthy adults in a Chinese population: the MJ Health Screening Center Study in Taiwan. Int J Obes 2001;25:1189-1195.
[35] Bakx JC, Van den Hoogen HJM, Deurenberg P, van Doremalen J, van den Bosch WJ. Changes in serum cholesterol levels over 18 years in a cohort of men and women: the Nijmegen cohort study. Prev Med 2000;30:138-145.
[36] Clinical reality of coronary prevention guidelines: a comparison of EUROASPIRE I and II in nine countries. EUROASPIRE I and II Group. European Action on Secondary Prevention by Intervention to Reduce Events. Lancet 2001;35:995-1001.

[37] Kotseva K, Wood D, De Backer G, De Bacquer D, Pyörälä K, Keil U et al. EUROASPIRE III: a survey on the lifestyle, risk factors and use of cardioprotective drug therapies in coronary patients from 22 European countries. Eur J Cardiovasc Prev Rehabil 2009;16(2):121-37.

[38] MONICA monograph: World's largest study of heart disease, stroke, risk factors, and population trends 1979-2002. Tundstall-Pendoe H edition. WHO, Geneva, 2003.

[39] Davey-Smith G, Song F, Sheldon TA. Cholesterol lowering and mortality: the importance of considering initial level of risk. Br Med J 1993;306:1367-1373.

[40] Roe MT,Ou FS, Alexander KP, Newby LK, Foody JM, Gibler WB, Boden WE et al. Patterns and prognostic implications of low high-density lipoprotein levels in patients with non-ST-segment elevation Acute coronary syndromes. Eur Heart J 2008;29(20):2480-2488.

[41] deGoma EM, Leeper NJ, Heidenreich PA. Clinical significance of high-density lipoprotein cholesterol in patients with low low-density lipoprotein cholesterol. J Am Coll Cardiol 2008;51:49-55.

[42] Lehto S, Palomaki P, Miettinen H et al. Serum cholesterol and high density lipoprotein cholesterol distribution in patients with acute myocardial infarction and in the general population of Kuopio province, eastern Finland. J Intern Med 1993;233(2):179-185.

Chapter 8

Lipoproteins and Cardiovascular Diseases

Adebowale Saba and Olayinka Oridupa

Additional information is available at the end of the chapter

http://dx.doi.org/10.5772/48132

1. Introduction

1.1. What are lipids?

Lipids consists of a broad group of naturally occurring molecules that include fats, waxes, sterols including cholesterol, fat-soluble vitamins (such as vitamins A, D, E, and K), monoglycerides, diglycerides, triglycerides, phospholipids, and others. Lipids were previously known as sources of energy storage and the building blocks for cell membrane. Lipids are now known to play several key roles in intracellular signalling, membrane trafficking, hormonal regulation, blood clotting (Muller-Roeber and Pical, 2002; Vance and Vance, 2002; Fahy *et al.*, 2009). All lipids may be defined as hydrophobic or amphiphilic small molecules. The amphiphilic nature of some lipids allows them to form structures such as vesicles, liposomes, or membranes in an aqueous environment. Biological lipids originate entirely or in part from two distinct types of biochemical subunits, which are ketoacyl and isoprene groups (Fahy *et al.*, 2009).

Lipids typically do not travel alone in the blood. Instead, it binds to a protein that transports it to its destination in the body. The complex formed by the binding of lipid to protein i.e. lipoprotein, makes lipids water soluble, which enables its transportation in blood. The lipoprotein particle is composed of an outer shell of phospholipids, which renders the particle soluble in water; a core of fats called lipid, including cholesterol and a surface apoprotein (apolipoprotein). Ideally, the lipoprotein aggregates should be described in terms of the different protein components (apolipoprotein) because this determines the overall structures and metabolism of the lipoprotein, and the interactions with receptor molecules in liver and peripheral tissues. The apolipoprotein molecule enables tissues to recognize and take up the lipoprotein particle. However, lipoproteins are classified based on their characteristic density on ultracentrifugation, which has been used to segregate the different lipoprotein classes. Lipoproteins are broadly classified as high density lipoprotein (HDL), low density lipoprotein (LDL), intermediate density lipoprotein (IDL), very low density lipoprotein (VLDL) and chylomicrons (CM). Each of these particles perform different functions and can be detrimental (VLDL, IDL, LDL) or beneficial (HDL) to the cardiovascular system.

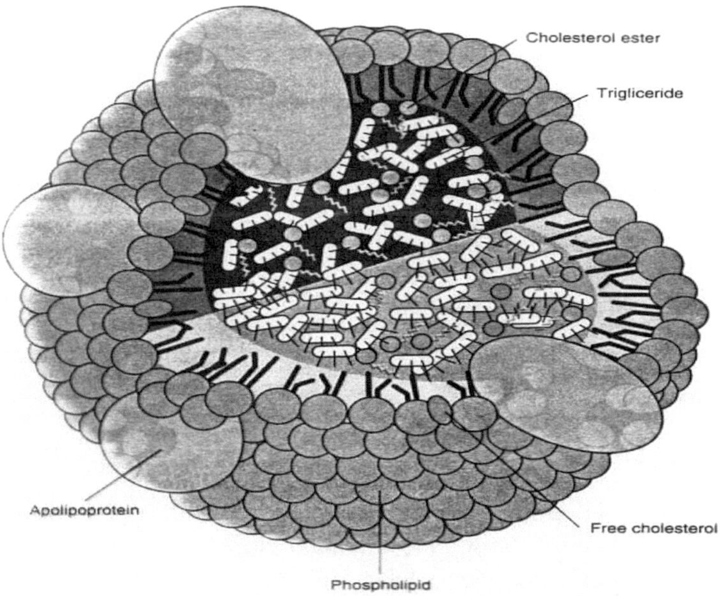

Figure 1. Structure of Lipoprotein available from http://www.campbell.edu

	CM	VLDL	IDL	HDL
Density (g/ml)	< 0.94	0.94 – 1.006	1.006 – 1.063	1.063 -1.210
Diameter (Á)	6000 - 2000	600	250	70-120
Total lipid (wt %) *	99	91	80	44
Triacylglycerols	85	55	10	6
Cholesterol esters	3	18	50	40
Cholesterol	2	7	11	7
Phospholipids	8	20	29	46

*Most of the remaining materials comprise the various apolipoproteins

Table 1. Physical properties and lipid compositions of lipoprotein classes

1.2. Role of cholesterol in membrane dynamics

It is relevant to establish the important of cholesterol in the body to be able to relate the various metabolic events associated with cholesterol and its homeostasis. Mammalian cell membranes contain varying proportions of cholesterol depending on organelle and cell type. These levels are tightly controlled by lipid transfer, through both vesicular and protein-bound pathways. With its rigid sterol backbone, cholesterol preferentially locates among saturated membrane lipids that have straight, elongated hydrocarbon chains rather than among kinked, unsaturated species. The presence of cholesterol in the membrane increases lateral ordering of lipids, reducing permeability and fluidity and potentially restricting diffusion of membrane proteins. Its distribution is not uniform within a membrane: regions of high cholesterol and corresponding low fluidity are termed lipid rafts. These areas act as platforms for the assembly of signalling complexes within the membrane and have been implicated in the development of numerous disease processes, notably arteriosclerosis and cancer (Di Vizio *et al.*, 2008; Ikonen, 2008).

2. Cardiovascular disease and risk factors

Elevated plasma levels of low density lipoprotein (LDL) and low levels of high density lipoprotein (HDL) poses a major risk of development of cardiovascular diseases (Grundy *et al.*, 1999). A dietary intake of saturated fat and a sedentary lifestyle has been associated with about 31% of coronary heart disease and 11% of stroke in humans. According to the Framingham Heart Study and other studies (Wilson *et al.*, 1998), the major and independent risk factors for coronary heart disease (CHD) are cigarette smoking of any amount, elevated blood pressure, elevated serum total cholesterol and low-density lipoprotein cholesterol (LDL-C), low serum high-density lipoprotein cholesterol (HDL-C), diabetes mellitus, and advancing age. More recently, a review by Patrick and Uzick (2001) documented new risk factors for CHD which included levels of circulating homocysteine, fibrinogen, C-reactive protein (CRP), endogenous tissue plasminogen-activator, plasminogen-activator inhibitor type I, lipoprotein(a), factor VII and certain infections such as *Chlamydia pneumonia*. These studies showed that the total risk of an individual is the summation of all major risk factors.

Other factors contributing to the total risk for CHD are categorized as conditional risk factors and predisposing risk factors. The conditional risk factors are associated with increased risk for CHD, although their causative, independent, and quantitative contributions to CHD have not been well documented. The predisposing risk factors are those that worsen the independent risk factors. Two of these risk factors; obesity and physical inactivity, are designated major risk factors by the American Heart Association (AHA) (Fletcher *et al.*, 1996; Eckel, 1997). The adverse effects of obesity are worsened when it is expressed as abdominal obesity, an indicator of insulin resistance. These risk factors apply before clinical manifestation of coronary atherosclerotic diseases. The clinical significance of these risk assessment is to identify high-risk patients who require attention, motivate patients to adhere to risk-reduction therapies and modify the intensity of risk reduction effort required in potential patients (Grundy *et al.*, 1999).

2.1. Lipoproteins, cholesterol and atherosclerosis

Cholesterol is a building block of the outer layer of cell membranes. Cholesterol is a waxy steroid of fat that is produced in the liver or intestines. It is used to produce hormones and cell membranes and is transported in the blood plasma of all mammals (Leah, 2009). As an essential structural component of mammalian cell membranes, it is required to establish proper membrane permeability and fluidity. In addition, cholesterol is an important component for the manufacture of bile acids, steroid hormones, and vitamin D. Cholesterol is the principal sterol synthesized by animals; however, small quantities can be synthesized in other eukaryotes such as plants and fungi. It is almost completely absent among prokaryotes including bacteria (Pearson *et al.*, 2003).

Owing to its limited solubility in water, cholesterol is transported in blood in lipoproteins. The lipoprotein outer layer is formed of amphiphilic cholesterol and phospholipid molecules, studded with proteins, surrounding a hydrophobic core of triglycerides and cholesterol esters. Lipoproteins are specifically targeted to cells by distinct apolipoproteins on their surface that bind to specific receptors. Low density lipoprotein (LDL) contains the highest level of cholesterol. LDL receptors in peripheral tissues bind LDL, triggering its endocytosis, lysosomal targeting and hydrolysis. When cells have abundant cholesterol, LDL receptor synthesis is inhibited by the sterol regulatory element binding proteins (SREBP) pathway (Wang *et al.*, 1993; Yokoyama *et al.*, 1993; Brown and Goldstein, 2009).

The biosynthesis of cholesterol is intensely regulated in the body with negative feedback of plasma cholesterol levels. The molecular basis of this regulation was set out by Michael Brown and Joseph Goldstein, earning them the Nobel Prize in Physiology and Medicine in 1985 (Leah, 2009). A key irreversible step of cholesterol synthesis is catalyzed by HMG-CoA reductase. Transcription of the HMG gene is controlled by SREBPs, transcription factors that bind sterol regulatory elements. SREBPs are only able to enter the nucleus when cholesterol levels fall. At other times they are tied up in a complex that includes Scap (SREBP-cleavage activating protein), an escort protein with a cholesterol-binding motif that senses cellular cholesterol levels. The SREBP pathway is now implicated in multiple regulatory aspects of lipid formation and metabolism (Brown and Goldstein, 2009).

2.2. Atherogenicity of lipoprotein sub-fractions

The first stages of cholesterol build up in the blood vessels (atherosclerosis) occur when LDL particles circulating in the blood penetrate through the inner lining of blood vessels and become trapped in the artery wall. The normal function of LDL is to deliver cholesterol to cells, where it is used in membranes or for the synthesis of steroid hormones. Cells take up cholesterol by receptor-mediated endocytosis. LDL binds to a specific LDL receptor and is internalized in an endocytic vesicle. Receptors are recycled to the cell surface, while hydrolysis in an endolysosome releases cholesterol for use in the cell. The liver removes LDL and other lipoproteins from the circulation by receptor-mediated endocytosis.

Deregulation of cholesterol levels results in the existence of more LDL in the blood than can be taken up by LDL receptors. Excess LDL is oxidized and taken up by macrophages, forming foam cells that can become trapped in the walls of blood vessels along with cells of inflammation (Zioncheck *et al.*, 1991; Young and McEneny, 2001). Fatty streaks, consisting of subendothelial collection of foam cells are initially formed in blood vessels. Small, dense LDL particles are more atherogenic than large, buoyant LDL particles, and oxidation of LDL also increases its atherogenicity. In addition, LDL belongs to the group of lipoproteins that contain apolipoprotein (apo) B-100. Some of the particles in this highly heterogeneous group contain other apolipoproteins, such as apo C-II, apo C-III, and apo E. Furthermore, some particles are larger and rich in triglycerides (large VLDL), whereas others are smaller and rich in cholesteryl esters (small VLDL, IDL). It is now known that remnant lipoproteins containing apo C-III are highly atherogenic and may be more specific measures of coronary heart disease (CHD) risk assessment than plasma triglycerides (Carmena *et al.*, 2004).

The end result is the formation of an atherosclerotic plaque which occludes the endothelial lumen and impedes blood flow, leading to myocardial infarction, the major cause of heart attacks and strokes. Although LDL levels correlate with heart attack risk, high density lipoprotein (HDL) has an inverse ratio of risk because HDL particles transport cholesterol to the liver for excretion. Modern cholesterol tests distinguish the LDL/HDL ratio as well as the overall level (Barter *et al.*, 2007). Other sub-fractions of lipoproteins such as chylomicrons, IDL and VLDL may enter the endothelial spaces due to their sizes, thus contribute substantially to development of atherosclerotic plaques. They may also increase prothrombotic factors, triggering cardiovascular diseases (Brunzell *et al.*, 2008).

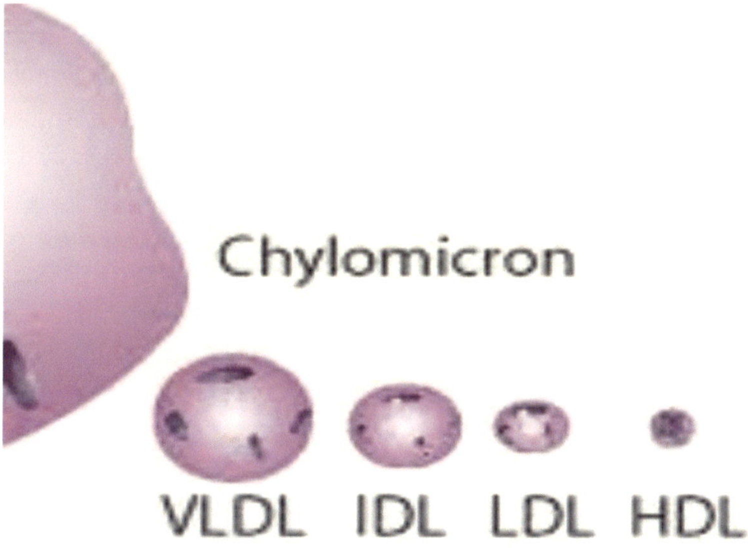

Figure 2. Sizes of Lipoproteins available at http://www.sigmaaldrich.com/european-export.html

Triglyceride-rich lipoproteins comprise a great variety of nascent and metabolically modified lipoprotein particles differing in size, density, and lipid and apolipoprotein composition. Studies have shown an inverse relationship between the size of lipoproteins and their ability to cross the endothelial barrier to enter the arterial intima. Chylomicrons and large VLDLs are probably not capable of entering the arterial wall. On the other hand, small VLDL and IDL can enter the arterial intima. Therefore, certain triglyceride rich lipoproteins are atherogenic, whereas others are not. A large body of evidence suggests that small VLDLs and IDLs are independently associated with atherosclerosis (Carmena *et al.*, 2004).

2.3. Apolipoproteins in lipoproteins

Apolipoproteins are the carrier proteins for lipoproteins and they consist of a single polypeptide chain often with relatively little tertiary structure. They are required to solubilise the non-polar lipids in the circulation and in some instances to recognise specific receptors. They are classified as Apo A1, A2, A4, A5, B48, B100, C1, C2, C3, D, E, H, J, L, M and Apo (a). Most apolipoproteins are synthesised by the liver and intestine.

Apolipoprotein	Molecular weight	Lipoprotein	Function
Apo AI	29,100	HDL	Lecithin: cholesterol acyltransferase (LCAT) activation. Main structural protein. Binds ABCA1 on macrophages
Apo AII	17,400	HDL	Enhances hepatic lipase activity
Apo AIII	46,000	CM	
Apo AIV	46,000	HDL, CM	Inhibits food intake in CNS
Apo AV	39,000	HDL	Enhances triacylglycerol uptake
Apo B48	241,000	CM	Derived from Apo B100 gene by RNA editing, lacks the LDL receptor binding site
Apo B100	512,000	LDL, IDL, VLDL	Binds to LDL receptor
Apo CI	7,600	VLDL, CM	Activates LCAT
Apo CII	8,900	VLDL, CM	Activates lipoprotein lipase
Apo CIII	8,750	VLDL, CM	Inhibits lipoprotein lipase
Apo D	33,000	HDL	Closely associated with LCAT, progesterone binding
Apo E	34,000	HDL	At least 3 forms. Binds to LDL receptor
Apo(a)	300,000 – 800,000	LDL, Lp(a)	Linked by disulfide bonds to apo B100 and similar to plasminogen, associated with premature coronary artery disease and stroke
Apo H	50,000	Chylomicrons	Involved with triacylglycerol metabolism
Apo M		HDL	Transports sphingosine-1-phosphate

Table 2. Classes of apolipoproteins, their molecular weight and functions

2.3.1. *Apolipoprotein A*

Apo A are subdivided into apo AI, AII, AIII, AIV and AV. Apo AI and AII are the major apolipoproteins in the HDL particle. HDL is primarily saddled with the responsibility of removing excess cholesterol from peripheral tissues and delivering it to the liver for excretion in bile as bile acids in a process known as reverse cholesterol transport. Apo AI is synthesised mainly by the liver and also by the intestine. The protein consists of 243 amino acids arranged as eight α-helical segments of 22 amino acids which have 11 –mer repeats, no disulfide bonds or glycosylations. The helices are believed to be amphipathic in nature with both hydrophobic and hydrophilic faces. This nature enhances its interaction with the lipid and aqueous phases. Apo AII on the other hand is found as a twin chain of 77 amino acids each, linked by disulfide bonds. It enhances the activity of hepatic lipase, thus increasing lipoprotein metabolism. Apo III is found in chylomicrons.

2.3.2. *Apolipoprotein E*

Three isoforms of this apolipoprotein exist and they are all synthesised mainly by the liver and also by several tissues such as arterial wall, brain and adipose tissue. They are important for homeostasis of lipid and lipoproteins in blood circulation as well as their metabolism in these tissues. Apo E is required for the clearance of VLDL remnant (IDL) from circulation in the liver. Other suggestions on its involvement with immune response and inflammation have also been put forward.

2.3.3. *Apolipoprotein B*

Two types of apolipoprotein B are synthesised from the intestine and liver as apo B100 which has the full length of 4536 amino acid residues and a truncated form with 48% of the full length known as apo B48. These proteins which are synthesised on ribosomes, an organelle located on the surface of rough endoplasmic reticulum are translocated through the reticular membrane into the lumen of endoplasmic reticulum. Assembly of VLDL occurs here by accretion of lipids to the core of apo B particle. This occurs in three distinct stages as the apo B grows bigger, forming the pre-VLDL to VLDL 2 which eventually grows to become the triacylglycerol-rich VLDL 1 or chylomicrons, the most energy dense substances in the body. VLDL 2 is a triacylglycerol-poor version of VLDL and its assembly occurs in golgi bodies. This is transported to basolateral membrane of the intestinal cells where final assembly of VLDL 1 or chylomicrons occur and these are secreted into the lamina propria of the intestinal cells by reverse exocytosis.

Apo B100 and B48 are large and water-insoluble, and are the only non-exchangeable apolipoproteins. They are major components of chylomicrons and VLDL, and usually remain with their lipid aggregates throughout their passage in plasma and several metabolic changes occurring during their circulation in plasma. Chylomicrons are usually transported through the intestinal lymphatic system and flow into blood circulation via the left subclavian vein. As apo B carries the VLDL or chylomicrons through the blood stream, their triacylglycerol content are removed by peripheral tissues via enzymatic activity of lipoprotein lipase, located in the endothelial wall. This makes free fatty acids available for energy production in muscle and some are stored in adipose tissue. Apo B48 remains with the remnant of the lipoprotein particle, along with dietary cholesterol and apo E. the lipoprotein remnant is eventually cleared from circulation in the liver by an apo E dependent receptor-mediated reaction.

ApoB-100 is one of the largest monomeric proteins known and it is the major apolipoprotein component common to the atherogenic lipoproteins [VLDL, LDL, IDL and Lp(a)] (Boerwinkle *et al.*, 1992; Carmena *et al.*, 2004). Apo B100 differs from apo B48 by the presence of LDL receptor site on the apo B100 molecule. Apo B-100 is encoded for by the Apo B gene and mutations in this gene cause familial hypercholesterolemia, an autosomal hereditary metabolic disorder. The level of apo B-100 is a stronger predictor of risk than LDL in humans. Apo B-100 is speculated to mediate delivery of the cholesterol content of lipoproteins to cells via an unknown mechanism. It is well established that Apo B-100 is associated with atherogenic lipoproteins, thus the number of Apo B-100 can be used to determine the risk of atherosclerosis and CHD in individuals. Apo B-100/apo A-I ratio are strongly and positively related to increased risk of fatal myocardial infarction. Apolipoprotein A-I (apo A-I) is the major apolipoprotein in the HDL structure. The Apo B-100/apo A-I ratio is especially valuable in patients with normal or low LDL concentrations, a frequent observation in type 2 diabetes mellitus which may present with hypertriglyceridemia and hyper-apo B concentrations (Carmena *et al.*, 2004).

2.3.4. Apolipoprotein C

Apolipoprotein C is subdivided into three and each has its own distinct function. Apo CI is involved with activation of Lecithin: cholesterol acyltransferase (LCAT) along with apo AI. This enzyme converts free cholesterol into cholesterol ester, which enhances the incorporation of cholesterol into the lipid core of a lipoprotein particle, particularly in assembly of HDL particle. The enzyme is mainly bound to HDL and LDL in plasma. Apo CII and CIII have antagonistic activity to each other which are required in regulation of lipoprotein lipase activity. Apo CII is required for activation of lipoprotein lipase, while CIII inhibits lipoprotein lipase activity. Apo CII is believed to open a lid-like region of the enzyme which allows the active site to hydrolyse the fatty acid ester bonds of triacylglycerols. In addition to inhibition of lipoprotein lipase, apo CIII also inhibits the binding of lipoproteins to receptors at the cell surface, thereby decreasing hydrolysis of triacylglycerols. High levels of apo CIII have been associated with elevated serum levels of triacylglycerols (hypertriglyceridemia).

2.3.5. Transfer of apolipoproteins in lipoprotein homeostasis

Lipids enter blood circulation bound to apolipoproteins as chylomicrons or VLDL which as secreted into the blood stream from the intestines. Chylomicrons or VLDL consist mainly of apo B100 and B48, but also consist of some apo AI, along with other apolipoprotein which will be discussed. These lipoproteins carry triacylglycerol-rich cholesterols to the peripheral tissues to provide sources of energy and for storage, while HDL carries excess cholesterol from peripheral tissue to the liver for excretion in bile acids. Immediately chylomicrons enter blood circulation, an exchange of apolipoproteins occurs between chylomicrons and HDL. The apo AI content of chylomicrons is exchanged for the apo C and E content of HDL. Apo C content is required for activation and inhibition of lipoprotein lipase which hydrolyses the triacylglycerol content of chylomicrons and VLDL, while apo E is needed for the receptor mediated clearance from circulation. Circulation of VLDL and chylomicrons in the blood stream exposes the particles to enzymatic release of triacylglycerols from the lipoprotein core and excess cholesterol is removed from cells. The triacylglycerol- poor and cholesterol rich LDL remnant produced is potentially toxic to the body and needed to be safely cleared from blood circulation. The main concern for this lipoprotein is its toxic effect on the cardiovascular system. The liver scavenges and disposes chylomicrons remnant more effectively than the LDL particles, a mechanism put in place by the body to get rid of the more atherogenic particle of the two; chylomicrons remnant. LDL particles are mostly removed by other mechanisms involving HDL.

2.3.6. *Apolipoprotein(a)*

Apolipoprotein(a) itself is a large glycoprotein that exhibits size heterogeneity among individuals with isoforms that range between 180 – 700kDa in size. Apo(a) genotypes were determined using a newly developed pulsed-field gel electrophoresis method which distinguished 19 different genotypes at the apo(a) locus. The apo(a) gene itself was found to account for virtually all the genetic variability in plasma Lp(a) levels (Boerwinkle *et al.*, 1992). The apo(a) cDNA contains multiple tandem copies of a sequence that encodes a cysteine-rich protein motif called a kringle. The particular repeated kringle in apo(a) is designated kringle 4 because it closely resembles the fourth kringle in plasminogen, with the protease domain of apo(a) containing 88% amino-acid identity to plasminogen. McLean *et al.* (1992) proposed that the apo(a) isoforms are of different sizes because of variations in the numbers of kringle 4-encoding repeats in the apo(a) gene. The molecular mass of apo(a) protein varies from 187 kDa for an apo(a) that contains 12 kringle 4 domains, to 662 kDa for an apo(a) that contains 50 kringle 4 domains (Carmena *et al.*, 2004).

2.4. Lipoprotein A [Lp(a)]

Lipoprotein [Lp(a)] is a variant of LDL with an additional apolipoprotein in the structure. Lp(a) is essentially an LDL particle with a large glycoprotein, apolipoprotein (a) [apo(a)] attached to it (McLean *et al.*, 1987; Loscalzo *et al.*, 1990; Boerwinkle *et al.*, 1992; Palabrica *et al.*, 1995). Lp(a) resembles low density lipoprotein (LDL) in lipid composition, but it is distinguished by the presence of apo(a) which is bound by a disulfide linkage to apolipoprotein B-100, a ligand in the LDL molecule by which LDL binds to its receptor. Lp(a) levels has been demonstrated to have a clear association with development of atherosclerosis and other cardiovascular diseases (Zenker *et al.*, 1986; Danesh *et al.*, 2000; Berglund and Anuurad, 2008; Danik *et al.*, 2008). The postulated atherogenicity of Lp(a) is probably due to the presence of apo(a) component of the Lp(a) molecule. A study showed that the removal of apo(a) from Lp(a) particles result in a lipoprotein with greatly enhanced affinity for the LDL receptor (Armstrong *et al.*, 1985).

2.4.1. *Synthesis of lipoprotein(a)*

Apo(a) is expressed by liver cells (hepatocytes), and the assembly of apo(a) and LDL particles seems to take place at the outer hepatocyte surface. The half-life of Lp(a) in the circulation is about 3 to 4 days (Rader *et al.*, 1993) and this particle varies in blood concentration from one individual to another from <0.2 - >200 mg/dL. Ethnicity is a factor, with those of Asian and African origin averaging the highest concentrations. Within ethnic groups, individual elevation of Lp(a) is directly associated with increased risk of cardiovascular diseases (Sandholzer *et al.*, 1991; Chien *et al.*, 2008). Lp(a) is usually unaffected by factors like age, blood pressure, and total cholesterol.

2.4.2. *Similarity between lipoprotein(a) and plasminogen*

The structure of Lp(a) is similar to plasminogen, a naturally occurring glycoprotein that participates in dissolving of clots that form in the bloodstream, and tissue plasminogen activator (tPA). Lp(a) competes with plasminogen for its binding site, leading to reduced fibrinolysis (Loscalzo *et al.*, 1990; Palabrica *et al.*, 1995). Also because Lp(a) stimulates secretion of PAI-1 it leads to thrombogenesis. In addition, because of LDL cholesterol content, Lp(a) contributes to atherosclerosis (Schreiner *et al.*, 1993; Sotiriou *et al.*, 2006) and ultimately a cardiovascular risk factor (Berglund and Ramakrishnan, 2004).

2.4.3. *Correlation between apolipoprotein size and Lp(a) concentration*

There is a general inverse correlation between the size of the apo(a) isoform and the Lp(a) plasma concentration (Bowden *et al.*, 1994; Kraft *et al.*, 1996) which is caused by a variable rate of degradation before the apo(a) protein has matured for Lp(a) assembly (White *et al.*, 1994). The plasma concentration of Lp(a) is unaffected by many physiological, pharmacological, and environmental factors that affect the levels of other plasma lipoproteins (Albers *et al.*, 1977). A genetic determination of plasma Lp(a) levels was strongly suggested due to this lack of environmental and physiological influences. Consistent with this formulation, early genetic studies suggested that the presence of Lp(a) in plasma was inherited as a single autosomal dominant trait (Berg and Mohr, 1963; Iselius *et al.*, 1981), with an estimated heritability level ranging from 0.75 to 0.98 (Boerwinkle *et al.*, 1992).

Plasma Lp(a) concentrations vary 1000-fold between individuals and represent a continuous quantitative genetic trait with a skewed distribution in Caucasian populations (Utermann, 1989). A study was conducted by Lackner *et al.* (1991) in which the apo(a) gene of members of 12 Caucasian families were segregated. It was found that within a given family, sibling pairs with identical apo(a) genotypes tended to have very similar plasma Lp(a) levels (Lackner *et al.*, 1991). However, individuals with the same apo(a) genotypes who were members of different families often had significantly different plasma concentrations of

Lp(a). Taken together, these observations suggest that the apo(a) gene is the major determinant of plasma Lp(a) levels and that cis-acting DNA sequences at or near the apo(a) locus, other than the number of kringle 4 repeats, contribute importantly to plasma Lp(a) concentrations (Boerwinkle *et al.*, 1992). Variation in the hypervariable apo(a) gene on chromosome 6q2.6-q2.7 and interaction of apo(a) alleles with defective LDL-receptor genes explain a large fraction of the variability of plasma Lp(a) concentrations (Utermann, 1989).

Furthermore, the size of the apo(a) glycoprotein varies in individuals and this size is inversely correlated with the plasma level of Lp(a). The reason for the inverse correlation between the size of the apo(a) gene and level of plasma Lp(a) is not known, but a variation of length within the kringle 4-encoding region of the apo(a) gene may account for a greater proportion of the inter-individual variation in plasma Lp(a) concentrations. Also, the number of kringle 4 repeats in the gene may not have a direct effect on plasma Lp(a) concentration (Boerwinkle *et al.*, 1992; Brunner *et al.*, 1996). A study conducted on apo(a) gene of mamorset monkeys showed a plasma Lp(a) concentration of a very wide range of over a 100-fold, but only one apo(a) isoform (Guo *et al.*, 1991). This may be explained by the differences in the composition of kringle 4 sequence of apo(a) genes in which individuals may have same sizes of apo(a) alleles but different plasma Lp(a) concentrations. The frequency of recombination activity in this locus may be responsible for the variation in their kringle 4 composition and number which may have marked effect on synthesis and /or degradation of Lp(a) (Boerwinkle *et al.*, 1992).

2.5. Role of oxidation in atherogenesis

Oxidative stress, especially LDL oxidation has been suggested for almost three decades as the most probable aetiology of atherosclerosis (Steinbrecher *et al.*, 1984). Markers of LDL oxidation in plasma, particularly circulating oxidized LDL and auto-antibodies against oxidized LDL, could be used to assess the development of atherosclerosis in patients (Carmena *et al.*, 2004). Circulating oxidized LDL is additive to the global risk score based on age, sex, total and HDL cholesterol, diabetes mellitus, hypertension, and smoking as a useful marker for identifying persons at risk for CAD (Holvoet *et al.*, 2001; Toshima *et al.*, 2000).

A study has associated circulating oxidized LDL with both subclinical atherosclerosis (clinically silent ultrasound assessed atherosclerotic changes in the carotid and femoral arteries) and inflammatory variables (C-reactive protein and the inflammatory cytokines interleukin-6 and tumor necrosis factor-α). This conclusion supports the concept that oxidatively modified LDL may play a major role in development of atherosclerosis (Hulthe and Fagerberg, 2002). It has been proposed that, because of the antigenic properties of oxidized LDL, the anti-oxidized LDL antibody titer could represent a useful index of in vivo LDL oxidation. Autoantibodies against oxidized LDL have been reported to be associated with atherosclerosis, but existing reports are still conflicting. Some studies have reported a positive relationship between autoantibodies against oxidized LDL and CHD (Sherer *et al.*, 2001) whereas another did not (Leinonen *et al.*, 1998). There is a strong cross-reactivity between autoantibodies against oxidized LDL and anticardiolipin antibodies, which have been positively associated with CHD (Erkkila *et al.*, 2000).

2.6. Relationship between insulin resistance, diabetes, and small, dense LDL

Cardiovascular heart disease risk is usually significantly increased when elevated levels of small, dense LDL accompanied by hypertriglyceridemia, reduced HDL-cholesterol levels, abdominal obesity, and insulin resistance. Results from the Que´bec Cardiovascular Study have indicated that persons displaying elevated plasma concentrations of insulin and apo B together with small, dense LDL particles showed a remarkable increase in CHD risk (Lamarche *et al.*, 1999).

Sensitivity to insulin in diabetic and non-diabetic individuals was assessed using nuclear magnetic resonance (NMR) spectroscopy. Insulin resistance had profound effects on lipoprotein size and an increase in serum triglycerides. The lipid profile revealed a 2- to 3-fold increase in concentrations of large VLDL particles with no change in medium or small VLDL, increase in overall LDL particle concentration with more small LDL particle size and reduced large LDL. In type 2 diabetes, these alterations could be attributed primarily to the underlying insulin resistance. These changes in the NMR lipoprotein subclass profile predictably increased the risk of cardiovascular disease but were not fully apparent in the conventional lipid profile (Garvey *et al.*, 2003). The Diabetes Atherosclerosis Intervention Study (DAIS) (Vakkilainen *et al.*, 2003) showed that lipid-modifying treatment decreased the angiographic progression of coronary atherosclerosis in subjects with type 2 diabetes. This effect was related in part to the correction of lipoprotein abnormalities. Compared with placebo, fenofibrate treatment significantly increased LDL particle size and HDL cholesterol and decreased plasma total cholesterol, LDL cholesterol, and triglyceride concentrations. The final LDL particle size was inversely correlated with the increase in percentage diameter stenosis (Vakkilainen *et al.*, 2003).

The Pittsburgh Epidemiology of Diabetes Complications Study on whether NMR lipoprotein spectroscopy improves the prediction of coronary artery disease (CAD) in patients with childhood-onset type 1 diabetes, independently of conventional lipid and other risk factors showed that both lipid mass and particle concentrations (NMR spectroscopy) of all VLDL subclasses, small LDL, medium LDL, and medium HDL were increased in CAD cases compared with controls, whereas large HDL was decreased. Mean LDL and HDL particle sizes were also less in CAD cases (Soedamah-Muthu *et al.*, 2003; Carmena *et al.*, 2004).

2.7. Dyslipoproteinaemia

Dyslipoproteinaemia is a term broadly used for derangement in lipid and lipoprotein metabolism, which may either be hyperlipoproteinaemia or hypolipoproteinaemia. Dyslipoproteinaemias are generally classified as familial (primary) or acquired (secondary). This chapter discusses hyperlipoproteinaemia with its close relevance to development of cardiovascular diseases. Primary hyperlipoproteinaemias are of genetic origin and may be due to a mutation in a receptor protein which presents as inborn errors of lipid metabolism, and includes common hypercholesterolemia, combined familiar hyperlipidemia, familiar hypercholesterolemia, familiar hypertriglyceridemia, VLDL remnants hyperlipidemia and primary chylomicronaemia (Garmendia, 2003). Primary hyperlipoproteinaemia was first classified by Fredrickson and Lees (1965) and this classification was adopted by World Health Organization (WHO). They divided primary hyperlipoproteinaemias into four types and details are show in Table 3 below.

The secondary hyperlipoproteinemias also mimic primary types and may present with similar symptoms. Secondary dyslipoproteinaemias are usually due to other underlying causes that lead to alterations in plasma lipid and lipoprotein metabolism, including hypothyroidism, diabetes mellitus, nephrotic syndrome, chronic biliary obstruction, renal insufficiency. Some drugs modify lipid metabolism and these include alcohol, beta-adrenergic blockers, diuretics, progestagens, corticosteroids (Garmendia, 2003). Treatment of the underlying cause or discontinuation of offending drug may resolve the dyslipoproteinaemia. Lipid and lipoprotein abnormalities are common observations and are regarded as modifiable risk factors for development of cardiovascular diseases.

Hyperlipo-proteinaemia	Sub-type	Classification	Defect	Lipoprotein increased	Treatment
Type I	a, c	Familial hyperchylo-micronemia	↓ Lipoprotein lipase (LpL)	Chylomicrons	Diet control
	b	apoprotein CII deficiency	Altered Apo CII		
Type II	a	Familial hypercholesterolemia	LDL receptor deficiency	LDL	Bile acid sequestrants, statins,niacin
	b	Familial combined hyperlipidemia	↓ LDL receptor or and ↑Apo B	LDL and VLDL	Statins, niacin,fibrate
Type III		Familial dysbetalipo-proteinaemia	Apo E2 synthesis	IDL	Fibrate, statins
Type IV		Familial hypertriglyce-ridaemia	↑VLDL and ↓LpL	VLDL	Fibrate, niacin, statin
Type V			↑VLDL and ↓LpL	VLDL and chylomicrons	Niacin, fibrate

Adapted from Fredrickson classification of hyperdyslipoproteinaemia (Fredrickson and Lees, 1965).

Table 3.

2.8. Theories of atherogenesis

Arteries are blood vessels that carry oxygenated blood from the heart to all tissues of the body. The arterial wall is composed of three layers, namely the intima (inner lining), media and adventitia. A single layer of endothelial cells line the inner surface of the intima, forming a barrier to blood cells and plasma flowing within the blood vessel. Atherosclerosis is characterized by lesions in the intima of arteries, seen as raised fibrous plagues ranging in colour from pearly gray to yellowish gray. The cellular components of the plaque include a cell similar to the adjacent endothelial cell, macrophages, fibrinogen from which fibrin is formed and white blood cells intersparsed between dense connective tissue which consist majorly of collagen fibers. The cells within and around the plaque are usually lipid ladened. Atherosclerosis poses a high risk not just because it can close up an artery, slowing down or entirely restricting blood flow, but may also lead to thrombus formation. A thrombus is a complex aggregation of platelets, red and white blood cells in a fibrin network. Several theories have emanated, suggesting the actual pathogenesis of atherosclerosis.

Schoenhagen (2006) documented the different theories that have been postulated in the course of history. In 1851, a scientist, Rokitansky suggested the encrustation theory or thrombogenesis in which it is said that atherosclerosis began in the intima of arteries with the deposition of thrombus. This is followed by the organisation of the thrombus through infiltration of fibroblast, secondarily followed by deposition of lipid. The German pathologist, Rudolf Virchow postulated the insudation or inflammation theory in 1856, a different initiation of atherosclerosis. It was suggested that infiltration of fatty substances from the blood stream into the arterial wall leads to deposition of cholesterol which acts as an irritant, causing inflammation and the proliferation of cells. The cholesterol deposits act as irritant in the arterial intima, initiating inflammatory process as macrophages are incriminated as key role players in the phases of the disease. This theory was further supported by the work of N.N. Anitschkow in 1933 where he discovered that a disease resembling human atherosclerosis could be reproduced in rabbits with high serum cholesterol or LDL levels. He thus stated this occurrence may be as a result of defects in metabolism of lipids and lipids Schoenhagen, 2006).

The flow theory relates the circulation of blood in vessels to its effect on arterial wall. It stated that lesions occurred more often at curved, branching, or bifurcated sites, generally at regions of perturbed blood flow. Other hypotheses that arose from the flow theory include the stagnation point hypothesis by Fox and Hugh (1966), high wall shear stress hypothesis proposed by Fry (1968), low wall shear stress hypothesis by Caro *et al.* (1969), diminished lateral pressure hypothesis by Texon (1980) and the convection-diffusion hypothesis.

All the theories above were established based on three methods. Atherosclerotic plaques from autopsy findings from individuals of both sexes, various ages and race with different diseases which included hyperlipidaemia, diabetes and hypertension were considered. Epidemiological studies of factors which promote or prevent development of atherosclerosis, and finally experimental pathology which established the sequence of lesion development or regression were considered. None of these theories entirely explains the pathogenesis of atherosclerosis, but each has explained an aspect of this process.

2.9. Prevention and treatment of cardiovascular diseases

The contributing factors to development of cardiovascular diseases are numerous as mentioned in the risk factors above. Lowering of plasma cholesterol levels is usually the first line of intervention for prevention and treatment of cardiovascular diseases. Dramatic successes have been recorded with cholesterol-lowering therapy which may suggest that maintenance of low cholesterol levels is sufficient to prevent development of atherosclerosis or reversing an established disease condition (Brunzell *et al.*, 2008). Different approaches have been used for prevention and treatment of this condition, some are enumerated below.

2.9.1. Role of High density lipoprotein-cholesterol (HDL-C)

High density lipoprotein-cholesterol (HDL-C) is the smallest of the lipoprotein sub-fractions. It is however, the most complicated and diverse of the lipoproteins. It is the major lipoprotein which transports excess cholesterol from the plasma to the liver for excretion or utilization in the liver and other hormone producing regions of the body. Excess cholesterol is eliminated from the body via the liver, which secretes cholesterol in bile or converts it to bile salts (Toth, 2005; Tall, 2008). Also, its anti-inflammatory property protects LDL from oxidation and limits the concentrations of oxidized components, which may pose as atherogenic treats. HDL-C have been associated with reduced risk of cardiovascular events (Duffy and Rader, 2009; Khera *et al.*, 2011). HDL-C plays a key role in the reverse transportation of cholesterol by accepting cholesterol from lipid-laden macrophages (Lehrke *et al.*, 2007). In the study conducted by Khera *et al.* (2011), the ability of HDL to promote cholesterol efflux from macrophages was strongly and inversely associated with both subclinical atherosclerosis and obstructive coronary artery disease. It was also discovered that the associations persisted after adjustment for traditional cardiovascular risk factors, including the levels of HDL cholesterol and apolipoprotein A-I. HDL has several protein constituents which are exchangeable with other lipoproteins, and it acquires different apolipoproteins in the process of maturation such as apo AII, AIV, AV, CI, CII, CIII and E which results in generation of diverse HDL particles with various metabolic functions. In addition to being carrier proteins for the HDL particle, these proteins have protective roles which they play against cardiovascular diseases, such as by acting as anti-inflammatory regulators to limit the activity of pro-inflammatory cytokines.

Nascent HDL is synthesised and secreted by the liver and small intestine. It travels in the circulation where it gathers cholesterol to form mature HDL, which then returns the cholesterol to the liver via various pathways. Apolipoprotein A-I (ApoA-I) is the major protein component of high density lipoprotein (HDL) in plasma. The protein is encoded for by APOAI gene (Breslow *et al.*, 1982, Arinami *et al.*, 1990). Defects in this gene have been associated with HDL deficiencies (HUGO Gene Nomenclature Committee, 2011). This protein increases the efflux of cholesterol from tissue to liver where it is excreted. A few individuals were reported to produce a HDL ApoA-I protein variant called ApoA-I Milano, an abnormal and apparently more efficient apolipoprotein. It has low measured HDL-C levels yet very low rates of cardiovascular events even with high blood cholesterol values

(Franceschini *et al.*, 1981). Apo CI, CII and apo E also accumulate in the nascent HDL particle, which serves as a store for these apolipoproteins in circulation.

Phospholipids are transferred from macrophages by a specific transporter molecule known as ATP-binding cassette transport protein A1 (ABCA-1) into the core of the lipoprotein, and cholesterols are extracted from the cells by a transporter protein derived from macrophages in the sub-endothelial spaces of tissues; ABCG-1 transporter. The eventual maturation of the HDL particle is dependent on the lecithin: cholesterol acyltransferase (LCAT), an enzyme activated by apoAI, which catalyses the formation of cholesterol esters from cholesterol. Mobilization of free cholesterol and phospholipids from IDL and LDL continues until a matured, spherical HDL particle is formed. Endocytosis of the matured HDL into hepatocytes occurs and the cholesterol and cholesterol esters are transported via a facilitated transfer to distinct pools within the cell. The modified HDL particles are secreted back into circulation where they can further acquire cholesterol before they re-circulate to the liver. The complete reverse cholesterol transport occurs with the addition of apo E to the HDL particle which facilitates their uptake and catabolism.

2.9.2. Lipoprotein Lipase as an anti-atherogenic agent

Activation of lipoprotein lipase (LpL) activity has been reported to have anti-atherogenic activity. Lipoprotein lipase (LpL) is a rate-limiting enzyme found on the surface of endothelial cells. It is polypeptide with 839 amino acids and an extracellular domain which binds to apo B100 and apo E. LpL catalyses the hydrolysis of the triacylglycerol (TAG) component of circulating chylomicrons and very low density lipoproteins (VLDL). The enzyme digests the TAG to fatty acids and monoglycerides. This provides non-esterified fatty acids and 2-monoacylglycerol which can be utilised immediately by cells for energy production or synthesis of other lipids. Unutilized fatty acids may be bound to circulating albumin and released slowly to meet future cellular requirements. Glycerol produced from LpL activity is transported back to the liver and kidneys, where it is converted to dihydroxyacetone phosphate in the alternative glycolytic pathway. The fatty acids from LpL activity in the muscle may diffuse into cells to be oxidized to two-carbon units or used to re-synthesis TAG which are stored in adipose cells (Clee *et al.*, 2000; Tsutsumi, 2003). Significant LpL activity occurs in muscle, adipose tissue and lactating mammary glands. Accumulation of VLDL remnants (IDL with apo B100 and apo E are converted to LDL with further loss of triacylglycerols. Both carrier proteins are necessary for recognition of IDL and LDL by the LDL receptors in the liver, after which they are taken up into hepatocytes by endocytosis and catabolized.

Research carried out over the past two decades have not only established a central role for LpL in the overall lipid metabolism and transport but have also identified additional, non-catalytic functions of the enzyme. Furthermore, abnormalities in LpL function have been found to be associated with a number of pathophysiological conditions, including atherosclerosis, chylomicronaemia, obesity, Alzheimer's disease, and dyslipidaemia associated with diabetes, insulin resistance, and infection (Mead *et al.*, 2002).

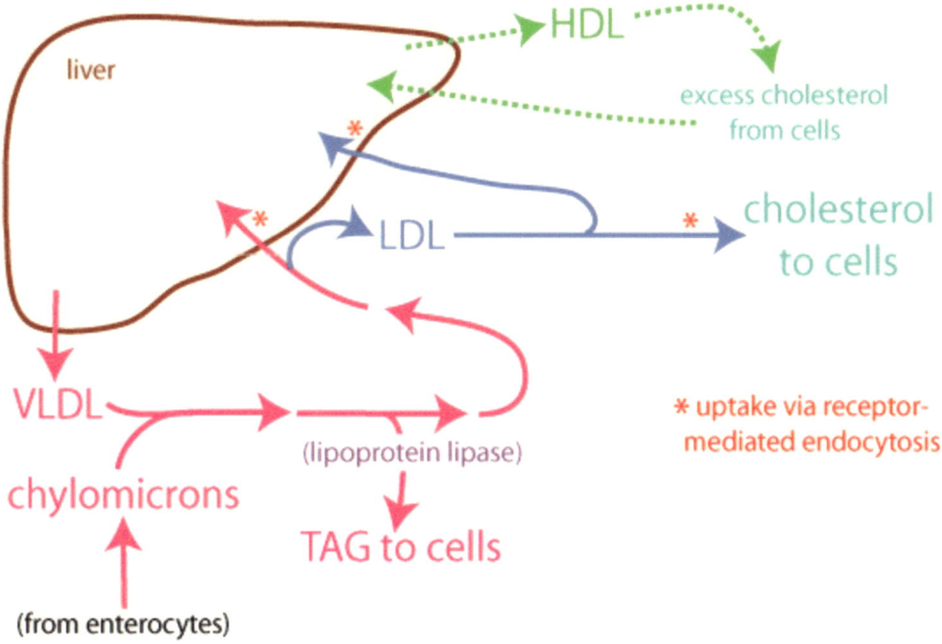

Figure 3. Summary of the fate of Lipoprotein sub-fractions (Adapted from http://courses.washington.edu/conj/bess/cholesterol/liver.html)

LpL encodes lipoprotein lipase, which is expressed in heart, muscle, and adipose tissue. LpL functions as a homodimer, and has the dual functions of triglyceride hydrolysis and ligand/bridging factor for receptor-mediated lipoprotein uptake. Through catalysis, VLDL is converted to IDL and then to LDL. Severe mutations that cause LpL deficiency result in type I hyperlipoproteinemia, while less extreme mutations in LpL are linked to many disorders of lipoprotein metabolism.

LpL isozymes are regulated differently depending on the tissue. For example, insulin is known to activate LpL in adipocytes and its placement in the capillary endothelium. By contrast, insulin has been shown to decrease expression of muscle LpL (Kiens *et al.*, 1989). The form that is in adipocytes is activated by insulin, whereas that in muscle and myocardium is activated by glucagon and epinepherine. This helps to explain why during fasting, LpL activity increases in muscle tissue and decreases in adipose tissue. After feasting, the opposite occurs (Braun and Severson, 1992; Mead *et al.*, 2002).

The concentration of LpL displayed on endothelial cell surface cannot be regulated by endothelial cells, as they neither synthesize nor degrade LpL. Instead, this regulation occurs by managing the flux of LpL arriving at the lipolytic site and being released into circulation attached to lipoproteins (Braun and Severson, 1992; Goldberg, 1996). The typical concentration of LpL in plasma is in the nanomolar range. Lipoprotein lipase deficiency leads to hypertriglyceridemia (elevated levels of triglycerides in the bloodstream) and decreased high density lipoprotein activity (Clee *et al.*, 2000; Tsutsumi, 2003; Okubo *et al.*, 2007). Diets high in refined carbohydrates have been shown to cause tissue-specific overexpression of LpL. This has been implicated in tissue-specific insulin resistance and consequent development of type 2 diabetes mellitus & obesity.

2.9.3. Influence of Hormones on plasma LDL-C and Lp(a) levels

Several studies have reported conflicting reports on the effect of hormonal replacement therapy on plasma LDL-C and Lp(a) levels (Taskinen *et al.*, 1996; Shlipak *et al.*, 2000; Vigna *et al.*, 2002). In a cohort study conducted by Danik *et al.* (2008), the effect of hormone replacement therapy (HT) on Lp(a) and cardiovascular risk was investigated. It was reported that the relationship of high Lp(a) levels with increased cardiovascular disease is modified by hormonal therapy. These data suggest that the predictive utility of Lp(a) is markedly attenuated among women taking HT and may inform clinicians' interpretation of Lp(a) values in such patients. It was noteworthy that the effect of hormonal therapy was observed only in women with high LDL cholesterol levels, in agreement with previous studies suggesting an interaction between Lp(a) and LDL cholesterol (Berglung and Anuurad, 2008).

3. Summary

Elevated serum LDL-C and low levels of HDL-C are known as major and independent risk factors for CHD. Small, dense lipoprotein sub-fractions have been reported to have atherogenic potentials, with particular reference to Lipoprotein(a) [Lp(a)], a variant of low density lipoprotein (LDL). Other atherogenic sub-fractions of lipoproteins are VLDL, LDL, and IDL. These sub-fractions are characterized by the presence of apolipoprotein B-100, with an additional apolipoprotein known as apo(a) in the Lp(a) structure. Apo(a) is structurally and functionally similar to plasminogen and it accounts for virtually all the genetic variability in plasma Lp(a) levels. Variation of length within the kringle 4-encoding region of the apo(a) gene may account for a greater proportion of the inter-individual variation in plasma Lp(a) concentrations, with a strong genetic involvement inherited as a single autosomal dominant trait.

In the course of the pathogenesis of atherosclerosis, oxidized LDL is taken up by macrophages and into endothelial cells. This leads to formation of atherosclerotic plaques which precedes development of CHD. Oxidized LDL is antigenic and titres of auto-antibodies against oxidized LDL in plasma can be used as indicator of a positive association with CHD. A positive association has also been established between plasma level of LDL, specifically oxidized LDL and other risk factors contributing to development of CHD. Such risk factors were identified as hypertriglyceridemia, reduced HDL-C levels, abdominal obesity and insulin resistance. Treatment of CHD can be achieved by lowering of plasma cholesterol levels which has been achieved by cholesterol-lowering therapy, suggesting that maintenance of low cholesterol levels may sufficiently prevent or reverse an established atherosclerosis. Increasing plasma levels of HDL-C has been reported to also be of benefit.

HDL-C has anti-inflammatory activity which may prevent oxidation of LDL and it plays a key role in reverse transportation of cholesterol from lipid-laden macrophages.

The enzyme Lipoprotein lipase (LpL) may be useful in chemotherapy or prophylaxis of CHD. LpL has anti-atherogenic activity by its dual functions of triglyceride hydrolysis and ligand/brigding factor for receptor-mediated lipoprotein uptake. Activation of the enzyme is dependent on the tissue, resulting in variability of its activity. Hormonal replacement therapy may also be of benefit to patients with CHD and related diseases, but reported on current findings are conflicting.

Author details

Adebowale Saba and Olayinka Oridupa
University of Ibadan, Nigeria

4. References

Albers, J. J.; Adolphson J. L. & Hazzard W. R. (1977). Radio-immunoassay of human plasma Lp(a) lipoprotein. *J. Lipid Res.* Vol 18, pp 331-338. ISSN 0022-2275

Arinami, T.; Hirano, T.; Kobayashi, K.; Yamanouchi, Y. & Hamaguchi, H. (1990). Assignment of the apolipoprotein A-I gene to 11q23 based on RFLP in a case with a partial deletion of chromosome 11, del(11)(q23.3----qter). *Hum. Genet.* Vol. 85, No. 1, pp 39–40. PMID 1972696

Armstrong, V.W., Walli, A.K. & Seidel, D. 1985. Isolation, characterization, and uptake in human fibroblasts of an apo (a)-free lipoprotein obtained on reduction of lipoprotein (a). *J Lipid Res*, Vol 26, pp 1314-1323. ISSN 0022-2275

Barter, P., Gotto, A.M., LaRosa, J.C., Maroni, J., Szarek, M., Grundy, M.S.S.M., Kastelein, J.J.P., Vera Bittner, V. & Fruchart J. (2007). HDL Cholesterol, Very Low Levels of LDL Cholesterol, and Cardiovascular Events. *N. Engl. J. Med.* Vol 357, pp 1301-1310 (2007). /doi:10.1056/NEJMoa064278

Berg, K. & Mohr, J. (1963). Genetics of the Lp system. *Acta Genet.* Vol. 13, pp 349-360. doi:10.1159/000151817

Berglund, L. & Ramakrishnan, R. (2004). Lipoprotein(a): an elusive cardiovascular risk factor. *Arterioscler. Thromb. Vasc. Biol.* Vol. 24, No.12, pp 2219–26. doi:10.1161/ 01.ATV.0000144010.55563.63. PMID 15345512

Berglund, L. & Anuurad, E. (2008). Role of lipoprotein(a) in cardiovascular diseases: Current and future perspectives. *J Am Coll Cardiol*, Vol 52, pp 132-134, doi:10.1016/j.jacc.2008.04.008.

Boerwinkle, E., Leffert, C.C., Lin, J., Lackner, C., Chiesa, G. & Hobbs, H.H. (1992). Apolipoprotein (a) gene accounts fro greater than 90% of the variation in plasm lipoprotein (a) concentrations. *J Clin Invest* Vol 90, pp 52-60. 0021-9738/92/07/0052/09

Bowden, J.F., Pritchard, P.H., Hill, J.S. & Frohlich, J.J. (1994). Lp(a) concentration and apo(a) isoform size. Relation to the presence of coronary artery disease in familial hypercholesterolemia. *Arterioscler Thromb,* Vol 14, pp 1561. doi: 10.1161/ 01.ATV.14.10.1561

Braun, J.E. & Severson, D.L. (1992). Regulation of the synthesis, processing and translocation of lipoprotein lipase. *Biochem J* Vol. 287, No.2, (October, 1992) pp 337–47. PMC1133170

Breslow, J.L., Ross, D., McPherson, J., Williams, H., Kurnit, D., Nussbaum, A.L., Karathanasis, S.K. & Zannis, V.I. (1982). Isolation and characterization of cDNA clones for human apolipoprotein A-I. *Proc. Natl. Acad. Sci. U.S.A.* Vol. 79, No. 22, pp 6861–5. doi:10.1073/pnas.79.22.6861. PMC 347233. PMID 6294659. http://www.pubmedcentral.nih.gov/articlerender.fcgi?tool=pmcentrez&artid=347233.

Brown, M.S. & Goldstein, J.L. (2009). Cholesterol feedback: from Schenheimer's bottle to Scap's MELADL. *J. Lipid Res.* (April 2009), Vol. 50, S15-S27. doi:10.1194/jlr.R800054-JLR200

Brunner, C., Lobentanz, E.M., Pethö-Schramm, A., Ernst, A., Kang, C., Dieplinger, H., Müller, H.J. & Utermann, G. (1996). The number of identical kringle IV repeats in apolipoprotein(a) affects its processing and secretion by HepG2 cells. *J. Biol. Chem.* Vol. 271, No. 50, pp 32403–10. doi:10.1074/jbc.271.50.32403. PMID 8943305.

Brunzell, J.D., Davidson, M., Furberg, C.D., Goldberg, R.B., Howard, B.V., Stein, J.H. & Witztum, J.L. (2008). Lipoprotein management in patients with cardiometabolic risk: Consensus conference report from the American Diabetes Association and the American College of Cardiology Foundation. *J Am Coll Cardiol* Vol. 51, pp 1512-1524. doi:10.1016/j.jacc.2008.02.034. Available from http://content.onlinejacc.org/cgi/content/full/51/15/1512

Carmena, R., Duriez, P. & Fruchart, J.C. (2004). Atherogenic lipoprotein particles in atherosclerosis. *Circulation* Vol. 109, pp 2-7. doi: 10.1161/01.CIR.0000131511.50734.44. Available from http://circ.ahajournals.org/content/109/23_suppl_1/III-2

Chien, K.L., Hsu, H.C., Su, T.C., Sung, F.C., Chen, M.F. & Lee, Y.T. (2008). Lipoprotein(a) and Cardiovascular Disease in Ethnic Chinese: The Chin-Shan Community Cardiovascular Cohort Study. *Clinical Chemistry* Vol. 54, pp 285-291, 2008. doi:10.1373/clinchem.2007.090969

Clee, S.M., Bissada, N., Miao, F., Miao, L., Marais, A.D., Henderson, H.E., Steures, P., McManus, J., McManus, MCManus, B., LeBoeuf, R.C., Kastelein, J.J.P. & Hayden, M.R. (2000). Plasma and vessel wall lipoprotein lipase have different roles in atherosclerosis. *J Lipid Res* Vol. 41, pp 521-531. Available from http://www.jlr.org/content/41/4/521.abstract

Danesh, J., Collins, R. & Peto, R.. (2000). Lipoprotein(a) and Coronary Heart Disease: Meta-Analysis of Prospective Studies. *Circulation.* Vol. 102, pp 1082-1085 doi: 10.1161/01.CIR.102.10.1082

Danik, S., Rifai, N., Buring, J.E. & Ridker P.M. (2008). Lipoprotein(a), Hormone Replacement Therapy, and Risk of Future Cardiovascular Events. *J Am Coll Cardiol,* Vol. 52, pp 124-131, doi:10.1016/j.jacc.2008.04.009. ISSN 0735-1097/08.

Di Vizio, D., Solomon, K.R. & Freeman, M.R. (2008). Cholesterol and cholesterol-rich membranes in prostate cancer: an update. *Tumori* Vol 5, pp 633-639. Available from http://www.tumorionline.it/allegati/00386_2008_05/fulltext/1%20-%20Di%20Vizio%20%28633-639%29.pdf

Duffy, D. & Rader, D.J. (2009). Update on strategies to increase HDL quantity and function. Nat Rev Cardiol; Vol. 6, pp 455-463. Available from www.ncbi.nlm.nih.gov

Eckel, R.H. (1997). Obesity and heart disease: a statement for healthcare professionals from the Nutrition Committee, American Heart Association. *Circulation*.Vol. 96, pp 3248 – 3250. doi: 10.1161/01.CIR.96.9.3248. Available from http://circ.ahajournals.org/content/96/9/3248.full

Fahy, E., Subramaniam, S., Murphy, R., Nishijima, M., Raetz, C., Shimizu, T., Spener, F., Van Meer, G., Wakelam, M. & Dennis, E.A. (2009). Update of the LIPID MAPS comprehensive classification system for lipids. *Journal of Lipid Research* Vol. 50, S9–S14. doi:10.1194/jlr.R800095-JLR200. PMID 19098281

Fredrickson, D.S., Lees, R.S. (1965). A system for phenotyping hyperlipoproteinaemia. *Circulation* Vol 31, No. 3, pp 321-327. Doi: 10.1161/01.CIR.31.3.321. PMID 14262568

Fletcher, G.F., Balady, G., Blair, S.N., Blumenthal, J., Caspersen, C., Chaitman, B., Epstein, S., Froelicher, E.S.S., Froelicher, V.F., Pina, I.L. & Pollock, M.L. (1996). Statement on exercise: benefits and recommendations for physical activity programs for all Americans: a statement for health professionals by the Committee on Exercise and Cardiac Rehabilitation of the Councilon Clinical Cardiology, American Heart Association. *Circulation,* Vol. 94, pp 857– 862. doi: 10.1161/01.CIR.94.4.857. Available from http://circ.ahajournals.org/content/94/4/857.long

Franceschini, G., Sirtori, M., Gianfranceschi, G. & Sirtori, C.R. (1981). Relation between the HDL apoproteins and AI isoproteins in subjects with the AIMilano abnormality. *Metab. Clin. Exp*. Vol 30, No. 5, pp 502–9. doi:10.1016/0026-0495(81)90188-8. PMID 6785551.

Garmendia, F. (2003). Advances in the knowledge and treatment of dyslipoproteinnemias. *An. Fac. Med.* Vol 64, No. 2, pp 101-106.

Garvey, W.T., Kwon, S., Zheng, D., Shaughnessy, S., Wallace, P., Hutto, A., Pugh, K., Jenkins, A.J., Klein, R.L. & Liao Y. (2003). Effects of insulin resistance and type 2 diabetes on lipoprotein subclass particle size and concentration determined by nuclear magnetic resonance. *Diabetes*. Vol. 52, pp 453–462. Available from http://diabetes.diabetesjournals.org/content/52/2/453.full.pdf

Goldberg, I.J. (1996). Lipoprotein lipase and lipolysis: central roles in lipoprotein metabolism and atherogenesis. J *Lipid Res* Vol. 37, No. 4, pp 693–707. Available from http://www.ncbi.nlm.nih.gov/pubmed/8732771

Guo, H.C., Michel, J.B., Blouquit, Y. & Chapman, M. J. (1991). Lipoprotein(a) and apolipoprotein(a) in a new world monkey, the common marmoset (callithrix jacchus): Association of variable plasma lipoprotein(a) levels with a single apolipoprotein(a) isoform. *Arterioscler. Thromb*. Vol. 11, pp 1030-1041. doi:10.1161/01.ATV.11.4.1030 Available from http://atvb.ahajournals.org/content/11/4/1030

Grundy, S.M., Paternak, R., Greenland, P., Smith, S. & Fuster, V. (1999). Assessment of cardiovascular risk by use of multiple-risk-factor assessment equations. *Circulation* Vol. 100, pp 1481-1492. doi: 10.1161/01.CIR.100.13.1481. Available from http://circ.ahajournals.org/content/100/13/1481

Holvoet, P., Mertens, A., Verhamme, P., Bogaerts, K., Beyens, G., Verhaeghe, R., Collen, D., Muls, E. & Van de Werf, F. (2001). Circulating oxidized LDL is a useful marker for identifying patients with coronary artery disease. *Arterioscler Thromb Vasc Biol,*. Vol. 21, pp 844–848. doi: 10.1161/01.ATV.21.5.844. Available from http://atvb.ahajournals.org/content/21/5/844.full

HUGO Gene Nomenclature Committee (HGNC), (2011). APOA1 apolipoprotein A-I (*Homo sapiens*). Gene ID: 335, protein coding. http://www.ncbi.nlm.nih.gov/sites/entrez?Db=gene&Cmd=ShowDetailView&TermToSearch=335 Accessed on 9 November, 2011.

Hulthe, J. & Fagerberg, B. (2002). Circulating oxidized LDL is associated with subclinical atherosclerosis development and inflammatory cytokines (AIR Study). *Arterioscler Thromb Vasc Biol,* Vol. 22, pp 1162–1167. doi: 10.1161/01.ATV.0000021150.63480.CD. Available from http://atvb.ahajournals.org/content/22/7/1162.abstract?ijkey=80aaca3f2daac9fbc15623e1618c678b0a8a2db5&keytype2=tf_ipsecsha

Ikonen, E. (2008). Cellular cholesterol trafficking and compartmentalization. *Nature Rev. Mol. Cell Biol.* Vol. 9, pp 125-138. doi:10.1038/nrm2336

Iselius, L., Dahlen, G. H., De Faire, U. & Lundman, T. (1981). Complex segregation analysis of the Lp(a)/pre-.l3-lipoprotein trait. *Clin. Genet.*Vol. 20, pp 147-151. DOI: 10.1111/j.1399-0004.1981.tb01820.x. available from http://onlinelibrary.wiley.com/doi/10.1111/j.1399-0004.1981.tb01820.x/abstract

Kiens, B., Lithell, H., Mikines, K.J. & Richter, E.A. (1989). Effects of insulin and exercise on muscle lipoprotein lipase activity in man and its relation to insulin action. *J. Clin. Invest.* Vol. 84, No. 4, pp 1124–9. 0021-9738/89/10/1124/06. Available from http://www.ncbi.nlm.nih.gov/pmc/articles/PMC329768/pdf/jcinvest00485-0080.pdf

Khera, A.V., Cuchel, M., de la Llera-Moya, M., Rodrigues, A., Burke, M.F., Jafri, K., French, B.C., Phillips, J.A., Mucksavage, M.L., Wilensky, R.L., Mohler, E.R., Rothblat, G.H. & Rader, D.J. (2011). Cholesterol Efflux Capacity, High-Density Lipoprotein Function, and Atherosclerosis. *N Engl J Med,* Vol. 364, pp127-135. PMID 21226578. Available from www.ncbi.nlm.nih.gov/pubmed/21226578

Kraft, H.G., Lingenhel, A., Köchl, S., Hoppichler, F., Kronenberg, F., Abe, A., Mühlberger, V., Schönitzer, D. & Utermann, G. (1996). Apolipoprotein(a) kringle IV repeat number predicts risk for coronary heart disease. *Arterioscler Thromb Vasc Biol,* Vol. 16, pp 713. PMID 8640397. Available from http://www.ncbi.nlm.nih.gov/pubmed/8640397

Lackner, C., E. Boerwinkle, Leffert, Rahmig, T. & Hobbs, H. H. (1991). Molecular basis of sapolipoprotein(a) isoform size heterogeneity as revealed by pulsed-field gel electrophoresis. *J. Clin. Invest.,* Vol. 87, pp 2077-2086. PMID 1645755. Available from http://www.ncbi.nlm.nih.gov/pubmed/1645755

Lamarche, B., Lemieux, I. & Despres, J.P. (1999). The small, dense LDL phenotype and the risk of coronary heart disease: epidemiology, pathophysiology and therapeutic aspects. *Diabetes Metab.* Vol. 25, pp 199–211. PMID 10499189. Available from http://www.ncbi.nlm.nih.gov/pubmed/10499189?dopt=Abstract

Leah, E. (2009). Cholesterol. Lipidomics Gateway. doi:10.1038/lipidmaps.2009.3. Available from http://www.lipidmaps.org/update/2009/090501/full/lipidmaps.2009.3.html.

Lehrke, M., Millington, S.C., Lefterova, M., Cumaranatunge, R.G., Szapary, P., Wilensky, R., Rader, D.J., Lazar, M.A. & Reilly, M.P. (2007). CXCL16 is a marker of inflammation, atherosclerosis, and acute coronary syndromes in humans. *J Am Coll Cardiol*, Vol. 49, pp 442-449. doi:10.1016/j.jacc.2006.09.034. Available from http://content.onlinejacc.org/cgi/content/full/49/4/442

Leinonen, J.S., Rantalaiho, V., Laippala, P., Wirta O, Pasternack A, Alho H, Jaakkola O, Yla-Herttuala S, Koivula T & Lehtimaki T. (1998). The level of autoantibodies against oxidized LDL is not associated with the presence of coronary heart disease or diabetic kidney disease in patients with non-insulindependent diabetes mellitus. *Free Radic Res.*, Vol. 29, pp 137–141. PMID 9790515. Available from http://www.ncbi.nlm.nih.gov/pubmed/9790515?dopt=Abstract

Loscalzo, J., Weinfeld, M., Fless, G.M. & Scanu, A.M. (1990). Lipoprotein(a), fibrin binding, and plasminogen activation. *Arteriosclerosis*, Vol. 10, (March/April, 1990), pp 240-245. Available from atvb.ahajournals.org/content/10/2/240.full.pdf

McLean, J.W., Tomlinson, J.E., Kuang, W.J., Eaton, D.L., Chen, E.Y., Fless, G.M., Scanu, A.M. & Lawn, R.M. (1987). cDNA sequence of human apolipoprotein(a) is homologous to plasminogen. *Nature* Vol. 300, No. 12, (November, 1987) pp 132-137.

Mead, J.R., Irvine, S.A. & Ramji, D.P. (2002). Lipoprotein lipase: structure, function, regulation, and role in disease. *J Mol Med*, Vol. 80, No. 12, pp 753–69. PMID 12483461. Available from http://www.ncbi.nlm.nih.gov/pubmed/12483461

Muller-Roeber, B. & Pical, C. (2002). Inositol Phospholipid Metabolism in Arabidopsis. Characterized and Putative Isoforms of Inositol Phospholipid Kinase and Phosphoinositide-Specific Phospholipase C. *Plant Physiol.*, Vol. 130, No. 1, pp 22-46 doi: 10.1104/pp.004770

Okubo, M., Horinishi, A., Saito, M., Ebara, T., Endo, Y., Kaku, K., Murase, T. & Eto, M.A. (2007). A novel complex deletion-insertion mutation mediated by Alu repetitive elements leads to lipoprotein lipase deficiency. *Mol. Genet. Metab.*, Vol. 92, No. 3, pp 229–33. PMID 17706445. Available from http://www.ncbi.nlm.nih.gov/pubmed/17706445

Palabrica, T.M., Liu, A.C., Aronovitz, M.J., Furie, B., Lawn, R.M. & Furie, B.C. (1995). Antifibrinolytic activity of apolipoprotein(a) in vivo: human apolipoprotein(a) transgenic mice are resistant to tissue plasminogen activator-mediated thrombolysis. *Nat Med*, Vol. 1, pp 256. Available from http://www.nature.com/naturemedicine

Patrick, L. & Uzick, M. (2001). Cardiovascular disease: C-reactive protein abd the inflammatory disease paradigm: HMG-CoA reductase inhibitors, alpha-tocopherol, red yeast rice and olive oil polyphenols. A review of the literature. *Altern Med Rev* Vol. 6, No. 3, pp 248-271. Available from http://www.thorne.com/altmedrev/.fulltext/6/3/248.pdf

Pearson, A., Budin, M. & Brocks, J.J. (2003). Phylogenetic and biochemical evidence for sterol synthesis in the bacterium Gemmata obscuriglobus. *Proc. Natl. Acad. Sci. U.S.A.* Vol. 100, No. 26, pp 15352–7. doi:10.1073/pnas.2536559100. PMC 307571. PMID 14660793.

Rader, D.J., Cain, W., Zech, L.A., Usher, D. & Brewer, H.B. (1993). Variation in lipoprotein(a) concentrations among individuals with the same apolipoprotein (a) isoform is determined by the rate of lipoprotein(a) production. *J. Clin. Invest.* Vol. 91, No. 2, pp 443–7. doi:10.1172/JCI116221. PMC 287951. PMID 8432853

Sandholzer, C., Hallman, D.M., Saha, N., Sigurdsson, G., Lackner, C., Császár, A., Boerwinkle, E. & Utermann, G. (1991). Effects of the apolipoprotein(a) size polymorphism on the lipoprotein(a) concentration in 7 ethnic groups. *Hum. Genet.* Vol. 86, No. 6, pp 607–14. doi:10.1007/BF00201550. PMID 2026424.

Schreiner, P.J., Morrisett, J.D., Sharrett, A.R., Patsch, W., Tyroler, H.A., Wu, K. & Heiss, G. (1993). Lipoprotein(a) as a risk factor for preclinical atherosclerosis. *Arterioscler. Thromb.*, Vol. 13, No. 6, pp 826–33. doi:10.1161/01.ATV.13.6.826. PMID 8499402

Schoenhagen, M. (2006). Current developments in atherosclerosis research. Misra Schoenhagen (editor). Published by Nova Science Inc., New York. ISBN 1-59454-493-X. Available from http://www.novapublishers.com

Sherer, Y., Tenenbaum, A., Praprotnik, S., Shemesh J, Blank M, Fisman EZ, Harats D, George J, Levy Y, Peter JB, Motro M, Shoenfield Y. (2001). Coronary artery disease but not coronary calcification is associated with elevated levels of cardiolipin, beta-2-glycoprotein-I, and oxidized LDL antibodies. *Cardiology*, Vol. 95, pp 20–24. PMID 11385187. Available from http://www.ncbi.nlm.nih.gov/pubmed/11385187?dopt=Abstract

Shlipak, M.G., Simon, J.A., Vittinghoff, E., Lin, F., Barrett-Connor, E., Knopp, R.H., Levy, R.I. & Hulley, S.B. (2000). Estrogen and progestin, lipoprotein(a), and the risk of recurrent coronary heart disease events after menopause. *JAMA*, Vol. 283, pp 1845-1852. URL: http://jama.ama-assn.org/cgi/content/abstract/283/14/1845

Soedamah-Muthu, S.S., Chang, Y.F., Otvos, J., Evans, R.W. & Orchard, T.J. (2003). Lipoprotein subclass measurements by nuclear magnetic resonance spectroscopy improve the prediction of coronary artery disease in type 1 diabetes: a prospective report from the Pittsburgh Epidemiology of Diabetes Complications Study. *Diabetologia*, Vol. 46, pp 674–682. PMID 12743701. Available from http://www.ncbi.nlm.nih.gov/pubmed/12743701?dopt=Abstract

Sotiriou, S.N., Orlova, V.V., Al-Fakhri, N., Ihanus, E., Economopoulou, M., Isermann, B., Bdeir, K., Nawroth, P.P., Preissner, K.T., Gahmberg, C.G., Koschinsky, M.L. & Chavakis, T. (2006). Lipoprotein(a) in atherosclerotic plaques recruits inflammatory cells through interaction with Mac-1 integrin". *FASEB J.* Vol. 20, No 3, pp 559–61. PMID 16403785. Available from http://www.ncbi.nlm.nih.gov/pubmed/16403785

Steinbrecher, U.P., Parthasarathy, S., Leake, D.S., Witztum, J.L. & Steinberg, D. (1984). Modification of lowdensity lipoprotein by endothelial cells involves lipid peroxidation and degradation of low density lipoprotein phospholipids. *Proc Natl Acad Sci U S A.*, Vol. 81, pp 3883–3887. Available from http://www.pnas.org/content/81/12/3883.abstract?ijkey=fea45b830edf328270f68e98a3cc03d2bb008eff&keytype2=tf_ipsecsha

Tall, A.R. (2008). Cholesterol efflux pathways and other potential mechanisms involved in the athero-protective effect of high density lipoproteins. *J Intern Med*, Vol. 263, No. 3, pp 256-273. ISSN 0954-6820

Taskinen, M.R., Puolakka, J., Pyorala, T., Luotola, H., Bjorn, M., Kaariainen, J., Lahdenpera, S. & Ehnholm, C. (1996). Hormone replacement therapy lowers plasma Lp(a) concentrations. Comparison of cyclic transdermal and continuous estrogen-progestin regimens. *Arterioscler Thromb Vasc Biol*, Vol. 16, pp 1215-1221. doi: 10.1161/ 01.ATV.16.10.1215. Available from http://atvb.ahajournals.org/content/16/10/1215.full

Toshima, S., Hasegawa, A., Kurabayashi, M., Itabe, H., Takano, T., Sugano, J., Shimamura, K., Kimura, J., Michishita, I., Suzuki, T. & Nagai, R. (2000). Circulating oxidized low density lipoprotein levels: a biochemical risk marker for coronary heart disease. *Arterioscler Thromb Vasc Biol.*, Vol. 20, pp 2243–2247. doi: 10.1161/01.ATV.20.10.2243. Available from http://atvb.ahajournals.org/content/20/10/2243.abstract?ijkey=4a5cafa55c6ce067f39ca5eb803d7801adb5f339&keytype2=tf_ipsecsha

Toth, P. (2005). The good cholesterol High-Density Lipoprotein. *Circulation*, Vol. 111, No. 5, pp e89-e91. Available from http://circ.ahajournals.org/cgi/content/full/111/5/e89.

Tsutsumi, K. (2003). Lipoprotein lipase and atherosclerosis. *Curr Vasc Pharm*, Vol. 1, pp 11-17. ISSN 1570-1611/03

Utermann, G. (1989). The mysteries of lipoprotein(a). *Science*, Vol. 246, No. 4932, pp 904-10. Available from www.sciencemag.org

Vakkilainen, J., Steiner, G., Ansquer, J.C., Aubin, F., Rattier, S., Foucher, C., Hamsten, A &, Taskinen, M. (2003). Relationships between low-density lipoprotein particle size, plasma lipoproteins, and progression of coronary artery disease: the Diabetes Atherosclerosis Intervention Study (DAIS). *Circulation.*, Vol. 107, pp 1733–1737. doi: 10.1161/01.CIR.0000057982.50167.6E. Available from http://circ.ahajournals.org/content/107/13/1733.abstract?ijkey=401352270488f9ca9f3caafaf43179b7a352181c&keytype2=tf_ipsecsha

Vance, D.E. & Vance, J.E. (2002). Biochemistry of lipids, lipoproteins and membranes. Elsevier, Amsterdam. ISBN 978-0-444-53219-0.

Vigna, G.B., Donega, P., Zanca, R., Barban, A., Passaro, A., Pansini, F., Bonaccorsi, G., Mollica, G. & Fellin, R. (2002). Simvastatin, transdermal patch, and oral estrogen-progestogen preparation in early-postmenopausal hypercholesterolemic women: a randomized, placebo-controlled clinical trial *Metabolism*, Vol. 51, pp 1463-1470.

Wang, X., Briggs, M. R., Hua, X., Yokoyama, C., Goldstein, J. L. & Brown. M. S. (1993). Nuclear protein that binds sterol regulatory element of LDL receptor promoter: II. Purification and characterization. *J. Biol. Chem.* Vol. 268, pp 14497–14504.

White, A.L., Rainwater, D.L., Hixson, J.E., Estlack, L.E. & Lanford, R.E. (1994). Intracellular processing of apo(a) in primary baboon hepatocytes. *Chem. Phys. Lipids*, Vol. 67-68, pp 123–33. doi:10.1016/0009-3084(94)90131-7. PMID 8187206.

Wilson, P.W., D'Agostino, R.B., Levy, D., Belanger, A.M., Silbershatz, H. & Kannel, W.B. (1998). Prediction of coronary heart disease using risk factor categories. *Circulation.*, Vol. 97, pp 1837–1847. PMID 9603539. Available from http://www.ncbi.nlm.nih.gov/pubmed/9603539

Young, I.S. & McEneny, J. (2001). Lipoprotein oxidation and atherosclerosis. Biochem Soc Trans 29 (2): 358-362. PMID 11356183. Available from http://www.ncbi.nlm.nih.gov/pubmed/11356183

Yokoyama, C., Wang, X., Briggs, M. R., Admon, A. , Wu, J., Hua, X., Goldstein, J. L. & Brown, M. S. (1993). SREBP-1, a basic helix-loop-helix leucine zipper protein that controls transcription of the LDL receptor gene. *Cell*. Vol. 75, pp 187–197. PMID 8402897. Available from http://www.ncbi.nlm.nih.gov/pubmed/8402897

Zenker, G., Koltringer, P., Bone, G., Niederkorn, K., Pfeiffer, K. & Jurgens, G. (1986). Lipoprotein(a) as a strong indicator for cerebrovascular disease. *Stroke*, Vol. 17, pp 942-945. doi: 10.1161/01.STR.17.5.942. Available from http://stroke.ahajournals.org/content/17/5/942.

Zioncheck, T.F., Powell, L.M., Rice, G.C., Eaton, D.L. & Lawn, R.M. (1991). Interaction of recombinant apolipoprotein(a) and lipoprotein(a) with macrophages. *J Clin Invest*, Vol. 87, pp 767. 0021-9738/91/03/0767/05

Chapter 9

Linking the Pathobiology of Hypercholesterolemia with the Neutrophil Mechanotransduction

Xiaoyan Zhang and Hainsworth Y. Shin

Additional information is available at the end of the chapter

http://dx.doi.org/10.5772/47867

1. Introduction

This chapter focuses on the potential contributions of the blood-borne neutrophils to hypercholesterolemia-related pathophysiology (e.g., thrombus formation, embolism, heart attack, stroke, etc.). Neutrophils are immersed in the cholesterol-abundant plasma of blood and play critical roles in the acute inflammatory response of the body to infection or tissue damage. Because of their high degree of sensitivity to inflammatory agonists and their arsenal of potent microbicidal and tissue degradative agents, a number of redundant cellular mechanisms exist to control or "turn-off" the inflammatory processes by these cells under physiological (non-pathological) conditions. Failure of these mechanisms leads to sustained levels of cell activity that contribute to a chronic inflammatory phenotype with the continuous release of proteases and cytokines as well as the potential to elicit non-specific damage to host tissues. Alternatively, chronic neutrophil activity may impair tissue perfusion via its effects on the rheological flow behavior of blood particularly in terms of the ability of leukocytes to transit the microcirculation[1]. Such potential damage mechanisms are thought to govern an increasing number of human pathological scenarios (e.g., Alzheimer's, diabetes, vascular disease) that correlate with a chronic inflammatory state. In this regard, chronic inflammation has gained recognition in the scientific community and even in the mainstream national media (e.g., Time[2] and Newsweek[3] magazines) as a common denominator for human diseases. The question is whether the dysregulation of neutrophil activity is a significant component of this potential disease mechanism.

We address this issue from a mechanobiological perspective by presenting evidence that supports a role of impaired neutrophil mechanotransduction of hemodynamics-derived fluid flow in the pathogenesis of hypercholesterolemia-linked diseases. For this purpose, we will first discuss the link between chronic inflammation and hypercholesterolemia and then highlight the neutrophil involvement in the pathophysiology of related cardiovascular diseases, e.g. atherosclerosis and microvascular dysfunction. To further exemplify this link, we will explain the potential mechanism(s) by which cholesterol in blood may impact the biochemical regulation of neutrophil activity at the cellular level. Finally, we will introduce our own evidence as well as those of others pointing to dysregulated neutrophil mechanotransduction as an important component of hypercholesterolemic pathologies.

2. Hypercholesterolemia and chronic inflammation

Hypercholesterolemia is the dominant risk factor for atherosclerosis and its downstream complications (e.g., heart attack, stroke, etc.). Over the past two decades, a wealth of insight has pointed to the development of atherosclerotic lesions in the large vessels (e.g., aorta, carotid, femoral artery, etc.) as occurring at the interface between hypercholesterolemia and inflammation (see reviews[4, 5]). According to the current paradigm, at atheroma-prone sites, inflamed endothelial cells (due to damage or dysregulation) initiate the invasion of blood leukocytes (predominantly, monocytes) and smooth muscle cells (SMCs) into the subendothelial (e.g., the intimal) layer of the vascular wall contributing to atherosclerotic tissue remodeling, thrombosis, and finally embolus formation. The main lipid species that appear to dominate this inflammatory process are modified low-density lipoprotein (LDL) particles, particularly oxidized LDL (oxLDL), which act as potent proatherogenic and proinflammatory factors responsible for not only loading monocyte-derived macrophages with cholesterol but also directly stimulating leukocytes and other vascular wall cells (for a more complete explanation, see reviews[5, 6]).

Hypercholesterolemia also induces chronic inflammation in the microcirculation[7]. Phenotypic changes in the microvasculature are observed long before the appearance of fatty streak lesions in the large arteries of animals placed on high fat (HFD), i.e. proatherogenic, diet[8, 9]. The inflammatory phenotype of the microvessels in hypercholesterolemic animals results in increased basal levels of rolling, adherent, and emigrating leukocytes in the postcapillary venules, predominantly neutrophils, as well as enhanced production of reactive oxygen species (ROS). Hypercholesterolemia also exaggerates microvascular responses to a range of proinflammatory stimuli. For example, the postcapillary venules of LDL receptor deficient (LDLr-/-) mice, a murine model of modest hypercholesterolemia (with 3-fold higher levels of plasma cholesterol compared to their wild-type (WT) counterparts), exhibit enhanced leukocyte adhesion and albumin leakage in response to experimental ischemia/reperfusion injury as compared to those of WT mice[10]. Interestingly, similar phenotypic changes can be observed in the microvasculature of normocholesterolemic animals administered oxLDL[11, 12], suggesting that oxLDL participates in hypercholesterolemia-related microvascular dysfunction.

Although the mechanisms responsible for the induction of inflammation by hypercholesterolemia in both microvessels and larger arteries remain unclear, it appears that they both begin with endothelial dysfunction characterized by reduced vasodilation, a proinflammatory state, and enhanced permeability to macromolecules (e.g., lipids). However, the original triggers for this endothelial dysfunction are still controversial. In this regard, oxLDL, based on its potent proinflammatory effects, has been considered as a candidate that initiates the inflammatory responses. In fact, production and release of ROS and myeloperoxidase (MPO)[13], which play critical roles in the oxidation of LDL-cholesterol conjugates and are tightly controlled under the physiological non-inflamed conditions, increase in response to hypercholesterolemia. The cellular basis for the causality between oxLDL formation and chronic inflammation, however, remains elusive.

Interestingly, the preferential formation of atherosclerotic lesions at bifurcations, severe curvatures, and stenoses in the arterial circulation strongly suggests that the hemodynamic flow environment is an important determinant in atherogenesis. Fluid flow-derived frictional (i.e., shear) stresses imposed on the surfaces of endothelium lining the vascular wall have been shown to serve an atheroprotective function when blood flow is laminar (i.e., smooth and ordered; for a more comprehensive discussion, see review[14]). For example, laminar fluid flow stimulates endothelial production of nitric oxide (NO), with vasodilatory and anti-inflammatory actions[15]. In contrast, oscillatory shear stresses enhance production and release of ROS[16]. In addition, disturbed flows lead to the upregulation of adhesion molecules on the endothelial surface (e.g., intercellular adhesion molecule-1 or ICAM-1) responsible for recruiting leukocytes to the vascular wall[17]. In effect, generation of complex distributions of fluid shear stresses on the vascular wall, such as at sites of bifurcations and branch points, appears to shift the endothelial phenotype from atheroprotective to proatherogenic.

However, complex flow fields are not sufficient for the onset/progression of hypercholesterolemia-related atherosclerosis since, for example, we are born with bifurcations and curvatures but do not develop atherosclerosis from birth. It is, therefore, clear that cardiovascular pathobiology due to hypercholesterolemia occurs at the intersection between vascular cell biology and the surrounding fluid stress environment. In this regard, it may be the sensitivity (i.e., responsiveness) of vascular cells to fluid shear stress that is altered in the face of hypercholesterolemia leading to a proinflammatory and a proatherogenic phenotype. Moreover, the endothelial cells are not the only cells in the vasculature. Neutrophils also exist in the cholesterol-enriched, fluid flow environment of the circulation and are critical for initiating acute inflammation. Recently, a growing body of evidence supporting the involvement of neutrophils has emerged.

3. The neutrophil involvement in hypercholesterolemia-related vascular dysfunction

Neutrophils make up the majority of the nucleated leukocytes in human blood with the remaining being monocytes and lymphocytes. As the principal gatekeepers of the acute inflammatory response of the body's immune system, neutrophils are extremely sensitive to inflammatory stimuli allowing them to rapidly (i.e., on the order of milliseconds) transition from an inactivated to an activated state. Upon activation, the upregulation of cell-cell adhesion molecules (e.g., selectins, integrins) enables neutrophils to roll and adhere onto the endothelium prior to their transmigration (via diapedesis) across the vascular wall and to the target tissues (i.e., sites of infection and tissue injury) where they release a potent array of biochemicals including proinflammatory mediators, ROS, and proteases to fight infection and orchestrate tissue repair. During this process, neutrophils also undergo changes in their physical attributes such as their size, deformability, and adhesiveness. It is these features that cause neutrophils to be major players in the pathobiology of hypercholesterolemia-related cardiovascular diseases for both the macro- and micro- circulations.

3.1. Potential roles of neutrophils in atherosclerosis

While it has long been appreciated that monocytes and their descendants, along with T lymphocytes, mast cells, and platelets, contribute to the development and destabilization of atherosclerotic lesions, only recently has the neutrophil been seriously considered as a contributing factor for disease onset and/or progression. Direct evidence comes from the identification of neutrophils in different locations of atherosclerotic lesions present in hypercholesterolemic mice and humans using antibodies to neutrophil-specific antigens including Ly6G, CD177, and CD66b[18-21]. Neutrophils, in fact, have been reported to accumulate in atheroprone arteries preceding plaque formation in hypercholesterolemic murine models of atherosclerosis[18, 20]. Further evidence of a neutrophilic component in the early stages of atherosclerosis is the positive correlation between the number of circulating neutrophils and lesion sizes[18]. Experimental data also point to neutrophil infiltration into the highly inflamed areas of atherosclerotic arteries during late disease stages[19] with contribution to lesion destabilization and thromboembolus formation[21, 22].

One way chronically activated neutrophils may enter atheroprone regions in the macrocirculation are at sites of disturbed flow and recirculation[23] where their enhanced residence times promote capture at the vascular wall[24]. Alternatively, activated neutrophils may disrupt vascular (i.e., adventitial or medial) wall perfusion in the vasa vasorum (or microcirculation) of large vessels (e.g., aorta) leading to vessel tissue injury followed by atherogenesis (from within the vessel wall to the luminal surface)[25-27]. In these ways, neutrophils may initiate or exacerbate atherosclerosis at different stages via their capability to release vast amounts of ROS and proteins stored in their cytosolic granules[28]. For example, while MPO released by activated neutrophils can reduce the bioavailability of NO[29, 30] and contribute to the onset of endothelial dysfunction, a number of granule proteins, such as LL-37, azurocidin, cathepsin G, and α-defensins, exert direct chemotactic activity for monocytes[31, 32]. Moreover, neutrophil secretory products, e.g., α-defensins, also promote macrophage maturation and activation, contributing to the uptake of oxLDL and the formation of foam cells[28]. Finally, neutrophil-derived proteolytic enzymes, particularly matrix metalloproteinase (MMP) -2 and -9[28], play critical roles in plaque destabilization and eventual rupture.

3.2. Effects of neutrophils on microvascular dysfunction

Similar to situations in large arteries, activated neutrophils promote microvascular dysfunction through the sustained release of proinflammatory, cytotoxic, and degradative agents. However, while leukocytes have no significant effect on the macrovascular flow properties of blood (which is dominated by the substantially greater numbers of erythrocytes), these cells, particularly the neutrophils, influence blood flow in the microcirculation (Figure 1). This is because vessel diameters in the microcirculation are in the range of 6 – 100 μm, which are comparable in size to the diameters of leukocytes.

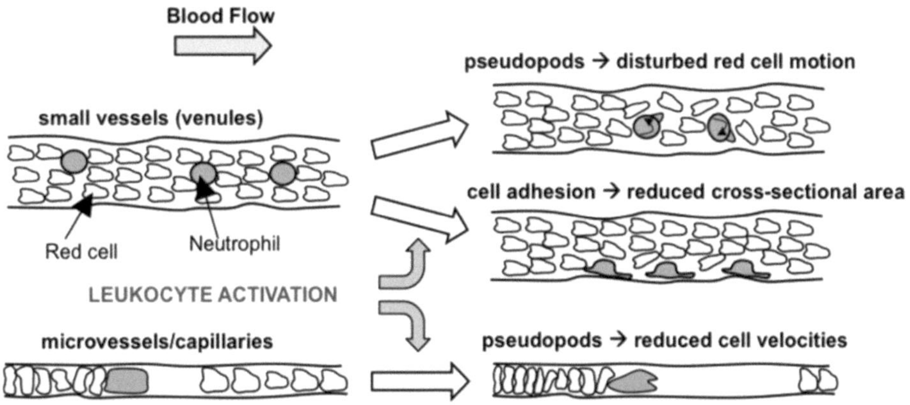

Figure 1. Potential rheological effects of leukocyte activation on the blood flow in the microcirculation. Sustained activation, e.g., due to proinflammatory stimuli, hinders leukocyte passage through the small vessels either by promoting pseudopod projections or through enabling cell adhesion to the vascular wall. Ultimately, these may elevate peripheral resistance and contribute to microvascular dysfunctions (adapted from Shin, H.Y., et al., 2011[79]).

The relatively comparable size scales of the neutrophil and vessel diameters are important to note since the quiescent neutrophil under physiological (i.e., non-inflamed) conditions is capable of efficient transit through the microvessels due to their inherently round, deformable, and non-adhesive state. On the other hand, cell activation physically hinders the passage of neutrophils through the small vessels of the microcirculation[1]. Pseudopods projected by activated neutrophils, while enabling cells to attach to other cells (e.g., endothelial cells, other blood cells) or phagocytose particles, also contribute to a reduction in cell deformability due to their enriched content of F-actin, and increases in geometric size and irregularity[1, 33], all of which serve to increase leukocyte transit time or enhance leukocyte retention in the microvasculature[34, 35]. In turn, activated leukocytes disrupt the motion of erythrocytes, leading to increases in the apparent viscosity of blood and microvascular resistance[1, 34, 36, 37]. Moreover, activated neutrophils are hyperadhesive and exhibit extensive interactions with other leukocytes or platelets, e.g. during hypercholesterolemia[7], which may also enhance the apparent viscosity of blood. Finally, once neutrophils adhere to endothelium, they further increase flow resistance by reducing microvessel diameters (resistance $\propto 1/[\text{diameter}]^4$)[34, 37]. The state of neutrophil activation is, thus, a critical determinant of tissue blood flow and perfusion.

In summary, as a result of their arsenal of noxious agents and their effects on microvascular blood flow, it is evident that tight regulation of neutrophil activity is an essential requirement for a healthy circulation. A failure to either prevent or "turn-off" cell activity, e.g., due to hypercholesterolemia, leads to sustained neutrophil activation which has potential impacts not only in terms of the initiation and progression of atherosclerosis in large arteries but also as it relates to microvascular blood flow and downstream tissue perfusion.

4. The influence of cholesterol on neutrophil activity

One way hypercholesterolemia may influence the activation state of neutrophils is to modify the lipid composition of biological membranes. Cholesterol is an essential component of mammalian cell membranes. Approximately 90% of the free (i.e. unesterified) cholesterol in cells resides in the plasma membrane[38]. These sterol molecules not only maintain the integrity of cell membranes, but also play an important role in regulating membrane properties (e.g., microviscosity) and functions (e.g., via their influence on membrane-bound signaling components). In addition to de novo biosynthesis, mammalian cells can take up cholesterol from the extracellular milieu. Exposure to elevated cholesterol levels both in vivo and in vitro enhances cholesterol abundance within the plasma membrane of neutrophils and other blood cells[39-42]. These findings, in conjunction with the wealth of evidence demonstrating the influence of the extracellular cholesterol levels on neutrophil activity, point to membrane cholesterol enrichment as a potential link between hypercholesterolemia and chronic neutrophil activity. To better understand this link, we next describe the possible cholesterol uptake pathways, the influence of cholesterol on physicochemical properties of the cell membrane, and lastly, the influence of cholesterol on neutrophil activity.

4.1. Transport of extracellular cholesterol into the plasma membrane

Due to its insolubility in aqueous media, cholesterol must be transported complexed to carrier molecules, i.e. within the hydrophobic cores of lipoproteins[43]. Lipoproteins (e.g., LDL) in the blood plasma are positioned in close proximity to the circulating blood cells. Conditions that elevate cholesterol-enriched lipoprotein levels may thus favor cholesterol transport into the membranes of these blood cells[39, 42]. In the laboratory, cyclodextrin derivatives (e.g., methyl-β-cyclodextrin or MβCD), synthetic cholesterol carrier molecules, are commonly used to alter membrane cholesterol abundance. Such treatments elicit acute changes in membrane cholesterol levels and downstream cell activity indicating the existence of mechanism(s) that permit rapid transport of cholesterol into nearby cell membranes. Cholesterol uptake may occur by either receptor dependent/independent endocytosis followed by rapid membrane mobilization[44] or direct exchange between the hydrophobic environments of carrier molecules and the lipid bilayer(Figure 2).

4.1.1. Receptor-dependent endocytosis

For a variety of cell types (including leukocytes)[45], LDL-cholesterol is taken up in vivo mainly through LDL receptor (LDLr)-mediated endocytosis. LDLr expression is subject to feedback regulation and, thus, is unlikely to contribute significantly to the overaccumulation of cellular cholesterol[43]. Cellular uptake of cholesterol can also occur via endocytosis mediated by other receptors[43]. Potentially, these pathways can load cholesterol continuously into the cell leading to cholesterol elevations in the plasma membrane[43]. For example, a family of scavenger receptors have been identified on monocytes, macrophages, and SMCs[6] that, by binding to modified LDL (e.g., oxidized and acetylated LDL) with high affinity, account for the majority of cholesterol uptake by these cells[6]. To our knowledge, such scavenger receptors have not yet been identified for neutrophils.

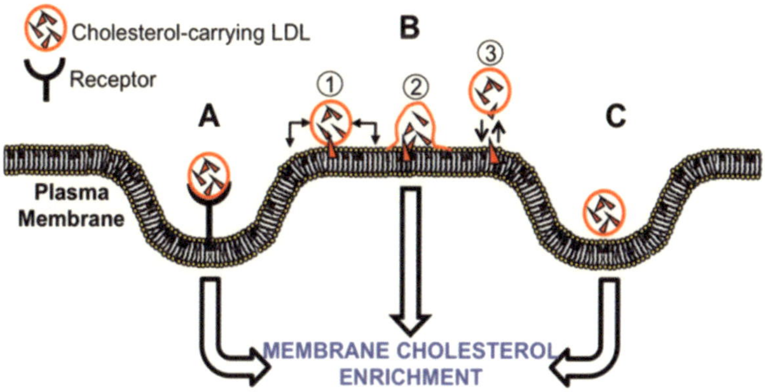

Figure 2. Schematic representation of three possible modes of cholesterol uptake. A: Receptor-mediated endocytosis; B: Direct surface exchange of cholesterol between extracellular carrier molecules and plasma membrane which may occur due to the formation of a transient collision complex without (①) or with (②) membrane fusion or resulting from diffusion across the aqueous phase (③); C: Receptor-independent endocytosis.

4.1.2. Receptor-independent endocytosis

The entire LDL particle can be internalized as a result of fluid or bulk endocytosis without receptor-mediated LDL binding to the cell surface[46]. It is taken up at a rate strictly proportional to its concentration in the extracellular milieu[43]. Alternatively, some LDL, e.g., cationized LDL, can also be taken up by the cell through a non-specific low affinity adsorptive endocytotic process. In this case, endocytosis occurs after cationized LDL binds to the negatively charged membrane surface[47]. For both of these modes of endocytosis, cholesterol transport is not influenced by intracellular cholesterol levels and thus may lead to progressive cholesterol uptake[43].

4.1.3. Cholesterol surface exchange

Cholesterol may also directly enter or exit the plasma membrane[48]. In this case, free cholesterol is exchanged between the hydrophobic cores of the plasma membrane and extracellular carriers (e.g., lipoproteins). The direction of net flux of cholesterol is governed by its concentration gradient between the lipid bilayer and the carrier molecules. Two mechanisms for this surface transfer have been proposed: 1) formation of transient collision complexes with/without membrane fusion and 2) direct diffusion across the aqueous phase. In principle, these transport modes follow similar kinetics with transfer rates depending on the concentrations and structures of both donor (e.g., extracellular LDL) and acceptor (e.g., cells) particles (for more details, see review[48]). This level of complexity contributes, in part, to the diverse half-times ranging from seconds to hours measured for the uptake of cholesterol by human erythrocytes[49]. Finally, this pathway is not under feedback control.

4.2. The influence of cholesterol on the plasma membrane dynamics

In the lipid bilayer, cholesterol orients with its polar hydroxyl group encountering the aqueous phase and the hydrophobic steroid ring parallel to and buried in the hydrocarbon chains of the phospholipids[50]. This unique orientation allows cholesterol to interact with membrane phospholipids and sphingolipids and thus influence their physicochemistry. Along these lines, cholesterol influences the physical and biological properties of the lipid bilayer and, in doing so, impacts the functions of membrane signaling molecules.

4.2.1. Effects of cholesterol on the physical properties of cellular membranes

A key function of cholesterol is to modulate the fluidity (i.e., the inverse of microviscosity) of the lipid bilayer. The close inter-positioning of sterols (i.e., cholesterol) between neighboring membrane phospholipids imposes a degree of immobility on the carbon atoms nearest the membrane surface, while increasing the freedom of motion deep within the hydrophobic core of the membrane[51]. In this regard, membrane fluidity reflects the temperature-dependent influence of cholesterol on the gel to liquid-crystal (i.e., solid-like to fluid-like) phase transition of the lipid bilayer[51]. Under physiological conditions (i.e., 37°C), biological membranes adopt a liquid crystalline state whereby increases and decreases in cholesterol content reduce and enhance membrane fluidity, respectively[42, 52, 53].

Operationally, membrane fluidity refers to the ensemble of physical properties that govern the motion of the phospholipid molecules in a membrane, including segmental, rotational, lateral, and translational motions[54]. In this fashion, lipid bilayer fluidity can physically influence the dynamics of membrane-associated molecules including proteins that drive downstream cell functions. Two mechanisms have been proposed to explain how membrane fluidity alters the activities and functions of membrane proteins. One mechanism occurs through effects on protein mobility, particularly lateral diffusion which impacts collisional encounters[54]. This effect appears to be of physiological significance particularly for diffusion-controlled processes that are mediated by large membrane proteins. For example, modulation of membrane fluidity influences Ca++-dependent cAMP signaling through changes in protein mobility that governs the coupling between hormone receptors and the adenylate cyclase catalytic unit[55, 56]. Alternatively, membrane fluidity can also influence the structural flexibility of membrane proteins and, thus, their ability to adopt an optimal conformation for activity[57]. Membrane fluidity has, in fact, been reported to impact the conformation-dependent activation of G-protein coupled receptors (GPCRs) on endothelial cells by affecting changes in the protein tertiary structure[58].

4.2.2. Role of cholesterol in lipid raft structure and function

Lipid rafts are nano-scale microdomains that are abundant in the plasma membrane. Structurally, lipid rafts consist of dynamic assemblies of cholesterol and sphingolipids in the outer leaflet of the phospholipid bilayer[59]. The preponderance of saturated hydrocarbon chains in raft sphingolipids renders lipid rafts with a distinct liquid-ordered (i.e., solid-like) phase that is dispersed in the liquid-disordered matrix of the lipid bilayer[60]. One important property of lipid rafts is that they include or exclude proteins to variable extents depending on the raft affinity of proteins[59]. Once individual rafts cluster, they spatially facilitate interactions between raft proteins and expose them to a new membrane environment that is enriched in accessory enzymes and/or second-messenger molecules. In doing so, lipid rafts serve to efficiently initiate and/or amplify signaling cascades. As such, lipid rafts act as signaling platforms that orchestrate outside-in and inside-out signal transduction. Interestingly, these cholesterol-rich microdomains have also been implicated as mechanotransduction centers such as caveolae, a subtype of lipid rafts that reportedly play a role in endothelial mechanotransduction of shear stress and pressure[61-63].

Cholesterol is required to support the formation of lipid rafts and maintain their functionality. It condenses the packing of sphingolipids in the exoplasmic leaflet by occupying the spaces between their saturated hydrocarbon chains near the hydrophilic polar head groups. In this way, cholesterol content and organization influence the stability of lipid rafts with an impact on their capacity to interact with target proteins. Removal or depletion of cholesterol from the plasma membrane using MβCD has been widely used to disrupt rafts and disperse raft proteins into the liquid-disordered matrix of the cell membrane[59]. Treatment of cells with cholesterol-sequestering agents (e.g., filipin or nystatin) or inhibition of cholesterol biosynthesis (e.g., lovastatin) as well as addition of exogenous cholesterol into cell membranes also disrupts raft structure leading to an impact on the functions of raft proteins[59]. As a consequence of these lipid raft-related perturbations, neutrophil functions (e.g., chemokine-induced calcium signaling, extracellular regulated kinase activity, cell polarization, shape change, adhesion, migration, integrin expression, and actin polymerization) are altered[64-68].

4.3. Effects of elevated cholesterol environments on neutrophil activity

Up to this point, we have described how perturbations in extracellular cholesterol levels modify the membrane physicochemistry and the mode by which these modifications may influence membrane protein-related signaling in the neutrophil. The altered cell signaling capacity of membrane-bound proteins is followed by changes in cell behavior that contribute to the principal role of the neutrophil as the first responder to tissue damage and infection. In this way, the influence of membrane cholesterol on the ability of the neutrophil to sense its environment extends to basic cell functions including cell adhesion and migration, phagocytosis, ROS production, and degranulation. We will discuss the effects of cholesterol on these cell functions in order to illustrate the link between the lipid bilayer properties and the control of neutrophil activation.

4.3.1. Expression of membrane adhesion molecules

Upon agonist stimulation, neutrophils exhibit upregulated expression of adhesion molecules that facilitate their recruitment to sites of inflammation by enabling their binding to other cells (e.g., leukocytes, platelets, endothelium)[33]. Two classes of adhesion molecules govern leukocyte interactions with other cells: selectins and integrins. In addition to the ligands for platelet (P)- or endothelial (E)- selectins, neutrophils constitutively express leukocyte (L)-selectins and β_2 (i.e., CD18) integrins, which participate in their initial capture and firm adhesion to other cells, respectively[33]. Currently, the impact of hypercholesterolemia on expression of the selectin family of adhesion molecules is unclear since neutrophils in a cholesterol-rich environment have been reported to exhibit both elevated surface expression[69] and cleavage of L-selectins[70]. In the case of the integrins, surface levels of CD18, particularly Mac-1 (CD11b/CD18), are elevated on neutrophils exposed to a hypercholesterolemic environment both in vitro and in vivo[29, 69, 70]. Notably, surface expression of Mac-1 by neutrophils in hypercholesterolemic patients positively correlates with serum cholesterol levels[29]. But, cholesterol enrichment does not appear to alter the expression of LFA-1 (CD11a/CD18)[69], another CD18 subtype. Thus, the influence of extracellular cholesterol levels on neutrophil adhesion molecule expression is receptor-specific.

Moreover, neutrophils exposed to elevated cholesterol levels undergo increased adhesive interactions with other cells. For example, neutrophils with increased membrane cholesterol exhibit enhanced tethering and firm arrest on activated endothelial cell monolayers[41, 71, 72]. Moreover, neutrophils exposed to hypercholesterolemia display increased heterotypic adhesion to platelets[73] as well as increased homotypic aggregation in response to 10 μM N-formyl-Met-Leu-Phe (fMLP)[74]. These studies confirm that cholesterol-dependent modulation of adhesion molecule expression has an impact on neutrophil adhesion to other leukocytes, platelets, or the endothelium lining the blood vessel lumen.

4.3.2. ROS production

Neutrophil-derived ROS includes superoxide (O^{2-}), hydrogen peroxide (H_2O_2), hydroxyl radicals (^{-}OH), and NO-related oxidants. Notably, total production of ROS by neutrophils in a hyperlipidemic environment positively correlates with levels of triglycerides and LDL but not with total amount of cholesterol in the plasma[75]. Recent studies, however, did demonstrate a positive correlation between O^{2-} release rate and plasma cholesterol levels[29]. Interestingly, enhanced O^{2-} release by neutrophils was detected in other clinical states associated with cardiovascular complications, namely hypertension and diabetes[76,

77], which are usually accompanied by hyperlipidemia. In fact, elevations in extracellular cholesterol levels have been shown to enhance neutrophil respiratory burst in response to agonist stimulation. Moreover, plasma activity of superoxide dismutase (SOD), which scavenges ROS, decreases with increases in total cholesterol[75].

4.3.3. Degranulation

Neutrophils contain four main types of granules: primary, secondary, and tertiary granules as well as secretory vesicles. These granules contain a multitude of cytokines (e.g., interleukins, tumor necrosis factor-α, etc.), enzymes (e.g., MPO, etc.), and proteases (e.g., cathepsins, MMPs, etc.). Upon activation, neutrophils degranulate and release these bioactive mediators into the extracellular milieu. Interestingly, although neutrophils from hyperlipidemic patients contain significantly lower levels of intracellular MPO, sera from these patients exhibit significantly higher levels of MPO[29]. These results point to a degranulation process that further links the activation state of neutrophils with the cholesterol levels in the blood environment.

5. Membrane cholesterol and the neutrophil mechanosensitivity to shear stress

In addition to the presence of inflammatory stimuli (e.g., oxLDL), elevated neutrophil activity in hypercholesterolemia may result from defects in their mechanotransduction of fluid shear stress, a control mechanism to prevent spontaneous neutrophil activity under physiological conditions[78, 79]. In this regard, the mechanosensitivity of neutrophils may serve as a key regulator of the inflammatory status of the circulation. We will first define the leukocyte mechanosensitivity to shear followed by a brief discussion of cellular mechanisms that link the extracellular flow environment to downstream neutrophil functions. Interestingly, such mechanotransduction processes occur across the plasma membrane that plays a critical role in regulating the activity of membrane proteins as well as the transmembrane movement of bioactive molecules. The direct contact of cell membrane with the extracellular flow environment makes it a likely target of local environmental factors (e.g., enhanced cholesterol abundance) that influence the neutrophil responsiveness to mechanical stimuli.

5.1. Regulation of neutrophil activity by fluid flow-derived shear stress

Neutrophils, either freely suspended in the bloodstream or adhered to/migrating on vascular endothelium, sense and respond to fluid shear stress[80-82]. Fluid shear stress (ranging from approximately 1 to 10 dyn/cm^2) minimizes neutrophil activity levels[78]. The most obvious manifestation of the cell-inactivating effects of shear exposure on cell activity is the retraction of existing pseudopodia by non-cytokine-stimulated human neutrophils adhered to a surface and subjected to a non-uniform flow field imposed by a micropipette with a tip of diameter in the range of 4 – 8 μm[82] (Figure 3A). This situation models brief and spontaneous periods of blood stasis followed by reperfusion, a typical scenario in the microvessels. Under this condition, neutrophils sediment, attach, extend pseudopods, and migrate on the vascular endothelium. Upon reintroduction of fluid flow, these cells retract pseudopods and detach into the flow field in a mechanobiological fashion. Such a scenario has been documented using intravital microscopy of microvascular networks of rodents (e.g., mesentery, spinotrapezius muscle, cremaster muscle)[81-83]. The ability of shear stress to minimize pseudopod activity has been further confirmed for non-adherent heterogeneous leukocyte populations[84] exposed to a constant shear field (5 dyn/cm^2) in a cone-plate viscometer (Figure 3B).

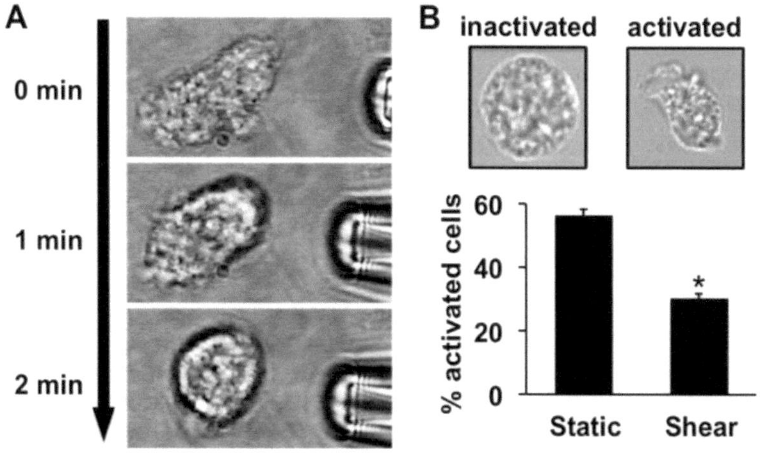

Figure 3. Deactivation of neutrophils under flow stimulation. A: A migrating/adherent neutrophil exposed to a micropipette flow (~ 2 dyn/cm^2) for 2 min. B: Non-adherent neutrophils in suspension exposed to cone-plate shear (5 dyn/cm^2) for 10 min. Bars are mean percentage of activated cells with pseudopods (see image insets) in each population tested ± SEM; *$p < 0.05$ compared to static condition using paired Student's t-test.

Notably, impairment of shear-induced pseudopod retraction by treating neutrophils with cell agonists above threshold concentrations, e.g. fMLP (>10^{-8} M), commits these cells to an activated (inflamed) phenotype and leads to their microvascular entrapment due to increases in adhesivity, size, and stiffness[35, 81, 82]. Thus, during inflammation, the biochemical milieu of the neutrophil overrides mechanobiological deactivation. Exposure to shear of magnitudes typically found in the macro- and micro- circulations is also associated with other attributes of neutrophil deactivation such as decreased surface expression of integrin receptors (i.e., CD18), depolymerization of the F-actin cytoskeleton, cell detachment, and attenuated phagocytic activity[81, 82]. Moreover, shear stress exposure enhances caspase 3-dependent apoptosis[85], in line with the relatively short lifespan (18 to 24 hrs) of these cells when they are passively circulating in the physiologic bloodstream. These observations support the key role of fluid flow-related shear stress as a biophysical stimulus that promotes neutrophil inactivation when cell activity is below a threshold level. As such, the mechanical influence of fluid flow serves an anti-inflammatory role.

5.2. Shear stress mechanotransduction at the neutrophil surface

An understanding of the fluid flow mechanoregulation of neutrophil activity in the circulation reveals clues regarding how impaired mechanosensitivity to flow may be a mitigating factor for hypercholesterolemic disorders. Membrane detachment during pseudopod retraction by migrating neutrophils in response to fluid shear stress points to two fundamental requirements that must be fulfilled by the cell signaling apparatus: 1) depolymerization of the F-actin cytoskeleton that serves as a structural and a signaling scaffold for neutrophil motility and 2) rapid disengagement of adhesion receptors that anchor the pseudopod to the underlying substrates. For suspended neutrophils, similar events are needed but, in this case, mechanisms must be in place to prevent the expression of adhesive proteins or interfere with engagement of adhesion molecule with substrates (e.g. foreign surfaces, other cells) presenting counter-receptors. These fundamental requirements point to the neutrophil surface components as critical players in mechanotransduction since the cell must sense the extracellular flow environment and remediate its interactions with the cellular microenvironment (e.g., the surrounding matrix and cells).

5.2.1. GPCRs and shear stress control of neutrophil pseudopod activity

Shear stress-induced pseudopod retraction by neutrophils occurs in parallel with a rapid decrease in F-actin content[86, 87]. Typically, remodeling of the F-actin cytoskeleton in leukocytes is controlled by the Ras superfamily of small guanine triphosphate (GTP)-binding proteins, particularly the small GTP-binding phosphatases (GTPases) including Rac1, Rac2, cdc42 and members of the Rho family (as reviewed in the literature[88-90]). Rather than stimulating the activity of molecules that coordinate pseudopod retraction (e.g., RhoA, MLCK), fluid shear stress appears to either inhibit (e.g., possibly through release of an inhibitor) or interfere with the ability of neutrophils to form and sustain pseudopod projections via reducing cytosolic activity of the key small GTPases (e.g., Rac1, Rac2) involved in actin polymerization[83]. These reported effects point to the actions of fluid shear stress on G protein signaling downstream of GPCRs that regulate neutrophil chemotaxis, such as the formyl peptide receptor (FPR).

Notably, fMLP, a ligand for FPR, dose-dependently impairs neutrophil pseudopod retraction responses to shear stimulation[81]. Along this line, HL-60-derived neutrophils subjected to shear stress exhibit reduced activity of $G_{\alpha i}$ downstream of FPR[91]. A critical piece of evidence pointing to FPR as a mechanosensory regulator of pseudopod retraction is the observation that transfection of FPR expression plasmid in undifferentiated HL-60 cells not only confers expression of this receptor but imparts on these cells the ability to form pseudopods that retract under the influence of fluid shear stress[91]. Furthermore, HL-60 promyelocytes differentiated into neutrophils and subsequently transfected with siRNA to silence FPR expression exhibit an attenuated pseudopod retraction response to shear exposure, despite the fact that these cells retain the ability to project pseudopods because of the presence of other cytokine-related GPCRs[91]. Together, these observations point to a role of fluid flow in regulating the activity of membrane-associated receptors by establishing the importance of membrane-bound GPCRs, specifically FPR, in the neutrophil pseudopod retraction response to shear stress. In conjunction with the dependence of GPCR activity on the membrane cholesterol content, it is conceivable that the influence of shear stress on GPCR activity is impacted by perturbations in extracellular cholesterol abundance and their effects on the cell membrane properties.

5.2.2. Cell surface CD18 integrins and shear stress regulation of neutrophil adhesion

Pseudopod retraction by migrating neutrophils subjected to fluid flow depends on their expression levels of CD18 integrins[92], consistent with the requirement of these receptors for cyclical pseudopod projection and retraction[93]. In addition to modulating CD18 interactions with their ligands (e.g. ICAM-1) during inflammation[94, 95], fluid shear stress appears to regulate integrin dynamics on the neutrophil surface under conditions that mimic low activation states by redistributing these receptors from areas of maximal shear stress to regions where shear is minimal, i.e. at focal adhesions. Moreover, shear exposure reduces CD18 levels on the surfaces of migrating, and also non-adherent, neutrophils even in the presence of inflammatory mediators, e.g. fMLP[81, 96]. Considering the role of CD18 in strengthening neutrophil attachment to the vascular wall, shear-mediated reductions in CD18 likely diminish the ability of cells to maintain adhesive attachments[97]. In this way, shear-mediated reductions in CD18 serve an anti-inflammatory role that ensures neutrophils in a non-inflamed environment remain in a non-adhesive state.

The mechanism underlying shear-induced reductions in CD18 surface levels involves proteolysis that occurs on the surfaces of migrating and suspended neutrophils. Proteolysis modulates the levels of a wide variety of transmembrane receptors on the neutrophil surface including L-selectin (involved in rolling interactions with endothelium)[98] and CD43, an anti-adhesive mucin-like molecule[99]. CD18 integrins also undergo cleavage of the intracellular domain by calpain to promote detachment of the cell uropod during neutrophil migration[100]. But shear-induced truncation of CD18 integrins differs from calpain-mediated cleavage in that the former involves lysosomal cysteine proteases (e.g., cathepsin B) that exert extracellular activity[96, 97]. Notably, the cell membrane is critically positioned between the intracellular levels, and the extracellular actions, of these proteases.

Additionally, cleavage of CD18 integrins under fluid flow also requires conformational changes in their extracellular domains[96]. Conformational activity of CD18 integrins involves shifts in the protein tertiary structure from a closed-bent to an open-extended configuration[96]. In the case of cytokine stimulation, this conformational change exposes ligand binding sites[101] that promote cell capture onto the vessel wall[95, 102]. Another consequence of CD18 conformational changes, which occur upon shear stress exposure, is to expose proteolytic cleavage sites[96]. With this evidence in mind, it is apparent that the physicochemical state of the cell membrane is a key factor in neutrophil mechanosensitivity that directly or indirectly affects the ability of shear stress to unfold the CD18 ectodomain.

5.2.3. RNS and ROS in shear mechanotransduction

Reactive nitrogen species (RNS; e.g., NO) and ROS are multi-functional free radical mediators of acute inflammation serving not only as anti-microbial agents but also as biological second messengers that influence leukocyte functions (e.g., chemotaxis, phagocytosis, etc.)[103, 104]. NO from exogenous and endogenous sources (such as membrane-associated NO synthase) inhibits neutrophil recruitment out of the microvasculature during acute inflammation[105, 106]. Interestingly, NO also enhances neutrophil pseudopod retraction in response to shear stress and counteracts the blocking effects of cell agonists (e.g. fMLP and platelet-activating factor)[81]. In contrast, ROS, particularly O^{2-}, interferes with the neutrophil shear response and is thought to contribute to the blocking effects of cell agonist, e.g. fMLP, on flow-induced pseudopod retraction[84].

Notably, the fact that inhibition of NO synthase activity in neutrophils has no effect on shear-induced pseudopod retraction[81] points to an exogenous source and an extracellular role for NO. This finding leaves open the possibility that the facultative effects of NO on the neutrophil shear response (i.e., pseudopod retraction) result from its ability to scavenge O^{2-} [103] and, in this way, mediate cell pseudopod activity[107, 108]. In support of this, SOD (an O^{2-} scavenger) also enhances the shear responses of fMLP-stimulated neutrophils[84]. Thus, O^{2-} is a critical mediator for neutrophil shear response. Since the cell membrane, particularly cholesterol-enriched lipid rafts, plays an important role in regulating the production/release of O^{2-}[109], its state may indirectly influence neutrophil mechanosensitivity to shear stress.

5.3. Neutrophil mechanosensitivity and cardiovascular disease

The accumulated evidence reported in the vascular mechanotransduction literature (see reviews[23, 78]) points to the following general paradigm. Exposure of vascular cells to physiological flows under normal (i.e., non-diseased, non-inflamed) conditions correlates with quiescence (i.e., baseline activity). This paradigm resulted from a multitude of studies that selectively examined the activity of various signaling pathways and putative force sensors in response to applied mechanical stresses. They, however, overlooked a subtle, but equally important, factor: mechanosensitivity or the degree to which cells respond to mechanical stresses. Just as biochemical perturbations (e.g. pathogens, inflammatory agonists) temporally and dose-dependently alter vascular cell activity leading to pathogenesis, so must changes in cell mechanosensitivity impact circulatory health.

Neutrophils experience wide variations in fluid stresses as they pass through the circulation and, thus function "normally" under a diverse array of mechanical stress distributions and magnitudes. In other words, aberrant mechanical stresses are unlikely to be a cause of cell dysfunction. What may change and contribute to "abnormal" behavior is their sensitivity to the surrounding fluid flow mechanoenvironment with a negative impact on the ability of fluid shear stress to deactivate the neutrophils. Along this line, the work of Geert Schmid-Schönbein at the University of California, San Diego has demonstrated that attenuated neutrophil shear responses contribute to the microvascular pathobiology observed in spontaneously hypertensive rats (SHRs)[110] and, in doing so, illustrated the potential impact of impaired shear stress mechanotransduction on cardiovascular health.

5.3.1. *Impaired fluid shear responses and downstream effects on vascular pathophysiology*

Significant features of the blood from SHRs are elevated numbers of circulating neutrophils, suppressed expression of adhesion molecules (e.g., selectins, CD18), and an activated phenotype[111-113]. Although the increased activity of neutrophils is not associated with increased adhesion to microvascular endothelium[78], their increased numbers raise peripheral vascular resistance[110]. One possible explanation is that circulating activated neutrophils in SHRs release vasoactive substances that constrict the small arteries and arterioles; this has been documented for atherosclerosis[114-116]. Extensive evidence, however, points to a hemorheological effect of leukocyte activation on microvascular resistance[1, 34, 36]. Specifically, the disturbed motion of white blood cells, due to pseudopod projection, significantly reduces erythrocyte velocities in the microcirculation increasing hemodynamic resistance and upstream blood pressures[36, 110] (see Figure 1).

The key evidence for the involvement of fluid flow mechanotransduction in microvascular abnormalities due to hypertension is that neutrophils from SHRs lack the ability to retract pseudopods in response to shear stress; in some cases, cells extend cellular projections under flow stimulation[110]. The underlying mechanism associated with the blockade and possible reversal of the pseudopod retraction response to shear stress reportedly involves the dependence of blood pressure in SHRs on the plasma level of glucocorticoid-related steroid hormones and the density of glucocorticoid receptors on the neutrophil surface[117, 118]. In line with this, glucocorticoid-treated[119] rats, like SHRs, exhibit elevated peripheral resistance in parallel with elevated numbers of neutrophils that lack a pseudopod retraction response to shear stress. Taken together, leukocyte shear mechanotransduction appears to be critical for the maintenance of a healthy circulation, particularly the microcirculation. Failure of this regulatory mechanism, e.g., due to impaired cell mechanosensitivity resulting from a pathological blood environment, may not only lead to sustained neutrophil activation but also result in disturbed blood flow. In this way, aberrant neutrophil mechanotransduction may contribute to microvascular damage that exacerbates ischemia-reperfusion injury or leads to peripheral vascular disease and downstream organ/tissue injury.

Studies on spontaneous hypertension also reveal a key point. Factors that drive phenotypic changes in neutrophils (e.g., from an inactivated to an activated state) dramatically alter their ability to sense the surrounding flow environment (i.e., mechanosensitivity) leading to the development of pathological behavior, including immune suppression. Intuitively, cell mechanosensitivity depends on the number and activity of proteins "moonlighting" as putative mechanosensors embedded in the cell membrane positioned at the interface between the intra- and extra- cellular milieu. These studies further strengthen the argument that the plasma membrane is a critical determinant of neutrophil mechanosensitivity.

5.3.2. *The plasma membrane and shear stress mechanosensitivity*

The fact that shear stress-induced neutrophil deactivation (e.g., FPR deactivation, G protein signaling, CD18 cleavage, pseudopod retraction, etc.) occurs in the absence of any passive cell deformation due to flow[120] substantiates the presence of a cell surface component(s) that transduces flow stimulation. Interestingly, neutrophils retract pseudopods independently of the fluid shear stress distribution imposed on the cell surface[82]. Thus, membrane properties appear to outweigh the location of mechanosensors on the cell surface. Moreover, non-adherent neutrophils respond to shear stress further emphasizing the importance of cell membrane-mediated over cell deformation-based (e.g., cytoskeleton-related, cell adhesion-dependent) neutrophil mechanotransduction.

The membrane itself may act as a mechanotransducer either via stress-induced changes in its fluidity[121-123] or through lipid rafts[62, 63, 124]. However, the concept that the membrane serves as a fluid stress sensor lacks the specificity that explains the diversity of cell type-specific responses to shear. An alternative, more plausible, viewpoint is that the cell membrane serves as mechanotransduction center for the cell. Along this line, the specificity associated with mechanotransduction depends on the specific mechanoreceptor(s) expressed by the cell. In this regard, a multitude of cell transmembrane proteins including various GPCRs[58, 91, 125], tyrosine kinase receptors[126-130], ion channels[131], NO synthases, and integrin-associated focal adhesions[132, 133] have been implicated as fluid shear stress transducers for a variety of cells (e.g., endothelial cells, osteoblasts, neutrophils) and microorganisms (e.g. dino-flagella)[134].

One potential action of fluid shear stress on transmembrane mechanosensors (e.g., FPR) is to alter their surface levels. In the case of GPCRs, exposing migrating neutrophil-like cells to parallel plate flow redistributes surface-associated FPRs to a perinuclear compartment in the cytosol[135]. These results suggest that internalization of FPRs under fluid shear stimulation leads to pseudopod retraction by counteracting their constitutive activity which drives pseudopod extension. It should be noted, however, that intact FPR must be present since cleavage of FPR is linked to an impaired ability of fluid shear stress to promote retraction of neutrophil pseudopods[136]. Since receptor internalization occurs across the lipid bilayer, shear-induced changes in mechanoreceptor surface levels may thus be a mechanosensitive neutrophil response influenced by properties of the cell membrane.

It is also feasible that the ability of shear stress to alter protein tertiary structure is a function of membrane properties. In addition to evidence regarding the influence of shear stress on the conformation of FPR and CD18 integrins, fluid flow also alters the structure of other membrane-bound GPCRs in other cell types including the bradykinin B_2 receptor for endothelial cells and the type I parathyroid hormone receptor for osteoblasts[58, 125]. Interestingly, physiologically relevant magnitudes of mechanical stresses are capable of physically altering the conformation of proteins[132, 133, 137]. Since these proteins are embedded in the cell membrane, it is possible that membrane properties influence flow-related perturbations of protein structure.

In the end, the physicochemical properties (e.g., fluidity, lipid rafts) of the cell membrane, with their influence on the ability of surface mechanosensors to adopt structural shifts under shear, come to the forefront in terms of how hypercholesterolemia modifies neutrophil mechanosensitivity. This is the topic of the next section.

5.3.3. Membrane cholesterol versus membrane fluidity in hypercholesterolemic impairment of neutrophil mechanosensitivity

Hypercholesterolemia is associated with chronic neutrophil activation and elevated blood cholesterol as well as cholesterol enrichment in the plasma membranes of blood cells. Based on the intimate relationship between protein dynamics (e.g., surface expression, conformational activity) and the cell membrane (as described in the previous section), the chemical and mechanical properties of the lipid bilayer may be critical determinants of the ability of neutrophils to sense fluid shear stress. Along this line, hypercholesterolemia-related membrane perturbations may reduce the neutrophil responsiveness to shear stress by interfering with critical mechanotransduction events, e.g. GPCR and CD18 conformational activity, protease release, and/or production of ROS, that must bidirectionally transmit biological activity across the cell membrane (Figure 4).

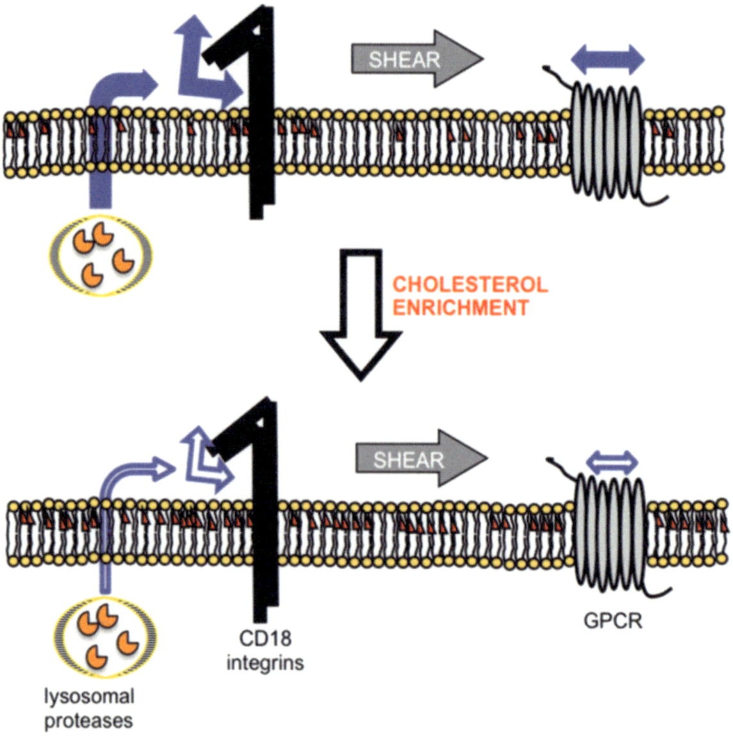

Figure 4. Effects of cholesterol abundance on neutrophil mechanotransduction. Elevations in extracellular cholesterol lead to membrane cholesterol enrichment which may alter cell mechanosensitivity either by influencing shear-induced structural changes of surface sensors, or by interfering with shear-induced release of lysosomal proteases. The cell membrane may also influence contributions from ROS/RNS (not shown).

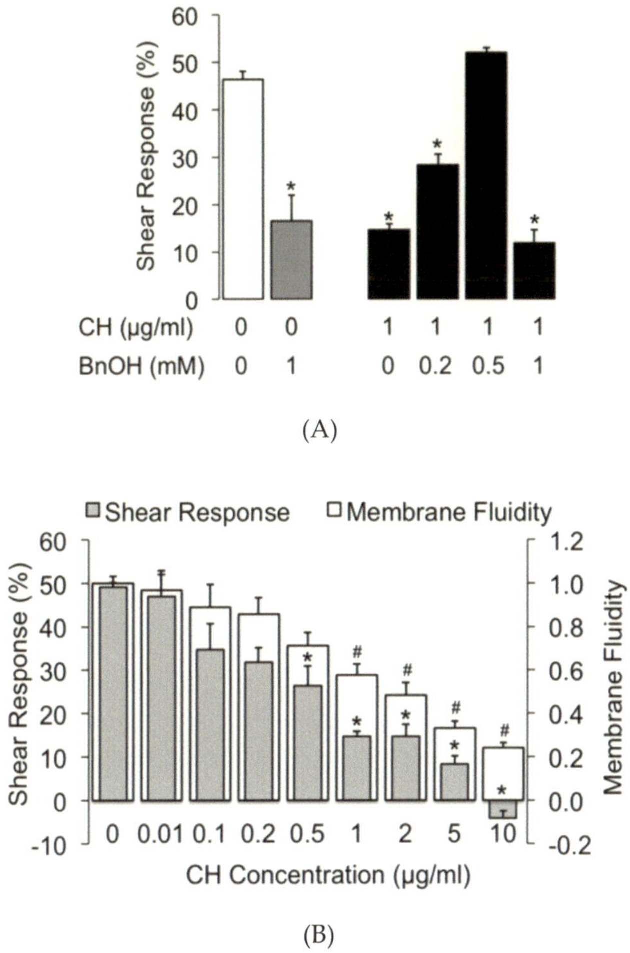

Figure 5. Relationship between membrane cholesterol-dependent fluidity and neutrophil shear responses. A: Recovery effects of benzyl alcohol (BnOH; a membrane fluidizer) on the shear response by neutrophils treated with cholesterol-enhancing agents (CH). B: Does-dependent effects of cholesterol enrichment on neutrophil shear response and membrane fluidity. Cone-plate shear: 5 dyn/cm^2 for 10 min. Bars are mean percentage of reductions in activated cells by shear ± SEM. *, #$p < 0.001$ compared to untreated cells using Student's t-test with Bonferrroni's adjustment.

Recently, we reported that neutrophil deactivation by shear stress depends on the cholesterol-dependent physicochemical properties (i.e., fluidity) of the cell membrane[40]. Fundamentally, we showed that the deactivating actions of fluid shear require a cell membrane containing an optimal level of cholesterol. Shear stress mechanotransduction is impaired if there is too much or too little cholesterol. Moreover, the membrane must be capable of supporting the formation of lipid rafts. But, the critical evidence from this work are our observations[40] that membrane fluidizer, benzyl alcohol, was capable of counteracting the rigidifying effects of membrane cholesterol enhancement (with cyclodextrin-cholesterol conjugates) and that the concentration of benzyl alcohol to achieve this depended on the amount of cholesterol loaded into the neutrophil membranes (Figure 5A). Thus, there is also an optimal membrane fluidity level permissive for shear-induced neutrophil deactivation. This was confirmed by regression analysis[40] which revealed a linear relationship (Figure 5B) between membrane cholesterol-related fluidity and the degree to which neutrophils within a population are inactivated by fluid flow. Membrane cholesterol enrichment therefore impairs neutrophil mechanosensitivity, at least in part, through its impact on membrane fluidity.

Interestingly, neutrophils from LDLr-/- mice fed a HFD exhibit a reduced and even reversed shear stress response relative to cells from similar mice maintained on a regular chow (i.e., normal) diet (ND)[40]. These observations were consistent with our in vitro data correlating membrane cholesterol levels with neutrophil mechanosensitivity[40]. In fact, the shear sensitivity of neutrophils from hypercholesterolemic mice tracks negatively with time-dependent increases in blood levels of cholesterol, particularly of the free form (Figure 6). Presumably, the gradual loading of cholesterol into the neutrophil membrane resulting from the progressive increases in the cholesterol concentration gradient across the outer leaflet of the cell membrane is responsible for the time-dependent decrease in shear mechanosensitivity. Impairment of neutrophil shear responses by membrane cholesterol enrichment may thus underlie the pathogenesis of hypercholesterolemic disorders via an effect on cell membrane fluidity which governs the ability of protein sensors to initiate a sufficient degree of mechanotransduction at the cell surface. As such, a chronic inflammatory state may develop.

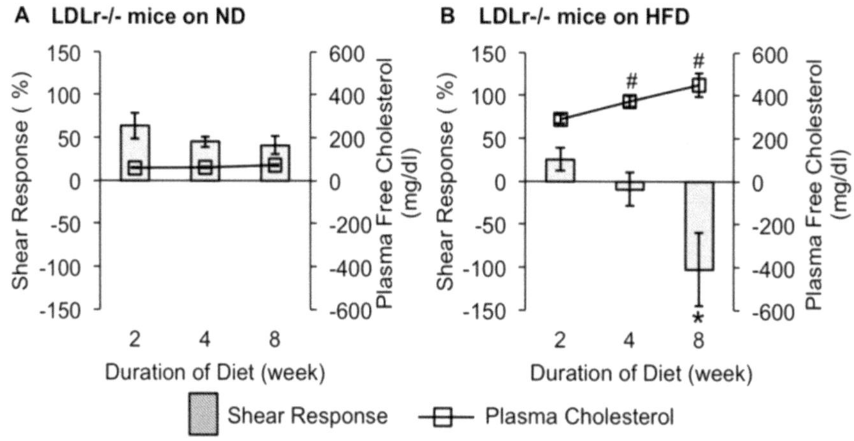

Figure 6. Correlation between neutrophil shear responses and serum levels of free cholesterol. A: LDLr-/- mice on normal diet (ND); B: LDLr-/- mice on high fat diet (HFD). Cone-plate shear: 5 dyn/cm^2 for 10 min. Bars and square dots are mean ± SEM. *, #$p < 0.02$ compared to 2-week using Student's t-test with Bonferrroni's adjustment.

6. Future directions

To date, the accumulated evidence strongly points to shear stress mechanotransduction as an important negative control mechanism for neutrophils flowing in blood under non-inflamed conditions and, thus, an important mediator of circulatory homeostasis. For the most part, the pathobiology of hypercholesterolemia is a process that takes decades to develop into a serious, life-threatening condition and tracks with gradual elevations in blood cholesterol levels. In addition, hypercholesterolemia is characterized by a chronic inflammatory phenotype associated with elevated levels of neutrophil activity in the blood. The question is how these two factors may be related or linked?

Based on the evidence presented in this chapter, the possibility that elevations in blood cholesterol levels impair the neutrophil-deactivating effects of fluid shear stress further suggests that vascular mechanotransduction is an important aspect of cardiovascular physiology and that the pathobiology of hypercholesterolemia may result, at least in part, from a putative disruption of this mechanotransducing function. This statement applies not only to neutrophils, but also to other cells in the circulation including the other white cells and the endothelium. Moreover, the presented evidence hints at the need to shift focus on the study of vascular mechanobiology from characterizing mechanotransduction (i.e., identifying mechanobiological signaling) in disease to actively investigating the influence of mechanosensitivity (i.e. the degree to which cells transduce fluid stresses) on vascular pathogenesis. In our case, we linked altered neutrophil mechanosensitivity with the gradual changes in blood cholesterol levels and leukocyte membranes during the development of hypercholesterolemia in LDLr-/- mice fed a fat-enriched diet. In light of our own evidence and those of others[3, 4, 8] showing that shear stress is anti-inflammatory for neutrophils, it is possible that a putative source of vascular dysfunction causal for hypercholesterolemic pathobiology is the aberrant neutrophil mechanosensitivity.

Despite recognition that vascular mechanotransduction is critical for circulatory homeostasis, there are no markers currently in use or, to our knowledge, in development that account for mechanosensitivity to predict vascular inflammatory status. Current indicators of inflammation include C-reactive protein (CRP; >3 mg/L is at cardiovascular risk) and serum amyloid protein A (SAA; >10 mg/L is at cardiovascular risk). But even though these two biochemical markers are the gold standard measures of inflammatory activity for blood[138, 139], they are upregulated when leukocyte activity levels are already elevated. It is thus not clear whether these molecules are viable "predictors" or just indicators of chronic inflammatory disorders. As such, understanding, characterizing, and formulating measures of neutrophil mechanosensitivity may prove useful in revealing earlier clues regarding the state of inflammation in blood.

In the end, the likelihood that a cholesterol-dependent loss of neutrophil sensitivity to fluid flow stimuli leads to pathological situations implicates a wide range of cardiovascular (and non-cardiovascular) diseases that correlate with both chronic inflammation and an altered cholesterol environment, e.g. hypercholesterolemia and diabetes[74, 140]. The critical issues are to increase efforts to define the link between chronic inflammation and impaired neutrophil mechanotransduction and to determine if chronic inflammation precedes or results from an impairment of vascular mechanotransduction. Further work is, therefore, needed to determine mechanistic-level connections between the cell surface, the flow sensors, the extracellular flow environment, and the influence of a hypercholesterolemic environment on these. The hope is that by fully defining the role of fluid mechanics in the physiological regulation of leukocytes, particularly the neutrophils, one may gain a better understanding of their role in the pathogenesis of cardiovascular disease.

Author details

Xiaoyan Zhang and Hainsworth Y. Shin*

Center for Biomedical Engineering, University of Kentucky, Lexington, KY, USA

7. References

[1] Mazzoni, M.C. and G.W. Schmid-Schönbein, *Mechanisms and consequences of cell activation in the microcirculation.* Cardiovasc Res, 1996. 32(4): p. 709-19.

[2] Gorman C, P.K.D.A., and Cray D, *The Fires Within.* Time, 2004. 163: p. 30-46.

[3] Underwood, A., *Quieting a Body's Defenses.* Newsweek, 2005.

[4] Libby, P., P.M. Ridker, and A. Maseri, *Inflammation and atherosclerosis.* Circulation, 2002. 105(9): p. 1135-43.

[5] Steinberg, D., *Atherogenesis in perspective: hypercholesterolemia and inflammation as partners in crime.* Nat Med, 2002. 8(11): p. 1211-7.

[6] Boullier, A., et al., *Scavenger receptors, oxidized LDL, and atherosclerosis.* Ann N Y Acad Sci, 2001. 947: p. 214-22; discussion 222-3.

[7] Stokes, K.Y., et al., *Hypercholesterolemia promotes inflammation and microvascular dysfunction: role of nitric oxide and superoxide.* Free Radic Biol Med, 2002. 33(8): p. 1026-36.

[8] Scalia, R., J.Z. Appel, 3rd, and A.M. Lefer, *Leukocyte-endothelium interaction during the early stages of hypercholesterolemia in the rabbit: role of P-selectin, ICAM-1, and VCAM-1.* Arterioscler Thromb Vasc Biol, 1998. 18(7): p. 1093-100.

[9] Stokes, K.Y., et al., *NAD(P)H oxidase-derived superoxide mediates hypercholesterolemia-induced leukocyte-endothelial cell adhesion.* Circ Res, 2001. 88(5): p. 499-505.

[10] Mori, N., et al., *Ischemia-reperfusion induced microvascular responses in LDL-receptor -/- mice.* Am J Physiol, 1999. 276(5 Pt 2): p. H1647-54.

[11] Lehr, H.A., et al., *P-selectin mediates the interaction of circulating leukocytes with platelets and microvascular endothelium in response to oxidized lipoprotein in vivo.* Lab Invest, 1994. 71(3): p. 380-6.

[12] Vink, H., A.A. Constantinescu, and J.A. Spaan, *Oxidized lipoproteins degrade the endothelial surface layer : implications for platelet-endothelial cell adhesion.* Circulation, 2000. 101(13): p. 1500-2.

---

* Corresponding Author

[13] Parthasarathy, S., D. Steinberg, and J.L. Witztum, *The role of oxidized low-density lipoproteins in the pathogenesis of atherosclerosis.* Annu Rev Med, 1992. 43: p. 219-25.

[14] Cunningham, K.S. and A.I. Gotlieb, *The role of shear stress in the pathogenesis of atherosclerosis.* Lab Invest, 2005. 85(1): p. 9-23.

[15] De Caterina, R., et al., *Nitric oxide decreases cytokine-induced endothelial activation. Nitric oxide selectively reduces endothelial expression of adhesion molecules and proinflammatory cytokines.* J Clin Invest, 1995. 96(1): p. 60-8.

[16] De Keulenaer, G.W., et al., *Oscillatory and steady laminar shear stress differentially affect human endothelial redox state: role of a superoxide-producing NADH oxidase.* Circ Res, 1998. 82(10): p. 1094-101.

[17] Nagel, T., et al., *Shear stress selectively upregulates intercellular adhesion molecule-1 expression in cultured human vascular endothelial cells.* J Clin Invest, 1994. 94(2): p. 885-91.

[18] Drechsler, M., et al., *Hyperlipidemia-triggered neutrophilia promotes early atherosclerosis.* Circulation, 2010. 122(18): p. 1837-45.

[19] Rotzius, P., et al., *Distinct infiltration of neutrophils in lesion shoulders in ApoE-/- mice.* Am J Pathol, 2010. 177(1): p. 493-500.

[20] van Leeuwen, M., et al., *Accumulation of myeloperoxidase-positive neutrophils in atherosclerotic lesions in LDLR-/- mice.* Arterioscler Thromb Vasc Biol, 2008. 28(1): p. 84-9.

[21] Ionita, M.G., et al., *High neutrophil numbers in human carotid atherosclerotic plaques are associated with characteristics of rupture-prone lesions.* Arterioscler Thromb Vasc Biol, 2010. 30(9): p. 1842-8.

[22] Naruko, T., et al., *Neutrophil infiltration of culprit lesions in acute coronary syndromes.* Circulation, 2002. 106(23): p. 2894-900.

[23] Davies, P.F., *Hemodynamic shear stress and the endothelium in cardiovascular pathophysiology.* Nat Clin Pract Cardiovasc Med, 2009. 6(1): p. 16-26.

[24] Burns, M.P. and N. DePaola, *Flow-conditioned HUVECs support clustered leukocyte adhesion by coexpressing ICAM-1 and E-selectin.* Am J Physiol Heart Circ Physiol, 2005. 288(1): p. H194-204.

[25] Maiellaro, K. and W.R. Taylor, *The role of the adventitia in vascular inflammation.* Cardiovasc Res, 2007. 75(4): p. 640-8.

[26] Mulligan-Kehoe, M.J., *The vasa vasorum in diseased and nondiseased arteries.* Am J Physiol Heart Circ Physiol, 2010. 298(2): p. H295-305.

[27] Ritman, E.L. and A. Lerman, *The dynamic vasa vasorum.* Cardiovasc Res, 2007. 75(4): p. 649-58.

[28] Soehnlein, O., *Multiple roles for neutrophils in atherosclerosis.* Circ Res, 2012. 110(6): p. 875-88.

[29] Mazor, R., et al., *Primed polymorphonuclear leukocytes constitute a possible link between inflammation and oxidative stress in hyperlipidemic patients.* Atherosclerosis, 2008. 197(2): p. 937-43.

[30] Nicholls, S.J. and S.L. Hazen, *Myeloperoxidase and cardiovascular disease.* Arterioscler Thromb Vasc Biol, 2005. 25(6): p. 1102-11.

[31] Chertov, O., et al., *Identification of human neutrophil-derived cathepsin G and azurocidin/CAP37 as chemoattractants for mononuclear cells and neutrophils.* J Exp Med, 1997. 186(5): p. 739-47.

[32] Soehnlein, O., et al., *Neutrophil secretion products pave the way for inflammatory monocytes.* Blood, 2008. 112(4): p. 1461-71.

[33] Schmid-Schönbein, G.W., *Analysis of inflammation.* Annu Rev Biomed Eng, 2006. 8: p. 93-131.

[34] Eppihimer, M.J. and H.H. Lipowsky, *Effects of leukocyte-capillary plugging on the resistance to flow in the microvasculature of cremaster muscle for normal and activated leukocytes.* Microvasc Res, 1996. 51(2): p. 187-201.

[35] Worthen, G.S., et al., *Mechanics of stimulated neutrophils: cell stiffening induces retention in capillaries.* Science, 1989. 245(4914): p. 183-6.

[36] Helmke, B.P., et al., *A mechanism for erythrocyte-mediated elevation of apparent viscosity by leukocytes in vivo without adhesion to the endothelium.* Biorheology, 1998. 35(6): p. 437-48.

[37] Lipowsky, H.H., *Microvascular rheology and hemodynamics.* Microcirculation, 2005. 12(1): p. 5-15.

[38] Lange, Y., et al., *Plasma membranes contain half the phospholipid and 90% of the cholesterol and sphingomyelin in cultured human fibroblasts.* J Biol Chem, 1989. 264(7): p. 3786-93.

[39] Day, A.P., et al., *Effect of simvastatin therapy on cell membrane cholesterol content and membrane function as assessed by polymorphonuclear cell NADPH oxidase activity.* Ann Clin Biochem, 1997. 34 (Pt 3): p. 269-75.

[40] Zhang, X., et al., *Membrane cholesterol modulates the fluid shear stress response of polymorphonuclear leukocytes via its effects on membrane fluidity.* Am J Physiol Cell Physiol, 2011. 301(2): p. C451-60.

[41] Oh, H., et al., *Membrane cholesterol is a biomechanical regulator of neutrophil adhesion.* Arterioscler Thromb Vasc Biol, 2009. 29(9): p. 1290-7.

[42] Cooper, R.A., *Influence of increased membrane cholesterol on membrane fluidity and cell function in human red blood cells.* J Supramol Struct, 1978. 8(4): p. 413-30.

[43] Goldstein, J.L. and M.S. Brown, *The low-density lipoprotein pathway and its relation to atherosclerosis.* Annu Rev Biochem, 1977. 46: p. 897-930.

[44] Brasaemle, D.L. and A.D. Attie, *Rapid intracellular transport of LDL-derived cholesterol to the plasma membrane in cultured fibroblasts.* J Lipid Res, 1990. 31(1): p. 103-12.

[45] Lara, L.L., et al., *Low density lipoprotein receptor expression and function in human polymorphonuclear leucocytes.* Clin Exp Immunol, 1997. 107(1): p. 205-12.

[46] Goldstein, J.L. and M.S. Brown, *Binding and degradation of low density lipoproteins by cultured human fibroblasts. Comparison of cells from a normal subject and from a patient with homozygous familial hypercholesterolemia.* J Biol Chem, 1974. 249(16): p. 5153-62.

[47] Basu, S.K., et al., *Degradation of cationized low density lipoprotein and regulation of cholesterol metabolism in homozygous familial hypercholesterolemia fibroblasts.* Proc Natl Acad Sci U S A, 1976. 73(9): p. 3178-82.

[48] Phillips, M.C., W.J. Johnson, and G.H. Rothblat, *Mechanisms and consequences of cellular cholesterol exchange and transfer*. Biochim Biophys Acta, 1987. 906(2): p. 223-76.

[49] Brasaemle, D.L., A.D. Robertson, and A.D. Attie, *Transbilayer movement of cholesterol in the human erythrocyte membrane*. J Lipid Res, 1988. 29(4): p. 481-9.

[50] Ohvo-Rekila, H., et al., *Cholesterol interactions with phospholipids in membranes*. Prog Lipid Res, 2002. 41(1): p. 66-97.

[51] Rothman, J.E. and D.M. Engelman, *Molecular mechanism for the interaction of phospholipid with cholesterol*. Nat New Biol, 1972. 237(71): p. 42-4.

[52] Chabanel, A., et al., *Influence of cholesterol content on red cell membrane viscoelasticity and fluidity*. Biophys J, 1983. 44(2): p. 171-6.

[53] Coderch, L., et al., *Influence of cholesterol on liposome fluidity by EPR. Relationship with percutaneous absorption*. J Control Release, 2000. 68(1): p. 85-95.

[54] Lenaz, G., *Lipid fluidity and membrane protein dynamics*. Biosci Rep, 1987. 7(11): p. 823-37.

[55] Rimon, G., et al., *Mode of coupling between hormone receptors and adenylate cyclase elucidated by modulation of membrane fluidity*. Nature, 1978. 276(5686): p. 394-6.

[56] Schramm, M., *Transfer of glucagon receptor from liver membranes to a foreign adenylate cyclase by a membrane fusion procedure*. Proc Natl Acad Sci U S A, 1979. 76(3): p. 1174-8.

[57] Lenaz, G.a.P.C., G., *Structure and Properties of Cell Membranes*, G. Benga, Editor 1985, CRC Press: Boca Raton, FLA. p. 73-136.

[58] Chachisvilis, M., Y.L. Zhang, and J.A. Frangos, *G protein-coupled receptors sense fluid shear stress in endothelial cells*. Proc Natl Acad Sci U S A, 2006. 103(42): p. 15463-8.

[59] Simons, K. and D. Toomre, *Lipid rafts and signal transduction*. Nat Rev Mol Cell Biol, 2000. 1(1): p. 31-9.

[60] Brown, D.A. and E. London, *Functions of lipid rafts in biological membranes*. Annu Rev Cell Dev Biol, 1998. 14: p. 111-36.

[61] Radel, C., M. Carlile-Klusacek, and V. Rizzo, *Participation of caveolae in beta1 integrin-mediated mechanotransduction*. Biochem Biophys Res Commun, 2007. 358(2): p. 626-31.

[62] Rizzo, V., et al., *In situ flow activates endothelial nitric oxide synthase in luminal caveolae of endothelium with rapid caveolin dissociation and calmodulin association*. J Biol Chem, 1998. 273(52): p. 34724-9.

[63] Rizzo, V., et al., *Rapid mechanotransduction in situ at the luminal cell surface of vascular endothelium and its caveolae*. J Biol Chem, 1998. 273(41): p. 26323-9.

[64] Marwali, M.R., et al., *Membrane cholesterol regulates LFA-1 function and lipid raft heterogeneity*. Blood, 2003. 102(1): p. 215-22.

[65] Niggli, V., et al., *Impact of cholesterol depletion on shape changes, actin reorganization, and signal transduction in neutrophil-like HL-60 cells*. Exp Cell Res, 2004. 296(2): p. 358-68.

[66] Pierini, L.M., et al., *Membrane lipid organization is critical for human neutrophil polarization*. J Biol Chem, 2003. 278(12): p. 10831-41.

[67] Seely, A.J., J.L. Pascual, and N.V. Christou, *Science review: Cell membrane expression (connectivity) regulates neutrophil delivery, function and clearance*. Crit Care, 2003. 7(4): p. 291-307.

[68] Tuluc, F., J. Meshki, and S.P. Kunapuli, *Membrane lipid microdomains differentially regulate intracellular signaling events in human neutrophils.* Int Immunopharmacol, 2003. 3(13-14): p. 1775-90.
[69] Stulc, T., et al., *Leukocyte and endothelial adhesion molecules in patients with hypercholesterolemia: the effect of atorvastatin treatment.* Physiol Res, 2008. 57(2): p. 184-94.
[70] Lehr, H.A., et al., *In vitro effects of oxidized low density lipoprotein on CD11b/CD18 and L-selectin presentation on neutrophils and monocytes with relevance for the in vivo situation.* Am J Pathol, 1995. 146(1): p. 218-27.
[71] Sugano, R., et al., *Polymorphonuclear leukocytes may impair endothelial function: results of crossover randomized study of lipid-lowering therapies.* Arterioscler Thromb Vasc Biol, 2005. 25(6): p. 1262-7.
[72] Furlow, M. and S.L. Diamond, *Interplay between membrane cholesterol and ethanol differentially regulates neutrophil tether mechanics and rolling dynamics.* Biorheology, 2011. 48(1): p. 49-64.
[73] Tailor, A. and D.N. Granger, *Hypercholesterolemia promotes leukocyte-dependent platelet adhesion in murine postcapillary venules.* Microcirculation, 2004. 11(7): p. 597-603.
[74] Lechi, C., et al., *Increased leukocyte aggregation in patients with hypercholesterolaemia.* Clin Chim Acta, 1984. 144(1): p. 11-6.
[75] Araujo, F.B., et al., *Evaluation of oxidative stress in patients with hyperlipidemia.* Atherosclerosis, 1995. 117(1): p. 61-71.
[76] Kristal, B., et al., *Participation of peripheral polymorphonuclear leukocytes in the oxidative stress and inflammation in patients with essential hypertension.* Am J Hypertens, 1998. 11(8 Pt 1): p. 921-8.
[77] Shurtz-Swirski, R., et al., *Involvement of peripheral polymorphonuclear leukocytes in oxidative stress and inflammation in type 2 diabetic patients.* Diabetes Care, 2001. 24(1): p. 104-10.
[78] Makino, A., et al., *Mechanotransduction in leukocyte activation: a review.* Biorheology, 2007. 44(4): p. 221-49.
[79] Shin, H.Y., et al., *Mechanobiological Evidence for the Control of Neutrophil Activity by Fluid Shear Stress* in *Mechanobiology Handbook*, J. Nagatomi, Editor 2011, CRC Press: Boca Raton, FL, USA. p. 139-75.
[80] Fukuda, S. and G.W. Schmid-Schönbein, *Centrifugation attenuates the fluid shear response of circulating leukocytes.* J Leukoc Biol, 2002. 72(1): p. 133-9.
[81] Fukuda, S., et al., *Mechanisms for regulation of fluid shear stress response in circulating leukocytes.* Circ Res, 2000. 86(1): p. E13-8.
[82] Moazzam, F., et al., *The leukocyte response to fluid stress.* Proc Natl Acad Sci U S A, 1997. 94(10): p. 5338-43.
[83] Makino, A., et al., *Control of neutrophil pseudopods by fluid shear: role of Rho family GTPases.* Am J Physiol Cell Physiol, 2005. 288(4): p. C863-71.
[84] Komai, Y. and G.W. Schmid-Schönbein, *De-activation of neutrophils in suspension by fluid shear stress: a requirement for erythrocytes.* Ann Biomed Eng, 2005. 33(10): p. 1375-86.

[85] Shive, M.S., W.G. Brodbeck, and J.M. Anderson, *Activation of caspase 3 during shear stress-induced neutrophil apoptosis on biomaterials.* J Biomed Mater Res, 2002. 62(2): p. 163-8.
[86] Shive, M.S., M.L. Salloum, and J.M. Anderson, *Shear stress-induced apoptosis of adherent neutrophils: a mechanism for persistence of cardiovascular device infections.* Proc Natl Acad Sci U S A, 2000. 97(12): p. 6710-5.
[87] Chen, H.Q., et al., *Effect of steady and oscillatory shear stress on F-actin content and distribution in neutrophils.* Biorheology, 2004. 41(5): p. 655-64.
[88] Cicchetti, G., P.G. Allen, and M. Glogauer, *Chemotactic signaling pathways in neutrophils: from receptor to actin assembly.* Crit Rev Oral Biol Med, 2002. 13(3): p. 220-8.
[89] Niggli, V., *Signaling to migration in neutrophils: importance of localized pathways.* Int J Biochem Cell Biol, 2003. 35(12): p. 1619-38.
[90] Tybulewicz, V.L. and R.B. Henderson, *Rho family GTPases and their regulators in lymphocytes.* Nat Rev Immunol, 2009. 9(9): p. 630-44.
[91] Makino, A., et al., *G protein-coupled receptors serve as mechanosensors for fluid shear stress in neutrophils.* Am J Physiol Cell Physiol, 2006. 290(6): p. C1633-9.
[92] Marschel, P. and G.W. Schmid-Schönbein, *Control of fluid shear response in circulating leukocytes by integrins.* Ann Biomed Eng, 2002. 30(3): p. 333-43.
[93] Anderson, S.I., et al., *Linked regulation of motility and integrin function in activated migrating neutrophils revealed by interference in remodelling of the cytoskeleton.* Cell Motil Cytoskeleton, 2003. 54(2): p. 135-46.
[94] Simon, S.I. and C.E. Green, *Molecular mechanics and dynamics of leukocyte recruitment during inflammation.* Annu Rev Biomed Eng, 2005. 7: p. 151-85.
[95] Simon, S.I. and H.L. Goldsmith, *Leukocyte adhesion dynamics in shear flow.* Ann Biomed Eng, 2002. 30(3): p. 315-32.
[96] Shin, H.Y., S.I. Simon, and G.W. Schmid-Schönbein, *Fluid shear-induced activation and cleavage of CD18 during pseudopod retraction by human neutrophils.* J Cell Physiol, 2008. 214(2): p. 528-36.
[97] Fukuda, S. and G.W. Schmid-Schönbein, *Regulation of CD18 expression on neutrophils in response to fluid shear stress.* Proc Natl Acad Sci U S A, 2003. 100(23): p. 13152-7.
[98] Walcheck, B., et al., *Neutrophil rolling altered by inhibition of L-selectin shedding in vitro.* Nature, 1996. 380(6576): p. 720-3.
[99] Carney, D.F., et al., *Effect of serine proteinase inhibitors on neutrophil function: alpha-1-proteinase inhibitor, antichymotrypsin, and a recombinant hybrid mutant of antichymotrypsin (LEX032) modulate neutrophil adhesion interactions.* J Leukoc Biol, 1998. 63(1): p. 75-82.
[100] Pfaff, M., X. Du, and M.H. Ginsberg, *Calpain cleavage of integrin beta cytoplasmic domains.* FEBS Lett, 1999. 460(1): p. 17-22.
[101] Arnaout, M.A., *Structure and function of the leukocyte adhesion molecules CD11/CD18.* Blood, 1990. 75(5): p. 1037-50.
[102] Radi, Z.A., M.E. Kehrli, Jr., and M.R. Ackermann, *Cell adhesion molecules, leukocyte trafficking, and strategies to reduce leukocyte infiltration.* J Vet Intern Med, 2001. 15(6): p. 516-29.

[103] Fialkow, L., Y. Wang, and G.P. Downey, *Reactive oxygen and nitrogen species as signaling molecules regulating neutrophil function.* Free Radic Biol Med, 2007. 42(2): p. 153-64.

[104] Guzik, T.J., R. Korbut, and T. Adamek-Guzik, *Nitric oxide and superoxide in inflammation and immune regulation.* J Physiol Pharmacol, 2003. 54(4): p. 469-87.

[105] Dal Secco, D., et al., *Nitric oxide inhibits neutrophil migration by a mechanism dependent on ICAM-1: role of soluble guanylate cyclase.* Nitric Oxide, 2006. 15(1): p. 77-86.

[106] Kubes, P., M. Suzuki, and D.N. Granger, *Nitric oxide: an endogenous modulator of leukocyte adhesion.* Proc Natl Acad Sci U S A, 1991. 88(11): p. 4651-5.

[107] Kubes, P., et al., *Nitric oxide synthesis inhibition induces leukocyte adhesion via superoxide and mast cells.* Faseb J, 1993. 7(13): p. 1293-9.

[108] Gaboury, J., et al., *Nitric oxide prevents leukocyte adherence: role of superoxide.* Am J Physiol, 1993. 265(3 Pt 2): p. H862-7.

[109] Vilhardt, F. and B. van Deurs, *The phagocyte NADPH oxidase depends on cholesterol-enriched membrane microdomains for assembly.* EMBO J, 2004. 23(4): p. 739-48.

[110] Fukuda, S., et al., *Contribution of fluid shear response in leukocytes to hemodynamic resistance in the spontaneously hypertensive rat.* Circ Res, 2004. 95(1): p. 100-8.

[111] Suzuki, H., et al., *Impaired leukocyte-endothelial cell interaction in spontaneously hypertensive rats.* Hypertension, 1994. 24(6): p. 719-27.

[112] Arndt, H., C.W. Smith, and D.N. Granger, *Leukocyte-endothelial cell adhesion in spontaneously hypertensive and normotensive rats.* Hypertension, 1993. 21(5): p. 667-73.

[113] Suematsu, M., et al., *The inflammatory aspect of the microcirculation in hypertension: oxidative stress, leukocytes/endothelial interaction, apoptosis.* Microcirculation, 2002. 9(4): p. 259-76.

[114] Kaul, S., R.C. Padgett, and D.D. Heistad, *Role of platelets and leukocytes in modulation of vascular tone.* Ann N Y Acad Sci, 1994. 714: p. 122-35.

[115] Mugge, A., et al., *Activation of leukocytes with complement C5a is associated with prostanoid-dependent constriction of large arteries in atherosclerotic monkeys in vivo.* Atherosclerosis, 1992. 95(2-3): p. 211-22.

[116] Faraci, F.M., et al., *Effect of atherosclerosis on cerebral vascular responses to activation of leukocytes and platelets in monkeys.* Stroke, 1991. 22(6): p. 790-6.

[117] DeLano, F.A. and G.W. Schmid-Schönbein, *Enhancement of glucocorticoid and mineralocorticoid receptor density in the microcirculation of the spontaneously hypertensive rat.* Microcirculation, 2004. 11(1): p. 69-78.

[118] Sutanto, W., et al., *Corticosteroid receptor plasticity in the central nervous system of various rat models.* Endocr Regul, 1992. 26(3): p. 111-8.

[119] Fukuda, S., H. Mitsuoka, and G.W. Schmid-Schönbein, *Leukocyte fluid shear response in the presence of glucocorticoid.* J Leukoc Biol, 2004. 75(4): p. 664-70.

[120] Sugihara-Seki, M. and G.W. Schmid-Schönbein, *The fluid shear stress distribution on the membrane of leukocytes in the microcirculation.* J Biomech Eng, 2003. 125(5): p. 628-38.

[121] Haidekker, M.A., N. L'Heureux, and J.A. Frangos, *Fluid shear stress increases membrane fluidity in endothelial cells: a study with DCVJ fluorescence.* Am J Physiol Heart Circ Physiol, 2000. 278(4): p. H1401-6.

[122] Butler, P.J., et al., *Rate sensitivity of shear-induced changes in the lateral diffusion of endothelial cell membrane lipids: a role for membrane perturbation in shear-induced MAPK activation.* Faseb J, 2002. 16(2): p. 216-8.

[123] Butler, P.J., et al., *Shear stress induces a time- and position-dependent increase in endothelial cell membrane fluidity.* Am J Physiol Cell Physiol, 2001. 280(4): p. C962-9.

[124] Ferraro, J.T., et al., *Depletion of plasma membrane cholesterol dampens hydrostatic pressure and shear stress-induced mechanotransduction pathways in osteoblast cultures.* Am J Physiol Cell Physiol, 2004. 286(4): p. C831-9.

[125] Zhang, Y.L., J.A. Frangos, and M. Chachisvilis, *Mechanical stimulus alters conformation of type 1 parathyroid hormone receptor in bone cells.* Am J Physiol Cell Physiol, 2009. 296(6): p. C1391-9.

[126] Chen, K.D., et al., *Mechanotransduction in response to shear stress. Roles of receptor tyrosine kinases, integrins, and Shc.* J Biol Chem, 1999. 274(26): p. 18393-400.

[127] Jin, Z.G., et al., *Ligand-independent activation of vascular endothelial growth factor receptor 2 by fluid shear stress regulates activation of endothelial nitric oxide synthase.* Circ Res, 2003. 93(4): p. 354-63.

[128] Lee, H.J. and G.Y. Koh, *Shear stress activates Tie2 receptor tyrosine kinase in human endothelial cells.* Biochem Biophys Res Commun, 2003. 304(2): p. 399-404.

[129] Milkiewicz, M., et al., *HIF-1alpha and HIF-2alpha play a central role in stretch-induced but not shear-stress-induced angiogenesis in rat skeletal muscle.* J Physiol, 2007. 583(Pt 2): p. 753-66.

[130] Shay-Salit, A., et al., *VEGF receptor 2 and the adherens junction as a mechanical transducer in vascular endothelial cells.* Proc Natl Acad Sci U S A, 2002. 99(14): p. 9462-7.

[131] Tarbell, J.M., S. Weinbaum, and R.D. Kamm, *Cellular fluid mechanics and mechanotransduction.* Ann Biomed Eng, 2005. 33(12): p. 1719-23.

[132] Kamm, R.D. and M.R. Kaazempur-Mofrad, *On the molecular basis for mechanotransduction.* Mech Chem Biosyst, 2004. 1(3): p. 201-9.

[133] Lee, S.E., R.D. Kamm, and M.R. Mofrad, *Force-induced activation of talin and its possible role in focal adhesion mechanotransduction.* J Biomech, 2007. 40(9): p. 2096-106.

[134] Chen, A.K., et al., *Evidence for the role of G-proteins in flow stimulation of dinoflagellate bioluminescence.* Am J Physiol Regul Integr Comp Physiol, 2007. 292(5): p. R2020-7.

[135] Su, S.S. and G.W. Schmid-Schönbein, *Internalization of Formyl Peptide Receptor in Leukocytes Subject to Fluid Stresses.* Cell Mol Bioeng, 2010. 3(1): p. 20-29.

[136] Chen, A.Y., et al., *Receptor cleavage reduces the fluid shear response in neutrophils of the spontaneously hypertensive rat.* Am J Physiol Cell Physiol, 2010. 299(6): p. C1441-9.

[137] Mofrad, M.R., et al., *Force-induced unfolding of the focal adhesion targeting domain and the influence of paxillin binding.* Mech Chem Biosyst, 2004. 1(4): p. 253-65.

[138] Gillmore, J.D., et al., *Amyloid load and clinical outcome in AA amyloidosis in relation to circulating concentration of serum amyloid A protein.* Lancet, 2001. 358(9275): p. 24-9.

[139] Johnson, B.D., et al., *Serum amyloid A as a predictor of coronary artery disease and cardiovascular outcome in women: the National Heart, Lung, and Blood Institute-Sponsored Women's Ischemia Syndrome Evaluation (WISE).* Circulation, 2004. 109(6): p. 726-32.

[140] Tomida, K., et al., *Hypercholesterolemia induces leukocyte entrapment in the retinal microcirculation of rats.* Curr Eye Res, 2001. 23(1): p. 38-43.

Section 4

Management of Hyper and Dyslipoproteinemias

Chapter 10

The Confounding Factor of Apolipoprotein E on Response to Chemotherapy and Hormone Regulation Altering Long-Term Cognition Outcomes

Summer F. Acevedo

Additional information is available at the end of the chapter

http://dx.doi.org/10.5772/48115

1. Introduction

One player in health cognitive functioning is shown to be lipoproteins, essential in the metabolism and redistribution of lipids: cholesterol, phospholipids and triacylglycerol. There are several classes of lipoproteins which are used to transport lipids throughout the body and range in density (protein/lipid ratio); chylomicrons (contain dietary lipids), intermediate low density lipoproteins (IDL), very low density lipoproteins (VLDL, bad cholesterol), low density lipoproteins (LDLs) and the high density lipoproteins (HDLs, good cholesterol). The Apolipoprotein/Apoprotein gene family of proteins is part of the lipoprotein complexes that function as regulators of binding between lipoproteins and receptors. These proteins act as enzyme co-factors during lipid metabolism, helping to stabilize lipoproteins during transportation from cell or tissue to its destination [1].

Apolipoprotein E (ApoE), initially termed the "arginine-rich apoprotein", was first identified as a part of the VLDL complexes. ApoE is synthesized principally in the liver, but has also been found in other tissues such as the brain, ovaries, lungs, adrenals, spleen, muscle cells, and macrophages [2]. The three most common alleles of *ApoE* are *ApoE2, ApoE3, ApoE4* [3] found in the nervous system are primarily produced in astroglia and microglia. The three major isoforms differ at position 112 (*ApoE2*/*ApoE3* Cysteine, *ApoE4* Arginine) and 158 (*ApoE2* Cysteine, *ApoE3*/*ApoE4* Arginine), the amino acid substitutions at position 112 affect salt bridge formation within the protein, which ultimately impacts on lipoprotein preference, stability of the protein and on receptor binding activities of the isoforms [4]. Being an *ApoE4* carrier or having the *ApoE2*/*ApoE3* genotype is associated with higher triglyercide levels, higher VLDL levels, higher total cholesterol, higher total lipoproteins levels and elevated LDL or "bad cholesterol" levels all of which contribute to hypertension and diabetes confounding factors that have to be considered when designing a chemotherapy treatment regiment [5].

In rodent models, the lack of *ApoE* or ApoE4 protein expression leads to destabilization of cell membranes, increased apoptosis, and heightened sensitivity to neuronal trauma; whereas, ApoE3 and ApoE2 protein expression allow for healthy cell functioning and neuroprotection [4]. ApoE4 has detrimental effects in transgenic mice, including behavioral abnormalities, such as deficits in spatial learning and memory using Morris water maze (MWM) [6], as well as significant alterations in the hippocampus and cortex [4,7]. The studies in mice are consistent with clinical studies indicating reduced spatial learning and memory in those who carry the *ApoE4* allele [8]. Experiments have also demonstrated that the three isoforms of human *ApoE* gene have different effects on the development of neurodegenerative diseases. Those individuals that are *ApoE4* carriers have an increased risk of age-related mild cognitive impairments (MCI) and the development of Alzheimer's disease (AD) particularly in females [9].

In ovarian cancer, ApoE protein levels act as a potential tumor-associated marker as found in serous carcinomas, but not in serious borderline or normal ovarian surface epithelium cells [10]. Up-regulation of ApoE protein levels is also seen in breast carcinomas, pancreatic cancer, stomach carcinomas, colon carcinomas and prostate carcinomas [10]. Blockage of ApoE expression in the serous carcinoma cell lines leads to cell cycle arrest and apoptosis. Women infused with ApoE protein at the time of diagnosis showed significantly higher survival rates [10]. This data suggest that upregulation of ApoE expression may be a defense mechanism to help body fight carcinomas.

2. Chemotherapy, *ApoE* and memory

Over the last 20 years it has become apparent that chemotherapy drugs not only attack cancer cells, but also cross the blood brain barrier (BBB) leading to negative effects on cognitive processing which is known as Chemo-brain or Chem-fog [11,12]. Methotrexate, 5-Flourouracil (5-FU) the most common chemotherapy drugs used to treat breast, colorectal, head and neck cancers been shown in both rodent models and clinical studies to lead to neurocognitive deficits in a variety of domains including visual memory and visuospatial functioning [11,13,14]. In clinical studies, there is considerable variability between studies as to the extent and frequency of such impairments heretofore mentioned. However, rodent studies are clearly show that the drugs in the CMF (cyclophosphamide, methotrexate, 5-FU) regiment lead to decreased hippocampus cell proliferation and induce MWM memory impairments [15,16]. In addition, cytarabine (cytosine arabinoside) and ifosfamide among other chemotherapy have been shown to lead to memory impairments, hemiparesis, aphasia and progressive dementia [17]. However, there is significant lack of pre-clinical testing of most chemotherapeutic agents and their long-term effects on memory.

Currently, only one study has examined if *ApoE4* carrier status as a potential genetic risk factor in breast or lymphoma survivors. They have found that even after 8 years, *ApoE4* carriers displayed impairments, specifically in visual memory and spatial ability [18]. This suggests that the *ApoE* genotype may be a confounding factor to consider when examining post-chemotherapy neurocognitive status. More studies are necessary to confirm this result.

Studies in women treated with CMF regiment for breast cancer found increased total cholesterol, LDL, HDL cholesterol and Apolipoprotein A-1 (ApoA-1) in those who developed permanent amenorrhea (loss of menstrual cycle, induced menopause) [19]. Studies report that around 30% of patients who have gone through chemotherapy develop permanent amenorrhea [20]. There is evidence that *ApoE4* carriers have an earlier onset of natural menopause [21]. Age of menopause and being an *ApoE4* carrier are both risk factors for age related diseases including AD and coronary artery disease (CAD). There is also a direct connection between ApoE mRNA levels and estrogen in various tissues, including the brain [22] leading to regulation of neurite outgrowth [7]. The data suggests that ApoE is a critical intermediary in the estrogen related neuroplasticity [7]. Lack of estrogen along with expression of ApoE4 protein which has reduced ApoE functioning is one potential cause of impaired cognitive performance in some women that have undergone chemotherapy. Studies do not take this factor into account or narrowly examine menopausal status at the time of testing.

3. Apolipoproteins effects on secondary drug response

3.1. Tamoxifen

Other confounding factors that can affect cognitive status after chemotherapy include other medications patients are taking to control comorbid conditions or to treat the tumor itself. Tamoxifen is used as an estrogen receptor modulator (SERM) in estrogen receptor (ER) positive breast cancer carcinomas. Tamoxifen is a pure estrogen receptor blocker. As with other drugs, tamoxifen and its metabolites can cross the BBB affecting ER in various brain regions including the cerebral cortex, hippocampus and amygdala [23]. In combination with chemotherapy, tamoxifen appears to intensify the cognitive impairments, particularly in visual memory, verbal working memory and visuospatial ability [24]. The Anatrozole, Tamoxifen Combined (ATAC) trial, also found verbal memory and processing speed impairments post-chemotherapy treated only with tamoxifen compared to women only on a combined ATAC treatment (Table 1) [25]. This suggests that tamoxifen has confounding effects when given as part of the chemotherapy regiment. Tamoxifen also appears to have both agonist and antagonist properties in the brain with reported up-regulation of pro-inflammatory cytokines shown to be related to cognitive dysfunction [26]. Positron emission tomography (PET) imaging of survivors does show higher hypometabolism with dual chemotherapy and tamoxifen, not seen in women treated only with tamoxifen [27]. Animal models using repeated tamoxifen or combinations of methotrexate and 5-FU injections both produced deficits in acquisition and retention in an operant learning paradigm (Table 1)

[28]. These studies were conducted when women were still in treatment, leaving the question of potential long-term effects. There is a study that examined women who used tamoxifen for <4 years compared to >6 years of exposure. The study found that the current exposure led to greater memory deficits compared to non-users (Table 1) [29]. This suggests that while on therapy, patients may have acute memory impairments and that alternative drugs should be seriously considered.

In breast cancer survivors undergoing tamoxifen treatment, their total cholesterol, VLDL, high density lipoproteins (HDL) and Apolipoprotein B (ApoB) protein levels have been shown to decrease in both *ApoE4* carriers and non-*ApoE4* carriers (Table 1) [30,31]. Breast cancer patients who are *ApoE4* carriers have higher plasma triglyceride levels and altered ApoA-1/ApoB ratio. Both are risk factors for cardiovascular events, after tamoxifen treatment (Table 1) [30,31]. In non-*ApoE4* carriers, there were lower levels of lipoprotein (a) after treatment, but no effect on triglycerides or the ApoA-1/ApoB ratio (Table 1). This suggests that non-*ApoE4* carriers have a more positive response to tamoxifen with respect to lipid profiles and risk for cardiovascular complications [31]. Considering tamoxifen and *ApoE4* both have a deleterious effect on cognition and tamoxifen has an *ApoE* genotype dependent effect on lipid profiles, suggest further investigations are warranted to understand the mechanistic relationship.

3.2. Anatrozole

Anatrozole also known as arimidex is an aromatase inhibitor that lowers estrogen levels and is used as a treatment in estrogen positive breast cancer patients post-surgery. Results of the ATAC trial of 9399 women indicated that those only on arimidex had better clinical outcomes including vascular events and gynecological problems compared to the tamoxifen group with no differences seen in cognitive outcomes [25,32]. There data suggests that arimidex was the preferred initial treatment by women treated for breast cancer [32]. However, other studies indicate that women on arimidex treatment had greater cognitive decline than tamoxifen treatment in verbal and visual memory (Table 1) [33,34]. At this point no effects have been seen on cholesterol, lipoproteins or apolipoprotein levels in both animal models and clinical studies (Table 1) [35,36]. Although, there are not alternations in lipids levels, the cognitive side effects are of concern with this medication and should be examined with respect to *ApoE* genotype.

3.3. Letrozole

Letrozole, a potent aromatase interfering with adrenal steroid biosynthesis, has also been assessed as a replacement for tamoxifen or as secondary maintenance treatment after tamoxifen as part of the Breast International Group (BIG 1-98) trial [37,38]. In the BIG 1-98 study, better overall cognitive outcomes were seen in women on letrozole treatment compared to those on a tamoxifen treatment (Table 1) [37,38]. In the tamoxifen only group, increased endometrial cancer and vaginal bleeding where found [38]. Participants given letrozole only did displayed more incidences of skeletal and cardiac events and hypercholesterolemia compared to the tamoxifen group [37]. After letrozole treatment, unfavorable effects have been seen including increased serum total cholesterol, LDL and ApoB with atherogenic ratio risk of total cholesterol/HDL and LDL/HDL levels (Table 1) [39]. Therefore, the better cognitive outcomes may not out-weigh negative effects on lipoprotein levels, particularly in those with or at risk for hypertension and/or diabetes. The potential confounding factor of *ApoE* genotype has not been reported.

3.4. Exemestane

There is another alternative for estrogen suppression therapy exemestane, an aromatase inhibitor, general used after tamoxifen is not working in post-menopausal women. Exemestane has been examined as part of the randomized Tamoxifen and Exemestane Adjunctive Multinational (TEAM) trial [24]. The preliminary data from the TEAM trial found that tamoxifen is associated with lower verbal and executive function, while exemestane did not seem to alter cognitive performance levels (Table 1) [24]. Analysis of data from 72 patients as part of the EORTC trail 10958 indicates that treatment with exemestane resulted in reduced triglyceride levels and tamoxifen treatment increased triglyceride levels [40]. All other lipid parameters including HDL, ApoA-1, ApoB or Lip (a) levels at 8, 24 and 48 weeks were unchanged by either treatment [40]. Further studies will be needed to determine if stable cognitive performance and lipid levels are effect of *ApoE* genotype.

3.5. Raloxifene

Raloxifene, also a SERM and is used as a hormone replacement therapy, has the positive effects of estrogen on the skeletal system and is an antagonist of estrogen in breast or endometrial tissues [41]. Raloxifene treatment also appears to be less detrimental to cognitive function assessed by Modified Mini-Mental State (3MS) compared to tamoxifen [42]. Used to prevent osteoporosis, it has been shown that after three years of treatment raloxifene did not affect overall cognitive scores [43]. The Multiple Outcomes of Raloxifene Evaluation (MORE) study of 7478 women, reported finding that raloxifene treatment lowered the risk of cognitive decline in word list recall test and there was no overall effect on cognitive function (Table 1) [44]. A subset of the National Surgical Adjuvant Breast and Bowel Project (NSABP) Study of Tamoxifen and Raloxifene (STAR), the CoSTAR study for women at high risk for breast cancer did not find any cognitive effects of either drug (Table 1) [45]. Together the evidence supports that raloxifene treatment does not impair cognitive function the way that tamoxifen treatment and in fact may even lower the risk of cognitive decline [34].

With respect to lipid and lipoprotein levels post hoc analysis of 2659 women in the MORE study, found that raloxifene treatment in women with or without high triglycerides lead to reduced cholesterol levels with healthier lipoprotein parameters (Table 1) [46]. Studies in Greek women, found that LDL cholesterol levels were lower in women treated with raloxifene [47]. Raloxifene also appears to raise HDL levels and ApoA-1 while decreasing ApoB protein levels and improving ratios of total cholesterol to lipoproteins (Table 1). After one year of treatment with raloxifene women had reduced fat mass and trunk and central regions along with decreased adiposity in their truck and abdominal regions (Table 1) [48]. Overall, raloxifene treatment improves cholesterol health and alters fat distribution in a positive manner to help prevent obesity, making it a better candidate for overall health compared to tamoxifen. Its effects in relation to *ApoE* genotype have not been reported.

3.6. Estradiol

In post-menopausal estradiol (also known as 17β-estradiol or oestradiol) treatment is used for estrogen replacement therapy. Healthy post-menopausal women given estradiol display improved visuospatial abilities measured by a mental rotation task (Table 1) [49]. Other non-randomized studies in women with surgically induced amenorrhea or those with AD indicate that estrogen replacement treatment may help to improve or minimize cognitive deficits [50]. Even in men those given estradiol performed better on visual memory after treatment (Table 1) [51]. These results are consistent with improved memory in mice given other replacement estrogens treatments [52]. Over half of randomized clinical studies find significant improvements in cognition and attention after estrogen replacement therapy (Table 1) [53]. Estradiol has been shown to increase levels of ApoE in the brain, proposed to be beneficial for neuronal reorganization and repair [7]. In a health study of 3,393 women, results suggest that estrogen replacement reduces the risk of age-related cognitive decline in non-*ApoE4* women, but not in *ApoE4* carriers (Table 1) [54]. Another study with 181 post-menopausal women, also found the best learning and memory performance after estrogen replacement is seen in non-*ApoE4* carriers [55]. This suggests that knowing *ApoE* genotype may be helpful to assess potential response to estrogen replacement therapy.

Treatment	Function	Effects on Lipids/Apolipoproteins	Effects on Cognition
Tamoxifen	Estrogen receptor modulator (SERM) lowers estrogen function	Decreased total cholesterol, VLDL, HDL and ApoB protein levels.	Leads to impairments in visual memory, verbal working memory and visuospatial ability.
Anatrozole	Aromatase inhibitor lowers estrogen function	No effects seen in rodent or clinical studies.	Reduced verbal and visual memory compared to tamoxifen. Deficits

Treatment	Function	Effects on Lipids/Apolipoproteins	Effects on Cognition
			in rodent model operant learning paradigm.
Letrozole	Aromatase inhibitor lowers estrogen function.	Increased serum total cholesterol, LDL and ApoB increase risk of cardiovascular events.	Better overall cognitive outcomes compared to tamoxifen treatment.
Exemestane	Aromatase inhibitor lowers estrogen function.	Reduces triglyercide levels.	No effect on cognitive performance compared to Tamoxifen group that had lower verbal and executive functioning.
Raloxifene	SERM, lowers estrogen function in reproductive tissue and used as estrogen replacement therapy in non-reproductive tissues.	Lower cholesterol, LDL, ApoB protein levels and increases ApoA-1, HDL level leading to better cardiovascular health. Also shown to reduced adiposity and fat mass.	Lowered the risk of cognitive decline or has no effect.
Estradiol	Estrogen replacement therapy	Increase ApoE protein levels in brain. Reduces cognitive decline in only non-*ApoE4* carriers.	Treatment improves visuospatial abilities and visual memory.
Tibolone	Estrogen replacement therapy	Reduces total cholesterol, triglyceride levels, HDL and ApoA-1 levels.	Decreased anxiety, improved quality of like and semantic memory.
Cetrorelix	Used to reduce gonadotrophins and sex steroids	Increases ApoA-1 and HDL levels.	Anxiolytic, anti-depressive and improved beta-amyloid 25-35 associated memory consolidation impairments.

Table 1. Hormone treatments effects of lipids/apolipoproteins levels and cognition.

3.7. Tibolone

Tibolone is another drug used in hormonal replacement therapy having estrogenic, progestogenic, and androgenic effects. Long-term treatment does appear to decrease anxiety, improve semantic memory and overall quality of life; however, one study reported that those in treatment did score worse on attention task compared to women not on treatment (Table 1) [56,57]. Tibolone appears to be the most beneficial with respect to reducing total cholesterol, triglyercide, HDL and ApoA-1 levels compared to raloxifene and estradiol (Table 1) [47,58-60]. This suggests that hormone replacement therapy with medications such as tibolone in post-menopausal women are beneficial to cognitive health and lipid profiles.

3.8. Cetrorelix

Cetrorelix an antagonist of hypothalamic luteinizing hormone-releasing hormone (LHRH), is used in treatment of prostate carcinoma, benign prostatic hyperplasia, and ovarian cancer to reduce gonadotrophins and sex steroids [61]. In mice, a study suggests that it is anxiolytic, anti-depressive and able to correct beta-amyloid 25-35 associated memory consolidation impairments (Table 1) [61]. Injection of cetrorelix into *ApoE* deficient mice ($ApoE^{-/-}$) mice suggests that the associated suppression of testosterone leads to increased atherosclerosis despite lower cholesterol levels in the male mice [62]. In female $ApoE^{-/-}$ mice, the reduction in testosterone also leads to reduction in estradiol, insulin and HDL levels without effects on atherosclerosis [62]. In a pilot study conducted in men, treatment with cetrorelix resulted in increased ApoA-1, HDL, insulin and leptin consistently (Table 1) [63]. Therefore, when this drug is used within a chemotherapy treatment regiment it is important to carefully monitor lipid levels. Additional studies are needed to examine if *ApoE* genotype has any effect on response and potential long-term cognitive side effects of this drug.

4. Recommendations

1. Determine if an *ApoE* genotype can help assess what is most treatment useful including how to properly maintain lipid levels during chemotherapy.
2. Find out lipid levels, track and maintain determined treatment to reduce risk of post-chemotherapy cognitive impairments.
3. Examine *ApoE* genotype before selecting pharmacotherapy options pre or post-chemotherapy treatment.

5. Conclusion

Overall of the studies SERMs/aromatase inhibitors raloxifene or exemestane may be better alternatives to tamoxifen or letrozole treatment in terms of effects on cognitive deficits and overall health risk in women treated with chemotherapy. In addition, *ApoE* genotype and cholesterol levels need to be taken into account when examining efficacy of these drugs and in hormone replacement therapies as efficacy is dependent on *ApoE* genotype. Knowing these issues will help doctors to address them early for improving quality of life, reducing services used and saving millions of dollars in unneeded medical expenses. Realistically, the wonder drug that can cure all cancer and has no side effects will not be found. What is needed is to reduce the impact and intensity of cognitive side-effects as much as possible, taking into account an individual's physiology and genetics. The more we know about cognitive status across ages, ethnicity, lipid levels, and genetic status the better we can treat mind and body.

Author details

Summer F. Acevedo*
Department of Physiology, Pharmacology, and Toxicology, Psychology Program,
Ponce School of Medicine and Health Sciences, Ponce, Puerto Rico

Acknowledgement

We acknowledge the support of Tirtsa Porrata-Doria and the Ponce School of Medicine and Health Sciences (PSMHS) Research Center for Minority Institutions (RCMI), Molecular Biology Core Lab (G12 RR003050). Special thanks go to Robert Ritchie from the PSMHS/RCMI Publications Office (G12 RR003050/8G12MD007579-27). Additionally, would like to acknowledge Dr. Jacob Raber at Oregon Health Science University for mentoring and sparking my interest in apolipoproteins and cancer research.

6. References

[1] Han, X. (2004) The role of apolipoprotein E in lipid metabolism in the central nervous system. Cell Mol. Life Sci. 61: 1896-1906.

[2] Mahley, R. W., Y. Huang, and K. H. Weisgraber (2006) Putting cholesterol in its place: apoE and reverse cholesterol transport. J. Clin. Invest. 116: 1226-1229.

[3] Mahley, R. W. (1988) Apolipoprotein E: cholesterol transport protein with expanding role in cell biology. Science 240: 622-630.

[4] Mahley, R. W., K. H. Weisgraber, and Y. Huang (2009) Apolipoprotein E: structure determines function, from atherosclerosis to Alzheimer's disease to AIDS. J. Lipid Res. 50 Suppl: S183-188.

[5] Tso, T. K., J. T. Snook, R. A. Lozano, and W. B. Zipf (2001) Risk factors for coronary heart disease in type 1 diabetic children: the influence of apoE phenotype and glycemic regulation. Diabetes Res. Clin. Pract. 54: 165-171.

[6] Morris, R. (1984) Developments of a water-maze procedure for studying spatial learning in the rat. J. Neurosci. Method. 11: 47-60.

* Corresponding Author

[7] Struble, R. G., C. Cady, B. P. Nathan, and M. McAsey (2008) Apolipoprotein E may be a critical factor in hormone therapy neuroprotection. Front. Biosci. 13: 5387-5405.
[8] Berteau-Pavy, F., B. Park, and J. Raber (2007) Effects of sex and APOE epsilon4 on object recognition and spatial navigation in the elderly. Neuroscience 147: 6-17.
[9] Mahley, R. W., and Y. Huang (2006) Apolipoprotein (apo) E4 and Alzheimer's disease: unique conformational and biophysical properties of apoE4 can modulate neuropathology. Acta. Neurol. Scand. Suppl. 185: 8-14.
[10] Chen, Y. C., G. Pohl, T. L. Wang, P. J. Morin, B. Risberg, G. B. Kristensen, A. Yu, B. Davidson, and M. Shih Ie (2005) Apolipoprotein E is required for cell proliferation and survival in ovarian cancer. Cancer Res. 65: 331-337.
[11] Ahles, T. A., and A. Saykin (2001) Cognitive effects of standard-dose chemotherapy in patients with cancer. Cancer Invest. 19: 812-820.
[12] Meyers, C. A. (2008) How chemotherapy damages the central nervous system. J. Biol. 7: 11.
[13] Janelsins, M. C., S. Kohli, S. G. Mohile, K. Usuki, T. A. Ahles, and G. R. Morrow (2011) An update on cancer- and chemotherapy-related cognitive dysfunction: current status. Semin. Oncol. 38: 431-438.
[14] Nelson, C. J., N. Nandy, and A. J. Roth (2007) Chemotherapy and cognitive deficits: mechanisms, findings, and potential interventions. Palliat. Support. Care 5: 273-280.
[15] Seigers, R., S. B. Schagen, C. M. Coppens, P. J. van der Most, F. S. van Dam, J. M. Koolhaas, and B. Buwalda (2009) Methotrexate decreases hippocampal cell proliferation and induces memory deficits in rats. Behav. Brain Res. 201: 279-284.
[16] Winocur, G., J. Vardy, M. A. Binns, L. Kerr, and I. Tannock (2006) The effects of the anti-cancer drugs, methotrexate and 5-fluorouracil, on cognitive function in mice. Pharmacol. Biochem. Behav. 85: 66-75.
[17] Verstappen, C. C., J. J. Heimans, k. Hoekman, and T. J. Postma (2003) Neurotoxic complications of chemotherapy in patients with cancer: clinical signs and optimal management. Drugs 63: 1549-1563.
[18] Ahles, T. A., A. J. Saykin, W. W. Noll, C. T. Furstenberg, S. Guerin, B. Cole, and L. A. Mott (2003) The relationship of APOE genotype to neuropsychological performance in long-term cancer survivors treated with standard dose chemotherapy. Psycho-oncology 12: 612-619.
[19] Saarto, T., C. Blomqvist, C. Ehnholm, M. R. Taskinen, and I. Elomaa (1996) Effects of chemotherapy-induced castration on serum lipids and apoproteins in premenopausal women with node-positive breast cancer. J. Clin. Endocrinol. Metab. 81: 4453-4457.
[20] Phillips, K. A., and J. Bernhard (2003) Adjuvant breast cancer treatment and cognitive function: current knowledge and research directions. J. Nat. Cancer Inst. 95: 190-197.
[21] Koochmeshgi, J., S. M. Hosseini-Mazinani, S. Morteza Seifati, N. Hosein-Pur-Nobari, and L. Teimoori-Toolabi (2004) Apolipoprotein E genotype and age at menopause. Ann. N. Y. Acad. Sci. 1019: 564-567.
[22] Srivastava, R. A., N. Srivastava, M. Averna, R. C. Lin, K. S. Korach, D. B. Lubahn, and G. Schonfeld (1997) Estrogen up-regulates apolipoprotein E (ApoE) gene expression by

increasing ApoE mRNA in the translating pool via the estrogen receptor alpha-mediated pathway. J. Biol. Chem. 272: 33360-33366.

[23] Ciocca, D. R., and L. M. Roig (1995) Estrogen receptors in human nontarget tissues: biological and clinical implications. Endocr. Rev. 16: 35-62.

[24] Schilder, C. M., C. Seynaeve, L. V. Beex, W. Boogerd, S. C. Linn, C. M. Gundy, H. M. Huizenga, J. W. Nortier, C. J. van de Velde, F. S. van Dam, and S. B. Schagen (2010) Effects of tamoxifen and exemestane on cognitive functioning of postmenopausal patients with breast cancer: results from the neuropsychological side study of the tamoxifen and exemestane adjuvant multinational trial. J. Clin. Oncol. 28: 1294-1300.

[25] Shilling, V., V. Jenkins, L. Fallowfield, and T. Howell (2003) The effects of hormone therapy on cognition in breast cancer. J. Steroid Biochem. Mol. Biol. 86: 405-412.

[26] Wefel, J. S., A. E. Kayl, and C. A. Meyers (2004) Neuropsychological dysfunction associated with cancer and cancer therapies: a conceptual review of an emerging target. Br. J. Cancer 90: 1691-1696.

[27] Silverman, D. H., C. J. Dy, S. A. Castellon, J. Lai, B. S. Pio, L. Abraham, K. Waddell, L. Petersen, M. E. Phelps, and P. A. Ganz (2007) Altered frontocortical, cerebellar, and basal ganglia activity in adjuvant-treated breast cancer survivors 5-10 years after chemotherapy. Breast Cancer Res. Treat. 103: 303-311.

[28] Walker, E. A., J. J. Foley, R. Clark-Vetri, and R. B. Raffa (2011) Effects of repeated administration of chemotherapeutic agents tamoxifen, methotrexate, and 5-fluorouracil on the acquisition and retention of a learned response in mice. Psychopharmacology 217: 539-548.

[29] Paganini-Hill, A., and L. J. Clark (2000) Preliminary assessment of cognitive function in breast cancer patients treated with tamoxifen. Breast Cancer Res. Treat. 64: 165-176.

[30] Chang, N. W., F. N. Chen, C. T. Wu, C. F. Lin, and D. R. Chen (2009) Apolipoprotein E4 allele influences the response of plasma triglyceride levels to tamoxifen in breast cancer patients. Clin. Chim. Acta. 401: 144-147.

[31] Liberopoulos, E., S. A. Karabina, A. Tselepis, E. Bairaktari, C. Nicolaides, N. Pavlidis, and M. Elisaf (2002) Are the effects of tamoxifen on the serum lipid profile modified by apolipoprotein E phenotypes? Oncology 62: 115-120.

[32] Howell, A., J. Cuzick, M. Baum, A. Buzdar, M. Dowsett, J. F. Forbes, G. Hoctin-Boes, J. Houghton, G. Y. Locker, and J. S. Tobias (2005) Results of the ATAC (Arimidex, tamoxifen, Alone or in Combination) trial after completion of 5 years' adjuvant treatment for breast cancer. Lancet 365: 60-62.

[33] Collins, B., J. Mackenzie, A. Stewart, C. Bielajew, and S. Verma (2009) Cognitive effects of hormonal therapy in early stage breast cancer patients: a prospective study. Psycho-oncology 18: 811-821.

[34] Agrawal, K., S. Onami, J. E. Mortimer, and S. K. Pal (2010) Cognitive changes associated with endocrine therapy for breast cancer. Maturitas 67: 209-214.

[35] Lew, R., P. Komesaroff, M. Williams, T. Dawood, and K. Sudhir (2003) Endogenous estrogens influence endothelial function in young men. Circ. Res. 93: 1127-1133.

[36] Sadlonova, V., P. Kubatka, K. Kajo, D. Ostatnikova, G. Nosalova, K. Adamicova, and J. Sadlonova (2009) Side effects of anastrozole in the experimental pre-menopausal mammary carcinogenesis. Neoplasma 56: 124-129.

[37] Phillips, K. A., K. Ribi, Z. Sun, A. Stephens, A. Thompson, V. Harvey, B. Thurlimann, F. Cardoso, O. Pagani, A. S. Coates, A. Goldhirsch, K. N. Price, R. D. Gelber, and J. Bernhard (2010) Cognitive function in postmenopausal women receiving adjuvant letrozole or tamoxifen for breast cancer in the BIG 1-98 randomized trial. Breast 19: 388-395.

[38] Thurlimann, B., A. Keshaviah, A. S. Coates, H. Mouridsen, L. Mauriac, J. F. Forbes, R. Paridaens, M. Castiglione-Gertsch, R. D. Gelber, M. Rabaglio, I. Smith, A. Wardley, K. N. Price, and A. Goldhirsch (2005) A comparison of letrozole and tamoxifen in postmenopausal women with early breast cancer. N. Engl. J. Med. 353: 2747-2757.

[39] Elisaf, M. S., E. T. Bairaktari, C. Nicolaides, B. Kakaidi, C. S. Tzallas, A. Katsaraki, and N. A. Pavlidis (2001) Effect of letrozole on the lipid profile in postmenopausal women with breast cancer. Euro. J. Cancer 37: 1510-1513.

[40] Atalay, G., L. Dirix, L. Biganzoli, L. Beex, M. Nooij, D. Cameron, C. Lohrisch, T. Cufer, J. P. Lobelle, M. R. Mattiaci, M. Piccart, and R. Paridaens (2004) The effect of exemestane on serum lipid profile in postmenopausal women with metastatic breast cancer: a companion study to EORTC Trial 10951, 'Randomized phase II study in first line hormonal treatment for metastatic breast cancer with exemestane or tamoxifen in postmenopausal patients'. Ann. Oncol. 15: 211-217.

[41] Rey, J. R., E. V. Cervino, M. L. Rentero, E. C. Crespo, A. O. Alvaro, and M. Casillas (2009) Raloxifene: mechanism of action, effects on bone tissue, and applicability in clinical traumatology practice. Open. Orthop. J. 3: 14-21.

[42] Espeland, M. A., S. A. Shumaker, M. Limacher, S. R. Rapp, T. B. Bevers, D. H. Barad, L. H. Coker, S. A. Gaussoin, M. L. Stefanick, D. S. Lane, P. M. Maki, and S. M. Resnick (2010) Relative effects of tamoxifen, raloxifene, and conjugated equine estrogens on cognition. J. Women's Health (Larchmt). 19: 371-379.

[43] Yaffe, K., K. Krueger, S. Sarkar, D. Grady, E. Barrett-Connor, D. A. Cox, and T. Nickelsen (2001) Cognitive function in postmenopausal women treated with raloxifene. N. Engl. J. Med. 344: 1207-1213.

[44] Alejandre-Gomez, M., L. M. Garcia-Segura, and I. Gonzalez-Burgos (2007) Administration of an inhibitor of estrogen biosynthesis facilitates working memory acquisition in male rats. Neurosci. Res. 58: 272-277.

[45] Legault, C., P. M. Maki, S. M. Resnick, L. Coker, P. Hogan, T. B. Bevers, and S. A. Shumaker (2009) Effects of tamoxifen and raloxifene on memory and other cognitive abilities: cognition in the study of tamoxifen and raloxifene. J. Clin. Oncol. 27: 5144-5152.

[46] Dayspring, T., Y. Qu, and C. Keech (2006) Effects of raloxifene on lipid and lipoprotein levels in postmenopausal osteoporotic women with and without hypertriglyceridemia. Metabolism 55: 972-979.

[47] Christodoulakos, G. E., I. V. Lambrinoudaki, C. P. Panoulis, C. A. Papadias, E. E. Kouskouni, and G. C. Creatsas (2004) Effect of hormone replacement therapy, tibolone

and raloxifene on serum lipids, apolipoprotein A1, apolipoprotein B and lipoprotein(a) in Greek postmenopausal women. Gynecol. Endocrinol. 18: 244-257.

[48] Francucci, C. M., P. Daniele, N. Iori, A. Camilletti, F. Massi, and M. Boscaro (2005) Effects of raloxifene on body fat distribution and lipid profile in healthy post-menopausal women. J. Endocrinol. Invest. 28: 623-631.

[49] Duka, T., R. Tasker, and J. F. McGowan (2000) The effects of 3-week estrogen hormone replacement on cognition in elderly healthy females. Psychopharmacology 149: 129-139.

[50] Henderson, V. W., A. Paganini-Hill, C. K. Emanuel, M. E. Dunn, and J. G. Buckwalter (1994) Estrogen replacement therapy in older women. Comparisons between Alzheimer's disease cases and nondemented control subjects. Arch. Neurol. 51: 896-900.

[51] Kampen, D. L., and B. B. Sherwin (1996) Estradiol is related to visual memory in healthy young men. Behav. Neurosci. 110: 613-617.

[52] Liu, F., M. Day, L. C. Muniz, D. Bitran, R. Arias, R. Revilla-Sanchez, S. Grauer, G. Zhang, C. Kelley, V. Pulito, A. Sung, R. F. Mervis, R. Navarra, W. D. Hirst, P. H. Reinhart, K. L. Marquis, S. J. Moss, M. N. Pangalos, and N. J. Brandon (2008) Activation of estrogen receptor-beta regulates hippocampal synaptic plasticity and improves memory. Nature Neurosci. 11: 334-343.

[53] Vearncombe, K. J., and N. A. Pachana (2009) Is cognitive functioning detrimentally affected after early, induced menopause? Menopause 16: 188-198.

[54] Yaffe, K., M. Haan, A. Byers, C. Tangen, and L. Kuller (2000) Estrogen use, APOE, and cognitive decline: evidence of gene-environment interaction. Neurology 54: 1949-1954.

[55] Burkhardt, M. S., J. K. Foster, S. M. Laws, L. D. Baker, S. Craft, S. E. Gandy, B. G. Stuckey, R. Clarnette, D. Nolan, B. Hewson-Bower, and R. N. Martins (2004) Oestrogen replacement therapy may improve memory functioning in the absence of APOE epsilon4. J. Alzheimer's Dis. 6: 221-228.

[56] Fluck, E., S. E. File, and J. Rymer (2002) Cognitive effects of 10 years of hormone-replacement therapy with tibolone. J. Clin. Psychopharmacol. 22: 62-67.

[57] Gulseren, L., D. Kalafat, H. Mandaci, S. Gulseren, and L. Camli (2005) Effects of tibolone on the quality of life, anxiety-depression levels and cognitive functions in natural menopause: an observational follow-up study. Aust. N. Z. J. Obstet. Gynaecol. 45: 71-73.

[58] Creatsas, G., G. Christodoulakos, I. Lambrinoudaki, C. Panoulis, C. Chondros, and P. Patramanis (2003) Serum lipids and apolipoproteins in Greek postmenopausal women: association with estrogen, estrogen-progestin, tibolone and raloxifene therapy. J. Endocrinol. Invest. 26: 545-551.

[59] von Eckardstein, A., D. Crook, J. Elbers, J. Ragoobir, B. Ezeh, F. Helmond, N. Miller, H. Dieplinger, H. C. Bennink, and G. Assmann (2003) Tibolone lowers high density lipoprotein cholesterol by increasing hepatic lipase activity but does not impair cholesterol efflux. Clin. Endocrinol. (Oxf) 58: 49-58.

[60] Garefalakis, M., and M. Hickey (2008) Role of androgens, progestins and tibolone in the treatment of menopausal symptoms: a review of the clinical evidence. Clin. Interv. Aging 3: 1-8.

[61] Telegdy, G., M. Tanaka, and A. V. Schally (2009) Effects of the LHRH antagonist Cetrorelix on the brain function in mice. Neuropeptides 43: 229-234.

[62] von Dehn, G., O. von Dehn, W. Volker, C. Langer, G. F. Weinbauer, H. M. Behre, E. Nieschlag, G. Assmann, and A. von Eckardstein (2001) Atherosclerosis in apolipoprotein E-deficient mice is decreased by the suppression of endogenous sex hormones. Horm. Metab. Res. 33: 110-114.

[63] Buchter, D., H. M. Behre, S. Kliesch, A. Chirazi, E. Nieschlag, G. Assmann, and A. von Eckardstein (1999) Effects of testosterone suppression in young men by the gonadotropin releasing hormone antagonist cetrorelix on plasma lipids, lipolytic enzymes, lipid transfer proteins, insulin, and leptin. Exp. Clin. Endocrinol. Diabetes 107: 522-529.

Chapter 11

Anticholesterolemic and Antiatherogenic Effects of Taurine Supplementation is Model Dependent

Karl-Erik Eilertsen, Rune Larsen,
Hanne K. Mæhre, Ida-Johanne Jensen and Edel O. Elvevoll

Additional information is available at the end of the chapter

http://dx.doi.org/10.5772/47865

1. Introduction

Taurine (2-aminoethanesulfonic acid) is a sulphur-containing compound characterized as an amino acid. The presence of a sulfonic group, as opposed to a carboxyl group in other amino acids, gives taurine a pKa value of 1.5 and it is the most acidic amino acid. It is an exclusively free amino acid, i.e. it is not incorporated into proteins, but still widely distributed in most body tissues.

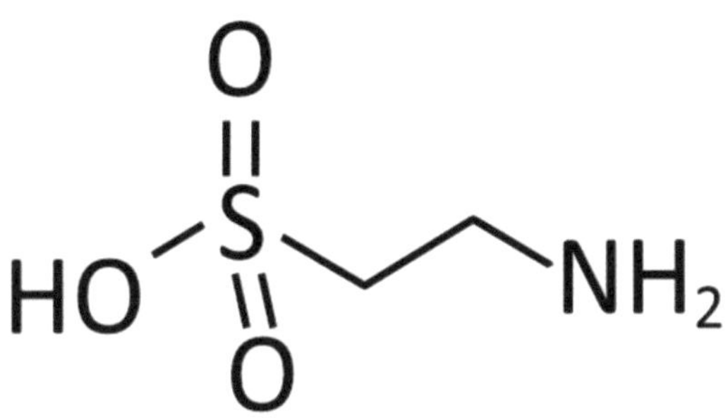

Figure 1. The structure of taurine

Taurine was identified almost two centuries ago and was named after the ox, *Bos taurus*, since it was first isolated from the bile of ox [1]. After its discovery, taurine was considered non-essential and biologically inert, however a multitude of functions have now been identified. Yet, all its physiological roles have not been fully elucidated. The phylogenetically oldest and best documented function of taurine is conjugation with bile acids in bile salt synthesis [2, 3]. In addition, taurine is involved in a variety of physiological processes as extensively reviewed [2], including neuromodulation in the central nervous system [4], energy production [5], protection against oxidation [6, 7] and immunomodulation

[8, 9]. An osmoregulatory role of taurine has also been established, playing a pivotal role in Central nervous system (CNS) cell volume regulation [10-12].

In felines taurine is considered indispensable and dietary deficiency leads to several clinical problems, including retinal degeneration and developmental abnormalities [13]. In humans it is regarded as a conditionally essential amino acid due to a limited ability to synthesize it [14, 15]. Taurine is now thought to play a more important role in human nutrition, and an increased dietary intake of taurine has been linked to several beneficial health outcomes in various diseases and medical conditions [16-18].

2. Taurine and nutrition

Estimates of dietary intake of taurine vary greatly. Although taurine content have been analysed in a variety of foods, it is usually excluded in food and nutrition data banks. Therefore, it is difficult to assess the dietary intake. A diet high in meat and especially seafood will provide a higher intake than a vegetarian diet which will provide very little taurine [19]. Mean ± SE dietary intake of taurine of 58 ± 19.5 mg/d was reported in omnivores [20], while it was not detected in a vegan diet. Laidlaw et al. [21] analysed taurine content in foods and calculated a taurine intake of less than 200 mg/d for individuals consuming a diet high in meat.

2.1. Taurine biosynthesis

Taurine is the most abundant intracellular free amino acid in the human body, the average amount being approximately 560 mmol (70g). The main organs of distribution are the retina, along with white blood cells, platelets, spleen, heart, muscle and brain [22].

As a product in the metabolism of sulphur-containing amino acids, taurine can be synthesised from its precursors methionine and cysteine, as shown in figure 2. The first step of the synthesis is methionine's reversible conversion to homocysteine by transmethylation and remethylation processes. Homocysteine can then be converted irreversibly to cysteine through the transsulfuration pathway catalyzed by cystathionine β-synthase and cystathionine γ-lyase [23]. Cysteine is, in turn, the origin of several biologically important molecules, including glutathione, inorganic sulphur and taurine [24]. Taurine can be synthesised from cysteine through several pathways, most commonly via cysteine sulfinic acid and hypotaurine, involving the enzymes cysteine dioxygenase (CDO) and cysteine sulfinic decarboxylase (CSAD) mainly present in the liver and brain. The activities of the enzymes involved, in particular the activity of CSAD, are both species and age dependent [25-27], being high in rodents and absent in cats. In addition, taurine synthesis is dependent on an adequate cysteine concentration, as production of glutathione is favoured when cysteine concentration is limited [28].

In humans the CSAD activity is low and the average daily synthesis of taurine ranges from 0.4 to 1.0 mmol (50-125mg). Excretion of taurine is very variable (0.22-1.85 mmol day-1) and affected by several factors such as genetics, age, gender, dietary intake, kidney function and health status [22]. The taurine body pool size is however regulated by the kidneys through renal absorption by the proximal tubule [14, 26, 29, 30].

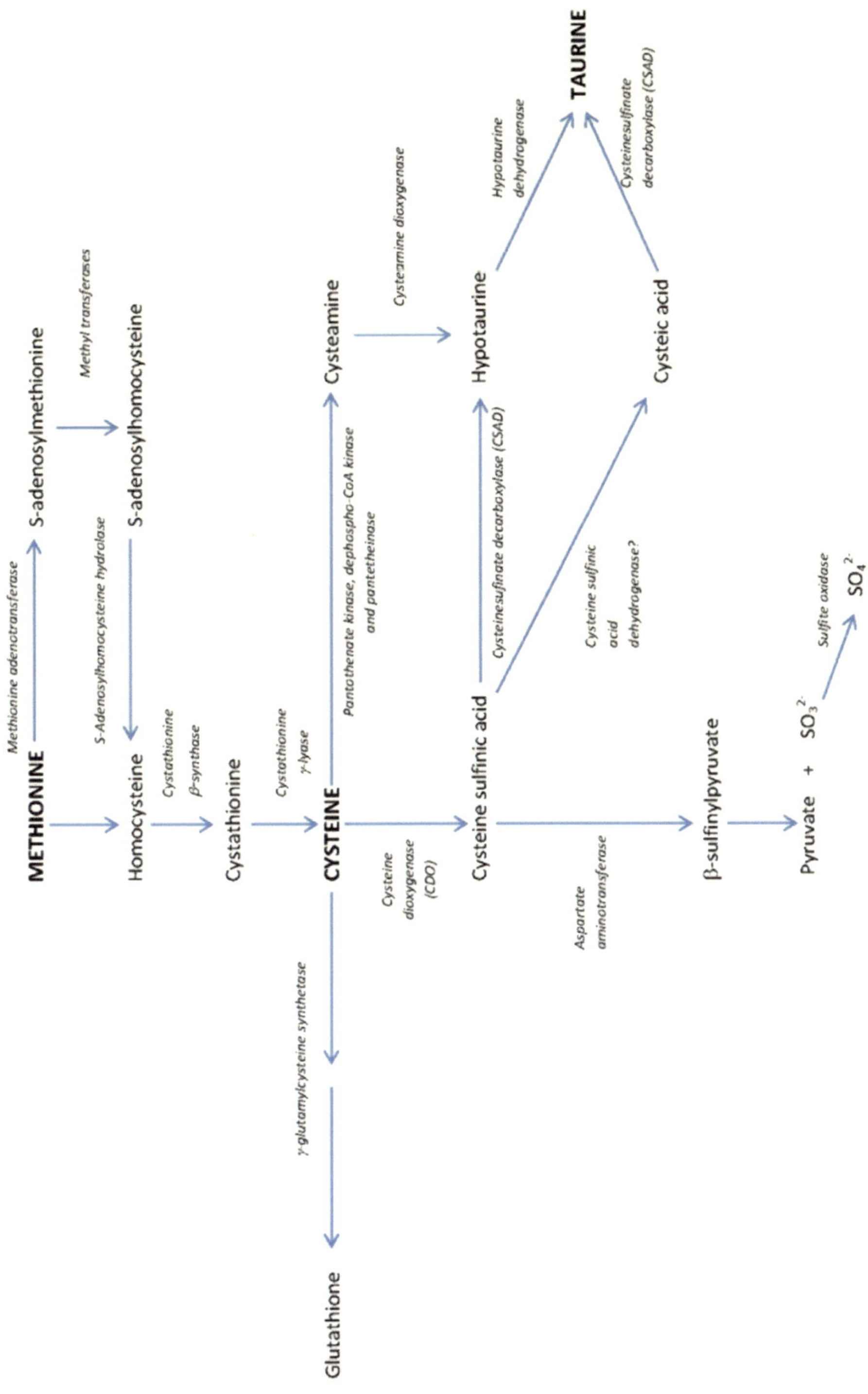

Figure 2. Taurine biosynthesis

2.2. Dietary sources

Taurine is found in most meats used for human consumption, whereas plants including grains, legumes, fruits and vegetables are devoid or contain only negligible amounts [21]. An exception is algae, mainly red algae (*Rhodophyta*), where notable amounts have been found [31, 32].

Taurine concentration has been investigated in a wide range of food products and it varies substantially between different marine and non-marine food items [17, 33]. A comparison of taurine concentrations in various foods is presented in table 1 [21, 34-43]. It is evident that seafood, and especially molluscs are high in taurine. Taurine is a key osmolyte in marine molluscs [44] and the highest taurine concentrations are found in marine bivalves and univalves [45]. Scallops and blue mussels are reported to have a respective taurine content of 827±15 and 510±12 mg per 100g raw muscle [21, 34]. In fact, the univalve abalone was already early in the last century 1918 exploited for preparation of taurine in large quantities [46].

There is also a tendency of taurine being more abundant in fish than in terrestrial animals. Taurine concentrations (mg per 100g raw fillets) of entire muscle of farmed Atlantic salmon (94 ± 16 mg), cod (120 ± 21 mg), saithe (162 ± 25 mg) and haddock (57 ± 6 mg) are reported to be intermediate [34]. Taurine content varies greatly between white and red muscle both in fish, poultry and mammals, with significantly higher levels being present in red muscle [21, 39, 40, 43], probably due to the increased vascularisation of these tissues.

Several studies investigating the retention and losses of taurine during food processing and preparation have been conducted [34, 36, 47-49]. Results indicate that taurine is susceptible to leaching losses similar to or even more than other free amino acids. Data on the oxidative and heat stability of taurine in foods is scarce. In milk, taurine losses seemed to proceed with the same degradation rate as lysine due to browning reactions [50].

2.3. Taurine supplementation

Taurine is maybe most famous for being an ingredient that is added to energy drinks, the concentration being approximately 4.0 g/L. Its physiological effect has been debated, with manufacturers, backed by studies, claiming that taurine in combination with other active ingredients may improve cognitive and muscular performance [51, 52]. The safety of taurine intake has also been investigated, especially in conjunction with its use in energy drinks. The European Food Safety Authority (EFSA) have concluded that taurine do not present any safety concerns with the levels currently used in energy drinks. The no observable adverse effect level (NOAEL) was at least 1000 mg/kg bw/day for pathological and behavioural changes, being much higher than an extreme consumer would be exposed to [53]. In their risk assessment, Shao and Hatchcock [54], found that absence of adverse effects was strong for taurine at supplemental intakes up to 3 g per day.

Food source	Taurine content (mg/100g wet weight) reported range ± SEM	References
Meat		
Beef, round	36	[38]
Beef (Bos taurus)	43 ± 8	[21]
Chicken, light meat	18 ± 3	[21]
Chicken, dark meat	169 ± 37	[21]
Turkey, light meat	30 ± 7	[21]
Turkey, dark meat	306 ± 69	[21]
Pork (loin)	61 ± 11	[21, 38]
Pork (loin)	50 ± 11	[21, 38]
Lamb, leg	45 ± 4	[38]
Veal	40 ± 13	[21]
Reindeer (loin)	62,4 ± 12	[42]
Red deer (loin)	28 ± 13	[37]
Fish		
Plaice	146 ± 5	[[35]
Cod fillet	120 ± 21	[34, 48]
Cod roe	365	[34]
Saithe fillet	162 ± 25	[34]
Haddock fillet	57 ± 6	[34]
African catfish	201 ± 32	[36]
Salmon fillet	94 ± 16	[34]
Mackerel	78	[35]
Bigeye tuna, white muscle	26	[41]
Bigeye tuna, dark muscle	270	[41]
Yellowfin tuna, white muscle	42	[41]
Yellowfin tuna, dark muscle	964	[41]
Bluefin tuna, white muscle	88	[41]
Bluefin tuna, dark muscle	195	[41]
Pacific saury, white muscle	223	[41]
Pacific saury, dark muscle	248	[41]
Milkfish, white muscle	95,7	[39]
Milkfish, dark muscle	309	[39]
Octopus	388 ± 13	[43]
Squid	356 ± 95	[21]
Fresh water fish		
Rainbow trout, white muscle	17	[40]
Rainbow trout, dark muscle	206	[40]

Coho salmon, white muscle	23	[40]
Coho salmon, dark muscle	275	[40]
Eel, white muscle	7	[40]
Eel, dark muscle	65	[40]
Catfish, white muscle	193	[40]
Catfish, dark muscle	465	[40]
Tilapia, white muscle	75	[40]
Tilapia, dark muscle	649	[40]
Carp, white muscle	129	[40]
Carp, dark muscle	579	[40]
Char, white muscle	15	[40]
Char, dark muscle	190	[40]
Sweet smelt, white muscle	137	[40]
Sweet smelt, dark muscle	294	[40]
Shellfish		
Peeled shrimps (Northern)	220 ± 2	[34]
Blue mussels	510 ± 12	[34]
Mussel	655 ± 72	[21]
Mussel	349	[43]
Clams	520 ± 97	[21, 43]
Scallops	827 ± 15	[21]
Scallop	332	[43]
Oysters	396 ± 29	[21]

Table 1. Taurine content in various food sources

Another food item where taurine is supplemented is in infant formulas. This practice started in the early 1980s after recognizing that preterm infants fed infant formulas had lower urine and plasma concentrations than infants fed pooled human milk [55]. The necessity of this supplementation remains disputed as clinical studies have not provided evidence of any clinical effects of growth and development in preterm or low birth weight infants [56]. High concentrations of taurine in the developing brain [57], as well as results from various animal studies clearly indicate the importance of taurine in neurodevelopment [58, 59].

2.4. Taurine and associated health benefits

An increased dietary intake of taurine has been associated with multiple beneficial health outcomes. Epidemiological data and animal studies suggests that dietary intake of taurine has beneficial effects on cardiovascular disease (CVD) [33, 60-62]. Perhaps the best characterized attribution of taurine is the antihypertensive effect although there are still questions about the exact mechanisms of action [63-66]. A long term effect of hypertension is the development of hypertrophy of the left ventricle, in which Angiotensin II (Ang II) plays an important role. Several studies have shown that taurine reverses these actions of Ang II [67, 68]. Animal studies have also indicated that taurine may reduce insulin resistance [69, 70], but most of the clinical studies have failed to prove the beneficial role of taurine in insulin resistance and diabetic complications [71, 72]. Taurine have also been found to ameliorate alcoholic steatohepatitis [73-75] in rats. In addition, some evidence have been brought forward of a potential therapeutic use of taurine in nonalcoholic fatty liver disease [76]. Despite taurine being linked to beneficial health outcomes in an increasing number of diseases and medical conditions, the number of studies is relatively small. The effects of taurine on cholesterol and CVD are most studied and documented.

3. Taurine and cholesterol metabolism

Perhaps the best studied function of taurine is its role in cholesterol metabolism. Cholesterol is metabolized and broken down to cholic acids, conjugated to taurine or glycine, and excreted in the bile [77].

3.1. Effects of taurine on circulating cholesterol levels

High blood cholesterol levels is the most pronounced risk factor for developing atherosclerosis, vascular inflammation and hardening of the arteries associated with excess cholesterol deposition in the vasculature. Taurine has generally been associated with a beneficial effect on blood cholesterol levels. Cholesterol is metabolized and broken down to cholic acids, conjugated to taurine or glycine, and excreted in the bile [77]. The conjugation pattern varies considerably across species. In dog and rat bile acids are entirely conjugated to taurine, whereas rabbits have all their bile acids conjugated to glycine. Species where glycine-conjugated bile acids dominate have higher blood cholesterol levels and are more susceptible to dietary induced hypercholesterolemia, and based on these observations it was hypothesized that dietary taurine might counteract dietary induced increase in blood cholesterol [2].

3.1.1. Effects of taurine on cholesterol levels in mice

Several studies have investigated the effect of dietary intake of taurine on lipids in different mice strains. Six months administration of 1% taurine (w/v) to the drinking water given to C57BL7/6J mice fed a high-fat diet resulted in reduced serum LDL and VLDL cholesterol and increased serum HDL cholesterol [78]. Similar results were obtained in a small study using the same mouse strain, where 1% taurine (w/w) added to a high cholesterol diet reduced serum triglycerides, total cholesterol and VLDL+LDL cholesterol levels already after 4 weeks treatment [79]. Cholesterol-fed and streptozotocin (STZ)-induced diabetic male ICR mice were given a diet enriched with 2% cholesterol (w/w) and 0.5% cholate (w/w) for 10 weeks [80]. In addition, mice received a daily dose of saline or taurine (50 or 100 mg/kg p.o.). Both taurine-treated groups had lower serum total and LDL cholesterol compared to the STZ/saline group.

In more extreme models such as apolipoprotein E-deficient (apo$E^{-/-}$) mice fed a normal rodent chow supplemented with 2% taurine (w/w) for 12 weeks, serum VLDL, LDL and total cholesterol levels increased compared to mice without taurine supplementation [81]. Similar results has been reported for extreme spontaneously hyperlipidemic mice (SHL; KOR-*Apoe*^{shl}), where 12 weeks treatment with 1% taurine (w/v) added to the drinking water increased serum HDL-cholesterol but did not affect serum total cholesterol or VLDL+LDL cholesterol levels [82].

3.1.2. *Effects of taurine on cholesterol levels in rats*

A large number of studies have investigated the effect of dietary taurine on dietary hypercholesterolemia in various rat models [65]. Rats fed high-fat diets seem to be the model with most consistent antihypercholesterolemic effects of dietary taurine. Rats have a relatively low taurine content in skeletal muscles (10-25 mg/100 g muscle) [83, 84].

Rats are generally not suitable for pharmacologically cholesterol and lipoprotein studies due to their substantially different lipid profile compared to humans.

Male wistar rats. Taurine supplementation does not alter plasma lipids in male wistar rats fed normal chow [85]. However, when these rats were fed high cholesterol diet containing 2% cholesterol and 1% cholic acid, dietary supplementation of 4% taurine significantly counteracted the observed increase in serum cholesterol by 44% [86]. This observation has been confirmed be several studies [87-90]. In Wistar male rats fed a cholesterol-containing diet (0.5% cholesterol w/w) for 40 days, serum cholesterol increased 5 fold compared to chow fed rats. Oral supplementation of 470 mg/kg/day taurine (0.5% w/v) in water lowered the increased serum cholesterol (54%) [91]. When these rats were fed a high-fat diet (11% coconut oil w/w) for 6 months, a daily oral supplementation of 1 mg taurine lowered serum cholesterol (37%), LDL cholesterol (34%), and triglycerides (95%), compared to the high-fat control diet [88]. Already after 14 days intervention, 5% dietary taurine (w/w) supplementation has been indicated to lower high cholesterol (1% cholesterol, 2.5% cholate) induced serum cholesterol (-42%) [90]. In wistar rats, the taurine effect has been indicated to be caused by an increased faecal bile acid excretion, increased hepatic cholesterol 7α-hydroxylase expression and activity [89, 90]. The rapid effect of taurine supplementation on serum cholesterol has been confirmed recently [92]. When fed a diet containing 60.7% sucrose, 9.0% lard, and 0.5% cholesterol for 14 days, serum cholesteryl ester and free cholesterol were reduced by 39% and 53% compared to rats fed control diet without taurine, respectively. Rats fed taurine also had smaller livers compared to control-fed rats. Hepatic cholesteryl esters were also reduced by approximately 20% in the taurine supplemented rats. This hypocholesterolemic effect was ascribed to a lower hepatic secretion of cholesteryl esters.

Streptozotocin-induced diabetic rats. Male Wistar rats injected with STZ are also used as a diabetic model. In these rats dietary taurine supplementation markedly reduced serum total cholesterol (-50%) induced by cholesterol-containing diet (1% cholesterol w/w) for 4 weeks [93].

Fructose-induced rat insulin resistance model. Male wistar rats fed a diet containing 60% fructose developed impaired glucose tolerance and insulin resistance [85]. Taurine administration (300 mg/kg/day *i.p.*) counteracted the fructose induced plasma total cholesterol, LDL cholesterol, and triglycerides by 11%, 21%, and 23%, respectively.

Spontaneously hypertensive rats (SHR). The effect of taurine supplementation on blood pressure in SHR rats was investigated already in the 1970ies [94]. How dietary administration of taurine affects cholesterol metabolism has not been reported in these studies. However, in a stroke prone substrain of the SHR rats, taurine supplementation has been indicated to prevent high-fat/high cholesterol induced elevation of serum cholesterol in SHR rats [95].

Sprague-Dawley rats. Also in male Sprague-Dawley rats fed a high-fat, high cholesterol diet (HFCD; 10% corn oil, 1.5% cholesterol) supplemented with taurine (1.5% w/w) plasma cholesterol was lowered by 31% compared to HFCD control rats [87]. LDL+VLDL cholesterol (-38%) and triglycerides (-43%) were also lower in rats supplemented with taurine compared to HFCD control rats. These results have been confirmed with an identical experimental setup for 5 weeks reporting a 20% and 25% reduction in serum total cholesterol and triglycerides, respectively [96]. In this model, plasma total cholesterol, LDL cholesterol and triglycerides were reduced in rats fed taurine supplemented cholesterol free diet compared the cholesterol free control diet [87].

3.1.3. *Effects of taurine on cholesterol levels in rabbits*

Different rabbit strains have been used to investigate the effects of dietary taurine supplementation on dietary induced hypercholesterolemia [86, 97, 98]. The results from the administration of taurine to rabbits have been ambiguous. In *Male New Zealand white rabbits,* fed a high cholesterol (1% w/w) diet, addition of 2.5% taurine (w/w) for 2.5 month reduced the serum total cholesterol and triglyceride levels by 22% and 38%, respectively, compared to high cholesterol diet alone [99]. In this study similar reductions were observed for hepatic and aorta lipid levels in these rabbits. However, when the same rabbit strain were given normal chow supplemented with 0.5% cholesterol (w/w) for 4 weeks no effect of dietary taurine (2.5% w/w) supplementation was observed [100]. Also when given a normal diet supplemented with 2% cholesterol (w/w), taurine added to the drinking water (0.1 or 0.5% w/v) for 14 weeks had no influence on serum cholesterol and triglycerides [97].

3.1.4. *Effects of taurine on cholesterol levels in hamsters*

Hamsters (Male Golden Syrian hamsters) have also been used as model for studying cholesterol metabolism. The rationale for this is that hamsters and humans have comparable blood cholesterol levels, hamsters use both taurine and glycine for bile acid conjugation and the lipoprotein profile in response to dietary cholesterol is comparable [101]. When Male Golden Syrian hamsters were fed a normal chow supplemented with 0.05% cholesterol or a 10% coconut oil, high-fat diet (0.05% cholesterol) for two weeks, taurine dissolved in drinking water (1% w/v) reduced serum total cholesterol in chow- (15% reduction) as well as high-fat diet-fed (42% reduction) hamsters [102]. A similar effect was observed for non-HDL (LDL+VLDL) cholesterol.

Recently, lipid metabolism has been closely studied in Male Golden Syrian Hamsters fed different diets with or without taurine for 4 weeks [103]. The groups received a high fat diet (chow mixed with 7% butter [w/w] and 0.2% cholesterol [w/w]) and drinking water without or supplemented with either 0.35% or 0.7% taurine (w/v). Hamsters given taurine was smaller, had less visceral fat and smaller livers after 4 weeks. Both taurine concentrations resulted in significant lower serum triglycerides, total cholesterol, and LDL+VLVL cholesterol. Up-regulated gene expression of the low-density lipoprotein receptor and CYP7A1 genes, paralleled by increased faecal cholesterol and bile acid concentrations in the taurine treated hamsters, indicated that the taurine effect on the cholesterol and lipid profiles is due to increased cholesterol metabolism.

3.1.5. Effects of taurine on cholesterol levels in humans

Historically, taurine has been believed to decrease blood cholesterol levels in adults. Only a limited number of studies have investigated the effect of oral taurine supplementation on blood cholesterol or lipoprotein levels in humans and ambiguous results have arisen from these. Early studies found no effect on serum cholesterol after incidental treatment of patients with 1.5 to 3 g taurine/day for up to 2 months [77, 104, 105]. To our knowledge there has been no well-designed random controlled clinical trial assessing the dose-response effect of oral taurine supplementation on blood lipids in healthy humans. However, the effect of taurine in relation to development of CVD has been documented through a human clinical trial. Results of a 7 week human intervention trial revealed that supplementation with 0.4 g taurine/day in combination with omega-3 fatty acids (1 g EPA+DHA/day) significantly improved the lipid profiles by reducing serum total and LDL cholesterol levels compared to supplementation with omega-3 fatty acids alone [106]. In another study the effects of oral supplementation with taurine (3 g/day) or placebo for 7 weeks was assessed in young obese healthy subjects [107]. In this study, taurine had no effect on serum cholesterol, but triglycerides and bodyweight was significantly reduced compared to placebo effect. Finally, a daily 6 g taurine supplementation to human healthy volunteers receiving a cholesterol-inducing diet for 3 weeks attenuated the expected increase of serum total cholesterol and LDL-cholesterol, whereas serum VLDL-cholesterol and triglyceride levels compared to the control group [108]. In insulin-dependent diabetes mellitus patients intake of taurine (1 g/day) reduced serum triglyceride levels, but no effect was observed on serum cholesterol [109]. Finally, in a randomized, double-blinded, crossover intervention, overweight non-diabetic men given a daily dose of 1.5 g taurine or placebo, no effect was reported on blood lipids [71]. In summary, results from oral taurine supplementation to humans are ambiguous, and further adequately designed interventions are warranted to further investigate the potential of taurine as a hypocholesterolemic agent.

4. Effects of taurine on atherogenesis/development of atherosclerosis

High blood cholesterol levels is the most pronounced risk factor for developing atherosclerosis, vascular inflammation and hardening of the arteries associated with excess cholesterol deposition in the vasculature. Apart from humans and monkeys, wild animals normally do not develop substantial atherosclerosis. There is however, an array of laboratory animal models in common use for studying the effects of pharmacological substances and dietary modifications on lesion formation. The effect of taurine has been investigated in several of these models.

4.1. Effects of taurine on atherosclerosis in mice

In the hyperlipidemic apoE$^{-/-}$ mice, taurine has been reported to delay atherogenesis by decreasing oxidized substances that cause inflammation, as well as increasing HDL-cholesterol [82]. Also in apoE$^{-/-}$ mice fed a normal rodent chow supplemented with 2% taurine (w/w) for 12 weeks, formation of atherosclerotic lesions were significantly reduced [81]. However this effect was independent of serum cholesterol as VLDL, LDL, and total cholesterol were increased.

Moreover, in spontaneously hyperlipidemic mice, taurine (1% w/v) provided through drinking water, was reported to suppress the development of lesion formation without affecting the levels of serum VLDL and LDL [82].

In our lab, apoE$^{-/-}$-mice were given Western diets (WD) containing 20% fat (w/w), 0.2% cholesterol (w/w) for 13 weeks [110]. The mice received WD, WD supplemented with 0.5% taurine (w/w) or WD supplemented with 0.5% taurine (w/w) in combination with a daily dose of marine long-chain omega-3 polyunsaturated fatty acids (n-3 PUFA) recommend in the dietary guidelines for humans. In these studies, taurine did not affect serum cholesterol or triglyceride levels alone or combination with n-3 PUFA. This may indicate that a larger supplementary dose of taurine is needed to prevent dietary induced hypercholesterolemia in apoE$^{-/-}$-mice.

4.2. Effects of taurine on lipid lesion formation in rats

Rats are generally not a suitable animal model for atherosclerosis as they do not develop lesion deposits resembling the early phase of human atherogenesis. However, in a rat model of balloon induced vascular neointima formation, supplementation with taurine (3% in drinking water) from 2 days before the surgical procedure and 14 days after, reduced vascular smooth muscle cell proliferation [111]. This key step in the initiation of atherogenesis was reduced by 28% compared to control fed rats. Taurine was located immunohistochemically mainly to the surface of the exposed media and adventitia of the injured carotid artery and higher levels were observed in the taurine treated rats. This corresponded to a lower vascular production of superoxide anion compared to the control animals. From these experiments it was concluded that the preventive effect of taurine towards neointima formation was attributable to anti-oxidative effects.

4.3. Effects of taurine on atherosclerosis in rabbits

The effect of dietary taurine on development of atherosclerosis has been investigated in different rabbit strains. Taurine has been indicated to prevent progression of atherosclerotic lesions in rabbits without affecting serum cholesterol in two different models. In New-Zealand white male rabbits given a diet containing 2% cholesterol (w/w), taurine added to the drinking water (0.1 or 0.5% w/v) for 14 weeks reduced the aortic deposition of fat [97]. This so-called anti-atherosclerotic effect was only significant for the highest taurine dose tested. Recently, it was indicated that the taurine antiatherosclerotic effect was evident in these rabbits after only 4 weeks on the atherogenic diet [100]. *Watanabe heritable hyperlipidemic (WHHL) rabbits* carries an inheritable mutation in the LDL receptor and is hence a typical genetically hyperlipidemic animal model. When WHHL rabbits were given drinking water containing 1% taurine (w/v) for 6 months they developed significantly less atherosclerotic lesion formation compared to rabbits not supplemented with taurine [112].

4.4. Effects of taurine on atherosclerosis in humans

Results on the effects of dietary taurine in humans are mainly from prospective studies. It is evident that individuals with high urinary excretion of taurine and high dietary intake of food high in taurine in general have fewer incidences of cardiovascular diseases compared to individuals with low dietary intake of taurine [33]. In addition, increased dietary intake of taurine either alone or in the combination with omega-3 fatty acids, has also been suggested to reduce MCP-1, an important risk factor of CVD [106]. No further randomised clinical trials on the effects of dietary supplementation of taurine on CVD disease markers has been reported.

5. Conclusion

Taurine appears to be able to prevent hypercholesterolemia and hepatic steatosis induced by high-fat and high-cholesterol diets in most animal models. The major mechanism by which taurine lowers serum cholesterol levels is by increased utilization of cholesterol for bile acid synthesis. In mice, rats, and hamsters, dietary intake of taurine cause reduction in diet-induced serum cholesterol accompanied by enhanced mRNA expression and enzymatic activity of 7a-hydroxylase, the rate-limiting enzyme of bile acid synthesis. In normal diets taurine does not appear to modify serum and liver cholesterol levels.

Dietary supplementation with taurine is indicated to have cardiovascular benefits. The effect on atherosclerosis appears to be highly dose- and model-dependent. In animal experiments using high-fat diets to induce increased levels of lipids, taurine has been demonstrated to significantly alleviate atherosclerotic lesions. The effects of taurine appear to be related to increased degradation and excretion of cholesterol as bile in the feces and the most common feature is that taurine increases expression and activity of cholesterol 7α-hydroxylase. Only a few studies have evaluated the effects of taurine in human subjects.

From the available data it is not possible to conclude about the proposed antihyperlipidemic and antiatherosclerotic, therefore more basic and clinical research on the effects of taurine supplementation on hypercholesterolemic and atherosclerotic effects are warranted. Randomized clinical trials of dietary taurine and taurine sources may provide further knowledge about the potential hypocholesterolemic and antiatherogenic effects of long-term dietary taurine supplementation in healthy volunteers and humans with hyperlipidemia, metabolic syndrome and cardiovascular diseases.

Author details

Karl-Erik Eilertsen, Rune Larsen,
Hanne K. Mæhre, Ida-Johanne Jensen and Edel O. Elvevoll
University of Tromsø, Norway

6. References

[1] Demarcay H. Ueber die Natur der Galle. Annalen der Pharmacie. 1938;27(3):270-291.
[2] Huxtable RJ. Physiological actions of taurine. Physiological Reviews. 1992;72(1):101-163.
[3] Jacobsen JG, Smith LH. Biochemistry and physiology of taurine and taurine derivatives. Physiological Reviews. 1968;48(2):424-.
[4] Belluzzi O, Puopolo M, Benedusi M, Kratskin I. Selective neuroinhibitory effects of taurine in slices of rat main olfactory bulb. Neuroscience. 2004;124(4):929-944.
[5] Giehl TJ, Qoronfleh MW, Wilkinson BJ. Transport, nutritional and metabolic studies of taurine in staphylococci. Journal of General Microbiology. 1987;133:849-856.
[6] Atmaca G. Antioxidant effects of sulfur-containing amino acids. Yonsei Medical Journal. 2004;45(5):776-788.
[7] Schaffer S, Azuma J, Takahashi K, Mozaffari M. Why is taurine cytoprotective? In: Lombardini JB, Schaffer SW, Azuma J, editors. Taurine 5: Beginning the 21st Century2003. p. 307-321.
[8] Schuller-Levis GB, Park E. Taurine and its chloramine: Modulators of immunity. Neurochemical Research. 2004;29(1):117-126.
[9] Wojtecka-Lukasik E, Czuprynska K, Maslinska D, Gajewski M, Gujski M, Maslinski S. Taurine - chloramine is a potent antiinflammatory substance. Inflammation Research. 2006;55:17-18.
[10] Lambert IH. Regulation of the cellular content of the organic osmolyte taurine in mammalian cells. Neurochemical Research. 2004;29(1):27-63.
[11] Oja SS, Saransaari P. Taurine as osmoregulator and neuromodulator in the brain. Metab Brain Dis. 1996;11(2):153-164.
[12] Olson JE, Martinho E. Regulation of taurine transport in rat hippocampal neurons by hypo-osmotic swelling. Journal of Neurochemistry. 2006;96(5):1375-89.
[13] Markwell PJ, Earle KE. Taurine - an essential nutrient for the cat - a brief review of the biochemistry of its requirement and the clinical consequences of deficiency. Nutrition Research. 1995;15(1):53-8.

[14] Schuller-Levis G, Park E. Is taurine a biomarker? Advances in Clinical Chemistry. 2006;41:1-21.
[15] Stapleton PP, Charles RP, Redmond HP, BouchierHayes DJ. Taurine and human nutrition. Clinical Nutrition. 1997;16(3):103-108.
[16] Bouckenooghe T, Remacle C, Reusens B. Is taurine a functional nutrient? Current Opinion in Clinical Nutrition and Metabolic Care. 2006;9(6):728-733.
[17] Wojcik OP, Koenig KL, Zeleniuch-Jacquotte A, Costa M, Chen Y. The potential protective effects of taurine on coronary heart disease. Atherosclerosis. 2010;208(1):19-25.
[18] Zulli A. Taurine in cardiovascular disease. Current Opinion in Clinical Nutrition and Metabolic Care. 2011;14(1):57-60.
[19] Laidlaw SA, Shultz TD, Cecchino JT, Kopple JD. Plasma and urine taurine levels in vegans. American Journal of Clinical Nutrition. 1988;47(4):660-663. Epub 1988/04/01.
[20] Rana SK, Sanders TA. Taurine concentrations in the diet, plasma, urine and breast milk of vegans compared with omnivores. British Journal of Nutrition. 1986;56(1):17-27. Epub 1986/07/01.
[21] Laidlaw SA, Grosvenor M, Kopple JD. The taurine content of common foodstuffs. Journal of Parenteral and Enteral Nutrition. 1990;14(2):183-188.
[22] Lourenco R, Camilo ME. Taurine: a conditionally essential amino acid in humans? An overview in health and disease. 2002;17:262-270.
[23] Brosnan JT, Brosnan ME. The sulfur-containing amino acids: An overview. Journal of Nutrition. 2006;136(6):1636S-1640S.
[24] Stipanuk MH, Dominy JE, Lee JI, Coloso RM. Mammalian cysteine metabolism: New insights into regulation of cysteine metabolism. Journal of Nutrition. 2006;136(6):1652S-1659S.
[25] Delarosa J, Stipanuk MH. Evidence for a rate-limiting role of cysteinesulfinate decarboxylase activity in taurine biosynthesis invivo. Comparative Biochemistry and Physiology B-Biochemistry & Molecular Biology. 1985;81(3):565-571.
[26] Jacobsen JG, Smith LH. Comparison of decarboxylation of cysteine sulphinic acid-1-14C and cysteic acid-1-14C by human, dot, and rat liver and brain. Nature. 1963;200(490):575-577.
[27] Worden JA, Stipanuk MH. A comparison by species, age and sex of cysteinesulfinate decarboxylase activity and taurine concentration in liver and brain of animals. Comparative Biochemistry and Physiology B-Biochemistry & Molecular Biology. 1985;82(2):233-239.
[28] Stipanuk MH. Role of the liver in regulation of body cysteine and taurine levels: A brief review. Neurochemical Research. 2004;29(1):105-110.
[29] Lourenco R, Camilo ME. Taurine: a conditionally essential amino acid in humans? An overview in health and disease. Nutricion hospitalaria : organo oficial de la Sociedad Espanola de Nutricion Parenteral y Enteral. 2002;17(6):262-270. Epub 2003/01/08.
[30] Tappaz ML. Taurine biosynthetic enzymes and taurine transporter: Molecular identification and regulations. Neurochemical Research. 2004;29(1):83-96.
[31] Dawczynski C, Schubert R, Jahreis G. Amino acids, fatty acids, and dietary fibre in edible seaweed products. Food Chemistry. 2007;103(3):891-839.

[32] Kataoka H, Ohnishi N. Occurrence of Taurine in Plants. Agricultural and Biological Chemistry. 1986;50(7):1887-1888.
[33] Yamori Y, Taguchi T, Hamada A, Kunimasa K, Mori H, Mori M. Taurine in health and diseases: consistent evidence from experimental and epidemiological studies. Journal of Biomedical Science. 2010;17.S6
[34] Dragnes BT, Larsen R, Ernstsen MH, Maehre H, Elvevoll EO. Impact of processing on the taurine content in processed seafood and their corresponding unprocessed raw materials. International Journal of Food Sciences and Nutrition. 2009;60(2):143-152.
[35] Gormley TR, Neumann T, Fagan JD, Brunton NP. Taurine content of raw and processed fish fillets/portions. European Food Research and Technology. 2007;225(5-6):837-842.
[36] Mierke-Klemeyer S, Larsen R, Oehlenschlager J, Maehre H, Elvevoll EO, Bandarra NM, et al. Retention of health-related beneficial components during household preparation of selenium-enriched African catfish (Clarias gariepinus) fillets. European Food Research and Technology. 2008;227(3):827-833.
[37] Purchas RW, Triumf EC, Egelandsdal B. Quality characteristics and composition of the longissimus muscle in the short-loin from male and female farmed red deer in New Zealand. Meat Science. 2010;86(2):505-510.
[38] Roe DA, Weston MO. Potential significance of free taurine in the diet. Nature. 1965;205:287-288.
[39] Shiau CY, Pong YJ, Chiou TK, Chai TY. Free amino acids and nucleotide-related compounds in milkfish (Chanos chanos) muscles and viscera. Journal of Agricultural and Food Chemistry. 1996;44(9):2650-2553.
[40] Suzuki T, Hirano T, Shirai T. Distribution of Extractive Nitrogenous Constituents in White and Dark Muscles of Fresh-Water Fish. Comparative Biochemistry and Physiology B-Biochemistry & Molecular Biology. 1990;96(1):107-111.
[41] Suzuki T, Hirano T, Suyama M. Free Imidazole Compounds in White and Dark Muscles of Migratory Marine Fish. Comparative Biochemistry and Physiology B-Biochemistry & Molecular Biology. 1987;87(3):615-619.
[42] Triumf EC, Purchas RW, Mielnik M, Maehre HK, Elvevoll E, Slinde E, et al. Composition and some quality characteristics of the longissimus muscle of reindeer in Norway compared to farmed New Zealand red deer. Meat Science. 2012;90(1):122-129.
[43] Zhao XH, Jia JB, Lin Y. Taurine content in Chinese food and daily taurine intake of Chinese men. Taurine 3. 1998;442:501-505.
[44] Lange R. Osmotic function of amino acids and taurine in mussel, mytilus edulis. Comparative Biochemistry and Physiology. 1963;10(2):173-179.
[45] Bradley HC. The occurence of taurin in invertebrate muscle. Science. 1904;20:25.
[46] Schmidt CLA, Watson T. A method for the preparation of taurin in large quantities. Journal of Biological Chemistry. 1918;33(3):499-500.
[47] Larsen R, Elvevoll EO. Water uptake, drip losses and retention of free amino acids and minerals in cod (Gadus morhua) fillet immersed in NaCl or KCl. Food Chemistry. 2008;107(1):369-376.
[48] Larsen R, Stormo SK, Dragnes BT, Elvevoll EO. Losses of taurine, creatine, glycine and alanine from cod (Gadus morhua L.) fillet during processing. Journal of Food Composition and Analysis. 2007;20(5):396-402.

[49] Purchas RW, Busboom JR, Wilkinson BHP. Changes in the forms of iron and in concentrations of taurine, carnosine, coenzyme Q(10), and creatine in beef longissimus muscle with cooking and simulated stomach and duodenal digestion. Meat Science. 2006;74(3):443-449.
[50] Saidi B, Warthesen JJ. Analysis and Heat-Stability of Taurine in Milk. Journal of Dairy Science. 1990;73(7):1700-1706.
[51] Geiss KR, Jester I, Falke W, Hamm M, Waag KL. The Effect of a Taurine-Containing Drink on Performance in 10 Endurance-Athletes. Amino Acids. 1994;7(1):45-56.
[52] Seidl R, Peyrl A, Nicham R, Hauser E. A taurine and caffeine-containing drink stimulates cognitive performance and well-being. Amino Acids. 2000;19(3-4):635-642.
[53] Authority EFS. Scientific Opinion of the Panel on Food Additives and Nutrient Sources added to Food on a request from the Commission on the use of taurine and D-glucurono-g-lactone as constituents of the so-called "energy" drinks. 2009;935: 1-31. The EFSA Journal. 2009;935:1-31.
[54] Shao A, Hathcock JN. Risk assessment for the amino acids taurine, l-glutamine and l-arginine. Regulatory Toxicology and Pharmacology. 2008;50(3):376-399.
[55] Chesney RW. Taurine - Is It Required for Infant Nutrition. Journal of Nutrition. 1988;118(1):6-10.
[56] Verner A, Craig S, McGuire W. Effect of taurine supplementation on growth and development in preterm or low birth weight infants. Cochrane Database of Systematic Reviews. 2007(4).
[57] Sturman JA, Gaull GE. Taurine in the Brain and Liver of Developing Human and Monkey. Journal of Neurochemistry. 1975;25(6):831-835.
[58] Chapman GE, Greenwood CE. Taurine in nutrition and brain development. Nutrition Research. 1988;8(8):955-968.
[59] Gaull GE. Taurine in pediatric nutrition: review and update. Pediatrics. 1989;83(3):433-442. Epub 1989/03/01.
[60] Yamauchitakihara K, Azuma J, Kishimoto S. Taurine protection against experimental arterial calcinosis in mice. Biochemical and Biophysical Research Communications. 1986;140(2):679-683.
[61] Yamori Y, Liu L, Mizushima S, Ikeda K, Nara Y, Group CS. Male cardiovascular mortality and dietary markers in 25 population samples of 16 countries. Journal of Hypertension. 2006;24(8):1499-1505.
[62] Yamori Y, Liu LJ, Ikeda K, Miura A, Mizushima S, Miki T, et al. Distribution of twenty-four hour urinary taurine excretion and association with ischemic heart disease mortality in 24 populations of 16 countries: Results from the WHO-CARDIAC Study. Hypertension Research. 2001;24(4):453-457.
[63] Hagar HH, El Etter E, Arafa M. Taurine attenuates hypertension and renal dysfunction induced by cyclosporine A in rats. Clinical and Experimental Pharmacology and Physiology. 2006;33(3):189-196.
[64] Hu JM, Xu XL, Yang JC, Wu GF, Sun CM, Lv QF. Antihypertensive Effect of Taurine in Rat. In: Azuma J, Schaffer SW, Ito T, editors. Advances in Experimental and Medical Biololgy. 2009. p. 75-84.
[65] Militante JD, Lombardini JB. Treatment of hypertension with oral taurine: experimental and clinical studies. Amino Acids. 2002;23(4):381-393.

[66] Nandhini AT, Thirunavukkarasu V, Anuradha CV. Potential role of kinins in the effects of taurine in high-fructose-fed rats. Canadian Journal of Physiology and Pharmacology. 2004;82(1):1-8.
[67] Azuma M, Takahashi K, Fukuda T, Ohyabu Y, Yamamoto I, Kim S, Iwao H, Schaffer SW, Azuma J. Taurine attenuates hypertrophy induced by angiotensin II in cultured neonatal rat cardiac myocytes. European Journal of Pharmacology. 2000;403(3):181-188.
[68] Rao MR, Tao L. Effects of taurine on signal transduction steps induced during hypertrophy of rat heart myocytes. In: Schaffer S, Lombardini JB, Huxtable RJ, editors. Taurine 3: Cellular and Regulatory Mechanisms1998. p. 137-143.
[69] Franconi F, Di Leo MAS, Bennardini F, Ghirlanda G. Is taurine beneficial in reducing risk factors for diabetes mellitus? Neurochemical Research. 2004;29(1):143-150.
[70] Franconi F, Loizzo A, Ghirlanda G, Seghieri G. Taurine supplementation and diabetes mellitus. Current Opinion in Clinical Nutrition and Metabolic Care. 2006;9(1):32-36.
[71] Brons C, Spohr C, Storgaard H, Dyerberg J, Vaag A. Effect of taurine treatment on insulin secretion and action, and on serum lipid levels in overweight men with a genetic predisposition for type II diabetes mellitus. European Journal of Clinical Nutrition. 2004;58(9):1239-1247.
[72] Ito T, Schaffer SW, Azuma J. The potential usefulness of taurine on diabetes mellitus and its complications. Amino Acids. 2012;42(5):1529-1539.
[73] Fang YJ, Chiu CH, Chang YY, Chou CH, Lin HW, Chen MF, Chen YC. Taurine ameliorates alcoholic steatohepatitis via enhancing self-antioxidant capacity and alcohol metabolism. Food Research International. 2011;44(9):3105-3110.
[74] Kerai MDJ, Waterfield CJ, Kenyon SH, Asker DS, Timbrell JA. Reversal of ethanol-induced hepatic steatosis and lipid peroxidation by taurine: A study in rats. Alcohol and Alcoholism. 1999;34(4):529-541.
[75] Wu GF, Yang JC, Sun CM, Luan XH, Shi J, Hu JM. Effect of Taurine on Alcoholic Liver Disease in Rats. Advances in Experimental and Medical Biology. 2009;643:313-22.
[76] Gentile CL, Nivala AM, Gonzales JC, Pfaffenbach KT, Wang D, Wei YR, Jiang H, Orlicky DJ, Petersen DR, Pagliassotti MJ, Maclean KN. Experimental evidence for therapeutic potential of taurine in the treatment of nonalcoholic fatty liver disease. American Journal of Physiology-Regulatory, Integrative andComparative Physiology. 2011;301(6):R1710-R722.
[77] Truswell AS, Mcveigh S, Mitchell WD, Brontest.B. Effect in Man of Feeding Taurine on Bile Acid Conjugation and Serum Cholesterol Levels. Journal of Atherosclerosis Research. 1965;5(5):526-529.
[78] Murakami S, Kondo-Ohta Y, Tomisawa K. Improvement in cholesterol metabolism in mice given chronic treatment of taurine and fed a high-fat diet. Life Sciences. 1998;64(1):83-91.
[79] Chen W, Matuda K, Nishimura N, Yokogoshi H. The effect of taurine on cholesterol degradation in mice fed a high-cholesterol diet. Life Sciences. 2004;74(15):1889-1898.
[80] Kamata K, Sugiura M, Kojima S, Kasuya Y. Restoration of endothelium-dependent relaxation in both hypercholesterolemia and diabetes by chronic taurine. European Journal of Pharmacology. 1996;303(1-2):47-53.

[81] Kondo Y, Toda Y, Kitajima H, Oda H, Nagate T, Kameo K, Murakami S. Taurine inhibits development of atherosclerotic lesions in apolipoprotein E-deficient mice. Clinical and Experimental Pharmacology and Physiology. 2001;28(10):809-815.

[82] Matsushima Y, Sekine T, Kondo Y, Sakurai T, Kameo K, Tachibana M, Murakami S. Effects of taurine on serum cholesterol levels and development of atherosclerosis in spontaneously hyperlipidaemic mice. Clinical and Experimental Pharmacology and Physiology. 2003;30(4):295-299.

[83] Matsuzaki Y, Miyazaki T, Miyakawa S, Bouscarel B, Ikegami T, Tanaka N. Decreased taurine concentration in skeletal muscles after exercise for various durations. Medicine and Science in Sports and Exercise. 2002;34(5):793-797.

[84] Yatabe Y, Miyakawa S, Miyazaki T, Matsuzaki Y, Ochiai N. Effects of taurine administration in rat skeletal muscles on exercise. Journal of Orthopaedic Science 2003;8(3):415-419.

[85] El Mesallamy HO, El-Demerdash E, Hammad LN, El Magdoub HM. Effect of taurine supplementation on hyperhomocysteinemia and markers of oxidative stress in high fructose diet induced insulin resistance. Diabetology & Metabolic Syndrome. 2010;2:46. Epub 2010/07/02.

[86] Herrmann RG. Effect of taurine, glycine and beta-sitosterols on serum and tissue cholesterol in the rat and rabbit. Circulation Research. 1959;7(2):224-227.

[87] Park T, Lee K, Um Y. Dietary taurine supplementation reduces plasma and liver cholesterol and triglyceride concentrations in rats fed a high-cholesterol diet. Nutrition Research. 1998;18(9):1559-1571.

[88] Sethupathy S, Elanchezhiyan C, Vasudevan K, Rajagopal G. Antiatherogenic effect of taurine in high fat diet fed rats. Indian Journal of Experimental Biology. Indian Journal of Experimental Biology. 2002;40(10):1169-1172.

[89] Sugiyama K, Ohishi A, Muramatsu K. Comparison between the plasma cholesterol-elevating effects of caffeine and methionine in rats on a high cholesterol diet. Agricultural and Biological Chemistry. 1989;53(11):3101-3103.

[90] Yokogoshi H, Mochizuki H, Nanami K, Hida Y, Miyachi F, Oda H. Dietary taurine enhances cholesterol degradation and reduces serum and liver cholesterol concentrations in rats fed a high-cholesterol diet. Journal of Nutrition. 1999;129(9):1705-1712.

[91] Masuda M, Horisaka K. Effect of taurine and homotaurine on bile acid metabolism in dietary hyperlipidemic rats. Journal of pharmacobio-dynamics. 1986;9(11):934-940. Epub 1986/11/01.

[92] Fukuda N, Yoshitama A, Sugita S, Fujita M, Murakami S. Dietary Taurine Reduces Hepatic Secretion of Cholesteryl Ester and Enhances Fatty Acid Oxidation in Rats Fed a High-Cholesterol Diet. Journal of Nutritional Science and Vitaminology. 2011;57(2):144-149.

[93] Mochizuki H, Takido J, Oda H, Yokogoshi H. Improving effect of dietary taurine on marked hypercholesterolemia induced by a high-cholesterol diet in streptozotocin-induced diabetic rats. Bioscience Biotechnology and Biochemistry. 1999;63(11):1984-1987.

[94] Nara Y, Yamori Y, Lovenberg W. Effect of Dietary Taurine on Blood-Pressure in Spontaneously Hypertensive Rats. Biochemical Pharmacology. 1978;27(23):2689-2692.

[95] Murakami S, Yamagishi I, Asami Y, Ohta Y, Toda Y, Nara Y, Yamori Y. Hypolipidemic effect of taurine in stroke-prone spontaneously hypertensive rats. Pharmacology. 1996;52(5):303-313.

[96] Choi MJ, Kim JH, Chang KJ. The effect of dietary taurine supplementation on plasma and liver lipid concentrations and free amino acid concentrations in rats fed a high-cholesterol diet. In: Oja SS, Saransaari P, editors. Taurine 62006. p. 235-242.

[97] Petty MA, Kintz J, Difrancesco GF. The effects of taurine on atherosclerosis development in cholesterol-fed rabbits. European Journal of Pharmacology. 1990;180(1):119 127.

[98] Takenaga T, Imada K, Otomo S. Hypolipidemic effect of taurine in golden Syrian hamsters. Taurine 4: Taurine and Excitable Tissues. 2000;483:187-192.

[99] Balkan J, Kanbagli O, Hatipoglu A, Kucuk M, Cevikbas U, Aykac-Toker G, Uysal M. Improving effect of dietary taurine supplementation on the oxidative stress and lipid levels in the plasma, liver and aorta of rabbits fed on a high-cholesterol diet. Bioscience Biotechnology and Biochemistry. 2002;66(8):1755-1758.

[100] Zulli A, Lau E, Wijaya BPP, Jin X, Sutarga K, Schwartz GD, Learmont J, Wookey PJ, Zinellu A, Carru C, Hare DL. High Dietary Taurine Reduces Apoptosis and Atherosclerosis in the Left Main Coronary Artery Association With Reduced CCAAT/Enhancer Binding Protein Homologous Protein and Total Plasma Homocysteine but not Lipidemia. Hypertension. 2009;53(6):1017-1022.

[101] Nistor A, Bulla A, Filip DA, Radu A. The hyperlipidemic hamster as a model of experimental atherosclerosis. Atherosclerosis. 1987;68(1-2):159-173.

[102] Murakami S, Kondo Y, Toda Y, Kitajima H, Kameo K, Sakono M, Fukuda N. Effect of taurine on cholesterol metabolism in hamsters: Up-regulation of low density lipoprotein (LDL) receptor by taurine. Life Sciences. 2002;70(20):2355-2366.

[103] Chang YY, Chou CH, Chiu CH, Yang KT, Lin YL, Weng WL, Chen YC. Preventive Effects of Taurine on Development of Hepatic Steatosis Induced by a High-Fat/Cholesterol Dietary Habit. Journal of Agricultural and Food Chemistry. 2011;59(1):450-457.

[104] Failey RB, Childress RH. Effect of Para-Aminobenzoic Acid on Serum Cholesterol Level in Man. American Journal of Clinical Nutrition. 1962;10(2):158-162.

[105] Hellstrom K, Sjovall J. Conjugation of Bile Acids in Patients with Hypothyroidism (Bile Acids and Steroids, 105). Journal of Atherosclerosis Research. 1961;1(3):205-210.

[106] Elvevoll EO, Eilertsen KE, Brox J, Dragnes BT, Falkenberg P, Olsen JO, Kirkhus B, Lamglait A, Østerud B. Seafood diets: Hypolipidemic and antiatherogenic effects of taurine and n-3 fatty acids. Atherosclerosis. 2008;200(2):396-402.

[107] Zhang M, Bi LF, Fang JH, Su XL, Da GL, Kuwamori T, Kagamimori S. Beneficial effects of taurine on serum lipids in overweight or obese non-diabetic subjects. Amino Acids. 2004;26(3):267-271.

[108] Mizushima S, Nara Y, Sawamura M, Yamori Y. Effects of oral taurine supplementation on lipids and sympathetic nerve tone. Taurine 2. 1996;403:615-622.

[109] Elizarova EP, Nedosugova LV. First experiments in taurine administration for diabetes mellitus - The effect on erythrocyte membranes. In: Huxtable RJ, Azuma J, Kuriyama K, Nakagawa M, Baba A, editors. Taurine 2: Basic and Clinical Aspects1996. p. 583-588.

[110] Eilertsen KE. Manuscript in preparation.
[111] Murakami S, Sakurai T, Toda Y, Morito A, Sakono M, Fukuda N. Prevention of neointima formation by taurine ingestion after carotid balloon injury. Vascular Pharmacology. 2010;53(3-4):177-184.
[112] Murakami S, Kondo Y, Sakurai T, Kitajima H, Nagate T. Taurine suppresses development of atherosclerosis in Watanabe heritable hyperlipidemic (WHHL) rabbits. Atherosclerosis. 2002;163(1):79-87.

Chapter 12

Endoscopic Treatment of Metabolic Syndrome

Eduardo Guimarães Hourneaux de Moura, Ivan Roberto Bonotto Orso,
Bruno da Costa Martins and Guilherme Sauniti Lopes

Additional information is available at the end of the chapter

http://dx.doi.org/10.5772/45989

1. Introduction

For a long time, obesity has been known as a risk factor for cardiovascular disease, which is one of the main causes of death in developed countries. The prevalence of obesity (defined as having a body mass index [BMI] of 30 kg/m2 or more) is increasing in both developing and developed countries. A 5-kg/m2 increase in body mass index (BMI) increases the risk of cardiac complications by 29% [1]. This risk is due to the coexistence of other factors associated with obesity, such as hypertension, dyslipidemia, nonalcoholic fatty liver disease and abnormalities in glycemic metabolism. Resistances to insulin and lipid abnormalities are commonly found among obese patients with type 2 diabetes mellitus (T2DM) and are strongly related to an increase of cardiovascular risk. Resistance to insulin and consequent compensatory hyperinsulinemia significantly increase the risk of death due to cardiovascular diseases [2–5].

To identify the patients with metabolic syndrome, insulin resistance, and greater cardiovascular risk, there are criteria, established by the Adult Treatment Panel III (NCEP ATP III), that include the presence of three of more of the following: central obesity (abdominal circumference above 102 cm in men and 88 cm in women), increased triglycerides (greater than or equal to 150 mg/dl) or use of a lipid-lowering agent, reduced HDL cholesterol (lower than 40 mg/dl among men and lower than 50mg/dl among women) or use of a lipid lowering agent, hypertension (systolic arterial pressure greater than or equal to 130 mmHg or diastolic pressure greater than or equal to 90mmHg) or use of an antihypertensive agent, andglucose levels greater than or equal to 100 mg/dl or use of an oral hypoglycemic agent and/or insulin [6].

Despite this, recent studies propose that the use of the TG/HDL (Triglicerides/High Density Lipoprotein) ratio may be a more practical way to estimate insulin resistance. It is believed that the greater the ratio, the greater the insulin resistance of the patient. This ratio provides an estimate of the sensitivity to insulin and is as accurate as the criteria for the metabolic syndrome defined by the ATP III, the concentration of plasma insulin when fasting, or other estimates that measure the amount of glucose and the plasma concentration of insulin in order to evaluate its action [5, 7 -9]. Some studies suggest that the increase of the TG/HDL ratio may better predict the risk of cardiovascular diseases than do conventional risk factors such as hypertension, tobacco use, and physical activity [10].

To control the obesity and insulin resistance the initial steps are lifestyle changes aimed at controlling diet and increasing activity with the goal of reducing body weight, followed by the addition of orally active pharmacologic agents and insulin to the treatment regimen.

However, dietary modification and pharmaceutical therapy offer limited potential for sustained weight loss, effective in fewer than 5% of cases [2]. In a meta-analysis of pharmacotherapy for obesity, the percentages of patients achieving 5% and 10% weight loss thresholds by using anti obesity drugs were 54% and 18%, respectively, but a lack of adherence to treatment limited the efficacy and effectiveness [11].

Weight loss surgery, in contrast, has been shown to effect a more durable response . In addition, it can induce reversal of obesity-associated comorbidities [2,12].

Of interest is the observation that obese patients with diabetes who undergo certain gastric bypass procedures demonstrate improvement in glycemia, often within days of surgery and before significant weight loss. The exact mechanism responsible for this dramatic effect of surgical procedures for obesity on diabetes improvement is not fully understood; however, the surgical rearrangement of the anatomy of the gastrointestinal (GI) tract changes the location where partially digested nutrients first contact the intestine, suggesting that correction of dysfunctional homeostatic mechanisms may contribute to the glycemic improvement. Whether it is a pure effect of weight reduction or bypass of the hormonally active foregut has a primary effect remains a controversy. Hypothesis includes weight reduction, decreased caloric intake, and bypass of the hormonally active foregut [12-14].

However, weight loss surgery is associated with complications such as anastomotic leak and ulcer presenting a mortality rate estimates range between 0.1 and 2.0% [2].

Endoscopic weight loss therapies may provide some of the benefits of weight loss surgery while being reversible, with a lower risk profile, and being available to patients who do not qualify for surgery or are poor candidates for surgery. Those endoscopic solutions for weight loss are also applicable as metabolic procedures to address comorbidities as type 2 diabetes, dyslipidemia and nonalcoholic fatty liver disease.

2. Physiopathology of metabolic improvement after metabolic surgery

Several studies have shown a significant diabetes improvement in obese patients who undergo certain gastric bypass procedures. The improvement of glycemia is observed, often within days of surgery and before significant weight loss. The mechanism responsible for this improvement is not fully understood; however, the surgical rearrangement of the anatomy of the gastrointestinal tract may contribute to the glycemic improvement. If the glicemic improvement is a pure effect of weight reduction or bypass of the hormonally active foregut has a primary effect remains a controversy. Hypothesis to explain those effects includes weight reduction, decreased caloric intake, and bypass of the hormonally active.foregut. [12-14]

Two hypothesis have been proposed to explain the effect of duodenal jejunal bypass on type 2 diabetes. The "hindgut hypothesis" holds that diabetes control results from the expedited delivery of nutrient chyme to the distal intestine, enhancing a physiologic signal that improves glucose metabolism. A potential candidate mediator of this effect is glucagon-like peptide 1 (GLP-1). This incretin hormone is secreted by L cells of the distal bowel in response to intestinal nutrients. It stimulates insulin secretion and exerts proliferative and antiapoptotic effects on pancreatic *beta* cells.16 If proven true, the hindgut hypothesis would spur further research on methods to enhance signaling by GLP-1 (or other distal gut peptides) to treat type 2 diabetes [14-16]. An alternative hypothesis is that the effect of selected bariatric operations on diabetes depends on exclusion of the duodenum and proximal jejunum from the transit of nutrients, possibly preventing secretion of a putative signal that promotes insulin resistance and type 2 diabetes ("foregut hypothesis"). Although no obvious candidate molecules can be identified with current knowledge, if proven true, this hypothesis might open new avenues in the search for the cause and cures of diabetes [14,17,18].

Several reports supported that the duodeno jejunal exclusion (foregut hypothesis) owes to a direct effect of the bypass of the hormonally active foregut. Rubino and Marescaux showed in their study in animal model that bypassing a short segment of proximal intestine directly ameliorates type 2 diabetes, independently of effects on food intake, body weight, malabsorption, or nutrient delivery to the hindgut [14].

However,in previous studies, it was observed that a strict calorie restriction, as performed in the first weeks after bariatric surgery, could bring itself to a normalization of plasma glucose and insulin levels before body weight decrease [19,20].

Wei-Jei Lee in a study comparing the band (restrictive) with the gastric bypass (duodeno jejunal bypass), with a longer follow-up, showed that the gastric banding group had a similar improvement of insulin resistance to the bypass group while similar weight reduction was achieved. The gastric banding group had similar result at postoperative 6 months compared to the gastric bypass group at the first postoperative month. Also, 3 months to 1 year and 6 months to 2 years were compatible. Suggesting that for a long-term effect of resolution of insulin resistance, sustained weight reduction plays the key mechanism [12].

Improvement in the glycemic control, insulin resistance and metabolic syndrome after bariatric surgery is regulated by a complex mechanism and still there is no certainty whether it is a pure effect of calorie restriction and weight reduction or it is caused by the bypass of the hormonally active foregut .

3. Metabolic improvements with the intragastric baloon

The intragastric baloon is a spherical silicone elastomer balloon that is resistant to degradation by gastric acid for approximately 6 months. It can be placed endoscopicallyand filled with 400 to 700 ml of saline and methylene blue dye, which changes the color of the urine in the event of balloon rupture.

Balloon insertion and removal are performed under conscious sedation or general anesthesia. Before the insertion, an upper gastrointestinal endoscopy is performed to detect possible contraindications to the procedure. The baloon placement device is inserted through the mouth into the stomach. Then the balloon is positioned in the fundus under endoscopic control, and inflated by injecting saline mixed with 10-ml methylene blue into the catheter. Finally, once the desired volume has been injected, the balloon is released by a short pull on the catheter. The baloon should be removed after a maximum of 6 months because beyond this period, the risk of spontaneous balloon deflation significantly increases.

A meta-analysis by Imaz et al. [21] of 15 studies comprising 3698 patients estimated 14.7 kg weight loss, 32.1% excess weight loss (EWL), and 5.7 kg/m2 decrease in BMI after 6 months.

In a review including 22 studies with a total of 4371 patients implanted with the intragastric baloon, demonstrated a mean weight loss of 17.8 Kg, with extremes of the means of 4.9–28.5 kg and higher absolute values observed in higher BMI categories. [22].

A prospective study, evaluating the effect of the baloon on weight, insulin resistance, and liver steatosis in obese patients showed that 76% of the patients had a BMI decrease of 3.5 Kg/m^2 or more. The mean (SD) weight loss with respect to baseline values was 16.4 (8.2) kg with a corresponding mean (SD) BMI reduction of 6.4 (3.2) kg/m2. The absolute percentage of participants with glycemia levels of 100 mg/dL or higher decreased from 50% to 12%, those with triglyceridemia 150 mg/dL or higher from 58% to 19%, and those with abnormal ALT level from 38% to 7% [23].

Two studies (one randomized, one uncontrolled) totaling 143 patients have reported that, one year after BIB removal, patients had regained 41% and 28% (mean values, respectively) of the absolute weight loss observed at BIB removal [24,25]. Another study following 88 patients for a median of 22 month after baloon withdrawal, observed that (50%) regained some weight, 34 (39%) maintained their weight, and the remaining 10 (11%) continued to lose weight [23].

It is also important to consider that 20–40% of patients fail to achieve a significant weight loss (often defined as ≥10% baseline weight or ≥25% excess weight). Such failures may be related to the request of early baloon removal by patients who present a digestive or psychological intolerance to the baloon, to the early vanishing of anticipated effects on hunger and early satiety, or to patient's adaptation of food intake [23].

In conclusion, the BIB strategy may be an alternative to current management of obesity focused on lifestyle changes, drug therapy, and treating associated metabolic complications.

Although the baloon has not yet proved to be a convincing means of primary long term weight loss, it holds some promise for improving co-morbidities and quality of life in nonmorbidly obese patients or those who are unwilling to undergo bariatric surgery. New perspectives are also beginning to show its potential value in specific patient groups especially, for example, those preparing for surgery.

4. Metabolic improvements with duodeno jejunal bypass liner

A totally different concept, that of mimicking principles of bariatric surgery, has recently been applied in the development of the endoscopic duodenal-jejunal bypass liner. In addition to early satiety and delayed gastric emptying, the intraluminal sleeve aims at creating a duodenojejunal exclusion.

The DJBL is a sterilized, single-use endoscopic device, which is minimally invasive and employed under radioscopic control. It is composed of a nitinol anchoring with tiny lateral barbs for fixation and an impermeable plastic conduit made of a fluorine polymer 62 cm in length, which impedes contact of the chyme with bile–pancreatic secretions prior to the proximal segments of the jejunum. FIGURE 1

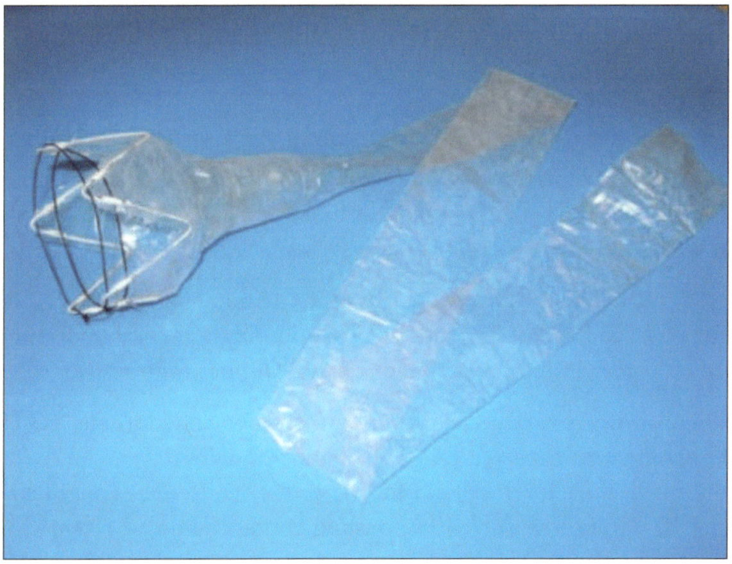

Figure 1. Impermeable plastic conduit and anchor system.

Endoscopic implantation is performed under general anesthesia. The device is introduced over a guidewire that has been previously positioned in the duodenal bulb with endoscopic assistance. The plastic conduit is stretched to overlay the duodenum and the proximal region of the jejunum. After the correct positioning of the plastic conduit, the anchoring system is freed, setting the device in the duodenal bulb. The infusion of a contrast agent is performed to verify the correct positioning of the prosthesis and the absence of obstructions within the plastic conduit. FIGURE 2

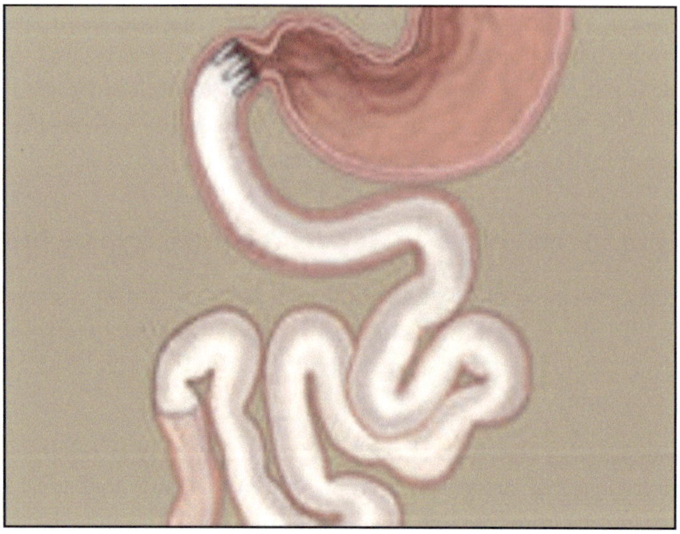

Figure 2. Implanted DJBL.

Previously studies with obese patients that used the duodenojejunal bypass liner (DJBL) demonstrated a significant weight loss. In addition, an improvement in the control of T2DM was observed, which was statistically greater than that of the group treated with a low-calorie diet [26-33].

Our group in the Gastrointestinal Endoscopy Unit of the University of São Paulo School of Medicine performed a prospective study to evaluates the effectiveness of this method over the control of hyperlipidemia, improvement of insulin resistance, metabolic syndrome, and in the potential benefit in the reduction of cardiovascular risk [34].

The inclusion criteria of the study were ages between 18 and 65 years, BMI≥35 kg/m2, T2DM with or without other comorbidities, and a triglyceride/high-density lipoprotein cholesterol ratio (TG/HDL) ratio greater than or equal to 3.5, indicating insulin resistance.

To identify patients with resistance to insulin and metabolic syndrome, TG/HDL ratio has proved to be an excellent practical indicator. This ratio estimates the resistance to insulin and is as accurate in terms of the clinical criteria for metabolic syndrome as specified by the Adult Treatment Panel III [6], the measure of the fasting concentration of plasma insulin,or other estimates that measure glycemia and the plasma concentration of insulin in order to identify individuals with insulin resistance [5, 7,8]. In addition, the TG/HDL ratio has proved to be an independent factor that was correlated with the risk of cardiovascular events [5].

A low TG/HDL ratio indicates large particles and a lower atherogenic potential of LDL cholesterol, while a high TG/ HDL ratio indicates a large population of small, dense, and pro-atherogenic particles of LDL cholesterol [7]. The lipid disorder consisting of the increase of plasma triglycerides and the reduction of HDL cholesterol, known as atherogenic dyslipidemia, is directly associated with insulin resistance and is also an independent risk factor for cardiovascular diseases [3–5, 7].

Eighty-one patients were selected for implantation of the device. Of these, 78 successfully received the implant and three were not given the implant due to anatomic factors (short bulb). Among the 78 patients, one was excluded from the analysis for not having performed the laboratory measures. Of the 77 remaining patients, we calculated the initial TG/ HDL ratio (at the time of the implant), identifying 54 patients (70%) with a ratio greater than or equal to 3.5, indicating the presence of insulin resistance and metabolic syndrome. These patients were included in the study and were monitored in order to evaluate whether or not an improvement occurred in this ratio during the period in which they had the implanted device.

We compared the TG/HDL ratio at the time of the implant with the ratio obtained after 6 months to evaluate whether there was an improvement in insulin resistance. We divided the patients into two groups: those who demonstrated a ratio below 3.5 in the end of the study, considered as control of the insulin resistance, and those that did not demonstrate values lower than 3.5. In the two groups, we evaluated whether control of T2DM and weight loss occurred during this period, and we correlated the influence of the control of diabetes and weight loss with the improvement of TG/HDL. We considered a significant weight loss to be a reduction of at least 10% of initial body weight and control of T2DM as an HbA1c level lower than 7%. The patients who presented a reduction of HbA1c levels greater than 1.5% yet did not obtain values lower than 7% were considered to have partial control over DM2.

The overall initial average of the TG/HDL ratio was 5.75 and presented a significant reduction down to 4.36 at the end of the 6 months ($p=<0.001$), indicating an improvement of insulin resistance (Table 1). Of these patients, 23 (42.6%) presented control of the TG/HDL ratio with values lower than 3.5 at the end of the study. This group presented a significant improvement in the ratio, which decreased from 5.15 to 2.85 ($p<0.001$). Thirty-one patients did not show a controlled TG/HDL ratio but rather a discrete improvement, with an initial average of 6.2 and a final of 5.47, with no statistical difference ($p=0.1641$).

	Patients N	Initial average TG/HDL ratio	Final average TG/HDL ratio	p
Baseline non controlled TG/HDL	54	5,75	4,36	0,001
Controlled TG/HDL at the end	23	5,15	2,85	<0,001
Not controlled TG/HDL at the end	31	6,2	5,47	0,1641

Table 1. Improvement on TG/HDL ratio.

In order to identify the differences between the group that presented an improvement in the ratio and the group that did not, we evaluated the control of T2DM (HbA1c improvement) and the success of weight loss (reduction >10% of initial weight).

In the evaluation of T2DM control (Table 2), we observed that all patients presented a significant improvement in the levels of HbA1c ($p<0.001$). In the group that controlled the

TG/HDL ratio, three patients already had diabetes controlled at the beginning of the study, with an initial average of HbA1c of 6.4% and a final of 5.83% (p=0.023). Fifteen patients did not have diabetes under control, presenting an initial average of 7.8%, and then developing control of T2DM, with a final average of 6.1% (p<0.001). Five patients did not have diabetes undercontrol and obtained partial control after the intervention. These patients presented higher initial levels of HbA1c than those of the group that controlled diabetes, with an initial average of 10.34%, and presented a significant reduction of the final average of 8.88% (p=0.03).

In the group that did not control the TG/HDL ratio, five patients already had T2DM under control, with an initial average of 6.6%mg/dl and a final average of 6.2% (p=0.037). Fifteen patients did not have diabetes under control and were able to bring it under control with initial and final averages of 8.5% and 6.4%, respectively (p=0.001). Eleven patients did not have diabetes under control and obtained partial control. These patients also presented a higher level of HbA1c than that of the patients who had controlled T2DM, with an initial average of 9.9%, reaching a significant reduction at the end of the study with an average of 7.7% (p=0.003), which is very close to the level required for T2DM control.

An association was not observed between the control of T2DM and an improvement in the TG/HDL ratio (Table 3).

Diabetes improvement on TG/HDL controlled patients				
Diabetes	**Patients N**	**Inial HbA1c average**	**Final HbA1c average**	**p**
Already controled and improved	3	6,4	5,38	0,023
Not controled who controlled	15	7,8	6,1	<0,001
Not controlled with partial control	11	10,34	8,88	0,03
HbA1c Worsening	0	-	-	.
Diabetes improvement on TG/HDL not controlled patients				
Diabetes	**Patients**	**Inial HbA1c average**	**Final HbA1c average**	**p**
Already controled and improved	5	6,6	6,2	0,037
Not controled who controlled	15	8,5	6,4	0,001
Not controlled with partial control	11	9,9	7,7	0,003
HbA1c Worsening	0	-	-	.

Legend: Diabetes evolution on TG/HDL controlled and not controlled patients. Glycemic improvement was statistically significant in all groups of patients.

Table 2. Diabetes improvement.

	Diabetes Improvement – HbA1c			
	Already controled and improved	Not controlled who controlled	Not controlled with partial control	HbA1c Worsening
TG/HDL controlled (n = 23)	3	15	5	0
TG/HDL not controlled (n= 31)	5	15	11	0

Legend: Relationship between improvement of diabetes and TG/HDL control. Patients improved diabetes regardless of having controlled the TG/HDL ratio (p 0,35).

Table 3. Relation between TG/HDL ratio control and HbA1C control.

In relation to weight loss (Table 4), the patients lost on average 12.6% of their initial weight. Among the 23 patients who controlled their TG/HDL ratio, 19 (82.6%) lost more than 10% of their initial weight. The average initial weight of these patients was 116.5 kg and the average final weight was 97 kg, constituting an average loss of 16.7% of initial weight. Four patients did not lose more than 10% of their weight. The average initial weight of these patients was 94.4 kg, and the average final weight was 87.47 kg, marking an average loss of 7.4%. In the group that did not control the TG/HDL ratio, 15 lost more than 10% of their initial weight (48%), with an average initial weight of 123.9 kg and an average final weight of 105.7 kg (loss of 14.6% in initial weight). Sixteen patients did not lose more than 10% of their weight, presenting an average initial weight of 111.9 kg and an average final weight of 103.5 kg, with an average loss of 7.5% in initial weight.

Weight loss on TG/HDL controlled patients				
Weight loss	Patients N (%)	Inicial average weight - Kg	Final average weight - Kg	Percentage of loss
Over than 10% of initial weight	19 (82,6)	116,5	97	16,7
Less than 10% of initial weight	4 (17,4)	94,4	87,47	7,4
Weight loss on TG/HDL not controlled patients				
Weight loss	**Patients N (%)**	**Inicial weight average**	**Final weight average**	**Percentage of loss**
Over than 10% of initial weight	15 (48)	123,9	105,7	14,6
Less than 10% of initial weight	16 (52)	111,9	103,5	7,5

Table 4. Weight loss

Comparing the patients who lost weight with the patients who controlled their TG/HDL ratio, an association can be observed between a weight loss greater than 10% of initial weight and control of the TG/HDL ratio ($p<0.01$), with an odds ratio of 5.06 (Table 5).

	Weight loss		
	Over than 10% of inicial weight	Less than 10% of inicial weight	Total
TG/HDL controlled - N (average weight loss %)	19 (16,7)	4 (7,4)	23
TG/HDL not controlled N (average weight loss %)	15 (14,6)	16 (7,5)	31

Legend: Relationship between weight loss and TG/HDL control. Control of TG/HDL ratio is related to weight loss greater than 10% of initial weight. (p 0,01- OR 5,06).

Table 5. Relation between TG/HDL ratio control and weight loss.

Of the 54 patients included in this 6 months study, 38 have completed (26 completed 24 weeks, 12 completed 20 weeks). Among the 16 patients left, 12 had the device removed at 16 weeks, 2 at 12 weeks, and 2 had the implant for just 4 weeks. The early implant removals occurred due to migration of the device in nine patients, the observation of a free device anchor during endoscopic exam in four patients, the presence of bleeding without migration in one patient, subject request in one case, and due to the decision of the researcher in one case.

In resume, all patients implanted with the device presented a statistically significant reduction of the levels of HbA1c, and the majority of these patients (70.3%) presented values lower than 7% at the end of the study and were therefore considered to be controlled diabetics. In addition, all patients presented a statistically significant reduction of initial weight, with an average general loss of 12.6% of initial weight. Regarding the improvement of insulin resistance and metabolic syndrome, there was a significant reduction of the TG/HDL ratio from 5.75 to 4.36 ($p=0.0001$). Of these patients, 42.6% controlled their insulin resistance, presenting a TG/HDL ratio value lower than 3.5 at the end of the study.

Among the patients who controlled the TG/HDL ratio, the reduction of the ratio went from 6.8 at the beginning of the study to 2.8 at the end ($p<0.001$).

In other study conducted by our group, twenty two implanted patients were followed during a period of 1 year [13]. In the full analysis population, the mean percentage excess weight loss was 35.5% ($P < 0.0001$). The reduction in excess body weight was reflected by reductions in BMI and waist circumference of 6.7 kg/m2 and 13.0 cm, respectively.

The improvement in glycemic control is convincingly demonstrated by the results with a percentage of subjects with HbA1c < 7% at baseline improved from 4.5% to 73.0% at final study assessment. Statistically significant reductions in fasting blood glucose (- 30.3 ± 10.2 mg/dL), fasting insulin (- 7.3 ± 2.6 lU/mL), and HbA1c (- 2.1 ± 0.3%) were observed.

Blood levels of total cholesterol, low-density lipoprotein cholesterol, and triglycerides also were significantly reduced during the study.

On this one year series, thirteen subjects completed the 52-week period, and 18 subjects completed at least 24 weeks. The mean duration of the implant period for all subjects was 41.9 ± 3.2 weeks. The reasons for early removal of the device were migration or rotation of the device (n = 3; 36, 36, and 48 weeks post implantation), GI bleeding (n = 1; 4 weeks post implantation), abdominal pain (n = 2; 21 and 30 weeks post-implantation), and principal investigator request due to subject's non compliance with study visits (n = 2; 20 and 32 weeks post implantation). The device was removed from one subject who presented an abdominal tumor not related to the device.

Sixteen subjects had HbA1c measured 3 and/or 6 months after explantation of the DJBL. These subjects demonstrated a mean decrease in HbA1c during the original 52-week study of - 2.3 ± 0.4%. Three and 6 months after removal of the device, their mean changes from baseline were - 2.3 ± 0.3% (n = 15) and - 1.7 – 0.7% (n = 11), respectively.

The DJBL offers a new non-surgical therapeutic possibility, positioned between pharmacological drugs and the various techniques employed in bariatric surgery. This technology platform may be employed prior to bariatric surgery to help control T2DM, in order to promote weight loss and a reduction of visceral fat, lipid control, a reduction of insulin resistance, and of cardiovascular risk,minimizing the risk of per operative clinical complications, accustoming the patient to a restricted diet that will be necessary in the post-operative period and can even be used as a substitution for bariatric surgery as a less invasive technique in selected cases.

5. Conclusion

As the prevalence of obesity increases, less invasive methods will be needed to obtain a sustained weight loss. Some new endoscopic tools and methods are being investigated and they could be applied as first-line therapy for obesity, to control the metabolic comorbidities, to reduce the operatory risk prior bariatric and metabolic surgery and as substitution of surgery in selected cases.

The intragastric baloon and the Duodeno Jejunal Bypass Liner are tools with promising results in the endoscopic treatment of obesity. They are still subject of research with a great potential for improvement. Although outcomes from the use of the intragastric baloon and the DJBL are not comparable to those of surgery with regard to weight loss and late results, these new techniques have showed an excelent result in ameliorating health status, in the control of the metabolic syndrome as well as improving the quality of life for a well selected group of patients.

Author details

Eduardo Guimarães Hourneaux de Moura*,
Bruno da Costa Martins, Guilherme Sauniti Lopes
Department of Gastroenterology, Gastrointestinal Endoscopy Unit,
Hospital das Clínicas - University of São Paulo School of Medicine, São Paulo, Brazil

Ivan Roberto Bonotto Orso
Department of Gastroenterology, Gastrointestinal Endoscopy Unit,
Hospital das Clínicas - University of São Paulo School of Medicine. São Paulo, Brazil
Department of Surgery, School of Medicine of the Assis Gurgacz Faculty,
Gastroclínica Cascavel, Brazil

6. References

[1] Bogers RP, Bemelmans WJ, Hoogenveen RT, et al. Association of overweight with increased risk of coronary heart disease partly independent of blood pressure and cholesterol levels: a metaanalysis of 21 cohort studies including more than 300 000 persons. Arch Intern Med. 2007;167(16):1720–8.
[2] Kumar N, Thompson CC. Endoscopic solutions for weight loss. Current Opinion in Gastroenterology 2011, 27:407–41
[3] McLaughlin T, Abbasi F, Cheal K, et al. Use of metabolic markers to identify overweight individuals who are insulin resistant. Ann Intern Med. 2003;139:802–9.
[4] Shishehbor MH, Hoogwef BJ, Lauer MS. Association of triglyceride to HDL cholesterol ratio with heart rate recovery. Diabetes Care. 2004;27(4):936–41.
[5] McLaughlin T, Reaven G, Abbasi F, et al. Is there a simple way to identify insulin resistant individuals at increased risk of cardiovascular disease? Am J Cardiol. 2005;96(3):399–404.
[6] Tong PC, Kong AP, SoWY, et al. The usefulness of the international diabetes federation and the national cholesterol education program's Adult Treatment Panel III definitions of the metabolic syndrome in predicting coronary heart disease in subjects with type 2 diabetes. Diabetes Care. 2007;30(5):1206–11.
[7] Hadaegh F, Dhalili D, Ghasemi A, et al. Triglyceride/HDLcholesterol ratio is an independent predictor for coronary heart disease in a population of Iranian men. Nutr Metab CardiovascDis. 2009;19(6):401–8.
[8] Vasques ACJ, Rosado LEFPM, Rosado GP, et al. Indicadores do perfil lipídico plasmático relacionados à resistência á insulina. Rev Assoc Méd Bras. 2009;55(3):342–6.
[9] Quizada Z, Paoli M, Zerpa Y, et al. The triglyceride/HDL cholesterol ratio as a marker of cardiovascular risk in obesechildren; association with traditional and emergent risk factors. Pediatr Diabetes. 2008;9(5):464–71.
[10] Kannel WB, Vasan RS, Keyes MJ, et al. Usefulness of the triglyceride high density lipoprotein versus the cholesterol high density lipoprotein ratio for predicting insulin

*Corresponding Author

resistance and cardiometabolic risk (from the Framingham Offspring Cohort). Am J Cardiol. 2008;101(4):497–501.

[11] Rucker D, Padwal R, Li SK, et al. Long term pharmacotherapy for obesity and overweight: update and meta-analysis. Br Med J 2007;335: 1194-9.)

[12] Lee WJ, Lee YC, Ser KH, Chen JC, Chen SC. Improvement of Insulin Resistance After Obesity Surgery: A Comparison of Gastric Banding and Bypass Procedures Obes Surg 2008; 18:1119–1125

[13] Moura,EGH, Martins BC, Lopes GS, Orso IR, Oliveira SL, Galva˜o Neto MP, et al. Metabolic Improvements in Obese Type 2 Diabetes Subjects Implanted for 1 Year with an Endoscopically Deployed Duodenal–Jejunal Bypass Liner. Diabetes Technol Ther 2012;14(2). DOI: 10.1089/dia.2011.0152

[14] Rubino F, Forgione A, Cummings DE, Vix M, Gnuli D, Mingrone G et al, The Mechanism of Diabetes Control After Gastrointestinal Bypass Surgery Reveals a Role of the Proximal Small Intestine in the Pathophysiology of Type 2. Ann Surg 2006;244: 741–749

[15] Mason EE. The mechanism of surgical treatment of type 2 diabetes. Obes Surg. 2005;15:459–461.

[16] Patriti A, Facchiano E, Sanna A, et al. The enteroinsular axis and the recovery from type 2 diabetes after bariatric surgery. Obes Surg. 2004; 14:840–848.

[17] Pories WJ, Albrecht RJ. Etiology of type II diabetes mellitus: role of the foregut. World J Surg. 2001;25:527–531.

[18] Rubino F, Gagner M, Gentileschi P, et al. The early effect of the Roux-en-Y gastric bypass on hormones involved in body weight regulation and glucose metabolism. Ann Surg. 2004;240:236 –242.

[19] Kelly DE, Wing R, Buonocore C, et al. Relative effect soft calorie restriction and weight loss in non-insulin-dependent diabetes mellitus. J Clin Endocrinol Metab. 1993;77:1287–93. 21.

[20] Ash S, Reeves MM, Yeo S, et al. Effect of intensive dietetic interventions on weight and glycemic control in overweight men with type II diabetes: a randomized trial. Int J Obes. 2003;27: 797–802.)

[21] Imaz I, Martı´nez-Cervell C, Garcı´a-Alvarez EE, et al. Safety and effectiveness of the intragastric balloon for obesity. A meta-analysis. Obes Surg 2008;7:841–846.

[22] Dumonceau JM. Evidence-based Review of the Bioenterics Intragastric Balloon for Weight Loss. Obes Surg (2008) 18:1611–1617

[23] Forlano R, Ippolito AM, Iacobellis A, et al. Effect of the BioEnterics intragastric balloon on weight, insulin resistance, and liver steatosis in obese patients. Gastrointest Endosc 2010; 71:927–933.

[24] Herve J, Wahlen CH, Schaeken A, et al. What becomes of patients one year after the intragastric balloon has been removed? Obes Surg. 2005;15:864–70.

[25] Mathus-Vliegen EM, Tytgat GN. Intragastric balloon for treatment- resistant obesity: safety, tolerance, and efficacy of 1-year balloon treatment followed by a 1-year balloon-free follow-up. Gastrointest Endosc. 2005;61:19–27.

[26] Gersin KS, Keller JE, Stefanidis D, et al. Duodenal–jejunal bypass sleeve: a totally endoscopic device for the treatment of morbid obesity. Surg Innov. 2007;14(4):275–8.

[27] Schauer P, Chand B, Brethauer S. New applications for endoscopy: the emerging field of endoluminal and transgastric bariatric surgery. Surg Endosc. 2007;21:347–56.

[28] Tarnoff M, Shikora S, Lembo A, et al. Chronic in-vivo experience with an endoscopically delivered and retrieved duodenal–jejunal bypass sleeve in a porcine model. SurG Endosc. 2008;22(4):1023–8.

[29] Tarnoff M, Shikora S, Lembo A. Acute technical feasibility of an endoscopic duodenal–jejunal bypass sleeve in a porcine model: a potentially novel treatment for obesity and type 2 diabetes. Surg Endosc. 2008;22(3):772–6.

[30] Tarnoff M, Rodriguez L, Escalona A, et al. Open label, prospective, randomized controlled trial of an endoscopic duodenal–jejunal bypass sleeve versus low calorie diet for pre-operative weight loss in bariatric surgery. Surg Endosc. 2009;23(3):650–6.

[31] Rodriguez-Grunert L, Galvao Neto MP, Alamo M, et al. First human experience with endoscopically delivered and retrieved duodenal– jejunal bypass sleeve. Surg Obes Relat Dis. 2008;4(1):55–9.

[32] Rodriguez L, Reyes E, Fagalde P, et al. Pilot clinical study of an endoscopic, removable duodenal–jejunal bypass liner for the treatment of type 2 diabetes. Diabetes Technol Ther. 2009;11 (11):725–32.

[33] Schouten R, Rijs CS, Bouvy ND, et al. A multicenter, randomized efficacy study of the EndoBarrier Gastrointestinal Liner for presurgical weight loss prior to bariatric surgery. Ann Surg. 2010;251(2):236–43

[34] Moura EGH, Orso IR, Martins BC, Lopes GS, Oliveira SL, et al. Improvement of Insulin Resistance and Reduction of Cardiovascular Risk Among Obese Patients with Type 2 Diabetes with the Duodenojejunal Bypass Liner. Obes Surg (2011) 21:941–947

Chapter 13

Nutritional Management of Disturbances in Lipoprotein Concentrations

Somayeh Hosseinpour-Niazi, Parvin Mirmiran and Fereidoun Azizi

Additional information is available at the end of the chapter

http://dx.doi.org/10.5772/47868

1. Introduction

Atherogenic dyslipidemia includes increase in blood concentrations of LDL cholesterol, total cholesterol, triglycerides and decrease in high-density lipoprotein cholesterol, both of which are frequently associated with the development of cardiovascular diseases (CVDs) [1,2] .Treatment of dyslipidemia can reduce the risk of CVDs [3]. In both industrialized and non-industrialized countries, the prevalence of dyslipidemia is increasing (4-7), therefore management of dyslipidemia has become a mainstay of routine clinical practice for both public health and clinicians. Although the benefits of lipid-lowering therapy have been demonstrated most conclusively, the role of diet determinants in dyslipidemia needs to be further considered [8]. Diet plays an important role in the concentrations of lipoprotein and is the primary intervention for patients with dyslipidemia. Understanding the relationships between dietary determinants and dyslipidemia and the effect of diet on lipoprotein concentrations may help to identify the dietary changes needed to reduce health risks [2]. Dietary changes, including reduced intakes of saturated fat and cholesterol, increased intakes of polyunsaturated fatty acids, fish, fruits and vegetables, and reduced energy intakes may have beneficial effects on lipoprotein concentrations [9-12]. One important aspect of diet is dietary patterns that address the effect of the diet as a whole and thus may provide insight beyond the effects described for single nutrients or foods [13]. The effects of some dietary patterns including the Mediterranean diet, the dietary stop to hypertension (DASH) and traditional dietary patterns, on lipoprotein particles need to be discussed [14, 15, 16]. In addition, other aspects of diet, including herbal, phytochemical, and dietary supplement (plant stanols and sterols) also play important roles in the prevention and treatment of dyslipidemia and may improve lipoprotein concentrations [17]. We searched the medical literature for studies of the effects of diet and its component including macronutrient, dietary food groups, dietary patterns and herbal on disturbances of lipoprotein concentrations. The purpose of this chapter is to update current knowledge on the role of the following dietary determinants in lipoprotein concentrations and dyslipidemia: including 1) macronutrients (total fat, saturated fatty acids, trans fatty acids, n-6 polyunsaturated fatty acids, n-3 polyunsaturated fatty acids, dietary cholesterol, carbohydrate and protein), 2) food groups (grains and cereal, fruit and vegetables, dairy products, nuts, beans and legumes, and meat, fish, poultry and eggs), 3) dietary patterns (Mediterranean diet, Dietary to Stop Hypertension, western **diet** and healthy diet), and therapeutic life style change (TLC), 4) dietary **supplements**, (plant stanols and sterols), herbal and phytochemicals.

2. Diet and lipoprotein

Lipoprotein concentrations are affected by both genetic and environmental factors. Among environmental factors such as physical activity and smoking, diet is an important component in preventing and improving dyslipidemia. Diet intervention is recommended by the National Cholesterol Education Program (NCEP) guidelines as first-line therapy for the management of disturbances in lipoprotein concentrations. Also the Third report of the NCEP recommended that if dietary therapy do not improve disturbances in lipoprotein concentrations, non-pharmacologic therapeutic factors such as viscous fiber and plant stanols and sterols should be recommended prior to advancing to drug therapy[18].

3. Macronutrient and lipoprotein

3.1. Total fat

The Nutrition Committee of the American Heart Association (AHA) emphasises on that diets providing up to 40% of dietary energy as primarily unsaturated fat (20% MUFA, 10% SFA, 10% PUFA and 1% TFA) were as heart healthy as low-fat diets (<30% of dietary energy) [19]. The effects of different dietary fatty acids on lipid profiles should be considered in the evaluation of strategies for controlling of disturbances in lipoprotein concentrations. Changes in dietary fat composition are clearly associated with changes in lipoprotein concentrations. Types of dietary fatty acids include saturated fatty acids (SFAs), monounsaturated fatty acid (MUFAs), polyunsaturated fatty acid (PUFAs) and dietary cholesterol, the effects of which on lipoprotein concentrations will be discussed.

3.2. Dietary saturated fatty acids (SFAs)

Among the dietary fatty acids only dietary SFAs and *trans* fatty acids increase LDL cholesterol concentrations [18]. The major sources of dietary SFAs are fast foods, processed foods, high-fat dairy products (whole milk, cheese, butter, ice cream, and cream), high-fat red meats, tropical oils such as palm oil, coconut oil, and palm kernel oil, baked products and mixed dishes containing dairy fats, shortening, and tropical oils. Dietary SFAs increase LDL and total cholesterol concentrations, in comparison with all dietary fatty acids except *trans* fatty acids [20-21], by inhibiting LDL receptor activity and enhancing apolipoprotein (apo) β-containing lipoprotein production [22]. Every 1 percent increase of total energy from dietary SFAs raises the serum LDL cholesterol about 2 percent. Conversely, a 1 percent reduction in saturated fatty acids will reduce serum cholesterol by about 2 percent [23,24]. The LDL cholesterol–raising effect of dietary SFAs depends on the intake of dietary cholesterol and PUFAs. In high intakes of dietary cholesterol, dietary SFAs decreased LDL receptor activity and increased plasma LDL concentrations [25]. However, in the adequate of dietary PUFAs (5–10% of total energy), dietary SFAs have no effect on LDL clearance [22]. In addition different dietary SFAs have different effects on lipoprotein concentrations [29]. Short chain SFAs have been shown to have a stronger LDL cholesterol raising effect, such that lauric acid (12:0) raised LDL cholesterol the most, followed by myristic (14:0) and palmitic (16:0) acids. In contrast, stearic acid (18:0), as a long chain SFA, has no effect on LDL and HDL cholesterol or the TC: HDL cholesterol ratio, and even lowers serum cholesterol [27,28]. Finally, the effects of dietary SFAs can be modulated by the foods in which they are contained. Cheeses may have smaller effects on LDL cholesterol concentrations than butter, and fermented dairy foods, such as yogurt, have been associated with LDL reductions [29]. Reduced intakes of dietary SFAs and cholesterol are first steps for the purpose of achieving the LDL cholesterol goal (<100 mg/dl). To maximize LDL cholesterol lowering by reducing dietary SFAs, it will be necessary to lower intakes of dietary SFAs approximately to <7 percent of total energy [18]. However the replacement of dietary SFAs with other macronutrients is important. Although replacement of dietary SFAs with carbohydrate decrease total, LDL, and HDL cholesterol, it also increases triglycerides [20]; however replacement of dietary SFAs by PUFAs decreases concentrations of total, LDL, and the LDL/HDL cholesterol ratio by decreasing LDL cholesterol production and increasing LDL clearance [30]. Although replacement of dietary SFAs with PUFAs has been shown to decrease HDL cholesterol, it decreases LDL cholesterol even more substantially; thus, the HDL:LDL ratio is increased [23] and the TC:HDL cholesterol ratio is decreased [26]. Replacement of 5% of total energy from SFAs with PUFAs reduces CHD risk by 42% [31]. Replacement of dietary SFAs with MUFAs has also been associated with improving lipoprotein concentrations, although this effect is slightly less than when PUFAs are the replacement dietary fatty acid [23]. Replacement of dietary SFAs with both MUFAs and carbohydrate decrease LDL cholesterol; however replacement with MUFA was associated with lower reductions in HDL cholesterol and lower arises in triglyceride concentrations [32].

3.3. Trans fatty acids

Trans fatty acids contain at least one double bond in the *trans* configuration [40] and were the most harmful macronutrient that increase disturbances in lipoprotein concentrations [26,33,34]. Dietary *trans* fatty acids, produced during the hydrogenation of either vegetable or fish oils (industrial TFA), are found in manufacturing products such as cookies, pastries, and salad dressings; *trans* fatty acids are also formed during anaerobic bacterial fermentation of unsaturated fatty acids that occurs in the rumen of polygastric animals such as cattle, sheep, and goats (natural trans fatty acids), and hence found in dairy products derived from the animals' milk and meat [33,35]. Industrial and natural *trans* fatty acids contain similar types of these fatty acids, but in different proportions. Industrial *trans* fatty acids contain trans isomers of oleic acid, the major ones being C18:1 trans-9 (elaidic acid) and C18: 1 trans-10 [35]. Consumption of industrial *trans* fatty acids increases total, LDL cholesterol, and total to HDL cholesterol ratio and the LDL to HDL cholesterol ratio [33, 35-37] and decrease HDL cholesterol [40]. Data on the effects of natural *trans* fatty acids on plasma lipoproteins in humans are inconsistent. An equivalent of 1% natural *trans* fatty acids of daily energy, has no significant effect on total cholesterol, LDL cholesterol, apo B, triglyceride concentrations but may be associated with a reduction in plasma HDL cholesterol concentrations [38]. However high intakes of natural *trans* fatty acids, but not low intakes, have adverse effects [39]. Therefore both natural and industrial *trans* fatty acids have detrimental effects on lipoprotein concentrations and their intakes should be limited [40]. The effects of *trans* fatty acids on lipid profiles are also variable, depending on their chain length; long chain *trans* fatty acids may have more adverse effect on lipid profiles. Partially hydrogenated fish oil or *trans* alpha-linolenic acid had more detrimental effect on lipoprotein compared with isocaloric amount of partially hydrogenated soy bean oil [37,41]. Effect of *trans* fatty acids on lipoprotein concentrations is a current topic of debate. *Trans* fatty acid intake increases lipoprotein a and triglycerides when substituted for dietary SFAs [42,43]. Issues related to the potential change in lipoprotein a levels induced by *trans* fatty acid intake and risk for disease need to be clarified.

Dietary guidelines for American 2010 emphasize that consumption of *trans* fatty acids should be reduced as much as possible by limiting foods that contain sources of these fatty acids [43]. On the basis of these data, it should be attempts to substitute unhydrogenated oil for hydrogenated or SFAs in diet.

3.4. Monounsaturated fatty acids

Monounsaturated fatty acids have received increased attention as being potentially beneficial for their association with low rates of CHD in olive-oil consuming populations of the Mediterranean style diet [18]. The most common form of dietary MUFAs is oleic acid (18:1 n-9), which occurs in the cis form. Olive oil, canola oil, and sunflower oil are the main sources of dietary MUFAs. Oleic acid is an effective hypocholesterolemic factor when substituted for dietary SFAs. MUFA-rich oil consumption has been one of the strategies recommended for modulating the plasma lipid profile in humans. Diets containing high MUFA-rich foods reduce plasma total and LDL cholesterol levels and enrich LDL particles with cholesteryl oleate, a change in LDL particle composition that has been shown to confer atherogenicity [23, 45-48]. Also compared with diets rich in saturated fat, MUFA-rich diets lower apolipoprotein β concentrations along with declines in LDL cholesterol level [49,50]. Consumption of MUFA-rich diets also induces lower triglycerides and higher HDL cholesterol concentrations compared with low-fat, high-carbohydrate diets [51]. Long term MUFA-rich diets result in an earlier postprandial peak in plasma triglyceride and apo β-48 concentrations [52,53]; this mechanism is not clear, however oleic acid has been shown to be preferentially esterified into triglycerides in the enterocyte [54], which may be result a faster entry rate of chylomicrons into the circulation, reflecting accelerated rates of digestion and absorption or upregulation of chylomicron synthesis and secretion [55]. However MUFA-rich diets increase clearance of plasma triglycerides compared with isocaloric SFA-rich or high complex carbohydrate diets and therefore decrease triglyceride concentrations [51,56,57]. MUFA substitution for dietary SFAs suggest an effective dietary strategy for improving disturbances of lipoprotein concentrations, which currently recommended in most national and international dietary guidelines [18].

3.5. N-6 Polyunsaturated fatty acids

Dietary n-6 PUFAs such as linoleic acid (18:2) are widely found in a variety of vegetables and vegetables oils [58]. Conjugated linoleic acid (CLA), a group of naturally occurring fatty acids that are mainly present in foods from ruminant sources, is a collective term used to describe positional and geometric derivatives of linoleic acid containing conjugated double bonds [59].

CLA have beneficial effects on lipoprotein disturbances. CLA reduced total, LDL and VLDL cholesterol, especially atherogenic apolipoprotein β-rich lipoproteins and triglycerides concentrations [60,61]. CLA increases the excretion of sterols and consequently decreases serum cholesterol concentration [86].

3.6. N-3 Polyunsaturated fatty acids

Dietary sources of n-3 PUFAs are limited. The shorter chain n-3 PUFAs FA, α-linolenic acid (ALA), is found in many plants, but the longer chains eicosapentaenoic acid (EPA) and docosahexaenoic acid (DHA) are produced almost exclusively by cold water algae, which are, in turn, ingested by fish. Humans cannot synthesize the n-3 double bond, but they do have the elongase and desaturase enzymes to convert ALA to EPA and DHA, a conversion, which however is an inefficient process. The conversion of ALA to EPA may be further reduced as a result of large amounts of n-6 FA in the diet, which compete for the same enzymes. Some studies however have found that ALA, irrespective of n-6 PUFAs, has a beneficial effect of lipid profiles [63]. Mechanism of actions of the medium- and long chain n-3 fatty acids appears to be independent. ALA exerts most of its effects by modulating lipoproteins, while EPA and DHA may reduce triglyceride synthesis [64]. Experts currently recommend the consumption of EPA and DHA, rather than ALA, to meet dietary goals for dietary n-3 PUFA [65]. Long-chain n-3 PUFA reduce triglyceride concentrations. An intake of 4 g EPA and DHA per day results in a 25–30% decrease of fasting triglyceride concentrations in both normolipidaemic and hypertriacylglycerolaemic subjects (66). Compares EPA and DHA, EPA-ethyl ester shows no change in triglyceride concentrations, suggesting that DHA is the active agent in fish oil, that decreases triglyceride concentrations. Therefore among long chain n-3 PUFAs, EPA may produce favourable effects on triglyceride and HDL cholesterol concentrations [67,68]. The hypotriglyceridaemic effect of long chain n-3 PUFAs, mediated by several mechanisms such as enhanced hepatic fatty acid oxidation [69], inhibition of fatty acid and triglyceride synthesis, reduced assembly and secretion of VLDL triglyceride concentrations [70], facilitates triglyceride rich lipoprotein removal through enhanced LPL activity in plasma [71]. Significant increases in HDL have been observed after DHA supplementation [67,68,72,73]; it may be related to decreased cholesteryl ester transfer protein activity that reduces the exchange from HDL cholesterol ester and VLDL, resulting in larger, more cholesterol-rich HDL cholesterol particles [74,75].

Inconsistent effects of DHA on total and LDL cholesterol levels have been shown; some investigators found a LDL cholesterol-raising effect [68,76] or no significant changes in total cholesterol or LDL cholesterol concentration [77,78]. After supplementation with n-3 Long chain PUFA, limited amounts of triglycerides are available for packaging into VLDL, which results in VLDL particles with low triglycerides that are readily converted to LDL, increases LDL cholesterol concentrations [79]. N-3 PUFAs could increase production of LDL via conversion of VLDL to LDL by increased lipolysis of VLDL and/or increased lipolytic activity or decreased clearance of LDL, by decreases in LDL receptor binding activity or reduced LDL receptor expression [80]. ALA, an n-3 polyunsaturated fatty acid found mainly in plant sources, including flaxseed oil, canola oil, and walnuts, is a metabolic precursor of DHA and EPA and any risk reduction may be mediated through conversion to these fatty acids; ALA cannot be synthesized by humans, and therefore, it is an essential fatty acid in diet [58]. Although evidence indicates that consumption of long chain n-3 PUFAs from seafood reduces the risk factors of cardiovascular disease, the effect of ALA intake in these risk factors is less well established. Daily supplementation with ALA-rich flaxseed is reported to reduce total cholesterol, LDL-cholesterol [81,82]. Weight of the evidence favors recommendations for modest dietary consumption of ALA (2 to 3 g per day) for primary and secondary prevention of CHD [58]. The relationship between ALA intake and CHD risk was seen among participants who consumed very little seafood; among men with limited seafood intake, each 1 g per day ALA intake was associated with 50% lower risk of CVDs; in contrast among subjects with some seafood intake, ALA intake was not associated with CHD risk. If benefits of ALA are greatest when EPA and DHA intakes are very low, the consumption of plant sources of n-3 fatty acids may be particularly important for CHD prevention among individuals who do not regularly consume fish [58].

3.7. Dietary cholesterol

The main source of dietary cholesterol is eggs, which contribute about one-third of the cholesterol in the diet; intake of dietary cholesterol has increased in recent year. Other sources of dietary cholesterol include animal products, dairy, meats, poultry, and shellfish [83]. High cholesterol intakes increase LDL cholesterol and the degree of rise varies from person to person. On average, the response of serum cholesterol to dietary cholesterol as revealed is approximately 10 mg/dL per 100 mg dietary cholesterol per 1000 kcal [84,85]. A recent meta-analysis showed that dietary cholesterol raises the ratio of total to HDL cholesterol, adversely affecting the serum cholesterol profile [86].

3.8. Carbohydrate

Recommendations to decrease fat and increase carbohydrate intake have come under scrutiny. Diets low in fat necessarily has a high proportion of carbohydrates, and high carbohydrate diet increase triglycerides, reduce HDL cholesterol concentrations, and increase LDL cholesterol concentrations [87]. In addition to carbohydrate intake, the type of carbohydrate, according to glycemic index, most likely influences lipid profiles [88]. Glycemic index refers to the value obtained by feeding a carbohydrate load and measuring the level of blood glucose. Using the glycemic index, carbohydrates with a low glycemic index may decrease triglyceride concentrations and increase HDL cholesterol [89]. Also substituting low-GI foods for high-GI foods lowers triglyceride concentrations by 15 to 25% [138]. High-carbohydrate diets increase triglyceride concentrations, compared to high-fat diets [91] via enhance hepatic lipogenesis [92] and decrease the synthesis of lipoprotein lipase [93]. A high carbohydrate diet also increases glucose and insulin concentrations, the latter increasing lipogenesis, leading to increases in triglyceride concentrations, triglyceride-enriched VLDL particles, and increases the LDL cholesterol concentrations [94]. Therefore reductions in dietary carbohydrate have been associated with reduced concentrations of LDL cholesterol [95] and increase means LDL particle size [96].

High carbohydrate diets (>60 percent of total energy) are associated with lipoprotein disturbances; reduction in the content of carbohydrate have beneficial effects on lipid profiles. However substitution of carbohydrate with other macronutrients is important. When carbohydrates are substituted for SFAs, the fall in LDL cholesterol levels equals that with monounsaturated fatty acids, and however, compared with MUFAs, this substitution frequently causes a fall in HDL cholesterol and a rise in triglycerides [23,97]. When dietary carbohydrate is consumed along with high-fiber diets, however, the rise in triglycerides or fall in HDL cholesterol has been reported to be reduced [98,99]. Addition of n-3 PUFA to low-fat, high-carbohydrate diets decreases the adverse effects of carbohydrate on blood lipids [51,100]. Also refined- and whole grains, as sources of carbohydrate, have an essential role in the metabolism of lipid profiles, that will be discussed in the section on food groups. In a relatively short period of time, dietary consumption of fructose has increased several fold above the amount present in natural foods, because of the use of high fructose corn sweeteners and sucrose in manufactured foods [101]. In human diets approximately one-third of dietary fructose comes from fruit, vegetables, and other natural sources and two-thirds is added to beverages and food in the diet (e.g. soft drinks, fruit-flavored drinks, candies, jams, syrups, and bakery products). Although there is little evidence that modest amounts of fructose have detrimental effects on carbohydrate and lipid metabolism, larger doses have been associated with numerous metabolic abnormalities, suggesting that high fructose consumption adversely affects health. High levels of plasma triacylglycerols are a well-established consequence of dietary fructose intake [101]. Numerous mechanisms have been suggested to explain this phenomenon [102,103], e.g. enhanced hepatic lipogenesis, and therefore overproduction of VLDL [102,104].

3.9. Protein

Plant sources of protein are predominantly legumes, dry beans, nuts, and, to a lesser extent, grain products and vegetables, which are low in saturated fats and cholesterol. Animal sources of protein include dairy products, egg whites, fish, poultry, and meats. Dietary protein in general has little effect on lipoprotein profiles. However, substituting plant protein including wheat gluten, soy proteins for animal protein decrease serum cholesterol [104,105]. Advice on the use of soy foods to displace animal products is consistent with the AHA advisory on soy [107], which states that 50 g/d soy protein consumption reduces approximate 3% LDL-C with no apparent dose-response effect [108]. Maximum reduction in LDL cholesterol was achieved when ~50 g of soy protein when was replaced meat or dairy protein [109]. Soy is a complex protein with a globulin fraction to which its cholesterol-lowering effect has been attributed; this fraction digested to peptides with inhibitory effects on cholesterol synthesis [110]. Isoflavones or the saponins found in soy, are also responsible for the cholesterol-lowering effect of soy [111,112]. Soy and other vegetable proteins also reduce oxidized LDL due to antioxidant activity [112,113].

4. Dietary food groups and lipoprotein

4.1. Grains and cereal

Based on evidence from both population and intervention studies, the recommended intake of whole grains of the 2005 Dietary Guidelines for Americans, is at least three ounces per day [114]. The Dietary Guidelines Advisory Committee (DGAC) 2010 Report emphasizes fiber-rich carbohydrate foods such as whole grains and vegetables, fruits, and cooked dry beans and peas, it specifically recommends that half of the grains consumed be whole grains, hence some whole grains should replace refined grains [115]. Similar recommendations are made by the American Heart Association [116] and the American Diabetes Association [117]. Whole grains are referred to as "complex" or "high-quality" carbohydrates, mainly due to their dietary fiber content [118], which has a beneficial effect on body weight, and lipid profiles because they are usually less energy-dense and more satiating than refined-grain foods [119] may be due to their high fiber content. Among whole grains, oat and barley have an advantage over wheat and brown rice in lowering serum lipids [120,121,122], contain viscous fibres, including β-glucacon [118] that lower serum cholesterol; 3.5 g of β-glucan from oats reduces LDL-C by 5% [123,124]. β-glucan interferes with reabsorption of bile acids and cholesterol by binding to bile acids, leading to increase bile acid excretion and lowering the bile acid levels in the liver and thereby increasing the conversion rate of cholesterol to bile acids. A viscous fiber intake of 10–25 g/d is recommended by the National Cholesterol Education Program's Adult Treatment Panel III as an additional diet option to decrease LDL cholesterol; an intake of 5–10 g/d lowers LDL-C by about 5% [126].

4.2. Fruit and vegetables

The 2010 Dietary Guidelines for Americans, recommend consuming sufficient amounts (5-13 servings, depending on energy needs) and a varieties of fruits and vegetables to reduce the risk of developing chronic diseases [115]; fruits, vegetables, or both should be emphasized at each meal, being major sources of vitamins C, E, and A, beta-carotene, other vitamins, fiber, flavonoids, and some minerals. Snacks and desserts that contain fruits and/or vegetables can be low in saturated fat, total fat, and cholesterol, and are very nutritious [18]. Fruits and vegetable intakes do not significantly change HDL cholesterol concentrations, but do decrease total and LDL cholesterol [9,127-132]. The protective effect of fruit and vegetables against CVDs is from their water-soluble and also viscous fibers (e.g. pectins) [133]. Viscous fiber increases fecal bile acid losses [134] and chenodeoxycholic acid synthesis [135].

4.3. Dairy products

Dairy products are important sources of protein, calcium, phosphorus, and vitamin D. The recommendation for intakes of dairy products is 2-3 serving per day; fat-free milk or 1 percent fat milk, fat-free or low-fat cheese (e.g., ≤3g per 1 oz serving), 1 percent fat cottage cheese or imitation cheeses made from vegetable oils, and fat-free or low-fat yogurt are good choices. Fat-free milk and other fat-free or low-fat dairy products provide as much or more calcium and protein than whole milk dairy products, with little or no saturated fat [18].

Recent studies confirm that milk products were associated with lower small dense LDL, and triglyceride concentrations, and higher HDL cholesterol [136]. In the CARDIA study, obese subjects with more frequent consumption of dairy products showed a trend towards lower risk of dyslipidaemia [137]. Minerals (calcium, magnesium), protein (casein and whey) and vitamins (riboflavin and vitamin B-12) have the hypocholesterolaemic effect of dairy product. The possible hypolipidaemic mechanism of calcium includes decreased intestinal absorption of cholesterol, bile acids, or fat [138], decreased fatty acid synthesis, increasing lipolysis, all of which lead to decreased triacylglycerol stores [139]. Milk proteins (whey) [140] or peptides [141] may also play a role. Whey may act independently or synergistically with the calcium; attenuate lipogenesis, and accelerate lipolysis [142]. Dairy products contain SFAs that could affect the blood lipid profile. A recent meta-analysis of 21 prospective cohort studies showed that the harmful effects of SFAs on CHD are still controversial [143]. An inverse association was shown between milk-specific fatty acids in serum cholesterol esters with serum cholesterol and apolipoprotein β levels [144]. Consumption of fat-free dairy products might decrease plasma cholesterol levels, while whole milk has neither a hypo- nor hypercholesterolaemic effect [139]. SFAs in dairy products can adversely influence CHD, although the effect of SFAs on CHD risk depends on the source of calories by which it is substituted to maintain energy balance [145]. Different dairy products have different effects on the lipid profiles. The LDL-C-raising effect of cheese was less than that of butter at comparable intakes of total fat and saturated fat [146,147]. Butter fat may increase total and LDL cholesterol by down-regulation of LDL removal from the circulation [148]. Fermented dairy products may have a favourite effect on lipid profiles. The protective effect of yogurt [139,149], a fermented dairy product, was shown to reduce absorption of cholesterol and therefore prevent dyslipidemia; it is thought to increase calcium bioavailability through its high acidity [149]. Fermented milk products may decrease cholesterol levels more than non-fermented products [149-151]. Probiotic yogurt decreased total cholesterol by 4% and LDL cholesterol by 5% [149]. A meta-analysis of fermented dairy products has shown a possible cholesterol lowering property, through the high content of probiotic bacteria [152].

4.4. Nuts

Although nuts are high in fat, in most nuts the predominant fats are unsaturated. Studies over the last decade have demonstrated favourable effects of nuts in modifying lipid risk factors for CHD [153]. However, their use is not yet part of standard advice for patients with hyperlipidemia, despite recognized health benefits for the general population. Intake of nuts fits well with current American Heart Association guidelines [19] to replace dietary SFAs with unsaturated fats and with the National Cholesterol Education Program (NCEP) guidelines to increase intake of dietary MUFAs [153]. Less atherogenic plasma lipid profiles associated with long-term consumption of nuts [154,155]. Addition of nuts to the habitual diet of both normocholesterolemic and hypercholesterolemic subjects results in a significant reduction in plasma total and LDL cholesterol, whereas HDL remains unchanged or increases [155-158]. One-percent reductions in LDL cholesterol would be achieved with daily intakes of 4-11 g of walnuts, pecans, peanuts, macadamias, and pistachios [50,155,157-161]. There are several components in nuts i.e. high MUFA, high PUFAs : SFAs ratio, proteins (specially high arginin), plant sterols, fiber, and associated phenolic substances, which may all contribute to the cardioprotective effect of nuts [154,162]. Also replacement of dietary SFAs with MUFAs due to the high MUFA content of nuts and high content of vitamin E in nuts reduce susceptibility of LDL to oxidation, a key event in the development of CVDs [233]. Consumption of almonds, either as the whole nut or the oil, lower total and LDL cholesterol concentrations. Addition of 100 g of almonds to the diets reduces total cholesterol by 9-16% and LDL cholesterol by 12-19 % in hypercholesterolemic subjects [164]; in one study almond consumption also reduced fasting triglyceride concentrations by 14%, compared with baseline [165]. Macadamia is another nut that improve lipid disturbances, and its inclusion as part of a healthy diet favourably altered the plasma lipid profile, despite the nuts being high in fat; their consumption reduced plasma total and LDL cholesterol concentrations and increase HDL cholesterol without any change in the triglyceride concentrations [166]. These changes could contribute to high MUFA intake and lower intake of PUFA and SFA consumption of macadamia nuts. Of nuts, walnuts are unique in improving dyslipidemia because they are a rich source of PUFAs, especially α-linolenic acid and linoleic acid; 100 g of walnuts contain 65.2 g fat; mainly from PUFAs (47.2 g) including α-linolenic acid (9.1 g) and linoleic acid (38.1 g) [167]. In a meta-analysis, consumption of walnuts resulted in decrease in total and LDL cholesterol concentrations, whereas HDL cholesterol and triglycerides were not affected [168]. Despite favourable effects of nuts on dyslipidemia, the intake of nuts should fit within the calorie and fat goal [18].

4.5. Beans and legumes

Legumes include a variety of beans such as navy, pinto, kidney, garbanzo, lima beans and peas such as split green peas or lentils. The Dietary Guidelines for Americans suggest consuming 3 cups of legumes per week [18, 169]. Legumes are a rich source of soluble dietary fiber and vegetable protein and have long been known to be hypercholesterolaemic foods [170,171]. One-half cup of cooked beans or peas can provide a range of dietary fiber from 4.6 g in fava beans up to 9.6 g fiber in navy beans, with a half cup of chick peas providing 6.2 g of total fiber, and 1.3 grams soluble dietary fiber [169]. In a meta-analysis both total and LDL cholesterol decreased, while HDL cholesterol did not change significantly, when diets uses supplemented with non-soy legumes [169]. The hypocholesterolaemic property of legumes is associated with the water-soluble fibre. Dietary fiber in legumes is not digested in the small intestine but be fermented in the colon and produces short chain fatty acids such as acetate, propionate and butyrate [172,173]; that inhibits hydroxy-3-methylglutaryl-CoA reductase, the limiting enzyme for cholesterol synthesis. Dietary fiber also decrease LDL cholesterol concentration by partially interrupting the enterohepatic circulation of bile acids via binding to bile acids in the intestines and preventing their re-absorption [174]. Consequently, an increase in the production of bile acids decreases the liver pool of cholesterol and increases uptake of serum cholesterol by the liver, decreasing thereby circulating cholesterol in the blood [175]. Another hypercholesterolemic component of legume is phytochemicals, which has been shown to reduce blood cholesterol levels and is present in small to moderate amounts in many types of legumes, such as chickpeas [176]. Dietary modification strategies that target the reduction of risk factors for CVDs should include an increase in legume consumption in addition to other strategies which have been of proven benefit [169].

4.6. Meat, fish, poultry and eggs

Recommendation for intakes of meat, fish and poultry are up to 5 oz per day from lean meats (beef, pork, and lamb), poultry, and fish [18]. To achieve NCEP dietary goals, individuals are often counselled to reduce the amount and frequency of red meat consumption because of its hypercholestromia effects [177-179]. Cholesterol raising effects of red meats appears to result from high contents of SFAs [177,179]. Therefore, lean red meats that provide small amounts of these fatty acids do not adversely influence the blood lipid profile, compared with lean white meats. In isoenergetic low-fat diets, lean meat, fish and, poultry had similar effects on blood lipid response in both hypercholesterolemic and normocholesterolemic subjects [178,180,181]. Data available suggest that meat protein, per se, is not hypercholesterolemic [177,181,182]. The blood cholesterol-raising potential of meat products appears to be a function of their SFA fat and cholesterol contents. Therefore, substituting lean for higher fat red meat should favourably influence serum total cholesterol and LDL-C levels. Incorporating lean beef, fish, or poultry into the AHA diet can be beneficial in lower disturbances of lipid profile in patients with hypercholesterolemia [178,183]. Therefore the hypercholesterolemic subjects known to be at high risk for CVDs, could be advised to include lean fish as well as lean beef or poultry without skin in an AHA diet to reduce their lipoprotein disturbances [184,185]; normolipidemic subjects can also incorporate lean fish in an AHA diet [184], althought it is not necessary to eliminate or drastically reduce intake of lean red meat consumption because it is a rich source of iron, zinc and vitamin B12. One of the dietary recommendations in the prevention of CVDs is to limit egg consumption, because they have been shown to be a major source of dietary cholesterol (One egg contains 200 mg/cholesterol) that increases both serum total and LDL-cholesterol concentrations [21,86,186]. Several epidemiologic studies however found no relation between egg consumption and risk of coronary heart disease [187,188], may be because dietary cholesterol increases not only concentrations of total and LDL cholesterol but also concentrations of HDL cholesterol [21,186,189,190]. Egg intake has been also shown to promote the formation of large LDL particles, which is less atherogenic [191]. Therefore dietary recommendations aimed at restricting egg consumption should not be generalized to include all individuals [191].

4.7. Dietary pattern

Using single nutrients or dietary food groups have some limitations in assessing their effect on lipid profiles separately because nutrients and foods are consumed in combination. To date, dietary patterns consider how foods are consumed in combination, and are used to evaluate the effects of overall nutritional habits on health status. There are two dietary patterns that demonstrate the beneficial effect on disturbances of lipoprotein concentrations; there include the dietary to stop hypertension (DASH) and the Mediterranean diet. The DASH dietary pattern, rich in fruits, vegetables, and low-fat dairy foods, emphasizes fish, poultry, and whole grains, and is reduced in total fat, SFAs and cholesterol, red meat, sweets, and sweetened beverages [192,193]; it lowers total, LDL and HDL cholesterols, without any adverse effects on triglyceride concentrations [194]; all of these coupled with decrease in blood pressure, reduce 10-year coronary heart disease risk of approximately 12% [194]. The Mediterranean dietary pattern consists of: (a) daily consumption: of non refined cereals and products (whole grain bread, pasta, brown rice, etc), vegetables (2 – 3 servings/ day), fruits (6 servings/day), olive oil (as the main added lipid) and dairy products (1 – 2 servings/day), (b) weekly consumption: of fish (4–5 servings/week), poultry (3 – 4 servings/week), olives, pulses, and nuts (3 servings/ week), potatoes, eggs and sweets (3 – 4 servings/week) and monthly consumption: of red meat and meat products (4 – 5 servings/month). It is also characterized by moderate consumption of wine (1 – 2 wineglasses/day). Mediterranean diet is a diet poor in SFAs and PUFAs but rich in MUFA (oleic acid) provided by the olive oil. The ratio of MUFAs : SFAs fat ratio is high > 2 [195]. This diet pattern is associated with reduction in total and LDL-cholesterol, and also a significant effect on triglycerides and VLDL concentrations, and a small positive or no effect on HDL-cholesterol [196-199] and improves dyslipidemia in dislipidemic patients [200]. This diet also includes antioxidant vitamins and phenolic compounds, and therefore reduces levels of circulating oxidized LDL and increases total antioxidant capacity [201]. Beside these two dietary patterns, other dietary pattern such as the western, and healthy dietary patterns affect lipoprotein profiles. The western pattern is characterized by high consumption of food such as refined grains, french fries, and red meats that have detrimental effects on lipid profiles. The healthy pattern included non-hydrogenated fat, vegetables, eggs, and fish and was negatively associated with lipoprotein disturbances [202-205]. In addition of dietary patterns, therapeutic lifestyle change is another dietary approach that ATP III recommends to reduce risks for CHD. This dietary approach includes the following: 1) Reduced intakes of dietary SFAs (<7% of total calories) and cholesterol (<200 mg/d), 2) weight reduction, 3) increased physical activity, and 4) therapeutic options for enhancing LDL lowering such as plant stanols/ sterols (2 g/d) and increased viscous (soluble) fiber (10-25 g/d) [18].

5. Dietary supplement

5.1. Plant stanols and sterols

Dyslipidemia may be treated with dietary interventions, including the daily consumption of foods with added plant stanols or plant sterols. Plant sterols are isolated from soybean and tall pine-tree oils. Also some foods such as macadamia nuts are a rich source (1.28 mg/g lipid) of plant sterols. Plant sterols can be esterified to unsaturated fatty acids, creating sterol esters, to increase lipid solubility. Hydrogenating of sterols produces plant stanols. Plant stanols and sterols are available in commercial margarines. Daily consumption of 2 g plant stanols or plant sterols, expressed as free plant stanol or plant sterol equivalents improves dyslipidemia [18]. FDA confirms a daily dose of plant sterols and stanols of 2 g per day as safe, a dose which reduces LDL cholesterol by 10% [206], with little or no change in HDL cholesterol or triglyceride levels. There were no apparent added benefits at higher doses of palnts stanols and sterols. Plant stanols and sterols compete with absorption of dietary cholesterol and bile acid [8]. The consumption of plant stanols and sterols is an effective LDL cholesterol lowering strategy for patients who are undergoing statin therapy. The lipid-lowering response to combined plant stanols and sterols/statin therapy target both intestinal and hepatic cholesterol metabolism. Consumption of plant stanols and sterols reduces intestinal cholesterol absorption and reduces hepatic cholesterol synthesis. Consumption of statins simultaneously with plant stanols and sterols inhibit hepatic cholesterol synthesis and therefore reduce in LDL cholesterol concentrations [8]. Plant sterols/stanols reduce absorption of dietary carotenoids, and decrease levels of plasma betacarotene; therefore increased intakes of fruits and vegetables are recommended with consumption of plant stanols/sterols[18].

5.2. Herbal

There is a need to identify additional non-pharmacologic therapeutic options for cholesterol lowering. There is also a need to find products that are more practical for the consumer than viscous fiber and plant stanols and sterols to permit widespread adoption.

5.2.1. Flavonoid

Flavonoids have 2 aromatic rings that are bound by an oxygenated heterocyclic ring. On the basis of their chemical structure, they are divided into several subclasses: flavones, flavonols, flavanones, flavan-3-ols, anthocyanins and isoflavones. Flavones and flavonols are found in leaf vegetables and onion. Flavanones are mainly found in grapefruits and citrus fruits. Tea and cocoa are the richest sources of flavan-3-ols. Soy and soy products such as tofu, and miso are the main sources of isoflavones [207,208]. Although increased resistance of LDL to oxidation was observed after treatment with various synthetic pharmaceutical agents, an effort is made to identify natural food products which can offer antioxidant defense against LDL oxidation. Polyphenolic flavonoids are powerful antioxidants and their antioxidative capacity is related to their chemical structure [209].

Incubation of LDL with flavonoids protects the lipoprotein against oxidation [210]. Certain flavonoids such as querectin could have a potentially protective role in suppression of LDL oxidation, regardless of the effect of antioxidant vitamins [211] via scavenging radicals and reduce total and LDL cholesterol concentrations, by reducing the hepatic lipogenesis [212].

5.2.2. *Tea*

The effect of tea on lipid profiles is uncertain. Although some studies have found no lipid-lowering effects from green or black tea consumption, most showed hypolipidemic effects for tea [213-218]. The association between tea drinking and lipid profile concentrations was linear for up to 10 cups per day, beyond which the association disappeared [219]. Daily consumption of 10 cups of green tea was associated with a reduction of approximately 2% in serum total cholesterol [219]. Tea also is a major source of flavonoids, the predominant ones in green tea being catechins. Theaflavins are polyphenol pigments present in black tea, formed from the polymerization of catechins during fermentation of green tea [220]. Catechins reduce intestinal cholesterol absorption [221], reduce hepatic cholesterol content [222] and increase fecal excretion of total fatty acids, neutral sterols, and acidic sterols [223] and up-regulate the LDL receptor in liver cells [224]. Polyphenol in black tea also increases fecal excretion of total lipids and cholesterol [225].

5.2.3. *Chocolate*

The beneficial effects of chocolate on healthy humans have been widely addressed in recent years. Supplementation of cocoa products affects lipid profiles in subjects with cardiovascular-related diseases such as hypercholesterolemia, glucose intolerance, and hypertension as well as healthy individuals [226-228]. Consumption of cocoa and dark chocolate increase the concentration of HDL cholesterol [229] and plasma antioxidant capacity, decrease the formation of lipid oxidation products, and inhibit the oxidation of LDL [230]. In a meta-analysis study, cocoa was associated with small decreases in total and LDL cholesterol, but not HDL cholesterol concentrations [231]. Cocoa products contain more polyphenols than teas. A particular group of flavonoids, namely, the flavan-3-ols was found in chocolate (flavanols) [232]. Moderate consumption of cocoa or dark chocolate, have potential health benefits [231], however, a high dose of polyphenols has been shown to exert cytotoxic effects on liver cells [233] and higher polyphenol supplementation may counteract its beneficial biological effects on lipid metabolism [234].

5.2.4. *Fenugreek*

Fenugreek (Trigonella foenum-graecum), an annual medicinal plant of the Fabaceae family is well documented for its pharmacological properties. Fenugreek seeds have been historically used for the treatment of various chronic diseases such as diabetes, dyslipidemia, and obesity [235,236]. The seeds of Fenugreek contain many nutrients including protein, carbohydrates, fat, vitamin, and minerals, fiber, saponins, choline and trigonelline, polyphenolic flavonoids, steroid saponins, polysaccharides mainly galactomannans and 4-hydroxyisoleucine [237,239], the fiber and saponin components of the seeds have been shown to have hypocholesterolemic effect [240], and the beneficial effect of raw fenugreek seeds on elevated serum cholesterol levels has been well established [241]. Raw fenugreek seeds reduce serum total cholesterol, LDL cholesterol, VLDL cholesterol and triglyceride concentrations, without altering the HDL fraction [242]; intakes of 20–25 g in three divided doses yielded maximum benefit in the control of cholesterol concentrations [243]. Its use as a dietary adjunct however is limited because of its bitterness. Soaking and washing of fenugreek seeds in water overnight removes the bitterness to a certain extent and makes then edible [243,244].

5.2.5. *Ginseng*

The beneficial metabolic effects of ginseng on lipid profiles as a hypolipidemic agent were reported over 20 years ago [245-247]. Ginseng leads to reduction of cholesterol and triglyceride concentrations in liver and serum. Administration of red ginseng powder and extract reduces plasma total cholesterol, triglycerides, FFA, and increased HDL-C [248,249]. Ginseng saponins may decrease blood cholesterol concentrations by increasing cholesterol excretion through bile acid formation [249,250]. Ginsenoside, one of active components of ginseng saponins, may accelerate serum cholesterol turnover by increased cholesterol degradation and excretion in the feaces notwithstanding increased hepatic cholesterogenesis [250,251]. Ginseng saponins as ginsenosides increase LDL receptors by promoting the synthesis of LDL receptors[252].

5.2.6. *Ginger*

Ginger has been listed in the "Generally Recognized as Safe" by FDA [338]; fresh ginger rhizome contains polyphenolic compounds such as gingerols; zingerone, which is the major active component and gingerol, is one of the most abundant constituents in the gingerol series and also responsible for its characteristic pungent taste [253,254]. Ginger oleo-resin and dried ginger rhizome reduce hypercholesterolaemia. The speculated mechanism for these compounds is by disrupting cholesterol absorption from the gastro-intestinal tract [255], which may be due to the presence of niacin in ginger, and it causes increased clearance of VLDL, lowers triglyceride levels, increases hepatic uptake of LDL and inhibition of cholesterogenesis [256]. Ginger powder significantly reduces the extent of lipid peroxidation and improves plasma antioxidant capacity, which decreases plasma-free radicals [257]. Moreover, polyphenolic flavonoids present in ginger may prevent coronary artery disease by reducing plasma cholesterol levels or by inhibiting LDL oxidation [258]. Reduction in serum triglycerides is dose dependent; doses of 200 and 400 mg/kg of ginger are more effective as antihypercholesterolaemics than atorvastatin when given for 4 weeks and are equivalent to it when given for shorter period under the same conditions of diet and life style for the treatment of the same pathologic condition. The triglyceride lowering effect of ginger may be due to ginger's ability to enhance lipase activity [255].

5.2.7. Licorice

Licorice root, derived from the plant Glycyrrhiza glabra is used widely in Asia as a sweetener or a spice, contains flavonoids from the flavan and chalcone subclasses, and has a antioxidative properties [259]. Licorice-derived glabridin binds to the LDL particle and protects it from oxidation by its capacity to scavenge free radicals and its property to reduce the LDL aggregation [260,261].

6. Conclusion

Diet therapy is the initial recommended intervention for prevention of and managing disturbances of lipoprotein concentrations, prior to advancing to drug therapy. Further research on the association between dietary components and lipoprotein disturbances is recommended.

Author details

Parvin Mirmiran*
Department of Clinical Nutrition and Dietetics, Faculty of Nutrition Sciences and Food Technology, National Nutrition and Food Technology Research Institute, Shahid Beheshti University of Medical Sciences, Tehran, Iran

Somayeh Hosseinpour-Niazi
Nutrition Related Non-Communicable Disease, Research Institute for Endocrine Sciences, Shahid Beheshti University of Medical Sciences, Tehran, Iran

Fereidoun Azizi
Endocrine Research Center, Research Institute for Endocrine Sciences, Shahid Beheshti University of Medical Sciences, Tehran, Iran

Acknowledgement

The authors would like to acknowledge Ms N. Shiva for language editing of the manuscript and to express their appreciation to Emad Yuzbashian for his valuable help. This study was funded by a grant from the Research Institute of Endocrine Sciences, Shadid Beheshti University of Medical Sciences, Tehran, Iran. None of the authors had any personal or financial conflicts of interest.

7. References

[1] Robins SJ, Lyass A, Zachariah JP, Massaro JM, Vasan RS (2011) Insulin resistance and the relationship of a dyslipidemia to coronary heart disease: the Framingham Heart Study. Arterioscler Thromb Vasc Biol. 31: 1208-1214.

---

* Corresponding Author

[2] Musunuru K (2010) Atherogenic dyslipidemia: cardiovascular risk and dietary intervention. Lipids. 45:907-914.
[3] Goff DC Jr, Bertoni AG, Kramer H, Bonds D, Blumenthal RS, Tsai MY, eat al (2006) Dyslipidemia prevalence, treatment, and control in the Multi-Ethnic Study of Atherosclerosis (MESA): gender, ethnicity, and coronary artery calcium. Circulation. 113:647-656.
[4] Hosseini-Esfahani F, Mousavi Nasl Khameneh A, Mirmiran P, Ghanbarian A, Azizi F (2011) Trends in risk factors for cardiovascular disease among Iranian adolescents: the tehran lipid and glucose study, 1999-2008. J Epidemiol. 21:319-328.
[5] Esteghamati A, Meysamie A, Khalilzadeh O, Rashidi A, Haghazali M, Asgari F, et al (2009) Third national Surveillance of Risk Factors of Non-Communicable Diseases (SuRFNCD-2007) in Iran: methods and results on prevalence of diabetes, hypertension, obesity, central obesity, and dyslipidemia. BMC Public Health.9:167.
[6] Goodman SG, Langer A, Bastien NR, McPherson R, Francis GA, Genest JJ Jr, et all (2010) DYSIS Canadian Investigators. Prevalence of dyslipidemia in statin-treated patients in Canada: results of the DYSlipidemia International Study (DYSIS). Can J Cardiol. 26:e330-335.
[7] Aguilar-Salinas CA, Gómez-Pérez FJ, Rull J, Villalpando S, Barquera S, Rojas R (2010) Prevalence of dyslipidemias in the Mexican National Health and Nutrition Survey 2006. Salud Publica Mex. 52:44-53.
[8] Rideout TC, Harding SV, Marinangeli CP, Jones PJ (2010) Combination drug-diet therapies for dyslipidemia. Transl Res. 155:220-227.
[9] Mirmiran P, Noori N, Zavareh MB, Azizi F (2009) Fruit and vegetable consumption and risk factors for cardiovascular disease. Metabolism. 58:460-8.
[10] Mirmiran P, Ramezankhani A, Azizi F (2009) Combined effects of saturated fat and cholesterol intakes on serum lipids: Tehran Lipid and Glucose Study. Nutrition. 25:526-531
[11] Mirmiran P, Mirbolooki M, Heydarian P, Salehi P, Azizi F (2008) Intrafamilial associations of lipid profiles and the role of nutrition: the Tehran lipid and glucose study. Ann Nutr Metab. 52: 68-73.
[12] Azadbakht L, Mirmiran P, Hedayati M, Esmaillzadeh A, Shiva N, Azizi F(2007) Particle size of LDL is affected by the National Cholesterol Education Program (NCEP) step II diet in dyslipidaemic adolescents. Br J Nutr. 98: 134-139.
[13] Hu FB (2002) Dietary pattern analysis: a new direction in nutritional epidemiology. Curr Opin Lipidol. 13:3–9.
[14] Azadbakht L, Mirmiran P, Esmaillzadeh A, Azizi T, Azizi F (2005) Beneficial effects of a Dietary Approaches to Stop Hypertension eating plan on features of the metabolic syndrome. Diabetes Care. 28: 2823-31.
[15] Azadbakht L, Mirmiran P, Esmaillzadeh A, Azizi F (2006) Dietary diversity score and cardiovascular risk factors in Tehranian adults. Public Health Nutr. 9: 728-736.
[16] Hosseini-Esfahani F, Jessri M, Mirmiran P, Bastan S, Azizi F (2010) Adherence to dietary recommendations and risk of metabolic syndrome: Tehran Lipid and Glucose Study. Metabolism. 59: 1833-1842.

[17] Bahadoran Z, Mirmiran P, Hosseinpanah F, Hedayati M, Hosseinpour-Niazi S, Azizi F (2011) Broccoli sprouts reduce oxidative stress in type 2 diabetes: a randomized double-blind clinical trial. ur J Clin Nutr. 65: 972-977.

[18] National Cholesterol Education Program (NCEP) Expert Panel on Detection, Evaluation, and Treatment of High Blood Cholesterol in Adults (Adult Treatment Panel III) (2002) Third Report of the National Cholesterol Education Program (NCEP) Expert Panel on Detection, Evaluation, and Treatment of High Blood Cholesterol in Adults (Adult Treatment Panel III) final report. Circulation.106: 3143-3421.

[19] Krauss RM, Eckel RH, Howard B, et al. (2000) AHA Dietary Guidelines: revision 2000: A statement for healthcare professionals from the Nutrition Committee of the American Heart Association. Circulation 102, 2284–2299.

[20] Siri-Tarino PW, Sun Q, Hu FB, Krauss RM (2010) Saturated fat, carbohydrate, and cardiovascular disease. Am J Clin Nutr. 91:502–9.

[21] Clarke R, Frost C, Collins R, Appleby P, Peto R (1997) Dietary lipids and blood cholesterol: quantitative meta-analysis of metabolic ward studies. BMJ. 314:112–117.

[22] Dietschy JM (1998) Dietary fatty acids and the regulation of plasma low density lipoprotein cholesterol concentrations. J Nutr. 128: 444–448.

[23] Mensink RP, Katan MB (1992) Effect of dietary fatty acids on serum lipids and lipoproteins. A meta-analysis of 27 trials. Arterioscler Thromb. 12:911–919.

[24] Kris-Etherton PM, Yu S (1997) Individual fatty acid effects on plasma lipids and lipoproteins: human studies. Am J Clin Nutr;65: 1628-1644.

[25] Hayes KC, Khosla P, Hajri T, Pronczuk A (1997) Saturated fatty acids and LDL receptor modulation in humans and monkeys. Prostaglandins Leukot Essent Fatty Acids. 57: 411–418.

[26] Mensink RP, Zock PL, Kester AD, Katan MB (2003) Effects of dietary fatty acids and carbohydrates on the ratio of serum total to HDL cholesterol and on serum lipids and apolipoproteins: a metaanalysis of 60 controlled trials. Am J Clin Nutr. 77:1146–1155.

[27] Bonanome A & Grundy SM (1988) Effect of dietary stearic acid on plasma cholesterol and lipoprotein levels. N Engl J Med. 318, 1244–1248.

[28] Grande F, Anderson JT & Keys A (1970) Comparison of effects of palmitic and stearic acids in the diet on serum cholesterol in man. Am J Clin Nutr. 23, 1184–1193.

[29] German JB, Gibson RA, Krauss RM, et al (2009) A reappraisal of the impact of dairy foods and milk fat on cardiovascular disease risk. Eur J Nutr, 48:191–203.

[30] National Cholesterol Education Program (1994) Second report of the expert panel on detection, evaluation, and treatment of high blood cholesterol in adults (Adult Treatment Panel II). Circulation. 89:1333-445.

[31] Siri-Tarino PW, Sun Q, Hu FB, Krauss RM (2010) Saturated fatty acids and risk of coronary heart disease: modulation by replacement nutrients. Curr Atheroscler Rep. 12: 384-390.

[32] Berglund L, Lefevre M, Ginsberg HN, et al (2007) Comparison of monounsaturated fat with carbohydrates as a replacement for saturated fat in subjects with a high metabolic risk profile: studies in the fasting and postprandial states. Am J Clin Nutr. 86:1611–1620

[33] Lichtenstein AH (1997) Trans fatty acids, plasma lipid levels, and risk of developing cardiovascular disease. A statement for healthcare professionals from the American Heart Association. Circulation. 95: 2588-2590.

[34] Sun Q, Ma J, Campos H, et al (2007) A prospective study of trans fatty acids in erythrocytes and risk of coronary heart disease. Circulation. 115: 1858–1865

[35] Brouwer IA, Wanders AJ, Katan MB (2010) Effect of animal and industrial trans fatty acids on HDL and LDL cholesterol levels in humans--a quantitative review. PLoS One. 5: 9434.

[36] Almendingen K, Jordal O, Kierulf P, Sandstad B, Pedersen JI (1995) Effects of partially hydrogenated fish oil, partially hydrogenated soybean oil, and butter on serum lipoproteins and Lp[a] in men. J Lipid Res. 36: 1370–1384.

[37] Vermunt SH, Beaufrere B, Riemersma RA, Sebedio JL, Chardigny JM, et al (2001) Dietary trans alpha-linolenic acid from deodorised rapeseed oil and plasma lipids and lipoproteins in healthy men: the TransLinE Study. Br J Nutr. 85: 387–392.

[38] Lacroix E, Charest A, Cyr A, Baril-Gravel L, Lebeuf Y, Paquin P, et al (2012) Randomized controlled study of the effect of a butter naturally enriched in trans fatty acids on blood lipids in healthy women. Am J Clin Nutr. 95: 318-25.

[39] Motard-Belanger A, Charest A, Grenier G, Paquin P, Chouinard Y, et al (2008) Study of the effect of trans fatty acids from ruminants on blood lipids and other risk factors for cardiovascular disease. Am J Clin Nutr. 87: 593–599.

[40] Mozaffarian D, Katan MB, Ascherio A, Stampfer MJ, Willett WC (2006) Trans fatty acids and cardiovascular disease. N Engl J Med. 354: 1601–1613.

[41] Almendingen K, Jordal O, Kierulf P, Sandstad B, Pedersen JI (1995) Effects of partially hydrogenated fish oil, partially hydrogenated soybean oil, and butter on serum lipoproteins and Lp[a] in men. J Lipid Res. 36: 1370–1384.

[42] Zock PL, Mensink RP (1996) Dietary *trans*-fatty acids and serum lipoproteins in humans. *Curr Opin Lipidol.* 7: 34-37

[43] United States Department of Agriculture, United States Department of Health and Human Services: Report of the Dietary Guidelines Advisory Committee on the Dietary Guidelines for Americans. Washington, DC: Government Printing Office; 2010

[44] Heyden, S (1994) Polyunsaturated and monounsaturated fatty acids in the diet to prevent coronary heart disease via cholesterol reduction. Ann. Nutr. Metab. 38: 117–122.

[45] Gustafsson I-B, Vessby B, Ohrvall M, Nydahl M (1994) A diet rich in monounsaturated rapeseed oil reduces the lipoprotein cholesterol concentration and increases the relative content of n23 fatty acids in serum in hyperlipemic subjects. Am J Clin Nutr. 59:667–74

[46] Mensink RP (1994) Dietary monounsaturated fatty acids and serum lipoprotein levels in healthy subjects. Atherosclerosis. 110. 65–68.

[47] Roche, H. M., Zampelas, A. & Knapper, J.M.E (1998) Effect of longterm olive oil dietary intervention on postprandial triacylglycerol and factor VII metabolism. Am. J. Clin. Nutr. 68: 552–560.

[48] Ginsberg, H. N., Barr, S. L., Gilbert, A., Karmally,W., Deckelbaum, R., Kaplan, K., Ramakrishnan, R., Holleran, S. & Dell, R. B. (1990) Reduction of plasma cholesterol

levels in normal men on an American Heart Association Step 1 diet or a Step 1 diet with added monounsaturated fat. N. Engl. J. Med. 322: 574–579.
[49] Allman-Farinelli MA, Gomes K, Favaloro EJ, Petocz P (2005) A diet rich in high-oleic-acid sunflower oil favorably alters low-density lipoprotein cholesterol, triglycerides, and factor VII coagulant activity. J Am Diet Assoc. 105: 1071-1079.
[50] Rajaram S, Burke K, Connell B, Myint T, Sabate J (2001) A monounsaturated fatty-acid pecan-enriched diet favourably alters the serum lipid profile of healthy men and women. J Nutr. 131:2275-2279.
[51] Jiménez-Gómez Y, Marín C, Peérez-Martínez P, Hartwich J, Malczewska-Malec M, Golabek I, et al (2010) A low-fat, high-complex carbohydrate diet supplemented with long-chain (n-3) fatty acids alters the postprandial lipoprotein profile in patients with metabolic syndrome. J Nutr. 140: 1595-1601.
[52] Roche HM, Zampelas A, Knapper JM,Webb D, Brooks C, Jackson KG, et al (1998) Effect of long-term olive oil dietary intervention on postprandial triacylglycerol and factor VII metabolism. Am J Clin Nutr. 68: 552–560.
[53] Roche, H. M., Zampelas, A., Jackson, K. G., Williams, C. M., Gibney, M. J. (1998) The effect of test meal monounsaturated fatty acid:saturated fatty acid ratio on postprandial lipid metabolism. Br. J. Nutr. 79: 419–424.
[54] Dashti N, Smith EA, Alaupovic P (1990) Increased production of apolipoprotein B and its lipoproteins by oleic acid in Caco-2 cells. J Lipid Res. 31:113–23.
[55] Silva KD, Kelly CN, Jones AE, Smith RD, Wootton SA, Miller GJ, et al (2003). Chylomicron particle size and number, factor VII activation and dietary monounsaturated fatty acids. Atherosclerosis. 166: 73–84.
[56] Rajaram S, Burke K, Connell B, et al (2001) A monounsaturated fatty acid-rich pecan-enriched diet favorably alters the serum lipid profile of healthy men and women. J Nutr. 131:2275–2279
[57] Curb DJ, Wergowske G, Dobbs JC, Abbot RD, Huang B (2000) Serum lipid effects of a high-monounsaturated fat diet based on macadamia nuts. Arch Int Med. 160: 1154-1158.
[58] Mozaffarian D (2005)Does alpha-linolenic acid intake reduce the risk of coronary heart disease? A review of the evidence. Altern Ther Health Med. 11: 24-30.
[59] Wandders AJ, Brouwer IA, Siebelink E, Katan MB (2010) Effect of a high intake of conjugated linoleic acid on lipoprotein levels in healthy human subjects. PLoS One. 5: 9000.
[60] Kostogrys RB, Pisulewski PM (2010) Effect of conjugated linoleic acid (CLA) on lipid profile and liver histology in laboratory rats fed high-fructose diet. Environ Toxicol Pharmacol. 30: 245-250.
[61] Nicolosi RJ, Rogers EJ, Kritchevsky D, Scimeca JA, Huth PJ (1997) Dietary conjugated linoleic acid reduces plasma lipoproteins and early aortic atherosclerosis in hypercholesterolemic hamsters. Artery. 22: 26.
[62] Yang L, Yeung SY, Huang Y, Wang HQ, Chen ZY. referential incorporation of trans, trans-conjugated linoleic acid isomers into the liver of suckling rats. Br J Nutr. 2002 Mar;87(3):253-60.

[63] Mirmiran P, Hosseinpour-Niazi S, Naderi Z, Bahadoran Z, Sadeghi M, Azizi F (2012) Association between interaction and ratio of ω-3 and ω-6 polyunsaturated fatty acid and the metabolic syndrome in adults. Nutrition. 27.

[64] Poudyal H, Panchal SK, Diwan V, Brown L (2011) Omega-3 fatty acids and metabolic syndrome: effects and emerging mechanisms of action. Prog Lipid Res. 50: 372-387.

[65] Kris-Etherton PM, Grieger JA, Etherton TD (2009) Dietary reference intakes for DHA and EPA. Prostaglandins Leukot Essent Fatty Acids. 81:99–104.

[66] Harris WS, Miller M, Tighe AP, Davidson MH, Schaefer EJ (2008) Omega-3 fatty acids and coronary heart disease risk: clinical and mechanistic perspectives. Atherosclerosis. 197:12–24.

[67] Egert S, Kannenberg F, Somoza V, Erbersdobler H,Wahrburg U (2009) Dietary alpha-linolenic acid, EPA, and DHA have differential effects on LDL fatty acid composition but similar effects on serum lipid profiles in normolipidemic humans. J Nutr. 139:861–8

[68] Geppert J, Kraft V, Demmelmair H, Koletzko B (2006) Microalgal docosahexaenoic acid decreases plasma triacylglycerol in normolipidaemic vegetarians: a randomized trial. Br J Nutr. 95: 779–86.

[69] Clarke SD, Jump D (1997) Polyunsaturated fatty acids regulate lipogenic and peroxisomal gene expression by independent mechanisms. Prostaglandins Leukot Essent Fatty Acids. 57:65–9.

[70] Nestel PJ (2000) Fish oil and cardiovascular disease: lipids and arterial function. Am J Clin Nutr. 71, 228–231.

[71] Harris WS, Lu G, Rambjor GS,Walen AI, Ontko JA, Cheng Q, Windsor SL (1997) Influence of n-3 fatty acid supplementation on the endogenous activities of plasma lipases. Am J Clin Nutr. 66:254–260.

[72] Maki KC, McKenney JM, Reeves MS, Lubin BC, Dicklin MR (2008) Effects of adding prescription omega-3 acid ethyl esters to simvastatin (20 mg/day) on lipids and lipoprotein particles in men and women with mixed dyslipidemia. Am J Cardiol.102:429–33.

[73] Neff LM, Culiner J, Cunningham-Rundles S, Seidman C, Meehan D, Maturi J, Wittkowski KM, et al (2011) Algal docosahexaenoic acid affects plasma lipoprotein particle size distribution in overweight and obese adults. J Nutr. 141: 207-213.

[74] Abbey M, Clifton P, Kestin M, Belling B & Nestel P (1990) Effect of fish oil on lipoproteins, lecithin:cholesterol acyltransferase, and lipid transfer protein activity in humans. Arteriosclerosis. 10: 85–94.

[75] Buckley R, Shewring B, Turner R, Yaqoob P, Minihane AM (2004) Circulating triacylglycerol and apoE levels in response to EPA and docosahexaenoic acid supplementation in adult human subjects. Br J Nutr. 92: 477–483.

[76] Theobald HE, Chowienczyk PJ,Whittall R, Humphries SE, Sanders TA (2004) LDL cholesterol-raising effect of low-dose docosahexaenoic acid in middle-aged men and women. Am J Clin Nutr. 79: 558–563.

[77] Conquer JA, Holub BJ (1998) Effect of supplementation with different doses of DHA on the levels of circulating DHA as non-esterified fatty acid in subjects of Asian Indian background. J Lipid Res. 39: 286–292.

[78] Nestel P, Shige H, Pomeroy M, Cehun M, Abbey M, Raederstorff D (2002) The n-3 fatty acids eicosapentaenoic acid and docosahexaenoic acid increase systemic arterial compliance in humans. Am J Clin Nutr 76: 326–330

[79] Griffin BA (2001) The effect of n-3 fatty acids on low density lipoprotein subfractions. Lipids. 36, 91–97.

[80] Lu G, Windsor SL, Harris WS (1999) Omega-3 fatty acids alter lipoprotein subfraction distributions and the in vitro conversion of very low density lipoproteins to low density lipoproteins. J Nutr Biochem. 10:151–158.

[81] Cunnane SC, Hamadeh MJ, Liede AC, Thompson LU, Wolever TM, Jenkins DJ (1995) Nutritional attributes of traditional flaxseed in healthy young adults. Am J Clin Nutr. 61:62–8.

[82] Lucas EA, Wild RD, Hammond LJ, Khalil DA, Juma S, Daggy BP, et al (2002) Flaxseed improves lipid profile without altering biomarkers of bone metabolism in postmenopausal women. J Clin Endocrinol Metab. 87:1527–1532.

[83] Putnam J, Gerrior S (1999) Trends in the U.S. food supply, 1970-97. In: America's eating habits: changes and consequences. Washington, D.C.: United States Department of Agriculture, Economic Research Service. 133-60.

[84] Grundy SM, Barrett-Connor E, Rudel LL, Miettinen T, Spector AA (1988) Workshop on the impact of dietary cholesterol on plasma lipoproteins and atherogenesis. Arteriosclerosis. 8:95-101

[85] National Research Council. Diet and health: implications for reducing chronic disease risk. Washington, D.C.: National Academy Press, 1989: 171-201

[86] Weggemans RM, Zock PL, Katan MB (2001) Dietary cholesterol from eggs increases the ratio of total cholesterol to high-density lipoprotein cholesterol in humans: a meta-analysis. Am J Clin Nutr. 73: 885-91.

[87] Schaefer EJ, Gleason Jam (2009) Dietary fructose and glucose differentially affect lipid and glucose homeostasis J Nutr, 139:1257S–1262S

[88] Barclay AW, Petocz P, McMillan-Price J, et al (2008) Glycemic index, glycemic load, and chronic disease risk—a meta—analysis of observational studies. Am J Clin Nutr. 87:627–637.

[89] Barclay AW, Petocz P, McMillan-Price J, et al (2008) Glycemic index, glycemic load, and chronic disease risk–a meta-analysis of observational studies. Am J Clin Nutr. 87:627–37.

[90] Pelkman CL (2001) Effects of the glycemic index of foods on serum concentrations of high-density lipoprotein cholesterol and triglycerides. Curr Atheroscler Rep. 3:456–461.

[91] Sacks FM, Katan M (2002) Randomized clinical trials on the effects of dietary fat and carbohydrate on plasma lipoproteins and cardiovascular disease. Am J Med.;113 Suppl 9 : 13–24.

[92] Aarsland A, Chinkes D, Wolfe RR (1997) Hepatic and whole-body fat synthesis in humans during carbohydrate overfeeding. Am J Clin Nutr. 65:1774–82

[93] McNeel RL, Mersmann HJ (2005) Low- and high-carbohydrate diets: body composition differences in rats. Obes Res. 13:1651–60.

[94] Volek JS, Phinney SD, Forsythe CE, Quann EE, Wood RJ, Puglisi MJ, et al (2009) Carbohydrate restriction has a more favorable impact on the metabolic syndrome than a low fat diet. Lipids. 44: 297-309.

[95] Krauss RM, Blanche PJ, Rawlings RS, et al (2006) Separate effects of reduced carbohydrate intake and weight loss on atherogenic dyslipidemia. Am J Clin Nutr. 83:1025–1031

[96] Sharman MJ, Kraemer WJ, Love DM, Avery NG, Gomez AL, Scheett TP, Volek JS (2002) A ketogenic diet favorably affects serum biomarkers for cardiovascular disease in normal-weight men. J Nutr. 132: 1879–1885.

[97] Turley ML, Skeaff CM, Mann JI, Cox B (1998) The effect of a low-fat, high-carbohydrate diet on serum high density lipoprotein cholesterol and triglyceride. Eur J Clin Nutr. 52:728-32.

[98] Vuksan V, Sievenpiper JL, Owen R, Swilley JA, Spadafora P, Jenkins DJA (2000) Beneficial effects of viscous dietary fiber from Konjac-mannan in subjects with the insulin resistance syndrome: results of a controlled metabolic trial. Diabetes Care.23:9-14.

[99] Grundy SM, Florentin L, Nix D, et al. (1988) Comparison of monounsaturated fatty acids and carbohydrates for reducing raised levels of plasma cholesterol in man. Am J Clin Nutr 47: 965–969.

[100] Jiménez-Gómez Y, Marín C, Peérez-Martínez P, Hartwich J, Malczewska-Malec M (2010) A low-fat, high-complex carbohydrate diet supplemented with long-chain (n-3) fatty acids alters the postprandial lipoprotein profile in patients with metabolic syndrome. J Nutr. 140: 1595-1601.

[101] Basciano H, Federico L, Adeli K (2005) Fructose, insulin resistance, and metabolic dyslipidemia. Nutr Metab (Lond). 2: 5.

[102] Busserolles J, Zimowska W, Rock E, Rayssiguier Y, Mazur A (2002) ts fed a high sucrose diet have altered heart antioxidant enzyme activity and gene expression Life Sci. 71: 1303-1312.

[103] Girard A, Madani S, Boukortt F, Cherkaoui-Malki M, Belleville J, Prost J (2006) Fructose-enriched diet modifies antioxidant status and lipid metabolism in spontaneously hypertensive rats. Nutrition. 22: 758-766.

[104] Elliott SS, Keim NL, Stern JS, Teff K, Havel PJ (2002) Fructose, weight gain, and the insulin resistance syndrome. Am J Clin Nutr.76: 911-922.

[105] Rukmini C, Raghurm TC (1991) Nutritional and biochemical aspects of the hypolipidemic action of rice bran oil: a review. J Am Coll Nutr. 10: 593–601.

[106] Turnbull WH, Leeds AR, Edwards DG (1992) Mycoprotein reduces blood lipids in free-living subjects. Am J Clin Nutr. 55: 415–419.

[107] Erdman JW Jr (2000) AHA Science Advisory: soy protein and cardiovascular disease: a statement for healthcare professionals from the Nutrition Committee of the AHA. Circulation. 102: 2555–2559.

[108] Sacks FM, Lichtenstein A, Van Horn L, Harris W, Kris-Etherton P, Winston M (2006) Soy protein, isoflavones, and cardiovascular health: an American Heart Association Science Advisory for professionals from the Nutrition Committee. Circulation. 113: 1034–1044.

[109] Jenkins DJ, Mirrahimi A, Srichaikul K, Berryman CE, Wang L, Carleton A, et al (2010) Soy protein reduces serum cholesterol by both intrinsic and food displacement mechanisms. J Nutr. 140: 2302-2311.

[110] Lovati MR, Manzoni C, Gianazza E, Arnoldi A, Kurowska E, Carroll KK, Sirtori CR (2000) Soy protein peptides regulate cholesterol homeostasis in Hep G2 cells. J Nutr. 130: 2543–2549.

[111] Zhuo XG, Melby MK, Watanabe S (2004) Soy isoflavone intake lowers serum LDL cholesterol: a meta-analysis of 8 randomized controlled trials in humans. J Nutr. 134: 2395–2400.

[112] Wiseman H, O'Reilly JD, Adlercreutz H, Mallet AI, Bowey EA, Rowland IR, et al (2000) Isoflavone phytoestrogens consumed in soy decrease F(2)-isoprostane concentrations and increase resistance of low-density lipoprotein to oxidation in humans. Am J Clin Nutr. 72: 395–400

[113] Jenkins DJ, Kendall CW, Vidgen E, Augustin LS, van Erk M, Geelen A, et al (2001) High-protein diets in hyperlipidemia: effect of wheat gluten on serum lipids, uric acid, and renal function. Am J Clin Nutr. 74:57–63.

[114] United States Department of Agriculture, United States Department of Health and Human Services: Dietary Guidelines for Americans, 6th edn. Washington, DC: Government Printing Office; 2005.

[115] United States Department of Agriculture, United States Department of Health and Human Services: Report of the Dietary Guidelines Advisory Committee on the Dietary Guidelines for Americans. Washington, DC: Government Printing Office; 2010

[116] Lichtenstein AH, Appel LJ, Brands M, Carnethon M, Daniels S, Franch HA, et al (2006) Summary of American Heart Association Diet and Lifestyle Recommendations revision 2006. Arterioscler Thromb Vasc Biol. 26:2186–2191

[117] American Diabetes Association (2007) Nutrition recommendations and interventions for diabetes: a position statement of the American Diabetes Association. Diabetes Care 30: 48–65.

[118] Harris KA, Kris-Etherton PM (2010) Effects of whole grains on coronary heart disease risk. Curr Atheroscler Rep. 12: 368-376.

[119] Koh-Banerjee P, Rimm EB (2003) Whole grain consumption and weight gain: a review of the epidemiological evidence, potential mechanisms and opportunities for future research. Proc Nutr Soc. 62: 25–29

[120] Kashtan H, Stern HS, Jenkins DJ, Jenkins AL, Hay K, Marcon N, et al (1992) Wheat-bran and oat-bran supplements' effects on blood lipids and lipoproteins. Am J Clin Nutr.55: 976-980.

[121] Shimizu C, Kihara M, Aoe S, Araki S, Ito K, Hayashi K, et al (2008) Effect of high beta-glucan barley on serum cholesterol concentrations and visceral fat area in Japanese men—a randomized, double-blinded, placebo-controlled trial. Plant Foods Hum Nutr. 63: 21–25

[122] Kelly SA, Summerbell CD, Brynes A, Whittaker V, Frost G (2007) Wholegrain cereals for coronary heart disease. Cochrane Database Syst Rev, 2:CD005051.

[123] US FDA. Food labeling: health claims: soluble fiber from whole oats and risk of coronary heart disease. Docket 95P–0197. Washington, DC: US FDA; 2001. p. 15343–4.

[124] Brown L, Rosner B, Willett WW, Sacks FM (1999) Cholesterol-lowering effects of dietary fiber: a meta-analysis. Am J Clin Nutr. 69:30–42.

[125] Papathanasopoulos A, Camilleri M (2010) Dietary fiber supplements: effects in obesity and metabolic syndrome and relationship to gastrointestinal functions. Gastroenterology. 138:65–72.

[126] National Heart Lung and Blood Institute: Third Report of the Expert Panel on Detection, Evaluation, and Treatment of High Blood Cholesterol in Adults (Adult Treatment Panel III). Edited by National Cholesterol Education Program. Bethesda, MD: National Institutes of Health; 2002.

[127] Dragsted LO, Krath B, Ravn-Haren G, Vogel UB, Vinggaard AM, Bo Jensen P, et al (2006) Biological effects of fruit and vegetables. Proc Nutr Soc. ;65: 61-7.

[128] Jenkins DJ, Kendall CW, Popovich DG, Vidgen E, Mehling CC, Vuksan V, et al (2001) Effect of a very-high-fiber vegetable, fruit, and nut diet on serum lipids and colonic function. Metabolism. 50:494–503

[129] Djousse L, Arnett DK, Coon H, Province MA, Moore LL, Ellison RC (2004) Fruit and vegetable consumption and LDL cholesterol: the national heart, lung, and blood institute family heart study. Am J Clin Nutr.79: 213-217

[130] Ballesteros MN, Cabrera RM, Saucedo MS, Yepiz-Plascencia GM, Ortega MI, Valencia ME (2001) Dietary fiber and lifestyle influence serum lipids in free living adult men. J Am Coll Nutr. 20:649–655.

[131] Stone NJ (2001) Lowering low-density cholesterol with diet: the important role of functional foods as adjuncts. Coron Artery Dis. 12:547–552.

[132] Fornes NS, Martins IS, Hernan M, Velasquez-Melendez G, Ascherio A (2000) Frequency of food consumption and lipoprotein serum levels in the population of an urban area, Brazil. Rev Saude Publica. 34: 380–387

[133] Pereira MA, O'Reilly E, Augustsson K, Fraser GE, Goldbourt U, Heitmann BL, Hallmans G, et al (2004) Dietary fiber and risk of coronary heart disease: a pooled analysis of cohort studies. Arch Intern Med. 164: 370–376.

[134] Kritchevsky D, Story JA (1974) Binding of bile salts in vitro by nonnutritive fiber. J Nutr. 104:458-462.

[135] Everson GT, Daggy BP, McKinley C, Story JA (1992) Effects of psyllium hydrophilic mucilloid on LDL-cholesterol and bile acid synthesis in hypercholesterolemic men. J Lipid Res. 33:1183-1192.

[136] Sjogren P, Rosell M, Skoglund-Andersson C, Zdravkovic S, Vessby B, de Faire U, et al (2004) Milkderived fatty acids are associated with a more favorable LDL particle size distribution in healthy men. J Nutr. 134: 1729– 1735

[137] Pereira MA, Jacobs DR Jr, Van Horn L, Slattery ML, Kartashov AI, Ludwig DS (2002) Dairy consumption, obesity, and the insulin resistance syndrome in young adults: the CARDIA Study. JAMA. 287: 2081–2089.

[138] Shahkhalili Y, Murset C, Meirim I, Duruz E, Guinchard S, Cavadini C, et al (2001) Calcium supplementation of chocolate: effect on cocoa butter digestibility and blood lipids in humans. Am J Clin Nutr. 73: 246–252.

[139] Pfeuffer M, Schrezenmeir J (2007) Milk and the metabolic syndrome. Obes Rev.8:109 – 118.

[140] Takeuchi T, Shimizu H, Ando K, Harada E (2004) Bovine lactoferrin reduces plasma triacylglycerol and NEFA accompanied by decreased hepatic cholesterol and triacylglycerol contents in rodents. Br J Nutr. 91: 533–538.

[141] Nagaoka S, Futamura Y, Miwa K, Awano T, Yamauchi K, Kanamaru Y, et al (2001) Identification of novel hypocholesterolemic peptides derived from bovine milk beta lactoglobulin. Biochem Biophys Res Commun. 281: 11–17.

[142] Shah, H. (2000) Effects of milk-derived bioactives: an overview. Br. J. Nutr. 84: 3–10.

[143] Siri-Tarino PW, Sun Q, Hu FB, Krauss RM (2010) Meta-analysis of prospective cohort studies evaluating the association of saturated fat with cardiovascular disease. Am J Clin Nutr. 91: 535-46

[144] Samuelson G, Bratteby LE, Mohsen R, Vessby B (2001) Dietary intake in healthy adolescents: inverse relationship between the estimated intake of saturated fatty acids and serum cholesterol. Br J Nutr. 85: 333–341.

[145] Jakobsen MU, O'Reilly EJ, Heitmann BL, Pereira MA, Bälter K, Fraser GE, et al (2009) Major types of dietary fat and risk of coronary heart disease: a pooled analysis of 11 cohort studies. Am J Clin Nutr. 89:1425–1432.

[146] Nestel PJ, Chronopulos A, Cehun M (2005) dairy fat in cheese raises LDL cholesterol less than that in butter in mildly hypercholesterolaemic subjects. Eur J Clin Nutr.59: 1059-1063.

[147] Biong AS, Müller H, Seljeflot I, Veierød MB, Pedersen JI, et al (2004) A comparison of the effects of cheese and butter on serum lipids, haemostatic variables and homocysteine. Br J Nutr. 92: 791–797

[148] Matthan NR, Welty FK, Barrett PH, Harausz C, Dolnikowski GG, Parks JS, et al (2004) Dietary hydrogenated fat increases high-density lipoprotein apoA-I catabolism and decreases low-density lipoprotein apoB-100 catabolism in hypercholesterolemic women. Arterioscler Thromb Vasc Biol. 24:1092–1097

[149] Pfeuffer M, Schrezenmeir J (2000) Bioactive substances in milk with properties decreasing risk of cardiovascular diseases. Br J Nutr. 84: 155–159

[150] Agerhol-Larsen L, Bell ML, Grunwald GK, Astrup A (2000) The effect of a probiotic milk product on plasma cholesterol: a metaanalysis of short-term intervention studies. Eur J Clin Nutr. 54: 856–860.

[151] Xiao JZ, Kondo S, Takahashi N, Miyaji K, Oshida K, Hiramatsu A, et al (2003) Effects of milk products fermented by Bifidobacterium longum on blood lipids in rats and healthy adult male volunteers. J Dairy Sci. 86: 2452– 2461

[152] Agerholm-Larsen L, Bell ML, Grunwald GK, Astrup A (2000) The effect of probiotic milk product on plasma cholesterol: a metaanalysis of short-term intervention studies. Eur. J. Clin. Nutr. 54: 856–860.

[153] Expert Panel on Detection, Evaluation, and Treatment of High Blood Cholesterol in Adults. Executive Summary of the Third Report of the National Cholesterol Education Program (NCEP) Expert Panel on Detection, Evaluation, and Treatment of High Blood Cholesterol in Adults (Adult Treatment Panel III). JAMA. 2001;285:2486–2497
[154] Fraser GE (1999) Nut consumption, lipids, and risk of a coronary event. Clin Cardiol. 22:1–5.
[155] Kris-Etherton PM, Pearson TA, Wan Y, Hargrove RL, Moriarty K, Fishell V, et al (1999) High-monounsaturated fatty acid diets lower both plasma cholesterol and triglyceride concentrations. Am J Clin Nutr. 70:1009–1015
[156] Abbey M, Noakes M, Belling GB, Nestel PJ (1994) Partial replacement of saturated fatty acids with almonds or walnuts lowers total plasma cholesterol and low-density-lipoprotein cholesterol. Am J Clin Nutr.. 59: 995–999
[157] Morgan WA, Clayshulte BJ (2000) Pecans lower low-density lipoprotein cholesterol in people with normal lipid levels. J Am Diet Assoc. 100: 312–318.
[158] Zambón D, Sabaté J, Muñoz S, Campero B, Casals E, Merlos M, et al (2000) Substituting walnuts for monounsaturated fat improves the serum lipid profile of hypercholesterolemic men and women: a randomized crossover trial. Ann Intern Med. 132: 538–546
[159] Sabaté J, Fraser GE, Burke K, Knutsen SF, Bennett H, Lindsted KD (1993) Effects of walnuts on serum lipid levels and blood pressure in normal men. N Engl J Med. 328:603- 607.
[160] Edwards K, Kwaw I, Matud J, Kurtz I (1999) Effect of pistachio nuts on serum lipid levels in patients with moderate hypercholesterolemia. J Am Coll Nutr.18:229–232
[161] O'Byrne DJ, Knauft DA, Shireman RB (1997) Low fat-monounsaturated rich diets containing high-oleic peanuts improve serum lipoprotein profiles. Lipids. 32:687–695
[162] Schaefer EJ, Lichtenstein AH, Lamon-Fava S, Contois JH, Li Z, Rasmussen H, McNamara JR, et al (1995) Efficacy of a National Cholesterol Education Program Step 2 diet in normolipidemic and hypercholesterolemic middle-aged and elderly men and women. Arterioscler Thromb Vasc Biol. 15:1079–1085.
[163] Reaven P, Parthasarathy S, Grasse BJ, Miller E, Steinberg D, Witztum JL (1993) Effects of oleate-rich and linoleate-rich diets on the susceptibility of low density lipoprotein to oxidative modification in mildly hypercholesterolemic subjects. J Clin Invest. 91:668–676
[164] Spiller, G. A., Jenkins, D.A.J., Bosello, O., Gates, J. E., Cragen, L. N., Bruce B (1998) Nuts and plasma lipids: an almond-based diet lowers LDL-C while preserving HDL-C. J. Am. Coll. Nutr. 17: 285–290.
[165] Hyson DA, Schneeman BO, Davis PA (2002) Almonds and almond oil have similar effects on plasma lipids and LDL oxidation in healthy men and women. Almonds and almond oil have similar effects on plasma lipids and LDL oxidation in healthy men and women. J Nutr. 132: 703-707
[166] Garg ML, Blake RJ, Wills RB (2003) Macadamia nut consumption lowers plasma total and LDL cholesterol levels in hypercholesterolemic men. J Nutr. 133:1060-1063.

[167] Li L, Tsao R, Yang R, Kramer JK, Hernandez M (2007) Fatty acid profiles, tocopherol contents, and antioxidant activities of heartnut (Juglans ailanthifolia var. cordiformis) and Persian walnut (Juglans regia L.). J Agric Food Chem. 55:1164-1169.

[168] Banel DK, Hu FB (2009) Effects of walnut consumption on blood lipids and other cardiovascular risk factors: a meta-analysis and systematic review. Am J Clin Nutr. 90: 56-63

[169] Bazzano LA, Thompson AM, Tees MT, Nguyen CH, Winham DM (2011) Non-soy legume consumption lowers cholesterol levels: a meta-analysis of randomized controlled trials. Nutr Metab Cardiovasc Dis. 21: 94-103.

[170] Anderson JW, Major AW (2002) Pulses and lipaemia, short- and long term effect: potential in the prevention of cardiovascular disease. Br J Nutr. 88:263-271

[171] Duranti M (2006) Grain legume proteins and nutraceutical properties. Fitoterapia. 77: 67-82

[172] Mallillin AC, Trinidad TP, Raterta R, Dagbay K, Loyola AS (2008) Dietary fiber and fermentability characteristics of root crops and legumes. Br J Nutr. 100: 485–488.

[173] Roberfroid M (1997) Health benefits of non-digestible oligosaccharides. In Dietary Fiber in Health and Disease (Advances in Experimental Biology), p. 427 [D Kritchevsky and C Bonfield, editors]. New York: Plenum Press.

[174] Duane WC (1997) Effects of legume consumption on serum cholesterol, biliary lipids, and sterol metabolism in humans. J Lipid Res 38: 1120–1128

[175] Galisteo M, Duarte J, Zarzuelo A (2008) Effects of dietary fibers on disturbances clustered in the metabolic syndrome. J Nutr Biochem. 19: 71-84

[176] Rochfort S, Panozzo J (2007) Phytochemicals for health, the role of pulses. J Agric Food Chem. 55:7981-7994

[177] Denke M (1994) Role of beef tallow, an enriched source of stearic acid, in a cholesterol lowering diet. Am J Clin Nutr. 60: 1044S-1049S

[178] Scott LW, Dunn JK, Pownall HJ, Brauchi DJ, McMann MC, Herd JA, et al (1994) Effects of beef and chicken consumption on plasma lipid levels in hypercholesterolemic men. Arch Intern Med. 154: 1261-1267.

[179] Wolmarans P, Benadé AJS, Kotze TJvW, Daubitzer AK, Marais MP, Laubscher R (1991) Plasma lipoprotein response to substituting fish for red meat in the diet. Am J Clin Nutr. 53:1171–1176

[180] Davidson MH, Hunninghake D, Maki KC, Kwiterovich PO, Kafonek S (1999) Comparison of the effects of lean red meat vs lean white meat on serum lipid levels among free-living persons with hypercholesterolemia. Arch Intern Med. 159:1331–1338

[181] Lankinen M, Schwab U, Erkkilä A, Seppänen-Laakso T, Hannila ML, Mussalo H, et al (1991) Effects of a lean beef diet and of a chicken and fish diet on lipoprotein profiles. Nutr Metab Cardiovasc Dis. 1: 25-30.

[182] Morgan S, Sinclair A, O'Dea K (1993) Effect on serum lipids of addition of safflower oil or olive oil to very-low-fat diets rich in lean beef. J Am Diet Assoc. 93: 644- 648.

[183] Grundy SM. Cholesterol and atherosclerosis. Diagnosis and treatment. New York: Gower Medical Publishing, 1990

[184] Lacaille B, Julien P, Deshaies Y, Lavigne C, Brun L-D, Jacques H (2000) Responses of plasma lipoproteins and sex hormones to the consumption of lean fish incorporated in a prudent-type diet in normolipidemic men. J Am Coll Nutr. 19: 745–753

[185] Gascon A, Jacques H, Moorjani S, Deshaies Y, Brun L-D, Julien P (1996) Plasma lipoprotein profile and lipolytic activities in response to the substitution of lean white fish for other animal protein sources in premenopausal women. Am J Clin Nutr. 63: 315–321

[186] Howell WH, McNamara DJ, Tosca MA, Smith BT, Gaines JA (1997) Plasma lipid and lipoprotein responses to dietary fat and cholesterol: a meta-analysis. Am J Clin Nutr.65:1747–1764.

[187] Dawber TR, Nickerson RJ, Brand FN, Pool J (1982) Eggs, serum cholesterol, and coronary heart disease. Am J Clin Nutr. 36:617–625

[188] Hu FB, Stampfer MJ, Rimm EB, Manson JE, Ascherio A, Colditz GA, et al (1999) A prospective study of egg consumption and risk of cardiovascular disease in men and women. JAMA. 281:1387–1394.

[189] Mayurasakorn K, Srisura W, Sitphahul P, Hongto PO (2008) High-density lipoprotein cholesterol changes after continuous egg consumption in healthy adults. J Med Assoc Thai. Mar. 91: 400-407.

[190] Mutungi G, Ratliff J, Puglisi M, Torres-Gonzalez M, Vaishnav U, Leite JO (2008) Dietary cholesterol from eggs increases plasma HDL cholesterol in overweight men consuming a carbohydrate-restricted diet. J Nutr. 138:272-176

[191] Fernandez ML (2006) Dietary cholesterol provided by eggs and plasma lipoproteins in healthy populations. Curr Opin Clin Nutr Metab Care. 9: 8-12.

[192] Harsha DW, Sacks FM, Obarzanek E, Svetkey LP, Lin PH, Bray GA (2004) Effect of dietary sodium intake on blood lipids: results from the DASH-sodium trial. Hypertension. 43: 393-398.

[193] Miller ER 3rd, Erlinger TP, Appel LJ (2006) The effects of macronutrients on blood pressure and lipids: an overview of the DASH and OmniHeart trials. Curr Atheroscler Rep. 8: 460-465.

[194] Obarzanek E, Sacks FM, Vollmer WM, Bray GA, Miller ER, Lin PH (2001) Effects on blood lipids of a blood pressure lowering diet: the Dietary Approaches to Stop Hypertension (DASH) Trial. Am J Clin Nutr. 74:80–89.

[195] Willett WC, Sacks F, Trichopoulou A, Drescher G, Ferro-Luzzi A, Helsing E, et al (1995) Mediterranean diet pyramid: a cultural model for healthy eating. Am J Clin Nutr. 6: 1402-1406

[196] Tzima N, Pitsavos C, Panagiotakos DB, Skoumas J, Zampelas A, Chrysohoou C, et al (2007) Mediterranean diet and insulin sensitivity, lipid profile and blood pressure levels, in overweight and obese people; the Attica study. Lipids Health Dis. 6:22.

[197] Salen P, de Lorgeril M (1999) [Hyperlipidemias. Concern with the Mediterranean diet]. Presse Med.28:2018-2024.

[198] Demarin V, Lisak M, Morović S (2011) Mediterranean diet in healthy lifestyle and prevention of stroke. Acta Clin Croat. 50: 67-77.

[199] Willett WC. Public Health Nutr 2006;9:105

[200] Mekki K, Bouzidi-bekada N, Kaddous A, Bouchenak M (2010) Mediterranean diet improves dyslipidemia and biomarkers in chronic renal failure patients. Food Funct.1: 10-5.

[201] Pitsavos, C., Panagiotakos, D.B., Tzima, N., Chrysohoou, C., Economou, M., Zampelas, A, et al (2005) Adherence to the Mediterranean diet is associated with total antioxidant capacity in healthy adults: the ATTICA study. Am. J. Clin. Nutr. 82: 694–699.

[202] Bouchard-Mercier A, Paradis AM, Godin G, Lamarche B, Pérusse L, Vohl MC (2010) Associations between dietary patterns and LDL peak particle diameter: a cross-sectional study. J Am Coll Nutr.29: 630-67.

[203] Ganguli D, Das N, Saha I, Biswas P, Datta S, Mukhopadhyay B (2011) Major dietary patterns and their associations with cardiovascular risk factors among women in West Bengal, India. Br J Nutr. 105:1520-1529

[204] Lim JH, Lee YS, Chang HC, Moon MK, Song Y (2011) Association between dietary patterns and blood lipid profiles in Korean adults with type 2 diabetes. J Korean Med Sci. 26: 1201-1208.

[205] Ambrosini GL, Huang RC, Mori TA, Hands BP, O'Sullivan TA, de Klerk NH, et al (2010) Dietary patterns and markers for the metabolic syndrome in Australian adolescents. Nutr Metab Cardiovasc Dis.20:274-283.

[206] FDA Talk Paper. FDA authorizes new coronary heart disease health claim for plant sterol and plant stanol esters. 5 September 2000. Washington (DC). Available from: http://www.fda.gov/Food/Labeling.Nutrition/LabelClaims/HealthClaimsMeetingSignificantScientificAgreementSSA/ucm074747.htm

[207] Manach C, Scalbert A, Morand C, Remesy C, Jimenez L (2004) Polyphenols: food sources and bioavailability. Am J Clin Nutr. 79:727–747

[208] Otaki N, Kimira M, Katsumata S, Uehara M, Watanabe S, Suzuki K (2009) Distribution and major sources of flavonoid intakes in the middle-aged Japanese women. J Clin Biochem Nutr. 44:231-238.

[209] Higgins JPT, Green S, eds. Highly sensitive search strategies for identifying reports of randomized controlled trials in MEDLINE. Cochrane handbook for systematic reviews of interventions 4.2.5 (updated May 2005); Appendix 5b. The Cochrane Library, Issue 3, 2005. Chichester, UK: Wiley & Sons, Ltd. 2005.

[210] Frankel, E.N., Kanner, J., German, J.B., Parks, E., Kinsella, J.E (1993) Inhibition of oxidation of human low-density lipoprotein by phenolic substances in red wine. Lancet. 341:454–457.

[211] Otaki N, Kimira M, Katsumata S, Uehara M, Watanabe S, Suzuki K (2009) Distribution and major sources of flavonoid intakes in the middle-aged Japanese women. J Clin Biochem Nutr. 44: 231-8.

[212] Odbayar TO, Badamhand D, Kimura T, Takashi Y, Tsushida T, Ide T (2006) Comparative studies of some phenolic compounds quercetin, rutin, and ferulic acid) affecting hepatic fatty acid synthesis in mice. J Agric Food Chem. 54:8261–8265

[213] Princen HM, van Duyvenvoorde W, Buytenhek R, Blonk C, Tijburg LB, Langius JA, et al (1998) No effect of consumption of green and black tea on plasma lipid and

antioxidant levels and on LDL oxidation in smokers. Arterioscler Thromb Vasc Biol. 18:833–841

[214] Bingham, S. A., Vorster, H., Jerling, J. C., Magee, E., Mulligan, A., Runswick, S. A., Cummings, J. H. (1997) Effect of black tea drinking on blood lipids, blood pressure and aspects of bowel habit. Brit. J. Nutr. 78: 41–55

[215] van het Hof KH, de Boer HS, Wiseman SA, Lien N, Westrate JA, Tijburg LB (1997) Consumption of green or black tea does not increase resistance of low-density lipoprotein to oxidation in humans. Am J Clin Nutr. 66:1125-1132

[216] McAnlis GT, McEneny J, Pearce J, Young IS (1998) Black tea consumption does not protect low density lipoprotein from oxidative modification. Eur J Clin Nutr. 52:202-206

[217] Duffy SJ, Vita JA, Holbrook M, Swerdloff PL, Keaney JF (2001) Effect of acute and chronic tea consumption on platelet aggregation in patients with coronary artery disease. Arterioscler Thromb Vasc Biol. 21: 1084-1089

[218] Maron DJ, Lu GP, Cai NS, Wu ZG, Li YH, Chen H, Zhu JQ, et al (2003) Cholesterol-lowering effect of a theaflavin-enriched green tea extract: a randomized controlled trial. Arch Intern Med. 163:1448-1453.

[219] Tokunaga S, White IR, Frost C, Tanaka K, Kono S, Tokudome S (2002) Green tea consumption and serum lipids and lipoproteins in a population of healthy workers in Japan. Ann Epidemiol.12: 157-165.

[220] Kris-Etherton PM, Keen CL (2002) Evidence that the antioxidant flavonoids in tea and cocoa are beneficial for cardiovascular health. Curr Opin Lipidol. 13:41-49.

[221] Ikeda I, Imasato Y, Sasaki E, Nakayama M, Nagao H, Takeo T, et al (1992) Tea catechins decrease micellar solubility and intestinal absorption of cholesterol in rats. Biochim Biophys Acta. 1127:141– 146.

[222] Yang TT, Koo MW (2000) Chinese green tea lowers cholesterol level through an increase in fecal lipid excretion. Life Sci. 66:411-423

[223] Chan PT, Fong WP, Cheung YL, Huang Y, Ho WK, Chen ZY (1999) Jasmine green tea epicatechins are hypolipidemic in hamsters (Mesocricetus auratus) fed a high fat diet. J Nutr. 129:1094-1101.

[224] Bursill C, Roach PD, Bottema CD, Pal S (2001) Green tea upregulates the low-density lipoprotein receptor through the sterol-regulated element binding protein in HepG2 liver cells. J Agric Food Chem. 49: 5639-5645

[225] Matsumoto N, Okushio K, Hara Y (1998) Effect of black tea polyphenols on plasma lipids in cholesterol-fed rats. J Nutr Sci Vitaminol. 44:337-342

[226] Crews WD Jr, Harrison DW, Wright JW (2008) A double-blind, placebocontrolled, randomized trial of the effects of dark chocolate and cocoa on variables associated with neuropsychological functioning and cardiovascular health: Clinical findings from a sample of healthy, cognitively intact older adults. Am J Clin Nutr. 87:872–880.

[227] Farouque HM, Leung M, Hope SA, Baldi M, Schechter C, Cameron JD, Meredith IT (2006) Acute and chronic effects of flavanol-rich cocoa on vascular function in subjects with coronary artery disease: a randomized double-blind placebo-controlled study. Clin Sci. 111: 71–80.

[228] Lecumberri E, Goya L, Mateos R, Alía M, Ramos S, Izquierdo-Pulido M, et al (2007) A diet rich in dietary fiber from cocoa improves lipid profile and reduces malondialdehyde in hypercholesterolemic rats. Nutrition. 23:332–41.

[229] Rein D, Lotito S, Holt RR, Keen CL, Schmitz HH, Fraga CG (2000) Epicatechin in human plasma. In vivo determination and effect of chocolate consumption on plasma oxidation status. J Nutr. 130: 2109–2114.

[230] Mathur S, Devaraj S, Grundy SM, Jialal I (2002) Cocoa products decrease low density lipoprotein oxidative susceptibility but do not affect biomarkers of inflammation in humans. J Nutr. 132:3663-3667.

[231] Jia L, Liu X, Bai YY, Li SH, Sun K, He C, et al (2010) Short-term effect of cocoa product consumption on lipid profile: a meta-analysis of randomized controlled trials. Am J Clin Nutr. 92:218–225.

[232] Keen CL, Holt RR, Oteiza PI, Fraga CG, Schmitz HH (2005) Cocoa antioxidants and cardiovascular health. Am J Clin Nutr. 81:298-303.

[233] Schmidt M, Schmitz HJ, Baumgart A, Guédon D, Netsch MI, Kreuter MH, Schmidlin CB, et al (2005) Toxicity of green tea extracts and their constituents in rat hepatocytes in primary culture. Food Chem Toxicol. 43: 307-14.

[234] Vinson JA, Proch J, Bose P, Muchler S, Taffera P, Shuta D, et al (2006) Chocolate is a powerful ex vivo and in vivo antioxidant, an antiatherosclerotic agent in an animal model, and a significant contributor to antioxidants in the European and American Diets. J Agric Food Chem. 54:8071–6.

[235] Basch E, Ulbricht C, Kuo G, Szapary P, Smith M (2003) Therapeutic applications of fenugreek. Altern Med Rev. 8:20–27.

[236] Handa T, Yamaguchi K, Sono Y, Yazawa K (2005) Effects of fenugreek seed extract in obese mice fed a high-fat diet. Biosci Biotechnol Biochem. 69: 1186–1188.

[237] Gupta R, Nair S (1999) Antioxidant flavonoids in common Indian diet. South Asian J Prev Cardio. 3:83-94.

[238] Petit PR, Sauvaire YD, Hillaire-Buys DM, Leconte OM, Baissac YG, Ponsin GR (1995) Steroid saponins from fenugreek seeds: extraction, purification, and pharmacological investigation on feeding behavior and plasma cholesterol. Steroids. 60:674-80

[239] Broca C, Breil V, Cruciani-Guglielmacci C, Manteghetti M, Rouault C, Derouet M, et al (2004) Insulinotropic agent ID-1101 (4-hydroxyisoleucine) activates insulin signaling in rat. Am J Physiol Endocrinol Metab. 287:463-71

[240] Dixit PP, Misar A, Mujumdar AM, Ghaskadbi S (2010) Pre-treatment of Syndrex protects mice from becoming diabetic after streptozotocin injection. Fitoterapia. 81(5):403-12

[241] Udayasekhara Rao P, Sesikeran B, Srinivasa Rao P, Nadamnui A, Vikas Rao V, Ramachandra RP (1996) Short term Nutritional and safety evaluation of Fenugreek. Nutr Res 16(9): 1495–1505

[242] Praveen KB, Dasgupta DJ, Prashar BS, Kaushal SS (1987) Preliminary Report: Effective reduction of LDL cholesterol by indigenous plant products. Current Science. 56(12): 80–81

[243] Saibaba A, Raghuram TC (1997) Fenugreek – The wonder Seed. Nutrition 31(2): 21–25.

[244] Neeraja A, Rajyalakshmi P (1996) Hypoglycemic effect of processed fenugreek seeds in humans. J Food Sci Technol. 33: 427–430.
[245] Muwalla MM, Abuirmmeileh NM (1991) Suppression of avian hepatic cholesterogenesis by dietary ginseng. J Nutr Biochem. 1:518– 521
[246] Zheng X, Yan Y (1991) The effect of ginsenosides of ginseng stem and leaf (GSL) on the lipid regulation and lipid peroxidation in chronic hyperlipidemic rabbits. Zhongguo Yaolixue Tonbao. 7:110–116
[247] Kim SH, Park KS (2003) Effects of Panax ginseng extract on lipid metabolism in humans. Pharmacol Res. 48:511-513.
[248] Kim SH, Park KS (2003) Effects of Panax ginseng extract on lipid metabolism in humans. Pharmacol Res. 48:511–513.
[249] Yamamoto M, Kumagai A (1984) Long term ginseng effects on hyperlipidemia in man with further study of its actions on atherogenesis and fatty liver rats. In: Proceedings of the 4th International Ginseng Symposium at Korea Ginseng & Tobaco Research Institute. p. 13–20.
[250] Joo CN. The preventive effect of Korean ginseng saponins on aortic atheroma formation in prolonged cholesterol fed rabbits. In: Proceedings of the 3rd International Ginseng Symposium at Korea Ginseng & Tobaco Research Institute; 1980. p. 27–36
[251] Kang DG, Yun YG, Ryoo JH, Lee HS (2002) Anti-hypertensive effect of water extract of danshen on renovascular hypertension through inhibition of the renin angiotensin system. Am J Chin Med. 30:87–93
[252] Yokozawa T, Kobayashi T, Kawai A, Oura H, Kawashima Y (1985) Hyperlipidemia-improving effects of ginseoside-Rb2in cholesterolfed rats. Chem Pharm Bull. 33:722–729.
[253] Langner E, Greifenberg S, Gruenwald J (1998) Ginger: history and use. Adv Ther 15:25–44
[254] Gruenwald J, Brendler T, Jaenicke C (2000) PDR for herbal medicines, 2nd edn. Medical Economics Company, Inc, Montvale, NJ
[255] Bhandari U, Kanojiah R, Pillai KK (2005) Effect of ethanol extract of Zingiber officinale on dyslipidaemia in diabetic rats. J Ethnopharmacol 97: 227–230
[256] Mary JM, John PK (2000) Agents used in hyperlipidaemia. In: Katzung BG (ed) Basic and clinical pharmacology, 8th edn. McGraw Hill Comp, New York, pp 581–595
[257] Afshari AT, Shirpoor A, Farshid A, Saadatian R, Rasmi Y, Saboory E et al (2007) The effect of ginger on diabetic nephropathy, plasma antioxidant capacity and lipid peroxidation in rats. Food Chem 101:148–153
[258] Belinky PA, Aviram M, Fuhrman B, Rosenblat M, Vaya J (1998) The antioxidative effects of the isoflavan glabridin on endogenous constituents of LDL during its oxidation. Atherosclerosis. 137:49–61
[259] Demizu S, Kajiyama K, Takahashi K, Hiraga Y, Yamamoto S, Tamura Y, et al (1988) Antioxidant and antimicrobial constituents of licorice: isolation and structure elucidation of a new benzofuran derivative. Chem Pharm Bull (Tokyo). 36: 3474-3479.

[260] Belinky PA, Aviram M, Fuhrman B, Rosenblat M, Vaya J (1998) The antioxidative effects of the isoflavan glabridin on endogenous constituents of LDL during its oxidation. Atherosclerosis. 137: 49

[261] Fuhrman B, Volkova N, Kaplan M, Presser D, Attias J, Hayek T, et al (2002) Antiatherosclerotic effects of licorice extract supplementation on hypercholesterolemic patients: increased resistance of LDL to atherogenic modifications, reduced plasma lipid levels, and decreased systolic blood pressure. Nutrition. 18: 268-273.

Chapter 14

Cyclosporin A-Induced Hyperlipidemia

Maaike Kockx and Leonard Kritharides

Additional information is available at the end of the chapter

http://dx.doi.org/10.5772/47866

1. Introduction

Cyclosporin A (CsA) is an immunosuppressant drug widely used in organ transplant recipients and patients with auto-immune disorders. Long-term treatment with CsA is associated with hyperlipidemia and an increased risk of atherosclerosis. The mechanisms by which cyclosporin A causes hyperlipidemia are unclear. Cell and animal studies have pointed to various mechanisms that may mediate CsA-induced hyperlipidemia. In this review we will give an overview of CsA-induced hyperlipidemia, with a focus on the data available that might explain the underlying mechanism(s) and describe the available treatment regimes used to treat hyperlipidemia induced by immunosuppressant drugs.

2. Hyperlipidemia in humans after solid organ transplantation

Hyperlipidemia is observed in about 60% of kidney, liver, cardiac and bone marrow transplants after treatment with CsA (for review see [1,2]. There are multiple factors potentially contributing to hyperlipidemia in these patients, such as post-transplantation obesity, multiple drug therapy and diabetes. The concurrent use of steroids in particular, makes it hard to establish a direct contribution of CsA to dyslipidemia in humans, as corticosteroids are known to exacerbate hyperlipidemia in transplant recipients [3,4].

Studies investigating plasma lipids after CsA monotherapy are limited [4,5,6,7,8,9] and only a few studies have directly compared the combination of CsA therapy with low dose prednisolone with other immune suppressing strategies in combination with low dose steroids [10,11]. In general, these studies indicate that CsA treatment can independently lead to elevated plasma triglyceride and cholesterol levels in humans and that these effects are reversible upon cessation of immunosuppression therapy (Table 1). Animal studies (reviewed in [12]), where the effect of CsA can be studied in a more controlled background, indicate that CsA directly raises plasma lipid levels in rats, mice, guinea pigs and rabbits, and have proven that animals are valuable models to study mechanisms of CsA-induced hyperlipidemia.

Treatment	*Patients*	*Patient number*	*Duration*	*Lipid effects*	*Reference*
Monotherapy	Amyotrophic lateral sclerosis	36	2 mnths	TC ↑(21%) LDL-C ↑(31%) apoB ↑(12%) TG = HDL =	[5]
Monotherapy	Autologous bone marrow transplants	13	32 days	TC↑ (26%) LDL-C ↑ HDL-C ↓ TG = VLDL-C =	[13]
Monotherapy	Renal transplants	59	3-6 and 12 mnths	TC = LDL-C = apoB ↑ TG ↑ HDL-C ↓ apoA-I ↓	[8]
Monotherapy	Renal transplants	58	>1 yr	TC ↑ LDL-C ↑ apoB ↑ TG ↑ VLDL-C = HDL-C ↓ HDL2-C = HDL3-C ↓	[14]
Monotherapy and CsA/pred	Bone marrow transplants	180	100 days	TC ↑ LDL-C ↑ apoB ↑ TG ↑ VLDL-trig ↑ VLDL-C = HDL ↓ HDL2 ↓ HDL3 = apoA-I ↓	[4]
Monotherapy	Psoriasis	15	3 mnths	TC ↑ (22%) LDL-C ↑ (35%) TG = VLDL-C = HDL-C =	[9]

Treatment	*Patients*	*Patient number*	*Duration*	*Lipid effects*	*Reference*
ALG/aza/cort v CsA/ALG/aza/cort	Renal transplants	702	52 wks	TC ↑(20%) LDL-C ↑ TG ↑ HDL-C =	[7]
Aza/pred v CsA v CsA/pred	Renal transplants	9	3 mnths	TC ↑ LDL-C ↑ (45%) TG = VLDL-C = HDL-C =	[6]
Aza/pred v CsA/pred	Renal transplants	20	7.7 yrs	TC ↑ LDL-C ↑ apoB ↑ TG ↑ VLDL-C ↑ HDL-C ↓	[10]

ALG, Minnesota antilymphocyte globulin; aza, azathioprine; cort, corticosteroids; pred, prednisolone
TC, total cholesterol; TG, total triglyceride; LDL, low density lipoprotein; VLDL, very low density lipoprotein; HDL, high density lipoprotein; apo, apolipoprotein;

Table 1. Effect of CsA on plasma lipid parameters in humans

2.1. Plasma VLDL

Triglyceride-containing VLDL particles are produced in the liver via lipidation of apolipoprotein B (apoB) by microsomal triglyceride transfer protein (MTP), generating triglyceride-poor (VLDL2) as well as triglyceride-rich VLDL (VLDL1) particles, both of which can be secreted [15]. In plasma, VLDL is converted to intermediate-density lipoprotein (IDL) by lipoprotein lipase (LPL). IDL can be further hydrolyzed by lipases to low density lipoprotein (LDL). CsA increases plasma VLDL levels in transplant recipients and a concomitant increase in plasma apoB levels is observed [4,10,11]. It is unclear whether both plasma VLDL1 and VLDL2 levels are elevated. In contrast to LDL levels, plasma triglyceride and VLDL levels appear to increase only after long-term treatment with CsA (Table 1 and [8])

Hypertriglyceridemia in transplant patients is associated with increased plasma apolipoprotein CIII (apoCIII) levels [16,17,18] and decreased lipase activity (see below). As apoCIII inhibits LPL and hepatic lipase (HL) as well as uptake of triglyceride lipoprotein in liver, the increase of apoCIII may be an important contributor to hypertriglyceridemia found in transplant patients.

2.2. Plasma LDL

Plasma LDL levels appear to be consistently elevated by CsA [4,5,6,7,9,10,13,14] even in patients where plasma VLDL levels are not altered [5,6,9,13]. A correlation between CsA levels and plasma LDL-C has been described in some studies [19], but was not observed in others [5,20]. Regulation of plasma LDL levels is complex, depending on hepatic VLDL production, subsequent lipolysis of VLDL, clearance of LDL via the LDL receptor (LDLr) in the liver and conversion into bile. CsA may affect LDL metabolism at several levels (section 3.2).

2.3. Plasma HDL

Total plasma HDL levels are inversely correlated with the risk of cardiovascular disease [21]. HDL particles are however heterogeneous in size and composition, and occur as HDL2a, HDL2b, HDL3a, HDL3b and HDL3c which are progressively smaller in diameter and contain higher protein to lipid ratios. The precise contribution of various HDL subclasses to cardiovascular disease is currently unclear [21,22]. Plasma HDL cholesterol levels are determined by production of nascent HDL particles in the liver and intestine, by plasma transfer reactions of lipids between HDL and lipolysed triglyceride lipoproteins such as VLDL or chylomicrons, hepatic uptake of HDL lipids via the scavenger receptor class B1 (SRB1) HDL receptor in the liver, and renal clearance of small, lipid-poor apoA-I particles. Nascent HDL particles are formed by lipidation of apolipoprotein A-I (apoA-I) via the ATP-binding cassette transporter-1 (ABCA1) located in cellular membranes, although ABCA1-independent pathways of apoA-I lipidation also exist [23]. The formed lipid-poor HDL particles acquire more lipid after interaction with ABCG1 and mature by the subsequent esterification of cholesterol by lecithin-acyl transferase (LCAT). Further remodeling occurs by phospholipid transfer protein (PLTP) generating HDL2. HDL2 can be converted into HDL3 by hydrolysis via lipases and by transfer of cholesteryl esters to triglyceride-containing lipoproteins with the reciprocal exchange for triglycerides, which is mediated by cholesteryl ester transfer protein (CETP).

Immunosuppressive therapy has been reported to increase, decrease or leave HDL levels unaffected [5,10,11,24]. Parallel changes in plasma apoA-I levels are usually observed. Increased HDL levels are observed in most transplant patients, but this is most likely related to the concomitant treatment with steroids, which are known to increase plasma HDL [3]. CsA may affect particular subclasses of HDL more than others. Independently of steroids, plasma HDL levels, especially the HDL3 subpopulation, were found to inversely relate to plasma CsA levels [19]. In a study of bone marrow transplant recipients CsA decreased total plasma HDL, and in particular HDL2 [4]. In rats, a similar decrease in plasma HDL and HDL2 levels was observed after CsA treatment [25]. A recent study performed in pediatric renal transplant recipients showed that although total plasma HDL levels were not changed with CsA treatment, the relative proportion of HDL2b decreased while the relative proportion of HDL3a, HDL3b and HDL3c increased [26]. This is important as decreased HLD2b with increased HDL3b is associated with an atherogenic lipoprotein phenotype characterized by increased triglycerides and small dense LDL [27]. This result also emphases that simple monitoring of total HDL cholesterol may be insufficient to understand the consequences of CsA on HDL biology.

2.4. Plasma lipoprotein (a)

Lipoprotein (a) [Lp(a)] is a LDL-like lipoprotein consisting of LDL with one molecule of apoB covalently linked to a molecule of apolipoprotein (a). Plasma Lp(a) levels, and especially certain genetic Lp(a) variants, are independently associated with an increased risk for CVD [28,29]. Elevated Lp(a) plasma levels have been observed in renal transplant studies [14,30] this was however, not observed by others [31]. Although some studies suggested normalization of elevated Lp(a) levels after successful transplantation due to improved kidney function [31,32], CsA treatment has been indicated to independently increase Lp(a) levels in renal transplant recipients [8,14,33]. The mechanisms by which CsA affect plasma Lp(a) levels are unexplored, but may involve similar mechanisms to that of elevation of plasma LDL levels. As the LDLr does not play a major role in the clearance of Lp(a), the mechanism however, is unlikely mediated via effects of CsA on the LDLr (see section 3.2.1).

2.5. Qualitative differences in lipoproteins

2.5.1. Particle changes

Elevated plasma triglyceride levels are associated with the formation of triglyceride rich LDL particles that are more atherogenic [34]. A high prevalence of smaller denser LDL particles is observed in transplant recipients [35] and appears to be associated with CsA therapy [26,36]. Inhibition of lipoprotein lipase (LPL) activity is associated with the formation of small dense LDL subclasses. As apoCIII inhibits lipase activity, increased plasma apoCIII levels observed with CsA-treatment may explain inhibited lipase activity and subsequent increase in small dense LDL particles [17]. In addition decreased lipase activity could contribute to decreased HDL2 subclasses observed, while effects on CETP by CsA may help explain increases in HDL3 subfractions (see section 2.3 and 3.1.2).

2.5.2. Interaction of CsA with plasma lipoproteins

In whole blood CsA is primarily transported bound to lipoproteins (33%) and erythrocytes (58%) and whole blood CsA levels correlate with lipoprotein levels [37,38]. *In vitro* and *in vivo* studies show that in serum from healthy patients 50-60% of CsA is bound to HDL, 20-30% to LDL, 10-25% to VLDL with 10-15% bound to the non-lipoprotein proteins [39,40,41,42]. However, the proportion of CsA bound to the LDL and VLDL fractions increases in hyperlipidemic serum, without changing the amount bound to free protein [40,41], indicating that the distribution of CsA between the lipoprotein classes will change as plasma lipoprotein concentrations change. The binding of CsA to lipoprotein particles may also depend on lipoprotein composition. For example, Wasan et al. [41] showed that high triglyceride content of HDL was associated with a decreased percentage of CsA recovered in the HDL fraction and an increased percentage recovered in the VLDL fraction. Interestingly, treatment of patients with lipid lowering agents, such as statins have been reported to increase the unbound fraction of CsA and clearance of CsA in plasma [43].

Concerns have been raised about changes to the bioavailabilty and activity of CsA resulting from its binding to lipoproteins, especially as decreased CsA activity and increased toxicity have been observed in patients with hyperlipidemia [42,44]. CsA levels are higher in hyperlipidemic patients due to decreased clearance which was reversed after lipid-lowering with fibrates (reviewed in [37]). *In vitro* studies using skin fibroblasts indicate that CsA bound to LDL does not affect binding to cells via the LDLr, but uptake of CsA is inhibited [45]. These studies were confirmed in HepG2 and Jurkat Tcells which showed decreased uptake of CsA in the presence of LDL [40]. In line with these findings, uptake of CsA in tissues from rats was reduced when CsA was co-injected with lipoproteins [46].

3. Mechanisms of CsA-Induced hyperlipidemia – What we learn from cell and animal studies

As the effects of CsA in humans are confounded by many factors such as other medication, obesity, insulin resistance and nutritional status, cell and animal studies are useful to elucidate the mechanism(s) of CsA-induced hyperlipidemia. Figure 1 depicts the reported CsA-effects on VLDL, LDL and HDL metabolism.

3.1. VLDL

3.1.1. Effects of CsA on VLDL synthesis and secretion

CsA decreased apoB translocation over the endoplasmic reticulum (ER) membrane in the human liver cell line HepG2 [47]. It was suggested that this was due to a reduction in the efficiency of lipid transfer by inhibition of MTP, however whether MTP activity is inhibited by CsA was not investigated. These findings are in line with the report from Kaptein et al. [48], which showed that CsA inhibits VLDL and apoB secretion from HepG2 cells, by post-translational mechanisms. In contrast, in mice, CsA increased the rate of hepatic VLDL secretion *in vivo*, while total apoB secretion was unaffected [49]. No effect of CsA on levels of VLDL receptors in either adipose tissue or skeletal muscle were found [50] suggesting that VLDL uptake may not be affected by CsA. There are no studies that we are aware of studying the effect of CsA on *in vivo* VLDL synthesis in humans.

3.1.2. VLDL metabolism

Inhibition of lipolysis by CsA could contribute to increased plasma VLDL and reduced HDL concentrations. Various studies have investigated lipase activity in patients, but results may be confounded by co-treatment with steroids. HL activity was increased in cardiac transplant patients and correlated with CsA dose while lipoprotein lipase (LPL) activity was decreased in these patients [51]. Others have shown decreased HL as well as LPL activity in kidney transplant recipients [52]. More directly, Tory et al [53] showed suppression of LPL activity in plasma from normolipidemic subjects treated with CsA, while in rats, CsA dose- and time-dependent decreased plasma LPL activity [24]. In addition, LPL abundance in skeletal muscle and adipose tissue was decreased in rats [50]. These latter studies suggested CsA can inhibit LPL activity independently of steroids. Although the precise mechanism of CsA-inhibited LPL activity is unknown, it helps to explain increased triglyceride levels observed after CsA treatment.

Some studies show reduced cholesteryl ester transfer protein (CETP) activity in transplant recipients [54]. In contrast, CsA directly added to human plasma *ex vivo* increased CETP activity [53]. These apparently anomalous results may relate to differences between the direct effects of CsA on CETP itself and indirect effects secondary to changes in the concentrations of other lipoproteins, but remain unexplained. Since CETP transfers cholesteryl ester from HDL to apoB-containing lipoproteins with reciprocal transfer of triglycerides, any effect of CsA on CETP activity could be expected to have major effects on plasma lipoprotein profiles.

3.2. LDL

3.2.1. LDL synthesis and catabolism

We have recently reviewed this literature in detail [55]. There appear to be conflicting conclusions arising from *in vitro* and *in vivo* studies. One of the key discrepancies is the role of LDLr expression and LDL clearance by the liver in mediating CsA-hyperlipidemia. In general, *in vitro* studies are consistent with a role for decreased LDL receptor expression or activity in liver cells after exposure to CsA [48,56]. *In vivo* studies however, show mixed effects, with no effect or an increase in hepatic LDLr protein or mRNA levels [49,50]. Similarly 3-hydroxy-3-methyl-glutaryl-CoA reductase (HMG-CoAr), the rate limiting enzyme in cholesterol synthesis, mRNA levels were upregulated in HepG2 cells and mouse liver after CsA, but hepatic HMG-CoA reductase protein levels in rat liver were unaffected by CsA treatment [49,50,57]. In rats, CsA decreased the fractional catabolic rate of LDL [58]. One very important consideration is the difference in concentrations of CsA used in *in vitro* studies relative to those achieved *in vivo* under normal transplant immunosuppression. *In vitro* studies commonly use concentrations of 10 μg/ml whereas plasma levels of CsA in humans and in animal studies are typically in the order of 100 ng/ml. This apparent 10-fold difference in concentration may underestimate the difference in effective concentrations tested *in vivo* and *in vitro* studies because of the complicating effects of *in vivo* hyperlipidemia, which under some circumstances can lessen the effective concentration of CsA delivered to some tissues [46].

3.3. HDL

CsA effects on plasma HDL and HDL subclasses may be mediated by effects on the synthesis and/or formation of HDL as well as by effecting remodeling of HDL through changes in lipase and/or CETP activity (see 3.1.2)

3.3.1. Effect of CsA on HDL synthesis and formation

In vitro studies have indicated that CsA potently inhibits ABCA1 activity thereby inhibiting apoA-I lipidation, the first step in HDL formation [59,60,61]. This was associated with decreased ABCA1 turnover and an increase in total and cell-surface levels of ABCA1 [59]. Uptake, Internalization and re-secreton of apoA-I were however decreased by CsA, suggesting that ABCA1 trapped at the plasma membrane is dysfunctional [59,60]. *In vivo* studies using wild type C57Bl6 mice corroborated these *in vitro* findings. CsA lowered plasma HDL levels after 6 days of treatment [59]. A lowering in plasma HDL in mice was however not observed by others after long-term treatment of mice with CsA combined with a high fat diet [62]. As many aspects of lipid metabolism can be affected by CsA, it may be difficult to determine a causal effect on HDL levels via ABCA1 inhibition in an *in vivo* whole body system NB.

Direct effects of CsA on the expression of ABCA1 and apoA-I have also been reported and may contribute to the changes in HDL formation. The target of immunosuppression by CsA, Nuclear Factor of activated T-cells, cytoplasmic 2 (NFATc2), was found to bind the mouse ABCA1 promoter and mediate CsA-inhibition of ABCA1 expression by inflammatory stimuli [63]. In addition CsA has been found to inhibit apoA-I gene expression in human HepG2 cells and rats [64]. A recent proteomic study in HepG2 cells showed that CsA decreased secretion levels of apoA-I suggesting that the transcriptional effects of CsA on apoA-I expression may lead to decreased amounts of secreted apoA-I [65].

3.3.2. Effects on HDL metabolism

As mentioned above (section 3.1.2), CsA directly suppresses LPL activity and increases CETP activity in human plasma and animals (section 3.1.2). LPL activity is strongly associated with plasma HDL2 concentrations [66], and decreased LPL levels in CsA treatment may therefore contribute to decreased HDL2 levels [4,25]. On the other hand, increased CETP activity will generate triglyceride-rich HDL, which is converted to smaller HDL3 particles by HL [66].

3.4. Effects on bile acid synthesis and secretion

3.4.1. Effects on bile synthesis

In liver, cholesterol is converted to bile acids by 7α-hydroxylase (CYP7α) or 27-hydroxylase (CYP27A1) [67]. In healthy humans, CYP7α is considered the predominantly pathway while CYP27A1 accounts for 10% of bile acid synthesis and subsequent formation of chenodeoxycholate. However inhibition of Cyp7α can increase the contribution of the CYP27A1 pathway [68]. *In vitro* studies show that CsA inhibits both CYP27A1 activity and subsequent formation of chenodeoxycholate in human and animal liver extracts and in primary hepatocyte cultures [57,69,70,71]. A CsA responsive element has been mapped on the CYP27A1 promoter [72], indicating that CsA affects transcription of the CYP27A1 gene directly. In most of the *in vitro* studies, CYP7α activity was not affected by CsA [69,70]. *In vivo,* in rat however, CsA decreased CYP7α protein levels [50], indicating that the predominant bile acid synthesis pathway may also be affected by CsA. The inhibitory effect of CsA on bile synthesis is suggested to contribute to increased plasma lipid concentrations in transplant recipients. Radioisotope studies performed in children after liver transplantation demonstrated that CsA treatment significantly inhibits bile salts synthesis rates, especially that of chenodeoxycholate and that bile acid synthesis rate inversely correlates with plasma cholesterol and triglyceride levels [73].

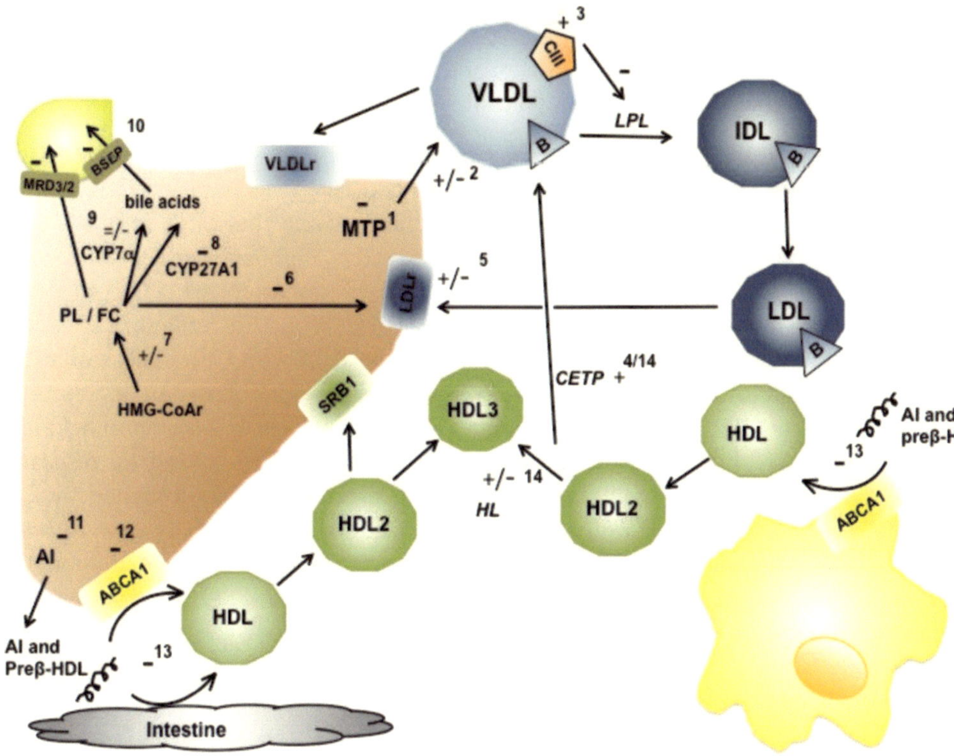

Figure 1. Mechanisms of CsA-mediated hyperlipidemia. Figure only displays pathways that are reported to be affected by CsA. 1) Inhibition of VLDL formation via inhibition of MTP, 2) Increased and decreased secretion of VLDL particles have been reported, 3) Decreased lipolysis of VLDL due to increased apoCIII and subsequent inhibition of LPL, 4) hypertriglyceridemia by increased CETP activity, 5) Increased LDL due to decreased LDLr expression as well as activity, 6) Increased liver FC content leading to decreased LDLr levels, 7) Increased and decreased levels of HMG-CoAr affecting cholesterol synthesis, 8/9) Inhibition of bile acid conversion via CYP27A1 or CYP7α leading to increased liver FC levels, however in most studies Cyp7α is not affected by CsA. NB: decreased CYP27A1 activity can increase HMG-CoAr levels via negative feedback, 10) Decreased flow of bile salts, cholesterol and phospholipids into bile, 11) Decreased expression and secretion of apoA-I, 12) Inhibition of ABCA1 expression, 13) inhibition of apoA-I lipidation via inhibition of ABCA1 activity 14) Stimulation of HL and CETP leads to increased formation of HDL2 to HDL3, however decreased HL activity has also been reported. VLDL, very low density lipoprotein; IDL, intermediate density lipoprotein; LDL, low density lipoprotein; HDL, high density lipoprotein; AI, apolipoprotein A-I, B, apolipoprotein B; CIII, apolipoprotein CIII; MTP, microsomal triglyceride transfer protein; LPL, lipoprotein lipase; HL, hepatic lipase; CETP, cholesteryl ester transfer protein; ABCA1, ATP-binding cassette transporter-1; SRB1, scavenger receptor class B1; LDLr, LDLreceptor; VLDLr, VLDLreceptor; PL, phospholipid; FC, free cholesterol; HMG-CoAr, 3-hydroxy-3-methyl-glutaryl-CoA reductase; CYP7α, 7α-hydroxylase; CYP27A1, 27-hydroxylase; MRD, multidrug resistance protein; BSEP, bile salt export protein.

The effects of CsA on CYP27A1 may relate to effects of CsA on cholesterol metabolism. 27-hydroxycholesterol is a potent negative feedback regulator of HMG-CoA reductase [74] and decreased CYP27A1 activity may therefore explain increased HMG-CoA reductase mRNA and cholesterol levels [57]. Although important in macrophages, it should be noted however that it is not clear whether such a feedback loop exists in liver cells [75]. Increased cholesterol synthesis could subsequently lead to downregulation of LDLr levels as observed in some CsA studies, also contributing to increased plasma cholesterol levels (see section 3.2.1).

Besides effects on bile acid synthesis CsA may affect bile flow. CsA treatment is associated with increased plasma bile acid concentrations and cholestasis in humans as well as in animal models [9,52,76]. Studies in rat indicate that bile flow and the secretion of bile salts, proteins and lipids into the bile are dose-dependently inhibited by CsA [52,76,77]. Interestingly, the changes in serum levels of bile acids are consistent with CsA-mediated inhibition of hepatocellular uptake of individual bile acids [78,79]. The inhibitory effect was greater for phosholipid secretion than that for cholesterol [80] and in some studies no inhibition of cholesterol excretion was observed [81], suggesting differential effects on transport mechanisms. Transport pumps involved in bile synthesis and secretion belong to the family of the ATP-binding cassette transporters which include, multidrug resistance proteins (MDR) and P-glycoprotein, and most of which are effectively inhibited by CsA [79,82]. Interestingly, comparison of the bile salt export pump (BSEP) activity from different species, showed that CsA inhibits bile salt transport with species and bile salt specific variation [83]. Rat BSEP was for example more effectively inhibited than mouse BSEP. Biliary cholesterol secretion is mediated via ABCG5 and ABCG8 [84]. Although both members of the ATP-binding cassette family, it has not been investigated whether CsA inhibits ABCG5/8 activity. As phospholipids are transported via MDR3, it is likely that differences in efficacy of CsA between inhibition of MRD3 and ABCG5/8 exist. It is clear that CsA can affect bile flow and secretion in cultured cells and animal models. It should be noted however, that in humans no inhibitory effect of CsA on secretion of bile acids and lipids or on bile composition after liver transplantation was observed [85]. Others have shown that although cholate synthesis was reduced by CsA, compensatory increased intestinal absorption counteracted this decrease [86]. It remains therefore unclear to what extent inhibition of bile flow and secretion by CsA are contributing to hypercholesteremia *in vivo*.

4. Therapies to address hyperlipidemia

Hyperlipidemia is associated with significant morbidity and mortality rates in transplant recipients [87]. Many strategies have been investigated to target dyslipidemia in transplant patients. A number of excellent comprehensive reviews have been published on the clinical management of hyperlipidemia and its risks (eg [88,89]). We will therefore restrict our comments to a very brief summary of this area.

4.1. Statins

Statins inhibit HMG-CoA reductase, the rate limiting enzyme in the cholesterol synthesis pathway and are world-wide the drug of choice to lower plasma LDL-C levels. Various statins have been tested in transplant patients and all show significant lowering of plasma cholesterol, LDL-C and apoB levels with some indicating improved survival rates (for review see [88,89,90]). A randomized trial, investigating the safety and efficacy of statins in renal transplant patients, the Assessment of LEscol in Renal Transplantation (ALERT) study, showed that fluvastatin effectively lowered LDL-C by 32% and reduced cardiac death and non-fatal myocardial infarction incidence significantly [91]. Importantly, statins may provide beneficial effects other then their lipid-lowering properties [92]. Wissing et al [93] reported improved flow mediated brachial artery vasodilatation by atorvastatin in kidney transplant patients and significant reductions in acute rejections have been observed in cardiac transplant patients [94].

Rhabdomyolysis, one of the few serious side effects of statins, is more common with high dose statin treatment. The risk is elevated in patients with renal disease and in patients taking drugs affecting statin metabolism, especaily in those taking CsA [88,89]. All statins have the potential to interact with CsA, as CsA substantially increases plasma levels of all statins. Although this is most notable for those metabolized via the Cyp3A4 pathway, statins not metabolized via the Cyp3A4 pathway [95] such as pravastatin and fluvastatin are also affected [95], suggesting that the interaction of CsA and statins may involve other mechanisms such as inhibition of drug transporters. Simvastatin poses the highest risk of myopathy, and particular care must be taken with higher doses of this agent, with recommendations that doses of 10mg/d are not exceeded in transplant patients [89]. Because statin therapy has been associated with mortality benefit after transplantation, correction of hyperlipidemia using lower doses of statins is mandatory after transplantation. Therefore careful clinical monitoring of patients as well as measurement of creatine kinase levels to detect muscle injury is advised, and the use of statins that are not metabolized via CYP3A4, such as fluvastatin or pravastatin may be preferential [95].

4.2. Fibrates

Fibrates lower plasma triglyceride levels via activation of the Peroxisome Proliferator Activated Receptor alpha (PPARα) and may be useful in transplant patients with elevated plasma triglycerides especially in combination with statin treatment to lower plasma cholesterol levels. Gemfibrozil was found to significantly lower plasma triglyceride levels in heart transplant patients and increase long term survival [96,97]. Fenofibrate is less well studied in transplant patients and may be associated with increased nephrotoxicity [88,98]. Care must be taken administering fibrates with CsA, particularly in combination with statins as drug-drug interactions exist via CYP3A4 as well as the hepatic uptake transporter the organic anion transporting polypeptide 1B1 (OAT1B1).

4.3. Ezetimibe

Inhibition of intestinal cholesterol absorption to lower high plasma cholesterol levels may be used when statins or fibrates are ineffective or are not tolerated. Ezetimibe proved to be an effective drug lowering plasma LDL-C levels significantly by blocking cholesterol absorption in the small intestine [99]. To that point though, various studies showed effective LDL-C lowering in liver, cardiac and renal transplant recipients [99]. Although, drug-drug interaction between CsA and ezetimibe were suggested (See [88]), CsA levels in studied transplant patients were not affected by combined ezetimibe use (reviewed in [99]). Co-administration of ezetimibe with (low-dose) statins has been found to effectively reduce high plasma cholesterol levels in transplant recipients and may be useful in patients that resistant to high-dose statin or where target plasma lipid levels can not be achieved by statin therapy alone [100,101].

5. Conclusions

CsA-induced hyperlipidemia is well established and remains a significant clinical issue. CsA potentially affects many aspects of lipid and lipoprotein metabolism and the precise underlying mechanism(s) causing dyslipidemia are still unclear. Further mechanistic studies may lead to the generation immunosuppressants that do not cause hyperlipidemia or may help to develop strategies to effectively target CsA-induced hyperlipidemia.

Author details

Maaike Kockx
Macrophage Biology Group, Centre for Vascular Research, University of New South Wales, Sydney, Australia

Leonard Kritharides*
Macrophage Biology Group, Centre for Vascular Research, University of New South Wales, Sydney, Australia
Department of Cardiology, Concord Repatriation General Hospital, University of Sydney, Sydney, Australia

6. References

[1] Kobashigawa JA, Kasiske BL (1997) Hyperlipidemia in solid organ transplantation. Transplantation 63: 331-338.
[2] Miller LW (2002) Cardiovascular toxicities of immunosuppressive agents. Am J Transplant 2: 807-818.
[3] Strohmayer EA, Krakoff LR (2011) Glucocorticoids and cardiovascular risk factors. Endocrinol Metab Clin North Am 40: 409-417, ix.
[4] Lopez-Miranda J, Perez-Jimenez F, Torres A, Espino-Montoro A, Gomez P, et al. (1992) Effect of cyclosporin on plasma lipoproteins in bone marrow transplantation patients. Clin Biochem 25: 379-386.
[5] Ballantyne CM, Podet EJ, Patsch WP, Harati Y, Appel V, et al. (1989) Effects of cyclosporine therapy on plasma lipoprotein levels. JAMA 262: 53-56.

* Corresponding Author

[6] Raine AE, Carter R, Mann JI, Morris PJ (1988) Adverse effect of cyclosporin on plasma cholesterol in renal transplant recipients. Nephrol Dial Transplant 3: 458-463.

[7] Kasiske BL, Tortorice KL, Heim-Duthoy KL, Awni WM, Rao KV (1991) The adverse impact of cyclosporine on serum lipids in renal transplant recipients. Am J Kidney Dis 17: 700-707.

[8] Hilbrands LB, Demacker PN, Hoitsma AJ, Stalenhoef AF, Koene RA (1995) The effects of cyclosporine and prednisone on serum lipid and (apo)lipoprotein levels in renal transplant recipients. J Am Soc Nephrol 5: 2073-2081.

[9] Edwards BD, Bhatnagar D, Mackness MI, Gokal R, Ballardie FW, et al. (1995) Effect of low-dose cyclosporin on plasma lipoproteins and markers of cholestasis in patients with psoriasis. QJM 88: 109-113.

[10] Schorn TF, Kliem V, Bojanovski M, Bojanovski D, Repp H, et al. (1991) Impact of long-term immunosuppression with cyclosporin A on serum lipids in stable renal transplant recipients. Transpl Int 4: 92-95.

[11] Ichimaru N, Takahara S, Kokado Y, Wang JD, Hatori M, et al. (2001) Changes in lipid metabolism and effect of simvastatin in renal transplant recipients induced by cyclosporine or tacrolimus. Atherosclerosis 158: 417-423.

[12] Kockx M, Guo DL, Traini M, Gaus K, Kay J, et al. (2009) Cyclosporin A decreases apolipoprotein E secretion from human macrophages via a protein phosphatase 2B-dependent and ATP-binding cassette transporter A1 (ABCA1)-independent pathway. J Biol Chem 284: 24144-24154.

[13] Luke DR, Beck JE, Vadiei K, Yousefpour M, LeMaistre CF, et al. (1990) Longitudinal study of cyclosporine and lipids in patients undergoing bone marrow transplantation. J Clin Pharmacol 30: 163-169.

[14] Brown JH, Anwar N, Short CD, Bhatnager D, Mackness MI, et al. (1993) Serum lipoprotein (a) in renal transplant recipients receiving cyclosporin monotherapy. Nephrol Dial Transplant 8: 863-867.

[15] Adiels M, Olofsson SO, Taskinen MR, Boren J (2008) Overproduction of very low-density lipoproteins is the hallmark of the dyslipidemia in the metabolic syndrome. Arterioscler Thromb Vasc Biol 28: 1225-1236.

[16] Kimak E, Solski J, Baranowicz-Gaszczyk I, Ksiazek A (2006) A long-term study of dyslipidemia and dyslipoproteinemia in stable post-renal transplant patients. Ren Fail 28: 483-486.

[17] Badiou S, Garrigue V, Dupuy AM, Chong G, Cristol JP, et al. (2006) Small dense low-density lipoprotein in renal transplant recipients: a potential target for prevention of cardiovascular complications? Transplant Proc 38: 2314-2316.

[18] Tur MD, Garrigue V, Vela C, Dupuy AM, Descomps B, et al. (2000) Apolipoprotein CIII is upregulated by anticalcineurins and rapamycin: implications in transplantation-induced dyslipidemia. Transplant Proc 32: 2783-2784.

[19] Kuster GM, Drexel H, Bleisch JA, Rentsch K, Pei P, et al. (1994) Relation of cyclosporine blood levels to adverse effects on lipoproteins. Transplantation 57: 1479-1483.

[20] Ramezani M, Einollahi B, Ahmadzad-Asl M, Nafar M, Pourfarziani V, et al. (2007) Hyperlipidemia after renal transplantation and its relation to graft and patient survival. Transplant Proc 39: 1044-1047.

[21] Kontush A, Chapman MJ (2006) Functionally defective high-density lipoprotein: a new therapeutic target at the crossroads of dyslipidemia, inflammation, and atherosclerosis. Pharmacol Rev 58: 342-374.
[22] Rosenson RS, Brewer HB, Jr., Chapman MJ, Fazio S, Hussain MM, et al. (2011) HDL measures, particle heterogeneity, proposed nomenclature, and relation to atherosclerotic cardiovascular events. Clin Chem 57: 392-410.
[23] Zheng H, Kiss RS, Franklin V, Wang MD, Haidar B, et al. (2005) ApoA-I lipidation in primary mouse hepatocytes. Separate controls for phospholipid and cholesterol transfers. J Biol Chem 280: 21612-21621.
[24] Lopez-Miranda J, Perez-Jimenez F, Gomez-Gerique JA, Espino-Montoro A, Hidalgo-Rojas L, et al. (1992) Effect of cyclosporin on plasma lipoprotein lipase activity in rats. Clin Biochem 25: 387-394.
[25] Espino A, Lopez-Miranda J, Blanco-Cerrada J, Zambrana JL, Aumente MA, et al. (1995) The effect of cyclosporine and methylprednisolone on plasma lipoprotein levels in rats. J Lab Clin Med 125: 222-227.
[26] Zeljkovic A, Vekic J, Spasojevic-Kalimanovska V, Jelic-Ivanovic Z, Peco-Antic A, et al. (2011) Characteristics of low-density and high-density lipoprotein subclasses in pediatric renal transplant recipients. Transpl Int 24: 1094-1102.
[27] Berneis KK, Krauss RM (2002) Metabolic origins and clinical significance of LDL heterogeneity. J Lipid Res 43: 1363-1379.
[28] Erqou S, Kaptoge S, Perry PL, Di Angelantonio E, Thompson A, et al. (2009) Lipoprotein(a) concentration and the risk of coronary heart disease, stroke, and nonvascular mortality. JAMA 302: 412-423.
[29] Clarke R, Peden JF, Hopewell JC, Kyriakou T, Goel A, et al. (2009) Genetic variants associated with Lp(a) lipoprotein level and coronary disease. N Engl J Med 361: 2518-2528.
[30] Fonseca I, Queiros J, Costa S, Santos MJ, Henriques AC, et al. (2002) Lipoprotein(A) in renal transplant recipients. Transplant Proc 34: 370-372.
[31] Innocenti M, Lorenzetti M, Naldi F, Paleologo G, Pasquariello A, et al. (1998) Evaluation of lipoprotein A in renal transplant recipients. Transplant Proc 30: 2048.
[32] Black IW, Wilcken DE (1992) Decreases in apolipoprotein(a) after renal transplantation: implications for lipoprotein(a) metabolism. Clin Chem 38: 353-357.
[33] Webb AT, Reaveley DA, O'Donnell M, O'Connor B, Seed M, et al. (1993) Does cyclosporin increase lipoprotein(a) concentrations in renal transplant recipients? Lancet 341: 268-270.
[34] Packard CJ, Shepherd J (1997) Lipoprotein heterogeneity and apolipoprotein B metabolism. Arterioscler Thromb Vasc Biol 17: 3542-3556.
[35] Rajman I, Harper L, McPake D, Kendall MJ, Wheeler DC (1998) Low-density lipoprotein subfraction profiles in chronic renal failure. Nephrol Dial Transplant 13: 2281-2287.
[36] Quaschning T, Mainka T, Nauck M, Rump LC, Wanner C, et al. (1999) Immunosuppression enhances atherogenicity of lipid profile after transplantation. Kidney Int Suppl 71: S235-237.

[37] Akhlaghi F, Trull AK (2002) Distribution of cyclosporin in organ transplant recipients. Clin Pharmacokinet 41: 615-637.
[38] Gardier AM, Mathe D, Guedeney X, Barre J, Benvenutti C, et al. (1993) Effects of plasma lipid levels on blood distribution and pharmacokinetics of cyclosporin A. Ther Drug Monit 15: 274-280.
[39] Sgoutas D, MacMahon W, Love A, Jerkunica I (1986) Interaction of cyclosporin A with human lipoproteins. J Pharm Pharmacol 38: 583-588.
[40] Rifai N, Chao FF, Pham Q, Thiessen J, Soldin SJ (1996) The role of lipoproteins in the transport and uptake of cyclosporine and dihydro-tacrolimus into HepG2 and JURKAT cell lines. Clin Biochem 29: 149-155.
[41] Wasan KM, Pritchard PH, Ramaswamy M, Wong W, Donnachie EM, et al. (1997) Differences in lipoprotein lipid concentration and composition modify the plasma distribution of cyclosporine. Pharm Res 14: 1613-1620.
[42] De Klippel N, Sennesael J, Lamote J, Ebinger G, de Keyser J (1992) Cyclosporin leukoencephalopathy induced by intravenous lipid solution. Lancet 339: 1114.
[43] Akhlaghi F, McLachlan AJ, Keogh AM, Brown KF (1997) Effect of simvastatin on cyclosporine unbound fraction and apparent blood clearance in heart transplant recipients. Br J Clin Pharmacol 44: 537-542.
[44] de Groen PC, Aksamit AJ, Rakela J, Forbes GS, Krom RA (1987) Central nervous system toxicity after liver transplantation. The role of cyclosporine and cholesterol. N Engl J Med 317: 861-866.
[45] Wasan KM, Ramaswamy M, Kwong M, Boulanger KD (2002) Role of plasma lipoproteins in modifying the toxic effects of water-insoluble drugs: studies with cyclosporine A. AAPS PharmSci 4: E30.
[46] Lemaire M, Pardridge WM, Chaudhuri G (1988) Influence of blood components on the tissue uptake indices of cyclosporin in rats. J Pharmacol Exp Ther 244: 740-743.
[47] Macri J, Adeli K (1997) Studies on intracellular translocation of apolipoprotein B in a permeabilized HepG2 system. J Biol Chem 272: 7328-7337.
[48] Kaptein A, de Wit EC, Princen HM (1994) Cotranslational inhibition of apoB-100 synthesis by cyclosporin A in the human hepatoma cell line HepG2. Arterioscler Thromb 14: 780-789.
[49] Wu J, Zhu YH, Patel SB (1999) Cyclosporin-induced dyslipoproteinemia is associated with selective activation of SREBP-2. Am J Physiol 277: E1087-1094.
[50] Vaziri ND, Liang K, Azad H (2000) Effect of cyclosporine on HMG-CoA reductase, cholesterol 7alpha-hydroxylase, LDL receptor, HDL receptor, VLDL receptor, and lipoprotein lipase expressions. J Pharmacol Exp Ther 294: 778-783.
[51] Superko HR, Haskell WL, Di Ricco CD (1990) Lipoprotein and hepatic lipase activity and high-density lipoprotein subclasses after cardiac transplantation. Am J Cardiol 66: 1131-1134.
[52] Deters M, Kirchner G, Koal T, Resch K, Kaever V (2004) Everolimus/cyclosporine interactions on bile flow and biliary excretion of bile salts and cholesterol in rats. Dig Dis Sci 49: 30-37.

[53] Tory R, Sachs-Barrable K, Hill JS, Wasan KM (2008) Cyclosporine A and Rapamycin induce in vitro cholesteryl ester transfer protein activity, and suppress lipoprotein lipase activity in human plasma. Int J Pharm 358: 219-223.

[54] Atger V, Leclerc T, Cambillau M, Guillemain R, Marti C, et al. (1993) Elevated high density lipoprotein concentrations in heart transplant recipients are related to impaired plasma cholesteryl ester transfer and hepatic lipase activity. Atherosclerosis 103: 29-41.

[55] Kockx M, Jessup W, Kritharides L (2010) Cyclosporin A and atherosclerosis--cellular pathways in atherogenesis. Pharmacol Ther 128: 106-118.

[56] Rayyes OA, Wallmark A, Floren CH (1996) Cyclosporine inhibits catabolism of low-density lipoproteins in HepG2 cells by about 25%. Hepatology 24: 613-619.

[57] Gueguen Y, Ferrari L, Souidi M, Batt AM, Lutton C, et al. (2007) Compared effect of immunosuppressive drugs cyclosporine A and rapamycin on cholesterol homeostasis key enzymes CYP27A1 and HMG-CoA reductase. Basic Clin Pharmacol Toxicol 100: 392-397.

[58] Lopez-Miranda J, Vilella E, Perez-Jimenez F, Espino A, Jimenez-Pereperez JA, et al. (1993) Low-density lipoprotein metabolism in rats treated with cyclosporine. Metabolism 42: 678-683.

[59] Le Goff W, Peng DQ, Settle M, Brubaker G, Morton RE, et al. (2004) Cyclosporin A traps ABCA1 at the plasma membrane and inhibits ABCA1-mediated lipid efflux to apolipoprotein A-I. Arterioscler Thromb Vasc Biol 24: 2155-2161.

[60] Lorenzi I, von Eckardstein A, Cavelier C, Radosavljevic S, Rohrer L (2008) Apolipoprotein A-I but not high-density lipoproteins are internalised by RAW macrophages: roles of ATP-binding cassette transporter A1 and scavenger receptor BI. J Mol Med 86: 171-183.

[61] Karwatsky J, Ma L, Dong F, Zha X (2009) Cholesterol efflux to apoA-I in ABCA1-expressing cells is regulated by Ca2+ dependent-calcineurin signaling. J Lipid Res.

[62] Emeson EE, Shen ML (1993) Accelerated atherosclerosis in hyperlipidemic C57BL/6 mice treated with cyclosporin A. Am J Pathol 142: 1906-1915.

[63] Maitra U, Parks JS, Li L (2009) An innate immunity signaling process suppresses macrophage ABCA1 expression through IRAK-1-mediated downregulation of retinoic acid receptor alpha and NFATc2. Mol Cell Biol 29: 5989-5997.

[64] Zheng XL, Wong NC (2006) Cyclosporin A inhibits apolipoprotein AI gene expression. J Mol Endocrinol 37: 367-373.

[65] Van Summeren A, Renes J, Bouwman FG, Noben JP, van Delft JH, et al. (2011) Proteomics investigations of drug-induced hepatotoxicity in HepG2 cells. Toxicol Sci 120: 109-122.

[66] von Eckardstein A, Huang Y, Assmann G (1994) Physiological role and clinical relevance of high-density lipoprotein subclasses. Curr Opin Lipidol 5: 404-416.

[67] Anderson KE, Kok E, Javitt NB (1972) Bile acid synthesis in man: metabolism of 7 -hydroxycholesterol- 14 C and 26-hydroxycholesterol- 3 H. J Clin Invest 51: 112-117.

[68] Duane WC, Javitt NB (1999) 27-hydroxycholesterol: production rates in normal human subjects. J Lipid Res 40: 1194-1199.

[69] Souidi M, Parquet M, Ferezou J, Lutton C (1999) Modulation of cholesterol 7alpha-hydroxylase and sterol 27-hydroxylase activities by steroids and physiological conditions in hamster. Life Sci 64: 1585-1593.
[70] Princen HM, Meijer P, Wolthers BG, Vonk RJ, Kuipers F (1991) Cyclosporin A blocks bile acid synthesis in cultured hepatocytes by specific inhibition of chenodeoxycholic acid synthesis. Biochem J 275 (Pt 2): 501-505.
[71] Winegar DA, Salisbury JA, Sundseth SS, Hawke RL (1996) Effects of cyclosporin on cholesterol 27-hydroxylation and LDL receptor activity in HepG2 cells. J Lipid Res 37: 179-191.
[72] Segev H, Honigman A, Rosen H, Leitersdorf E (2001) Transcriptional regulation of the human sterol 27-hydroxylase gene (CYP27) and promoter mapping. Atherosclerosis 156: 339-347.
[73] Hulzebos CV, Bijleveld CM, Stellaard F, Kuipers F, Fidler V, et al. (2004) Cyclosporine A-induced reduction of bile salt synthesis associated with increased plasma lipids in children after liver transplantation. Liver Transpl 10: 872-880.
[74] Esterman AL, Baum H, Javitt NB, Darlington GJ (1983) 26-hydroxycholesterol: regulation of hydroxymethylglutaryl-CoA reductase activity in Chinese hamster ovary cell culture. J Lipid Res 24: 1304-1309.
[75] Javitt NB (2002) 25R,26-Hydroxycholesterol revisited: synthesis, metabolism, and biologic roles. J Lipid Res 43: 665-670.
[76] Stone BG, Udani M, Sanghvi A, Warty V, Plocki K, et al. (1987) Cyclosporin A-induced cholestasis. The mechanism in a rat model. Gastroenterology 93: 344-351.
[77] Roman ID, Monte MJ, Gonzalez-Buitrago JM, Esteller A, Jimenez R (1990) Inhibition of hepatocytary vesicular transport by cyclosporin A in the rat: relationship with cholestasis and hyperbilirubinemia. Hepatology 12: 83-91.
[78] Azer SA, Stacey NH (1994) Differential effects of cyclosporin A on transport of bile acids by rat hepatocytes: relationship to individual serum bile acid levels. Toxicol Appl Pharmacol 124: 302-309.
[79] Stieger B, Fattinger K, Madon J, Kullak-Ublick GA, Meier PJ (2000) Drug- and estrogen-induced cholestasis through inhibition of the hepatocellular bile salt export pump (Bsep) of rat liver. Gastroenterology 118: 422-430.
[80] Galan AI, Roman ID, Munoz ME, Cava F, Gonzalez-Buitrago JM, et al. (1992) Inhibition of biliary lipid and protein secretion by cyclosporine A in the rat. Biochem Pharmacol 44: 1105-1113.
[81] Chan FK, Shaffer EA (1997) Cholestatic effects of cyclosporine in the rat. Transplantation 63: 1574-1578.
[82] Bohme M, Jedlitschky G, Leier I, Buchler M, Keppler D (1994) ATP-dependent export pumps and their inhibition by cyclosporins. Adv Enzyme Regul 34: 371-380.
[83] Kis E, Ioja E, Nagy T, Szente L, Heredi-Szabo K, et al. (2009) Effect of membrane cholesterol on BSEP/Bsep activity: species specificity studies for substrates and inhibitors. Drug Metab Dispos 37: 1878-1886.
[84] Yu L, Gupta S, Xu F, Liverman AD, Moschetta A, et al. (2005) Expression of ABCG5 and ABCG8 is required for regulation of biliary cholesterol secretion. J Biol Chem 280: 8742-8747.

[85] Baiocchi L, Angelico M, De Luca L, Ombres D, Anselmo A, et al. (2006) Cyclosporine A versus tacrolimus monotherapy. Comparison on bile lipids in the first 3 months after liver transplant in humans. Transpl Int 19: 389-395.
[86] Hulzebos CV, Wolters H, Plosch T, Kramer W, Stengelin S, et al. (2003) Cyclosporin a and enterohepatic circulation of bile salts in rats: decreased cholate synthesis but increased intestinal reabsorption. J Pharmacol Exp Ther 304: 356-363.
[87] Toussaint C, Kinnaert P, Vereerstraeten P (1988) Late mortality and morbidity five to eighteen years after kidney transplantation. Transplantation 45: 554-558.
[88] Bilchick KC, Henrikson CA, Skojec D, Kasper EK, Blumenthal RS (2004) Treatment of hyperlipidemia in cardiac transplant recipients. Am Heart J 148: 200-210.
[89] Ballantyne CM, Corsini A, Davidson MH, Holdaas H, Jacobson TA, et al. (2003) Risk for myopathy with statin therapy in high-risk patients. Arch Intern Med 163: 553-564.
[90] Ojo AO (2006) Cardiovascular complications after renal transplantation and their prevention. Transplantation 82: 603-611.
[91] Holdaas H, Fellstrom B, Jardine AG, Holme I, Nyberg G, et al. (2003) Effect of fluvastatin on cardiac outcomes in renal transplant recipients: a multicentre, randomised, placebo-controlled trial. Lancet 361: 2024-2031.
[92] Blum A, Shamburek R (2009) The pleiotropic effects of statins on endothelial function, vascular inflammation, immunomodulation and thrombogenesis. Atherosclerosis 203: 325-330.
[93] Wissing KM, Unger P, Ghisdal L, Broeders N, Berkenboom G, et al. (2006) Effect of atorvastatin therapy and conversion to tacrolimus on hypercholesterolemia and endothelial dysfunction after renal transplantation. Transplantation 82: 771-778.
[94] Kobashigawa JA, Katznelson S, Laks H, Johnson JA, Yeatman L, et al. (1995) Effect of pravastatin on outcomes after cardiac transplantation. N Engl J Med 333: 621-627.
[95] Asberg A (2003) Interactions between cyclosporin and lipid-lowering drugs: implications for organ transplant recipients. Drugs 63: 367-378.
[96] Pflugfelder PW, Huff M, Oskalns R, Rudas L, Kostuk WJ (1995) Cholesterol-lowering therapy after heart transplantation: a 12-month randomized trial. J Heart Lung Transplant 14: 613-622.
[97] Stapleton DD, Mehra MR, Dumas D, Smart FW, Milani RV, et al. (1997) Lipid-lowering therapy and long-term survival in heart transplantation. Am J Cardiol 80: 802-805.
[98] Boissonnat P, Salen P, Guidollet J, Ferrera R, Dureau G, et al. (1994) The long-term effects of the lipid-lowering agent fenofibrate in hyperlipidemic heart transplant recipients. Transplantation 58: 245-247.
[99] Suchy D, Labuzek K, Stadnicki A, Okopien B Ezetimibe--a new approach in hypercholesterolemia management. Pharmacol Rep 63: 1335-1348.
[100] Yoon HE, Song JC, Hyoung BJ, Hwang HS, Lee SY, et al. (2009) The efficacy and safety of ezetimibe and low-dose simvastatin as a primary treatment for dyslipidemia in renal transplant recipients. Korean J Intern Med 24: 233-237.
[101] Lopez V, Gutierrez C, Gutierrez E, Sola E, Cabello M, et al. (2008) Treatment with ezetimibe in kidney transplant recipients with uncontrolled dyslipidemia. Transplant Proc 40: 2925-2926.

Chapter 15

Lipoproteins and Apolipoproteins of the Ageing Eye

Edward Loane

Additional information is available at the end of the chapter

http://dx.doi.org/10.5772/45985

1. Introduction

In this chapter, we outline the structure of the retina and the aetiopathogenesis of the major age-related eye disease: age-related macular degeneration (AMD). We then discuss the role that lipoproteins and apolipoproteins play in the ageing eye and in the development of AMD.

2. The macula and retina

The macula is the central part of the retina, the neurosensory portion of the eye, and it is responsible for detailed central and colour vision due to its high concentration of cone photoreceptors. Anatomically, the macula is centred on the foveola, and has a ganglion cell layer of more than one cell in thickness. The macula has a diameter of approximately 5.5 mm. The macula is characterised by a yellowish colour (hence the term *macula lutea,* which is Latin for 'yellow spot'), attributable to the presence of macular pigment (MP).[1] The concentration of MP peaks at the centre of the macula, where the appearance of the 'yellow spot' may be clearly evident on clinical examination or fundus photography [Figure 1]. MP is optically undetectable outside the macula.[2] Within the layer structure of the retina, the highest concentration of MP is seen in the receptor axon layer and the inner plexiform layer.[1]

The retina consists of a neurosensory portion comprised of nine individual layers, and an external retinal pigment epithelium (RPE). The RPE plays an important physiological role in the maintenance of neurosensory retinal health, through functions including Vitamin A metabolism, phagocytosis of photoreceptor outer segments, maintenance of the outer blood-retina barrier, heat exchange, and the active transport of substances in and out of the RPE.[3] The blood supply of the retina is derived from the inner retinal vasculature and the outer choriocapillaris. Non-pathological changes that occur in the RPE with age include an increase in cellular pleomorphism and a decrease in cell number, with migration of peripheral RPE cells towards the macula, reduced melanin composition, and an accumulation of the age-pigment lipofuscin.[4;5] These changes may lead to a reduction in the metabolic activity of the RPE, with subsequent apoptosis, which pre-dates pathological change.[5;6] The RPE is separated from the choriocapillaris by Bruch's membrane (BrM). BrM is a semipermeable filtration barrier, comprised of five individual layers.[7;8] Disruption of BrM may result in alteration of its filtration properties, impacting on the function of the RPE and the neurosensory retina.[9] Changes that occur in BrM with age include an increase in its overall thickness, with a reconfiguration of associated lipids and proteins and the accumulation of debris.[10;11] When this debris accumulates between BrM and the RPE, it is referred to as a basal laminar deposit (BlamD) and is not specifically pathological in nature.[12] However, when deposits accumulate within the inner collagenous layer of BrM, they are referred to as basal linear deposits (BlinDs) and are a histopathological hallmark of AMD.[13] These deposits (BlamDs and BlinDs) contain a wide range of constituents including collagen, inflammatory proteins and lipoproteins. When sufficient debris accumulates in BlinDs, they are visible clinically as drusen.[14;15]

3. Age-related macular degeneration

Age-related macular degeneration (AMD) is the leading cause of blindness in people over 50 years of age in the developed world, and it results in loss of central and colour vision if not treated, or if not amenable to treatment.[16-18] The loss of central vision impacts greatly on the individual, as their ability to perform simple daily tasks, such as reading, watching television, driving and recognizing people's faces becomes increasingly difficult. Thus, their quality of life and their ability to lead an independent life diminish significantly as the disease progresses. The peripheral retina is not affected in individuals with AMD, regardless of stage, such that, in the absence of other ocular pathology, peripheral (navigational) vision remains unchanged.

It is currently estimated that late AMD affects 513,000 people in the United Kingdom (2.4% of those over the age of 50), and that this number will increase to 679,000 by the year 2020.[19] Prevalence data from the United States in 2004 estimated that more than 1.75 million individuals were affected by the disease, with this latter figure expected to rise to almost 3 million by the year 2020.[20] The prevalence of this condition is likely to increase dramatically in the future, as a result of increasing life-expectancy and the resultant increasing senescence of society.[21] Data from the National Eye Institute in the United States in 2004 indicated that the prevalence of advanced AMD in people over 40 years of age was 1.47%, rising to 15% in white females aged over 80 years. Beyond its impact on the individual sufferer,[22] the predicted increase in longevity (Figure 2), coupled with the predicted growth in world population (Figure 3) will significantly increase the socio-economic burden that AMD places on countries and their health-care systems.[23-26]

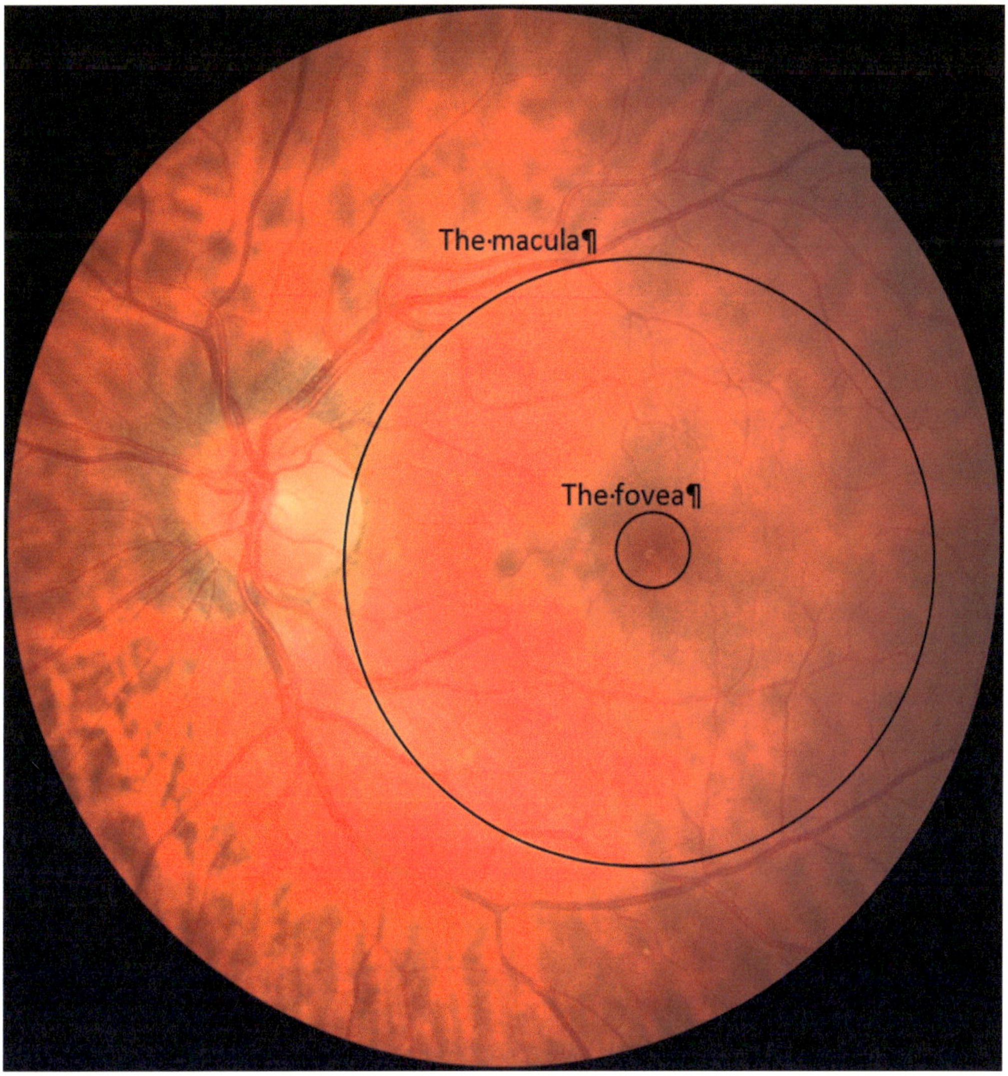

Figure 1. Colour fundus photograph showing the macula, surrounding the fovea, which is centred on the foveola (not marked, but evident as the 'yellow spot') of a left eye.

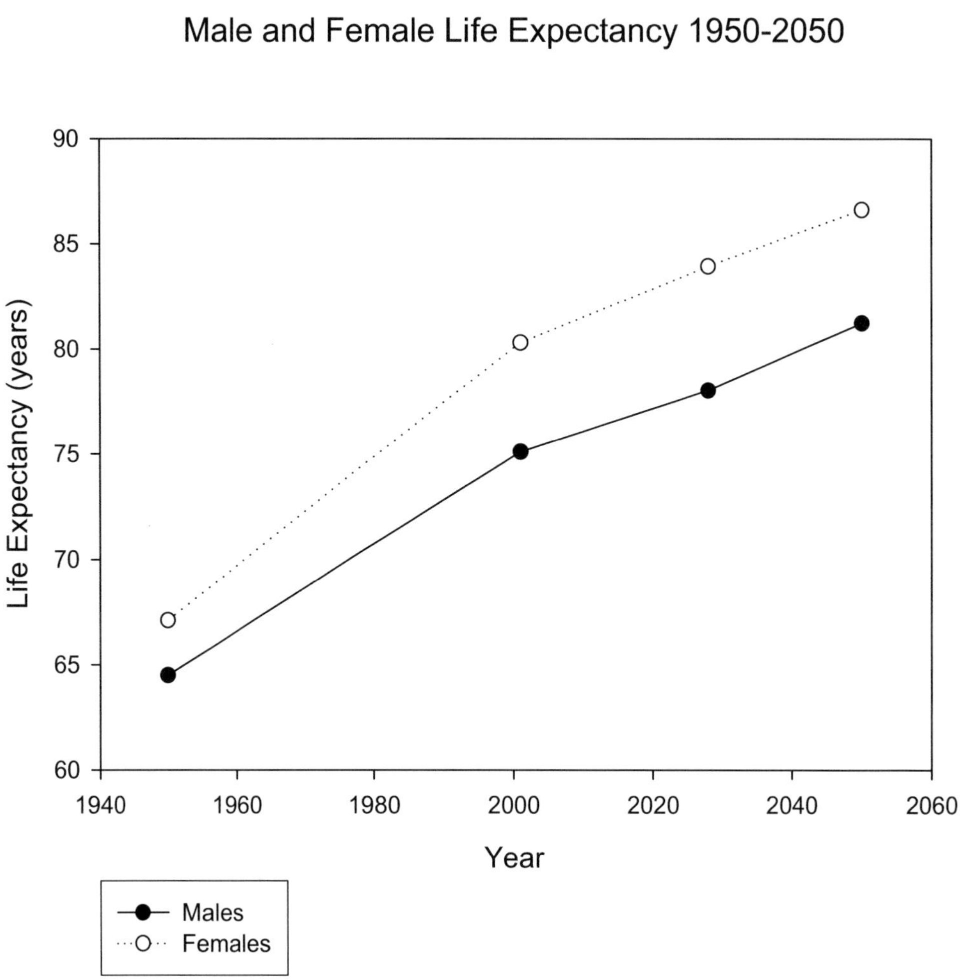

Figure 2. Male and female life expectancy 1950-2050.
* Figures from 1950 and 2001: Irish Department of Health and Children data;
Projected figures for 2028 and 2050: USA data.

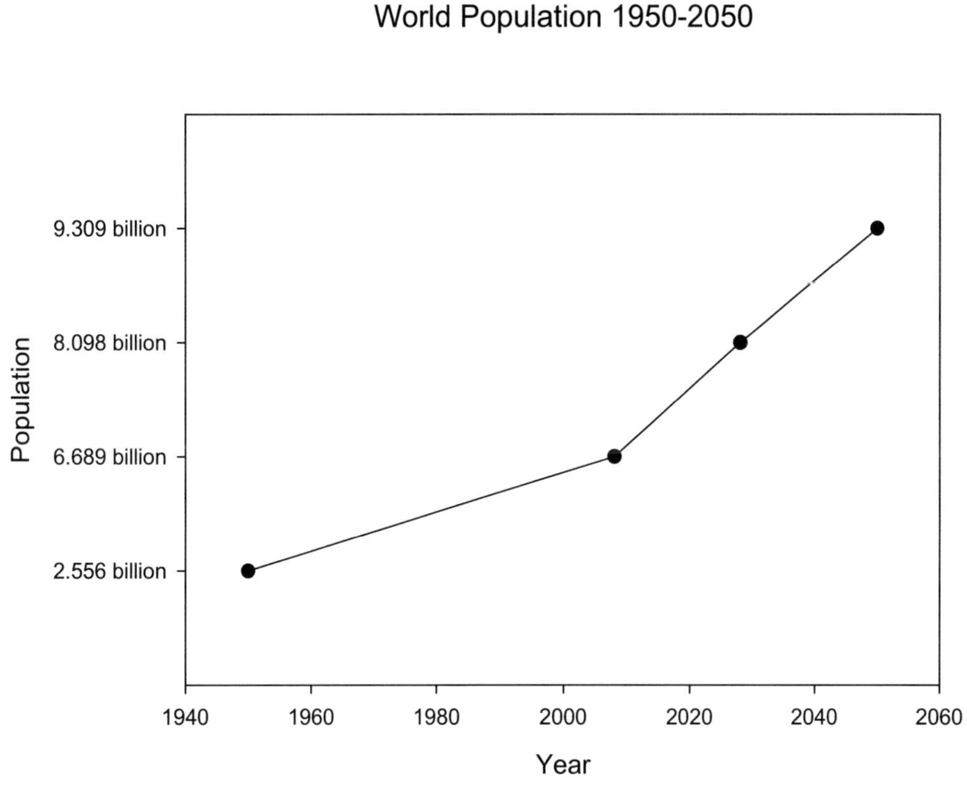

Figure 3. World population 1950-2050 (predicted).

4. Classification of AMD

In 1995, the International Age-Related Maculopathy Epidemiological Study Group clarified the definition and core grading system used to detect and define AMD.[27] This was done to homogenize the systems used to identify and classify this disease in all future clinical and epidemiological studies. This current classification system defines AMD primarily on the basis of morphological changes, without reference to visual acuity.

AMD is defined as a disorder of the macular area, most often clinically apparent after 50 years of age, and characterised by any of the following findings, which are not patently due to another disorder:

1. Soft drusen ≥ 63 μm in diameter. Drusen are whitish-yellow spots that lie external to the neurosensory retina or the RPE (Figure 4). Drusen may be soft and confluent, soft distinct, or soft indistinct. Hard drusen do not, of themselves, characterize AMD.
2. Hyperpigmentation in the outer retina or choroid associated with drusen.
3. Hypopigmentation of the RPE, most often more sharply demarcated than drusen, without any visible choroidal vessels associated with drusen.

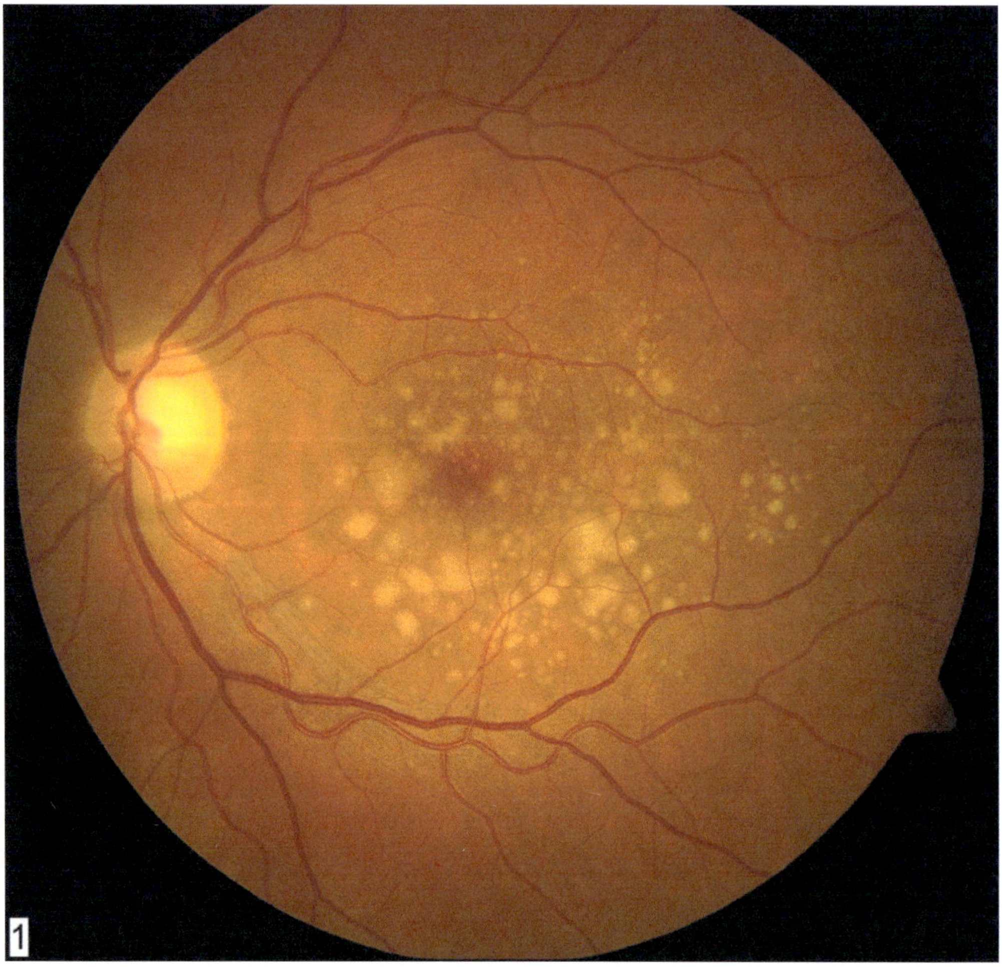

Figure 4. Macular soft drusen of a left eye.

These age-related pathological changes, which are associated with progressive accumulation of debris under the retina, predispose to the late stage of AMD.[28;29] Late AMD is classified as either geographic atrophy (atrophic AMD) or neovascular AMD (choroidal neovascularisation, also referred to as 'exudative AMD' or 'disciform AMD').

Geographic atrophy (GA) is characterised by the following, which is not patently due to another disorder:

1. Any sharply delineated area of hypopigmentation, or depigmentation, or apparent absence of the RPE, in which the choroidal vasculature is more visible than in the surrounding area. The area of atrophy must be ≥ 175 μm in diameter (Figure 5).

Neovascular AMD is characterised by any of the following, which are not patently due to another disorder:

1. RPE detachment(s), which may be associated with neurosensory retinal detachment.
2. Subretinal or sub-RPE neovascularisation.
3. Epiretinal, intraretinal, subretinal, or sub-RPE glial tissue or fibrin-like deposits.
4. Subretinal haemorrhage (Figure 6).
5. Hard exudates (lipids) within the macular area, related to any of the above, in the absence of other retinal vascular disease.

Rarely, neovascular AMD may develop in an area of GA. If this happens, the affected eye is re-classified as having neovascular AMD.

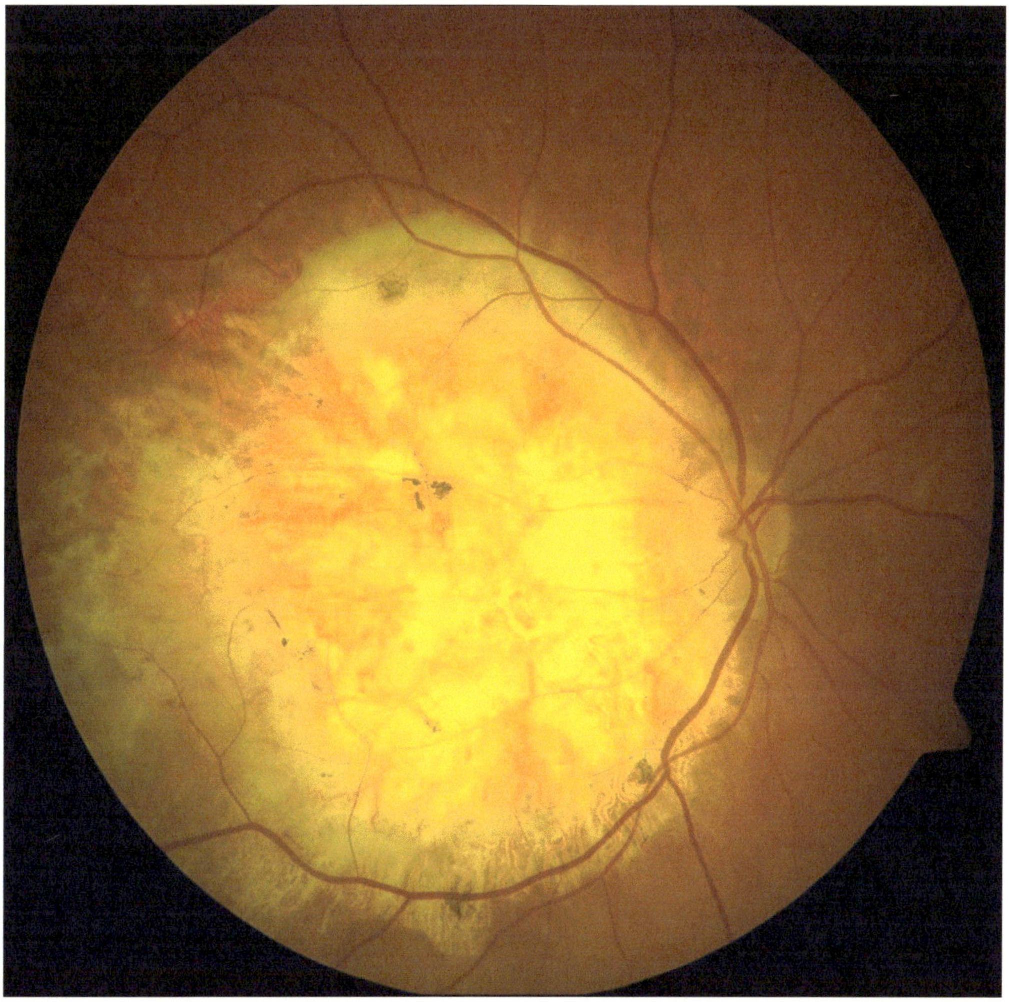

Figure 5. Geographic atrophy, affecting the entire macula of a right eye.

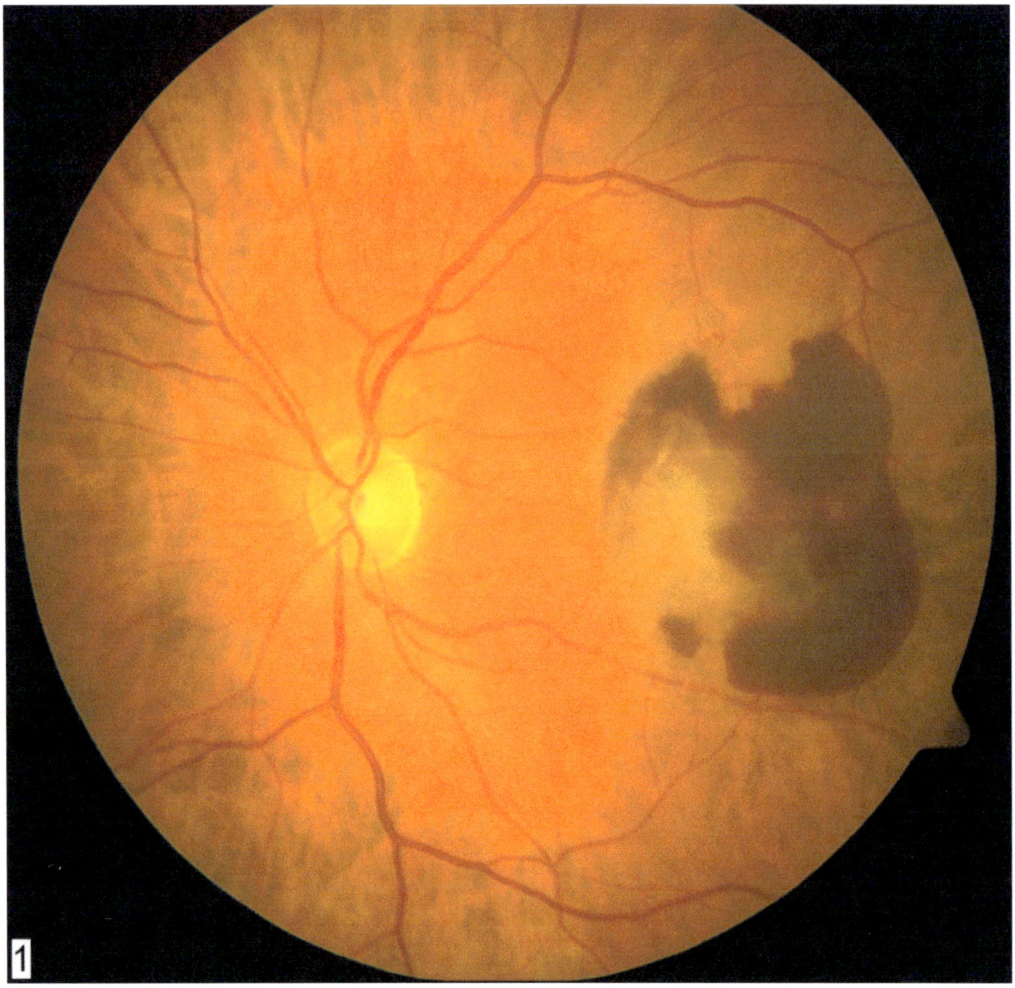

Figure 6. Neovascular AMD, showing sub-retinal haemorrhage in a left eye.

5. Pathogenesis of AMD

AMD has a multi-factorial pathogenesis.[30;31] Therefore, the development of AMD is dependent on a complex interaction between an individual's genetic composition (genotype) and lifestyle (or environmental) factors. This interaction is complex and incompletely understood; however, certain factors have been well established as representing risk for this condition, whereas others are known as putative risk factors, according to our current understanding of this disease. The well-established risk factors for the development of AMD are: increasing age, a positive family history of AMD (including specific genotypes), and tobacco smoking.[30;32;33] Therefore, tobacco smoking is the only proven environmental/lifestyle risk factor for this disease.[34;35] Putative risk factors include: obesity,[36;37] hypertension,[38] light iris colour,[39] cumulative sunlight exposure,[40] and a diet low in anti-oxidant fruits and vegetables,[41] particularly those containing the hydroxy-carotenoids: lutein and zeaxanthin.[42] Although the pathogenesis of AMD remains incompletely understood, there is a growing consensus that one or more of the following processes contribute to this condition: inflammation; oxidative stress; cumulative blue light damage; RPE cell and BrM dysfunction; reduced foveolar choroidal circulation.

6. Macular pigment

Macular pigment (MP) is composed of the hydroxy-carotenoids lutein (L), zeaxanthin (Z), and *meso*-zeaxanthin (*meso*-Z). L and Z are of dietary origin and are not synthesized *de novo* in humans, whereas *meso*-Z is not found in a conventional western diet, but is understood to be primarily formed in the retina following conversion from L.[43;44] Interestingly, it has been shown that L is the dominant carotenoid in the diet,[45] whereas Z/*meso*-Z have been shown to be the dominant carotenoids at the central macula.[46;47] MP is found in highest concentration at the central macula, where it functions as a powerful antioxidant and acts as a filter of actinic short wavelength blue light, thus limiting (photo-)oxidative damage to retinal cells.[48] These properties of MP are believed to be the mechanism whereby it may protect against the development, and/or progression, of AMD.

Although MP is entirely of dietary origin, it is also subject to heritability, as reported in 2005 by Liew *et al.* in a classic twin study.[49] In that study of 76 monozygotic and 74 dizygotic female twin pairs, they estimated that heritability accounted for between 67% and 85% of an individual's MP level. However, to date a direct significant association between MP levels and the major risk genes for AMD has not been shown.[50]

MP can be measured *in vivo* by non-invasive psychophysical means, resulting in an MP optical density measurement.[51;52]

7. Lipoproteins

Circulating lipoproteins consist of a complex of triglycerides, phospholipids and cholesterol, and one or more specific proteins, referred to as apolipoproteins. The association of lipoproteins with high affinity receptors on cell surfaces regulates lipid metabolism and transport in the body.[53] Lipoproteins are classified into the following six groups: chylomicrons; chylomicron remnants; very low density lipoproteins (VLDL); intermediate density lipoproteins (IDL); low density lipoproteins (LDL); high density lipoproteins (HDL).[53]

Chylomicrons are synthesised by the intestine and deliver dietary triglycerides to muscle and adipose tissue, and dietary cholesterol to the liver. Lipoprotein lipase, located at capillary endothelial cell surfaces, hydrolyses the triglyceride core of the chylomicron, thus liberating fatty acids and glycerol, which are used as energy sources by various cells, or are taken up by adipocytes and stored as triglycerides. Chylomicron remnants, which are rich in cholesterol, result from chylomicron metabolism, and are rapidly cleared by the liver.[53]

Subsequently, the liver synthesises a second class of triglyceride-rich lipoprotein, referred to as VLDL, which, upon secretion, functions as a transporter of lipids and cholesterol. In the bloodstream, VLDL undergoes progressive removal of triglycerides from its core by lipoprotein lipase, in a similar way to chylomicrons. The VLDL particles thus become increasingly smaller, leading to the formation of IDL, and LDL. LDL are the final metabolic products of VLDL and are responsible for most of the cholesterol transport in serum.[53]

HDL are the smallest lipoproteins, arising from several sources including the intestine and liver. HDL are involved in a process known as 'reverse cholesterol transport', whereby HDL acquire cholesterol from cells and deliver it to the liver.[53] This is a particularly important mechanism in humans, as the quantities of cholesterol transported out of the gut and liver far exceed the quantities converted to steroid hormones, or those lost through the skin in sebum. Thus, unless the requirement for cell membrane repair or synthesis is high, excess cholesterol must be returned to the liver for excretion.[54]

8. Association of carotenoids with plasma lipoproteins

The majority of plasma carotenoids are transported on LDL, with 55% of total carotenoids associated with this lipoprotein, whereas HDL is associated with 33%, and VLDL is associated with 10-19%, of the total carotenoids.[55] However, in the case of the hydroxy-carotenoids, L and Z, some studies have reported that they are relatively equally distributed between LDL and HDL molecules, but other studies have reported that HDL is the preferential carrier of the MP carotenoids in plasma.[56;57]

MP is inversely related to percentage body fat.[58] Interestingly, Viroonudomphol *et al.* have demonstrated lower levels of HDL in overweight and obese subjects, consistent with the possibility that a relative lack of HDL may impair transport and/or retinal capture of the carotenoids.[59] Furthermore, Seddon and co-workers have demonstrated a significantly increased risk of AMD in association with obesity.[33] These findings have prompted the suggestion that an individual's lipoprotein, and apolipoprotein, profile may influence the transport and delivery of these carotenoids to the retina, with a consequential impact on MP.

A recent study, designed to investigate the respective relationships between lipoprotein profile, MP optical density and serum concentrations of L and Z, was conducted in 302 healthy adult subjects.[60] This study found that there was a statistically significant inverse association between serum triglyceride concentration and MP optical density, and an inverse association between serum triglyceride concentration and serum L concentration in subjects with a positive family history of AMD. There have been no previous reports on the association between serum triglyceride concentration and either MP optical density or serum concentrations of L and/or Z. Elevated serum triglyceride concentration is an element of an undesirable lipoprotein profile and represents risk for cardiovascular disease.[61;62] Since there is an inverse association between serum triglyceride concentration and serum HDL concentration,[62] one could expect an inverse association between serum triglyceride concentration and serum L, since HDL appears to be the most important lipoprotein involved in the transport of L in serum. This expected inverse association was observed in subjects with a positive family history of AMD. In this study sample there was a positive and significant association between serum HDL concentration (and serum cholesterol concentration) and serum L and Z concentrations. Of note, there was no significant association observed between MP optical density and either serum cholesterol concentration or serum HDL concentration. There was also no association between serum LDL concentration and MP optical density (or serum concentrations of its constituent carotenoids). These findings suggest that a desirable lipoprotein profile (higher serum HDL, lower serum LDL and lower serum triglyceride concentrations) is associated with greater serum L concentration. However, the impact of lipoprotein profile on the capture and/or stabilization of these carotenoids at the macula, where they comprise MP, is less clear from this data.

In this study, the lipoprotein particle-concentration of L and/or Z in serum was not directly measured, nor were lipoprotein subspecies measured, as performed by Goulinet *et al.*[57] In their study, they fractionated HDL and LDL subspecies on the basis of their hydrated density by gradient ultracentrifugation, and they found that serum L and Z (combined) were relatively equally distributed between HDL and LDL; but more importantly, they found that there was a progressive decrease in the concentration of these carotenoids with increasing density (and decreasing lipoprotein particle size) from light to dense LDL. They also found that the majority of macular carotenoid transport by LDL was accounted for by the most abundant subspecies, LDL3 (intermediate LDL) and LDL4 (dense LDL). This is highly relevant to the transport of L and Z in serum, as LDL3 and LDL4, despite being the most abundant subspecies of LDL in that study, had reduced particle-concentrations of these carotenoids compared to less dense LDL subspecies, making them more vulnerable to oxidation.[63] LDL is the primary component of total cholesterol,[62] and has previously been reported in various studies to transport between 22-44% of L and Z in serum.[55;57;64-66] Of note, it has been shown that there is no significant difference in the transport of L and Z by lipoproteins between subjects with and without AMD.[65]

The findings of Goulinet *et al* in relation to HDL were similar to that of LDL, in that there was a progressive and marked decrease in HDL particle concentration of L and Z, with maximal carotenoid concentration evident in the lightest, largest HDL subspecies (HDL2-1), and minimal concentration in the densest HDL. Certainly, the findings of Goulinet *et al* with respect to HDL, in concert with our findings, are consistent with the view that HDL plays an important role in the transport of L and Z in human serum, and are provocative given that AMD and cardiovascular disease share certain antecedants.[32;57;60;67-70] Furthermore, and again consistent with a shared pathogenesis between AMD and cardiovascular disease, the finding of an inverse association between serum triglyceride concentration and MP optical density (and between serum triglyceride concentration and serum L concentration) in subjects with a positive family history of AMD, is noteworthy.[60] Since AMD has been shown to be associated with low serum concentrations of L,[71] and given that risk factors for AMD are associated with a relative lack of MP,[31] our observations are yet another example of how AMD and cardiovascular disease share risk factors.[32;60-62;67-70]

In 2007, Connor *et al* reported on the role that HDL plays in the transport of L and Z in serum in a study involving WHAM chicks.[64] WHAM chickens have a recessive sex-linked mutation in the *ABCA1* transporter gene that results in very low circulating HDL concentration, with normal, or increased, concentrations of other plasma lipoproteins, particularly LDL. The analogous mutation in humans results in Tangier disease, which is characterized by a similar deficiency in circulating HDL concentration.[72] In their study, involving 24 WHAM chicks and 24 control chicks, Connor *et al* found that one-day old WHAM chicks had only 9% of the L concentration in plasma when compared with control chicks, and only 6% of the retinal concentration of controls (the corresponding concentrations of Z were 6% and 9%, respectively). Following a high-L diet for 28 days, there was a significant increase in the plasma and retinal concentrations of L in WHAM chicks and controls, but the increases were still greatly inferior in the WHAM chicks when compared with control chicks and, furthermore, still did not reach the concentrations observed in the one-day old control chicks. The observations of Connor *et al* suggest an important role for HDL in the transport of L and Z in serum and/or their incorporation into the retina, and are consistent with our findings.[60;64]

Interestingly, although all subjects in our study were healthy volunteers with no evidence of ocular pathology, it is notable that, on average, subjects with a positive family history of AMD had a higher serum concentration of L than subjects with a negative family history of AMD, yet MP optical density levels in both groups were comparable, as were serum concentrations of HDL.[60] As was shown in this study, and as has previously been documented,[73] serum concentrations of L and Z generally correlate positively with MP optical density. Therefore, it is plausible to suggest that in the subjects in this study with a positive family history of AMD, the delivery to, and/or uptake by, the retina of the macular carotenoids is defective when compared to subjects without such a family history.[60] Indeed, although MP optical density levels were comparable between subjects with and without a family history of AMD, subjects with a positive family history of this disease also had higher serum L concentrations. This is consistent with the observations of Nolan *et al*, where a relative lack of MP was seen in association with a positive family history of AMD in 828 healthy subjects, but where dietary and serum concentrations of L and Z were comparable for subjects with and without a family history of this condition, suggesting defective retinal capture of circulating L and/or Z in persons who are genetically predisposed to AMD.[31] Mechanisms governing the retinal capture and/or stabilization of L and/or Z may be subject to influence by HDL subspecies profile, by affecting receptor-mediated uptake of these carotenoids from serum. Indeed, apolipoprotein profile is probably a determinant of retinal uptake of the macular carotenoids from serum, reflected in our recently reported finding that individuals with at least one Apo ε4 allele exhibit significantly higher MP optical density than individuals without this protective allele, despite statistically comparable serum concentrations of L and Z.[74] Interestingly, the lack of an association between MP optical density and either serum cholesterol concentration or serum HDL concentration in our study would suggest that our observations are more likely due to impaired uptake and/or stabilization of circulating L and/or Z by the macula than due to any impact the HDL subspecies profile may have on the transport of the macular carotenoids in serum.

Another recent study has shown somewhat conflicting evidence regarding the association between circulating lipoprotein levels and MP levels in serum and in the macula.[75] These differences may be attributable to differences in the methods used to measure serum lipoproteins, although it should be noted that this study also found a positive association between serum L and serum HDL levels, underscoring the importance of HDL as a transporter of L in serum. However, it should be emphasised that a notable paucity of data still remains regarding the mechanism(s) whereby L and Z accumulate in the liver, are repackaged into lipoproteins, and transported via the circulatory system to specific target tissues such as the retina.

9. Apolipoproteins

Plasma lipoproteins include one or more protein constituents, known as apolipoproteins. Apolipoproteins have been classified into several subgroups, including apolipoprotein A (ApoA), apolipoprotein B (ApoB), apolipoprotein C (ApoC), and apolipoprotein E (ApoE). These subgroups are themselves further sub-classified, for example: ApoA-I, ApoA-II etc. Each lipoprotein class is associated with certain apolipoproteins, for example: chylomicrons and VLDL are associated with ApoB; chylomicrons, VLDL and HDL are associated with ApoE.[76] The primary role of apolipoproteins is the transport and redistribution of lipids amongst various tissues in the body. Specific apolipoproteins are recognised by cell surface receptors, and this facilitates the high affinity binding required for delivery to target tissues. Certain apolipoproteins also act as cofactors of enzymes involved in lipoprotein metabolic pathways, including those of lipoprotein lipase and lecithin-cholesterol acyl transferase (LCAT), which catalyse the formation of cholesterol esters. Another role of specific apolipoproteins is the maintenance of the structure of lipoproteins, by stabilizing their micellar structure, and by providing a hydrophilic surface in association with phospholipids.[53] The function of apolipoproteins has provoked interest in their possible role in a range of degenerative conditions. In particular, several investigators have suggested an association between ApoE and various diseases, including Alzheimer's disease, atherosclerosis and AMD.[77-80]

Abalain *et al.* investigated the association between AMD and serum levels of lipoproteins and lipoparticles.[78] They found that there was no difference in serum ApoA-I and ApoB levels between AMD patients and controls. However, they found that serum ApoE levels were higher, and that serum ApoC-III levels were lower, in AMD patients compared with controls. The higher level of serum ApoE in AMD patients is consistent with the findings of Boerwinkle and Utermann, who found that the Apo ε4 allele is associated with lower serum ApoE levels, and that the Apo ε2 allele is associated with higher serum levels of ApoE.[79] ApoC-III interferes with lipoprotein metabolism and, when associated with ApoB as a lipoparticle, it has been shown to be involved in atherogenesis.[80] Abalain *et al.* found no difference in the levels of this particular lipoparticle between AMD patients and controls.[78] The evidence to date suggests that, of the apolipoproteins, ApoE has the strongest association with AMD.

10. Apolipoprotein E

ApoE is a structural component of plasma chylomicrons, VLDL, and a subclass of HDL. It is a 299 amino-acid protein, and is synthesised in a large number of tissues including the spleen, kidneys, lungs, adrenal glands, liver, brain and retinal Müller cells.[81] ApoE is polymorphic, with three common isoforms: E2, E3 and E4, which are coded for by three separate alleles: Apo ε2, Apo ε3 and Apo ε4. These alleles are differentiated on the basis of cysteine-arginine residue interchanges at sites 112 and 158 in the amino acid sequence.[82] As a result of this polymorphism, six common phenotypes exist: three homozygous phenotypes (ε3ε3, ε2ε2, ε4ε4) and three heterozygous phenotypes (ε2ε3, ε2ε4, ε3ε4). ApoE is crucial to many processes, including: cholesterol transport and metabolism; receptor-mediated uptake of specific lipoproteins; heparin binding; formation of cholesteryl-ester-rich particles; lipolytic processing of type IIIβ-VLDL; inhibition of mitogenic stimulation of lymphocytes; transport of lipids within the brain.[53]

ApoE is an important regulator of cholesterol metabolism because of its affinity for ApoE-specific receptors in the liver, and its affinity for LDL receptors in the liver and other peripheral tissues requiring cholesterol.[53] ApoE-specific receptors are present on the membranes of hepatic parenchymal cells, and have a high binding affinity for chylomicron remnants, IDL and a sub-class of HDL. ApoE also regulates the activity of several lipid-metabolising enzymes, including lipoprotein lipase, and LCAT.

ApoE is found in greatest concentrations in the liver. However, it is also the predominant apolipoprotein in the brain, and is responsible for lipid transport and cholesterol regulation within the central nervous system (CNS). ApoE is a major component of plasma and cerebrospinal fluid, and plays a fundamental role after CNS injury, where it appears to regulate the transport of cholesterol and phospholipids during the early and intermediate phases of the reinnervation process.[83;84]

ApoE polymorphisms result in differences in the metabolism of ApoE-containing lipoprotein particles.[85] For example, it is possible that certain ApoE polymorphisms affect their ability to interact with lipoprotein lipase in the conversion of VLDL to LDL.[86] Indeed, ApoE polymorphism influences plasma lipid levels both in sedentary states and in their response to exercise, and it is therefore believed to be related to risk for coronary artery disease. In general, carriers of the Apo ε4 allele have higher levels of total cholesterol and LDL-cholesterol than those with the Apo ε3 allele. ApoE polymorphism also appears to play a role in the responsiveness of blood lipids to dietary and lipid-lowering drug interventions. Thus, the ApoE gene-environmental interactions contribute to population variance in blood lipid-lipoprotein levels.[87]

ApoE receptors also play an important role in lipoprotein metabolism. The primary physiological role of ApoE is to facilitate the binding of lipoproteins to LDL receptors, thereby regulating the uptake of cholesterol required by the cell. For instance, large amounts of lipids are released from degenerating cell membranes after nerve cell loss, thus stimulating astrocytes to synthesise ApoE, which binds these excess lipids and distributes them appropriately for reuse in cell membrane biosynthesis.[88] This observation prompted Klaver *et al.* to speculate that a high degree of ApoE biosynthesis is required to support the high rate of photoreceptor renewal at the macula.[88] Indeed, it has been demonstrated that mice which were fed a high-fat diet, or which were deficient in ApoE, exhibit an increase in the thickness of BrM, which is seen in association with ageing and with AMD.[89]

Ishida *et al.* identified the presence of ApoE and lipids at the inner aspect of the RPE, and proposed that both compounds may be secreted by the RPE.[90] The role of ApoE in reverse cholesterol transport prompted the authors to suggest that this apolipoprotein may also facilitate the efflux of lipids from the RPE into the adjacent BrM, and they proposed a possible pathway for RPE cell-secreted lipids to cross BrM, where partially digested or undigested photoreceptor outer segments are secreted across the basal surface in association with ApoE. Subsequent binding with HDL at BrM may then facilitate desorption of the lipid particles into the circulation.[90]

In the retina ApoE is synthesised in Müller cells and in the RPE, and the presence of ApoE has been demonstrated in drusen.[81;91;92] It has been suggested, therefore, that age and/or disease-related disruption of normal ApoE function may result in the accumulation of lipoproteins at the interface between the RPE and BrM, consistent with observations that lipid deposits in drusen are largely composed of cholesteryl esters and unsaturated fatty acids.

These findings are consistent with the view that ApoE plays an important physiological role in the maintenance of macular health, and that an impaired ApoE system may affect the functional integrity of BrM. Furthermore, there is a biologically plausible rationale whereby the ApoE profile might influence the transport, capture, and stabilization of key compounds, such as L and Z, at the macula.

11. Lipoproteins, apolipoproteins and the retina

As noted previously, the ageing retina features changes in the RPE and BrM, which include changes in the lipoprotein and apolipoprotein composition of both structures. These changes may progress to the disease state of AMD. In recent times, evidence accrued from light microscopy, ultrastructural studies, lipid histochemistry, isolated lipoprotein assays, and gene expression analysis had led to the identification of many of the constituents that deposit in the RPE and BrM with age and AMD.[93] One of the universal changes that occurs with age is the development of BlamDs between the RPE and BrM.[11;12] This process may progress to the development of a 'lipid wall', mainly composed of neutral lipid deposits, decreasing the permeability of BrM and hindering metabolic activity between the RPE and BrM, preceding pathological changes associated with AMD.[10;93;94] When these deposits accumulate within the inner collagenous layer of BrM, they are referred to as basal linear deposits (BlinDs) and are a histopathological hallmark of AMD, which, when sufficiently large, can be recognised clinically as drusen.[13-15;95]

Much of the debris that accumulates in BrM in the form of BlinDs is composed of lipoproteins and lipoprotein particles.[14] It has been found that almost 60% of the total cholesterol within these lipoproteins is esterified cholesterol.[96] Furthermore, the esterified cholesterol within BrM was enriched between 16 and 40-fold compared to plasma. If these extracellular lipid deposits had been derived from plasma, more than 90% of the phospholipid would be phosphatidylcholine, whereas in actual fact, these lipoproteins are comprised of less than 50% phosphatidylcholine.[96] Indeed, the composition of drusen, which are essentially large BlinDs, has been shown to include esterified and unesterified cholesterol, and multiple apolipoproteins, including apolipoproteins B, A-I, C-I, C-II, and E, appearing with frequencies ranging from 100% (ApoE) to approximately 60% (A-I).[88;91;97;98] Interestingly, ApoC-III, although abundant in plasma, is present in fewer drusen (16.6%) than ApoC-I (93.1%), which is not present in plasma in large quantities, indicating either a specific retention of plasma-derived apolipoproteins within drusen, or an intraocular source for these apolipoproteins.[93] It is now understood that the majority of lipoproteins in BrM have undergone intracellular processing within the RPE prior to secretion as neutral lipids, mainly esterified cholesterol.[99;100] The RPE origin has been definitively shown by two groups using metabolic labelling and immunoprecipitation in rat-derived and human-derived RPE cell lines that were shown to secrete full-length ApoB.[101;102] This evidence is further strengthened by the finding of microsomal triglyceride transfer protein within native human RPE, indicating that the RPE is capable of secreting lipoprotein particles.[102] The pattern of lipid deposition in BrM with age, in which debris appears firstly in the elastic layer and then fills in towards the RPE, is also consistent with this lipid being primarily of RPE origin.[103]

The hydrophobic nature of the age-related thickening of BrM has been implicated in the aetiopathogenesis of AMD. In the case of Apo E, it is noteworthy that ApoE4 presents a positive charge relative to both ApoE2 and ApoE3. ApoE4 possesses arginine at residue 112 of the amino acid sequence, whereas ApoE3 possesses cysteine at this position, and in the case of ApoE2, the most frequent variant has cysteine instead of the normally occurring arginine at residue 158. Thus, ApoE3 presents a neutral charge, and ApoE2 a negative charge, relative to ApoE4.[53] Souied *et al.* suggested that this difference in charges between the ApoE isoforms may also contribute to differences in the clearance of debris through BrM.[104]

It appears that Müller cells are the most prominent biosynthetic sources of ApoE in the neural retina, and RPE cells are the most prominent sources in the RPE/choroid.[91] However, it remains unclear whether the concentration of ApoE in the cytoplasm of some RPE cells, especially those in close proximity to drusen, is the result of biosynthesis or selective accumulation. It has been shown that, in both the central and peripheral nervous systems, ApoE expression by astrocytes is up-regulated in response to neuronal injury and neuro-degenerative disease.[84;105;106] Indeed, there is evidence for ApoE up-regulation by Müller cells in degenerating human retina, where increased ApoE immuno-reactivity is found in the sub-retinal space of detached retinas[107] and in the Müller cells of retinas affected by glaucoma or AMD.[108] Furthermore, the relatively high levels of ApoE mRNA detected in the retina, especially in the eyes of older donors and in an individual with documented AMD, support the view that up-regulation by retinal glia may be responsible for the observed increase in ApoE expression.[91]

12. Apo ε4 allele status and AMD

ApoE gene status is believed to be a determinant of AMD risk.[88;104;109-111] The *ApoE* gene has three separate alleles: Apo ε2, Apo ε3 and Apo ε4, resulting in six common phenotypes: three homozygous (ε3ε3, ε2ε2, ε4ε4) and three heterozygous (ε2ε3, ε2ε4, ε3ε4) phenotypes. The ε4 allele has been found to be associated with a reduced risk of AMD, whereas the ε2 allele has been associated with an increased risk of developing this disease.[88;104;109-113]

Due to the lack of cysteine residues at positions 112 and 158, preventing the formation of disulphide bridges with ApoA-II or other peptide components, the Apo ε4 allele has an inability to form dimers. It has been suggested that this inability of the Apo ε4 allele to form dimers, when compared with the Apo ε2 and Apo ε3 alleles, favours easier transport of lipids through BrM because of the smaller sized lipid particles, thus protecting against a loss of permeability of BrM.[104]

In the same way, it is possible that the neurosensory retina and the RPE respond to conditions of high oxidative injury by up-regulation of ApoE synthesis and/or accumulation, with implications for selective capture and stabilisation of L and Z in the retina.[91] It has been demonstrated that there is selective binding of certain receptors within the CNS to HDL particles enriched with ApoE, and that there is a lack of binding of these receptors to HDL particles deficient in ApoE.[114] Should this selectivity of the uptake mechanism be dependent on the ApoE polymorphism of the transporting lipoproteins, and given that the Apo ε4 allele is putatively protective for AMD, it is tempting to hypothesise that retinal capture of L and Z may be related to apolipoprotein profile. In other words, the apolipoprotein composition as well as the lipoprotein profile, may play an important role in the transport and delivery of L and Z, and their subsequent accumulation and stabilisation within the retina.[115] Therefore, it is possible that the putative protective effect of the Apo ε4 allele against AMD is attributable, at least in part, to the role its phenotypic expression (ApoE4) plays in the transport and delivery of the macular carotenoids to the retina, and to their stabilisation within the retina. Furthermore, recent research has shown an association between possession of at least one Apo ε4 allele and higher levels of MP across the macula, which is consistent with the view that apolipoprotein profile influences the transport and/or retinal capture of the macular carotenoids.[74]

13. Conclusion

In conclusion, the role that lipoproteins and apolipoproteins play in the ageing eye and in the aetiopathogenesis of AMD is complex and, as yet, incompletely understood. Lipoproteins and apolipoproteins play an important role in the delivery of potentially protective nutrients from the digestive tract to the eye. The local ocular metabolic activity, centred on the RPE and BrM, involves an exchange of nutrients from the choroidal circulatory system via BrM to the RPE and retina, with a reverse process whereby waste products are removed from the retina by the RPE through BrM in association with locally produced lipoproteins and apolipoproteins (particularly ApoB and ApoE). Unfortunately, over time it appears that these lipoproteins and apolipoproteins can accumulate between the RPE and BrM, and within BrM, leading to degradation in the metabolic efficiency between these two structures and the choroidal circulation. This deposition has been described as a 'lipid wall' and precedes the development of AMD.[93;94] Methods to detect and arrest or delay this process before it becomes clinically apparent and visually consequential to the patient have yet to be developed. Recent advances in our understanding of the lipoprotein and apolipoprotein molecular biology of the ageing and AMD-affected eye will help to direct future treatment strategies.[100]

Author details

Edward Loane
Department of Ophthalmology, Mater Misericordiae University Hospital, Dublin, Ireland

14. References

[1] Snodderly DM, Brown PK, Delori FC, Auran JD. The Macular Pigment .1. Absorbance Spectra, Localization, and Discrimination from Other Yellow Pigments in Primate Retinas. Investigative Ophthalmology & Visual Science 1984;25(6):660-73.
[2] Werner JS, Donnelly SK, Kliegl R. Aging and human macular pigment density. Appended with translations from the work of Max Schultze and Ewald Hering. Vision Research 1987;27:275-68.
[3] Bok D. The retinal pigment epithelium: a versatile partner in vision. J Cell Sci Suppl 1993;17:189-95.:189-95.
[4] Boulton M, yhaw-Barker P. The role of the retinal pigment epithelium: topographical variation and ageing changes. Eye (Lond) 2001 Jun;15(Pt 3):384-9.
[5] Del Priore LV, Kuo YH, Tezel TH. Age-related changes in human RPE cell density and apoptosis proportion in situ. Invest Ophthalmol Vis Sci 2002 Oct;43(10):3312-8.
[6] Dunaief JL, Dentchev T, Ying GS, Milam AH. The role of apoptosis in age-related macular degeneration. Arch Ophthalmol 2002 Nov;120(11):1435-42.
[7] American Academy of Ophthalmology. Basic and Clinical Science Course, Section 2: Fundamentals and Principles of Ophthalmology. 2011.
[8] Snell RS, Lemp MA. Clinical Anatomy of the Eye. Second ed. Wiley-Blackwell; 1998.
[9] Marshall J. The ageing retina: physiology or pathology. Eye (Lond) 1987;1(Pt 2):282-95.
[10] Zarbin MA. Current concepts in the pathogenesis of age-related macular degeneration. Arch Ophthalmol 2004;122(4):598-614.
[11] Pauleikhoff D, Harper CA, Marshall J, Bird AC. Aging changes in Bruch's membrane. A histochemical and morphologic study. Ophthalmology 1990 Feb;97(2):171-8.

[12] van der Schaft TL, de Bruijn WC, Mooy CM, de Jong PT. Basal laminar deposit in the aging peripheral human retina. Graefes Arch Clin Exp Ophthalmol 1993 Aug;231(8):470-5.
[13] Curcio CA, Millican CL. Basal linear deposit and large drusen are specific for early age-related maculopathy. Arch Ophthalmol 1999 Mar;117(3):329-39.
[14] Curcio CA, Presley JB, Millican CL, Medeiros NE. Basal deposits and drusen in eyes with age-related maculopathy: evidence for solid lipid particles. Exp Eye Res 2005 Jun;80(6):761-75.
[15] Lommatzsch A, Hermans P, Muller KD, Bornfeld N, Bird AC, Pauleikhoff D. Are low inflammatory reactions involved in exudative age-related macular degeneration? Morphological and immunhistochemical analysis of AMD associated with basal deposits. Graefes Arch Clin Exp Ophthalmol 2008 Jun;246(6):803-10.
[16] Bressler NM. Age-related macular degeneration is the leading cause of blindness. JAMA 2004 Apr 21;291(15):1900-1.
[17] Congdon NG, Friedman DS, Lietman T. Important causes of visual impairment in the world today. JAMA 2003 Oct 15;290(15):2057-60.
[18] Klein R, Wang Q, Klein BEK, Moss SE, Meuer SM. The Relationship of Age-Related Maculopathy, Cataract, and Glaucoma to Visual-Acuity. Investigative Ophthalmology & Visual Science 1995;36(1):182-91.
[19] Owen CG, Jarrar Z, Wormald R, Cook DG, Fletcher AE, Rudnicka AR. The estimated prevalence and incidence of late stage age related macular degeneration in the UK. Br J Ophthalmol 2012 Feb 13.
[20] Friedman DS, O'Colmain BJ, Munoz B, Tomany SC, McCarty C, de Jong PT, et al. Prevalence of age-related macular degeneration in the United States. Arch Ophthalmol 2004 Apr;122(4):564-72.
[21] van Leeuwen R, Klaver CC, Vingerling JR, Hofman A, de Jong PT. Epidemiology of age-related maculopathy: a review. Eur J Epidemiol 2003;18(9):845-54.
[22] Augustin A, Sahel JA, Bandello F, Dardennes R, Maurel F, Negrini C, et al. Anxiety and depression prevalence rates in age-related macular degeneration. Invest Ophthalmol Vis Sci 2007 Apr;48(4):1498-503.
[23] Gupta OP, Brown GC, Brown MM. Age-related macular degeneration: the costs to society and the patient. Curr Opin Ophthalmol 2007 May;18(3):201-5.
[24] Owen CG, Fletcher AE, Donoghue M, Rudnicka AR. How big is the burden of visual loss caused by age related macular degeneration in the United Kingdom? Br J Ophthalmol 2003 Mar 1;87(3):312-7.
[25] Bandello F, Lafuma A, Berdeaux G. Public health impact of neovascular age-related macular degeneration treatments extrapolated from visual acuity. Invest Ophthalmol Vis Sci 2007 Jan;48(1):96-103.
[26] Cruess AF, Zlateva G, Xu X, Soubrane G, Pauleikhoff D, Lotery A, et al. Economic burden of bilateral neovascular age-related macular degeneration: multi-country observational study. Pharmacoeconomics 2008;26(1):57-73.
[27] Bird AC, Bressler NM, Bressler SB, Chisholm IH, Coscas G, Davis DM, et al. An international classification and grading system for age-related maculopathy and age-

related macular degeneration. The International ARM Epidemiological Study Group. Survey of Ophthalmology 1995;39(5):367-74.
[28] Gass JD. Pathogenesis of disciform detachment of the neuroepithelium. Am J Ophthalmol 1967 Mar;63(3):Suppl-139.
[29] Sarks SH. Council Lecture. Drusen and their relationship to senile macular degeneration. Aust J Ophthalmol 1980 May;8(2):117-30.
[30] Tomany SC, Wang HJ, van Leeuwen R, Klein R, Mitchell P, Vingerling JR, et al. Risk factors for incident age-related macular degeneration - Pooled findings from 3 continents. Ophthalmology 2004;111(7):1280-7.
[31] Nolan JM, Stack J, O'Donovan O, Loane E, Beatty S. Risk factors for age-related maculopathy are associated with a relative lack of macular pigment. Exp Eye Res 2007 Jan;84(1):61-74.
[32] Delcourt C, Michel F, Colvez A, Lacroux A, Delage M, Vernet MH, et al. Associations of cardiovascular disease and its risk factors with age-related macular degeneration: the POLA study. Ophthalmic Epidemiology 2001 Sep;8(4):237-49.
[33] Seddon JM, Cote J, Davis N, Rosner B. Progression of age-related macular degeneration: associated with body mass index, waist circumference, and waist-hip ratio. Arch Ophthalmol 2003;121:785-92.
[34] Tomany SC, Cruickshanks KJ, Klein R, Klein BEK, Knudtson MD. Sunlight and the 10-year incidence of age-related maculopathy - The Beaver Dam eye study. Arch Ophthalmol 2004;122(5):750-7.
[35] SanGiovanni JP, Chew EY, Clemons TE, Ferris FL, III, Gensler G, Lindblad AS, et al. The relationship of dietary carotenoid and vitamin A, E, and C intake with age-related macular degeneration in a case-control study: AREDS Report No. 22. Arch Ophthalmol 2007 Sep;125(9):1225-32.
[36] Hammond BR, Johnson MA. The Age-related Eye Disease Study (AREDS). Nutrition Reviews 2002;60(9):283-8.
[37] Klein R, Klein BEK, Franke T. The Relationship of Cardiovascular-Disease and Its Risk-Factors to Age-Related Maculopathy - the Beaver Dam Eye Study. Ophthalmology 1993;100(3):406-14.
[38] Hyman L, Schachat AP, He QM, Leske MC. Hypertension, cardiovascular disease, and age-related macular degeneration. Arch Ophthalmol 2000;118(3):351-8.
[39] Tomany SC, Klein R, Klein BEK. The relationship between iris color, hair color, and skin sun sensitivity and the 10-year incidence of age-related maculopathy - The beaver dam eye study. Ophthalmology 2003;110(8):1526-33.
[40] Klein R, Tomany SC, Cruickshanks KJ, Klein BEK. Sunlight and the 10-year incidence of age-related maculopathy. The Beaver Dam Eye Study. Arch Ophthalmol 2004 May;122(5):750-7.
[41] Delcourt C, Carriere I, Delage M, Barberger-Gateau P, Schalch W. Plasma lutein and zeaxanthin and other carotenoids as modifiable risk factors for age-related maculopathy and cataract: the POLA Study. Invest Ophthalmol Vis Sci 2006 Jun;47(6):2329-35.

[42] Sommerburg O, Keunen JEE, Bird AC, van Kuijk FJGM. Fruits and vegetables that are sources for lutein and zeaxanthin: the macular pigment in human eyes. Br J Ophthalmol 1998;82(8):907-10.

[43] Bone RA, Landrum JT, Hime GW, Cains A, Zamor J. Stereochemistry of the Human Macular Carotenoids. Investigative Ophthalmology & Visual Science 1993;34(6):2033-40.

[44] Johnson EJ, Neuringer M, Russell RM, Schalch W, Snodderly DM. Nutritional manipulation of primate retinas, III: effects of lutein or zeaxanthin supplementation on adipose tissue and retina of xanthophyll-free monkeys. Investigative Ophthalmology Visual Science 2005 Feb 1;46(2):692-702.

[45] Bialostosky K, Wright JD, Kennedy-Stephenson J, McDowell M, Johnson CL. Dietary intake of macronutrients, micronutrients, and other dietary constituents: United States 1988-94. Vital Health Stat 11 2002 Jul;(245):1-158.

[46] Bone RA, Landrum JT, Fernandez L, Tarsis SL. Analysis of the macular pigment by HPLC - Retinal distribution and age study. Investigative Ophthalmology & Visual Science 1988;29(6):843-9.

[47] Snodderly DM, Handelman GJ, Adler AJ. Distribution of individual macular pigment carotenoids in central retina of macaque and squirrel monkeys. Investigative Ophthalmology & Visual Science 1991;32(2):268-79.

[48] Snodderly DM. Evidence for Protection Against Age-Related Macular Degeneration by Carotenoids and Antioxidant Vitamins. Am J Clin Nutr 1995;62(6):S1448-S1461.

[49] Liew SHM, Gilbert C, Spector TD, Mellerio J, Marshall J, van Kuijk FJGM, et al. Heritability of Macular Pigment: a Twin Study. Investigative Ophthalmology & Visual Science 2005;46(12):4430-6.

[50] Loane E, Nolan JM, McKay GJ, Beatty S. The association between macular pigment optical density and CFH, ARMS2, C2/BF, and C3 genotype. Exp Eye Res 2011 Nov;93(5):592-8.

[51] Loane E, Stack J, Beatty S, Nolan JM. Measurement of macular pigment optical density using two different heterochromatic flicker photometers. Curr Eye Res 2007 Jun;32(6):555-64.

[52] Wooten BR, Hammond BR, Land RI, Snodderly DM. A practical method for measuring macular pigment optical density. Investigative Ophthalmology & Visual Science 1999;40(11):2481-9.

[53] Mahley RW, Innerarity TL, Rall SC, Jr., Weisgraber KH. Plasma lipoproteins: apolipoprotein structure and function. J Lipid Res 1984 Dec 1;25(12):1277-94.

[54] Durrington PN. Lipoproteins and their metabolism. In: Durrington PN, editor. Hyperlipidaemia: Diagnosis and Management. Butterworth-Heinemann Ltd; 1989.

[55] Clevidence BA, Bieri JG. Association of carotenoids with human plasma lipoproteins. Methods in Enzymology 1993;214:33-46.

[56] Erdman JW, Jr., Bierer TL, Gugger ET. Absorption and transport of carotenoids. Ann N Y Acad Sci 1993 Dec 31;691:76-85.

[57] Goulinet S, Chapman MJ. Plasma LDL and HDL subspecies are heterogenous in particle content of tocopherols oxygenated and hydrocarbon carotenoids - Relevance to

oxidative resistance and atherogenesis. Arteriosclerosis Thrombosis and Vascular Biology 1997;17(4):786-96.
[58] Nolan J, O'Donovan O, Kavanagh H, Stack J, Harrison M, Muldoon A, et al. Macular pigment and percentage of body fat. Investigative Ophthalmology Visual Science 2004 Nov 1;45(11):3940-50.
[59] Viroonudomphol D, Pongpaew P, Tungtrongchitr R, Changbumrung S, Tungtrongchitr A, Phonrat B, et al. The relationships between anthropometric measurements, serum vitamin A and E concentrations and lipid profiles in overweight and obese subjects. Asia Pacific Journal of Clinical Nutrition 2003;12(1):73-9.
[60] Loane E, Nolan JM, Beatty S. The respective relationships between lipoprotein profile, macular pigment optical density, and serum concentrations of lutein and zeaxanthin. Invest Ophthalmol Vis Sci 2010 Nov;51(11):5897-905.
[61] Hokanson JE, Austin MA. Plasma triglyceride level is a risk factor for cardiovascular disease independent of high-density lipoprotein cholesterol level: a meta-analysis of population-based prospective studies. J Cardiovasc Risk 1996 Apr;3(2):213-9.
[62] Morrison A, Hokanson JE. The independent relationship between triglycerides and coronary heart disease. Vasc Health Risk Manag 2009;5(1):89-95.
[63] Dejager S, Bruckert E, Chapman MJ. Dense low density lipoprotein subspecies with diminished oxidative resistance predominate in combined hyperlipidemia. J Lipid Res 1993 Feb 1;34(2):295-308.
[64] Connor WE, Duell PB, Kean R, Wang Y. The Prime Role of HDL to Transport Lutein into the Retina: Evidence from HDL-Deficient WHAM Chicks Having a Mutant ABCA1 Transporter. Investigative Ophthalmology & Visual Science 2007 Sep 1;48(9):4226-31.
[65] Wang W, Connor SL, Johnson EJ, Klein ML, Hughes S, Connor WE. Effect of dietary lutein and zeaxanthin on plasma carotenoids and their transport in lipoproteins in age-related macular degeneration. Am J Clin Nutr 2007 Mar;85(3):762-9.
[66] Cardinault N, Abalain JH, Sairafi B, Coudray C, Grolier P, Rambeau M, et al. Lycopene but not lutein nor zeaxanthin decreases in serum and lipoproteins in age-related macular degeneration patients. Clin Chim Acta 2005 Jul 1;357(1):34-42.
[67] Klein R, Klein BEK, Tomany SC, Cruickshanks KJ. The association of cardiovascular disease with the long-term incidence of age-related maculopathy - The Beaver Dam Eye Study. Ophthalmology 2003;110(4):636-43.
[68] Rizzo M, Berneis K. Low-density lipoprotein size and cardiovascular risk assessment. QJM 2006 Jan;99(1):1-14.
[69] Snow KK, Seddon JM. Do age-related macular degeneration and cardiovascular disease share common antecedents? Ophthalmic Epidemiology 1999;6:125-43.
[70] Klein R, Deng Y, Klein BE, Hyman L, Seddon J, Frank RN, et al. Cardiovascular disease, its risk factors and treatment, and age-related macular degeneration: Women's Health Initiative Sight Exam ancillary study. Am J Ophthalmol 2007 Mar;143(3):473-83.
[71] Gale CR, Hall NF, Phillips DIW, Martyn CN. Lutein and zeaxanthin status and risk of age-related macular degeneration. Investigative Ophthalmology & Visual Science 2003;44(6):2461-5.

[72] Brooks-Wilson A, Marcil M, Clee SM, Zhang LH, Roomp K, van DM, et al. Mutations in ABC1 in Tangier disease and familial high-density lipoprotein deficiency. Nat Genet 1999 Aug;22(4):336-45.
[73] Beatty S, Nolan J, Kavanagh H, O'Donovan O. Macular pigment optical density and its relationship with serum and dietary levels of lutein and zeaxanthin. Archives of Biochemistry and Biophysics 2004;430(1):70-6.
[74] Loane E, McKay GJ, Nolan JM, Beatty S. Apolipoprotein E genotype is associated with macular pigment optical density. Invest Ophthalmol Vis Sci 2010 May;51(5):2636-43.
[75] Renzi LM, Hammond BR, Jr., Dengler M, Roberts R. The relation between serum lipids and lutein and zeaxanthin in the serum and retina: results from cross-sectional, case-control and case study designs. Lipids Health Dis 2012 Feb 29;11:33.:33.
[76] Mahley RW, Innerarity TL. Lipoprotein receptors and cholesterol homeostasis. Biochim Biophys Acta 1983 May 24;737(2):197-222.
[77] Corder EH, Saunders AM, Strittmatter WJ, Schmechel DE, Gaskell PC, Small GW, et al. Gene dose of apolipoprotein E type 4 allele and the risk of Alzheimer's disease in late onset families. Science 1993 Aug 13;261(5123):921-3.
[78] Abalain JH, Carre JL, Leglise D, Robinet A, Legall F, Meskar A, et al. Is age-related macular degeneration associated with serum lipoprotein and lipoparticle levels? Clinica Chimica Acta 2002 Dec;326(1-2):97-104.
[79] Boerwinkle E, Utermann G. Simultaneous effects of the apolipoprotein E polymorphism on apolipoprotein E, apolipoprotein B, and cholesterol metabolism. Am J Hum Genet 1988 Jan;42(1):104-12.
[80] Parra HJ, Arveiler D, Evans AE, Cambou JP, Amouyel P, Bingham A, et al. A case-control study of lipoprotein particles in two populations at contrasting risk for coronary heart disease. The ECTIM Study. Arterioscler Thromb 1992 Jun;12(6):701-7.
[81] Shanmugaratnam J, Berg E, Kimerer L, Johnson RJ, Amaratunga A, Schreiber BM, et al. Retinal muller glia secrete apolipoproteins E and J which are efficiently assembled into lipoprotein particles. Brain Res Mol Brain Res 1997 Oct 15;50(1-2):113-20.
[82] Utermann G, Langenbeck U, Beisiegel U, Weber W. Genetics of the apolipoprotein E system in man. Am J Hum Genet 1980 May;32(3):339-47.
[83] Boyles JK, Zoellner CD, Anderson LJ, Kosik LM, Pitas RE, Weisgraber KH, et al. A role for apolipoprotein E, apolipoprotein A-I, and low density lipoprotein receptors in cholesterol transport during regeneration and remyelination of the rat sciatic nerve. J Clin Invest 1989 Mar;83(3):1015-31.
[84] Poirier J. Apolipoprotein E in animal models of CNS injury and in Alzheimer's disease. Trends Neurosci 1994 Dec;17(12):525-30.
[85] Gregg RE, Zech LA, Schaefer EJ, Stark D, Wilson D, Brewer HB, Jr. Abnormal in vivo metabolism of apolipoprotein E4 in humans. J Clin Invest 1986 Sep;78(3):815-21.
[86] Ehnholm C, Mahley RW, Chappell DA, Weisgraber KH, Ludwig E, Witztum JL. Role of apolipoprotein E in the lipolytic conversion of β-very low density lipoproteins to low density lipoproteins in type III hyperlipoproteinemia. PNAS 1984 Sep 1;81(17):5566-70.

[87] Leon AS, Togashi K, Rankinen T, Despres JP, Rao DC, Skinner JS, et al. Association of apolipoprotein E polymorphism with blood lipids and maximal oxygen uptake in the sedentary state and after exercise training in the HERITAGE family study. Metabolism 2004 Jan;53(1):108-16.

[88] Klaver CC, Kliffen M, van Duijn CM, Hofman A, Cruts M, Grobbee DE, et al. Genetic association of apolipoprotein E with age-related macular degeneration. Am J Hum Genet 1998 Jul;63(1):200-6.

[89] Ong JM, Zorapapel NC, Rich KA, Wagstaff RE, Lambert RW, Rosenberg SE, et al. Effects of cholesterol and apolipoprotein E on retinal abnormalities in apoE-deficient mice. Investigative Ophthalmology & Visual Science 2001 Jul 1;42(8):1891-900.

[90] Ishida BY, Bailey KR, Duncan KG, Chalkley RJ, Burlingame AL, Kane JP, et al. Regulated expression of apolipoprotein E by human retinal pigment epithelial cells. J Lipid Res 2004 Feb 1;45(2):263-71.

[91] Anderson DH, Ozaki S, Nealon M, Neitz J, Mullins RF, Hageman GS, et al. Local cellular sources of apolipoprotein E in the human retina and retinal pigmented epithelium: implications for the process of drusen formation. American Journal of Ophthalmology 2001 Jun;131(6):767-81.

[92] Dentchev T, Milam AH, Lee VM, Trojanowski JQ, Dunaief JL. Amyloid-beta is found in drusen from some age-related macular degeneration retinas, but not in drusen from normal retinas. Mol Vis 2003 May 14;9:184-90.

[93] Curcio CA, Johnson M, Huang JD, Rudolf M. Apolipoprotein B-containing lipoproteins in retinal aging and age-related macular degeneration. J Lipid Res 2010 Mar;51(3):451-67.

[94] Ruberti JW, Curcio CA, Millican CL, Menco BP, Huang JD, Johnson M. Quick-freeze/deep-etch visualization of age-related lipid accumulation in Bruch's membrane. Invest Ophthalmol Vis Sci 2003 Apr;44(4):1753-9.

[95] Sarks S, Cherepanoff S, Killingsworth M, Sarks J. Relationship of Basal laminar deposit and membranous debris to the clinical presentation of early age-related macular degeneration. Invest Ophthalmol Vis Sci 2007 Mar;48(3):968-77.

[96] Curcio CA, Millican CL, Bailey T, Kruth HS. Accumulation of cholesterol with age in human Bruch's membrane. Invest Ophthalmol Vis Sci 2001 Jan;42(1):265-74.

[97] Malek G, Li CM, Guidry C, Medeiros NE, Curcio CA. Apolipoprotein B in Cholesterol-Containing Drusen and Basal Deposits of Human Eyes with Age-Related Maculopathy. Am J Pathol 2003 Feb 1;162(2):413-25.

[98] Li CM, Clark ME, Chimento MF, Curcio CA. Apolipoprotein localization in isolated drusen and retinal apolipoprotein gene expression. Invest Ophthalmol Vis Sci 2006 Jul;47(7):3119-28.

[99] Ebrahimi KB, Handa JT. Lipids, lipoproteins, and age-related macular degeneration. J Lipids 2011;2011:802059. Epub;%2011 Jul 28.:802059.

[100] Curcio CA, Johnson M, Rudolf M, Huang JD. The oil spill in ageing Bruch membrane. Br J Ophthalmol 2011 Dec;95(12):1638-45.

[101] Wu T, Fujihara M, Tian J, Jovanovic M, Grayson C, Cano M, et al. Apolipoprotein B100 secretion by cultured ARPE-19 cells is modulated by alteration of cholesterol levels. J Neurochem 2010 Sep;114(6):1734-44.
[102] Li CM, Presley JB, Zhang X, Dashti N, Chung BH, Medeiros NE, et al. Retina expresses microsomal triglyceride transfer protein: implications for age-related maculopathy. J Lipid Res 2005 Apr;46(4):628-40.
[103] Huang JD, Presley JB, Chimento MF, Curcio CA, Johnson M. Age-related changes in human macular Bruch's membrane as seen by quick-freeze/deep-etch. Exp Eye Res 2007 Aug;85(2):202-18.
[104] Souied EH, Benlian P, Amouyel P, Feingold J, Lagarde JP, Munnich A, et al. The epsilon4 allele of the apolipoprotein E gene as a potential protective factor for exudative age-related macular degeneration. Am J Ophthalmol 1998 Mar;125(3):353-9.
[105] Mouchel Y, Lefrancois T, Fages C, Tardy M. Apolipoprotein E gene expression in astrocytes: developmental pattern and regulation. Neuroreport 1995 Dec 29;7(1): 205-8.
[106] Snipes GJ, McGuire CB, Norden JJ, Freeman JA. Nerve injury stimulates the secretion of apolipoprotein E by nonneuronal cells. PNAS 1986 Feb 15;83(4):1130-4.
[107] Schneeberger SA, Iwahashi CK, Hjelmeland LM, Davis PA, Morse LS. Apolipoprotein E in the subretinal fluid of rhegmatogenous and exudative retinal detachments. Retina 1997;17(1):38-43.
[108] Kuhrt H, Hartig W, Grimm D, Faude F, Kasper M, Reichenbach A. Changes in CD44 and ApoE immunoreactivities due to retinal pathology of man and rat. J Hirnforsch 1997;38(2):223-9.
[109] Baird PN, Guida E, Chu DT, Vu HTV, Guymer RH. The ε 2 and ε4 alleles of the apolipoprotein gene are associated with age-related macular degeneration. Investigative Ophthalmology & Visual Science 2004 May 1;45(5):1311-5.
[110] Simonelli F, Margaglione M, Testa F, Cappucci G, Manitto MP, Brancato R, et al. Apolipoprotein E polymorphisms in age-related macular degeneration in an Italian population. Ophthalmic Res 2001 Nov;33(6):325-8.
[111] Zareparsi S, Reddick AC, Branham KEH, Moore KB, Jessup L, Thoms S, et al. Association of apolipoprotein E alleles with susceptibility to age-related macular degeneration in a large cohort from a single center. Investigative Ophthalmology & Visual Science 2004;45(5):1306-10.
[112] Bojanowski CM, Shen D, Chew EY, Ning B, Csaky KG, Green WR, et al. An apolipoprotein E variant may protect against age-related macular degeneration through cytokine regulation. Environ Mol Mutagen 2006 Oct;47(8):594-602.
[113] Fritsche LG, Freitag-Wolf S, Bettecken T, Meitinger T, Keilhauer CN, Krawczak M, et al. Age-related macular degeneration and functional promoter and coding variants of the apolipoprotein E gene. Hum Mutat 2008 Dec 18.
[114] Stewart JE, Skinner ER, Best PV. Receptor binding of an apolipoprotein E-rich subfraction of high density lipoprotein to rat and human brain membranes. The International Journal of Biochemistry & Cell Biology 1998 Mar 1;30(3):407-15.

[115] Schneider WJ, Kovanen PT, Brown MS, Goldstein JL, Utermann G, Weber W, et al. Familial dysbetalipoproteinemia. Abnormal binding of mutant apoprotein E to low density lipoprotein receptors of human fibroblasts and membranes from liver and adrenal of rats, rabbits, and cows. J Clin Invest 1981 Oct;68(4):1075-85.

Section 5

Lipid Oxidation and Anti-Oxidants

Chapter 16

Pathophysiology of Lipoprotein Oxidation

Vikram Jairam, Koji Uchida and Vasanthy Narayanaswami

Additional information is available at the end of the chapter

http://dx.doi.org/10.5772/50622

1. Introduction

Lipoproteins are large lipid/protein complexes that play a major role in transport of lipids and lipophilic molecules in the plasma and central nervous system (CNS). Plasma lipoproteins represent a dynamic continuum of particles that are constantly undergoing re-modulation in their lipid and protein components leading to re-structuring of the particle under normal physiological conditions. The re-modulation is the result of lipid transfer and metabolism mediated by non-enzymatic and enzymatic processes, and of lipid association/dissociation behavior of apolipoproteins. Variations in protein and lipid components and composition arise because of changes in the feeding, metabolic and hormonal states, and due to differences in age, gender and disease states.

In this chapter, we will focus on oxidation of lipoproteins, paying attention to sources of oxidative stress, oxidation products of specific lipid and protein components, and the pathophysiology of oxidized lipoproteins in various disease states. The disease states that are addressed in this chapter include cardiovascular disease (CVD)/atherosclerosis, Alzheimer's disease and diabetes.

2. Lipoproteins

2.1. Classes of lipoproteins in plasma and CNS

Lipoproteins are classified based on their density, ranging from high density lipoproteins (HDL), low density lipoproteins (LDL), intermediate density lipoproteins (IDL), very low density lipoproteins (VLDL) and, chylomicrons (CM), Figure 1. They can also be classified based on their mobility during electrophoresis on an agarose gel as α-, pre-β- and β-lipoproteins, which correspond to HDL, VLDL and LDL, respectively. Lipoproteins vary significantly in particle diameter and density, protein and lipid components and composition. In general, the particle diameter is inversely related to the density.

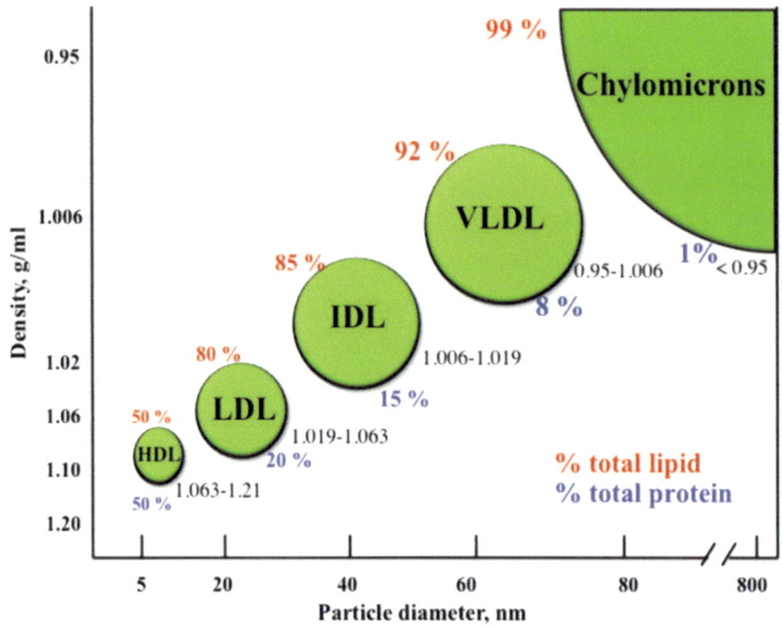

Figure 1. Lipoprotein classes. The classification of the major types of lipoproteins is based on their densities obtained by flotation ultracentrifugation analysis. The density range for each class is shown, in addition to the lipid (red) and protein (blue) content. The diagram is not to scale.

One of the main functions of lipoproteins is to transport hydrophobic factors in the highly aqueous vascular system. The CM is composed of lipids of dietary origin and is synthesized by the intestines. The VLDL is synthesized and secreted by the liver, and plays a role in distribution of lipids synthesized by the liver to the peripheral tissues. In the blood stream, the VLDL undergoes particle re-modulation to IDL and LDL, during which process the LDL become enriched in cholesterylesters. LDL plays a predominant role in delivery of cholesterol to the peripheral tissues and to the liver, with the cell surface-localized LDL receptor (LDLr) family of proteins playing a role in the cellular uptake and internalization of lipoproteins in target cells. The HDL may be synthesized by the liver and intestines or derived from other lipoproteins. In addition, peripheral tissues such as the macrophages are an important source of HDL, which are formed from cellular cholesterol efflux, and eventually transported to the liver. This process is called reverse cholesterol transport (RCT), which is an important mechanism for removal of peripheral cholesterol to the liver for eventual disposal.

Much less is known about lipoprotein metabolism in the CNS. Studies on lipoproteins secreted by astrocytes and those isolated from the cerebrospinal fluid (CSF) indicate the presence of only HDL-sized particles; large triglyceride-containing lipoprotein particles have not been detected in the CNS. The CNS maintains autonomy in terms of cholesterol synthesis and metabolism, right from the time when the blood brain barrier is established during development. One of the main functions of HDL secreted by astrocytes appears to be cholesterol delivery to the neurons via the LDLr family of proteins, for eventual use in the process of synaptogenesis.

2.2. Lipid and protein components of lipoproteins

In general lipoproteins are spherical in shape with a monolayer of amphipathic lipids (for example phospholipids, cholesterol and sphingolipids) and proteins encircling a core of neutral lipids (such as triglycerides and cholesterylesters), **Figure 2**. The lipid composition and content vary significantly in the different lipoproteins: in general, the larger lipoproteins (CM, VLDL and IDL) are enriched in triglycerides, while the smaller lipoproteins (LDL and HDL) are enriched in cholesterylesters and cholesterol. Unesterified or free cholesterol is esterified to cholesterylester by the action of lecithin-cholesterol acyltransferase (LCAT) on HDL. The phospholipids serve as the donor of the fatty acyl chains (especially 18:1 or 18:2 fatty acids) utilized in the esterification process. The fatty acid composition of triglycerides in the fasted state is dominated by 16:0 and 18:1 fatty acids.

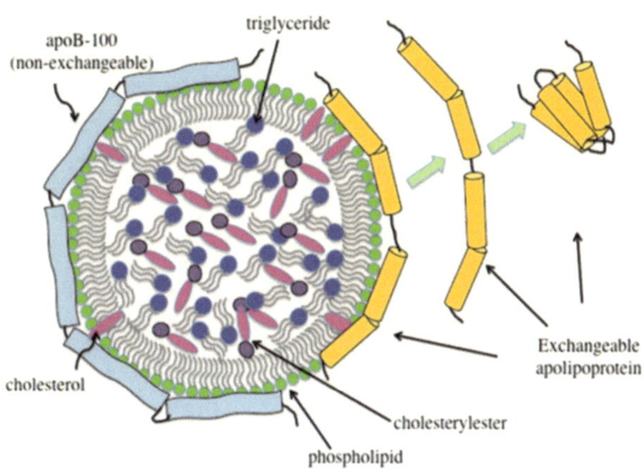

Figure 2. Schematic representation of a generic lipoprotein particle. Lipoproteins have a spherical geometry with a monolayer of amphipathic lipids and proteins encircling a core of neutral lipids. ApoB-100 (pale blue) is a single large polypeptide and is the non-exchangeable component of lipoproteins such as VLDL and LDL. The molecular mass of exchangeable lipoproteins (yellow cylinders) varies from 8-50 kDa. The exchangeable apolipoproteins have the ability to exist in lipid-free and lipid-bound states.

The protein composition and content of the lipoproteins also vary significantly from one particle to another. There are two types of apolipoproteins: non-exchangeable and exchangeable. The non-exchangeable apolipoproteins, apolipoprotein B-100 (apoB-100) and apoB-48, are present as a single copy per lipoprotein particle: apoB-100 on VLDL, IDL or LDL and apoB-48 on CM and CM remnants. ApoB-100 is 4536 residues long that is synthesized in the liver, while apoB-48, represents the N-terminal 48% of apoB-100, and is synthesized in the intestines. Both apoB-100 and apoB-48 are integral to the structure and stability of the lipoprotein particle. There are different types of exchangeable apolipoproteins, Table 1, which undergo a reversible association with lipoproteins depending on the metabolic state.

Apolipoprotein	Lipoprotein	Function
ApoAI	HDL	activates LCAT; promotes ABCA1-mediated cholesterol efflux in RCT
ApoAII	HDL	Inhibits LCAT
ApoAIV	Chylomicrons, HDL	Activates LCAT, cholesterol clearance and transport
ApoB-48	Chylomicron and chylomicron remnants	Cholesterol clearance and transport; lacks LDLr binding sites
ApoB-100	VLDL, IDL and LDL	Binds to LDLr
ApoCI	VLDL, HDL	Activates LCAT
ApoCII	VLDL, IDL, chylomicrons	Activates lipoprotein lipase
ApoCIII	VLDL, IDL, chylomicrons	Inhibits lipoprotein lipase
ApoD	HDL	Carrier proteins family (lipocalins)
ApoE2	VLDL, IDL, chylomicrons, chylomicron remnants	Poor LDLr binding activity; associated with Type III hyperlipoproteinemia and CVD
ApoE3	HDL, VLDL, IDL, chylomicron remnants (higher binding preference for HDL over VLDL)	Binds to LDLr family of proteins with high affinity; significant role in cholesterol efflux and RCT in atherosclerosis
ApoE4	VLDL, HDL, IDL, chylomicron remnants (higher binding preference for VLDL over HDL)	Binds to LDLr family of proteins with high affinity; associated with CVD and Alzheimer's disease
ApoM	HDL	Transports sphingosine-1-phosphate
Apo(a)	Lipoprotein(a) (Lp(a))	Linked to apoB-100 via disulfide bond; similar to plasminogen

[1] Adapted with permission from the AOCS Lipid Library, "Plasma Lipoproteins: Composition, Structure and Biochemistry," Table 3, 'The main properties of apoproteins, in http://lipidlibrary.aocs.org/lipids/lipoprot/index.htm, accessed June 12, 2012

Table 1. Major apolipoprotein components of lipoproteins and associated functions[1]

The exchangeable apolipoproteins have the ability to exist in lipid-free and lipid-bound states, and undergo a large conformational change upon transitioning from one state to the other. Exchangeable apolipoproteins are characterized by an abundance of amphipathic α-helices that are folded into a helix bundle. For example, in the case of apoE, the protein is composed of a series of α-helices that are folded into an N-terminal (NT) and a C-terminal (CT) domain. The NT domain is comprised of a 4-helix bundle bearing the LDLr binding sites on helix 4, which has an abundance of basic residues, while the CT domain harbors high-affinity lipid binding sites and apoE self-association sites. A similar arrangement was noted for apoAI; however, apoAI does not have the capability to interact with LDLr.

In summary, lipoprotein particles offer several targets that are vulnerable to attack by oxidative species. Oxidative modification is expected to compromise the structure and function of the protein and lipid components.

3. Oxidative stress

Reactive oxygen species (ROS) generally refer to oxygen free radicals with one or more unpaired electrons, and other highly reactive oxygen-containing molecules. They may be generated by the cells as products of normal cellular metabolism, or derived from exogenous sources including environmental pollutants, which could in turn, trigger further release of ROS. Whereas the ROS may be beneficial at low concentrations and play key physiological roles, their deleterious effect at high concentrations contributes to the etiology of several disease states. Organisms have developed an exquisite arsenal of defense mechanisms to combat the harmful effects of ROS, thereby maintaining a redox homeostasis. However, when the ROS overcome the cellular defense mechanism, there is a dysregulation of the redox balance, which leads to the state of oxidative stress (Esterbauer et al., 1991).

3.1. ROS

The high reactivity of free radicals is attributed to the unpaired electron(s) on the oxygen molecule. The primary ROS is the superoxide anion, which is generated by the addition of one electron to molecular oxygen during mitochondrial electron transport (Equation 1). This in turn gives rise to hydrogen peroxide following the action of antioxidant enzymes, superoxide dismutases (SOD) (Equation 2). The superoxide anions also give rise to the hydroxyl radicals in the presence of trace amounts of Fe^{2+} (Fenton reaction) (Equation 3); the hydroxyl radical is an extremely deleterious species that is capable of causing indiscriminate damage in the immediate vicinity of its formation.

$$O_2 + e^- \longrightarrow \bullet O_2^- \qquad \text{Superoxide anion} \qquad (1)$$

$$\bullet O_2^- + 2H^+ \xrightarrow{SOD} H_2O_2 + O_2 \qquad \text{Hydrogen peroxide} \qquad (2)$$

$$\bullet O_2^- + H_2O_2 \xrightarrow{Fe^{2+}} \bullet OH + OH^- + O_2 \qquad \text{Hydroxyl radical} \qquad (3)$$

SOD catalyzes the dismutation of superoxide anion into oxygen and hydrogen peroxide, thereby affording protection to the cell. Three forms of SOD exist in humans and other mammals: SOD1 is located in the cytoplasm, SOD2 in the mitochondria, while SOD3 is extracellular. Mutations in SOD lead to diseases commonly associated with high oxidative stress such as familial amyotrophic lateral sclerosis, Parkinson's Disease, and cardiovascular disease (Fukai et al., 2002; Noor et al., 2002; Tórsdóttir et al., 2006). Overexpression of SOD inhibits LDL oxidation by endothelial cells (Fang et al., 1998). Higher SOD levels were demonstrated to be protective against LDL oxidation *in vitro* (Laukkanen et al., 2000).

Red blood cells (RBC) are particularly susceptible to ROS and free radical stress due to its constant interactions with oxygen. Despite their lack of mitochondria, RBC are under tremendous oxidative stress, due to the abundance of hemoglobin. Hemoglobin is prone to oxidation through either exogenous or endogenous sources, which results in superoxide production. Therefore, the RBC contains many SOD enzymes to convert the superoxides to hydrogen peroxide. Furthermore, hydrogen peroxide can combine with hemoglobin to form ferrylhemoglobin, a strong oxidizing agent (Rifkind et al, 2002). The RBCs are also equipped with other antioxidant enzymes such as catalase and glutathione peroxidase to overcome the harmful effects of ROS.

3.2. Reactive aldehydes

It is well established that polyunsaturated fatty acids (PUFA) undergo lipid peroxidation that is initiated by ROS, which generate a 'spectrum' of reactive aldehyde species. The reactive aldehydes formed in biological systems are more complex than those formed in simple systems like purified lipid preparations and their physiological concentrations are believed to be lower. They play a crucial role in amplifying the free radical-initiated reaction by generating a complex array of toxic end products. Although considered as end products of oxidative damage of lipids, the aldehydic products display reactivity with a wide variety of biological molecules under cellular conditions, thereby enhancing the pathogenesis of the diseases. They are therefore considered as toxic second messengers of oxidative stress and lipid peroxidation. Protein-bound aldehydes have been proposed as potential markers of oxidative stress as evidenced by immunohistochemical analysis of atherosclerotic lesions (Uchida et al., 1998a). Numerous α,β-unsaturated aldehydes have been reported in literature: we will focus on acrolein, 4-hydroxynonenal (4-HNE) and malondialdehyde (MDA), **Figure 3.**

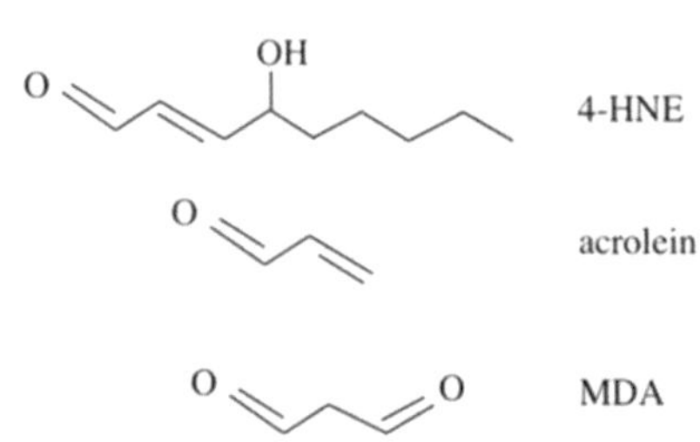

Figure 3. Major aldehydic products of lipid peroxidation

3.2.1. Acrolein

Acrolein is an acrid smelling environmental pollutant that is formed during combustion of organic and plastic substances (Beauchamp et al., 1985). It is present as one of the major components in the gaseous phase of tobacco smoke (up to 140 μg/cigarette) (Witz, 1989). It is also generated as a natural metabolite during oxidative stress mediated lipid peroxidation

(Uchida et al., 1998b). In addition, oxidation of threonines by myeloperoxidase (MPO) gives rise to acrolein (Anderson, et al., 1997; Savenkova et al., 1994). Acrolein is the most reactive of all α,β-unsaturated aldehydes; it causes oxidative modification of proteins by reacting with the sulfhydryls of cysteines, ε-amino groups of lysines and imidazole group of histidines (Witz, 1989; Esterbauer et al., 1991). Some possible products formed upon reaction of acrolein with lysine side chains include aldimine adducts (Schiff base formation), propanal adducts (Michel addition) and N^{ε}-(3-formyl-3,4-dehydropiperidino)lysine (FDP-lysine), **Figure 4**.

3.2.2. 4-HNE

Discovered more than 50 years ago, alkenals have been under intense scrutiny in terms of their chemistry, biochemistry, toxicology and as oxidative stress agents. Amongst these alkenals, 4-HNE was determined to be the most toxic. It was discovered to be a product of lipid peroxidation, particularly of n-6 PUFA such as linoleic acid and arachidonic acid. Since then, 4-HNE has been implicated in a number of diseases, including atherosclerosis, Alzheimer's Disease, Parkinson's Disease, liver cirrhosis, and cancer (Zarkovic, 2003).

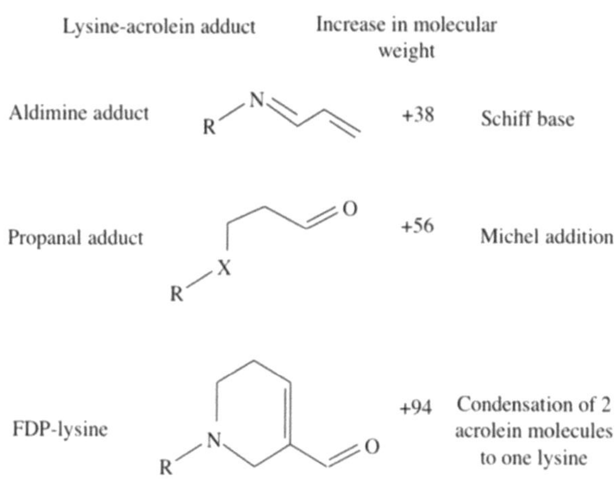

Figure 4. Modifications of lysine side chains by acrolein. The major lysine modification products of acrolein are shown, along with the expected increments in molecular weight during mass spectral analysis. Adapted from Furuhata et al, 2003.

From a chemical perspective, the reactivity of 4-HNE is conferred by the carbonyl group at C1 position, and a C=C bond between the second and third carbons which provide a partial positive charge to C3 (Schaur, 2003). This results in 4-HNE being highly reactive towards thiol and amino groups of proteins. 4-HNE reacts with thiols via a Michael addition, in which a nucleophile such as the sulfhydryl of cysteine or glutathione reacts with the C3 of HNE to form a covalent adduct. In addition, amino compounds such as lysine, ethanolamine, guanine, and the imidazole group of histidine are capable of forming Michael addition adducts with 4-HNE. 4-HNE can also undergo reactions involving a reduction or an epoxidation of the double bond. 4-HNE has been shown to react *in vivo* with common biomolecules, and can often have an inhibitory role with enzymes (Schaur, 2003; Korotchkina et al, 2001). Cross-reactivity of anti-4-HNE antibody with oxidized LDL (Ox-LDL) suggests that 4-HNE may contribute to the overall process of atherosclerotic plaque formation (Uchida et al., 1993).

3.2.3. MDA

In fresh samples, MDA is formed mainly from PUFA such as arachidonic acid (Esterbauer et al., 1991). Initially determined by the thiobarbituric acid (TBA) assay as TBA reactive substances (TBARS), the presence of pre-existing MDA or protein-bound MDA needs to be confirmed by other sensitive assays as well. Under physiological conditions, MDA readily modifies protein side chains forming stable cross-linked adducts with the ε-amino groups of lysines. Other side chains that may be modified include those of histidine, tyrosine, arginine and methionine.

While the levels of acrolein, 4-HNE, MDA and other aldehydic compounds *per se* are thought to be too low for detection, their role in mediating oxidative damage has been confirmed by the presence of autoantibodies against aldehyde-modified proteins. It provides evidence for the *in vivo* occurrence of lipid peroxidation products and their role in disease progression (Steinberg et al, 1989). We have a better understanding of the association between lipid peroxidative products, oxidative stress and disease progression, in part due to the development of antibodies directed against oxidized lipids and protein-bound aldehydes in the past two decades, Table 2. Currently, we have a repertoire of antibodies directed against different types of oxidatively modified lipids and proteins, **Table 2**, that serve as powerful tools for ELISA, immunohistochemical and immunoblotting analyses.

3.3. Sources of ROS

In addition to the mitochondrial electron transport, there are enzymatic sources of ROS that contribute to the redox status in biological systems. Located in the blood stream, they are of relevance to lipoprotein oxidation.

3.3.1. MPO

MPO is an enzyme found abundantly in neutrophil granulocytes and to some extent in macrophages. It is a lysosomal protein stored in the azurophilic granules in neutrophils and can be identified by its characteristic green heme pigment. During the respiratory burst of the neutrophil, MPO catalyses the production of hypochlorous acid (HOCl) from hydrogen peroxide and chloride anion (Equation 4). MPO is also responsible for generation of 3-chlorotyrosine and 3-nitrotyrosine; individuals with coronary artery disease show elevated levels of these two products in their blood and HDL. MPO has been shown to play a role in promoting atherosclerotic lesions through oxidative modification of apoAI (Shao et al., 2012) (described under apoAI).

$$H_2O_2 + Cl^- + H^+ \rightarrow HOCl + H_2O \tag{4}$$

Lipid peroxidation products	mAb	Epitope
Oxidized lipids		
[a]9-Hydroxy-10E,12Z-octadecadienoic acid (9-HODE)	9H2	9-HODE
[b]13-Hydroxy-9Z,11E-octadecadienoic acid (13-HODE)	13H1	13-HODE
[a]12-Hydroxyeicosatetraenoic aci (12-HETE)	12H8	12-HETE
[a]Leukotoxin (epoxylinoleic acid)	21D1	Leukotoxin
[a]7-Ketocholesterol (7-KC)	35A-8	7-KC
[c]15-deoxy-$\Delta^{12,14}$-prostaglandin J_2 (15d-PGJ_2)	11G2	15d-PGJ_2
Protein-bound aldehydes		
[d]Acrolein (ACR)	5F6	ACR-lysine
[e]Crotonaldehyde (CRA)	82D3	CRA-lysine
[a]2-Hexenal (HE)	CT5	HE-lysine
[f]2-Nonenal (NE)	27Q4	NE-lysine
[g]Malondialdehyde (MDA)	1F83	MDA-lysine
[h]4-Hydroxy-2-hexenal (HHE)	53	HHE-histidine
[i]2-Hydroxyheptanal (2HH)	3C8	2-HH-lysine
[j]4-Hydroxy-2-nonenal (HNE)	HNEJ2	HNE-histidine
[k]4-Hydroxy-2-nonenal (HNE)	2C12	HNE-lysine
[l]4*R*-4-Hydroxy-2-nonenal ((*R*)-HNE)	R310	(*R*)-HNE-histidine
[m]4*S*-4-Hydroxy-2-nonenal ((*S*)-HNE)	S412	(*S*)-HNE-histidine
[n]4-Hydroperoxy-2-nonenal (HPNE)	PM9	HPNE-lysine
[o]4-Oxo-2-nonenal (ONE)	9K3	ONE-lysine
Protein-bound core aldehyde		
[p]9-Oxononanoylcholesterol (9-ONC)	2A81	9-ONC-lysine

[a]Unpublished
[b]Shibata et al. (2009) Acta Histochem. Cytochem. 42, 197; [c]Shibata et al. (2011) JBC 277, 10459; [d]Uchida et al. (1998) PNAS 95, 4882; [e]Ichihashi et al. (2002) JBC 276, 23903; [f]Ishino et al. (2010) JBC 285, 15302; [g]Yamada et al. (2001) JLR 42, 1187; [h]Yamada et al. (2004) JLR 45, 626; [i]Itakura et al. (2003) BBRC 308, 452; [j]Toyokuni et al. (1995) FEBS Lett. 359, 189; [k]Itakura et al. (2000) FEBS Lett. 473, 249; [l]Hashimoto et al. (2003) JBC 278, 5044; [m]Hashimoto et al. (1995) JBC 278, 5044; [n]Shimozu et al. (2011) JBC 286, 29313; [o]Shibata et al. (2011) JBC 286, 19943; [p]Kawai et al. (2003) JBC 278, 21040; [q]Kawai et al. (2003) JBC 278, 50346

Table 2. Lipid peroxidation-specific monoclonal antibodies

3.3.2. NADPH oxidase

NADPH oxidase is an enzyme complex that is normally found in the plasma membranes of neutrophils and monocytes. Like MPO, it is activated during the respiratory burst of the neutrophil. Its main role is to generate superoxide by transferring electrons from NADPH to molecular oxygen (Equation 5). The superoxide is then used to destroy phagocytosed bacteria or pathogens. NADPH oxidase and its product, superoxide, are major contributory factors for foam cell formation in atherosclerosis (Meyer & Schmitt, 2000).

$$2O_2 + NADPH \rightarrow 2 \bullet O_2^- + NADP^+ + H^+ \quad (5)$$

3.3.3. Lipoxygenase

Lipoxygenases are iron-containing enzymes that catalyze the dioxygenation of PUFA in lipids. Specifically, they form hydroperoxides from fatty acids and molecular oxygen. 15-lipoxygenase is the main protein in this family; it is involved in the metabolism of eicosanoids (prostaglandins, leukotrienes), which function as secondary messengers. 15-lipoxygenase has been shown to be involved in LDL oxidation (Bailey et al., 1995), with considerably greater 15-lipoxygenase activities in atherosclerotic compared to normal aortas (Hiltunen et al., 1995).

4. Lipoprotein oxidation

We now turn to the specifics of lipoprotein oxidation with reference to the different lipid and protein components.

4.1. Oxidation of lipid components

Of the different lipid components that may be potentially oxidized in various lipoprotein particles, we will focus on the oxidation of fatty acyl chains, including those on phospholipids, cholesterylester and triglycerides, and on cholesterol derivatives. The reader is referred to a comprehensive review by Subbaiah and colleagues (Levitan et al., 2010) for an introduction to the role of ceramides in Ox-LDL and of sphingosine 1-phosphate in HDL

4.1.1. Oxidation of fatty acyl chain

One group of biological targets that are highly vulnerable to attack by ROS and aldehydes are the lipids: their abundance in lipoproteins and the ease with which their unsaturated bonds are oxidatively modified make them susceptible to damage. Products of lipid peroxidation have been associated with the pathophysiology of numerous disease states, including atherosclerosis, diabetes and cancer. Although found in low concentrations in normal healthy tissues, they are found to be enriched in pathological cells and tissues, including macrophage foam cells and atherosclerotic lesions (Olkkonnen, 2008; Brown & Jessup, 1999; Olkkonen & Lehto, 2004; Javitt, 2008; Tsimikas et al., 2005; Berliner & Watson, 2005). PUFA peroxidation products cause further damage to proteins by oxidative modification of amino acid side chains and formation of protein carbonyl groups (Refsgaard et al., 2000).

Briefly, lipid peroxidation is initiated by the hydroxyl radical abstracting a hydrogen from the methylene group adjacent to a double bond of fatty acids, **Figure 5**. The fatty acid may also be part of the phospholipid at the *sn*-2 position or esterified to the –OH group of cholesterol in a lipoprotein. This process gives rise to an unstable lipid radical, which undergoes rearrangement of double bonds and addition of oxygen to form a peroxyl radical.

The lipid radical or the peroxyl radical could react with a neighbouring fatty acyl chain as well, thereby propagating the peroxidation process. The chain reaction may be terminated by eventual formation of a lipid hydroperoxide.

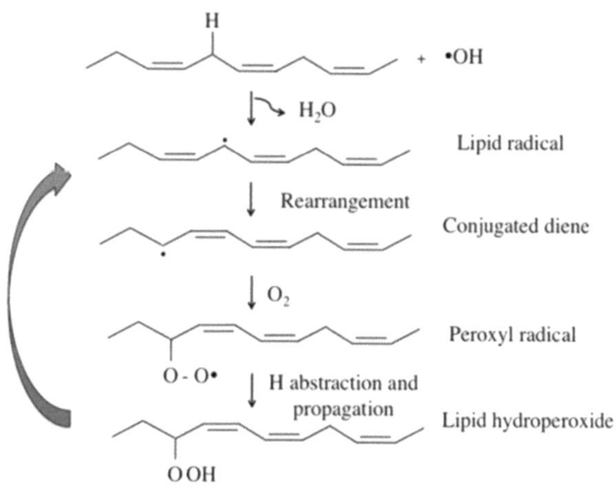

Figure 5. Lipid peroxidation. The lipid peroxidation process may be divided into the initiation, propagation and termination steps; only the double bond containing segment of a fatty acid is shown.

In the context of a lipoprotein, it was postulated that LDL oxidation was initiated through cell-generated ROS formation, with the involvement of the lipoxygenase pathway (Cyrus et al, 1999). Both apoE-null and LDLr-null mice genetically deficient in the 12/15-lipoxygenase were found to have significantly less atherosclerosis. Other groups developed mice overexpressing the 12/15-lipoxygenase, and found spontaneous aortic fatty streak lesions on a chow diet (Reilly et al., 2004). Finally, it was shown that the endothelial cells within the vessels required 12/15-lipoxygenase to generate oxidized phospholipids. It was also shown that the MPO pathway contributed to Ox-LDL in human atherosclerotic lesions (Savenkova et al., 1994).

LDL oxidation may occur within the arterial endothelial cells, which have high levels of precursor molecules like linoleic acid and arachidonic acid - fatty acids that are involved in producing eicosanoids. Hydroperoxide reaction with linoleic acid produces 13(*S*)-hydroperoxy-octadecadienoic acid (13(*S*)-HPODE), and reaction with arachidonic acid produces 15(*S*)-hydroperoxy-eicosatetraenoic acid (15(*S*)-HPETE). These molecules are present as part of LDL surface phospholipids and trigger further oxidation of phospholipids with arachidonic acid. These are early events occurring prior to apoB-100 modification and constitute the 'minimally-modified' LDL (Navab et al., 2001). Another consequence of phospholipid oxidation is fragmentation of the fatty acyl group at *sn*-2 position resulting in short chain fatty acids, which structurally and functionally resemble platelet activating factors with chemotactic activity.

Although less understood, *in vitro* oxidation of HDL is also shown to generate oxidized lipids and proteins. It is likely that HDL lipids may be oxidized initially even before LDL lipids, upon exposure of human plasma to peroxyl radicals, Cu^{2+} ions or lipoxygenase

(Garner et al., 1998). These studies also show that Met residues are oxidized in apoAI and apoAII. Further, ^{1}H and ^{31}P NMR analysis indicate a loss of the unsaturated system with appearance of epoxides on fatty acyl chains and 5,6-epoxide derivatives of cholesterol indicating significant modification of HDL (Bradamante et al., 1992).

4.1.2. Oxysterols

Oxysterols are molecules that are formed from the oxidation of cholesterol, which can occur at several sites (Vaya & Schipper, 2007). They display higher water solubility compared to cholesterol. The major oxysterols isolated from plasma LDL include 7α-OH and 7β-OH cholesterol. In addition, 7-keto cholesterol, which impairs cholesterol efflux and reduces cell membrane fluidity, and 5,6-epoxide derivatives of cholesterol, have also been identified in Ox-LDL (Levitan et al., 2010). Other types of oxysterols have been localized in atherosclerotic lesions, including those wherein the side chains of cholesterol are oxidized (for example, 27-hydroxycholesterol). The cytochrome P-450 system is largely responsible for generating the hydroxylated derivatives. They also function as transcriptional effectors and attenuate the Liver X Receptor (LXR). A majority of PUFA in LDL is esterified as cholesterylester; thus, hydroperoxide and hydroxide derivatives of cholesterylester are found in abundance in human atherosclerotic lesions.

In the CNS, the neurons convert cholesterol to 24S-hydroxycholesterol (also known as cerebrosterol), **Figure 6**; the conversion facilitates its movement out of the CNS, since cholesterol as such does not cross the blood brain barrier. Almost all circulating 24S-hydroxycholesterol originates from the brain (Lutjohann et al., 1996; Lutjohann et al., 2000; Bjorkhem et al., 1998) and may reflect CNS cholesterol turnover.

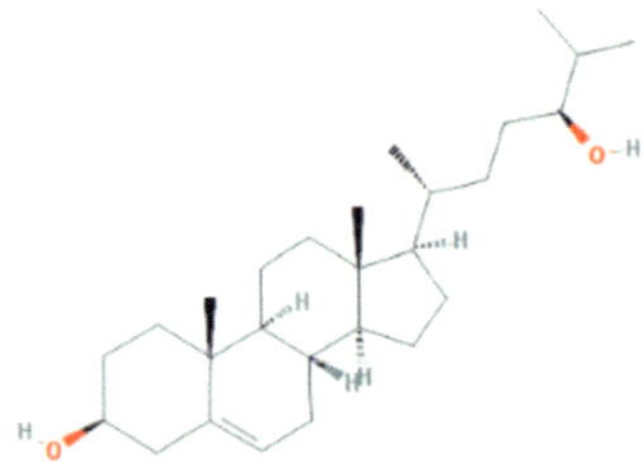

Figure 6. 24S-hydroxycholesterol (obtained from: http://pubchem.ncbi.nlm.nih.gov)

4.2. Oxidation of protein components

4.2.1. ApoB-100

From a historical perspective, apoB-100 was the earliest apolipoprotein to be identified as being a target for oxidative modification on lipoproteins (Steinberg et al., 1989). It is implied in the discussions involving Ox-LDL hypothesis, along with oxidation of the lipid moieties in LDL. The term Ox-LDL is used to identify LDL that has been modified to an extent that it is not recognized by the LDLr anymore; instead it becomes a ligand for the scavenger receptors family of proteins. On the other hand, the term 'minimally-modified' LDL has been adopted to encompass the different preparations that have been modified enough to be chemically distinguishable from, but recapitulates the LDLr binding feature of, unmodified LDL. ApoB-100 is one of the oxidizable targets on lipoproteins: one of the earliest *in vitro* studies demonstrate that incubation of human LDL with 4-HNE (Haberland et al., 1984; Jurgens et al., 1986), results in modification of 45 lysines, 7 histidines, 23 serines and 51 tyrosine residues on apoB-100.

4-HNE forms covalent adducts with lysine residues on apoB-100, thereby blocking its ability to recognize the macrophage LDLr. Evidence for *in vivo* LDL oxidation was provided by immunocytochemical staining and immunoblot analysis of extracts from atherosclerotic lesions of LDLr-deficient rabbits using antibodies against Ox-LDL, MDA-lysine or 4-HNE-lysine, (Palinski et al., 1990). This study also demonstrated higher titers of autoantibodies against MDA-conjugated LDL in the human and rabbit antisera. 4-HNE-modified LDL is an efficient ligand for scavenger receptors (Hoff & O'Neil, et al, 1993; Rosenfeld et al., 1990). ApoB-100 can also be modified by MPO, which leads to formation of chlorotyrosine and nitrotyrosine derivatives.

4.2.2. ApoE

ApoE is a 299 residue, 34 kDa protein that is commonly associated with VLDL, CM remnants and a sub-class of HDL. It is a major cholesterol transport protein in the plasma and the CNS (Hatters et al., 2006). In humans, apoE is polymorphic; variation in the *APOE* gene results in three major alleles, ε2, ε3 and ε4, occurring at frequencies of 8%, 77% and 15%, respectively in the population. The products of the three alleles are the isoforms, apoE2, apoE3 and apoE4, which differ in the amino acids at positions 112 and 158: apoE2 has Cys while apoE4 has Arg at these locations; apoE3 has a Cys and Arg at these locations, respectively. ApoE3 is considered an anti-atherogenic protein; individuals homozygous for the *APOE* ε2 allele are prone to developing familial type III hyperlipoproteinemia and premature atherosclerosis. The inheritance of one or more of the *APOE* ε4 alleles predisposes the bearer to hypercholesterolemia, as well as Alzheimer's disease, affecting both the age of onset and the severity of these diseases.

In vitro oxidative modification of the receptor-binding domain of apoE3 by acrolein generates epitopes recognized by an antibody specific for acrolein-lysine adducts, mAb5F6 (Tamamizu-Kato et al., 2007) with formation of both intra- and inter-molecular cross-linked products. This modification resulted in severe impairment of three major functions of apoE3: (i) its ability to interact with the LDLr, a function mediated by specific lysines and arginines located on helix 4 of the receptor-binding domain; (ii) its ability to bind heparin, which is facilitated predominantly by two specific lysines (K143 and K146), **Figure 7**; (iii) its ability to bind lipids. These studies indicate that acrolein either directly modifies the lysines in helix 4 that are involved in LDLr and heparin binding or that modification of lysines elsewhere on apoE3 alters the conformation of lysines in helix 4, thereby disrupting its binding. Further evidence was provided by direct exposure of VLDL isolated from human plasma to acrolein or Cu^{2+}, which disrupted its ability to bind and internalize the lipoprotein particle via LDLr, LDLr-related protein or HSPG on hepatocytes. (Arai et al., 1999; 2005). Taken together, oxidative modification appears to compromise the functional integrity of apoE3.

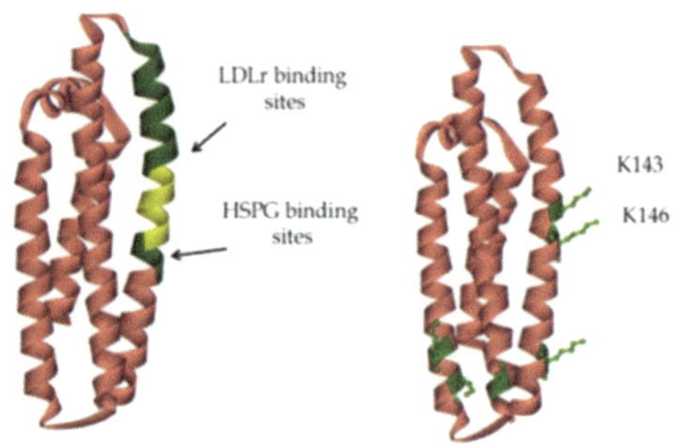

Figure 7. Lysines relevant in LDLr and HSPG binding function of apoE3. Ribbon diagram of apoE3 receptor-binding domain is shown with the LDLr- and HSPG-binding sites represented in green and yellow, respectively.

From a physiological perspective, the loss of apoE function by oxidative modification has direct implications in CVD: (i) decreased uptake and internalization of apoE-containing lipoproteins leading to their accumulation in the blood; (ii) decreased ability of apoE to interact with heparan sulfate proteoglycans (HSPG) lining the blood vessels and cell surfaces, where apoE is believed to be stored; and (iii) impaired ability of apoE to interact with lipoproteins, which is an essential prerequisite for the receptor-binding domain to elicit LDLr binding.

In the CNS, where apoE is the predominant apolipoprotein that has been identified so far, oxidative modification is expected to have serious implications in progression of neurological diseases such as Alzheimer's disease in an isoform-specific manner. About 65% of individuals with late-onset familial and sporadic Alzheimer's disease bear the *APOE* ε4 allele (Huang et al., 2004). The precise mechanism by which apoE4 is associated with Alzheimer's disease remains a contentious issue. While the role of apoE4 in aggravating the beta amyloid toxicity has received widespread attention, the inherent propensity of apoE4 to misfold noted under *in vitro* physiological conditions requires further scrutiny *in vivo*. Further, since oxidative damage plays a significant role in the pathogenesis associated with Alzheimer's disease (Perry et al., 2002), it is likely that oxidative modification of apoE4 further exacerbates its role in the etiology of the disease. Indeed, cerebrospinal fluid obtained by lumbar puncture in a limited number of Alzheimer's disease patients homozygous for *APOE3* or *APOE4*, and age-matched controls with or without dementia display a 50 kDa apoE-immunoreactive protein co-migrating with proteins immunoreactive for 4-HNE and MDA adducts (Montine et al., 1996; Bassett et al., 1999).

A similar 50 kDa apoE-immunoreactive protein was also reported in P19 neuroglial cultures differentiated into neurons and astrocytes subjected to oxidative stress. Interestingly, apoE3 appeared to be cross-linked to a greater extent than apoE4. The cross-linking has been attributed to the susceptible site provided by apoE3 in the form of Cys112, and the known reactivity of 4-HNE with sulfhydryl groups (Esterbauer et al., 1991). *In vitro* modification of apoE3 or apoE4 (isolated from plasma) by 4-HNE yielded cross-linked products with apparent molecular weights corresponding to dimeric and trimeric apoE (Montine et al., 1996). A similar trend was noted with MDA. It is possible that the greater susceptibility of apoE3 to cross-linking and oxidation than apoE4 is a reflection of its greater potency as an antioxidant.

4.2.3. *ApoAI*

ApoAI is a 243 residue, 28 kDa exchangeable apolipoprotein that is a major component of HDL. Like apoE, it is composed predominantly of amphipathic α helices, with an N-terminal domain 4-helix bundle. Under normal physiological conditions, apoAI plays a critical role in promoting ATP Binding Cassette Transporter A1 (ABCA1)-mediated cholesterol efflux from macrophages. This aids in mobilizing cholesterol and phospholipids from peripheral tissues to the liver by the RCT process, for eventual disposal by biliary secretion. In atherosclerotic lesions, apoE plays a dominant role in RCT by virtue of the fact that cholesterol-laden macrophages secrete lipid-poor apoE, which in turn in promotes ABCA1-mediated cholesterol efflux (Huang et al, 1995).

With both apoAI and apoE, a nascent discoidal form of HDL is generated that is composed of a bilayer of phospholipids and cholesterol circumscribed by the α-helices of the protein. The discoidal HDL is an excellent substrate for LCAT, the enzyme that catalyzes the transfer of a fatty acyl chain from phospholipids to the free hydroxyl group of cholesterol to form cholesterylesters. The conversion of the amphipathic free cholesterol to the hydrophobic cholesterylesters promotes its transition to the core of the lipoprotein particle, thereby generating a spherical HDL containing a cholesterylester core. The HDL is targeted to the liver and steroidogenic tissues where they are recognized by the scavenger receptor class B Type 1 (SR-B1), which mediate selective uptake of cholesterylesters into the cells.

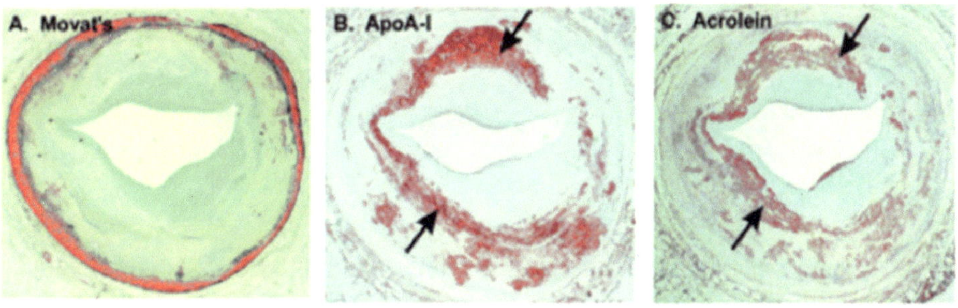

Figure 8. Immunohistochemical co-localization of apoAI and acrolein adducts in human atherosclerotic lesions. This research was originally published in The Journal of Biological Chemistry. Shao, B., Fu, X., McDonald, T. O., Green, P. S., Uchida, K., O'Brien, K. D., Oram, J. F. & Heinecke, J. W. Acrolein impairs ATP Binding Cassette Transporter A1-dependent cholesterol export from cells through site-specific modification of apolipoprotein A-I. J. Biol. Chem. (2005) Vol. 280, No. 43, pp. 36386-36396 © the American Society for Biochemistry and Molecular Biology.

There is strong evidence that MPO oxidizes HDL *in vivo* (Daugherty et al., 1994; Bergt et al., 2004; Pennathur et al., 2004; Zheng et al., 2004). In addition, aldehyde modification of HDL is also associated with the loss in ability of HDL to activate LCAT (McCall et al., 1995). Acrolein modifies apoAI specifically at Lys226 in helix 10 converting it to N^{ε}-(3-methylpyridinium)lysine (MP-Lys). A corresponding decrease in ABCA1-mediated cholesterol efflux was also noted. In addition, immunohistochemical analysis demonstrated co-localization of acrolein-adducts and apoAI in human atherosclerotic lesions, **Figure 8**, confirming previous studies (Uchida et al., 1998b). Further chlorination of Tyr192 and oxidation of specific Met residues in apoAI via the MPO pathway impairs its ability to promote cholesterol efflux (Shao et al., 2010). Taken together, it appears that oxidative modification of HDL apoAI may be one of the contributory factors to atherogenesis.

Figure 9 provides a simplified overview of the roles of apoB-100, apoAI and apoE in lipid distribution between liver and peripheral tissues. It also shows potential functional sites that are likely to be affected because of oxidative modification of lipoproteins.

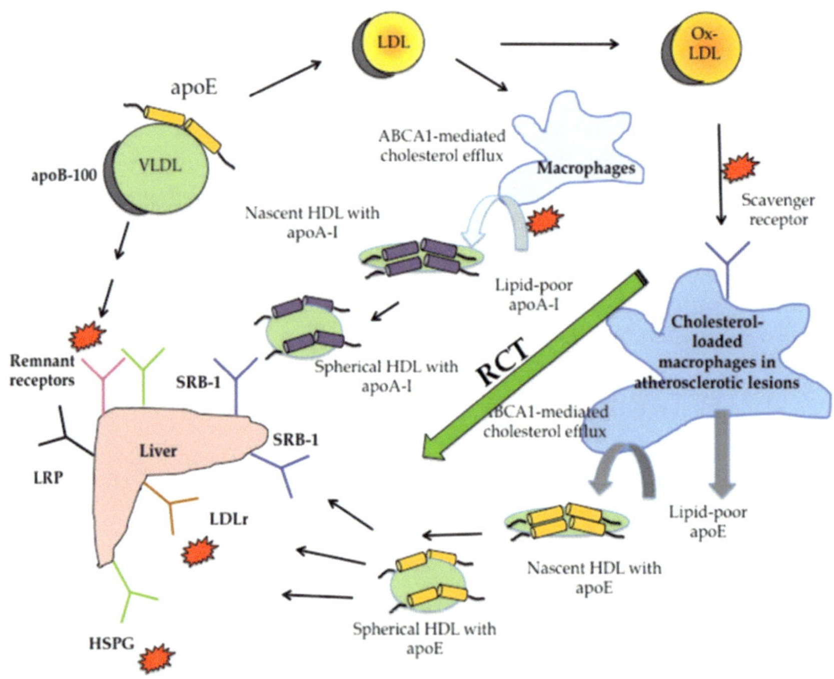

Figure 9. Distribution cholesterol mediated by apoB-100, apoE and apoAI. The green block arrow shows the general direction of the RCT process. The red stars draw attention to processes that are affected by oxidative modification of protein and lipid components of lipoproteins.

5. Lipoprotein oxidation and disease states

5.1. Atherosclerosis

Atherosclerosis is one of the leading causes of death worldwide, and is commonly associated with coronary and cerebrovascular diseases (Rocha & Libby, 2009; Moore & Tabas, 2011). Originally considered purely a lipid-storage disease, it is now recognized that atherosclerosis is intrinsically linked to inflammation, particularly with respect to involvement of innate and adaptive immunity.

Atherosclerosis is a progressive disorder marked by several stages with varying extent of lesions marking each stage. It is initiated by accumulation of lipoproteins in the sub-endothelium, typically at arterial branch points, which tend to have impaired laminar flow. The physiological response to this retention includes: (i) chemical modification (due to the production of ROS and other oxidative factors from intimal macrophages) and aggregation of the lipoproteins, (ii) early inflammatory response such as T cell recruitment, (iii) cytokine secretion, and, (iv) endothelial alterations. In response to the chemokines, the circulating monocytes enter the arterial wall and eventually differentiate to macrophages under the influence of macrophage colony stimulating factors.

The macrophages internalize oxidized and modified lipoproteins via several types of scavenger receptors, including class A and B scavenger receptors (e.g. CD36) and lectin-like oxidized LDLr-1 (LOX-1). Mice lacking CD36, platelet-activating-factor receptor, and toll-like receptors 2 and 4 show decreased atherosclerosis. The internalized cholesterol is stored in the cytoplasm as cholesterylester lipid droplets, surrounded by a monolayer of phospholipids. When viewed by electron microscopy, these cells have a foamy appearance, and are therefore called foam cells. Foam cell formation marks the earliest pathological lesion in atherosclerosis called 'fatty streaks'.

As the fatty streaks progress, they induce migration of smooth muscle cells from media to the intima of the arterial wall, which is accompanied by secretion of collagen and matrix proteins and macrophage proliferation. As lipid accumulation continues to occur in macrophages, smooth muscle cells also take up lipids. Collectively, these processes give rise to fibrous lesions. As the lesion progresses, foam cells die and release lipids, which aggregate with lipoproteins trapped in the matrix, eventually leading to advanced lesions with calcification and hemorrhage.

Ox-LDL also plays a role in modulating smooth muscle function, by increasing their adhesion to macrophages and foam cells in the plaques. At low concentrations, Ox-LDL stimulates proliferation of smooth muscle cells, whereas at higher concentrations, they cause smooth muscle apoptosis by up-regulating levels of the pro-apoptotic lipid ceramide. During this process, a necrotic core is formed in the vessel lumen. The necrotic plaque is unstable and subject to disintegration or rupture, leading to formation of a thrombus. Ox-LDL are key stimulators of the coagulation pathway, and promote the secretion of tissue factor from endothelial cells, which is required to form a clot. At various stages of the lesions unstable angina, heart attack or stroke may occur.

In summary, our understanding of the atherogenesis process has progressed rapidly since the deleterious nature of Ox-LDL was first pointed out about 3 decades back. We now recognize it as a multifactorial disease with inflammation and oxidative stress playing key roles. The last decade has seen the additional role of HDL in mitigating the severity of atherosclerosis; in this context, the role of HDL (not HDL levels *per se*, but the robustness of HDL function) is currently under intense scrutiny.

5.2. Alzheimer's disease

Alzheimer's disease is a neurodegenerative disorder that is clinically characterized by progressive cognitive decline and dementia. The major neuropathological hallmarks of this age-related disease are the extra cellular accumulation of amyloid beta peptide (Aβ), the intracellular presence of neurofibrillary tangles composed of hyperphosphorylated tau and the loss of cholinergic neurons in the brain.

Several hypotheses have been put forward to explain the pathogenesis of Alzheimer's disease, of which the oxidative stress hypothesis is of relevance here. In general, the brain of Alzheimer's disease subjects is believed to be in a heightened state of oxidative stress and is characterized by higher levels of lipid peroxidation products (acrolein, 4-HNE and MDA) (Arlt et al., 2002; Butterfield et al., 2010). The brain is particularly susceptible to oxidative stress due to the high oxygen and metal content, lipid levels, and paucity of antioxidants (such as vitamins C and E) compared to normal tissues (Schippling et al., 2000). The Alzheimer's disease brain is also rich in isoprostanes (a stable marker of peroxidation of arachidonic acid) and neuroprostanes (Lovell et al., 2001).

Accumulation of soluble and insoluble assemblies of Aβ (a peptide composed of 39-43 residues) in the brain parenchyma and in the cerebral vasculature is considered the primary event in Alzheimer's disease pathogenesis by the amyloid hypothesis. Aβ is derived from the ubiquitously expressed transmembrane protein amyloid precursor protein by regulated intramembranous proteolysis. The oligomeric form of Aβ is considered as the most toxic species; one of the reasons for its toxicity is its ability to act as an oxidative stress agent in a lipid environment. The oxidative nature of Aβ has been attributed to the presence of Met35, which is capable of undergoing one-electron oxidation to form a sulfuranyl radical cation (S•+). This in turn is able to abstract a H-atom from a fatty acyl chain as shown in **Figure 7**, and initiate a series of free radical-mediated events in a lipid milieu such as a membrane bilayer or a lipoprotein particle (Butterfield et al., 2005).

In the CNS, apoE is localized on HDL-sized spherical and discoidal particles. Lipoproteins isolated from CSF of Alzheimer's disease patients were shown to have a higher sensitivity to *in vitro* oxidation compared to those from normal subjects. It has also been shown that CSF lipoproteins can be oxidized through transition metal ions such as Cu^{2+} or Fe^{3+}. These metal ions are present in CSF as complexes with metal-binding proteins like ceruloplasmin or transferrin (Schippling et al., 2000); under pathological conditions, they are released in a catalytically active form. Furthermore, Aβ itself can induce oxidation through interactions with metal ions. Aβ contains three histidines and one tyrosine, all of which can chelate transition metal ions. This promotes Aβ aggregation and a pro-oxidative state of Aβ through reduction of the transition metal (Artl, et al., 2002). The peptide can then produce ROS or induce lipid peroxidation, both of which strongly affect lipoprotein stability.

Conversion of cholesterol to 24S-hydroxycholesterol is believed to be a mechanism by which the brain maintains cholesterol homeostasis. Plasma concentrations of 24Shydroxycholesterol are utilized as a biomarker and a diagnostic tool for neurological disorders. Neuronal damage is accompanied by destruction of neuronal membranes, which causes more cholesterol to be converted into 24S-hydroxycholesterol. In agreement significantly higher peripheral concentrations of 24S-hydroxycholesterol were found in Alzheimer's disease and vascular demented patients (Lutjohann at al., 2000). Importantly, the latter study showed that the apoE genotype does not contribute significantly to the elevated plasma levels of 24S-hydroxycholesterol in Alzheimer's disease patients (Bretillon, 2000).

The CSF levels of 24S-hydroxycholesterol appears to be sensitive to changes in the brain (possibly because they are not affected by hepatic clearance rates of this oxysterol) and may represent better markers both for neurodegenerative diseases and for disturbances in the blood brain barrier (Leoni & Caccia, 2011). In early stages of Alzheimer's disease, there are significantly higher CSF concentrations of 24S-hydroxycholesterol suggesting increased cholesterol turnover in the CNS during degeneration (Leoni & Caccia, 2011). These levels decrease as the disease advances, possibly reflecting the loss of cells expressing cholesterol 24S-hydroxylase, the enzyme responsible for the conversion of brain cholesterol into 24S-hydroxycholesterol. In Alzheimer's disease and mild cognitive impairment, but not in normal individuals, the levels of 24S-hydroxycholesterol significantly correlate with CSF levels of apoE (Shafaati et al., 2007).

The elevation of 24S-hydroxycholesterol in CSF is consistent with a significant role for this oxysterol as a signaling molecule during neuronal degeneration. It has been shown that 24S-hydroxycholesterol is able to induce expression of apoE and ABC transporters in astrocytes through activation of LXR, and to stimulate cellular cholesterol efflux (Baldan et al., 2009).

5.3. Diabetes

Diabetes mellitus is a disease characterized by hyperglycemia and insufficiency or resistance to insulin. In general, diabetes is associated with increased generation of free radicals, heightened state of oxidative stress, and attenuated antioxidant response. Studies have revealed greater levels of TBARS in the plasma of diabetic patients compared to controls (Kawamura et al., 1994). Plasma lipoproteins isolated from diabetic rats have been shown to be cytotoxic *in vitro* in cultured cells, suggesting they may have been oxidatively modified *in vivo* (Morel & Chisolm, 1989). Other studies have shown that glucose autoxidation and protein glycation can result in the formation of radicals like superoxide anions that promote lipoprotein oxidation.

Hyperglycemia promotes glycation of the protein and lipid components of lipoproteins leading to generation of advanced glycation end-products (AGE) (Sun et al., 2009). Glycation is the process whereby glucose attaches to the ε-amino group of lysines or the α-amino group of an N-terminal amino acid in a non-enzymatic manner (Maritim et al., 2003). Biochemically, glucose attachment to the protein results in the formation of an unstable Schiff base, which then rearranges to an Amadori product. These Amadori products then undergo dehydration reactions and rearrange themselves to finally form AGE, which are responsible for many of the irreversible pathological effects seen in diabetes. The receptors for AGE (RAGE) are abundantly expressed on vascular endothelial cells, smooth muscle cells and macrophages, which are enhanced in atherosclerotic lesions in diabetes. Since Ox-LDL has AGE epitopes, it binds RAGE on macrophages and enhances macrophage proliferation and oxidative stress.

Like oxidation, glycation of LDL prevents LDLr-mediated cellular uptake of lipoproteins and promotes scavenger receptor-mediated uptake. *In vivo*, small, dense, which is more atherogenic than large buoyant LDL, appears to be preferentially glycated; also, *in vitro* studies suggest that it is more susceptible to glycation (Soran & Durrington, 2011). Diabetic individuals display higher plasma concentrations of glycated LDL than non-diabetic individuals.

Type 1 diabetes subjects have lipid disorders (diabetic dyslipidemia), with a pro-atherogenic lipid profile: increased concentration of TG and LDL cholesterol, low HDL levels (Verges, 2009). In addition, they display increased cholesterol-triglyceride ratio within their VLDL, increased triglyceride in their LDL and HDL, glycation of apolipoproteins, increased oxidation of LDL and an increase in small dense LDL (relatively more atherogenic). HDL from Type 1 diabetes subjects is less effective in promoting cholesterol efflux and has reduced antioxidant properties.

Subjects with Type 2 diabetes also have a proatherogenic lipid profile with quantitative and qualitative differences in their lipoproteins (Verges, 2005). Typically, they have increased triglyceride levels, increased VLDL production and decreased VLDL catabolism, and decreased HDL cholesterol levels. They have large VLDL particles that are richer in triglyceride, small dense LDL particles, increase in triglyceride content of LDL and HDL, Ox- and glycated-LDL, glycation of apolipoproteins and increased susceptibility of LDL to oxidation.

6. Concluding remarks

In conclusion, this chapter has provided a broad overview of the role of oxidative stress, ROS, and lipoprotein oxidation in the pathophysiology of disease states such as atherosclerosis, Alzheimer's disease and diabetes. While Ox-LDL and inflammation seem to be bona fide factors in the development of atherosclerosis, the role of dysfunctional HDL in these disease states is not known at this point. There are several points of uncertainty regarding the *in vivo* source of ROS and oxidative stress, the physiological behavior of oxidized lipoproteins, particularly in Alzheimer's disease and diabetes, and the line-up of antioxidants and autoantibodies in response to the oxidized factors. It is anticipated that the next decade will provide more insights into the molecular and mechanistic basis of the effect of oxidative damage on lipoprotein in disease states. This would pave the way for new therapeutic options for preventing and treating these diseases.

Author details

Vikram Jairam
Yale University School of Medicine, USA

Koji Uchida
Nagoya University, Japan

Vasanthy Narayanaswami
California State University Long Beach, USA

Acknowledgement

The authors were funded by the Tobacco Related Disease Research Program (TRDRP 17RT-0165), the Alzheimer's Association, the National Institutes of Health (NIH-HL096365), CSUPERB Faculty Development Grant and the Drake Family Trust.

7. References

Anderson, M. M., Hazen, S. L., Hsu, F. F. & Heinecke, J. W. (1997). Human neutrophils employ the myeloperoxidase-hydrogen peroxide-chloride system to convert hydroxy-amino acids into glycolaldehyde, 2-hydroxypropanal, and acrolein. A mechanism for the generation of highly reactive alpha-hydroxy and alpha,beta-unsaturated aldehydes by phagocytes at sites of inflammation. *J. Clin. Invest.*, Vol. 99, No. 3, (Feb 1997), pp. 424-432.

Arai, H., Kashiwagi, S., Nagasaka, Y., Uchida, K., Hoshii, Y. & Nakamura, K. (1999). Oxidative modification of apolipoprotein E in human very-low-density lipoprotein and its inhibition by glycosaminoglycans. *Arch. Biochem. Biophys.*, Vol. 367, No. 1, (Jul 1999), pp. 1-8.

Arai, H., Uchida, K. & Nakamura, K. (2005). Effect of ascorbate on acrolein modification of very low density lipoprotein and uptake of oxidized apolipoprotein e by hepatocytes. *Biosci., Biotechnol. Biochem.*, Vol. 69, No. 9, (Sep 2005), pp. 1760-1762.

Arlt, S., Beisiegel, U. & Kontush, A. (2002). Lipid Peroxidation in Neurodegeneration: New Insights into Alzheimer's Disease. *Curr. Opin. Lipidol.*, Vol. 13, No. 3, (June 2002), pp. 289-294, 0957-9672

Bailey, J. M., Makheja, A. M., Lee, R. & Simon, T. H. (1995). Systemic Activation of 15-lipoxygenase in Heart, Lung, and Vascular Tissues by Hypercholesterolemia: Relationship to Lipoprotein Oxidation and Atherogenesis. *Atherosclerosis*, Vol. 113, No. 2, (Mar 1995), pp. 247-258.

Baldán, Á., Bojanic, D. D., & Edwards, P. A. (2009) The ABCs of sterol transport. *J Lipid Res.* 2009 April; 50(Supplement): S80–S85

Bassett, C. N., Neely, M. D., Sidell, K. R., Markesbery, W. R., Swift, L. L. & Montine, T. J. (1999). Cerebrospinal Fluid Lipoproteins Are More Vulnerable to Oxidation in

Alzheimer's Disease and Are Neurotoxic When Oxidized Ex Vivo. *Lipids,* Vol. 34, No. 12, (Dec 1999), pp. 1273-1280.

Beauchamp, R. O., Jr., Andjelkovich, D. A., Kligerman, A. D., Morgan, K. T. & Heck, H. D. (1985). A critical review of the literature on acrolein toxicity. *Crit. Rev. Toxicol.,* Vol. 14, No. 4, (1985), pp. 309-380.

Bergt, C., Pennathur, S., Fu, X., Byun, J., O'Brien, K., McDonald, T. O., Singh, P., Anantharamaiah, G. M., Chait, A., Brunzell, J., Geary, R. L., Oram, J. F. & Heinecke, J. W. (2004). The myeloperoxidase product hypochlorous acid oxidizes HDL in the human artery wall and impairs ABCA1-dependent cholesterol transport. *Proc. Natl. Acad. Sci. (USA),* Vol. 101, No. 35, (Aug 2004), pp. 13032-13037.

Berliner, J. A. & Watson, A. D. (2005). A Role for Oxidized Phospholipids in Atherosclerosis. *New Engl. J. Med.,* Vol. 353, No. 1, (Jul 2005), pp. 9-11.

Björkhem, I., Lütjohann, D., Diczfalusy, U., Ståhle, L., Ahlborg, G. & Wahren, J. (1998). Cholesterol homeostasis in human brain: turnover of 24S-hydroxycholesterol and evidence for a cerebral origin of most of this oxysterol in the circulation. *J. Lipid Res.,* Vol. 39, No. 8, (Aug 1998), pp. 1594-1600.

Bretillon, L., Sidén, A., Wahlund, L. O., Lütjohann, D., Minthon, L., Crisby, M., Hillert, J., Groth, C. G., Diczfalusy, U. & Björkhem, I. (2000). Plasma levels of 24S-hydroxycholesterol in patients with neurological diseases. *Neurosci. Lett.,* Vol. 293, No. 2, (Oct 2000), pp. 87-90.

Bradamante, S., Barenghi, L., Giudici, G. A. & Vergani, C. (1992). Free radicals promote modifications in plasma high-density lipoprotein: nuclear magnetic resonance analysis. *Free Rad. Biol. Med.,* Vol. 12, No. 3, (1992), pp. 193-203.

Brown, A. J. & Jessup, W. (1999). Oxysterols and atherosclerosis. *Atherosclerosis,* Vol. 142, No. 1, (Jan 1999), pp. 1-28.

Butterfield, D. A. & Boyd-Kimball, D. (2005). The critical role of methionine 35 in Alzheimer's amyloid beta-peptide (1-42)-induced oxidative stress and neurotoxicity. *Biochim. Biophys. Acta,* Vol. 1703, No. 2, (Jan 2005), pp. 149-156.

Butterfield, D. A., Bader Lange, M. L. & Sultana, R. (2010). Involvements of the lipid peroxidation product, HNE, in the pathogenesis and progression of Alzheimer's disease. *Biochim. Biophys. Acta,* Vol. 1801, No. 8, (Aug 2010), pp. 924-929.

Cyrus, T., Witztum, J. L., Rader, D. J., Tangirala, R., Fazio, S., Linton, M. F. & Funk, C. D. (1999). Disruption of the 12/15-lipoxygenase gene diminishes atherosclerosis in apo E-deficient mice. *J. Clin. Invest.,* Vol. 103, No. 11, (Jun 1999), pp. 1597-1604.

Daugherty, A., Dunn, J. L., Rateri, D. L. & Heinecke, J. W. (1994). Myeloperoxidase, a catalyst for lipoprotein oxidation, is expressed in human atherosclerotic lesions. *J. Clin. Invest.,* Vol. 94, No. 1, (Jul 1994), pp. 437-444.

Esterbauer, H., Schaur, R. & Zollner, H. (1991). Chemistry and Biochemistry of 4-hydroxynonenal, Malonaldehyde and Related Aldehydes. *Free Rad. Biol. Med.,* Vol. 11, No. 1, pp. 81-128.

Fang, X., Weintraub, N. L., Rios, C. D., Chappell, D. A., Zwacka, R. M., Engelhardt, J. F., Oberley, L. W., Yan, T., Heistad, D. D. & Spector, A.A. (1998). Overexpression of

Human Superoxide Dismutase Inhibits Oxidation of Low-Density Lipoprotein by Endothelial Cells. *Circ. Res.*, Vol. 82, No. 12, pp. 1289-1297, 1524-4571.

Furuhata, A., Ishii, T., Kumazawa, S., Yamada, T., Nakayama, T. & Uchida, K. (2003). N(epsilon)-(3-methylpyridinium)lysine, a major antigenic adduct generated in acrolein-modified protein. *J. Biol. Chem.*, Vol. 278, No. 49, (Dec 2003), pp. 48658-48665.

Fukai, T., Folz, R. J., Landmesser, U. & Harrison, D. G. (2002). Extracellular superoxide dismutase and cardiovascular disease. *Cardiovasc. Res.*, Vol. 55, No. 2, (Aug 2002), pp. 239-249.

Garner, B., Witting, P. K., Waldeck, A. R, Christison, J. K, Raftery, M. & Stocker, R. (1998). Oxidation of high density lipoproteins. I. Formation of methionine sulfoxide in apolipoproteins AI and AII is an early event that accompanies lipid peroxidation and can be enhanced by alpha-tocopherol. *J. Biol. Chem.*, Vol. 273, No. 11, (Mar 1998), pp. 6080-6087.

Haberland, M. E., Olch, C. L., Folgelman, A. M. (1984) Role of lysines in mediating interaction of modified low density lipoproteins with the scavenger receptor of human monocyte macrophages. *J. Biol. Chem.* 1984 Sep 25;259(18):11305-11311.

Hatters, D.M., Peters-Libeu, C.A., & Weisgraber, K. H. (2006). Apolipoprotein E structure: insights into function. *Trends Biochem. Sci.* 31, 445-454.

Hiltunen, T., Luoma, J., Nikkari, T. & Ylä-Herttuala, S. (1995). Induction of 15-lipoxygenase mRNA and protein in early atherosclerotic lesions. *Circulation*, Vol. 92, No. 11, (Dec 1995), pp. 3297-3303.

Hoff, H. F. & O'Neil, J. (1993). Structural and Functional Changes in LDL after Modification with Both 4-hydroxynonenal and Malondialdehyde. *J. Lipid Res.*, Vol. 34, No. 7, (Jul 1993), pp. 1209-1217.

Huang, Y., von Eckardstein, A., Wu, S., Assmann, G. (1995) Effects of the apolipoprotein E polymorphism on uptake and transfer of cell-derived cholesterol in plasma. *J. Clin. Invest.* 96, 2693-701.

Huang, Y., Weisgraber, K. H., Mucke, L. & Mahley, R. W. (2004) Apolipoprotein E: diversity of cellular origins, structural and biophysical properties, and effects in Alzheimer's disease. *J. Mol. Neurosci.* 23, 189-204.

Javitt, N. B. (2008). Oxysterols: novel biologic roles for the 21st century. *Steroids*, Vol. 73, No. 2, (Feb 2008), pp. 149-157.

Jürgens, G., Hoff, H. F., Chisolm, G. M. 3rd, & Esterbauer, H. (1987) *Chem. Phys. Lipids*. Vol. 45 (2-4, (Nov-Dec 1987), 315-336.

Kawamura, M., Heinecke, J. W. & Chait, A. (1994). Pathophysiological Concentrations of Glucose Promote Oxidative Modification of Low Density Lipoprotein by a Superoxide-dependent Pathway. *J. Clin. Invest.*, Vol. 94, No. 2, (Aug 1994), pp. 771-778.

Korotchkina, L. G., Yang, H., Tirosh, O., Packer, L. & Patel, MS. (2001). Protection by Thiols of the Mitochondrial Complexes from 4-hydroxy-2-nonenal. *Free Rad. Biol. Med.*, Vol. 30, No. 9, (May 2001), pp. 992-999.

Laukkanen, M. O., Lehtolainen, P., Turunen, P., Aittomäki, S., Oikari, P., Marklund, S. L., Ylä-Herttuala, S. (2000). Rabbit extracellular superoxide dismutase: expression and effect on LDL oxidation. *Gene*, Vol. 254, No. 1-2, (Aug 2000), pp. 173-179.

Leoni, V. & Caccia, C. (2011). Oxysterols as Biomarkers in Neurodegenerative Diseases. *Chem. Phys. Lipids*, Vol. 164, No. 6, (Sept 2011), pp. 515-524.

Levitan, I., Volkov, S. & Subbaiah, P. V. (2010). Oxidized LDL: diversity, patterns of recognition, and pathophysiology. *Antioxid. & Redox Signaling*, Vol. 13, No. 1, (Jul 2010), pp. 39-75.

Lovell, M. A., Xie, C. & Markesbery, W. R. (2001). Acrolein is increased in Alzheimer's disease brain and is toxic to primary hippocampal cultures. *Neurobiol. Aging*, Vol. 22, No. 2, (Mar-Apr 2001), pp. 187-194.

Lütjohann, D., Breuer, O., Ahlborg, G., Nennesmo, I., Sidén, A., Diczfalusy, U. & Björkhem, I. (1996). Cholesterol homeostasis in human brain: evidence for an age-dependent flux of 24S-hydroxycholesterol from the brain into the circulation. *Proc. Natl. Acad. Sci. (USA)*, Vol. 93, No. 18, (Sept 1996), pp. 9799-9804.

Maritim, A. C., Sanders, R. A., & Watkins, J. B. (2003). Diabetes, Oxidative Stress, and Antioxidants: A Review. *J. Biochem. Mol. Toxicol.*, Vol. 17, No. 1, (Feb 2003), pp. 24-38.

McCall, M. R., Tang, J. Y., Bielicki, J. K. & Forte, T. M. (1995). Inhibition of Lecithin-Cholesterol Acyltransferase and Modification of HDL Apolipoproteins by Aldehydes. *Arterioscl. Thromb. Vasc. Biol.*, Vol. 15, No. 10, (Oct 1995), pp. 1599-1606.

Meyer, J. & Schmitt, M.E. (2000). A Central Role for the Endothelial NADPH Oxidase in Atherosclerosis. *FEBS Lett.*, Vol. 472, No. 1, (Apr 2000), pp. 1-4.

Montine, T. J., Huang, D. Y., Valentine, W. M., Amarnath, V., Saunders, A., Weisgraber, K. H., Graham, D. G. & Strittmatter, W. J. (1996). Crosslinking of apolipoprotein E by products of lipid peroxidation. *J. Neuropathol. Exp. Neurol.*, Vol. 55, No. 2, (Feb 1996), pp. 202-210.

Moore, K. J. & Tabas, I. (2011) Macrophages in the pathogenesis of atherosclerosis. *Cell*, 2011 Vol. 145, No 3, (Apr 29), pp. 341-355.

Morel, D. W. & Chisolm, G. M. (1989). Antioxidant treatment of diabetic rats inhibits lipoprotein oxidation and cytotoxicity. *J. Lipid Res.*, Vol. 30, No. 12, (Dec 1989), pp. 1827-1834.

Navab, M., Berliner, J. A., Subbanagounder, G., Hama, S., Lusis, A. J., Castellani, L. W., Reddy, S., Shih, D., Shi, W., Watson, A. D., Van Lenten, B. J., Vora, D. & Fogelman, A. M. (2001). HDL and the inflammatory response induced by LDL-derived oxidized phospholipids. *Arterioscl. Thromb. Vasc. Biol.*, Vol. 21, No. 4, (Apr 2001), pp. 481-488.

Noor, R., Mittal, S. & Iqbal, J. (2002). Superoxide dismutase--applications and relevance to human diseases. *Medical Science Monitor: International Medical Journal of Experimental and Clinical Research*, Vol. 8, No. 9, (Sep 2002), pp. 210-215.

Olkkonen, V. M. & Lehto, M. (2004). Oxysterols and oxysterol binding proteins: role in lipid metabolism and atherosclerosis. *Ann. Med.*, Vol. 36, No. 8, (2004), pp. 562-572.

Palinski, W., Ylä-Herttuala, S., Rosenfeld, M. E., Butler, S. W., Socher, S. A., Parthasarathy, S., Curtiss, L. K., Witztum, J. L. (1990) Antisera and monoclonal antibodies specific for epitopes generated during oxidative modification of low density lipoprotein, *Arteriosclerosis*, Vol. 10, No 10 (3) (May-Jun 1990), 325-335.

Perry, G., Cash, A. D. & Smith, M. A. (2002). Alzheimer Disease and Oxidative Stress. *J. Biomed. Biotech.*, Vol. 2, No. 3, (2002), pp. 120-123.

Refsgaard, H. H., Tsai, L. & Stadtman, E. R. (2000). Modifications of proteins by polyunsaturated fatty acid peroxidation products. *Proc. Natl. Acad. Sci. (USA)*, Vol. 97, No. 2, (Jan 2000), pp. 611-616.

Reilly, K. B., Srinivasan, S., Hatley, M. E., Patricia, M. K., Lannigan, J., Bolick, D. T., Vandenhoff, G., Pei, H., Natarajan, R., Nadler, J. L. & Hedrick, C. C. (2004). 12/15-Lipoxygenase activity mediates inflammatory monocyte/endothelial interactions and atherosclerosis in vivo. *J. Biol. Chem.*, Vol. 279, No. 10, (Mar 2004), pp. 9440-9450.

Rifkind, J. M., Abugo, O. O., Nagababu, E., Ramasamy, S., Demehin, A., Jayakumar, R. (2002). Aging and the red cell. *Adv. Cell Aging & Geront.*, Vol. 11, (2002), pp. 283-307.

Rocha, V. Z. & Libby, P. (2009). Obesity, Inflammation, and Atherosclerosis. *Nature Reviews Cardiology*, Vol. 6, No. 6, (June 2009), pp. 399-409.

Rosenfeld, M. E., Palinski, W., Ylä-Herttuala, S., Butler, S., & Witztum, J. L. (1990) Distribution of oxidation specific lipid-protein adducts and apolipoprotein B in atherosclerotic lesions of varying severity from WHHL rabbits, *Arteriosclerosis*, Vol. 10, No. 10 (3) (May-Jun 1990), 336-349

Savenkova, M. L., Mueller, D. M. & Heinecke, J. W. (1994). Tyrosyl radical generated by myeloperoxidase is a physiological catalyst for the initiation of lipid peroxidation in low density lipoprotein. *J. Biol. Chem.*, Vol. 269, No. 32, (Aug 1994), pp. 20394-20400.

Schaur, R. (2003). Basic Aspects of the Biochemical Reactivity of 4-hydroxynonenal. *Mol. Asp. Med.*, Vol. 24, No. 4-5 (Aug-Oct 2003), pp. 149-159.

Schippling, S., Kontush, A., Arlt, S., Buhmann, C., Sturenberg, H., Mann, U., Muller-Thomsen, T. & Beisiegel, U. (2000). Increased Lipoprotein Oxidation in Alzheimer's Disease. *Free Rad. Biol. Med.*, Vol. 28, No. 3, (Feb 2000), pp. 351-360.

Shao, B., Fu, X., McDonald, T. O., green, P. S., Uchida, K., O'Brien, K. D., Oram, J. F. & Heinecke, J. W. (2005). Acrolein impairs ATP Binding Cassette Transporter A1-dependent cholesterol export from cells through site-specific modification of apolipoprotein A-I. *J. Biol. Chem.*, Vol. 280, No. 43, (Oct 2005), pp. 36386-36396.

Shao, B., Pennathur, S. & Heinecke, J. W. (2012). Myeloperoxidase Targets apoA-I, the Major High Density Lipoprotein Protein, for Site-Specific Oxidation in Human Atherosclerotic Lesions. *J. Biol. Chem.*, Vol. 287, No. 9, (Feb 2012), pp. 6375-6386.

Shafaati, M., Solomon, A., Kivipelto, M., Björkhem, I. & Leoni, V. (2007). Levels of ApoE in cerebrospinal fluid are correlated with Tau and 24S-hydroxycholesterol in patients with cognitive disorders. *Neurosci. Lett.*, Vol. 425, No. 2, (Sep 2007), pp. 78-82.

Soran, H. & Durrington, P. N. (2011). Susceptibility of LDL and its subfractions to glycation. *Curr. Opinion in Lipidology*, Vol. 22, No. 4, (Aug 2011), pp. 254-261.

Steinberg, D., Parthasarathy, S., Carew, T. E., Khoo, J. C. & Witztum, J. L. (1989). Beyond cholesterol. Modifications of low-density lipoprotein that increase its atherogenicity. *New Eng. J. Med.*, Vol. 320, No. 14, (Apr 1989), pp. 915-924.

Sun, L., Ishida, T., Yasuda, T., Kojima, Y., Honjo, T., Yamamoto, Y., Yamamoto, H., Ishibashi, S., Hirata, K. & Hayashi, Y. (2009). RAGE mediates oxidized LDL-induced pro-inflammatory effects and atherosclerosis in non-diabetic LDL receptor-deficient mice. *Cardiovasc. Res.*, Vol. 82, No. 2, (May 2009), pp. 371-381.

Tamamizu-Kato, S., Wong, J. Y., Jairam, V., Uchida, K., Raussens, V., Kato, H., Ruysschaert, J. M, Narayanaswami, V. (2007). Modification by acrolein, a component of tobacco smoke and age-related oxidative stress, mediates functional impairment of human apolipoprotein E. *Biochemistry*, Vol. 46, No. 28, (Jul 2007), pp. 8392-8400.

Tórsdóttir, G., Sveinbjörnsdóttir, S., Kristinsson, J., Snaedal, J. & Jóhannesson, T. (2006). Ceruloplasmin and superoxide dismutase (SOD1) in Parkinson's disease: a follow-up study. *J. Neurol. Sci.* ., Vol. 241, No. 1-2, (Feb 2006), pp. 53-58.

Uchida, K.,Szweda, L. I., Chae, H-Z., & Stadtman, E. R. (1993), Immunochemical detection of 4-hydroxynonenal protein adducts in oxidized hepatocytes. *Proc. Natl. Acad. Sci. (USA),* Vol. 90, No. 18, (Sept 1993), pp. 8742-8746.

Uchida, K., Kanematsu, M., Sakai, K., Matsuda, T., Hattori, N., Mizuno, Y., Suzuki, D., Miyata, T., Noguchi, N., Niki, E. & Osawa, T. (1998a). Protein-bound acrolein: potential markers for oxidative stress. *Proc. Natl. Acad. Sci. (USA),* Vol. 95, No. 9, (Apr 1998), pp. 4882-4887.

Uchida, K., Kanematsu, M., Morimitsu, Y., Osawa, T., Noguchi, N. & Niki, E. (1998b). Acrolein is a product of lipid peroxidation reaction. Formation of free acrolein and its conjugate with lysine residues in oxidized low density lipoproteins. *J. Biol. Chem.*, Vol. 273, No. 26, (Jun 1998), pp. 16058-16066.

Vaya, J. & Schipper, H. M. (2007). Oxysterols, Cholesterol Homeostasis, and Alzheimer Disease. *J. Neurochem.*, Vol. 102, No. 6, (Sep 2007), pp. 1727-1737.

Vergès B. (2005). New insight into the pathophysiology of lipid abnormalities in type 2 diabetes. *Diab. Metab.,* Vol. 31, No. 5, (Nov 2005), pp. 429-439.

Vergès B. (2009). Lipid disorders in type 1 diabetes. *Diab. Metab.*, Vol. 35, No. 5, (Nov 2009a), pp. 353-360.

Witz, G. (1989). Biological interactions of alpha, beta-unsaturated aldehydes. *Free Rad. Biol. Med.*, Vol. 7, No. 3, (1989), pp. 333-349.

Zarkovic, N. (2003). 4-Hydroxynonenal as a bioactive marker of pathophysiological processes. *Mol. Asp. Med.*, Vol. 24, No. 4-5, pp. 281-291.

Zheng, L., Nukuna, B., Brennan, M. L., Sun, M., Goormastic, M., Settle, M., Schmitt, D., Fu, X., Thomson, L., Fox, P. L., Ischiropoulos, H., Smith, J. D., Kinter, M. & Hazen, S. L. (2004). Apolipoprotein A-I is a selective target for myeloperoxidase-catalyzed oxidation and functional impairment in subjects with cardiovascular disease. *J. Clin. Invest.,* Vol. 114, No. 4 (Aug 2004), pp. 529-541.

Chapter 17

Oxidized Phospholipids: Introduction and Biological Significance

Mohammad Z. Ashraf and Swati Srivastava

Additional information is available at the end of the chapter

http://dx.doi.org/10.5772/50461

1. Introduction

Phospholipids containing polyunsaturated fatty acids are highly prone to modification by reactive oxygen species. They tend to undergo lipid peroxidation to form OxPLs which induce cytotoxicity and apoptosis and plays a significant role in inflammation. There are reports that provide insights for involvement of OxPLs in interleukin transcription, phenotype switching of smooth muscle cells and apoptotic mechanisms of the modified phospholipids. Thus peroxidation greatly alters the physiochemical properties of membrane lipid bilayers and consequently induces signaling depending upon the formation or reorganization of membrane domains or specific molecular binding (Deigner et al, 2008). Distinct OxPLs species may interact with specific binding sites and receptors leading to the activation of individual signaling pathways. The most prevalent human coronary atherosclerosis is a chronic inflammatory disease that occurs due to lipid abnormalities. Pro-inflammatory oxidized low-density lipoprotein (OxLDL) has been suggested to be a link between lipid accumulation and inflammation in vessel walls. Increased levels of phospholipids' oxidation products have been detected in different organs and pathological states, including atherosclerotic vessels (Watson et al 1997, Subbanagounder et al 2000), inflamed lung (Yoshimi et al 2005, Nakamura et al 1998), non-alcoholic liver disease (Ikura et al 2006), plasma of patients with coronary artery disease (Tsimikas et al 2005), as well as in apoptotic cells (Huber et al 2002, Chang et al 2004), virus-infected cells (Van Lenten et al 2004) and cells stimulated with inflammatory agonists (Subbanagounder et al 2002). Moreover, studies have been done on two HDL-associated enzymes, serum paraoxonase (PON1) and PAF-acetylhydrolase (PAF-AH), which are responsible for hydrolysis of plasma oxidized phospholipids (Forte et al 2002) thereby providing evidence for their role in atherosclerosis. Another important marker of oxidative stress is the association of OxPLs with the apolipoprotein B-100 particle (OxPLs/apoB) of

LDL. Increased levels of OxPLs/apoB are implicated in coronary artery disease, progression of carotid and femoral atherosclerosis and the prediction of cardiovascular events (Tsimikas et al 2005).

2. Formation of OxPLs

OxPLs are generated by the oxidation of polyunsaturated fatty acid residues, which are usually present in the phospholipids at the *sn*-2 position. Oxidation of phospholipids is initiated either enzymatically by lipoxygenases or by reactive oxygen species and propagates *via* the classical mechanism of lipid peroxidation chain reaction. This implies that the production of OxPLs cannot be regulated by adjusting the amount or activity of enzymes. Hence there is a probability of the uncontrolled generation of OxPLs during oxidative stress. Several evidences suggest that OxPLs are formed from Poly Unsaturated Fatty Acids (PUFAs) at the sn-2 position (Bochkov et al 2007, Podrez et al 2002). Bioactive oxidized phospholipids may contain fragmentation products of PUFA, such as 1-palmitoyl-2-oxovaleroyl-*sn*-glycero-3-phosphorylcholine and 9-keto-10-dodecendioic acid ester of 2-lyso-phosphatidyl choline (KOdiA-PC); prostaglandins, such as 15 deoxy-delta 12, 14 prostaglandin I2 (PGI2) and 1-palmitoyl-2-(5,6-epoxyisoprostane E2)-*sn*-glycero-3-phosphoryl choline (PEIPC); and levuglandins. These molecules exhibit different biological activities. Chromatographic separation of many products formed by oxidation of 1-palmitoyl-2-arachidonoyl-*sn*-glycero-3-phosphorylcholine (PAPC) led to the identification of 1-palmitoyl-2-(5-oxovaleroyl)-*sn*-glycero-3-phosphatidylcholine (POVPC), 1-palmitoyl-2-glutaroyl-*sn*-glycero-3-phosphatidylcholine (PGPC) and 1-palmitoyl-2-(5,6-epoxyisopropane E_2)-*sn*-glycero-3-phosphatidylcholine (PEIPC) as potent lipid mediators of inflammation. High structural variation may explain why OxPLs demonstrate a remarkable variety of biological activities (FIGURE-1).

Enzymatic and non-enzymatic reactions, free-radical, and radical-free processes are capable of initiating wide spectrum of reactions causing oxidation of PUFAs. Majority of these reactions produce identical primary oxidation products (i.e., peroxyl radicals and hydroperoxides). Subsequent oxidation of OxPLs is an enzyme-independent stochastic process producing a wide spectrum of OxPLs. Peroxidation products thus generated proceeds according to several mechanisms such as oxidation of PUFA residue, cyclization of peroxyl radical or oxidative fragmentation of esterified PUFAs generating either full-length residues incorporating several oxygen atoms, or shortened fatty acid residues. Introduction of additional oxygen atoms into PUFAs is a common mechanism that increases complexity of OxPLs mixtures however biological activities of poly-oxygenated PLs are still not characterized. On the other hand, cyclization of peroxyl radical produces cyclic peroxide, which undergoes re-arrangements yielding bicyclic endoperoxide, or oxidation introducing additional non-cyclic or cyclic peroxide group. Cyclization of peroxyl radical is only possible for FAs having three or more double bonds (Salomon et al 2005).

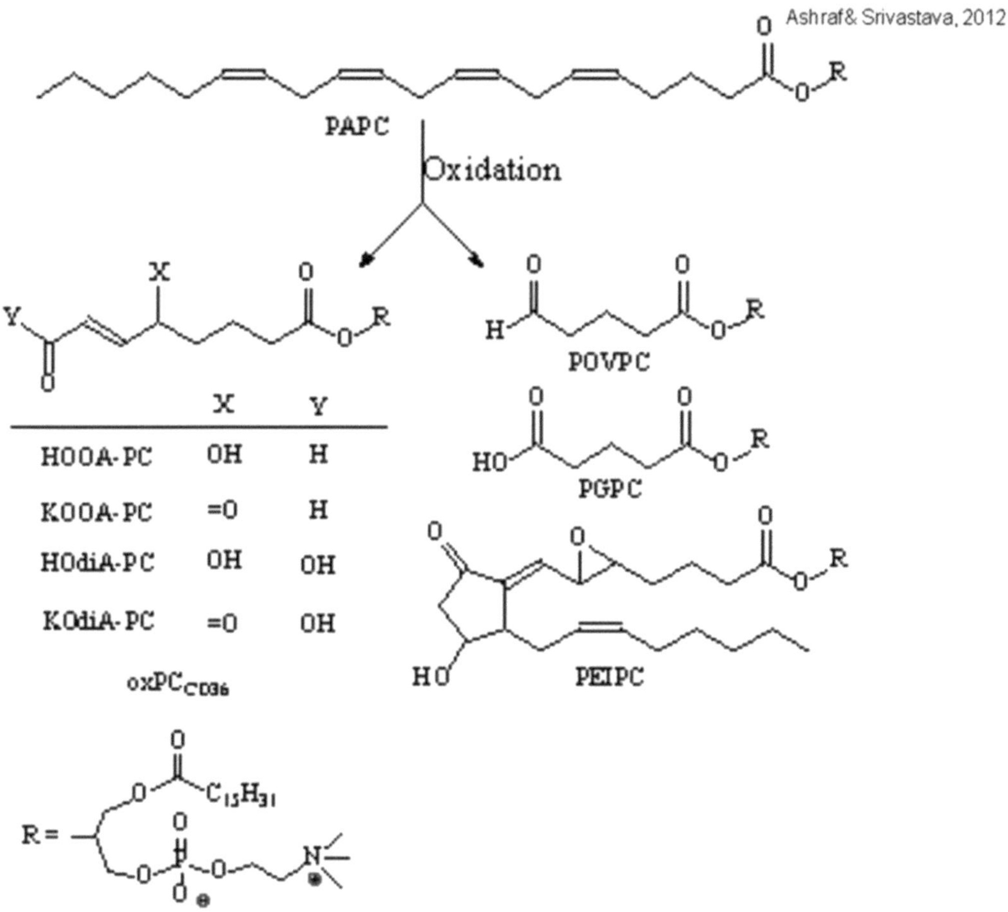

Figure 1. Representative chemical structures of oxidized phospholipids formed during oxidation of PAPC.

2.1. Oxidative cleavage and generation of fragmented OxPLs species

Peroxides/ peroxyls are transformed into advanced oxidation products by fragmentation of hydroperoxides. γ-Hydroxy (or oxo) a,b-unsaturated PLs with terminal aldehyde groups are produced from hydroperoxides via oxidation/fragmentation or polymerization/cleavage. Oxidative fragmentation of hydroperoxides occurs via several mechanisms including b-scission, Hock rearrangement, or cyclization of alkoxy radical produced from hydroperoxide (Gugiu et al 2006). γ-Hydroxy (or oxo)-α,β-unsaturated aldehyde PLs are highly reactive compounds, that are able to covalently link to amino groups of proteins, as well as thiol groups of biomolecules (Hoff et al 2003). On the other hand, peroxyl radical can cross-react with double bonds present in hydroperoxides yielding peroxydimers, these are unstable products and spontaneously break down forming either new radicals or α,β-unsaturated aldehydes (Schneider et al 2008). In addition to these products, saturated fragmented species containing terminal carbonyl groups are produced by oxidative fragmentation of PUFA-PLs, most common amongst which are oxononanoate and azelaoate formed from linoleic acid, oxovaleroate, and glutaroate generated from arachidonic acid, or oxobutyrate and succinate produced from docosahexaenoic acid (Gu et al 2003, Podrez et al 2002). Saturated fragmented OxPLs can be formed by further oxidation of γ-hydroxy (or oxo)-α, β-unsaturated PLs in addition to direct formation from hydroperoxides, (Podrez et al 2002). Saturated fragmented OxPLs l lack double bonds and hence they are resistant to further oxidation as the absence of double bonds within fragmented chains results in reduced reactivity of aldehyde containing saturated OxPLs as compared to α,β-unsaturated fragmented OxPLs.

2.2. Non-enzymatic oxidation of PL-PUFAs

This process is initiated by free radicals or non-radical reactive oxygen species (ROS). Free radical-mediated chain reaction is initiated by the formation of carbon-centered radicals and/or hydroperoxides of PUFAs (peroxidation of PUFAs). Due to the presence of methylene groups located between double bonds (bisallylic methylene groups), PUFAs are more susceptible to oxidation as compared to saturated FAs. As a result they are characterized by weakened hydrogen-carbon bonds. Free radicals can abstract hydrogen from bisallylic methylene leading to the formation of carbon-centered radicals within PUFAs. Now occurs the initiation step of lipid peroxidation, Carbon-centered radicals rapidly react with molecular oxygen, producing peroxyl radicals. These Peroxyl radicals react with bisallylic methylene groups in other PUFA molecules, leading to the transformation of peroxyl radicals to hydroperoxides and generation of new carbon-centered radicals. Thus, additional cycles of peroxidation are initiated. PUFA hydroperoxides in turn produce reactive alkoxyl and hydroxyl radicals via iron or copper-catalyzed Fenton-like reactions, further propagating the chain reaction (Bochkov et al 2010).

2.3. Enzymatic oxidation of PL-PUFAs

1, 4-pentadiene motifs are recognized within unsaturated fatty acids by lipoxygenases (LOXs) and molecular oxygen with high stereoselectivity is introduced. The majority of lipoxygenases oxidize only unesterified PUFAs. Only one group (12/15-LOX) amongst all known LOXs is capable of oxidizing PL-esterified fatty acids. This class of enzymes is present in different biological species and includes mouse, rat, rabbit, bovine, and porcine leukocyte-type 12- LOX, rabbit and human reticulocyte-type 15-LOX, and soybean LOX (Huang et al 2008, Wittwer et al 2007). Switching of activity of electron transport in mitochondria to peroxidation by cytochrome c (cyt c) has been suggested by Kagan et al (2005). This transformation begins when cyt c binds to negatively charged cardiolipin (CL), leading to conformational changes and subsequent release of PL-protein complex from mitochondria into cytosol. The complex of cyt c with CL activated by traces of PUFA-OOH or H_2O_2 acquires the ability to oxidize CL, PS, or PI, with formation of PL-OOH (Kagan et al 2009).

Alternatively, OxPLs are also generated by re-esterification of free oxidized PUFAs into lyso-PLs. Several types of OxPLs have been found to be generated by this mechanism both *in vivo* and *in vitro* (Arai et al 1997, Birkle et al 1984).

2.4. Detoxification of reactive OxPLs

Detoxification of OxPLs comprises the mechanisms that terminate peroxidation chain reaction and inactivate chemically reactive toxic groups produced by oxidation. Hydroxides are characterized by significantly lower chemical reactivity and therefore are considered to be stable and non-toxic compared to hydroperoxides (Spiteller et al 1997). Most commonly, the enzyme catalyzing the reduction of hydroperoxides to hydroxides is glutathione peroxidase (GPx). Lipid hydroperoxides are reduced in a reaction that involve selenocysteine residue of GPx and glutathione thus generating lipid hydroxide and oxidized glutathione. With respect to membrane-bound hydroperoxides of PL esterified PUFAs, PL glutathione peroxidase (GPx4) has the highest activity amongst GPx enzymes (Savaskan et al 2007).

A variety of products containing aldehyde and keto functional groups are formed upon oxidation of OxPLs which are further reduced by aldo-keto reductases to respective hydroxyl groups. Apart from playing physiological role in metabolism of sugar aldehydes, aldo-keto reductases also play a role in detoxification of toxic phospholipid aldehydes (Jin et al 2007).

Another aspect of detoxification is OxPLs cleavage. Platelet activating factor acetylhydrolase (PAF-AH) has been recognized for its ability to cleave and thus inactivate PAF (McIntyre et al 2009).The enzyme was shown to hydrolyze fragmented saturated OxPLs (Stremler et al 1991), as well as long-chain OxPLs, including esterified F2-isoprostanes, PC-hydroperoxides and PEIPC (Kriska et al 2007, Davis et al 2008).

3. Mechanism of action

Specific receptor binding of OxPLs is the subject of an ongoing debate. Available evidence suggests that OxPLs interact with various signal transduction receptors and pattern recognition receptors present on the cell surface. Most commonly known receptors include CD36, SRB1, EP2, VEGFR2 and the PAF receptor (Bochkov et al 2007, Zimman et al 2007). It has been demonstrated that when present in vesicles, truncated oxidized fatty acids at the sn-2 position move from the hydrophobic interior to the aqueous exterior of the vesicle. this would allow their recognition by cell surface receptors Earlier models of isoprostane-containing phospholipids have suggested that they are highly twisted and may distort membrane areas in which they are present (Morrow et al 1992). Moumtzi et al (2007) have shown that phospholipid oxidation products can integrate into lipid membranes of cells and lipoproteins; they can either act as ligands or may cause local membrane disruption. Besides, peroxidation of phospholipids leads to the accumulation of lysoforms as a result of both non-enzymatic decylation and enzymatic hydrolysis reactions catalyzed to a large extent by lipoprotein-as-associated phospholipase A_2 (also known as PAF acetylhydrolase), which has high substrate selectivity toward polar phospholipids, including the oxidized forms (Zalewski et al 2005). Some lysophospholipids bind and activate G protein-coupled receptors (GPCR). Parhami et al (1993 & 1995) explained that oxidized phospholipids act by binding to a G protein-coupled receptor. These authors demonstrated that minimally modified LDL stimulated a putative Gs-coupled receptor, thus increasing cyclic AMP (cAMP) levels in endothelial cells. Lysophosphatidylcholine and lysophosphatidic acid triggered the activity of G2A and LPA1-LPA4 receptors respectively (Tomura et al 2005, Anliker et al 2004). In addition to GPCR, OxPLs also activate other classes of receptors such as peroxisome proliferator-activated receptors (PPAR). Thus, phospholipid peroxidation may induce the generation of lysophospholipids that are known to accumulate in LDL (OxLDL) and atherosclerotic lesions (Siess et al 2004, Tselepis et al 2002).

Prostaglandin receptors have been recently implicated into OxPLs-induced inflammation. OXPAPC and its component lipid PEIPC are able to stimulate prostaglandin E_2 and D_2 receptors (EP2 and DP respectively) and to compete with receptor binding of radio labeled prostaglandin E_2 (Li et al 2006). Previously, it was observed that POVPC binds to human macrophages via the PAF receptor (PAF-R). Occupancy of the PAF-R by the OxPLs modifies the transcription levels of pro-inflammatory genes such as IL-8 (Pegorier et al 2006).

Some effects of OxPLs are probably not mediated by signal transducing receptors. Modulation of cellular cholesterol depots has been suggested as a non-receptor mediated mechanism of OxPLs sensing by cells. It is well illustrated that OxPAPC induces depletion and re-distribution of cellular cholesterol reserves finally leading to the activation of a transcription factor SREBP, a well recognized sensor for cellular cholesterol contents. In turn, SREBP activates IL-8 production (Yeh et al 2004).The human aortic EC gene expression was found to be stimulated by PAPC. Furthermore, OxPAPC may bind to a 37KDa glycosylphosphatidylinositol anchored protein, which interacts with TLR4 to induce interleukin-8 (IL-8) transcription (Walton et al 2003). Leitinger et al (2003) and Watson et al (1997) have described a possible role of toll-like receptors (TLRs) in OxPLs-induced inflammation. Studies have confirmed that Asp299Gly-TLR4 polymorphism plays a protective role in attenuation of atherosclerosis.

Mitogen activated protein kinase phosphatase-1 (MKP-1) was reported to be involved in OxPAPC-induced MCP-1 production. Also activation of eNOS by OxPAPC is regulated via a phosphatidylinositol-3-kinase/Akt-mediated mechanism, OxPAPC-induced SREBP activation is significantly reduced with eNOS inhibition (Berliner and Gharavi, 2008).

Chen et al (2007) reported that LDL-associated phosphatidylcholine esterified with *sn*-2-azelaic acid at the *sn*-2 position is readily taken up by cells. This compound, one of the main phospholipid oxidation products in LDL, induces apoptosis of HL60 cells at low micromolar concentrations. Since the intact phospholipid is required for signaling, this effect can be prevented by over-expression of PAF acetyl hydrolase known for oxidizing phospholipids with polar residues at the *sn*-2 position.

Another biologically active phospholipid described is platelet activating factor (PAF) having various inflammatory actions such as platelet aggregation, hypotension, anaphylactic shock and increased vascular permeability (Prescott et al 2000). PAF is structurally identified as 1-0-alkyl-2-acetyl-*sn*-glycero-3-phosphocholine. Atherogenic effects are also induced by PAF by activating monocytes and stimulating smooth muscle cell growth. In contrast to the tightly regulated physiological generation of PAF, uncontrolled processes of free radical oxidation generate analogs of PAF *in vivo* and *in vitro*. As a result of this uncontrolled chemical reaction, fragmentation of the residue at sn-2 position occurs and these oxidatively generated PAF mimetics stimulate monocytes, leukocytes and platelets. They are found in atherosclerotic lesions and even in blood from individuals exposed to cigarette smoke (Heery et al 1995).

Other oxidized phospholipids such as POVPC and PGPC have also been shown to play major roles in activation of endothelial cells and induction of leukocyte binding. They are identified as abundant products in oxidized LDL. The effect of POVPC is protein kinase-A dependent leading to the stimulation of the cAMP-mediated pathway (Berliner and Gharavi, 2008).

OxPLs also induces autocrine mediators such as vascular endothelial growth factor (VEGF), which works through activation of transcription factor-4 (ATF4) (Oskolkova et al 2008).

4. OxPLs receptors

It has been shown that OxPLs stimulate a number of signal-transducing receptors located on the cell surface or in the nucleus, including G protein-coupled receptors, receptor tyrosine kinases, Toll-like receptors, receptors coupled to endocytosis, and nuclear ligand-activated transcription factors such as PPARs.

4.1. Prostaglandin receptors

OxPCs containing esterified PEIPC activate receptors recognizing prostaglandins E2 and D respectively (Li et al 2006). Activation of EP2 receptor on ECs results in activation of integrins and increased binding of monocytes.

4.2. Scavenger receptors

OxPLs comprise a major group of ligands for scavenger receptors. Different classes of Scavenger receptors range from Class A, B, D, E and F depending upon the nature and type of ligand (FIGURE-2). CD36 have been described as the major receptor expressed on macrophages and involved in the process of atherogenesis and apoptosis. The role of CD36 has been shown to be responsible for recognition of free oxidized phospholipids (Boullier et al 2000, Podrez et al 2000). Also Boullier et al (2000) and Watson et al (1997) have pointed out that oxidized phospholipid is covalently linked to apolipoprotein B-100 in extensively oxidized LDL (e.g. Cu^{2+} -oxLDL) and serve as ligand for CD36. Scavenger receptor- ligand interaction initiates signaling cascades that regulate macrophage activation, lipid metabolism and inflammatory pathways which may influence the development and stability of atherosclerotic plaque. Recent studies have demonstrated the expression of scavenger receptors especially CD36 and SR-BI on platelets suggesting their critical role in platelet hyper-reactivity in dyslipidemia and atheroprogression.

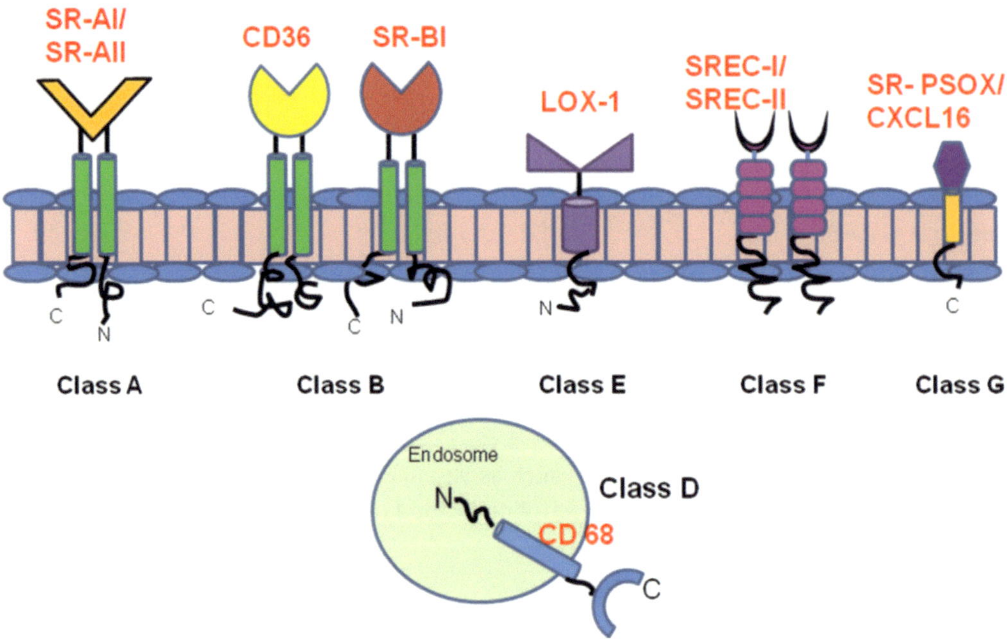

Ashraf & Srivastava, 2012

Figure 2. Schematic representations of different class of scavenger receptors involved in OxPLs binding.

4.3. PAF receptors

OxPLs initiate activation of receptor specific for PAF, which act as an important lipid mediator of inflammation and platelet aggregation. It recognizes alkyl-acyl-phosphatidylcholines specifically and contains an ether bond at the sn-1 position in combination with unusually short sn-2 acetyl residue. Oxidative fragmentation of sn-2 PUFAs in alkyl-PCs generates products such as 1-alkyl-2-butenoyl and 1-alkyl-2-butanoyl that are recognized by PAF receptor (Androulakis et al 2005, Marathe et al 1999). However, the role of the PAF receptor in the overall biological activity of OxPCs is not characterized.

4.4. VEGF receptors

It has been demonstrated that phosphorylation (activation) of VEGFR2 is enhanced within the first minutes of incubation with OxPAPC (Zimman et al 2007). They hypothesized that trans-activation of VEGFR2 in OxPAPC-treated cells was mediated by c-SRC.

4.5. Sphingosine-1-phosphate (S1P) receptor 1

It has shown that OxPAPC stimulates the recruitment of S1P1 to caveolin-enriched membrane microdomains, and induces its phosphorylation (activation) by AKT. Transactivation of S1P1 by OxPAPC plays a role in barrier-protective function of OxPLs.

4.6. Toll-like receptor 4

TLR4 plays a role in OxPAPC-mediated induction of IL-8 in HeLa cells. OxPAPC also induces lung injury and IL-6 production by mouse lung macrophages via the TLR4–TRIF–TRAF6 pathway (Imai et al 2008). On the other hand various classes of OxPLs do not influence the basal levels of E-selectin, ICAM-1, VCAM-1, TNFa, IL-6, IL-1a, IL-1b, and COX-2 in whole blood or individual cell types, including human umbilical vein ECs, blood monocytes, macrophage cell line, or fibroblasts (Bochkov et al 2002, Erridge et al 2008).

4.7. PPARα and PPARγ

Peroxisome proliferator-activated receptors (PPARs) are intracellular ligand-activated transcription factors. Diacyl-OxPLs stimulated a PPAR response element-driven reporter construct in transfected HAECs and the effect of OxPAPC, POVPC, and PGPC was mediated by PPARα as indicated by the activation of the ligand binding domain of PPARα, but not PPARγ or PPARδ (Lee et al 2000).

Second messengers up-regulated by OxPLs: Apart from the above described receptors , minimally modified Low Density Lipoproteins (MM-LDL) also induces elevation of Ca^{2+} in ECs (Honda et al 1999) and also OxPAPC was shown to induce rapid and reversible Ca2þ-responses in ECs (Bochkov et al 2002). MM-LDL causes a saturable dose-dependent increase in cAMP levels in aortic ECs that may arise due to activation of Gs and inhibition of Gi heterotrimeric G-protein complexes (Parhami et al 1995).

5. Biological function

Many cellular events are initiated and modulated by biologically active oxidized phospholipids. OxPLs were initially characterized as an active principle of minimally modified LDL (MM-LDL), responsible for its ability to stimulate EC to bind the leukocytes (Watson et al 1995). MM-LDL and OxPLs has the characteristic feature of inflammatory agonist i.e., their ability to activate binding of monocytes but not neutrophils (Watson et al 1997). In contrast to lipopolysaccride (LPS), tumor necrosis factor α (TNFα), or interleukin 1 (IL-1), MM-LDL does not up-regulate the expression of ICAM-1, VCAM-1 and E-Selectin on EC (Kim et al 1994), but promotes surface deposition of CS-1-containing variant of fibronectin (CS-1 FN) serving as ligand for the $\alpha4\beta1$ (VLA-4) integrin expressed on the surface of monocytes (Shih et al 1999). Similar to MM-LDL, OxPLs selectively stimulate adhesion of monocytes by CS-1 FN-dependent mechanism. Likewise other inflammatory agonists, OxPLs also stimulate the production of cyto- and chemokines. OxPLs are known to up-regulate expression IL-6, IL-8, MCP-1, GROα, MIP-1α, MIP-1β and CXCL3 (Subbanagounder et al 2002, Furnkranz et al 2005, Lee et al 2000, Reddy et al 2002, Kadl et al 2002, Gargalovic et al 2006, Huo et al 2001).

Expression of a number of genes related to angiogenesis, atherosclerosis, inflammation and wound healing are modulated by oxidized phospholipids in human aortic endothelial cells (Berliner and Gharavi, 2008; Gargalovic et al., 2006). Bochkov and colleagues (2002, 2007) have made known that OxPLs counteract the lipopolysaccride (LPS) pathway. Considering anti-inflammatory role of OxPLs, they reported that oxidized 1-palmitoyl-2arachidonoyl-*sn*-glycero-3-phosphocholine (OxPAPC) interfered with the ability of LPS to bind to the LPS-binding protein (LBP) and to CD-14, thus suppressing LPS-induced nuclear factor-κB (NF-κB)-mediated up-regulation of inflammatory genes.

Knapp and coworkers (2007) found that OxPAPC inhibits the interaction of LPS with LPS-binding protein and CD14. This also reduces phagocytotic activity of neutrophils and macrophages by a CD-14-independent mechanism. However, in these experiments, administration of OxPAPC rendered mice highly susceptible to *Escherichia coli* peritonitis, which may cause mortality during gram-negative sepsis *in vivo*. Thus the overall harmful profile of phospholipid oxidation products includes the impairment of host response to bacterial infections.

Recently, Gharavi and colleagues (2007) have reported the activation of JAK2/STAT3 pathway by phospholipids and implicated their role in atherogenesis. 1-Palmitoyl-2-epoxyisoprostane-*sn*-glycero-3-phosphocholine, an oxidation product of -1palmitoyl-2-arachidonoyl-*sn*-glycero-3-phosphocholine, induces c-Src kinase-dependent activation of JAK2 in endothelial cells and synthesis of chemotactic factors, such as interleukin (IL)-8. In turn, STAT3 activation and regulation of IL-8 transcription is dependent on JAK2 leading to the enhanced levels of STAT3 activity in inflammatory regions of human atherosclerotic lesions. Since STAT3 activation is involved in other chronic inflammatory diseases such as rheumatoid arthritis, psoriasis etc, it has been suggested that STAT3 activation by oxidized phospholipids could be an important interventional target for atherosclerosis and other diseases with inflammatory components.

5.1. Regulation of vascular cell function

OxPLs have multiple effects on endothelial cells. After 4h treatment with 50μg/ml of OxPAPC ~1000 genes are regulated amongst which ~600 are up-regulated and ~400 are down-regulated (Gargalovic et al 2006). Also, a major difference in responsiveness to specific effects of Ox-PAPC of endothelial cells from different human donors has been documented (Gargalovic et al 2006). The atherogenic pathways which were found to be upregulated include inflammation, cholesterol synthesis, coagulation and decrease in cell division. Some important effects of OxPAPC on endothelial cell function independent of gene regulation have been reported. OxPAPC has been shown to increase monocytes but not neutrophils binding by activating β-1 integrin (Berlin et al 2008, Leitinger et al 2005).

Many effects of OxPLs are mediated by its interaction with CD36. Several studies have indicated that LDL supplemented with OxPAPC or vesicles supplemented with fragmented α/β unsaturated fatty acids at the sn-2 position, such as KOdiA or HODA PC, bind to CD36 (Podrez et al 2002, Greenberg et al 2006). Another important phagocytic function of macrophages is the uptake of apoptotic cells, which are abundant in atherosclerotic plaques. OxPLs including oxidized phosphatidyl serine and phosphatidyl choline derivatives were shown to serve as ligands for macrophage uptake of apoptotic cells (Chou et al 2008, Greenberg et al 2006).

OxPLs also interact and bind with other recognition receptors in macrophages such as TLRs, CD14, LPS binding protein and C-reactive protein competing with negative ligands (Bochkov et al 2007, Bochkov et al 2002, Erridge et al 2008, Miller et al 2003). Thus, the formation of OxPLs during inflammation may represent an important feedback mechanism to limit further tissue damage. OxPLs have also been shown to activate macrophages. Currently conducted studies have revealed the role of OxPAPC in inducing lung injury and cytokine production by lung macrophages (Imai et al 2008).

The role of OxPLs in adaptive immune response can't be overlooked where they modulate the maturation process of dendritic cells (DCs). OxPLs also regulate innate immunity in human leprosy (Cruz et al 2008). In addition to the effects on DCs, OxPLs have also been shown to affect and induce T-cells (Seyerl et al 2008).

Phenotypic switching of smooth muscle cells (SMCs) involving increased proliferation; enhanced migration and down-regulation of SMC differentiation marker genes play a critical role in atherogenesis. Many studies have shown that OxPLs stimulate differentiation and cell division of SMCs (Heery et al 1995, Pidkovka et al 2007) while others have shown activation of apoptotic signaling pathways (Fruhwirth et al 2008).

5.2. Gene expression

OxPLs have profound effect on gene expression. OxPAPC have been shown to modulate the expression of approximately 1000 genes in human aortic ECs which include both up-regulated and down-regulated mRNAs (Gargalovic et al 2006). OxPLs regulate genes related to inflammation, lipid metabolism, cellular stress, proliferation, and differentiation. These include VEGF-A and IL-8, which are induced by OxPLs independent of their transcription factors.

5.3. Pathophysiological functions

Pathophysiologically OxPLs are involved in various proinflammatory and cardiovascular disorder; details are being described below (FIGURE-3).

5.4. Atherosclerosis

Quantification of OxPLs using liquid chromatography coupled with mass spectrometry has indicated that atherosclerotic vessels contain high concentrations of OxPCs. Different species of OxPCs were detected in atherosclerotic vessels including PL-hydroperoxides and hydroxides (Waddington et al 2001). In addition to elevated levels of OxPLs, atherosclerotic vessels express high amounts of proteins known to be induced by OxPLs in vitro. The latter includes MKP-1 (Reddy et al 2004), ATF3, ATF4 (Gargalovic et al 2006), SREBP-1 (Yeh et al 2004), HO-1 and IL-8 (Cheng et al 2009), MCP-1 and COX-2 (Ma et al 2008). OxPLs act on all major cell types involved in atherogenesis including monocytes, endothelial and vascular smooth muscle cells, lymphocytes, and platelets.

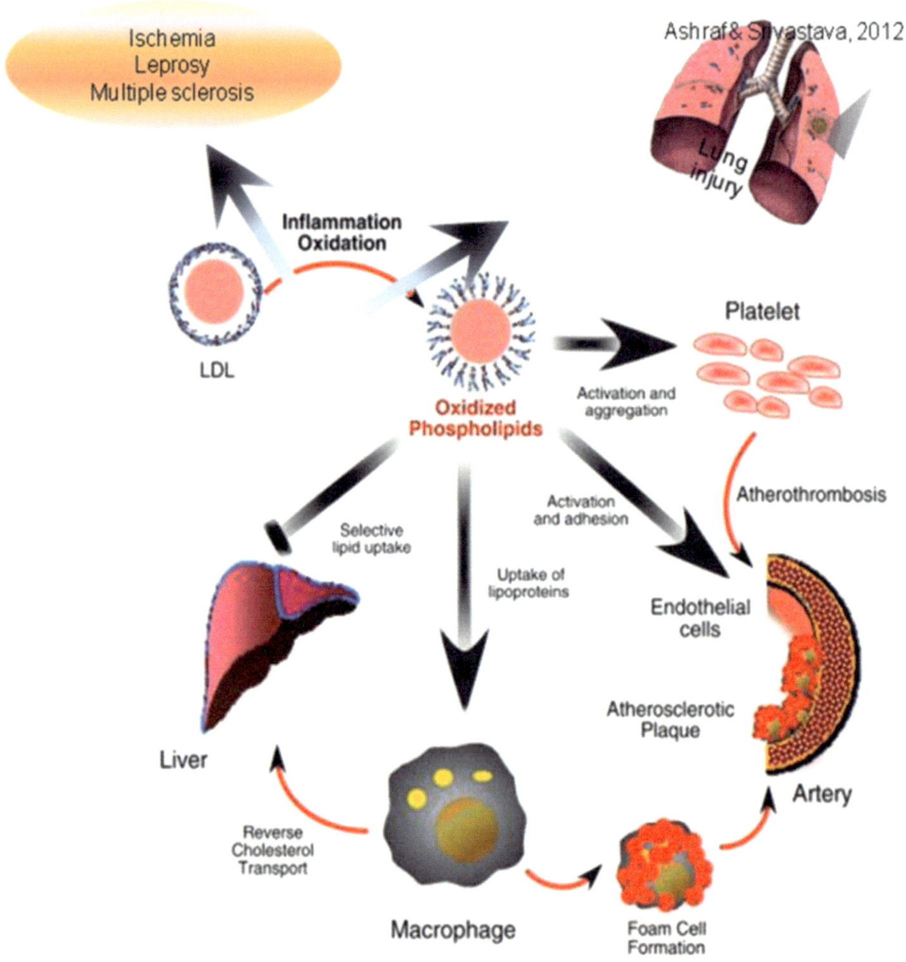

Figure 3. Oxidized phospholipids present in oxidized LDL induce various Diseases.

5.5. Lung injury

The epithelial lining pulmonary surfactant is permanently exposed to high concentrations of oxygen and other oxidants present in the air. Ozone gas also plays a role in generating oxidatively truncated PLs (Uhlson et al 2002). Under normal healthy conditions surfactant is protected from oxidation by maintaining low contents of PUFAs, antioxidant action of glutathione present in the lining fluid and surfactant proteins A and D (Kuzmenko et al 2004). However, the accumulation of biologically active OxPLs products occurs in pathological states due to the oxidation of surfactant PCs, membrane lipids and apoptosis of bronchial cells. Studies conducted with animal models have shown that OxPLs protect lungs from acute lung injury. Ma et al. (2004) showed that OxPAPC inhibits elevation of TNFa in mice upon intratracheal or systemic administration of LPS or CpG DNA. Hence the available data shows that OxPLs may induce either beneficial or detrimental effects on lungs. The action of OxPLs on the lungs may depend upon their concentrations, lower levels of OxPLs protect endothelial barrier whereas high concentrations of the same OxPLs induce disruptive effects (Birukov et al 2004, DeMaio et al 2006).

5.6. Ischemia

Ischemia/reperfusion results in elevated levels of OxPLs both in tissues and systemic levels. PAF like (alkyl-acyl) OxPLs were detected within the first minutes after reperfusion of kidneys after warm ischemia (Lloberas et al 2002). Plasma concentrations of fragmented OxPCs were increased in patients during the reperfusion period after coronary surgery with cardiopulmonary bypass (Frey et al 2000). Hence available data shows that ischemia/reperfusion is a pathological state characterized by elevated local and circulating levels of OxPLs.

5.7. Inflammation

Inflammation is characterized by a massive production of ROS. The elevation of circulating levels of OxLDL in response to inflammatory stimuli has already been shown. The OxPLs production in response to inflammation is induced by different cell types including leukocytes. Phorbol ester-stimulated neutrophils and monocytes incubated with PUFA-PCs produced mono- and bishydroperoxides of PC, as well as isoP–PC, thus suggesting that activated phagocytes can oxidize lipids in the surrounding medium (Jerlich et al 2003).

5.8. Radiation stress

Formation of OxPLs can be activated by visual and UV-light. OxPLs accumulating in retinas serve as ligands for CD36-dependent phagocytosis of shed photoreceptor outer segments by retinal pigment epithelium; this process is necessary for normal function of the retina (Sun et al 2006). Generation of OxPLs by light exposure has also been shown in skin cells. UVA-1-irradiated PAPC containing several OxPLs species induced expression of antioxidant and anti-inflammatory enzyme heme oxygenase-1 in dermal fibroblasts, keratinocytes, and in a three-dimensional epidermal equivalent model (Gruber et al 2007). Therefore, OxPLs are likely to play a protective role in UVA irradiated skin by inducing HO-1.

5.9. Leprosy

Oxidized PCs have been detected in lepromatous (disseminated) leprosy lesions, but not in tuberculous leprosy characterized by stronger host immune response and self-contained infection (Cruz et al 2008). Lepromatous leprosy lesions are characterized by the accumulation of OxPLs, which can counteract innate and specific immune responses, thereby promoting survival.

5.10. Multiple sclerosis

Multiple sclerosis (MS) is an autoimmune disease of the brain that causes neurodegeneration. Role of OxPLs in MS is supported by Qin et al. (2007), demonstrating the presence of OxPLs (alone and conjugated to a 15 KDa protein) in extracts of MS lesions directly by Western blot analyses using the E06 antibody. OxPLs might be promoting the inflammatory process in MS lesions.

6. Medical relevance

Increasing number of studies suggest the role of oxidized phospholipids in development of atherosclerosis by interacting with specific receptors as well as through their reactive groups that can bind covalently to proteins, forming lipid-protein adducts that become dysfunctional. It is a challenge to determine if therapeutic inhibition of the OxPLs interaction with vessel wall cells can inhibit atherosclerosis. Also it will be interesting to identify the lipid oxidation products that activate each response in the various cell types and the receptors or binding molecules and signal transduction pathways activated by these lipids.

Pro-inflammatory oxidized phospholipids are significant predictors of the presence of carotid and femoral atherosclerosis, development of new lesions and increased risk of cardiovascular events (Ashraf et al 2009). Hence oxidized phospholipids could serve as biomarker for diagnosis of coronary artery disease and they could also be used as potential targets for therapeutic intervention.

7. Conclusions

The inflammatory profile of OxPLs combines both pro- and anti-inflammatory effects. OxPLs may show detrimental as well as beneficial cellular effects. OxPLs exert pro-inflammatory effects on different cell types such as endothelium where they induce a shift from antithrombotic and anti-inflammatory state to procoagulant and inflammatory phenotype of EC. Although OxPLs stimulate a number of classical inflammation mechanisms, they are not capable of activating many signaling and adhesion events characteristic of acute inflammation, such as activation of the NFκB pathway, expression of ICAM-1 and E-selectin or adhesion of granulocytes. Several studies have provided evidence that OxPLs play an important role in atherosclerosis. In addition, OxPLs also up-regulate monocytes-specific chemokines and stimulate EC to bind monocytes, thus initiating monocytic inflammation. Thus it can be concluded that OxPLs can stimulate and inhibit inflammation depending upon the biological situation. Advancement in this field can be expected from studies that are based on well defined synthetic and labeled OxPLs species and the modern techniques of system biology. Also advances in the knowledge of signaling pathways and the interaction partners of oxidized phospholipid will increase our understanding of inflammatory processes and molecular mechanisms of various diseases such as atherosclerosis. These studies may also help in playing important role in future therapeutic diagnostics.

Abbreviations

Oxidized phospholipids (OxPL)
Oxidized low-density lipoprotein (OxLDL)
Serum paraoxonase (PON1)
PAF-acetylhydrolase (PAF-AH)

9-keto-10-dodecendioic acid ester of 2-lyso-phosphatidyl choline (KOdiA-PC)
15 deoxy-delta 12, 14 prostaglandin I2 (PGI2)
1-palmitoyl-2-(5,6-epoxyisoprostane E2)-*sn*-glycero-3-phosphoryl choline (PEIPC)
Reactive oxygen species (ROS)
Lipoxygenases (LOXs)
Glutathione peroxidase (GPx)
G protein-coupled receptors (GPCR)
Peroxisome proliferator-activated receptors (PPAR)
Toll-like receptors (TLRs)
Vascular endothelial growth factor (VEGF)
Lipopolysaccride (LPS)
Tumor necrosis factor α (TNFα)
Dendritic cells (DCs)
Smooth muscle cells (SMCs)
Multiple sclerosis (MS)

Author details

Mohammad Z. Ashraf and Swati Srivastava
Genomics Group, Defence Institute of Physiology & Allied Sciences, India

8. References

Androulakis N, Durand H, Ninio E, and Tsoukatos DC. Molecular and mechanistic characterization of platelet activating factor-like bioactivity produced upon LDL oxidation. J Lipid Res 2005; 46: 1923–1932.

Anliker B, Chun J. Cell surface receptors in lysophospholipids signaling. Semin cell Dev Biol 2004;92:1086-1094.

Arai M, Imai H, Metori A, and Nakagawa Y. Preferential esterification of endogenously formed 5-hydroxyeicosatetraenoic acid to phospholipids in activated polymorphonuclear leukocytes. Eur J Biochem 1997; 244: 513–519.

Ashraf M.Z., Kar N. S., Podrez E.A. Oxidized phospholipids: Biomarker for cardiovascular diseases. Int J Biochem Cell Biol 2009; 41: 1241-1244.

Berliner JA and NM Gharavi. Endothelial cell regulation by phospholipid oxidation products. Free Radic Biol Med 2008; 45:119-123.

Birkle DL and Bazan NG. Effect of Kþ depolarization on the synthesis of prostaglandins and hydroxyeicosatetra- (5,8,11,14) enoic acids (HETE) in the rat retina. Evidence for esterification of 1 2-HETE in lipids. Biochim Biophys Acta 1984; 795: 564–573.

Birukov KG, Bochkov VN, Birukova AA, Kawkitinarong K, Rios A, Leitner A, Verin AD, Bokoch GM, Leitinger N, and Garcia JG. Epoxycyclopentenone-containing oxidized phospholipids restore endothelial barrier function via cdc42 and Rac. Circ Res 2004; 95: 892–901.

Bluml S, Rosc B, Lorincz A, Seyerl M, Kirchberger S, Oskolkova O, Bochkov VN, Majdic O, Ligeti E, and Stockl J. The oxidation state of phospholipids controls the oxidative burst in neutrophil granulocytes. J Immunol 2008; 181: 4347–4353.

Bochkov VN, Kadl A, Huber J, et al. Protective role of phospholipids oxidation products in endotoxin-induced tissue damage. Nature 2002; 419:77-81.

Bochkov VN, Mechtcheriakova D, Lucerna M, Huber J, Malli R, Graier WF, Hofer E, Binder BR, and Leitinger N. Oxidized phospholipids stimulate tissue factor expression in human endothelial cells via activation of ERK/EGR-1 and Ca(++)/NFAT. Blood 2002; 99: 199–206.

Bochkov VN, Oskolkova OV, Birukov KG, Levonen AL, CJ Binder, Stockl J. Generation and Biological Activities of Oxidized Phospholipids. Antioxidants & Redox Signaling. 2010; 12: , 1009-59.

Bochkov VN, Philippova M, Oskolkova O, Kadl A, Furnkranz A, Karabeg E, Afonyushkin T, Gruber F, Breuss J, Minchenko A, Mechtcheriakova D, Hohensinner P, Rychli K, Wojta J, Resink T, Erne P, Binder BR, and Leitinger N. Oxidized phospholipids stimulate angiogenesis via autocrine mechanisms, implicating a novel role for lipid oxidation in the evolution of atherosclerotic lesions. Circ Res 2006; 99: 900–908.

Bochkov VN. Inflammatory profile of oxidized phospholipids. Thromb Haemost 2007; 97:348-354.

Boullier A, Gillotte KL, Horkko S, Green SR, Friedman P, Dennis EA et al. The binding of oxidized low density lipoprotein to mouse CD36 is mediated in part by oxidized phospholipids that are associated with both the lipid and protein moieties of the lipoprotein. J Biol chem. 2000; 275:9163-9169.

Byun J, Mueller DM, Fabjan JS, and Heinecke JW. Nitrogen dioxide radical generated by the myeloperoxidasehydrogen peroxide-nitrite system promotes lipid peroxidation of low density lipoprotein. FEBS Lett 1999; 455: 243–246.

Carr AC, Winterbourn CC, and van den Berg JJ. Peroxidasemediated bromination of unsaturated fatty acids to form bromohydrins. Arch Biochem Biophys 1996; 327: 227–233.

Chang MK, Binder CJ, Miller YI, et al. Apoptotic cells with oxidized-specific epitopes are immunogenic and proinflammatory. J Exp Med 2004; 200: 1359-1370.

Chen R, Yang L, MsIntyre TM. Cytotoxic phospholipid oxidation products. Cell death from mitochondrial damage and the intrinsic caspase cascade. J Biol Chem 2007; 282:24842-24850.

Cheng C, Noordeloos AM, Jeney V, Soares MP, Moll F, Pasterkamp G, Serruys PW, and Duckers HJ. Heme oxygenase 1 determines atherosclerotic lesion progression into a vulnerable plaque. Circulation 2009; 119: 3017–3027.

Chou MY, K Hartvigsen, LF Hansen, L Fogelstrand, PX Shaw, A Boullier, CJ Binder and JL Witztum. Oxidation-specific epitopes are important targets of innate immunity. J Intern Med 2008; 263: 479-488.

Cruz D, AD Watson, CS Miller, D Montoya, MT Ochoa, PA Seiling, MA Gutierrez, M Navab, ST Reddy, JL Witztum, et al. Host derived oxidized phospholipids and HDL regulate innate immunity in human leprosy. J Clin Invest; 2008; 118:2917-2928.

Davis B, Koster G, Douet LJ, Scigelova M, Woffendin G, Ward JM, Smith A, Humphries J, Burnand KG, Macphee CH, and Postle AD. Electrospray ionization mass spectrometry identifies substrates and products of lipoproteinassociated phospholipase A2 in oxidized human low density lipoprotein. J Biol Chem 2008; 283: 6428–6437.

DeMaio L, Rouhanizadeh M, Reddy S, Sevanian A, Hwang J, and Hsiai TK. Oxidized phospholipids mediate occluding expression and phosphorylation in vascular endothelial cells. Am J Physiol Heart Circ Physiol 2006; 290: H674–H683.

Erridge C, S Kennedy, CM Spickett and DJ Webb. Oxidized phospholipid inhibition of toll-like receptor (TLR) signaling is restricted to TLR2 and TLR4: roles for CD14, LPS-binding protein and MD2 as targets for specificty of inhibition. J Biol Chem 2008; 283:24748-24759.

Forte TM, Subbanagounder G, Berliner JA, Blanche PJ, Clermont AO, Jia Z et al. Altered activities of anti-atherogenic enzymes LCAT, paraoxonase and platelet activating factor acetylhydrolase in atherosclerosis-susceptible mice. J Lipid Res 2002; 43: 477-485.

Frey B, Haupt R, Alms S, Holzmann G, Konig T, Kern H, Kox W, Rustow B, and Schlame M. Increase in fragmented phosphatidylcholine in blood plasma by oxidative stress. J Lipid Res 41: 1145–1153, 2000.

Fruhwirth GO and a Hermetter. Mediation of apoptosis by oxidized phospholipids. Subcell Biochem 2008; 49:351-367.

Furnkranz A, Schober A, Bochkov VN, et al. Oxidized phospholipids trigger atherogenic inflammation in murine arteries. Arterioscler Thromb Vasc Biol 2005; 25: 633–638.

Gargalovic PS, Gharavi NM, Clark MJ, et al. The unfolded protein response is an important regulator of inflammatory genes in endothelial cells. Arterioscler Thromb Vasc Biol 2006; 26: 2490–2496.

Gargalovic PS, Gharavi NM, Clark MJ, Pagnon J, Yang WP, He A, Truong A, Baruch-Oren T, Berliner JA, Kirchgessner TG, and Lusis AJ. The unfolded protein response is an important regulator of inflammatory genes in endothelial . Arterioscler Thromb Vasc Biol 2006; 26: 2490–2496.

Gargalovic PS, M Imura, B Zhang, NM Gharavi, MJ clark, J Pagnon, WP Yang, A He, A Troung, S Patel et al. Identification of imflammatory gene modules based on variations of human endothelial cell responses to oxidized lipids. Proc Natl Acad Sci USA 2006; 103:12741-12746.

Gharavi NM, Alva JA, Mouillesseaux KP et al. Role of the JAK/STAT pathway in the regulation of interleukin-8 transcription by oxidized phospholipids *in vitro* and in atherosclerosis *in vivo*. J Biol Chem 2007; 282: 31460-31468.

Girotti AW and Kriska T. Role of lipid hydroperoxides in photo-oxidative stress signaling. Antioxid Redox Signal 2004; 6: 301–310.

Greenberg ME, M Sun, R Zhang, M Febbraio, R. Silverstein and SL Hazen. Oxidized phosphatidylserine-CD36 interactions play an essential role in macrophage-dependent phagocytosis of apoptotic cells. J Exp Med 2006; 203:2613-2625.

Gruber F, Oskolkova O, Leitner A, Mildner M, Mlitz V, Lengauer B, Kadl A, Mrass P, Kronke G, Binder BR, Bochkov VN, Leitinger N, and Tschachler E. Photooxidation generates biologically active phospholipids that induce heme oxygenase-1 in skin cells. J Biol Chem 2007; 282: 16934–16941.

Gu X, Sun M, Gugiu B, Hazen S, Crabb JW, and Salomon RG. Oxidatively truncated docosahexaenoate phospholipids: Total synthesis, generation, and peptide adduction chemistry. J Org Chem 2003; 68: 3749–3761.

Gugiu BG, Mesaros CA, Sun M, Gu X, Crabb JW, and Salomon RG. Identification of oxidatively truncated ethanolamine phospholipids in retina and their generation from polyunsaturated phosphatidylethanolamines. Chem Res Toxicol 2006; 19: 262–271.

Heery JM, Kozak M, Stafforini DM, Jones DA, Zimmerman GA, McIntyre TM et al. Oxidatively modified LDL contains phospholipids with platelet-activating factor-like acitivity and stimulates the growth of smooth muscle cells. J clin Invest 1995; 96: 2322-2330.

Hoff HF, O'Neil J, Wu Z, Hoppe G, and Salomon RL. Phospholipid hydroxyalkenals: Biological and chemical properties of specific oxidized lipids present in atherosclerotic lesions. Arterioscler Thromb Vasc Biol 2003; 23: 275–282.

Honda HM, Leitinger N, Frankel M, Goldhaber JI, Natarajan R, Nadler JL, Weiss JN, and Berliner JA. Induction of monocyte binding to endothelial cells by MM-LDL: role of lipoxygenase metabolites. Arterioscler Thromb Vasc Biol 1999; 19: 680–686.

Huang LS, Kang JS, Kim MR, and Sok DE. Oxygenation of arachidonoyl lysophospholipids by lipoxygenases from soybean, porcine leukocyte, or rabbit reticulocyte. J Agric Food Chem 2008; 56: 1224–1232.

Huber J, Valves A, Mitulovic G, et al. Oxidized membrane vesicles and blebs from apoptotic cells contain biologically active oxidized phospholipids that induce monocyte-endothelial interactions. Arterioscler Thromb Vasc Biol 2002; 22: 101-107.

Huo Y, Weber C, Forlow SB, et al. The chemokine KC, but not monocyte chemoattractant protein-1, triggers monocyte arrest on early atherosclerotic endothelium. J Clin Invest 2001; 108: 1307–1314.

Ikura Y, Ohsawa M, Suekane T, et al. Localization of oxidized phosphatidylcholine in nonalcoholic fatty liver disease: impact on disease progression. Hepatology 2006; 43: 506-514.

Imai Y, Kuba K, Neely GG, Yaghubian-Malhami R, Perkmann T, van LG, Ermolaeva M, Veldhuizen R, Leung YH, Wang H, Liu H, Sun Y, Pasparakis M, Kopf M, Mech C, Bavari S, Peiris JS, Slutsky AS, Akira S, Hultqvist M, Holmdahl R, Nicholls J, Jiang C, Binder CJ, and Penninger JM. Identification of oxidative stress and Toll-like receptor 4 signaling as a key pathway of acute lung injury. Cell 2008; 133: 235–249.

Jerlich A, Schaur RJ, Pitt AR, and Spickett CM. The formation of phosphatidylcholine oxidation products by stimulated phagocytes. Free Radic Res 2003; 37: 645–653.

Jin Y and Penning TM. Aldo-keto reductases and bioactivation/detoxication. Annu Rev Pharmacol Toxicol 2007; 47: 263–292.

Kadl A, Huber J, Gruber F, et al. Analysis of inflammatory gene induction by oxidized phospholipids *in vivo* by quantitative real-time RT-PCR in comparison with effects of LPS. Vascul Pharmacol 2002; 38: 219–27.

Kagan VE, Bayir HA, Belikova NA, Kapralov O, Tyurina YY, Tyurin VA, Jiang J, Stoyanovsky DA, Wipf P, Kochanek PM, Greenberger JS, Pitt B, Shvedova AA, and Borisenko G. Cytochrome c/cardiolipin relations in mitochondria: A kiss of death. Free Radic Biol Med 2009; 46: 1439–1453.

Kagan VE, Tyurin VA, Jiang J, Tyurina YY, Ritov VB, Amoscato AA, Osipov AN, Belikova NA, Kapralov AA, Kini V, Vlasova II, Zhao Q, Zou M, Di P, Svistunenko DA, Kurnikov IV, and Borisenko GG. Cytochrome c acts as a cardiolipin oxygenase required for release of proapoptotic factors. Nat Chem Biol 2005; 1: 223–232.

Kim JA, Territo MC, Wayner E, et al. Partial characterization of leukocyte binding molecules on endothelial cells induced by minimally oxidized LDL. Arterioscler Thromb 1994; 14: 427–433.

Knapp S, Matt U, Leitinger N, et al. Oxidized phospholipids inhibit phagocytosis and impair outcome in gram-negative sepsis *in vivo.* J Immunol 2007; 178: 993-1001.

Kriska T, Marathe GK, Schmidt JC, McIntyre TM, and Girotti AW. Phospholipase action of platelet-activating factor acetylhydrolase, but not paraoxonase-1, on long fatty acyl chain phospholipid hydroperoxides. J Biol Chem 2007; 282: 100–108.

Kuzmenko AI, Wu H, Bridges JP, and McCormack FX. Surfactant lipid peroxidation damages surfactant protein A and inhibits interactions with phospholipid vesicles. J Lipid Res 2004; 45: 1061–1068.

Lee H, Shi W, Tontonoz P, et al. Role for peroxisome proliferator-activated receptor alpha in oxidized phospholipid-induced synthesis of monocyte chemotactic protein-1 and interleukin-8 by endothelial cells. Circ Res 2000; 87: 516–521.

Leitinger N, Tyner TR, Oslund L, et al. Structurally similar oxidized phospholipids differentially regulate endothelial binding of monocytes and neutrophils. Proc Natl Acad Sci USA 1999; 96: 12010–12015.

Leitinger N. Oxidized phospholipid as triggers of inflammation in atherosclerosis. Mol Nutr Food Res 2005; 49:1063-1071.

Leitinger N. Oxidized phospholipids as modulators of inflammation in atherosclerosis. Curr Opin Lipidol 2003; 14: 421-430.

Li R, Mouillesseaux KP, Montoya D, Cruz D, Gharavi N, Dun M, Koroniak L, and Berliner JA. Identification of prostaglandin E2 receptor subtype 2 as a receptor activated by OxPAPC. Circ Res 2006; 98: 642–650.

Lloberas N, Torras J, Herrero-Fresneda I, Cruzado JM, Riera M, Hurtado I, and Grinyo JM. Postischemic renal oxidative stress induces inflammatory response through PAF and oxidized phospholipids. Prevention by antioxidant treatment. FASEB J 2002; 16: 908–910.

Ma Y, Malbon CC, Williams DL, and Thorngate FE. Altered gene expression in early atherosclerosis is blocked by low level apolipoprotein E. PLoS One2008; 3: e2503.

Ma Z, Li J, Yang L, Mu Y, Xie W, Pitt B, and Li S. Inhibition of LPS- and CpG DNA-induced TN F-alpha response by oxidized phospholipids. Am J Physiol Lung Cell Mol Physiol 2004; 286: L808–L816.

Marathe GK, Davies SS, Harrison KA, Silva AR, Murphy RC, Castro-Faria Neto H, Prescott SM, Zimmerman GA, and McIntyre TM. Inflammatory platelet-activating factorlike phospholipids in oxidized low density lipoproteins are fragmented alkyl phosphatidylcholines. J Biol Chem 1999; 274: 28395–28404.

McEver RP and Cummings RD. Perspectives series: Cell adhesion in vascular biology. Role of PSGL-1 binding to selectins in leukocyte recruitment. J Clin Invest 1997; 100: 485– 491.

McIntyre TM, Prescott SM, and Stafforini DM. The emerging roles of PAF acetylhydrolase. J Lipid Res 2009; 50: S255– S259.

Miller YI, S Viriyakosol, CJ Binder, JR Feramisco, TN Kirkland and JL Witztum. Minimally modified LDL binds to CD14, induces macrophage spreading via TLR4/MD-2 and inhibits phagocytosis of apoptotic cells. J Biol Chem 2003; 278: 1561-1568.

Morrow JD, JA Awad, HJ Boss, IA Blair and LJ Roberts 2nd. Non-cyclooxygenase-derives prostanoids (F2-isoprostanes) are formed in situ on phospholipids. Proc Natl Acad Sci, USA 1992; 89:10721-10725.

Moumtzi A, M Trenker, K Flicker, E Zenzmaier, R Saf, and A Hermetter. Import and fate of fluorescent analogs of oxidized phospholipids in vascular smooth muscle cells. J Lipid Res 2007; 48:565-582.

Nakamura T, Henson PM, Murphy RC. Occurrence of oxidized metabolites of arachidonic acid esterified to phospholipids in murine lung tissue. Annal Biochem 1998; 262: 23-32.

O'Donnell VB, Eiserich JP, Chumley PH, Jablonsky MJ, Krishna NR, Kirk M, Barnes S, Darley–Usmar VM, and Freeman BA. Nitration of unsaturated fatty acids by nitric oxide-derived reactive nitrogen species peroxynitrite, nitrous acid, nitrogen dioxide, and nitronium ion. Chem Res Toxicol 1999; 12: 83–92.

Oskolkova OV, Afonyushkin T, Leitner A, von Schlieffen E, Gargalovic PS, Lusis AJ, et al. ATF4-dependent transcription is a key mechanism in VEGF up-regulation by oxidized phospholipids: critical role of oxidized *sn*-2 residues in activation of unfolded protein response. Blood 2008; 112: 330-339.

Parhami F, Fang ZT, Fogelman AM, et al. Minimally modified low density lipoprotein-induced inflammatory responses in endothelial cells are mediated by cyclic adenosine monophosphate. J Clin Invest 1993; 92: 471-478.

Parhami F, Fang ZT, Yang B, et al. Stimulation of Gs and inhibition of Gi protein functions by minimally oxidized LDL. Arteioscler Thromb Vasc Biol 1995; 15: 2019-2024.

Pegorier S, Stengel D, Durand H, et al. Oxidized phospholipid: POVPC binds to platelet-activating-factor on hman macrophages. Implications in atherosclerosis. Atherosclerosis 2006; 188: 433-443.

Pidkovka NA, OA Cherepanova, T Yoshida, MR Alexander, RA Deaton, JA Thomas, N Leitinger, and GK Owens. Oxidized phospholipids induce phenotypic swithcing of vascular smooth muscle cells in vivo and in vitro Circ Res 2007; 101:792-801.

Podrez E.A, E. Poliakov, Z Shen, R. Zhang, Y. Deng, M Sun, P J Finton, L. Shan, M Febbraio, D P Hajjar et al. A novel family of atherogenic oxidized phospholipids promotes macrophage foam cell formation via the scavenger receptor CD36 and is enriched in atherosclerotic lesions. J Biol Chem 2002; 277:38517-38523.

Podrez EA, Febbraio M, Sheibani N, Schmitt D, Silverstein RL, Hajjar DP, et al. Macrophage scavenger receptor CD36 is the major receptor for LDL modified by monocyte-generated reactive nitrogen species. J Clin Invest 2000; 105: 1095-1108.

Podrez EA, Poliakov E, Shen Z, Zhang R, Deng Y, Sun M, Finton PJ, Shan L, Gugiu B, Fox PL, Hoff HF, Salomon RG, and Hazen SL. Identification of a novel family of oxidized phospholipids that serve as ligands for the macrophage scavenger receptor CD36. J Biol Chem 2002; 277: 38503–38516.

Podrez EA, Poliakov E, Shen Z, Zhang R, Deng Y, Sun M, Finton PJ, Shan L, Gugiu B, Fox PL, Hoff HF, Salomon RG, and Hazen SL. Identification of a novel family of oxidized phospholipids that serve as ligands for the macrophage scavenger receptor CD36. J Biol Chem 2002; 277: 38503–38516.

Prescott SM, Zimmerman GA, Stafforini DM, MsIntyre TM. Platelet-activating factor and related lipid mediators. Annu Rev Biochem 2000; 69: 419-445.

Qin J, Goswami R, Balabanov R, and Dawson G. Oxidized phosphatidylcholine is a marker for neuroinflammation in multiple sclerosis brain. J Neurosci Res 2007; 85: 977–984.

Reddy ST, Grijalva V, Ng C, et al. Identification of genes induced by oxidized phospholipids in human aortic endothelial cells. Vascul Pharmacol 2002; 38: 211–218.

Reddy ST, Nguyen JT, Grijalva V, Hough G, Hama S, Navab M, and Fogelman AM. Potential role for mitogenactivated protein kinase phosphatase-1 in the development of atherosclerotic lesions in mouse models. Arterioscler Thromb Vasc Biol 2004; 24: 1676–1681.

Salomon RG. Levuglandins and isolevuglandins: Stealthy toxins of oxidative injury. Antioxid Redox Signal 2005; 7: 185 201.

Savaskan NE, Ufer C, Kuhn H, and Borchert A. Molecular biology of glutathione peroxidase 4: From genomic structure to developmental expression and neural function. Biol Chem 388: 1007–1017, 2007.

Schneider C, Porter NA, and Brash AR. Routes to 4- hydroxynonenal: fundamental issues in the mechanisms of lipid peroxidation. J Biol Chem 2008; 283: 15539–15543.

Seiss W, Tigyi G. Thrombogenic and artherogenic activities of lysophosphatidic acid. J Cell Biochem 2004; 92: 1086-1094.

Seyerl M, S Bluml, S. Kirchberger, VN Bochkov, O Oskolkova, O Majdic and J stockl. Oxidized phospholipid induce enery in human peripheral T cells. Eur J Immunol 2008; 38:778-787.

Shih PT, Elices MJ, Fang ZT, et al. Minimally modified low-density lipoprotein induces monocyte adhesion to endothelial connecting segment-1 by activating beta1 integrin. J Clin Invest 1999; 103: 613–625.

Spiteller P and Spiteller G. 9-Hydroxy-10,12-octadecadienoic acid (9-HODE) and 13-hydroxy-9,11-octadecadienoic acid (13-HODE): Excellent markers for lipid peroxidation. Chemistry and Physics of Lipids 1997; 89: 131–139.

Stremler KE, Stafforini DM, Prescott SM, and McIntyre TM. Human plasma platelet-activating factor acetylhydrolase. Oxidatively fragmented phospholipids as substrates. J Biol Chem 1991; 266: 11095–11103.

Subbanagounder G, Deng Y, Borromeo C, et al. Hydroxy alkenal phospholipids regulate inflammatory functions of endothelial cells. Vascul Pharmacol 2002; 38: 201–209.

Subbanagounder G, Leitinger N, Schwenke DC et al. Determinants of bioactivity of oxidized phospholipids. Specific oxidized fatty acyl groups at the sn-2 position. Arterioscler Thromb Vasc Biol 2000; 20: 2248-2254.

Subbanagounder G, Wong JW, Lee H, et al. Epoxyisoprostane and epoxycyclopentenone phospholipids regulate monocyte chemotactic protein-1 and interleukin-8 synthesis. Formation of these oxidized phospholipids in response to interleukin-1 beta. J Biol Chem 2002; 277: 7271-7281.

Sun M, Finnemann SC, Febbraio M, Shan L, Annangudi SP, Podrez EA, Hoppe G, Darrow R, Organisciak DT, Salomon RG, Silverstein RL, and Hazen SL. Light- induced oxidation of photoreceptor outer segment phospholipids generates ligands for CD36-mediated phagocytosis by retinal pigment epithelium: A potential mechanism for modulating outer segment phagocytosis under oxidant stress conditions. J Biol Chem 2006; 281: 4222–4230.

Tomura H, Mogi C, Sato K et al. proton-sensing and lysolipid-sensitive G-protein coupled receptors: a novel type of multi-functional receptors. Cell signal 2005; 17: 1466-1467.

Tselepis AD, John Chapman M, Inflammation, bioactive lipids and atherosclerosis: potential roles of a lipoprotein-associated phospholipase A2, platelet activating factor-acetylhydrolase. Atheroscler Suppl 2002; 3: 57-68.

Tsimikas S, Brilakis ES, Miller ER, et al. Oxidized phospholipid, Lp(a) lipoprotein, and coronary artery disease. N Engl J Med 2005; 353: 46-57.

Uhlson C, Harrison K, Allen CB, Ahmad S, White CW, and Murphy RC. Oxidized phospholipids derived from ozonetreated lung surfactant extract reduce macrophage and epithelial cell viability. Chem Res Toxicol 2002; 15: 896–906.

Van Lenten BJ, Wagner AC, Navab M, et al. D-4F, an apolipoprotein A-1 mimetic peptide, inhibits the inflammatory response induced by influenza A infection of human type II pneumocytes. Circulation 2004; 110:3252-3258.

Waddington E, Sienuarine K, Puddey I, and Croft K. Identification and quantitation of unique fatty acid oxidation products in human atherosclerotic plaque using highperformance liquid chromatography. Anal Biochem 2001;292: 234–244.

Walton KA, Hsieh X, Gharavi N, Wang S, Wang G, Yeh M, Cole AL, and Berliner JA. Receptors involved in the oxidized 1-palmitoyl 2-arachidonoyl-sn-glycero-3-phosphorylcholine-mediated synthesis of interleukin-8. A role for Toll-like receptor 4 and a glycosylphosphatidylinositol anchored protein. J Biol Chem 2003; 278: 29661–29666.

Watson AD, Berliner JA, Hama SY, et al. Protective effect of high density lipoprotein associated paraoxonase. Inhibition of the biological activity of minimally oxidized low density lipoprotein. J Clin Invest 1995; 96: 2882–2891.

Watson AD, Leitinger N, Navab M et al. Structural identification by mass spectrometry of oxidized phospholipids in minimally oxidized low density lipoprotein that induced monocyte/endothelial interactions and evidence for their presence *in vivo*. J Biol chem 1997; 272: 13597-13607.

Winterbourn CC, van den Berg JJ, Roitman E, and Kuypers FA. Chlorohydrin formation from unsaturated fatty acids reacted with hypochlorous acid. Arch Biochem Biophys 1992; 296: 547–555.

Wittwer J and Hersberger M. The two faces of the 15- lipoxygenase in atherosclerosis. Prostaglandins Leukot Essent Fatty Acids 2007;77: 67–77.

Yeh M, Cole AL, Choi J, Liu Y, Tulchinsky D, Qiao JH, Fishbein MC, Dooley AN, Hovnanian T, Mouilleseaux K, Vora DK, Yang WP, Gargalovic P, Kirchgessner T, Shyy JY, and Berliner JA. Role for sterol regulatory element-binding protein in activation of endothelial cells by phospholipid oxidation products. Circ Res 2004; 95: 780–788.

Yoshimi N, Ikura Y, Sugama Y et al. Oxidized phosphatidylcholine in alveolar macrophages in idiopathic interstitial pneumonias. Lung 2005; 183: 109-121.

Zalewski A, Macphee C. Role of lipoprotein-associated phospholipase A2 in atherosclerosis: biology, epidemiology and possible therapeutic target. Arterioscler Thromb Vasc Biol 2005; 25: 923-931.

Zimman A, K P Mouillesseaux, T Le, N M Gharavi, A Ryvkin, T G Graeber, T T Chen, A D Watson and J A Berliner. Vascular endothelial growth factor receptor 2 plays a role in the activation of aortic endothelial cells by oxidized phospholipids. Arterioscler Thromb Vasc Biol 2007; 27: 332-338.

Chapter 18

HDL-Associated Paraoxonase 1 Gene Polymorphisms as a Genetic Markers for Wide Spread Diseases

Ivana Pejin-Grubiša

Additional information is available at the end of the chapter

http://dx.doi.org/10.5772/48104

1. Introduction

The story of paraoxonase 1 (PON1) begins in 1946, when Abraham Mazur reported the presence of an enzyme in human and rabbit tissues which was able to hydrolyse organophosphate compounds [1]. In 1950s, enzyme was named "paraoxonase" according to its ability to hydrolyse paraoxon, the toxic metabolite of the organophosphate insecticide parathion [2,3]. Later it was discovered that it exhibits a broad spectrum of activities and has diverse substrates. Mackness and colleagues linked PON1 to cardiovascular diseases in 1991 and demonstrated that PON1 could prevent the accumulation of oxidized lipids in low-density lipoprotein (LDL) [4]. However, despite intensive research over sixty years the exact physiological function of PON1 is still unclear.

2. Body

2.1. Paraoxonase 1

The paraoxonase (PON) family of the enzymes consists of three members, PON1, PON2 and PON3 that share approximately 65% similarity at the amino acid level. These were named in order of their discovery, but according to the structural homology and predicted evolutionary distance between them it seems that PON2 is the oldest and PON1 is the youngest family member [5].

PON1 and PON3 enzymes are secreted from liver cells and associate with HDL in the circulation [6]. Low levels of PON1 may be expressed in a number of tissues, primarily in epithelia. PON2 in humans is more widely expressed and is found in nearly every human tissue including heart, kidney, liver, lung, placenta, small intestine, spleen stomach, testis [6-

8]. Also, human PON2 mRNA is detected in the cells of the artery wall, including endothelial cells, smooth muscle cells and macrophages and is undetectable in HDL, LDL or the media of cultured cells [6, 7]. Of the three PON proteins, PON3 is the most recently identified and the least characterized.

PON1 is the most studied and best understood. This calcium-dependent esterase is consisting of 354 amino acids with a molecular mass od approximately 45 kDa [9, 10] and requires calcium ions for structural stability and enzymatic activities [11]. It is capable of hydrolyzing organophosphates such as oxon metabolites of a insecticides parathion, diazinon and chlorpyrihos and nerve agents sarin and soman, aromatic esters such is phenyl acetate (arylesterase activity) and a variety of aromatic and aliphatic lactones (lactonase activity) [12-18]. Beside its protective role against dietary and environmental lactones, PON1 also catalyzes the reaction of lactonization of γ- and δ-hydroxycarboxylic acids [18]. It is capable of hydrolyse the oxidised lipid derivates 5-hydroxy-eicosatetraenoic acid lactone acid (5-HETEL) and 4-hydroxy-docosahexaeonic acid (4-HDoHE) which are potent triggers of an inflammatory response and therefore determinants of atherosclerotic disease [16,19].

2.2. Paraoxonase 1 gene polymorphisms

Genes that code for three PON proteins (*pon1, pon2 and pon3*) are located to each other on the long arm of chromosome 7 in humans (7q21.3-22.1) and share approximately 70% similarity at the nucleotide level [8]. Earlier studies on different human populations showed that the hydrolytic activity of serum PON1 was polymorphically distributed [20-22] and a number of research demonstrated that the molecular basis of these differences were Q192R, L55M and C(-107)T polymorphisms in *pon1* gene.

The *pon1* gene contains functional polymorphisms in both the coding and promoter regions. In the coding region, two common polymorphism are a glutamine (Q) to arginine (R) substitution at codon 192 (Q192R) and a leucine (L) to methionine (M) substitution at position 55 (L55M). In Q192R polymorphism the exchange of codon CAA to CGA in exon 6 of *pon1* gene determines isoforms of the enzyme which differ greatly in the rate of hydrolysis a number of substrates. Paraoxon is hydrolysed at a far greater rate by the R192 isoform compared to the Q192, but some organophospates and lactones are hydrolyzed faster by Q192 [12,13]. Recent study showed that R isoform of the enzyme has higher lactonase activity and increased antiatherogenic potential [23].

L55M polymorphism (exchange of codon TTG to ATG in exon 3) is correlated with blood enzyme level with isoform L55 associated with higher serum enzymatic activity. Still, it is not clear whether this is because of a decreased stability of the M55 alloenzyme [24] and/or because of the linkage disequilibrium with -107/-108 T allele [25, 26]. Isoform M55 showed lower stability and loses activity more rapidly and to a greater extent than the L isoform [24]. This is due to the key role of L55 in packing in the propeller's central tunnel, and of its neighboring residues which ligate calcium iones [27].

At least five polymorphisms have been detected in the human *pon1* gene promoter region: C(-107/-108)T, G(-126)C, G(-162)A, G(-832)A and G(-909)C, but only C(-107T) apear to affects expression level of PON1 enzyme [25,26]. This single nucleotide polymorphism (SNP) is within stimulating protein-1 (Sp1) binding site with allele T that disrupts the recognition sequence for Sp1 and results in decreased affinity for it [28].

The frequencies of alleles of Q192R, L55M and C(-107)T polymorphisms are different among populations worldwide (Table 1). Data for European population showed predominance of Q192 and -107C alleles over R192 and -107T alleles. Spanish and Serbian populations showed higher frequency of the -107T allele. For codon 55 polymorphism, populations worldwide show predominance of L55 over M55 allele. In Asia allele Q192 is more frequent only in Indian Punjabis and Iranians and allele -107C is predominant among examined populations. Afro-Americans and Amerindian tribes showed higher frequency of allele R compared with allele Q and predominance of allele-107C. Only Mexicans showed higher frequency of -107T allele. There is a very little data from „black" continent and it concerns only Q192R polymorphism frequency with higher frequency of allele R only in Beninese (Table 1).

More than 200 single nucleotide polymorphisms (SNPs) have been identified in the human *pon1* gene but only these three have been associated with a number of pathophysiological conditions.

2.3. Pon1 variants and oxidative stress-related disorders

The central role of HDL is in the process of reverse cholesterol transport (RHC). Also it has antioxidative, antiinflammatory and antifibrinolytic functions that contribute to its antiatherosclerotic effects. Mackness and coworkers were the first that showed that HDL acted at a specific point in the oxidation cascade: it metabolises oxidized phospholipids on LDL [29]. Although several other HDL-associated proteins such as apo AI, lecithin:cholesterol acyltransferase (LCAT) and platelet-activating factor acetyltransferase (PAFAH) also have antioxidant properties, PON1 seems to be the predominant antioxidant enzyme [4, 29-31]. HDL isolated from the blood of PON1 knock-out mice or from avian species which naturally lack PON1, has at best, no effect on LDL-oxidation and at worst promotes LDL-oxidation [32,33]. Conversely, HDL isolated from mice overexpressing human PON1 completely abolishes LDL-oxidation [34]. Several human studies have shown an inverse linear relationship between the concentration of oxidised-LDL in the circulation and PON1 activity, strongly implicating PON1 in the metabolism of oxidised-LDL *in vivo* [35,36].

Enzymatic and nonenzymatic systems of antioxidative protection are included in scavenging free radicals and their metabolic products and in maintaining normal cellular physiology. Increased level of free radicals and impairment of antioxidant status are processes underlying pathophysiologic mechanisms in a variety of diseases including atherosclerosis, diabetes mellitus, cancer, chronic liver impairment, several neurological diseases, many infectious diseases and association studies have identified links between *pon1* gene polymorphisms and susceptibility and outcome of these diseases.

***pon1* polymorphisms**	**Q192R**		**L55M**		**C(-107)T**		**References**
	Q	**R**	**L**	**M**	**C**	**T**	
Populations of Europe							
Finnish	0.69	0.31	0.67	0.33	-	-	76
Dutch	0.68	0.32	0.63	0.37	-	-	77
Spanish	0.7	0.3	0.63	0.37	0.46	0.54	78
Italians	0.65	0.35	0.66	0.34	0.57	0.43	79
English	0.78	0.22	0.7	0.3	0.52	0.48	80
Turkish	0.69	0.31	0.7	0.3	-	-	81
Croatian	0.77	0.23	0.66	0.34	0.54	0.46	82
Czecs	0.54	0.46	0.69	0.31	0.59	0.41	83
Serbian	0.77	0.23	0.68	0.32	-	-	84
Populations of Asia							
Asian Indians Punjabis	0.74	0.26	0.81	0.19	0.52	0.48	85
Japanese	0.4	0.6	0.94	0.06	0.48	0.52	86
Koreans	0.38	0.620	0.94	0.06	-	-	87
Chinese	0.42	0.58	0.95	0.05	0.57	0.43	88, 89
Iranian	0.69	0.31	0.59	0.41	-	-	90
Populations of America							
Caucasian-Americans	0.73	0.27	0.64	0.36	0.5	0.5	26
Canadians	0.73	0.27	0.64	0.36	0.48	0.52	91, 28
African-Americans	0.37	0.63	0.79	0.21	0.85	0.15	92
Amazonian Amerindian tribes	0.27	0.730	0.967	0.033	-	-	93
Caribean-Hispanics	0.540	0.460	0.71	0.29	0.65	0.35	92
Mexicans	0.510	0.490	0.84	0.16	0.45	0.55	94
Peruvians	0.539	0.461	-	-	0.61	0.39	95
Populations of Africa							
Beninese	0.388	0.612	-	-	-	-	96
Ethiopians	0.592	0.408	-	-	-	-	96
Egyptians	0.67	0.33	-	-	-	-	97

Table 1. The allele frequencies of *pon1* gene polymorphisms Q192R, L55M and C(-107)T in populations worldwide

According to World Health Organization (WHO data for 2010), 95% of mortality in Serbia is caused by chronic noncontagious diseases, wherefrom 58% of it is caused by cardiovascular diseases (CVD) [37]. Although patients with CVD commonly have at least one identifiable risk factor, many ischemic events occur in the absence of any of it [38]. Atherogenesis, one of the main risk factors for CVD, is initiated by oxidation of the low-density lipoprotein (LDL) and by impairment in oxidative stress-antioxidant balance.

Enhanced oxidative stress such as in diabetes, leads to the development of accelerated atherosclerosis. Atherosclerosis in patients with diabetes tends to occur earlier and be more aggressive. People with type 2 diabetes have a 3–4 fold increased risk of developing atherosclerosis compared to people without type 2 diabetes. Serbia falls into the group of European countries with the highest diabetes mortality rates where diabetes is the fifth leading cause of death and the fifth cause of the burden of disease [39]. At least a half of the persons with non-insulin dependent diabetes mellitus (NIDDM) have not been diagnosed and are not aware of their disease [40,41].

Due to the abovementioned, there has been a marked interest in discovering additional markers of oxidative stress, including gene variants, which may have a role in predicting wide spread diseases risk. Because controversial results have been reported so far, the aim of studies performed in our laboratory was to evaluate possible interactions between *pon1* gene polymorphisms and clinical manifestations of atherosclerosis and diabetes mellitus type 2 in our population.

Allele and genotype frequencies for Q192R, L55M and C(-107)T did not show significant difference between cases with clinical manifestations of atherosclerosis (60 subjects) and controls (100 subjects) ($P>0.05$). Although the M allele (L55M) has shown a somewhat higher risk ($OR=1.23$) and the T allele (-107C/T) has shown a 1.49 times lower risk of occurence of the disease ($OR=0.67$) the difference did not reach statistical significance, most likely due to low number of subjects (Grubisa et al., unpublished data).

Also, we investigated the association between these polymorphisms and atherosclerosis in patients with type 2 diabetes mellitus (140 subjects). Our results have shown that R allele is a risk factor for atherosclerosis in these patients ($OR=2.22$, $P<0.0001$). Although M allele has shown a little higher risk ($OR=1.26$) and allele T has shown a slightly lower risk ($OR=0.85$) the results obtained do not support an association between these *pon1* gene variants and atherosclerosis in NIDDM patients (Grubisa et al., unpublished data).

Lactones are hydrolyzed preferentially by either PON1 Q or R isoformes, depending of their structure. R192 is more efficient at hydrolyzing homocysteine thiolactone, while δ-valerolactone and 2-coumaranone are more rapidly hydrolyzed by PON1Q192 [12]. In 1990's the results obtained indicated that the Q192R polymorphism may play the role in coronay heart disease (CHD) etiology because this genotype is associated with LDL oxidation; the PON1-192 R isoform is less effective at hydrolysing lipid peroxides than the Q isoform [42,43]. It have been shown that position 192 is involved in HDL binding as a part of amphipathic helix H2 of active site [27]. Gaidukov and coworkers reported from *in vitro* and sera tests that the PON1-192Q izoform binds HDL with a 3-fold lower affinity than the R isozyme and consequently exhibits significantly reduced stability, lipolactonase activity, and macrophage cholesterol efflux [27]. The higher lactonase activity is manifested by increased antiatherogenic potency: the observed rate of HDL-mediated cholesterol efflux from macrophages is 2.2-fold higher for the 192R [27]. Also it was shown that the affinity and stability of the PON1 on HDL was lower in sera of individuals with the Q192 variant than in individuals with the 192R variant [27]. Low levels of HDL particles is one of the strongest risk factors for coronary heart disease and one of the characteristic features of diabetic dyslipidemia and it seems that proteins on HDL play a major role in the protection against atherosclerois-based cardiovascular diseases. HDL carryng apolipoprotein A-I binds PON1 with high affinity, stabilizes the enzyme and stimulates its lipolactonase activity [44].

PON1 is also an extracellular homocysteine-thiolactonase (Hcy-thiolactonase). Hcy-thiolactone is a toxic metabolite linked to immune activation and thrombogenesis in human cardiovascular diseases and is elevated under conditions predisposing atherosclerosis [45-47]. A small fraction of Hcy, a sulfur-containing amino acid, is metabolized to a Hcy-thiolactone in an error-editing reaction in protein biosynthesis when Hcy is mistakenly selected instead of dietary methonine (Met) [48]. Hcy-thiolactone is neutral at physiological pH and can diffuse out of the cell and accumulate in the extracellular fluids where is hydrolyzed to Hcy by extracellular Hcy-thiolactonase-paraoxonase 1 [49]

Hcy-thiolactonase activity is strongly associated with *pon1* genotype in diverse human populations [15]. High Hcy-thiolactonase activity is associated with L55 and R192 alleles, more frequent in blacks than in whites and low activity is associated with M55 and Q192 alleles, more frequent in whites than in blacks [15]. Despite the impact of *pon1* genotype on Hcy-thiolactonase activity, these genetic variations are not asociated with atherosclerosis-based cardiovascular diseases. It seems that PON1 phenotype is better predictor [16,50].

Human clinical studies suggest that PON1 phenotype, i.e., paraoxonase activity is a much stronger predictor of cardiovascular disease status than PON1 genetic polymorphisms [51-55] a finding that has been confirmed in other studies [52,55]. Bhattacharyya and colleagues demonstrated that both the *pon1* Q192R polymorphism and serum PON1 activity are associated with prevalent coronary artery disease and incident adverse cardiovascular events [56]. This study complemented the study of Gaidukov, demonstrating that individuals with the arginine (R) at position 192 have higher serum levels of PON1 activity, lower systemic indices of systemic oxidative stress and corresponding reductions in both prevalent coronary artery disease and prospective cardiac events [56]. Plasma PON1 activity can vary up to 40 to 50-fold, and differences in PON1 protein levels up to 13–15-fold are also present within a single PON1 Q192R genotype in adults [57,58]. A number of studies indicated that measurement of an individual's PON1 function (serum activity) takes into account all polymorphism and other factors that might affect PON1 activity or expression. However, modulation of PON1 by alcohol, smoking, drugs, diet, certain physiological and pathological conditions should also be considered. These factors can increase or decrease PON1 activity [59] as well as HDL status.

However, PON1 activity is partially inactivated during the detoxification of lipid hydroperoxides [60]. This effect can be possibly related to displacement of calcium ions or inhibition through free radicals directly. It has been suggested that other antioxidant enzymes might prevent this inhibition of PON1 activity. Antioxidant enzymes, all show co-activity and might work in a collaboration against oxidative stress and elevation in oxidative stress might inhibit these enzymes [61].

Paraoxonases are important detoxifying and anti-oxidative enzymes, which establishes their role in organophosphate poisoning, diabetes, obesity, cardiovascular diseases, and innate immunity [62, 63] Consequently, PON2 has been the focus of a great deal of research in recent years. Both PON1 and PON2 protect against atherosclerosis development and share ability to hydrolyze lactones with both overlaping and distinct substrate specificities [19]. Although PON1 is associated with circulating serum HDL and reduces oxidative stress in lipoproteins, macrophages in arterial walls and in atherosclerotic lesion by its ability to hydrolyze specific oxidized lipids, PON2 acts as an intracelullar antioxidant [7,64-68] associated with plasma membrane [6-8]. The mechanism how PON2 modulates oxidative stress is still unknown, although Altenhöfer demonstrated that PON2 prevents superoxide generation, but was ineffective against existing radicals [69]. Oxidative stress affects PON2 expression too, but additional studies are needed to highlight the PON2 expression level under oxidative stress since controversial results both from *in vivo* and *in vitro* experiments have been reported [6,7,66,70-74].

3. Conclusion

Paraoxonase 1 is found to be associated with HDL particles within circulation and therefore promotes some of HDL's functions. There is no consistent evidence for involvement of *pon1* genotypes in atherosclerosis and diabetes mellitus type 2. Studies analyzed the role of *pon1* polymorphisms in oxidative stress-based diseases showed a great variation in ethnics, environmental background, age and gender of case and control groups. Allele frequencies appeared to be dependent on geographic locations, perhaps also due to genetic drift. Probably the effect of each polymorphism alone of he so called oxidative stress-associated genes is not strong enough to affect initiation and progression of atherosclerosis as well as PON1 enzyme status (activity levels and catalytic efficiency specified by the Q192R polymorphism) [75].

Author details

Ivana Pejin-Grubiša

Department of Human Genetics and Prenatal Diagnostics, Zvezdara University Medical Center, Belgrade, Serbia

4. References

[1] Mazur A (1946) An Enzyme in Animal Tissue Capable of Hydrolyzing the Phosphorus-fluorine Bond of Alkyl Fluorophosphates. J. Biol. Chem. 164:271–289.

[2] Aldridge WN (1953a) Serum Esterases I. Two Types of Esterase (A and B) Hydrolysing p-nitrophenyl Acetate, Propionate and Butyrate and a Method for Their Determination. Biochem. J. 53:110–117.

[3] Aldridge WN (1953b) Serum Esterases II. An Enzyme Hydrolysing Diethyl p-nitrophenyl Acetate (E600) and Its Identity With the A-esterase of Mammalian Sera. Biochem. J. 53:117–124.

[4] Mackness M, Arrol S, Durrington PN (1991) Paraoxonase Prevents Accumulation of Lipoperoxides in Low-Density Lipoprotein. FEBS Lett. 286:152–154.

[5] Draganov DI, La Du BN (2004) Pharmacogenetics of Paraoxonses, a Brief Review. Naunyn-Schmiedeberg's Arch. Pharmacol. 369:78–88.

[6] Ng CJ, Shih DM, Hama SY, Villa B, Navab M, Reddy ST (2005) The Paraoxonase Gene Family and Atherosclerosis. Free Radic. Biol. Med. 38:153–163.

[7] Ng CJ, Wadleigh DJ, Gangopadhyay A, Hama S, Grijalva VR, Navab M, Fogelman AM, Reddy ST (2001) Paraoxonase-2 is a Ubiquitously Expressed Protein With Antioxidant Properties and is Capable of Preventing Cell-mediated Oxidative Modification of Low Density Lipoprotein. J. Biol. Chem. 276(48): 44444–44449.

[8] Primo-Parmo SL, Sorenson RC, Teiber J, La Du BN, (1996) The Human Serum Paraoxonase/Arylesterase Gene (PON1) is One Member of a Multigene Family. Genomics 33:498–507.

[9] Furlong CE, Richter RJ, Seidel SL, Costa LG, Motulsky AG (1989) Spectrophotometric Assays for the Enzymatic Hydrolysis of the Active Metabolites of Chlorpyrifos and Parathion by Plasma Paraoxonase/Arylesterase. Anal. Biochem. 180: 242–247.

[10] Hassett C, Richter RJ, Humbert R, Chapline C, Crabb JW, Omiecinski CJ, Furlong CE (1991) Characterization of cDNA Clones Encoding Rabbit and Human Serum Paraoxonase: the Mature Protein Retains its Signal Sequence. Biochemistry 30:10141–10149.

[11] Kuo CL, La Du BN (1998) Calcium Binding by Human and Rabbit Serum Paraoxonases. Structural Stability and Enzymatic Activity. Drug Metab. Dispos. 26:653–660.

[12] Billecke S, Draganov D, Counsell R, Stetson P, Watson C, Hsu C, La Du BN (2000) Human Serum Paraoxonase (PON1) Isozymes Q and R Hydrolyze Lactones and Cyclic Carbonate Esters. Drug Metab. Dispos. 28:1335–1342.

[13] Davies HG, Richter RJ, Keifer M, Broomfield CA, Sowalla J, and Furlong CE (1996) The Effect of the Human Serum Paraoxonase Polymorphism is Reversed With Diazoxon, Soman and Sarin. Nat. Genet. 14:334–336.

[14] Furlong CE, Li WF, Brophy VH, Jarvik GP, Richter RJ, Shih DM, Lusis AJ, Costa LG (2000) The PON1 Gene and Detoxication. Neurotoxicology 21:581–587.

[15] Jakubowski H, Ambrosius WT, Pratt JH (2001) Genetic Determinants of Homocysteine Thiolactonase Activity in Humans: Implications for Atherosclerosis. FEBS Lett. 491:35–39.

[16] Khersonsky O, Tawfik DS (2005) Structure-Reactivity Studies of Serum Paraoxonase PON1 Suggest That its Native Activity is Lactonase. Biochemistry 44:6371–6382.

[17] Rodrigo L, Mackness B, Durrington PN, Hernandez A, Mackness MI (2001) Hydrolysis of Platelet-Activating Factor by Human Serum Paraoxonase. Biochem. J. 354:1–7.

[18] Teiber JF, Draganov DI, La Du BN (2003) Lactonase and Lactonizing Activities of Human Serum Paraoxonase (PON1) and Rabbit Serum PON3. Biochem. Pharmacol. 66:887–896.

[19] Draganov DI, Teiber JF, Speelman A, Osawa Y, Sunahara R, La Du BN (2005) Human Paraoxonases (PON1, PON2, and PON3) are Lactonases With Overlapping and Distinct Substrate Specificities. J. Lipid Res. 46:1239–1247.

[20] Eckerson HW, Wyte CM, LaDu BN (1983) The Human Serum Paraoxonase/Arylesterase Polymorphism. Am. J. Hum. Genet. 35:1126–1138.

[21] Mueller RF, Hornung S, Furlong CE, Anderson J, Giblett ER, Motulsky AG (1983) Plasma Paraoxonase Polymorphism: a New Enzyme Assay, Population, Family Biochemical and Linkage Studies. Am. J. Hum. Genet. 35:393–408.

[22] Playfer JR, Eze LC, Bullen MF, Evans DA (1976) Genetic Polymorphism and Interethnic Variability of Plasma Paraoxonase Activity. J. Med. Genet. 13:337–342.

[23] Gaidukov L, Rosenblat M, Aviram M, Tawfik DS (2006) The 192R/Q Polymorphs of Serum Paraoxonase PON1 Differ in HDL Binding, Lipolactonase Stimulation, and Cholesterol Efflux: J. Lipid Res. 47:2492–2502.

[24] Leviev I, Deakin S, James RW (2001) Decreased Stability of the M54 Isoform of Paraoxonase as a Contributory Factor to Variations in Human Serum Paraoxonase Concentrations. J. Lipid Res. 42: 528–535.

[25] Leviev I, James RW (2000) Promoter Polymorphisms of Human Paraoxonase PON1 Gene and Serum Paraoxonase Activities and Concentrations. Arterioscler. Thromb.Vasc. Biol. 20:516–521.

[26] Brophy VH, Jampsa RL, Clendenning JB, McKinstry LA, Jarvik GP, Furlong CE (2001) Effects of 5′ Regulatory-Region Polymorphisms on Paraoxonase-Gene (PON1) Expression. Am. J. Hum. Genet. 68:1428–1436.

[27] Harel M, Aharoni A, Gaidukov L, Brumshtein B, Khersonsky O, Meged R, Dvir H, Ravelli RBG, McCarthy A, Toker L, Silman I, Sussman JL, Tawfik DS (2004) Structure and Evolution of the Serum Paraoxonase Family of Detoxifying and Anti-Atherosclerotic Enzymes. Nat. Struct. Mol. Biol. 11:412–419.

[28] Deakin S, Leviev I, Brulhart-Meynet MC, James RW (2003) Paraoxonase-1 Promoter Haplotypes and Serum Paraoxonase: a Predominant Role for Polymorphic Position -107, Implicating the Sp1 Transcription Factor. Biochem. J. 372:643–649.

[29] Mackness MI, Arrol S, Abbott CA, Durrington PN (1993) Protection of Low-Density Lipoprotein Against Oxidative Modification by High-Density Lipoprotein Associated Paraoxonase. Atherosclerosis. 104:129–135.

[30] Ahmed Z, Ravandi A, Maguire GF, Emili A, Draganov D, La Du BN, Kuksis A, Connelly PW (2001) Apolipoprotein AI Promotes the Formation of Phosphatidylcholine Core Aldehydes That are Hydrolysed by Paraoxonase (PON1) During High Density Lipoprotein Oxidation With a Peroxynitrite Donor. J. Biol. Chem. 276:24473–24481.

[31] Watson AD, Berliner JA, Hama SY, La Du BN, Fault KF, Fogelman AM, Navab M (1995) Protective Effect of High Density Lipoprotein Associated Paraoxonase-Inhibition of the Biological Activity of Minimally Oxidised Low-Density Lipoprotein. J. Clin. Invest. 96: 2882–2891.

[32] Shih DM, Gu L, Xia YR, Navab M, Li WF, Hama S, Castellani LW, Furlong CE, Costa LG, Fogelman AM, Lusis AJ (1998) Mice Lacking Serum Paraoxonase are Susceptible to Organophosphate Toxicity and Atherosclerosis. Nature. 394:284–287.

[33] Mackness B, Mackness M.I, Arrol S, Turkie W, Durrington PN (1998). Effect of the Human Serum Paraoxonase 55 and 192 Genetic Polymorphisms on the Protection by High Density Lipoprotein Against Low Density Lipoprotein Oxidative Modification. FEBS Letts. 423:57–60.

[34] Tward A, Xia YR, Wang XP, Shi YS, Park C, Castellani LW, Lusis A, Shih DH (2002) Decreased Atherosclerotic Lesion Formation in Human Serum Paraoxonase Transgenic Mice. Circulation 106:484–490.

[35] Sampson MJ, Braschi S, Willis G, Astley SB (2005) Paraoxonase-1 (PON1) Genotype and Activity and *in vivo* Oxidised, Plasma Low-Density Lipoprotein in Type II Diabetes. Clin. Sci 109:189–197.

[36] Tsuzura S, Ikeda Y, Suehiro T, Ota K, Osaki F, Arii K, Kumon Y, Hashimoto K (2004) Correlation of Plasma Oxidized Low-Density Lipoprotein Levels to Vascular Complications and Human Serum Paraoxonase in Patients With Type 2 Diabetes. Metabolism 53:297–302.

[37] World Health Organization (WHO) (2011) Noncommunicable Diseases Country Profiles 2011. Available: http://www.who.int/nmh/publications/ncd_profiles2011/en/

[38] Futterman LG, Lemberg L (1998) Fifty Percent of Patients With Coronary Artery Disease Do Not Have Any of the Conventional Risk Factors. Am J Crit Care. 7:240-4.

[39] Incidence and Mortality of Diabetes in Serbia 2010. Serbian Diabetes Registry. Report No5. Institute of Public Health of Serbia „Dr Milan Jovanović Batut" 2011.

[40] Ford ES (2005) Risks for All-Cause Mortality, Cardiovascular Disease, and Diabetes Associated With the Metabolic Syndrome: a Summary of the Evidence. Diabet. Care 28:1769–1778.

[41] McEwan P, Williams JE, Griffiths Bagust A, Peters JR, Hopkinson P, Currie CJ (2004) Evaluating the Performance of the Framingham Risk Equations in a Population With Diabetes. Diabet. Med. 21:318–323.

[42] Aviram M, Billecke S, Sorenson R, Bisgaier C, Newton R, Rosenblat M, Erogul J, Hsu C, Dunlop C, La Du BN (1998) Paraoxonase Active Site Required for Protection Against LDL Oxidation Involves its Free Sulphydryl Group and is Different From That Required for its Arylesterase/Paraoxonase Activities: Selective Action of Human Paraoxonase Alloenzymes Q and R. Arterioscl. Thromb. Vasc. Biol. 10:1617–1624.

[43] Mackness B, Durrington PN, Mackness MI (1999) Polymorphisms of Paraoxonase Genes and Low-Density Lipoprotein Lipid Peroxidation: Lancet 353:468–469.

[44] Gaidukov L, Tawfik DS (2005) High Affinity, Stability, and Lactonase Activity of Serum Paraoxonase PON1 Anchored on HDL With ApoA-I. Biochemistry 44:11843–11854.

[45] Jakubowski H (1997) Metabolism of Homocysteine Thiolactone in Human Cell Cultures. Possible Mechanism for Pathological Consequences of Elevated Homocysteine Levels. J. Biol. Chem. 272:1935–1942.

[46] Jakubowski H, Zhang L, Bardeguez A, Aviv A (2000) Homocysteine Thiolactone and Protein Homocysteinylation in Human Endothelial Cells: Implications for Atherosclerosis. Circ. Res. 87: 45–51.
[47] Jakubowski H (2006) Pathophysiological Consequences of Homocysteine Excess. J. Nutr. 136:1741S–1749S.
[48] Jakubowski H (2004) Molecular Basis of Homocysteine Toxicity in Humans. Cell Mol. Life Sci. 61:470–487.
[49] Chwatko G, Jakubowski H (2005) The Determination of Homocysteine-Thiolactone in Human Plasma. Anal. Biochem. 337:271–277.
[50] Domagała TB, Łacinski M, Trzeciak WH, Mackness B, Mackness MI, Jakubowski H (2006) The Correlation of Homocysteine-Thiolactonase Activity of the Paraoxonase (PON1) Protein With Coronary Heart Disease Status. Cell Mol. Biol. (Noisy-le-grand) 52:3–9.
[51] Jarvik GP, Rozek LS, Brophy VH, Hatsukami TS, Richter RJ, Schellenberg GD, Furlong CE (2000) Paraoxonase (PON1) Phenotype is a Better Predictor of Vascular Disease Than is PON1192 or PON155 Genotype. Arterioscler. Thromb. Vasc. Biol. 20:2441-2447.
[52] Jarvik GP, Jampsa R, Richter RJ, Carlson CS, Rieder MJ, Nickerson DA, Furlong CE (2003) Novel Paraoxonase (PON1) Nonsense and Missense Mutations Predicted by Functional Genomic Assay of PON1 Status. Pharmacogenetics. 13:291-295.
[53] Mackness B, Davies GK, Turkie W, Lee E, Roberts DH, Hill E, Roberts C, Durrington PN, Mackness MI (2001) Paraoxonase Status in Coronary Heart Disease: Are Activity and Concentration More Important Than Genotype? Arterioscler. Thromb. Vasc. Biol. 21:1451–1457.
[54] Mackness B, Durrington P, McElduff P, Yarnell J, Azam N, Watt M, Mackness M (2003) Low Paraoxonase Activity Predicts Coronary Events in the Caerphilly Prospective Study. Circulation. 107:2775–2779.
[55] Mackness MI, Durrington PN, Mackness B (2004) The Role of Paraoxonase 1 Activity in Cardiovascular Disease: Potential for Therapeutic Intervention. Am. J. Cardiovasc. Drugs. 4:211–217.
[56] Bhattacharyya T, Nicholls SJ, Topol EJ, Zhang R, Yang X, Schmitt D, Fu X, Shao M, Brennan DM, Ellis SG, Allayee H, Lusis AJ, Hazen SL (2008) Relationship of Paraoxonase 1 (PON1) Gene Polymorphisms and Functional Activity With Systemic Oxidative Stress and Cardiovascular Risk. JAMA 299:1265–1276.
[57] Costa LG, Cole TB, Jarvik GP, Furlong CE (2003) Functional Genomics of the Paraoxonase (PON1) Polymorphisms: Effect on Pesticide Sensitivity, Cardiovascular Disease, and Drug Metabolism. Annu Rev. Med. 54:371–392.
[58] Richter RJ, Jarvik GP, Furlong CE (1999) Determination of Paraoxonase 1 (PON1) Status Without the Use of Toxic Organophosphate Substrates. Circ Cardiovasc Genet. 1:147–152.
[59] Costa LG, Cole TB, Furlong CE. (2005) Paraoxonase (PON1): From Toxicology to Cardiovascular Medicine. Acta Biomed. Suppl 2:50-57.

[60] Karabina SA, Lehner AN, Frank E, Parthasarathy S, Santanam N(2005) Oxidative Inactivation of Paraoxonase-Implications in Diabetes Mellitus and Atherosclerosis. Biochim. Biophys. Acta. 1725:213–221.

[61] Sozmen EY, Sagin FG, Kayikcioglu M, Sozmen B (2008) Oxidative Stress & Antioxidants and PON1 in Health and Disease. In: Mackness B, Mackness M, Aviram M, Paragh G, editors. The Paraoxonases: Their Role in Disease Development and Xenobiotic Metabolism. Dordrecht:Springer.pp 61-73.

[62] Camps J, Marsillach J, Joven J (2009) The Paraoxonases: Role in Human Diseases and Methodological Difficulties in Measurement.Crit. Rev. Clin. Lab. Sci. 46:83–106.

[63] Shih DM, Lusis AJ (2009) The Roles of PON1 and PON2 in Cardiovascular Disease and Innate Immunity. Curr. Opin. Lipidol. 20: 288–292.

[64] Ng CJ, N. Bourquard N, Grijalva V, Hama S, Shih DM, Navab M, Fogelman AM, Lusis AJ, Young S, Reddy ST (2006) Paraoxonase-2 Deficiency Aggravates Atherosclerosis in Mice Despite Lower Apolipoprotein-B-containing Lipoproteins: Anti-atherogenic Role for Paraoxonase-2. J. Biol. Chem. 281:29491–29500.

[65] Ng CJ, Hama SY, Bourquard N, Navab M, Reddy ST (2006) Adenovirus Mediated Expression of Human Paraoxonase 2 Protects Against the Development of Atherosclerosis in Apolipoprotein E-deficient mice. Mol. Genet. Metab. 89:368–373.

[66] Fortunato G, Di Taranto MD, Bracale UM, Del Guercio L, Carbone F, Mazzaccara C, Morgante A, D'Armiento FP, D'Armiento M, Porcellini M, sacchetti L, Bracale G, Salvatore F (2008) Decreased Paraoxonase-2 Expression in Human Carotids During the Progression of Atherosclerosis. Arterioscler. Thromb. Vasc. Biol. 28:594–600.

[67] Horke S, Witte I, Wilgenbus P, Kruger M, Strand D, Forstermann U (2007) Paraoxonase-2 Reduces Oxidative Stress in Vascular Cells and Decreases Endoplasmic Reticulum Stress-induced Caspase Activation. Circulation. 115:2055–2064.

[68] Devarajan A, Bourquard N, Hama S, Navab M, grijalva VR, Morvardi S, Clarke CF, Vergnes L, Reue K, Teiber JF, Reddy ST (2011) Paraoxonase 2 Deficiency Alters Mitochondrial Function and Exacerbates the Development of Atherosclerosis. Antioxid. Redox Signal. 14:341-351.

[69] Altenhöfer S, Witte I, Teiber JF, Wilgenbus P, Pautz A, Li H, Daiber A, Witan H, Clement AM, Förstermann U, Horke S (2010) One Enzyme, Two Functions: PON2 Prevents Mitochondrial Superoxide Formation and Apoptosis Indenpendent From its Lactonase Activity. J. Biol. Chem. 285:24398-24403.

[70] Rosenblat M, Draganov D, Watson CE, Bisgaier CL, La Du BN, Aviram M (2003) Mouse Macrophage Paraoxonase 2 Activity is Increased Whereas Cellular Paraoxonase 3 Activity is Decreased Under Oxidative Stress. Arterioscler. Thromb. Vasc. Biol. 23:468–474.

[71] Shiner M., B. Fuhrman B, Aviram M (2004) Paraoxonase 2 (PON2) Expression is Upregulated via a Reduced-nicotinamide-adenine-dinucleotide-phosphate (NADPH)-oxidase-dependent Mechanism During Monocytes Differentiation Into Macrophages. Free Radic. Biol. Med. 37: 2052–2063.

[72] Shiner M, Fuhrman B, Aviram M (2006) A Biphasic U-shape Effect of Cellular Oxidative Stress on the Macrophage Anti-oxidant Paraoxonase 2 (PON2) Enzymatic Activity. Biochem. Biophys. Res. Commun. 349:1094–1099.

[73] Rosenblat M, Hayek T, Hussein K, Aviram M (2004) Decreased Macrophage Paraoxonase 2 Expression in Patients with Hypercholesterolemia is the Result of Their Increased Cellular Cholesterol Content: Effect of Atorvastatin Therapy. Arterioscler. Thromb. Vasc. Biol. 24:175–180.

[74] Levy E, Trudel K, Bendayan M, Seidman EG, Delvin E, Lavoie JC, Precourt LP, Amre D, Sinnett D (2007) Biological Role, Protein Expression, Subcellular Localization and Oxidative Stress Response of Paraoxonase 2 in the Intestine of Humans and Rats. Am. J. Physiol. Gastrointest. Liver Physiol. 293:G1252–G1261.

[75] Li WF., Costa LG, Furlong CE (1993) Serum Paraoxonase Status: a Major Factor in Determining Resistance to Organophosphates. J. Toxicol. Environ. Health. 40:337–346.

[76] Clarimon J, Eerola J, Hellstörm O, Tienari PJ, Singleton A (2004) Paraoxonase 1 (*PON1*) Gene Polymorphisms and Parkinson's Disease in a Finnish Population. Neurosci. Lett. 367:168–170.

[77] Leus FR, Zwart M, Kastelein JJP, Voorbij HAM (2001) PON2 Gene Variants are Associated With Clinical Manifestations of Cardiovascular Disease in Familial Hypercholesterolemia Patients. Atherosclerosis. 154:641–649.

[78] Parra S, Alonso-Villaverde C, Coll B, Ferré N, Marsillach J, Aragonès G, Mackness M, Mackness B, Masana L, Joven J, Camps J (2007) Serum Paraoxonase-1 Activity and Concentration are Influenced by Human Immunodeficiency Virus Infection. Atherosclerosis. 194:175-181.

[79] Sardo MA, Campo S, Bonaiuto M, Bonaiuto A, Saitta C, Trimarchi G, Castaldo M, Bitto A, Cinquegrani M, Saitta A (2005) Antioxidant Effect of Atorvastatin is Independent of PON1 Gene T(–107)C, Q192R and L55M Polymorphisms in Hypercholesterolaemic Patients. Curr. Med. Res. Opin. 21:777–784.

[80] O'Leary KA, Edwards RJ, Town MM, Boobis AR (2005) Genetic and Other Sources of Variation in the Activity of Serum Paraoxonase/Diazoxonase in Humans: Consequences for Risk From Exposure to Diazinon. Pharmacogenet. Genomics 15:51-60.

[81] Aynacioglu AS, Cascorbi I, Mrozikiewich PM, Nacak M, Tapanyigit EE, Roots I (1999) Paraoxonase 1 Mutations in a Turkish Population. Toxico.L. Appl. Pharmacol. 157:174–177.

[82] Grdić M, Barišić K, Rumora L, Salamunić I, Tadijanović M, Žanić-Grubišić T, Pšikalova R, Fleger-Meštrić Z, Juretić D (2008) Genetic Frequencies of Paraoxonase 1 Gene Polymorphisms in Croatian Population. Croat. Chem. Acta 81:105-111.

[83] Flekač M, Škrha J, Zídková K, Lacinová Z, Hilgertová J (2008) Paraoxonase 1 Gene Polymorphisms and Enzyme Activities in Diabetes Mellitus. Physiol Res. 57:717-726.

[84] Pejin-Grubiša I, Buzadžic I, Jankovic-Oreščanin B, Barjaktarović-Vučinić N (2010) Distribution of Paraoxonase 1 Coding Region Polymorphisms in Serbian Population. Genetika 42:235-247.

[85] Gupta N, Singh S, Maturu N, Sharma YP, Gill KD (2011) Paraoxonase 1 Polymorphisms, Haplotypes and Activity in Predicting CAD Risk in North-West Indian Punjabis. PloS One 6(5). Available:http://www.plosone.org/article/info%3Adoi%2F10.1371%2Fjournal.pone.0017805. Accessed 2011 May 24

[86] Suehiro T, Nakamura T, Inoue M, Shiinoki T, Ikeda Y, Kumon Y, Shindo M, Tanaka H, Hashimoto K (2000) A Polymorphism Upstream From the Human Paraoxonase (PON1) Gene and its Asociation With PON1 Expression. Atherosclerosis. 150:295-298.

[87] Hong SH, Song J, Min WK, Kim JQ (2001) Genetic Variations of the Paraoxonase Gene in Patients With Coronary Artery Disease. Clin. Biochem. 34:475-481.

[88] Mohamed Ali S, Chia SE (2008) Interethnic Variability of Plasma Paraoxonase (PON1) Activity Towards Organophosphates and PON1 Polymorphism Among Asian Populations-a Short Review. Ind. Health 46:309-317.

[89] Zhang F, Liu HW, Fan P, Bai H, Song Q (2011) The -108 C/T Polymorphism in Paraoxonase 1 Gene in Chinese Patients With Polycystic Ovary Syndrome. Sichuan Da Xue Xue Bao Yi Xue Ban. 42:24-28.

[90] Sepahvand F, Rahimi-Moghaddam P, Shafiei M, Ghaffari SM, Rostam-Shirazi M, Mahmoudian M (2007) Frequency of Paraoxonase 192/55 Polymorphism in an Iranian Population. J. Toxicol. Environ. Health 70:1125–1129.

[91] McKeown-Eyssen C, Baines C, Cole DEC, Riley N, Tyndale RF, Marshall L, Jazmaji V (2004) Case-Control Study of Genotypes in Multiple Chemical Sensitivity: CYP2D6, NAT1, NAT2, PON1, PON2 and MTHFR. Int. J. Epidem. 33:1–8.

[92] Chen J, Kumar M, Chen W, Berkowitz G, wetmur JG (2003) Increased Influence of Genetic Variation on PON1 Activity in Neonates. Environ. Health Perspect. 111:1403–1409.

[93] Santos NPC, Santos AKCR, Santos SEB (2005) Frequency of the Q192R and L55M Polymorphisms of the Human Serum Paraoxonase Gene (PON1) in Ten Amazonian Amerindian Tribes. Genet. Mol. Biol. 28:36-39.

[94] Rojas-Garcia AE, Solis-Heredia MJ, Pina-Guzman B, Vega L, LopezCarrillo L, Quintanilla-Vega B (2005) Genetic Polymorphisms and Activity of PON1 in a Mexican Population. Toxicol. Appl. Pharmacol. 205:282–289.

[95] Cataño HC, Cueva JL, Cardenas AM, Izaguirre V, Zavaleta AI, Carranca E, Hernández AF (2006) Distribution of Paraoxonase-1 Gene Polymorphisms and Enzyme Activity in a Peruvian Population. Environ. Molecul. Mutagen. 47:699-706.

[96] Scacchi R, Corbo RM, Rickards O, De Stefano GF (2003) New Data on the World Distribution of Paraoxonase (PON1Gln192→Arg) Gene Frequencies. Hum. Biol. 75:365-373.

[97] El-Fasakhany FM, El-Segeaya O, Alahwal L, Abu Al-Nooman S (2007) Paraoxonase 1 Activity and Paraoxonase 192 Gene Polymorphism in Non Insulin Dependent Diabetes Mellitus Patients Among Egyptian Population. Tanta. Med. Scien. J. 2:68-77.

Chapter 19

Antioxidant Complexes and Lipoprotein Metabolism – Experience of Grape Extracts Application Under Metabolic Syndrome and Neurogenic Stress

Andriy L. Zagayko, Anna B. Kravchenko,
Mykhaylo V. Voloshchenko and Oxana A. Krasilnikova

Additional information is available at the end of the chapter

http://dx.doi.org/10.5772/48083

1. Introduction

The oxidative hypothesis of atherosclerosis states that peroxide modification of LDL (or other lipoproteins) is important and probably required for the pathogenesis of arterial sclerotic disease; thus, there is an assumption that inhibition of LDL oxidation would increase or prevent atherosclerosis and its clinical consequences [1]. It is believed that the basis for the atherosclerotic plaque development is the foam cell formation from oxidized low-density lipoproteins (LDL) captured by monocytes and macrophages via scavenger-receptors.

Oxidation of LDL is also important for the healthy vessel functioning. High LDL concentrations can suppress the function of arteries in relation to release of nitric oxide from the endothelium, and many of such effects are mediated by the products of lipid oxidation [2]. Moreover, oxidized LDL inhibit the endothelium-dependent nitric oxide mediated relaxations in a rabbit isolated coronary arteries. Oxidized LDL induce apoptosis in the vascular cells, including macrophages, and this is prevented by nitric oxide [3].

One of the most important mechanisms of the inflammation proatherogenic effect is development of the systemic oxidative stress, and, as a consequence of proatherogenic abnormalities of the blood lipoprotein metabolism, there is appearance of antibodies to them, alterations of the main artery wall structure [4].

At the same time, on the one hand, a high atherogenicity of strongly oxidized LDL, especially tiny subfractions, has been confirmed; on the other hand, the oxidative stress is one of the causes of endothelial dysfunction.

Endothelium vascular wall cells are involved into the interaction with the pathogenic LDL [5]. While macrophages are being overloaded with esterified cholesterol, oxysterols and other biologically active substances, including powerful enzymes with a wide spectrum of action, a foam cell is formed from the macrophage. Yet so far to its apoptosis the foam cell secrets a wide complex of interleukins, enzymes, mediators. Many of them induce a local inflammatory process, destruction of the surrounding intercellular substance, damage of the fibrous structures and separate cells.

Many factors are considered as the most important factors for atherosclerosis development risk. Among such factors an important role belongs to the so-called proatherogenic states, including chronic stress and metabolic syndrome (MS) [6]. The proatherogenic character of stress is connected, first of all, with the activation of free radical oxidation and hyperlipidemia development. One of the principal statements of all contemporary conceptions of the atherosclerosis pathogenesis is thought to be the destruction of the cell membrane structure, which universal damage factor is peroxide oxidation of lipids (POL) [7].

It is well-known that free-radical processes play the leading role in atherosclerosis pathogenesis. So the antioxidants using in correction of proatherogenic states is fully explicable especially when we speak about natural antioxidants. Thus, the investigation of their biological effects under stress and metabolic syndrome is of grate interest and may be a perspective direction of research.

At the same time it is known that the enzymes associated with HDL, paraoxonase and PAF-acetyl hydrolase can hydrolyse biologically active lipids of mm-LDL, destroy monocyte aggregates and decrease the endothelial activation of mm-LDL [8]. HDL also contain a high concentration of tocopherol due to which they can be free radical scavengers as well.

Antioxidants protect LDL from peroxide oxidation and consequently from intensive uptake of LDL by macrophages decreasing the foam cell formation, the endothelium damage and possibility for lipids to infiltrate the intima. This condition supports the actuality of searching medicines for treating atherosclerosis, in which inhibition of the POL process plays an important part in the mechanism of their action [9]. Tocopherol, carotene, probucol, a number of plant medicines containing flavonoids are proposed as antioxidants.

The overwhelming majority of antioxidant substances used in pharmacotherapy are xenobiotics and so substrates of CYP system actvating ROS formation. Moreover some of them, such as probucol, leade to HDL-C decreasing.

Therefore, the substances of natural, in particular, plant origins that possess a complex activity draw attention of researchers.

Phenolic compounds are widely present in the world of plants; they are the most widespread product of the plant metabolism. Participation of polyphenols in redox processes to produce stable quinone structures by their phenolic forms reveals an antiradical direction of their action which provides their direct antioxidant activity. At present it has been proven that polyphenols as antiradical agents not only hinder the initiation of free radical oxidation, but also interrupt the chain of lipoperoxidation [10]. A great variety of studies carried out both *in vitro* and *in vivo* supports the ability of polyphenols to inactivate ("to bind", "to scavenge") the radicals that initiate chains of oxidation. First of all, it relates to the primary ROS - $^{\cdot}O_2^{-}$ and $^{\cdot}OH$ [11].

There are some data that such natural polyphenols as catechins and procyanidins exposed to the human blood plasma produce certain complexes primarily with ApoA-1, i.e. with HDL.

One of the richest sources of polyphenols is *Vitis vinifera* and products of its processing, in particular wine.

Phenolic substances of grapes, including flavonoids and other polyphenols of grape, wine and grape seeds, are of a great interest due to their antioxidant properties and the ability to scavenge free radicals [12].

Studies *in vitro* have shown that grape, wine and grape seeds inhibit the oxidation of LDL. The activity of those substances as oxidation inhibitors in wine diluted 1,000 times markedly exceeded the analogous values for vitamins C and E [13]. It has been experimentally proven that red wine polyphenols slow down LDL oxidation processes and prevent platelet aggregation, thus preventing coronary heart diseases [14].

However, there is not a lot of research in this field yet. Arguments for anti-atherogenic properties of antioxidants are not enough. Results of convincing research are needed in order to decisively recommend antioxidants for treatment and prophylaxis of atherosclerosis.

2. Actuality

Taking into account the leading role of the free-radical processes in atherosclerosis pathogenesis one can make a conclusion about expediency of using natural antioxidants in prophylaxis and correction of this disease. [15]. Consequently, the study of the antioxidant influence on the development of stress-reactions and metabolic syndrome (MS) with the purpose of prevention of harmful complications for the cardiovascular system is of undoubted interest.

A number of studies also confirm the ability of a natural antioxidant α-tocopherol to reduce the risk of cardiovascular system diseases developed in patients with MS. It has been found that administration of α-tocopherol limits oxidation and cytotoxicity of LDL in the blood plasma significantly, supports the vascular endothelial function and reduces the intensity of systemic inflammation in the conditions of MS. The inhibiting effect of this antioxidant on aggregation and adhesion of platelets, adhesion of monocytes to endothelial cells and the smooth muscle cell proliferation has also been shown. However, numbers of experimental studies confirm that the single use of α-tocopherol is not enough for prevention of cardiovascular diseases in patients with MS [16]. It has been determined that the use of α-tocopherol in combination with ascorbic acid and aspirin (as a thrombolytic drug) is more effective [8]. In the ASAP study, the combination of vitamins E plus C was also tested, and this significantly decreased the intima-to-media progression rates in human. The ATBC clinical study used a combination of vitamin E and β-carotene in human as a secondary prevention strategy; however, no benefit on major coronary events has been found. The large MRC/BHF Heart Protection Study (HPS) for secondary prevention also examined the benefit of the antioxidant combination (vitamins E and C and β-carotene).

A strong dose-dependent effect of α-tocopherol administration is one of the unwanted effects. It is known that even a slight increase of the α-tocopherol dose could affect lipoprotein oxidation, the endothelium function and the degree of systemic inflammation [12].

The results of Cambridge Heart Antioxidant Study (CHAOS) of using antioxidants in cardiology published in 1996 give the opportunity to say that in patients with true (confirmed by angiography) coronary atherosclerosis vitamin E administration (a daily dose of 544-1088 mg (400-800 MU) reduces the risk of non-fatal myocardial infarction. The overall mortality from cardiovascular diseases in this case does not decrease. A favourable effect is revealed only after one-year administration of tocopherol.

At the same time in the Heart Outcomes Prevention Evaluation (HOPE) study, which was devoted to the study of the action of both ramipril and vitamin E (400 MU/daily dose), it was found that use of this antioxidant during approximately 4.5 years did not cause any effect on either the primary (myocardial infarction, insult and death from cardiovascular diseases) or any other end points of research. In another large-scale study on the primary prophylaxis of atherosclerotic diseases in people at least with one risk factors (hypertension, hypercholesterolemia, obesity, preliminary MI of the closest relative or advanced age) vitamin E (300 ME/daily dose) was used during 3.6 years and did not reveal any effect on any of the end points (the incidence of cardiovascular events and death). The vitamin E effectiveness was also not confirmed for various other cases (hypercholesterolemia, the level of sportsmen training, sexual potency, retardation of aging processes, etc.).

Empirically vitamin E is used in various diseases; however, the majority of the reports about tocopherol effectiveness is based on the single clinical observation and experiment data. Nowadays there are no reliable results on the role of vitamin E in prevention of tumour diseases, though the ability to reduce formation of nitrosamines (potentially carcinogenic substances being formed in the stomach), to decrease the formation of free radicals and have antitoxic effects when using chemotherapeutic remedies is well-known. In addition, the long-term intake of vitamin E in the doses from 11 to 800 mg does not cause side effects.

In HDL Atherosclerosis Treatment Study (HATS) there was the treatment of atherosclerosis depending on the high density lipoprotein cholesterol (HDL-C) level; in 160 patients with coronary heart disease with the confirmed coronary artery stenosis and the low HDL-C level the higher (800 MU/day) dose of vitamin E than in HOPE was used. The treatment combination also included 1000 mg of vitamin C, 25 mg of β-carotene and 100 mg of selenium. The study lasted 3 years and revealed that antioxidants had no influence on the HDL-C level, but in combination with hypocholesterolemic drugs they reduced their effect on LDL-C and especially – on HDL-C.

The dominant carotenoid revealed in blood and various tissues (such as liver, kidneys, adrenal glands, ovaries and prostate) is lycopene. Due to its structure and mechanism of action lycopene belongs to the group of antioxidants; a lycopene molecule contains 13 double bonds, which can interact with free radicals. Like β-carotene lycopene can serve as a precursor of vitamin A. However, the lycopene antioxidant activity is two times stronger than that of vitamin A.

Lycopene is recommended as an adjuvant in the treatment of the following diseases: idiopatic male infertility, chronic prostatitis, preeclampsia and intrauterine growth retardation (IUGR), mastopathia, diabetes mellitus, cardiovascular diseases, leucoplakia, age-related degeneration of yellow spots and cataract [17]. As with other oxidants, lycopene is administered in immunodeficiency states against chronic infections and to reduce the harmful action of unfavourable environmental factors.

The most representative evidence of the antioxidants' positive role in cardiovascular diseases prophylaxis was obtained in the multicultural European community multicentre study on antioxidants, myocardial infarction, and breast cancer (EURAMIC), during which the relationship between the antioxidant status and acute myocardial infarction in patients from 10 European countries was determined. The protective action was proven only for lycopene. In the Kuopio Ischemic Heart Disease Risk Factor Study (KIHD) the high level of blood plasma lycopene is associated with decreased risk of acute coronary syndrome and insult. In Erasmus Rotterdam Health Study (ERGO, also called "Rotterdam Study") it has been proven that lycopene prevents development and progression of atherosclerosis.

The meta-analysis of 72 epidemiological studies conducted concerning the connection between the tomato intake and cancer has determined the associative feed-back between the blood plasma lycopene level and the risk of cancer in 57 studies and 35 from 57 obtained associations were statistically significant [18].

Probucol (phenbutol) is a hypolipidemic medicine and belongs to butyl phenol derivatives. Probucol is the medicine that is similar in structure to hydroxytoluene – the compound with the potent antioxidant properties.

The hypolipidemic effect of Probucol is caused by activation of non-receptor ways of LDL extraction from the blood. It is believed that the prominent antioxidant activity of probucol prevents LDL oxidation.

Probucol decreases the total cholesterol content in plasma due to intensification of the LDL catabolism at the final stage of cholesterol elimination from the organism. It also inhibits the cholesterol biosynthesis at early stages and to a small extent slows down the food cholesterol absorption. It does not influence the triacylglycerol and the VLDL content, but significantly decreases the antiatherogenic HDL level in the blood. It is believed that decrease of the HDL-C level reflects improvement of cholesterol esters transfer with HDL on acceptor lipoproteins due to increase of the cholesteryl ester transfer protein (CETP) activity.

In spite of undesired decrease in the HDL-C concentration probucol causes regression of xanthelasma; this effect is revealed best of all in patients with the most dramatic HDL-C decrease. This important observation demonstrates that the low HDL-C content is not undoubtedly a negative phenomenon. The data obtained in experiments with animals indicate that probucol due to its antioxidant properties prevents lipid peroxidation and thereby inhibits the LDL uptake by macrophages, therefore, it inhibits atherogenesis. This allows suggesting that the therapeutic effect of the medicine may not be connected with its ability to decrease the LDL level. There is no clinical evidence of this hypothesis at the moment.

The medicine is absorbed slowly when taken internally, it is readily soluble in the adipose tissue releasing gradually into the bloodstream, and so its action is kept for a long time (up to 6 months after discontinuation of the treatment).

When using probucol in MultiVitamins and Probucol (MVP) research the renewal of the endothelium function in patients with IHD, decrease of restenosis cases after coronary angioplasty (when taking it at least 4 weeks before the procedure and further treatment during 6 months) was observed. Other antioxidants (α-tocopherol in high doses (700 mg per day), β-carotene and vitamin C) turned out to be ineffective.

Combined application of the endogenous antiradical antioxidants is of particular interest. In HPS (Heart Protection Study) along with the study of the simvastatin effectiveness the prophylactic action of antioxidants was investigated. The use of the vitamin complex (600 mg of vitamin E, 250 mg of vitamin C and 20 mg of β-carotene per day) lasted in average 5.5 years and did not reveal any differences in placebo groups and groups taking vitamins. Moreover, if the tendency exists, it reflects increasing of vascular events in the antioxidant intent-to-treat group. The action of antioxidants was compared with the effect of the combined use of simvastatin and nicotinic acid (niacin). Moreover, one of the groups received simvastatine+niacin and antioxidants. Angiographic and clinical data of this study were also disappointing with respect to the use of antioxidants.

Unfortunately, a great part of the compounds synthesized, which are used for pharmacocorrection of these states, are xenobiotics, so they can activate the free-radical formation process. Synthetic antioxidants, in particular probucol, can not be recommended for patient use because they decrease the HDL-C level.

The lack of antioxidant medicines popularity and the absence of traditions of their common use in practical medicine are caused a number of reasons: unsatisfactory previous study of this issue, complexity of adequate estimation of oxidation state parameters in the organism and the absence of the effective medicines with the antioxidant activity that are able to quickly reduce the consequences of the oxidative stress.

Therefore, the main indications for using antioxidants are excessively activated free-radical oxidation processes accompanying different pathologies. The choice of specific medicines, correct indications and contraindications for their use has not been developed yet and require further research.

3. Experiment design

In our experiments we studied the indicators of lipid and lipoprotein metabolism in the blood plasma and the liver under the experimental metabolic syndrome (MS) in Syrian hamsters of different sex and age.

In the experiments purebred male rats with 180-220 g of the body weight were used. The animals were kept in vivarium on a balanced diet. During 21 days the animals were given low alcoholic beverages from grapes of red and white grades *per os* daily. These beverages were introduced in the maximum effective doses of 9 mg of polyphenols/100 g the body weight. Taking into account the fact that the polyphenol content in the beverages investigated was quite low, the effective dose was introduced 3 times a day by 2 ml of liquids per 100 g of the animal's body weight. Control animals were introduced the corresponding volume of the saline solution. Ethanol was given in the corresponding dose.

Stress was caused by immobilization on the abdomen for 3 hours [19]. Animals were decapitated 3 hours after the immobilization. The blood was collected to get the serum. The liver was perfused by the cold extraction medium (0.25 M sucrose in 0.025 M tris-HCl, pH 7.5), homogenized in the Potter homogenizer with 2 ml of the extraction medium per 1 g of the liver. All manipulations with animals were held under chloralose-urethane anaesthesia.

To distribute the plasma lipoproteins the samples were centrifuged at 65,000 rpm (342,000 *g*) for 4 h at 4°C in the Optima XL-100K ultracentrifuge (Beckman Coulter) set at slow acceleration and deceleration [20]. Samples were fractionated within 1 h of centrifugation.

Lipids were extracted with chloroform and methanol (1:2 v/v) twice, as described by Bligh et al [21], and the supernatant was collected for determination of TG and FFA. TG and FFA were determined by enzymatic colorimetric methods with commercial kits (Zhongsheng, Beijing, China). The total cholesterol content was detected with the help of standard enzymatic cholesteroloxidase kits of "Boehringer Mannheim GmbH diagnostica" firm (Germany). The total lipid concentration was determined with the help of a standard kit "Eagle Diagnostics" (USA) – the reaction with vanillin reagent.

Determination of the lipid peroxide product quantity was performed in heptane-isopropanol extracts [22]. The optical density was measured at the wavelength of 220 nm (for compounds with isolated double bonds), 232 nm (for diene conjugates) and 278 nm – for ketodienes and conjugate trienes.

The TBA content was determined on the spectrophotometer with the help of the reaction with thiobarbituric acid [23].

A modified version of the high performance liquid chromatography (HPLC) procedure developed by Stacewicz-Sapuntzakis et al. [24] was used to measure vitamins E in the plasma. The HPLC system included a 150 × 3.9 mm Nova-pak C18 (4 microns) column with a guard-pak pre-column (both from Waters, Milford, MA), Waters Millipore TCM column heater, Waters 490 multi-wavelength detector, Hitachi 655–61 processor, Hitachi 655A-11 liquid chromatography, and BioRad autosampler AS-100.

The serum ascorbic acid concentrations were measured as described by using HPLC [25] with salicylsalicylic acid as a deproteinizing agent, metaphosphoric acid as a stabilizer.

The serum PON1 activity was measured by the rate of generation of p-nitrophenol determined at 405 nm according to MacKness B et al. [26].

The plasma cholesterol ester transfer protein (CETP) activity was examined using the modifications of Khosla et al. [44]. The CETP activity in duplicate10-μL aliquots of the plasma was determined after incubations with ^{3}H-cholesterol ester (CE)-labeled HDL_3 and LDL. Radioactivity transferred from ^{3}H-HDL_3 to LDL (measured in the supernatant after precipitation with heparin/$MnCl^{2+}$) was used to calculate the CETP activity (expressed as the percentage of radioactivity transferred from ^{3}H-HDL_3 to LDL per 16 h of incubation).

To measure endothelium-bound LPL, the perfusion solution was changed to buffer containing 1% fatty acid–free BSA and heparin (5 units/ml). The coronary effluent was collected in timed fractions over 10 min and assayed for the LPL activity by measuring the hydrolysis of a sonicated [3H]triolein substrate emulsion [27].

The plasma LCAT activity was measured by determination of the amount of radioactivity in each spot calculating the free cholesterol/ total cholesterol ratio in each plasma sample before and after the LCAT reaction and thus estimating the esterification rate [28]. The fractional esterification rate (% . h′) expressed as the percentage of the free cholesterol esterified in the plasma sample per hour.

The HL activity was evaluated using the glycerol-stabilized emulsion of triolein and egg phosphatidylcholine containing glycerol-tri[9,10(n)-^{3}H] oleate by determination of the radioactivity amount during incubation [29].

Statistical analysis. All data were analyzed for statistical significance with SPSS 13.0 software. The data were presented as means ± standard deviation. Statistical analysis used one-way ANOVA. $P<0.05$ was considered to be statistically significant.

4. Discussion

The results of our studies suggest the existence of significant changes in the lipid metabolism, as well as sex and age differences in the lipid and lipoprotein metabolism both in healthy animals and in animals with MS.

In male hamsters fed with a high-calorie diet atherogenic dyslipidemia develops independently of age (Table 1). As it can be seen from the data obtained, increase of the total lipid content in the animal blood plasma is caused by increasing of the ApoB-containing lipoprotein (ApoB-LP) level since the HDL content is not changed. At the same time it has been found that the plasma TAG level in young (47%) and in adult animals (30%) increased in comparison with the intact group.

Increase of the TAG blood content in conditions of MS is considered to be a key factor for development of atherogenic dyslipidemia that is typical for this pathology [30]. A strong correlation between hypertriacylglycerolemia plus the HDL-C level decrease and accumulation of LDL*B* in the blood plasma has been demonstrated in many experiments and clinical studies [18].

It is assumed that atherogenic alterations occur as a result of lipoprotein disbalance in the blood plasma, i.e. because of predominance of the LDL and VLDL fractions over the antiatherogenic HDL fraction (especially when the values of the LDL+VLDL/HDL index are higher than 3.5).

Age	Group	Parameters			
		TAG, g/L	Total cholesterol, mmol/L	ApoB-LP, g/L	HDL, g/L
4 weeks	Intact	1.06±0.07	2.93±0.19	4.72±0.23	1.11±0.05
	MS	1.56±0.09*	3.56±0.10*	6.68±0.15*	0.98±0.07
20 weeks	Intact	1.57±0.22	2.84±0.15	5.66±0.34	1.01±0.02
	MS	2.00±0.13*	3.71±0.18*	6.68±0.21*	0.85±0.08
1 year	Intact	1.50±0.10	2.73±0.02	5.21±0.06	1.74±0.13
	MS	2.27±0.13*	3.15±0.08*	7.00±0.22*	2.32±0.13*

The data presented as mean±SD
* –p≤0.05 versus intact animals

Table 1. Some plasma lipid values in male Syrian golden hamsters with MS (in each group n=10).

As it is known, there are 2 phenotypes of LDL: LDL*A* and LDL*B* that differ by size, density, the lipid content and the atherogencity coefficient. LDL*B* have less size (d 25.5-25.75) comparing to LDL*A* (d > 25.75) and are characterized by a lower content of polar lipids, as well as a higher content of cholesterol esters. Lipoproteins of this subfraction are slowly removed from the bloodstream that is caused by their low affinity to B/E-receptors for LDL, higher sensitivity to glycosylation and oxidative damage [31]; they also have a high affinity to scavenger-receptors of macrophages [32].

All this features explain a high atherogenicity of LDL*B* subfraction. Numerous clinical and epidemiological studies have confirmed that accumulation of LDL*B* in the blood is an independent risk factor for atherosclerosis occurrence [33].

Normally, there are predominantly LDL*A* in the blood plasma, and LDL*B* are present in a small percent of the total LDL, but in MS and insulin resistance the LDL*B* content increases significantly.

It is well-known that in MS the key factor for TAG and ApoB-LP accumulation in the blood is the VLDL hyperproduction by the liver [34]. According to our data, accumulation of ApoB-LP in the blood occurs parallelly with increase in the content of this lipoprotein fraction in the liver (Table 2).

These results allow us to make a suggestion that VLDL formation is activated in the liver of the animals fed with a high-calorie diet in our experiment.

The mechanisms of the VLDL hyperproduction by the liver in the conditions of FFA intensive supply to hepatocytes have remained still unclear. The stimulation of VLDL formation can occur both by using the elevated uptake of the blood FFA and via activation of fatty acid biosynthesis *de novo* because of hyperglycemia.

Age	Group	Parameters				
		Total lipids, mg/g liver	ApoB-LP, mg/g liver	HDL, mg/g liver	G6PDH, nmol/mg protein/min	Lysosomal lipase, nmol/mg protein/min
Week 4	Intact	104.24±2.52	11.46±0.37	1.25±0.14	3.74±0.33	0.67±0.03
	MS	124.16±2.05*	15.16±0.54*	1.11±0.07	2.80±0.17*	1.09±0.07*
Week 20	Intact	112.62±2.66	13.03±0.50	0.94±0.10	4.44±0.28	0.54±0.03
	MS	143.59±2.65*	15.69±0.36*	1.10±0.20	3.13±0.28*	1.27±0.09*

The data presented as mean±SD
* –p≤0.05 versus intact animals

Table 2. Some liver lipid metabolism values in male Syrian golden hamsters with MS used in the current study (in the crude tissue, in each group n=10).

It is known that in insulin resistance FFA that come to hepatocytes from the blood are primarily used for the TAG re-synthesis. It leads to increase in the intracellular TAG content and correlates with the increase of the VLDL secretion rate into the bloodstream. The VLDL morphology, which is specified predominantly at the second stage of their formation, depends significantly on the intracellular TAG content and hepatocyte sensitivity to insulin [35]. More active phospholipase D-dependent pre-VLDL lipidation takes place in the elevated intercellular TAG content and insulin resistance of hepatocytes [36]. Insulin blocks the VLDL1 formation in the liver. In the conditions of insulin resistance this effect and the elevated intercellular TAG content stimulate formation and secretion predominantly of VLDL1 by the liver [37].

The VLDL1 secretion increase leads to significant changes in the lipid and lipoprotein metabolism in the blood: the increased TAG content and accumulation of LDL*B* with high atherogenicity in the blood. These changes are typical for MS and considered to be separate risk factors for development of atherosclerosis.

Metabolism of ApoB-LP in the blood plasma is tightly connected with metabolism of HDL performing a reverse cholesterol transport from peripheral tissues to the liver. The leading factors in the process of transformation of VLDL into LDL in the bloodstream and determination of the LDL morphology are the rate of cholesterol esters transfer from HDL to ApoB-LP mediated by cholesteryl ester transfer protein (CETP), and the rate of TAG hydrolysis in the ApoB-LP composition mediated by lipoprotein lipase (LPL) and hepatic lipase (HL) [38].

According to data of many clinical studies, increase of the CETP activity of the HDL composition in most cases leads to decrease of the HDL-C level and accumulation of LDL*B* in the blood plasma. Moreover, a degree of these modifications correlates with the blood TAG level.

We observed significant changes in the cholesterol and HDL metabolism in the blood plasma in animals fed with a high-calorie diet. These changes have expressed the proatherogenic character and could be one of the causes for the LDL*B* accumulation in the blood.

Our results suggest that increase of the total blood cholesterol level in hamsters fed with a high-calorie diet is obviously connected with increase of the cholesterol content in the ApoB-LP composition as its level in the HDL composition decreases (Table 3).

Age (at the beginning of the experiment)	Group	Parameters			
		HDL-C, mkmol/L	HDL-CE, mkmol/L	LCAT, mkmol/l/h	CETP, mkmol/l/h
Week 4	Intact	174.17±18.99	1028.33±12.76	54.92±0.58	20.42±1.76
	MS	80.83±9.17*	810.00±22.78*	49.00±2.50	33.83±1.56*
Week 20	Intact	138.00±8.00	770.00±32.56	45.50±2.55	59.50±5.39
	MS	164.50±9.97	512.50±0.01*	20.25±2.28*	116.88±9.43*

The data presented as mean±SD
* –p≤0.05 versus intact animals, [a] –p≤0.05 versus intact animals 4 weeks.

Table 3. Plasma HDL-C and HDL-CE, cholesterol esterifying activity and CE transfer in Syrian golden hamsters with the experimental MS (in each group n=10).

Decrease in the HDL cholesterol level is apparently connected with increase of the transfer rate of cholesteryl esters from HDL to ApoB-LP. According to our data the rate of the cholesteryl esters transfer from HDL in the animals fed with a high-calorie diet grows to 166% and 199% compare to the values of young and adult intact animals, respectively (Table 3).

At the same time decrease in the free cholesterol and HDL esterified cholesterol levels was determined in young males, but in adult animals only the HDL esterified cholesterol content lowered. The cholesteryl ester transfer rate from HDL to ApoB-LP is activated when the TAG content increases in the blood, it is observed in the postprandial period, as well as in ApoB-LP metabolism abnormalities [39]. In both cases the cholesteryl ester transfer activation is a consequence of increasing the TAG-rich lipoproteins (TRL) in the bloodstream [40]. The latter is also confirmed by our data pertaining to the increase of the neutral lipids content in the ApoB-containing lipoproteins in hamsters with the experimental MS. These differences seem to be connected with the difference in the HDL free cholesterol esterification rate in males of various ages. This rate is primarily determined by the activity of LCAT – the enzyme associated with HDL [41].

The increase of the cholesteryl-ester transfer activity from HDL is mostly the result of the CETP activation. The increase of the CETP activity in MS was demonstrated in a great number of experiments [22]. It is known that the activation of CETP biosynthesis in the liver is primarily the cause for increasing the activity of this protein in the blood HDL composition, but mechanisms of CETP induction have been still unclear.

Thus, increase of the cholesteryl ester transfer rate from HDL on the background of hypertriacylglycerolemia, which is observed in our experiment in the animals fed with a high-calorie diet (Table 3), is atherogenic since the cholesteryl ester transfer predominantly to TAG-enriched lipoproteins leads to accumulation of CE-enriched VLDL1, which are major precursors of LDL*B*. Intensive TAG supply to HDL in exchange for cholesteryl esters results in accumulation of TAG-enriched HDL particles, which are the predominant substrate for hepatic lipase (HL), in the blood. So, HDL particles are rapidly removed from the bloodstream and it leads to decrease of the HDL-C content.

That is why changes in the enzymes activity, which hydrolyze lipoprotein lipids in the bloodstream, in particular – in LPL and HL activity, affect significantly the lipoprotein metabolism in MS.

TAG in the TAG-enriched lipoproteins (chylomicrons and VLDL) are the substrate for LPL. FFA, released after hydrolysis under the action of LPL, come to adipocytes and muscle cells where they are deposited as the TAG component or used as a source of energy. TAG hydrolysis in the VLDL composition increases availability of cholesterol for its transfer to HDL, therefore, in this way LPL mediates the reverse cholesterol transfer. The LPL activity is regulated by the influence on transcription, translation and enzyme transport from the cells. Insulin is known to activate LPL that results in decrease of the total blood TAG level and stimulation of cholesterol reverse transfer [35].

According to our data, the plasma LPL activity decreased in young male hamsters fed with a high-calorie diet (Table 4).

Age (at the beginning of the experiment)	Group	Parameters	
		LPL (U/ml)	HL (U/ml)
Week 4	Intact	8±2	51±4
	MS	4±1*	91±3*
Week 20	Intact	83±2	3±1
	MS	129±3*	2±1

The data presented as mean±SD
* –p≤0.05 versus intact animals

Table 4. Postheparin plasma lipase activities in Syrian golden hamsters with the experimental MS (in each group n=10).

The results obtained are in agreement with the literature data about the reduction of the LPL activity in obesity and insulin resistance [42]. The mechanisms of the LPL activity inhibition in these conditions are still unclear though a definite contribution could be made by development of insulin resistance.

The increase of the cholesteryl ester transfer rate from HDL on the background of hypertriacylglycerolemia, which was stated in our experiment both in animals fed with a high-calorie diet and in chronic stress, is an atherogenic factor for two reasons. Firstly, the cholesteryl ester transfer predominantly to the TAG-enriched lipoprotein fractions leads to accumulation of VLDL1 enriched with cholesteryl esters, which are the main LDL*B* precursors. Secondly, the intensive exchange of cholesteryl esters in HDL for TAGs results in accumulation of TAG-enriched HDL in the blood, which are the predominant substrates for HL, and they are rapidly removed from the bloodstream, and it, in turn, causes decrease in the HDL-C concentration. The activation of the lipoprotein secretion by the liver is also observed in the conditions of the acute chemical and emotional painful stress. This fact may be considered to be a sign of proatherogenesis since it is accompanied by hyperlipidemia development due to increase of atherogenic lipoprotein fractions.

As shown in our studies, decrease of the LPL activity in the blood plasma of young males fed with a high-calorie diet can be an additional factor for TAG accumulation in the blood and the HDL-C level reduction observed in our experiment.

HL mediates a selective transport of VLDL remnants to hepatocytes via LDL-receptors, takes part in reverse transport of cholesterol accelerating HDL coming into the liver via scavenger receptors (SRB1). Hydrolyzing TAG in the ApoB-LP composition HL plays a significant role in their re-modelling in the bloodstream. It is known that the HL activity specifies substantially the lipid composition, size and properties of LDL [43].

The HL activity is predominantly regulated at the transcriptional level under the influence of sex hormones, glucocorticoids and adipokines. The rate of the HL gene transcription is also dependent on the intercellular lipid content, primarily cholesterol in the hepatocytes [44].

In our experiment the blood plasma HL activity in male hamster fed with a high-calorie diet increased irrespective of age (Table 4), it corresponds to literature data. In a number of studies it has been shown that the HL activity increases in insulin resistance, obesity, and a high-calorie diet [45]. Moreover, it has been determined that increase of the HL mRNA content is observed when using a high-calorie diet; this is the evidence of the enzyme biosynthesis activation under these conditions. The authors associate this fact with the decrease in the blood plasma adiponectin level, which can inhibit the HL synthesis in hepatocytes.

Considering these data, as well as the data obtained in our studies about adiponectin decrease in the blood plasma in obesity (Fig. 1), we may suppose that one of the causes for the HL activity increasing when taking a high-calorie diet in our experiment is decrease of adiponectin secretion by the adipose tissue.

The HL activity increase is considered to be one of the key factors for the atherogenic dyslipidemia development in obesity and MS. In a number of works a clear correlation between the HL activity and the LDL*B* content in the blood plasma was demonstrated [19]. It is believed that namely HL activation results in the increased LDL*B* formation [33]. The latter occurs with increase of the TAG-enriched VLDL1 content in the blood and the CETP activation. Furthermore, the HL activity increase leads to decrease of the HDL cholesterol level [46]. This is associated with the fact that hydrolysis of TAG in the HDL3 composition results in their transformation into HDL2, which are rapidly removed from the bloodstream by the liver. Thus, the HDL-C level decrease that we determined in our experiment (Table 3) can be a consequence of the HL activity increase.

In the current study we have found that the blood FFA level increase is accompanied by the ApoB-LP synthesis activation in the liver of Syrian male hamsters fed with a high-calorie diet irrespective of age. This causes increase of the TAG and ApoB-LP level in the blood. Decrease of the HDL-C level is a consequence of the rate of cholesteryl ester exchange between HDL and LDL due to activation of CETP and HL. As a result of these changes the atherogenic dyslipidemia development, which is typical for MS, is observed.

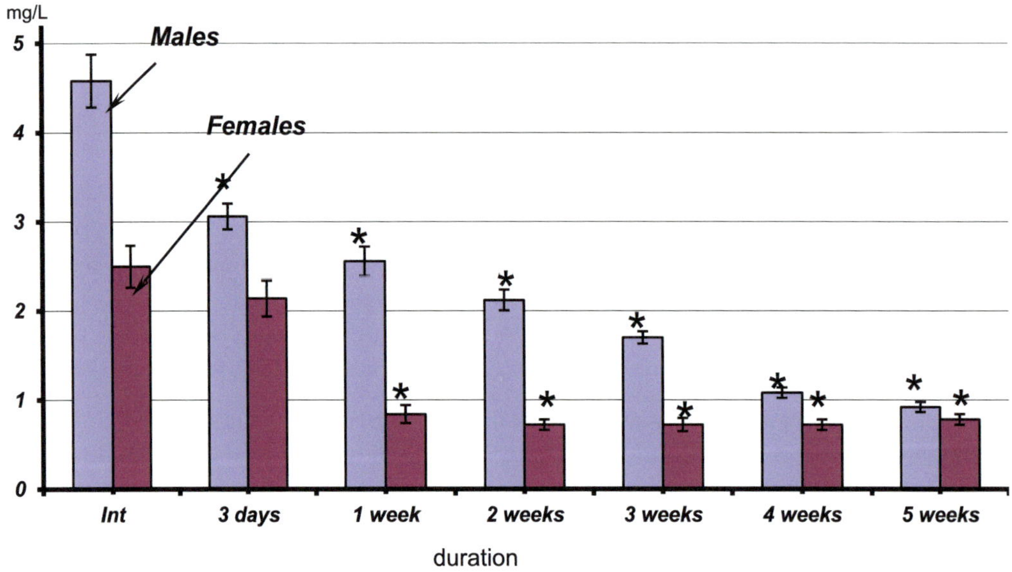

Figure 1. Plasma adiponectine level in Syrian golden hamsters with the experimental MS development (values are mean±SD; * –p≤0.05 versus intact animals, in each group n=10, * – p < 0.05 versus intact animals).

We have found the age differences of the lipid profile in the blood plasma of the normal male hamsters. So, in intact males (with age from 4 to 20 weeks) on the background of the constant content of total lipids and lipoproteins in the blood plasma there was increase of the FFA level (60% comparing to a 4-week intact), TAG (48%) and ApoB-LP (20%), and the HDL level showed a tendency to decrease. These results are the evidence of the lipidation increase with age. It has been also shown that in adult males the unesterified cholesterol and cholesteryl ester levels are lower than in young animals (20% and 25%, respectively), and the cholesteryl ester transfer rate from HDL in adult animals exceeds this index value in young animals (191%) (Table 3).

The data obtained correspond to the literature data about age-dependent changes in the lipid metabolism in males, which have the proatherogenic character [47]. It is known that with age the sex hormones level lowers in males and the glucocorticoid secretion level increases. The plasma lipid profile in males is determined, among other factors, by the secretion level of sex hormones possessing antiatherogenic properties. A lot of studies proved the presence of direct correlation between the blood testosterone plus the dehydrotestosterone level and the HDL-C content [48]. Moreover, the high level of sex hormones correlates with decrease of the TAG content and the total cholesterol in the blood. Thus, increase of the TAG level and decrease of the HDL cholesterol content in the blood plasma of males with age may be connected with reduction of the sex hormone secretion (Table 5). Changes of the lipid profile in the blood plasma of males with age may be also associated with increase of glucocorticoid secretion, which was observed in our experiment (Table 5).

Thus, with age the blood plasma lipid profile in males is subjected to unfavourable changes such as increase of the FFA and TAG content and decrease of HDL-C. The latter may be connected with decrease of the sex hormone level, and increase of the cortisol secretion.

However, despite the more favourable lipid profile in the blood plasma of young males comparing with normal adult animals, atherogenic dyslipidemia in obesity and insulin resistance develops irrespective of age.

In contrast with males, in females the atherogenic dyslipidemia development is significantly dependent on age (Table 6). In particular, while in males with age there are no significant changes in the liver ApoB-LP content, in females this index raises during maturation – in intact animals – to 20%, and in animals in the experimental MS – to 31%. That indicates intensification of lipolytic processes in the liver of females during ageing, and may serve as a manifestation of the lipid metabolism activation. The analogous changes in the total lipid content also proved this tendency (Table 6).

Sex	Parameter	Group	Age (at the beginning of the experiment)		
			4 weeks	20 weeks	1 year
Females	Estradiol, pmol/mL	Intact	0.55±0.05	0.64±0.06	0.54±0.05
		MS	0.63±0.06*	0.75±0.08	0.36±0.04*
	Cortisol, nmol/L	Intact	47.00±3.85	73.17±5.56	76.00±4.95
		MS	74.0±5.49*	87.5±4.45	117.00±2.63*
Males	Estradiol, pmol/mL	Intact	0.24±0.02	0.19±0.02	0.29±0.03
		MS	0.27±0.02	0.29±0.02*	0.20±0.03*
	Testosterone, pmol/mL	Intact	4.02±0.39	4.24±0.31	3.58±0.37
		MS	4.45±0.41*	3.51±0.40*	3.03±0.31*
	Cortisol, nmol/L	Intact	61.17±3.71	94.8±3.06	85.33±5.40
		MS	84.67±3.62*	132.00±7.88*	148.40±9.54*

The data presented as mean±SD
* –p≤0.05 versus intact animals

Table 5. Plasma sex hormones and cortisol levels in hamsters with the experimental MS (in each group n=16)

Age	Group	Parameters			
		Total lipids, mg/g	ApoB-LP, mg/g liver	HDL, mg/g	Lysosomal lipase, nmol/mg protein/min
4 weeks	Intact	117.67±4.72	8.87±0.24	1.27±0.08	0.34±0.03
	MS	144.34±5.00*	10.24±0.25*	0.65±0.05*	1.24±0.05*
10 weeks	Intact	137.54±3.91	10.65±0.46	0.89±0.07	0.83±0.04
	MS	179.22±3.44*	13.44±0.30*	0.46±0.06*	1.33±0.08*

The data presented as mean±SD
* –p≤0.05 versus intact animals

Table 6. Lipid metabolism parameters in the liver homogenate in Syrian golden female hamsters with MS (in the crude tissue, in each group n=16)

Oxidation of LDL and VLDL (i.e. ApoB-LP) is an alternative way of the lipoprotein catabolism, which leads to their uptake by macrophages via scavenger-receptors, and may lead to the transformation of these cells into "foam" ones. That is why it is one of the factors of atherogenesis in MS.

In our experiment we also observed the composition changes in lipoproteins and in particular HDL particle enrichment with lipids (Table 7). However, the cholesterol content of these lipoproteins decreased in contrast to the ApoB-LP cholesterol content that was increased.

Parameters	Group	
	Intact	MS
Total lipids, % of the total HDL composition	49.45±1.35	57.31±1.91*
Total cholesterol, % of the total HDL composition	14.97±0.23	11.21±0.76*
TAG, % % of the total HDL composition	1.75±0.07	3.08±0.15*
α-Tocopherol, mmol/L	8.02±0.39	5.70±0.35*
Isolated double bonds, U/ml	8.64±0.59	7.31±0.17*
Diene conjugates, mmol/L	18.88±2.10	31.68±1.65*
Ketodienes+conjugated trienes, U/ml	1.15±0.08	1.48±0.06*
Total hydroperoxides, mmol/L	69.04±3.46	78.31±1.33*

The data presented as mean±SD or percentage
* –p≤0.05 versus intact animals

Table 7. The plasma HDL composition in Syrian golden hamsters (1 year) with the experimental MS (in each group n=10).

There are several possible reasons for that phenomenon. One of them is a well-known fact that HDL contains high levels of both unsaturated fatty acids, which are rapidly utilized, and proteins, which hydrophilic properties compensate the lack of phospholipids, as well as α-tocopherol and enzymatic antioxidants, particularly paraoxonase, which protect these lipoproteins from peroxidation. There is no doubt that the changes in the cholesterol metabolism enzymes activity associated with HDL (CETP) are involved in this process (Table 3).

Nevertheless, the content decrease of compounds with isolated double bonds and accumulation of the lipoperoxidation products has been determined in the HDL fraction in MS (Table 7). Moreover, the data obtained have shown that the content of ketodienes and coupled trienes in the HDL fraction is 129% comparing to control; the content of diene conjugates – 168% and the content of the total hydroperoxides – 115%. It has been also found that there is decrease of the α-tocopherol content in HDL (41%) comparing to the control values (Table 7).

Thus, HDL can protect LDL from oxidation "providing" a cell with paraoxonase and PAF-acetyl hydrolase. However, this protective effect of HDL is reduced in response to induction of the stress acute phase in animal models [49].

As can be seen from our data (Table 8), the HDL-associated paraoxonase activity is generally decreased in experimental MS.

Sex	Age	Groups	Activity, nmol/mL/min
Males	4 week	Intact	80.78±3.69
		MS	67.06±3.70*
	20 week	Intact	62.19±2.63
		MS	37.29±3.33*
Females	4 week	Intact	104.41±2.95
		MS	75.45±2.21*
	20 week	Intact	127.27±2.95
		MS	121.93±3.05

The data presented as mean±SD
* –p≤0.05 versus intact animals

Table 8. The plasma HDL paraoxonase activity in Syrian golden hamsters with experimental MS (in each group n=10).

The data obtained show that application of antioxidant complexes for correction of unfavourable changes in proatherogenic states may be perspective since free radical oxidation activation is a common pathogenetic link of all those states; this link is not only involved in damage of cells and their components, but also as an alternative way of catabolism it accelerates the lipid recyclization.

At the same time, since a significant feature of proatherogenic states is the hormone status imbalance, polyphenolic antioxidants need special attention because these compounds along with the antioxidative activity also demonstrate phytoestrogen properties [50], and it may be an additional factor of the lipid metabolism regulation.

An important effect of flavonoids is scavenging of oxygen-derived free radicals. The experimental systems *in vitro* have also shown that flavonoids possess anti-inflammatory, antiallergic, antiviral, and anticarcinogenic properties. The so-called "Mediterranean diet" is thought to prevent cardiovascular diseases, as a consequence of its high content of antioxidants, which are crucial in ameliorating oxidative events implicated in many diseases. In addition to the antioxidant/antiradical activity, red wine polyphenols (RWPs) have been shown to possess many biological properties, including inhibition of platelet aggregation, the vasorelaxing activity, modulation of the lipid metabolism, and inhibition of the low-density lipoprotein oxidation.

In our research we have used wine, juice and polyphenolic extracts from grapes of different grades, and polyphenolic concentrates "Enoant" and "Polyphen" obtained from *Vitis Vinifera* grapes to correct the changes in the lipid metabolism in the conditions of the experimental metabolic syndrome, acute and chronical stress. All substances used in our research were developed in National Institute for Vine and Wine "Magarach" (Yalta, Ukraine). The studies carried out have specified that polyphenolic extracts and concentrates are quite active remedies that decrease negative effects in MS though the effectiveness of various substances administered are significantly different.

So, administration of any of the investigated substances has significantly decreased the total blood plasma lipoprotein content in hamsters with MS, but the use of "Cabernet" extract has the most pronounced effect (Fig. 2). The same tendency is observed in decreasing the ApoB-LP content, the total cholesterol and FFA level have also decreased, though practically no difference between the grape varieties investigated has found.

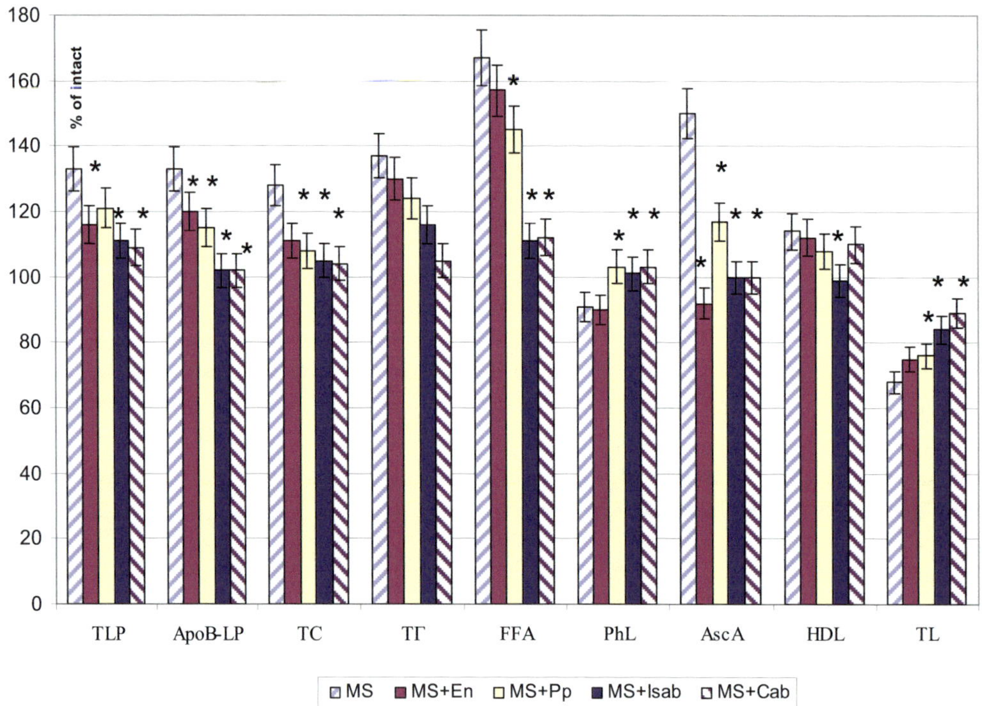

* – p≤0.05 versus intact animals.

Figure 2. The effect of *Vitis Vinifera* substances on some plasma lipid metabolism values in male Syrian golden hamsters (1 year old) with the experimental MS (in each group n= 7)

The non-enzyme antioxidant level (α-tocopherol, reduced glutathione and ascorbic acid) in the blood serum has also reached reference values under the influence of the polyphenolic extracts. This fact confirms the high antioxidant activity of the studied substances.

Normalization of the blood plasma phospholipid content under the influence of polyphenolic extracts arouses the interest. The phospholipid content returned to the intact level, which may be a result of their oxidation reduction, given that the unsaturated fatty acids in phospholipids are compounds that undergo oxidation by free radicals quickly and easily.

However, in spite of the quite favourable effect of "Isabella" extract, its administration also has negative consequences, particularly the HDL level decrease to the value observed in intact animals accompanied by the LDL content increase.

The investigated substances normalize also the blood lipoproteins composition. Thus, the total lipid and the total cholesterol content decrease in the ApoB-LP composition, moreover

"Enoant" lowers the cholesterol content in this atherogenic lipoprotein fraction even below the control level.

Generally, the TAG content is also normalized under the action of all the investigated substances, but taking into account the ratio – cholesterol/triacylglycerols, "Polyphen" has the most favourable effect.

The polyphenol extracts and concentrates have significantly improved the ApoB-LP oxidative status in animals with MS. The best results have been obtained when using "Cabernet", as well as for other indexes investigated (Table 9).

Parameter	Group				
	MS	MS +"Enoant"	MS +"Polyphen"	MS+ "Isabella"	MS+ "Cabernet"
Total lipids, % of the total ApoB-LP composition	88.87 ±0.71*	83.70 ±0.78 */**	83.12 ±0.37*/**	82.22 ±0.09*/**	81.00 ±0.19*
Total cholesterol, % of the total ApoB-LP composition	8.39 ±0.24	7.87 ±0.04*/**	8.13 ±0.04	8.06 ±0.12**	8.17 ±0.08*/**
TAG, % of the total ApoB-LP composition	37.55 ±1.89*/**	53.97 ±0.10*/**	53.02 ±0.14*/**	50.65 ±1.23*/**	49.57 ±0.40*/**
α - Tocopherol, mmol/L	2.68 ±0.08*	2.98 ±0.05*/**	3.06 ±0,04*/**	3.17 ±0,02*/**	3.19 ±0,05*/**
Isolated double bonds, U/ml	1.71 ±0.06*	1.84 ±0.03*/**	1.90 ±0.02*/**	1.97 ±0.03**	2.09 ±0.03**
Diene conjugates, mmol/L	37.25 ±1.50*	30.63 ±0.41*/**	29.54 ±0.34*/**	28.77 ±0.14*/**	26.68 ±1.94**
Ketodienes+conjugated trienes, U/ml	8.18 ±0.11*	7.49 ±0.04*/**	7.23 ±0.08*/**	7.17 ±0.08*/**	6.99 ±0.26**
Total hydroperoxides, mmol/L	108.25 ±1.39*	98.52 ±0.55*/**	94.97 ±0.15*/**	90.65 ±1.15*/**	89.30 ±1.06*/**

The data presented as mean±SD or percentage
* – p≤0.05 versus intact animals, ** – p≤0.05 versus model of MS

Table 9. The effect of *Vitis Vinifera* substances on the plasma ApoB-LP composition in male Syrian golden hamsters (1 year old) with the experimental MS (in each group n= 10)

The HDL composition in the blood is also affected by the substances studied. In these particles the total lipid content decreases and even reaches the level of intact animals when using "Cabernet" extract (Table 10).

The cholesterol level also changes: it decreases when using "Enoant" and increases under the action of "Isabella" and "Cabernet" extracts.

The HDL-C content decreases under the action of "Enoant" may occur due to peroxide processes inhibition since cholesterol accumulation in lipoprotein particles, as it was mentioned before, has a compensatory character in response to the phospholipid oxidation of the lipoprotein particle hydrophilic cover. The TAG content decreased under the action of all substances, and "Isabella" was the most effective substance. The TAG content decrease is probably mediated by the phytoestrogenic action of polyphenols directed to lipolysis inhibition in the adipose tissue.

Parameter	Group				
	MS	MS +"Enoant"	MS +"Polyphen"	MS+ "Isabella"	MS+ "Cabernet "
Total lipids, % of the total ApoB-LP composition	57.31 ±1.91*	54.91 ±0.21*/**	53.09 ±0.08*/**	51.74 ±0.74*/**	49.20 ±0.42**
Total cholesterol, % of the total ApoB-LP composition	11.21 ±0.76*	10.54 ±0.30*/**	11.14 ±0.04*	12.05 ±0.21*	12.45 ±0.34*/**
TAG, % of the total ApoB-LP composition	3.08 ±0.15*	2.90 ±0.09*/**	2.11 +0.12*/**	1.92 ±0.04*/**	1.96 ±0.03*/**
α - Tocopherol, mmol/L	5.70 ±0.35*	7.47 ±0.20*/**	7.19 ±0.17*/**	7.36 ±0.11*/**	8.13 ±0.06**
Isolated double bonds, U/ml	7.31 ±0.17*	7.67 ±0.08*/**	7.69 ±0.07*/**	7.99 ±0.05*/**	8.15 ±0.01*/**
Diene conjugates, mmol/L	31.68 ±1.65*	24.85 ±0.35*/**	23.44 ±0.40*/**	22.55 ±0.34*/**	21.88 ±0.23**
Ketodienes+conjugated trienes, U/ml	1.48 ±0.06*	1.24 ±0.03**	1.32 ±0.03*/**	1.25 ±0.03*/**	1.54 ±0.47*/**
Total hydroperoxides, mmol/L	78.31 ±1.33*	75.26 ±0.31*/**	75.62 ±0.54*/**	74.48 ±0.55*/**	73.41 ±0.39*/**

The data presented as mean±SD or percentage
* – p≤0.05 versus intact animals, ** – p≤0.05 versus model of MS

Table 10. The effect of *Vitis Vinifera* substances on the plasma HDL composition in male Syrian golden hamsters (1 year old) with the experimental MS (in each group n= 10)

The exact bimolecular mechanisms for this cardioprotection are unclear, but it is likely that actions mediated both through the estrogen receptors, such as the beneficial alteration in lipid profiles and upregulation of the low-density lipoprotein (LDL) receptor, and independently of the estrogen receptors, such as antioxidant action, contribute to the cardioprotective effects of phytoestrogens observed.

The potential role of phytoestrogens, including isoflavonoids, as cardioprotective agents has been extensively reviewed. The data obtained in our experiments showed that in male hamsters with the experimental MS the treatment with grape extracts reduced VLDL cholesterol (VLDL-C) and TG by 30 and 40 % compared with the control animals. Furthermore, golden Syrian hamsters fed with red wine phenolics had a significant decrease in the plasma apo B concentrations. Similar to our previous study, grape polyphenols may have altered hepatic secretion of TG-rich VLDL. This reduction is evident when observing the decreases in both plasma apo B and apo E concentrations. The significant decrease in apo E concentrations may have further reduced plasma TG concentrations. In general, apo E displaces apo C-II from the VLDL particle, thereby inhibiting the lipoprotein lipase (LPL) activity and overall lipolysis. Furthermore, Huang et al. [51] showed that adding apo C-II to transgenic apo-E3–enriched

VLDL increased the LPL activity in a dose-dependent manner. The reductions in apo E and TG concentrations suggest less displacement by apo E, thereby promoting the grape polyphenols activity and further reducing the TG concentrations in the plasma.

Due to decreases in TG concentrations, administration of "Cabernet" extract was shown to affect the overall lipoprotein metabolism. Decreased concentrations of the plasma TG altered substrate availability in the delipidation cascade, leading to the decrease observed in LDL-C concentrations. After a 3-week treatment period the grape polyphenols treatment induced a significant decrease in the cholesteryl ester transfer protein (CETP) activity as well. Such decrease in the CETP activity may be partially a result of the substantial decrease in substrate availability, including both the plasma TG and LDL-C.

It is evident that grape polyphenols modify the packaging of VLDL through alteration in the hepatic enzyme activity and apo B secretion. These modifications seem to decrease the overall secretion of the VLDL particles and therefore, decrease plasma TG and related apo concentrations. Due to decrease of the TG substrate, further modifications in the lipoprotein metabolism may occur.

The alteration in the TG metabolism may not be the single mechanism driving the hypocholesterolemic effects of grape polyphenols. When golden Syrian hamsters were treated with dealcoholized red wine, red wine, or grape juice, similar significant reductions in both TC and LDL-C concentrations were apparent in all treatment groups compared with the control [51]. Although there was a trend for decrease in TG concentrations in all treatment groups compared with the control, the differences were not significant. That study, along with others, suggests the presence of an additional mechanism by which grape polyphenols exert the cardioprotective effect. In Hep G-2 cells, dealcoholized red wine was shown to upregulate significantly the LDL receptor activity. This significant increase in activity was similar to the increase seen when Hep G-2 cells were treated with atorvastatin. Furthermore, when Hep G-2 cells were treated with increasing doses of red wine, LDL receptor mRNA abundance was significantly increased in a dose-responsive manner. The increase of the LDL receptor activity and abundance may be a result of the homeostatic intracellular cholesterol feedback loop. In general, decrease in the intracellular cholesterol will upregulate the LDL receptor expression and activity, whereas increase in the intracellular cholesterol will downregulate the receptor [48]. Grape polyphenols were shown to decrease hepatic cholesterol concentrations; therefore, the liver compensates for this deficiency by upregulating the LDL receptor and the overall decrease in the plasma LDL concentrations occurs.

One possible explanation of the anti-atherogenic activity of grape polyphenols is the well-known HDL cholesterol-increasing effect of polyphenols in various species, including transgenic mice [52].

In our experiments it has been found that the grape extract treatment induced slight (15%) increase in HDL cholesterol concentrations is possibly related to the significant decrease in the hepatic lipase activity (Table 11). The reductions observed in both hepatic and LPL activities by grape polyphenols treatment may prevent formation of small atherogenic VLDL*B* particles and may also decrease their uptake by the LDL receptor -related protein.

In addition to increases in HDL cholesterol concentrations, grape extracts also change the size and quality of HDL particles [53]. Although the mechanisms by which polyphenols influence the metabolism of HDL particles are not clear, changes in LPL and cholesteryl ester transfer protein (CETP) may play an important role.

Polyphenols treatment in humans is associated with decrease in the CETP content correlated with the concomitant increase in HDL cholesterol concentrations [54]. Consistent with our findings, grape extracts caused a significant increase in the postheparin LPL activity and HDL cholesterol concentrations in patients with moderate hypercholesterolemia and in hamsters [39]. However, the HDL cholesterol-increasing action of polyphenols in animals (mouse, hamster and rat) without CETP in some cases [52] suggests that this effect is may be independent of the CETP activity.

milliunits	Control (MS) (n=50)	Grape extract "Cabernet" (n=50)
LPL	356.0±53.2	258.6±57.3*
Hepatic lipase	232.6±25.9	216.2±34.7

*P<0.05 versus control animals.

Table 11. The plasma postheparin lipases activity in male hamsters with MS (in each group n=10)

The "Cabernet" extract appeared to be the most effective substance in relation to the HDL defence from peroxidation, though the other substances revealed the same but not so high activity. They decreased the content of products (diene conjugates, ketodienes+coupled trienes, total hydroperoxides) effectively and increased – substrates (compounds with isolated double bonds) of lipoperoxidation, prevented decrease of the antioxidant level (α-tocopherol).

It should be pointed out that the level of lipid peroxidation secondary products (ketodienes+coupled trienes) decreased more effectively under the influence of "Enoant" (to intact values).

Under the action of the studied substances the lipoprotein supply to the liver also decreases, evidenced by the decrease of the ApoB-LP content in the organ. Moreover, in the composition of these lipoproteins the TAG content normalizes, and it indicates normalization of the activity of lipases catalyzing the lipoprotein metabolism in the blood (Table 12).

The liver oxidative status is also improved: the antioxidant levels almost restore, the peroxidation products content decreases, the content of compounds with isolated double bonds increases (Tables 3, 5, 9, 12).

Testing of the "Enoant" action – one of the substances studied – in female hamsters of different age with the experimental MS proved the effectiveness of the antioxidant therapy of this pathology.

So, the total lipids, TAG and FFA contents decrease in the blood plasma of those animals under the action of "Enoant" (Table 12). In addition, in adult females "Enoant" causes decrease in the ApoB-LP and total cholesterol content, and it, in turn, reduces atherogenic changes in MS.

Parameter	Group				
	MS	MS + "Enoant"	MS + "Polyphen "	MS+ "Isabella"	MS+ "Cabernet"
Total cholesterol, % of the total ApoB-LP composition.	7.18±0.06*	8.23±0.26*/**	8.46±0.05**	8.76±0.05**	9.10±0.13**
TAG, % of the total ApoB-LP composition	42.00±1.29*	44.64±0.52**	44.42±0.43**	45.95±0.50**	45.41±0.73**
Isolated double bonds, U/g	2.13±0.06*	2.39±0.04*/**	2.71±0.03*/**	2.99±0.09**	2.77±0.16**
Total hydroperoxides, mmol/g	101.03±2.00*	90.55±1.54*/**	88.69±1.02*/**	80.46±0.77*/**	78.83±2.71*/**

The data presented as mean±SD or percentage
* – p≤0.05 versus intact animals, ** – p≤0.05 versus the model of MS

Table 12. The effect of *Vitis Vinifera* substances on the liver cytosol ApoB-LP composition in male Syrian golden hamsters (1 year old) with the experimental MS (in the crude tissue, in each group n= 10)

The increase of α-tocopherol (the main lipid-phase antioxidant) in the blood plasma of animals that received "Enoant" proved its antioxidant activity in our experiment (Table 13).

Furthermore, the significant decrease of the body weight was observed in hamsters that received "Enoant" along with a high-calorie diet compared to the animals on a high-calorie diet alone.

Based on these findings, we may conclude that introduction of grape polyphenolic extracts and concentrates in MS can prevent the increase of the total lipid and ApoB-LP content in the blood plasma, prevent the activation of free radical processes in the plasma lipoprotein particles, and normalize the liver lipid metabolism. The ability of the investigated substances to reduce negative consequences of MS such as atherosclerosis development has been proven.

The last suggestion is confirmed by our results concerning the aorta wall lipid composition in the experimental MS. The introduction of "Enoant" for prophylaxis and treatment reduces significantly atherogenesis manifestations in the aorta, decreasing the aorta media lipidation and the neutral lipid content (Fig. 3, 4).

Thus, from our data, we can conclude that antioxidants, particularly grape polyphenolic concentrates and extracts, which have pronounced antioxidant, phytoestrogenic and stress-protector properties, should be included into a complex therapy of MS to reduce its negative effects.

The next experiment was designed to investigate the action of grape wines and polyphenolic concentrates on development of proatherogenic effects of the emotional-painful stress. In our experiments we used purebred female rats because, as it was shown in previous studies, the acute stress response in females was more expressive than in males.

Age	Group	Parameter						
		Total lipids, mg/ml	ApoB-LP, mg/ml	Total cholesterol, mmol/L	TAG, mg/ml	FFA, mmol/L	Diene conjugates in ApoB-LP, nmol/ml	α - Tocopherol, nmol/ml
4 weeks	MS	4.52 ±0.17	4.00 ±0.16	2.04 ±0.08	1.08 ±0.49	1.17 ±0.06	24.82 ±1.46	6.67 ±0.22
	MS+ "Enoant"	3.54 ±0.16**	3.47 ±0.13	1.88 ±0.06	0.91 ±0.02**	0.95 ±0.02**	22.27 ±0.99	9.79 ±0.77**
20 weeks	MS	7.75 ±0.20	3.84 ±0.11	2.58 ±0.07	1.40 ±0.04	1.42 ±0.04	23.58 ±1.35	10.49 ±0.82
	MS+ "Enoant"	6.88 ±0.14**	3.30 ±0.08**	2.22 ±0.05**	1.18 ±0.03**	1.15 ±0.03**	21.10 ±1.14	12.87 ±0.36**

The data presented as mean±SD or percentage
** – p≤0.05 versus the model of MS

Table 13. The effect of polyphenol concentrate "Enoant" on some plasma lipid metabolism values in female Syrian golden hamsters with the experimental MS (in each group n= 10)

Age	Group	Parameters			
		Total lipids, mg/g	α - Tocopherol, nmol/g	Ascorbic acid, mkmol/g	TBA active substances, nmol/g
4 weeks	MS	140.75 ±9,15	21.01 ±1.47	3.73 ±0.14	1.97 ±0.06
	MS+ "Enoant"	111.53 ±4.08**	25.31 ±0.34**	4.08 ±0.10	1.74 ±0.09
20 weeks	MS	154.18 ±2.70	19.59 ±0.39	6.10 ±0.35	1.95 ±0.09
	MS+ "Enoant"	121.04 ±4.18**	26.11 ±1.03**	5.96 ±0.24	1.73 ±0.09

The data presented as mean±SD or percentage
** – p≤0.05 versus the model of MS

Table 14. The effect of polyphenol concentrate "Enoant" on some liver lipid metabolism values in female Syrian golden hamsters with the experimental MS (in each group n= 10)

During 21 days animals were daily given *per os* grape wines of "Cabernet" and "Rkatsiteli" grades in the doses that corresponded to 300 ml of wine for a human of 70 kg. Other animals were given alcohol in the dose corresponding to 30 ml of alcohol for a human of 70 kg, as well as polyphenolic concentrates "Enoant" and "Polyphen" in the doses of 0.05 ml/kg of the body weight. The grape wines and polyphenolic concentrates were produced by the National Institute of Grape and Wine "Magarach". Control animals were given the corresponding volume of the physiological solution.

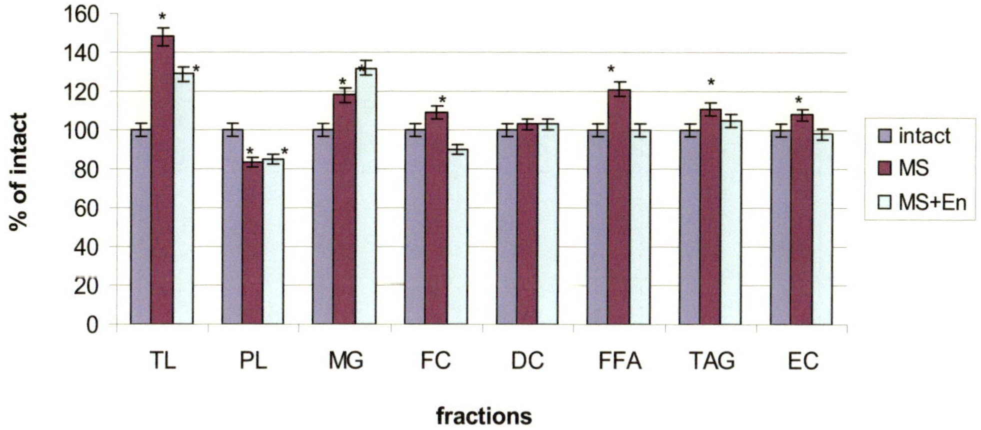

Figure 3. The lipid content in the aorta wall in male Syrian golden hamsters with the experimental MS and "Enoant" treatment (M±m, in each group n=10).

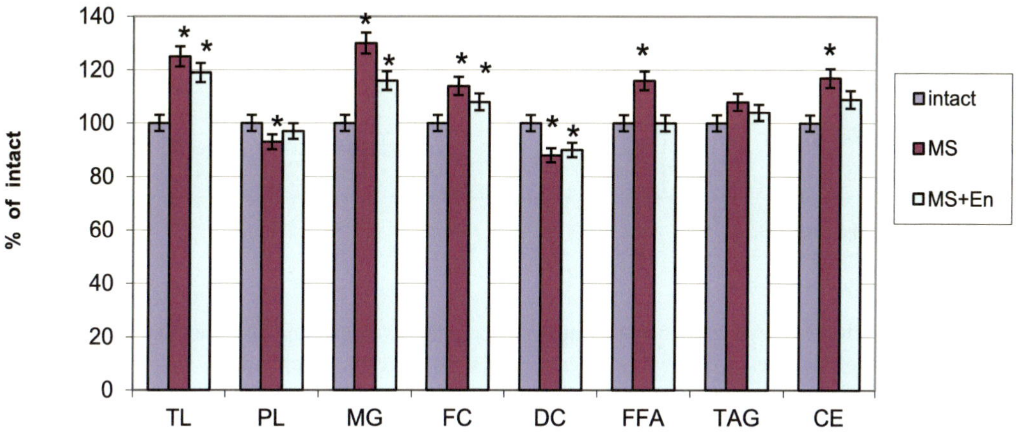

Figure 4. The lipid content in the aorta wall in female Syrian golden hamsters with the experimental MS and "Enoant" treatment (M±m, in each group n=10).

It was shown that all the substances investigated: polyphenolic concentrates "Enoant" and "Polyphen", 10% solution of ethanol, and grape wines "Cabernet" and "Rkatsiteli" possessed the stress-protective activity, which intensity was dependent on the substance used (Tables 15-17).

When introducing only "Enoant" and "Polyphen" to the animals these complexes did not cause any changes on the investigated indexes of the pro-oxidant and antioxidant status in the liver and it is an indication about safety of using these concentrates.

"Enoant" and "Polyphen" revealed the significant protective activity in the emotional-painful stress. It allows to use them as stress-protective, hepatoprotective and antiatherogenic remedies.

Parameter	Group									
	Stress + Enoant		Stress + Polyphen		Stress + ethanol		Stress + Cabernet		Stress + Rkatsiteli	
	str.	non-str.	str.	non-str.	str.	non-str.	str.	non-str.	str.	non-str.
TL, mg/ml	4.14 ±0.45*	3.98 ±0.31 **	3.81 ±0.03 */**	3.82 ±0.48 **	5.50 ±0.74 */**	5.58 ±0.37*	3.41 ±0.66 **	3.46 ±0.43 **	3.43 ±0.15 **	3.57 ±0.54 **
TAG, mg/ml	0.76 ±0.08 */**	0.78 ±0.09*	0.51 ±0.08	0.71 ±0.09*	0.66 ±0.12 */**	0.91 ±0.05 */**	0.50 ±0.05	0.58 ±0.04 */**	0.54 ±0.10	0.54 ±0.05 **
Total cholesterol, mg/ml	70.76 ±9.34*	54.90 ±6.92 */**	74.14 ±8.21*	82.93 ±9.48 **	85.71 ±5.71	69.10 ±7.37 **	56.71 ±7.32*	62.25 ±6.50 **	97.51 ±9.66 **	80.83 ±8.74 **
HDL, mg/ml	0.98 ±0.14	0.89 ±0.08	0.96 ±0.05	0.88 ±0.08	1.10 ±0.19	1.00 ±0.12	1.26 ±0.15	1.04 ±0.08	1.08 ±0.12	0.96 ±0.05
ApoB-LP, mg/ml	1.56 ±0.12 **	1.89 ±0.26*	1.59 ±0.16 **	1.54 ±0.22 **	1.85 ±0.19 **	1.67 ±0.18 **	2.07 ±0.29*	1.67 ±0.18 **	1.86 ±0.20 **	1.63 ±0.18 **
Corticosterone, nmol/l	5.50 ±0.99 */**	5.87 ±1.08 */**	7.10 ±0.80 */**	29.33 ±8.58	15.00 ±1.72 */**	38.75 ±5.64 **	32.75 ±5.17 **	26.33 ±4.40 */**	16.93 ±2.43 */**	40.60 ±6.38 **

The data presented as mean±SD
* - $p \le 0.05$ versus intact animals
** - $p \le 0.05$ versus stressed animals

Table 15. The effect of grape polyphenol complexes and grape wines on the lipid metabolism and the plasma corticosterone level in rats with the neurogenic stress (in each group n=10).

Moreover, we have found out that the stress-protective activity of grape wines is equal to the polyphenolic concentrates activity given in the similar dose.

Wines of "Cabernet" and "Rkatsiteli" grades normalized the total lipid content both in a liver homogenate and in the blood plasma in stress; in addition, TAG levels also reached the control values.

Grape wine components prevented the FFA content increase noted when introducing the solution of alcohol. This fact may prove the protective action of the components mediated by inhibition of fatty infiltration of organs.

The latter is confirmed by the absence of influence of grape wine introduction on the NADPH-generating dehydrogenases activity in the liver.

The cholesterol content decrease in the blood plasma when introducing grape wines has attracted our attention, as well as a favourable redistribution of cholesterol in the LP fractions – decrease of ApoB-containing lipoprotein level with the unchanged HDL content.

Parameter		Group									
		Stress + Enoant		Stress + Polyphen		Stress + ethanol		Stress + Cabernet		Stress + Rkatsiteli	
		str.	non-str.	str.	non-str.	str.	non-str.	str.	non-str.	str.	non-str.
PON, nmol/ml/min		240.25 ±20.86 */**	216.00 ±16.24 */**	231.33 ±15.53 */**	231.25 ±15.52 **	197.50 ±14.37 */**	150.00 ±19.24 */**	239.25 ±24.18 */**	246.00 ±16.56 **	222.75 ±13.98**	215.25 ±13.78 **
Ascorbic acid, mkmol/ml		49.99 ±4.41*	43.20 ±4.16*	53.76 ±4.51 **	45.08 ±6.08 */**	38.04 ±4.12 */**	34.83 ±4.43*	48.07 ±2.13*	48.10 ±2.49 */**	55.87 ±6.65 **	53.41 ±5.42 */**
α-Tocopherol, nmol/ml		10.70 ±1.34 **	10.62 ±0.67 **	10.39 ±1.12 **	10.8 ±1.26 **	6.07 ±0.76 */**	5.35 ±0.78 */**	9.45 ±1.11	9.99 ±0.48 **	10.26 ±0.48 **	9.32 ±1.04
Oxidized ApoB-LP	Isolated double bonds, U/ml	2.76 ±0.36*	2.23 ±0.24*	3.14 ±0.46 **	3.51 ±0.42 */**	3.13 ±0.37 **	2.35 ±0.24*	3.25 ±0.38 **	2.64 ±0.28*	4.13 ±0.43 **	3.11 ±0.41*
	Diene conjugates, mmol/ml	22.47 ±3.20 */**	25.93 ±2.98 */**	21.54 ±1.22 */**	25.93 ±1.76 */**	30.44 ±2.20*	26.28 ±1.24*	23.05 ±4.48*	26.99 ±1.38*	19.54 ±0.93 **	21.56 ±0.99 */**
	Ketodienes+conjugated trienes, U/ml	2.31 ±0.14	2.58 ±0.29 */**	2.20 ±0.11 **	2.37 ±0.10*	3.01 ±0.50*	2.97 ±0.42*	2.29 ±0.16	2.60 ±0.31*	2.22 ±0.11	2.23 ±0.06 **

The data presented as mean±SD
* - p≤0.05 versus intact animals
** - p≤0.05 versus stressed animals

Table 16. The effect of grape polyphenol complexes and grape wines on the plasma oxidant/antioxidant status in rats with the neurogenic stress (in each group n=10).

Paraoxonase activity was normalized in the animals given wines and the antioxidant content both in the blood plasma and in the liver was significantly higher than in the control animals.

These effects together with much lower level of ApoB-LP oxidation in the animals given grape wines prove the high antiatherogenic potential of the wines investigated.

In addition, the grape wines have revealed a rather high level of the stress-protective activity and it is indicated by a significant decrease of the corticosterone content in the blood plasma in stressed animals given wines.

Since grape wines have shown a high level of the stress-protective activity we investigated how the ratio of wine components – polyphenols and ethanol – can influence the stress-protective activity of this complex.

As was shown in our experiments "Enoant" administration even in the combination with ethanol does not reduce the stress-protective action of it, but on the contrary – it intensifies this action preventing unfavourable effects of the alcohol. At the same time the TAG and FFA level in the liver tissue of rats given ethanol together with "Enoant" decreases even when using the lowest dose investigated (0.01 ml per 100 g) (Table 18). Since the content of TAG and FFA increases apparently due to the lipogenesis activation when using ethanol, which might lead to fatty infiltration of the liver, then reduction of this process activity could protect the liver.

Parameter	Group									
	Stress + Enoant		Stress + Polyphen		Stress + ethanol		Stress + Cabernet		Stress + Rkatsiteli	
	str.	non-str.	str.	non-str.	str.	non-str.	str.	non-str.	str.	non-str.
Total lipids, mg/g	149.17 ±33.14	151.84 ±19.86 **	130.98 ±23.66	201.93 ±33.71 **	193.32 ±25.61 */**	220.54 ±23.22 */**	171.39 ±28.66 **	180.31 ±15.98 **	164.93 ±16.93 **	190.03 ±18.21**
TAG, mg/g	4.65 ±0.78	3.77 ±0.47	6.00 ±0.32	5.94 ±1.08 **	7.86 ±0.25 **	6.89 ±0.49 **	4.90 ±1.08	3.88 ±1.08	5.74 ±0.61	4.93 ±0.68
ApoB-LP, mg/g	4.58 ±0.06 **	4.43 ±0.07 */**	3.72 ±0.34*	3.15 ±0.44*	5.00 ±0.82	4.10 ±0.49 */**	3.06 ±0.16 */**	3.15 ±0.16 */**	2.87 ±0.30 */**	3.02 ±0.36*
FFA, mg/g	1.06 ±0.12	1.00 ±0.11	1.08 ±0.16	1.26 ±0.08*	1.35 ±0.15*	1.45 ±0.13*	1.05 ±0.22	1.22 ±0.20	1.15 ±0.13	1.40 ±0.31

The data presented as mean±SD
* - p≤0.05 versus intact animals
** - p≤0.05 versus stressed animals

Table 17. The effect of grape polyphenol complexes and grape wines on the liver lipid metabolism in rats with the neurogenic stress, in the crude tissue (in each group n=10).

Parameter	Group									
	Stress + Enoant		Stress + Polyphen		Stress + ethanol		Stress + Cabernet		Stress + Rkatsiteli	
	str.	non-str.	str.	non-str.	str.	non-str.	str.	non-str.	str.	non-str.
GSH, mkmol/g	6.02 ±0.27 */**	3.70 ±0.42*	5.12 ±0.46	5.48 ±0.34	4.40 ±0.34	2.87 ±0.29 */**	4.64 ±0.70	4.38 ±0.35	4.47 ±0.51	3.59 ±0.50*
α-Tocopherol, nmol/g	27.82 ±2.86 */**	34.01 ±3.56 **	34.34 ±4.14 **	25.39 ±1.18 */**	6.55 ±0.61 */**	5.59 ±0.54 */**	26.95 ±1.32 */**	24.64 ±0.81 */**	29.24 ±1.82 */**	28.9 ±3.15 **
Ascorbic acid, mkmol/g	1.31 ±0.13 **	1.29 ±0.10 */**	1.17 ±0.23*	1.25 ±0.15 */**	0.59 ±0.07 */**	0.81 ±0.06*	1.04 ±0.21*	1.21 ±0.11 */**	1.24 ±0.20 */**	1.24 ±0.17 */**
Isolated double bonds, U/g	18.63 ±1.88 */**	15.99 ±1.50 */**	20.40 ±1.34 */**	18.66 ±1.52 */**	16.98 ±0.66*	13.89 ±1.73*	24.40 ±2.92 */**	16.15 ±1.76 */**	26.67 ±1.62 */**	18.82 ±1.90 */**
Diene conjugates, mmol/g	13.58 ±0.74 */**	12.87 ±0.54 */**	13.48 ±0.62 */**	12.85 ±0.26 */**	13.61 ±0.08 */**	13.20 ±0.58 */**	10.90 ±1.10 **	15.00 ±2.12*	10.66 ±1.88 **	14.32 ±1.62*
Ketodienes+conjugated trienes, U/g	13.57 ±0.54*	13.39 ±1.67	11.44 ±1.56 **	11.84 ±1.88 */**	12.68 ±1.91	13.04 ±1.31 **	14.20 ±2.02*	15.74 ±1.98	17.15 ±1.62*	16.04 ±1.78
TBA-active products, nmol/mg protein	0.21 ±0.02*	0.19 ±0.02 **	0.28 ±0.03*	0.22 ±0.03 **	0.55 ±0.08 */**	0.68 ±0.06 */**	0.15 ±0.02	0.28 ±0.04*	0.18 ±0.02	0.17 ±0.02 **

The data presented as mean±SD
* - $p \le 0.05$ versus intact animals
** - $p \le 0.05$ versus stressed animals

Table 18. The effect of grape polyphenol complexes and grape wines on the liver tissue oxidant/antioxidant status in rats with the neurogenic stress (in the crude tissue, in each group n=10).

It should be noted that the effect of high doses of "Enoant" (0.1 and 0.15 ml/100 g) was ambiguous. On the one hand, it caused α-tocopherol accumulation in the liver that might be an indicator of their protective action, but on the other hand, it probably revealed some prooxidative effect initiating the increase in the content of the POL final products – thiobarbituric acid-active products, and also activating ApoB-containing lipoproteins oxidation. In this case the secondary oxidative stress developed. A tendency to decrease the lipid content in the liver and to increase it in the blood plasma testifies about it.

Such effect is typical for high doses of many antioxidants capable to reveal the prooxidant action, including α-tocopherol. These data indicate the necessity of reasonable attitude to antioxidants therapy, including "Enoant".

Parameter	Group						
	Stress	Stress+Ethanol+ Enoant, ml per 100 g of the body weight:					
	+Ethanol	0.01	0.03	0.05	0.07	0.1	0.15
Total lipids, mg/ml	5.89 ±0.08*	5.69 ±0.06*	5.30 ±0.05*	4.82 ±0.15*	3.51 ±0.08	3.44 ±0.09	3.58 ±0.15
TAG, mg/ml	0.91 ±0.07*	0.99 ±0.04*	0.73 ±0.04*	0.52 ±0.03	0.43 ±0.03	0.39 ±0.02*	0.50 ±0.02
Total cholesterol, mg/ml	0.49 ±0.07	0.46 ±0.05	0.54 ±0.04	0.40 ±0.01*	0.55 ±0.09	0.55 ±0.03	0.56 ±0.02
HDL, mg/ml	0.93 ±0.02	1.00 ±0.06	1.02 ±0.07	0.91 ±0.05	1.81 ±0.03*	1.23 ±0.06*	1.32 ±0.04*
ApoB-LP, mg/ml	1.73 ±0.05*	1.84 ±0.05*	1.48 ±0.06	1.44 ±0.03	1.18 ±0.05	1.41 ±0.02	1.64 ±0.04*
α-Tocopherol, nmol/мml	4.11 ±0.34*	4.88 ±0.17*	5.84 ±0.14*	6.68 ±0.24*	8.20 ±0.34*	9.23 ±0.35	8.80 ±0.46
Ascorbic acid, mkmol/L	33.81 ±1.73*	34.04 ±2.73*	39.04 ±1.60*	49.25 ±1.10*	55.01 ±1.67	56.98 ±2.03	49.96 ±3.43
Diene conjugates in ApoB-LP, mkmol/L	28.11 ±0.34*	28.99 ±0.14*	29.35 ±0.80*	28.81 ±1.30*	23.91 ±0.51	20.60 ±1.43	28,27 ±1.35*
TBA-active products, mkmol/L	2.39 ±0.55*	2.01 ±0.30*	1.48 ±0.16*	1.11 ±0.04	1.18 ±0.34	0.77 ±0.21	1.31 ±0.20*
Corticosterone, nmol/L	35.25 ±4.27	24.00 ±3.03	28.25 ±4.54	16.00 ±0.44*	25.50 ±0.50	17.50 ±2.33*	21.60 ±6.00

The data presented as mean±SD
* - p≤0.05 versus intact animals

Table 19. The effect of different doses of polyphenol concentrate "Enoant" in combination with ethanol on the plasma parameters of the stress response development in rats with the neurogenic stress (in each group n=10).

At the same time small doses of "Enoant" have a relatively low biological activity; they do not reduce negative effects of ethanol intake and do not inhibit the stress response significantly.

Therefore, we can conclude that the most effective doses of "Enoant" are 0.05-0.07 ml/100 g of the body weight because with these doses "Enoant" has not only high stress-protective, antiatherogenic and hepatoprotective activities, but practically neutralizes negative effects of ethanol.

Thus, our results suggest that grape wines have a high stress-protective, antiatherogenic and hepatoprotective activity that is equal to grape polyphenolic non-alcoholic concentrates characteristics, and the wine components in the doses studied have prevented negative effects of ethanol. Introduction of ethanol to animals in the human equivalent dose – 0.43 ml/kg of the body weight increases their tolerance to stress, but is an unfavourable factor that could result in MS development, fatty infiltration of organs and other pathologies. The polyphenolic concentrates "Enoant" and "Polyphen" in the human equivalent dose – 0.3 ml/kg of the body weight reveal a significant stress-protective, hepatoprotective and anti-atherogenic activity under the action of the emotional-painful stress. Grape wines from "Cabernet" and "Rkatsiteli" grades in the human equivalent dose – 4.3 ml/kg of the body weight also reveal a high stress-protective, antiatherogenic and hepatoprotective activity equal to grape polyphenolic non-alcoholic concentrates, and the wine components in the doses used prevented the negative effect of ethanol.

The highest activity has been shown by the combination of "Enoant" and ethanol that corresponds to the ratio of components in dry red wines, as well as the absence of significant difference in the protective effects of red and white wines, in spite of the difference in the polyphenol content [55]. Based on these results, in the second series of our experiments we decided to investigate "Cabernet" and "Rkatsiteli" wine effects on the development of stress-reaction proatherogenic consequences under the action of the emotional-painful stress in different periods of introduction.

It has been shown that "Cabernet" had a higher level of the anti-atherogenic activity than "Rkatsiteli"; in relation to the stress-protective activity the wines of these grades did not differ markedly. Such effect is likely connected with accumulation of polyphenols in the organism.

To examine the last supposition it was necessary to determine how different periods of introduction of the investigated wines influenced the stress-reaction development. We have carried out the study of wine intake influence on the development of proatherogenic consequences of the emotional and painful stress in different terms after consumption.

The data obtained in the experiments showed significant improvement of the antioxidant status both in the blood plasma and the liver tissue one day after the introduction of "Cabernet" wine (tables 20, 21).

At the same time "Rkatsiteli" wine did not reveal such activity. A similar condition persisted for 2-5 days of administration.

Periods of time		Parameter					
		Total lipids, mg/g	TAG, mg/g	GSH, mkmol/g	α-Tocopherol, nmol/g	Diene conjugates, nmol/g	TBA-active products, nmol/g
Day 1	C+Str	94.94±5.65*	3.80±0.11*	2.12±0.17*	18.88±0.79*	14.93±0.37*	1.68±0.10
	R+Str	103.16±4.63*	5.15±0.61*	2.37±0.28*	15.60±0.39*	15.94±0.39*	2.24±0.32
Day 2	C+Str	105.17±5.12*	3.58±0.23*	2.51±0.34*	26.04±2.17*	14.13±0.09*	1.07±0.24*
	R+Str	106.93±9.42*	5.67±0.34*	2.77±0.87*	15.65±0.92*	15.10±0.07*	1.91±0.15
Day 3	C+Str	115.38±11.65*	3.92±0.13*	3.82±0.37*	23.69±2.18*	13.85±0.48*	1.70±0.11
	R+Str	103.28±5.81*	6.47±0.28*	2.47±0.26*	16.89±0.71*	14.44±0.25*	2.29±0.20*
Day 5	C+Str	139.14±8.06*	5.57±0.31*	2.43±0.37*	26.96±2.12	13.19±0.34	1.35±0.15
	R+Str	116.66±3.60*	6.88±0.37*	2.43±0.42*	18.64±1.18*	14.22±0.16*	2.06±0.13
Day 8	C+Str	161.18±6.05*	7.27±0.15	4.11±0.21	32.12±0.85	12.28±0.52	1.64±0.13
	R+Str	122.20±7.07*	6.35±0.45	2.33±0.28*	23.08±2.08*	12.93±0.44	1.73±0.04
Day 10	C+Str	181.82±9.24	8.04±0.63	5.15±0.42	30.62±2.53	11.86±0.12	1.30±0.19*
	R+Str	153.99±5.30*	8.41±0.56	3.22±0.48*	26.31±1.26*	11.84±0.48	1.29±0.14*
Day 12	C+Str	174.72±6.15	8.34±0.55	3.74±0.25	32.55±5.58	12.42±0.36	1.47±0.27
	R+Str	164.4±8.03*	8.63±0.47	3.25±0.17*	28.28±2.26	11.41±0.22	1.45±0.31
Day 15	C+Str	162.16±12.81	7.50±0.43	4.61±0.22	38.16±2.06*	10.50±0.52*	1.34±0.06*
	R+Str	172.13±10.42	7.68±0.69	4.42±0.34	41.99±2.42	10.27±0.63*	1.24±0.17

The data presented as mean±SD
* - p≤0.05 versus intact animals

Table 20. The effect of prophylactic administration of grape wines of "Cabernet" (C) and "Rkatsiteli" (R) grades on the stress response development in the liver tissue in rats with the neurogenic stress in different periods of time, in the crude tissue (in each group n=10).

Periods of time		Parameter						
		Total lipids, mg/ml	TAG, mg/ml	Total cholesterol, mg/ml	ApoB-LP, mg/ml	α-Tocopherol, nmol/ml	Diene conjugates, nmol/ml	Corticosterone nmol/l
Day 1	C+Str	5.15±0.50*	0.78±0.07*	0.94±0.04	1.53±0.02*	8.52±0.27*	31.49±2.58*	75.00±8.66*
	R+Str	6.15±0.44*	0.93±0.07*	1.08±0.05#	1.68±0.03*	6.29±0.47*	38.35±1.56*	111.70±10.00
Day 2	C+Str	3.87±0.23	0.71±0.04*	0.89±0.03	1.52±0.02*	8.90±0.72*	31.65±1.92*	96.67±21.86*
	R+Str	4.85±0.16*	0.84±0.06*	0.86±0.04	1.70±0.04*	7.77±0.27	32.17±1.71*	81.54±16.50*
Day 3	C+Str	3.42±0.32	0.56±0.03	0.73±0.05*	1.54±0.04*	10.20±0.52	24.72±2.89	35.67±9.49
	R+Str	4.45±0.41	0.70±0.04*	0.89±0.06	1.66±0.04*	8.49±0.38*	22.24±1.84	43.00±8.50
Day 5	C+Str	3.54±0.27	0.49±0.05	0.72±0.02*	1.45±0.06	10.28±0.65	28.34±1.77*	41.50±18.50
	R+Str	3.56±0.23	0.58±0.05	0.88±0.02	1.66±0.04*	10.01±0.27	20.46±2.03	57.10±3.00*
Day 8	C+Str	3.47±0.39	0.49±0.08	0.77±0.02*	1.36±0.03	12.02±0.35	20.47±2.03	42.50±10.61
	R+Str	4.08±0.31	0.4±0.06	0.84±0.03	1.55±0.04	10.04±0.68	20.15±2.61	40.50±9.19
Day 10	C+Str	3.73±0.24	0.54±0.05	0.69±0.02*	1.33±0.02	11.03±0.89	22.68±2.46	34.00±18.38
	R+Str	3.60±0.35	0.55±0.09	0.76±0.04*	1.43±0.04	11.39±0.47	22.90±1.93	38.00±2.82
Day 12	C+Str	3.71±0.35	0.49±0.07	0.65±0.03*	1.26±0.04	11.18±1.01	21.72±1.57	25.50±2.12
	R+Str	3.55±0.45	0.054±0.06	0.75±0.03*	1.35±0.03	11.40±0.93	20.08±1.45	27.00±1.31
Day 15	C+Str	3.78±0.36	0.59±0.04	0.61±0.02*	1.33±0.02	11.04±1.32	20.99±0.92	34.00±7.00
	R+Str	4.22±0.57	0.51±0.09	0.69±0.03*	1.29±0.06	11.72±0.93	20.59±1.92	31.00±10.82

The data presented as mean±SD
* - p≤0.05 versus intact animals

Table 21. The effect of prophylactic administration of grape wines of "Cabernet" (C) and "Rkatsiteli" (R) grades on the plasma parameters of the stress response development in rats with the neurogenic stress (in each group n=10).

However, on day 8 of administration the antioxidant and stress-protective effects of these wines were almost similar, and on the day 10 –they practically did not differ.

On days 12 and 15 there were also no differences as to the antioxidant and stress-protective action of the wines studied, which significantly reduced activation of the free radical oxidation under the action of stress normalizing the most of the indexes investigated.

Thus, the investigated wines are characterized by the high level of the antioxidant and stress-protective activity, and in the first days of introduction "Cabernet" wine improved more effectively the antioxidant status in the blood and the liver tissue than "Rkatsiteli" wine, but by day 10 the effects of the studied wines had no substantial difference.

Probably, these results are dependent on polyphenol cumulation in the organism because it is known that the polyphenol content of "Cabernet" is 10 times more than of "Rkatsiteli".

Thus, the results suggest that "Cabernet" and "Rkatsiteli" wines have already revealed the high stress-protective, hepatoprotective and anti-atherogenic activity in the conditions of the emotional-painful stress on the 2-3 days after introduction, and practically normalized the oxidative status and the lipid metabolism under the action of stress in prophylactic administration within 10 days. This indicates that grape polyphenols possess a high total antioxidant activity. At the same time the last suggestion required further research.

In order to examine the effects of wine stocks and polyphenolic concentrates obtained from other grape grades on development of proatherogenic consequences of the emotional-painful stress we investigated the action of substances obtained from the grapes of hybrid grades "Krasen", "Golubok" and "Podarok Magaracha" produced by the National Institute of Grape and Wine "Magarach".

In the series of experiments we used purebred male rats that during 21 day were given daily, *per os,* table wine stocks of the grades "Podarok Magaracha", "Krasen" and "Golubok" in the human equivalent dose corresponding to 300 ml of wine for a human with 70 kg of the body weight. Other groups of animals were given ethanol in the human equivalent dose corresponding to 30 ml of ethanol for a human with 70 kg of the body weight taking into account the species sensitivity coefficients, as well as the table wine stocks of the grades mentioned in doses equivalent to the polyphenol content of the given wines calculated by the polyphenol content in active doses (AD – 9 mg of polyphenols/100 g of the body weight).

The results have demonstrated that not only polyphenolic concentrates, but the table wine stocks also revealed a substantial stress-protective activity to a different extent (Tables 22-25).

In fact, "Krasen" table wine stock revealed the highest activity; the stress-protective activity was almost 2.4 times more the ethanol activity in the dose studied. This product effectively prevented the activation of free radical oxidation both in the blood (increased the level of compounds with isolated double bonds in the atherogenic ApoB-LP, decreased the content of peroxidation products – diene conjugates – almost 3 times comparing to the stressed animals, and 15% - comparing to the intact animals), and the liver tissue (prevented the antioxidant content decrease, particularly the content of α-tocopherol and ascorbic acid returned practically to the intact level, and there was 40% decrease of the diene conjugates level). At the same time this table wine stock prevented hyperlipidemia and the shift of metabolism to the increased lipolysis, there was 60% decrease of the blood total lipid content comparing to the stressed animals, and 11% - comparing with the intact animals. At the same time the TAG content in the liver was equal to the intact level that also demonstrated the protective action of this table wine stock. Reduction of lipogenesis in the liver tissue under the action of this product is important, and it protects the organ from steatosis. It should be also mentioned that the given product normalized the cholesterol content in the blood plasma.

Group	Parameter				
	Total lipids, mg/g	TAG, mg/g	FFA, mmol/g	ApoB-LP, mg/g	Lysosomal lipase, nmol/mg protein/min
Str.+Con. Podarok Magaracha (AD)	149.03 ±2.59*,**	5.74 ±0.07**	4.46 ±0.04*	4.22 ±0.03*,**	0.45 ±0.03**
Str.+Con. Krasen (AD)	147.13 ±1.15*	4.27 ±0.02*,**	4.19 ±0.05*,**	4.61 ±0.01*,**	0.50 ±0.02
Srt.+Wine Podarok Magaracha	161.74 ±1.91*,**	6.33 ±0.05**	4.30 ±0.11*,**	4.43 ±0.03*,**	0.32 ±0.01*,**
Srt.+Wine Krasen	155.88 ±1.35*,**	6.04 ±0.16**	3.24 ±0.04**	4.50 ±0.07*,**	0.56 ±0.03**
Str.+Con. Podarok Magaracha (DW)	141.29 ±1.79*	4.95 ±0.15*,**	3.83 ±0.09**	3.18 ±0.03*	0.65 ±0.02**
Str.+Con. Krasen (DW)	145.87 ±3.19*	4.24 ±0.07*,**	4.23 ±0.07*,**	4.39 ±0.08*,**	0.55 ±0.03**
Wine Podarok Magaracha	182.4 ±3.08*	6.51 ±0.07**	2.70 ±0.08*,**	4.80 ±0.10*,**	0.37 ±0.01*
Wine Krasen	164.36 ±1.86	5.97 ±0.17	2.95 ±0.09*	4.93 ±0.18*	0.35 ±0.02*
Con. Podarok Magaracha (AD)	191.33 ±2.03*	6.47 ±0.04*	4.61 ±0.31*	4.33 ±0.08*	0.83 ±0.03*
Con. Krasen (AD)	170.4 ±2.09	6.15 ±0.14	3.22 ±0.05	4.97 ±0.11*	0.69 ±0.02#
Ethanol	229.76 ±3.39*	7.40 ±0.13*	4.57 ±0.13*	6.11 ±0.07*	0.38 ±0.02*

The data presented as mean±SD
* - p≤0.05 versus intact animals
** - p≤0.05 versus stressed animals

Table 22. The effect of grape polyphenol concentrates and grape wines on the liver lipid metabolism in rats with the neurogenic stress (in the crude tissue, in each group n=10).

It is also necessary to point out that the control intake of the investigated substances (Tables 22-25) did not reveal negative effects on the organisms of the experimental animals. Moreover, in addition to the antioxidant activity these substances revealed a significant hypocholesterolemic and anti-atherogenic action, which was more pronounced when using "Krasen" grade wine stock and the concentrate.

Group	Parameter				
	GSH, mkmol/g	α-Tocopherol, nmol/g	Ascorbic acid, mkmol/g	Diene conjugates, nmol/g	TBA-active products, nmol/ mg protein
Str.+Con. Podarok Magaracha (AD)	3.34 ±0.02*,**	24.01 ±0.47*,**	1.21 ±0.01*,**	14.96 ±0.22*,**	0.49 ±0,01
Str.+Con. Krasen (AD)	3.8 ±0.02*,**	26.71 ±0.46*,**	1.28 ±0.01*,**	15.03 ±0.11*,**	0,45 ±0.02
Srt.+Wine Podarok Magaracha	3.48 ±0.02*,**	26.71 ±0.34*,**	1.33 ±0.02*,**	13.81 ±0.18#,**	0.42 ±0.03**
Srt.+Wine Krasen	3.57 ±0.08*,**	28.46 ±0.75**	1.41 ±0.03*,**	13.36 ±0.40**	0.21 ±0.01*,**
Str.+Con. Podarok Magaracha (DW)	2.00 ±0.07*,**	21.72 ±0.51*,**	1.02 ±0.03*,**	16.05 ±0.11*,##	0.47 ±0.01
Str.+Con. Krasen (DW)	3.27 ±0.06*	21.12 ±0.39*,**	1.04 ±0.03*,**	16.72 ±0.16*,##	0.45 ±0.02
Wine Podarok Magaracha	4.68 ±0.10*,**	33.59 ±0.60	2.16 ±0.04*	10.42 ±0.53*	0.15 ±0.01*
Wine Krasen	4.56 ±0.24	35.48 ±0.78*	1.57 ±0.03	9.27 ±0.24*	0.13 ±0.01*
Con. Podarok Magaracha (AD)	4.46 ±0.13	35.29 ±0.45*	1.95 ±0.03*	10.87 ±0.41	0.20 ±0.01*
Con. Krasen (AD)	4.82 ±0.14*	27.84 ±0.39	2.00 ±0.04*	10.26 ±0.06*	0.15 ±0.01*
Ethanol	5.31 ±0.35*	24.16 ±1.40*	2.06 ±0.03*	12.62 ±0.60	0.46 ±0.02

The data presented as mean±SD
* - p≤0.05 versus intact animals
** - p≤0.05 versus stressed animals

Table 23. The effect of grape polyphenol concentrates and grape wines on the oxidant/antioxidant status in the liver tissue in rats with the neurogenic stress (in the crude tissue, in each group n=10).

Group	Parameter						
	Total lipides, mg/ml	TAG, g/ml	FFA, mmol/L	Total cholesterol, g/ml	HDL, mg/ml	ApoB-LP, mg/ml	Cortico-sterone, nmol/L
Str.+Con. Podarok Magaracha (AD)	3.89 ±0.08**	0.72 ±0.01*,**	1.40 ±0.02#,**	56.66 ±1.49*,**	0.79 ±0.02*,**	1.45 ±0.05*,**	47 ±2**
Str.+Con. Krasen (AD)	4.12 ±0.08#	0.67 ±0.01*,**	1.16 ±0.03*,**	58.70 ±1.77	0.82 ±0.03	0.,90 ±0.04*,**	47 ±3**
Srt.+Wine Podarok Magaracha	4.60 ±0.11*	0.65 ±0.02*,**	1.63 ±0.02*	63.32 ±1.01**	0.85 ±0.02**	1.21 ±0.02**	41 ±1*,**
Srt.+Wine Krasen	3.42 ±0.09*,**	0.52 ±0.03**	1.40 ±0.04**	64.69 ±1,70**	0.86 ±0.02	1.08 ±0.03#,**	42 ±1**
Str.+Con. Podarok Magaracha (DW)	5.25 ±0.07*	0.74 ±0.02*	0.74 ±0.02*,**	56.67 ±1.15*,**	0.87 ±0.03	1.75 ±0.07*,**	60 ±2
Str.+Con. Krasen (DW)	3.88 ±0.08**	0.67 ±0.01*,**	0.67 ±0.01*,**	55.58 ±1.21*,**	0.83 ±0.03**	1.06 ±0.03*,**	63 ±1*
Wine Podarok Magaracha	3.72 ±0.09	0.38 ±0.01*	1.14 ±0.02*	60.78 ±1.64	1.4 ±0.02*	1.11 ±0.02	54 ±2
Wine Krasen	3.76 ±0.03	0.36 ±0.01*	1.18 ±0.02*	57.21 ±0.81#	1.11 ±0.02*	1.14 ±0.01	44 ±4
Con. Podarok Magaracha (AD)	4.32 ±0.04*	0.56 ±0.02	1.23 ±0.01	68.70 ±1.22	1.04 ±0.02*	1.14 ±0.02	71 ±2*
Con. Krasen (AD)	3.78 ±0.04	0.39 ±0.02*	1.43 ±0.04*	60.89 ±1.67	1.16 ±0.02*	1.20 ±0.02	35 ±2*
Ethanol	4.05 ±0.09	0.74 ±0.03*	1.65 ±0.02*	55.80 ±1.46*	1.24 ±0.06*	1.32 ±0.02*	75 ±3*

The data presented as mean±SD
* - p≤0.05 versus intact animals
** - p≤0.05 versus stressed animals

Table 24. The effect of grape polyphenol concentrates and grape wines on the plasma lipid metabolism parameters and corticosterone level in rats with the neurogenic stress (in each group n=10).

Group	Parameter				
	PON, nmol/ ml×min	Ascorbic acid, mkmol/L	α-Tocopherol, nmol/ml	Isolated double bonds in ApoB-LP	Diene conjugates in ApoB-LP
Str.+Con. Podarok Magaracha (AD)	174 ±2*	37.26 ±1.60*,**	7.77 ±0.21*	2.07 ±0.09*	27.03 +1.30*,#
Str.+Con. Krasen (AD)	194 ±3*,**	41.79 ±0.44*,**	8.25 ±0.66*,**	2.35 ±0.15*	26.86 ±0.43*
Srt.+Wine Podarok Magaracha	173 ±3*	40.08 ±1.72*,**	8.91 ±0.29**	1.92 ±0.25*	18.45 ±0.18*,**
Srt.+Wine Krasen	189 ±2*,**	54.30 ±0.97*,**	10.42 ±0.16*,**	2.47 ±0.03*,**	14.50 ±0.35*,**
Str.+Con. Podarok Magaracha (DW)	159 ±1*	35.46 ±1.10*	7.04 ±0.07*	1.52 ±0.06*	25.73 ±0.68*,**
Str.+Con. Krasen (DW)	179 ±2*,**	48.59 ±0.96*,**	8.03 ±0.06*,**	2.21 ±0.07*	27.68 ±0.60*,**
Wine Podarok Magaracha	223 ±6	75.95 ±0.74*	10.03 ±1.72	5.43 ±0.12*	15.67 ±0.15*
Wine Krasen	215 ±3*	74.79 ±0.34*	11.52 ±0.24*	5.55 ±0.15*	15.81 ±0.15*
Con. Podarok Magaracha (AD)	221 ±4	67.04 ±1.40	9.49 ±0.24	4.83 ±0.11	16.87 ±0.21
Con. Krasen (AD)	255 ±3	70.61 ±2.31	9.97 ±0.16	5.39 ±0.04*	15.47 ±0.34*
Ethanol	236 ±3	79.49 ±2.18*	10.22 ±0.36	5.75 ±0.17*	17.82 ±0.24

The data presented as mean±SD
* - p≤0.05 versus intact animals
** - p≤0.05 versus stressed animals

Table 25. The effect of grape polyphenol concentrates and grape wines on the plasma oxidant/antioxidant status in rats with the emotional-painful stress (in each group n=10).

5. Conclusion

Based on our findings, it is possible to state that antioxidant complexes, particularly polyphenol extracts and the concentrates obtained from *Vitis Vinifera,* which are safe and reveal the potent antioxidant and stress-protective activity, should be used for reduction of proaterogenic states consequences in the complex prophylactic and treatment of atherosclerosis as effective stress-protective remedies.

Thus, administration of *Vitis Vinifera* substances can prevent the increase of the total lipoprotein and ApoB-LP content in the blood, and prevent the free radical process activation in the plasma lipoprotein particles, and, in general, normalize the lipid and lipoprotein metabolism in the liver in metabolic syndrome. These results have proven the ability of the investigated complexes to reduce such negative consequence of metabolic syndrome as development of atherosclerosis.

In addition, according to obtained research data the polyphenolic concentrates possess a potent protective activity both in acute and chronical neurogenic stress.

Our studies suggest that multicomponent active substances with antioxidant properties are more effective in correction of the proatherogenic states caused by stress and metabolic syndrome negative effects in comparison with individual antioxidants (particularly, α-tocopherol). The research data suggest that the increased plasma antioxidant activity alone does not result in decreased foam cell formation, at least in the studied animal model. Moreover, *in vitro* studies have shown that α-tocopherol can be pro-oxidative rather than protective for lipids in isolated LDL. Similarly with vitamin E, vitamin C additives do not offer consistent benefit against atherosclerosis in animals.

The occurrence of tocopherol-mediated peroxidation and the mode of its prevention predicts that the balance of α-tocopherol and available coantioxidants, rather than α-tocopherol alone, determines whether LDL lipid peroxidation occurs in biological systems. Inhibition of the free radical process with the polyphenolic complexes administration can be associated with their ability to increase the level of antioxidants – α-tocopherol, ascorbic acid and reduced glutathione in the test animal liver tissue compared with the group of the stressed animals. The complexes obtained from *Vitis Vinifera,* in particular, polyphenolic concentrates "Enoant" and "Polyphen", as well as grape wines (particularly "Cabernet") with their moderate use revealed the potent antioxidant activity. The preliminary results also suggest that coantioxidants inhibit lipoprotein lipid peroxidation *in vivo*. Thus, if LDL oxidation causes atherosclerosis, the requirement for coantioxidants may explain why supplementation with individual antioxidants, particular vitamin E alone, overall has yielded inconclusive results in the controlled human and animal intervention studies.

In conclusion, our research results may be used for the development of the atherosclerosis prophylaxis strategy, and treatment of diabetes mellitus and metabolic syndrome because recent studies proved insufficient effectiveness of α-tocopherol and advantages of multicomponent antioxidant complexes administration. The high effectiveness of the polyphenolic complexes obtained from *Vitis Vinifera,* including polyphenolic concentrates

"Enoant" (from grape of "Cabernet" grade) and "Polyphen" (from grape of "Rkatsiteli" grade) produced by the National Institute of Grape and Wine "Magarach" has been proven. Our results also confirmed the high effectiveness of the antioxidant complexes from grapes in the correction of the endothelial dysfunction, thus, including these extracts in the treatment schemes is very reasonable. As it would be expected from our observations, increasing the antioxidant oxidant defense by antioxidant supplementation has the ability to restore the endothelial vasomotor function.

An important question to be asked is whether the polyphenol antioxidants exerted their inhibitory effect on lesion progression only because of their antioxidant properties or, possibly, because of additional biological properties, in particular – the phytoestrogen activity.

However, further studies, especially in humans, are required to validate the role of these antioxidants in inhibiting LDL oxidation.

Nevertheless, there are some limitations in the use of the concentrates produced from red grade grapes because of the uric acid content changes.

Author details

Andriy L. Zagayko*, Anna B. Kravchenko,
Mykhaylo V. Voloshchenko and Oxana A. Krasilnikova
Biochemistry Department, National University of Pharmacy, Kharkiv, Ukraine

6. References

[1] Navab M, Anantharamaiah GM, Fogelman AM (2005) The role of high-density lipoprotein in inflammation. Trends Cardiovasc Med. 15:158-161.

[2] Carvalho MD, Vendrame CM, Ketelhuth DF, Yamashiro-Kanashiro EH, Goto H, Gidlund M (2010) High-density lipoprotein inhibits the uptake of modified low- density lipoprotein and the expression of CD36 and FcgammaRI. J. Atheroscler. Thromb. 17: 844-857.

[3] Mumby S, Koh TW, Pepper JR, Gutteridge JM (2001) Risk of iron overload is decreased in beating heart coronary artery surgery compared to conventional bypass. Biochim Biophys Acta. 1537: 204-210.

[4] Park D, Kyung J, Kim D, Hwang SY, Choi EK, Kim YB (2012) Anti-hypercholesterolemic and anti-atherosclerotic effects of polarized-light therapy in rabbits fed a high-cholesterol diet. Lab. Anim Res. 28: 39-46

[5] Kang MK, Chang HJ, Kim YJ, Park AR, Park S, Jang Y, Chung N (2012) Prevalence and determinants of coronary artery disease in first-degree relatives of premature coronary artery disease. Coron. Artery Dis. 2:167-173.

* Corresponding Author

[6] Miyazaki Y, Glass L, Triplitt C, Wajcberg E, Mandarino LJ, DeFronzo RA (2002) Abdominal fat distribution and peripheral and hepatic insulin resistance in type 2 diabetes mellitus. Am. J. Physiol. Endocrinol. Metab. 283: E1135- E1143.

[7] Mayr M. (2006) Oxidized low-density lipoprotein autoantibodies, chronic infections, and carotid atherosclerosis in a population-based study. J. Am. Coll. Cardiol. 47: 2436-2443.

[8] Stocker R, Keaney JF (2004) Role of oxidative modifications in atherosclerosis. Physiol. Rev. 84: 1381 – 1478.

[9] Beckman JA, Creager MA, Libby P (2002) Diabetes and Atherosclerosis: Epidemiology, Pathophysiology, and Management. JAMA. 287: 2570-2570.

[10] Masson D, Jiang XC, Lagrost L, Tall AR (2009) The role of plasma lipid transfer proteins in lipoprotein metabolism and atherogenesis. J. Lipid Res. 50: S201- S206.

[11] Sun AY, Wang Q, Simonyi A, Sun GY (2008) Botanical phenolics and brain health. Neuromolecular Med. 10: 259-274.

[12] Iriti M, Faoro F (2009) Bioactivity of grape chemicals for human health. Nat. Prod. Commun. 4: 611-634.

[13] Walzem RL (2008) Wine and health: state of proofs and research needs. Inflammopharmacology. 16: 265-271.

[14] Goupy P, Bautista-Ortin AB, Fulcrand H, Dangles O (2009) Antioxidant activity of wine pigments derived from anthocyanins: hydrogen transfer reactions to the dpph radical and inhibition of the heme-induced peroxidation of linoleic acid. Agric. Food Chem. 57: 5762-5770.

[15] Goupy P, Bautista-Ortin AB, Fulcrand H, Dangles O (2009) Antioxidant activity of wine pigments derived from anthocyanins: hydrogen transfer reactions to the dpph radical and inhibition of the heme-induced peroxidation of linoleic acid. Agric. Food Chem. 57: 5762-5770.

[16] Dohadwala MM, Vita JA (2009) Grapes and cardiovascular disease. J. Nutr. 139:1788S-17893S.

[17] Chorell E, Svensson MB, Moritz T, Antti H (2012) Physical fitness level is reflected by alterations in the human plasma metabolome. Mol. Biosyst. 8: 1187-1196.

[18] Yamaguchi N, Mezaki Y, Miura M, Imai K, Morii M, Hebiguchi T, Yoshikawa K (2011) Antiproliferative and proapoptotic effects of tocopherol and tocol on activated hepatic stellate cells. J. Nutr. Sci. Vitaminol. 57: 317-325.

[19] Raghavamenon A, Garelnabi M, Babu S, Aldrich A, Litvinov D, Parthasarathy S (2009) Alpha-tocopherolis ineffective in preventing the decomposition of preformed lipid peroxides and may promote the accumulation of toxic aldehydes: a potential explanation for the failure of antioxidants to affect human atherosclerosis. Antioxid. Redox. Signal. 11: 1237-1248.

[20] Mein JR, Lian F, Wang XD. (2008) Biological activity of lycopene metabolites: implications for cancer prevention. Nutr. Rev. 66: 667-683.

[21] Bligh EG, Dyer WJ (1959) A rapid method of total lipid extraction and purification. Can. J. Biochem. Physiol. 37: 911–917.

[22] Makarova OP, Saperova MA, Skurupiy VA (2010) Lipid peroxidation in the liver and lungs in SiO(2)-induced granulomatosis. Bull. Exp. Biol. Med. 149: 702-705.

[23] Gao H, Zhou YW (2005) Anti-lipid peroxidation and protection of liver mitochondria against injuries by picroside II. World J. Gastroenterol. 11: 3671-3674.

[24] Stacewicz-Sapuntzakis M, Bowen PE, Kikendall JW, Burgess M (1987) Simultaneous determination of serum retinol and various carotinoids: their distribution in middle-age men and women. J. Micronutr. Anal. 3: 27-33.

[25] Cahill L, Corey P, El-Sohemy A (2009) Vitamin C deficiency in a population of young Canadian adults. Am. J. Epidemiol. 170: 464–471.

[26] MacKness B, Mackness MI, Durrington PN, Arrol S, Evans AE, McMaster D, Ferrières J, Ruidavets JB, Williams NR, Howard AN (2000) Paraoxonase activity in two healthy populations with differing rates of coronary heart disease. Eur. J. Clin. Invest. 30: 4–10.

[27] Idris CA, Sundram K (2002) Effect of dietary cholesterol, trans and saturated fatty acids on serum lipoproteins in non-human primates. Asia Pac J. Clin. Nutr. 7:S408- S415.

[28] Pulinilkunnil T, Abrahani A, Varghese J, Chan N, Tang I, Ghosh S, Kulpa J, Allard M, Brownsey R, Rodrigues B (2003) Evidence for rapid "metabolic switching" through lipoprotein lipase occupation of endothelial-binding sites. J. Mol. Cell Cardiol. 35: 1093-1103.

[29] Basford JE, Wancata L, Hofmann SM, Silva RA, Davidson WS, Howles PN, Hui DY (2011) Hepatic deficiency of low density lipoprotein receptor-related protein-1 reduces high density lipoprotein secretion and plasma levels in mice. J. Biol. Chem. 286: 13079-13087.

[30] Kanda T, Brown JD, Orasanu G, Vogel S, Gonzalez FJ (2009) PPARgamma in the endothelium regulates metabolic responses to high-fat diet in mice. *J Clin Invest.*;119:110–124.

[31] Belaïd-Nouira Y, Bakhta H, Bouaziz M, Flehi-Slim I, Haouas Z, Ben Cheikh H (2012) Study of lipid profile and parieto-temporal lipid peroxidation in AlCl3 mediated neurotoxicity. modulatory effect of fenugreek seeds. Lipids Health Dis. 11: 16-26.

[32] Dubé JB, Boffa MB, Hegele RA, Koschinsky ML (2012) Lipoprotein(a): more interesting than ever after 50 years. Curr. Opin. Lipidol. 23:133-140.

[33] Tsimikas S, Miller YI (2011) Oxidative modification of lipoproteins: mechanisms, role in inflammation and potential clinical applications in cardiovascular disease. Curr. Pharm. Des. 17: 27-37.

[34] Tani S, Saito Y, Anazawa T, Kawamata H, Furuya S, Takahashi H, Iida K, Matsumoto M, Washio T, Kumabe N, Nagao K, Hirayama A (2011) Low-density lipoprotein cholesterol/apolipoprotein B ratio may be a useful index that differs in statin-treated patients with and without coronary artery disease: a case control study. Int Heart J. 52: 343-347.

[35] Perona JS, Avella M, Botham KM, Ruiz-Gutierrez V (2008) Differential modulation of hepatic very low-density lipoprotein secretion by triacylglycerol-rich lipoproteins derived from different oleic-acid rich dietary oils. Br. J. Nutr. 99: 29-36.

[36] Subramanian S, Chait A (2012) Hypertriglyceridemia secondary to obesity and diabetes. Biochim. Biophys. Acta. 1821: 819-825

[37] Dallinga-Thie GM, Franssen R, Mooij HL, Visser ME, Hassing HC, Peelman F, Kastelein JJ, Péterfy M, Nieuwdorp M (2010) The metabolism of triglyceride-rich lipoproteins revisited: new players, new insight. Atherosclerosis. 211: 1-8.

[38] Chang BH, Li L, Saha P, Chan L (2010) Absence of adipose differentiation related protein upregulates hepatic VLDL secretion, relieves hepatosteatosis, and improves whole body insulin resistance in leptin-deficient mice. J. Lipid Res. 51: 2132-2142.

[39] Annema W, Tietge U (2011) Role of hepatic lipase and endothelial lipase in high-density lipoprotein-mediated reverse cholesterol transport. J.Curr Atheroscler. Rep. 13: 257-265.

[40] Hansen MK, McVey MJ, White RF, Legos JJ, Brusq JM, Grillot DA, Issandou M, Barone FC (2010) Selective CETP inhibition and PPAR alpha agonism increase HDL cholesterol and reduce LDL cholesterol in human ApoB100/human CETP transgenic mice. J. Cardiovasc. Pharmacol. Ther. 15: 196-202.

[41] Salerno AG, Patrício PR, Berti JA, Oliveira HC (2009) Cholesteryl ester transfer protein (CETP) increases postprandial triglyceridaemia and delays triacylglycerol plasma clearance in transgenic mice. Biochem J. 419: 629-634.

[42] Swarbrick MM, Stanhope KL, Elliott SS, Graham JL, Krauss RM, Christiansen MP, Griffen SC, Keim NL, Havel PJ (2008) Consumption of fructose-sweetened beverages for 10 weeks increases postprandial triacylglycerol and apolipoprotein-B concentrations in overweight and obese women. Br. J. Nutr. 100: 947-952.

[43] Benn M, Stene MC, Nordestgaard BG, Jensen GB, Steffensen R, Tybjaerg-Hansen A (2008) Common and rare alleles in apolipoprotein B contribute to plasma levels of low-density lipoprotein cholesterol in the general population. J. Clin. Endocrinol. Metab. 93:1038-1045.

[44] Lewis RM, Hanson MA, Burdge GC (2011) Umbilical venous-arterial plasma composition differences suggest differential incorporation of fatty acids in NEFA and cholesteryl ester pools. Br. J. Nutr. 106: 463-467.

[45] Calabresi L, Franceschini G (2010) Lecithin:cholesterolacyltransferase, high-density lipoproteins, and atheroprotection in humans. Trends Cardiovasc. Med. 20: 50-53.

[46] Kolovou GD, Anagnostopoulou KK, Kostakou PM, Mikhailidis DP (2009) Cholesterol ester transfer protein (CETP), postprandial lipemia and hypolipidemic drugs..Curr Med Chem. 16: 4345-4360.

[47] Eu CH, Lim WY, Ton SH, bin Abdul Kadir K (2010) Glycyrrhizic acid improved lipoprotein lipase expression, insulin sensitivity, serum lipid and lipid deposition in high-fat diet-induced obese rats. Lipids Health. Dis. 29: 79-81.

[48] Storey SM, Atshaves BP, McIntosh AL, Landrock KK, Martin GG, Huang H, Ross Payne H, Johnson JD, Macfarlane RD, Kier AB, Schroeder F (2010) Effect of sterol carrier protein-2 gene ablation on HDL-mediated cholesterol efflux from cultured primary mouse hepatocytes. Am J Physiol Gastrointest Liver Physiol. 299: G244- G254.

[49] Sekiya M, Osuga J, Yahagi N, Okazaki H, Tamura Y, Igarashi M, Takase S, Harada K, Okazaki S, Iizuka Y, Ohashi K, Yagyu H, Okazaki M, Gotoda T, Nagai R, Kadowaki T, Shimano H, Yamada N, Ishibashi S (2008) Hormone-sensitive lipase is involved in hepatic cholesteryl ester hydrolysis. J. Lipid Res. 49: 1829-1838.

[50] Lindi V, Schwab U, Louheranta A, Vessby B, Hermansen K, Tapsell L, Riccardi G, Rivellese AA, Laakso M, Uusitupa MI (2008) The G-250A polymorphism in the hepaticlipase gene promoter is associated with changes in hepatic lipase activity and LDL cholesterol: The KANWU Study. KANWU Study Group. Nutr. Metab. Cardiovasc. Dis. 18: 88-95.

[51] Kolovou GD, Anagnostopoulou KK, Kostakou PM, Mikhailidis DP (2009) Cholesterol ester transfer protein (CETP), postprandial lipemia and hypolipidemic drugs. Curr. Med. Chem. 16: 4345-4360.

[52] Behre HM, Simoni M, Nieschlag E (1997) Strong association between serum levels of leptin and testosterone in men. Clin. Endocrinol. 47: 237–240.

[53] Huang Y, Liu XQ, Rall SC, Taylor JM, von Eckardstein A, Assmann G, Mahley RW (1998) Overexpression and accumulation of apolipoprotein E as a cause of hypertriglyceridemia. J. Biol. Chem. 273: 26388-26393.

[54] Xie X, Zhu Y (2007) Insights into cholesterol efflux in vascular endothelial cells. Cardiovasc Hematol Disord Drug Targets. 7: 127-134.

[55] Miida T, Seino U, Miyazaki O, Hanyu O, Hirayama S, Saito T, Ishikawa Y, Akamatsu S, Nakano T, Nakajima K, Okazaki M, Okada M (2008) Probucol markedly reduces HDL phospholipids and elevated prebeta1-HDL without delayed conversion into alpha-migrating HDL: putative role of angiopoietin-like protein 3 in probucol-induced HDL remodeling. Atherosclerosis. 200: 329-335.

Chapter 20

The Anti-Atherogenic Effects of Lycopene

Amany M. M. Basuny

Additional information is available at the end of the chapter

http://dx.doi.org/10.5772/48134

1. Introduction

Cardiovascular diseases (CVD) are one of the leading causes of death in world. Many epidemiological studies have concluded that a diet rich in fruits and vegetables reduces the incidence of heart disease in humans (Khachik *et al.*, 2002). Carotenoids are important photochemical those are considered to be responsible for the health protective effects of fruits and vegetables (Omoni & Aluko, 2005). The carotenoids are a group of over 600 fat soluble pigments that are responsible for the natural yellow, orange, and red colors of fruits and vegetables (Giovannucci, 2002). Lycopene is one of such carotenoids, and is the pigment principally responsible for the distinctive red color of ripe tomato (*Lycopersicon esculentum*) and tomato products (Shi, 2000). Several epidemiological studies have suggested that a high consumption of tomatoes and tomato products containing lycopene may protect against CVD (Wu *et al.*, 2003). These epidemiological leads have stimulated a number of animal model studies designed to test this hypothesis and to establish the beneficial effects of lycopene. Evidence from these studies suggests that lycopene has anti-atherogenic effects both in vitro and in vivo. The focus of this chapter is the anti-atherogenic effects of lycopene. This chapter will also highlight the chemical composition of lycopene, its sources and function, as well as potential impact an human health.

2. Sources and function of lycopene

Animals and humans do to not synthesize lycopene, and thus depend on dietary sources. Tomatoes and tomato products are the major dietary sources of lycopene. Other sources include watermelon, pink grapefruit, apricots, pink guava and papaya (Willis & Wians, 2003). Lycopene is the most abundant carotenoid in ripe tomatoes, comprising approximately 80-90% of the pigments present. The amount of lycopene in fresh tomatoes depends on the variety, maturity, and environmental conditions in which the fruit matures (Shi, 2000).

Source	Lycopene content (mg/100g wet basis)
Tomatoes fresh	0.72 – 20
Tomato juice	5.00 – 11.60
Tomato sauce	6.20
Tomato paste	5.40 – 15.00
Tomato soup	7.99
Ketchup	9.90 – 13.44
Pizza sauce	12.71
Watermelon	2.30 – 7.20
Pink guava	5.23 – 5.50
Pink grapefruit	0.35 – 3.36
Papaya	0.11 – 5.30
Carrot	0.65 – 0.78
Pumpkin	0.38 – 0.46
Sweet potato	0.02 – 0.11
Apricot	0.01- 0..05

Table 1. shows the lycopene content of tomatoes, some commonly consumed tomato products and other lycopene containing fruits and vegetables.

Lycopene is also widely distributed in the human body. It is one of the major carotenoids found in the human serum (between 21 and 43% of total carotenoids) with plasma levels ranging from 0.22 to 1.06 nmol/ml (Cohen, 2002). It is also found in various tissues throughout the body such as the liver, kidney, adrenal glands, tests, ovaries and the prostate gland (Basu & Imrhan, 2006). Unlike other carotenoids like α-and β-carotene, lycopene lacks the β.:onone rang structure common to other carotenooids (Agarwal & Rao, 2000). Although it lacks provitamine an activity, lycopene is known to be a potent antioxidant (Livny *et al.*, 2002). Reactive oxygen (ROS) species have been implicated in playing a major role in the causation and progression of several chronic diseases. These ROS are highly reactive oxidant molecules that are generated endogenously through regular metabolic activity. They react with cellular components, causing oxidative damage to such critical cellular biomolecules as lipids, proteins and DNA. Antioxidants are protective agents that inactive ROS and therefore, significantly delay or prevent oxidative damage associated with chronic disease risk. Lycopene is one of the most potent antioxidants among the dietary carotenoids and may help lower the risk of chronic diseases including cancer and heart disease.

3. Chemical composition of lycopene

Lycopene is a lipophelic, 40-carbon atom highly unsaturated, straight chain hydrocarbon containing 11 conjugated and 2 non-conjugated double bonds. The all-trans isomer of lycopene is the most predominant isomer in fresh tomatoes and is the most thermodynamically stable from (figure 1). The many conjugated double bonds of lycopene make it a potentially powerful antioxidant, a characteristic believed to be responsible for its beneficial effects. The antioxidant activity of lycopene is high light by its singlet oxygen-quenching property and its ability to trap peroxyl radicals. This singlet quenching ability of lycopene is twice as high as that of β-carotene and 10 times higher than that of α-tocopherol and butylated hydroxyl toluene.

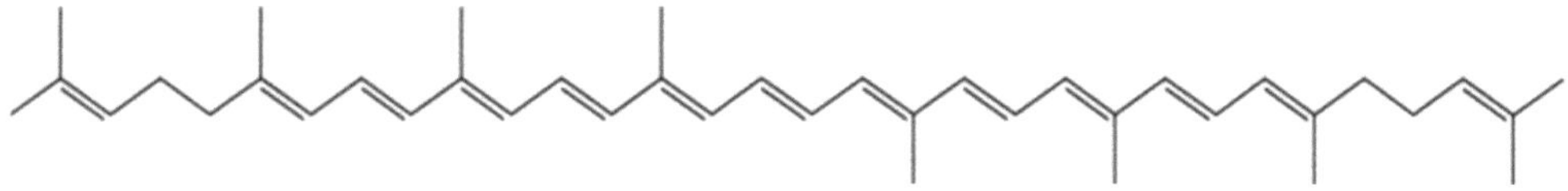

Figure 1. All-*trans* Lycopene.

As a result of the 11 conjugated carbon-carbon double bonds in its backbone, lycopene can theoretically assume 211 or 2048 geometrical configurations (Omani & Aluko, 2005).

However, it is now known that the biosynthesis in plants leads to the all-*trans*-form, and this is independent of its thermodynamic stability. In human plasma, lycopene is an isomeric mixture, containing at least 60% of the total lycopene as cis- isomers (Kim *et al.*, 2012).

All-*trans*, 5-*cis*, 9-*cis*, 13-*cis*, and 15-*cis* are the most commonly identified isomeric forms of lycopene with the stability sequence being 5-*cis*>all-*trans*>9-*cis*>13-*cis*>15- *cis*>7-*cis*>11-*cis*, (Agarwal & Rao, 2000) so that the 5-*cis*-form is thermodynamically more stable than the all-*trans*-isomer. Whereas a large number of geometrical isomers are theoretically possible for all-*trans* lycopene, according to only certain ethylenic groups of a lycopene molecule can participate in *cis-trans* isomerization because of steric hindrance. In fact, only about 72 lycopene *cis* isomers are structurally favorable. Figure 2 illustrates the structural distinctions of the predominant lycopene geometrical isomers.

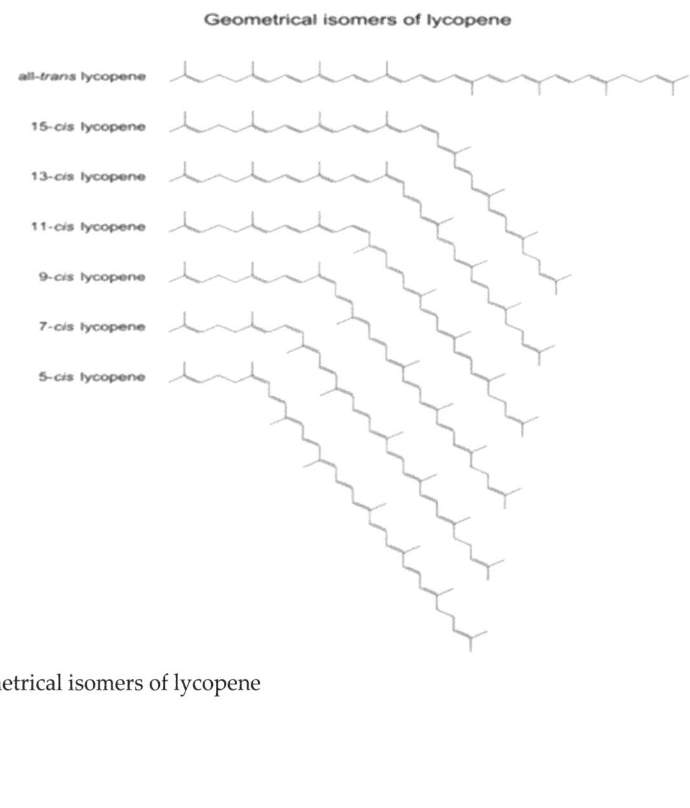

Figure 2. Geometrical isomers of lycopene

4. Mechanisms action of lycopene

A cellular and molecular study have shown lycopene to be one of the most potent antioxidants and has been suggested to prevent atherogenesis by protecting critical bimolecules such as DNA, proteins, lipids and low density lipoproteins (Pool-zobel *et al.*, 1997). Lycopene, because of its high number of conjugated double bonds, exhibits higher singlet oxygen quenching ability compared to β-carotene or α-tocopherol (Di-Mascio *et al.*, 1989). Cis lycopene has been shown to predominate in both benign and malignant prostate tissues, suggesting a possible beneficial effect of high cis-isomer concentrations, and also the involvement of tissue isomerases in vivo isomerization from all trans to cis form (Clinton *et al.*, 1996). Where as Levin *et al.*, (1997) have shown that 9- cis- β-carotene is a better antioxidant than its all-trans counterpart, no such mechanistic data have been reported in case of individual lycopene isomers. Handley *et al.*, (2003) reported a significant increase in 5-cis lycopene concentrations following a 1- week lycopene-restricted diet, and a subsequent reduction in 5-cis, and a concomitant increase in cis-β, cis-D and cis-E lycopene isomers during the 15-day dietary intervention with tomato products in healthy individuals. Although this study reported a decrease in LDL oxidizability due to the intervention with tomato lycopene, the individual antioxidant role of lycopene isomers and their inter conversions remain unclear. At a physiological concentration of 0.3 μmol/1, lycopene has been shown to inhibit growth of non-neoplastic human prostate epithelial cells in vitro, through cell cycle arrest which may be of significant implications in preventing benign prostate hyperplasia, a risk factor for prostate cancer (Obermuller-Jevic *et al.*, 2003). Lycopene has also been shown to significantly reduce LNCaP human prostate cancer cell survival in a dose-dependent manner, and this anti-neoplastic action may be explained by increased DNA damage at high lycopene concentrations (> 5μm), whereas lower levels of lycopene reduced malondialdehyde formation, with no effects on DNA (Hwang & Bowen, 2005). Physiologically attainable concentrations of lycopene have been shown to induce mitochondrial apoptosis in LNCaP human prostate cancer cells, although no effects were observed on cellular proliferation or necrosis (Hantz *et al.*, 2005). Lycopene has also been shown to interfere in lipid metabolism, lipid oxidation and corresponding development of atherosclerosis. Lycopene treatment has been shown to cause a 37% suppression of cellular cholesterol synthesis in J-774A.1 macrophage cell line, and augment the activity of macrophage LDL receptors (Fuhrman *et al.*, 1997). Oxidized LDLs are highly atherogenic as they stimulate cholesterol accumulation and foam cell formation, initiating the fatty streaks of atherosclerosis (Libby, 2006). LDL susceptibility to oxidative modifications is decrease by an acyl analog of platelet-activating (PAF), acyl-PAF, which experts its beneficial role during the initiation and progression of atherosclerosis. Purified lycopene in association with α-tocopherol or tomato lipophillic extracts has been shown to enhance acyl-PAF biosynthesis in endothelial cells during oxidative stress (Balestrieri *et al.*, 2004). Fuhrman *et al.*, (2000) further reported comparative data in which tomato oleoresin exhibited superior capacity to inhibit in vitro LDL oxidation in comparison with pure lycopene by up to fivefold. A combination of purified lycopene (5μmol/I) with α-toopherol in the concentration range of 1-10μmol/I resulted in a significant greater inhibition of in vitro LDL oxidation, than the expected additive individual inhibitions. In this study, purified lycopene was also shown to act synergistically with other natural antioxidants like the flavonoid glabridin, the phenolics rosmarinic acid and carnosic acid, and garlic acid in inhibiting LDL oxidation in vitro. These observations suggested a superior antiatherogeneic characteristic of tomato oleoresin over pure lycopene. The combination of lycopene with other natural antioxidants, as in tomatoes, may be more potent in inhibiting lipid peroxidation, than lycopene per se. The antiatherogenic effects of lycopene are generally believed to be due to its antioxidant properties. Dietary lycopene increases blood and tissue lycopene levels and acting as an antioxidant, lycopene traps reactive oxygen species and reduce the oxidative damage to lipids (lipoproteins and membrane lipids), proteins including important enzymes, and DNA, therapy lowering oxidative stress. This reduced oxidative stress then leads to a reduced risk for chronic diseases associated with oxidative stress such as cardiovascular disease (Omani & Aluko 2005). Alternatively, some non-oxidative mechanisms may be responsible for the beneficial effects of lycopene. The increased lycopene status in the body may regulate gene functions, improve intercellular communication, modulate hormone and immune response, or regulate metabolism, thus lowering the risk for chronic disease (Agarwl & Rao, 2000). A possible mechanism speculated for the protective role of lycopene in heart disease is via the inhibition of cellular HMGCoA reducate, the rate-limiting enzyme in cholesterol synthesis (Fuhrman *et al.*, 1997).

5. Lycopene stability

Being acyclic, lycopene possesses symmetrical planarity and has no vitamin A activity, and as a highly conjugated polyene, it is particularly susceptible to oxidative degradation. Physical and chemical factors known to degrade other carotenoids, including elevated temperature, exposure to light, oxygen, extremes in pH, and molecules with active surfaces that can destabilize the double bonds, apply to lycopene as well (Rao *et al.*, 2003).

In a study to determine the photoprotective potential of dietary antioxidants including lycopene carried out by Handley *et al.*, (2003) carotenoids were prepared in special nanoparticle formulations together with vitamin C and/or vitamin E. The presence of vitamin E in the formulation further increased the stability and cellular uptake of lycopene, which suggests that vitamin E in the nanoparticle, protects lycopene against oxidative transformation. Their findings suggest that lycopene stability may be improved by nanoparticle formulation and incorporation of vitamin E in the lycopene formulation.

Badimon *et al.*, 2010 studied the stability of lycopene during heating and illumination. They carried out various pretreatment steps to the all-trans lycopene standard, which included; dissolving the lycopene standard into hexane and evaporating to dryness under nitrogen in vials, after which a thin film formed at the bottom surface. The resulting lycopene was heated at 50, 100, and 150°C or illuminated at a distance of 30 cm with illumination intensity in the range of 2000–3000 lux (25°C) for varied lengths of time (up to100 hours for heating and 5 days for illumination). After analysis, the degradation of total lycopene (all-*trans* plus *cis* forms) during heating or illumination was found to fit a firstorder model. At 50°C, the isomerization dominated in the first 9 hours; however, degradation was favored afterwards. At 100 and 150°C, the degradation proceeded faster than the isomerization, whereas, during illumination, isomerization was the main reaction. The degradation rate constant (min−1) of lycopene was found to rise with increasing temperature with an activation energy calculated as 61.0 kJ/mol.

The stability of crystalline lycopene was determined under various temperature conditions (5, 25, and 35°C) while stored in airtight containers, sealed under inert gas, and protected from light. After 30 months of storage, crystalline lycopene remained stable when stored under the recommended conditions (Barros *et al.*, 2011).

Lycopene (synthetically prepared by the Wittig reaction) 5% TG (Tablet Grade) and lycopene 10% WS (Water Soluble) beadlet formulations tested for over 24 months of storage, and Lycopene 10% FS (Fluid Suspension) liquid formulation tested for over 12 months of storage under various temperature conditions (5 and 25°C), were all found to be stable.(25) For the 10% WS lycopene beadlet formulations, an important market application form, stability with respect to oxidation under ambient light conditions and room temperature for 12 months in beverages was found to be 93% of the initial content of the beverage lycopene (Pool-zobel *et al.*, 1997).

6. Dietary intake of lycopene

The human body is unable to synthesize carotenoids, which qualifies diet as the only source of these components in blood and tissues. At least 85% of our dietary lycopene comes from tomato fruit and tomato-based products, the remainder being obtained from other fruits such as watermelon, pink grapefruit, guava, and papaya, Tomatoes are an integral part of the human diet and are commonly consumed in fresh form or in processed form such as tomato juice, paste, puree, ketchup, soup, and sauce. Kim *et al.*, (2012) used a tomato products consumption frequency questionnaire to estimate the average daily consumption of different tomato products in the Canadian population.

Di-Mascio *et al.*, (1989) estimated that 50% of the dietary lycopene was obtained from fresh tomatoes, while the average daily intake of lycopene was estimated to be 25 mg in the Canadian population. In a British study conducted with elderly females, the daily consumption of lycopene-rich food, such as tomatoes and baked beans in tomato sauce (measured by weight of foods eaten), was equivalent to a daily lycopene intake of 1.03 mg per person (Omani & Aluko, 2005) developed a database from which the carotenoid intake of the German population, stratified by sex and age, was evaluated on the basis of the German National Food Consumption Survey (NVS). The mean total carotenoid intake amounted to 5.33 mg/day. The average intake of lycopene was 1.28 mg/day with tomatoes and tomato products providing most of the lycopene.

A study presenting data on dietary intake of specific carotenoids in The Netherlands, based on a food composition database for carotenoids, was done by Furhman *et al.*, (1997). Regularly eaten vegetables, the main dietary source of carotenoids, were sampled comprehensively and analyzed with modern analytic methods. The database was complemented with data from literature and information from food manufacturers. Intake of carotenoids was calculated for participants of the Dutch Cohort Study on diet and cancer, aged 55 to 69 in 1986, and the mean intake of lycopene was 1.0 mg/day for men and 1.3 mg/day for women.

6.1. Bioavailability of lycopene

Although 90% of the lycopene in dietary sources is found in the linear, all-trans conformation, human tissues (Particularly liver, adrenal, adipose tissue, testes and prostate) contain mainly cis-isomers. Hollowy *et al.*, (2002) reported that a dietary supplementation of tomato pure for 2 weeks in healthy volunteers led to a completely different isomer pattern of plasma lycopene in these volunteers, versus those present in tomato pure. 5-cis, 13-cis and 9-cis lycopene isomers, not detected in tomato puree, were predominant in the serum (Hollowary *et al.,* 2000).Analysis of plasma lycopene in male participants in the health professionals follow-up study revealed 12 distinct cis-isomers and the total cis-lycopene contributed about 60-80% of total lycopene concentrations (Wu *et al.,* 2003). Studies conducted with lymph cannulated ferrts have shown better absorption of cis-isomers and their subsequent enrichment in tissues (Boileau *et al.,* 1999). Physiochemical studies also suggest that cis-isomer geometry accounts for more efficient incorporation of lycopene into mixed micelles in the lumen of the intestine and into chylomicrons by the enterocyte. Cis-isomers are also preferentially incorporated by the liver into very low-density lipoprotein (VLDL) and get secreted into the blood (Britton, 1995). Research has shown convincing evidence regarding the isomerization of all trans-lycopene to cis-isomers, under acidic conditions of the gastric juice. Incubation of lycopene derived from capsules with simulated gastric juice for 1-min shown a 40% cis-lycopene content, whereas the levels did not exceed 20% even after 3h incubation with water as a control. However, when tomato puree was incubated for 3h with simulated gastric juice, the cis-lycopene content was only 18% versus 10% on incubation with water. Thus, gastric pH and food matrix influence isomerization and subsequent absorption and increased bioavailability of cis-lycopene (Re *et al.,* 2001).

The process of cooking which releases lycopene from the matrix into the lipid phase of the meal increases its bioavailability, and tomato paste and tomato puree are more bioavailable sources of lycopene than raw tomatoes (Gartner *et al.,* 1997 & Porrini *et al.,* 1998). Factors such as certain fibers, fat substituents, plant sterols and cholesterol-lowering drugs can interfere with the incorporation of lycopene into micelles, thus lowering its absorption (Boileau *et al.,* 2002). Several clinical trials have also shown the bioavailability of lycopene from processed tomato products (Table 2). Agarwal and Rao (1998), reported a significant increase in serum lycopene levels following a 1-week daily, consumption of spaghetti sauce (39mg of lycopene), tomato juice (50mg of lycopene) or tomato oleoresin (75 or 150 mg of lycopene), in comparison with the placebo, in healthy human volunteers. There was also indication that the lycopene levels increased in a dose-dependent manner in the case of tomato sauce and tomato oleoresin. Reboul *et al.,* (2005) further demonstrated that enrichment of tomato paste with 6% tomato peel increases lycopene bioavailability in men, thereby suggesting the beneficial effects of peel enrichment, which are usually eliminated during tomato processing. Richelle *et al.,* (2002) compared the bioavailability of lycopene from tomato paste and from lactolycopene formulation (Lycopene from tomato oleoresin embedded in a whey protein matrix), and reported similar bioavailability of lycopene from the two sources in healthy subjects. Dietary fat has been shown to promote lycopene absorption, principally via stimulating bile production for the formation of bile acid micelles. Consumption of tomato products with olive oil or sunflower oil has been shown to produce an identical bioavailability of lycopene, although plasma antioxidant activity improved with olive oil consumption, suggesting a favorable impact of monounsaturated fatty acids on lycopene absorption and its antioxidant mechanism (Lee *et al.,* 2000). In an interesting study Unlu *et al.,* (2005) reported the role of avocado lipids in enhancing lycopene absorption. In this study, in healthy, nonpregnant, nonsmoking adults, the addition of avocado oil (12 or 24g) to salsa (300g) enhanced lycopene absorption, resulting in 4.4 times the mean area under the concentration-versus-time curve after intake of avocado-free salsa. This study demonstrates the favorable impact of avocado consumption on lycopene absorption and has been attributed to the fatty acid distribution of avocados (66.00% oleic acid), which may facilitate the formation of chylomicrons. In a comparative study by Hoppe et al., (2003), both synthetic and tomato –based lycopene supplementation showed similar significant increases of serum total lycopene above baseline whereas no significant changes were found in the placebo group. In an attempt to study lycopene metabolism, Diwadkar-Navsariwala *et al.,* (2003) developed a physiological pharmcokinetic model to describe the disposition of lycopene, administered as a tomato beverage formulation at five graded does (10, 30, 60, 90, or 120 mg) in healthy men. Blood was collected before dose administration and at scheduled study intervals until 672h. The overall results of this study showed that independent of dose, 80% of the subjects absorbed less than 6mg of lycopene, suggesting a possible saturation of absorptive mechanisms. This may have important implications for planning clinical trials with pharmacological doses of lycopene in the control and prevention of chronic disease, if absorption saturation occurs at normally consumed levels of dietary lycopene.

6.2. The anti-atherogenic effects of lycopene

In a previous study (Basuny *et al.,* 2006 and 2009) was to study the effect of tomato lycopene on hypercholesterolemia. Lycopene of tomato wastes was extracted and determination. The level of tomato lycopene was 145.50ppm. An aliquots of the concentrated tomato lycopene, represent 100, 200, 400 and 800ppm; grade lycopene (200ppm) and butylated hydroxyl toluene (BHT, 200ppm) were investigated by 1,1-diphenyl-2-picrylhydrazyl (DPPH) free radical scavenging method. These compounds were administered to rats fed on hypercholestrolemic diet daily from 10 weeks by stomach tube. Serum lipid contents (total lipids, total cholesterol, high density lipoprotein cholesterol and low density lipoprotein cholesterol), oxidative biomarkers (glutathione peroxidase and malonaldhyde), the liver (aspartate aminotransferase, alanine aminotranseferase and alkaline phosphatase activities) and kidney (uric acid, urea and creatinine) function testes were measured to assess the safety limits of the lycopene in tomato wastes. The data of the aforementioned measurements indicated that the administration of tomato lycopene did not cause any changes in liver and kidney functions. On the contrary, rats fed on hypercholesterolemic diet induced significant increases in the enzymes activities and the serum levels of total lipids, total cholesterol and low and high density lipoproteins cholesterol and decreased levels of the glutathione peroxidase and malonaldhyde. In conclusion, presently available data from epidemiological and a number of animal studies have provided evidence to suggest that lycopene, the naturally present carotenoid in tomatoes and other fruits and vegetables, possesses anti-atherogenic effects. However, there is a need for more human dietary intervention studies in order to better understand the role of lycopene in human health.

Scientific evidence indicates that oxidation of low density lipoprotein (LDL), which carry cholesterol in the blood stream plays an important role in the development of atherosclerosis, the underlying disorder leading to heart attacks and ischemic strokes (Rao, 2002). Several studies indicate that consuming the antioxidant lycopene that is contained in tomatoes and tomato lycopene products can reduce the risk of cardiovascular diseases (CVD). Available evidence from the Kuopio Ischaemic Heart Disease Risk Factor (KIHD) study suggests that the thickness of the innermost wall of blood vessels and the risk of myocardial infarction reduced in persons with higher serum and adipose tissue concentrations of lycopene (Rissanen *et al.*, 2003). This finding suggests that the serum lycopene concentration may play a role in the early stages of atherosclerosis. A thick artery wall is a sign of early atherosclerosis, and increased thickness of the intima media has been shown to predict coronary events. Similarly, the relationship between plasma lycopene concentration and intima-media thickness of the common carotid artery wall (CCA-IMT) was investigation in 520 middle-aged men and women 45-69 years as parts of the Antioxidant Supplementation in Atherosclerosis Prevention (ASAP) study (Rissanen *et al.*, 2000). Low levels of plasma lycopene were associated with a 17.80% increment in CCA-IMT in men, while there was no significant difference among women. These findings also suggest that low plasma lycopene concentrations are associated with early atherosclerosis, evidenced by increased CCA-IMT in middle-aged men.

Findings from the Rotterdam Study (Klipstein-Grobusch *et al.*, 2000) showed modest inverse associations between levels of serum lycopene and atherosclerosis, assessed by the presence of calcified plaques in the abdominal aorta. Study population comprised of 108 cases of aortic atherosclerosis and 109 controls aged 55 years and over. The association between serum lycopene levels and atherosclerosis was most pronounced among subjects who were current and former smokers. No association with risk of aortic calcification for the serum carotenoids α-carotene, β-carotene, lutein and zeaxanthin was observed. These results suggest that lycopene may play a protective role in the development of atherosclerosis. Results from the European Study of Antioxidant, Myocardial Infarction, and Cancer of the breast (the EURAMIC study) also show that men with the highest concentration of lycopene in their adipose tissue biopsy had a 48% reduction in risk of myocardial information compared with men with the lowest adipose lycopene concentrations (Kohlmeir *et al.*, 1997). An increase in LDL oxidation is known to be associated with an increased risk of atherosclerosis and coronary heart disease (Parthasarathy, 1998). Agarwal and Rao (1998) investigated the effect of dietary supplementation of lycopene on LDL oxidation in 19 healthy human subjects. Dietary lycopene was provided using tomato juice, spaghetti sauce and tomato oleoresin for a period of 1 week each. Blood samples were collected at the end of each treatment, and TBARS and conjugated dienes were measured to estimate LDL oxidation. In addition to significantly increasing serum lycopene levels by a least twofold, lycopene supplementation significantly reduced serum lipid peroxidation and LDL oxidation. The average decrease of LDL –TBARS and LDL-conjugated diene for the tomato products treatment over placebo was 25 and 13%, respectively. These results suggest significance for lycopene in decreasing risk for coronary heart disease. Results from the ongoing Women's Health Study (WHS) showed that women with the highest intake of tomato-based foods rich in lycopene had a reduced risk for CVD compared to women with a low intake of those foods (Sesso *et al.,* 2003). Results showed that women who consumed seven servings or more of tomato based foods like tomato sauce and pizza each week had a nearly 30% risk reduction in total CVD compared to the group with intakes of less than one serving per week. The researchers also found out that women who ate more than 10 servings per week had an even more pronounced reduction in risk (65%) for specific CVD outcomes such as heart attack or stroke. Though not statistically significant, the strongest association of dietary lycopene with CVD protection was seen among women with a median dietary lycopene intake of 20.20 mg/day, who had a 33% reduction in risk of the disease when compared with women with the lowest dietary lycopene intake (3.3 mg/day).

Lycopene has also been shown to have a hypercholesterolemic effect both in vivo and in vitro. In a small dietary supplementation study, six healthy male subjects were fed 60 mg/day lycopene for 3 months. At the end of the treatment period, a significant 14% reduction in plasma LDL cholesterol levels was observed in vivo with no effect on HDL cholesterol concentration (Fuhrman *et al.,* 1997) & Lorenz *et al.,* 2012).

6.3. Safety of lycopene

The safety issue for carotenoids attracted much attention after the publication of the β-carotene supplementation trials, which yielded negative results. It is interesting that in thus studies an increased risk for lung cancer was related to a 12- and 16 fold increase in β-carotene plasma levels due to supplementation. β-carotene plasma levels increased from 0.32μml before supplementation up to 3.90 and 5.90 μm, respectively. Rao *et al.,* (2003), which showed no effect for β-carotene supplementation, only a 5-fold increase in the carotenoid serum level was achieved. Interestingly, the only study with positive results after supplementation with β-carotene was achieved in linxian, a chinese community with very low carotenoid levels (0.11μm) before the intervention (Jonker et al., 2003). Although supplementation caused an 11-fold increase in β-carotene level, the final concentration of β-carotene reached was a relatively low 1.5μm. Interestingly, reviewing many studies which measured serum levels of β-carotene and lycopene after supplementation suggests that β-carotene serum levels are significantly higher than those found for lycopene. Serum levels reached for β-carotene are around 3μm and may exceed 5μm after supplementation; on the other hand lycopene levels above 1.2μm are rarely seen even after long-term application. Moreover, the serum level achieved for lycopene was not directly correlated to the amount of the supplementation carotenoid (Nahum *et el.,* 2001). For example, supplemented as high as 75 mg/day did not increase lycopene serum levels more than 1μm (Agarwl & Rao 1998). In conclusion, by some unknown mechanism, lycopene plasma levels after supplementation remain relatively low, which may provide a safety value.

6.4. Lycopene relationship with other micronutrients

When reviewing data related to the chemoprevention of various diseases, it become evident that the use of a single carotenoid, or any other micronutrient which has been successful in vitro and animal models, does not prove as favorable in human intervention studies. That is, there is no magic bullet. In fact, accumulating evidence suggests that a concerted, synergistic action of various micronutrients is, more likely to be the basis of the disease-prevention activity of a diet rich in vegetables and fruits. Indeed, the sources of lycopene used in most of the human studies reviewed there were either prepared tomato products or tomato extracts containing lycopene and other tomato micronutrients and carotenoids in various proportions. Pure lycopene has not been tested as a single in human prevention studies. On the other hand, many studies showing the beneficial effect of lycopene in alleviating chronic conditions have been conducted in which the subjects were provided with tomato-based foods, or tomato extracts, but not with the pure compound. For example, the oleoresin preparation used in many of these studies also contained other tomato carotenoids such as phytoene, phytofluene and β-carotene (Amir *et al.*, 1999; Pastori *et al.*, 1998 & Stahl *et al.*, 1998). In a recent study (Bioleau *et al.*, 2003) that compared the potency of freeze-dried whole tomatoes (tomato powder) or pure lycopene in a rat model of prostate cancer. Rats were treated with the carcinogen (N-methyl1-N-nitrosourea) combined with androgens to stimulate prostate carcinogenesis, and the ability of these two preparations containing lycopene to enhance survival was compared. Mortality with prostate cancer was lower by 25 % (p- 0.09) for rats fed the tomato powder diet than for rats fed control feed. Prostate cancer morality of rats fed our lycopene was similar to that of the control group. The authors concluded that consumption of tomato powder but not pure lycopene inhibited prostate carcinogenesis, suggesting that tomato products contain other compounds, besides lycopene, that modify prostate carcinogenesis.

6.5. Epidemiologic studies: lycopene and cardiovascular diseases

Epidemiological observations also report an inverse association between plasma of tissue lycopene levels and the incidence of cardiovascular diseases. In the Kuopio Ischemic Heart Disease Risk Factor Study, lower levels of plasma lycopene were seen in men who had a coronary event compared with men who did not. In addition, a higher concentration of serum lycopene was inversely correlated with a decrease in the mean and maximal intima-mediated thickness of the common carotid artery (CCA-IMT) with lo lycopene, resulting in an 18% increase in CCA-IMT (Rissanen *et al.*, 2003). The European Multiccnter Case-Control Study on antioxidants, Myocardial Infarction and Breast Cancer Study (EURAMIC Study) reported that a higher lycopene concentration was independently protective against cardiovascular diseases (Basu & Imrhan 2006). The Women's Health Study further revealed that a decreased risk for developing cardiovascular diseases was more strongly associated with higher tomato intake than with lycopene intake (Sesso *et al.*, 2003). Processed tomato products definitely provide a bioavailability source of lycopene and have a positive correlation with plasma and tissue lycopene levels. However, these studies do not suggest a role of lycopene perse, in reducing the risks for cardiovascular diseases, as plasma level of lycopene, in epidemiologic studies, only reflects the consumption of tomato and tomato products.

7. Conclusion

Thus, it can be concluded that moderate amounts of whole food-based supplementation (2–4 servings) of tomato soup, tomato puree, tomato paste, tomato juice or other tomato beverages, consumed with dietary fats, such as olive oil or avocados, leads to increases in plasma carotenoids, particu- larly lycopene. The recommended daily intake of lycopene has been set at 35 mg that can be obtained by consuming two glasses of tomato juice or through a combination of tomato products (Rao and Agarwal, 2000). These foods may have both chemopreventive as well as chemotherapeutic values as outlined in Figure 3. In the light of recent clinical trials, a combination of naturally occurring carotenoids, including lycopene, in food sources and supplements, is a better approach to disease prevention and therapy, versus a single nutrient. Lycopene has shown distinct antioxidant and anticarcinogenic effects at cellular levels, and definitely contributes to the health benefits of consumption of tomato products. However, until further research establishes sig- nificant health benefits of lycopene supplementation per se, in humans, the conclusion may be drawn that consumption of naturally occurring carotenoid-rich fruits and vegetables, particularly processed tomato products containing lycopene, should be encouraged, with positive implications in health and disease.

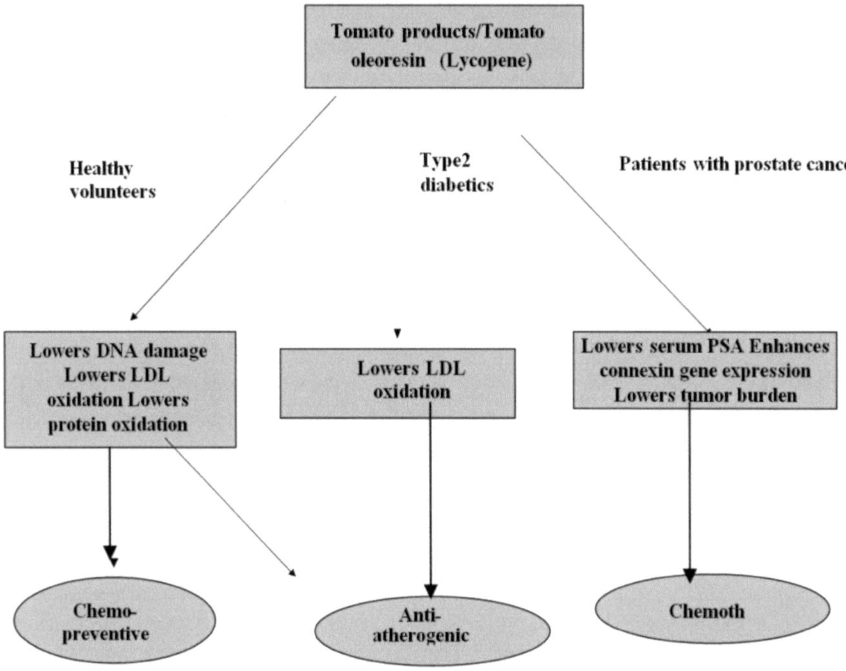

Figure 3. Summary of mechanisms of action of tomato products or tomato oleoresin supplementation, containing lycopene, in health and disease.

		Type and duration of lycopene supplementation	Effects on biomarkers of oxidative stress/ carcinogenesis	
Agarwal and Rao (1998)	19 healthy subjects (mean age 29 years, BMI 2472.8 kg/m^2)	0 mg lycopene (placebo), 39 mg lycopene (spaghetti sauce), 50 mg lycopene (tomato juice), or 75 mg lycopene (tomato oleoresin) per day for 1 week	25% decrease in LDL-TBARS 13% decrease in LDL-CD for all groups versus placebo (P<0.05)	Increase at 7 days in all groups versus placebo (P<0.05)
Riso et al. (1999)	10 healthy subjects (mean age 23.171.1 years, BMI 20.571.5 kg/m^2)	16.5 mg lycopene (60 g tomato puree), per day for 21 days	38% decrease in DNA damage in lymphocytes (P<0.05)	Increase at 21 days versus baseline (P<0.001)
Bub et al. (2000)	23 healthy volunteers (mean age 3474 years, BMI 2372 kg/m^2)	40 mg lycopene (330 ml tomato juice) for 2 weeks	12% decrease in plasma TBARS 18% increase in LDL lag time (P<0.05) no effects on water-soluble antioxidants, FRAP, glutathione peroxidase and reductase activities (P<0.05)	Increase at 2 weeks versus baseline (P<0.05)
Chopra et al. (2000)	34 healthy females (mean age 37.578.5 years, BMI 2473.5 kg/m^2)	440 mg lycopene (200 g tomato puree þ 100 g watermelon) per day for 7 days	Significant decrease in LDL oxidizability in nonsmokers (P<0.05); no effects in smokers (P<0.05)	Increase at 7 days versus baseline (P<0.05)
Porrini and Riso (2000)	9 healthy subjects (mean age 25.472.2years, BMI 20.371.5 kg/m^2)	7 mg lycopene (25 g tomato puree), per day for 14 days	50% decrease in DNA damage in lymphocytes (P<0.05)	Increase at 14 days versus baseline (P<0.001)
Upritchard et al. (2000)	15 well-controlled type II diabetics (mean age 6378years, BMI30.977 kg/m^2)	Tomato juice (500 ml) per day or placebo for 4 weeks	Decreased LDL oxidizability versus baseline (P<0.001)	Increase at 4 weeks versus baseline (P<0.001)
Hininger et al. (2001)	175 healthy volunteers (mean age 33.571 years, BMI-24.370.5 kg/m^2)	15 mg lycopene (natural tomato extract) or placebo per day for 12 weeks	No effects on LDL oxidation, reduced glutathione, protein SH groups and antioxidant metalloenzyme activities (P<0.05)	Increase at 12 weeks versus baseline (P<0.05)
Chen et al. (2001)	32 patients with localized prostate adenocarcinoma (mean age 63.776.1 years, BMI 28.074.9 kg/m^2)	30 mg lycopene (200 g spaghetti sauce) per day for 3 weeks before surgery or a reference group with no supplementation	Decreased leukocyte and prostate tissue oxidative DNA damage; decreased serum PSA levels (P<0.05)	Increase at 3 weeks versus baseline (P<0.001)
Kucuk et al. (2001)	26 patients with newly diagnosed, clinically localized prostate cancer (mean age 62.1571.85 years, BMI not reported)	15 mg lycopene (Lyc-O-Mato capsules) twice daily or no supplementation for 3 weeks before surgery	Decreased tumor growth in the intervention group versus control(P<0.05); decreased plasma PSA levels and increased expression of connexin43 in prostate tissue in the intervention group versus control (P<0.05);decreased plasma IGF-1 levels in intervention and control groups(P<0.05)	No effects at 3 weeks versus baseline (P<0.05)

		Type and duration of lycopene supplementation	Effects on biomarkers of oxidative stress/ carcinogenesis	
Porrini et al. (2002)	9 healthy subjects (mean age 25.272.2 years, BMI 20.271.6 kg/m^2)	7 mg lycopene (25 g tomato puree) with 150 g of spinach and 10 g of olive oil per day for 3 weeks	Decreased DNA oxidative damage ($P<0.05$)	Not reported

Table 2. Summary of clinical trials investigating the effects of supplementation of tomato products, tomato oleoresin or purified lycopene on biomarkers of oxidative stress and Carcinogenesis

Author details

Amany M. M. Basuny
Department of Fats & Oils, Food Technology Research Institute,
Agriculture Research Centre, Giza, Egypt

8. References

Agarwal, A. & Rao, A. (1998): Tomato lycopene and low density lipoprotein oxidation: a human dietary intervention study. Lipids, 33: 981-984.

Agrawal, S. & Rao, V. (2000): Tomato lycopene and its role in human health and chronic diseases. Canadian Medical Association Journal, 163: 739-744.

Amir, H.; Karas, M. & Giat, J. (1999): Lycopene 1, 25 di-hydroxy vitamin-D3 cooperate in the inhibition of cell cycle progression and induction of differentiation in Hl-60 leukemic cells. Nutrition Cancer, 33: 105-112.

Badiman, L.; Vilahur, G. & Padro, T. (2010): Nutraceuticals and atherosclerosis: Human trials. Cardiovascular Therabeutics, 28: 202-215.

Barros, L.; Carbrita, L.; Boas, M.; Carvaiho, A. & Ferreira, I. (2011): Chemical, biochemical and electrochemical assays to evaluate phytochemicals and antioxidant activity of wild plants. Food Chemistry, 127: 1600-1608.

Basu, A. & Imrhan, V. (2006): Tomato versus lycopene in oxidative stress and carcinogenesis: conclusions from clinical trials. European Journal OF Clinical Nurition, 1-9.

Basuny, A. M.; Mostafat, D. M. & Azouz, A. (20060: Supplementation of polyunsaturated oils with lycopene as natural antioxidant and antipolymerization during heating process. Minia Journal of Agricultural Research and Development, 26: 449-469.

Basuny, A. M.; Gaafar, A. M. & Arafat, S. M. (2009): Tomato lycopene is a natural antioxidant and cn alleviate hypercholesterolemia. African Journal of Biotechnology, 23: 6627-6633.

Boileau, T. W.; Liao, Z.; Kim, S.; Lemeshow, S.; Erdman, J. & Clinton, S. (2003): Prostate carcinogensis in N-methyl-N-nitrosourea (NMW-testosterone-treated rats fed tomato powder, lycopene, or energy-restricted diets. J Natl. Cancer Inst. 95: 1578-1586.

Boileau AC, Merchen NR, Wasson K, Atkinson CA, Erdman JW (1999). cis-Lycopene is more bioavailable than trans-lycopene in vitro and in vivo in lymph-cannulated ferrets. J Nutr 129, 1176–1181.

Boileau TWM, Boileau AC, Erdman JW (2002). Bioavailability of alltrans and cis-isomers of lycopene. Exp Biol Med 227, 914–919.

Britton, G. (1995): Structure and properties of carotenoids in relation to function. FASEB J 9, 1551–1558.

Briviba, K.; Schnabele, K.; Rechkemmer, G.; Bub, A. (2004): Supplementa- tion of a diet low in carotenoids with tomato or carrot juice does not affect lipid peroxidation in plasma and feces of healthy men. J Nutr 134, 1081–1083.

Bub, A.; Watzl, B.; Abrahamse, L.; Delincee, H.; Adam, S. & Wever, J. (2000): Moderate intervention with carotenoid-rich vegetable products reduces lipid peroxidation in men. J Nutr 130, 2200–2206.

Bub, A.; Barth, S. W.; Watzl, B.; Briviba, K. & Rechkemmer, G. (2005): Araoxonase 1 Q192R (PON1-192) polymorphism is associated with reduced lipid peroxidation in healthy young men on a low- carotenoid diet supplemented with tomato juice. Br J Nutr 93, 291–297.

Chen, L.; Stacewicz-Sapuntzakis, M.; Duncan, C.; Sharifi, R.; Ghosh, L. & Van Breemen, R. (2001): Oxidative DNA damage in prostate cancer patients consuming tomato sauce-based entrees as a whole-food intervention. J Natl Cancer Inst 93, 1872–1879.

Chopra, M.; O'Neill, M. E.; Keogh, N.; Wortley, G.; Southon, S. & Thurnham, D. I. (2000): Influence of increased fruit and vegetable intake on plasma and lipoprotein carotenoids and LDL oxidation in smokers and nonsmokers. Clin Chem 46, 1818–1829.

Clinton, S. K.; Emenhiser, C.; Schwartz, S. J.; Bostwick, D. G.; Williams, A. W. & Moore, B. J. (1996): Cis–trans lycopene isomers, carotenoids, and retinol in the human prostate. Cancer Epidemiol Biomarkers Prev 5, 823–833.

Cohen, L. (2002): A review of animal model studies of tomato carotenoids, lycopene and cancer chemoprevention. Experimental Biology and Medicine, 277: 864-868.

Di Mascio, P.; Kaiser, S. & Sies, H. (1989): Lycopene as the most efficient biological carotenoid singlet oxygen quencher. Arch Biochem Biophys 274, 532–538.

Diwadkar-Navsariwala, V.; Novotny, J. A.; Gustin, D. M.; Sosman, J. A.; Rodvold, K. A. & Crowell, J. A. (2003): A physiological pharmacokinetic model describing the disposition of lycopene in healthy men. J Lipid Res 44, 1927–1939.

Fuhrman, B.; Elis, A. & Aviram, M. (1997): Hydpocholesterolemic effect of lycopene and β-carotene is related to suppression of cholesterol synthesis and augmentation of LDL receptor activity in macrophages-Biochemical and Biophysical Research Communications, 233: 658-662.

Gartner, C.; Stahl, W. & Sies, H. (1997): Lycopene is more bioavailable from tomato paste than from fresh tomatoes. Am J Clin Nutr 66, 116–122.

Giovannucci, E. (2002): A review of epidemiologic studies of tomatoes, lycopene and prostate cancer. Experimental Biology and Medicine, 227: 852-859.

Hadley, C. W.; Clinton, S. K. & Schwartz, S. J. (2003): The consumption of processed tomato products enhances plasma lycopene concentrations in association with reduced lipoprotein sensitivity to oxidative damage. J Nutr 133, 727–732.

Hantz, H. L.; Young, L. F.; Martin, K. R. (2005): Physiologically attainable concentrations of lycopene induce mitochondrial apoptosis in LNCaP human prostate cancer cells. Exp Biol Med 230, 171–179.

Hininger, I. A.; Meyer-Wenger, A.; Moser, U.; Wright, A.; Southon, S. & Thurnham, D. (2001): No significant effects of lutein, lycopene or b-carotene supplementation on biological markers of oxidative stress and LDL oxidizability in healthy adult subjects. J Am Coll Nutr 20, 232–238.

Holloway, D. E.; Yang, M.; Paganga, G.; Rice-Evans, C. A. & Bramley, P. M. (2000): Isomerization of dietary lycopene during assimilation and transport in plasma. Free Radical Res 32, 93–102.

Hoppe, P. P.; Kramer, K.; Van den Berg, H.; Steenge, G. & Vliet, T. (2003): Synthetic and tomato-based lycopene have identical bioavailability in humans. Eur J Nutr 42, 272–278.

Hwang, E. S. & Bowen, P. E. (2005): Effects of lycopene and tomato paste extracts on DNA and lipid oxidation in LNCaP human prostate cancer cells. Biofactors 23, 97–105.

Jonker, D.; Kuper, C.; Fraile, N.; Estrella, A. & Otero, C. (2003): Ninety-day oral toxicity study of lycopene from Blakeslea trispora in rats. Regul Toxicol Pharmacology, 37: 396-406.

Kim, Y.; Park, Y.; Lee, K.; Jeon, S.; Gregor, R. & Choi, S. (2012): Dose dependent effects of lycopene enriched tomato wino on liver and adipose tissue in high fat diet fed rats. Food Chemistry, 130: 42-48.

Kiokias, S. & Gordon, M. H. (2003): Dietary supplementation with a natural carotenoid mixture decreases oxidative stress. Eur J Clin. Nutr 57, 1135–1140.

Klipstein-Grobusch, K.; Launer, L.; Geleijnse, J.; Boeing, H.; Hofman, A. & Wtteman, J. (2000): Serum caroteniods and atherosclerosis. The Rotterdam study. Atherosclerosis, 148: 49-56.

Khachik, F.; Carvalho, L.; Bernstein, P.S; Muir, G.; Zhao, D. & Katz, N. (2002): Chemistry, distribution and metabolism of tomato carotenoids and their impact on human health. Experimental Biology and Medicine, 227: 845-851.

Kohlmeir, L.; Kark, J.; Gomez-Garcia, E.; Martin, B.; Steck, S. & Kardinaal, A. (1997): Lycopene and myocardial infraction risk in the EURAMIC study. American Journal of Epidemiology, 146: 618-626.

Kucuk, O.; Sarkar, F. H.; Sakr, W.; Djurie, Z.; Pollak, M. N. & Khachik, F. (2001): Phase II randomized clinical trial of lycopene supplemen- tation before radical prostatectomy. Cancer Epidemiol Biomarkers Prev 10, 861–868.

Lee, A.; Thurnham, D. & Chopra, M. (2000): Consumption of tomato products with olive oil but not sunflower oil increases the antioxidant activity of plasma. Free Radical Biol Med 29, 1051–1055.

Levin, G.; Yeshurun, M. & Mokady, S. (1997): In vivo antiperoxidative effect of 9-cis b-carotene compared with that of the all-trans isomer. Nutr Cancer 27, 293–297.

Libby, P. (2006): Inflammation and cardiovascular disease mechanisms. Am J Clin Nutr 83, 456S–460S.

Liu, C.; Russell, R. M. & Wang, X. D. (2006): Lycopene supplementation prevents smoke-induced changes in p53, p53 phosphorylation, cell proliferation, and apoptosis in the gastric mucosa of ferrets. J Nutr 136, 106–111.

Lorenz, M.; Fechner, M.; Kalkowski, J.; Frohlich, K.; Trautman, A.; Bohm, V.; Liebisch, G.; Lehneis, S.; Schmitz, G.; Ludwing, A.; Baumann, G.; Stangl, K. & Stangle, V. (2012): Effects of lycopene on the initial state of atherosclerosis in New Zealand white rabbits. PLoS one 7: 1-8.

Livny, O.; Kaplan, I.; Reifen, R.; Polak, S.; Madar, Z. & Schwartz, B. (2002): Lycopene inhibits proliferation and enhances gap-junctional communication of KB-1 human oral tumor cells. Journal of Nutrition, 132: 3754-3759.

Nahum, A.; Hirsch, K. & Danilenko, M. (2000): Lycopene inhibition of cell cycle progression in breast and endometrial cancer cells in associated with reduction in cyclin D levels and retention of P 27 in the cyclin E- cdk 2 complexes. Oncogene, 26: 3428-3436.

Omoni, O. & Aluko, R. (2005): The anticarcinogenic and antiatherogenic effects of lycopene: a review. Trends in Food Science & Technology, 16: 344-350.

Parthasarathy, S. (1998): Mechanisms by which dietary antioxidants may prevent cardiovascular diseases. Journal of Medicinal Food, 1: 45-51.

Paster, M.; Fander, H.; Boscoboinik, D. & Azzi, A. (1998): Lycopene in association with α-tocopherol inhibits at physiological concentrations proliferation of prostate carcinoma cells. Biochemistry Biophysics Research communication, 35: 582-585.

Obermuller-Jevic, U. C.; Olano-Martin, E.; Corbacho, A. M.; Eiserich, J. P. Van der Vliet. A. & Valacchi, G. (2003): Lycopene inhibits the growth of normal human prostate epithelial cells in vitro. J Nutr 133, 3356–3360.

Pool-Zobel, B. L.; Bub, A.; Muller, H.; Wollowski, I. & Rechkemmer, G. (1997): Consumption of vegetables reduces genetic damage in humans: first result of a human intervention trial with carotenoid-rich foods. Carcinogenesis 18, 1847–1850.

Porrini, M.; Riso, P. & Testolin, G. (1998): Absorption of lycopene from single or daily portions of raw and processed tomato. Br J Nutr 80, 353–361.

Porrini, M. & Riso, P. (2000): Lymphocyte lycopene concentration and DNA protection from oxidative damage is increased in women after a short period of tomato consumption 130, 189–192.

Porrini, M.; Riso, P. & Oriani, G. (2002): Spinach and tomato consumption increases lymphocyte DNA resistance to oxidative stress but this is not related to cell carotenoid concentrations. Eur J Nutr 41, 95-100.

Porrini, M.; Riso, P.; Brusamolino, A.; Berti, C.; Guarnieri, S. & Visioli, F. (2005): Daily intake of a formulated tomato drink affects carotenoid plasma and lymphocyte concentrations and improves cellular antioxidant protection. Br J Nutr 93, 93–99.

Rao, A. V. & Shen, H. (2002): Effect of low dose lycopene intake on lycopene bioavailability and oxidative stress. Nutr Res 22, 1125-1131.

Rao, G.; Guns, E. & Rao, A. (2003): Lycopene: Its role in human health and disease. Agro Food Industry In Tech, 8: 25-30.

Rao, A. V. (2004): Processed tomato products as a source of dietary lycopene: bioavailability and antioxidant properties. Can J Diet Pract Res 65, 161–165.

Reboul, E. Borel, P.; Mikail, C.; Abou, L.; Charbonnier, M. & Caris-Veyrat. C. (2005): Enrichment of tomato paste with 6% tomato peel increases lycopene and b-carotene bioavailability in men. J Nutr 135, 790–794.

Re, R.; Fraser P. D.; Long M, Bramley P. M. & Rice-Evans C. (2001): Isomerization of lycopene in the gastric milieu. Biochem Biophys Res Commun 281, 576–581.

Richelle, M.; Bortlik, K.; Liardet, S.; Hager, C.; Lambelet, P. & Baur, M. (2002): A food-based formulation provides lycopene with the same bioavailability to humans as that from tomato paste. J Nutr 132, 404–408.

Riso, P.; Pinder, A.; Santangelo, A. & Porrini, M. (1999): Does tomato consumption effectively increase the resistance of lymphocyte DNA to oxidative damage? Am J Clin Nutr 69, 712–718.

Riso, P.; Visioli, F.; Erba, D.; Testolin, G. & Porrini, M. (2004): Lycopene and vitamin C concentrations increased in plasma and lymphocytes after tomato intake. Effects on cellular antioxidant protection. Eur J Clin Nutr 58, 1350–1358.

Rissanen, T.; Voutilainen, S.; Nyyssonen, K.; Salonen. & Salonen J. T. (2000): Low plasma lycopene concentrations is associated with increased intima-media thickness of the carotid artery wall. Arteisclerosis, Thrombosis and Vascular Biology, 20: 677-2681.

Rissanen, T.; Voutilainen, S.; Nyyssonen, K.; Salonon, J. Kaplan, G. & Salonen, J. (2003): Serum lycopene concentration and carotid atherosclerosis: the Kuopio Ischemic Heart Disease Risk Factor Study. Am J Clin Nutr 77, 133–138.

Sesson, H. D.; Liu, S.; Gaziano, M. & Buring, J. (2003): Dietary lycopene, tomato-based food products and cardiovascular disease in women. Journal of Nutrition, 133: 2336-341.

Shi, J. (2000): Lycopene in tomatoes: Chemical and physical properties affected by food processing. Critical Reviews In Food Science and Nutrition, 40: 1-42.

Stahl, W.; Junghans, A.; Boer, B.; Driomina, E.; Briviba, K. & Sies, H. (1998): Carotenoid mixtures protect multilamellar liposomes against oxidative damage: synergistic effects of lycopene and lutein. FEBS Lett. 42: 305-308.

Unlu, N. Z.; Bohn, T.; Clinton, S. K. & Schwartz, S. J. (2005): Carotenoid absorption from salad and salsa by humans is enhanced by the addition of avocado or avocado oil. J Nutr 135, 431–436.

Upritchard, J. E.; Sutherland, W. H. F.; Mann, J, I. (2000): Effect of supplementation with tomato juice, vitamin E, and vitamin C on LDL oxidation and products of inflammatory activity in Type 2 diabetes. Diabetes Care 23, 733–738.

Visioli, F.; Riso, P.; Grande, S.; Gall, C. & Porrini, M. (2003): Protective activity of tomato products on in vivo markers of lipid oxidation. Eur J Nutr 42, 201–206.

Willis, M. S. & Wiams, F. H. (2003): The role of nutrition in preventing prostate cancer: a review of the proposed mechanisms of action of various dietary substances. Clinica Cimica Acta, 330:57-83.

Wu, K.; Schwaz, S. J.; Platz, A.; Clinton, S.; Erdman, J. & Ferruzzi, M. (2003): Variations in plasma lycopene and specific isomers over time in a cohort of US men. Journal of Nutrition, 133; 1930-1936.

Zhao, X.; Aldini, G.; Johnson, E. J.; Rasmussen, H. Kraemer, K. & Woolf, H. (2006): Modification of lymphocyte DNA damage by carotenoid supplementation in postmenopausal women. Am J Clin Nutr 83, 163–169.

Section 6

Animal Models for Lipoprotein Research

Chapter 21

Animal Models as Tools for Translational Research: Focus on Atherosclerosis, Metabolic Syndrome and Type-II Diabetes Mellitus

Isaac Karimi

Additional information is available at the end of the chapter

http://dx.doi.org/10.5772/47769

1. Introduction

Close to one century ago, Joslin, an American diabetologist, proposed the link between diabetes and obesity [1]. He concluded that "diabetes is largely a penalty of obesity, and the greater the obesity, the more likely is Nature to enforce it". In the 1950s, Vague [2] described that central obesity predisposes not only to diabetes but also to atherosclerosis. In the 1970s, for the first time, Haller [3] used the term "metabolic syndrome" (MetS) for associations of obesity, diabetes mellitus (DM), hyperlipoproteinemia, hyperuricemia, and hepatic steatosis when describing the additive effects of risk factors on atherosclerosis. Phillips developed the concept of metabolic risk factors for myocardial infarction and described a cluster of abnormalities including glucose intolerance, hyperinsulinemia, hyperlipidemia, and hypertension [4,5]. In 1988, Reaven, an American endocrinologist, propounded that insulin resistance (IR) was the cause of glucose intolerance, hyperinsulinaemia, increased very-low-density lipoprotein cholesterol (VLDL-C), decreased high-density lipoprotein cholesterol (HDL-C) and hypertension and named the constellation of abnormalities "syndrome X" [6]. Reaven did not include abdominal obesity, which has also been hypothesized as the underlying factor, as part of the condition. In the late 1990s and the early 21st century, MetS was widely recognized as a leading risk factor for cardiovascular morbidity and mortality and variously defined by World Health Organization [7], International Diabetes Federation (IDF [8]), the European Group for the Study of Insulin Resistance [9] and the National Cholesterol Education Program Adult Treatment Panel III [10] based on the reference intervals of its components. Accordingly to these definitions, MetS is thought to represent a combination of cardiometabolic risk determinants, including obesity, glucose intolerance and IR, dyslipidemia (including hypertriglyceridaemia, increased free fatty acids (FFAs) and decreased HDL-C) and hypertension and more recently a growing list of clinical manifestations like polycystic ovarian syndrome (PCOS), atherosclerosis, proinflammatory state, oxidative stress and non-alcoholic fatty liver disease (NAFLD) has been associated to it.

The MetS is increasingly recognized as a strong predictor of patient risk for developing coronary artery disease. It is associated with an atherogenic dyslipidemia characterized by elevated levels of triglycerides (TGs), reduced levels of HDL-C and a preponderance of small dense low-density lipoprotein (LDL) particles [11]. An atherogenic dyslipidemia is an integral component of MetS, and a major contributor to the cardiovascular risks in patients. These alarming situations increase the priority for developing new methods and technologies to investigate and to fight the MetS and its related comorbidities. Translational physiology offers us specific animal models for investigating these conditions to help support biomedical research efforts towards finding the necessary cures. This chapter summarizes various types of animal models that used as a tool in lipoprotein clinical researches and critically evaluates the physiological fidelity of these animal models to the human condition. The animal models are used to investigate biological or pathobiological phenomena or employed to find therapeutic and/or toxic effects of a xenobiotic or food ingredients. The laboratory animal models are developed and used to study the cause, nature, and cure of human lipoprotein disorders. They may conveniently be categorized in one of the following two groups:

1. Experimental animal models of lipoprotein disorders
2. Spontaneous animal models of lipoprotein disorders

2. Experimental animal models of lipoprotein disorders

Experimental (induced) models are healthy animals in which the condition (usually disease) to be investigated is experimentally induced, for instance, the induction of DM with encephalomyocarditis virus or alloxan. Although homologous animal models that completely show symptoms and the course of the lipoprotein metabolic disorders are very rare, the most induced models are exploratory, helping to understand mechanisms operative in fundamental normal biology or mechanisms associated with an abnormal biological function. Generally, induced models of metabolic disorders are prepared by genetic manipulation, dietary intervention, surgery, applying xenobiotics (drugs or toxins), and a combination of mentioned methods (see review [12]). This chapter will focus mainly on diet-induced and spontaneous animal models commonly used to investigate lipoprotein metabolic disorders. Readers referred to chapter 22 to study transgenic models of lipoprotein disorders.

Nowadays, obesity, particularly visceral (or central) obesity, is accepted as network backbone of the other MetS components and their manifestations. It has been reported that the incidence of MetS and type 2 diabetes (T2D) increases with the severity of obesity [13]. In this context, increasing body mass index is positively associated with prevalence of both impaired glucose tolerance and T2D and also correlated with dyslipidemia component of MetS that characterized by (a) increased flux of free fatty acids (FFA), (b) raised TGs values,

(c) low HDL-C values, (d) increased small, dense low density lipoprotein particles (e) increased TC and LDL-C and (f) raised apolipoprotein (apo) B levels

The intake of high energy diet and sedentary behavior in developed countries has an accrual effect on the incidence of obesity. Although the association between visceral fat and MetS is strong, the mechanism is not fully elucidated. The adipose tissue is not an inert tissue and constitutively produces adipocytokines that involve in pathogenesis of MetS and IR (see review [14]). Diets play a fundamental role in inducing obesity-related diseases in human, and most animal models do use diet as a way to precipitate the obesity-related diseases. Today, most diet-driven animal disease models are generated using open source, purified ingredient diets. The "open source" nature of purified ingredient diets allows researchers to compare data from different studies, since the diet formulas are generally freely available to the public, while the "close source", chow diets are differently formulated. Purified ingredients, on the other hand, are highly refined and contain just a single nutrient (ie. fructose). These ingredients have little variability and therefore provide consistency between batches, and so help to minimize data variability. There are numerous differences between chows and purified diets, creating countless variables, thus making it difficult to interpret the results when these diets are used together in a study. Chow is a nonpurified diet composed of a mixture of intact feed. In contrast, purified diets provide macronutrients as purified ingredients. For example, carbohydrate in chow diets is derived from complex mixtures of corn and wheat flakes, wheat middlings, ground corn, and dried whey. In addition to carbohydrate, these ingredients provide variable amounts of protein, fat, vitamins, minerals, and various phytochemicals and other (anti)nutrients. Some of these compounds, in particular the phytoestrogens, may act as endocrine disruptors that alter endocrine milieu and disease progression and so are usually unwanted variables. Finally, purified ingredient diet formulas can be easily modified so that researchers can intentionally and specifically change one ingredient at a time, allowing them to study the effects of large or small changes in the nutritional quantity and quality of the diet. Because of these advantages, most metabolic disease animal research uses and requires purified ingredient diets. In addition to purified and chow diets, some scientists used what is known as the cafeteria diet (CAF) to induce obesity. In this model, animals are allowed free access to standard chow and water while concurrently offered highly pleasant, energy dense, unhealthy human junk foods including cookies, candy, cheese, and processed meats *ad libitum*. These foods contain a substantial amount of salt, sugar, and fat and are meant to simulate the human "Western diet". However, the nutritive and nonnutritive components of these foods are not well defined. In addition, the animal may choose a different selection of foods each day. In this section, I discuss how high diets influence the phenotypes of the obesity and/or MetS in translated animals.

2.1. The mouse models

The advantages of mouse models that made them suitable for translating human conditions include a well-known genome, relative ease of genetic manipulation, a short breeding span, access to physiological and invasive testing, short reproductive cycle, large litter size, much lower cost and possibility of conducting longitudinal studies using larger numbers of animals, rapid development of atherosclerotic plaques, only partial resemblance to humans, very high levels of blood lipids, useful for noninvasive imaging and large experience.

Normal mice have traditionally not been ideal models of cardiovascular disease research since they typically have very low levels of TC and LDL-C but high levels of HDL-C. This is in contrast to humans in whom the reverse is true because unlike humans and several other animals, mice do not possess plasma cholesteryl ester transfer protein (CETP) and, therefore, about 70% of the plasma TC is found in HDL particles. Mouse models have proved to be useful to study development and progression of atherosclerotic lesion, and several reviews have extensively discussed the different available models (see review [15]). The ability of mice to maintain their cholesterol profile even in the face of high-cholesterol diets means that very little actual atherosclerosis develops [16]. As wild-type mice are resistant to lesion development, the current mouse models for atherosclerosis are based on genetic modifications of lipoprotein metabolism with additional dietary changes. In order to 'force' the atherosclerosis phenotype on normal mice, it is usually necessary to combine high concentrations of dietary cholesterol with 0.25%-0.5% cholic acid which promotes fat and cholesterol absorption from the intestine [17]. However, cholic acid can also promote liver inflammation, decrease bile acid production, and alter circulating TG and HDL-C, it may independently affect the development of atherosclerosis [15]. Atherosclerosis is a complex multifactorial disease with different etiologies that synergistically promote lesion development. High-fat diets (HFDs) are used to model obesity, dyslipidaemia, atherosclerosis, IR and MetS in rodents (see reviews [18,19]. High-fat diet (HFD) feeding in mice increased systolic blood pressure and induced endothelial dysfunction [20] and some kind of nephropathy [21]. Different types of HFDs have been used with fat fractions ranging between 20% and 60% energy as fat as either animal-derived fats, such as lard or beef tallow, or plant oils such as olive or coconut oil [22]. Long-term feeding of rats (60% of energy) and mice (35% fat wt/wt) with HFD increased body weight compared to standard chow-fed controls [23]. Although the increase in body weight was significant after as little as 2 weeks, the diet-induced phenotype became apparent after more than 4 weeks of HFD feeding [23]. Long-term feeding with both animal and plant fat-enriched diets eventually led to moderate hyperglycaemia and impaired glucose tolerance in most rat and mouse strains [24]. Lard, coconut oil and olive oil (42% of energy content) increased body weight, deposition of liver TGs, plasma TGs and FFAs concentrations and plasma insulin concentrations [22]. Lard, coconut oil and olive oil caused hepatic steatosis with no signs of inflammation and fibrosis [22]. Although HFD induces most of the symptoms of human MetS in rodents, it does not resemble the diet causing MetS and associated complications, as the human diet is more complex than a HFD. Other major components of modern diets are refined carbohydrates and fructose. The epidemiologic data has proved that a significant correlation in the prevalence of diabetes with fat, carbohydrate, corn syrup (source of fructose), and total energy intakes. The striking features of these studies are the fact that intake of corn syrup was positively associated with T2D, while protein and fat were not (see review [25]). Most studies have utilized mice as animal models to define the role of carbohydrate enriched diets in formation of different aspects of MetS. The interested readers will have to go to the current literature in order to understand more fully the fidelity of mice for translation of similar conditions in humans. In this context, high-fat high-carbohydrate (HFHC) diet contains 55% fructose and 45% sucrose (wt/vol) in drinking water has been fed to nongenetically modified adult male C57Bl/6 mice for 16 weeks led to obesity and nonalcoholic steatohepatitis (NASH) [26]. HFHC has been used to induce hyperglycemia, glucose intolerance, IR, increased fat pad weight and adipocyte hypertrophy and commonly HFHC-fed mouse models used to screen therapeutic effects of various drugs and diets against MetS and its comorbidities (e.g., [27]). Charlton and colleagues recently proposed an obese mouse model of NASH that induced by feeding fast food (high SFs, cholesterol, and fructose) diet [28]. C57BL/6J mouse is T2D model by simply feeding HFD to nonobese, nondiabetic C57BL/6J mouse strain. It is characterized by marked obesity, hyperinsulinaemia, IR and glucose intolerance [29]. Diets contribute to T2D in mouse identical to human, and a HFD and sucrose administration appears to speed up the development of the disease in mice. In this context, impaired insulin secretion and/or impaired insulin action also contribute to the diabetic phenotype for these mice [30]. However, the mouse models are observed to develop diabetes in relation to profound obesity and do not display the same islet pathology as humans with T2D [31]. A large number of investigations also have been carried out within recent years, concerning the therapeutic action of various (bio)pharmaceutical and nutraceutical compounds on mouse models of lipoprotein disorders (e.g., [32]).

2.2. The rat models

Rats, like humans, showed different vulnerability to diet-induced obesity. At first, an animal model of diet-induced obesity is one introduced by Levin and coworkers [33] and developed into a purified diet model [34]. In this model, Sprague-Dawley (SD) rats fed a purified moderately high-fat (MHF) diet exhibit a bimodal pattern in body weight gain similar to that observed in humans. Approximately half of the rats gain weight rapidly compared with chow-fed rats (obesity prone [OP] or diet-induced obese [DIO]), whereas the other half gain BW at a rate similar to or lower than that of the chow-fed animals (obesity resistant [OR] or diet-resistant [DR]) [34-36]. Most rodents tend to become obese on HFD and very high-fat diet (VHFD), but there can be variable responses in glucose tolerance, IR, TGs, and other parameters depending on the strain and gender, and source of dietary fat [22,37]. When outbred SD and Wistar rats were placed on HFD (32 or 45 kcal% fat), there was a wide distribution in body weight gain and a subset of animals became obese, whereas others remained as lean as the animals fed with a low-fat diet (LFD) have shown that the rat model of diet-induced obesity develops mild hypertension accompanied by vascular and renal changes similar to those observed in obese hypertensive humans [35,38]. The MHF diet that they used contains 32% kcal fat, a value similar to the average Western diet, as opposed to many other models that have very high levels of fat [38]. All rats fed the MHF diet did not become obese and their body weight displayed a bimodal distribution. The increased body weight reflects an increase in the adipose mass in the OP rats versus the chow-fed rats [38].

Elevation of plasma TGs and FFAs was commonly observed in patients with diabetic dyslipidemia or obesity [39]. Evidence showed that hyperglycemia and hypertriglyceridemia had direct effects on arterial wall and induced endothelial dysfunction [40]. The elevation of TGs and fasting plasma glucose was noted in HFD studies [41]. However, the levels of TGs and TC in high-fat fed DR rats were no more than chow-fed control rats. The HDL-cholesterol level decreases in hypercholesterolemic and MHF-fed rats [38]. The TGs content of plasma, LDL, and VLDL has been increased in OP rats fed MHF diet after 3 weeks [38]. As opposed to cholesterol content, this difference is even greater after 10 weeks of the MHF diet therefore authors concluded that factors other than diet like reduced growth hormone secretion are also responsible for the high levels of TGs in OP rats [38,42]. The underlying mechanism is not known. Insight into the differences in endocrine and lipoprotein metabolism may provide further evidences. For example, Yang et al's study showed that DR rats had higher levels of plasma peptide YY, a gut-derived anorexigen, than DIO and the control groups. This indicates that a difference in appetite control is responsible for the lower caloric intake and weight gain in DR rats [43]. One of the common features of obesity in humans is dyslipidemia which occurs in rat model of diet-induced obesity and is frequently associated with hypertension [38]. I have decided to ignore molecular mechanisms of hypertension in OP rat because of limited space. However, hypertension developed in OP, but not OR, rats, is a multifactorial disorder and diet is not the major factor that causes the high blood pressure in this model.

According to Barker hypothesis, adult metabolic diseases are programmed during fetal life [44]. To investigate the mechanisms by which altered intrauterine *milieu* predisposes to later development of MetS, different animal models have been developed (see review [45]). Interestingly, offspring of rats fed high saturated fats during pregnancy have fetal IR [46], abnormal cholesterol metabolism [47] and raised adult blood pressure [48]. Furthermore, the outbred Sprague-Dawley DIO and DR rats have been selectively bred over time such that their future body weight response to a HFD is known *in utero*, allowing the researcher to look early in life (prior to the onset of obesity) for genetic traits that may later predispose them to their DIO or DR phenotypes [37,49].

The inbred obese Zucker diabetic fatty (ZDF) rat is high-fidelity model with close resemblance to human case in obesity and T2D. The males become obese and diabetic on a LFD, but HFD feeding promotes more robust disease. The female ZDF rat is unique in that while they are obese, they do not develop diabetes unless fed a diet (in this case, chow-based) containing 48 kcal% fat [50]. The female ZDF rat is also suitable model mimics pre-diabetic state in humans because she shows a prolonged period of insulin sensitivity prior to the onset of diet-induced diabetes [51]. The ZDF rats show profound dyslipoproteinaemia with increased TC and TGs levels and lower chylomicra disposal rates that mimics conditions occurred in human case of obesity [52]. Although normal rats are not ideal model of cardiovascular disease research since they typically have very low levels of TC and LDL-C but high levels of HDL-C, they are mild diet-responsive. The ability of rats to sustain their cholesterol profile even in the face of high-cholesterol diets means that very little actual atherosclerosis develops [16]. However, feeding Wistar rats a high calorie "Western diet"

(45% fat) for up to 48 weeks induces obesity and cardiac dysfunction, while a high fat diet (60% fat) induces obesity only [53]. The "Western diet" composed of a purified ingredient SF-rich HFD, and cholesterol (~0.2% by weight) can elevate TC and LDL-C and in turn cause atherosclerosis in certain rodent models and humans [54]. A mixture of high levels of dietary cholesterol with 0.25%-0.5% cholic acid has been used to induce atherosclerosis phenotype on normal rats and mice for many years ago [55]. More recently, Zaragosa and colleagues introduced various animal models of cardiovascular diseases (see review [56]). Surprisingly rat does not develop atheroma in the process of atherosclerosis (see review [56]). Generally rats are highly resistant to the development of atherosclerosis because they lack physiological resemblance on many aspects with humans that are pathophysiologically important [57]. For example, HDL is dominating lipoprotein in these animals and rat platelets are generally resistant in hyperlipidemic condition (see review [58]). Rats are potentially practical model for studying hypercholesterolemia along with hypertension (see review [58]). They exhibit augmented thrombotic response and develop coronary atherosclerotic lesions under hypertensive and hyperlipidemic conditions (e.g., [59]). Triglyceride-rich diets containing various amounts of cholesterol, with or without cholic acid have been used to induce hypercholesterolemia in rats. The fat sources vary from lard to soybean, canola or sunflower oils. Nevertheless, the question of the caloric value of the employed diets has not yet been considered properly since their high fat content, which is the strategy used in order to induce hypercholesterolemia, leads to lower ingestion by the animals and induces malnutrition. To overcome this shortcoming, Matos and colleagues [60] proposed a diet containing 25% soybean oil, 1.0% cholesterol, 13% fiber (cellulose) and 4,538.4 Kcal/Kg that led to an increase in LDL-C, a decrease in the HDL-C fraction and affected less the hepatic function of the rats during eight weeks. Roberts and colleagues presented a rat model of diet-induced syndrome X and they explored potential mechanisms of hypercholesterolemia in diet-induced syndrome X [61]. To induce syndrome X, female Fischer rats were fed a high-fat (primarily from lard plus a small amount of corn oil), refined-carbohydrate (sucrose) diet for 20 months [61]. Sampey and colleagues [62] have demonstrated that the CAF is a more robust model of MetS than lard-based HFD and that the rapid-onset of weight gain, obesity, multiorgan dysfunctions and pathologies observed in the CAF model more closely reflect the modern human condition of early onset obesity. However, they did not repot possible lipid-lipoprotein disorders that may be occurred in their model. Recently, Manting and colleagues [63] have shown that a combination of chronic stress and HFD (83.25% basal feed, 10% lard, 1.5% cholesterol, 0.2% sodium taurodeoxycholate, 5% sugar and 0.05% propylthiouracil) can induce lipid metabolism disorder in Wistar rats and they claimed that their multiple factor model better mimics the disease characteristics of human beings.

2.3. The hamster models

Hamsters are another animal model can be used to assess some aspects of MetS. Like rats and mice, HDL-C is predominant plasma cholesterol-rich lipoprotein in these animal, but in contrast, dietary cholesterol (~ 0.1%) can significantly elevate LDL-C and like humans, SF can increase these levels further [64]. The combination of high dietary SF and cholesterol is commonly used to promote atherosclerosis in these animals and atherosclerotic lesions similar to those found in humans can be found after prolonged feeding periods [65]. Actually, cholesterol itself may not always be necessary for this phenotype, since a purified diet with no cholesterol but high concentrations of SF can promote more aortic cholesterol accumulation compared to a diet with both cocoa butter and 0.15% cholesterol [66]. Cholesterol-fed hamsters have been used to screen therapeutic anti-atherosclerotic and hypolipidemic properties of (phyto)medicines (e.g., [67,68]). Hamsters have been proposed as an animal model to evaluate diet-induced atherosclerosis since the 1980s [69]. Relative to other normal rodent models, hamsters have a low rate of endogenous cholesterol synthesis, cholesteryl ester transfer protein (CETP) activity and tissue specific editing of apolipoprotein (apo) B mRNA and secretion of apo B-100 from the liver and apo B-48 from the small intestine. Hamsters, like humans, take up approximately 80% of LDL-C *via* the LDL receptor pathway. The morphology of aortic foam cells and lesions in hamsters fed atherogenic diets was reported to be similar to human lesions [70]. Recently, in a systematic review Dillard and colleagues concluded that the Golden-Syrian hamster does not appear to be a constructive model to determine the mechanism(s) of diet-induced development of atherosclerotic lesions (see review [71]) however the authors only concentrated on atherogenecity of cholesterol- and fat-rich diets in hamster models of atherosclerosis.

Leung and colleagues investigated intestinal lipoprotein production and the response to insulin sensitization in the high fat-fed Syrian Golden hamster for 5 weeks [72]. They concluded that Syrian Golden Hamsters were fed 60% fat is a good model of nutritionally-induced IR that intestinal overproduction of lipoproteins appear to contribute to the hypertriglyceridemia of IR in this animal model and insulin sensitization with rosiglitazone (an insulin sensitizer) ameliorates intestinal apoB48 particle overproduction in this model. An appropriate dyslipidemic animal model that has diabetes would provide an important tool for research on the treatment of diabetic dyslipidemia. Ten days of high fat feeding in golden Syrian hamsters resulted in a significant increase in IR and baseline serum lipid levels accompanied by a prominent dyslipidemia. Thirteen days of treatment with fenofibrate, a peroxisome proliferator-activated receptor alpha (PPAR alpha) selective agonist, produced a dose-dependent improvement in serum lipid levels characterized by lowered VLDL-C and LDL-C and raised HDL-C in a fashion similar to that seen in man [73]. Various diet formula, fat resources and time tables have been found to induce some aspects of MetS in the literature. For example, a diet consisted of 80 g of anhydrous butterfat, 100 g of corn oil, 20 g of Menhaden fish oil and 1.5 g of cholesterol has been used to encourage hypercholesterolemia in male golden Syrian hamsters for 21 days [68]. Male golden hamsters were given 15% HFD contained 100 g of lard and 50 g of soybean oil and 100 g of sucrose showed diabetic dyslipidemia for eight weeks [74]. F1B hamster is a genetically-defined hamster, derived from two highly inbred lines, namely by crossbreeding between Bio 87.20 female with a Bio 1.5 male. F1B hamster is an exciting animal model for hyperlipidemic-related applications. The F1B strain is very responsive to SF and cholesterol by increasing the non-HDL fraction to a greater extent than the HDL fraction [75]. Dietary fatty acid chain length, degree of saturation and *cis-trans* conformation have been shown to alter several metabolic pathways involving cholesterol throughout the body, the combined effect of which is reflected in plasma lipid and lipoprotein profiles (see review [76]). Interestingly, intake of *trans*-fatty acids in shortenings and margarines has been linked to increased risk of cardiovascular disease through effects on lipoprotein metabolism and substituting *trans*-fatty acids for either saturated or polyunsaturated fatty acids results in more deleterious lipid-lipoprotein profiles [77]. Hamsters are candidate model to investigate cardiometabolic risks of different fat resource and fat-rich diets [78]. Similarity with the human LDL receptor gene, makes hamster ideal to study LDL receptor antagonists and also useful for drugs which interfere with CETP activities and reverse cholesterol transport (RCT) from peripheral tissues to the liver for biliary and fecal excretion [79]. A considerable amount of experimental attention is currently directed at understanding the *in vivo* mechanisms of RCT. Although not established *in vivo*, RCT is thought to be impaired in patients with MetS, in which liver steatosis prevalence is relatively high. In this sense, Briand and coworkers [80] introduced a hamster model of MetS to study RCT. These scientists with the help of HFT diet containing 27% fat, 0.5% cholesterol, and 0.25% deoxycholate as well as 10% fructose in drinking water for 4 weeks induced promoted IR, dyslipidemia with significantly higher plasma non-HDL-C concentrations and CETP activity, and hepatic steatosis. *In vivo* RCT was assessed by intraperitoneally injecting (3)H-cholesterol labeled macrophages. Finally their results indicate a significant increase in macrophage-derived cholesterol fecal excretion, which may not compensate for the diet-induced dyslipidemia and liver steatosis [80]. One of the main target organs of MetS is liver, in which it manifests itself as NAFLD [81]. Bhathena and colleagues currently developed a triumphant BioF1B Golden Syrian hamster model of MetS that successfully manifested hyperlipidemia, IR and NAFLD [81]. They induced this model by feeding hamsters a high-fat, high-cholesterol, inadequate methionine- and choline-containing diet. In addition to F1B hamster strain from Biobreeders (Watertown, MA) that commonly used to study diet-induced metabolic disorders other three outbred strains are Charles River (CR), Sasco and Harlan (see review [71]). All these strains are derived from inbred or outbred Golden-Syrian hamster.

Similar to rats, hamsters fed high fructose diets (~60% of energy) may develop IR and hypertriglyceridemia in TG after only two weeks compared to diets low in fructose [72,73]. Interestingly, hamsters fed high-sucrose diets did not have elevated TG levels and developed only mild IR relative to those fed diets high in fructose [72]. Avramoglu and colleagues reviewed mechanisms of metabolic dyslipidemia in insulin resistant states (see review [82]). They developed an explanatory fructose-fed hamster model of insulin resistance to study hepatic lipid metabolism as its lipoprotein metabolism as described previously [83,84]. Hamsters exhibit obesity, hypertriglyceridemia, increase plasma FFAs concentration and IR if fed fructose-rich diet for a two week period. Fructose feeding induced a noteworthy increase in synthesis and secretion of total TGs as well as VLDL-TG by primary hamster hepatocytes [73]. The microsomal triglyceride transfer protein plays a pivotal role in VLDL assembly and its activity showed a striking 2.1-fold elevation in hepatocytes derived from fructose-fed versus control hamsters [73]. The apoB production also has been increased in the fructose-fed hamsters [73]. Fructose-fed hamster also has been introduced as an exploratory animal model to excavate role of intestinal lipoprotein overproduction in the dyslipidemia of insulin-resistant states [74]. The authors have shown that fructose feeding for 3 weeks increases secretion of apoB48-containing lipoproteins in the fasting state and during steady state fat feeding. Wang and coworkers [75] investigated the composition of plasma lipoproteins in hamsters fed high-carbohydrate diets of varying complexity (60% carbohydrate as chow, cornstarch, or fructose) for 2 weeks. They showed that hamsters fed the high-fructose diet showed significantly increased VLDL–triglyceride (92.3%), free cholesterol (68.6%), and phospholipid (95%), whereas apolipoprotein B levels remained unchanged. Fructose feeding induced a 42.5% increase LDL–triglyceride concurrent with a 20% reduction in LDL–cholesteryl ester. Compositional changes were associated with reduced LDL diameter. In contrast, fructose feeding caused elevations in all HDL fractions.

2.4. The guinea pig models

A number of seminal reviews on the details of the criteria that make guinea pigs suitable animal models for studying lipoprotein metabolism are available (see reviews [85-87]) and a summary will be presented here. Guinea pigs in contrast to other rodents have higher levels of plasma LDL-C compared to HDL-C. As humans, guinea pigs have higher concentrations of free compared to esterified cholesterol found in the liver and they show evidence of moderate rates of hepatic cholesterol synthesis and catabolism. Similar to humans, the binding domain for the LDL receptor of guinea pigs discriminates normal and familial binding defective apo B-100 and apo B mRNA editing in liver is scarce (< 1%) compared to 18 to 70% in other species [88]. The three important proteins involved in lipoprotein remodeling and RCT (CETP, lecithin:cholesterol acyltransferase (LCAT), and lipoprotein lipase (LPL) have been reported in guinea pigs.

Guinea pigs have been used as models to dissect the mechanisms by which various dietary fat resources influences plasma lipid-lipoprotein profiles. In contrast to hamsters they do not possess a fore-stomach fermentation which modifies dietary macronutrients before reaching the small intestine [89]. Guinea pigs are not only superior models for studying the mechanisms by which statins [90], cholestyramine [91], apical sodium bile acid transport inhibitors [92] and microsomal transfer protein inhibitors [93] lower plasma LDL-C but also are selected to investigate the mechanisms by which certain drugs or toxins affect lipid-lipoprotein metabolism (e.g.,[94]). Guinea pigs respond to dietary fat saturation, dietary cholesterol and dietary fiber by alterations in LDL-C (see review [87]). For example, the SF-rich diet will increase TC and LDL-C levels much more than polyunsaturated fat (PF)-rich diet in guinea pigs and cholesterol-rich diets can further increase TC and LDL-C levels [95].

The suitability of guinea pigs as models of atherosclerosis is augmented by an array of review and assessment features (see review [59]). However, guinea pigs do not develop advanced atherosclerotic lesions, and are not an entrenched model for atherosclerosis progression [96].

High plasma level of lipoprotein (a) (also called Lp(a)) is associated with coronary heart disease and other forms of atherosclerosis in humans (see review [97]), and as primates, hedgehogs [98] and guinea pigs possess Lp(a) among normal animal models [99]. Guinea pigs are practical model to study the role of oxidized LDL (oxLDL) in progression of atherosclerosis [100]. Initial atherosclerosis induced by various formula of HFD in guinea-pigs. Intake of HFD (guinea-pig pellet diet + 0.2% w/w cholesterol) can induce the onset of early atherosclerotic changes in coronary artery, aorta and major organs at least for one month [101]. High SF diet supplemented with high cholesterol (0.25%) will advance an atherosclerotic process for twelve weeks in guinea pigs [102]. Yang and colleagues [103] introduced a hyperlipidemic guinea pig model in a comparative investigation. They concluded that different response of TG metabolism to a HFD (0.1% cholesterol and 10% lard) in guinea pigs and rats suggests that Hartley guinea pigs could be a better hypertriglyceridemia animal model than rats for research on lipid metabolism disorders and hypolipidemic drugs. It seems that chronic dyslipidemia associated with hypertriglyceridemia may reduce auditory function. In this context guinea pigs fed a HFD used as an animal model to find the relationship between of sensorineural hearing loss and dyslipidemia [104].

As rats, guinea pigs are accepted models for studying fetal programming of cardiovascular diseases. Interestingly, adipogenesis in the guinea pig is very active during early postnatal life and was altered by a maternal HFD; thus, it is an adequate model for intrauterine fat deposition [105]. More studies are requested to explore lipid-lipoprotein profiles of guinea pigs that received a maternal HFD. Caillier and colleagues [106] currently generated a guinea pig model of MetS by 150-day exposure to diabetogenic high fat high sucrose or the high fat high fructose diets. To my knowledge, it would be early to consider guinea pig as an animal model of MetS since the literatures are scarce.

2.5. The rabbit models

A century ago, rabbits were used as translated animal models of atherosclerosis [107]. Since then, a number of animal models have been used to explain the relationship between disorders of lipid metabolism and atherogenesis (see reviews [108,109]). In this sense, dietary lipid manipulation and use of naturally defective animals, such as Watanabe heritable hyperlipidemia (WHHL) rabbits, have been the focus of most experimental settings (see chapter 22). Rabbits are appropriate animal models for studying lipoprotein metabolism and its disorders because they share with humans several aspects of lipoprotein metabolism, such as similarities in composition of apolipoprotein B containing lipoproteins, hepatic production of apo B 100-containing VLDL, plasma CETP activity, human-like apo B, low hepatic lipase activity and high absorption rate of dietary cholesterol. Unlike humans, rabbits are hepatic lipase–deficient and do not have an analogue of human apo A-II. Rabbits do not form spontaneous atherosclerotic lesions and therefore require very high cholesterol levels to induce more advanced disease (see review [110]). Rabbits also have significant differences in their lipid metabolism from humans, which can result in their development of "cholesterol storage syndrome" while on high-cholesterol diets (0.5–3%), with cholesterol deposited in their liver, adrenal cortex, and reticuloendothelial and genitourinary systems

[108]. We found that a high-cholesterol diet contained in 0.47% cholesterol would be tolerable for adult male rabbits for 4 weeks but our high-cholesterol diet was mildly atherogenic [94,111]. The atherosclerotic lesions of rabbits do not completely resemble those in humans [108] and the formed lesions are more fatty and macrophage rich than human [112]. Atherogenic diets are usually associated with hypercholesterolemia and the development of atherosclerotic lesions in the aortic arch and thoracic aorta rather than in the abdominal aorta that is almost always affected in humans. New Zealand White rabbits are the strain commonly used in atherosclerosis research. Although they have low plasma TC concentrations and HDL as dominant lipoprotein [113], βVLDL becomes the major class of plasma lipoproteins when exposed to cholesterol-rich diet (see review [59]). In conjunction with chylomicron remnants, βVLDL becomes highly atherogenic. Long-term experiments using diets high in cholesterol are discouraging in rabbits, because they cannot increase the excretion of sterols and resulting hepatotoxicity does not allow the animal to survive (see review [114]).

Various HFD and intervention period have been used to induce MetS in rabbits or its components like IR, visceral obesity, hypertension, dyslipidemia (e.g., [115,116]). Rabbits are suitable animals to investigate MetS-associated multiorgan dysfunctions. Helfenstein and colleagues recently proposed an experimental model of impaired glucose tolerance combined with hypercholesterolaemia induced by diets (high-fat/high-sucrose (10/40%) and cholesterol-enriched diet for 24 weeks) that gained weight, increased blood glucose, TC, LDL-C, TGs, and decreased HDL-C in New Zealand male rabbits [117]. Their cheap model reproduced several metabolic characteristics of human DM and promoted early signs of retinopathy. Corona and colleagues [118] reviewed relationships between hypogonadism and MetS emphasizing their possible interaction in the pathogenesis of cardiovascular diseases. However they concluded that the clinical significance of the MetS-associated hypogonadism needs further clarifications. Vignozzi and colleagues [119] described an animal model of MetS obtained by feeding male rabbits for 12 weeks. In their experiment, HFD-animals develop hypogonadism and all the MetS features like hyperglycemia, glucose intolerance, dyslipidemia, hypertension, and visceral obesity. A recently established rabbit model of HFD-induced MetS showed hypogonadism and the presence of prostate gland alterations, including inflammation, hypoxia and fibrosis [120]. Rabbits fed a cholesterol-rich diet (1% cholesterol) for 8 weeks and 12 weeks share several physiopathological aspects of NAFLD [121]. Because this model is not insulin resistant and obese, it may be useful for elucidating the mechanism of NAFLD related mainly to hyperlipidemia.

3. Spontaneous animal models of lipoprotein disorders

The pathophysiology of disorders of lipoprotein metabolism of humans cannot highly translated to wild-type rodents since they are very resistant to atherogenesis and have no similarity to human lipid and lipoprotein metabolism; further, they do not develop cardiovascular diseases identical to humans. Therefore search for more reliable model is still continuing. In this regard, pig, is a considered a very good model of human atherosclerosis, because it is similar to humans in terms of body size and other physiological features (see review [122]). Pigs spontaneously develop atherosclerosis even on a normal porcine diet and dietary modification lead to sever atherosclerosis [123]. As humans, pigs transport most cholesterol in LDL-C and dietary modification alters their plasma lipoproteins closely resemble those occurring in humans. In contrast to rodents, swine atherosclerosis, like the human illness, progresses to advanced stages (see review [122]). For example, mini-pigs fed fat-enriched food showed fatty streaks in their abdominal and thoracic aorta and coronary arteries during 18 months [124]. Cholesterol contents of diets also affects the extension and exacerbation of atherosclerotic changes in pigs (see review [122]). Johansen and colleagues [125] suggested an obese Göttingen minipig model of MetS that was highly responsive to a high fat high energy diet. Several swine models of T2D and IR have been proposed (see review [18]). However, spontaneity in development of MetS and IR, is not common in this species [58]. Dogs do not naturally show atherosclerosis and cholesterol- and SF-rich diets combined with thyroid suppression is required for atherosclerosis development [126]. Beagles and miniature Schnauzer dogs show useful similarities with human in cholesterol synthesis, and lipoproteins level [127,128]. Feline DM, in both spontaneous and inducible forms, therefore provides a reliable animal model of human T2D and may provide additional insights into the clinical, physiological, and pathological features of this disease (see review [129]). Considerably more studies must be forthcoming to establish firmly how lipoprotein profile participates in pathogenesis of atherosclerosis, DM and possibly MetS in pet animal since companion animal obesity would be a serious veterinary medical concern in near future [130]. The subject of suitability of other domesticated animal species such as pigs and sheep, as well as feral, migrating and hibernating species for studying lipid and lipoprotein metabolism has been concisely reviewed (see review [131]). For thorough coverage of this aspect of animal models, this work is recommended.

A number of other interesting wild rodents that are explanatory or exploratory animal models of different disorders of lipid and/or carbohydrate metabolism have been introduced. Recently, *Octodon degus* (degu) has been proposed as an animal model of diet-induced development of atherosclerosis [132]. To induce atherosclerosis, degus were fed for 16 weeks chow containing 0.25% cholesterol and 6% palm oil. Cholesterol-fed degus exhibited 4- to 5-fold increases in TC, principally in the VLDL-C and LDL-C fractions and developed cholesteryl ester-rich atherosclerotic lesions throughout the aorta [132]. Hedgehogs are homologous animal models for studying roles of Lp(a) in atherosclerosis [98]. Sand rat, Tuco-Tuco and spiny mouse are unusual models of diet-induced obesity and T2D (see review [12]. In laboratory condition, sand rat (*Psammomys obesus*) develops obesity and diabetes when fed on standard chow (high energy diet) instead of its usual energy-diluted vegetable diet composed mainly of saltbush *Atriplex* [133]. Surprisingly, sand rats are studied extensively and serve as more statable polygenic model for the study of diabesity syndrome [133,134]. Spiny mouse (*Acomys calirinus*) is another small rodent living in semiarid areas of eastern Mediterranean. They gain weight and exhibit marked impaired pancreatic beta cell when they are placed in captivity on high energy rodent lab chow [135]. *Ctenomis talarum* (Tuco-tuco) is another feral species which exhibit similar characteristic features of sand rat and spiny mice when fed on high energy rodent diet [136]. Brandt's vole

(*Lasiopodomys brandtii*) is another rodent model used in diet-induced obesity [137]. The Nile grass rat (NGR), *Arvicanthis niloticus,* is a herbivorous rodent inhabiting dry savanna, woodlands, and grasslands in Africa. Noda and colleagues [138] recently showed that the NGR is a precious, spontaneous model for exploring the etiology and pathophysiology of MetS as well as its various complications.

Avian models of human atherosclerosis include pigeon, chicken, Japanese quail, turkey and parrots. Although these avian models are not frequently used in studying atherosclerosis, it is worthy to note, that spontaneous atherosclerosis in the chicken was first described in 1914 [139]. Use of pigeon models of atherosclerosis has been extensively reviewed (see review [140]). Briefly, the most important key points supporting the use of pigeons as models for human atherosclerosis include: 1. Pigeons are hypercholesterolemic compared to humans. 2. Pigeons are primarily HDL-C carriers but βVLDL-C and LDL-C become major lipid carriers when these animals are fed cholesterol-rich diet. 3. Their lipid metabolism and lesion progression are similar to humans. 4. Pigeons also resemble humans in cellular and vascular dysfunctions involving in atherogenesis. 5. Pigeons are negative animal models to study relevance of apoE, apo B48, chylomicra or LDL receptor in atherosclerosis pathology. 6. Pigeons are susceptible to both spontaneous and diet-induced atherosclerosis. Parrots are exceptional animal models to assess the impacts of various risk factors include elevated cholesterol level, diet composition, social stress and inactivity (similar to sedentary behavior in humans) on occurrence and progression of atherosclerosis (see review [59]).

Phylogenetically, nonhuman primates are more similar to humans than other models in terms of lipid-lipoprotein profiles, pathophysiology of atherosclerosis, feeding habits, and genotype. It is demonstrated that, along with aging, some rhesus monkeys spontaneously develop diabesity (e.g. [141]). In this context, *Macaca nigra* is very valuable in studies focused on the interactions between atherosclerosis and diabetes [142]. Spontaneous diabetes has been documented in nonhuman primates include cynomolgus, rhesus, bonnet, Formosan rock, pig-tailed, celebes macaques, African green monkeys, and baboons (see reviews [18,143]). Diabetic nonhuman primates have detrimental changes in plasma lipid and lipoprotein concentrations and lipoprotein composition which may contribute to progression of atherosclerosis. As both the prediabetic condition (similar to MetS in humans) and overt diabetes become better translated in monkeys, their use in pharmacological studies is increasing. Monkeys can be categorized into hyperresponders and hyporesponders based on initiation, progression and severity of atherosclerotic lesions. Several nonhuman primates, such as squirrel monkeys, baboons, and wooly and spider monkeys, may develop spontaneous early stage (fatty streaks) atherosclerosis at different anatomical locations (see reviews [59,108]). Rare cases of LDL receptor deficiency in a rhesus monkey family associated with increased levels of LDL-C, Lp(a), and advanced atherosclerotic lesions in the aorta, and to a lesser extent in coronary arteries, were reported [144]. Nonhuman primates are more reliable model to study cardiovascular disease plus MetS rather than rodents, since they develop MetS and cardiovascular diseases as they age. They develop spontaneous (in some species) and high fat high cholesterol diet-induced atherosclerotic lesions [145]. Nonhuman primates are good model of hypertension and its harmful effect on atherosclerosis development. The close similarity of plasma lipoprotein-lipid level, plaque development and its calcification and mineralization with humans makes nonhuman primates practical model to explore the correlation between plasma lipids and plaque development (see review [59]). Kaufman and colleagues measured some anthropometric indices and metabolic parameters in 250 laboratory-born bonnet macaques living in social groups and maintained on commercial monkey chow [146]. Finally they concluded socially reared and housed bonnet macaques may provide a useful model for studying the pathogenesis, prevention, and treatment of the MetS. Recently, in an excellent investigation, Zhang and colleagues established a rhesus monkey model of spontaneous MetS using population screening approaches suitable to explore the pathogenesis of MetS in relation to cardiovascular disease and DM [147]. To sum, the inconsistency in anatomic location of atherosclerotic lesions, high cost of husbandry and veterinary services, limited animal availability, difficult handling, together with ethical queries are major obstacles in the use of monkeys as common animal models in studying MetS and its comorbidities.

4. Conclusion

The incidence of metabolic syndrome is increasing on a pandemic level. One of the major underlying cause and/or outcome of metabolic syndrome is dyslipidemia, which contribute greatly to the cardiovascular problems associated with the syndrome. The animal models have a vital role to play in extending our understanding of metabolic syndrome and its related comorbidities. Conventional laboratory animals such as mice, rats, hamsters, guinea pigs and rabbits have been examined to gain a better perceptive of the relationship between disorders of lipid metabolism and their clinical correlations. High-fat diets frequently used to induce different aspects of metabolic syndrome in rodent models. However, nonconventional animal models like pig, pigeon, and feral animals (e.g., spiny mice, sand rat, hedgehogs) can consider as spontaneous animal models suitable for studying both the pathogenesis and potential therapeutic agents in lipoprotein disorders. The attempts to find animal models relevant to the study of metabolic syndrome are continuing.

Author details

Isaac Karimi
Division of Biochemistry, Physiology and Pharmacology, Department of Basic Veterinary Sciences, School of Veterinary Medicine, Razi University, Kermanshah, Iran

5. References

[1] Joslin EP (1922) The prevention of diabetes mellitus. JAMA. 76(2):79-84.
[2] Vague J (1956) The degree of masculine differentiation of obesities: a factor determining predisposition to diabetes, atherosclerosis, gout, and uric calculous disease. Am J Clin Nutr 4: 20–34.

[3] Haller H (1977) Epidermiology and associated risk factors of hyperlipoproteinemia. Z Gesamte Inn Med. 15;32(8):124-8. [Article in German]
[4] Phillips GB (1978) Sex hormones, risk factors and cardiovascular disease. Am J Med. 65:7-11.
[5] Phillips GB (1977) Relationship between serum sex hormones and glucose, insulin, and lipid abnormalities in men with myocardial infarction. Proc Natl Acad Sci USA.74: 1729-1733.
[6] Reaven GM (1988) Banting lecture 1988. Role of insulin resistance in human disease. Diabetes 37: 1595–1607.
[7] World Health Organization (1999) WHO consultation: definition, diagnosis and clasification of diabetes mellitus and its complications. Geneva: WHO.
[8] Alberti KG, Zimmet P, Shaw J (2005) The metabolic syndrome: a new worldwide definition. Lancet. 24; 366:1059–62.
[9] Balkau B, Charles MA (1999) Comment on the provisional report from the WHO consultation. European Group for the Study of Insulin Resistance (EGIR). Diabet Med.16:442–3.
[10] Executive Summary of The Third Report of The National Cholesterol Education Program (NCEP) (2001) Expert Panel on Detection, Evaluation, And Treatment of High Blood Cholesterol In Adults (Adult Treatment Panel III). JAMA.285:2486–97.
[11] Cziraky MJ (2004) Management of dyslipidemia in patients with metabolic syndrome. J Am Pharm Assoc. 44: 478-488.
[12] Srinivasan K, Ramarao P (2007) Animal models in type 2 diabetes research: An overview. Indian J Med Res. 125: 451-472
[13] Weiss R, Dziura J, Burgert TS, Tamborlane WV, Taksali SE, Yeckel CW, Allen K, Lopes M, Savoye M, Morrison J, Sherwin RS, Caprio S (2004) Obesity and the metabolic syndrome in children and adolescents. N Engl J Med.350:2362–74.
[14] Bruce KD, Byrne CD (2009) The metabolic syndrome: common origins of a multifactorial disorder. Postgrad Med J. 85:614–621.
[15] Getz GS, Reardon CA (2006) Diet and murine atherosclerosis. Arterioscler Thromb Vasc Biol. 26:242-249.
[16] Maxwell KN, Soccio RE, Duncan EM, Sehayek E, Breslow JL (2003) Novel putative SREBP and LXR target genes identified by microarray analysis in liver of cholesterol-fed mice. J Lipid Res. 44:2109-2119.
[17] Nishina PM, Lowe S, Verstuyft J, Naggert JK, Kuypers FA, Paigen B (1993) Effects of dietary fats from animal and plant sources on diet-induced fatty streak lesions in C57BL/6J mice. J Lipid Res. 34:1413-1422.
[18] Cefalu WT (2006) Animal models of type 2 diabetes: clinical presentation and pathophysiological relevance to the human condition. ILAR J. 47(3).
[19] Panchal SK, Brown L (2011) Rodent models for metabolic syndrome research. J Biomed Biotechnol. 2011:351982.
[20] Kobayasi R, Akamine EH, Davel AP, Rodrigues MAM, Carvalho CRO and Rossoni LV (2010) "Oxidative stress and inflammatory mediators contribute to endothelial dysfunction in high-fat diet-induced obesity in mice," J Hypertension. 28: 2111–2119.

[21] Deji N, Kume S, Araki SI et al (2009) "Structural and functional changes in the kidneys of high-fat diet-induced obese mice,". Am J Physio. 296: F118–F126.

[22] Buettner R, Parhofer KG, Woenckhaus M, Wrede CE, Kunz-Schughart LA, Scholmerich J, Bollheimer LC (2006) Defining high-fat-diet rat models: metabolic and molecular effects of different fat types. J Mol. Endocrinol. 36:485-501.

[23] Sutherland LN, Capozzi LC, Turchinsky NJ, Bell RC, Wright DC (2008) "Time course of high-fat diet-induced reductions in adipose tissue mitochondrial proteins: potential mechanisms and the relationship to glucose intolerance," Am J Physio. 295: E1076–E1083.

[24] Sweazea KL, Lekic M, Walker BR (2010) "Comparison of mechanisms involved in impaired vascular reactivity between high sucrose and high fat diets in rats," Nutr Metabol. 7:48.

[25] Basciano H, Federico L, Adeli K (2005) Fructose, insulin resistance, and metabolic dyslipidemia. Nutr Metabol. 2:5 doi:10.1186/1743-7075-2-5.

[26] Kohli R, Kirby M, Xanthakos SA, Softic S, Feldstein AE, Saxena V, Tang PH, Miles L, Miles MV, Balistreri WF, Woods SC, Seeley RJ (2010) High-fructose, medium chain trans fat diet induces liver fibrosis and elevates plasma coenzyme Q9 in a novel murine model of obesity and nonalcoholic steatohepatitis. Hepatology. 52(3):934-44.

[27] Nascimento FA, Barbosa-da-Silva S, Fernandes-Santos C, Mandarim-de-Lacerda CA, Aguila MB (2010) Adipose tissue, liver and pancreas structural alterations in C57BL/6 mice fed high-fat-high-sucrose diet supplemented with fish oil (n-3 fatty acid rich oil). Exp Toxicol Pathol. 62(1):17-25.

[28] Charlton M, Krishnan A, Viker K, Sanderson S, Cazanave S, McConico A, Masuoko H, Gores G (2011) Fast food diet mouse: novel small animal model of NASH with ballooning, progressive fibrosis, and high physiological fidelity to the human condition. Am J Physiol Gastrointest Liver Physiol. 301(5):G825-34.

[29] Surwit RS, Kuhn CM, Cochrane C, McCubbin JA, Feinglos MN (1988) Diet-induced type II diabetes in C57BL/6J mice. Diabetes. 37: 1163-7.

[30] Ikegami H, Fujisawa T, Ogihara T (2004) Mouse models of type 1 and type 2 diabetes derived from the same closed colony: Genetic susceptibility shared between two types of diabetes. ILAR J. 45:268-277.

[31] Harmon JS, Gleason CE, Tanaka Y, Poitout V, Robertson RP (2001) Antecedent hyperglycemia, not hyperlipidemia, is associated with increased islet triacylglycerol content and decreased insulin gene mRNA level in Zucker diabetic fatty rats. Diabetes. 50:2481-2486.

[32] Hu Y, Davies GE (2010) Berberine inhibits adipogenesis in high-fat diet-induced obesity mice. Fitoterapia. 81: 358-366.

[33] Levin BE, Triscari J, Sullivan AC (1983) Relationship between sympathetic activity and diet-induced obesity in two rat strains. Am J Physiol. 245: R364-R371.

[34] Lauterio TJ, Bond JP, Ulman EA (1994) Development and characterization of a purified diet to identify obesity-susceptible and -resistant rat populations. J Nutr. 124:2172-2178.

[35] Chang S, Graham B, Yakubu F, Lin D, Peters JC, Hill JO (1990) Metabolic differences between obesity-prone and obesity-resistant rats. Am J Physiol 259:R1103-R1110.

[36] Levin BE, Dunn-Meynell AA (2006) Differential effects of exercise on body weight gain and adiposity in obesity-prone and –resistant rats. Int J Obes(Lond). 30:722-727.

[37] Levin BE, Keesey RE (1998) Defense of differing body weight set points in diet-induced obese and resistant rats. Am J Physiol. 274:R412-R419.

[38] Dobrian AD, Davies MJ, Prewitt RL, Lauterio TJ (2000) Development of Hypertension in a Rat Model of Diet-Induced Obesity. Hypertension.35:1009-1015.

[39] Krauss RM (2004) Lipids and lipoproteins in patients with type 2 diabetes. Diabetes Care.27:04–1496.

[40] Monti LD, Landoni C, Setola E, Galluccio E, Lucotti P, Sandoli EP, Origgi A, Lucignani G,Piatti P, Fazio F (2004) Myocardial insulin resistance associated with chronic hypertriglyceridemia and increased FFA levels in Type 2 diabetic patients. Am J Physiol Heart Circ Physiol. 287:H1225–31.

[41] Akiyama T, Tachibana I, Shirohara H, Watanabe N, Otsuki M (1996) High-fat hypercaloric diet induces obesity glucose intolerance and hyperlipidemia in normal adult male Wistar rat. Diabetes Res Clin Pract. 31: 27–35.

[42] Lauterio TJ, Barkan A, DeAngelo M, DeMott-Friberg R, Ramirez R (1998) Plasma growth hormone secretion is impaired in obesity-prone rats before onset of diet-induced obesity. Am J Physiol. 275:E6-E11.

[43] Yang N, Wang C, Xu M, Mao L, Liu L, Sun X (2005) Interaction of dietary composition and PYY gene expression in diet-induced obesity in rats. J Huazhong Univ Sci Technolog Med Sci.25:243–6.

[44] Barker DJP, Gluckman PD, Godfrey KM, Harding JE, Owens JA, Robinson JS (1993) Fetal nutrition and cardiovascular disease in adult life. Lancet 341: 938–941.

[45] Bertram CE, Hanson MA (2001) Animal models and programming of the metabolic syndrome. Brit Med Bull. 60:103–121

[46] Guo F, Jen KLC (1995) High fat feeding during pregnancy and lactation affects offspring metabolism in rats. Physiol Behav.57: 681–6

[47] Brown SA, Rogers LK, Dunn JK, Gotto AM, Jr, Patsch W (1990) Development of cholesterol homeostatic memory in the rat is influenced by maternal diets. Metab Clin Exp. 39: 468–73.

[48] Langley-Evans SC (1996) Intrauterine programming of hypertension in the rat: nutrient interactions. Comp Biochem Physiol. 114: 327–31

[49] Ricci MR, Levin BE (2003) Ontogeny of diet-induced obesity in selectively bred Sprague-Dawley rats. Am J Physiol Regul. Integr. Comp Physiol. 285:R610-R618.

[50] Corsetti JP, Sparks JD, Peterson RG, Smith RL, Sparks CE (2000) Effect of dietary fat on the development of non-insulin dependent diabetes mellitus in obese Zucker diabetic fatty male and female rats. Atherosclerosis. 148:231-241.

[51] Owens D (2006) Spontaneous, surgically and chemically induced models of disease. In The Laboratory Rat. Suckow MA, Weisbroth SH, Franklin CL, Eds. Elsevier Academic Press. 711-732 p.

[52] Blay M, Peinado-Onsurbe J, Julve J, Rodríguez V, Fernández-López JA, Remesar X, Alemany M (2001) Anomalous lipoproteins in obese Zucker rats. Diabetes Obes Metab. 3(4):259-70.

[53] Ballal K, Wilson CR, Harmancey R, Taegtmeyer H (2010) Obesogenic high fat western diet induces oxidative stress and apoptosis in rat heart. Mol Cell Biochem. 344(1-2):221-30.

[54] Hegsted DM, Ausman LM, Johnson JA, Dallal GE (1993) Dietary fat and serum lipids: an evaluation of the experimental data. Am J Clin Nutr. 57:875-883.

[55] Fillios LC, Andrus SB, Mann GV, Stare FJ (1956) Experimental production of gross atherosclerosis in the rat. J Exp Med. 104: 539.

[56] Zaragoza C, Gomez-Guerrero C, Martin-Ventura JL, Blanco-Colio L, Lavin B, Mallavia B, Tarin C, Mas S, Ortiz A, Egido J (2011) Animal models of cardiovascular diseases. J Biomed Biotechnol. doi:10.1155/2011/497841

[57] Russell JC, Proctor SD (2006) Small animal models of cardiovascular disease: tools for the study of the roles of metabolic syndrome, dyslipidemia, and atherosclerosis. Cardiovasc Pathol. 15: 318-30.

[58] Singh V, Tiwari RL, Dikshit M, Barthwal MK (2009) Models to study atherosclerosis: a mechanistic insight. Curr Vasc Pharmacol. 7: 75-109.

[59] Gomibuchi H, Okazaki M, Iwai S, Kumai T, Kobayashi S, Oguchi K (2007) Development of hyperfibrinogenemia in spontaneously hypertensive and hyperlipidemic rats: a potentially useful animal model as a complication of hypertension and hyperlipidemia. Exp Anim.56: 1-10.

[60] Matos SL, Paula Hd, Pedrosa ML, dos Santos RC, de Oliveira EL, Júnior DAC, Silva ME (2005) Dietary models for inducing hypercholesterolemia in rats. Braz Arch Biol Techn. 48:203-209.

[61] Roberts CK, Liang K, Barnard RJ, Kim CH,Vaziri ND (2004) HMG-CoA reductase, cholesterol 7a-hydroxylase, LDL receptor, SR-B1, and ACAT in diet-induced syndrome X. Kidney Int. 66:1503-1511.

[62] Sampey BP, Vanhoose AM, Winfield HM, Freemerman AJ, Muehlbauer MJ, Fueger PT, Newgard CB, Makowski L (2011) Cafeteria diet is a robust model of human metabolic syndrome with liver and adipose inflammation: comparison to high-fat diet. Obesity.19: 1109-1117.

[63] Manting L, Haihong Z, Jing L, Shaodong C, Yihua L (2011) The model of rat lipid metabolism disorder induced by chronic stress accompanying high-fat-diet. Lipids Health Dis. 10:153.

[64] Spady DK, Woollett LA, Dietschy JM (1993) Regulation of plasma LDL- cholesterol levels by dietary cholesterol and fatty acids. Annu Rev Nutr. 13:355-381.

[65] Pien CS, Davis WP, Marone AJ, Foxall TL (2006) Characterization of diet induced aortic atherosclerosis in Syrian F1B Hamsters. J Exp Anim Sci. 42:65-83, 2006

[66] Alexaki A, Wilson TA, Atallah MT, Handelman G, Nicolosi RJ (2004) Hamsters fed diets high in saturated fat have increased cholesterol accumulation and cytokine production in the aortic arch compared with cholesterol-fed hamsters with moderately elevated plasma non-HDL cholesterol concentrations. J Nutr. 134:410-415.

[67] Vinson JA, Mandarano M, Hirst M, Trevithick JR, Bose P (2003) Phenol antioxidant quantity and quality in foods: beers and the effect of two types of beer on an animal model of atherosclerosis. J Agric Food Chem. 51(18):5528-33.

[68] Rimando AM, Nagmani R, Feller DF, Yokoyama W (2005) Pterostilbene, a new agonist for the peroxisome proliferator-activated receptor r-isoform, lowers plasma lipoproteins and cholesterol in hypercholesterolemic hamsters. J Agric Food Chem. 53: 3403–3407

[69] Nistor A, Bulla A, Filip DA, Radu A (1987) The hyperlipidemic hamster as a model of experimental atherosclerosis. Atherosclerosis. 68:159-173.

[70] Kahlon T, Chow F, Irving D, Sayre R (1996) Cholesterol response and foam cell formation in hamsters fed two levels of saturated fat and various levels of cholesterol. Nutr Res. 16:1353-1368.

[71] Dillard A, Matthan NR, Lichtenstein AH (2010) Use of hamster as a model to study diet-induced Atherosclerosis. Nutr Metab. 7:89

[72] Leung N, Naples M, Uffelman K, Szeto L, Adeli K, Lewis GF (2004) Rosiglitazone improves intestinal lipoprotein overproduction in the fat-fed Syrian Golden hamster, an animal model of nutritionally-induced insulin resistance. Atherosclerosis. 174(2):235-41.

[73] Wang PR, Guo Q, Ippolito M, Wu M, Milot D, Ventre J, Doebber T, Wright SD, Chao YS (2001) High fat fed hamster, a unique animal model for treatment of diabetic dyslipidemia with peroxisome proliferator activated receptor alpha selective agonists. Eur J Pharmacol. 427(3):285-93.

[74] Li S-Y, Chang C-Q, Ma F-Y, Yu C-L (2009) Modulating effects of chlorogenic acid on lipids and glucose metabolism and expression of hepatic peroxisome proliferator-activated receptor-α in golden hamsters fed on high fat diet. Biomed Environ Sci. 22:122-129.

[75] Tyburczy C, Major C, Lock AL, Destaillats F, Lawrence P, Brenna JT, Salter AM, Bauman DE (2009) Individual trans octadecenoic acids and partially hyrdogenated vegetable oil differentially affect hepatic lipid and lipoprotein metabolism in golden Syrian hamsters. J Nutr.139:257–263.

[76] Kritchevsky D (2001) Diet and atherosclerosis. J Nutr Health Aging. 5:155–159.

[77] Dorfman SE, Laurent D, Gounarides JS, Li X, Mullarkey TL, Rocheford EC, Sari-Sarraf F, Hirsch EA, Hughes TE, Commerford SR (2009) Metabolic implications of dietary trans-fatty acids. Obesity.17:1200–1207.

[78] Costa RRS, Villela NR, Souza MdGS, Boa BCS, Cyrino FZGA, Silva SV, Lisboa PC, Moura EG, Barja-Fidalgo TC, Bouskela E (2011) High fat diet induces central obesity, insulin resistance and microvascular dysfunction in hamsters. Microvascular Res. 82:416-422.

[79] Kothari HV, Poirier KJ, Lee WH, Satoh Y (1997) Inhibition of cholesterol ester transfer protein CGS 25159 and changes in lipoproteins in hamsters. Atherosclerosis. 128: 59-66.

[80] Briand F, Thiéblemont Q, Muzotte E, Sulpice T (2012). High-Fat and Fructose Intake Induces Insulin Resistance, Dyslipidemia, and Liver Steatosis and Alters In Vivo Macrophage-to-Feces Reverse Cholesterol Transport in Hamsters. J Nutr. 142(4):704-9.

[81] Bhathena J, Kulamarva A, Martoni C, Malgorzata A, Urbanska, Malhotra M, Paul A, Prakash S (2011) Diet-induced metabolic hamster model of nonalcoholic fatty liver disease. Diabetes, Metabolic Syndrome and Obesity: Targets and Therapy.4:195–203

[82] Avramoglu RK, Qiu W, Adeli K (2003) Mechanisms of metabolic dyslipidemia in insulin resistant states: deregulation of hepatic and intestinal lipoprotein secretion. Frontiers in Bioscience. 8: 464-476.

[83] Sullivan MP, Cerda JJ, Robbins FL, Burgin CW, Beatty RJ (1993) The gerbil, hamster, and guinea pig as rodent models for hyperlipidemia. Lab Anim Sci. 43,575-578.

[84] Hoang VQ, Botham KM, Benson GM, Eldredge EE, Jackson B, Pearce N, Suckling KE (1993) Bile acid synthesis in hamster hepatocytes in primary culture: sources of cholesterol and comparison with other species. Biochim Biophys Acta. 1210:73-80.

[85] Fernandez ML(2001) Guinea pigs as models for cholesterol and lipoprotein metabolism. J Nutr.131:10-20.

[86] West KL, Fernandez ML (2004) Guinea pigs as models to study the hypocholesesterolemic effects of drugs. Cardiovasc Rev. 22:7-22.

[87] Fernandez ML, Volek JS (2006) Guinea pigs: a suitable animal model to study lipoprotein metabolism, atherosclerosis and inflammation. Nutr Metab (Lond). 3: 17.

[88] Greeve J, Altkemper I, Dietrich J-H, Greten H, Windler E (1993) Apolipoprotein mRNA editing in 12 different mammalian species: hepatic expression is reflected in low concentrations of apoB-containing plasma lipoproteins. J Lipid Res. 34:1367-1383.

[89] Fernandez ML, Yount NY, McNamara DJ (1990) Whole body cholesterol synthesis in the guinea pig. Effects of dietary fat quality. Biochim Biophys Acta. 1044:340-348.

[90] Madsen CS, Janovitz E, Zhang R, Nguyen-Tran V, Ryan CS, Yin X, Monshizadegan H, Chang M, D'Arienzo C, Scheer S, Setters R, Search D, Chen X, Zhuang S, Kunselman L, Peters A, Harrity T, Apedo A, Huang C, Cuff CA, Kowala MC, Blanar MA, Sun CQ, Robl JA, Stein PD (2008). The Guinea pig as a preclinical model for demonstrating the efficacy and safety of statins. J Pharmacol Exp Ther. 324(2):576-86.

[91] Fernandez ML, Roy S, Vergara-Jimenez M (2000) Resistant starch and cholestyramine have distinct effects on hepatic cholesterol metabolism in guinea pigs fed a hypercholesterolemic diet. Nutr Res.20:837-850.

[92] West K, Ramjiganesh T, Roy S, Keller BT, Fernandez ML (2002) SC-435, an ileal, apical sodium-dependent bile acid transporter inhibitor (ASBT) alters hepatic cholesterol metabolism and lowers plasma low-density-lipoprotein-cholesterol concentrations in guinea pigs. J Pharmacol Exp Theraup.303:291-299.

[93] Aggarwal D, West KL, Zern TL, Shrestha S, Vergara-Jimenez M, Fernandez ML (2005) JTT-130, a microsomal triglyceride transfer protein (MTP) inhibitor lowers plasma triglycerides and LDL cholesterol concentrations without increasing hepatic triglycerides in guinea pigs. BMC Cardiovasc Disord. 27:5:30.

[94] Karimi I, Hayatgheybi H, Shamspur T, Kamalak A, Pooyanmehr M, Marandi Y (2011) Chemical composition and effect of an essential oil of *Salix aegyptiaca* L. (Musk willow) in hypercholesterolemic rabbit model. Braz J Pharmacog. 21(3): 407-414.

[95] Lin ECK, Fernandez ML, Tosca MA, McNamara DJ (1994) Regulation of hepatic LDL metabolism in the guinea pig by dietary fat and cholesterol. J Lipid Res.35:446-57.

[96] Lynch SM, Gaziano JM, Frei B (1996) Ascorbic acid and atherosclerotic cardiovascular disease. Subcell Biochem. 25: 331-67.

[97] McCormick SPA (2004) Lipoprotein(a): biology and clinical importance. Clin Biochem Rev. 25(1): 69–80.

[98] Hrzenjak A, Frank S, Wo X, Zhou Y, Van Berkel T, Kostner GM (2003) Galactose-specific asialoglycoprotein receptor is involved in lipoprotein (a) catabolism. Biochem J. 376:765-71.

[99] Rath M, Pauling L (1990) Hypothesis: lipoprotein(a) is a surrogate for ascorbate. Proc Natl Acad Sci USA.87: 6204-7.

[100] Leite JO, DeOgburn R, Ratliff J, Su R, Smyth JA, Volek JS, McGrane MM, Dardik A, Fernandez ML (2010) Low-carbohydrate diets reduce lipid accumulation and arterial inflammation in guinea pigs fed a high-cholesterol diet. Atherosclerosis. 209(2):442-8.

[101] Mangathayaru K, Kuruvilla S, Balakrishna K, Venkhatesh J (2009) Modulatory effect of *Inula racemosa* Hook. f. (Asteraceae) on experimental atherosclerosis in guinea-pigs. J Pharm Pharmacol. 61(8):1111-8.

[102] Leite JO, Vaishnav U, Puglisi M, Fraser H, Trias J, Fernandez ML (2009) A-002 (Varespladib), a phospholipase A2 inhibitor, reduces atherosclerosis in guinea pigs. BMC Cardiovasc Disord. 17:7.

[103] Yang R, Guo P, Song X, Liu F, Gao N (2011) Hyperlipidemic guinea pig model: mechanisms of triglyceride metabolism disorder and comparison to rat. Biol. Pharm. Bull. 34(7): 1046-1051.

[104] Evans MB, Tonini R, Shope CD, Oghalai JS, Jerger JF, Insull W Jr, Brownell WE (2006) Dyslipidemia and auditory function. Otol Neurotol. 27(5):609-14.

[105] Castañeda-Gutiérrez E, Pouteau E, Pescia G, Moulin J, Aprikian O, Macé K (2011) The guinea pig as a model for metabolic programming of adiposity. Am J Clin Nutr. 94:1838S-1845S

[106] Caillier B, Pilote S, Patoine D, Levac X, Couture C, Daleau P, Simard C, Drolet B (2012) Metabolic syndrome potentiates the cardiac action potential-prolonging action of drugs: a possible 'anti-proarrhythmic' role for amlodipine. Pharmacol Res. 65(3):320-7.

[107] Ignatowski AC (1908) Influence of animal food on the organism of rabbits. Izv Imp Voyenno-Med Akad Peter. 16:154–173.

[108] Moghadasian MH, Frohlich JJ, McManus BM (2001) Advances in experimental dyslipidemia and atherosclerosis. Lab Invest. 81:1173– 1183.

[109] Xiangdong L, Yuanwu L, Hua Z, Liming R, Qiuyan L, Ning L (2011) Animal models for the atherosclerosis research: a review. Protein Cell. 2(3):189-201.

[110] Sider KL, Blaser MC, Simmons CA (2011) Animal models of calcific aortic valve disease. International Journal of Inflammation.ID 364310, 18 pages doi:10.4061/2011/364310.

[111] Karimi I, Hayatgheybi H, Razmjo M, Yousefi M, Dadyan A, Hadipour MM (2010) Anti-hyperlipidaemic effects of an essential oil of *Melissa officinalis*. L in cholesterol-fed rabbits. J Appl Biologl Sci. 4(1): 23-28.

[112] Badimon L (2001) Atherosclerosis and thrombosis: lessons from animal models. Thromb Haemost. 86: 356-65.

[113] Taylor JM, Fan J (1997) Transgenic rabbit models for the study of atherosclerosis. Front Biosci. 2: d298-308.

[114] Yanni AE (2004) The laboratory rabbit: an animal model of atherosclerosis research. Lab Anim. 38: 246–256.

[115] Morelli A, Vignozzi L, Maggi M, Adorini L (2011) Farnesoid X receptor activation improves erectile dysfunction in models of metabolic syndrome and diabetes. Biochim Biophys Acta. 1812(8):859-66.

[116] Mallidis C, Czerwiec A, Filippi S, O'Neill J, Maggi M, McClure N (2011) Spermatogenic and sperm quality differences in an experimental model of metabolic syndrome and hypogonadal hypogonadism. Reproduction. 142(1):63-71.

[117] Helfenstein T, Fonseca FA, Ihara SS, Bottós JM, Moreira FT, Pott H Jr, Farah ME, Martins MC, Izar MC (2011) Impaired glucose tolerance plus hyperlipidaemia induced by diet promotes retina microaneurysms in New Zeala6nd rabbits. Int J Exp Pathol. 92(1):40-9.

[118] Corona G, Rastrelli G, Morelli A, Vignozzi L, Mannucci E, Maggi M (2011) Hypogonadism and metabolic syndrome. J Endocrinol Invest. 34(7):557-67.

[119] Vignozzi L, Morelli A, Sarchielli E, Comeglio P, Filippi S, Cellai I, Maneschi E, Serni S, Gacci M, Carini M, Piccinni MP, Saad F, Adorini L, Vannelli GB, Maggi M (2012) Testosterone protects from metabolic syndrome-associated prostate inflammation: an experimental study in rabbit. J Endocrinol. 212(1):71-84.

[120] Morelli A, Comeglio P, Filippi S, Sarchielli E, Cellai I, Vignozzi L, Yehiely-Cohen R, Maneschi E, Gacci M, Carini M, Adorini L, Vannelli GB, Maggi M (2012) Testosterone and farnesoid X receptor agonist INT-747 counteract high fat diet-induced bladder alterations in a rabbit model of metabolic syndrome. J Steroid Biochem Mol Biol. DOI: 10.1016/j.jsbmb.2012.02.007

[121] Kainuma M, Fujimoto M, Sekiya N, Tsuneyama K, Cheng C, Takano Y, Terasawa K, Shimada Y (2006) Cholesterol-fed rabbit as a unique model of nonalcoholic, nonobese, non-insulin-resistant fatty liver disease with characteristic fibrosis. J Gastroenterol. 41:971-80.

[122] Pasławski R, Pasławska U, Szuba A, Nicpoń J (2011) Swine as a Model of Experimental Atherosclerosis. Adv Clin Exp Med. 20: 211–215.

[123] Royo T, Alfón J, Berrozpe M, Badimon L (2000) Effect of gemfibrozil on peripheral atherosclerosis and platelet activation in a pig model of hyperlipidemia. Eur J Clin Invest. 30(10): 843–852.

[124] Jacobsson L (1998) Experimental hyperlipidemia and atherosclerosis in mini-pigs; influence of certain drugs. Scand J Lab Anim Sci. 25: 85–91.

[125] Johansen T, Hansen HS, Richelsen B, Malmlof R (2001) The obese Göttingen minipig as a model of the metabolic syndrome: Dietary effects on obesity, insulin sensitivity, and growth hormone profile. Comp Med. 51:150-155.

[126] Narayanaswamy M, Wright KC, Kandarpa K (2000) Animal models for atherosclerosis, restenosis, and endovascular graft research. J Vasc Interv Radiol.11: 5-17.

[127] Berkhout TA, Simon HM, Jackson B, Yates J, Pearce N, Groot PH, Bentzen C, Niesor E, Kerns WD, Suckling KE (1997) SR-12813 lowers plasma cholesterol in beagle dogs by decreasing cholesterol biosynthesis. Atherosclerosis. 133: 203-12.

[128] Xenoulis PG, Suchodolski JS, Levinski MD, Steiner JM (2007) Investigation of hypertriglyceridemia in healthy Miniature Schnauzers. J Vet Intern Med. 21: 1224-30.

[129] Henson MS, O'Brien TD (2006) Feline models of type 2 diabetes mellitus. ILAR J. 47: 234-242.

[130] German AJ (2006) The growing problem of obesity in dogs and cats. J Nutr. 136:1940S-1946S.

[131] Clarke I (2008) Models of 'obesity' in large animals and birds. Obesity and Metabolism. 36: 107-117.

[132] Homan R, Hanselman JC, Bak-Mueller S, Washburn M, Lester P, Jensen HE, Pinkosky SL, Castle C, Taylor B (2010) Atherosclerosis in *Octodon degus* (degu) as a model for human disease. Atherosclerosis. 212(1):48-54.

[133] Shafrir E, Ziv E (1998) Cellular mechanism of nutrionally induced insulin resistance: the desert rodent *Psammomys obesus* and other animals in which insulin resistance leads to detrimental outcome. J Basic Clin Physiol Pharmacol. 9: 347-85.

[134] Kaiser N, Nesher R, Donath MY, Fraenkel M, Behar V, Magnan C (2005) *Psammomys obesus*, a model for environment-gene interactions in type 2 diabetes. Diabetes. 54: S137-44.

[135] Velasquez MT, Kimmel PL, Michaelis OE (1990) Animal models of spontaneous diabetic kidney disease. FASEB J. 4: 2850-9.

[136] Weir BJ (1974) The development of diabetes in the tuco-tuco (*Ctenomys talarum*). Proc R Soc Med. 67(9):843-6.

[137] Zhao Z-J, Chen J-F, Wang D-H (2010) Diet-induced obesity in the short-day-lean Brandt's vole. Physiol Behav. 99(1):47-53.

[138] Noda K, Melhorn MI, Zandi S, Frimmel S, Tayyari F, Hisatomi T, Almulki L, Pronczuk A, Hayes KC, Hafezi-Moghadam A (2010) An animal model of spontaneous metabolic syndrome: Nile grass rat. FASEB J. 24: 2443–2453.

[139] Roberts JC, Jr, Straus R (Eds) (1965) Comparative atherosclerosis; the morphology of spontaneous and induced atherosclerotic lesions in animals and its relation to human disease, Harper & Row, New York

[140] Anderson JL, Smith SC, Taylor Jr RL (2012) Spontaneous Atherosclerosis in Pigeons: A good model of human disease, atherogenesis, Prof. Sampath Parthasarathy (Ed.), ISBN: 978-953-307-992-9, InTech.

[141] Tigno XT, Gerzanich G, Hansen BC (2004) Age-related changes in metabolic parameters of nonhuman primates. J Gerontol A Biol Sci Med Sci. 59:1081–1088.

[142] Kojic ZZ (2003) Animal models in the study of atherosclerosis. Srp Arh Celok Lek. 131: 266-70.

[143] Wagner JD, Kavanagh K, Ward GM, Auerbach BJ, Harwood Jr HJ, Kaplan JR (2006) Old World Nonhuman Primate Models of Type 2 Diabetes Mellitus. ILAR J. 47: 259-271.

[144] Kusumi Y, Scanu AM, McGill HC, and Wissler RW (1993) Atherosclerosis in a rhesus monkey with genetic hypercholesterolemia and elevated plasma Lp(a). Atherosclerosis. 99:165–174.

[145] Pick R, Johnson PJ, Glick G (1974) Deleterious effects of hypertension on the development of aortic and coronary atherosclerosis in stumptail macaques (*Macaca speciosa*) on an atherogenic diet. Circ Res. 35: 472-82.

[146] Kaufman D, Smith ELP, Gohil BC, Banerji M-A, Coplan JD, Kral JG, Rosenblum LA (2005) Early Appearance of the metabolic syndrome in socially reared bonnet macaques. J Clin Endocrinol Meta. 90:404–408.

[147] Zhang X, Zhang R, Raab S, Zheng W, Wang J, Liu N, Zhu T, Xue L, Song Z, Mao J, Li K, Zhang H, Zhang Y, Han C, Ding Y, Wang H, Hou N, Liu Y, Shang S, Li C, Sebokova E, Cheng H, Huang PL (2011) Rhesus macaques develop metabolic syndrome with reversible vascular dysfunction responsive to pioglitazone. Circulation. 124(1):77-86.

Chapter 22

Genetically Modified Animal Models for Lipoprotein Research

Masashi Shiomi, Tomonari Koike
and Tatsuro Ishida

Additional information is available at the end of the chapter

http://dx.doi.org/10.5772/50258

1. Introduction

Coronary artery disease (CAD) is a leading cause of death in the world and one of the major risk factors for CAD is dyslipidemia. In understanding dyslipidemia and developing therapeutics, animal models, especially genetically modified animals, have played important roles and contributed greatly to progress in this field. Before the development of genetically modified animals, the Watanabe heritable hyperlipidemic (WHHL) rabbit, the first animal model for familial hypercholesterolemia, developed by Yoshio Watanabe in 1980 (Watanabe, 1980), helped to verify a low-density lipoprotein (LDL) receptor-pathway in vivo and to clarify lipoprotein metabolism in humans (Goldstein, 1983), in addition to the process by which atherosclerosis develops (Shiomi, 2009). Furthermore, WHHL rabbits have contributed to the development of hypocholesterolemic agents, statins, (Watanabe, 1981; Tsujita, 1986) and to clarifying anti-atherosclerotic effects (Watanabe, 1988; Shiomi, 1995; 2009). In the present, WHHL rabbits were improved by selective breeding to produce the WHHLMI strain, which suffers from severe and vulnerable coronary atheromatous plaques and myocardial infarction due to coronary occlusion with progression of atherosclerotic plaques (Shiomi, 2003). However, WHHL or WHHLMI rabbits were not suitable for studying the role of genes in lipid metabolism, because it is difficult to apply genetic modification techniques to rabbits.

The first transgenic mice were developed in 1982 (Gordon, 1982) and the first knockout mice in 1984 (Bradley, 1984). Genetically modified mice are commonly used to study lipoprotein metabolism and atherosclerosis. The first transgenic mice for lipoprotein metabolism were LDLR-overexpressing mice, developed in 1988 (Hofmann, 1988), and the first knockout (KO) mice for lipoprotein metabolism were apolipoprotein (apo) E-KO mice, developed in 1992 (Zhang, 1992). Thereafter, numerous genetically modified mouse models were produced and these mice have contributed to a better understanding of lipoprotein metabolism. However, we should recognize that the lipoprotein metabolism of genetically modified mice is not entirely the same as that of humans, despite hyperlipidemia or hypercholesterolemia. In addition, recent studies have demonstrated different phenotypes manifested in mice and rabbits after the same gene transfer (Fan, 2003). The first transgenic rabbit was developed in 1985 (Hammer, 1985) and the first transgenic rabbit for lipoprotein metabolism, the hepatic lipase-overexpressing rabbit, was developed in 1994 (Fan, 1994). The differences in phenotype following gene transfer between mice and rabbits may be due to species differences in lipoprotein metabolism. When using animal models in experimental research, one has to be careful interpreting the results. Since other chapters explain in detail the functions of enzymes, apolipoproteins, and receptors relating to lipoprotein metabolism, this chapter concentrates on introducing various genetically modified animals and species differences in phenotype expression after gene modification for researchers wishing to study lipoprotein metabolism.

2. Species differences in lipoprotein metabolism

2.1. Species differences in lipoprotein profiles

Fig. 1 shows lipoprotein profiles of mice, rabbits, and humans analysed with high performance liquid chromatography (HPLC). The main lipoprotein is LDL in human normal subjects but high-density lipoprotein (HDL) in wild-type mice. LDL is increased markedly in the plasma of patients with hypercholesterolemia (Yin, 2012) and WHHLMI rabbits, but lipoproteins that elute at the position of very low-density lipoprotein (VLDL) are increased in apoE-deficient mice, one of the most commonly used genetically modified mice (Piedrahita, 1992). Although serum lipid levels and the lipoprotein profile of apoE-KO mice vary depending on the colony, the lipoprotein profile of apoE-KO mice is similar to that of cholesterol-fed rabbits (Yin, 2012). In LDL receptor (LDLR)-deficient individuals, LDL increased markedly in patients (familial hypercholesterolemia) and rabbits (WHHLMI rabbits) despite a normal diet or chow, while plasma LDL levels were not so high in homozygous LDLR-KO mice fed standard chow as described by Ishibashi et al. (1993), although serum lipid levels and lipoprotein profiles of LDLR-KO mice vary in each colony similar to those of apoE-KO mice. After consumption of a cholesterol-enriched diet, plasma cholesterol levels increased markedly in LDLR-KO mice, similar to FH patients and WHHL rabbits, but the increased lipoprotein fraction eluted at the position of VLDL (Ishibashi, 1994) and the lipoprotein profile was similar to that of cholesterol-fed rabbits (Yin, 2012). These differences in lipoprotein profiles between mice and humans or rabbits are probably due to the species differences in lipoprotein metabolism.

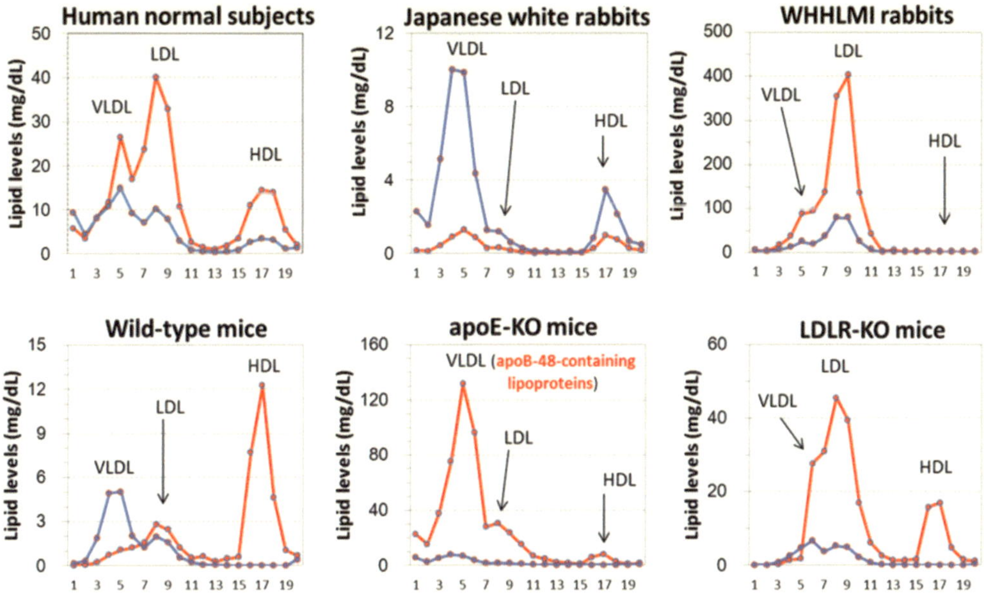

Figure 1. Lipoprotein profiles of a healthy human, rabbits, and mice. Lipoprotein profiles were analyzed with high performance liquid chromatography. Animals were fed standard chow. Red lines indicate cholesterol and blue lines indicate triglyceride

2.2. Lipoprotein metabolism in mice and rats

Fig. 2 shows a schematic diagram of lipoprotein metabolism in wild-type mice and rats. Dietary cholesterol is absorbed at the intestine mediated by Niemann-Pick C1-like 1 protein (NPC1L1) (Altmann, 2004) and ATP-binding cassette transporters (ABC) (Berge, 2000). Dietary fat is hydrolyzed to form monoglycerides and fatty acids in the intestine. These lipolytic products are translocated to the enterocyte membrane, and migrate to the endoplasmic reticulum. Overexpression of the tis7 gene in mice increases triglyceride absorption (Wang, 2009), and the transport of saturated fatty acids and cholesterol is decreased in fatty acid-binding protein (FABP)-KO mice (Newberry, 2009). Monoglycerides and fatty acids are re-esterified into triglycerides at the cytoplasmic surface of the endoplasmic reticulum. Synthesized triglycerides, cholesterol, and apolipoproteins are assembled in chylomicron particles. In the process of chylomicron assembly, MTP assists in binding lipids to apolipoproteins (Hussain, 2012). Since the apoB mRNA editing enzyme (apobec-1) functions at the intestinal wall, chylomicron particles contain apoB-48 as a major apolipoprotein but do not contain apoB-100. Chylomicron particles are released into the lacteal vessel. ApoB-48 is a marker of exogenous lipoprotein in humans and rabbits. At the capillary of peripheral tissue, chylomicron particles release free fatty acids (FFA) to adipose tissues mediated by lipoprotein lipase (LPL) and are converted to chylomicron remnants (Goldstein, 1983). Chylomicron remnants rapidly disappear from the circulation through apoE-receptors (apoER, remnant receptors) expressed in the liver. The metabolism of exogenous lipoproteins is preserved across species. In endogenous lipoprotein metabolism, cholesterol and other lipids synthesized in the liver are assembled in VLDL particles. Since apobec-1 is not expressed in the liver in humans and rabbits, VLDL particles contain apoB-100 but not apoB-48. However, VLDL particles of mice and rats contain both apoB-100 and apoB-48, because apobec-1 is expressed in the liver (Greeve, 1993). At the capillary in peripheral tissue, VLDL particles release FFA similar to chylomicron particles, which is then transformed into intermediate-density lipoprotein (IDL) (Goldstein, 1983). Part of the remaining VLDL particles and/or partially catalyzed VLDL particles bind to VLDL receptors expressed in peripheral tissue (Takahashi, 2004). Finally, apoB-48-containing VLDL and IDL in mice and rats disappear from the circulation through apoER expressed in the liver similar to chylomicron remnants. Some IDL particles containing apoB-100 bind to LDLRs in the liver, the rest release fatty acids via hepatic lipase (HL) and are transformed into LDL. LDL particles bind to LDLRs expressed at the surface of somatic cells. Therefore, a marker of endogenous lipoproteins is apoB-100 in humans and rabbits but endogenous lipoproteins of mice and rats contain both apoB-100 and apoB-48. Since the fractional catabolic rate for apoB-48-containing lipoproteins is very high compared to that for apoB-100-containing lipoproteins (Li, 1996), concentrations of VLDL and LDL are very low in wild-type mice and rats compared to humans. In reverse cholesterol transport from peripheral tissue to the liver, high-density lipoproteins (HDLs) receive free cholesterol from macrophages through scavenger receptor type B-I (SR-BI) and ABCs such as ABCA1 (Rohrer, 2009). The free cholesterol transported from macrophages is esterified by lecithin:cholesterol acyltransferase (LCAT) in plasma. The esterified cholesterol in HDL particles is transferred to VLDL, IDL, and LDL in plasma by cholesterol ester transfer protein (CETP) in humans and rabbits (Son, 1986), while mice and rats do not have CETP activity in plasma (Agellon, 1991). Therefore, the cholesterol ester in HDL particles is not transferred to apoB-containing lipoproteins in mice and rats. This is one of the major reasons why HDL is the predominant lipoprotein in these two species. Circulating HDL particles bind to SR-BI in the liver. Therefore, the large difference in lipoprotein metabolism between mice / rats and humans / rabbits is characterized by both the expression of apobec-1 in the liver and absence of CETP in the plasma of mice and rats. Dyslipidemia develops when lipoprotein metabolism is impaired. For example, LDLR deficiency causes familial hypercholesterolemia, and impaired LPL function causes hypertriglyceridemia. A number of animal models including transgenic animals, knockout animals, and spontaneous mutant animals can develop hypercholesterolemia, hypertriglyceridemia, and postprandial hypertriglyceridemia. However, the lipoprotein profile of these mice is greatly different from that of human patients.

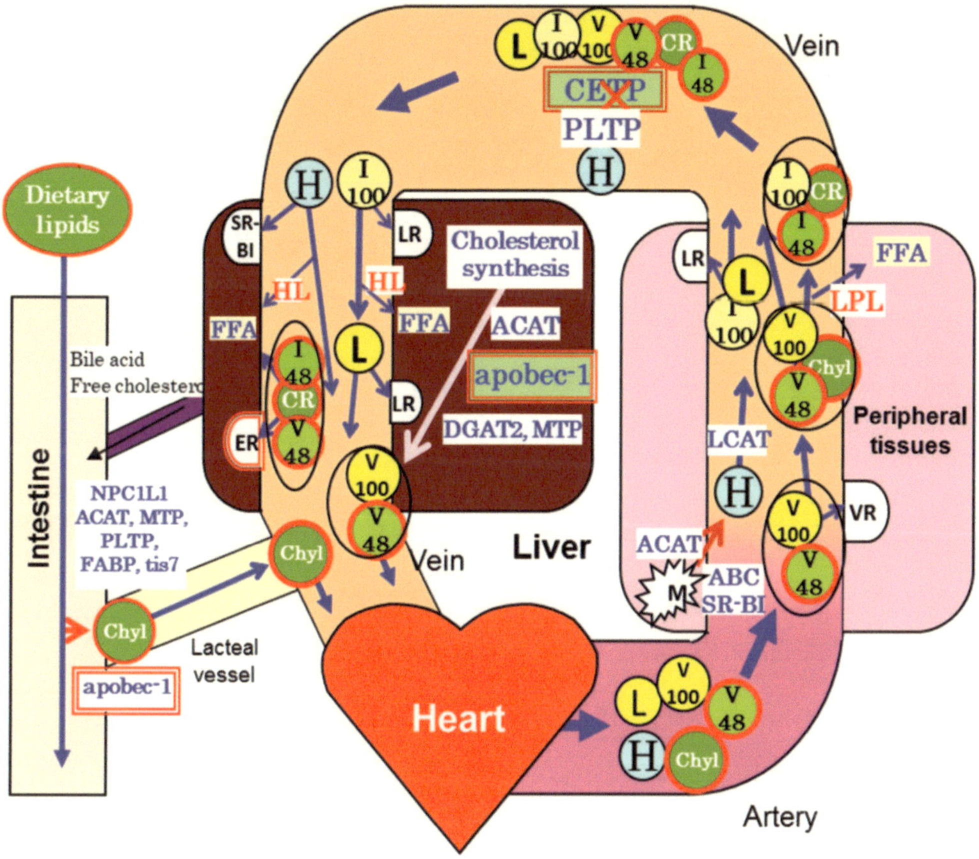

Figure 2. Lipoprotein metabolism in mice and rats (wild-type). Abbreviations: ACAT, acyl-CoA:cholesterol acyltransferase; apoB, apolipoprotein B; CETP, cholesterol ester transfer protein; Chyl, chylomicron; CR, chylomicron remnant; DGAT2, acyl-Co-A:diacylglycerol acyltransferase 2; ER, apoE receptor; FABP, fatty acids-binding protein; FFA, free fatty acids; H, high-density lipoprotein; HL, hepatic lipase; I48, intermediate-density lipoprotein (IDL) with apoB-48; I100, IDL with apoB-100; L, low-density lipoprotein (LDL); LCAT, lecithin:cholesterol acyltransferase; LPL, lipoprotein lipase; LR, LDL receptor; M, macrophage; MTP, microsomal triglyceride transfer protein; NPC1L1, Niemann-Pick C1-like 1; PLTP, phospholipid transfer protein; SR-BI, scavenger receptor type B-I; V48, very low-density lipoprotein (VLDL) with apoB-48; V100, VLDL with apoB-100; VR, VLDL receptor

3. Genetically modified animal models as tools for studying lipoprotein metabolism

Genetically modified animals for studying lipoprotein metabolism are summarized in Table 1. The list includes genetically modified mice, rats, rabbits, and chicken, but not double- or triple-modified animals. Several double- or triple-modified mice were developed by cross-breeding, for example, the apoE/LDLR-dKO mouse, the apoE/SRBI-dKO mouse, and others. Genetically modified animals can be good tools for clarifying the role of genes in lipoprotein metabolism and atherosclerosis if researchers take into consideration species differences.

	Transgenic animals			Knockout animals		
	mouse	rat	rabbit	mouse	rat	chick
Apolipoprotein						
apoA-I	Walsh, 1989	Swanson, 1992	Duverger, 1996	Plump, 1997		
apoA-II	Marzal-Casacub, 1996		Koike, 2009	Weng, 1996		
apoB100	Farese, 1995		Fan, 1995	Young, 1995		
apoC-I	Simonet, 1991			Gautier, 2002		
apoC-II	Shachter, 1994					
apoC-III	Aalto-Setala, 1992			Ding, 2011		
apoC-IV	Allan, 1996					
apoE	Shimano, 1992		Fan, 1998	Piedrahita, 1992		
apo(a)	Chiesa, 1992		Rouy, 1998			
apoM	Christoffersen, 2008		Christoffersen, 2008			
Cholesterol absorption in intestine						
NPC1L1				Altmann, 2004		
ABCA1				McNeish, 2003		Mulligan, 2003
ACAT1				Buhman, 2000		
Apobec 1				Kendrick, 2001		
MTP				Xie, 2006		
PLTP				Liu, 2007		
SR-BI				Mardones, 2001		
FABP				Newberry, 2009		
tis7	Wang, 2005					
VLDL secretion from liver						
DGAT2				Liu, 2008		
MTP				Raabe, 1998		
Apobec-1				Morrison, 1996		
Lipolytic enzyme						
LPL	Shimada, 1993		Fan, 2001	Coleman, 1995		
HL	Braschi, 1998		Fan, 1994	Gonzalez-Navarro, 2004		
EL	Ishida, 2003;			Ishida, 2003; Ma, 2003		
Lipoprotein metabolism						
LDLR	Hofmann, 1998			Ishibashi, 1993	Asahina, 2012	
PCSK9	Herbert, 2010			Rashid, 2005		
VLDLR				Frykman, 1995		
SR-type A				Suzuki, 1997		
Reverse cholesterol transport						
ABCA1	Vaisman, 2001			McNeish, 2000		
ABCG1	Kennedy, 2005			Kennedy, 2005		
SR-BI	Wang, 1998			Rigotti, 1997		
LCAT	Vaisman, 1995		Hoeg, 1996			
CETP	Aggellon, 1991	Herrera, 1999				
PLTP	Jiang, 1996		Masson, 2011	Jiang, 1999		

Table 1. List of genetically modified animals regarding lipoprotein metabolism.

3.1. Cholesterol absorption in the intestine

Recent studies using genetically modified animals have help to clarify the mechanism of cholesterol absorption in the jejunum. Dietary cholesterol forms micelles with bail acids in the lumen of the jejunum, which are then transported through NPC1L1. Thereafter, the free cholesterol is esterified by acyl coenzyme A:cholesterol acyltransferase 2 (ACAT2) and chylomicron particles are formed by the packaging of esterified cholesterol, triglyceride, and apolipoprotein B by MTP (Hussain, 2012). NPC1L1 was found by Altman et al (2002). NPC1L1, highly expressed in the jejunum and located on the surface of absorptive enterocytes, is critical for the intestinal absorption of dietary and biliary cholesterol (Altmann, 2004). NPC1L1 mediates cholesterol uptake through vesicular endocytosis. Davis et al (2004) produced NPC1L1 KO mice, which had substantially reduced intestinal uptake of cholesterol and sitosterol. NPC1L1-deficiency resulted in the up-regulation of intestinal hydroxymethylglutaryl-CoA synthase mRNA expression and an increase in intestinal cholesterol synthesis, the down-regulation of ABCA1 mRNA expression, and no change in ABCG5 and ABCG8 mRNA levels. Therefore, NPC1L1 is required for intestinal uptake of both cholesterol and phytosterols and plays a major role in cholesterol homeostasis. These findings in NPC1L1-KO mice were similar to results obtained with an inhibitor of cholesterol absorption, ezetimibe (Garcia-Calvo, 2005). Phospholipid transfer protein (PLTP) is also involved in cholesterol absorption in the intestine (Liu R, 2007). PLTP-KO mice absorb significantly less cholesterol than wild-type mice. In addition, mRNA levels of NPC1L1 and ABCA1 and MTP activity levels were significantly decreased in the small intestine of PLTP-KO mice. The free cholesterol taken up through NPC1L1 and PLTP is esterified by ACAT2. Experiments with ACAT2-KO mice demonstrated that a deficiency of ACAT2 activity inhibits cholesterol absorption in the intestine (Buhman, 2000). In the intestine, lipids absorbed are packaged with apolipoproteins and form chylomicron particles. The major structural apolipoprotein in chylomicron particles is apoB-48. ApoB-48 is produced by apobec-1, which inserts a stop codon. A deficiency of apobec-1 in the intestine resulted in reduction in the secretion and assembly of chylomicron particles (Kendrick, 2001). These results from apobec-1-KO mice suggested that apoB-48 is involved in the assembly of chylomicron particles (Lo, 2008). Finally, absorbed lipids and synthesized apolipoproteins are assembled by MTP. MTP-KO mice demonstrated a decrease in cholesterol absorption and chylomicron secretion, in addition to manifestations of steatorrhea (Xie, 2006). Although ABC and SR-BI were considered important to cholesterol absorption until the year 2000, SR-BI is not essential for intestinal cholesterol absorption (Mardones, 2001). Cholesterol absorption was independent of ABCA1 in KO mice (McNeish, 2000) and ABCA1-mutant chickens (Mulligan, 2003). Thus, studies with genetically modified animals have verified the mechanisms of dietary lipid absorption revealed by experiments in vitro.

3.2. Formation and secretion of VLDL particles from liver

The liver is the main organ in lipoprotein metabolism. Endogenous lipoprotein (VLDL) particles are produced in liver. The production and secretion of VLDL consist of the synthesis of cholesterol, triglyceride, phospholipids, and apolipoproteins, and assembly of these components. As described, apobec-1 is expressed in the liver in mice and rats, but not in humans and rabbits. Therefore, apoB-48-containing VLDL particles are secreted from the mouse and rat liver. Compared to those containing apoB-100, VLDL particles containing apoB-48 are rapidly cleared from circulation through apoER expressed on hepatocytes, similar to chylomicron remnants (Fig 2). To better approximate human lipoprotein metabolism, apobec-1-deficient mice were developed by gene targeting (Morrison, 1996). The LDL levels increased and HDL levels decreased in the circulation. However, overexpression of human apoB-100 showed different results between mice and rabbits. Plasma cholesterol levels decreased in apoB-100-overexpresing mice (Farese, 1996), although plasma cholesterol, triglyceride, and LDL levels increased and HDL levels decreased in apoB-100-overexpressing rabbits (Fan, 1995). This difference may be due to differences in CETP activity in the circulation between mice and rabbits. In addition, suppression of acyl-CoA: diacylglycerol acyltransferase 2 (DGAT2) expression by antisense treatment decreased VLDL secretion in mice (Liu, 2008). This suggests that suppression of triglyceride synthesis decreases VLDL secretion. Studies using inhibitors of MTP suggested that MTP plays a key role in the production of VLDL particles and inhibition of MTP activity decreases VLDL secretion in WHHL rabbits (Shiomi, 2001). Indeed, MTP+/- mice fed a high-fat diet demonstrated decreased levels of apoB-containing lipoproteins in plasma (Raabe, 1998). Furthermore, hepatocytes synthesize and/or secrete a lot of apolipoproteins, such as apoA, apoB, apoC, apoE, apoM, and apo(a). The function of these apolipoproteins has been clarified using genetically modified animals. However, influences of the overexpression of apoB, apoE, and apo(a) differ between mice and rabbits. ApoE overexpression resulted in a marked decrease in non-HDL cholesterol in mice (Shimano, 1992), while in rabbits, cholesterol of LDL and HDL increased and the fractional catabolic rate of chylomicron also increased (Fan, 1998). These differences between mice and rabbits may be due to CETP activity and the expression of abobec-1 in mouse liver. Lipoprotein (a), an atherogenic lipoprotein, is formed by the binding of apo(a) to LDL particles and is detected in plasma of only humans and monkeys. In human-apo(a) transgenic mice (Chiesa, 1992), apo(a) does not bind to mouse LDL particles, while human apo(a) binds to rabbit LDL particles and lipoprotein (a) is also atherogenic in rabbits (Rouy, 1998; Fan, 1999). Therefore, the role of endogenous apoB-containing lipoproteins (VLDL, IDL, and LDL) in the regulation of plasma lipid levels differs between genetically modified mice and rabbits or humans.

3.3. Lipolysis of apoB-containing lipoproteins

Lipoproteins are transporters in circulation that provide cholesterol as a material for steroid hormones and the cytoskeleton, and triglycerides (fatty acids) for energy to peripheral tissue. In the transportation of fatty acids, lipoprotein lipase (LPL), hepatic lipase (HL), and endothelial lipase (EL) function at capillaries and apoC affects lipolytic activities. LPL mediates the lipolysis of VLDL and chylomicrons, and these lipoprotein particles are transformed into IDL and chylomicron remnants, respectively. Although LPL-/- mice die within a day after birth because of dramatic hypertriglyceridemia, impaired fat tolerance, and hypoglycemia (Weinstock, 1995), these LPL-KO mice could be rescued by transient LPL expression induced by adenoviral-mediated gene transfer (Strauss, 2001). Rescued adult LPL-KO mice exhibit severe hypertriglyceridemia as patients with homozygous LPL-deficiency. LPL+/- mice showed increases in plasma triglyceride levels due to increases in the fraction of VLDL and chylomicrons in the circulation. Overexpression of LPL in mice (Shimada, 1993) and rabbits (Fan, 2001) caused decreases in plasma triglyceride, VLDL, and LDL levels, in addition to the suppression of atherosclerotic lesions (Shimada, 1996). ApoC-I modulates this metabolism. Although knockout of apoC-I gene did not affect serum lipid levels, expression of human CETP markedly increased levels of cholesterol ester in plasma, VLDL, and LDL in apoC-I-KO mice (Gautier, 2002). In contrast, in transgenic mice overexpressing human apoC-I, plasma triglyceride and total cholesterol levels were increased compared to those in wild-type mice. In addition, overexpression of apoC-I, apoC-II, apoC-III, and apoC-IV also increased plasma triglyceride and total cholesterol levels, and suppressed LPL activity in mice (Simonet, 1991; Shachter, 1994; Aalto-Setala, 1992; Allan, 1996) and rabbits (Ding, 2011). HL modulates the metabolism of both apoB-containing and apoA-containing lipoproteins. In apoE-KO mice, deficiency of HL showed a decrease in the fractional catabolic rate of apoB-48-containing VLDL, IDL and LDL despite no effects on apoB-100-containing LDL, in addition to increases in total cholesterol and triglyceride levels in apoB-containing and apoA-containing lipoproteins (Mezdour, 1997 & Gonzalez-Navarro H, 2004). However, development of atherosclerotic lesions was reduced in HL-KO mice. In HL-transgenic mice (Brashci, 1998) and rabbits (Fan, 1994), catabolism of both HDL and apoB-48-containing lipoproteins is enhanced, and plasma total cholesterol and triglyceride levels are decreased. Therefore, HL may be associated with catabolism of not only apoB-containing lipoproteins but HDL. EL is located in arterial endothelial cells and has phospholipase activity against phospholipids in HDL particles (Broedl, 2003; Ishida, 2003). EL hydrolyzes phospholipids on HDL particles and promotes catabolism of HDL. Overexpression of EL decreases in HDL cholesterol and apoA-I levels decreased in mice (Ishida, 2003; Jaye, 1999). By contrast, a deficiency of EL increases HDL levels (Ishida, 2003; Ma, 2003), in addition to atherogenic action (Ishida, 2004) and allergic asthma (Otera, 2009). Another study confirmed the high HDL-C levels in EL-/- mice but did not document an association with atherosclerosis (Ko, 2005). Thus, the role of EL in reverse cholesterol transport and atherosclerosis has not been fully elucidated. Further studies are required to clarify the function of EL in the metabolism of HDL and atherosclerosis. In studies about lipolysis, genetically modified animals have demonstrated no species differences, and are useful in this field.

3.4. Receptor-mediated catabolism of apoB-containing lipoproteins

3.4.1. LDL receptor

Lipoprotein receptors, such as LDLRs, VLDL receptors (VLDLRs), apoE receptors (remnant receptors, apoERs), and scavenger receptors (SRs), take up lipoprotein particles into parenchymal cells and/or phagocytes. LDLRs are expressed on the surface of parenchymal cells and bind to circulating LDL. The ligands are apoB-100 and apoE. A deficiency or the suppression of LDLRs results in the accumulation of LDL in the circulation, a condition known as human familial hypercholesterolemia. Several animal models for LDLR-deficiency have been developed. One of the better known models is the WHHL (Watanabe, 1980) or WHHLMI (Shiomi, 2003; 2009) rabbit. WHHL or WHHLMI rabbits show hypercholesterolemia due to LDLR deficiency even when fed standard chow. However, LDLR-KO mice (Ishibashi, 1993 & 1994) and LDLR-KO rats (Asahina, 2012) showed mildly increased serum cholesterol levels. Sanan et al. (1998) reported that LDLR-KO mice expressing human apoB-100 showed hypercholesterolemia due to the accumulation of LDL in the plasma even in chow feeding. In addition, Teng et al (1997) demonstrated that adenovirus-mediated gene delivery of apobec-1 reduced plasma apoB-100 levels, leading to the almost complete elimination of LDL particles and a reduction in LDL cholesterol in LDLR-KO mice. These studies suggests that the absence of any increase in plasma cholesterol levels in LDLR-KO mice is due to the expression of apobec-1 in the liver and apoB-100-containing LDL is a key player in LDLR deficiency to increased plasma cholesterol levels. Expression of LDLRs on the cell surface is regulated by proprotein convertase subtilisin/kexin type 9 (PCSK9). Recently, Huijgen et al. (2012) reported that plasma levels of PCSK9 were associated with LDL cholesterol levels in patients with familial hypercholesterolemia. In addition, overexpression of PCSK9 induced negative modulation of LDLR expression and decreased plasma LDL clearance, also promoting atherosclerosis (Herbert, 2010). In contrast, knockout of PCSK9 resulted in an increase in the LDL receptor protein (Mbikay, 2010) and a decrease in plasma cholesterol levels (Rashid, 2005). Furthermore, PCSK9 regulates the expression of not only LDLRs but VLDLRs and apoERs (Poirier, 2008). Therefore, PCSK9 can be considered a new target in the treatment of hypercholesterolemia.

3.4.2. VLDL receptor

VLDL particles are incorporated through VLDLRs. VLDLRs are expressed in heart, muscle, adipose tissues, and macrophages but not in liver in humans and rabbits. In mice, however, VLDLRs are not expressed in macrophages (Takahashi, 2011), suggesting the process of atherogenesis to be somewhat different between mice and humans or rabbits. Knockout of VLDLRs does not affect lipoprotein metabolism but decreases body weight, BMI, and epididymal fat in mice (Frykman, 1995). In addition, LPL activity is decreased by VLDLR-deficiency. These observations suggest VLDLRs to be associated with metabolic syndrome. Furthermore, a recent study suggests that the expression of VLDLRs is affected by PCSK9 (Roubtsova, 2011). Surprisingly, adipose tissues of apoE-KO mice did not express LDLR, VLDLR, and LDLR-related proteins (Huang, 2009), although wild-type mice developed these receptors in adipose tissue. Since the VLDLR has various functions, genetically modified animals may contribute to further studies.

3.4.3. *Remnant receptor*

Remnant receptors are mainly expressed in liver and contribute to the metabolism of exogenous lipoproteins, which contain apoB-48, in humans and rabbits. Therefore, down-regulation of remnant receptor function causes the accumulation of chylomicron remnants in plasma. As their ligand is apolipoprotein E, remnant receptors are also called apoE receptors (apoERs). In mice and rats, VLDL, IDL, and LDL contain apoB-48 due to the expression of apobec-1 in the liver. These apoB-48-containing lipoproteins bind to apoERs through interaction with the apoE ligand and disappear from the circulation rapidly. The fractional catabolic rate for apoB-48-containing VLDL is remarkably high compared to that for apoB-100-containing VLDL (Gonzalez-Navarro, 2004). This is one of the reasons why concentrations of VLDL and LDL are very low in plasma of mice and rats. In contrast, apoE-KO mice have very high VLDL concentration and the VLDL fraction consists of apoB-48 (Gonzalez-Navarro, 2004). Since apoE is a ligand of apoER, lipoproteins containing only apoB-48 cannot bind to apoERs in apoE-KO mice. This is the reason for the hypercholesterolemia in apoE-KO mice. Consequently, the hypercholesterolemia of apoE-KO mice due to the accumulation of apoB-48-containing lipoproteins is different from human hypercholesterolemia due to the accumulation of apoB-100-containing lipoproteins. This difference affects the development of hypocholesterolemic agents.

3.4.4. *Scavenger receptor type A*

Scavenger receptor type A (SR-A) is expressed on phagocytes and plays an important role in the removal of modified lipoproteins, such as oxidized-LDL, acetyl-LDL, and glycated-LDL. Therefore, SR-A plays an important role in atherogenesis. Knockout of SR-A decreases the uptake of modified LDL, but does not affect plasma lipid levels (Suzuki, 1997).

3.5. Reverse cholesterol transport

The plasma concentration of HDL is inversely related to the risk of atherosclerotic vascular diseases. HDL plays a key role in the reverse transport of cholesterol from peripheral tissue to liver. Recent studies suggest that HDL is also associated with anti-inflammation, anti-thrombosis, anti-oxidation, and the enhancement of endothelial function. Newly synthesized apoA-I binds to ABCs (particularly ABC-A1) or SR-BI of macrophages and takes up free cholesterol from macrophages. The complex of apoA-I and free cholesterol is transformed into discoidal nascent HDL (pre-β HDL). These nascent HDLs become HDL particles (HDL3) after esterification of the free cholesterol by LCAT in plasma. In the process of the transformation from discoidal HDL to HDL3, HDL takes up apoE and free cholesterol from macrophages mainly by ABCG1. Therefore, as the HDL matures, its cholesterol content increases. In humans and rabbits, CETP in plasma exchanges the cholesterol ester of HDL particles with triglyceride in apoB-containing lipoproteins. Therefore, peripheral cholesterol is transported by two pathways; an LDLR pathway mediated by CETP function and a SR-BI pathway. However, mice and rats do not have CETP activity in plasma.

Therefore, the pathway of reverse cholesterol transport is markedly different between mice / rats and humans / rabbits.

3.5.1. *Apolipoproteins of HDL particles*

HDL particles contain apoA, apoC, apoE, and apoM. ApoA, the main structural apolipoprotein of HDL particles, is mainly classified as apoA-I, apoA-II, and apoA-IV. ApoA and apoE play an important role in the efflux of cholesterol from macrophages to discoidal and small HDL, respectively. Recent studies suggested that apoM is related to the anti-oxidative function of HDL (Elsoe, 2012).

ApoA-I is a major structural apolipoprotein of HDL particles. ApoA is synthesized mainly in the liver and intestine, and from HDL particles hydrolyzed by HL. Humans and mice have two types of HDLs. One contains only apoA-I and the other, both apoA-I and apoA-II. However, rabbit HDLs are apoA-I particles (Chapman, 1980 & Koike, 2009). ApoA-I plays an important role in the reverse transport of cholesterol. Overexpression of apoA-I increases HDL-cholesterol levels in mice (Walsh, 1989), rats (Swanson, 1992), and rabbits (Duverger, 1996). In contrast, knockout of the apoA-I gene in mice decreases cholesterol levels in HDL, VLDL, and whole plasma (Plump, 1997). In addition, apoA-I-deficient HDL is a poor substrate for HL and LCAT. These studies using genetically modified animals indicate that apoA-I plays an important role in cholesterol reverse transport. Another major apolipoprotein of HDL is apoA-II. Overexpression of human apoA-II in mouse liver resulted in a decrease in plasma cholesterol levels due to a decrease in HDL cholesterol but an increase in plasma triglyceride levels (Marzal-Casacub, 1996). In addition, LCAT activity and mouse apoA-II levels in plasma were decreased in human apoA-II transgenic mice. Consequently, the changes in plasma lipid levels in human apoA-II transgenic mice may be due to a reduction in levels of mouse apoA-II. These results suggest species differences in apoA-II. In addition, apoA-II is dimer in human but monomer in mice. Knockout of apoA-II in mice resulted in a decrease in not only HDL-cholesterol but non-HDL cholesterol. In addition, the fractional catabolic rate for apoA-I was increased by a deficiency of apoA-II (Weng, 1996). Furthermore, the deficiency was associated with lower free fatty acid, glucose, and insulin levels, suggesting insulin hypersensitivity, while apoA-II does not relate to insulin sensitivity in humans. Therefore, the function of apoA-II is very confusing in mouse models. Conversely, rabbits overexpressing human apoA-II, which do not have apoA-II, lipid levels in plasma and non-HDL lipoproteins were increased but HDL-cholesterol levels and activities of LPL and HL were decreased (Koike, 2009). Therefore, effects of human apoA-II overexpression may be different between mice and rabbits. To clarify the function of apoA-II, more studies are required.

3.5.2. *Transfer of cholesterol from macrophages to HDL*

The start of the reverse cholesterol transport process is the transfer of cholesterol from macrophages to apoA-I, in which ABCs play important roles. Several strains of mice with genetically modified ABCA1 and ABCG1 have been produced. Overexpression of ABCA1 in mice increases cholesterol efflux from macrophages, in addition to increases in levels of cholesterol, apoA-I, and apoA-II in HDL (Vaisman, 2001). In contrast, ABCA1-KO mice showed a marked decrease in HDL-cholesterol, LDL-cholesterol, and plasma apoB levels, and an absence of apoA-I in plasma, but an increase in cholesterol absorption and the accumulation of lipid-laden macrophages (McNeish, 2000). Furthermore, in ABCG1-KO mice, cholesterol efflux from macrophages to HDL is decreased (Kennedy, 2005). After its efflux from macrophages to apoA-I and HDL, free cholesterol is esterified by LCAT in the plasma. Overexpression of LCAT in mice increases levels of cholesterol, apoA-I, apoA-II, and apoE in plasma, in addition to HDL cholesterol (Vaisman, 1995). LCAT-overexpressing rabbits showed an increase in HDL-cholesterol on a chow diet but non-HDL cholesterol was not increased on a cholesterol diet (Hoeg, 1996). Therefore, LCAT plays an important role in the esterification of free cholesterol in HDL. Studies using ABCA1-KO mice, ABCG1-KO mice, and ABCA1/ABCG1-dKO mice (Out, 2008) have elucidated how cholesterol is transported from macrophages to HDL. ApoE also promotes reverse cholesterol transport by enhancing the efflux of free cholesterol from peripheral macrophages to maturing HDL particles (Hayek, 1994). The efflux of free cholesterol to apoE-binding HDL is mediated by ABCG1/4 (Matsuura, 2006). The efflux from macrophages to apoE-containing HDL3 (small HDL particles) depends on apoE. HDL-cholesterol levels are markedly low in apoE-KO mice, despite being high HDL in wild-type mice (Zhang, 1992). Therefore, the low HDL-cholesterol levels in apoE-KO mice are due to a decrease in the efflux of cholesterol from macrophages to HDL. By contrast, the overexpression of apoE induced a marked decrease in apoB-containing lipoproteins in mice (Shimano, 1992), but human apoE3 overexpression increased cholesterol levels in not only HDL but LDL in rabbits (Fan, 1998). These differences between mice and rabbits are due to the activity of CETP in plasma, since no CETP activity is detected in mice but strong activity is found in rabbits.

3.5.3. *Transfer of lipids from HDL to apoB-containing lipoproteins*

One of the main courses of reverse cholesterol transport depends on CETP in plasma in human and rabbits (Son, 1986). However, mice and rats do not have CETP activity in plasma (Agellon, 1991). Therefore, in mice, the HDL cholesterol level is high but cholesterol levels of apoB-containing lipoproteins are markedly low. Overexpression of human CETP in mice induces a decrease in HDL-cholesterol but no changes in cholesterol levels in VLDL and LDL (Agellon, 1991). Similar findings were made in Dahl rats (Herrera, 1999). One of the reasons for no changes in cholesterol levels of non-HDL fraction is due to a rapid clearance of apoB-48-containing lipoproteins through apoER. These results suggest that CETP transfers cholesterol ester from HDL to apoB-containing lipoproteins. Furthermore, CETP expression led to atherosclerosis in Dahl rats fed a cholesterol diet. Therefore, CETP-overexpression can cause complex responses depending on diet. PLTP is a plasma protein, which transfers phospholipids from apoB-containing lipoproteins to HDL. Knockout of PLTP resulted in a decrease in cholesterol (Jiang, 1999; 2001). In PLTP transgenic mice, PLTP did not affect lipid levels of apoB-containing lipoproteins but increased phospholipid and cholesterol levels in HDL (Jiang, 1996). However, overexpression of PLTP in rabbits increased cholesterol levels in apoB-containing lipoproteins but had no effect on HDL lipids in a high-cholesterol deiet feeding (Masson, 2011). These differences in the function of PLTP between mice and rabbits may be due to fundamental differences in lipoprotein metabolism, such as CETP activity in the plasma and apobec-1 expression in the liver. Therefore, one has to be deliberate in interpreting results from gene modification studies.

3.5.4. HDL receptors

HDL particles are incorporated by SR-BI, a HDL receptor, expressed in liver. In humans and rabbits, cholesterol is transferred from peripheral macrophages to liver through two pathways, via CETP-LDLR and SR-BI, while in mice and rats, cholesterol is transported to liver via SR-BI expressed in liver. Overexpression of SR-BI in mice induces a decrease in plasma lipids and an increase in the fractional catabolic rate for HDL (Wang, 1998). In contrast, SR-BI-KO mice show increases in plasma cholesterol levels, HDL particle size, and levels of apoE and apoA-I in HDL particles (Rigotti, 1997). These results demonstrate the function of SR-BI in reverse cholesterol transport. However, these changes in plasma lipid levels reflect HDL lipid levels, because mice do not have CETP activity in the plasma.

4. Species differences in phenotypes between genetically modified animals

Table 2 shows species differences in phenotypes of lipoprotein metabolism between genetically modified animals when the same genes were modified. Overexpression of apoB-100 increased HDL-cholesterol but had no effect on non-HDL-cholesterol in mice fed a chow diet (Farese, 1996) but increased in cholesterol levels in plasma and LDL in rabbits (Fan, 1995). Overexpression of apoE decreased levels of apoB-containing lipoproteins in mice (Shimano, 1992) but increased cholesterol levels in plasma and LDL in rabbits (Fan, 1998). In addition, overexpression of PLTP increased HDL-cholesterol and apoA-I levels but did not affect LDL-cholesterol in mice (Jiang, 1996), while it increased LDL-cholesterol and did not affect HDL in rabbits (Masson, 2011). These differences may be due to high CETP activity in rabbits and no CETP activity in mice. Lipoprotein(a) is detected in humans and monkeys, and is an atherogenic lipoprotein. In mice overexpressing human apo(a) (Chiesa, 1992), apo(a) did not bind to mouse LDL particles, while in rabbits, human apo(a) can bound to rabbit LDL particles and formed Lp(a) (Rouy, 1998; Fan, 1999). Finally, LDLR-KO increased cholesterol levels mildly in plasma and LDL in mice (Ishibashi, 1993) and rats (Asahina, 2012), while spontaneous LDLR-deficient rabbits (WHHL or WHHLMI rabbits) show severe hypercholesterolemia due to the accumulation of LDL in plasma even on a normal diet (Goldstein, 1993; Shiomi, 2003; 2009). In humans, LDLR-deficiency produces severe hypercholesterolemia due to the accumulation of LDL in plasma. These differences in effects of LDLR between humans / rabbits and mice / rats may be due to the expression of apobec-1 in the liver and CETP activity in the plasma. Therefore, one has to consider species differences when using animal models. LDLR-KO mice on a high-fat diet showed dramatic hypercholesterolemia (Ishibashi, 1994), and the plasma cholesterol level is comparable with or higher than that of apoE-KO mice. In contrast, the degree of atherosclerosis is greater in apoE-KO mice than LDLR-KO mice. In addition, LDLR-KO mice on a chow diet did not show massive atherosclerotic lesions at the age of 12 months (Ishibashi, 1994), while apoE-KO mice on a chow diet for the same period showed massive atherosclerotic lesions (Zhang, 1992 and Reddick, 1994). These differences in plasma lipid profiles and atherosclerosis are not fully understood but are likely to be attributable to the quality or subtype of the circulating lipoproteins.

Genes modification	Mice and/or rats	Rabbits
Apo-B100 overexpression	Increase in HDL-cholesterol and no effect on non-HDL-cholesterol on chow feeding	Increase in plasma and LDL cholesterol
Apo-E overexpression	Decrease in non-HDL cholesterol	Increase in plasma and LDL cholesterol
Apo (a) overexpression	Not bind to mouse LDL	Binds to rabbit LDL and forms Lp(a)
PLTP overexpression	Increase in HDL-cholesterol and apoA-I but no effects on LDL-cholesterol	Increase in LDL-cholesterol but no effect on HDL
LDLR-deficiency	Mild increase in plasma cholesterol	Marked increase in plasma cholesterol

Table 2. Species differences in phenotype on overexpression of the same genes.

5. Genetically modified animal models for human dyslipidemia

Table 3 summarizes plasma lipid profiles of genetically modified animal models for human dyslipidemia. Plasma lipid and/or lipoprotein profiles of genetically modified mice resemble those for human diseases involving apoC-II deficiency, LPL deficiency, and CETP deficiency. However, plasma lipid levels and lipoprotein profiles of genetically deficient mice are markedly different from those of humans with a deficiency of ABCA1, apoE or LDLR. In ABCA1-KO mice, cholesterol levels markedly decreased in not only HDL but whole plasma, while patients with Tangier disease, who do not have ABCA1 and show very low levels of HDL-cholesterol, exhibit a mild decrease in plasma cholesterol levels. This difference between ABCA1-KO mice and patients with Tangier disease may be due to CETP activity in plasma. ApoE-KO mice fed normal chow show hypercholesterolemia and the increased lipoprotein fraction is VLDL, which contains apoB-48, although plasma triglyceride levels are almost normal. In addition, HDL-cholesterol levels are markedly low. However, patients with apoE deficiency show type III hyperlipidemia by the WHO classification. The increased lipoprotein is VLDL and IDL, and both cholesterol and triglyceride levels are increased. HDL cholesterol is almost normal (Mabuchi, 1989). These differences in lipoprotein metabolism between mice and patients may be due to the expression of apobec-1 in mouse liver and the triglyceride content of the VLDL fraction. Considering these observations, the hypercholesterolemia in apoE-KO mice may not reflect human hypercholesterolemia. In LDLR deficiency, although humans and rabbit models show marked hypercholesterolemia due to the accumulation of LDL in plasma despite a normal diet (Watanabe, 1980; Shiomi, 2009), the accumulation of LDL in plasma in homozygous LDLR-KO mice is mild (Ishibashi, 1993; 1994). These differences in plasma lipid levels and lipoprotein profiles between LDLR-KO mice and familial hypercholesterolemia or WHHL rabbits are due to the rapid clearance of apoB-48-containing VLDL, IDL, and LDL through apoER in mouse liver. Recently, LDLR-KO rats were developed (Asahina, 2012) . These animals have a similar lipoprotein profile to LDLR-KO mice. Studies demonstrate that both the expression of apobec-1 in liver and a deficiency of CETP in plasma greatly affect lipoprotein metabolism and plasma lipid levels in mice and rats. To solve these problems with LDLR-KO mice and apoE-KO mice, cross breeding with apobec-1-KO/CETP-expressing animals may be required in studies of lipoprotein metabolism. In the development of statins, potent anti-hyperlipidemic agents used by more than 40 million patients world-wide, no cholesterol-lowering effect was observed in mice and rats, although strong cholesterol-lowering effects were found in rabbits, dogs, monkeys and chickens (Tsujita, 1986). In addition, simvastatin, a statin, did not decrease serum cholesterol levels in LDLR-KO mice and CETP(+/-)LDLR(-/-)mice, and increased serum cholesterol levels in apoE-KO mice at a dose of 30 mg/kg/day (Yin, 2012), although an extremely high dose of statins (0.168% in diet, 200-300 mg/kg/day) decreased serum cholesterol levels in LDLR-KO mice (Krause, 1998). In contrast, WHHL rabbits, an animal model of familial hypercholesterolemia, have played important roles in studies of the hypocholesterolemic effects and anti-atherosclerotic effects of statins (Shiomi, 1995; 2009). The cholesterol-lowering effect of statins is mainly mediated by an increase in LDLR in liver. Therefore, the effect is weak when the contribution of LDLR to the regulation of plasma cholesterol levels is small, as in mice and rats. These studies suggest the need to select animal models based on study purposes.

Gene modification	Mice and/or rats	Human
ABCA1 deficiency	Marked decrease in cholesterol levels in both plasma and HDL	Tangier disease, Marked decrease in HDL cholesterol but mild decrease in plasma cholesterol
apoC-II deficiency	Hypertriglyceridemia	Hypertriglyceridemia
apoE-deficiency	Hypercholesterolemia Increase in VLDL and decrease in HDL	Combined hyperlipidemia Increase in IDL and no changes in HDL
LDLR-deficiency	Mild increase in plasma cholesterol Mild increase in LDL	Hypercholesterolemia Marked increase in LDL
LPL deficiency	Hypertriglyceridemia Lethal right after birth in homozygotes	Hypertriglyceridemia
CETP deficiency	High HDL cholesterol (wild-type mice)	High HDL cholesterol

Table 3. Plasma lipid profiles of genetically modified animal models for human dyslipidemia

6. Conclusion

In this chapter, the authors summarized achievements of studies using genetically modified animal models in lipoprotein research. The cross-breeding of genetically modified animals, such as double KO mice, triple KO mice, and others, has contributed to studies of lipoprotein metabolism. Studies using genetically modified mice have elucidated the mechanisms of cholesterol absorption in the intestine, lipolysis of apoB-containing lipoproteins, lipoprotein receptor function, and cholesterol efflux from macrophages to HDL. Although genetically modified animals are useful to elucidate the function of genes related to lipoprotein metabolism, we have to carefully select animal species to know the effect of these genes on lipid levels of whole plasma and lipoprotein profiles in humans. In addition, genetically modified mice have limitations in studies about the development of hypocholesterolemic agents, because of the expression of apobec-1 in liver and a deficiency of CETP activity in plasma. Consequently, these fundamental differences in lipoprotein metabolism between mice and humans affect the interpretation of results of gene modification about lipoprotein metabolism in mice. To solve these problems, genetically modified mice should be produced using CETP-transgenic/apobec-1-KO mice or animals having a background of no expression of apobec-1 in the liver and expression of CETP in plasma. We have to be careful in the interpretation of results obtained using genetically modified animals, and to select animal models in response to study purposes to extrapolate the results to humans. Recently, techniques of X-linked severe combined immunodeficiency (X-SCID) using zinc-finger nucleases, transcriptional activator-like effector nucleases (TALEN), and mutation using an N-ethyl-N-nitrosourea mutagenesis have become available for knockout gene expression. These techniques will be able to produce KO-animals other than mice. These animals will contribute further to studies of lipoprotein metabolism and lipid disorders in humans.

Author details

Masashi Shiomi, Tomonari Koike and Tatsuro Ishida
Kobe University Graduate School of Medicine, Japan

Acknowledgement

This work was supported in part by grants-in-aid for scientific research from the Ministry of Education, Culture, Sports and Technology, Japan (23300157).

7. References

Aalto-Setala, K. (1992). Transgenic animals in lipoprotein research. *Ann Med*, Vol.24, No.5, pp.405-409

Agellon, L.B.; Walsh, A.; Hayek, T.; Moulin, P.; Jiang, X.C.; Shelanski, S.A.; Breslow, J.L. & Tall, A.R. (1991). Reduced high density lipoprotein cholesterol in human cholesteryl ester transfer protein transgenic mice. *J Biol Chem*, Vol.266, No.17, pp.10796-10801

Allan, C.M. & Taylor, J.M. (1996). Expression of a novel human apolipoprotein (apoC-IV) causes hypertriglyceridemia in transgenic mice. *J Lipid Res*, Vol 37, No. 7, pp. 1510-1518

Altmann, S.W.; Davis, H.R.; Yao, X.; Laverty, M.; Compton, D.S.; Zhu, L.J.; Crona, J.H.; Caplen, M.A.; Hoos, L.M.; Tetzloff, G.; Priestley, T.; Burnett, D.A.; Strader, C.D. & Graziano, M.P. (2002). The identification of intestinal scavenger receptor class B, type I (SR-BI) by expression cloning and its role in cholesterol absorption. *Biochim Biophys Acta*, Vol.1580, No.1, pp. 77-93

Altmann, S. W.; Davis, H. R.; Zhu, L.; Yao, X.; Hoos, L. M.; Tetzloff, G.; Iyer S. N.; Maguire, M.; Golovko, A.; Zeng, M.; Wang, L.; Murgolo, N. & Graziano, M. P. (2004). Niemann-Pick C1 like 1 protein is critical for intestinal cholesterol absorption. *Science*, Vol.303, No.5661, pp.1201-1204

Asahina, M.; Mashimo, T.; Takeyama, M.; Tozawa, R.; Hashimoto, T.; Takizawa, A.; Ueda, M.; Aoto, T.; Kuramoto, K. & Serikawa, T. (2012). Hypercholesterolemia and atherosclerosis in low density lipoprotein receptor mutant rats. *Biochem Biophys Res Commun*, Vol. 418, No.3, pp. 553-558

Berge, K.E.; Tian, H.; Graf, G.A.; Yu, L.; Grishin, N.V.; Schultz, J.: Kwiterovich, P.; Shan, B.; Barnes, R. &, Hobbs, H.H. (2000). Accumulation of dietary cholesterol in sitosterolemia caused by mutations in adjacent ABC transporters. *Science*, Vol.290, No.5497, pp.1771-1775.

Bradley A, Evans M, Kaufman MH, & Robertson E. (1984). Formation of germ-line chimaeras from embryo-derived teratocarcinoma cell lines. *Nature*, Vol.309, No.5965, pp. 255-256

Braschi, S., Couture, N.; Gambarotta, A.; Gauthier, B.R.; Coffill, C.R.; Sparks, D.L.; Maeda, N. & Schultz, J.R. (1998). Hepatic lipase affects both HDL and ApoB-containing lipoprotein levels in the mouse. *Biochim Biophys Acta*, Vol.1392, No.2-3, pp.276-290

Broedl, U.C; Maugeais, C.; Marchadier, D.; Glick, J.M. & Rader, D.L. (2003). Effects of nonlipolytic ligand function of endothelial lipase on high density lipoprotein metabolism in vivo. J Biol Chem, Vol.278, No.42, pp.40688-40693

Buhman, K.K.; Accad, M.; Novak, S; Choi, R.S.; Wong, J.S.; Hamilton, R.L.; Turley, S. &, Farese, R.V. Jr. (2000). Resistance to diet-induced hypercholesterolemia and gallstone formation in ACAT2-deficient mice. *Nat Med*, Vol.6, No.12, pp.1341-1347

Chapman, M.J. (1980). Animal lipoproteins: Chemistry, structure, and comparative aspects. *J Lipid Res*. Vol.21, No., pp.789–853

Chiesa, G.; Hobbs, H.H.; Koschinsky, M.L.; Lawn R.M.; Maika, S.D. & Hammer, R.E. (1992). Reconstitution of lipoprotein(a) by infusion of human low density lipoprotein into transgenic mice expressing human apolipoprotein(a). *J Biol Chem*, Vol.267, No.34, pp. 24369-24374

Christoffersen, C.; Ahnström, J.; Axler, O.; Christensen, E.I.; Dahlbäck, B. & Nielsen, L.B. (2008). The signal peptide anchors apolipoprotein M in plasma lipoproteins and prevents rapid clearance of apolipoprotein M from plasma. J Biol Chem, Vol.283, No.27, pp.18765-18772

Christoffersen, C,; Jauhiainen, M.; Moser, M.; Porse, B.; Ehnholm, C.; Boesl, M.; Dahlbäck, B. & Nielsen, L.B. (2008). Effect of apolipoprotein M on high density lipoprotein metabolism and atherosclerosis in low density lipoprotein receptor knock-out mice. *J Biol Chem*, Vol.283, No.4, pp.1839-1847

Coleman, T.; Seip, R.L.; Gimble, J.M.; Kee, D.; Meda, N. & Semenkovich, C.F. (1995). COOH-terminal disruption of lipoprotein lipase in mice is lethal in homozygotes, but heterozygotes have, elevated triglycerides and impaired enzyme activity. *J Biol Chem*, Vol.270, No. 21, pp. 12518-12525

Davis, H.R. Jr.; Zhu, L.J.; Hoos, L.M.; Tetzloff, G.; Maguire, M.; Liu, J.; Yao, X.; Iyer, S.P.N.; Lam, M.H.; Lund, E.G.; Detmers, P.A.; Graziano, M.P. & Altman, S.W. (2004). Niemann-Pick C1 like 1 (NPC1L1) is the intestinal phytosterol and cholesterol transporter and a key modulator of whole-body cholesterol homeostasis. *J Biol Chem*, Vol. 279, No.32, pp. 33586-33592

Ding, Y.; Wang, Y.; Zhu, H.; Fan, J.; Liu, G. & Liu. E. (2011). Hypertriglyceridemia and delayed clearance if fat load in transgenic rabbits expressing human apolipoprotein CIII. *Transgenic Res*, Vol 20, No.4, pp.867-875

Duverger, N.; Kruth, H.; Emmanuel, F.; Caillaud, J.M.; Viglietta, C.; Castro, G.; Tailleux, A.; Fievet, C.; Fruchart, J.C.; Houdebine, L.M. & Denefle, P. (1996). Inhibition of atherosclerosis development in cholesterol-fed human apolipoprotein A-I-transgenic rabbits. *Circulation*. Vol.94, No.4, pp.713-717

Elsøe, S.; Ahnström, J.; Christoffersen, C.; Hoofnagle, A.N.; Plomgaard, P.; Heinecke, J.W.; Binder, C.J.; Björkbacka, H.; Dahlbäck, B. & Nielsen, L.B. (20129. Apolipoprotein M binds oxidized phospholipids and increases the antioxidant effect of HDL. *Atherosclerosis*, Vol.221, No.1, pp.91-97

Fan, J.; Araki, M.; Wu, L.; Challah, M.; Shimoyamada, H.; Lawn, R.M.; Kakuta, H.; Shikama, H. & Watanabe, T. (1999). Assembly of lipoprotein (a) in transgenic rabbits expressing human apolipoprotein (a). *Biochem Biophys Res Commun*, Vol.255, No.3, pp.639-644.

Fan, J.; Ji, Z.S.; Huang, Y.; de Silva, H.; Sanan, D.; Mahley, R.W.; Innerarity, T.L. & Taylor, J.M. (1998). Increased expression of apolipoprotein E in transgenic rabbits results in reduced levels of very low density lipoproteins and an accumulation of low density lipoproteins in plasma. *J Clin Invest*, Vol.101, No.10, pp.2151-2164

Fan, J.; McCormick, S.P.; Krauss, R.M.; Taylor, S.; Quan, R.; Taylor, J.M. & Young, S.G. (1995). Overexpression of human apolipoprotein B-100 in transgenic rabbits results in increased levels of LDL and decreased levels of HDL. *Arteroscler Thromb Vasc Biol*, Vol.15, No.11, pp.1889-1899

Fan, J.; Unoki, H.; Kojima, N.; Sun, H.; Shimoyamada, H.; Deng, H.; Okazaki, M.; Shikama, H.; Yamada, N. & Watanabe, T. (2001). Overexpression of lipoprotein lipase in transgenic rabbits inhibits diet-induced hypercholesterolemia and atherosclerosis. *J Biol Chem*, Vol.276, No.43, pp.40071-40079

Fan, J., Wang. J. Bensadoun, A., Lauer, S.J., Dang, Q., Mahley, R.W. & Taylor, J.M. (1994). Overexpression of hepatic lipase in transgenic rabbits leads to a marked reduction of plasma high density lipoprotein and intermediate density lipoproteins. *Proc Natl Acd Sci USA*,Vol. 91, No. 18, pp. 8724-8728

Fan, J. & Watanabe, T. (2003). Transgenic rabbits as therapeutic protein bioreactors and human disease models. *Pharmacology and Therapeutics*, Vol.99, No.3, pp. 261-282

Farese, R.V.Jr.; Cases, S.; Ruland, S.L.; Kayden, H.J.; Wong, J.S.; Young, S.G. & Hamilton, R.L. (1996). A novel function for apolipoprotein B: lipoprotein synthesis in the yolk sac is critical for maternal-fetal lipid transport in mice. *J Lipid Res*, Vol.37, No.2, pp.347-360

Farese, R.V.Jr.; Ruland, S.L.; Flynn, L.M.; Stokowski, R.P. & Young, S.G. (1995). Knockout of the mouse apolipoprotein B gene results in embryonic lethality in homozygotes and protection against diet-induced hypercholesterolemia in heterozygotes. *Proc Natl Acad Sci USA*, Vol .92, No.5, pp.1774-1778

Frykman, P.K.; Brown, M.S.; Yamamoto, T.; Goldstein, J.L. & Herz, J. (1995). Normal plasma lipoproteins and fertility in gene-targeted mice homozygous for a disruption in the gene encoding very low density lipoprotein receptor. *Proc Natl Acad Sci U S A*, Vol.92, Mo.18, pp.8453-8457

Garcia-Calvo, M.; Lisnock, J.; Bull, H.G.; Hawes, B.E.; Burnett, D.A.; Braun, M.P.; Crona, J.H.; Davis, H.R. Jr.; Dean. D.C.; Detmers, P.A.; Graziano, M.P.; Hughes, M.; Macintyre, D.E.; Ogawa, A.; O'neill, K.A.; Iyer, S.P.; Shevell, D.E.; Smith, M.M.; Tang, Y.S.; Makarewicz, A.M.; Ujjainwalla, F.; Altmann, S.W.; Chapman, K.T. & Thornberry, N.A. (2005). The target of ezetimibe is Niemann-Pick C1 like 1 (NPC1L1). *Proc Natl Acd Sci U S A*, Vol. 102, No. 23, pp. 8132-8137

Gautier, T.; Masson, D.; Jong, M.C.; Duverneuil, L.; Guern, N.L.; Deckert, V.; de Barros, J.P.; Dumont, L.; Bataille, A.; Zak, Z.; Jiang, X.; Tall, A.R.; Havekes, L.M. & Lagrost, L. (2002). Apolipoprotein CI deficiency markedly augments plasma lipoprotein changes mediated by human cholesterol ester transfer protein (CETP) in CETP transgenic/apoCI-knocked out mice. *J Biol Chem*, Vol 277, No.35, pp. 31354-31363

Goldstein, J.L.; Kita, T. & Brown, M.S. (1983). Defective lipoprotein receptors and atherosclerosis: Lessons from an animal counterpart of familial hypercholesterolemia. *N Engl J Med*, Vol. 309, No.5, pp. 288-296

Gonzalez-Navarro, H.; Nong, Z.; Amar, M.J.A.; Shamburek, R.D.; Najib-Fruchart, J.; Paigen, B.J.; Brewer, B. Jr. & Santamarina-Fojo, S. (2004). The ligand-binding function of hepatic lipase modulates the development of atherosclerosis in transgenic mice. *J Biol Chem*, Vol.279, No.44 , pp. 45312-45321

Gordon, J. W. & Ruddle, F.H. (1982). Germ line transmission in transgenic mice. *Progress in Clinical and Biological Research*, Vol.85, pp. 111-124.

Greeve, J.; Altkemper, I.; Dieterich, J.H.; Greten, H. & Windler, E. (1993). Apolipoprotein B mRNA editing in 12 different mammalian species: hepatic expression is reflected in low concentrations of apoB-containing plasma 1ipoproteins. *J Lipid Res*, Vol.34, No. 8, pp. 1367-1383.

Hammer, R.E., Pursel, V.G., Rexroad, C.E. Jr., Wall, R.J., Bolt, D.J., Ebert, K.M., Palmiter, R.D. & Brinster, R.L. (1985). Production of transgenic rabbits, sheep and pigs by microinjection. *Nature*, Vol.315, No.6021, pp. 680-683

Hayek, T.; Oiknine, J.; Brook, J.G. & Aviram, M. (1994). Role of HDL apolipoprotein E in cellular cholesterol efflux: studies in apo E knockout transgenic mice. *Biochem Biophys Res Commun*, Vol.205, No.2, pp.1072-11078

Herbert B, Patel D, Waddington SN, Eden ER, McAleenan A, Sun XM, Soutar AK. (2010). Increased secretion of lipoproteins in transgenic mice expressing human D374Y PCSK9 under physiological genetic control. *Arterioscler Thromb Vasc Biol*,Vol.30, No.7, pp.1333-1339

Herrera, V.L.; Makrides, S.C.; Xie, H.X.; Adari, H.; Krauss, R.M.; Ryan, U.S. & Ruiz-Opazo, N. (1999). Spontaneous combined hyperlipidemia, coronary heart disease and decreased survival in Dahl salt-sensitive hypertensive rats transgenic for human cholesteryl ester transfer protein. *Nat Med*, Vol.5, No.12, pp.1383-1389

Hoeg, J.M.; Santamarina-Fojo, S.; Bérard, A.M.; Cornhill, J.F.; Herderick, E.E.; Feldman, S.H.; Haudenschild, C.C.; Vaisman, B.L.; Hoyt, R.F.Jr.; Demosky, S.J.Jr.; Kauffman, R.D.; Hazel, C.M.; Marcovina, S.M. & Brewer, H.B. Jr. (1996). Overexpression of lecithin:cholesterol acyltransferase in transgenic rabbits prevents diet-induced atherosclerosis. *Proc Natl Acad Sci U S A*, Vol.93, No.21, pp.11448-11453

Hoeg, J. M.; Vaisman, B. L.; Demosky, S.J. Jr.; Meyn, S.M.; Talley, G.D.; Hoyt, R.F. Jr.; Feldman, S.; Berard, A.M.; Sakai, N.; Wood, D.; Brousseau, M.E.; Marcovina, S.; Brewer, H.B.Jr. & Santamarina-Hojo, S. (1996). "Lecithin:cholesterol acyltransferase overexpression generates hyperalpha-lipoproteinemia and a nonatherogenic lipoprotein pattern in transgenic rabbits. *J Biol Chem*, Vol.271, No.8, pp. 4396-4402

Hofmann, S.L.; Russell, D.W.; Brown, M.S.; Goldstein, J.L. & Hammer, R.E. (1988). Overexpression of low density lipoprotein (LDL) receptor eliminates LDL from plasma in transgenic mice. *Science*, Vol.239, No.4845, pp.1277-1281

Huang, Z.H.; Minshall, R.D. & Mazzone, T. (2009). Mechanism for endogenously expressed apoE modulation of adipocyte very low density lipoprotein metabolism. *J Biol Chem*, Vol.284, No.46, pp.31512-13522

Huijgen, R.; Fouchier, S.W.; Denoun, M.; Hutten, B.A.; Vissers, M.N.; Lambert, G. & Kastelein, J.J.P. (2012). Plasma levels of proprotein convertase subtilisin Kexin type 9 (PCSK9) and phenotypic variability in familial hypercholesterolemia. *J Lipid Res*, (in press)

Hussain, M.M.; Rava, P.; Walsh, M.; Rana, M. & Iqbal, J. (2012). Multiple functions of microsmal triglyceride transfer protein. *Nutr Metab*, Vol. 9, No. , pp. 14-

Ishibashi, S.; Brown, M.S.; Goldstein, J.L.; Gerard, R.D.; Hammer, R.E. & Herz, J. (1993). Hypercholesterolemia in low density lipoprotein receptor knockout mice and its reversal by adenovirus-mediated gene delivery. *J Clin Invest*, Vol.92, No.2, pp.883-893

Ishibashi, S., Goldstein, J.L., Brown, M.S., Herz, J. & Burns, D.K. (1994). Massive xanthomatosis and atherosclerosis in cholesterol-fed low density lipoprotein receptor-negative mice. *J Clin Invest*, Vol.93, No.5, pp.1885-1893

Ishida, T.; Choi, S.; Kundu, R.K.; Hirata, K.; Rubin, E.M.; Cooper ,A.D. & Quertermous, T. (2003). Endothelial lipase is a major determinant of HDL level. *J Clin Invest*, Vol.111, No.3, pp.347-355.

Ishida T, Choi SY, Kundu RK, Spin J, Yamashita T, Hirata K, Kojima Y, Yokoyama M, Cooper AD, Quertermous T. (2004). Endothelial lipase modulates susceptibility to atherosclerosis in apolipoprotein-E-deficient mice. J Biol Chem. Vol.279, No.43, pp45085-92.

Jiang, X.C.; Bruce, C.; Mar, J.; Lin, M.; Ji, Y.; Francone, O.L; & Tall, A.R. (1999). Targeted mutation of plasma phospholipid transfer protein gene markedly reduces high-density lipoprotein levels. *J Clin Invest*, Vol.103, No.6, pp.907-914

Jiang, X.; Francone, O.L.; Bruce, C.; Milne, R.; Mar, J.; Walsh, A.; Breslow, J.L. & Tall, A.R. (1996). Increased prebeta-high density lipoprotein, apolipoprotein AI, and phospholipid in mice expressing the human phospholipid transfer protein and human apolipoprotein AI transgenes. *J Clin Invest*, Vol.98, No.10, pp.2373-2380

Jiang, X.C.; Qin, S.; Qiao, C.; Kawano, K.; Lin, M.; Skold, A.; Xiao, X. & Tall, A.R. (2001). Apolipoprotein B secretion and atherosclerosis are decreased in mice with phospholipid-transfer protein deficiency. *Nat Med*, Vol.7, No.7, pp.847-852

Kennedy, M.A.; Barrera, G.C.; Nakamura, K.; Baldán, A.; Tarr, P.; Fishbein, M.C.; Frank, J.; Francone, O.L. & Edwards, P.A. (2005). ABCG1 has a critical role in mediating cholesterol efflux to HDL and preventing cellular lipid accumulation. *Cell Metab*, Vol.1, No.2, pp.121-131

Kendrick, J.S.; Chan, L. & Higgins, J.A. (2001). Superior role of apolipoprotein B48 over apolipoprotein B100 in chylomicron assembly and fat absorption: an investigation of apobec-1 knock-out and wild-type mice. *Biochem J*, Vol.356, No.3, pp. 821-827

Ko KW, Paul A, Ma K, Li L, Chan L. (2005). Endothelial lipase modulates HDL but has no effect on atherosclerosis development in apoE-/- and LDLR-/- mice. J Lipid Res. Vol.46, No.12. pp2586-94.

Koike, T.; Kitajima, S.; Yu, Y.; Li, Y.; Nishijima, K.; Liu, E.; Sun, H.; Wagar, A.B.; Shibata, N.; Inoue, T.; Wang, Y.; Zhang, B.; Kobayashi, J.; Morimoto, M.; Saku, K.; Watanabe, T. & Fan, J. (2009). Expression of human apoAII in transgenic rabbits leads to dyslipidemia: a new model for combined hyperlipidemia. *Arterioscler Thromb Vasc Biology*, Vol.29, No.12, pp. 2047-2053.

Koo, C.; Innerarity, T.L. & Mahley, R.W. (1985). Obligatory role of cholesterol and apolipoprotein E in the formation of large cholesterol-enriched and receptor-active high density lipoproteins. *J Biol Chem*, Vol.260, No.22, pp. 11934-11943

Krause, B.R. & Princen, H.M.G. (1998). Lack of predictablity of classical animal models for hypolipidemic activity: a ggod time for mice? Atherosclerosis, Vol 140, No.1 , pp.15-24

Li, X.; Catalina, F.; Grundy, S.M. & Patel, S. (1996). Method to measure apolipoprotein B-48 and B-100 secretion rates in an individual mouse: evidence for a very rapid turnover of VLDL and preferential removal of B-48- relative to B-100-containing lipoproteins. *J. Lipid Res.* Vol. 37, No.1, pp. 210-220.

Liu, R.; Iqbal, J.; Yeang, C.; Wang, D.Q.; Hussain, M.M. & Jiang, X.C. (2007). Phospholipid transfer protein-deficient protein-deficient mice absorb less cholesterol. *Arterioscler Thromb Vasc Biol*, Vol. 27, No.9, pp. 2014-2021

Liu, Y.; Millar, J.S.; Cronley, D.A,; Graham, M.; Crooke, R, Bilheimer, JT. & Rader, D.J. (2008). Knockout of acyl-Coa:diacylglycerol acyltransferase 2 with antisense oligonucleotide reduces VLDL TG and apoB secretion in mice. *Biochim Biophys Acta*, Vol.1781, No. 3, pp.97-104

Lo, C.M.; Nordskog, B.K.; Nauli, A.M.; Zheng, S.; Vonlehmden, S.B.; Yang, Q.; Lee, D.; Swift, L.L.; Davidson, N.O. & Tso, P. (2008). Why does the gut choose apolipoprotein B48 but not B100 for chylomicron formation? *Am J Physiol Gastrointest Liver Physio,* Vol.294, No.1, pp.G344-G352.

Ma K, Cilingiroglu M, Otvos JD, Ballantyne CM, Marian AJ, Chan L. (2003). Endothelial lipase is a major genetic determinant for high-density lipoprotein concentration, structure, and metabolism. Proc Natl Acad Sci U S A. Vol.100, No.2, pp2748-53.

Mabuchi, H.; Itoh, H.; Takeda, M.; Kajinami, K.; Wakasugi, T.; Koizumi, J.; Takeda, R. & Asagami, C. (1989). A young type III hyperlipoproteinemic patient associated with apolipoprotein E deficiency. *Metabolism,* Vol.38, No.2, pp.115-119.

Mardones, P.; Quiñones, V.; Amigo, L.; Moreno, M.; Miquel, J.F.; Schwarz, M.; Miettinen, H.E.; Trigatti, B.; Krieger, M.; VanPatten, S.; Cohen, D.E. & Rigotti, A. (2001). Hepatic cholesterol and bile acid metabolism and intestinal cholesterol absorption in scavenger receptor class B type I-deficient mice. *J Lipid Res,* Vol.42, No.2, pp.170-180

Marzal-Casacuberta, A.; Blanco-Vaca, F.; Ishida, B.Y.; Julve-Gil, J.; Shen, J.; Calvet-Márquez, S.; González-Sastre, F. & Chan, L. (1996). Functional lecithin:cholesterol acyltransferase deficiency and high density lipoprotein deficiency in transgenic mice overexpressing human apolipoprotein A-II. *J Biol Chem,* Vol.271, No.12, pp.6720-6728

Masson, D.; Deckert, V.; Gautier, T.; Klein, A.; Desrumaux, C.; Viglietta, C.; Pais de Barros, J.P.; Le Guern, N.; Grober, J.; Labbé, J.; Ménétrier, F.; Ripoll, P.J.; Leroux-Coyau, M.; Jolivet, G.; Houdebine, & Lagrost, L. (2011). Worsening of diet-induced atherosclerosis in a new model of transgenic rabbit expressing the human plasma phospholipid transfer protein. *Arterioscler Thromb Vasc Biol,* Vol.31,No.4, pp.766-774

Matsuura, F.; Wang, N.; Chen, W.; Jiang, X.C. & Tall, A.R. (2006). HDL from CETP-deficient subjects shows enhanced ability to promote cholesterol efflux from macrophages in an apoE- and ABCG1-dependent pathway. *J Clin Invest,* Vol.116, No.5, pp.1435-1442

Mbikay M, Sirois F, Mayne J, Wang GS, Chen A, Dewpura T, Prat A, Seidah NG, Chretien M, Scott FW. (2010). PCSK9-deficient mice exhibit impaired glucose tolerance and pancreatic islet abnormalities. *FEBS Lett,* Vol.584, No.4, pp.701-706

McNeish, J.; Aiello, R.J.; Guyot, D,; Turi, T.; Gabel, C.; Aldinger, C.; Hoppe, K.L.; Roach, M.L.; Royer, L.J.; de Wet, J.; Broccardo, C.; Chimini, G. & Francone, O.L. (2000). High density lipoprotein deficiency and foam cell accumulation in mice with targeted disruption of ATP-binding cassette transporter-1. *Proc Natl Acad Sci U S A,* Vol.97, No.8, pp.4245-4250

Mezdour, H.R.; Jones, R.; Dengremont, C.; Castro, G. & Maeda, N. (1997). Hepatic lipase deficiency increases plasma cholesterol but reduces susceptibility to atherosclerosis in apolipoprotein E-deficient mice. J. Biol. Chem., Vol.272, No.21, pp. 13570–13575Morrison, J.R.; Paszty, C.; Stevens, M.E.; Hughes, S.D.; Forte, T.; Scott, J. & Rubin, E.M. (1996). Apolipoprotein B RNA editing enzyme-deficient mice are viable despite alterations in lipoprotein metabolism. *Proc Natl Acad Sci USA,* Vol. 93, No.14, pp.7154-7159

Mulligan, J.D.; Flowers, M.T.; Tebon, A.; Bitgood, J.J.; Wellington, C.; Hayden, M.R. & Attie, A.D. (2003). ABCA1 is essential for efficient basolateral cholesterol efflux during the absorption of dietary cholesterol in chickens. *J Biol Chem,* Vol.278, No.15, pp.13356-3366

Newberry, E.P.; Kennedy, S.M.; Xie, Y.; Luo, J. & Davidson, N.O. (2009). Diet-induced alterations in intestinal and extrahepatic lipid metabolism in liver fatty acid binding protein knockout mice. *Mol Cell Biochem*, Vol 326, No. 1-2, pp. 79-86

Otera, H.; Ishida, T.; Nishiuma, T.; Kobayashi, K.; Kotani, Y.; Yasuda, T.; Kundu, R.K.; Quertermous, T.; Hirata, K. & Nishiuma, Y. (2009). Targeting inactivation of endothelial lipase attenuates lung allergic inflammation through raising plasma HDL level and inhibiting eosinophil infiltration. Am J Physiol Lung Cell Mol Physiol, Vol 296, No. 4, pp. L594-602

Out, R.; Hoekstra, M.; Habets, K.; Meurs, I.; de Waard, V.; Hildebrand, R.B.; Wang, Y.; Chimini, G.; Kuiper, J.; Van Berkel, T.J. & Van Eck, M. (2008). Combined deletion of macrophage ABCA1 and ABCG1 leads to massive lipid accumulation in tissue macrophages and distinct atherosclerosis at relatively low plasma cholesterol levels. *Arterioscler Thromb Vasc Biol*, Vol.28, No.2, pp.258-264

Piedrahita, J.A.; Zhang, S.H.; Hagaman, J.R.; Oliver, P.M. & Maeda N. (1992). Generation of mice carrying a mutant apolipoprotein E gene inactivated by gene targeting in embryonic stem cell. *Proc Natl Acd Sci USA*, Vol.89, No.10, pp. 4471-4475

Poirier, S.; Mayer, G.; Benjannet, S.; Bergeron, E.; Marcinkiewicz, J.; Nassoury, N.; Mayer, H.; Nimpf, J.; Prat, A. & Seidah, N.G. (2008). The proprotein convertase PCSK9 induces the degradation of low density lipoprotein receptor (LDLR) and its closest family members VLDLR and ApoER2. *J Biol Chem*, Vol.283, No.4, pp.2363-2372

Plump, A.S.; Azrolan, N.; Odaka, H.; Wu, L.; Jiang, X.; Tall, A.; Eisenberg, S. & Breslow, J.L. (1997). ApoA-I knockout mice: characterization of HDL metabolism in homozygotes and identification of a post-RNA mechanism of apoA-I up-regulation in heterozygotes. *J Lipid Res*, Vol.38, No.5, pp.1033-1047

Raabe, M.; Flynn, L.M.; Zlot, C.H.; Wong, J.S.; Veniant, M.M.; Hamilton, R.L. & Young, S.G. (1998). Knockout of the abetalipoproteinemia gene in mice: reduced lipoprotein secretion in heterozygotes and embryonic lethality in homozygotes. *Proc Natl Acd Sci USA*, Vol.95, No.15, pp.8686-8691

Rashid, S.; Curtis, D.E.; Garuti, R.; Anderson, N.N.; Bashmakov, Y. ; Ho, Y.K.; Hammer, R.E.; Moon, Y.A. & Horton, J.D. (2005). Decreased plasma cholesterol and hypersensitivity to statins in mice lacking Pcsk9. *Proc Natl Acad Sci USA*, Vol.102, No.15, pp.5374-5379

Reddick RL, Zhang SH, Maeda N. Atherosclerosis in mice lacking apo E. Evaluation of lesional development and progression. Arterioscler Thromb. 1994;14(1):141-7.

Rigotti, A.; Trigatti, B.L.; Penman, M.; Rayburn, H.; Herz, J. & Krieger, M. (1997). A targeted mutation in the murine gene encoding the high density lipoprotein (HDL) receptor scavenger receptor class B type I reveals its key role in HDL metabolism. *Proc Natl Acad Sci U S A*, Vol.94, No.23, pp.12610-12615

Rohrer, L.; Ohnsorg, R.M.; Landolt, F.; Rinninger, F. & von Eckardstein, A. (2009). High0density lipoprotein transport through aortic endothelial cells involves scavenger receptor BI and ATP-binding cassette transporter G1. *Cir Res*, Vol.104, No.10, pp.1142-1150

Rouy, D.; Duverger, N.; Lin, S.D.; Emmanuel, F.; Houdebine, L.M.; Denefle, P.; Viglietta, C.; Gong, E.; Rubin, E.M. & Hughes, S.D. (1998). Apolipoprotein(a) yeast artificial chromosome transgenic rabbits. Lipoprotein(a) assembly with human and rabbit apolipoprotein B. *J Biol Chem*, Vol. 273, No. 2, pp. 1247-1251

Sanan, D.A.; Newland, D.L.; Tao, R.; Marcovina, S.; Wang, J.; Mooser, V.; Hammer, R,E, & Hobbs, H.H. (1998). Low density lipoprotein receptor-negative mice expressing human apolipoprotein B-100 develop complex atherosclerotic lesions on a chow diet: No accentuation by apolipoprotein(a). *Proc Natl Acad Sci USA*; Vol.95, No. 8, pp. 4544-4549

Shachter, N.S.; Hayek, T.; Leff, T.; Smith, J.D.; Rosenberg, D.W.; Walsh, A.; Ramakrishnan, R.; Goldberg, I.J.; Ginsberg, H.N. & Breslow, J.L. (1994). Overexpression of apolipoprotein CII causes hypertriglyceridemia in transgenic mice. *J Clin Invest*, Vol.93, No.4, pp.1683-1690

Shimada, M.; Ishibashi, S.; Inaba, T.; Yagyu, H.; Harada, K.; Osuga, J.; Ohashi, K.; Yazaki, Y. & Yamada, N. (1996). Suppression of diet-induced atherosclerosis in low density lipoprotein receptor knockout mice overexpressing lipoprotein lipase. Proc Natl Acad Sci U S A, Vol.93, No.14, pp.7242-7246

Shimada, M.; Shimano, H.; Gotoda, T.; Yamamoto, K.; Kawamura, M.; Inaba, T.; Yazaki, Y. & Yamada, N. (1993). Overexpression of human lipoprotein lipase in transgenic mice. Resistance to diet-induced hypertriglyceridemia and hypercholesterolemia. *J Biol Chem*, Vol.268, No.24, pp.17924-17929

Shimano, H.; Yamada, N.; Katsuki, M.; Shimada, M.; Gotoda, T.; Harada, K.; Murase, T.; Fukazawa, C.; Takaku, F. & Yazaki, Y. (1992). Overexpression of apolipoprotein E in transgenic mice: marked reduction in plasma lipoproteins except high density lipoprotein and resistance against diet-induced hypercholesterolemia. *Proc Natl Acad Sci U S A*, Vol.89, No.5, pp.1750-1754

Shiomi, M. & Ito, T. (2001). MTP inhibitor decreases plasma cholesterol levels in LDL receptor-deficient WHHL rabbits by lowering the VLDL secretion. *Eur J Pharmacol* Vol.431, No.1, pp 127-131

Shiomi, M. & Ito, T. (2009). The Watanabe heritable hyperlipidemic (WHHL) rabbit, its characteristics and history of development: A tribute to the late Dr. Yoshio Watanabe. *Atherosclerosis*, Vol.207, No.1, pp. 1-7.

Shiomi, M.; Ito, T.; Tsukada, T.; Yata, T.; Watanabe, Y.; Tsujita, Y.; Fukami, M.; Fukushige, J.; Hosohawa, T. & Tamura, A. (1995). Reduction of serum cholesterol levels alters lesional composition of atherosclerotic plaques: Effect of pravastatin sodium on atherosclerosis in mature WHHL rabbits. *Arterioscler Theromb Vasc Biol*, Vol.15, No.11, pp.1938-1944

Shiomi, M.; Ito, T.; Yamada, S.; Kawashima, S. & Fan, J. (2003). Development of an animal model for spontaneous myocardial infarction (WHHLMI rabbit). *Arterioscler Thromb Vasc Biol*, Vol.23, No.7, pp. 1239-1244

Simonet, W.S.; Bucay, N.; Pitas, R.E.; Lauer, S.J. & Taylor, J.M. (1991). Mulyiple tissue-specific elements control the apolipoprotein E-C-I gene locus in transgenic mice. *J Biol Chem*, Vol.266, No. 14, pp. 8651-8654

Son, Y.S. & Zilversmit, D.B. (1989). Increased lipid transfer activities in hyperlipidemic rabbit plasma. *Arteriosclerosis*, Vol. 6, No.3, pp. 345-351

Strauss, J.G.; Frank, S.; Kratky, D.; Hammerle, G.; Hrzenjak, A.; Knipping, G.; von Eckardstein, A.; Kostner, G.M. & Zechner, R. (2001). Adenovirus-mediated rescueof lipoprotein lipase-deficient mice. Lipolysis of triglyceride-rich lipoproteins is essential for high density lipoprotein maturation in mice. *J Biol Chem,* Vol. 276, No.39, pp.36083–36090.

Suzuki, H.; Kurihara, Y.; Takeya, M.; Kamada, N.; Kataoka, M.; Jishage, K.; Ueda, O.; Sakaguchi, H.; Higashi, T.; Suzuki, T.; Takashima, Y.; Kawabe, Y.; Cynshi, O.; Wada, Y.; Honda, M.; Kurihara, H.; Aburatani, H.; Doi, T.; Matsumoto, A.; Azuma, S.; Noda, T.; Toyoda, Y.; Itakura, H.; Yazaki, Y.; Horiuchi, S.; Takahashi, K.; Kruijt, J.K.; van Berkel, T.J.C.; Steinbrecher, U.P.; Ishibashi, S.; Maeda, N.; Gordon, S. & Kodama, T. (1997). A role for macrophage scavenger receptors in atherosclerosis and susceptibility to infection. *Nature*, Vol.386, No.6622, pp. 292-296

Swanson, M. E.; Hughes, T. E.; Denny, I.S.; France, D.S.; Paterniti, J.R.Jr.; Tapparelli, C.; Gfeller, P. & Burki, K. (1992). High level expression of human apolipoprotein A-I in transgenic rats raises total serum high density lipoprotein cholesterol and lowers rat apolipoprotein A-I. *Transgenic Res*, Vol.1, No.3, pp. 142-147

Takahashi, S.; Ito, T.; Zenimaru, Y.; Suzuki, J.; Miyamori, I.; Takahashi, M.; Ishida, T.; Hirata, K.; Yamamoto, T.; Iwasaki. T.; Hattori, H. & Shiomi, M. (2011). Species differences of macrophage very low-density-lipoprotein (VLDL) receptor protein expression. *Biochem Biophys Res Commun.* Vol.407, No.4, pp.656-662

Takahashi, S.; Sakai, J.; Hattori, H.; Zenimaru, Y.; Suzuki, J.; Miyamori, I. & Yamamoto T. (2004). The very low-density lipoprotein (VLDL) receptor: characterization and function as a peripheral lipoprotein receptor. *J Atheroscler Thromb*, Vol.11, No.4, pp. 200-208

Teng, B.; Ishida, B.; Forte, T.M.; Blumenthal, S.; Song, L.Z.; Fotto, A.M. Jr, & Cham, L. (1997). Effective lowering of plasma, LDL, and esterified cholesterol in LDL receptor-knockout mice by adenovirus-mediated gene delivery of apoB mRNA editing enzyme (Apobec-1). *Arterioscler Thromb Vasc Biol,* Vol. 17, No. 5, pp. 889-897

Tsujita, Y.; Kuroda, M.; Shimada, Y.; Tanzawa, K.; Arai, M.; Kaneko, I.; Tanaka, M.; Masuda, H.; Tarumi, C.; Watanabe, Y. & Fujii, S. (1986). CS-514, a competitive inhibitor of 3-hydroxy-3-methylglutaryl coenzyme A reductase: Tissue-selective inhibition of sterol synthesis and hypolipidemic effect on various animal species. *Biochim Biophys Acta,* Vol.877, No.1, pp.50-60

Vaisman, B.L.; Klein, H.G.; Rouis, M.; Bérard, A.M.; Kindt, M.R.; Talley, G.D.; Meyn, S.M.; Hoyt, R.F.Jr.; Marcovina, S.M. ; Albers, J.J.; Hoeg, J.M.; Brewer, H.B.Jr. & Santamarin-Fojo, S. (1995). Overexpression of human lecithin cholesterol acyltransferase leads to hyperalphalipoproteinemia in transgenic mice. *J Biol Chem,* Vol.270, No.20, pp.12269-12275

Vaisman, B.L.; Lambert, G.; Amar, M.; Joyce, C.; Ito, T.; Shamburek, R.D.; Cain, W.J.; Fruchart-Najib, J.; Neufeld, E.B.; Remaley, A.T.; Brewer, H.B.J. & Santamarina-Fojo, S. (2001). ABCA1 overexpression leads to hyperalphalipoproteinemia and increased biliary cholesterol excretion in transgenic mice. *J Clin Invest*. Vol.108, No.2, pp.303–309

Walsh, A.; Ito, Y. & Breslow, J.L. (1989). High levels of human apolipoprotein A-I in transgenic mice result in increased plasma levels of small high density lipoprotein (HDL) particles comparable to human HDL3. *J Biol Chem,* Vol. 264, No.11, pp.6488-6494

Wang, N.; Arai, T.; Ji, Y.; Rinninger, F. & Tall. A.R. (1998). Liver-specific overexpression of scavenger receptor BI decreases levels of very low density lipoprotein ApoB, low density lipoprotein ApoB, and high density lipoprotein in transgenic mice. *J Biol Chem*, Vol.273, No.49, pp.32920-3296

Wang, Y.; Iordanov, H.; Swietlicki, E.A.; Wang, L.; Fritsch, C.; Coleman, T.; Semenkovich, C.F.; Levin, M.S. & Rubin, D.C. (2005). Targeted intestinal overexpression of the immediate early gene tis7 in transgenic mice increases triglyceride absorption and adiposity. *J Biol Chem*, Vol. 280, No. 41, pp. 34764-34775

Watanabe, Y. (1980). Serial inbreeding of rabbits with hereditary hyperlipidemia (WHHL rabbit): Incidence and development of atherosclerosis and xanthoma. *Atherosclerosis*, Vol.36, No.2, pp.261-268

Watanabe, Y.; Ito. T.; Saeki, M.; Kuroda, M.; Tanzawa, K.; Mochizuki M.; Tsujita Y. & Arai, M. (1981). Hypolipidemic effects of CS-500 (ML-236B) in WHHL-rabbit, a heritable animal model for hyperlipidemia. *Atherosclerosis*, Vol. 38, No.1-2, pp.27-31

Watanabe, Y.; Ito, T.; Shiomi, M.; Tsujita, Y.; Kuroda, M.; Arai, M.; Fukami, M. & Tamura, A. (1988). Preventive effect of pravastatin sodium, a potent inhibitor of 3-hydroxy-3-methylglutaryl coenzyme A reductase, on coronary atherosclerosis and xanthoma in WHHL rabbits. *Biochim Biophys Acta*, Vol. 960, No.3, pp.294-302

Weinstock PH, Bisgaier CL, Aalto-Setälä K, Radner H, Ramakrishnan R, Levak-Frank S, Essenburg AD, Zechner R, Breslow JL. (1995). Severe hypertriglyceridemia, reduced high density lipoprotein, and neonatal death in lipoprotein lipase knockout mice. Mild hypertriglyceridemia with impaired very low density lipoprotein clearance in heterozygotes. J Clin Invest. Vol.96, No.6, pp2555-68.

Weng, W. & Breslow, J.L. (1996). Dramatically decreased high density lipoprotein cholesterol, increased remnant clearance, and insulin hypersensitivity in apolipoprotein A-II knockout mice suggest a complex role for apolipoprotein A-II in atherosclerosis susceptibility. *Proc Natl Acad Sci USA*, Vol.93, No.25, pp. 14788-14794

Xie, Y.; Newberry, E.P.; Young, S.G.; Robine, S.; Hamilton, R.L.; Wong, J.S.; Luo, J.; Kennedy, S. & Davidson, N.O. (2006). Compensatory increase in hepatic lipogenesis in mice with conditional intestine-specific Mttp deficiency. *J Biol Chem*, Vol.281, No.7, pp.4075-86

Yagyu, H.; Lutz, E.P.; Kako, Y.; Marks, S.; Hu, Y.; Choi, S.Y.; Bensadoun, A. & Goldberg, I.J. (2002). Very low density lipoprotein (VLDL) receptor-deficient mice have reduced lipoprotein lipase activity. Possible causes of hypertriglyceridemia and reduced body mass with VLDL receptor deficiency. *J Biol Chem*, Vol.277, No.12, pp.10037-10043

Yin, W.; Carballo-Jane, E.; McLaren, D.G.; Mendoza, V.H.; Gagen, K.; Geoghagen, N.S.; McNamara, L.A.; Gorski, J.N.; Eiermann, G.J.; Petrov, A.; Wolf, M.; Tong, X.; Wilsie, L.C.; Akiyama, T.E.; Chen, J.; Thankappan, A.; Xue, J.; Ping, X.; Andrews, G.; Wickham, L.A.; Gai, C.L.; Trinh, T. Kulick, A.A.; Donnelly, M.J.; Voronin, G.O.; Rosa, R.; Cumiskey, A.M.; Bekkari, K.; Mitnaul, L.J.; Puig, O.; Chen, F.; Raubertas, R.; Wong, P.H.; Hansen, B.C.; Koblan, K.S.; Roddy, T.P.; Hubbard, B.K. & Strack, A.M. (2012). Plasma lipid profiling across species for the identification of optimal animal models of human dyslipidemia. *J Lipid Res*, Vol 53, No.1, pp. 51-65.

Young, S. G.; Cham, C.M.; Pittas , R.E.; Burri, B.J.; Connolly, A.; Flynn, L.; Pappu, A.S.; Wong, J.S.; Hamilton, R.L. & Farese, R.V.Jr. (1995). A genetic model for absent chylomicron formation: mice apolipoprotein B in the liver, but not in the intestine. *J Clin Invest*, Vol.96, No. 6, pp.2932-2946

Zhang, S.H.; Reddick, R.L.; Piedrahita, J.A. & Maeda, N. (1992). Spontaneous hypercholesterolemia and arterial lesions in mice lacking apolipoprotein E. *Science*, Vol.258, No.5081, pp.468-471

Section 7

Role of Lipoproteins in Neurodegenerative Diseases

Chapter 23

Plasma Lipoproteins in Brain Inflammatory and Neurodegenerative Diseases

Armando Sena, Carlos Capela, Camila Nóbrega,
Véronique Férret-Sena, Elisa Campos and Rui Pedrosa

Additional information is available at the end of the chapter

http://dx.doi.org/10.5772/51268

1. Introduction

Functions of the central nervous system (CNS) are mainly performed by neurons and glial cells (astrocytes, oligodendrocytes and microglia). Microglia or microcytes have macrophage-like immune related functions; oligodendrocytes are the myelinating cells in the CNS; and astrocytes have diverse roles in synaptogenesis, neurotransmission, myelination and reactive mechanisms to injury. CNS tissue is separated from blood circulation by specialized cell barriers, the most extensive being the endothelium of the so-called blood-brain barrier (BBB).

Brain cholesterol and lipid homeostasis is largely independent of plasma lipoproteins because the BBB restricts the transport of these molecules. In consequence, lipoprotein fractions and compositions in the CNS are different from those in the blood, and consist mainly of high-density lipoproteins (HDL)-like particles. Glial cells (in particular astrocytes) are the main source of cholesterol and HDL-like particles in the CNS [1]. This specialized scenario is reflected on the analysis of cerebrospinal fluid (CSF). The CSF contains apolipoproteins similar to those of plasma, including apoE, apoA-I and A-II, apoC-I, C-II and C-III, apoJ and apoD, but not apoB; apoE and apo A-I are the most abundant. Importantly, while HDL-cholesterol and apoA-I in blood influence its levels in the CSF, this is not the case for apoE, apoJ and apoD, which are synthetized by glial cells [1-2]. In vitro studies have suggested apoA-I expression by brain endothelium and that plasma HDL (containing apoA-I) is transcytosed across the BBB [3]. In consequence, the implications of plasma lipoprotein metabolism in brain physiology and pathological states have been controversial. Nevertheless, many CNS disorders are associated with disturbances of the plasma lipoprotein profile and there is increasingly evidence for pathogenic and clinical relevance of these alterations.

In this chapter we do not pretend to make an exhaustive review on the vast literature related to this theme. Rather, we intend to incorporate some relevant studies in a comprehensive framework addressed to open new avenues of research. With this purpose, we will mainly focus on two frequent and disabling conditions, multiple sclerosis (MS) and Alzheimer disease (AD), and discuss the involvement of plasma lipoproteins in brain inflammatory and neurodegenerative mechanisms. With this approach we expect that useful insights may emerge regarding the contribution of plasma lipoproteins in CNS physiology and pathological states.

2. Multiple sclerosis

MS is a demyelinating inflammatory and neurodegenerative disease of the CNS with heterogeneous pathology (see below) and clinical outcomes. More than 80% of MS patients present initially with acute attacks (relapses) of neurological dysfunction (follow by variable degree of recovery and periods of "remission"), characterizing the relapsing-remitting phenotype (RR-MS). Most of these patients develop a disabling progressive course independently of eventual relapses (secondary progressive MS). A small percentage of patients (10-15%) presents initially with a progressive disease course (primary progressive MS). Patients with clinical isolated syndrome (CIS) have an isolated episode suggestive of MS. The investigation of CIS patients is of special theoretical and practical interest because of their increased risk to develop the disease.

A possible involvement of plasma lipoprotein in MS pathogenesis was suggested in 1953 by the work of Swank [4]. This author presented evidence for a favorable disease course in patients taking a diet poor in animal fat [5]. Sinclair, in 1956, called attention for the importance of a deficiency in polyunsaturated fatty acids and remarked similar epidemiological aspects of MS and cerebrovascular disease [6]. A landmark work, providing a clear potential involvement of plasma lipoproteins in disease was published by Shore *et al*, in 1987 [7]. These authors studied the animal model of MS, experimental autoimmune encephalomyelitis (EAE) and concluded that "*major changes in apoE-containing lipoproteins are undoubtedly significant in the altered immune function in EAE*". Supporting this prediction, it was observed higher plasma apoE concentrations in MS patients during relapses in comparison to remission states and lower levels in patients under remission in comparison to normal controls [8-10]. Studying EAE induction in apoE-deficient female mice, Karussis in 2003, found that apoE deficiency might be connected with a defective neuronal repair mechanism and enhanced immune reactivity and worse course of the disease [11]. These results could indicate that plasma apoE may have an immunosuppressive role in MS [9]. In agreement with this concept, our group observed that lower levels of plasma apoE might promote immune reactivity in these patients [12].

In their work, Shore *et al* observed higher concentrations of total LDL and HDL cholesterol after onset of clinical symptoms. Giubilei *et al*, in 2002, studied plasma lipoproteins and magnetic resonance imaging (MRI) in patients with a first clinical episode suggestive of MS (CIS), supporting the findings in EAE [13]. These authors observed high total and HDL-cholesterol in these patients and a significant correlation between disease activity (as assessed by MRI) and both total and LDL-cholesterol levels. Jamroz-Wisniewska *et al* found high total cholesterol levels in patients (RR and progressive forms) and also higher LDL-cholesterol in RR patients in remission and in progressive forms than in healthy subjects [14]. Serum paraoxonase 1 (a HDL associated enzyme) activity in relapses was significantly lower in RR patients in comparison to other MS groups. An epidemiological survey based on almost 9000 patients with MS found that the presence of hypercholesterolemia, among other vascular co-morbidities, increased the risk of a more rapid disabling progression of the disease [15]. Recently, Weinstock-Guttman *et al* studied the serum lipid profiles in association with clinical disability and MRI measures in 492 MS patients [16]. They found that worsening disability was associated with higher total and LDL cholesterol, and triglycerides. Higher HDL levels were associated with lower probability for the presence of acute inflammatory lesions (assessed by MRI). Other authors have found higher HDL-cholesterol (and total blood homocysteine) levels in MS patients during a phase of clinical inactivity in comparison to normal controls [17].

The possible influence of apoE allele polymorphism in MS susceptibility and disease severity has been addressed in many studies. Overall, literature does not suggest a role of apoE alleles as risk factor of developing MS [18-19]. An association of apoE polymorphism with disease severity in MS patients has been more controversial. Using MRI methodology, some studies have shown an association between the apoE4 isoform and more severe brain tissue destruction in these patients [20-22]. However, an influence of this isoform on the clinical course of the disease is not established and the interaction with potential confounders should be considered. For example, it was suggested that an influence of apoE polymorphism on the clinical course, and even the risk of MS, could particularly exist in women [23]. Our group has provided evidence for an influence of cigarette smoking in apoE4-carriers, in modulating the clinical severity of RR-MS patients [24]. Some studies have suggested an association of apoE4 allele and apoA1 promoter polymorphism with cognitive impairment in these patients, which may occur very early in the clinical course of the disease [25-26].

As mentioned above, MS is a heterogeneous clinical entity. RR-MS has a higher prevalence in women (which is increasing) and the course of the disease is in general more disabling in men. It is not unreasonable to hypothesize that gender-related and other genetic influences could implicate different impacts of lipoprotein metabolism in MS. Few studies have analyzed the influence of MS therapies in plasma lipoproteins of these patients. However, these studies could provide useful insights on the pathogenic role of this metabolism. Our group first suggested that interferon beta therapy changes this metabolism in RR-MS patients. In particular, we found that at 12 month of therapy, lower apoA1 and higher apoE levels were associated with the presence of relapses and/or progression of the disease [27]. Others authors have found that MS therapy is associated with a decrease of plasma total cholesterol [28-29]. Overall, the reviewed data strongly support a role of plasma lipoproteins metabolism in the pathophysiology of the disease, as discussed below.

2.1. Pathophysiological mechanisms

A major link between plasma lipoproteins and MS concerns the immune system. It is well known that immune reactivity interacts with adaptive alterations of lipoprotein metabolism [30-31]. Recent reports have showed that distinct metabolic programs are essential for survival and functional specialization of different lymphocyte cell populations. For example, lipid oxidation is essential for Treg generation while Th1 differentiation and cytokine production by differentiated Th1, Th2 and Th17 cells are suppressed by lipids and require glucose metabolism [32]. Although the immunopathogenesis of MS lesions (demyelinating plaques) is heterogeneous and may differ in different patients, an imbalance favoring a Th1 effector cell activation is generally accepted [33]. Therefore, it would not be unexpected if an abnormal lipid modulation of immune functions could contribute for MS pathogenesis. However, a primary role of lymphocytes (T cells and B cells) in mediating CNS injury in this disease (at least in all patients) is controversial [32]. Myeloid cells play a pivotal role in the regulation of infiltrating lymphocyte cell activities and are involved in myelin breakdown and axonal injury [33-34]. Macrophages of M1 phenotype are characterized by high production of pro-inflammatory mediators and are crucial in Th1 cell response, while M2 phenotypes are associated with tissue remodeling/repair and expression of anti-inflammatory molecules [35]. In MS lesions, myelin phagocytosis by myeloid cells induces a foamy appearance. Foamy macrophages are originated from resident myeloid cells (microglia) and infiltrating monocytes and are suggested to be of M2-type macrophages and to contribute to the resolution of brain inflammation [36-37].

Macrophage polarization is modulated by different factors. For example, the M2 antinflammatory phenotype is induced by HDLs and apoE [35, 38], and fatty acid and phospholipid synthesis is essential for phagocytic differentiation of human monocytes [39]. ApoE is one ligand for the LDL-receptor-related-protein-1 (LRP1). Quite interesting, LRP1 mediates the downregulation of microglial inflammatory activity by apoE [40] and is essential for phagocytosis of degraded myelin in mice with EAE [41]. Moreover, LRP1 is also expressed in neurons and astrocytes and regulates BBB permeability [42]. This scenario is consistent with a reduction of inflammatory infiltrates and clinical disability by apoE-derived peptides in EAE [43] and immunosuppressive and neuroprotective effects of plasma apoE in EAE [11] and MS patients [8-10, 12].

Among the transcriptional factors regulating macrophage polarization, peroxisome proliferator-activated receptor (PPAR) γ is known to promote M2 macrophages [35]. This is of potential interest in the context of preliminary evidence implicating PPARs in MS pathogenesis and as therapeutic targets for the disease [44-45].

In brain, apoE is associated with HDL-like particles, also containing the second major apolipoprotein, apoA-I. These apolipoproteins are primarily located on separated lipoproteins particles [1]. Although apoE in the brain is predominantly synthetized by glial cells, plasma HDL/apoA-I may cross the BBB and influence its levels in the brain [2]. HDL effects include an inhibition of cytocine-induced expression of adhesion molecules in endothelial cells, which could further depress brain parenchyma immune reactivity [46]. As mentioned, higher levels of plasma HDL were found in CIS and RR-MS and were associated with a lower probability in development of acute inflammatory lesions in these patients [13, 16-17]. Recently, preliminary evidence from our group suggests that higher plasma HDL levels are associated with an increased intrathecal IgG synthesis in these patients [47]. Because low plasma HDL-cholesterol is associated with a predominance of pro-inflammatory phenotype of monocyte-derived macrophage [48], these findings suggest an immunosuppressive role of HDL in the development of MS lesions. This interpretation is further supported by the beneficial therapeutical effects of fingolimod in the disease [49].

Fingolimod (FTY720) is a structural analog of sphingosine, which down modulates sphingosine 1-phosphate (S1P) receptors. S1P is a major component of HDL, including in the CNS and induces an anti-inflammatory phenotype in macrophages. SP1 receptors are widespread in CNS cells and a defect of sphingolipid and phospholipid metabolism is observed early in normal appearing white and grey matter in MS patients. Moreover, S1P is reduced in affected white matter and is increased in CSF of these patients [49]. Importantly, FTY720 treatment has been shown to have neuroprotective effects independent of immunomodulatory mechanisms [50]. These data suggest a protective role of endogenous HDL components not only in the genesis of acute inflammatory lesions but also in the neurodegenerative process of MS.

An involvement of oxidative stress in MS, including of lipid peroxidation has recently received much support [51]. Newcombe *et al*, in 1994, demonstrated for the first time the presence of oxidized LDL (ox-LDL) and their peroxidative end-products in early and actively demyelinating plaques in post-mortem MS brain [36]. They suggested that plasma LDL enters (through a damaged BBB) the parenchyma and is oxidatively modified in the lesions. More recent data supports an important involvement of oxidative damage including oxidized phospholipids in myelin and axon injury in MS [52]. Several studies have also demonstrated that measures of oxidative stress and lipid peroxidation are consistently increased in the blood of these patients [51]. Our group reported increased levels of serum oxLDL in RR-MS patients in remission in comparison to normal controls and higher levels during relapses [53]. These findings are consistent with a contribution of plasma ox-LDL in promotion BBB permeability and acute inflammatory CNS lesions in the disease. However, increased plasma lipid peroxidation or oxidative stress is probably not associated with disability progression in these patients [54]. The pathophysiology of acute lesions (MS plaques) and disability progression are indeed thought to be mediated by different mechanisms. In fact, it was suggested that low oxygen radical formation in peripheral leukocytes may be associated with a increased severity of the disease [55]. These findings indicate that the role of oxidative stress in MS is complex. An oral formulation of dimethylfumarate (BG-12) activates the Nrf2 antioxidant pathway and was recently observed to be of clinical benefit in RR-MS patients, possibly in disease progression also [56]. These recent promising results should stimulate future research to clarify the involvement of lipid peroxidation in the disease. It should be noted that this involvement further supports a role of plasma HDL in disease pathogenesis, as discussed above. Plasma HDL-associated α-tocopherol is transcytosed across the BBB and may have antioxidant as well as anti-inflammatory effects [3].

Ludewig and Laman (2004) remarked the similarities that may exist between the atherosclerotic plaque development and MS lesions and suggested: "*Systematic comparison of these two diseases involving foam cells in chronic lesions may prove fruitful*" [57]. As we have reviewed, recent research clearly supports this prediction. Moreover, patients with MS have several vascular abnormalities and a higher risk for ischemic stroke [58]. In 2003, our group first reported a pilot trial suggesting a benefit of statin monotherapy in the pathogenesis process (assessed by MRI) and clinical activity of RR-MS patients [59]. These beneficial effects were confirmed by Vollmer *et al* trial in 2004 [60] and in a long-term follow-up of our patients [61]. Very recently, beneficial effects of statin monotherapy were reported in patients with a first clinical episode (CIS) suggestive of MS [62-63]. A synthesis of some shared pathophysiological factors involved in MS and atherosclerosis is presented on Table 1. As we will discuss below, the presence of similar mechanisms involving plasma lipoprotein metabolism in the pathogenesis of atherosclerosis/ischemic and demyelinating lesions may be extensive to other chronic inflammatory and neurodegenerative pathologies.

Pathophysiology	Comment	References
	Lesions	
Foam Cells Plaques	Macrophage lipid uptake in early lesion formation	[36-37]
	Lipoprotein Related	
Total and LDL cholesterol	Promotion of lesion formation and/or progression	[13-16]
HDL-Cholesterol	Protective of lesion formation and/or progression	[13, 16-17, 47]
ApoA-I	Immunosuppressive and protective	[26-27]
ApoE	Immunosuppressive and protective	[8-10, 12]
Sphingosine-1-phosphate	Modulation of immune reactivity and lesion formation	[49-50]
Oxidative stress and oxLDL	Lipid peroxidation in lesion formation	[51-53]
Secretory phospholipase A_2	Increased expression	[115]
	Immunopathogenesis	
M_2-Macrophages	Anti-inflammatory, phagocytic cells in lesions	[37]
T-Cells	Promoting lesion formation	[33]
Inflammatory cytokines	Promoting lesion formation	[33]
Interleukin-10	Protective of lesion formation	[116]

Pathophysiology	Comment	References
Adipocytokines	Leptin and Adiponectin involvement in immune dysfunction	[117-118]
MMP-9	Upregulation associated with lesion formation	[42]
	Others	
Statins	Pathological and clinical benefits	[59, 63]
PPARs	PPAR-gamma agonists protective	[45]
Estrogens	Protective of lesion formation and/or progression	[119]
Homocysteine	Increased levels associated with lesion formation	[17, 58]
Platelets	Increased adhesiveness and aggregation	[58, 120]
Smoking	Promotes lesion formation	[121]
Ischemic events	Increased risk associated with atherogenesis and MS pathogenesis	[58]

Table 1. Some pathogenic similarities between multiple sclerosis and atherosclerosis.

3. Alzheimer disease

Possession of the apoE4 allele is the major genetic risk factor for sporadic late-onset AD [64-65]. This observation led to a large body of research on cholesterol and lipid metabolism in patients and animal models of AD during the last two decades. However, the investigation of this metabolism in patients with clinical AD is not sufficient to clarify its role in the pathogenesis of the dementia. It is generally accepted that the pathogenic processes in AD begin many decades before the appearance of evident symptoms. More recently, a major focus of interest has been on longitudinal studies addressing the association between lipoprotein profiles and clinical evolution of cognitive normal subjects or patients with mild cognitive impairment (MCI). A large percentage of patients with the diagnosis of MCI by the 6th decade are known to develop AD later in life. Therefore prospective studies are crucial for development of efficient preventive or therapeutic measures.

Amyloid-β (Aβ) deposition in plaques (AP) (also known senile or neuritic plaques (NP)) and neurofibrillary tangles (NFT), characterized by hyperphosphorylated tau protein aggregates, are pathologic hallmarks of AD [66-67]. The association between plasma lipoprotein profiles and risk of development of clinical manifestations of dementia has been controversial. An association between high cholesterol levels in midlife and an increased risk for dementia in old age has been suggested by several publications [68], but it was not confirmed by a recent large population study [69]. Instead, this study found that low cholesterol levels in late life were predictive of subsequent dementia. Supporting this conclusion, another study in elderly individuals found that low HDL-cholesterol and low total and Non-HDL cholesterol were associated with higher AD risk [70]. These authors suggested a protective effect of late life total cholesterol level on the risk for mild cognitive impairment and AD. Low HDL-cholesterol levels were also associated with decline of memory in middle-aged adults [71]. Within this framework, decreased plasma apoA-I levels have also been found in AD as well as in vascular dementia, and higher apoA-I levels associated with decreased risk of dementia [72-73]. Few studies have investigated the association of lipid profiles with AD-related pathology. A recent work has found that high total cholesterol, LDL-cholesterol and non-HDL-cholesterol levels were associated with risk of development of AP, but not NFT [74]. However, as we will discuss below, the genesis of pathological hallmarks of AD is not invariably associated with clinical manifestations of cognitive impairment and dementia.

The apoE4 allele is an established risk factor for the development of sporadic AD; it is associated with an early age at onset of dementia in an allele dose-dependent manner; and with increased Aβ burden. Moreover, in MCI it predicts conversion to AD. In contrast, apoE2 allele is associated with delayed age of onset of AD [66]. Recent data have provided evidence for an important role of apoE protein levels, independently of the genotype. In one study, middle-aged offspring with familial history of AD were found to have lower plasma apoE levels when compared with offspring without familial history of AD, independent of APOE genotype [75]. In other study, plasma apoE levels were found to be lower in patients with AD and decreased with Aβ load [76].

Overall the reviewed data strongly support a role of plasma lipoprotein metabolism in the pathogenesis of AD, as discussed in more detail below.

3.1. Pathophysiological mechanisms

As already mentioned, NP and NFT are the hallmarks of AD pathology. However, these aggregates are present in a variable extend in about 30% of cognitively normal elderly subjects. In AD, synaptic structural and functional alterations also occur early and are more pronounced than in normal ageing individuals. ApoE-containing lipoproteins, mainly derived from astrocytes, may influence these pathogenic processes in several ways. Cholesterol associated with these lipoproteins is necessary for neurons and to stimulate axonal growth and synaptogenesis. Lipidate-apoE contributes for clearing out Aβ from the brain, a process mediated by apoE receptors (especially LRP1) present in glial cells, neurons and in endothelium of the BBB. Pathways for Aβ clearance also include proteolytic degradation and oligomerization in the aggregates of amyloid plaques, mechanisms also modulated by apoE. For all these processes the isoform apoE4 (which is in general associated with less secreted production of the protein) is less efficient and promotes synaptic dysfunction, toxicity of soluble Aβ and NP deposits. Moreover, it is suggested that apoE4 fragments induce mitochondrial dysfunction and neurotoxicity and that cholesterol levels may regulate Aβ production [65, 66]. Supporting this important role of apoE for AD and the harmful effects of Aβ on cognitive functions, cognitive performance in normal older adults was associated with Aβ load (PET), mainly in ε4 carriers [77].

The above findings do not exclude the contribution of other apolipoproteins for Aβ pathology. For example, apoJ and apoD (see below) also modulate Aβ deposition, a deficiency of apoA-I promotes cognitive impairment and polymorphisms of all these apolipoproteins were associated with risk for AD [1]. Interestingly, increased plasma levels of apoJ (clusterin) are not present before the development of AD but are indeed associated with the severity and progression of the disease, supporting a neuroprotective role [78].

The link between a lipoprotein dysregulation and tau pathology (NFT deposition), in contrast, is not well understood. Beyond the involvement of cholesterol and apolipoproteins, AD is associated with disturbances of sphingolipids and phospholipid metabolism that may contribute for its pathogenesis [67]. Moreover, cognitive impairment and dementia, including AD, are frequently associated with markers of systemic and brain inflammatory activity [79], vascular atherogenic [80] and white-matter (myelin) pathology [81]. An underlying dysregulation of lipoprotein metabolism could be linked to all these pathogenic pathways.

The scenario briefly described above is clearly consistent with the observations that low plasma apoE may be associated with increased risk of AD and correlates with Aβ load, as assessed by PET [75-76]. As remarked, the last studies emphasized the importance of total apoE levels, independently of the genotype. Supporting this concept, it was recently reported in AD mouse models a stimulation of Aβ clearance and cognitive function by inducing apoE expression [82]. After apoE-mediated transport through the BBB, plasma Aβ transport is accomplished by triglyceride-rich lipoproteins (TRL) rich in apoE, for uptake in liver [83]. These findings are also consistent with the risk conferred by low plasma apoE levels. Low plasma apoE levels could also promote systemic immune reactivity and atherogenic pathology in these patients.

Although no relation exists between plasma and brain apoE levels, a strong correlation was found between HDL-cholesterol and apoA-I in serum and in CSF lipoproteins (which are HDL-like particles) [2]. This scenario could contribute to the risk of cognitive impairment and AD conferred by low plasma HDL-cholesterol and apoA-I levels [70-73]. On one hand, these deficiencies could be linked to an increased systemic inflammatory and oxidative status and promotion of atherogenesis. On the other hand, low HDL and apoA-I levels would provide less neurotrophic and immunosuppressive abilities to the brain [84]. If high total and LDL or non-HDL cholesterol in plasma cannot influence its levels in the brain, how could they be associated in some studies with an increased AD risk and Aβ load (NPs)? Experimental studies suggest that plasma cholesterol levels do not normally regulate production of brain Aβ [85]. One possibility resides in the fact that high non-HDL cholesterol in these patients may be associated with low HDL, apoE and apoA-I levels, a pro-inflammatory systemic status and increased atherogenic/ischemic pathology. Supporting this hypothesis, in animal models, cognitive impairment following high fat diet consumption was associated with brain inflammation [86]. Among other markers of inflammation [79], serum levels of adipocytokines have been associated with cognitive impairment and progression of AD, as well as atherogenic/ischemic disease [87-88]. Metabolic syndrome [89] and insulin resistance and type 2 diabetes [90] are associated not only with higher risk of vascular disease but also with risk of dementia, including AD. All these conditions may promote the development of dementia also by affecting myelin integrity and white-matter connective functions.

It should be noted that clinical overt cognitive impairment and dementia do not depend solely on the severity of neurodegenerative and vascular pathologies. Human brain is provided with potential compensatory or plastic mechanism, which may mitigate the clinical impact of ageing-associated pathologies [91-92]. This means that in old age, risk factors for dementia may not have the same significance they have in previous decades. Those factors may include high total and non-HDL cholesterol plasma levels, which may have a major impact in promoting atherogenesis/ischemic/inflammatory processes and AD-related pathology in middle-life, but not in neuroplastic mechanisms increasingly required with advancing age. Lower total and LDL cholesterol have indeed been associated with a poor prognosis in the ischemic stroke [93] and in elderly individuals, as observed above, this profile may increase the risk for overt dementia. Increased body mass index (BMI) in middle life appears to be a risk factor for latter development of cognitive decline and AD, but in late life the burden of cerebral amyloid and tau is associated with lower BMI in cognitively normal and MCI subjects [94-95]. These facts could contribute to the inconsistent results regarding the benefits of statins on prevention and treatment of AD, despite in vitro and animal studies demonstrating an effect in decreasing Aβ formation [96].

In Figure 1 are presented some of the suggested implications of lipoproteins in the pathogenesis of AD and MS.

4. Other brain disorders

Results concerning an association between high serum cholesterol levels and the risk for **Parkinson disease (PD)** have been conflicting. However, this possible association may not exist in older subjects (≥55 years). As for AD, low serum total and LDL-cholesterol levels may increase the risk of PD with advancing age [97-99]. In this context, it is intriguing that hyperlipidemia probably has also a protective role on the neurodegenerative process of **amyotrophic lateral sclerosis (ALS)** [100-101]. PD and ALS also involve inflammatory processes and it has been noted some convergence in the mechanisms underlying neurodegeneration in these disorders, in AD and MS. [102]. An abnormal brain cholesterol homeostasis may also contribute to the pathophysiology of **Huntington's disease** [103]. In what concerns PD, and in contrast to AD, an involvement of apoE genotypes is not clarified. Several studies reported no influence of apoE4 allele in the development of PD or in dementia associated with the disease, which is in contrast to its established role in AD pathogenesis [104]. However, apoE and LRP1 were found to be increased in brain from PD patients, suggesting an involvement in the deposition of α-sinuclein aggregates (Lewy bodies) typical of this disease [105].

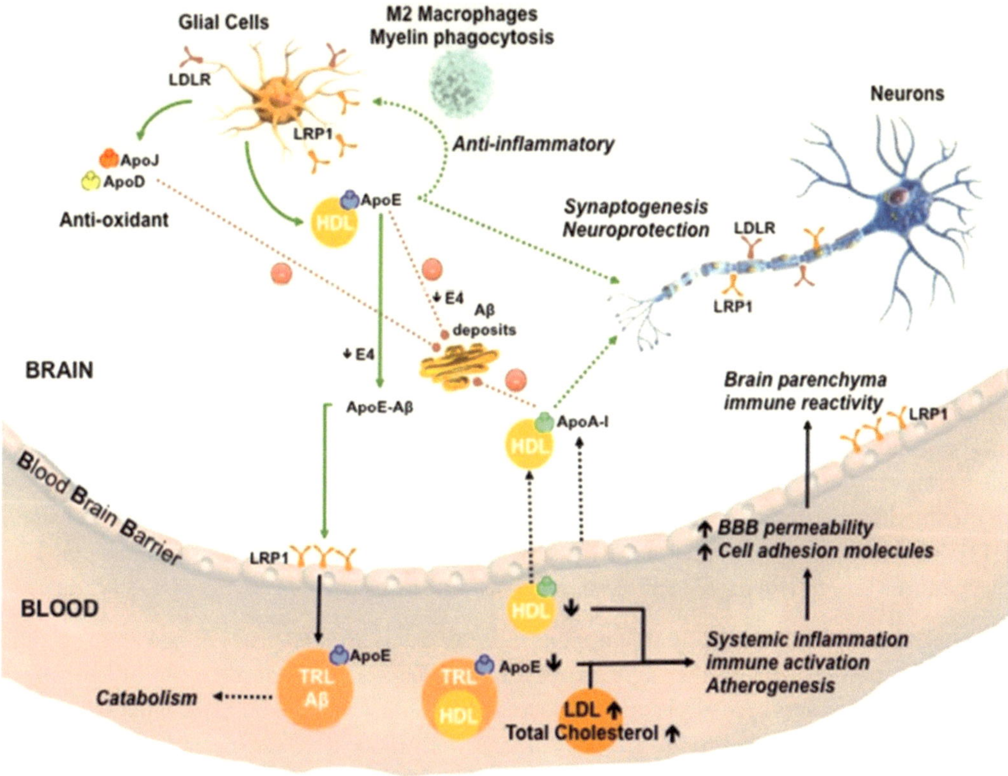

Figure 1. Some putative implications of lipoproteins for the pathogenesis of Multiple Sclerosis and Alzheimer disease (see text for interpretation).

As mentioned above, apolipoprotein D in the CNS is normally synthetized by glial cells (astrocytes and oligodendrocytes). Although present in the CSF in lower concentrations than apoE, A-I and J, some studies have suggested a possible neuroprotective role of apoD in neuropathological states [1,106]. ApoD is a member of the lipocalin family of proteins that are involved in the transport of small hydrophobic ligands. Among several proposed ligands (cholesterol, progesterone, pregnenolone, bilirubin), apoD can bind with high affinity arachidonic acid (AA). Inflammatory responses and oxidative stress associated with brain insults are known to mobilize AA from membranes. Therefore, apoD could have a neuroprotective role by controlling oxidative damage [106-107]. In fact, higher levels of apoD have been found in brain or CSF of AD and other neuropathologies [108]. Curiously, an increase of apoD has also been reported in plasma and certain brain regions of patients with **schizophrenia** and **bipolar disorder**. In these conditions, a disturbance of phospholipid metabolism has been proposed and apoD could represent a response addressed to stabilize membrane AA or bind free AA [106]. The fact that atypical antipsychotics such as clozapine up-regulate apoD expression supports neurotrophic effects of this protein [106]. It should be noted that other apolipoproteins have been implicated in these neuropsychiatric disorders, including apoE, and apoA-I [106, 109-110]. Interestingly, as observed for AD and MS, lower serum apoA-I levels were found in schizophrenia [110].

Overall, these data emphasizes the relevance of plasma lipoprotein metabolism in brain physiology and the convergence of similar dysfunctions of this metabolism associated with several neuropathologies.

5. Conclusion

This review has addressed MS and AD as a strategy to explore the potential relevance of plasma lipoproteins in CNS inflammatory and neurodegenerative disorders. Despite quite different in their demographics, clinical and pathological characteristics, some similarities in their inflammatory and neurodegenerative components have been noted previously [102].

In MS as in AD, the genesis of brain pathology is thought to begin many years before the clinical overt disease. Despite the occurrence of widespread lesions, brain plastic compensatory mechanisms may maintain those disorders clinically silent, delay their symptoms or modify their clinical evolution. Molecular mechanisms underlying grey and white matter plasticity are of outstanding neurobiological and medical importance and are currently poorly understood [111]. This review suggests that an involvement of lipoprotein metabolism in brain plasticity mechanisms is highly plausible and deserves much future research.

Clinical signs of MS very rarely first appear in individuals after 60 years of age and sporadic AD rarely manifest before that age. However, it is remarkable that a profile of low HDL-cholesterol, apoE and apoA-I plasma levels and elevated total and non-HDL cholesterol may promote the risk or progression of disability in both disorders. As discussed, this profile could be associated with both the genesis of lesions in the CNS and the systemic immune-related or metabolic alterations implicated in their pathophysiology (Table 1, Figure 1). It is to note that disturbances in brain cholesterol transport (that may occur in MS, AD and other neuropathologies) can lead to alterations in cholesterol uptake from plasma to brain and decrease plasma HDL levels (112). In MS as in AD, this lipoprotein profile may promote foam cell plaque formations. In young individuals genetically susceptible to MS, this profile may promote the genesis of demyelinating plaques; instead with advanced age, atheroma plaques formation prevails, contributing to AD, in genetically susceptible subjects. Supporting this speculation, MS pathogenesis may share many lipoprotein-related and inflammatory mechanisms underlying atherogenesis (Table1). In addition, with aging, this lipoprotein profile could have a convergent impact for the maintenance of the typical CNS lesions occurring in MS and AD. In fact, advanced ageing may be associated with lower recruitment of anti-inflammatory and phagocytic macrophages and other blood-derived factors to the CNS [113]. This situation, on one hand, favors lower capacity of β-amyloid clearance, oligodendrocyte toxicity and myelin lesions, early present in incipient AD. On the other hand, it restricts remyelination capacities in MS, which are more accentuated with advancing ageing in these patients. The presence of age-related changes in blood circulation has recently been noted of possible relevance for MS and AD [114]. These relevant age-related changes should comprise circulating lipoprotein metabolism.

Despite the similarities of lipoproteins involvement in these two disorders, including the neuroprotective, immunosuppressive and vascular/ischemic protective functions of HDL-cholesterol and associated apolipoproteins (Fig. 1), distinctive implications on their pathogenesis are expected. In MS, a participation of lymphocyte infiltration is certainly important while this is not the case for AD. For example, sphingosine-1-phosphate component of HDL could be special relevant for the immune dysfunction and the abnormal sphingosine metabolism associated with the genesis of demyelinating plaques and neurodegenerative processes in MS. In AD, triglyceride-rich plasma lipoproteins and apoE4 isoform are especially relevant in the clearance of Aβ and genesis of amyloid plaques. It should be emphasized that MS and AD are pathological and clinical heterogeneous diseases. For example, the immunopathogenesis of MS differ among patients even with similar clinical profiles and prominent atherosclerosis lesions are absent in some patients with AD. Therefore, the contribution of plasma lipoprotein metabolism for the pathogenesis of these disorders may be variable and this could explain discrepancies among some studies. Future work aimed to clarify the roles of plasma lipoproteins in these diseases should address clinical homogeneous patient populations, include concomitant pathological and immunological markers and consider potential environmental confounders. Ideally, laboratory data should be correlated with neuroimaging measures. Finally, MS and AD are clear examples of complex conditions for which multiple genetic risk factors for developing and progression are to be expected. Selected genetic typing of the study population is therefore convenient, because lipoprotein alterations may not have the same significance and the same therapeutical implications in different genetic backgrounds.

In sum, the available reviewed data suggest that plasma lipoproteins metabolism is a fruitful "window" to an improved understanding of MS and AD and other neurological diseases. Of outstanding interest, plasma lipoproteins may represent useful targets for discovering preventive and therapeutical strategies for these common disabling human conditions.

A very recent paper from Dr Lawrence Steinman group at Stanford University highlights the importance of lipids in the pathogenesis of MS and the therapeutic potential of lipid-based strategies for the disease (Science Transl Med 2012; 8 (137); E-pub 2012 6 Jun).

Author details

Armando Sena, Carlos Capela, Camila Nóbrega and Elisa Campos
Centro de Estudos de Doenças Crónicas (CEDOC), Faculdade de Ciências Médicas, UNL, Campo Mártires da Pátria, Lisboa, Portugal

Armando Sena, Carlos Capela, Camila Nóbrega and Rui Pedrosa
Serviço de Neurologia, Hospital dos Capuchos, Centro Hospitalar de Lisboa-Central, Lisboa, Portugal

Armando Sena, Camila Nóbrega and Véronique Férret-Sena
Interdisciplinary Centre of Research Egas Moniz (CRiEM), Cooperativa Egas Moniz, Monte da Caparica, Portugal

Acknowledgement

The authors thank to nurses Cristina Araújo and Ana Mendes (Neurology Service) for helping in research and dedicated assistance to our patients and to Merck-Serono, Biogen Idec, Bayer HealthCare, Lundbeck, Octapharma, Teva Pharma and Sanofi-Aventis for supporting our research.

6. References

[1] Hayashi H (2011) Lipid Metabolism and Glial Lipoproteins in the Central Nervous System. Biol. pharm. bull. 34: 453-461.

[2] Fagan AM, Younkin LH, Morris JC et al (2000) Differences in the Aβ40/ Aβ42 Ratio Associated with Cerebrospinal Fluid Lipoproteins as a Function of Apolipoprotein E Genotype. Ann. neurol. 48: 201-210.

[3] Balazs Z, Panzenboeck u, Hammer A et al (2004) Uptake and Transport of High-Density Lipoproteins (HDL) and HDL-Associated α-Tocopherol by an *in vitro* Blood-Brain Barrier Model. J. neurochem. 89: 939-950.

[4] Swank RL (1953) Treatment of Multiple Sclerosis with Low-Fat Diet. Arch. neurol. psych. 69: 91-103.

[5] Swank RL (1970) Multiple Sclerosis. Twenty Years on Low-Fat Diet. Arch. neurol. 23: 460-474

[6] Sinclair HM (1956) Deficiency of Essential Fatty Acids and Atherosclerosis, etcetera. Lancet, 270: 381-383.

[7] Shore VG, Smith ME, Perret V et al (1987) Alterations in Plasma Lipoproteins and Apolipoproteins in Experimental Allergic Encephalomyelitis. J. lipid res. 28: 119-129.

[8] Rifai N, Christenson RH, Gelman BB et al (1987) Changes in Cerebrospinal Fluid IgG and Apoliporotein E Indices in Patients with Multiple Sclerosis during Demyelination and Remyelination. Clin. chem. 33: 1155-1157.

[9] Gelman BB, Rifai N, Christenson RH, et al (1988) Cerebrospinal Fluid and Plasma Apolipoproteins in Patients with Multiple Sclerosis. Ann. clin. lab. science. 18: 46-52.

[10] Carlsson J, Armstrong VW, Reiber H et al (1991) Clinical Relevance of the Quantification of Apolipoprotien E in Cerebrospinal Fluid. Clin. chim. acta, 196: 167-176.

[11] Karussis D, Michaelson DM, Grigoriadis N et al (2003) Lack of Apolipoprotein-E Exarcebates Experimentally Allergic Encephalomyelitis. Mult. scler. 9: 476-480.

[12] Sena A, Bendtzen K, Cascais MJ et al (2010) Influence of Apolipoprotein E Plasma Levels and Tobacco Smoking on the Induction of Neutralising Antibodies to Interferon-Beta. J. neurol. 257: 1703-1707.

[13] Giubilei F, Antonioni G, Di Legge S et al (2002) Blood Cholesterol and MRI Activity in First Clinical Episode Suggestive of Multiple Sclerosis. Acta neurol. scand. 106: 109-112.

[14] Jamroz-Wisniewska A, Beltowski J, Stemasiak Z et al (2009) Paraoxonase 1 Acitivity in Different Types of Multiple Sclerosis. Mult.scler. 15: 399-402.

[15] Marrie RA, Rudick R, Horwitz R et al (2010) Vascular Comorbility is Associated with More Rapid Disability Progression in Multiple Sclerosis. Neurology 74: 1041-1047.

[16] Weinstock-Guttman B, Zivadinov R, Mahfooz N et al (2011) Serum Lipid Profiles are Associated with Disability and MRI Outcomes in Multiple Sclerosis. J. neuroinflammation 8:127-133

[17] Salemi G, Gueli MC, Vitale F, et al (2010) Blood Lipids, Homocysteine, Stress Factors, and Vitamins in Clinically Stable Multiple Sclerosis Patients. Lipids in health and disease 9:19-21.

[18] Burwick RM, Ramsay PP, Haines JL et al (2006) ApoE Epsilon Variation in Multiple Sclerosis Susceptibility and Disease Severity. Neurology 66: 1373-1383.

[19] Pinholt M, Frederiksen JL, Christiansen M (2006) The Association Between Apolipoprotein E and Multiple Sclerosis. Eur. j. neur. 13: 573-580.

[20] Enzinger C, Ropele S, Strasser-Fuchs S et al (2003) Lower Levels of N-Acetylaspartate in Multiple Sclerosis Patients with the Apolipoprotein E ε4 Allele. Arch. neurol. 60:65-70.

[21] Enzinger C, Ropele S, Smith S et al (2004) Accelerated Evolution of Brain Atrophy and "Black Holes" in MS Patients with ApoE- ε4. Ann. neurol. 55:563-569.

[22] De Stefano N, Bartolozzi ML, Nacmias B et al (2004) Influence of Apolipoprotein E ε4 Genotype on Brain Tissue Integrity in Relapsing-Remitting Multiple Sclerosis. Arch. neurol. 61: 536-540.

[23] Kantarci OH, Hebrink DD, Achenbach SJ et al (2004) Association of ApoE Polymorphisms with Disease Severity in MS is Limited to Women. Neurology 62: 811-814.

[24] Sena A, Couderc R, Ferret-Sena V et al (2009) Apolipoprotein E Polymorphisms Interacts with Cigarette Smoking in Progression of Multiple Sclerosis. Eur. j. neur. 16: 832-837.

[25] Shi J, Zhao CB, Vollmer TL et al (2008) ApoE ε4 Allele is Associated with Cognitive Impairment in Patients with Multiple Sclerosis. Neurology, 70: 185-190.

[26] Koutsis G, Panas M, Giogkaraki E et al (2009) An ApoAI Promoter Polymorphism is Associated with Cognitive Performance in Patientswith Multiple Sclerosis. Mult. scler. 15: 174-179.

[27] Sena A, Pedrosa R, Ferret-Sena V et al (2000) Interferon β1a Therapy Changes Lipoprotein Metabolism in Patients with Multiple Sclerosis. Clin. chem. lab. med. 38: 209-213.

[28] Morra BV, Coppola G, Orefice G et al (2004) Interferon β Treatment Decreases Cholesterol Plasma Levels in Multiple Sclerosis Patients. Neurology 62: 829-830.

[29] Coppola G, Lanzillo R, Florio C et al (2006) Long-Term Clinical Experience with Weekley Interferon β-1a in Relapsing Multiple Sclerosis. Eur. j. neurol. 13: 1014-1021.

[30] Hansson GK (2007) Light Hits the Liver. Science 316: 206-207.

[31] Glass CK, Saijo, K (2008) Oxysterols Hold T Cells in Check. Nature 455:40-41.

[32] Michalek RD, Gerriets VA, Jacobs SR et al (2011) Cutting Edge: Distinct Glycolytic and Lipid Oxidative Metabolic Programs Are Essential for Effector and Regulatory $CD4^+$ T Cells Subsets. J. immunol. 186: 3299-3303.

[33] Henderson APD, Barnett MH, Parratt JDE et al (2009) Multiple Sclerosis – Distribution of Inflammatory Cells in Newly Forming Lesions. Ann. neurol. 66: 739-753.

[34] Weber MS, Prod'homme T, Ypussef S. et al (2007) Type II Monocytes Modulate T Cell-Mediated Central Nervous System Autoimmune Disease. Nat. med. 13: 935-943.

[35] Chinetti-Gbaguidi G, Staels B (2011) Macrophage Polarization in Metabolic Disorders: Functions and Regulation. Curr. opin. lipidol. 22: 365-372.

[36] Newcombe J, Li H, Cuzner ML (1994) Low Density Lipoprotein Uptake by Macrophages in Multiple Sclerosis Plaques: Implications for Pathogenesis. Neuropathol. appl. neurobiol. 20: 152-162.

[37] Boven LA, Van Mars M, Van Zwam M et al (2006) Myelin-Laden Macrophages Are Anti-Inflammatory with Foam Cells in Multiple Sclerosis. Brain 129: 517-526.

[38] Baitsch D, Bock HH, Engel T et al (2011) Apolipoprotein E Induces Antiinflammatory Phenotype in Macrophages. Arterioscler. thromb. vasc. biol. 31: 1160-1168.

[39] Ecker J, Liebisch G, Englmaier M et al (2010) Induction of Fatty Acids Synthesis is a Key Requirement for Phagocytic Differentiation of Human Monocytes. Proc. natl. acad. sci. USA 107: 7817-7822.

[40] Pocivavsek A, Michailenko I, Strickland DK et al (2009), Microglial Low-Density Lipoprotein Receptor-Related Protein 1 Modulates c-Jun N-Terminal Kinase Activation. J. neuroimmunol. 214: 25-32.

[41] Gaultier A, Wu X, Le Moan N et al (2008) Low-Density Lipoprotein Receptor-Related Protein 1 Is An Essencial Receptor for Myelin Phagacytosis. J. cell sci. 122: 1155-1162.

[42] Yepes M, Sandkvist M, Moore EG et al (2003) Tissue-Type Plasminogen Activator Opening of the Blood-Brain Barrier Via the LDL Receptor-RelatedProtein. J. clin. invest. 112:1533-1540.

[43] Li F-Q, Sempowski GD, McKenna SE et al (2006) Apolipoprotein E-derived Peptides Ameliorate Clinical Disability and Inflammatory Infiltrates into the Spinal Cord in a Murine Model of Multiple Sclerosis. J pharmacol. exp. ther. 318: 956-965.

[44] Sena A, Tavares A, Ferret-Sena V et al (2008) Peroxisome Proliferator-Activated Receptors (PPARs) in Relapsing-Remitting Multiple Sclerosis Patients. Mult. scler. 14: S244.

[45] Shukla DK, Kaiser CC, Stebbins GT et al (2010) Effects of Poliglitazone on Diffusion Tensor Imaging Indices in Multiple Sclerosis Patients. Neuroscience Letters 472: 153-156.

[46] Cockerill GW, Rye KA, Gamble JR et al (1995) High-Density Lipoproteins Inhibit Cytocine-Induced Expression of Endothelial Cell Adhesion Molecules. Arterioscler. thromb. vasc. biol. 15: 1987-1994.

[47] Nobrega C, Capela C, Gorjon A et al (2011) Plasma Lipoproteins and Intrathecal Immunoglobulin Synthesis in Multiple Sclerosis. J. neurol. 258 (Suppl 1): S202.

[48] Sarov-Blat, L, Kiss RS, Haidar B et al (2007) Predominance of a Proinflammatory Phenotype in Monocyte-Derived Macrophages from Subjects with Low Plasma HDL-Cholesterol. Arterioscler. thromb. vasc. biol. 27: 1115-1122.

[49] Mehling M, Johnson TA, Antel J (2011) Clinical Immunology of the Sphongosine 1-Phosphate Receptor Modulator Fingolimod (FTY720) in Multiple Sclerosis. Neurology, 76 (Suppl 3): S20-S27.

[50] Norimatsu Y, Ohmori T, Kimura A et al (2012) FTY720 Improves Functional Recovery after Spinal Cord Injury by Primarily Nonimmunomodulatory Mechanisms. Am. J. pathol. 180: 1625-1635.

[51] Ferretti G, Bacchetti T (2011) Peroxidation of Lipoproteins in Multiple Sclerosis. J. neurol. sci. 311: 92-97

[52] Haider L, Fisher MT, Frischer JM et al (2011) Oxidative Damage in Multiple Sclerosis Lesions. Brain 134: 1914-1924.

[53] Sena A, Pedrosa R, Roque R et al (2006) Oxidised Low Density Lipoprotein in Serum of Relapsing-Remitting Multiple Sclerosis Patients. Mult. scler. 12 (Suppl 1): S168-S169.

[54] Koch M, Mostert J, Arutjunyan AV et al (2007) Plasma Lipid Peroxidation and Progression of Disability in Multiple Sclerosis. Eur. j. neurol. 14: 529-533.

[55] Mossberg N, Movitz C, Hellstrand K et al (2009) Oxygen Radical Production in Leukocytes and Disease Severity in Multiple Sclerosis. J. immunol. 213: 131-134.

[56] Linker RA, Lee D-H, Ryan S et al (2011) Fumaric Acid Esters Exert Neuroprotective Effects in Neuroinflammation Via Activation of the Nrf2 Aantioxidante Pathway. Brain 134: 678-692.

[57] Ludewig B, Laman JD (2004) The In and Out of Monocytes in Atherosclerotic Plaques: Balancing Inflammation through Migration. Proc. natl. acad. sci. USA 101: 11529-11530.

[58] D'haesseleer M, Cambron M, Vanopdenbosch L et al (2011) Vascular Aspects of Multiple Sclerosis. Lancet neurol. 10: 657-666.

[59] Sena A, Pedrosa R, Morais MG (2003) Therapeutical Potential of Lovastatin in Multiple Sclerosis. J. neurol. 250: 754-755.

[60] Vollmer T, Key L, Durkalski V et al (2004) Oral Simvastatin Treatment in Relapsing-Remitting Multiple Sclerosis. Lancet, 363: 1607-1608.

[61] Sena A, Pedrosa R, Morais MG (2007) Beneficial Effect of Statins in Multiple Sclerosis: Is It Dose-Dependent? Atherosclerosis, 191: 462.

[62] Tskiri A, Lakkenbach K, Fuglø D et al (2011) Simvastatin Improves Final Visual Outcome in Acute Optic Neuritis: a Randomized Study. Mult. scler. j. 18: 72-81.

[63] Waubant E, Pelletier D, Mass M et al (2012) Randomized Controlles Trial of Atorvastatin in Clinically Isolated Syndrome. Neurology 78:1171-1178.

[64] Strittmatter WJ, Saunders AM, Schmechel D et al (1993) Apolipoprotein E: High-Avidity Binding to β-Amyloid and Increased Frequency of Type 4 Allele in Late-Onset Alzheimer Disease. Proc. natl. acad. sci. USA 90: 1977-1981.

[65] Corder EH, Saunders AM, Strittmatter WJ (1993) Gene Dose of Apolipoprotein E Type 4 Allele and the Risk of Alzheimer's Disease in Late-Onset Families. Science 261:921-931.

[66] Bu G, (2009) Apolipoprotein E and its Receptors in Alzheimer's Disease: Pathways, Pathogenesis and Therapy. Nat. rev. neuroscience 10: 333-344.

[67] Di Paolo G, Kim T-W (2011) Linking Lipids to Alzheimer's Disease. Cholesterol and Beyond. Nat. rev. neuroscience 12: 284-296.

[68] Haan MN (2010) Midlife Cholesterol Level and Dementia 32 Years Later. Is There a Risk? Neurology 75: 1862-1863.

[69] Mielke MM, Zandi PP, Shao H (2010) The 32-Year Relationship Between Cholesterol and Dementia from Midlife to Late Life. Neurology 75: 1888-1895.

[70] Reitz C, Tang M-X, Schupf N et al (2010) Association of Higher Levels of High-Density Lipoprotein Cholesterol in Elderly Individuals and Lower Risk of Late-Onset Alzheimer Disease. Arch. neurol. 67: 1491-1497.

[71] Singh-Manoux A, Gimeno D, Kivimaki M et al (2008) Low HDL Cholesterol Is a Risk Factor for Deficit and Decline in Memory in Midlife. The Whitehall II Study. Arterioscler. thromb. vasc. biol. 28: 1556-1562.

[72] Merched A, Xia Y, Visvikis S et al (2000) The Relation Between Apolipoprotein AI and Dementia. The Honolulu-Asia Aging Study. Am. j. epidemiol. 165: 985-992.

[73] Kutiyama M, Takahashi K, Yamano T et al (1994) Low Levels of Serum Apolipoprotein AI and AII in Senile Dementia. Jpn j. psychiatry neurol. 48: 589-593.

[74] Matsuzaki T, Sasaki K, Hata J et al (2011) Association of Alzheimer Disease Pathology with Abnormal Lipid Metabolism. The Hisayama Study. Neurology 77: 1068-1075.

[75] Van Vliet p, Westendorp RGJ, Eikelenboom P et al (2009) Parental History of Alzheimer Disease Associated with Lower Plasma Apolipoprotein E Levels. Neurology 73: 681-687.

[76] Gupta VB, Laws SM, Villemagne VL et al (2011) Plasma Apolipoprotein E and Alzheimer Disease Risk. The AIBL Study of Aging. Neurology 76: 1091-1098.

[77] Kantarci K, Lowe V, Przybelski SA et al (2012) ApoE Modifies the Association Between Aβ Load and Cognition in Cognitively Normal Older Adults. Neurology 78: 232-240.

[78] Schrijvers EM, Koudstaal PJ, Hofman A et al (2011) Plasma Clusterin and the Risk of Alzheimer Disease. Jama 305: 1322-1326.

[79] Holmes C, Cunningham C, Zotova E et al (2009) Systemic Inflammation and Disease Progression in Alzheimer Disease. Neurology 73: 768-774.

[80] Iadecola C (2003) Atherosclerosis and Neurodegeneration. Unexpected Conspirators in Alzheimer's Dementia. Arterioscler. thromb. vasc. biol. 23: 1951-1953.

[81] Desai MK, Mastrangelo MA, Ryan DA et al (2010) Early Oligodendrocyte/Myelin Pathology in Alzheimer's Disease Mice Constitutes a Novel Therapeutic Target. Am. J. pathol. 177: 1422-1435.

[82] Cramer PE, Cirrito JR, Wesson DW et al (2012) ApoE-Directed Therapeutics Rapidly Clear β-Amyloid and Reverse Deficits in AD Mouse Models. Science 335: 1503-1506.

[83] Takechi, R, Galloway S, Pallebage-Gamarallage MMS et al (2008) Chylomicron Amyloid-Beta in the Aetiology of Alzheimer's Disease. Atherosclerosis (Suppl 9) 19-25.

[84] Kontush A, Chapman MJ (2008) HDL: Close to Our Memories? Arterioscler. thromb. vasc. biol. 28: 1418-1420.

[85] Elder GA, Cho JY, English DF et al (2007) Elevated Plasma Cholesterol. Does Not Affect Brain Aβ in Mice Lacking the Low-Density Lipoprotein Receptor. J. neurochem. 102: 1220-1231.

[86] Pistell PJ, Morrison CD, Gupta S et al (2010) Cognitive Impairment Following High Fat Diet Consumption Is Associated with Brain Inflammation. J. immunol. 219: 25-32.

[87] Lieb W, Beiser AS, Vasan RS et al (2009) Association of Plasma Leptin Levels with Incident Alzheimer Disease and MRI Measures of Brain Aging. Jama 302: 2565-2572.
[88] Une K, Takei A, Tomita N et al (2010) Adiponectin in Plasma and Cerebrospinal Fluid in MCI and Alzheimer's Disease. Eur. j. neurol. 18: 1006-1009.
[89] Yaffe K, Weston AL, Blackwell T et al (2009) The Metabolic Syndrome and Development of Cognitive Impairment Among Older Women. Arch. neurol. 66: 324-328.
[90] Craft S (2009) The Role of Metabolic Disorders in Alzheimer Disease and Vascular Dementia. Arch. neurol. 66: 300-305.
[91] Brayne C, Ince PG, Keage H et al (2010) Education, the Brain and Dementia: Neuroprotection or Compensation? Brain 133: 2210-2216.
[92] Belleville S, Clément F, Mellah S et al (2011) Training-Related Brain Plasticity in Subjects at Risk of Development Alzheimer's Disease. Brain 134: 1623-1634.
[93] Cuadrado-Godia E, Jiménez-Conde J, Ois A et al (2009) Sex Differences in the Prognostic Value of the Lipid Profile After the First Ischemic Stroke. J. neurol. 256: 989-995.
[94] Hughes TF, Borenstein AR, Schofield E et al (2009) Association Between Late-Life Body Mass Index and Dementia. The Kame Project. Neurology 72: 1741-1746.
[95] Vidoni ED, Townley RA, Honea RA et al (2011) Alzheimer Disease Biomarkers Are Associated with Body Mass Index. Neurology 77: 1913-1920.
[96] Shepardson NE, Shankar GM, Selkoe DJ (2011) Cholesterol Level and Statin Use in Alzheimer Disease. I. Review of Epidemiological and Preclinical Studies. Arch. neurol. 68: 1239-1244.
[97] De Lau LML, Koudstaal PJ, Hofman A et al (2006) Serum Cholesterol Levels and the Risk of Parkinson's Disease. Am. j. epidemiol. 164: 998-1002.
[98] Huang X, Chen H, Miller WC et al (2007) Lower Low-Density Lipoprotein Cholesterol Levels Are Associated with Parkinson's Disease. Mov. disord. 22: 377-381.
[99] Hu G, Antikainen R, Jousilahti P et al (2008) Total Cholesterol and the Risk of Parkinson Disease. Neurology 70: 1972-1979.
[100] Dupuis L, Corcia P, Fergani A et al (2008) Dyslipidemia Is a Protective Factor in Amyotrophic Lateral Sclerosis. Neurology 70: 1004-1009.
[101] Chiò A, Calvo A, Ilardi A et al (2009) Lower Serum Lipids Are Related to Respiratory Impairment in Patients with ALS. Neurology 73: 1681-1685.
[102] Glass CK; Saijo K, Winner B et al (2010) Mechanisms Underlying Inflammation in Neurodegeneration. Cell140: 918-934.
[103] Valenza M, Cattaneo E (2011) Emerging Roles for Cholesterol in Huntington's Disease. Trends in neurosci. 34: 474-486.
[104] Whitehead AS, Bertrandy S, Finnan F et al (1996) Frequency of the Apolipoprotein E ε4 Allele in a Case-Control Study of Early Onset Parkinson's Disease. J. neurol. neurosurg. psychiatry (1996) 61: 347-351.
[105] Wilhelmus MMM, Bol JGJM, Rozemuller AJM et al (2011) Apolipoprotein E and LRP1 Increase Early in Parkinson's Disease Pathogenesis. Am. J. pathol. 179: 2152-2156.

[106] Thomas EA, Sutcliffe JG (2002) The Neurobiology of Apoliporoteins in Psichiatric Disorders. Mol. Neurobiol. 26: 369-388.

[107] Muffat J, walker DW, Benzer S (2008) Human ApoD, an Up-Regulated in Neurodegenerative Diseases, Extends Lifespan and Increases Stress Resistance in *Drosophila*. Proc. natl. acad. sci. USA 105: 7088-7093.

[108] Terrise L, Poirier J, Bertrand P et al (1998) Increased Levels of Apolipoprotein D in Cerebrospinal Fluid and Hippocampus of Alzheimer's Patients. J. neurochem. 71: 1643-1650.

[109] Digney A, Keriakous D, Scarr E et al (2005) Differential Changes in Apolipoprotein E in Schizophrenia and Bipolar I Disorder. Biol. Psychiatry 57: 711-715.

[110] Huang JT-J, Wang L, Prabakaran S et al (2008) Independent Protein-Profiling Studies Show a Decrease in Apolipoprotein AI Levels in Schizophrenia CSF, Brain and Peripheral Tissues. Mol. psychiatry 13: 1118-1128.

[111] Zatorre RJ, Fields RD, Johansen-Berg H et al (2012) Plasticity in Gray and White: Neuroimaging Changes in Brain Structure During Learning. Nat. rev. neurosci. 15: 528-536.

[112] Karasinska JM, Rinninger F, Lütjohann et al (2009) Specific Loss of Brain ABCA1 Increases Brain Cholesterol Uptake and Influences Neuronal Structure and Function, J. neurosci., 29: 3579-3589.

[113] Ruckh JM, Zhao J-W, Shadrach JL et al (2012) Rejuvenation of Regeneration in the Aging Central Nervous System. Cell stem cell 10: 96-103.

[114] Redmond SA, Chan JR (2012) Revitalizing Remyelination–the Answer Is Circulating. Science 336: 161-162.

[115] Cunnigham TJ, Yao L, Oetinger M et al (2006) Secreted Phospholipase A2 Activity in Experimental Autoimmune Encephalomyelitis and Multiple Sclerosis. J. neuroinflammation 3: 26-33.

[116] Hesse D, Krakauer M, Lund H et al (2011) Disease Protection and Interleukin-10 Induction by Endogenous Interferon-β in Multiple Sclerosis? Eur. j. neurol. 18: 266-272.

[117] Matarese G, Procaccini C, De Rosa V, (2008) The Intricate Interface Between Imuune and Metabolic Regulation: a Role for Leptin in the Pathogenesis of Multiple Sclerosis? J. leukoc. biol. 84: 893-899.

[118] Hietaharju A, Kuusisto H, Nieminen R et al (2010) Elevated Cerebrospinal Fluid Adiponectin and Adipsin Levels in Patients with Multiple Sclerosis: a Finnish Co-Twin Study. Eur. j. neurol. 17: 332-334.

[119] Sena A, Couderc R, Vasconcelos JC et al (2012) Oral Contraceptive Use and Clinical Outcomes in Patients with Multiple Sclerosis. J. neurol. sci. 317(1-2): 47-51.

[120] Neu IS, Prosiegel M, Pfaffenrath V (1982) Platelet Aggregation and Multiple Sclerosis. Acta neurol. scandinav. 66:497-504.

[121] Hawkes CH, (2007) Smoking is a Risk Factor for Multiple Sclerosis: a metanalysis. Mult. scler. 13: 610-615.

Chapter 24

Genetics of Ischemic Stroke: Emphasis on Candidate-Gene Association Studies

Sanja Stankovic, Milika Asanin and Nada Majkic-Singh

Additional information is available at the end of the chapter

http://dx.doi.org/10.5772/50257

1. Introduction

Stroke is the leading cause of neurological disability and among the leading causes of death worldwide. It is a focal neurological deficit that results from events that decrease or stop cerebral blood flow. As the consequence neurons cease functioning and irreversible neuronal ischemia and injury occur.

Broadly, strokes are classified into two main types-ischemic and hemorrhagic. Ischemic stroke (IS) is characterized by blockage in blood flow to a focal area of the brain, until hemorrhagic stroke is caused by bleeding into the brain. Acute IS is more common than hemorrhagic stroke. Although according the previous literature data about 80% of strokes were ischemic, the retrospective review from a stroke center found that about 60% were ischemic [1]. Except their causes and pathophysiology ischemic and hemorrhagic types differ in their treatments and outcomes [2].

Based on the system of categorizing stroke developed in multicenter Trial of Org 10172 in Acute Stroke Treatment (TOAST), IS may be divided into the following major subtypes: large artery infarction, small-vessel (lacunar) infarction, and cardioembolic infarction. This classification on the basis of inferred origin of cerebrovascular occlusion [3] is the most frequently used. Other studies used systems based on clinical presentation or location and size of the lesion within the brain (such as the Oxfordshire Community Stroke Project system) [4]. It classifies patients in five infarct types: cerebral infarction, lacunar infarct, total anterior circulation infarct, partial anterior circulation infarct, and posterior circulation infarcts. Many other classifications have been proposed, such as those from the Lausanne Stroke Registry and the Étude du profil Génétique de l'Infarctus Cérébral (GÉNIC) study [5,6]. The first one included atherosclerosis with stenosis, atherosclerosis without stenosis, emboligenic heart disease, hypertensive arteriopathy, cerebrall hemorrhage, mixed causes and undetermined causes. The former included atherothrombotic stroke, cardioembolic stroke, lacunar stroke, arterial dissection, unknown causes stroke. Although stroke is often considered a disease of elderly persons, one third of strokes occur in persons younger than 65 years.

Risk factors for IS includes modifiable and non-modifiable etiologies. Non-modifiable risk factors include: age, sex, race, ethnicity, heredity, etc. Modifiable risk factors include the followings: hypertension, diabetes mellitus, hypercholesterolemia, atrial fibrillation, lifestyle factors, etc. Unfortunately, modifiable risk factors accounts for only approximately 60% of the population-attributable risk for stroke [7].

2. Genetic risk factors in stroke

Evidence continues to accumulate to suggest important roles for genetic factors in stroke. Genetic risk factors are particularly interesting, because they can offer a direct clue to the biological pathways involved. Genetic factors might affect stroke risk at various levels. They could act through conventional risk factors, interact with conventional and environmental risk factors, or contribute directly to an established stroke mechanism. They could further affect the latency of stroke or infarct size, and stroke outcome [8]. Stroke may be the outcome of a number of monogenic disorders or, more commonly, a polygenic multifactorial disease.

Evidence shows that genetic factors are more important in small- and large-vessel stroke than in cardioembolic stroke [9,10]. Some intermediate phenotypes also exhibit high heritability, such as carotid intima-medial wall thickness and white-matter lesions [8].

Genetic predisposition to stroke has been proven in animal models and in humans (twins, affected sibling pair, families). Several studies demonstrated higher rates of stroke among relatives of patients who died from stroke than among relatives of healthy control subjects. In a large study of stroke patients and age and sex matched controls, the odds ratios (ORs) of having a family history of stroke were 2.24 for large vessel-disease and 1.93 for small vessel disease [9]. Twin studies have confirmed a significant genetic component to stroke, with the stroke prevalence fivefold higher in monozygotic than in dizygotic twins [11]. Touze and Rothwell [12] in a meta–analysis based on 18 studies confirmed sex differences in heritability of IS; women with stroke were more likely than men to have a parental history of stroke, which is accounted for by an excess maternal history of stroke. Also, genetic predisposition could differ depending on age and IS subtype.

The initial expectancy to find only one or a few common mutations that substantially contribute to the risk of IS shifted toward the hypothesis of a large number of small-effect genetic variants with complex gene-gene and gene-environment interactions. The first approach used in identification of genetic variants contributing to stroke was linkage studies. Linkage analysis relies on the cosegregation of known polymorphic DNA marker with nearby, unknown disease-causing alleles in families. This approach was successful in monogenic diseases, but was less successful in the identification of genetic loci that contribute to the occurrence of polygenic stroke. The second approach was candidate gene approach.

2.1. Candidate-gene association studies of ischemic stroke

Until recently, candidate gene approach was the most common in genetic investigation of IS. A gene identified as a "candidate" is hypothesized to be involved in IS risk, and then, genetic variants, usually single nucleotide polymorphisms (SNPs), are identified within that gene. The SNPs are selected on the basis of their localization in genes which encode proteins with a known function in a biological pathway implicated in the pathophysiology of the disease. Then, the frequency of the SNPs is determined in a series of cases and controls and the obtained results are compared. They use a case-control study design. A gene variant that is more common in patients than in controls may cause stroke or be located close to the true causal variant.

Genes encoding products involved in lipid metabolism, thrombosis, and inflammation are believed to be potential genetic factors for IS [13-15]. Although a large group of candidate genes have been studied, most of the epidemiological results are conflicting. Especially great interest is shown in exploring potential links between polymorphisms in genes encoding proteins involved in lipid metabolism and the risk of IS.

This chapter summarize the results of meta-analyses and case-control studies assessing the linkage of specific candidate genes with the risk of IS and specific subtypes. Electronic databases (Medline (http://www.ncbi.nlm.nih. gov/pubmed/), Embase (http://www. embase.com/), Google Scholar (http://scholar.google.com/), Yahoo (http://www.yahoo.com/), Kobson (http://www.kobson. nb.rs/) were searched until March 2012 and the obtained results were included in the text.

It is very well known that individuals with higher levels of plasma cholesterol, decreased high-density lipoprotein (HDL) and increased low-density lipoprotein (LDL) have a higher risk of premature atherosclerosis. The phenotype may arise not only from single gene disorders, but also from a number of genetic and environmental factors, including polymorphic variants of genes encoding the apolipoproteins, lipoprotein receptors and the key enzymes of plasma lipoprotein metabolism.

Apolipoprotein E. One of the most intensively investigated candidate genes for IS that received widespread attention is the apolipoprotein (apo) E gene. It forms a cluster with certain apoC genes on the long arm of chromosome 19 (19q13.2). The human apoE gene is polymorphic, with three common alleles (ε2, ε3, ε4) coding for three isoforms (E2, E3, E4). The association studies of apoE gene polymorphisms with IS gave conflicting results based on 9 meta-analyses [16-24] and 77 case-control studies [25-101]. In small case-control or cross-sectional studies, both the ε2ε3 genotype and the ε4 allele have been over-represented in patients with IS. Other groups have examined the role of the apoE genotype in modulating the outcome of cerebral infarction as this lipoprotein appears to be an important regulator of lipid turnover within the brain and of neuronal membrane maintenance and repair. McCarron et al. [102] found a favorable effect of the ε4 allele on stroke outcome. Stankovic and colleagues [85] reviewed the conflicting results on the importance of the apoε alleles in predisposition to IS.

Seven meta–analyses [17-19,21-24] gave a positive association between the ε4 allele and IS. The first one [22], published in 1999, revealed a significantly higher apoε4 allele frequency in affected patients compared with controls (OR 1.68, 95% CI 1.36–2.09, P<0.001). In the next decade, five meta–analyses [17-19, 21,23] confirmed that ε4 allele carriers have a higher risk of IS compared with pooled ε2 and ε3 allele carriers in European populations, persons of non-European descent, Asians, Han Chinese and persons with early-onset IS. Performing large-scale meta-analysis (10674 cases/33430 controls) consisted of four meta-analysis [19,21-23] and 9 case-control studies [33,35,36,54,59,65,66,84,88], Hamzi et al. [24] calculated OR for the apoε4 allele to be 0.95 (95%CI 0.77-1.14, P=0.002).

Approximately half of all case–control studies [26,27,29,33,38,41,45,47,49,51,53,54,57,58, 60,64,67,69,71,73-76,78-80,82,84,85,89-91,93-96,99,100] showed an increased frequency of the ε4 allele in stroke patients, making it a highly probable risk factor for IS; in four, significant association with large-vessel IS was observed. Three groups described the ε2 allele as a risk factor for IS [76,85,94]. The status of the E2/3 genotype as a protective or risk factor is controversial. One report [100] demonstrated a protective role of the ε4 allele for small-vessel disease, and another [93] concluded that the E3/4 genotype could be a risk factor for lacunar stroke compared with the E3/3 genotype.

Several SNPs have been described in the 5′ regulatory region (c.491A>T, c.427T>C, c.219G>T, and c.113G>C), but current information is very preliminary. A higher risk of IS was associated with the G allele of the tightly linked c.219G>T and c113G>C promoter polymorphisms [96], and with the T allele of c.427T>C polymorphism [94]. One paper [94] reported the C allele of c.427T>C polymorphism as protective for IS.

Other apolipoproteins. Except apoE gene polymorphism that was frequently investigated polymorphism in patients with IS, another apolipoprotein genes have undergone intense investigation (apo AI/CIII, apoAIV, apoAV, apoB, apoH). The most published studies investigating the relationship between these polymorphisms and IS are small in sample size and inconclusive in their results.

Some authors have studied the association between IS and DNA polymorphisms in apoAI gene (*Sst*I (rs5128), *Msp*I, c.75G>A, c.84T>C), apoCIII gene (c.641C>A, c.482 C>T, c.455C>T, c.1100C>T, c3175C>G, c3206T>G), apoAIV (p.Thr347Ser, p.Gln360His), and apoH (c.1025G>C, c.341G>A), mainly with negative results [28,30,31,34,52,103,104].

The apoB gene is located on chromosome 2q23, spanning approximately 43 kb and has 29 exons and 28 introns. ApoB polymorphisms (T71I (c>t; rs17246849), A591V (c>t; rs17240681), *Bfa*I (P2712L; c>t; rs17240903), *Msp*I (R3611Q; g>a; rs17247291), *Eco*RI (E4154K; g>a; rs1042031), and *Eco57*I (N4311S; a>g; rs17240958), p.Arg3500Gln, c.4311A>G) were examined in patients with IS. Only two studies found that apoB polymorphisms [105,106] were associated with IS risk. Zhang et al. [107] found that C7673T polymorphism in apoB gene is associated with risk of ischemic cerebral infarction with family history in 47 Han Chinese patients. In Danish prospective study (the Copenhagen City Heart Study) [108] with 23-yr follow-up the E4154K KK homozygosity was associated with an 80% reduction in risk of IS (0.2 (0.1-0.7)) compared with non-carriers. The other SNPs or haplotypes examined in this study were not associated with risk of IS.

The most promising results in IS studies are connected with apoA5 and apo(a) gene polymorphisms. It is well konown that apoAV is a member of apoAI/CIII/AIV gene cluster. apoAV gene consists of 4 exons and codes 369 amino acids protein. The common variants within the apoAV gene are associated with plasma tryglicerides (TG) levels, by enhancing the intravascular triglyceride hydrolysis by activating lipoprotein lipase (LPL), or can decrease the serum concentration of triglycerides through the inhibition of the hepatic very low density lipoprotein (VLDL) production. Literature data suggest significant association between apoAV gene polymorphisms (c.1131T>C, c.12238T>C, c.553G>T) and IS risk [34,109-112]. The association of apoAV 56G allele was observed in the large-vessel associated stroke group compared to the healthy controls [113]. The same group of authors [114] examined three polymorphism in apoAV gene in small-vessel, large-vessel and mixed subgroups of 378 patients with stroke and healthy controls. They found that patients carriers of -1131C and IVS3+476A alleles confer risk for all IS types, In this study the T1259C variant was not associated with IS that is in agreement with previous study of Jeromi et al.[112]. Recently published study on Han Chinese population confirmed the previously found association between c.1131T>C polymorphism in apoAV gene and IS risk [115].

There is growing and convincing evidence that elevated lipoprotein (a) levels have a significant role in stroke. Genetic studies demonstrated that Lp(a) is an inherited trait determined almost entirely by the apo(a) gene locus. Variations at the apo(a) gene locus beyond the kringle IV-2 domain seem to influence Lp(a) concentrations [116]. The pentanucleotide TTTTA repeat (PNTR) polymorphism located at the 5′ untranslated region of the apo(a) gene accounts for 10% to 14% of the variation in plasma Lp(a) concentrations [117], and was reported to be inversely correlated with Lp(a) levels. Low numbers of apo(a) TTTTA VNTR were associated with IS in three studies [118-120] that were included with the only meta-analysis [19] that evaluated the association of apo(a) TTTTA VNTR polymorphism and IS.

The Precocious Coronary Artery Disease (PROCARDIS) study identified 2 single-nucleotide polymorphisms (SNPs) at the Lp(a) locus (LPA) on chromosome 6q26–27 (rs3798220 (T/C) and rs10455872(A/G)) that each was strongly and independently related to Lp(a) levels and risk of coronary disease [121]. Wang et al. [122] in meta-analysis of 3550 IS cases and 6560 controls showed no significant association of LPA variants previously associated with Lp(a) levels with IS (OR per allele 0.96, 95% CI 0.88-1.04, for rs1853021 and 0.95, 95% CI 0.88-1.03, for rs1800769). Also, there was the lack of evidence of an association of LPA score and prevalent or incident stroke in Heart Protection Study (1326 prevalent and 507 incident IS cases) [123]. It does not exclude the possibility that lowering Lp(a) could have beneficial effects on the risk of stroke or stroke subtypes. On the contrary, theWomen's Health Study (123 IS cases) suggested a positive association of rs3798220 with stroke [124].

Future studies are warranted to assess whether the analysis of previously mentioned polymorphisms may be useful for the clinical approach to evaluate risk factors for IS.

Cholestryl ester transfer protein (CETP). CETP participates in HDL metabolism by facilitating the transfer of cholesteryl esters from HDL to apoB-containing lipoproteins in exchange for triglycerides being transferred to HDL This glycoprotein is secreted mainly from the liver and circulates in plasma, bound mainly to HDL. A deficiency of CETP is connected with anti-atherogenic profile, with increased HDL and decreased LDL levels. The CETP gene is located on chromosome 16q21 and consists of 16 exons. Several polymorphisms have been described, including (Taq1 B in intron-1(rs708272), 405V and A373P (rs5880) in exon 12, R451Q (rs 1800777) in exon 15, and -629A/C (rs 1800775). Of these, the most widely studied is the *Taq*I B polymorphism which results from a nucleotide substitution at position 277 of the first intron (rs708272). CETP Taq1 B2B2 genotype is associated with decreased CETP activity, higher HDL-cholesterol concentrations [125,126], decreased risk of coronary artery disease [126,127], lower carotid intimal medial thickness and stenosis [128], lower incidence of microangiopathy in patients with type 2 diabetes [129], and atrial fibrillation [130].

The relationship between CETP polymorphisms and the risk of IS has been the subject of eight reports [28,30,34,131-135]. An association with CETP *Taq*1 B polymorphism was found in one study [133] but not in another [132]. Some isolated reports of a significant association relate to the rs12720922 and rs9939244 [134] and the rs5883 [135] polymorphisms. Clearly, more extensive investigations in this area are warranted.

ATP-binding cassette transporter I(ABCAI). ABCA1 is a transmembrane protein present on peripheral tissue cells, crucial in the initial step of HDL formation. It mediates the transfer of cellular phospholipids and cholesterol to acceptor apolipoproteins such as apolipoprotein A-I [136]. The ABCA1 locus is located on chromosome 9q22-q31, and is composed of 50 exons ranging in size from 33bp to 249bp. More than 100 common and rare variants have been described [137]. Several polymorphisms of the ABCA1 gene have been investigated for their association with IS.

The first published study in IS on 244 Hungarian patients [138] suggests a protective role for the *ABCA1*-R219K and V771M polymorphisms. Pasdar et al. [139] studied four common polymorphisms in ABCA1 gene: G/A-L158L, G/A-R219K, G/A-G316G and G/A-R1587K in 400 Caucasian IS patients. There was no significant difference in allele frequencies of all polymorphisms, as the haplotypes arrangement. This study did not support a major role for the ABCA1 gene as a risk factor for IS. Following a report of an association of -14C/T polymorphism in the promoter region of the ABCA1 gene with IS [140], extensive studies to confirm this association in different populations are essential.

Lipoprotein lipase (LPL). Lipoprotein lipase (LPL) is a member of the lipase gene family [141] that may play a central role in lipid metabolism. The major sources of LPL synthesis are skeletal and heart muscle as well as adipose tissue, from which the mature enzyme is then secreted and transported to the vascular endothelium, the physiological site of the enzyme's action [142]. The physiological action of LPL consists of the hydrolysis of the triacylglycerol component of triglycerides and VLDL, resulting in the production of chylomicron remnants, and in the case of VLDL, resulting in the production of smaller, intermediate-density lipoproteins [143]. LPL is also synthesized by macrophages and macrophage-derived foam cells in atherosclerotic lesions [144-146], and this fraction of the enzyme has been linked to LPL-related proatherogenic effects. LPL possess a noncatalytic activity on lipoproteins such as molecular bridging [147] and retention of LDL-C by proteoglycans of the subendothelial matrix occurs, thereby proposing LPL activity in the arterial wall to promote atherosclerosis.

The human LPL gene is localized to chromosome 8p22, spanning 35kb. It contains 10 exons. The gene locus is highly polymorphic and contains many single nucleotide polymorphisms (SNPs) in both coding and non-coding regions. Some cause loss of enzymatic activity and others have only mild detrimental effects on LPL function, or serve more as markers for genetic variation elsewhere in the genome [148].

Epidemiological evidence on the potential role of LPL in IS remains scarce and controversial. Two SNPs in the coding DNA (cSNPs) that have been studied extensively cause point mutations in exons 2 and 6, with substitution of an aspartic acid to an asparagine residue at position 9 (D9N, p.Asp9Asn), and an asparagine to a serine residue at position 291 (N291S, p.Asn291Ser), respectively. These mutations occur at high frequencies in the general population (up to 5%) and are associated with elevated TG, decreased HDL-cholesterol levels, and concomitantly with a higher incidence of cardiovascular disease compared with non-carriers [149]. Polymorphism Ser447Ter is a consequence of a C to G transversion at nucleotide 1595 in exon 9, which converts the serine 447 codon (TCA) to a premature termination codon (TGA). This polymorphism is associated with increased lipolytic function and beneficial effects on lipid homeostasis and atheroprotection [148]. *Hind*III polymorphisms of the LPL gene in intron 8, which identifies a two-allele polymorphism with restriction fragments of 6 kb (H1) and 11 kb (H2), is associated with elevated TG levels [150], low HDL-cholesterol levels [151], and was considered as a possible IS-associated polymorphism [152] Also, *Pvu* II polymorphism in intron 6 has been associated with high TG levels and coronary artery disease.

Four meta–analysis [16,153,154], and 17 case–control studies have been reported [28,30,34,72,88,94,132,152,155-163] about the association of LPL gene polymorphisms and IS. In a meta–analysis of six studies [153] the inverse association between LPL Ser447Ter polymorphism and IS risk was of borderline significance (OR=0.88, 95%CI 0.79–0.99, P=0.033). In recently published meta-analysis [154] of 4681 IS patients and 8516 controls from 13 studies LPL Ter447 variant was associated with a significantly reduced risk for IS (OR 0.79, 95%CI 0.68-0.93, P=0.005) in Causcasian and East-Asian population. According the data of four studies (387 cases/589 controls), this association was of great importance in atherosclerotic stroke (OR 0.44, 95%CI 0.32-0.62, P<0.00001). In the meta-analysis of same authors [154] that included 7 studies (3669 cases and 6693 controls) no significant association between Ser291 variant and IS stroke risk was found. This is in accordance with the conclusion of previously published meta-analysis of LPL Asn291Ser polymorphism and IS [16]. A positive association between S447X variant and stroke has been reported in specific subtypes, as in the study of Shimo-Nakanishi et al. [152], Zhao et al. [160], Guan et al. [161], and Xu et al. [163] which reported a relationship with atherosclerotic stroke, and in the prospective cohort study of Morrison et al. [72] who described a positive association between S447X and asymptomatic stroke lesions, and in the study of Kostulas et al. [162] where the protective role of G-allele of LPL S447X polymorphism had a lower frequency in males. Shimo-Nakanishi et al. [152] observed a protective role of H- H- and H-H+ genotypes vs. H+H+ (*Hind*III polymorphism), and Xu et al. [163] noted a protective role of the P allele (*Pvu*II polymorphism) for IS. In conclusion, there is evidence to support an association between LPL gene polymorphism and IS, but this notion needs to be strengthened by further investigations.

Hepatic lipase. Despite the numerous association studies of LPL gene polymorphisms and IS, and these have generated consistently negative results [28,30,164].

Paraoxonase(PON). Paraoxonase is a glycoprotein, HDL-associated esterase, that hydrolyzes products of lipid peroxidation and prevents the oxidation of LDL. It has antioxidant and anti-atherogenic properties [165,166]. The paraoxonase gene maps to chromosome 7q21.3. It codes three isoforms, PON1, PON2, and PON3, that share 60 to 65% homology at the amino acid level [167]. PON1 and PON3 reside on circulating HDL particles. PON2 is ubiquitously expressed and does not appear to be associated with HDL particles [168-170]. PON genes polymorphisms may affect the corresponding enzyme activity.

Two non-synonymous *PON1* polymorphisms with possible regulatory effects on enzyme activity [171], namely rs662 (c.575A>G or p.Gln192Arg) and rs854560 (c.163T>A or p.Leu55Met), have been extensively investigated as potential risk factors for atherosclerosis-related phenotypes, including coronary artery disease, peripheral arterial disease and IS [171-173]. Two previously published systematic reviews suggested that the G allele of rs662 is associated with a small increase (per-allele OR 1.12) in the risk of coronary artery disease, while no such association was found for rs854560 [172,173]. Inter-individual variability in PON1 levels is determined by the Q192R (Gln192Arg) and L55M (Leu55Met) coding region polymorphisms and by two described polymorphisms in the promoter of the PON1 gene, C(-107)T and G(-824)A. Five polymorphic sites were found in the promoter region of the PON1 gene: c.107C>T, c.126G>C, c.160G/A, c.824G>A, and c.907G>C. Specific polymorphisms are associated with the risk of acute IS.

According the literature data there are three meta–analysis [18,174,175] and 26 case–control studies [28,30,34,35,176-197] explored the association of PON1 polymorphisms and IS risk. A positive association of Gln192Arg PON1 polymorphism and IS was described in the meta–analyses and in five case-control studies [177,178,184,185,188], but this association was negative in all other reports. Only two studies in Turkish populations obtained evidence for a positive association of Leu55Met PON1 polymorphism and IS [188,193], in contrast to 12 where no evidence for this association was found [28,30,177-179,181,186,187,190-192,194].

Two recently published meta-analysis included the studies examined the association of two common polymorphisms in the coding region of PON1 gene (rs662 and rs854560) and the occurrence of IS. In meta-analysis [174] of 22 studies (7384 cases/11074 controls) PON1 polymorphism rs662 was associated with increased risk for IS (OR 1.10 per G allele copy, 95%CI 1.04–1.17, P=0.001), while no significant association of rs854560 was observed in meta-analysis of 16 eligible studies (OR 0.97 per T allele copy, 95% CI 0.90–1.04, P=0.37). The other meta-analysis [175] included 8 studies on rs854560 polymorphism and 9 studies on rs662 polymorphism. This analysis provides strong evidence that the rs662 polymorphism of PON1 gene is associated with IS (OR 1.21, 95%CI 1.02-1.43, P=0.03), and that the rs854560 gene polymorphism is not associated with IS (OR 1.12, 95%CI 0.96-1.31, P=0.13).

Man et al. [198] in 191 Han Chinese patients with acute IS, of whom 25% had concurrent stenosis found that genotype distributions of PON1 Q192R differed significantly between patients with stroke and controls, and that the presence of at least one R allele in PON1 Q192R was associated with concurrent stenosis.

Polymorphism c.107C>T is important because it contributes 23% of the variances in PON1 levels. Since the presence of T at position -107 of the PON1 gene disturbs a recognition sequence for stimulating protein-1 (Sp1), the TT genotype is associated with the lowest serum PON1 levels. Although the frequency of the T allele and TT genotype did not differ significantly between young adults with arterial IS and controls, the presence of the -107T allele was associated with an independent increase in the risk of arterial IS [197].

There are two common polymorphisms of the PON2 gene: A148G (Ala148Gly) and C311S (Ser311Cys)). Almost all research groups except one [192] agree that there is no significant association between IS and these polymorphisms [28,30,177,181,187,199] Four polymorphisms in the PON3 gene were examined in two studies on IS patients [181,187]. No evidence for an association was obtained. Whereas rs662 (c.575A>G or p.Gln192Arg) polymorphism of the PON1 gene could be regarded as a potential risk factor for IS, this does not seem to be the case for PON2 and PON3.

Although, Lazaros et al. [200] did not identified none of the PON polymorphisms (PON1(Q/R) 192, PON1(M/L) 55, and PON2(S/C) 311) as a risk factors for IS, they concluded that PON2 311C allele was significantly increased in patients with severe forms of IS and could be reviewed as a possible predisposing factor for severe cases of IS.

Large-scale multicenter-controlled prospective studies are warranted to further explore the effects of PON polymorphisms on stroke susceptibility and severity.

Low-density lipoprotein receptor (LDLR). LDLR is a cell surface receptor that plays an integral role in plasma lipoprotein metabolism, especially in cholesterol homeostasis. The LDLR gene is localized on chromosome 19, and comprises 45 kb with 18 exons. Mutations in this gene may lead to dysfunction of the receptor resulting in familial hypercholesterolemia and premature ischemic heart disease. The most frequently studied is A370T polymorphism (c.1171G>A in exon 8 that changes alanine to threonine at position 370 in the LDLR protein. The other described polymorphisms are *Nco*I, *Ava*II, c.1773C>T, and rs2738446 and rs2738450.

Only few studies explored the association of LDLR gene polymorphisms and IS. Guo et al. [201] investigated the relationship between *Nco*I and *Ava*II polymorphisms of the LDLR gene in Han Chinese patients with atherosclerotic cerebral infarction and concluded that the coexistence of A-A- and N+N+ genotypes significantly increases the risk of atherosclerotic cerebral infarction (RR 5.56, $p<0.001$). The data of Frikke-Schmidt et al. [202] support an association between c.370A>T polymorphism (370A allele) and increased risk of stroke. Two studies reported an association between rs2738450 and IS [135,203]. Recently published study [204], for the first time revealed the association of rs1122608 (located 58.7 kb upstream of the LDLR gene) and 530 IS patients in Chinese Han population.

Oxidized LDL that play a key role in the atherogenesis process exert most effects through the interaction with its major receptor lectin like oxidized low density lipoprotein receptor 1(LOX-1). LOX-1 is encoded by the lectin like oxidized low density lipoprotein receptor 1 (OLR1) gene, located in the p12.3–p13.2 region of human chromosome 12 and consists of 6 exons. Few SNPs located within introns 4, 5, and 3′ untranslated region, are associated with higher risk of developing acute myocardial infarction. Polymorphism (rs11053646, G501C) located in exon 4, leads to a change from a lysine to an asparagine at position 167 (K167N). As the consequence, reduced binding and internalization of the oxLDL was noticed. Only one paper [205] relates G501C polymorphisms of the OLR1 gene and IS, with negative results. Except LOX-1 full receptor, LOXIN as an isoform lacking part of the functional domen was identified and it has a protective role by blocking LOX-1 activation. One recently published study examined the prevalence of OLR1 gene polymorphisms, IVS4-14 A/G and IVS4-73 C/T, which regulate the expression of LOXIN, in 43 patients with ischemic cerebrovascular diseases (ICVD). Patients with G homozygosity for IVS4-14 polymorphism and T homozygosity for IVS4-73 polymorphism have higher risk to develop ICVD [206]. Man et al. [198] in 191 Han Chinese patients with acute IS, of whom 25% had concurrent stenosis examined whether oxidized low-density lipoprotein receptor (OLR) 3′ untranslated region (UTR) C > T (rs1050283) polymorphism and found that TT allele in OLR rs1050283 were associated with concurrent stenoses.

The association of LDLR and OLR1gene polymorphisms with IS should be further assessed in different populations and in wider series of patients.

Soluble epoxide hydrolase 2. Soluble (cytosolic) epoxide hydrolase (sEH) has two activities as epoxide hydrolase and phosphatase. It is an enyzme involved in conversion of epoxyeicosatrienoic acids (EETs) metabolites of arachidonic acid in less active corresponding diols. EETs functions as vasodilatators, have anti-inflammatory effects [207], and anti-trombotic effects [208,209]. EETs have been shown to regulate cerebral blood flow and, through their mitogenic properties, may contribute to angiogenesis in the brain. Hence, they may protect against IS [210-212]. It modifies blood pressure [213] or plasma lipid levels and composition of lipoprotein particles [214]. Soluble EH is encoded by EPHX2 gene located at chromosome 8 (8p21-p12). This gene contains 19 exons. It encodes 555 amino acids. Fourty four SNPs and one insertion/deletion polymoprhism [215] was identified in these gene. Substitutions Lys55Arg, Cys154Tyr and Glu470Gly resulted in an enzyme with increased epoxide hydrolase activity, until two other variants, the Arg287Gln substitution and the Ser402^{Argins} insertion resulted in enzymes with reduced epoxide hydrolase activity.

Genetic studies links polymorphisms in the human EPHX2 gene with modified risk of IS in a number of human populations [216-218]. In the Fornage's study, a positive association was observed between the Glu470Gly variant and the incidence of IS in African American cohort [216].

Zhang et al. [218] examined potential associations between *EPHX2* G860A polymorphism and IS risk in Chinese population. The G860A polymorphism results in an amino acid substitution (R287Q, Arg287Gln) that alters enzyme stability and reduces enzyme activity [215,219]. They concluded that the presence of at least one A allele at position 860 of EPHX2 was independently associated with a decreased risk of IS. Gschwendtner et al. [217] in Caucasians found significant association between rs751141, rs7357432, rs2291635 and IS. The haplotype containing the associated alleles of the three SNPs showed an odds ratio of 1.59 (1.06-2.37, *P*=0.022) in the large-vessel subgroup and an odds ratio of 1.54 (0.96-2.41, *P*=0.062) in the subgroup of patients with undetermined etiology. Lee et al. [220] did not find positive association of three polymorphism in *EPHX2* gene (R103C, R287Q, and $Arg^{402\text{-}403ins}$) and IS risk.

Fava's study [221] examined whether the EPHX2 missense K55R and R287Q, together with the –1452T>C (rs7003694) in the promoter region and the +1784A>G (rs1042032) in the 3'UTR polymorphisms, are associated with hypertension and with risk of cardiovascular events in middle-aged Swedes. They found no significant difference in the incidence of IS in carriers of different EPHX2 R287Q, EPHX2 –1452T>C genotypes, EPHX2 +1784A>G (P>0.05), until the higher incidence of IS was evident in male EPHX2 R-homozygotes versus male K-allele carriers.

2.2. Genome-wide association studies (GWAS) in ischemic stroke

The completion of the Human genome project, together with rapid improvements in laboratory techniques in this field, has enabled investigators to examine multiple genetic variants simultaneously in large study populations and it can be used for unlocking the genetic basis of complex human diseases [222,223]. The genetic variants that can be identified by GWAS are common SNP and have low effect size. By introducing GWAS a major limitation of the candidate gene study was overcame and candidate gene studies have now been largely superseded by the GWAS technique.

To date, GWAS of IS has been performed in 6 cohorts, resulting in 7 publications with somewhat inconsistent results. The initial step in a genome-wide genotyping study in patients with IS was performed in 2007 [224]. The analysis which compared 408,803 unique SNPs in 249 white patients with IS and 268 white neurologically normal controls in five US stroke centers do not suggest any single common genetic variant exerting a major risk for IS. The other recently published genome-wide association study [225] found a significant association between two SNPs rs11833579 and rs12425791on chromosome 12p13 with total, ischemic, and atherothrombotic stroke in white persons. The SNPs are located closed to the gene Ninjurin2 (nerve injury-induced protein 2-NINJ2) and WNK1- serine-threonine kinase that regulate ion channels involved in sodium and potassium transport. Finally, SNPs in paired-like homeodomain transcription factor 2 (*PITX2*) and zinc finger homeobox 3

(*ZFHX3*) were observed to be associated with cardioembolic stroke and atrial fibrillation in Icelandic population [226,227].

Three GWAS were performed in Japanese populations. Kubo et al. [228] found significant association of non-synonymous SNP (1425 G/A) in protein kinase C-eta (PRKCH) with lacunar infarction in the pathogenesis of IS. Hata et al. [229] found that SNP in the 5'-flanking region of angiotensin receptor like-1 (AGTRL1) gene (rs9943582, - 154G/A) to have a significant association with brain infarction. Also, rs9615362 of cell surface receptor CELSR1 (cadherin epidermal growth factor laminin A seven-pass G-type receptor 1) was associated with IS [230].

3. Conclusion

Genetics of IS represents a unique challenge. Among the most examined candidate genes in IS are those associated with lipid metabolism. Unfortunately, the results are complex and far from clear-cut. According the literature review in this chapter it can be consluded that genes (polymorphisms) that are the most likely to be associated with IS are: apoE (apo ε2/ε3/ε4) and PON1 gene (p.Gln192Arg). Insufficient or inconsistent data that neither supported nor excluded an association of some genes polymorphisms with IS apoAV (c.1131T>C), LPA (rs3798220), LPL (S447X), LDLR (c.370A>T), OLR1(IIVS4-14A/G, IVS4-73C/T) and EPHX2 (G860A). For other genes/polymorphisms that were reviewed in this paper, we are reasonably confident that an association with IS can be ruled out.

The reasons for contradictory results in the studies may be limited sample size, heterogeneity of study designs and endpoints, differences in inclusion and exclusion criteria, ethnically different patient populations, selection of control population, different stroke subtypes and age of stroke onset, type of statistical evaluation, covariates, correction for multiple testing etc. One of the limitations of multiple non-reproducible candidate gene studies was the restriction to a single or rather few genetic variants tested for association with disease in examined gene. Further, genetic variants of candidate genes with strong effects at the transcriptional level or others affecting the functionality of the protein may have escaped the test for association with disease risk. Thus, in retrospect, it is not surprising that the candidate gene approach resulted in only limited success in the elucidation of IS stroke genes.

Research in the field of IS should be directed towards facilitation of the characterization of IS pathogenesis at the molecular level and the development of genetic markers' panels for assessment of IS risk. Technological developments such as GWAS, NGS technology, transcription profiling and proteomics will provide huge amounts of genetic information and allow investigators to identify variants in patients with specific stroke subtype and to identify how they exert their effects at the molecular level. The replication in an independent study, in large and well-characterized groups of patients of different ethnic origin, is required to confirm previously obtained results. On the basis of genetic or genomic information the therapeutic outcome or side effects in stroke patients could be predicted, as the effectiveness and safety of applied therapy. Also, this approach may help in stroke prevention by identification of presymptomatic at-risk individuals, resulting in minimizing patients' morbidity and mortality and reducing health care costs associated with stroke.

Author details

Sanja Stankovic*, Milika Asanin and Nada Majkic-Singh
Clinical Center of Serbia, University of Belgrade, Serbia

4. References

[1] Shiber JR, Fontane E, Adewale A. A Stroke registry: hemorrhagic vs ischemic strokes. Am J Emerg Med 2010; 28(3):331-333.
[2] Baird AE. Genetics and genomics of stroke. J Am Coll Cardiol 2010;56(4):245-253.
[3] Adams HP, Bendixen BH, Kappelle LJ, Biller J, Love BB, Gordon DL, Marsh EE. Classification of subtype of acute ischemic stroke: definitions for use in a multicenter clinical trial. Stroke 1993;24(1):35-41.
[4] Bamford J, Sandercock PA, Dennis MS, Burn J, Warlow CP: Classification and natural history of clinically identifiable subtypes of brain infarction. Lancet 1991;337(8756):1521-1526.
[5] Bogousslavsky J, Van Melle G, Regli F: The Lausanne Stroke Registry: analysis of 1,000 consecutive patients with first stroke. Stroke 1988;19(9):1083-1092.
[6] Touboul PJ, Elbaz A, Koller C, Lucas C, Adrai V, Chedru F, Amarenco P, GENIC Investigators: Common carotid artery intima-media thickness and ischemic stroke subtypes: the GENIC case-control study. Circulation 2000;102(3):313-318.
[7] Whisnant JP. Modeling of risk factors for ischemic stroke. The Willis Lecture. Stroke 1997;28(9):1840-1844.
[8] Dichgans M. Genetics of ischaemic stroke. Lancet Neurol 2007;6(2):149-161.
[9] Schulz UG, Flossmann E, Rothwell PM. Heritability of ischemic stroke in relation to age, vascular risk factors, and subtypes of incident stroke in population-based studies. Stroke 2004;35(4):819-824.
[10] Jerrard-Dunne P, Cloud G, Hassan A, Markus HS. Evaluating the genetic component of ischemic stroke subtypes: a family history study. Stroke 2003;34(6):1364-1369.
[11] Brass LM, Isaacsohn JL, Merikangas AR. A study of twins and stroke. Stroke 1992;23(2):221-223.
[12] Touzé E, Rothwell PM. Sex differences in heritability of ischemic stroke: a systematic review and meta-analysis. Stroke 2008;39(1):16-23.
[13] Hassan A, Markus HS. Genetics and ischaemic stroke. Brain 2000;123 (Pt 9):1784-1812.
[14] Bersano A, Ballabio E, Bresolin N, Candelise L. Genetic polymorphisms for the study of multifactorial stroke.Hum Mutat 2008;29(6):776-795.

* Corresponding Author

[15] Stankovic S, Majkic-Singh N. Genetic aspects of ischemic stroke: coagulation, homocysteine, and lipoprotein metabolism as potential risk factors. Crit Rev Clin Lab Sci 2010;47(2):72-123.

[16] Casas JP, Hingorani AD, Bautista LE, Sharma P. Meta-analysis of genetic studies in ischemic stroke: thirty-two genes involving approximately 18,000 cases and 58,000 controls. Arch Neurol 2004; 61:1652-1662.

[17] Xin XY, Song YY, Ma JF, Fan CN, Ding JQ, Yang GY, Chen SD. Gene polymorphisms and risk of adult early-onset ischemic stroke: A meta-analysis. Thromb Res 2009;124(5):619-624.

[18] Xu X, Li J, Sheng W, Liu L. Meta-analysis of genetic studies from journals published in China of ischemic stroke in the Han Chinese population. Cerebrovasc Dis 2008;26(1):48-62.

[19] Ariyaratnam R, Casas JP, Whittaker J, Smeeth L, Hingorani AD, Sharma P. Genetics of ischaemic stroke among persons of non-European descent: a meta-analysis of eight genes involving approximately 32,500 Individuals. PLoS Med 2007;4:e131.

[20] Rao R, Tah V, Casas JP, Hingorani A, Whittaker J, Smeeth L, Sharma P. Ischaemic stroke subtypes and their genetic basis: a comprehensive meta-analysis of small and large vessel stroke. Eur Neurol 2009;61(2):76-86.

[21] Banerjee I, Veena Gupta V, Ganesh S. Association of gene polymorphism with genetic susceptibility to stroke in Asian populations: a meta-analysis. J Hum Genet 2007;52(3):205-219.

[22] McCarron MO, Delong D, Alberts MJ. ApoE genotype as a risk factor for ischemic cerebrovascular disease: a meta-analysis. Neurology 1999;53(6):1308-1311.

[23] Sudlow C, Martínez González NA, Kim J, Clark C. Does apolipoprotein E genotype influence the risk of ischemic stroke, intracerebral hemorrhage, or subarachnoid hemorrhage? Systematic review and meta-analyses of 31 studies among 5961 cases and 17,965 controls. Stroke 2006;37(2):364-370.

[24] Hamzi K, Tazzite A, Nadifi S. Large-scale meta-anlysis of genetic studies in ischemic stroke: Five genes involving 152797 individuals. Indian J Hum Genet 2011;17(3):212-217.

[25] Karttunen V, Alfthan G, Hiltunen L, Rasi V, Kervinen K, Kesaniemi YA, Hillbom M. Risk factors for cryptogenic ischaemic stroke. Eur J Neurol 2002;9(6):625-632.

[26] Szolnoki Z, Somogyvári F, Kondacs A, Szabó M, Fodor L. Evaluation of the interactions of common genetic mutations in stroke subtypes. J Neurol 2002;249(10):1391-1397.

[27] Szolnoki Z, Somogyvári F, Kondacs A, Szabó M, Fodor L, Bene J, Melegh B. Evaluation of the modifying effects of unfavourable genotypes on classical clinical risk factors for ischaemic stroke. J Neurol Neurosurg Psychiatry 2003;74(12):1615-1620.

[28] Zee RYL, Cook NR, Cheng S, Reynolds R, Erlich HA, Lindpaintner K, Ridker PM. Polymorphism in the P-selectin and interleukin-4 genes as determinants of stroke: a population-based, prospective genetic analysis. Hum Mol Genet 2004;13(4):389-396.

[29] Pezzini A, Grassi M, Del Zotto E, Archetti S, Spezi R, Vergani V, Assanelli D, Caimi L, Padovani A. Cumulative effect of predisposing genotypes and their interaction with modifiable factors on the risk of ischemic stroke in young adults. Stroke 2005;36(3):533-539.

[30] Lalouschek W, Endler G, Schillinger M, Hsieh K, Lang W, Cheng S, Bauer P, Wagner O, Mannhalter C. Candidate genetic risk factors of stroke: results of a multilocus genotyping assay. Clin Chem 2007;53(4):600-605.

[31] Berger K, Stögbauer F, Stoll M, Wellmann J, Huge A, Cheng S, Kessler C, John U, Assmann G, Ringelstein EB, Funke H. The glu298asp polymorphism in the nitric oxide synthase 3 gene is associated with the risk of ischemic stroke in two large independent case-control studies. Hum Genet 2007;121(2):169-178.

[32] Gao X, Yang H, ZhiPing T. Association studies of genetic polymorphism, environmental factors and their interaction in ischemic stroke. Neurosci Lett 2006;398(3):172-177.

[33] Kessler C, Spitzer C, Stauske D, Mende S, Stadlmüller J, Walther R, Rettig R. The apolipoprotein E and beta-fibrinogen G/A-455 gene polymorphisms are associated with ischemic stroke involving large-vessel disease. Arterioscler Thromb Vasc Biol 1997;17(11):2880-2884.

[34] Yamada Y, Metoki N, Yoshida H, Satoh K, Ichihara S, Kato K, Kameyama T, Yokoi K, Matsuo H, Segawa T, Watanabe S, Nozawa Y. Genetic risk for ischemic and hemorrhagic stroke. Arterioscler Thromb Vasc Biol 2006;26(8):1920-1925.

[35] Topić E, Šimundić AM, Štefanović M, Demarin V, Vuković V, Lovrenčić-Huzjan A, Žuntar I. Polymorphism of apoprotein E (APOE), methylenetetrahydrofolte reductase (MTHFR) and paraoxonase (PON1) genes in patients with cerebrovascular disease. Clin Chem Lab Med 2001;39(4):346-350.

[36] McIlroy SP, Dynan KB, Lawson JT, Patterson CC, Passmore AP. Moderately elevated plasma homocysteine, methylenetetrahydrofolate reductase genotype, and risk for stroke, vascular dementia, and Alzheimer disease in Northern Ireland. Stroke 2002;33(10):2351-2356.

[37] Mahieux F, Bailleul S, Fenelon R, Couderc R, Laruelle P, Gunel M. Prevalence of apolipoprotein E phenotypes in patients with acute ischemic stroke. Stroke 1990;21:I-115.

[38] Pedro-Botet J, Sentí M, Nogues X, Rubiés-Prat J, Roquer J, D'Olhaberriague L, Olivé J. Lipoprotein and apolipoprotein profile in men with ischemic stroke. Role of lipoprotein (a), triglyceride-rich lipoproteins, and apolipoprotein E polymorphism. Stroke 1992;23(11):1556-1562.

[39] Couderc R, Mahieux F, Bailleul S, Fenelon G, Mary R, Fermanian J. Prevalence of apolipoprotein E phenotypes in ischemic cerebrovascular disease. A case-control study. Stroke 1993;24(5):661-664.

[40] Coria F, Rubio I, Nuñez E, Sempere AP, SantaEngarcia N, Bayón C, Cuadrado N. Apolipoprotein E variants in ischemic stroke. Stroke 1995;26(12):2375-2376.

[41] De Andrade M, Thandi I, Brown S, Gotto A Jr, Patsch W, Boerwinkle E. Relationship of the apolipoprotein E polymorphism with carotid artery atherosclerosis. Am J Hum Genet 1995;56(6): 1379-1390.

[42] Kuusisto J, Mykkänen L, Kervinen K, Kesäniemi YA, Laakso M. Apolipoprotein E4 phenotype is not an important risk factor for coronary heart disease or stroke in elderly subjects. Arterioscler Thromb Vasc Biol 1995;15(9):1280-1286.

[43] Basun H, Corder EH, Guo Z, Lannfelt L, Corder LS, Manton KG, Winblad B, Viitanen M. Apolipoprotein E polymorphism and stroke in a population sample aged 75 years or more. Stroke 1996;27(8):1310-1315.

[44] Hachinski V, Graffagnino C, Beaudry M, Bernier G, Buck C, Donner A, Spence JD, Doig G, Wolfe BM. Lipids and stroke: a paradox resolved. Arch Neurol 1996;53(4):303-308.

[45] Ferrucci L, Guralnik JM, Pahor M, Harris T, Corti MC, Hyman BT, Wallace RB, Havlik RJ. Apolipoprotein E epsilon 2 allele and risk of stroke in the older population. Stroke 1997;28(12):2410-2416.

[46] Nakata Y, Katsuya T, Rakugi H, Takami S, Sato N, Kamide K, Ohishi M, Miki T, Higaki J, Ogihara T. Polymorphism of angiotensin converting enzyme, angiotensinogen, and apolipoprotein E genes in a Japanese population with cerebrovascular disease. Am J Hypertens 1997;10(12Pt1):1391-1395.

[47] Schmidt R, Schmidt H, Fazekas F, Schumacher M, Niederkorn K, Kapeller P, Weinrauch V, Kostner GM. Apolipoprotein E polymorphism and silent microangiopathy-related cerebral damage. Results of the Austrian Stroke Prevention Study. Stroke 1997;28(5):951-956.

[48] Yang G, Jinjin G, Jianfei N. The relationship between polymorphisms of apolipoprotein E gene and atherosclerotic cerebral infarction. Zhonghua Shen Jing Ge Za Zhi 1997;30:236–239.

[49] Wang DS, Jiang L, Dai YM. Primary study of ApoE gene polymorphism in patients with cerebral infarction. Zhong Feng Yu Shen Jing Ji Bing Za Zhi 1997;14:71–74.

[50] Zhu TB, Zhao SP, You XK. Effect of apolipoprotein E gene on plasma levels of lipids, lipoprotein, apolipoprotein and relation to cerebral infarction. Hu Nan Yi Xue 1997;14:265–266.

[51] Yan SK, Zhou X, Li XL. Relationship between gene polymorphism of apolipoprotein E and serum lipids, lipoproteins, and apolipoproteins in Chinese patients with atherothrombotic brain infarction. Zhong Guo Shen Jing Mian Yi Xue He Shen Jing Bing Xue Za Zhi 1997;4:16–21.

[52] Aalto-Setälä K, Palomäki H, Miettinen H, Vuorio A, Kuusi T, Raininko R, Salonen O, Kaste M, Kontula K. Genetic risk factors and ischaemic cerebrovascular disease: role of common variation of the genes encoding apolipoproteins and angiotensin-converting enzyme. Ann Med 1998;30(2):224–233.

[53] Ji Y, Urakami K, Adachi Y, Maeda M, Isoe K, Nakashima K. Apolipoprotein E polymorphism in patients with Alzheimer's disease, vascular dementia and ischemic cerebrovascular disease. Dement Geriatr Cogn Disord 1998;9(5):243-245.

[54] Margaglione M, Seripa D, Gravina C, Grandone E, Vecchione G, Cappucci G, Merla G, Papa S, Postiglione A, Di Minno G, Fazio VM. Prevalence of apolipoprotein E alleles in healthy subjects and survivors of ischemic stroke: an Italian case-control study. Stroke 1998;29(2):399-403.

[55] Cao W, Chen F, Teng L, Wang S, Fu S, Zhang G. The relationship between apolipoprotein E gene polymorphism and coronary heart disease and arteriosclerotic cerebral infarction. Zhonghua Yi Xue Yi Chuan Xue Za Zhi 1999;16:249–251.

[56] Peng DQ, Zhao SP, Wang JL. Lipoprotein (a), and apolipoprotein E4 as independent risk factors for ischemic stroke. J Cardiovasc Risk 1999;6(1):1-6.
[57] Liu WG, Li ZH. The relationship between polymorphisms of apolipoprotein E gene and atherosclerotic cerebral infarction in middle-aged and young adults. Lin Chuang Shen Jing Bing Xue Za Zhi 1999;12:134–136.
[58] Peng DQ, Zhao SP. Comparison of apolipoprotein E genotype distribution in two types of stroke. Zhong Guo Dong Mai Ying Hua Za Zhi 1999;7:34–36.
[59] Catto AJ, McCormack LJ, Mansfield MW, Carter AM, Bamford JM, Robinson P, Grant PJ. Apolipoprotein E polymorphism in cerebrovascular disease. Acta Neurol Scand 2000;101(6):399-404.
[60] Kokubo Y, Chowdhury AH, Date C, Yokoyama T, Sobue H, Tanaka H. Age-dependent association of apolipoprotein E genotypes with stroke subtypes in a Japanese rural population. Stroke 2000;31(6):1299-1306.
[61] McCarron MO, Muir KW, Nicoll JA, Stewart J, Currie Y, Brown K, Bone I. Prospective study of apolipoprotein E genotype and functional outcome following ischemic stroke. Arch Neurol 2000;57(10):1480-1484.
[62] Ding J, Zhu WB, Fan W. Association between apolipoprotein E polymorphisms and cerebral stroke. Zhong Guo Shen Jing Jing Shen Ji Bing Za Zhi 2000;26:371–372.
[63] Wang TG, He ZY, Li YQ. The relation between apolipoprotein E gene polymorphism and atherosclerotic cerebral infarction. Yi Chuan 2000;22:4–6.
[64] Chowdhury AH, Yokoyama T, Kokubo Y, Zaman MM, Haque A, Tanaka H. Apolipoprotein E genetic polymorphism and stroke subtypes in a Bangladeshi hospital-based study. J Epidemiol 2001;11(3):131-138.
[65] Frikke-Schmidt R, Nordestgaard BG, Thudium D, Moes Grønholdt ML, Tybjaerg-Hansen A. ApoE genotype predicts AD and other dementia but not ischemic cerebrovascular disease. Neurology 2001;56(2):194-200.
[66] MacLeod MJ, De Lange RP, Breen G, Meiklejohn D, Lemmon H, Clair DS. Lack of association between apolipoprotein E genoype and ischaemic stroke in a Scottish population. Eur J Clin Invest 2001;31(7):570-573.
[67] Serteser M, Visvikis S, Ozben T, Herbeth B, Balkan S, Siest G. Lipid profile and apolipoprotein E genotyping in stroke: a case-control study. Neuroscience-Net 2001;3, article 10015.
[68] Slooter AJC, Bots ML, Havekes LM, del Sol AI, Cruts M, Grobbee DE, Hofman A, Van Broeckhoven C, Witteman JC, van Dujn CM. Apolipoprotein E and carotid artery atherosclerosis: the Rotterdam study. Stroke 2001;32(9):1947-1952.
[69] Li YW, He X, Zhao LX. The relationship between polymorphisms of apolipoprotein E gene and cerebrovascular disorder. Xin Nao Xue Guan Bing Fang Zhi 2001;1:17–19.
[70] Li ZH, LiuWG, Zhao XY, Chen YQ. Risk factor for stroke and ApoE polymorphism in the young and middle-aged. Cu Zhong Yu Shen Jing Ji Bing 2001;8:326–329.
[71] Luthra K, Prasad K, Kumar P, Dwivedi M, Pandey RM, Das N. Apolipoprotein E gene polymorphism in cerebrovascular disease: a case-control study. Clin Genet 2002;62(1):39-44.

[72] Morrison AC, Ballantyne CM, Bray M, Chambless LE, Sharrett AR, Boerwinkle E. LPL polymorphism predicts stroke risk in men. Genet Epidemiol 2002;22(3):233-242.

[73] Shen LH, Ke KF, Li ZH. Research on apolipoprotein E gene polymorphism in patients with atherosclerotic cerebral infarction. Jiao Tong Yi Xue 2002;16:504-505.

[74] Xia Y, Li HL, Wang JL. Association between apolipoprotein E polymorphism and lipid metabolism in patients with cerebral infarction. Zhong Guo Bing Li Sheng Li Za Zhi 2002;18:826-829.

[75] Zhu L, Cui TP. The relation of apolipoprotein E gene polymorphism and cerebral infarction. Xue Shuan Yu Zhi Xue Xue 2002;8:14-15.

[76] Kolovou GD, Daskalova DCh, Hatzivassiliou M, Yiannakouris N, Pilatis ND, Elisaf M, Mikhailidis DP, Cariolou MA, Cokkinos DV. The epsilon 2 and 4 alleles of apolipoprotein E and ischemic vascular events in the Greek population-implications for the interpretation of similar studies. Angiology 2003;54(1):51-58.

[77] Slowik A, Iskra T, Turaj W, Hartwich J, Dembinska-Kiec A, Szczudlik A. LDL phenotype B and other lipid abnormalities in patients with large vessel disease and small vessel disease. J Neurol Sci 2003;214(1-2):11-16.

[78] Souza DR, Campos BF, de Arruda EF, Yamamoto LJ, Trinidane DM, Tognola WA. Influence of the polymorphism of apolipoprotein E in cerebral vascular disease. Arq Neuropsiquiatr 2003; 61(1):7-13.

[79] Um JY, Kim HM, Park HS, Joo JC, Kim KY, Kim YK, Hong SH. Candidate genes of cerebral infarction and traditional classification in Koreans with cerebral infarction. Int J Neurosci 2005;115(6):743-756.

[80] Wang XT, Huang HJ, Ju K. Apolipoprotein E gene polymorphism in people with cerebrovascular disease in the south of the Zhejiang province. Shen Jing Ji Bing Yu Jing Shen Wei Sheng 2003;3:17-19.

[81] Duzenli S, Pirim I, Gepdiremen A, Deniz O. Apolipoprotein E polymorphism and stroke in a population from eastern Turkey. J Neurogenet 2004;18(1):365-375.

[82] Jin ZQ, Fan YS, Ding J, Chen M, Fan W, Zhang GJ, Zhang BH, Yu SJ, Zhang YS, Ji WF, Zhang JG. Association of apolipoprotein E 4 polymorphism with cerebral infarction in Chinese Han population. Acta Pharmacol Sin 2004;25(3):352-356.

[83] Lin HF, Lai CL, Tai CT, Lin RT, Liu CK. Apolipoprotein E polymorpshim in ischemic cerebrovascular diseases and vascular dementia patients in Taiwan. Neuroepidemiology 2004;23(3):129-134.

[84] Pezzini A, Grassi M, Zotto ED, Bazzoli E, Archetti S, Assanelli D, Akkawi NM, Albertini A, Padovani A. Synergistic effect of apolipoprotein E polymorphisms and cigarette smoking on risk of ischemic stroke in young adults. Stroke 2004;35(2):438-442.

[85] Stanković S, Jovanović-Marković Z, Majkić-Singh N, Stanković A, Glišić S, Živković M, Kostic V, Alavantic D. Apolipoprotein E gene polymorphism as a risk factor for ischemic cerebrovascular disease. Jugoslov Med Biohem 2004;23(3):255-264.

[86] Cerrato P, Baima C, Grasso M, Lentini A, Bosco G, Cassader M, Gambino R, Cavallo Perin P, Pagano G, Fornengo P, Imperiale D, Bergamasco B, Bruno G. Apolipoprotein E polymorphism and stroke subtypes in an Italian cohort. Cerebrovasc Dis 2005;20(4):264-269.

[87] Zhou J, Xue YL, Guan YX, Yang YD, Fu SB, Zhang JC. Association study of apolipoprotein e gene polymorphism and cerebral infarction in type 2 diabetic patients. Yi Chuan 2005;27:35-38.

[88] Baum L, Ng HK, Wong KS, Tomlinson B, Rainer TH, Chen X, Cheung WS, Tang J, Tam WWS, Goggins W, Tong CSW, Kam D, Chan Y, Thomas GN, Chook P, Woo KS. Associations of apolipoprotein E exon 4 and lipoprotein lipase S447X polymorphisms with acute ischemic stroke and myocardial infarction. Clin Chem Lab Med 2006;44(3):274-281.

[89] Kang SY, Lee WI. Apolipoprotein e polymorphism in ischemic stroke patients with different pathogenetic origins. Korean J Lab Med 2006;26(3):210-216.

[90] Jiang ZQ, Liu H, Zhang GZ. Relationship between polymorphism of apolipoprotein E gene and atherosclerotic cerebral infarction, hypertensive intracerebral hemorrhage in the youth. J Gannan Med Univ 2006;26:331-334.

[91] Wang JH, Ning XJ, Lu HY. The study on apolipoprotein E gene polymorphism characteristics of cerebral infarction and intracerebral hemorrhage. Zhong Guo Man Xing Bing Yu Fang Yu Kong Zhi 2006;14:21-23.

[92] Giassakis G, Veletza S, Papanas N, Heliopoulos I, Piperidou H. Apolipoprotein E and first-ever ischaemic stroke in Greek hospitalized patients. J Int Med Res 2007;35(1):127-133.

[93] Lai CL, Liu CK, Lin RT, Tai CT. Association of apolipoprotein E polymorphism with ischemic stroke subtypes in Taiwan. Kaohsiung J Med Sci 2007;23(10):491-497.

[94] Parfenov MG, Nikolaeva TY, Sudomoina MA, Fedorova SA, Guekht AB, Gusev EI, Favorova OO. Polymorphism of apolipoprotein E (APOE) and lipoprotein lipase (LPL) genes and ischaemic stroke in individuals of Yakut ethnicity. J Neurol Sci 2007;255(1-2):42–49.

[95] Saidi S, Slamia LB, Ammou SB, Mahjoub T, Almawi WY. Association of apolipoprotein E gene polymorphism with ischemic stroke involving large-vessel disease and its relation to serum lipid levels. J Stroke Cerebrovasc Dis 2007;16(4):160-166.

[96] Abboud S, Viiri LE, Lütjohann D, Goebeler S, Luoto T, Friedrichs S, Desfontaines P, Gazagnes MD, Laloux P, Peeters A, Seeldrayers P, Lehtimaki T, Karhunen P, Pandolfo M, Laaksonen R. Associations of apolipoprotein E gene with ischemic stroke and intracranial atherosclerosis. Eur J Hum Genet 2008;16(8):955-960.

[97] Artieda M, Gañán A, Cenarro A, García-Otín AL, Jericó I, Civeira F, Pocoví M. Association and linkage disequilibrium analyses of APOE polymorphisms in atherosclerosis. Dis Markers 2008;24(2):65-72

[98] Tasdemir N, Tamam Y, Toprak R, Tamam B, Tasdemir MS. Association of apolipoprotein E genotype and cerebrovascular disease risk factors in a Turkish population. Int J Neurosci 2008;118(8):1109-1129.

[99] Wang B, Zhao H, Zhou L, Dai X, Wang D, Cao J, Niu W. Association of genetic variation in apolipoprotein E and low density lipoprotein receptor with ischemic stroke in Northern Han Chinese. J Neurol Sci 2009;276(1-2):118-122.

[100] Saidi S, Zammiti W, Slamia LB, Ammou SB, Almawi WY, Mahjoub T. Interaction of angiotensin-converting enzyme and apolipoprotein E gene polymorphisms in ischemic stroke involving large-vesssel disease. J Thromb Thrombolysis 2009;27(1):68-74.

[101] Tascilar N, Dursun A, Ankarali H, Mungan G, Sumbuloglu V, Ekem S, Bozdogan S, Baris S, Aciman E, Cabuk F. Relationship of apoE polymorphism with lipoprotein(a), apoA, apoB and lipid levels in atherosclerotic infarct. J Neurol Sci 2009;277(1-2):17-21.
[102] McCarron MO, Muir KW, Weir CJ, Dyker AG, Bone I, Nicoll JA, Lees KR. The apolipoprotein E epsilon4 allele and outcome in cerebrovascular disease. Stroke 1998;29(9):1882-1887.
[103] Wang L, Gu Y, Wu G, Wang W, Liu J, Liu J, Wu Z. A case control study on the distribution of apolipoprotein AI gene polymorphisms in the survivors of atherosclerosis cerebral infarction. Zhonghua Liu Xing Bing Xue Za Zhi 2000;21:22–25.
[104] Xia J, Yang Q, Yang Q, Xu H, Zhang L. The relationship of apolipoprotein H G1025C (Try316Ser) polymorphism with stroke and its effect on plasma lipid levels in Changsha Hans. Zhonghua Yi Xue Yi Chuan Xue Za Zhi 2003;20:114–118.
[105] Wang L, Gu Y, Wu G. The relation between polymorphisms of apolipoprotein B gene and atherosclerotic cerebral infarction. Zhonghua Yi Xue Za Zhi 1999;79:603–606.
[106] Stanković A, Stanković S, Jovanović-Marković Z, Zivković M, Djurić T, Glišić-Milosavljević S, Alavantić D. Apolipoprotein B gene polymorphisms in patients from Serbia with ischemic cerebrovascular disease. Arch Biol Sci 2007;59(4):303–309.
[107] Zhang L, Zeng Y, Ma M, Yang Q, Hu Z, Du X. Association study between C7673T polymorphism in apolipoprotein B gene and cerebral infarction with family history in a Chinese population. Neurol India 2009;57(5):584-588.
[108] Benn M, Nordestgaard BG, Jensen JS, Tybjaerg-Hansen A. Polymorphisms in apolipoprotein B and risk of ischemic stroke. J Clin Endocrinol Metab 2007;92(9):3611-3617.
[109] Havasi V, Szolnoki Z, Talián G, Bene J, Komlósi K, Maász A, Somogyvári F, Kondacs A, Szabó M, Fodor L, Bodor A, Melegh B. Apolipoprotein A5 gene promoter region T-1131C polymorphism associates with elevated circulating triglyceride levels and confers susceptibility of ischemic stroke. J Mol Neurosci 2006;29(2):177-183.
[110] Li J, Xu, Zhu XY. Association of APOA5 gene polymorphism with levels of lipids and atherosclerotic cerebral infarction in Chinese. Zhonghua Yi Xue Yi Chuam Xue Za Zhi 2007;24:576–578.
[111] Zhang K, Qiu F, Li L, Gu GY, Tao Y, Wang L, Luo XY, Xia YQ. The associated study on apolipoprotein A5 gene polymorphisms with carotid artherosclerosis in patients with cerebral infartion. Zhonghua Yi Xue Yi Chuan Xue Za Zhi 2008;25:284-288.
[112] Járomi L, Csöngei V, Polgár N, Szolnoki Z, Maász A, Horvatovich K, Faragó B, Sipeky C, Sáfrány E, Magyari L, Kisfali P, Mohás M, Janicsek I, Lakner L, Melegh B. Functional variants of glucokinase regulatory protein and apolipoprotein A5 genes in ischemic stroke. J Mol Neurosci 2010;41(1):121-128.
[113] Maász A, Kisfali P, Szolnoki Z, Hadarits F, Melegh B. Apolipoprotein A5 gene C56G variant confers risk for the development of large-vessel associated ischemic stroke. J Neurol 2008;255(5):649-654.
[114] Maasz A, Kisfali P, Jaromi L, Horvatovich K, Szolnoki Z, Csongei V, Safrany E, Sipeky C, Hadarits F, Melegh B. Apolipoprotein A5 gene IVS3+G476A allelic variant confers susceptibility for development of ischemic stroke. Circ J 2008;72(7):1065-1070.

[115] Li X, Su D, Zhang X, Zhang C. Association of apolipoprotein A5 gene promoter region -1131T>C with risk of stroke in Han Chinese. Eur J Intern Med 2011;22(1):99-102.
[116] Ogorelkova M, Kraft HG, Ehnholm C, Utermann G. Single nucleotide polymorphisms in exons of the apo(a) kringles IV types 6 to 10 domain affect Lp(a) plasma concentrations and have different patterns in Africans and Caucasians. Hum Mol Genet 2001;10(8):815-824.
[117] Trommsdorff M, Köchl S, Lingenhel A, Kronenberg F, Delport R, Vermaak H, Lemming L, Klausen IC, Faergeman O, Utermann G, Kraft HG. A pentanucleotide repeat polymorphism in the 5′ control region of the apolipoprotein (a) gene is associated with lipoprotein (a) plasma concentrations in Caucasians. J Clin Invest 1995;96(1):150-157.
[118] Hu B, Zhou X, Shao H. Relationship between pentanucleotide repeat polymorphism of apolipoprotein (a) gene and atherosclerosis cerebral infarction in Han nationality. Zhonghua Shen Jing Ge Za Zhi 2000;33:172-175.
[119] Liu X, Sun L, Li Z, Gao Y, Hui R. Relation of pentanucleotide repeat polymorphism of apolipoprotein (a) gene to plasma lipoprotein (a) level among Chinese patients with myocardial infarction and cerebral infarction. Zhonghua Yi Xue Za Zhi 2002;82:1396-1400.
[120] Sun L, Li Z, Zhang H, Ma A, Liao Y, Wang D, Zhao B, Zhu Z, Zhao J, Zhang Z, Wang W, Hui R. Pentanucleotide TTTTA repeat polymorphism of apolipoprotein(a) gene and plasma lipoprotein(a) are associated with ischemic and hemorrhagic stroke in Chinese: a multicenter case-control study in China. Stroke 2003;34(7):1617-1622.
[121] Clarke R, Peden JF, Hopewell JC, Kyriakou T, Goel A, Heath SC, Parish S, Barlera S, Franzosi MG, Rust S, Bennett D, Silveira A, Malarstig A, Green FR, Lathrop M, Gigante B, Leander K, de Faire U, Seedorf U, Hamsten A, Collins R, Watkins H, Farrall M. Genetic variants associated with Lp(a) lipoprotein level and coronary disease. N Engl J Med 2009;361(26):2518-2528.
[122] Wang X, Cheng S, Brophy VH, Erlich HA, Mannhalter C, Berger K, Lalouschek W, Browner WS, Shi Y, Ringelstein EB, Kessler C, Luedemann J, Lindpaintner K, Liu L, Ridker PM, Zee RY, Cook NR. A meta-analysis of candidate gene polymorphisms and ischemic stroke in 6 study populations: association of lymphotoxin-alpha in nonhypertensive patients. Stroke 2009;40(3):683-695.
[123] Hopewell JC, Clarke R, Parish S, Armitage J, Lathrop M, Hager J, Collins R; Heart Protection Study Collaborative Group. Lipoprotein(a) genetic variants associated with coronary and peripheral vascular disease but not with stroke risk in the Heart Protection Study. Circ Cardiovasc Genet 2011;4(1):68-73.
[124] Chasman DI, Shiffman D, Zee RY, Louie JZ, Luke MM, Rowland CM, Catanese JJ, Buring JE, Devlin JJ, Ridker PM. Polymorphism in the apolipoprotein(a) gene, plasma lipoprotein(a), cardiovascular disease, and low-dose aspirin therapy. Atherosclerosis 2009;203(2):371-376.
[125] Ordovas JM, Cupples LA, Corella D, Otvos JD, Osgood D, Martinez A, Lahoz C, Coltell O, Wilson PW, Schaefer EJ. Association of cholesteryl ester transfer protein-TaqIB polymorphism with variations in lipoprotein subclasses and coronary heart disease risk: the Framingham study. Arterioscler Thromb Vasc Biol 2000;20(5):1323-1329.

[126] Brousseau ME, O'Connor JJ Jr, Ordovas JM, Collins D, Otvos JD, Massov T, McNamara JR, Rubins HB, Robins SJ, Schaefer EJ. Cholesteryl ester transfer protein TaqI B2B2 genotype is associated with higher HDL cholesterol levels and lower risk of coronary heart disease end points in men with HDL deficiency: Veterans Affairs HDL Cholesterol Intervention Trial. Arterioscler Thromb Vasc Biol 2002;22(7):1148-1154.
[127] Kuivenhoven JA, Jukema JW, Zwinderman AH, de Knijff P, McPherson R, Bruschke AV, Lie KI, Kastelein JJ. The role of a common variant of the cholesteryl ester transfer protein gene in the progression of coronary atherosclerosis. The Regression Growth Evaluation Statin Study Group. N Engl J Med 1998;338(2):86-93.
[128] Elosua R, Cupples LA, Fox CS, Polak JF, D'Agostino RA Sr, Wolf PA, O'Donnell CJ, Ordovas JM. Association between well-characterized lipoprotein-related genetic variants and carotid intimal medial thickness and stenosis: The Framingham Heart Study. Atherosclerosis 2006;189(1):222-228.
[129] Meguro S, Takei I, Murata M, Hirose H, Takei N, Mitsuyoshi Y, ishii K, Oguchi S, Shinohara J, Takeshita E, Watanabe K, Saruta T. Cholesteryl ester transfer protein polymorphism associated with macroangiopathy in Japanese patients with type 2 diabetes. Atherosclerosis 2001;156(1):151–156.
[130] Asselbergs FW, Moore JH, van den Berg MP, Rimm EB, de Boer RA, Dullaart RP, Navis G, van Gilst WH. A role for CETP TaqIB polymorphism in determining susceptibility to atrial fibrillation: a nested case control study. BMC Med Genet 2006;7:39.
[131] Zhuang Y, Wang J, Qiang H, Li Y, Liu X, Li L, Chen G. Cholesteryl ester transfer protein levels and gene deficiency in Chinese patients with cardio-cerebrovascular diseases. Chin Med J (Engl) 2002;115(3):371-374.
[132] Fidani L, Hatzitolios AI, Goulas A, Savopoulos C, Basayannis C, Kotsis A. Cholesterlyl ester transfer protein TaqI B and lipoprotein lipase Ser447Ter gene polymorphisms are not associated with ischaemic stroke in Greek patients. Neurosci Lett 2005;384(1-2):102-105.
[133] Quarta G, Stanzione R, Evangelista A, Zanda B, Sciarretta S, Di Angelantonio E, Marchitti S, Di Murro D, Volpe M, Rubattu S. A protective role of a cholesteryl ester transfer protein gene variant towards ischaemic stroke in Sardinians. J Int Med 2007;262(5):555-561.
[134] Enquobahrie DA, Smith NL, Bis JC, Carty CL, Rice KM, Lumley T, Hindorff LA, Lemaitre RN, Williams MA, Siscovick DS, Heckbert SR, Psaty BM. Cholesterol ester transfer protein, interleukin-8, peroxisome proliferator activator receptor alpha, and toll-like receptor 4 genetic variations and risk of incident nonfatal myocardial infarction and ischemic stroke. Am J Cardiol 2008;101(12):1683-1688.
[135] Hindorff LA, Lemaitre RN, Smith NL, Bis JC, Marciante KD, Rice KM, Lumley T, Enquobahrie DA, Li G, Heckbert SR, Psaty BM. Common genetic variation in six lipid-related and statin-related genes, statin use and risk of incident nonfatal myocardial infarction and stroke. Pharmacogenet Genomics 2008;18(8):677-682.
[136] Von Eckardstein A, Nofer JR, Assman G. High density lipoproteins and atherosclerosis. Role of cholesterol efflux and reverse cholesterol transport. Arterioscler Thromb Vasc Biol 2001;21(1):13-27.

[137] Braunham LR, Singaraja RR, Hayden MR. Variations on a gene: rare and common variants in ABCA1 and their impact on HDL cholesterol levels and atherosclerosis. Annu Rev Nutr 2006;26:105-129.

[138] Andrikovics H, Pongrácz E, Kalina E, Szilvási A, Aslanidis C, Schmitz G, Tordai I. Decreased frequencies of ABCA1 polymorphisms R219K and V771M in Hungarian patients with cerebrovascular and cardiovascular diseases. Cerebrovasc Dis 2006;21(4):254-259.

[139] Pasdar A, Yadegarfar G, Cumming A, Whalley L, St Clair D, MacLeod MJ. The effect of ABCA1 gene polymorphisms on ischaemic stroke risk and relationship with lipid profile. BMC Med Genetics 2007;8:30-36.

[140] Yamada Y, Metoki N, Yoshida H, Satoh K, Kato K, Hibino T, Yokoi K, Watanabe S, Ichihara S, Aoyagi Y, Yasunaga A, Park H, Tanaka M, Nozawa Y. Genetic factors for ischemic and hemorrhagic stroke in Japanese individuals. Stroke 2008;39(8):2211-2218.

[141] Hide WA, Chan L, Li WH. Structure and evolution of the lipase superfamily. J Lipid Res 1992;33(2):167-178.

[142] Mead JR, Irvine SA, Ramji DP. Lipoprotein lipase: structure, function, regulation, and role in disease. J Mol Med 2002;80(12):753-769.

[143] Goldberg IJ. Lipoprotein lipase and lipolysis: central roles in lipoprotein metabolism and atherogenesis. J Lipid Res 1996;37(4):693-707.

[144] Yla-Herttuala S, Lipton BA, Rosenfeld ME, Goldberg IJ, Steinberg D, Witztum JL. Macrophages and smooth muscle cells express lipoprotein lipase in human and rabbit atherosclerotic lesions. Proc Natl Acad Sci USA 1991;88(22):10143-10147.

[145] O'Brien KD, Gordon D, Deeb S, Ferguson M, Chait A. Lipoprotein lipase is synthesized by macrophage-derived foam cells in human coronary atherosclerotic plaques. J Clin Invest 1992;89(5):1544-1550.

[146] Lindqvist P, Ostlund-Lindqvist AM, Witztum JL, Steinberg D, Little JA. The role of lipoprotein lipase in the metabolism of triglyceride-rich lipoproteins by macrophages. J Biol Chem 1983;258(15):9086-9092.

[147] Mead JR, Ramji DP. The pivotal role of lipoprotein lipase in atherosclerosis. Cardiovasc Res 2002;55(2):261-269.

[148] Wittrup HH, Tybjaerg-Hansen A, Nordestgaard BG. Lipoprotein lipase mutations, plasma lipids and lipoproteins, and risk of ischemic heart disease. A meta-analysis. Circulation 1999;99(22):2901-2907.

[149] Kastelein JJ, Ordovas JM, Wittekoek ME, Pimstone SN, Wilson WF, Gagné SE, Larson MG, Schaefer EJ, Boer JM, Gerdes C, Hayden MR. Two common mutations (D9N, N291S) in lipoprotein lipase: a cumulative analysis of their influence on plasma lipids and lipoproteins in men and women. Clin Genet 1999;56(4):297-305.

[150] Chamberlain JC, Thorn JA, Oka K, Galton DJ, Stocks J. DNA polymorphisms at the lipoprotein lipase gene: associations in normal and hypertriglyceridaemic subjects. Atherosclerosis 1989;79(1):85-91.

[151] Gerdes C, Gerdes LU, Hansen PS, Faergeman O. Polymorphisms in the lipoprotein lipase gene and their associations with plasma lipid concentrations in 40-year-old Danish men. Circulation 1995;92(7):1765-1769.

[152] Shimo-Nakanishi Y, Urabe T, Hattori N, Watanabe Y, Nagao T, Yokochi M, Hamamoto M, Mizuno Y. Polymorphism of the lipoprotein lipase gene and risk of atherothrombotic cerebral infarction in the Japanese. Stroke 2001;32(7):1481-1486.

[153] Wang X, Cheng S, Brophy VH, Erlich HA, Mannhalter C, Berger K, Lalouschek W, Browner WS, Shi Y, Ringelstein EB, Kessler C, Luedemann J, Lindpaintner K, Liu L, Ridker PM, Zee RY, Cook NR. A meta-analysis of candidate gene polymorphisms and ischemic stroke in 6 study populations: association of lymphotoxin-alpha in nonhypertensive patients. Stroke 2009;40(3):683-95.

[154] Wang C, Sun T, Li H, Bai J, Li Y. Lipoprotein lipase Ser447Ter polymorphism associated with the risk of ischemic stroke: A meta-analysis. Thromb Res 2011;128(5):e107-e112.

[155] Cummings SR, Nevitt MC, Browner WS, Stone K, Fox KM, Ensrud KE, Cauley J, Black D, Vogt TM. Risk factors for hip fracture in white women. Study of Osteoporotic Fractures Research Group. N Engl J Med 1995;332(12):767-773.

[156] Zhao Y, Ma LY, Liu YX, Wang XY, Liu LS, Lindpaintner K. Relationship between alpha-ENaC gene Thr663Ala polymorphism and ischemic stroke. Zhongguo Yi Xue Ke Xue Yuan Xue Bao 2001;23:499-501.

[157] Huang P, Kostulas K, Huang WX, Crisby M, Kostulas V, Hillert J. Lipoprotein lipase gene polymorphisms in ischaemic stroke and carotid stenosis. Eur J Clin Invest 1997;27(9):740-742.

[158] Wittrup HH, Nordestgaard BG, Sillesen H, Schnohr P, Tybjaerg-Hansen A. A common mutation in lipoprotein lipase confers a 2-fold increase in risk of ischemic cerebrovascular disease in women but not in men. Circulation 2000;101(20):2393-2397.

[159] Myllykangas L, Polvikoski T, Sulkava R, Notkola IL, Rastas S, Verkkoniemi A, Tienari PJ, Niinistö L, Hardy J, Pérez-Tur J, Kontula K, Haltia M. Association of lipoprotein lipase Ser447Ter polymorphism with brain infarction: a population-based neuropathological study. Ann Med 2001;33(7):486-492.

[160] Zhao SP, Tong QG, Xiao ZJ, Cheng YC, Zhou HN, Nie S. The lipoprotein lipase Ser447Ter mutation and risk of stroke in the Chinese. Clin Chim Acta 2003;330(1-2):161-164.

[161] Guan GD, Xu E, Wang XJ, Xu YH, Qiu SD. Associations between Ser447Ter gene polymorphism of lipoprotein lipase and atherosclerotic cerebral infarction. Zhonghua Yi Xue Yi Chuan Xue Za Zhi 2006;23:519-522.

[162] Kostulas K, Brophy VH, Moraitis K, Manolescu A, Kostulas V, Gretarsdottir S, Cheng S, Hillert J. Genetic profile of ischemic cerebrovascular disease and carotid stenosis. Acta Neurol Scand 2008;118(3):146-152.

[163] Xu E, Li W, Zhan L, Guan G, Wang X, Chen S, Shi Y. Polymorphisms of the lipoprotein lipase gene are associated with atherosclerotic cerebral infarction in the Chinese. Neuroscience 2008;155(2):403-408.

[164] Tang X, Zhu YP, Li N, Chen DF, Zhang ZX, Dou HD, Hu YH. Genetic epidemiological study on discordant sib pairs of ischemic stroke in Beijing Fangshan District. Beijing Da Xue Xue Bao 2007;39:119-125.

[165] Aviram M, Billecke S, Sorenson R, Bisgaier C, Newton R, Rosenblat M, Erogul J, Hsu C, Dunlop C, La Du B. Paraoxonase active site required for protection against LDL

oxidation involves its free sulfhydryl group and is different from that required for Its arylesterase/paraoxonase activities: selective action of human paraoxonase allozymes Q and R. Arterioscler Thromb Vasc Biol 1998;18(10):1617-1624.

[166] Salonen JT, Ylä-Herttuala S, Yamamoto R, Butler S, Korpela H, Salonen R, Nyyssönen K, Palinski W, Witztum JL. Autoantibody against oxidised LDL and progression of carotid atherosclerosis. Lancet 1992;339(8798):883-887.

[167] Primo-Parmo SL, Sorenson RC, Teiber J, La Du BN. The human serum paraoxonase/arylesterase gene (PON1) is one member of a multigene family. Genomics 1996;33(3):498-507.

[168] Watson AD, Berliner JA, Hama SY, La Du BN, Faull KF, Fogelman AM, Navab M. Protective effect of high density lipoprotein associated paraoxonase. Inhibition of the biological activity of minimally oxidized low density lipoprotein. J Clin Invest 1995;96(6):2882-2891.

[169] Reddy ST, Wadleigh DJ, Grijalva V, Ng C, Hama S, Gangopadhyay A, Shih DM, Lusis AJ, Navab M, Fogelman AM. Human paraoxonase-3 is an HDL-associated enzyme with biological activity similar to paraoxonase-1 protein but is not regulated by oxidized lipids. Arterioscler Thromb Vasc Biol 2001;21(4):542-547.

[170] Ng CJ, Wadleigh DJ, Gangopadhyay A, Hama S, Grijalva VR, Navab M, Fogelman AM, Reddy ST. Paraoxonase-2 is a ubiquitously expressed protein with antioxidant properties and is capable of preventing cell-mediated oxidative modification of low density lipoprotein. J Biol Chem 2001;276(48):44444-44449.

[171] Mackness M, Mackness B. Paraoxonase 1 and atherosclerosis: is the gene or the protein more important? Free Radic Biol Med 2004;37(9):1317-1323.

[172] Wheeler JG, Keavney BD, Watkins H, Collins R, Danesh J. Four paraoxonase gene polymorphisms in 11212 cases of coronary heart disease and 12786 controls: meta-analysis of 43 studies. Lancet 2004;363(9410):689-695.

[173] Lawlor DA, Day IN, Gaunt TR, Hinks LJ, Briggs PJ, Kiessling M, Timpson N, Smith GD, Ebrahim S. The association of the PON1 Q192R polymorphism with coronary heart disease: findings from the British Women's Heart and Health cohort study and a meta-analysis. BMC Genet 2004;5:17.

[174] Dahabreh IJ, Kitsios GD, Kent DM, Trikalinos TA. Paraoxonase 1 polymorphisms and ischemic stroke risk: A systematic review and meta-analysis. Genet Med 2010;12(10):606-615.

[175] Banerjee I. Relationship between Paraoxonase 1 (PON1) gene polymorphisms and susceptibility of stroke: a meta-analysis. Eur J Epidemiol 2010;25(7):449-458.

[176] Cao H, Girard-Globa A, Serusclat A, Bernard S, Bondon P, Picard S, Berthezene F, Moulin P. Lack of association between carotid intima-media thickness and paraoxonase gene polymorphism in non-insulin dependent diabetes mellitus. Atherosclerosis 1998;138(2):361-366.

[177] Imai Y, Morita H, Kurihara H, Sugiyama T, Kato N, Ebihara A, Hamada C, Kurihara Y, Shindo T, Oh-hashi Y, Yazaki Y. Evidence for association between paraoxonase gene polymorphisms and atherosclerotic diseases. Atherosclerosis 2000;149(2):435-442.

[178] Voetsch B, Benke KS, Damasceno BP, Siqueira LH, Loscalzo J. Paraoxonase 192 Gln-->Arg polymorphism: an independent risk factor for nonfatal arterial ischemic stroke among young adults. Stroke 2002;33(6):1459-1464.

[179] Ueno T, Shimazaki E, Matsumoto T, Watanabe H, Tsunemi A, Takahashi Y, Mori M, Hamano R, Fujioka T, Soma M, Matsumoto K, Kanmatsuse K. Paraoxonase1 polymorphism Leu-Met55 is associated with cerebral infarction in Japanese populations. Med Sci Monit 2003;9(6):CR208-CR212.

[180] Chen JH, Zeng QX. Relationship between the paraoxonase gene 192 polymorphism and atherosclerotic cerebral infarction. Zhong Guo Lin Chuang Kang Fu 2003;7:3036-3037.

[181] Ranade K, Kirchgessner TG, Iakoubova OA, Devlin JJ, DelMonte T, Vishnupad P, Hui L, Tsuchihashi Z, Sacks FM, Sabatine MS, Braunwald E, White TJ, Shaw PM, Dracopoli NC. Evaluation of the paraoxonases as candidate genes for stroke: Gln192Arg polymorphism in the paraoxonse 1 gene is associated with increased risk of stroke. Stroke 2005;36(11):2346-2350.

[182] Huang Q, Liu YH, Yang Q. The association of PON1 Q192R gene polymorphism with atherosclerotic cerebral infarction. Zhong Hua Shen Jing Ke Za Zhi 2005;38:454-455.

[183] Wu J, Zhao SP, Tan LM. The relationship between PON1-192 polymorphism and type of cerebral infarction. Nao Yu Shen Jing Ji Bing Za Zhi 2005;13:253-255.

[184] Yu LT, Yu DC, Li L. The relationship between paraoxonase gene 192Gln/Arg polymorphism and ischemic cerebrovascular disease. Zhong Hua Lao Nian Xin Nao Xue Guan Bing Za Zhi 2005;7:254-256.

[185] Baum L, Ng HK, Woo KS, Tomlinson B, Rainer TH, Chen X, Cheung WS, Chan DK, Thomas GN, Tong CS, Wong KS. Paraoxonase 1 gene Q192R polymorphism affects stroke and myocardial infarction risk. Clin Biochem 2006;39(3):191-195.

[186] Huang Q, Liu YH, Yang QD, Xiao B, Ge L, Zhang N, Xia J, Zhang L, Liu ZJ. Human serum paraoxonase gene polymorphisms, Q192R and L55M, are not associated with the risk of cerebral infarction in Chinese Han population. Neurol Res 2006;28(5):549-554.

[187] Pasdar A, Ross-Adams H, Cumming A, Cheung J, Whalley L, St Clair D, MacLeod MJ. Paraoxonase gene polymorphism and haplotype analysis in a stroke population. BMC Medical Genetics 2006;7:28-33.

[188] Aydin M, Gencer M, Cetinkaya Y, Ozkok E, Ozbek Z, Kilic G, Orken C, Tireli H, Kara I. PON1 55/192 polymorphism, oxidative stress, type, prognosis and severity of stroke. IUBMB Life 2006;58(3):165-172.

[189] Chen WR, Xiao ZJ, Zhao SQ. The relationship between the gene polymorphism in paraoxonase and lacunar infarction. Cu Zhong Yu Shen Jing Ji Bing 2006;13:75-78.

[190] Schiavon R, Turazzini M, De Fanti E, Battaglia P, Targa L, Del Colle R, Fasolin A, Silvestri M, Biasioli S, Guidi G. PON1 activity and genotype in patients with arterial ischemic stroke and in healthy individuals. Acta Neurol Scand 2007;116(1):26-30.

[191] Shin BS, Oh SY, Kim YS, Kim KW. The paraoxonase gene polymorphism in stroke patients and lipid profile. Acta Neurol Scand 2008;117(4):237-243.

[192] Slowik A, Wloch D, Szermer P, Wolkow P, Malecki M, Pera J, Turaj W, Dziedzic T, Klimkowicz-Mrowiec A, Kopec G, Figlewicz DA, Szczudlik A. Paraoxonase 2 gene

C311S polymorphism is associated with a risk of large vessel disese stroke in a Polish population. Cerebrovasc Dis 2007;23(5-6):395-400.
[193] Can Demirdöğen B, Türkanoğlu A, Bek S, Sanisoğlu Y, Demirkaya S, Vural O, Arinç E, Adali O. Paraoxonase/arylesterase ratio, PON1 192Q/R polymorphism and PON1 status are associated with increased risk of ischemic stroke. Clin Biochem 2008;41(1-2):1-9.
[194] Demirdöğen BC, Demirkaya S, Türkanoğlu A, Bek S, Arınç E, Adali O. Analysis of paraoxonase 1 (PON1) genetic polymorphisms and activities as risk factors for ischemic stroke in Turkish population. Cell Biochem Funct 2009;27(8):558-567.
[195] Xiao ZJ, Chen J, Sun Y, Zheng ZJ. Lack of association between the paraoxonase 1 Q/R192 single nucleotide polymorphism and stroke in a Chinese cohort. Acta Neurol Belg 2009;109(3):205-209.
[196] Schmidt R, Schmidt H, Fazekas F, Kapeller P, Roob G, Lechner A, Kostner GM, Hartung HP. MRI cerebral white matter lesions and paraoxonase PON1 polymorphisms: three-year follow-up of the Austrian Stroke Prevention Study. Arterioscler Thromb Vasc Biol 2000;20(7):1811-1816.
[197] Voetsch B, Benke KS, Panhuysen CI, Damasceno BP, Loscalzo J. The combined effect of paraoxonase promoter and coding region polymorphisms on the risk of arterial ischemic stroke among young adults. Arch Neurol 2004;61(3):351-356.
[198] Man BL, Baum L, Fu YP, Chan YY, Lam W, Hui CF, Leung WH, Wong KS. Genetic polymorphisms of Chinese patients with ischemic stroke and concurrent stenoses of extracranial and intracranial vessels. J Clin Neurosci 2010;17(10):1244-1247.
[199] Xu HW, Zhao Z, Yuan N, Xiao B, Yang XS, Tang BS. Relationship between single nucleotide polymorphisms of paraoxonase 2 and stroke. Zhonghua Yi Xue Yi Chuan Xue Za Zhi 2007;24:328-330.
[200] Lazaros L, Markoula S, Kyritsis A, Georgiou I. Paraoxonase gene polymorphisms and stroke severity. Eur J Neurol 2010;17(5):757-759.
[201] Guo Y, Guo J, Zheng D, Pan L, Li Q, Ruan G. Relationship between the Nco I, Ava II polymorphism of low density lipoprotein receptor gene and atherosclerotic cerebral infarction. Zhonghua Yi Xue Yi Chuan Xue Za Zhi 2002;19:209-212.
[202] Frikke-Schmidt R, Nordestgaard BG, Schnohr P, Tybjærg-Hansen A. Single nucleotide polymorphism in the low-density lipoprotein receptor is associated with a threefold risk of stroke. A case-control and prospective study. Eur Heart J 2004;25(11):943-951.
[203] Lee JD, Lin YH, Hsu HL, Huang YC, Wu CY, Ryu SJ, Lee M, Huang YC, Hsiao MC, Chang YJ, Chang CH, Lee TH. Genetic polymorphisms of low density lipoprotein receptor can modify stroke presentation. Neurol Res. 2010;32(5):535-540.
[204] Yang XC, Zhang Q, Li SJ, Wan XH, Zhong GZ, Hu WL, Li L, Yu SZ, Jin L, Wang XF. Association study between three polymorphisms and myocardial infarction and ischemic stroke in Chinese Han population. Thromb Res 2010;126(4):292-294.
[205] Hattori H, Sonoda A, Sato H, Ito D, Tanahashi N, Murata M, Saito I, Watanabe K, Suzuki N. G501C polymorphism of oxidized LDL receptor geen (OLR1) and ischemic stroke. Brain Res 2006;1121(1):246-249.

[206] Vietri MT, Molinari AM, Boggia M, Parisi M, Cioffi M.. IVS4-14 A/G and IVS4-73 C/T polymorphisms in OLR1 gene in patients with ischemic cerebrovascular diseases. Genet Test Mol Biomarkers 2010;14(1):9-11.

[207] Imig JD, Navar LG, Roman RJ, Reddy KK, Falck JR. Actions of epoxygenase metabolites on the preglomerular vasculature. J Am Soc Nephrol 1996;7(11):2364-2370.

[208] Heizer ML, McKinney JS, Ellis EF. 14,15-Epoxyeicosatrienoic acid inhibits platelet aggregation in mouse cerebral arterioles. Stroke 1991;22(11):1389-1393.

[209] Krötz F, Riexinger T, Buerkle MA, Nithipatikom K, Gloe T, Sohn H, Campbell WB, Pohl U. Membrane-potential-dependent inhibition of platelet adhesion to endothelial cells by epoxyeicosatrienoic acids. Arterioscler Thromb Vasc Biol 2004;24(3):595-600.

[210] Zhang W, Otsuka T, Sugo N, Ardeshiri A, Alhadid YK, Iliff JJ, DeBarber AE, Koop DR, Alkayed NJ. Soluble epoxide hydrolase gene deletion is protective against experimental cerebral ischemia. Stroke 2008;39(7): 2073-2078.

[211] Zhang L, Ding H, Yan J, Hui R, Wang W, Kissling GE, Zeldin DC, Wang DW. Genetic variation in cytochrome P450 2J2 and soluble epoxide hydrolase and risk of ischemic stroke in a Chinese population. Pharmacogenet Genomics 2008;18(1):45-51.

[212] Fornage M, Lee CR, Doris PA, Bray MS, Heiss G, Zeldin DC, Boerwinkle E. The soluble epoxide hydrolase gene harbors sequence variation associated with susceptibility to and protection from incident ischemic stroke. Hum Mol Genet 2005;14(19):2829-2837.

[213] Newman JW, Morisseau C, Hammock BD. Epoxide hydrolases: their roles and interactions with lipid metabolism. Prog Lipid Res 2005;44(1):1-51.

[214] Sato K, Emi M, Ezura Y, Fujita Y, Takada D, Ishigami T, Umemura S, Xin Y, Wu LL, Larrinaga-Shum S, Stephenson SH, Hunt SC, Hopkins PN. Soluble epoxide hydrolase variant (Glu287Arg) modifies plasma total cholesterol and triglyceride phenotype in familial hypercholesterolemia: intrafamilial association study in an eight-generation hyperlipidemic kindred. J Hum Genet 2004;49(1):29-34.

[215] Przybyla-Zawislak BD, Srivastava PK, Vázquez-Matiás H, et al. Polymorphism in human soluble epoxide hydrolase. J Mol Pharmacol 2003;64(2):482-490.

[216] Fornage M, Lee CR, Doris PA, Bray MS, Heiss G, Zeldin DC, Boerwinkle E. The soluble epoxide hydrolase gene harbors sequence variation associated with susceptibility to and protection from incident ischemic stroke. Hum Mol Genet 2005;14(19):2829-2837.

[217] Gschwendtner A, Ripke S, Freilinger T, Lichtner P, Müller-Myhsok B, Wichmann H, Meitinger T, Dichgans M. Genetic variation in soluble epoxide hydrolase (EPHX2) is associated with an increased risk of ischemic stroke in white Europeans. Stroke 2008;39(5):1593-1596.

[218] Zhang L, Ding H, Yan J, Hui R, Wang W, Kissling GE, Zeldin DC, Wang DW. Genetic variation in cytochrome P450 2J2 and soluble epoxide hydrolase and risk of ischemic stroke in a Chinese population. Pharmacogenet Genomics 2008;18(1):45-51.

[219] Sandberg M, Hassett C, Adman ET, Meijer J, Omiecinski CJ. Identification and functional characterization of human soluble epoxide hydrolase genetic polymorphisms. J Biol Chem 2000;275(37):28873-28881.

[220] Lee J, Dahl M, Grande P, Tybjaerg-Hansen A, Nordestgaard BG. Genetically reduced soluble epoxide hydrolase activity and risk of stroke and other cardiovascular disease. Stroke 2010,41(1):27-33.

[221] Fava C, Montagnana M, Danese E, Almgren P, Hedblad B, Engström G, Berglund G, Minuz P, Melander O. Homozygosity for the EPHX2 K55R polymorphism increases the long-term risk of ischemic stroke in men: a study in Swedes. Pharmacogenet Genomics 2010;20(2):94-103.

[222] Wang WYS, Barratt BJ, Clayton DG, Todd JA. Genome-wide association studies: theoretical and practical concerns. Nat Rev Genet 2005;6(2):109-118.

[223] Wellcome Trust Case Control Consortium. 2007. Genomewide association study of 14,000 cases of seven common diseases and 3,000 shared controls. Nature 2007;447(7145):661-678.

[224] Matarin M, Brown WM, Scholz S, Simon-Sanchez J, Fung HC, Hernandez D, Gibbs JR, De Vrieze FW, Crews C, Britton A, Langefeld CD, Brott TG, Brown RD Jr, Worrall BB, Frankel M, Silliman S, Case LD, Singleton A, Hardy JA, Rich SS, Meschia JF. A genome-wide genotyping study in patients with ischaemic stroke: Initial analysis and data release. Lancet Neurol 2007;6(5):414–420.

[225] Ikram MA, Seshadri S, Bis JC, Fornage M, DeStefano AL, Aulchenko YS, Debette S, Lumley T, Folsom AR, van den Herik EG, Bos MJ, Beiser A, Cushman M, Launer LJ, Shahar E, Struchalin M, Du Y, Glazer NL, Rosamond WD, Rivadeneira F, Kelly-Hayes M, Lopez OL, Coresh J, Hofman A, DeCarli C, Heckbert SR, Koudstaal PJ, Yang Q, Smith NL, Kase CS, Rice K, Haritunians T, Roks G, de Kort PL, Taylor KD, de Lau LM, Oostra BA, Uitterlinden AG, Rotter JI, Boerwinkle E, Psaty BM, Mosley TH, van Duijn CM, Breteler MM, Longstreth WT Jr, Wolf PA. Genomewide association studies of stroke. N Engl J Med 2009;360(17):1718-1728.

[226] Gretarsdottir S, Thorleifsson G, Manolescu A, Styrkarsdottir U, Helgadottir A, Gschwendtner A, Kostulas K, Kuhlenbaumer G, Bevan S, Jonsdottir T, Bjarnason H, Saemundsdottir J, Palsson S, Arnar DO, Holm H, Thorgeirsson G, Valdimarsson EM, Sveinbjornsdottir S, Gieger C, Berger K, Wichmann HE, Hillert J, Markus H, Gulcher JR, Ringelstein EB, Kong A, Dichgans M, Gudbjartsson DF, Thorsteinsdottir U, Stefansson K. Risk variants for atrial fibrillation on chromosome 4q25 associate with ischemic stroke. Ann Neurol 2008;64(4):402-409.

[227] Gudbjartsson DF, Holm H, Gretarsdottir S, Thorleifsson G, Walters GB, Thorgeirsson G, Gulcher J, Mathiesen EB, Njolstad I, Nyrnes A, Wilsgaard T, Hald EM, Hveem K, Stoltenberg C, Kucera G, Stubblefield T, Carter S, Roden D, Ng MC, Baum L, So WY, Wong KS, Chan JC, Gieger C, Wichmann HE, Gschwendtner A, Dichgans M, Kuhlenbaumer G, Berger K, Ringelstein EB, Bevan S, Markus HS, Kostulas K, Hillert J, Sveinbjornsdottir S, Valdimarsson EM, Lochen ML, Ma RC, Darbar D, Kong A, Arnar DO, Thorsteinsdottir U, Stefansson K. A sequence variant in zfhx3 on 16q22 associates with atrial fibrillation and ischemic stroke. Nat Genet 2009;41(8):876-878.

[228] Kubo M, Hata J, Ninomiya T, Matsuda K, Yonemoto K, Nakano T, Matsushita T, Yamazaki K, Ohnishi Y, Saito S, Kitazono T, Ibayashi S, Sueishi K, Iida M, Nakamura Y,

Kiyohara Y. A nonsynonymous snp in prkch (protein kinase c eta) increases the risk of cerebral infarction. Nat Genet 2007;39(2):212-217.

[229] Hata J, Matsuda K, Ninomiya T, Yonemoto K, Matsushita T, Ohnishi Y, Saito S, Kitazono T, Ibayashi S, Iida M, Kiyohara Y, Nakamura Y, Kubo M. Functional snp in an sp1-binding site of agtrl1 gene is associated with susceptibility to brain infarction. Hum Mol Genet 2007;16(6):630-639.

[230] Yamada Y, Fuku N, Tanaka M, Aoyagi Y, Sawabe M, Metoki N, Yoshida H, Satoh K, Kato K, Watanabe S, Nozawa Y, Hasegawa A, Kojima T. Identification of celsr1 as a susceptibility gene for ischemic stroke in japanese individuals by a genome-wide association study. Atherosclerosis 2009;207(1):144-149.

Chapter 25

Lipoproteins and Apolipoproteins in Alzheimer's Disease

Etsuro Matsubara

Additional information is available at the end of the chapter

http://dx.doi.org/10.5772/45977

1. Introduction

Alzheimer's disease (AD) represents the so-called "storage disorder" of amyloid β (Aβ). The AD brain contains soluble and insoluble Aβ, both of which have been hypothesized to underlie the development of cognitive deficits or dementia (1-3). The steady-state level of Aβ is controlled by the generation of Aβ from its precursor, the degradation of Aβ within the brain, and transport of Aβ out of the brain. The imbalance among three metabolic pathways results in excessive accumulation and deposition of Aβ in the brain, which may trigger a complex downstream cascade (e.g., primary amyloid plaque formation or secondary tauopathy and neurodegeneration) leading to memory loss or dementia in AD. Accumulated lines of evidence indicate that such a memory loss represents a synaptic failure caused directly by soluble Aβ oligomers (AβOs) (4-6), whereas amyloid fibrils may cause neuronal injury indirectly via microglial activation (7). Many attentions are paid to understand the mechanism underlying the neurotoxic action of AβOs so far. However, the exact metabolic conditions controlling the *in vivo* generation of soluble AβOs has been out of attention.

Several lines of evidence indicated that lipidic environments in the central nervous system (CNS) represent one of the prevailing metabolic conditions. We then hypothesized that an alteration of the lipoprotein-soluble Aβ interaction in the CNS is capable of initiating and/or accelerating the cascade favoring Aβ assembly (8). We found that dissociation of Aβ42 from lipoprotein in the cerebrospinal fluid from AD accelerates Aβ42 assembly (9). Thus, lipoprotein is a key molecule to maintain monomeric soluble Aβ42 in CNS.

In this chapter, we review the issue regarding how lipoprotein and apolipoproteins contribute to physiological metabolic conditions. Then, we focus on how they constitute the

AD-related metabolic conditions in the CNS. We are certain that these points of view introduce a novel approach to find a therapeutic intervention for AD.

2. Lipoproteins, apolipoproteins, and Aβ metabolism in the CNS

In the CNS, we need to be aware that cholesterol metabolism is quite different from that in systemic circulation. Lipidic environments in the CNS were regulated by HDL-like lipoproteins, mainly lipidated apolipoprotein E (apoE), which is in charge of cholesterol transport to and from neurons (10, 11). This is also the case in lipidated apolipoprotein J (apoJ) (12). In addition to lipid trafficking, apoE or apoJ as a form of HDL-like lipoprotein plays a major role in Aβ metabolism in the CNS. Both apolipoproteins are well known as major carrier proteins for Aβ (13-17). Interestingly, transgenic mouse models of AD (apoE$^{-/-}$/apoJ$^{-/-}$) revealed that both apolipoproteins regulate in a cooperative manner the clearance and the deposition of Aβ in brain (18). The hypothetical pathways involved in the clearance of CNS Aß are efflux of Aß into the plasma via blood-brain barrier (BBB). Two lipoprotein-receptors, LRP-1 and LRP-2, seem to be responsible for efflux of lipoprotein-free or lipoprotein-associated (apoJ-associated) Aß from the brain to blood, respectively (19). *In vivo* relevance of LRP-1-mediated Aß transport has been confirmed in transgenic mice expressing low LRP-1-receptor and APP, which develops extensive Aß accumulation much faster than transgenic mice expressing high level of APP (20). Reduced expression of brain endotherial LRP-1 was also observed in AD patients, which was associated with impaired Aß clearance and cerebrovascular accumulation. LPR-2 appeared to function bi-directionally (influx vs efflux) at BBB. In contrast to LRP2-mediated influx (21), LPR2-mediated efflux of brain Aß was actively operated under physiological concentration of either Aß or apoJ (19). Interestingly, a recent study shows that apoE4 binding to Aß redirects its clearance from LRP-1 to VLDLR, which resulted in slower efflux of brain Aß than LRP-1 (22). In contrast, apoE2-Aß and apoE3-Aß complexes are cleared at BBB via both LPR-1 and VLDLR at a substantially faster rate than apoE4-Aß complexes(22). Impairment of the above-mentioned receptor-mediated clearance at BBB could contribute to the pathogenesis of AD. Alternatively, ApoE4-HDL shows less cholesterol exchange between lipid particles and the neuronal membrane as compared with apoE3-HDL (23), leading to altered membrane functions, e.g., signal transduction, enzyme activities, ion channel properties, and conformation of sAß peptides, which contribute to the disease-related metabolic conditions. Furthermore, when the generation of HDL-like lipoproteins in the AD mouse model is suppressed or overexpressed via the specific regulation of ATP-binding cassette A1 (ABCA1), Aβ deposition exhibits augmentation or reduction, respectively, which depends on the degree of ABCA1-mediated lipidation of apoE in the CNS (24, 25). From these points of view, lipidic environments in the CNS represent one of the prevailing metabolic conditions. We hypothesized that an alteration of the lipoprotein-sAβ interaction in the CNS is capable of initiating and/or accelerating the cascade favoring Aß assembly. Thus, we postulate that lipoproteins or apolipoproteins may regulate the metabolic conditions controlling the *in vivo* generation of soluble Aβ Os.

3. Aß is present in either lipoprotein-free or lipoprotein-associated form in brain parenchyma

To assess the above-mentioned issue, we examined whether the dissociation of sAß from lipoprotein-particles occurs in the brain. The combination of size exclusion chromatography (SEC) and enzyme-linked immunosorbent assay (ELISA) revealed that the dissociation of sAß from lipoprotein-particles occurs in brain parenchyma and the presence of soluble dimeric lipoprotein-free Aß in AD brains (8). These findings may support the hypothesis that functionally declined lipoproteins may be major determinants in the production of metabolic conditions leading to higher levels of soluble dimeric SDS-resistant form of Aβ in AD brains (8, 26). At this moment, it remains undetermined whether dissociation of Aβ from lipoprotein or less association of Aβ to lipoproteins accounts for such a metabolic conditions. To further verify this hypothesis, we focused on the entorhinal cortex (EC), followed by biochemical analyses using an anti-oligomer specific antibody, namely 2C3 (9, 27). Fifty brains obtained from healthy elderly are composed of three Braak NFT stages; Braak NFT stages I-II (n=35, normal control); Braak NFT stages III-IV (n=13, MCI stage); Braak NFT stages IV-V (n=2, AD stages). Immunoblot analysis of the delipidated EC employing monoclonal 2C3 revealed that the accumulation of soluble 12-mers precedes the appearance of neuronal loss or cognitive impairment, and is enhanced as the Braak neurofibrially tangle (NFT) stages progress, indicating that the ECs of AD patients indeed bear metabolic conditions that accelerate Aβ assembly.

4. Aß is present in either lipoprotein-free or lipoprotein-associated form in cerebrospinal fluid (CSF)[9]

The presence of lipoprotein-free sAβOs in CSF was also assessed in age-matched normal controls (NCs) and patients with Alzheimer's disease (AD) by SEC and ELISA specific for either AβOs or AβMs. The SEC experiment using pooled CSF revealed that the dissociation of sAβMs from lipoprotein particles indeed occurs in CSF, which was lower in AD than in NCs. Furthermore, the SEC experiment using lipoprotein-depleted pooled CSF (LPD-CSF) confirmed the presence of oligomeric 2C3 conformers (4- to 35-mers), which appeared to be higher in AD patients than in NCs. To address the issue on the presence of any metabolic conditions favoring Aβ assembly, we compared the levels of lipoprotein-free sAβMs and sAβOs in LPD-CSF from the 12 sporadic AD patients and 13 NCs to evaluate the AβOs/AβMs ratio (the O/M index). The levels of 2C3 oligomeric conformers composed of Aβ42 are significantly higher in AD patients than in NCs. The O/M index for either Aβ42 or Aβ40 is also significantly higher in AD patients than in NCs. Of note, the relative amounts of total lipoprotein-associated sAβMs (~70%) versus lipoprotein-free sAβMs (~30%) remained

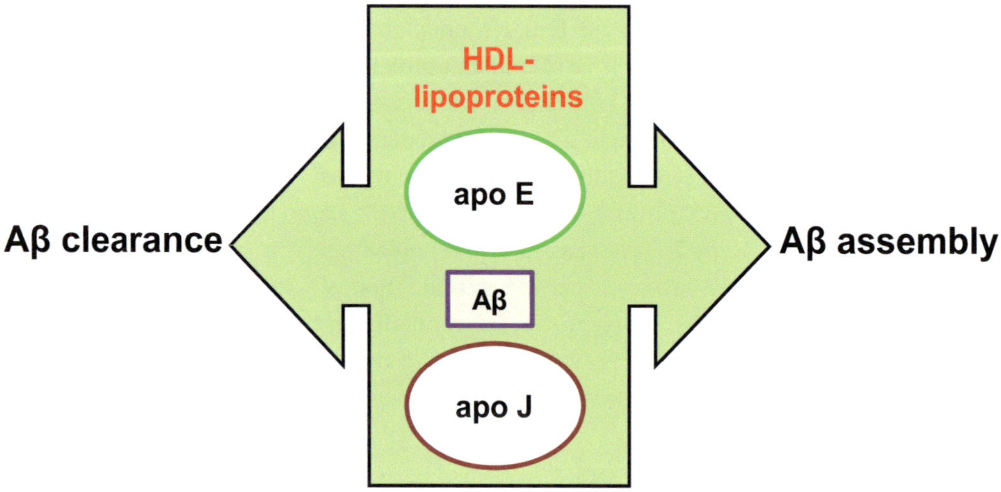

Figure 1. Hypothetical metabolic conditions favoring Aβ assembly. Functionally declined lipoproteins may accelerate the generation of metabolic conditions leading to higher levels of soluble Aβassembly in the CNS.

essentially unchanged in sporadic AD patients as compared with NCs. However, the relative amounts of lipoprotein-free Aβ42 was significantly lower in the sporadic AD patients (9.3 ± 3.9 %) than in NCs (13.2 ± 4.5 %), which is in accordance with our above-mentioned finding that the level of oligomeric 2C3 conformers composed of Aβ42 was significantly elevated in AD patients. Thus, it is likely that the conversion of lipoprotein-free monomeric soluble Aβ42 into oligomeric assembly preferentially occurs in AD CSF, mirroring the disease-related metabolic conditions in the brain parenchyma.

5. Summary

We previously reported that ~90% of sAβMs that circulate in normal plasma is associated with lipoprotein particles (27). From the above data, it is plausible to assume that about 70% of CSF sAβMs is normally associated with lipoprotein particles, indicating that CNS constitutes a risky environment where the lipoproteins-sAβMs interaction is impaired, leading to Aβ assembly. From this point of view, a key molecule to maintain monomeric sAβ42 metabolism in CNS appears to be HDL-like lipoprotein particles. In this sense, the dissociation of sAβ42 from or the lack of association with HDL-like lipoprotein particles not only constitutes a potential mechanism to initiate and/or accelerate the cascade favoring Aβ42 assembly in the brain, but also results in a reduced clearance of physiological lipoprotein-associated sAβ42 peptides in the brain. Thus, above-mentioned CNS environments may strongly affect conformation of sAβ peptides, resulting in the conversion of sAβ42 monomers into sAβ42 assembly. The findings suggest that functionally declined lipoproteins may accelerate the generation of metabolic conditions leading to higher levels of sAβ42 assembly in the CNS.

Author details

Etsuro Matsubara
Department of Neurology, Institute of Brain Science, Hirosaki Graduate School of Medicine, Japan

6. References

[1] Hardy J, Allsop D: Amyloid deposition as the central event in the aetiology of Alzheimer's disease. *Trends Pharmacol Sci* 1991, 12: 383-388.
[2] Lue LF, Kuo YM, Roher AE, Brachova L, Shen Y, Sue L, Beach T, Kurth JH, Rydel RE, Rogers J: Soluble amyloid β peptide concentration as a predictor of synaptic change in Alzheimer's disease. *Am J Pathol* 1999, 155: 853-862.
[3] McLean CA, Cherny RA, Fraser FW, Fuller SJ, Smith MJ, Beyreuther K, Bush AI, Masters CL: Soluble pool of Abeta amyloid as a determinant of severity of neurodegeneration in Alzheimer's disease. *Ann Neurol* 1999, 46: 860-866.

[4] Klein WL, Krafft GA, Finch CE: Targeting small Abeta oligomers: the solution to an Alzheimer's disease conundrum? *Trends Neurosci* 2001, 24: 219-224.

[5] Selkoe DJ: Alzheimer's disease is a synaptic failure. *Science* 2002, 298: 789-791.

[6] Hass C, SelkoeDJ: Soluble protein oligomers in neurodegeneration: lessons from the Alzheimer's amyloid β-peptide. *Nat Rev Mol Cell Biol* 2007, 8: 101-112.

[7] Akiyama H, Barger S, Barnum S, Bradt B, Bauer J, Cole GM, Cooper NR, Eikelenboom P, Emmerling M, Fiebich BL, Finch CE, Frautschy S, Griffin WS, Hampel H, Hull M, Landreth G, Lue L, Mrak R, Mackenzie IR, McGeer PL, O'Banion MK, Pachter J, Pasinetti G, Plata-Salaman C, Rogers J, Rydel R, Shen Y, Streit W, Strohmeyer R, Tooyoma I, Van Muiswinkel FL, Veerhuis R, Walker D, Webster S, Wegrzyniak B, Wenk G, Wyss-Coray T: Inflammation and Alzheimer's disease. *Neurobiol Aging* 2000, 21: 383-421.

[8] Matsubara E, Sekijima Y, Tokuda T, Urakami K, Amari M, Shizuka-Ikeda M, Tomidokoro Y, Ikeda M, Kawarabayashi T, Harigaya Y, Ikeda S, Murakami T, Abe K, Otomo E, Hirai S, Frangione B, Ghiso J, Shoji M. 2004. Soluble Abeta homeostasis in AD and DS: impairment of anti-amyloidogenic protection by lipoproteins. *Neurobiol Aging* 25:833-841.

[9] Takamura A, Kawarabayashi T, Yokoseki T, Shibata M, Morishima-Kawashima M, Saito Y, Murayama S, Ihara Y, Abe K, Shoji M, Michikawa M, Matsubara E. The Dissociation of Aβ from Lipoprotein in Cerebrospinal Fluid from Alzheimer's Disease accelerates Aβ42 assembly. J Neurosci Res. 2011;89(6):815-821.

[10] Michikawa M, Gong JS, Fan QW, Sawamura N, Yanagisawa K. 2001. A novel action of alzheimer's amyloid beta-protein (Abeta): oligomeric Abeta promotes lipid release. *J Neurosci* 21:7226-7235.

[11] Gong JS, Sawamura N, Zou K, Sakai J, Yanagisawa K, Michikawa M. 2002. Amyloid beta-protein affects cholesterol metabolism in cultured neurons: implications for pivotal role of cholesterol in the amyloid cascade. *J Neurosci Res* 70:438-46.

[12] DeMattos RB, Brendza RP, Heuser JE, Kierson M, Cirrito JR, Fryer J, Sullivan PM, Fagan AM, Han X, Holtzman DM. Purification and characterization of astrocyte-secreted apolipoprotein E and J-containing lipoproteins from wild-type and human apoE transgenic mice. Neurochem Int. 2001, 39(5-6):415-25.

[13] Ghiso J, Matsubara E, Koudinov A, Choi-Miura NH, Tomita M, Wisniewski T, Frangione B. The cerebrospinal-fluid soluble form of Alzheimer's amyloid ß is complexed to SP-40,40 (apolipoprotein J), an inhbitor of the complement membrane-attack complex. Biochem J.,1993;293:27-30.

[14] Wisniewski T, Golabek A, Matsubara E, Ghiso J, Frangione B. Apolipoprotein E: binding to soluble Alzheimer's ß –amyloid. Biochem Biophys Res Commun., 1993;192:359-365.

[15] Koudinov A, Matsubara E, Frangione B, Ghiso J. The soluble form of Alzheimer's amyloid ß protein is complexed to high density lipoprotein 3 and very high density

lipoprotein in normal human plasma. Biochem Biophys Res Commun., 1994;205:1164-1171.

[16] Matsubara E, Frangione B, Ghiso J. Characterization of apolipoprotein J-Alzheimer's A beta interaction. *J Biol Chem.* 1995, 270:7563-7567.

[17] Matsubara E, Soto C, Governale S, Frangione B, Ghisom J. Apolipoprotein J and Alzheimer's amyloid ß solubility. Biochem J, 1996;316:671-679.

[18] DeMattos RB, Bales KR, Cummins DJ, Paul SM, Holtzman DM. Brain to plasma amyloid-beta efflux: a measure of brain amyloid burden in a mouse model of Alzheimer's disease. Science 295: 2264-2267, 2002

[19] Bell RD, Sagare AP, Friedman AE, Bedi GS, Holtzman DM, Deane R, Zlokovic BV. Transport pathways for clearance of human Alzheimer's amyloid beta-peptide and apolipoproteins E and J in the mouse central nervous system. J Cereb Blood Flow Metab. 27:909-918, 2007.

[20] Shibata M, Yamada S, Kumar SR, Calero M, Bading J, Frangione B, Holtzman DM, Miller CA, Strickland DK, Ghiso J, Zlokovic BV. Clearance of Alzheimer's amyloid-ss(1-40) peptide from brain by LDL receptor-related protein-1 at the blood-brain barrier. J Clin Invest. 2000 Dec;106(12):1489-99.

[21] Shayo M, McLay RN, Kastin AJ, Banks WA. The putative blood-brain barrier transporter for the beta-amyloid binding protein apolipoprotein J is saturated at physiological concentration. Life Sci 60: 115-118, 1997.

[22] Deane R, Sagare A, Hamm K, Parisi M, Lane S, Finn MB, Holtzman DM, Zlokovic BV. apoE isoform-specific disruption of amyloid beta peptide clearance from mouse brain. J Clin Invest. 2008 Dec;118(12):4002-13.

[23] Zou K, Gong JS, Yanagisawa K, Michikawa M. 2002. A novel function of monomeric amyloid beta-protein serving as an antioxidant molecule against metal-induced oxidative damage. J Neurosci 22:4833-4841.

[24] Wahrle SE, Jiang H, Parsadanian M, Hartman RE, Bales KR, Paul SM, Holtzman DM. 2005. Deletion of Abca1 increases Abeta deposition in the PDAPP transgenic mouse model of Alzheimer disease. J Biol Chem 280:43236-43242.

[25] Wahrle SE, Jiang H, Parsadanian M, Kim J, Li A, Knoten A, Jain S, Hirsch-Reinshagen V, Wellington CL, Bales KR, Paul SM, Holtzman DM. 2008. Overexpression of ABCA1 reduces amyloid deposition in the PDAPP mouse model of Alzheimer disease. J Clin Invest 118:671-682.

[26] Matsubara E, Ghiso J, Frangione B, Amari M, Tomidokoro Y, Ikeda Y, Harigaya Y, Okamoto K, Shoji M. 1999. Lipoprotein-free amyloidogenic peptides in plasma are elevated in patients with sporadic Alzheimer's disease and Down's syndrome.Ann Neurol 45:537-541.

[27] Takamura A, Okamoto Y, Kawarabayashi T, Yokoseki T, Shibata M, Mouri A, Nabeshima T, Sun H, Abe K, Shoji M, Yanagisawa K, Michikawa M, Matsubara E.

Extracellular and Intraneuronal HMW-AbetaOs Represent a Molecular Basis of Memory Loss in Alzheimer's Disease Model Mouse. Mol Neurodegener. 2011;20: 6.

Section 8

Lipoproteins and Cancer

Chapter 26

Lipoproteins and Cancer

Caryl J. Antalis and Kimberly K. Buhman

Additional information is available at the end of the chapter

http://dx.doi.org/10.5772/50989

1. Introduction

Circulating lipoproteins perform vital functions, including the transport of fatty acids and cholesterol from intestine and liver throughout the body. However, in well-fed Western societies, elevated concentrations of lipoproteins in blood have long been recognized to convey increased risk for cardiovascular disease. High fat diets, obesity, and heredity can all contribute to hyperlipidemia. More recently, there has been concern for the possible effects of hyperlipidemia on risk for or progression of cancers, which have a far greater demand for lipids than normal tissues. For example, obesity is now an established risk factor for certain types of cancer and is also found to affect the prognosis for cancer patients (Calle and Kaaks 2004; Cleary and Grossmann 2009). While the association of obesity with cancer is complex, higher circulating lipids may be a contributing element. Similarly type 2 diabetes, a condition of multiple co-morbidities including hyperlipidemia, is associated with the incidence of and mortality from cancer (Faulds and Dahlman-Wright 2012).

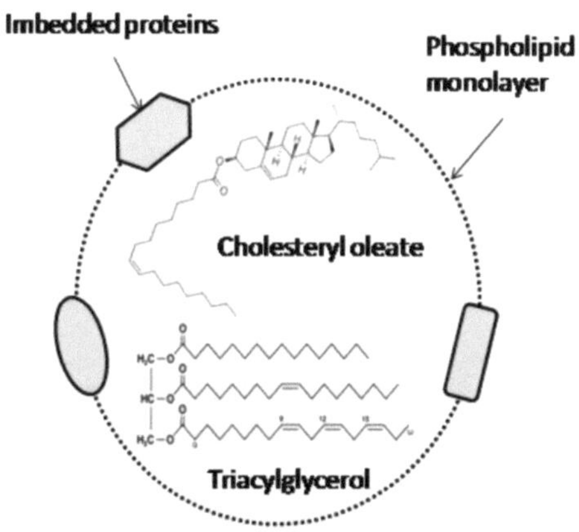

Figure 1. Cytoplasmic lipid droplets consist of an oily core of TAG and CE surrounded by a phospholipid monolayer, specific coat proteins, and other proteins.

The association of hyperlipidemia with cancer began with early observations of an accumulation of cholesterol in tumors (reviewed, (Mulas, Abete et al. 2011)). Higher levels of cholesterol and cholesteryl esters (CE) in malignant compared to less malignant tumors and normal tissues were first measured chemically (Yasuda and Bloor 1932). The accumulation of lipids in tumors was subsequently noted in tumor sections through histological examination and staining for lipid droplets (also called lipid bodies) (Freitas, Pontiggia et al. 1990). Lipid droplets are cellular organelles that store neutral lipids triacylglycerol (TAG) and CE (**Fig. 1**). Adipocytes store lipids in a single, large lipid droplet. Most other cell types have fewer, smaller lipid droplets except under pathological conditions when increased numbers and amounts of lipid may be present (Bozza and Viola 2010). Lipid droplets were detected *in vivo* in tumors with proton magnetic resonance (Delikatny, Chawla et al. 2011), and more recently, *in vivo* and *in vitro* with coherent anti-Stokes Raman scattering microscopy (Le, Huff et al. 2009). Unlike adipocyte lipid droplets, tumor cell lipid droplets contain significant quantities of CE (Tosi and Tugnoli 2005); therefore as these tumors grow and accumulate cholesterol, they may be expected to affect whole body cholesterol homeostasis and circulating cholesterol levels.

The observation of changes in plasma cholesterol in cancer patients constitutes the second line of evidence in the association of lipoproteins with cancer. It appeared in multiple studies over many years that lower plasma cholesterol was associated with a higher risk of cancer (Rose and Shipley 1980). This was a concern because lowering plasma cholesterol is a goal in cardiovascular disease prevention. The relationship between plasma cholesterol and cancer was examined in many population-based studies. Although total plasma cholesterol (total-C) measurements were used in many studies, determinations of individual lipoprotein cholesterol fractions were increasingly included. Plasma cholesterol resides primarily in low density lipoproteins (LDL) and high density lipoproteins (HDL), the lipoproteins that transport cholesterol to cells and collect excess cholesterol from cells, respectively. High HDL-C is a protective factor against atherosclerosis, while high LDL-C is positively associated with risk of atherosclerosis.

Two trends ultimately emerged from the data. First, total-C concentrations were lower two to six years prior to a cancer diagnosis, suggesting reverse causation: i.e., the early stages of the tumor led to lower circulating cholesterol (Sharp and Pocock 1997). Second, the plasma cholesterol fraction associated with tumor-caused decreases was primarily HDL-C, although the trend was detectable in total-C values also (Ahn, Lim et al. 2009). These conclusions were supported by data showing an increase in HDL-C when the patient was in remission (Dessi, Batetta et al. 1995).

The observations above suggest that in some types of cancer, tumor cells accumulate cholesterol as CE in lipid droplets and efflux less cholesterol to HDL, resulting in lower circulating HDL-C, detectable even before the tumor can be diagnosed. There is also some indication that low HDL-C levels may contribute to the development of cancer (Mondul, Weinstein et al. 2011). HDL has antioxidant and anti-inflammatory properties in addition to its role in reverse cholesterol transport (Kwiterovich 2000), and low HDL-C is a defining characteristic of the metabolic syndrome which has already been linked to cancer risk (Faulds and Dahlman-Wright 2012). Although lower HDL-C can have multiple etiologies, it can be one indicator of the presence of a tumor. If some tumors accumulate cholesterol, then it might be reasonable to ask if LDL-C fuels the development of this type of tumor.

In this chapter, we will review the evidence that LDL-C, which is usually highly correlated to total-C, is positively associated with the risk of some types of cancer. We will also review the growing body of data on what mechanisms may be involved in tumor cholesterol accumulation and what markers may be useful to identify tumors that are stimulated by cholesterol. We will address the questions: does higher circulating cholesterol increase the risk of or prognosis for certain cancers, and should lowering LDL-C be a goal in the prevention or management of some types of cancer?

2. Clinical and epidemiological evidence for an association of LDL with cancer

The presence of cancer can affect whole body cholesterol homeostasis, leading to the observation of low plasma HDL-C in cancer patients as described above. Plasma LDL-C levels in cancer may be confounded by the increased catabolism of LDL by a known or undiagnosed tumor, leading to an apparent association of low LDL-C with some types of cancer (Vitols, Gahrton et al. 1985). These apparent interactions of synchronous lipoprotein levels with cancer make it difficult to distinguish a tumor-promoting effect of lipoproteins from a tumor-induced effect on lipoproteins. Prospective studies that include a baseline measurement of blood cholesterol levels and a sufficient follow-up period could reveal if there was a positive association of hypercholesterolemia with the incidence of cancer, or in cancer patients, with prognosis or survival. Such studies have been conducted and the results have been somewhat inconsistent, which may be partially explained by the fact that tumors vary greatly by tissue of origin and even by sub-types of tumor arising from the same tissue.

Additional insight has been gained from studies of statins and statin users. Statins (inhibitors of 3-hydroxy-3-methylglutaryl-CoA reductase (HMGCR)), the rate limiting step in cholesterol biosynthesis) are considered to have pleiotropic effects against cancer due to the multiple biosynthetic products downstream of HMGCR (Gazzerro, Proto et al. 2012). However, pharmacokinetic data suggests that the peripheral tissues do not have access to high enough concentrations of therapeutic statins to effect other pathways and that the major effect of statins is through the reduction of cholesterol biosynthesis in the liver (Solomon and Freeman 2008). Statins lower plasma total-C, which reflects a large reduction in LDL-C (up to 50% or more), a lesser reduction of VLDL-C and minor effects on HDL-C. The reduction of circulating LDL-C, a major consequence of statin use, is likely the primary anti-cancer action of statins.

The largest prospective study to date on cholesterol and cancer was done in Korean adults enrolled in the Korean National Health Insurance Corporation (NHIC); participants (n = > one million) underwent biennial medical evaluations where a baseline fasting total-C measurement was obtained and follow-up data was collected for up to 14 years (Kitahara, Berrington de Gonzalez et al. 2011). The study identified cancer types that had a positive trend with quintiles of total-C in men (prostate, P = 0.002, and colon, P = 0.05) and women (breast, P = 0.003, and colon, P = 0.004), as well as those that had a negative trend in men (esophageal, stomach, liver, and lung) and women (liver). The results were adjusted for multiple factors including BMI, and excluded cancers diagnosed in the first 5 years of follow-up. This study identified the hormone-related cancers and colon cancer as having the greatest association with total-C. These cancers are also the most heavily studied with respect to the effects of total-C, statins, or dietary fat.

Prostate cancer. Early stage prostate cancer (PrC) is stimulated by circulating testosterone through over-expression of the androgen receptor (AR). AR signaling regulates the expression of the PrC marker prostate specific antigen (PSA); androgen-deprivation (castration) therapies block AR signaling, providing an effective treatment and reducing PSA levels. However, over time advanced PrC emerges which is resistant to castration therapies (androgen-independent), although the AR may still play a role in tumor progression (Taplin and Balk 2004). Testosterone is synthesized from cholesterol in the testes, but also in advanced prostate tumor cells, providing a rationale for an effect of cholesterol availability on prostate tumorigenesis (Mostaghel, Solomon et al. 2012).

Several large prospective studies in the USA showed an association between higher baseline plasma total-C and the development of high-grade (Gleason sum ≥ 7), but not total or low-grade PrC. In the Health Professionals Follow-Up Study, 18,018 men provided a baseline blood sample and were followed for up to 7 years (Platz, Clinton et al. 2008). Men with low total-C had a reduced incidence of high-grade PrC (odds ratio (OR) = 0.61, 95% CI, 0.39-0.98), and the association persisted after excluding men who were diagnosed within 2 years of blood draw. In the Prostate Cancer Prevention Trial (7 years), 5586 men in the placebo arm with a lower baseline total-C measurement had a reduced incidence of Gleason 8-10 PrC (OR = 0.41, 95% CI, 0.22-0.77) (Platz, Till et al. 2009). In the CLUE II study, 6816 men in Washington County, Maryland were followed for a mean of 12 years (Mondul, Clipp et al. 2010). Those with a baseline total-C in the desirable or borderline range had a reduced incidence of high grade PrC (hazard ratio (HR) = 0.68, 95% CI, 0.40-1.18), which was more pronounced in men with a higher BMI (HR = 0.36, 95% CI, 0.16-0.79). Excluding users of cholesterol-lowering drugs or cases diagnosed within two years of follow-up did not change the results.

The differential effects of total-C on high-grade PrC were supported in several studies conducted outside the USA. In the Alpha Tocopherol, Beta Carotene Cancer Prevention Study cohort, baseline fasting total-C and HDL-C were obtained for >29,000 Finnish male smokers who were enrolled between 1985 and 1988. After long-term follow-up (still ongoing) in 2006, and excluding the first 10 years from baseline, it was found that men with higher total-C had increased risk of overall (HR = 1.22, 95% CI, 1.03-1.44) and advanced (HR = 1.85, 95% CI, 1.13-3.03) PrC (Mondul, Weinstein et al. 2011). The Midspan studies (begun in the 1960s and 1970s in Scotland, UK) had a median follow-up period of 24 years after a baseline plasma total-C measurement (Shafique, McLoone et al. 2012). In 12,926 men diagnosed with PrC >5 years after entry into the study (n = 650), the HR for the risk of high-grade disease (Gleason score ≥ 8) in those with cholesterol levels in the second highest quintile or the highest two quintiles combined compared to the lowest quintile was 1.75 (95% CI, 1.03-2.97) and 1.88 (95% CI, 1.08-3.27), respectively. The use of statins was not available. The Nijmegen Biomedical Study in the Netherlands reported that among 2118 men followed for a median period of 6.7 years who had never used cholesterol-lowering drugs (and excluding those diagnosed in the first year), those with higher baseline total-C had increased risk for PrC (HR = 1.39, 95% CI, 1.03 1.88) and aggressive PrC (HR = 1.65, 95% CI, 1.10–2.47 (Kok, van Roermund et al. 2011). An even stronger association was seen for LDL-C levels and PrC (HR = 1.42, 95% CI, 1.00–2.02) and aggressive PrC (HR = 1.83, 95% CI, 1.15–2.90).

Some studies did not support a role for cholesterol in PrC. No association of baseline plasma total-C or HDL-C with incident, advanced, or fatal PrC was found in the HUNT 2 study where a cohort of 29,364 Norwegian men were followed for a mean 9.3 years (Martin, Vatten et al. 2009). A stated limitation of the study was the small number of advanced or fatal cases. Similarly, no association of total-C with incidence of PrC was found in the Apolipoprotein MOrtality RISk (AMORIS) study, which followed 200,660 Swedish men for a mean of 8 years (Van Hemelrijck, Garmo et al. 2011). In this study no information was available on tumor severity, precluding a finding of a differential effect based on tumor grade.

Other types of studies have contributed evidence for the effects of blood cholesterol on PrC. In a cross-sectional cohort study of 531 American men, the incidence of benign prostate hyperplasia was 4-fold greater in those with diabetes who were in the highest compared to the lowest quartile of LDL-C; this effect was not seen in those without diabetes (Parsons, Bergstrom et al. 2008). A positive diagnosis of PrC in African-American (AA) men (n = 521), but not non-AA men (n = 451), undergoing biopsy was >3-fold higher for those in the highest quartile of LDL-C compared to the lowest (Moses, Abd et al. 2009). In a case-control study in 1294 Italian men <75 years of age with incident PrC compared to 1451 men hospitalized with acute, non-neoplastic conditions, the odd ratio (OR) for prostate cancer was 1.54 (95% CI, 1.26-1.89) for those with hypercholesterolemia (Pelucchi, Serraino et al. 2011). A post hoc analysis of the REDUCE study (which evaluated the anti-testosterone dutasteride in men with high prostate specific antigen (PSA) values but no PrC) examined the association of coronary artery disease (CAD) with PrC risk (Thomas, Gerber et al. 2012). In 6729 men who underwent at least one biopsy, those with CAD had an increased risk of PrC diagnosis (OR = 1.35, 95% CI, 1.08–1.67), suggesting common risk factors.

The benefit of statins in PrC prevention or treatment is still under evaluation, but observational studies have demonstrated reduced risk of PrC in statin users (reviewed, (Solomon and Freeman 2008; Marcella, David et al. 2011)). Statin use was recently shown to reduce the risk if death from PrC in a case-control study; cases were residents of New Jersey, USA ages 55 to 79 years who died from PrC between 1997 and 2000 (n = 380) and controls from the population were matched by 5-year age group and race. The unadjusted OR for death from PrC was 0.49 (95% CI, 0.34-0.70) for any exposure to statins and decreased to 0.37 ($P < .0001$) after multivariate adjustment (Marcella, David et al. 2011). Users of high-potency statins had about 2.5 times more protection compared with users of low-potency statins; the authors suggest that this points to cholesterol-lowering as the mechanism of protection. A positive association between LDL-C and PSA was demonstrated in a longitudinal study of 1214 American veterans undergoing statin treatment between 1990 and 2006 (Hamilton, Goldberg et al. 2008). After a relatively short period of statin use (< 1 year), there was a near-linear relationship between changes in LDL-C and changes in PSA values. After adjustment for multiple factors, for every 10% change in LDL-C, PSA changed by 1.64% (95% CI, 0.64% to 2.65%, $P = .001$). This relationship held over increases or decreases in the values, although the mean and median changes in LDL and PSA were -26% and -4.1%, respectively (Hamilton, Goldberg et al. 2008). A subsequent study showed that statin use dose-dependently lowered the risk of a PSA recurrence in men who underwent a radical prostectomy (n = 1319) (30% lower risk of PSA recurrence (HR = 0.70, 95% CI, 0.50-0.97) (Hamilton, Banez et al. 2010). Median follow-up time was 24 months for statin users (n = 236, 18%), 36 months for non-users.

Breast cancer. Epidemiological studies showing a higher incidence of breast cancer (BrC) in Westernized countries led to a focus on the role of dietary fat in BrC risk (Kelsey 1993). Although dietary fat may affect circulating cholesterol levels, the specific contribution of plasma lipoproteins to BrC has received less attention. In addition, the relationship between circulating cholesterol and BrC risk may be complicated by the fact that, as for testosterone, cholesterol is a biosynthetic estrogen precursor and structurally similar to estrogen. Estrogen lowers plasma LDL by increasing the expression of the LDLR (Kovanen, Brown et al. 1979; Hulley, Grady et al. 1998), but stimulates breast tumor growth through over-expression of estrogen receptor alpha (herein referred to as ER). Obesity and menopausal status can affect circulating lipids, estrogen levels, and BrC risk.

The Nurses' Health Study of >70,000 female, married, American nurses used self-reported serum cholesterol levels to analyze the association of blood cholesterol with risk of invasive BrC during up to 12 years of follow-up (Eliassen, Colditz et al. 2005). In that study, BrC incidence was not affected by cholesterol levels or use of statins or other lipid-lowering drugs. In a 10-year follow-up of postmenopausal Korean women (n = 170,374), a positive trend for quartiles of baseline fasting serum total-C and BrC incidence was found (HR = 1.31, 95% CI, 1.06-1.61); however, after adjustment for BMI the trend was no longer significant (Ha, Sung et al. 2009). In contrast, 157 of 5865 peri/postmenopausal Swedish women in the Malmö Preventive Project developed BrC over a mean of 6.6 years; relative risk was increased by quartiles of baseline fasting total-C (P for trend, 0.05) (Manjer, Kaaks et al. 2001). This effect was not seen among the 112/3873 premenopausal women who developed BrC over a mean of 9.6 yrs. BMI was not a factor in the risk of BrC in either group.

Because BrC has multiple types with distinct and recognizable patterns of gene expression, different treatments and prognoses, it may be more useful to examine BrC types separately

(Hu, Fan et al. 2006). Expression of the ER is an important discriminating factor among BrC types, with ER- BrC having fewer treatment options and a worse prognosis. A number of studies have shown differences in cholesterol metabolism between ER+ and ER- BrC. LDLR and ER content were determined (by ligand binding) in tumors from 72 Swedish patients who had undergone mastectomy (Rudling, Stahle et al. 1986). Interesting, LDLR content was negatively, while ER content was positively correlated with survival in months. LDLR content strongly and independently predicted a worse prognosis in these patients (Rudling, Stahle et al. 1986). This finding is consistent with more recent data on tumor gene expression, where LDLR mRNA expression was generally higher in ER- as compared to ER+ human breast tumors in multiple studies ($P < 0.05$, oncomine.org).

Circulating cholesterol may affect severity, recurrence, or outcome of BrC. In a prospective study of Canadian women diagnosed with early stage BrC (n = 520) and followed for a median period of 8.7 years, a trend toward higher risk of recurrence was seen in women with a higher fasting baseline total-C or LDL-C (Bahl, Ennis et al. 2005). Unfortunately, women with preexisting hyperlipidemia were excluded from the study, leaving a population with a smaller range of cholesterol levels in the evaluation. In 24,329 Norwegian women, a higher baseline non-fasting total-C level was not associated with BrC incidence (Vatten and Foss 1990), but those in the highest quartile did have an increased the risk of death from BrC (HR = 2.0, 95% CI, 1.1 – 3.7) (Vatten, Foss et al. 1991). In the Women's Intervention Nutrition Study (WINS), women with BrC counseled for a low-fat diet (20% of calories) and followed for a median period of 5 years had a 24% lower risk of recurrence (n = 96/975, HR = 0.76, 95% CI, 0.60 to 0.98) as compared to the control group (n = 181/1462); interestingly, the effect was even stronger in those whose tumor was ER- (n = 28/205, HR = 0.58, 95% CI, 0.37 to 0.91) as compared to those whose tumor was ER+ (n = 59/273) (Chlebowski, Blackburn et al. 2006). Although neither total-C nor LDL-C were reported, serum fatty acid analysis showed a reduction in saturated fats in the diet group, and saturated fats are known to increase circulating cholesterol levels (Blackburn and Wang 2007).

A number of clinical trials are underway to evaluate statins for the prevention or treatment of breast cancer. Large scale prospective studies on the association of statin use with risk of breast cancer have had mixed results (Cauley, McTiernan et al. 2006; Jacobs, Newton et al. 2011), but beneficial effects of statins on disease recurrence have been documented. In a prospective cohort study of all female residents in Denmark diagnosed with stage I-III invasive BrC between 1996 and 2003 (n = 18,769), users of simvastatin (a lipophilic statin) had a 10% lower risk of recurrence (95% CI, -11% to -8%) as compared with nonusers of statins (Ahern, Pedersen et al. 2011). No reduced risk was observed in users of hydrophilic statins. In 703 American women treated for stage II/III breast cancer between 1999 and 2005 and followed until 2008, users of statins (n = 156) had a reduced risk of recurrence in multivariate analysis (HR = 0.40, 95% CI, 0.24–0.67) (Chae, Valsecchi et al. 2011). No effect was seen on overall survival. Interestingly, a retrospective analysis of BrC patients in the Kaiser Permanente Cancer Registry in California (n = 2141) found that those who had used statins for one year or more had fewer aggressive ER-/PR- tumors and were more likely to have low grade and less invasive tumors (Kumar, Benz et al. 2008).

In a small study of women with newly diagnosed BrC (chemotherapy and radiotherapy naïve, n = 17) who were postmenopausal and normal weight, it was found that oxidized LDL (oxLDL) (P < 0.001), total-C (P = 0.001) and LDL-C (P = 0.001) were higher compared to a matched control group (n = 30) (Delimaris, Faviou et al. 2007). While LDL-C may contribute to cancer risk or prognosis, as in cardiovascular disease oxLDL may also play a role. OxLDL is present as a small percentage of total LDL in normal individuals, but the percentage of oxLDL may increase in pathological states (Holvoet, Lee et al. 2008; Mello, da Silva et al. 2011). An oxLDL receptor (OLR1) and was recently identified experimentally as part of gene signature responsible for transformation, tumor growth, and proliferation in multiple cancer cell lines (Hirsch, Iliopoulos et al. 2010). There is evidence that oxLDL is higher in hypercholesterolemic subjects, and that lowering total LDL with statins will result in lower oxLDL (Stojakovic, Claudel et al. 2010; Tavridou, Efthimiadis et al. 2010).

Ovarian cancer. Ovarian cancer (OvC) has a much lower incidence than BrC, but is more deadly as most tumors are highly advanced at diagnosis. OvC is not stimulated by estrogen, but there is some evidence that circulating cholesterol affects outcomes. In a prospective study of 132 American women with stage III or IV OvC, serum banked at the time of diagnostic surgery was analyzed for total-C, HDL-C, and TAG (LDL was calculated; statin users were excluded) (Li, Elmore et al. 2010). Disease-specific survival was longer in patients with normal LDL as compared to those with elevated LDL-C (59 and 51 months, respectively, P = 0.04). In another study at the same site, statin use was found to be an independent positive prognostic factor in 126 women with stage III/IV OvC, 17 of whom were taking statins at the time of initial surgery (Elmore, Ioffe et al. 2008). Mean progression-free survival, as well as overall survival, was longer for statin users (24 months compared to 16 months, P = 0.007) as compared to statin non-users (62 months compared to 46 months, P = 0.04). Serum was not available to determine actual levels of lipoproteins. In a small study, women with OvC (n = 15) compared to a matched control group (n = 30) had higher oxLDL (P = 0.006) and there was a trend toward higher LDL-C (P = 0.076) (Delimaris, Faviou et al. 2007). The women had not yet received any chemotherapy or radiotherapy at the time of blood collection.

Colorectal cancer. Colon cancer risk was associated with baseline total-C in the Korean NHIC data (Kitahara, Berrington de Gonzalez et al. 2011). Other studies have had mixed results. In the European Prospective Investigation into Cancer and Nutrition, 1238 incident cases of colorectal cancer (CRC) and matched controls were analyzed for an association of CRC risk with serum lipoproteins (van Duijnhoven, Bueno-De-Mesquita et al. 2011). No significant trend for quintiles of total-C or LDL-C with CRC incidence was detected; a negative trend for HDL-C with colon cancer was seen, even when excluding the first two years of follow-up. No correction for the use of statins, aspirin or other medications was possible in this study. In the Japan Collaborative Cohort Study for Evaluation of Cancer Risk, the association of oxLDL and autoantibodies to oxLDL (oLAB) with the incidence of CRC was examined (Suzuki, Ito et al. 2004). A positive trend was found for oxLDL and CRC, even after multiple adjustments (P = 0.038, n = 119 cases, 316 controls); the trend for oLAB was not significant. The adjusted OR for the highest compared to the lowest quartile of oxLDL was 3.10, 95% CI, 1.04-9.23. Although total-C was not different between cases and controls, oxLDL was strongly associated with total-C ($P < 0.001$, $n = 304$).

Plasma cholesterol may affect the progression of colon cancer to a more aggressive disease. The fasting lipid profiles of Italian men and women with metastatic CRC ($n = 22$) had higher synchronous total cholesterol, LDL-cholesterol and LDL/HDL ratios compared to those without metastases ($n = 62$) ($P = 0.03$, 0.01, and 0.002, respectively) (Notarnicola, Altomare et al. 2005). These results were independent of BMI. The authors hypothesized that LDL is beneficial for the proliferation and invasion steps of tumor progression. The effect of statin use on CRC incidence is unsettled due to mixed results from several retrospective analyses (Poynter, Gruber et al. 2005; Flick, Habel et al. 2009; Singh, Mahmud et al. 2009). There is hopeful data that statins may lower the recurrence rate of CRC , and a large-scale clinical trial is currently examining the potential of statin therapy to reduce the relapse rate in colon cancer in patients who have had surgery for early stage colon cancer (Hede 2011).

Other cancers. There is little consistent evidence to date from large prospective studies for the positive association of total-C or LDL-C with the incidence of other cancers. However, retrospective case control and observational studies showing a reduced risk of cancer in statin users are suggestive that lowering LDL-C may be an effective preventative strategy for a wider range of cancer types. For example, renal clear cell carcinoma (the most prevalent renal cell carcinoma) is known to accumulate large amounts of CE (Gebhard, Clayman et al. 1987), and a large case control study in American veterans ($n = 1446$ cases) found a 48% reduction in risk for this cancer in statin users (Khurana, Caldito et al. 2008). In the same population, a 55% reduction in the incidence of lung cancer in statin users compared to nonusers was found ($n = 7280$ cases) (Khurana, Bejjanki et al. 2007).

The evidence cited in this section suggests that higher circulating cholesterol can have the strongest effects on more advanced tumors. The question of whether more advanced or aggressive tumors accumulate more cholesterol as compared to early stage tumors *in vivo* has not been specifically addressed, although there is some evidence to suggest that this is the case (Tosi and Tugnoli 2005). Experimental data in the next section provide more support for the association of exogenous cholesterol with more aggressive cancer, as well as insight into how and why cancer cells accumulate cholesterol against normal homeostatic mechanisms.

3. Experimental and mechanistic evidence for role of LDL in cancer

Cholesterol homeostasis. If cholesterol homeostasis is altered in cancer cells to meet a greater demand for cholesterol, an understanding of the mechanisms involved will open up new targets against cancer. In normal cells, free cholesterol in cells is closely regulated to maintain adequate membrane cholesterol but prevent free cholesterol toxicity. Excess cholesterol is stored in the form of neutral cholesteryl esters (CE) that are available to the cell through the CE cycle (Brown, Ho et al. 1980), or is effluxed to circulating HDL for transport back to the liver (Fielding and Fielding 2001). In cholesterol-accumulating tumors, there is more CE storage and less efflux of cholesterol to HDL. Is this cholesterol newly synthesized

Study (Country)	Years of follow-up	n (n for cases)	Sex	Type of cancer	Association with risk of cancer for: Total-C	LDL-C	HDL-C	Reference
National Health Insurance Corp. enrollees (South Korea)	Up to 14	1,189,719 (M:53,944 F: 24,475)	M,F	All	Positive for PrC (M), BrC (F), CRC (M,F); negative for stomach, liver (M,F), lung (M)	Not measured	Not measured	{Kitahara, 2011}
Health Professionals Follow-Up (USA)	Up to 7	18,018 (698)	M	PrC	Positive for high-grade PrC	Not measured	Not measured	{Platz, 2008}
Prostate Cancer Prevention Trial (USA)	Up to 7	5,586 (1,251)	M	PrC	Positive for high-grade PrC			{Platz, 2009}
CLUE II (USA)	Mean of 11.9	6,816 (438)	M	PrC	Positive for high-grade PrC	Not measured	Not measured	{Mondul, 2010}
Alpha-Tocopherol, Beta-Carotene Cancer Prevention (smokers, Finland)	>10	29,093 (2,041)	M	PrC	Positive for aggressive and advanced PrC	Not measured	Negative trend	{Mondul, 2011}
Midspan (Scotland, UK)	Up to 37	12,926 (650)	M	PrC	Positive for high-grade PrC	Not measured	Not measured	{Shafique, 2012}
Nijmegen Biomedical (Netherlands)	Mean of 6.6	2,118 (43)	M	PrC	Positive for total and aggressive PrC	Positive for total and aggressive PrC	Positive for non-aggressive PrC	{Kok, 2011}
HUNT 2 (Norway)	Mean of 9.3	29,364 (687)	M	PrC	None	Not measured	None	{Martin, 2009}
Apolipoprotein MOrtality RISk (Sweden)	Mean of 7.0 - 8.3	200,660 (5,112)	M	PrC	None			{Van Hemelrijck, 2011}
Nurses' Health (self-reported serum chol-esterol) (USA)	6 - 12	79,994 (3177)	F	BrC	None	Not measured	Not measured	{Eliassen, 2005}
Postmenopausal public servants (South Korea)	Up to 10	170,374 (714)	F	BrC	Positive trend	Not measured	Not measured	{Ha, 2009}
Malmö Preventive Project (Sweden)	Up to 20	9,738 (269)	F	BrC	Positive for postmenopausal; none for premenopausal	Not measured	Not measured	{Manjer, 2001}
National Health Screening Service (Norway)	11 - 14	24,329 (242)	F	BrC	Negative (pre-menopausal); none (post-menopausal)	Not measured	Not measured	{Vatten, 1990}
EPIC and Nutrition (nested case-control)	Mean of 3.8	521,448 (1238)	M,F	CRC	None	Not measured	Positive for colon cancer	{van Duijn-hoven, 2011}

Table 1. Large, prospective studies with a baseline total cholesterol measurement and long-term follow-up for cancer incidence. M, male; F, female; PrC, prostate cancer; BrC, breast cancer; CRC, colorectal cancer.

or obtained from LDL, and what determines this? Normal cells obtain cholesterol primarily through endocytosis of circulating LDL through the LDLR, but have the capacity for endogenous synthesis via the mevalonate pathway; both mechanisms are tightly controlled for cholesterol homeostasis (Goldstein, DeBose-Boyd et al. 2006). The expressions of both LDLR and HMGCR are regulated by the transcription factors sterol response element binding proteins (SREBP1/2), whose processing and maturation proceed in response to decreased intracellular cholesterol (Brown and Goldstein 1997). The observed accumulation of CE in some tumors, the positive association of total-C with the risk of some types of cancer, and the demand for cholesterol for membrane building in growing cells, all suggest that the expression of these proteins and other components of the cholesterol homeostatic response system are altered in cancer.

Cholesterol biosynthesis in cancer. In order to obtain sufficient cholesterol, proliferating cells may accelerate the rate of cholesterol biosynthesis. Oncogenes that transform cells and dysregulate growth activate anabolic and biosynthetic pathways leading to *de novo* cholesterol and fatty acid synthesis. This is accomplished by a greatly increased flux of glucose into cells and through the glycolytic pathway to produce energy, and transport of TCA cycle citrate from the mitochondria to the cytosol for lipid biosynthesis (Vander Heiden, Cantley et al. 2009). The cytosolic enzyme ATP citrate lyase converts citrate to acetyl-CoA, the basic building block for both fatty acids and cholesterol. Growth factor activation of tyrosine kinase receptors and downstream PI3K/AKT and MAP-kinase signaling pathways increase expression and activation of the SREBPs (Kotzka, Muller-Wieland et al. 2000; Porstmann, Griffiths et al. 2005; Krycer, Sharpe et al. 2010), which control many lipid biosynthetic enzymes. Interesting, it was recently demonstrated that a mutated form of the cell cycle regulator p53, common in many tumors, bound to the promoter regions of the SREBPs and increased the expression of mevalonate pathway genes in BrC cells (Freed-Pastor, Mizuno et al. 2012).

A high enough rate of *de novo* biosynthesis may not always be possible; for example in solid tumors, expansion and insufficient vascularization may limit the delivery of glucose and oxygen. If oxygen is limited, activation of the hypoxia inducible factor 1 (HIF1) pathway can increase survival but divert pyruvate to lactate, reducing production of citrate (Gordan, Thompson et al. 2007). If glucose is limited, reducing ATP production, the AMP activated protein kinase (AMPK) pathway can inactivate key biosynthetic enzymes by phosphorylation (Shackelford and Shaw 2009). If biosynthesis becomes constrained, cells would have an advantage by being able to obtain lipids exogenously from circulating lipoproteins.

Cholesterol uptake in cancer. Uptake of cholesterol from LDL is primarily through the LDLR, although several scavenger receptors may also contribute. Over-expression of LDLR without feedback regulation by cholesterol has been observed in many types of cancer cells (Chen, Li et al. 1988; Hirakawa, Maruyama et al. 1991; Chen and Hughes-Fulford 2001; Antalis, Uchida et al. 2011). Although the role of SREBPs in feedback regulation of LDLR expression is well understood (Goldstein, DeBose-Boyd et al. 2006), there is evidence that cell signaling pathways also contribute to LDLR up-regulation in cancer. In BrC cells, LDLR mRNA expression was 3-5-fold higher in ER- as compared to ER+ cell lines; PKC activation was strongly associated with increased LDLR expression in ER+ BrC cells, and to a lesser extent, even in ER- cells (Stranzl, Schmidt et al. 1997). Activation of the p42/44 (MAPK) cascade was sufficient to induce LDLR transcription in human hepatoma HepG2 cells expressing oncogenic Raf-1 kinase (Kapoor, Atkins et al. 2002). In glioblastoma cells, chronic activation of the EGF receptor tyrosine kinase, or other mechanisms which ultimately activated the PI3K/AKT pathway, led to increased expression of SREBP1 and the LDLR and to LDL-responsive proliferation (Guo, Reinitz et al. 2011).

Increased dietary cholesterol has been shown to promote tumorigenesis in animal models. A Western-type high cholesterol diet compared to a chow diet increased tumor incidence and metastasis in a mouse model of PrC (Llaverias, Danilo et al. 2010). The same group, using similar diets, showed an increase in tumor formation and more aggressive tumors in a mouse model of BrC (Llaverias, Danilo et al. 2011). In both studies, plasma total-C was reduced following tumor development, suggesting utilization of circulating cholesterol by the tumor and similarity to what is observed in people with cancer.

Role of cholesterol esterification. Whether tumor cells obtain the needed cholesterol endogenously or exogenously, it would be imperative to have a way to manage the increased flux of cholesterol so as to meet the dual goals of ensuring a ready supply and avoiding toxicity. Cholesterol toxicity is prevented by effluxing the excess free cholesterol to an extracellular acceptor or converting free cholesterol to non-toxic esters of fatty acids. The observed low HDL-C in cancer patients, combined with the observed increased cholesterol content in tumors suggest that efflux mechanisms are reduced and esterification is increased. Synthesis and storage of CE in lipid droplets not only reduces toxicity but provides an accessible depot of cholesterol for future cell needs.

The enzyme responsible for cholesterol esterification is acyl-CoA:cholesterol acyltransferase 1 (ACAT1/SOAT1), a constitutive resident of the endoplasmic reticulum. ACAT1 esterifies cholesterol obtained from LDL and also from endogenous synthesis (Chang, Li et al. 2009). ACAT1 is frequently found to be over-expressed in cancer vs. normal tissues in human tumor gene expression analyses, including cancers of brain, breast, cervix, esophagus, head and neck, kidney, and testis ($P < 0.05$, oncomine.org). Over-expression of ACAT1 has been specifically associated with cholesterol accumulation in renal clear cell carcinoma, a tumor type characterized by 35-fold more CE as compared to normal kidney (Gebhard, Clayman et al. 1987).

ACAT activity has been associated with proliferation in cancer cells. The CE content of lymphocytes from patients with acute or chronic lymphocytic leukemia ($n = 30$) was 6-fold higher as compared to lymphocytes from healthy age-matched controls ($n = 15$), and plasma HDL was >40% reduced in the leukemia patients compared to the controls (Mulas, Abete et al. 2011). Phytohemaglutinin (PHA)-stimulated proliferation of the isolated leukemic cells was positively correlated to esterification of oleate to cholesterol, and inhibition of ACAT greatly reduced PHA-induced proliferation (Mulas, Abete et al. 2011). Cholesterol esterification and ACAT1 expression were also studied in leukemia cell lines. Cells with a greater ability to esterify cholesterol and with lower cholesterol efflux (CEM) had a higher rate of proliferation as compared to cells with a greater ability to synthesize cholesterol *de novo* (MOLT4) (Dessi, Batetta et al. 1997). Further work demonstrated that the faster-growing CEM cells expressed more ACAT1 and less HMGCR mRNA as compared to the slower-growing MOLT4 cells (Batetta, Pani et al. 1999).

In BrC, we showed that more aggressive basal-like ER- BrC cells had more lipid droplets and a much higher ratio of CE to TAG in stored neutral lipids as compared to less aggressive ER+ BrC cells; this was associated with higher expression of ACAT1 (Antalis, Arnold et al. 2010). The cell line differences were mirrored in gene expression analyses of human breast tumors, where higher expression of ACAT1/SOAT1 is characteristic of basal-like ER- tumors (Antalis, Arnold et al. 2010). We further showed that ER- cells took up more LDL as compared to ER+ cells, and that LDL dose-responsively increased proliferation only of ER- cells and in an ACAT-sensitive manner. In a follow-up study, we examined the effect of lipoprotein deprivation on chemotactic migration of the highly motile basal-like ER- cell line MDA-MB-231. We showed that lipid droplets were depleted and migration was reduced 85% when cells were grown in medium without lipoproteins, and that adding back LDL or fatty acids restored migration in an ACAT-sensitive manner (Antalis, Uchida et al. 2011). In addition, LDLR expression in these cells was not affected by exogenous LDL but was reduced 75% in the presence of an ACAT inhibitor, suggesting that high ACAT1 expression permitted continued high expression of the LDLR.

What mediates the over-expression of ACAT1 in cancer is not completely understood. Although ACAT1 is a critical component of intracellular cholesterol homeostasis, its expression is not known to be regulated by the SREBPs (Goldstein, DeBose-Boyd et al. 2006). In monocytes and macrophages, ACAT1 expression was up-regulated by interferon γ and all-*trans*-retinoic acid via STAT1 (Yang, Duan et al. 2001) and by dexamethasone via a glucocorticoid response element in its promoter (Yang, Yang et al. 2004). ACAT1 has also been shown to have an NFκB binding element in its proximal promoter and to be up-regulated in response to TNFα signaling through NFκB (Lei, Xiong et al. 2009). Cholesterol acts as an allosteric activator of ACAT1 activity (Liu, Chang et al. 2005).

The LXR pathway. The transcription factor LXR is a major regulator of fatty acid and cholesterol metabolism in cells. When cellular free cholesterol levels are high, some cholesterol is oxidized to form oxysterols, which act as endogenous ligands for LXR; thus LXRs are considered "cholesterol sensors"(Tontonoz 2011). LXR has an absolute requirement for RXRα as a dimerization partner. RXRα expression is highly regulated by both transcription and protein degradation (Boudjelal, Wang et al. 2000; Lefebvre, Benomar et al. 2010). RXRα availability is also affected by competition with its other binding partners, including PPAR, RAR, VDR, TR and FXR. LXR/RXRα is a permissive heterodimer, being stimulated by agonists of either partner (Tontonoz 2011).

LXR signaling is known to have dual roles: up-regulation of genes of fatty acid biosynthesis (including fatty acid synthase and stearoyl-CoA desaturase 1/2) and repression of NFκB controlled inflammatory genes (including IL-6, COX-2, and nitric oxide synthase) (Joseph, Castrillo et al. 2003). In addition, LXR/RXRα controls the transcription of key genes in cholesterol homeostasis: MYLIP/IDOL, the E3-ligase that ubiquitinates the LDLR leading to its degradation, ABCA1 and ABCG1, transporters involved in cholesterol efflux to APOA1 and HDL, and others (Tontonoz 2011). The demonstrated control of ACAT1 by NFκB suggests that its transcription could be antagonized by LXR activity. LXR signaling may have the ability to mediate the balance between lipid biosynthesis/efflux mechanisms and uptake/storage mechanisms. **Fig. 2** and **Fig. 3** illustrate how key factors in cellular cholesterol homeostasis may be affected by the activity of LXR and its target genes.

The uptake of exogenous LDL through LDLR leads to increased cellular free cholesterol, reduced maturation of SREBPs and reduced transcription of LDLR. When LXR/RXRα is active (**Fig. 2**), LDLR protein is degraded by MYLIP and cholesterol efflux mechanisms are increased (Beltowski 2008). ACAT1 transcription may be reduced by the inhibitory effect of LXR/RXRα on NFκB transactivation activity, blocking cholesterol accumulation. Similarly ApoA1, the apolipoprotein acceptor for cholesterol efflux, which under some conditions is repressed by NFκB, could be increased (Mogilenko, Dizhe et al. 2009). As a result, normal cellular cholesterol homeostasis is enforced.

When LXR/RXRα is less active (**Fig. 3**), and under the influence of cytokines, a different pattern of gene expression predominates. Cholesterol efflux is reduced and thus free cholesterol is maintained at a high enough level in bilayer membranes that maturation of SREBPs is not triggered. More free cholesterol is esterified and stored in lipid droplets, due to a possible induction of ACAT1. LDLR protein degradation is reduced, allowing the cell to maintain high LDLR expression and unrestrained uptake of LDL. In this way, cellular cholesterol homeostasis is perturbed in the direction of LDL uptake and cholesterol accumulation.

The pathways described in **Figs. 2 and 3** are hypothesized to explain the observed cholesterol accumulation in some tumors and cancer cell lines. LDLR is placed at the center of the process of LDL uptake and accumulation, with LXR pathway inactivation being the key factor allowing cholesterol accumulation. No doubt the situation is more complicated than shown, as it does not account for scavenger receptor participation. However, the central role of LXR makes it a potential target in cancer.

LXR agonists have been tested in experimental models of cancer. In glioblastoma cells over-expressing the EGFR, EGF stimulated PI3K/Akt-driven up-regulation of SREBP1 and LDLR (Guo, Reinitz et al. 2011). An LXR agonist induced MYLIP/IDOL-mediated degradation of LDLR, ABCA1-mediated cholesterol efflux, and cell death both *in vitro* and in an animal model. In OvC cells, oxLDL stimulated proliferation and secretion of the cytokine cardiotrophin 1 (Scoles, Xu et al. 2010). An LXR agonist blocked both the cytokine secretion and the proliferation induced by oxLDL; the authors attribute the response to increased cholesterol efflux and decreased inflammatory effects of the LXR agonist. In an athymic model of PrC, progression of androgen-dependent tumors to androgen-independent tumors after castration was accompanied by decreases in expression of LXR target genes in the tumor, and treatment with an LXR agonist delayed the progression for about 4 weeks (Chuu, Hiipakka et al. 2006).

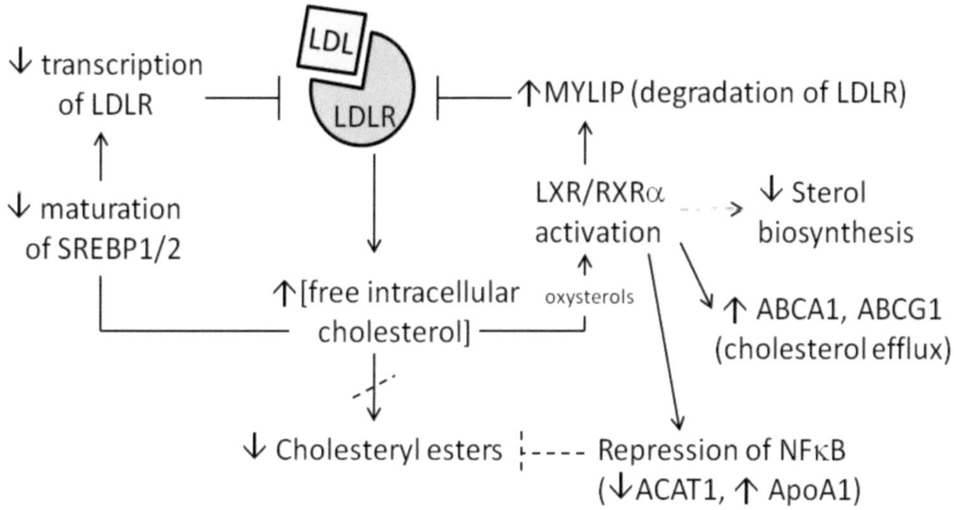

Figure 2. LXR transcriptional targets control intracellular cholesterol concentrations. Dotted line indicates pathways not proven.

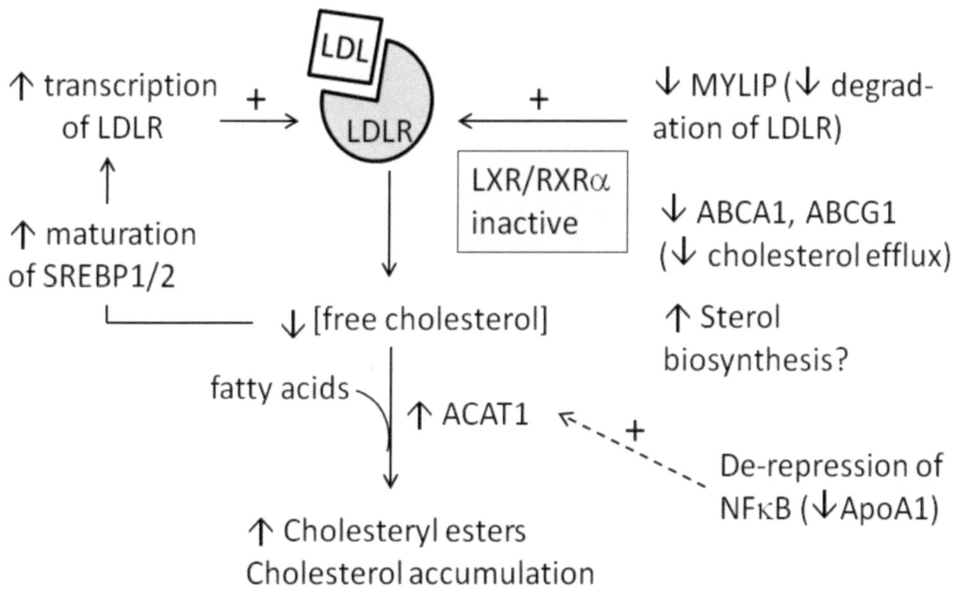

Figure 3. Reduced LXR signaling allows increased LDL uptake and intracellular cholesterol accumulation. Dotted line indicates pathways not proven.

Cholesterol and tumorigenesis. The question remains as to the role that CEs may play in the survival, proliferation and metastasis of cancer cells. We and others have proposed that accumulation of CE spares energy needed for *de novo* sterol synthesis, allowing greater proliferation and migration and perhaps a quicker return to growth after a period of stasis (Batetta, Pani et al. 1999; Antalis, Arnold et al. 2010; Antalis, Uchida et al. 2011). The process of cholesterol esterification was linked to proliferation in multiple studies in different cancer cell lines (Batetta, Pani et al. 1999; Peiretti, Dessi et al. 2007; Paillasse, de Medina et al. 2009; Antalis, Arnold et al. 2010; Mulas, Abete et al. 2011), implying a complex network of signaling pathways and gene expression that ties cholesterol accretion to tumorigenesis. However, the exact role of CE in tumorigenesis remains to be determined.

PrC is a unique case considering the slow growth characteristics of this malignancy. The lipid raft concept has been proposed to account for the tumorigenic effects of cholesterol (Freeman, Cinar et al. 2007), and a higher level of cholesterol in PrC cells has been linked to membrane lipid raft-induced oncogenic cell signaling (Hager, Solomon et al. 2006). A connection between LXR signaling and lipid raft-associated signaling was demonstrated in androgen-responsive LnCAP cells, where an LXR agonist down-regulated Akt signaling in a cholesterol- and lipid raft-dependent manner, resulting in apoptosis of cells and xenograft tumors (Pommier, Alves et al. 2010). In addition, a relationship between androgens and cholesterol metabolism was demonstrated in PrC cells. It was first noted that androgen stimulation caused a dramatic increase in lipid droplets in LNCap cells. The induced neutral lipids included both TAG (33-fold) and CE (7-fold increase), most of which originated from new lipid synthesis (Swinnen, Van Veldhoven et al. 1996). This was later found to be due to an up-regulation of the SREBPs and lipid biosynthetic genes (Nelson, Clegg et al. 2002). The androgen-independent PC-3 cells had a higher content of CE and but not higher ACAT1 activity or expression as compared to LNCap cells (Locke, Wasan et al. 2008). In both an androgen-independent cell line and a mouse xenograft model of PrC progression, changes in cholesterol metabolism and homeostasis were associated with initiation of tumoral androgen production and expression of the AR and PSA (Locke, Wasan et al. 2008; Leon, Locke et al. 2010). These data, along with the clinical data cited in **Section 2,** suggest that in PrC cholesterol accumulation may be important for androgen synthesis, which is closely involved with PrC progression even under castration therapy.

Another function of LDL and other lipoproteins is the provision of essential fatty acids. Mammalian cells are not able to make polyunsaturated fatty acids; the essential n-6 and n-3 fatty acids are derived from the diet and carried to cells by lipoproteins. Human glioma, one of the deadliest types of cancer, was found to contain up to 100-fold more CE compared to control tissue, and the fatty acid composition of the tumor CEs indicated an LDL origin (Nygren, von Holst et al. 1997). The n-6 fatty acid arachidonic acid is necessary for synthesis of second messengers such as the prostaglandin PGE_2, a tumor promoter (Wang and Dubois 2006). In androgen-independent PrC PC-3 cells, PGE_2 production increased >3-fold in response to LDL (Chen and Hughes-Fulford 2001). Thus the fatty acids esterified to cholesterol and other lipids may be important for the effect of LDL on cancer cells.

Finally, although lower plasma HDL-C in cancer patients may be due to reduced efflux of cholesterol to HDL from the tumor, there is evidence that some cancer cells can take up CE from circulating HDL, providing another explanation for low HDL. Recent investigations with the CEM-CCRF lymphoblastic cell line into the source of intracellular CE showed that

HDL-CE were taken up and stored without hydrolysis and re-esterification, while LDL-CE were hydrolyzed and re-esterified (Uda, Accossu et al. 2012). Although the mechanism was not clear, the data implied that HDL as well as LDL could be a source of CE for leukemic cells. A previous study in BrC cells showed that either HDL or LDL dose-dependently stimulated proliferation of ER- cell lines, but only HDL had the effect on ER+ cells lines (Rotheneder and Kostner 1989). In an animal model of PrC, a diet high in fat and cholesterol resulted in increased tumor incidence and increased tumor expression of scavenger receptor B1, the receptor responsible for selective uptake of HDL-C (the major form of circulating cholesterol in mice) by cells (Llaverias, Danilo et al. 2010). The question of whether HDL can supply cholesterol to tumor cells *in vivo* in humans remains open.

4. Conclusions and future directions

The heterogeneous nature of cancer and the changes that accompany tumor progression make it very difficult to draw overall conclusions about the effects of circulating cholesterol on cancer incidence or progression. However, large scale prospective studies have shown that higher plasma total-C and LDL-C can increase the risk for some cancers, with the hormone-related cancers in men and women being especially affected. Data also point to a more potent effect of exogenous cholesterol on more aggressive cancers. These conclusions are supported by data on the effect of statins, which have been shown to reduce both the risk and the progression of some cancers. As more clinical trial data emerges, we will have a clearer picture of the usefulness of cholesterol reduction and statins in cancer and what types of cancer respond to these therapies.

Individualized approaches are the future for cancer therapy. Gene and protein expressions may serve as biomarkers to identify tumors that are stimulated by LDL. The genes/proteins expected to be more expressed as a result of LXR/RXRα pathway activation, i.e. MYLIP and ABCA1, and those expected to be more expressed as a result of LXR/RXRα pathway inactivation, i.e. ACAT1/SOAT1 and LDLR, may be used to distinguish tumors that are cholesterol-accumulating. The cholesterol and CE content of tumor biopsies determined by chemical or enzymatic methods could also be used as biomarkers. Imaging methods such as magnetic resonance (Delikatny, Chawla et al. 2011) and coherent anti-Stokes Raman scattering (Le, Huff et al. 2009) have the potential to allow *in vivo* visualization of lipids in tumors. These kinds of data will help to substantiate and clarify the association of CE accumulation with types of cancer.

If it can be shown that a tumor has the markers of higher cholesterol uptake and accumulation, treatments to lower circulating lipids and affect intracellular cholesterol homeostasis are available. Existing drugs developed for prevention or treatment of cardiovascular disease or metabolic syndrome, such as statins and metformin (an AMPK activator), are being "repurposed" for the treatment of cancer. ACAT inhibitors that did not have the expected result of reducing atherosclerotic plaques in clinical trials may find a new use in cholesterol-accumulating cancers. A new ACAT1-specific inhibitor was effective in killing glioma cells in *in vitro* studies (Bemlih, Poirier et al. 2010). LXR pathway modulators that can increase cholesterol efflux and HDL-C levels without stimulating lipid biosynthesis in the liver, needed to treat cardiovascular disease and metabolic syndrome, may also be useful in cancer (Ratni, Blum-Kaelin et al. 2009). Dietary regimens targeting fat and cholesterol reduction in those with hyperlipidemia, with known benefits in preventing and treating heart disease, may be recommended to decrease the risk or recurrence of some types of cancer.

Author details

Caryl J. Antalis and Kimberly K. Buhman
Indiana University School of Medicine & Purdue University, USA

5. References

Ahern, T. P., L. Pedersen, et al. (2011). "Statin prescriptions and breast cancer recurrence risk: a Danish nationwide prospective cohort study." *J Natl Cancer Inst* 103(19): 1461-1468.

Ahn, J., U. Lim, et al. (2009). "Prediagnostic total and high-density lipoprotein cholesterol and risk of cancer." *Cancer Epidemiol Biomarkers Prev* 18(11): 2814-2821.

Antalis, C. J., T. Arnold, et al. (2010). "High ACAT1 expression in estrogen receptor negative basal-like breast cancer cells is associated with LDL-induced proliferation." *Breast Cancer Res Treat* 122(3): 661-670.

Antalis, C. J., A. Uchida, et al. (2011). "Migration of MDA-MB-231 breast cancer cells depends on the availability of exogenous lipids and cholesterol esterification." *Clin Exp Metastasis* 28(8): 733-741.

Bahl, M., M. Ennis, et al. (2005). "Serum lipids and outcome of early-stage breast cancer: results of a prospective cohort study." *Breast Cancer Res Treat* 94(2): 135-144.

Batetta, B., A. Pani, et al. (1999). "Correlation between cholesterol esterification, MDR1 gene expression and rate of cell proliferation in CEM and MOLT4 cell lines." *Cell Prolif* 32(1): 49-61.

Beltowski, J. (2008). "Liver X receptors (LXR) as therapeutic targets in dyslipidemia." *Cardiovasc Ther* 26(4): 297-316.

Bemlih, S., M. D. Poirier, et al. (2010). "Acyl-coenzyme A: cholesterol acyltransferase inhibitor Avasimibe affect survival and proliferation of glioma tumor cell lines." *Cancer Biol Ther* 9(12): 1025-1032.

Bjorge, T., A. Lukanova, et al. (2011). "Metabolic risk factors and ovarian cancer in the Metabolic Syndrome and Cancer project." *Int J Epidemiol* 40(6): 1667-1677.

Blackburn, G. L. and K. A. Wang (2007). "Dietary fat reduction and breast cancer outcome: results from the Women's Intervention Nutrition Study (WINS)." *Am J Clin Nutr* 86(3): s878-881.

Boudjelal, M., Z. Wang, et al. (2000). "Ubiquitin/proteasome pathway regulates levels of retinoic acid receptor gamma and retinoid X receptor alpha in human keratinocytes." *Cancer Res* 60(8): 2247-2252.

Bozza, P. T. and J. P. Viola (2010). "Lipid droplets in inflammation and cancer." *Prostaglandins Leukot Essent Fatty Acids* 82(4-6): 243-250.

Brown, M. S. and J. L. Goldstein (1997). "The SREBP pathway: regulation of cholesterol metabolism by proteolysis of a membrane-bound transcription factor." *Cell* 89(3): 331-340.

Brown, M. S., Y. K. Ho, et al. (1980). "The cholesteryl ester cycle in macrophage foam cells. Continual hydrolysis and re-esterification of cytoplasmic cholesteryl esters." *J Biol Chem* 255(19): 9344-9352.

Calle, E. E. and R. Kaaks (2004). "Overweight, obesity and cancer: epidemiological evidence and proposed mechanisms." *Nat Rev Cancer* 4(8): 579-591.

Cauley, J. A., A. McTiernan, et al. (2006). "Statin use and breast cancer: prospective results from the Women's Health Initiative." *J Natl Cancer Inst* 98(10): 700-707.

Chae, Y. K., M. E. Valsecchi, et al. (2011). "Reduced risk of breast cancer recurrence in patients using ACE inhibitors, ARBs, and/or statins." *Cancer Invest* 29(9): 585-593.

Chang, T. Y., B. L. Li, et al. (2009). "Acyl-coenzyme A:cholesterol acyltransferases." *Am J Physiol Endocrinol Metab* 297(1): E1-9.

Chen, J. K., L. Li, et al. (1988). "Altered low density lipoprotein receptor regulation is associated with cholesteryl ester accumulation in Simian virus 40 transformed rodent fibroblast cell lines." *In Vitro Cell Dev Biol* 24(4): 353-358.

Chen, Y. and M. Hughes-Fulford (2001). "Human prostate cancer cells lack feedback regulation of low-density lipoprotein receptor and its regulator, SREBP2." *Int J Cancer* 91(1): 41-45.

Chlebowski, R. T., G. L. Blackburn, et al. (2006). "Dietary fat reduction and breast cancer outcome: interim efficacy results from the Women's Intervention Nutrition Study." *J Natl Cancer Inst* 98(24): 1767-1776.

Chuu, C. P., R. A. Hiipakka, et al. (2006). "Inhibition of tumor growth and progression of LNCaP prostate cancer cells in athymic mice by androgen and liver X receptor agonist." *Cancer Res* 66(13): 6482-6486.

Cleary, M. P. and M. E. Grossmann (2009). "Minireview: Obesity and breast cancer: the estrogen connection." *Endocrinology* 150(6): 2537-2542.

Delikatny, E. J., S. Chawla, et al. (2011). "MR-visible lipids and the tumor microenvironment." *NMR Biomed* 24(6): 592-611.

Delimaris, I., E. Faviou, et al. (2007). "Oxidized LDL, serum oxidizability and serum lipid levels in patients with breast or ovarian cancer." *Clin Biochem* 40(15): 1129-1134.

Dessi, S., B. Batetta, et al. (1997). "Role of cholesterol synthesis and esterification in the growth of CEM and MOLT4 lymphoblastic cells." *Biochem J* 321 (Pt 3): 603-608.

Dessi, S., B. Batetta, et al. (1995). "Clinical remission is associated with restoration of normal high-density lipoprotein cholesterol levels in children with malignancies." *Clin Sci (Lond)* 89(5): 505-510.

Eliassen, A. H., G. A. Colditz, et al. (2005). "Serum lipids, lipid-lowering drugs, and the risk of breast cancer." *Arch Intern Med* 165(19): 2264-2271.

Elmore, R. G., Y. Ioffe, et al. (2008). "Impact of statin therapy on survival in epithelial ovarian cancer." *Gynecol Oncol* 111(1): 102-105.

Faulds, M. H. and K. Dahlman-Wright (2012). "Metabolic diseases and cancer risk." *Curr Opin Oncol* 24(1): 58-61.

Fielding, C. J. and P. E. Fielding (2001). "Cellular cholesterol efflux." *Biochim Biophys Acta* 1533(3): 175-189.

Flick, E. D., L. A. Habel, et al. (2009). "Statin use and risk of colorectal cancer in a cohort of middle-aged men in the US: a prospective cohort study." *Drugs* 69(11): 1445-1457.

Freed-Pastor, W. A., H. Mizuno, et al. (2012). "Mutant p53 disrupts mammary tissue architecture via the mevalonate pathway." *Cell* 148(1-2): 244-258.

Freeman, M. R., B. Cinar, et al. (2007). "Transit of hormonal and EGF receptor-dependent signals through cholesterol-rich membranes." *Steroids* 72(2): 210-217.

Freitas, I., P. Pontiggia, et al. (1990). "Histochemical probes for the detection of hypoxic tumour cells." *Anticancer Res* 10(3): 613-622.

Gazzerro, P., M. C. Proto, et al. (2012). "Pharmacological actions of statins: a critical appraisal in the management of cancer." *Pharmacol Rev* 64(1): 102-146.

Gebhard, R. L., R. V. Clayman, et al. (1987). "Abnormal cholesterol metabolism in renal clear cell carcinoma." *J Lipid Res* 28(10): 1177-1184.

Goldstein, J. L., R. A. DeBose-Boyd, et al. (2006). "Protein sensors for membrane sterols." *Cell* 124(1): 35-46.

Gordan, J. D., C. B. Thompson, et al. (2007). "HIF and c-Myc: sibling rivals for control of cancer cell metabolism and proliferation." *Cancer Cell* 12(2): 108-113.

Guo, D., F. Reinitz, et al. (2011). "An LXR agonist promotes GBM cell death through inhibition of an EGFR/AKT/SREBP-1/LDLR-dependent pathway." *Cancer Discov* 1(5): 442-456.

Ha, M., J. Sung, et al. (2009). "Serum total cholesterol and the risk of breast cancer in postmenopausal Korean women." *Cancer Causes Control* 20(7): 1055-1060.

Hager, M. H., K. R. Solomon, et al. (2006). "The role of cholesterol in prostate cancer." *Curr Opin Clin Nutr Metab Care* 9(4): 379-385.

Hamilton, R. J., L. L. Banez, et al. (2010). "Statin medication use and the risk of biochemical recurrence after radical prostatectomy: results from the Shared Equal Access Regional Cancer Hospital (SEARCH) Database." *Cancer* 116(14): 3389-3398.

Hamilton, R. J., K. C. Goldberg, et al. (2008). "The influence of statin medications on prostate-specific antigen levels." *J Natl Cancer Inst* 100(21): 1511-1518.

Hede, K. (2011). "Hints that statins reduce colon cancer risk finally being put to the test." *J Natl Cancer Inst* 103(5): 364-366.

Hirakawa, T., K. Maruyama, et al. (1991). "Massive accumulation of neutral lipids in cells conditionally transformed by an activated H-ras oncogene." *Oncogene* 6(2): 289-295.

Hirsch, H. A., D. Iliopoulos, et al. (2010). "A transcriptional signature and common gene networks link cancer with lipid metabolism and diverse human diseases." *Cancer Cell* 17(4): 348-361.

Holvoet, P., D. H. Lee, et al. (2008). "Association between circulating oxidized low-density lipoprotein and incidence of the metabolic syndrome." *JAMA* 299(19): 2287-2293.

Hu, Z., C. Fan, et al. (2006). "The molecular portraits of breast tumors are conserved across microarray platforms." *BMC Genomics* 7: 96.

Hulley, S., D. Grady, et al. (1998). "Randomized trial of estrogen plus progestin for secondary prevention of coronary heart disease in postmenopausal women. Heart and Estrogen/progestin Replacement Study (HERS) Research Group." *JAMA* 280(7): 605-613.

Jacobs, E. J., C. C. Newton, et al. (2011). "Long-term use of cholesterol-lowering drugs and cancer incidence in a large United States cohort." *Cancer Res* 71(5): 1763-1771.

Joseph, S. B., A. Castrillo, et al. (2003). "Reciprocal regulation of inflammation and lipid metabolism by liver X receptors." *Nat Med* 9(2): 213-219.

Kapoor, G. S., B. A. Atkins, et al. (2002). "Activation of Raf-1/MEK-1/2/p42/44(MAPK) cascade alone is sufficient to uncouple LDL receptor expression from cell growth." *Mol Cell Biochem* 236(1-2): 13-22.

Kelsey, J. L. (1993). "Breast cancer epidemiology: summary and future directions." *Epidemiol Rev* 15(1): 256-263.

Khurana, V., H. R. Bejjanki, et al. (2007). "Statins reduce the risk of lung cancer in humans: a large case-control study of US veterans." *Chest* 131(5): 1282-1288.

Khurana, V., G. Caldito, et al. (2008). "Statins might reduce risk of renal cell carcinoma in humans: case-control study of 500,000 veterans." *Urology* 71(1): 118-122.

Kitahara, C. M., A. Berrington de Gonzalez, et al. (2011). "Total cholesterol and cancer risk in a large prospective study in Korea." *J Clin Oncol* 29(12): 1592-1598.

Kok, D. E., J. G. van Roermund, et al. (2011). "Blood lipid levels and prostate cancer risk; a cohort study." *Prostate Cancer Prostatic Dis*.

Kotzka, J., D. Muller-Wieland, et al. (2000). "Sterol regulatory element binding proteins (SREBP)-1a and SREBP-2 are linked to the MAP-kinase cascade." *J Lipid Res* 41(1): 99-108.

Kovanen, P. T., M. S. Brown, et al. (1979). "Increased binding of low density lipoprotein to liver membranes from rats treated with 17 alpha-ethinyl estradiol." *J Biol Chem* 254(22): 11367-11373.

Krycer, J. R., L. J. Sharpe, et al. (2010). "The Akt-SREBP nexus: cell signaling meets lipid metabolism." *Trends Endocrinol Metab* 21(5): 268-276.

Kumar, A. S., C. C. Benz, et al. (2008). "Estrogen receptor-negative breast cancer is less likely to arise among lipophilic statin users." *Cancer Epidemiol Biomarkers Prev* 17(5): 1028-1033.

Kwiterovich, P. O., Jr. (2000). "The metabolic pathways of high-density lipoprotein, low-density lipoprotein, and triglycerides: a current review." *Am J Cardiol* 86(12A): 5L-10L.

Le, T. T., T. B. Huff, et al. (2009). "Coherent anti-Stokes Raman scattering imaging of lipids in cancer metastasis." *BMC Cancer* 9: 42.

Lefebvre, B., Y. Benomar, et al. (2010). "Proteasomal degradation of retinoid X receptor alpha reprograms transcriptional activity of PPARgamma in obese mice and humans." *J Clin Invest* 120(5): 1454-1468.

Lei, L., Y. Xiong, et al. (2009). "TNF-alpha stimulates the ACAT1 expression in differentiating monocytes to promote the CE-laden cell formation." *J Lipid Res* 50(6): 1057-1067.

Leon, C. G., J. A. Locke, et al. (2010). "Alterations in cholesterol regulation contribute to the production of intratumoral androgens during progression to castration-resistant prostate cancer in a mouse xenograft model." *Prostate* 70(4): 390-400.

Li, A. J., R. G. Elmore, et al. (2010). "Serum low-density lipoprotein levels correlate with survival in advanced stage epithelial ovarian cancers." *Gynecol Oncol* 116(1): 78-81.

Liu, J., C. C. Chang, et al. (2005). "Investigating the allosterism of acyl-CoA:cholesterol acyltransferase (ACAT) by using various sterols: in vitro and intact cell studies." *Biochem J* 391(Pt 2): 389-397.

Llaverias, G., C. Danilo, et al. (2011). "Role of cholesterol in the development and progression of breast cancer." *Am J Pathol* 178(1): 402-412.

Llaverias, G., C. Danilo, et al. (2010). "A Western-type diet accelerates tumor progression in an autochthonous mouse model of prostate cancer." *Am J Pathol* 177(6): 3180-3191.

Locke, J. A., K. M. Wasan, et al. (2008). "Androgen-mediated cholesterol metabolism in LNCaP and PC-3 cell lines is regulated through two different isoforms of acyl-coenzyme A:Cholesterol Acyltransferase (ACAT)." *Prostate* 68(1): 20-33.

Manjer, J., R. Kaaks, et al. (2001). "Risk of breast cancer in relation to anthropometry, blood pressure, blood lipids and glucose metabolism: a prospective study within the Malmo Preventive Project." *Eur J Cancer Prev* 10(1): 33-42.

Marcella, S. W., A. David, et al. (2011). "Statin use and fatal prostate cancer: A matched case-control study." *Cancer*.

Martin, R. M., L. Vatten, et al. (2009). "Components of the metabolic syndrome and risk of prostate cancer: the HUNT 2 cohort, Norway." *Cancer Causes Control* 20(7): 1181-1192.

Mello, A. P., I. T. da Silva, et al. (2011). "Electronegative low-density lipoprotein: origin and impact on health and disease." *Atherosclerosis* 215(2): 257-265.

Mogilenko, D. A., E. B. Dizhe, et al. (2009). "Role of the nuclear receptors HNF4 alpha, PPAR alpha, and LXRs in the TNF alpha-mediated inhibition of human apolipoprotein A-I gene expression in HepG2 cells." *Biochemistry* 48(50): 11950-11960.

Mondul, A. M., S. L. Clipp, et al. (2010). "Association between plasma total cholesterol concentration and incident prostate cancer in the CLUE II cohort." *Cancer Causes Control* 21(1): 61-68.

Mondul, A. M., S. J. Weinstein, et al. (2011). "Serum total and HDL cholesterol and risk of prostate cancer." *Cancer Causes Control* 22(11): 1545-1552.

Moses, K. A., T. T. Abd, et al. (2009). "Increased low density lipoprotein and increased likelihood of positive prostate biopsy in black americans." *J Urol* 182(5): 2219-2225.

Mostaghel, E. A., K. R. Solomon, et al. (2012). "Impact of circulating cholesterol levels on growth and intratumoral androgen concentration of prostate tumors." *PLoS One* 7(1): e30062.

Mulas, M. F., C. Abete, et al. (2011). "Cholesterol esters as growth regulators of lymphocytic leukaemia cells." *Cell Prolif* 44(4): 360-371.

Nelson, P. S., N. Clegg, et al. (2002). "The program of androgen-responsive genes in neoplastic prostate epithelium." *Proc Natl Acad Sci U S A* 99(18): 11890-11895.

Notarnicola, M., D. F. Altomare, et al. (2005). "Serum lipid profile in colorectal cancer patients with and without synchronous distant metastases." *Oncology* 68(4-6): 371-374.

Nygren, C., H. von Holst, et al. (1997). "Increased levels of cholesterol esters in glioma tissue and surrounding areas of human brain." *Br J Neurosurg* 11(3): 216-220.

Paillasse, M. R., P. de Medina, et al. (2009). "Signaling through cholesterol esterification: a new pathway for the cholecystokinin 2 receptor involved in cell growth and invasion." *J Lipid Res* 50(11): 2203-2211.

Parsons, J. K., J. Bergstrom, et al. (2008). "Lipids, lipoproteins and the risk of benign prostatic hyperplasia in community-dwelling men." *BJU Int* 101(3): 313-318.

Peiretti, E., S. Dessi, et al. (2007). "Modulation of cholesterol homeostasis by antiproliferative drugs in human pterygium fibroblasts." *Invest Ophthalmol Vis Sci* 48(8): 3450-3458.

Pelucchi, C., D. Serraino, et al. (2011). "The metabolic syndrome and risk of prostate cancer in Italy." *Ann Epidemiol* 21(11): 835-841.

Platz, E. A., S. K. Clinton, et al. (2008). "Association between plasma cholesterol and prostate cancer in the PSA era." *Int J Cancer* 123(7): 1693-1698.

Platz, E. A., C. Till, et al. (2009). "Men with low serum cholesterol have a lower risk of high-grade prostate cancer in the placebo arm of the prostate cancer prevention trial." *Cancer Epidemiol Biomarkers Prev* 18(11): 2807-2813.

Pommier, A. J., G. Alves, et al. (2010). "Liver X Receptor activation downregulates AKT survival signaling in lipid rafts and induces apoptosis of prostate cancer cells." *Oncogene* 29(18): 2712-2723.

Porstmann, T., B. Griffiths, et al. (2005). "PKB/Akt induces transcription of enzymes involved in cholesterol and fatty acid biosynthesis via activation of SREBP." *Oncogene* 24(43): 6465-6481.

Poynter, J. N., S. B. Gruber, et al. (2005). "Statins and the risk of colorectal cancer." *N Engl J Med* 352(21): 2184-2192.

Ratni, H., D. Blum-Kaelin, et al. (2009). "Discovery of tetrahydro-cyclopenta[b]indole as selective LXRs modulator." *Bioorg Med Chem Lett* 19(6): 1654-1657.

Rose, G. and M. J. Shipley (1980). "Plasma lipids and mortality: a source of error." *Lancet* 1(8167): 523-526.

Rotheneder, M. and G. M. Kostner (1989). "Effects of low- and high-density lipoproteins on the proliferation of human breast cancer cells in vitro: differences between hormone-dependent and hormone-independent cell lines." *Int J Cancer* 43(5): 875-879.

Rudling, M. J., L. Stahle, et al. (1986). "Content of low density lipoprotein receptors in breast cancer tissue related to survival of patients." *Br Med J (Clin Res Ed)* 292(6520): 580-582.

Scoles, D. R., X. Xu, et al. (2010). "Liver X receptor agonist inhibits proliferation of ovarian carcinoma cells stimulated by oxidized low density lipoprotein." *Gynecol Oncol* 116(1): 109-116.

Shackelford, D. B. and R. J. Shaw (2009). "The LKB1-AMPK pathway: metabolism and growth control in tumour suppression." *Nat Rev Cancer* 9(8): 563-575.

Shafique, K., P. McLoone, et al. (2012). "Cholesterol and the risk of grade-specific prostate cancer incidence: evidence from two large prospective cohort studies with up to 37 years' follow up." *BMC Cancer* 12: 25.

Sharp, S. J. and S. J. Pocock (1997). "Time trends in serum cholesterol before cancer death." *Epidemiology* 8(2): 132-136.

Singh, H., S. M. Mahmud, et al. (2009). "Long-term use of statins and risk of colorectal cancer: a population-based study." *Am J Gastroenterol* 104(12): 3015-3023.

Solomon, K. R. and M. R. Freeman (2008). "Do the cholesterol-lowering properties of statins affect cancer risk?" *Trends Endocrinol Metab* 19(4): 113-121.

Stojakovic, T., T. Claudel, et al. (2010). "Low-dose atorvastatin improves dyslipidemia and vascular function in patients with primary biliary cirrhosis after one year of treatment." *Atherosclerosis* 209(1): 178-183.

Stranzl, A., H. Schmidt, et al. (1997). "Low-density lipoprotein receptor mRNA in human breast cancer cells: influence by PKC modulators." *Breast Cancer Res Treat* 42(3): 195-205.

Suzuki, K., Y. Ito, et al. (2004). "Serum oxidized low-density lipoprotein levels and risk of colorectal cancer: a case-control study nested in the Japan Collaborative Cohort Study." *Cancer Epidemiol Biomarkers Prev* 13(11 Pt 1): 1781-1787.

Swinnen, J. V., P. P. Van Veldhoven, et al. (1996). "Androgens markedly stimulate the accumulation of neutral lipids in the human prostatic adenocarcinoma cell line LNCaP." *Endocrinology* 137(10): 4468-4474.

Taplin, M. E. and S. P. Balk (2004). "Androgen receptor: a key molecule in the progression of prostate cancer to hormone independence." *J Cell Biochem* 91(3): 483-490.

Tavridou, A., A. Efthimiadis, et al. (2010). "Simvastatin-induced changes in circulating oxidized low-density lipoprotein in different types of dyslipidemia." *Heart Vessels* 25(4): 288-293.

Thomas, J. A., 2nd, L. Gerber, et al. (2012). "Prostate Cancer Risk in Men with Baseline History of Coronary Artery Disease: Results from the REDUCE Study." *Cancer Epidemiol Biomarkers Prev.*

Tontonoz, P. (2011). "Transcriptional and Posttranscriptional Control of Cholesterol Homeostasis by Liver X Receptors." *Cold Spring Harb Symp Quant Biol.*

Tosi, M. R. and V. Tugnoli (2005). "Cholesteryl esters in malignancy." *Clin Chim Acta* 359(1-2): 27-45.

Uda, S., S. Accossu, et al. (2012). "A lipoprotein source of cholesteryl esters is essential for proliferation of CEM-CCRF lymphoblastic cell line." *Tumour Biol* 33(2): 443-453.

van Duijnhoven, F. J., H. B. Bueno-De-Mesquita, et al. (2011). "Blood lipid and lipoprotein concentrations and colorectal cancer risk in the European Prospective Investigation into Cancer and Nutrition." *Gut* 60(8): 1094-1102.

Van Hemelrijck, M., H. Garmo, et al. (2011). "Prostate cancer risk in the Swedish AMORIS study: the interplay among triglycerides, total cholesterol, and glucose." *Cancer* 117(10): 2086-2095.

Vander Heiden, M. G., L. C. Cantley, et al. (2009). "Understanding the Warburg effect: the metabolic requirements of cell proliferation." *Science* 324(5930): 1029-1033.

Vatten, L. J. and O. P. Foss (1990). "Total serum cholesterol and triglycerides and risk of breast cancer: a prospective study of 24,329 Norwegian women." *Cancer Res* 50(8): 2341-2346.

Vatten, L. J., O. P. Foss, et al. (1991). "Overall survival of breast cancer patients in relation to preclinically determined total serum cholesterol, body mass index, height and cigarette smoking: a population-based study." *Eur J Cancer* 27(5): 641-646.

Vitols, S., G. Gahrton, et al. (1985). "Hypocholesterolaemia in malignancy due to elevated low-density-lipoprotein-receptor activity in tumour cells: evidence from studies in patients with leukaemia." *Lancet* 2(8465): 1150-1154.

Wang, D. and R. N. Dubois (2006). "Prostaglandins and cancer." *Gut* 55(1): 115-122.

Yang, J. B., Z. J. Duan, et al. (2001). "Synergistic transcriptional activation of human Acyl-coenzyme A: cholesterol acyltransterase-1 gene by interferon-gamma and all-trans-retinoic acid THP-1 cells." *J Biol Chem* 276(24): 20989-20998.

Yang, L., J. B. Yang, et al. (2004). "Enhancement of human ACAT1 gene expression to promote the macrophage-derived foam cell formation by dexamethasone." *Cell Res* 14(4): 315-323.

Yasuda, M. and W. R. Bloor (1932). "Lipid Content of Tumors." *J Clin Invest* 11(4): 677-682.

Chapter 27

Role of Lipoproteins in Carcinogenesis and in Chemoprevention

Adebowale Bernard Saba and Temitayo Ajibade

Additional information is available at the end of the chapter

http://dx.doi.org/10.5772/46065

1. Introduction

Lipoproteins are complex aggregates of lipids and proteins that render endogenous lipids compatible with the aqueous environment of body fluids (Brown, 2007). The major physiological role of lipoproteins is to transport water-insoluble lipids from their point of origin to their respective destinations. Lipoproteins are synthesised mainly in the liver and intestines. They are in a state of constant flux, in circulation, changing in composition and physical structure as the peripheral tissues take up the various components before the remnants are returned to the liver. The most abundant lipid constituents of lipoproteins are free cholesterol, cholesterol esters, triacylglycerols and phospholipids, though fat-soluble vitamins and anti-oxidants are also found in lipoproteins (Kwiterovich, 2000).

Classification of lipoproteins

Lipoproteins are classified as chylomicrons (CM), very-low-density lipoproteins (VLDL), low-density lipoproteins (LDL) and high-density lipoproteins (HDL), based on the relative densities of the aggregates on ultracentrifugation. These classes are further refined by improved separation procedures, and intermediate-density lipoproteins (IDL) and subdivisions of the HDL (e.g. HDL_1, HDL_2, HDL_3 etc) are often defined. Density of lipoproteins is determined largely by the relative concentrations of triacylglycerols and proteins and by the diameters of the broadly spherical particles, which vary from about 6000Å in CM to 100Å or less in the smallest HDL. An alternative nomenclature is based on the relative mobilities on electrophoresis on agarose gels. Thus, α, pre-β and β lipoproteins correspond to HDL, VLDL and LDL, respectively (Lacko *et al.*, 2007).

Chylomicrons

Chylomicrons, the largest and least dense of the lipoproteins are formed in the intestinal cell walls from dietary fat and cholesterol. Their main task is to carry triglycerides from the intestine to the tissues where they are needed as a source of energy. In the circulation, triglycerides are removed from chylomicrons via the action of lipoprotein lipase (LPL), an enzyme present in the capillaries of many tissues. If present in large amounts, such as after a fatty meal, chylomicrons cause the plasma to appear milky.

Very low density lipoproteins

Very low density lipoproteins (VLDLs) are synthesised in the liver. Much like chylomicrons, they function primarily to distribute triglycerides to target sites such as adipose tissue and skeletal muscle where they are used for storage and energy. The manner in which triglycerides are removed from the circulation is the same as that of chylomicrons. Gradually with removal of triglycerides and protein, VLDLs are converted to LDL. High plasma levels of VLDL are associated with familial hypertriglyceridaemia, diabetes mellitus and underactive thyroid.

Low density lipoproteins

Low density lipoproteins are cholesterol-rich particles. About 70% of plasma cholesterol occurs in this form. LDLs are chiefly involved in the transport of the cholesterol manufactured in the liver to the tissues, where it is used. Uptake of cholesterol into cells occurs when lipoprotein binds to LDL receptors on the cell surface. LDL is then taken into the cell and broken down into free cholesterol and amino acids. Disorders involving a defect in or lack of LDL receptors are usually characterised by high plasma cholesterol levels. The cholesterol cannot be cleared efficiently from the blood and therefore accumulates.

High density lipoproteins

The high-density lipoproteins (HDLs) are small, dense, and spherical lipid-protein complexes which are normally considered to consist of those plasma lipoprotein particles which fall into the density range of 1.063–1.210 g/mL. HDL particles are composed of an outer layer containing free cholesterol, phospholipid, and various apolipoproteins (Apo), which covers a hydrophobic core consisting primarily of triglycerides and cholesterol esters (Barter *et al.*, 2003). The major proteins are Apo A-I (Mr 28,000) and Apo A-II (Mr 17,000). Apo A-I, the primary protein constituent of these particles, accounts for about 60% of the protein content of HDL. Apo A-I is synthesized in the intestines and liver and is thought to be largely responsible for the antiatherogenic effects of HDL. Some HDL particles carry only Apo A-I, whereas others contain both Apo A-I and Apo A-II (Shah *et al.*, 2001). Other apolipoprotein species found in HDL particles include Apo A-IV, Apo C (C-I, C-II, and C-III), and Apo E.

High density lipoprotein subtypes

Several subtypes of HDL particles have been identified on the basis of density, electrophoretic mobility, particle size, and apolipoprotein composition (Albers *et al.*, 1984). Differences in particle size are mainly the result of the number of apolipoprotein particles and the volume of the cholesterol ester in the core of the particle. HDL can also be classified into larger, less dense HDL2 or smaller, denser HDL3 which falls within the density ranges 1.063–1.125 and 1.125–1.210 g/mL, respectively. Although the major proportion of HDL is normally present in HDL3, individual variability in HDL levels in human populations usually reflects different amounts of HDL2 (Skinner, 1994). HDL2 is richer in particles containing Apo A-I without Apo A-II, whereas HDL3 is richer in particles containing both Apo A-I and Apo A-II (Gotto, 2001).

Synthesis, Metabolism and Regulation of plasma lipoprotein concentration

Normal metabolism and homeostasis of carbohydrates, amino acids and lipids *in vivo* depend on integrated liver function. Most plasma apolipoproteins and endogenous lipids and lipoproteins, including apolipoprotein(a) (apo(a)) and lipoprotein(a) (Lp(a)), are synthesized in the liver. The apolipoprotein(a) is a high molecular weight glycoprotein, of 250-838 kD (Bowden, 1994) and a total of 34 different apo(a) isoforms have been identified in populations (Marcovina *et al.*, 1993). The core components of Lp(a) are neutral lipid and an apoB-100 molecule, which are covalently connected by a disulfide-bond bridge and surrounded by hydrophilic apo(a) (Byrne, 1994). The heterogeneity of apo(a) determines the changes in plasma Lp(a) concentrations, and there is a negative correlation between the molecular weight of apo(a) and the plasma Lp(a) concentration (Wade, 1993)

Lp(a) has a simple Mendelian dominant inheritance, which is controlled by the alleles Lpa and Lp0 (Utermann *et al.*, 1988). Plasma Lp(a) concentration is controlled by three alleles., i.e., LpA, Lpa and Lp0 (Hasstedt *et al.* 1986). Pedigree analysis indicated that the size polymorphism of Lp(a) is controlled by a series of alleles of a single point (Utermann, 1988). The apo(a) gene which is located in q26-27 of chromosome 6 in humans has a linkage to the plasminogen (PGN) gene, and is inherited in a codominant Mendelian model (Amemiya, 1996). Apolipoprotein(a) mRNA (14 kb) encodes for a mature protein of 4529 amino acid residues in the presence of a signal peptide with 19 amino acid residues (McLean, 1987) while a high-degree homology exists between the molecular structures of the apo(a) gene and the PGN gene. The high-degree homology between the apo(a) gene and the PGN gene determines the biological actions of Lp(a) (Romics *et al.*, 1996).

2. Role of lipoprotein in cancer

Cancerous cells generally have high requirements for cholesterol as they are rapidly dividing cells. The Low density lipoproteins (LDL) which are cholesterol-rich particles have been especially found to play significant role in the pathogenesis of a large number of cancers. For instance, increased LDL requirement and receptor activity have been reported in cancer of the prostate gland (Chen and Hughes-Fulford, 2001); colon (Niendorf *et al.*, 1995); adrenal gland (Nakagawa *et al.*, 1995); hormone unresponsive breast tumors (Stranzl *et al.*, 1997), cancers of gynecological origin, tumors of lung tissues (Vitols *et al.*, 1992), leukemia (Tatidis *et al.*, 2002), and malignant brain tumors (Rudling *et al.*, 1990). In contrast, high density lipoproteins (HDL) have been reported significantly lowered in patients with primary or metastatic liver cancer (Moorman *et al.*, 1998). Hoyer and Engohm (1992) observed an inverse association between serum HDL-cholesterol and risk of breast cancer in a cohort of 5,207 Danish women, who participated in the Glostrup population studies of breast cancer.

Lymphoma patients often exhibit abnormal lipid metabolism. Numerous clinical studies of lymphoma patients have reported lipid abnormalities that are similar to the dyslipidemia observed in inflammatory and infectious diseases that are believed to develop secondary to circulating cytokines and the accompanying acute-phase response (Blackman *et al.*, 1993). Spiegel *et al.* (1989) investigated plasma lipids and lipoproteins at presentations in 25 patients with acute leukemia and non-Hodgkin's lymphoma and reported that all patients demonstrated an abnormality in at least one plasma lipid fraction and most exhibited a predictable pattern of lipid alterations that consisted of extremely low levels of HDL-cholesterol, elevated triglyceride, and elevated very low density lipoprotein (VLDL). The degree of lipid abnormality was directly related to the underlying tumor burden and particularly to the presence of bone marrow involvement. Therefore, low levels of circulating HDL-cholesterol in lymphoma patients may occur before the clinical onset of cancer and may serve as a marker for inflammation-induced lymphomagenesis, rather than a consequence of lymphoma-induced acute-phase responses.

2.1. Role of lipoprotein(a) in pathogenesis of cancer

Lipoprotein[a] is an intriguing molecule consisting of a low-density lipoprotein core and a covalently bound apolipoprotein[a]. Apolipoprotein[a] possess an inactive protease domain which is a single copy of the plasminogen kringle 5 and multiple repeats of domains homologous to the plasminogen kringle 4. The plasminogen kringle 5 (K5) domain, which is distinct from angiostatin, possesses potent anti-angiogenic properties on its own, which can be exploited in cancer therapy. The angiostatic effect and novel proinflammatory role of the K5 protein is via its ability to recruit tumor-associated neutrophils and NKT lymphocytes, leading to a potent antitumor response (Perri *et al.*, 2007).

Recently, anti-angiogenic agents have been found to promote leucocyte-vessel wall interaction as part of their anti-tumorigenic effects. Studies on animal models have indicated that the proteolytic break-down products of apolipoprotein[a] may posses anti-angiogenic and anti-tumorigenic effects both *in vitro* and *in vivo*. This is a convenient premise to develop novel therapeutic modalities which may efficiently suppress tumor growth and metastasis (Giuseppe *et al.*, 2007). Significant decrease in Lipoprotein[a] levels have been reported in liver cancer patients by Samonakis *et al.*, 2004. Although the liver plays an important role in lipid metabolism, several non-hepatic factors such as hormones, cytokines, genetics and nutrition are also involved in different ways. For example, several inflammatory and tumoral diseases are characterized by the production and delivery of cytokines influencing serum Lipoprotein[a] levels.

The mechanisms by which cancers induce cachexia involve inflammatory cytokine production; which is responsible for a wide number of metabolic disorders, essentially involving lipid metabolism (Langstein and Norton, 1991) and serum Lipoprotein[a] level changes during inflammatory disease. Liver damage has been linked to reduce Lipoprotein[a] serum levels (Malaguarnera *et al.*, 1996). Geiss *et al.*(1996) observed marked increase in Lipoprotein[a] concentration from 7 mg/dl in acute stage to 32 mg/dl in convalescence in hepatitis patients.The lipoprotein[a] half-life is short *in vivo,* 3.3~3.9 d, (Krempler *et al.,* 1983), and influenced early by liver function alterations (Malaguarnera *et al.,* 1994), The high involvement of Lipoprotein[a] in lipid and protein metabolism have been suggested to be a sensitive and early marker of liver malfunction, therefore Lipoprotein[a] may supply useful additional information for a more complete assessment and monitoring of the liver function in patients with hepatocellular carcinoma and liver cirrhosis (Uccello *et al.,* 2011).

2.1.1. *Role of high density lipoprotein (HDL) in the pathogenesis of cancer*

The origin and fate of HDL are less well understood than other lipoproteins. HDL may be formed both in the intestine and in the liver. It is also formed during lipolysis of Triglycerides-rich lipoproteins. Apolipoprotein AI (ApoAI) and apolipoprotein AII (apoAII) are the major apolipoproteins of HDL and the production rate of apoA-I is an important determinant of the variability of plasma HDL concentrations. Production rate of apoA-I is however, influenced by many factors and apoA-I transcriptional regulation has an impact on plasma HDL concentrations.

A large number of cellular lipid transporters and receptors including a spectrum of HDL and intermediates of HDL participate in the transport of excess cholesterol from peripheral cells to the liver. At one end of the spectrum are lipid-free or lipid-poor apoA-I particles, referred to as pre-β HDL. These particles are secreted by the liver and small intestine or generated from surface material from partially lipolyzed chylomicrons or from HDL_2 in the periphery by the action of cholesteryl ester transfer protein (CETP), hepatic lipase, or phospholipid transfer protein (PTP) (Dullaart *et al.,* 2001). Lipolyzed chylomicrons are HDL precursors that accept unesterified cholesterol and phospholipids transferred by ATP-binding cassette transporter A1 (ABCA1) on peripheral cells, giving rise to discoidal lipoproteins containing apoA-I (Yokoyama, 2005). The acquired unesterified cholesterol of the apoA-I is esterified by the plasma enzyme lecithin:cholesterol acyltransferase (LCAT) to form cholesteryl ester which is packaged into the hydrophobic core of the discoidal particle, converting it to spherical HDL_3. HDL_3 can continue to accept unesterified cholesterol and phospholipids from the class B, type I scavenger receptor (SR-BI); through continued action of LCAT (Fig. 1). This causes the hydrophobic core of the discoidal lipoprotein particle to expand and the size increases, thereby forming HDL_2.

High density lipoprotein (HDL) plays a key role in the reverse cholesterol transport pathway (RCTP) (Genest *et al.,* 1990) and various fractions of HDL have been shown to offer a new approach to study liver diseases (Cooper *et al.,* 1996). Plasma HDL-C, HDL-PL and HDL-C/HDL-PL have been reported to be lower in hepatocellular carcinoma patients than those in normal patients (Li *et al.,* 1993). In a study on 40 patients with hepatocellular carcinoma, LDL-C level was found to be significantly lower in the hepatocellular carcinoma patients than in the controls, but HDL-C did not show a statistically significant difference to the controls (Motta *et al.,* 2001). HDL-C itself has also been reported significantly decreased in patients with primary or metastatic liver cancer (Kanel *et al.* 1983). Therefore, variations in the level of plasma lipids and lipoproteins may assist in describing the nature of cirrhosis and hepatocellular carcinoma (Ahaneku *et al.,* 1992).

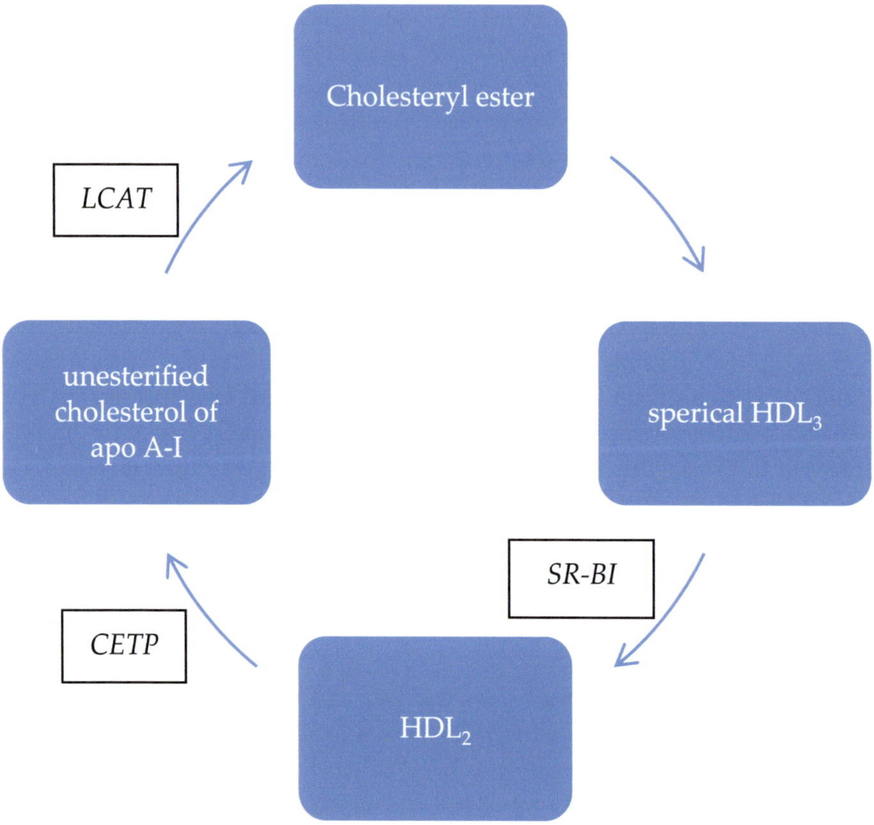

Figure 1. Formation of HDL_2 from unesterified cholesterol

Furthermore, an inverse and significant association exists between levels of HDL cholesterol and the risk of incident cancer, according to the results of a recent study (Jafri *et al.*, 2010). Exceptionally, a few cancer risk factors are associated with increased levels of HDL. For instance, many breast cancer risk factors are associated with high HDL-C and the relationship between breast cancer and HDL-C is independent of other risk factors (Moorman *et al.*, 1998).

HDL could play a role in carcinogenesis through its influence on cell cycle entry, via a mitogen-activated protein kinase-dependent pathway or regulation of apoptosis. Specifically, an inverse association exists between serum HDL and risk of breast cancer, and several studies have reported lower levels of HDL in breast cancer patients. Further studies have also shown that tumor progression from localized to metastatic disease is associated with declining HDL levels. At least two population-based screening surveys involving Norwegian women have established low HDL, as part of the metabolic syndrome associated with increased post-menopausal breast cancer risk. The risk of post-menopausal breast cancer among overweight and obese women in the highest serum HDL-cholesterol quartile was one-third the risk of women in the lowest serum HDL-cholesterol quartile (Furberg *et al.*, 2004).

2.2. Serum lipid profile and incidence of cancer

Cancer patients often present altered serum lipid profile including changes of HDL level. Case-control studies of newly diagnosed lung cancer have shown that HDL levels are reduced in lung cancer cases. Patients with advanced nonresectable lung cancer also often have decreased serum HDL. Although the biological mechanisms that might link low plasma levels of HDL with cancer are not well understood, the HDL regulation of cell cycle entry through a mitogen activated protein kinase-dependent pathway and apoptosis, modulation of cytokine production, and antioxidative function (Perletti *et al.*, 1996) has been suggested biologically plausible.

The association between higher HDL and lower overall cancer incidence observed in the ATBC cohort is biologically plausible, as HDL has anti-inflammatory properties (Perletti *et al.*, 1996). However, it is also plausible that this association reflects the effect of factors such as inflammation, which are associated with both HDL and risk of cancer.

2.3. Effect of inflammation on HDL levels in the body

Inflammation reduces HDL and likely increases risk of lung cancer (Mendez *et al.*, 1991). Studies have shown that chronic inflammation is known to reduce both serum HDL-cholesterol levels and its anti-inflammatory properties. The lipoprotein abnormalities seen in patients with inflammatory diseases are thought to develop secondary to circulating cytokines and the accompanying acute-phase response (Blackman *et al.*, 1993). Low HDL-cholesterol may therefore be a marker for the severity of systemic inflammation and inflammation-induced non-Hodgkin's lymphoma risk. Conversely, high HDL-cholesterol itself may be protective against non-Hodgkin's lymphoma. High-density lipoprotein-cholesterol seems to modulate inflammatory responses independent of non-HDL cholesterol levels by suppressing chemotactic activity of monocytes and lymphocytes and inhibiting cytokine-induced expression of endothelial cell adhesion molecules.

2.4. Association between antioxidant enzymes and HDL

There are a number of enzymes associated with HDL that have antioxidant properties, including paraoxonase, platelet-activating factor acetylhydrolase, and glutathione peroxidase. Paraoxonase-1 (PON1), the enzyme primarily responsible for HDL's antioxidant function, is closely bound to the HDL particle. PON1's enzymatic activity is highly regulated by environmental factors such as diet and physical activity, by certain drugs, and by genetic factors, especially certain genetic polymorphisms in the paraoxonase-1 gene, *PON1*. PON1 is synthesized in the mammalian liver and circulates in blood bound to HDL apolipoprotein (apo) A-1 and apo J. There are 2 other proteins in the same family as PON1 that probably also have antioxidant actions. These are PON2 and PON3. PON2 is ubiquitously expressed within cells, whereas PON3 exhibits a basal constitutive antioxidant activity and is essentially bound to HDL (Reddy *et al.*, 2001; Ng *et al.*, 2001). These enzymes have the ability to prevent the formation of proinflammatory oxidized phospholipids and to block the activity of those already formed; however, these oxidized lipids negatively regulate the activities of the HDL-associated enzymes. During an acute-phase response in rabbits, mice, and humans, there seems to be an increase in the formation of these oxidized lipids that results in the inhibition of the HDL-associated enzymes and an association of acute-phase proteins with HDL that renders HDL pro-inflammatory rather than being anti-inflammatory.

2.5. Role of lipoproteins in cancer chemotherapy

The efficacy of cancer chemotherapy is often limited by severe cytotoxic effects induced by anticancer drugs, on healthy and cancerous cells (Sehouli *et al.*, 2002). In addition, most of the available dosage forms perform with less than optimal efficiency because of poor solubility and limited accessibility to target tissues (Rosen and Abribat, 2005). Furthermore, it is difficult to eradicate cancer cells *in vivo* because they share the same biochemical machinery with normal cells.

In spite of the overwhelmingly large number of anticancer drugs that have been developed, none is completely selective for cancer cells. Consequently, all anticancer drugs presently in use induce significant dose-limiting toxic side effects. For this reason, there has been increased emphasis on selective delivery of drugs to tumours in ways that bypass normal body tissues.

2.5.1. Lipoproteins as special anticancer drug delivery agent

The cytotoxic effects of cancer chemotherapeutic drugs on healthy organs can be significantly diminished by employing special drug delivery systems targeted specifically to cancer cells (Minko *et al., 2004*). Targeting is especially important in circumstances where a localized tumor is removed surgically, and chemotherapy is prescribed as a follow-up preventive against potential metastases (Dharap *et al.*, 2005). Among the vehicles that can be used for special anticancer drug delivery are lipoproteins.

Lipoproteins have been considered appropriate drug-delivery vehicles for anticancer drugs (Braschi *et al.*, 1999), owing to their structural features, biocompatibility and targeting capability via receptor mediated mechanisms (Nikanjam *et al.*, 2007). The basic structure of lipoproteins, which comprises of an outer protein–phospholipid shell with a lipophilic surface and an interior hydrophobic compartment, positions them as ideal transporters of hydrophobic drugs, including anticancer agents. Due to their biocompatible, lipoproteins have considerable advantages over the conventional carrier systems currently used in cancer chemotherapy in that they provide the opportunity for targeted delivery of the anticancer drug they carry through endocytosis by receptor mediated uptake or by selective uptake of core components (Pathania, *et al.*, 2003).

The main advantages of lipoprotein-based formulations are their biocompatible components, their relative stability in the blood circulation and their track record of having already been safely injected into human subjects (Bisoendial *et al.*, 2002). In addition, lipoprotein-based formulations have vast targeting potential via receptor-mediated mechanisms that are overexpressed in cancer cells compared with normal cells. It has been shown that anticancer drugs can be targeted to cancer cells that generally express a high level of lipoprotein receptors, by encapsulation into reconstituted high density lipoprotein nanoparticles which are similar to the endogenous lipoprotein particles responsible for shuttling of hydrophobic molecules to different parts of the body.

Lipoprotein complexes are ideal for loading and targeted delivery of cancer therapeutic and diagnostic agents because they can mimic the shape and structure of endogenous lipoproteins, and as such, remain in circulation for an extended period of time, while largely evading the reticuloendothelial cells in the body's defenses. The small size (less than 30 nm) of the low-density and high-density classes of lipoproteins allows them to maneuver deeply into tumors. Furthermore, lipoproteins can be targeted to their endogenous receptors, especially, when the receptors are implicated in cancer. Although the lipophilic character of certain pharmaceuticals may be a disadvantage during intravenous therapy, this can be advantageous in anticancer drug deliveries as the highly lipid compounds are ideally suited for incorporation into lipoproteins. Most of the 'orphaned' anticancer drugs that have failed primarily due to their poor water solubility can also be made to progress faster through development process by incorporating them into lipoproteins.

2.5.2. *Low density lipoprotein in cancer chemotherapy*

Low density lipoprotein (LDL) has been found to represent a suitable carrier for cytotoxic drugs that may target them to cancer (Kader and Pater, 2002). This is because the low-density lipoprotein receptor (LDLR) has been found to be over-expressed in numerous cancers. The upregulated levels of low-density receptor in these cancers are believed to provide the cancer cells with the necessary lipid substrates needed for active membrane synthesis. In fact, sequestration of plasma LDL cholesterol in cancers has been suggested to explain the low levels of circulating total and LDL cholesterol observed in patients with malignancies. These findings have led many researchers to investigate the possibility of exploiting LDL as a delivery vehicle for cancer diagnostics and therapeutics (Corbin and Zheng, 2007).

The high requirement of LDL by cancer cells and thus the overexpression of LDL receptor can be utilized for developing a novel targeted drug delivery system. This can be achieved by targeting of the LDL particle and allowing the anticancer drugs to be transferred to the natural LDL inside of the body. LDL will function as a secondary carrier of anticancer molecules and deliver these molecules selectively to cancerous cells via elevated LDL receptors. This approach requires the anticancer molecules to have affinity for the LDL particle endogenously and to have certain special physicochemical properties.

There are at least three different ways in which diagnostic or anticancer agents can be incorporated into LDL. The first of these is protein labeling in which the anticancer drug is covalently attached to the amino acid residues of apolipoprotein (apo) B-100 protein of LDL; the second is surface labeling which involves intercalation of the diagnostic or anticancer agent into the phospholipid monolayer of LDL; the third is reconstitution core loading via substitution of anticancer agents into the lipid core of LDL. A large number of researchers have shown that various anticancer agents could be actively incorporated into LDL through intercalation or reconstitution methods. Moreover, these novel LDL–drug complexes were shown to be more efficacious against cancer cells than their conventional counterparts.

Coupling of doxorubicin to human LDL to form a LDL-doxorubicin complex injected to mice resulted in greater accumulation of LDL-doxorubicin, in the liver, than free doxorubicin. In contrast, LDL- doxorubicin was less accumulated in heart than free doxorubicin. This suggests LDL could be used as carriers to conjugate anti-cancer drugs.

2.5.3. Problems limiting the use of LDL as delivery vehicle

Although a lot of studies have produced promising results, progress towards utilizing LDL as a delivery vehicle in the clinical setting has been impeded by the need to isolate LDL from fresh donor plasma. Relying on donor plasma to acquire LDL is problematic because LDL samples vary from batch to batch, methods for isolating LDL are lengthy and large quantities of LDL are difficult to attain. Furthermore, isolated LDL can only be stored for finite periods before aggregation and degradation processes compromise the integrity of the LDL sample. As a result of these limitations, attempts have been made to prepare synthetic LDL-like particles consisting of phospholipid/cholesterol ester microemulsions and apoB-100 (the LDLR-binding component of LDL). Difficulties also plagued this endeavor owing to the size and complexity of the apoB-100 protein which is one of the largest monomeric proteins known consisting of over 4500 amino acids with a molecular weight of 550 kDa. Furthermore, apoB-100 is highly insoluble in aqueous solutions, making it difficult to work with. In addition, these problems are compounded still by the difficulties of having to isolate apoB-100 from donor plasma and other approaches to working with apoB-100 are therefore needed.

2.5.4. High-density Lipoprotein in cancer chemotherapy

The targeted delivery of anticancer agents via lipoprotein carriers is based on the concept that cancer cells have a higher expression of lipoprotein receptors (Lacko *et al.*, 2002; Cao *et al.*, 2004) due to their increased need for cholesterol to promote rapid proliferation. Clinical studies have shown that HDL cholesterol levels, like LDL levels, are lower in cancer patients, including those with haematological malignancies (Fiorenza *et al.*, 2000). Unfortunately, the targeting of chemotherapeutic agents via HDL is daunting because overwhelming efforts are required for the isolation HDL from human plasma and a lot of biosafety concerns are attached with the injection of human-blood-derived products (Adams *et al.*, 2003). Consequently, the focus of future studies is likely to be on synthetic/reconstituted lipoproteins with favourable drug-carrying capacity, and the exploitation of their potential for targeting tumour cells and tissues.

HDL transports cholesterol to liver cells, where they are recognized and taken up via specific receptors. Cholesteryl esters within HDL are selectively uptaken by hepatocytes via the scavenger receptor class B type I (SR-BI). An interesting feature of SR-BI is that the receptor selectively translocates HDL-cholesteryl esters from the lipoprotein particle to the cytosol of the liver parenchymal cells without a parallel uptake of the apolipoproteins and this property may allow for the delivery of its loaded drugs avoiding lysosomal degradation (Lou *et al.*, 2005). The high affinity of cancer cells for HDL has made them useful as carriers for delivering anticancer drugs into hepatoma cells to treat HCC. Anti-cancer drug-HDL complexes work as efficient drug delivery vehicles due to the ability of cancer cells to acquire HDL core components (Wasan *et al.*, 1996). Complexing of anti-cancer drugs with HDL does not influence characteristics of the anticancer drugs (Kader *et al.*, 2002) and administration of anti-cancer drug-HDL complex may reduce toxic side-effects during the chemotherapy(Lacko, *et al.*,2002).

In a cell culture system, cellular uptake of recombinant HDL-aclacinomycin by the SMMC-7721 hepatoma cells was significantly higher than that of free aclacinomycin at the concentration range of 0.5–10 μg/mL. Cytotoxicity of recombinant HDL- aclacinomycin to the hepatoma cells was significantly higher than that of free aclacinomycin at concentration range of less than 5 μg/mL just as IC50 of recombinant HDL-aclacinomycin was lower than IC50 of free aclacinomycin (Lou *et al.*, 2005). These results strongly suggest that HDL could be used as carriers to conjugate water-insoluble anti-cancer drugs in order to achieve higher therapeutic concentrations of the drugs in the microenvironment of the cancer cells.

3. Conclusion

Lipoproteins are complex endogenous aggregates of lipids and proteins that function primarily for the transport of water insoluble lipids from their point of origin to their respective destinations. Lipoproteins are classified as chylomicrons, very low density lipoproteins, low density lipoproteins and high density lipoproteins, based on the relative densities of the aggregates on ultracentrifugation.

Lipoproteins, which are cholesterol-rich particles, have been especially found to play significant roles in the pathogenesis of a large number of cancers because rapidly dividing cancerous cells generally have high requirements for cholesterol. This is exemplified by the increased LDL-requirement associated with a number of cancers including cancers of the prostate, colon, adrenal gland etc.

The efficacy of cancer chemotherapy is often limited by severe deleterious effects induced by anticancer drugs, on healthy and cancerous cells because of lack of specificity for the cancerous cells. In addition most available anticancer drugs do not perform optimally because of limited accessibility to target tissues. The deleterious effects of anticancer drugs on healthy organs can be markedly diminished by employing special drug delivery systems that specifically target cancer cells, using lipoproteins as carriers.

The main advantages of lipoproteins as anti-cancer drug carriers are: (1) lipoproteins are spherical particles consisting of a core of apolar lipids surrounded by a phospholipid monolayer, in which cholesterol and apoproteins are embedded. Therefore, highly lipophilic drugs can be incorporated into the apolar core without affecting lipoprotein receptor recognition; (2) lipoproteins are completely bio-degradable, do not trigger immunological responses, escape from recognition and elimination by the reticuloendothelial system, and have a relatively long half-life in the circulation; (3) lipoproteins can be recognized and taken up via specific receptors, and can mediate cellular uptake of the carried drugs ; and (4) many cancer cells show a high ability of lipoprotein uptake and therefore high therapeutic levels of the conjugated drugs can be rapidly attained at the target site(s).

Author details

Adebowale Bernard Saba and Temitayo Ajibade
Department of Veterinary Physiology,
Biochemistry and Pharmacology, University of Ibadan, Ibadan, Nigeria

4. References

Adams, K. M.; Lamber, N.C.; Heimfeld S.; Tylee, T. S. & Pang J. M. (2003). Male DNA in female donor apheresis and CD34-enriched products. *Blood* 102: 3845-3847. *doi:10.1182/blood-2003-05-1570.*

Ahaneku, J.E.; Taylor, G.O.; Olubuyide, I.O. & Agbedana, E.O. (1992). Abnormal lipid and lipoprotein patterns in liver cirrhosis with and without hepatocellular carcinoma. *J Pak Med Assoc*, 42(11):260-263. PMID:1336073

Albers, J. J.; Tollefson J. H.;Chen, C. H. & Steinmetz A. (1984) "Isolation and characterization of human plasma lipid transfer proteins," *Arteriosclerosis*; 4 (1): 49–58. PMID:6691846

Amemiya, H.; Arinami, T.; Kikuchi S.; Yamakawa-Kobayashi K.; Li L.; Fujiwara H. (1996). Apolipoprotein(a) and pentanucleotide repeat polymorphisms are associated with the degree of atherosclerosis in coronary heart disease. *Atherosclerosis*. 123:181-191. PMID:8782849

Barter, P.J.; Kastelein, & Kastelein J. (2003). "High density lipoproteins (HDLs) and atherosclerosis; the unanswered questions," *Atherosclerosis*. 168 (2) 195–211. PMID:12801602

Bisoendial, R.J.; Hovingh, G.K & de Groot, E. (2002). Measurement of subclinical atherosclerosis: beyond risk factor assessment. *Curr Opin Lipidol* 13:595-603. PMID:12441883

Blackman, J.D.; Cabana, V.G. & Mazzone, T. (1993). The acute-phase response and associated lipoprotein abnormalities accompanying lymphoma. *J Intern Med*. 233(2):201-4. PMID: 8433082

Bowden, J.F.; Pritchard, P.H.; Hill, J.S. & Frohlich, J.J. (1994). Lp(a) concentration and apo(a) isoform size. Relation to the presence of coronary artery disease in familial hypercholesterolemia. *Arterioscler Thromb* 14:1561-1568. PMID:7918305

Braschi, S.; Neville, T.A.; Vohl, M.C.& Sparks, D.L. (1999). Apolipoprotein A-I charge and conformation regulate the clearance of reconstituted high density lipoprotein *in vivo*. *J Lipid Res*.4 0(3):522–532. PMID:10064741

Brown, W.V. (2007). High-density lipoprotein and transport of cholesterol and triglyceride in blood. *J. Clin. Lipidology*, 1, 7-19. PMID: 21291664

Byrne, C.D. & Lawn, R.M. (1994). Studies on the structure and function of the apolipoprotein(a) gene. *Clin Genet*; 46:34-41. PMID:7988075

Cao, W.M.; Murao, K. & Imachi H. (2004). A mutant high-density lipoprotein receptor inhibits proliferation of human breast cancer cells. *Cancer Res*; 64: 1515-1521. PMID:14973113

Chen, Y. & Hughes-Fulford M. (2001). Human prostate cancer cells lack feedback regulation of low-density lipoprotein receptor and its regulator, SREBP2. *Int J Cancer*. 2001;91:41–45. PMID:11149418

Cooper, M.E., Akdeniz A. & Hardy K.J. (1996): Effects of liver transplantation and resection on lipid parameters: a longitudinal study. *Aust N Z J Surg*; 66(11):743-746. PMID:8918381

Corbin, I.R. & Zheng G. (2007). Mimicking Nature's Nanocarrier: Synthetic Low-density Lipoprotein-like Nanoparticles for Cancer-drug Delivery. *Nanomedicine* ;2(3):375-380. PMID:17716181

Dharap, S.S.; Wang, Y.; Chandna, P.; Khandare, J.J.; Qiu, B; Gunaseelan, S.; Sinko, P.J.; Stein, S; Farmanfarmaian, A. & Minko, T. (2005). Tumor-specific targeting of an anticancer drug delivery system by LHRH peptide. *PNAS* 102 36 12962-12967. PMID:16123131

Dullaart, R.P.F. & van Tol, A. (2001). Role of phospholipid transfer protein and pre β-high density lipoproteins in maintaining cholesterol efflux from Fu5AH cells to plasma from insulin-resistant subjects. *Scand. J. Clin. Lab. Invest.*;61:69–74. PMID:11300613

Fiorenza, A.M.; Branchi, A. & Sommariva, D. (2000). Serum lipoprotein profile in patients with cancer. A comparison with non-cancer subjects. *Int J Clin Lab Res*; 30:141-145. PMID:11196072

Furberg, A.; Marit, B.V.; Wilsgaard, T.; Bernstein, L.& Thune, .I (2004). Serum High-Density Lipoprotein Cholesterol, Metabolic Profile, and Breast Cancer Risk. *JNCI J Natl Cancer Inst* 96 (15): 1152-1160. PMID: 15292387

Geiss, H.C.; Ritter, M.M.; Richter, W.O.; Schwandt, P.& Zachoval, R. (1996). Low lipoprotein (a) levels during acute viral hepatitis. Hepatology.;24(6):1334–1337. PMID:8938156

Genest, J.J.; McNamara, J.R.; Ordovas, J.M.; Martin-Munley, S.; Jenner, J.L.; Millar, J.;Salem, D.N. & Schaefer, E.J. (1990): Effect of elective hospitalization on plasma lipoprotein cholesterol and apolipoproteins A-I, B and Lp(a). *Am J Cardiol*, 65(9):677-679. PMID:2106773

Giuseppe, L.; Massimo F.; Gian L.S. & Gian CG (2007). Lipoprotein[a] and cancer: Anti-neoplastic effect besides its cardiovascular potency. *Cancer Treatment Reviews* 33, 427-436. doi:10.1016/j.ctrv.2007.02.006

Gotto, A.M. (2001). "Low high-density lipoprotein cholesterol as a risk factor in coronary heart disease: a working group report," *Circulation*; 103 (17) 2213–2218. PMID:11331265

Hasstedt, S.J.; Ash, K.O. & Williams R.R. (1986). A re-examination of major locus hypotheses for high density lipoprotein cholesterol level using 2,170 persons screened in 55 Utah pedigrees. *Am J Med Genet*;24:57-67. PMID:3706413

Hoyer, A. P.; & Engholm G. (1992). "Serum lipids and breast cancer risk: a cohort study of 5,207 Danish women," Cancer Causes and Control; 3(5) 403–408. PMID:1525320

Jafri, H.; Alsheikh-Ali, A.A. & Karas, R.H. (2010). Baseline and on-treatment high-density lipoprotein cholesterol and the risk of cancer in randomized controlled trials of lipid-altering therapy. *J Am Coll Cardiol*; 55:2846-54. PMID:21173414

Kader, A. & Pater, A. (2002). Loading anticancer drugs into HDL as well as LDL has little affect on properties of complexes and enhances cytotoxicity to human carcinoma cells. *J Control Release;* 80(1–3):29-44. PMID:11943385

Kanel, G.C.; Radvan, G. & Peters RL (1983). High-density lipoprotein cholesterol and liver disease. *Hepatology*; 3: 343-348. PMID:6840679

Krempler. F.; Kostner, G.M.; Roscher, A.; Haslauer, F.;Bolzano, K. &Sandhofer, F. (1983). Studies on the role of specific cell surface receptors in the removal of lipoprotein (a) in man. *J Clin Invest.*;71:1431–1441. PMID: 6304146

Kwiterovich, P.O. (2000). "The metabolic pathways of high-density lipoprotein, low-density lipoprotein, and triglycerides: a current review". *The American journal of cardiology* 86 (12A): 5L–10L. PMID: 11374859

Lacko, A.G.; Nair, M.; Paranjape, S.; Johnson, S. & McConathy, W.J. (2002). High density lipoprotein complexes as delivery vehicles for anticancer drugs. *Anticancer Res.* 22(4):2045-9. PMID:12174882

Lacko, A.G.; Nair, M.; Prokai, L.& McConathy, W.J. (2007).Prospects and challenges of the development of lipoprotein-based formulations for anti-cancer drugs. *Expert Opin Drug Deliv.*;4(6):665-75. PMID:17970668

Langstein, H.N. & Norton, J.A. (1991). Mechanisms of cancer cachexia. *Hematol Oncol Clin North Am.* 5:103–123. PMID: 2026566

Li, W.X. (1993). [Serum cholesterol and cancer mortality: eleven-year prospective cohort study on more han nine thousand persons]. *Zhonghua Liu Xing Bing Xue Za Zhi.* 14(1):6-9. PMID:8504456

Lou, B.; Liao, X.L.; Wu, M.P.; Cheng, P.F.; Yin, C.Y. & Fei, Z. (2005). High-density lipoprotein as a potential carrier for delivery of a lipophilic antitumoral drug into hepatoma cells. *World J Gastroenterol;* 11(7):954-959. PMID:15742395

Malaguarnera, M.; Trovato, G.; Restuccia, S.; Giugno, I., Franze, C.M., Receputo, G., Siciliano, R.; Motta, M. & Trovato, B.A. (1994). Treatment of nonresectable hepatocellular carcinoma: review of the literature and meta-analysis. *Adv Ther*; 11:303–319. PMID:10150270

Malaguarnera, M.; Giugno, I.; Trovato, B.A.; Panebianco, M.P.; Restuccia, N. & Ruello, P. (1996). Lipoprotein(a) in cirrhosis. A new index of liver functions? *Curr Med Res Opin*; 13:479–485. PMID:8840366

Marcovina, S.M.; Zhang, Z.H.; Gaur, V.P. & Albers, J.J. (1993). Identification of 34 apolipoprotein(a) isoforms: differential expression of apolipoprotein(a) alleles between American blacks and whites. *Biochem Biophys Res Commun*;191:1192-1196. PMID:8466495

McLean, J.W.; Tomlinson, J.E.; Kuang, W.J.; Eaton, D.L.; Chen, E.Y. & Fless, G.M. (1987). cDNA sequence of human apolipoprotein(a) is homologous to plasminogen. *Nature*;330:132-137. PMID:3670400

Mendez, A.J.; Oram, J.F.& Bierman, E.L. (1991). Protein kinase C as a mediator of high density lipoprotein receptor-dependent efflux of intracellular cholesterol. *J Biol Chem*; 266. PMID:1645339

Minko, T.; Dharap, S.S.; Pakunlu, R.I.& Wang, Y. (2004). Molecular targeting of drug delivery systems to cancer. *Curr. Drug Targets*; 5 389-406. PMID: 15134222

Moorman, P.G.; Hulka, B.S.; Hiatt, R.A.; Krieger, N.; Newman, B.; Vogelman, J.H. & Orentreich, N. (1998). Association between high-density lipoprotein cholesterol and breast cancer varies by menopausal status. *Cancer Epidemiol Biomarkers Prev*;7(6):483-8. PMID: 9641492

Motta, M.; Giugno, I.; Ruello, P.; Pistone, G.; Di Fazio, I. & Malaguarnera, M. (2001). Lipoprotein(a) behaviour in patients with hepatocellular carcinoma. *Minerva Med.*;92:301–305. PMID:11675573

Niendorf, A.; Nagele, H.; Gerding, D.; Meyer-Pannwitt, U. & Gebhardt, A (1995) Increased LDL receptor mRNA expression in colon cancer is correlated with a rise in plasma cholesterol levels after curative surgery. *Int J Cancer,* 61: 461–464. PMID:7759150

Ng, C.J.; Wadleigh, D.J.; Gangopadhyay, A.; Hama, S.; Grijalva, V.R. & Navab, M. (2001). Paraoxonase-2 is a ubiquitously expressed protein with antioxidant properties and is capable of preventing cell-mediated oxidative modification of low density lipoprotein. *J Biol Chem* ; 276:44444-9. PMID:11579088.

Nikanjam, M.; Gibbs, A.R.; Hunt, C.A.; Budinger, T.F. & Forte, T.M. (2007). Synthetic nano-LDL with paclitaxel oleate as a targeted drug delivery vehicle for glioblastoma multiforme. *J Control Release, 124:163–171.* PMID:17964677

Pathania, D.; Millard, M. & Neamati N (2009). Opportunities in discovery and delivery of anticancer drugs targeting mitochondria and cancer cell metabolism. *Adv Drug Deliv.*,30;61(14):1250-75. PMID:19716393

Perletti, G.; Tessitore, L.; Sesca, E.; Pani, P.; Dianzani, M.U. & Piccinini, F. (1996). Epsilon PKC acts like a marker of progressive malignancy in rat liver, but fails to enhance tumorigenesis in rat hepatoma cells in culture. *Biochem Biophys Res Commun*; 221: 688-691. PMID: 8630022

Perri, S.R.; Martineau, D.; François, M.; Bisson, L.; Durocher, Y. & Galipeau, J. (2007). Plasminogen Kringle 5 blocks tumor progression by antiangiogenic and proinflammatory pathways. *Mol Cancer Ther*;6 (2):441–9. doi:10.1158/1535-7163.MCT-06-0434

Reddy, S.T.; Wadleigh, D.J.; Grijalva, V., Ng, C., Hama, S. & Gangopadhyay A (2001). Human paraoxonase-3 is an HDL-associated enzyme with biological activity similar to paraoxonase-1 protein but is not regulated by oxidized lipids. *Arterioscler Thromb Vasc Biol*;21:542-7. PMID:11304470

Romics, L.; Nemesánszky, E.; Szalay, F.; Császár, A.; Tresch, J. & Karádi, I. (1996). Lipoprotein(a) concentration and phenotypes in primary biliary cirrhosis. *Clin Chim Acta*;255:165-171. PMID:8937759

Rosen, H. & Abribat T (2005). The rise and rise of drug delivery. *Nat Rev Drug Discov.* ;4:381–5. PMID: 15864267

Rudling, M.J.; Angelin, B.; Peterson, C.O. & Collins, V.P. (1990). Low density lipoprotein receptor activity in human intracranial tumors and its relation to the cholesterol requirement. *Cancer Res.* 50, 483-487. Available from http://cancerres.aacrjournals.org/content/50/3/483#related-urls

Samonakis, D.N.; Koutroubaki, I.E.; Sfiridaki, A; Malliaraki, N.; Antoniou, P.& Romanos J (2004). Hypercoagulable states in patients with hepatocellular carcinoma. *Dig Dis Sci*; 49:854-858. PMID: 15259509

Sehouli, J.; Stengel, D.; Elling, D.; Ortmann, O.; Blohmer, J. Riess, H. &Lichtenegger W (2002) *Gynecol. Oncol.*, 85 , 321-326. PMID: 11972395

Shah, P.K.; Kaul, S.; Nilsson, J. & Cercek, B. (2001). "Exploiting the vascular protective effects of high-density lipoprotein and its apolipoproteins: an idea whose time for testing is coming, part I," *Circulation,* 104 (19). 2376–2383. PMID:11696481

Skinner, E. R. (1994). "High-density lipoprotein subclasses," *Current Opinion in Lipidology*, 5 (3). 241–247. doi:10.1155/2011/496925

Spiegel, D.; Bloom, J.R.; Kramer, H.C. & Gottheil, E. (1989). Effect of treatment on the survival of patients with metastasic breast cancer. *Lancet*, 2, 888–891. PMID: 9885092

Uccello, M.; Malaguarnera, G.; Pelligra, E.M.; Biondi, A.; Basile, F. & Motta M (2011). Lipoprotein(a) as a potential marker of residual liver function in hepatocellular carcinoma. *Indian J Med Paediatr Oncol*.; 32(2): 71–75. PMID:22174493

Utermann, G; Duba, C.& Menzel, H.J. (1988). Genetics of the quantitative Lp(a) lipoprotein trait. II. Inheritance of Lp(a) glycoprotein phenotypes. *Hum Genet*;78:47-50. PMCID: PMC287411

Vitols, S.; Peterson, C.; Larsson, 0.; Holm, P. & Aberg B (1992). Elevated uptake of low density lipoproteins by human lung cancer tissue in vivo. *Cancer Res*., 52, 6244-6247. 0007-0920/95

Wade, D.P.; Clarke, J.G.; Lindahl, G.E.; Liu, A.C.; Zysow, B.R.& Meer, K. (1993) l. 5' control regions of the apolipoprotein(a) gene and members of the related plasminogen gene family. *Proc Natl Acad Sci* U S A;90:1369-1373. PMID:7679504

Wasan, K.M. & Morton RE (1996): Differences in lipoprotein concentration and composition modify the plasma distribution of free and liposomal annamycin. *Pharm Res.*, 13(3):462-468. DOI: 10.1021/js960495j

Yokoyama, S. (2005). Assembly of high density lipoprotein by the ABCA1/apolipoprotein pathway. *Curr. Opin. Lipidol.*;16:269–279. Downloaded from www.jlr.org

Chapter 28

Structural Origin of ELOA Toxicity – Implication for HAMLET-Type Protein Complexes with Oleic Acid

Vladana Vukojević and Ludmilla A. Morozova-Roche

Additional information is available at the end of the chapter

http://dx.doi.org/10.5772/52522

1. Introduction

Self-assembled proteinaceous complexes with oleic acid (OA) acquire distinct properties that are not characteristic of the native protein. Most notably, the newly obtained features include the ability to specifically kill tumor cells while sparing the healthy, normally functioning ones, as it is the case with human or bovine α-lactalbumin made lethal to tumor cells (HAMLET or BAMLET) [1,2] or to indiscriminately induce cell death in all tested cell lines, as it is the case with equine lysozyme (EL) complex with oleic acid (ELOA) [3,4]. While extensive information has been accumulated on the structural, functional and therapeutic properties of protein complexes with OA, many questions remain still unanswered, such as what is the structural origin of their toxicity, what are the specific targets at the cell surface and/or the cellular interior, what are the mechanisms of cellular uptake?

In this chapter, we summarize our current understanding of the structure and function of HAMLET-type protein complexes with oleic acid, using ELOA as an example.

2. Origin of HAMLET - Human α-lactalbumin made lethal for tumor cells

Complexes of human α-lactalbumin with OA were discovered by Catharina Svanborg and co-workers about two decades ago [5]. Initially, Håkansson et al. [5,6] and Svensson et al. [7] discovered that a multimeric human α-lactalbumin derivative isolated from the casein fraction of milk was a potent Ca^{2+}-elevating and apoptosis-inducing agent with a broad, yet selective cytotoxic activity. It was found that the apoptosis-inducing fraction of α-lactalbumin contained oligomers of α-lactalbumin that have undergone a conformational change towards a molten globule-like state [7]. Oligomerization appeared to have conserved α-lactalbumin in a state with molten globule-like properties under physiological conditions. Multimeric α-lactalbumin was shown to bind to the cell surface, enter the cytoplasm and accumulate in cell nuclei [7]. Multimeric α-lactalbumin was also shown to increase the rate of respiration in isolated mitochondria by exerting an uncoupling effect, which was abolished completely by bovine serum albumin. Multimeric α-lactalbumin accumulated in the nuclei of sensitive cells rather than in the cytoplasm, vesicular fraction, or ER-Golgi complex [6]. Nuclear uptake was shown to occur rapidly in cells that are susceptible to an apoptosis-inducing effect, but not in nuclei of resistant cells. Nuclear uptake was shown to proceed through the nuclear pore complex and was critical for the induction of DNA fragmentation. Ca^{2+} was required for induction of DNA fragmentation by multimeric α-lactalbumin but nuclear uptake of multimeric α-lactalbumin is independent of Ca^{2+}.

Similar cytotoxic activity was observed by in vitro produced HAMLET complexes, in which human α-lactalbumin was converted into the apoptosis-inducing tumoricidal folding variant by binding OA [8-10]. The formation of HAMLET was carried out in chromatography ion exchange columns preconditioned with fatty acids. It was also identified that HAMLET formation is governed by stereo-specific lipid-protein interactions and that only unsaturated C16-C20 fatty acids in cis conformation, but not other fatty acids could induce HAMLET [11]. Among such complexes, only HAMLET complex with OA and cis vaccenic acid complexes were shown to kill tumor cells efficiently, while the C16 or C20 cis fatty acid complexes with α-lactalbumin showed low or intermediate activity [11].

HAMLET's remarkable tumor-selective cytotoxicity correlated with the conformational change of the protein that has taken place upon complex formation, i.e. conversion to molten globule-like state. However, α-lactalbumin in a molten globule state without OA does not possess such activity per se, indicating that the presence of both components is required. As partially unfolded α-lactalbumin can revert easily to its native state upon Ca^{2+} binding in natural cell culture media or within cells, the D87A Ca^{2+}- binding site mutant of α-lactalbumin was produced [12], which was lacking Ca^{2+}-binding property and remained partially unfolded at physiological conditions. Such mutant formed a tumoricidal HAMLET-like complex with OA, but the partially unfolded protein alone did not kill tumor cells. Another non-native α-lactalbumin variant with all amino acids building disulfide bridges substituted by Ala residues also did not exhibit cytotoxic activity in the absence of OA, while its HAMLET-like form displayed strong tumoricidal activity against lymphoma and carcinoma cell lines [13]. Together, these experiments consistently confirmed that both molten globule like protein conformation and specific fatty acids are required for the tumoricidal activity of the investigated complexes.

It has been suggested that naturally occurring HAMLET may have a protective function. In the stomach of nursing children low pH can induce the release of Ca^{2+} from the high-affinity Ca^{2+}-binding site of α-lactalbumin and activate lipases hydrolyzing free fatty acids from milk triglycerides, thereby providing naturally occurring conditions that favor the formation of α-lactalbumin lethal to tumors [14]. This could be important for lowering the incidence of cancer in breast-fed children by purging tumor cells from the gut of the neonate.

3. Equine lysozyme (EL) as a structural homologue of α-lactalbumin

The protein component of ELOA is equine lysozyme (EL), a protein that is abundant in mare milk and kumys (a fermented beverage produced from mare milk that is widely used in Middle Asia). EL belongs to an important calcium-binding sub-family within the extended family of lysozymes, *i.e.* in contrast to common c-type lysozyme EL possesses high affinity calcium binding site, resembling with this regards α-lactalbumins. Lysozymes and α-lactalbumins are characterized by not more than 35-40% in sequence homology, but share remarkably similar tertiary folds. EL serves as an evolutionary bridge between lysozymes and α-lactalbumins, combining the structural and folding properties of both. These are rather small molecules of about 14.6 kDa, consisting of two sub-domains – α-helical and β-sheet rich domains separated by a deep cleft. Lysozyme active site is located in this cleft (absent in α-lactalbumins). Calcium is coordinated by a loop positioned at the bottom of the cleft and important for the structural integrity of the protein, yet the physiological function of calcium binding to EL and other calcium-binding lysozymes is still unclear. The calcium-binding usually increases the protein stability against denaturing treatments, however in the case of EL, the significantly lower stability and cooperatively was observed compared to non-calcium-binding lysozymes even in its holo-form, while in the apo-form its thermodynamic stability is closer of α-lactalbumins than to c-type lysozymes [15,16]. EL forms a wide range of partially folded states under equilibrium conditions similar to these of α-lactalbumins [16,3,17,18]. However, EL molten globule is much more structured compared to the "classical" molten globules of α-lactalbumins, possessing an extended native-like hydrophobic core stabilised by interactions between three major α-helices (A, B and D-helices) in the α-domain [17,18]. Like c-type lysozymes, during refolding kinetics EL forms an ensemble of well-defined transient kinetic intermediates, possessing very persistent structures [19]. Importantly, the rapidly formed kinetic intermediate of EL (2.5 ms refolding time) is characterised by the same extended core structure as its equilibrium molten globule analogues populated under acidic conditions, indicating that the hydrophobic collapse into molten globule-like state is an essential step in protein folding. Given its distinct structural properties, EL may be used as an invaluable research object in revealing the general mechanism and role of intermediate states in protein folding.

4. Controlled ELOA production using ion-exchange chromatography

Similar to HAMLET, ELOA was produced at the solid-liquid interface in an ion-exchange chromatography column preconditioned with OA (Figure 1).

ELOA was eluted as a strong peak by using a 0-1.5 M NaCl gradient. In the absence of OA, free EL was eluted as a narrow peak at a low NaCl concentration [4]. EL was subjected to column chromatography without decalcification as it has been performed with human α-lactalbumin during original HAMLET production, indicating a difference in the generic properties of EL and α-lactalbumin. ELOA complex remains stable in its lyophilized form suitable for long storage as well as it can be kept in solution for up to a week. It is also important to note, that co-incubation of a 50 fold excess of OA mixed with EL in solution at room temperature did not lead to ELOA formation as was evident from the lack of characteristic ELOA conformational transitions monitored by near-UV CD [4]. Thus, the application of a solid-liquid interface facilitating protein self-assembly and protein-OA interactions proved to be an efficient approach in production of both ELOA and HAMLET complexes. By comparison, the complex of hen egg white lysozyme with oleic acid was also produced under the same conditions, but it was very low populated, unstable and OA can be easily depleted from its structure. Hen egg white lysozyme is much more stable than EL and it is evident that the hydrophobic interface in the column chromatography is not sufficient to cause its partial unfolding and interactions with OA molecules.

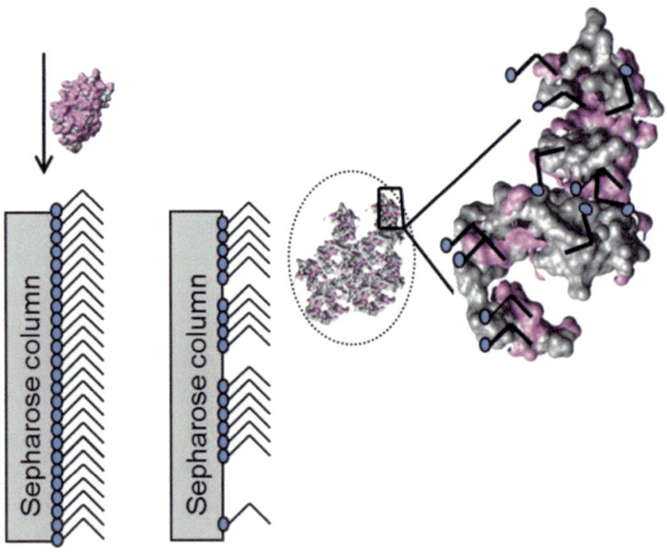

Figure 1. Schematic presentation of ELOA formation at the solid–liquid interface in a Sepharose chromatography column. The positively charged Sepharose matrix is preloaded with oleic acid (the hydrophilic carboxyl group is denoted by a blue circle and the aliphatic chain by a gray line with a "kink" at the position of the double bond). When folded, EL molecules (shown in space-filling representation, with exposed hydrophilic residues in purple and buried hydrophobic residues in grey) are added to the column some hydrophobic residues become exposed and interact with oleic acid molecules forming ELOA.

Indeed, hydrophobic and charged surfaces often facilitate the self-assembly processes by recruiting proteins and modifying their interactions [20]. Within the ion-exchange matrix bound OA molecules constitute an extended surface, facilitating both charged and hydrophobic interactions with EL molecules, while in solution OA, like many other small aliphatic molecules, would be present as a micelle. In addition, the solid-liquid interface may induce EL partial unfolding and expose its hydrophobic surfaces buried in the native state; this can also be critical for ELOA complex formation. It is worth noting, that hydrophobic interactions within the column chromatography may effectively model the interactions, which can take place at the hydrophobic and charged surfaces in biological systems. For example, the interactions with cell lipid membranes may be able to induce protein-ligand complexation otherwise not occurring in solution.

5. EL conformation in ELOA

Similar to human α-lactalbumin in the HAMLET complex, EL in ELOA acquires a partially folded state as evident from spectroscopic and NMR measurements [4]. The ELOA near-UV CD spectrum at room temperature shows the presence of less structure than in the native holo-state and even in the EL molten globule at 57 °C, i.e. the characteristic CD peaks are largely overlapped and the magnitude of the ellipticity is diminished at all wavelengths. ELOA spectra in the far-UV CD region recorded at both 25 °and 57 °C exhibit the same shape as the EL molten globule spectrum at 57 °C, which together with the near UV CD data demonstrate disordering of the tertiary interactions, but preservation of the secondary structure.

Consistently with molten globule conformation, the 1D ^{1}H NMR spectrum of ELOA at pH 9.0 exhibit very broad aromatic and aliphatic resonances, indicative of conformational mobility in a millisecond time scale, and a complete absence of resolved methyl peaks in the up-field region of 2.5 - 0.5 ppm [4,21]. This is in contrast to the NMR spectrum of native EL characterized by well-dispersed resonances in both the aromatic and aliphatic regions. Examination of the 1D ^{1}H NMR spectrum of ELOA showed up-field shifts of the resonance of bound OA compared with the resonances of free OA. This unequivocally demonstrates that OA molecules are integrated in ELOA. Specifically, OAs interact directly with EL aromatic residues as manifested in the presence of cross-peaks between the protons of aromatic residues and OA observed in the ^{1}H NOESY spectrum of ELOA. Due to the poor chemical shift dispersion of the ELOA spectrum, it is impossible to assign the positive NOE cross-peaks to specific aromatic residues, nonetheless this is an absolutely clear indication that EL aromatic residues are directly involved in OA binding.

Similar to typical molten globule states, ELOA binds the hydrophobic dye ANS, which is commonly used to examine the partially folded protein conformations. Interaction with ELOA resulted in ca. 10-fold increase in dye fluorescence compared with free ANS in solution. A shorter wavelength shift of the spectrum maximum of ANS (from 515 to 495 nm) indicates that ANS in its bound form is involved into a more hydrophobic environment. These results demonstrate that the ELOA complex is characterized by some exposed hydrophobic surfaces, which attract hydrophobic ANS molecules.

The surface dynamics and exposure of aromatic residues of ELOA were also probed by photochemically induced dynamic nuclear polarization (photo-CIDNP) spectroscopy [21]. CIDNP method evaluates the surface structure of proteins and complexes by means of a laser induced photochemical reaction, which takes place only if the aromatic side-chains of histidine, tyrosine and tryptophan residues are accessible to a photosensitizer [22]. The ELOA CIDNP spectra were compared with those of holo EL and EL molten globule.

The CIDNP spectra of the native EL at several pH 4.5, 6.9 and 9.0 are well-resolved and were assigned by comparison with NMR chemical shifts [17,18,22]. ELOA and EL molten display less resolved CIDNP spectra, consistent with their millisecond conformational fluctuations, although it is still straight forward to distinguish tyrosine and tryptophan/histidine residues based on their emissive (negative) and absorptive (positive) polarizations, respectively. In the EL molten globule state the characteristic emissive peak corresponding to the Tyr ε protons is the dominant feature, while the other peaks observed for the native state are not present, indicating that these residues are not surface-accessible. The ELOA spectrum contains the same emissive Tyr peak (≈ 6.75 ppm) as seen for the molten globule state. In addition, at ca. 7.7 ppm narrow absorptive photo-CIDNP signals assigned to either His 114 or Trp 63 or potentially both are present in ELOA spectrum, but not in the spectrum of EL molten globule. Both of these residues occur close to the EL inter-domain cleft, indicating limited conformational mobility in this region compared to the rest of the protein and hence that this region is affected by the presence of OA and may be an OA-binding site. Occupation of this cleft by OA may induce further large-scale changes in the relative positions of EL α- and β-domains, possibly lowering the affinity of the calcium-binding site. Hence, although there are clear similarities with EL molten globule, ELOA is characterized by some more structured regions arising from OA binding. These structural changes may be also related to ELOA functional activity as exposed hydrophobic residues in these regions may promote the ELOA interactions with the hydrophobic environment in lipid bilayers and cell membranes.

The thermal unfolding transition of ELOA, monitored by far-UV CD ellipticity at 222 nm, was manifested in an overall decrease of the CD signal and occurred over a very board range of temperatures from 30 °C and up to 80 °C. In EL alone, dissolved at both pH 9.0 and pH 7.0, two unfolding transitions were observed over the same temperature range, however, these transitions were not distinguished in ELOA. This indicates that the conformational changes in ELOA and EL may have different structural origin. It is interesting to note, that HAMLET is less stable towards thermal denaturation than holo human α-lactalbumin, while exhibiting the same stability towards urea denaturation [23]. This demonstrates that OA may produce some destabilizing effect on proteinaceous compounds in both ELOA and HAMLET, but to different degree and with different manifestation in their thermal unfolding transitions.

6. ELOA stoichiometry and comparison with EL amyloid oligomers

The question which is still debated concerns how many protein and OA molecules can be involved in the HAMLET-type complex formation and which conditions can affect this process. Firstly, in the case of ELOA the analysis of 1D ^{1}H NMR spectrum enabled us to determine the amount of bound OA per protein molecule by comparing the peak areas of the bound OA, reflecting the contribution of 2 olefinic protons, with the peaks corresponding to EL aromatic proton resonances [4]. This value can vary from 4 to 48 OA molecules per EL molecule, depending on the specific chromatographic conditions during ELOA formation. In general, increasing saturation of the column with OA resulted in the formation of ELOA with higher OA content. Secondly, the number of EL molecules in ELOA was determined by pulsed-field gradient NMR diffusion measurements and estimated to be 4–9 in most cases [4]. Thus, both number of OA and protein molecules can vary significantly within the ELOA complexes and the largest ELOA lies at the upper scale among the HAMLET-type complexes.

At the same time, the size of ELOA complexes tends to decrease upon dilution to micromolar concentration range. Their dimensions were characterized by AFM using the volume measurements of the round-shaped particles naturally attached to the mica surface, under the assumption that they acquire a shape of spherical cup due to their interactions with mica [24,22]. At the concentrations used in this study (< 1.5 μM) ELOA was predominantly present in the form of low molecular weight complexes - monomers to pentamers, while some larger aggregates were observed in a lower quantity. This finding was corroborated by fluorescence correlation spectroscopy (FCS). By comparing the average residence time of fluorescently labeled ELOA (*ca.* 120 μs) to the residence time of the reference fluorescent dye (*ca.* 35 μs), FCS indicated that ELOA is predominantly present as a low molecular weight complex (20-30 kDa) in diluted solutions [22].

It is interesting to draw comparison between ELOA and EL amyloid oligomers since they display some common properties. Amyloid oligomers attracted particular attention among protein self-assembled complexes due to their critical involvement in several amyloid and conformational diseases [25-27]. Oligomerisation precedes the amyloid fibril formation and oligomers may serve as nuclei for fibrillar growth. It has also been suggested that oligomers, rather than the apparently inert amyloid fibrils are major cytotoxic agents in amyloid diseases. Both α-lactalbumins and lysozymes form amyloids *in vitro* [28,29] and the lysozymes amyloid formation is associated with systemic amyloidosis in the body [30]. Under EL self-assembly both amyloid oligomers and ELOA become well populated, providing a unique opportunity to compare them directly.

Both ELOA and EL amyloid oligomers exhibited very similar stochiometry with 4 to 20-30 EL molecules involved [4,22,24]. Both ELOA and the amyloid species of corresponding size display the cytotoxic apoptotic activity, clearly absent in EL itself. ELOA and EL amyloid oligomers were characterized by spherical morphology examined by AFM and both tended to self-assemble into donut-like circular structures with very similar diameters of ca. 30 nm as measured by AFM [4,31]. In addition, ELOA and EL amyloid oligomers possess characteristic amyloid tinctorial properties such as binding of thioflavin-T dye, which is known as an amyloid specific marker. Thus, ELOA has some common structural and cytotoxic features with both HAMLET and amyloid oligomers and their further studies may shed light on both these phenomena and potential link between them. It is important that ELOA complexes are stable enough to be amenable for structural characterization at atomic resolution, whereas the amyloid oligomers are often transient in nature and tend to associate into larger aggregates or split into monomers. Amyloid oligomers are also not well populated and attempts have been made to stabilize them by using fatty acids and surfactants [32-35], which extend further the comparison between HAMLET-type complexes and amyloid species.

7. Live cell study of ELOA interaction with the plasma membrane

Molecular mechanisms of protein complexes interaction with living cells and their primary targets at the cell surface remain largely unknown and disputed [20,36]. Methods with single molecule sensitivity, Fluorescence Correlation Spectroscopy (FCS) and Confocal Laser Scanning Microscopy (CLSM) imaging by avalanche photodiodes (APD), so called APD imaging [37], which enable quantitative and nondestructive studies of molecular interactions and mobility in living cells, revealed that ELOA primarily acts on the plasma membrane of PC12 cells, inflicting damage and eventually causing plasma membrane rupture (Figure 2 A and B) followed by a rapid influx and distribution of ELOA inside the already dead cell (Figure 2 C) [21].

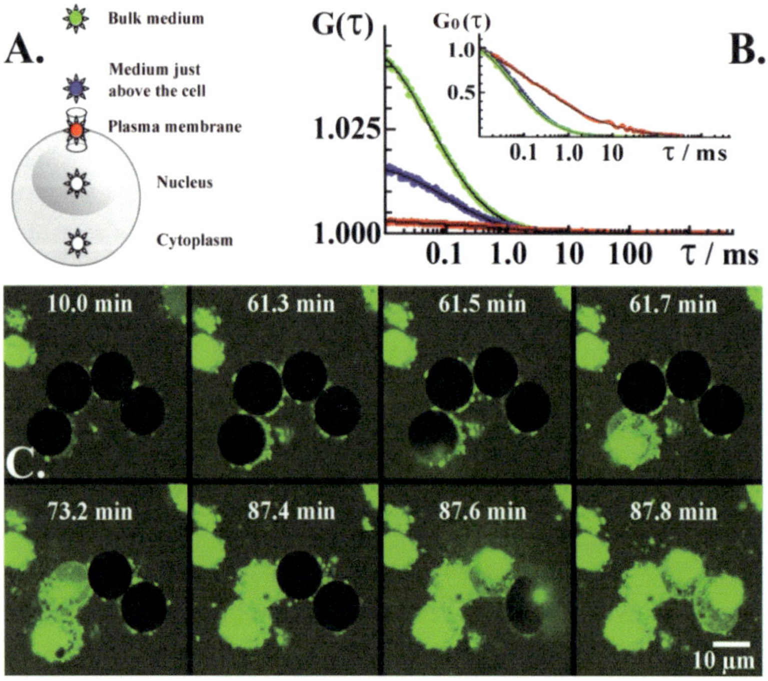

Figure 2. ELOA interaction with live PC12 cells studied by Fluorescence Correlation Spectroscopy (FCS) and APD imaging. (A) Schematic presentation of different locations – the cell culturing medium, PC12 cell plasma membrane and nucleus, at which FCS measurements were performed. **(B)** FCS measurements show that the concentration of ELOA in the bulk medium (C_{medium}^{bulk} = 240 nM) is lower than in the immediate vicinity of the cell ($c_{medium}^{cell\ surr.}$ = 670 nM) and the plasma membrane ($c_{medium}^{plasma\ membrane.}$ = 2.5 µM). **Insert:** Autocorrelation curves normalized to the same amplitude show that lateral mobility of ELOA in the plasma membrane (red) is significantly slower than in the medium (blue and green), as evident from the shift of the autocorrelation curve recorded at the plasma membrane (red) towards longer characteristic times. ELOA was neither detected in the cell nucleus nor in the cytoplasm. FCS measurements were taken 40-45 min after exposing PC12 cells to fluorescently labeled ELOA. **(C)** APD imaging shows that fluorescently labeled ELOA complexes are not gradually taken up by PC12 cells. Instead, the concentration of ELOA complexes in the immediate cellular surroundings progressively increases, reaching a local concentration that is several times higher than the concentration in the bulk medium. At a critical time-point (61.5 min), the plasma membrane ruptures. Only then the ELOA complexes "stream in" and swiftly distribute in the cellular interior, preferring particularly the cell nucleus. The scale bar is 10 µm.

8. Putative mechanism of ELOA-induced cellular toxicity

Local rearrangements of lipid organization in the plasma membrane of PC12 cells (Figure 3) observed using a general lipophilic marker that differently partitions between the ordered and disordered phase of the lipid bilayer (1,1'-dioctadecyl-3,3,3',3'-tetramethylind ocarbocyanine perchlorate dye, DiIC18(5)) [21,38], are consistent with the hypothesis that ELOA may form transient pores in the plasma membrane.

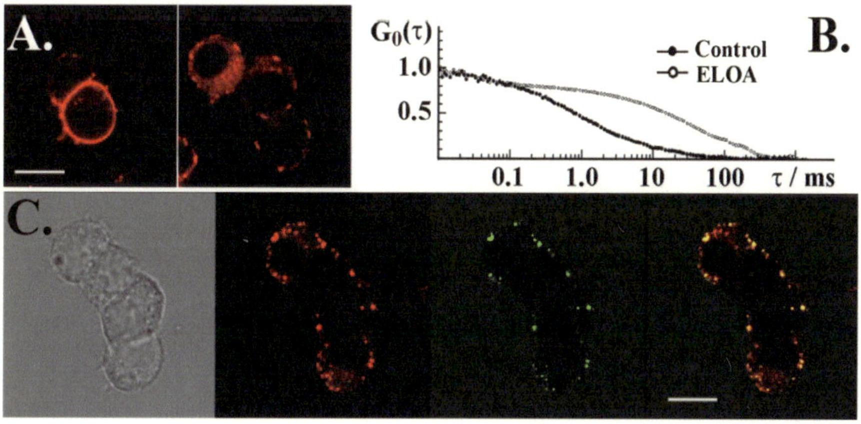

Figure 3. Lipid marker distribution in the plasma membrane of live PC12 treated with ELOA. (A) Representative image showing uniform distribution of the fluorescent lipid marker $DiIC_{18}(5)$ in PC12 cells not exposed to ELOA (left). In cells exposed to ELOA the distribution of $DiIC_{18}(5)$ becomes patchy, and regions of local accumulation could be observed (right). **(B)** $DiIC_{18}(5)$ partitioning between different regions in the plasma membrane is also affected, as evident from the shifting of the autocorrelation curve to longer characteristic times in cells treated with ELOA. (C) Transmitted-light and APD images of PC12 cells taken 40 min after exposure to ELOA show that the plasma membrane marker $DiIC_{18}(5)$ (red) colocalizes with the fluorescently labeled ELOA complexes (green). The scale bar is 10 μm.

9. Future development and prospective applications

Recently, complexes of bovine β-lactoglobulin and pike parvalbumin with OA were produced and classified as HAMLET-type complexes [39]. These proteins are neither structurally related to α-lactalbumins nor to lysozymes. Nevertheless, their complexes with OA displayed cytotoxic activity that bears a resemblance to the cytotoxic activity of HAMLET [39]. This suggests that protein self-assembly may be mediated by oleic acid and more oleic acid-protein complexes can be discovered in future. Their putative ability to eliminate specifically rapidly divided cells, such as cancer cells, has a significant therapeutic potential. The mechanisms of their toxic activity are still debated. Our research provides first insight at a single cell level that ELOA interactions with the cellular membrane play critical role in cytotoxicity, leading to membrane permeability and even rupture. There are obvious differences in the composition and structure of protein-oleic acid complexes arising due to differences in the structure and dynamics of the protein component and differences in the conditions of complex formation. The common feature of these complexes can be related to the fact that they all serve as cargo vessels delivering oleic acid to the cells and facilitating its penetration into cell membrane and cell interior. HAMLET is the first example of proteinaceous complexes with oleic acid effectively used in combating various cancer conditions and other complexes can be also potentially used for this purpose if their properties will be well-understood and controlled.

Author details

Vladana Vukojević
Department of Clinical Neuroscience, Karolinska Institute, Stockholm, Sweden

Ludmilla A. Morozova-Roche
Department of Medical Biochemistry and Biophysics, Umeå University, Umeå, Sweden

Acknowledgement

Support from the Swedish Medical Research Council, Insamlingsstiftelsen, Umeå and the Kempe foundation is gratefully acknowledged.

10. References

[1] Mossberg AK, Hun Mok K, Morozova-Roche LA, Svanborg C (2010) Structure and function of human α-lactalbumin made lethal to tumor cells (HAMLET)-type complexes. FEBS J. 277:4614-4625.
[2] Lisková K, Kelly AL, O'Brien N, Brodkorb A (2010) Effect of denaturation of alpha-lactalbumin on the formation of BAMLET (bovine alpha-lactalbumin made lethal to tumor cells). J Agric Food Chem. 58:4421-4427.
[3] Morozova-Roche LA (2007) Equine lysozyme: the molecular basis of folding, self-assembly and innate amyloid toxicity. FEBS Lett. 581: 2587-2592.
[4] Wilhelm K, Darinskas A, Noppe W, Duchardt E, Mok KH, Vukojević V, Schleucher J, Morozova-Roche LA (2009) Protein oligomerization induced by oleic acid at the solid-liquid interface--equine lysozyme cytotoxic complexes. FEBS J. 276:3975-3989.
[5] Håkansson A, Zhivotovsky B, Orrenius S, Sabharwal H, Svanborg C (1995) Apoptosis induced by a human milk protein. Proc Natl Acad Sci U S A. 92:8064-8068.
[6] Håkansson A, Andréasson J, Zhivotovsky B, Karpman D, Orrenius S, Svanborg C (1999) Multimeric alpha-lactalbumin from human milk induces apoptosis through a direct effect on cell nuclei. Exp Cell Res. 246:451-460.
[7] Svensson M, Sabharwal H, Håkansson A, Mossberg AK, Lipniunas P, Leffler H, Svanborg C, Linse S (1999) Molecular characterization of alpha-lactalbumin folding variants that induce apoptosis in tumor cells. J Biol Chem. 274:6388-6396.
[8] Svensson M, Håkansson A, Mossberg AK, Linse S, Svanborg C (2000) Conversion of alpha-lactalbumin to a protein inducing apoptosis. Proc Natl Acad Sci U S A. 97:4221-4226.
[9] Svanborg C, Agerstam H, Aronson A, Bjerkvig R, Düringer C, Fischer W, Gustafsson L, Hallgren O, Leijonhuvud I, Linse S, Mossberg AK, Nilsson H, Pettersson J, Svensson M

(2003) HAMLET kills tumor cells by an apoptosis-like mechanism--cellular, molecular, and therapeutic aspects. Adv Cancer Res. 88:1-29.
[10] Gustafsson L, Hallgren O, Mossberg AK, Pettersson J, Fischer W, Aronsson A, Svanborg C (2005) HAMLET kills tumor cells by apoptosis: structure, cellular mechanisms, and therapy. J Nutr. 135:1299-1303.
[11] Svensson M, Mossberg AK, Pettersson J, Linse S, Svanborg C (2003) Lipids as cofactors in protein folding: stereo-specific lipid-protein interactions are required to form HAMLET (human alpha-lactalbumin made lethal to tumor cells). Protein Sci. 12:2805-2814.
[12] Svensson M, Fast J, Mossberg AK, Düringer C, Gustafsson L, Hallgren O, Brooks CL, Berliner L, Linse S, Svanborg C (2003) Alpha-lactalbumin unfolding is not sufficient to cause apoptosis, but is required for the conversion to HAMLET (human alpha-lactalbumin made lethal to tumor cells). Protein Sci. 12:2794-2804.
[13] Pettersson-Kastberg J, Aits S, Gustafsson L, Mossberg A, Storm P, Trulsson M, Persson F, Mok KH, Svanborg C (2009) Can misfolded proteins be beneficial? The HAMLET case. Ann Med. 41:162-176.
[14] Svensson M, Håkansson A, Mossberg AK, Linse S, Svanborg C (2000) Conversion of alpha-lactalbumin to a protein inducing apoptosis. Proc Natl Acad Sci U S A. 97:4221-4226.
[15] Morozova L, Haezebrouck P, Van Cauwelaert F (1991) Stability of equine lysozyme. I. Thermal unfolding behaviour. Biophys Chem. 41:185-191.
[16] Van Dael H, Haezebrouck P, Morozova L, Arico-Muendel C, Dobson CM (1993) Partially folded states of equine lysozyme. Structural characterization and significance for protein folding. Biochemistry. 32:11886-11894.
[17] Morozova LA, Haynie DT, Arico-Muendel C, Van Dael H, Dobson CM (1995) Structural basis of the stability of a lysozyme molten globule. Nature Struct. Biol. 10:171- 175.
[18] Morozova-Roche LA, Arico-Muendel C, Haynie DT, Emelyanenko VI, Van Dael H, Dobson CM (1997) Structural characterisation and comparison of the native and A-states of equine lysozyme. J. Mol. Biol. 268: 903-921.
[19] Morozova-Roche LA, Jones JA, Noppe W, Dobson CM (1999) Independent nucleation and heterogeneous assembly of structure during folding of equine lysozyme. J. Mol. Biol. 289: 1055-1073.
[20] Stefani M (2007) Generic cell dysfunction in neurodegenerative disorders: role of surfaces in early protein misfolding, aggregation, and aggregate cytotoxicity. Neuroscientist. 13:519-531.
[21] Vukojević V, Bowen AM, Wilhelm K, Ming Y, Ce Z, Schleucher J, Hore PJ, Terenius L, Morozova-Roche LA (2010) Lipoprotein complex of equine lysozyme with oleic acid (ELOA) interactions with the plasma membrane of live cells. Langmuir. 26:14782-14787.
[22] Mok KH, Hore PJ (2004) Photo-CIDNP NMR methods for studying protein folding. Methods. 34:75-87.
[23] Fast J, Mossberg AK, Svanborg C, Linse S (2005) Stability of HAMLET-a kinetically trapped alpha-lactalbumin oleic acid complex. Protein Sci. 14:329-340.
[24] Malisauskas M, Ostman J, Darinskas A, Zamotin V, Liutkevicius E, Lundgren E, Morozova-Roche LA (2005) Does the cytotoxic effect of transient amyloid oligomers from common equine lysozyme in vitro imply innate amyloid toxicity? J Biol Chem. 280:6269-6275.

[25] Campioni S, Mannini B, Zampagni M, Pensalfini A, Parrini C, Evangelisti E, Relini A, Stefani M, Dobson CM, Cecchi C, Chiti F (2010) A causative link between the structure of aberrant protein oligomers and their toxicity. Nat Chem Biol. 6:140-147.
[26] Chiti F, Dobson CM (2006) Protein misfolding, functional amyloid, and human disease. Annu. Rev. Biochem. 75:333-366.
[27] Morozova-Roche LA, Malisauskas M (2007) A false paradise - mixed blessings in the protein universe: the amyloid as a new challenge in drug development. Curr. Med. Chem. 14:1221–1230.
[28] Morozova-Roche LA, Zurdo J, Spencer A, Noppe W, Receveur V, Archer DB, Joniau M, Dobson CM (2000) Amyloid fibril formation and seeding by wild-type human lysozyme and its disease-related mutational variants. J Struct Biol. 130:339-351.
[29] Goers J, Permyakov SE, Permyakov EA, Uversky VN, Fink AL (2002) Conformational prerequisites for alpha-lactalbumin fibrillation. Biochemistry. 41:12546–12551.
[30] Harrison RF, Hawkins PN, Roche WR, MacMahon RF, Hubscher SG, Buckels JA (1996) 'Fragile' liver and massive hepatic haemorrhage due to hereditary amyloidosis. Gut. 38:151-152.
[31] Malisauskas M, Zamotin V, Jass J, Noppe W, Dobson CM, Morozova-Roche LA (2003) Amyloid protofilaments from the calcium-binding protein equine lysozyme: formation of ring and linear structures depends on pH and metal ion concentration. J Mol Biol. 330:879-890.
[32] Nagarajan S, Ramalingam K, Neelakanta Reddy P, Cereghetti DM, Padma Malar EJ, Rajadas J (2008) Lipid-induced conformational transition of the amyloid core fragment Abeta(28-35) and its A30G and A30I mutants. FEBS J. 275:2415-2427.
[33] Otzen DE, Sehgal P, Westh P (2009) Alpha-Lactalbumin is unfolded by all classes of surfactants but by different mechanisms. J Colloid Interface Sci. 329:273-283.
[34] Sharon R, Bar-Joseph I, Frosch MP, Walsh DM, Hamilton JA, Selkoe DJ (2003) The formation of highly soluble oligomers of alpha-synuclein is regulated by fatty acids and enhanced in Parkinson's disease. Neuron. 37:583–595.
[35] Otzen DE, Nesgaard LW, Andersen KK, Hansen JH, Christiansen G, Doe H, Sehgal P (2008) Aggregation of S6 in a quasi-native state by sub-micellar SDS. Biochim. Biophys. Acta 1784:400–414.
[36] Cecchi C, Baglioni S, Fiorillo C, Pensalfini A, Liguri G, Nosi D, Rigacci S, Bucciantini M, Stefani M (2005) J. Cell Sci. 118:3459-3470.
[37] Vukojević V, Heidkamp M, Ming Y, Johansson B, Terenius L, Rigler R (2008) Quantitative single-molecule imaging by confocal laser scanning microscopy. Proc Natl Acad Sci U S A. 105:18176-1081.
[38] Loura LM, Fedorov A, Prieto M (2000) Partition of membrane probes in a gel/fluid two-component lipid system: a fluorescence resonance energy transfer study. Biochim Biophys Acta. 1467:101-112.
[39] Permyakov SE, Knyazeva EL, Khasanova LM, Fadeev RS, Zhadan AP, Roche-Hakansson H, Håkansson AP, Akatov VS, Permyakov EA (2012) Oleic acid is a key cytotoxic component of HAMLET-like complexes. Biol Chem. 393:85-92.

Section 9

Lipoproteins in Inflammatory and Infectious Diseases

Chapter 29

Adiponectin: A Perspective Adipose Tissue Marker with Antiinflammatory and Antiaterogenic Potencial

Dalibor Novotný, Helena Vaverková and David Karásek

Additional information is available at the end of the chapter

http://dx.doi.org/10.5772/47754

1. Introduction

Adipose tissue as a substantial part of the human body contains about 10 % of body mass. It serves both as a reservoir of the energy storage and the active endocrine tissue producing many proactive substances including adipokines. These molecules have many important metabolic effects [1]. Adiponectin is an adipose tissue-derived adipokine which circulates at relatively high concentrations in blood. It has protective role in the initiation and progression of atherosclerosis through its antiinflammatory and antiatherogenic effects. Adiponectin serum levels are decreased in obesity, type 2 diabetes, and patients with coronary artery disease, etc [2]. The level of circulating adiponectin correlates positively with HDL cholesterol, and negatively with inflammatory markers, markers of insulin resistance, triglyceride-rich lipoprotein particles, and other adipokines. Adiponectin disposes of protective actions on development of various obesity-linked diseases. The antiinflammatory properties may be the major component of its beneficial effects on cardiovascular and metabolic disorders including atherosclerosis and insulin resistance. In addition, adiponectin displays a direct biological activity through the induction of a classical pathway of complement activation.

2. Adiponectin and atherosclerosis

Human adiponectin is a protein containing 244 amino acids. It is produced by apM1 cDNA transcripts. Adiponectin consists of two structurally distinct domains and the C-terminal part is likely to be involved in protection against atherosclerosis (Figure 1).

As a member of the soluble collagen superfamily, adiponectin has a structural homology with collagen type VIII, X, complement C1q and tumor necrosis factor alpha family [2]. In human plasma, adiponectin is present in a variety of heterogeneous isoforms, from large multimeric structures of high molecular weight to trimeric isoform. Monomeric one is present only in adipose tissue. The biological activity of various multimeric isoforms are not fully known yet, but it appears different isoforms have varying effects in different diseases. Although some studies have proposed that the ratio of high molecular weight (HMW) form to the other forms may serve as a better indicator of metabolic disorders, the majority of studies that have linked adiponectin to metabolic diseases have used assays for total adiponectin.

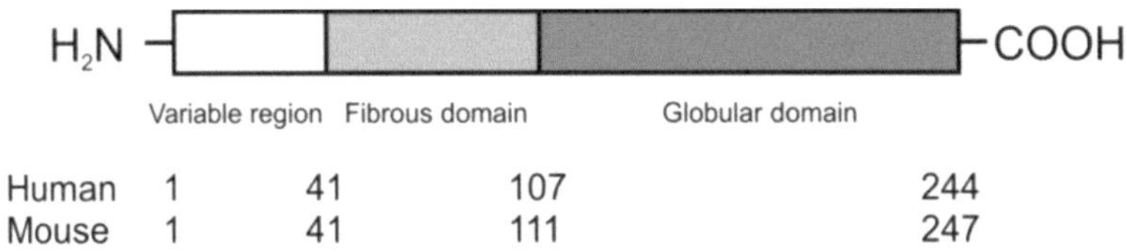

Figure 1. Schematic presentation of adiponectin structure (adapted from [2]).

Adiponectin gene is located on chromosome 3q27 and contains 3 exons and 2 introns. In 2003, DNA sequences encoding two receptors for adiponectin Adipo R1 and Adipo R2 were identified [3]. They are localized on chromosome 1 (1p36.13-q41) and 12 (12p13.31), respectively, with expression in most organs (AdipoR1 in skeletal muscle, AdipoR2 in the liver, in particular). Adiponectin gene is polymorphic, located in the region that contains susceptible loci for type 2 diabetes mellitus and metabolic syndrome. A number of single nucleotide polymorphisms (SNPs) and missense mutations were observed, especially in exons 2, 3 and the gene promoter.

Figure 2 schematically depicts some of the antiatherogenic properties of adiponectin towards different types of cells that have been established in experimental models. Adiponectin negatively regulates the expression of TNF alpha and C-reactive protein (CRP) in adipose tissue. On the contrary, its expression is negatively regulated by TNF alpha and interleukin 6 (IL 6). Adiponectin reduces expression of vascular and intracellular adhesion molecules (VCAM 1, ICAM 1), E-selectin, interleukin 8, and monocyte adhesion to human aortal endothelial cells after their stimulation with TNF alpha [4]. The proliferation and migration of smooth muscle cells induced by platelet growth factor (PDGF) is abolished or diminished by adiponectin action as inhibition of activation of nuclear factor kappa B in endothelial cells. This effect is partially mediated by its ability to support the action of cyclic adenosine monophosphate - proteinkinase A system (cAMP-PKA).

In endothelial cells, adiponectin inhibits the production of reactive oxygen species (ROS) induced by high levels of glucose via above mentioned the cAMP-PKA system. Adiponectin inhibits macrophage transformation to foam cells and reduces the intracellular content of cholesterol esters via suppression of expression of scavenger receptors, class A (SR-A). In these cells, adiponectin reduces lipopolysaccharides stimulated TNF alpha production. Recent clinical trials show a positive correlation of plasma levels of adiponectin and IL 10 [5]. In accordance with these findings, adiponectin has an antiatherogenic properties in mice models. Adenovirus-mediated supplementation of adiponectin inhibits the formation of atherosclerotic lesions and reduces the levels of mRNA of SR-A, TNF alpha and VCAM 1 in the vascular wall [6]. It is interesting that in these models, adiponectin has no effect on glucose and lipid parameters. The authors conclude that adiponectin affects atherogenesis through a series of antiinflammatory effects on macrophages and vascular endothelium.

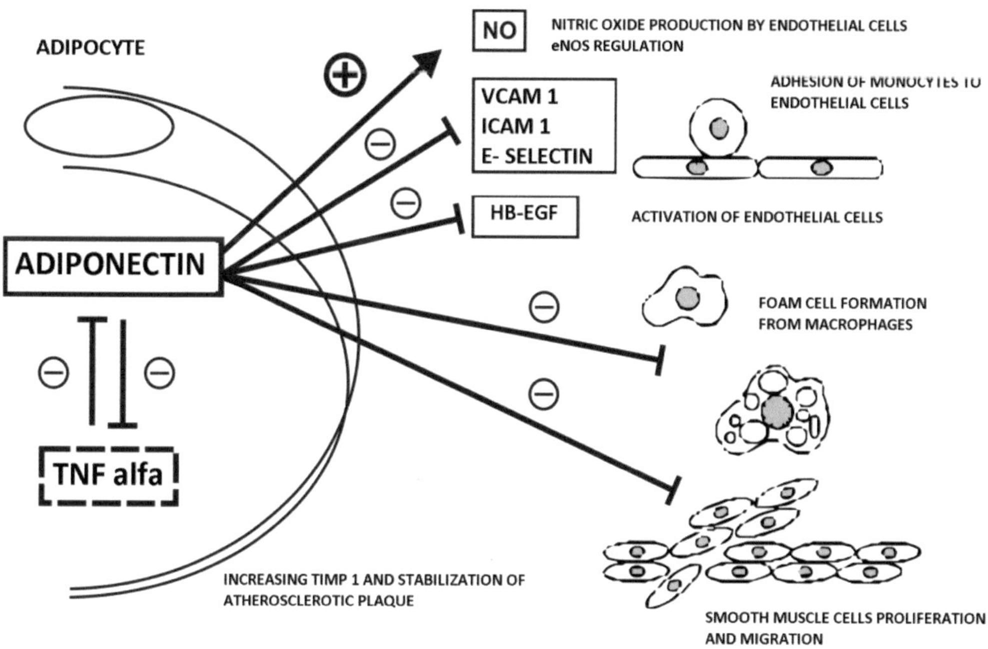

Figure 2. Some of protective actions of adiponectin (adapted from [2]).

A very important finding was observed in recent work describing the relationship of adiponectin-immune system. Adiponectin is able to bind to a number of target molecules, including the damaged endothelium and the surface of apoptotic cells. The significance of this phenomenon is not entirely clear. The study describes in vitro binding of purified C1q complement to recombinant adiponectin and dependence on calcium and magnesium ions. It was found that this binding stimulates the classical pathway of complement activation. Adiponectin does serve as an antiinflammatory factor, but may also induce biological activity through activation of complement. The authors hypotesize the binding of C1q leads to conformational changes in the adiponectin molecule, which induces the classical pathway of complement activation. Adiponectin may play an important role in immunity by its direct biological effect [7]. There is also evidence of adiponection accumulation on injured vascular arterial wall (but not in healthy one). This may lead to the hypothesis of the "consumption" of circulating adiponectin in patients with ischemic heart disease.

In some recent studies, adiponectin has a positive effect in endothelial homeostasis. It acts as a regulator of the enzyme endothelial nitric oxide synthase (eNOS), which is a key determinant of endothelial function and angiogenesis (the production of NO inhibits the inflammatory response in the arteries), and also promotes phosphorylation of eNOS in endothelial cells, increases its expression and induces NO production after suppression of its activity caused by the effect of oxidized low-density lipoproteins (oxLDL) [8].

Adiponectin promotes cyclooxygenase 2 (COX-2) expression and prostaglandin E2 (PGE2) synthesis in cardiac cells. It also has an antiapoptotic properties in vitro, as in endothelial cells. In the heart tissue adiponectin thus acts as a regulator of cardiac damage through its antiinflammatory effect and as a factor preventing the reconstruction of cardiac tissue. In order to become a useful biomarker of cardiovascular risk, it is necessary to determine which of its isoforms exhibit cardioprotectivity, and to clarify mechanism of their action in various pathophysiological conditions [7].

There is an increasing number of papers on experimental models point to the fact that adiponectin plays an important protective role in the development of insulin resistance and diabetes. Severe insulin resistance was seen in adiponectin-deficient knockout mice (KO-AD) after administration of high fat and/or carbohydrates diets. Administration of adiponectin led to reduced hyperglycemia in the diabetic mice without affecting insulin levels. In another study, increased muscle fatty acid oxidation and reduction of plasma glucose, free fatty acids and triglycerides were observed. Studies on experimental animal models revealed the administration of adiponectin has a beneficial action against the development of obesity and atherosclerosis. It seems that adiponectin acts not only as a factor increasing insulin sensitivity, and the protective effect may result from its ability to suppress production of proinflammatory cytokine [4].

2.1. Adiponectin and its relationship to obesity and metabolic syndrome

There has been a growing evidence of significantly reduced levels of adiponectin in obese individuals compared to subjects with normal body mass index (BMI) [9]. An inverse relationship with BMI was observed in both men and women, as well as negative correlation of adiponectin with visceral fat accumulation. It is obvious that hypoadiponectinemia (levels typically less than 4 mg/l) is associated with the development of insulin resistance and type 2 diabetes mellitus, independently of BMI and metabolic syndrome. Low adiponectinemia are considered the independent risk factor for developing hypertension. Kern et al. measured adiponectin plasma concentrations and mRNA levels in adipose tissue in nondiabetic subjects with varying degree of obesity and IR. They found a strong correlation of these two parameters. The obese individuals had significantly lower plasma adiponectin. When BMI was less than 30 kg/m2, women had twice more the body fat than men, but adiponectin levels were higher on average of 65% than in men (14.2 mg/l vs. 8.6 mg/l). Individuals with the highest levels of mRNA secreted the lowest levels of TNF alpha in adipose tissue. The authors conclude that adiponectin expression in adipose tissue is highest in lean subjects and women, and correlates with higher index of insulin sensitivity and lower TNF alpha expression [9]. Another study found that expression of adiponectin mRNA in adipose tissue may reflect short-term energy changes in some obese subjects. Expression of adiponectin and insulin sensitivity may be influenced by genetic variations in the adiponectin gene in response to acute energy fluctuations [10].

Metabolic syndrome characterized by abdominal obesity, dyslipidemia, hypertension and hyperglycemia, is a general risk for the development of atherosclerotic vascular disease. The study with 661 Japanese individuals investigated possible application of adiponectin as a biomarker of the metabolic syndrome [11]. Its plasma levels negatively correlated with waist circumference, visceral fat, TGL concentration, glucose and fasting insulinemia, systolic and diastolic blood pressure, and positively with HDL cholesterol. With decreasing levels of adiponectin, on the contrary, the number of components of metabolic syndrome increased. Total of 52% men and of 38% women with levels below 4.0 mg/l met criteria for MS. The authors suppose hypoadiponectinemia is closely associated with the clinical phenotype and its measurement could be useful in the MS treatment. Saely et al. observed a group of patients undergoing coronary angiography. Low adiponectin levels were independently associated with both metabolic syndrome and angiographically confirmed coronary atherosclerosis [12]. The highest levels of adiponectin were seen in subjects without MS and heart disease (12.1 ± 8.3 mg/l), whereas the lowest levels in patients with MS and presence of heart disease (6.7 ± 3.8 mg/l). Another study then identified a link between serum adipokines and cholesterol metabolism in individuals with MS. In 58 subjects with impaired glucose tolerance or elevated fasting glucose and signs of MS the markers of cholesterol synthesis were measured (determined by the ratio of non-cholesterol sterols to cholesterol and dietary cholesterol portion), in relation to adipokines and ultrasensitive CRP (hsCRP). It was found that adiponectin, leptin and CRP were associated with cholesterol metabolites (variations) and the high ratio of cholesterol synthesis to its absorption is characterized by high levels of serum leptin and low adiponectin [13].

2.2. Adiponectin and its relationship to heart disease

The hypoadiponectinemia was also found in patients with angiographically documented coronary atherosclerosis or acute coronary syndrome. In men, plasma adiponectin significantly predicted the extent of coronary atherosclerosis [14]. A prospective study of patients with end-stage renal disease showed an inverse relationship between cardiovascular events and adiponectinemia. Higher adiponectin levels represent a low risk of myocardial infarction in healthy men individuals and moderately reduced risk of coronary heart artery disease in diabetic men patients [4]. In contrast, adiponectin concentrations did not correlate significantly with the risk of heart disease in American Indians or the British women. A large prospective study involving British men with heart disease combined with a meta-analysis of seven previously published studies found only a weak association of adiponectin with the disease [4]. This inconsistent data could be due to differences in study populations (ethnicity, gender, type of disease etc.). In any cases, it remains unclear whether hypoadiponectinemia is a reliable indicator of heart disease.

2.3. Adiponection and inflammation

CRP is known to be an independent predictor of future risk for cardiovascular events and risk factor for developing MS. Its positive association with BMI is considered as a useful biomarker for chronic inflammation linked to obesity. Plasma levels of CRP correlated negatively with adiponectin levels [15] which was confirmed by various studies. Since CRP mRNA in humans is expressed in adipose tissue, adiponectin can apparently participate in influencing CRP levels in plasma by regulation of its expression. The regulation of CRP synthesis in the liver is also influenced by proinflammatory adipokines IL 6 and TNF alpha. On the one side, adiponectin expression is regulated by proinflammatory cytokines, on the other side, adiponectin modulates the activity and the production of TNF alpha in different tissues. Several studies have found links between hypoadiponectinemia and elevated serum IL 6. So far, there is no evidence of a link between adiponectin and TNF alpha in plasma in humans. Nishida et al. describe the action of IL 6, adiponectin, CRP and metabolic syndrome in subclinical atherosclerosis [16]. The relative influence of these parameters on a group of healthy subjects was observed. In 714 men and 364 women aged 40 to 59 years, thickness of the intima-media complex (IMT), pulse blood flow velocity and components of MS were measured. IL 6 levels correlated with IMT parameter, while adiponectin correlated negatively with IMT only in men. Individuals with either high IL 6 or CRP, or low levels of adiponectin, had increased IMT in the presence of MS. Increasing number of MS components was expressed more strongly in women than in men. The authors speculate IL 6 and adiponectin are important risk factors for premature arterial alterations in men.

In another study, the relationship of adiponectin to markers of inflammation, atherogenic dyslipidemia and heart disease was investigated in patients with coronary artery disease [17]. Study participants were in a rehabilitation program to reduce the cardiovascular risk factors. After adjusting for age and sex, adiponectin was associated positively with HDL cholesterol and N- terminal propeptide of B natriuretic peptide (NT-proBNP), while the association was negative for triglycerides. In this study, no relationship was found with markers of inflammation. The same results were obtained after next adjustment for other parameters; BMI, alcohol intake, smoking, presence of diabetes and/or hypertension and lipid-lowering therapy, and fasting glucose. The authors conclude serum adiponectin is associated with the presence of the atherogenic dyslipidemia and NT-proBNP levels, but not with markers of systemic inflammation (IL 6, CRP) in patients with manifest coronary heart disease. Atherogenic dyslipidemia may be a link between adiponectin and progression of atherosclerosis. The role of systemic inflammation as part of the adiponectin-atherosclerosis relationship may decrease during the course of the disease, and could be more amplified in the earlier stage of disease development.

3. Adiponectin and gene polymorphisms

As mentioned above, the adiponectin gene is located on chromosome 3q27, containing 3 exons and 2 introns. This region also encopasses the susceptibile loci for type 2 diabetes and metabolic syndrome. The sequence polymorphism was found in the form of several single nucleotide polymorphisms (SNPs) and a number of missense mutations. Sequence analysis of the gene for adiponectin in Japanese and Caucasian populations found more than 10 SNPs, some of which are associated with BMI, metabolic syndrome, insulin sensitivity, hyperglycemia, type 2 diabetes, levels of plasma adiponectin, etc. The results of studies, however, are inconsistent, providing conflicting results. In many cases, the haplotype analysis was performed from a combination of alleles of individual SNPs.

Kondo et al. analyzed a cohort of Japanese patients with type 2 diabetes and nondiabetic controls to detect mutations in the gene for adiponectin [18]. Four missense mutations in the globular domain (I+164T, R+112 C, H+241P, R+221S) were identified. The frequency of one mutation, the substitution of I+164T, was significantly higher in patients than in controls of comparable age and body weight. Mutation carriers had lower adiponectin concentrations in plasma and also showed the presence of a feature characteristic of the metabolic syndrome (hypertension, hyperlipidemia, diabetes, atherosclerosis). Hypoadiponectinemia was already evident at the same time in heterozygotes I+164 T mutation carriers and also in R +112 C, but this was the case of only 3 patients. The authors suggest I+164 T variant is associated with low adiponectin levels in plasma and type 2 diabetes mellitus.

Another study has examined the adiponectin gene locus as a candidate site for coronary artery disease [19]. 383 Japanese patients with angiographically confirmed disease and 318 individuals adjusted for age and BMI were the subjects of this study. Analyses of SNPs were performed using real time polymerase chain reaction (rtPCR) and restriction fragment length polymorphism (RFLP). In patients, the higher incidence of T+164 mutation and lower adiponectin levels in plasma were seen, independently of BMI. Subjects with the mutation showed a clinical phenotype of metabolic syndrome. According to the authors, the I+164T polymorphism is associated with metabolic syndrome and coronary artery disease in Japanese population.

Hara et al. examined the relationship between two SNPs located at exon 2 of adiponectin gene (T+45G and G+276T) and type 2 diabetes in the Japanese population [20]. Subjects with the GG genotype at position +45 or +276 had an increased risk of DM compared to TT genotypes. GG +276 homozygotes showed higher insulin resistance index and the presence of G allele at position 276 was characterized by lower levels of plasma adiponectin in subjects with higher BMI (GG: 10.4 mg/l, TT: 16.6 mg/l). The different results showed the study focused on the relationship between haplotypes of the adiponectin gene with obesity and other signs of metabolic syndrome in nondiabetic Caucasian population [21]. Both polymorphisms, T+45G and G+276T, separately significantly correlated with IR. The common haplotype was also closely associated with a number of components of metabolic syndrome. Homozygotes for middle-risk haplotype TG (i.e. individuals with +45 TT variant and +276 GG variant) had higher body weight, waist circumference, blood pressure, fasting glucose, insulin, cholesterol/ HDLcholesterol ratio and lower adiponectin levels, after adjustment for age, sex and body weight. However, in the second group (614 Caucasian individuals with type 2 DM) the risk haplotype was associated with increased body weight, not with DM. It is hypothesized the variability of the adiponectin gene is connected with obesity and other features of insulin resistance, but the risk haplotype is probably a marker of linkage disequilibrium with a polymorphism yet unidentified that directly affects the plasma levels of adiponectin and insulin sensitivity. Moreover, Fillipi et al. found no association of SNP T+45G with insulin resistance [22]. The T+276 G polymorphism was associated with higher BMI, lower insulin and adiponectin, but, unlike previous study, in the TT genotype. In discussion the authors analyzed possible causes of these results and conclude the same mentioned above. There is the high probability of the existence of further SNPs or gene mutations, which is in linkage disequilibrium with SNP +276 and which determines its effect. Variations in the adiponectin gene and risk for subsequent type 2 diabetes in women has been of interest in the study of Hu et al. [23].

A prospective study focused on the determination of SNPs participation in the development of IR in French population found that variations in the adiponectin gene affects weight gain, body fat distribution and the development and the onset of hyperglycemia, as well as serum adiponectin [24]. At the start of a three-year study, the normoglycemic individuals with no signs of diabetes or impaired glucose tolerance were influenced mainly by two SNPs: G-11391A and T+45G.

An interesting work was published in 2006 in Clinical Chemistry by Hegener et al. [24]; the prospective study monitoring the risk of atherothrombotic disease in individuals with no signs of diabetes. Five SNPs in the gene for adiponectin were investigated in 600 Caucasian men with subsequent aterotrombotic events (myocardial infarction or stroke) and 600 controls. After adjustment for potential risk factors, regression analysis then revealed two variants with a decreased risk of stroke (C-11377G and G-11066A). This study has provided evidence of links of specific adiponectin gene variants with reduced risk of stroke.

3.1. Relationship between G+276T single nucleotide polymorphism of adiponectin gene and markers of insulin resistance in dyslipidemic patients

In many recent studies, the adiponectin gene has been proposed as a potential candidate gene for insulin resistance but only a few of them have confirmed this relationship. Insulin resistance is considered the key factor in the patogenesis of common disorders, such as atherosclerosis, metabolic syndrome and diabetes mellitus. The genetic backround is likely to be polygenic but the genes involved are mostly unknown.

In our work, we have studied the possible relationship between single nucleotide polymorphism G+276T and IR markers, including lipid and lipoprotein profiles and adiponectin plasma levels in 355 dyslipidemic patients and their first-degree relatives.

3.2. Subjects

The group consisted of 355 patients attending Lipid Center of 3rd Medical Clinic, Faculty Hospital Olomouc, and their first-degree relatives. Patients had the first examination between January 2004 and January 2006. All patients were examinated by a physician and the family history were collected and medical history with physical examinations were performed. All individuals were tested for secondary hyperlipidemia, especially on the presence of diabetes mellitus, hypothyroidism, hepatic and renal failure and nephrotic syndrome. Violation of the following criteria led to exclusion from the study: hypolipidemic treatment in the previous 6 weeks, the presence of secondary hyperlipidemia, acute infection, acute cardiovascular or cerebrovascular attack within the past 3 months, cardiac disease (NYHA III and IV). Participiants were also divided into three groups. Group G1 included the presence of individuals with clinically manifest atherosclerosis, the group G2 individuals with dyslipidemia defined by Sniderman [25] (apolipoprotein B > 1.2 g/l and/or triglycerides > 1.5 mmol/l) but without clinical signs of the presence of atherosclerosis. Group 3 consisted of healthy individuals with the apolipoprotein B < 1.2 g/l and triglycerides < 1.5 mmol/l. The participants signed informed consent before taking a blood sample for DNA testing. The study was approved by the Ethical Committee of the Faculty of Medicine, Faculty Hospital Olomouc.

3.3. Materials and methods

Venous blood for biochemical tests were collected after 12-hour fasting. Total cholesterol, HDL cholesterol and triglycerides were determined enzymatically using an analyzer Modular SWA (Roche, Switzerland), as well as other routine biochemical analyses. LDL cholesterol was calculated using the Friedewald equation for specimens with TG < 4.5 mmol/l (available for 242 subjects). Concentrations of apolipoproteins AI and B were determined by immunoturbidimetric method, as well as C-reactive protein levels, established by highly sensitive method (all Roche, Switzerland). Insulin was determined by IRMA (Immunotech, France). HOMA parameter (homeostatic model) was calculated from the formula: fasting glucose x fasting insulin / 22.5. C-peptide and proinsulin were determined by commercially available kits (Immunotech, France, DRG Instruments GmbH, Germany, respectively). Serum levels of soluble adhesion molecules ICAM 1 and VCAM 1 were analyzed by immunoenzymatic technique (Immunotech, France). Adiponectin determination was performed by the ELISA method (BioVendor, Czech Republic). The following markers of endothelial dysfunction were examined: plasminogen activator inhibitor-1 (PAI-1) and tissue plasminogen activator (tPA), both determined by ELISA methods (Technoclone, Vienna, Austria). Concentrations of adhesion molecules, insulin, proinsulin, C-peptide and adiponectin were measured on samples frozen at - 80 ° C until analysis.

G+276T adiponectin gene SNP was detected by real time polymerase chain reaction with fluorescent hybridization probes (FRET) on the Light Cycler instrument, v.2.0 (Roche), according Fillipi et al [22]. Genotyping was performed after the isolation of DNA from peripheral blood samples using phenol method [26]. DNA isolates were then stored at - 20 °C until analysis. The primer and probe synthesis was made at the in TibMolbiol (Germany). The sequence of oligonucleotides for the detection of SNP +276 G> T were as follows:

Primers:

5'- GGC CTC TTT CAT CAC AGA CC -3'

5'- AGA TGC AGC AAA GCC AAA GT -3'

Probes:

5'- AAG CTT TGC TTT CTC CCT GTG TCT A--FL

5'- LCRed640- GCC TTA GTT AAT AAT GAA TGC CTT—PH

Individual genotypes were determined by melting curve analysis after the amplification process. The fluorescence signal was converted and delivered to the graph as the dependency of negative fluorescence change with temperature (y axis) on temperature (x axis). As the result, creation of the characteristic peaks representing the melting temperature of the product and allow to distinguish the genotypes GG, GT and TT was performed. Example of analysis is shown in Figure 2.

3.4. Statistical analysis

Quantitative data were expressed as a mean ± standard deviation. Parameters with abnormal distribution were logarithmically transformed before statistical analysis. Differences between genotypes in continuous variables were determined by using ANOVA after adjustment for age, gender and waist circumference (SPSS 12.0 statistical package, SPSS Inc., USA). Furthermore, the calculation of frequency of alleles (G and T) and genotypes (GG, GT and TT) in individual groups and subgroups were performed.

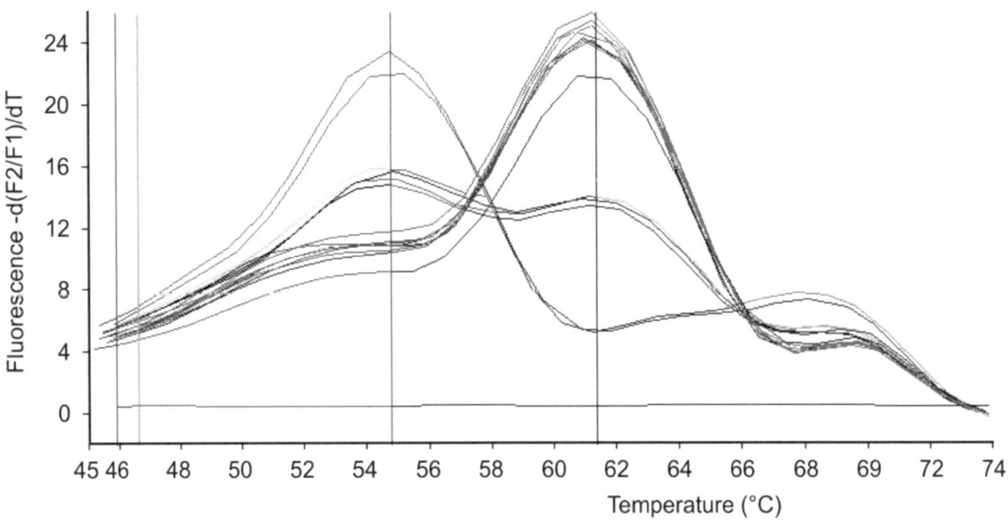

Figure 3. An example of the melting curve analyses for G+276T polymorphism of adiponectine gene. (melting temperature for T and G alelles: $T_{m(T)} = 54.8 \pm 1.5$ ºC, $T_{m(G)} = 61.3 \pm 1.5$ ºC).

3.5. Results

In Table 1 the clinical and laboratory characteristics of the groups of dyslipidemic patients divided according to genotypes at position +276 of the gene for adiponectin are shown. Table 2 presents the results of laboratory parameters that differed significantly between each of groups determined by genotype at position +276. The data are adjusted for age, gender and waist circumference. The results show that the GG genotype carriers had significantly higher levels of total cholesterol (GG: 6.54 ± 1.74 mmol/l, GT: 6.18 ± 1.45 mmol/l, TT: 6.25 ± 1.64 mmol/l, $p < 0.05$) and LDL cholesterol (GG: 4.12 ± 1.49 mmol/l, GT: 3.78 ± 1.31 mmol/l, TT: 3.70 ± 1.34 mmol/l, $p < 0.05$) than T allele carriers. In heterozygotes, however, the presence of T allele at position +276 was associated with higher concentrations of PAI-1 (GG: 71.50 ± 41.0 μg/l, GT: 81.0 ± 38.7 μg/l, TT: 70.14 ± 44.4 μg/l, $p < 0.05$). We did not find any significant association with other markers of IR, such as BMI, blood glucose, insulin, or serum adiponectin. Table 3 depicts the frequencies of genotypes and alleles at position +276, Table 4 then presents the distribution of genotypes in groups according to triglyceride levels (cut-off value of TGL = 1.5 mmol/l).

	GG	GT	TT
Number	188	144	23
BMI (kg/m^2)	26 ± 4	26 ± 4	26 ± 5
Systolic blood pressure (mmHg)	130 ± 18	132 ± 18	129 ± 14
Diastolic blood pressure (mmHg)	80 ± 10	83 ± 9	80 ± 7
Total cholesterol (mmol/l) *	6.47 ± 1.73	6.18 ± 1.44	6.26 ± 1.64
Triglycerides (mmol/l)	2.38 ± 2.97	2.43 ± 2.23	2.30 ± 3.44
HDL cholesterol (mmol/l)	1.47 ± 0.44	1.44 ± 0.44	1.46 ± 0.39
LDL cholesterol (mmol/l) *	4.07 ± 1.47	3.79 ± 1.31	3.70 ± 1.34
Apolipoprotein AI (g/l)	1.55 ± 0.30	1.57 ± 0.34	1.60 ± 0.29
Apolipoprotein B (g/l)	1.21 ± 0.39	1.15 ± 0.32	1.15 ± 0.32
hsCRP (mg/l)	1.99 ± 1.94	2.20 ± 1.93	2.02 ± 1.69
tPA (μg/l)	4.08 ± 4.81	4.31 ± 4.46	4.03 ± 3.89
PAI-I (μg/l) *	69.7 ± 40.7	79.9 ± 39.0	72.3 ± 44.4
VCAM 1(μg/l)	808 ± 247	823 ± 287	743 ± 184
ICAM 1 (μg/l)	563 ± 140	592 ± 165	585 ± 209
Fasting glucose (mmol/l)	5.09 ± 0.91	5.25 ± 1.23	4.82 ± 0.68
Insulin (U/l)	8.33 ± 5.55	7.99 ± 4.84	7.77 ± 4.54
HOMA IR	1.93 ± 1.45	1.92 ± 1.33	1.84 ± 1.18
C- peptide (μg/l)	2.38 ± 1.25	2.40 ± 1.12	2.41 ± 1.30
Adiponectin (mg/l)	12.9 ± 7.6	13.0 ± 7.0	12.0 ± 5.7

* GG vs. GT+TT, $p < 0.05$

Table 1. Clinical and laboratory characteristics according to adiponectin genotypes at position +276 (G+276T).

	GG	GT	TT	p
Number	188	144	23	
Total cholesterol (mmol/l)	6.54 ± 1.74	6.18 ± 1.45	6.25 ± 1.64	< 0.05
LDL cholesterol (mmol/l)*	4.12 ± 1.49	3.78 ± 1.31	3.70 ± 1.34	< 0.05
PAI-I (μg/l)	71.5 ± 41.0	81.0 ± 38.7	70.14 ± 44.4	< 0.05

*only 242 patients included

Table 2. Laboratory characteristics according to adiponectin genotypes at position +276 (G+276T) with significant differences between groups (GG vs. GT+TT, after adjustment for sex, age and BMI).

APM1 G+276T	Patients (n = 355)
Genotype	
GG	188 (53 %)
GT	144 (41 %)
TT	23 (6 %)
Allele	
G	520 (73 %)
T	190 (27 %)

Table 3. Genotype and allele frequencies for G+276T polymorphism in dyslipidemic patients.

Genotype	GG	GT	TT
TG ≤ 1.5 (n = 225)	119 (53 %)	94 (42 %)	12 (5 %)
TG > 1.5 (n = 148)	85 (57 %)	52 (35 %)	11 (8 %)

Chi-square 1.981, p = 0.37

Table 4. Genotype frequencies for G+276T polymorphism in dyslipidemic patients according to level of triglycerides (mmol/l).

3.6. Discussion

Insulin resistance is considered the key factor in the pathogenesis of complex diseases such as atherosclerosis, metabolic syndrome and diabetes mellitus. Genetic background IR is probably multifactorial but the participating genes are largely unknown.

In this study, the relationship of polymorphism G+276T of adiponectin gene and markers of insulin resistance was investigated. We found an association between genotype GT and one marker of IR, PAI-I. However, we found no association with serum adiponectin, insulin, HOMA and BMI. Our work did not confirm the preliminary findings from 2005, where the relationship between the adhesion molecules ICAM 1 and TT genotype was observed [27].

Possible association between SNPs and dyslipidemic phenotypes defined by Sniderman classification, based on serum TGL and apo B, was not seen. We found no linkage (data not specified), even in a situation where the only criterion was TGL alone. The genotype distribution in this case was comparable in both groups.

As shown in Table 1, GG genotype was associated with higher levels of total cholesterol and LDL cholesterol compared with GT and TT genotypes. This was found in our previous study as well [27].

Table 3 displays the fact that distribution of genotypes at position 276 is comparable with those published in previous works [20, 21, 22].

3.7. Conclusions

In summary, our study found only a weak association of adiponectin gene SNP G+276T with IR markers. The relationship of GG genotype and selected quantitative lipid parameters were confirmed, in accordance with several studies. Based on some recent literature we suggest the gene variant G+276T may be marker of one or more haplotypes containing a causal polymorphism determining IR or diabetes mellitus. Differences among populations on the linkage disequilibrium structure may result in association on the disease haplotype with different SNP alleles in different population. More studies will be necessary to perform for evaluation of the influence of G+276T SNP on insulin resistance.

4. Adiponectin and its relationship to endothelial dysfunction

In vitro experiments revealed the physiological concentrations of adiponectin inhibited TNF alpha induced expression of VCAM 1 and ICAM 1 on the endothelium and exhibited other antiatherogenic effects. In 2008 Vaverková et al. published a study concerning the relationship between adiponectin and serum concentrations of soluble adhesive molecules VCAM 1 and ICAM 1 as well as with markers of insulin resistance and inflammation in patients with cardiovascular disease and in dyslipidemic patients at high risk of cardiovascular disease [28].

The aim of the study was to evaluate the relationship of adiponectin to soluble forms of vascular cell adhesion molecule 1 (VCAM 1) and intercellular cell adhesion molecule 1 (ICAM 1) in patients with cardiovascular disease or dyslipidemia.

The data from experimental research in animals support the hypothesis of antiaterogenic properties of adiponectin. Adiponectin accumulates in the arterial wall of injured arteries [29]. In adenovirus-treated animals the increase of adiponectin significantly reduced progression of atherosclerotic lesions [6]. In vitro experiments revealed the fact that physiological concentrations of adiponectin inhibited TNF alfa induced expression of VCAM 1 and ICAM 1 on the endothelium [29] and exhibited other antiatherogenic effects.

We have investigated the relationship between adiponectin and serum concentrations of VCAM 1 and ICAM 1 as well as with markers of insulin resistance and inflammation in patients with cardiovascular disease and in dyslipidemic patients at high risk of CVD.

4.1. Subjects

264 patients of Lipid Center at Faculty Hospital Olomouc were included in the study. All patients were examined by a physician and the following information were obtained:

medical history, physical examination and NYHA classification. Subjects were tested for secondary hyperlipidemia. Patients were divided into three groups, those with the presence of clinically manifest atherosclerosis (G1), those with dyslipidemia defined according to Sniderman, but without clinically manifest atherosclerosis (G2), and healthy individuals (G3).

4.2. Results

The characteristics of the three subgroups of the studied cohort are shown in Table 5. Participants with CVD (G1) had comparable lipid, lipoprotein and apolipoprotein profile to the dyslipidemic subjects without CVD (G2) but were more insulin resistant. These differences persisted after adjustment for age, sex and BMI. The G1 had also the highest soluble ICAM 1, the difference in VCAM 1 was not statistically significant. Subjects with dyslipidemia (G2) had significantly lower adiponectin levels and higher levels of ICAM 1 compared with G3. Lower adiponectin levels in patients with CVD did not reach statistical significance, possibly due to a small number of patients. Adiponectin correlated with many lipid and nonlipid markers of insulin resistance. Adiponectin did not correlate with ICAM 1, but there was a strong positive association of adiponectin with VCAM 1. While ICAM 1 and VCAM 1 were strongly intercorrelated, they showed different association pattern with other risk factors. ICAM 1 correlated strongly with many markers of insulin resistance and hsCRP, while VCAM 1 were negatively associated with apo AI and apo B, and positively with adiponectin. Association of adiponectin with VCAM 1 was most prominent in group G1 and G2, but was not significant with G3. Results of multiple backward stepwise regression analysis confirmed these observations. Adiponectin levels were independently positively associated with sex (higher in women), HDL cholesterol and VCAM 1, and negatively with hsCRP. In multiple stewise regression analysis with VCAM 1 as the dependent variable, VCAM 1 was independently associated with ICAM 1 ($p < 0.0001$), adiponectin ($p < 0.0001$), HDL cholesterol ($p = 0.0208$) and triglycerides ($p = 0.0091$). On the other hand, ICAM 1 was independentely associated with VCAM 1 ($p < 0.0001$), atherogenic index ($p < 0.0001$), hsCRP ($p = 0.0001$) and HOMA ($p = 0.0307$). (More detailed results are given in lit. [28].)

4.3. Discussion

Our study confirms the previously described correlations of adiponectin with many lipid and nonlipid markers of IR as well as its relationships with HDL cholesterol, sex and hsCRP [30, 31, 32]. The unexpected finding was the significant independent positive association of adiponectin with VCAM 1 but not with ICAM 1 serum concentrations in patients with or at risk for CVD. Their expression results in adhesion of circulating leukocytes to the endothelial cells and their subsequent transendothelial migration- an important step in initiation and progression of atherosclerosis. VCAM 1 and ICAM 1 have different expression pattern and probably different roles in atherogenesis [33]. Soluble forms of these molecules can be measured in peripheral circulation. The origins of circulating soluble cell adhesion molecules are not entirely clear, but they may derive from shedding or proteolytic cleavage from endothelial cell.

	G1 (CVD+, DLP+/-)	G2 (CVD-, DLP+)	G3 (CVD-, DLP-)
Number	29 (M 18/F 11)	173 (M 97/F 76)	62 (M 19/ F 43)
Age (years)	60.0 ± 9.1	44.9 ± 13.8	36.4 ± 14.5
BMI (kg/m²)	27.5 ± 3.7	26.3 ± 5.7	23.6 ± 4.3
Waist (cm)	92.3 ± 13.1	88.4 ± 11.4	77.4 ± 10.8
Systolic blood pressure (mm Hg)	143 ± 15	131 ± 17	120 ± 13
Diastolic blood pressure (mm Hg)	86.3 ± 9	83 ± 8	75.6 ± 9.8
Total cholesterol (mmol/l)	6.8 ± 1.2	6.7 ± 1.4	4.7 ± 0.7
Triglycerides (mmol/l)	3.0 ± 2.1	2.7 ± 2.3	0.9 ± 0.2
AIP: log (TGL/HDLchol)	0.29 ± 0.38	0.24 ± 1.16	-0.2 ± 0.2
HDL cholesterol (mmol/l)	1.32 ± 0.43	1.34 ± 0.37	1.56 ± 0.36
LDL cholesterol (mmol/l)	4.2 ± 1.1	4.3 ± 1.3	2.8 ± 0.6
Apolipoprotein AI (g/l)	1.52 ± 0.28	1.51 ± 0.29	1.60 ± 0.30
Apolipoprotein B (g/l)	1.29 ± 0.3	1.33 ± 0.33	0.84 ± 0.17
hsCRP (mg/l)	3.4 ± 4.9	2.68 ± 3.6	2.69 ± 5.6
VCAM 1(μg/l)	885 ± 261	800 ± 285	860 ± 265
ICAM 1 (μg/l)	673 ± 202	601 ± 164	538 ± 114
Fasting glucose (mmol/l)	5.8 ± 1.8	5.1 ± 0.8	4.8 ± 0.6
Insulin (mIU/l)	8.8 ± 4.8	8.9 ± 5.3	6.6 ± 3.4
HOMA $_{IR}$	2.3 ± 1.6	2.1 ± 1.3	1.4 ± 0.8
C- peptide (μg/l)	3.2 ± 1.2	2.6 ± 1.0	1.9 ± 0.7
Proinsulin (mIU/l)	17.0 ± 8.0	15.6 ± 9.9	11.3 ± 5.2
Adiponectin (mg/l)	15.5 ± 8.0	12.3 ± 6.6	16.1 ± 6.8

Table 5. The demographic, clinical and laboratory characteristics of the study population

The expression pattern of adhesion molecules may explain why VCAM 1 is a marker of increased risk for future coronary events only in patients with atherosclerosis [34]. Patients with stable CAD have moderately increased and in several studies even normal levels of soluble VCAM 1 in comparison with healthy controls. The highest level of VCAM 1 was noted in patients with acute myocardial infarction [35]. In another study, VCAM 1 was a useful marker for predicting future ischemic events in the 6 months after presentation with unstable angina pectoris or nonQ myocardial infarction [36]. In our cohort, levels of VCAM 1 in the CVD patients were not significantly higher than in controls. This is in agreement with several other works.

4.4. Conclusions

Many studies, including experiments in vitro, animal models and studies in human, have shown that adiponectin has antiatherogenic and antiinflammatory properties. Low adiponectin levels were found in patients with CAD independently of other risk factors. Therefore, the finding of positive and independent association of adiponectin with the marker of endothelial dysfunction VCAM 1 was suprising. This positive association was present both in patients with CVD and dyslipidemic subjects without CVD, but it was not significant in healthy subjects without dyslipidemia. We hypothesize that adiponectin, which accumulates in the arterial wall only in place of endothelial injury and atherosclerotic plaques (that is the same places where VCAM 1 is expressed) may be involved in shedding of ectodomains of VCAM 1 from endothelial surface. This may represent a mechanism by which VCAM 1 effects on the cell surface can be downregulated. In this way, adiponectin could protect vascular wall from adhesion of leukocytes and thus from progression of atherosclerosis.

5. Adiponectin and dyslipidemia: Relationship of adiponectin, fibroblast growth factor 21 and adipocyte fatty acid binding protein levels to dyslipidemic phenotypes – Pilot study

5.1. Background

Adipose tissue is an important place of many metabolic and inflammatory processes. Adipokines are considered to be the mediators of these pathways.

Adiponectin (ADP, AdipoQ, apM1, GBP28) is a "favourable" adipokine of fat tissue circulating at relatively high concentrations in human plasma. Adiponectin has the protective effects in early stages and during progression of atherosclerosis probably by its antiinflammatory and antiatherogenic actions.

Fibroblast growth factor 21 (FGF 21) is also a "favourable" cytokine of adipose tissue considered as a new metabolic regulator of non insulin dependent glucose transport in cells [37]. Systematic administration of FGF 21 decreases plasma levels both of glucose and triglycerides, and leads to improving of lipoprotein profiles in genetic compromised FGF transgenic mice and primates [38]. Increased levels of FGF 21 and a negative correlation with HDL and adiponectin were found in patients with metabolic syndrome [39].

Adipocyte fatty acid binding protein (A-FABP) is a „unfavourable" adipokine, probably a new marker and/or predictor of metabolic syndrome [40]. A-FABP is a dominant cytoplasmic protein of mature adipocytes and a regulator of lipid and glucose metabolism, present also in macrophages of fat tissue. Its expression is induced by oxidated LDL [41]. Higher levels of A-FABP are associated with increased fasting glucose, triglycerides, insulin BMI and waist circumference, and decreased HDL in patients with metabolic syndrome. Inhibition of A-FABP action is associated with reversion of atherosclerosis (improving of diabetic and lipoprotein parameters).

The aim of our study was to evaluate the relationship between adiponectin, FGF 21 and A-FABP levels and dyslipidemic phenotypes defined on the basis of concentrations of triglycerides and apolipoprotein B [25].

5.2. Subjects, material and methods

119 pacients of Lipid Center at Faculty Hospital Olomouc were included on the pilot scheme. Routine serum biochemical parameters were analyzed on Modular SWA (Roche, Switzerland) in the day of blood collection. Levels of ADP, FGF 21 and A-FABP were determinated by imunochemical Elisa methods (BioVendor, Czech Republic). The analytical characteristics from data sheets were verificated according to laboratory protocol for all procedures.

119 individuals were divided into four dyslipidemic phenotypes (DLP) according to Sniderman classification- see Table 6.

	TGL (mmol/l)	Apo B (g/l)
DLP1	< 1.5	< 1.2
DLP2	≥ 1.5	< 1.2
DLP3	< 1.5	≥ 1.2
DLP4	≥ 1.5	≥ 1.2

Table 6. Classification of dyslipidemic phenotypes

5.3. Results

Basic clinical characteristics are shown in Table 7. Concentrations of adipokines and other biochemical parameters are given in Table 8.

	Number (n)	F/M	Age (y)	Waist (cm)	SBP (mg Hg)	DBP (mm Hg)	Smoking (n)	Manifestation of ATS
DLP1	32	16/16	41 ± 10.0	85 ± 9.3	129 ± 12	77 ± 9	4	3
DLP2	38	20/18	47.1 ± 10.1	96 ± 12	130 ± 19	78 ± 11	7	3
DLP3	13	3/10	47.8 ± 10.5	88 ± 8.0	125 ± 18	75 ± 7	2	0
DLP4	36	22/15	49.9 ± 10.7	92 ± 9.0	126 ± 15	75 ± 9	9	4

Table 7. Basic clinical characteristics of DLP groups

	ADP (mg/l)	FGF 21 (ng/l)	A-FABP (ug/l)	CHOL (mmol/l)	TGL (mmol/l)	HDLchol (mmol/l)	LDLchol (mmol/l)	Apo AI (g/l)	Apo B (g/l)	GLU (mmol/l)	BMI (kg/m²)
DLP1	10.6 ± 6.0	186 ± 100	22.5 ± 10.3	5.68 ± 0.8	1.04 ± 0.26	1.74 ± 0.41	3.48 ± 0.75	1.74 ± 0.41	0.91 ± 0.16	5.16 ± 0.66	25.9 ± 5.2
DLP2	8.0 ± 5.1	333 ± 360*	33.9 ± 29.0*	6.23 ± 1.92	5.29 ± 8.0	1.22 ± 0.31*	3.14 ± 0.93	1.51 ± 0.30	0.98 ± 0.14	5.57 ± 1.19*	28.3 ± 4.7*
DLP3	8.6 ± 4.9	165 ± 104	14.4 ± 4.6	7.47 ± 1.14	1.11 ± 0.2	1.45 ± 0.33	5.52 ± 1.27	1.51 ± 0.25	1.41 ± 0.25	5.03 ± 0.52	25.2 ± 3.1
DLP4	9.0 ± 5.9	384 ± 347**	29.2 ± 18.4**	8.43 ± 2.0	3.5 ± 2.2	1.27 ± 0.45**	5.59 ± 2.0	1.50 ± 0.43	1.62 ± 0.39	5.36 ± 1.27***	27.2 ± 5.0**

Differences between groups were analyzed with ANOVA. Parameters with skewed distribution (TGL, ADP, FGF 21, A-FABP) were log transformed to normalize their distributions before statistical analyses.

* DLP2 vs. DLP1 and DLP3, $p < 0.01$, ** DLP4 vs. DLP1 and DLP3, $p < 0.01$, *** DLP4 vs. DLP1 and DLP3, $p < 0.05$

Table 8. Adipokines and other biochemical parameters in connection with DLP

The highest levels of ADP were observed in DLP1 (no significance). Suprisingly, there was seen no negative association between adiponectin levels and DLP2 (DLP4). FGF 21 and A-FABP were significantly increased in the groups with the most important atherogenic potential (DLP2, DLP4). These two parameters correlated with higher levels of triglycerides, fasting glucose, BMI and lower HDL cholesterol, both in DLP2 and DLP4.

5.4. Conclusions

No association was found between ADP levels and other adipokines in DLP groups in our study. There was the correlation between FGF 21 and A-FABP in groups with TGL > 1.5 mmol/l. Increased levels of both parameters were associated with increased glucose, BMI and decreased HDL cholesterol levels (in accordance with lit.[40]). The increase of FGF 21 concentrations are probably due to the compensatory response to higher A-FABP that is considered the predictor of metabolic syndrome. In individuals with MS, the determination of A-FABP could be considered as a parameter with the independent metabolic effects [42]. The clinical potential especially of A-FABP in diagnostics and prediction of metabolic syndrome should be continue to observe.

6. Adiponectin in members of families with familial combined hyperlipidemia

Familial combined hyperlipidemia (FCH) is the most common genetic hyperlipidemia which affects 1.0% to 2.0% of the population. The lipids and lipoprotein levels are, however, only moderately elevated and do not fully explain the increased risk of cardiovascular disease. The aim of the study of Karásek et al. [43] was to evaluate plasma levels of adiponectin in asymptomatic, nonsmoking members of families with FCH. We also investigated the association between adiponectin and selected risk factors of atherosclerosis and markers of insulin resistance and chronic inflammation. Furthermore, we investigated the relationship between adiponectin and the intima-media thickness of the CCA (IMT), a recognized morphologic marker of early atherosclerosis.

6.1. Subjects and methods

The study was carried out with 82 members of 29 FCH families. A family with FCH was defined by a proband exhibiting plasma cholesterol and triglycerides concentrations above 90th percetile, adjusted for age and sex, based on data from the Czech population. At least one first-degree relative of the proband should have plasma cholesterol and/or triglycerides above 90th percentile, adjusted for age and sex, or level of apo B more than 1.25 g/l. Secondary hyperlipidemia was excluded by additional testing. Other exclusion criteria were a history of clinically manifest atherosclerosis, heart failure, cerebrovascular ischemic disease, peripheral vascular disease, smoking, hypolipidemic therapy in the previous 8 weeks, hormone therapy with estrogens and acute infection or trauma. Members of FCH were divided into 2 groups: HL (hyperlipidemic members of FCH families, i.e. probands and their hyperlipidemic first-degree relatives) and NL (normolipidemic first-degree relatives). The control groups, C-HL and C-NL, were sex and age matched to groups. Control groups consisted of healthy individuals with a negative family history of hyperlipidemia and early manifestation of atherosclerosis. Nobody was treated for hypertension.

Laboratory parameters were analyzed by routine methods described above. Ultrasound scanning was performed with a 10 MHz linear array transducer (Hewlett-Packard, Image Point, M2410A). All measurements were performed with the subjects in a supine position. Three video records were made of common carotid artery (CCA). IMT measurements were processed off-line using software Image-Pro Plus (v. 4.0, Media-Cybernetics, Silver Spring). The average of the IMT of 3 frozen images of both sides was chosen as the outcome variable. Subjects with an atherosclerotic plaque in the evaluated region were not included in the study. The measurement of IMT was made without knowledge of laboratory results.

6.2. Results

In comparison to sex and age matched controls, HL subjects had significantly higher diastolic blood pressure (DBP), BMI, insulin resistance and elevated levels of C-peptide and proinsulin. They had higher IMT, hsCRP and ICAM 1 as well. By definition, the FCH subjects showed higher plasma cholesterol and triglycerides concentrations compared with controls and normolipidemic relatives. They had a more atherogenic lipid and lipoprotein profile as reflected by increased LDL cholesterol and apo B concentrations. Normolipidemic relatives had significantly higher DBP, TGL and proinsulin concentrations compared with their sex and age matched controls. There was no difference in other measured anthropometric and biochemical parameters.

Compared with healthy controls, HL subjects had lower levels of adiponectin (13.02 ± 4.58 mg/l vs. 16.19 ± 5.39 mg/l, $p < 0.05$). In the NL relatives, there was no significant differences in adiponectin (15.77 ± 2.95 mg/l vs. 16.53 ± 4.26 mg/l). In all FCH families, a significant negative correlation was found between adiponectin and TGL ($r = -0.35$, $p < 0.01$), proinsulin ($r = -0.26$, $p < 0.05$), hsCRP ($r = -0.24$, $p < 0.05$), BMI ($r = -0.27$, $p < 0.05$) and waist circumference ($r = -0.32$, $p < 0.01$). Levels of adiponectin did not correlate with IMT, in members of FCH families or in controls. By using regression model in HL subjects, levels of adiponectin were predicted by apo B ($p < 0.05$) and hsCRP ($p < 0.05$). (More detailed results are given in lit. [43].)

6.3. Discussion and conclusions

This study reported decreased adiponectin levels in asmyptomatic hyperlipidemic members of FCH families. There was no difference in serum adiponectin levels between their first-degree normolipidemic relatives and healthy controls. A negative correlation between adiponectin and markers of insulin resistance, chronic inflammation and visceral obesity was found in FCH families. The results were consistent with previous findings and support an insulin-sensitizing effect of adiponectin. In hyperlipidemic individuals, the levels of plasma adiponectin were predicted by apolipoprotein B and high sensitive CRP, independent of insuline resistance and visceral obesity. Authors conclude low adiponectin levels are associated with proinflammatory status and insulin resistance, and could partially explain the increased risk of coronary heart disease, even if the lipids and lipoprotein levels are only moderately elevated.

The study did not confirm any correlation between adiponectin levels and IMT, a marker of subclinical atherosclerosis, in FCH subjects. Publications regarding the relationship between these parameters are not entirely consistent. Similar results were observed in other work published by Karásek et al [44] where IMT proved to correlate with age, lipid parameters, markers of insulin resistance and that of visceral obesity and blood pressure. These parameters seem to be risk factors instead of adiponectin. The lack of correlation between adiponectin and IMT does not argue for adiponectin as an independent predictor for next cardiovascular events in clinically asymptomatic, dyslipidemic individuals.

7. Conclusion of chapter

Adiponectin is another promising parameter of the metabolic syndrome, atherosclerosis and associated syndromes. Its effect should be studied in many other situations. Nowadays, its determination in plasma provides valuable information, for example in patients with angiographically documented coronary artery disease, even if not all studies confirm this relationship. We can rely on the fact that its levels show no or very little circadian variability, its concentration is independent of fasting, it has low intraindividual variability, it is present in high concentrations in plasma and its levels can be influenced by diet, lifestyle or medication. Probably the most effective way to increase adiponectin levels in plasma and thus to reduce cardiovascular risk in obese individuals is a reduction in body weight. Beneficial effect of thiazolidinediones are also used to treat patients with type 2 diabetes to increase adiponectin production and plasma levels.

On the other hand, although adiponectin is associated with many of the traditional cardiovascular risk factors and further evidence has shown that hypoadiponectinemia is associated with atherosclerotic cardiovascular events such as myocardial infarction and brain infarction [45, 46], recent epidemiologic studies have shown contradictory results. Some of them revealed that hyperadiponectinemia rather than hypoadiponectinemia is associated with liver cirrhosis, rheumatoid arthritis, inflammatory bowel disease and systemic lupus erythematosus, all of which are conditions predisposed to wasting . Release of adiponectin from fat tissue is increased under conditions of malnutrition and plasma adiponectin concentration rises in the inflammatory state. Therefore, adiponectin can act as a mirror reflecting the degree of systemic wasting, and thus can predict death [47].

We can speculate about the real impact of high adiponectin levels on atherosclerosis: are they protective of harmful? In healthy subjects without clinically important signs of atherosclerosis, adiponectin has the protective effects especially due to its tissue-insulin senzitizing action. However, in individuals with advanced atherosclerosis and/or inflammatory disease, the positive association of adiponectin levels with markers of endothelial dysfunction/hemostasis (VCAM 1, but also with thrombomodulin and von Willebrand factor) could explain the increased total and cardiovascular mortality and the one associated with high adiponectin levels. It could be also the case of other populations, such as elderly people, patients with heart failure, patients with chronic kidney diseases, patients with type 1 diabetes mellitus etc. In recent studies, adiponectin effects should be evaluated in these populations.

Author details

Dalibor Novotný
Department of Clinical Biochemistry, Faculty Hospital Olomouc, Czech Republic

Helena Vaverková and David Karásek
3rd Medical Clinic, Faculty Hospital Olomouc, Czech Republic

Acknowledgement

This work was supported by the grants IGA MZ CR NR/9068-3 and IGA MZ CR NS/10284-3.

8. References

[1] Funahashi T, Nakamura T, Shimomura I, et al (1999) Role of adipocytokines on the pathogenesis of atherosclerosis in visceral obesity. Intern. Med. 38: 202-206.
[2] Shimada K, Miyazaki T, Hiroyuki D (2004) Adiponectin and atherosclerotic disease. Clin. Chim. Acta 344: 1-12.
[3] Yamamouchi T, Kamon J, Ito Y, et al (2003) Cloning of adiponectin receptors that mediate antidiabetic metabolic effects. Nature 423: 762-769.
[4] Ouchi N, Walsh K (2007) Adiponectin as an anti-inflammatory factor. Clin. Chim. Acta 380: 24-30.
[5] Choi KM, Ryu OH, Lee KW, et al (2007): Serum adiponectin, interleukin-10 levels and inflammatory markers in the metabolic syndrome. Diabetes Res. Clin. Pract. 75: 235-240.
[6] Okamoto Z, Kihara S, Ouchi N, et al (2002) Adiponectin reduces atherosclerosis in apolipoprotein E-deficient mouse. Circulation 106: 2767-2770.
[7] Peake PW, Shen Y, Walther A, Charlesworth JA (2008) Adiponectin binds C1q and activates the classical pathway of complement. Biochem. Biophys. Res. Commun. doi: 10.1016/j.bbrc.2007.12.161.
[8] Motoshima H, Wu X, Mahadev K, Goldstein BJ (2004) Adiponectin supresses proliferation and superoxide generation and enhances eNOS activity in endothelial cells treated with oxidized LDL. Biochem. Biophys. Res. Commun. 315: 264-271.
[9] Kern PA, Di Gregorio GB, Lu T, Rassouli N, Ranganathan G (2003) Adiponectin expression from human adipose tissue. Relation to obesity, insulin resistance and tumor necrosis alfa expression. Diabetes 52: 1779-1785.

[10] Liu YM, Lacorte JM, Viguerie N, et al (2003) Adiponectin gene expression in subcutaneous adipose tissue of obese women in response to short-term low calorie diet and refeeding. J. Clin. Endocrinol. Metab. 88: 5881-5886.

[11] Ryo M, Nakamura T, Kihara S, et al (2004) Adiponectin as a biomarker of the metabolic syndrome. Circ. J. 68: 975-981.

[12] Saely ChH, Risch L, Hoefle G, et al (2007) Low serum adiponectin is independently associated with both the metabolic syndrome and angiographically determined coronary atherosclerosis. Clin. Chim. Acta 383: 97-102.

[13] Hallikainen M, Kolehmainen M, Schwab U, et al (2007): Serum adipokines are associated with cholesterol metabolism in the metabolic syndrome. Clin. Chim. Acta 383: 126-132.

[14] von Eynatten M, Schneider JG, Humpert PM, et al (2006) Serum adiponectin levels are an independent predictor of the extent of coronary artery disease in men. J. Am. Coll. Cardiol. 47: 2124-2126.

[15] Ouchi N, Kihara S, Funahashi T, et al (2003) Reciprocial association of C-reactive protein with adiponectin in blood stream and adipose tissue. Circulation 107: 671-674.

[16] Nishida M, Moriyama T, Ishii K, et al (2007) Effects of IL-6, adiponectin, CRP and metabolic syndrome on subclinical atherosclerosis. Clin. Chim. Acta 384: 99-104.

[17] von Eynatten M, Hamann A, Twardella D, et al (2006) Relathionship of adiponectin with markers of systemic inflammation, atherogenic dyslipidemia, and heart failure in patients with coronary heart disease. Clin. Chem. 52: 853-859.

[18] Kondo H, Shinomura I, Matsukawa Y, et al (2002) Association of adiponectin mutation with type 2 diabetes. Diabetes 51: 2325-2328.

[19] Ohashi K, Ouchi N, Kihara S, et al (2004) Adiponectin I164T mutation is associated with the metabolic syndrome and coronary artery disease. J. Am. Coll. Cardiol. 43: 1195-2000.

[20] Hara K, Boutin P, Mori Y, et al (2002) Genetic variation in the gene encoding adiponectin is associated with an increased risk of type 2 diabetes in the Japanese population. Diabetes 51: 536-540 .

[21] Menzaghi C, Ercolino T, Di Paola R, et al (2002) A haplotype locus at the adiponectin locus is associated with obesity and other features of the insulin resistance syndrome. Diabetes 51: 2306-2312 .

[22] Filippi E, Sentinelli F, Trischitta V, et al (2003) Association of the human adiponectin gene and insulin resistance. E. J. Hum. Genet. doi:10.1038/sj.ejhg.5201120.

[23] Hu FB, Doria A, Li T, et al (2004) Genetic variation at the adiponectin locus and risk of type 2 diabetes in women. Diabetes 53: 209-213.

[24] Hegener HH, Lee IM, Cook NR, et al (2006) Association of adiponectin gene variations with risk of incident myocardial infarction and ischemic stroke: a nested case-control study. Clin. Chem. 52: 2021-2027.

[25] Sniderman AD (2004) Applying apo B to the diagnosis and therapy of the atherogenic dyslipoproteinemias: a clinical diagnostics algorithm. Curr. Opin. Lipidol. 15: 433-438.

[26] John SW, Weitzner G, Rozen R, Scriver CR (1991) A rapid procedure for extracting genomic DNA from leukocytes. Nucleic Acids Res. 9: 408.

[27] Novotny D, Vaverkova H, Karasek D, Halenka M (2005) Relationship between +276 G-T single nucleotide polymorphism (SNP) of adiponectin gene and markers of insulin resistance in dyslipidemic patients. 75th EAS Congress, Prague, Supplement of book of abstracts: 7.

[28] Vaverková H, Karásek D, Novotný D, et al (2008) Positive association of adiponectin with soluble vascular cell adhesion molecule sVCAM-1 levels in patients with vascular disease or dyslipidemia. Atherosclerosis 197(2): 725-731.

[29] Okamoto Z, Arita Z, Nishida M, et al (2000) An adipocyte-derived plasma protein, adiponectin, adheres to injured vascular walls. Horm. Metab. Res. 32: 47-50.

[30] Pischon T, Rimm EB (2006) Adiponectin: a promissing marker for cardiovascular disease. Clin. Chem. 52: 797-799.

[31] Matsushita K, Yatsuya H, Tamakoshi K, et al (2006) Inverse association between adiponectin and C-reactive protein in substantially healthy Japanese men. Atherosclerosis 188: 184-189.

[32] Kazumi T, Kawaguchi A, Hirano T, Yoshino G (2004) Serum adiponectin is associated with high-density lipoprotein cholesterol, triglycerides, and low-density lipoprotein particle size in young healthy men. Metabolism 53: 589-593.

[33] Blankenberg S, Barbaux S, Tiret L (2003) Adhesion molecules and atherosclerosis. Atherosclerosis 170: 191-203.

[34] The AtheroGene Investigators (2001) Circulating cell adhesion molecules and death in patients with coronary artery disease. Circulation 104: 1336-1342.

[35] Guray U, Erbay AR, Guray Z, et al (2004) Levels of soluble adhesion molecules in various clinical presentations of coronary atherosclerosis. Int. J. Cardiol. 96: 235-240.

[36] Jager A, van Hinsbergh VWM, Kostense PJ, et al (2000) Increased levels of soluble vascular cell adhesion molecule 1 are associated with risk of cardiovascular mortality in type 2 diabetes, The Hoorn Study. Diabetes 49: 485-491.

[37] Spranger J, Kroke A, Mohling M, et al (2003) Adiponectin and protection against type 2 diabetes mellitus. Lancet 361: 226-228.

[38] Kharitonenkov A, Shiyanova TL, Koester A, et al (2005) FGF-21 as a novel metabolic regulator. J. Clin. Invest. 115(6): 1627-1635.

[39] Kharitonenkov A, Wroblewski V, Koester A, et al (2007) The metabolic state of diabetic monkeys is regulated by fibroblast growth factor 21. Endocrinology 148(2): 774-781.

[40] Zhang X, Yeung DCY, Karpisek M, et al (2008) Serum FGF21 levels are increased in obesity and are independently associated with metabolic syndrome in humans. Diabetes 57: 1246-1253.

[41] Stejskal D, Karpisek M (2006) Adipocyte fatty acid binding protein in a Caucasian population: a new marker of metabolic syndrome? Eur. J. Clin. Invest. 36(9): 621-625.

[42] Xu A, Wang Y, Xu JY, et al (2006) Adipocyte fatty acid binding protein is a plasma biomarker closely associated with obesity and metabolic syndrome. Clin.Chem. 52: 405-413.

[43] Novotný D, Karásek D, Vaverková H, Jackuliaková D, Malina P The relationship of adiponectin, fibroblast growth factor 21 and adipocyte fatty acid binding protein levels to dyslipidemic phenotypes- pilot study (2011). Clin. Chem. Lab. Med. 49: S394.

[44] Karásek D, Vaverková H, Halenka M, Jackuliakova D, Fryšák Z, Novotný D (2010) Adiponectin in members of families with familial combined hyperlipidemia. The Endocrinologist 20(3): 117-121.

[45] Karásek D, Vaverková H, Halenka M, Jackuliakova D, Frysak Z, Novotny D (2011) Total adiponectin levels in dyslipidemic individuals: relationship to metabolic parameters and intima-media thickness. Biomed. Pap. Med. Fac. Univ. Palacky. Olomouc 155(1): 55-62.

[46] Chen MP, Tsai JC, Chung FM, et al (2005) Hypoadiponectinemia is associated with ischemic cerebrovascular disease. Arterioscler Thromb Vasc Biol 25: 821-826.

[47] Wang WH, Yu WH, Dong XQ, et al (2011) Plasma adiponectin as an independent predictor of early death after acute intracerebral hemorrhagie. Clin. Chim. Acta 412: 1626-1631.

[48] Jernas M, Olsson B, Sjöholm K, et al (2009) Changes in adipose tissue gene expression and plasma levels of adipokines and acute-phase proteins in patients with critical illness. Metabolism 58:102-108.

Chapter 30

Dyslipoproteinemia in Chronic HCV Infection

Yoshio Aizawa, Hiroshi Abe, Kai Yoshizawa, Haruya Ishiguro, Yuta Aida, Noritomo Shimada and Akihito Tsubota

Additional information is available at the end of the chapter

http://dx.doi.org/10.5772/48117

1. Introduction

Hepatitis C virus (HCV) is a unique virus whose life cycle is closely associated with lipoprotein metabolism [1, 2, 3]. Assembly of HCV particles, formation of HCV-virions, is closely connected to the formation of lipid droplets in hepatic cells that may serve as an assembly platform [1, 4]. In addition, the production of HCV particles is tightly linked to the very low-density lipoprotein (VLDL) production pathway [5, 6]. HCV particles circulating in the blood during chronic HCV infection form lipo-viral particles (LVP) that are rich in triglycerides (TG), apoB-100 and apoE, with physiochemical similarity to VLDL particles and are highly infectious [7, 8]. In contrast, denser HCV particles are less infectious. These data strongly suggest that both viral particles and VLDL are integral components of LVPs with high infective capability. Although LVPs are thought to be assembled in liver cells by association with host lipoproteins prior to secretion, association between HCV and VLDL in the circulation after secretion from the liver cannot be ruled out.

Studies to date have indicated that the process of HCV assembly and secretion largely utilizes the VLDL pathway. Therefore, suppression of apoB-100 or apoE also inhibits secretion of HCV. Inhibition of microsomal triglyceride transfer protein (MTP), a critical protein for the initial step of VLDL assembly by co-translational lipidation of apoB-100 [9, 10], inhibits HCV secretion.

As HCV depends on VLDL pathways for its assembly and secretion, the lipid-rich environment of the liver cell combined with reduced VLDL secretion may be required for efficient assembly and secretion of HCV virions by ensuring the feasibility of co-assembly with VLDL. Hypobetalipoproteinemia, reduced activity of MTP with negative correlation to hepatic steatosis and viral load, is observed in HCV-G3 chronic infection [11]. Secretion of apoB-100 was reduced by HCV nonstructural proteins using the HCV subgenomic replicon expression system and interaction between the HCV NS5A and apoB-100 was observed [12].

In addition, there has been accumulating evidence that the HCV core protein is induced upon the redistribution of lipid droplets, affecting the assembly of both HCV and VLDL [13, 14]. The function of core protein in lipid metabolism has been widely examined including in models of steatosis involving HCV core protein transgenic mice [15]. Findings include reduction in activity of MTP [9] and Tyr164Phe substitution in relation to marked steatosis in HCV-G3 infection [16]. Although the participation of Arg70Gln/His substitution in steatosis of hepatocytes has been proposed [17], the precise connection between hepatic steatosis and HCV-G1b and/or HCV-G2 infection remains unclear.

As mentioned earlier, HCV particles in peripheral blood may associate not only with VLDL, but also with other lipoproteins, especially LDL, since circulating LVPs span a wide range of buoyant gravity and physicochemical characteristics [8, 18]. The association of lipoproteins with HCV particles may be beneficial for HCV through protection against anti-HCV neutralizing antibodies, as the antigenicity of HCV surface proteins is hidden beneath the associated lipoprotein particle [18]. Lipoprotein particles isolated from sera of HCV patients displayed differentially modulated lipid synthesis in human monocyte-derived macrophages in comparison to lipoproteins obtained from normal subjects, suggesting that HCV infection influences the biochemical composition of lipoproteins, thus revealing an alternative influence on lipid metabolism [19]. HCV entry into liver cells may occur through many receptors, including CD81 (direct binding to HCV E2 protein) and claudin-1, both of which act during the later steps of HCV entry [20]. The predominant role of LDL-receptors or remnant receptors is to catch VLDL-derived lipoprotein particles. Meanwhile, SR-BI (a receptor for HDL and oxidized LDL) directly binds HCV E2 protein [21]. Very recently, Nieman-Pick C1-like 1 cholesterol absorption receptor has been reported as a new factor for HCV entry to hepatic cells [22]. Interestingly, lipoprotein lipase, which hydrolyzes VLDL, is reported to increase the binding of LVP to hepatic cells while simultaneously decreasing infection levels of hepatic cells [23].

These findings suggest that examination of serum lipid profiles in chronic HCV infection may be important for understanding the biological features of HCV infection. Compared to normal subjects, low levels of TC, high-density lipoprotein cholesterol (HDL-C) and LDL-C was reported in chronic HCV-G3a infection [24]. However, lipoprotein profiles in infections of genotypes other than HCV-G3 have not been fully described and the data are somewhat conflicted. Moriya et al. indicated that TC levels and apoB, CII and CIII were reduced in HCV-G1b compared with chronic HCV-G2a or hepatitis B virus (HBV) infection [25], while others have not reported such a distinction. In addition, distortion of serum lipid levels has been widely observed in connection with virological outcome of IFN-based antiviral therapy, especially in HCV-G1 infection. Lower LDL-C, HDL-C, TC and/or TG was reported to be a possible predictor for unfavorable response to IFN-based therapy [24, 26, 27]. However, after the discovery of a genetic polymorphism near the human IL28B gene as the most potent predictor of the outcome of IFN-based therapy, the distortion of serum lipid levels is no longer thought to be an independent factor, but rather a confounding variable for predicting therapeutic efficacy [28].

In this chapter, we described the lipoprotein profiles in chronic HCV-G1b infection (the most common genotype in Japan) compared with that in chronic HCV-G2 infection (the second most common genotype in Japan). In addition, the influence of the genetic polymorphism near the human IL28B gene and aa substitutions in the core and NS5A regions of HCV on lipoprotein profiles in chronic HCV-G1b infection was determined.

To examine the serum lipid profiles of many patients, ultracentrifugation was unsuitable. We instead measured serum lipoprotein using HPLC in addition to a conventional laboratory method involving measurement of apolipoproteins. To examine serum LDL-C levels, the Friedwald equation can be used as an indirect calculation method yielding "total cholesterol minus HDL-C minus 0.2 × TG. However, the precision of this equation has not been determined in pathological conditions such as chronic HCV infection.

2. Methods

2.1. Patients and materials

Fasting sera of patients who were diagnosed as having chronic HCV infection were collected and stored at lower than -30 degrees centigrade until examination of apolipoproteins and/or lipoproteins. At the time of serum collection, TC, TG and HDL-C were measured using a routine laboratory kit. Serum LDL-C was measured by a direct assay using an LDL-cholesterol kit (Sekisui Medical, Japan). A good correlation between the direct assay and indirect measurement using the Friedwald equation (r=0.96) in 96 healthy adults whose TG level was lower than 400 mg/dl was observed.

Detection of HCV infection was made using a commercial real-time PCR Kit. Detection of HCV genotype was performed by a PCR method based on the 5′ non-coding sequence. Some patients diagnosed as having HCV-G1b infection, aa substitutions at core 70/91, interferon sensitivity determined region (ISDR) and/or IFN RBV resistance determining region (IRRDR) were examined. All patients examined were Japanese without ongoing treatment of IFN-based antiviral therapy.

All patients were confirmed not to be co-infected with HIV, HTLV, tuberculosis or other chronic bacterial infections. In addition, patients who were diagnosed with hepatic cirrhosis were excluded from the study. Study protocols were approved by the review board of each institution and written informed consent was obtained prior to study enrollment.

2.2. Detection of serum apolipoproteins

Serum levels of apolipoproteins (apoA1, apoA11, apoB, apoCII, apoCIII and apoE) were analyzed by immunonephelometry using an apolipoprotein detection kit (Sekisui Medical, Japan). Serum apoB-48 was assayed by chemiluminescent enzyme immunoassay [29] using an apoB-48 CLEIA kit (Fujirebio, Japan). Serum apoB-100 level was determined as apoB minus apoB-48.

2.3. Detection of lipoprotein fractions by HPLC based method

Fasting serum lipoprotein profiles were analyzed using an HPLC system with on-line enzymatic dual detection of cholesterol and TG as described previously (LipoSEARCH, Skylight Biotech, Japan) [30]. Briefly, 10 μl of whole serum sample was injected into two connected columns (300 × 7.8 mm) of TSKgel LipopropakXL (Tosoh, Japan) and eluted by TSKeluent Lp-1 (Tosoh). The eluent from the columns was divided by a micro splitter and continuously monitored at 550 nm after an online enzymatic reaction with a commercial kit, Determiner L TC (Kyowa Medex, Japan) and Determiner L TG (Kyowa Medex). Then, the cholesterol and TG concentrations were calculated by the computer program, which was designed to process complex chromatograms with a modified Gaussian curve fitting for resolving overlapping peaks by mathematical treatment.

Lipoprotein particles were fractionated into four major lipoproteins according to particle diameter as follows: >80 nm classified as chylomicrons; 30 to 80 nm as VLDL; 16 to 30 nm as LDL; and 8 to 16 nm as HDL. Next, the concentration of cholesterol and TG was measured in each major lipoprotein fraction and the ratio of cholesterol:TG concentration (C:T ratio) was calculated. This system has been successfully applied elsewhere in clinical research with excellent reproducibility [31]. Although freezing and thawing may affect the lipoprotein fraction, the influence of freezing and thawing on the measurement of cholesterol and TG concentration in healthy samples is fairly low, as described on homepage of Skylight Biotech (http://www.lipo-search.com).

Figure 1 illustrates chromatographic pattern of cholesterol and triglycerides derived LipoSEARCH was illustrated.

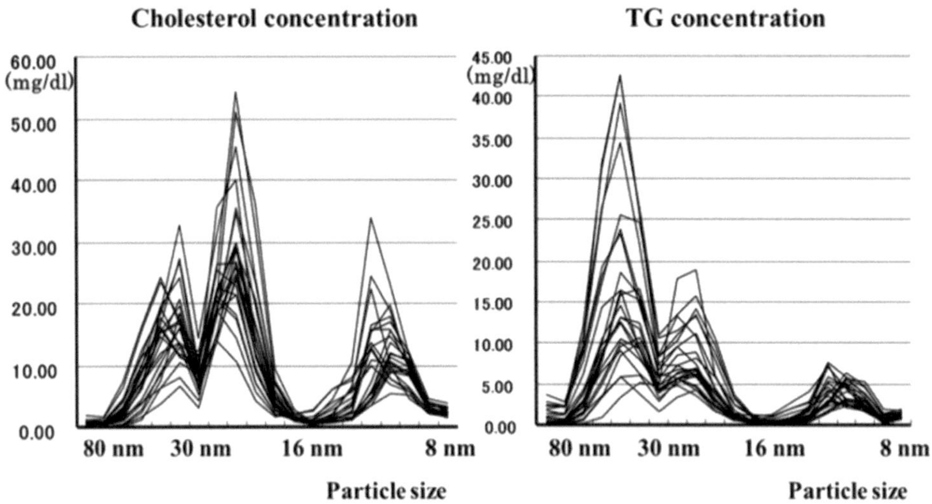

Figure 1. Chromatographic pattern of fasting serum cholesterol and TG derived HPLC-based method (LipoSEARCH).

2.4. Detection of amino acid substitutions at aa 70/91 in core region of HCV-G1b

HCV RNA was extracted from serum samples and reverse transcribed with random primers and MMLV reverse transcriptase (Takara Shuzo, Japan). Based on the method of Akuta et al. [32], nucleotide sequences of the core region were analyzed by direct sequencing after nested PCR. Subsequently, the aa substitutions at position 70 (arginine, Arg70; or glutamine/histidine, Gln70/His70) and at position 91 (leucine, Leu91; or methionine Met91) were determined.

2.5. Detection of amino acid substitutions related to ISDR in NS5A of HCV-G1b

Nucleotide sequences of NS5A-ISDR were analyzed by direct sequencing after double-round PCR according to Enomoto et al. [33]. Then, the aa sequence from 2209-2248, termed the ISDR region, in NS5A was determined and the numbers of aa substitutions defined.

2.6. Detection of amino acid substitutions related to IRRDR in NS5A of HCV-G1b

Nucleotide sequences of NS5A-IRRDR were analyzed by direct sequencing after double-round PCR according to El-Shamy et al. [34] and the numbers of aa substitutions within this region (2334-2376) determined. Moreover, the particular position of aa substitutions were evaluated in connection to dyslipoproteinemia, found in chronic HCV-G1b patients.

3. Results

Difference in lipoprotein profiles between patients with chronic HCV-G2 and HCV-G1b infection, between HCV-G1b patients with favorable and unfavorable response to PEG-IFN plus RBV combination therapy, and between minor (non-responder) genotypes and major (responder) genotypes of IL28B (rs8099917)

Serum lipoprotein profiles are clearly distorted in patients with chronic HCV-G3 infection and this is characterized by a decrease in apoB-100-related cholesterol [24]. However, serum lipoprotein disturbances in other genotypes are an issue that has been under discussion. In addition, disturbances in lipoprotein profiles have been reported in patients who were not responsive to IFN-based antiviral therapy in comparison to responsive patients with chronic HCV-G1 infection [26, 27]. However, whether or not dyslipoproteinemia is an independent factor affecting the efficacy of IFN-based therapy remains controversial. The latest interpretation seems to be that dyslipoproteinemia in HCV-G1 patients may be a confounding factor of the host genotype of IL28B that is the strongest predictor for virological outcome following PEG-IFN plus RBV therapy [28].

Initially, we compared serum levels of apolipoproteins in chronically HCV-G2- and G1b-infected patients paying special attention to virological outcome of PEG-IFN plus RBV therapy in the HCV-G1b patients. Of the pre-treatment fasting sera taken from 42 HCV-G1b patients, 23 achieved a sustained viral response (SVR; negative for HCV RNA at 6 months after standard 48 weeks of therapy); 8 had a transient viral response (TVR; HCV RNA-negative during therapy, but reappearing after therapy); and 11 had a non-viral response (NVR; HCV RNA-positive during therapy). 24 HCV-G2 patients were also examined.

There were no differences in the concentration of apoAI, apoAII, apoCIII, apoE and apoB-48 (138.2±27.4 mg/dl vs. 142.4±32.4 mg/dl, 30.4±6.5 mg/dl vs. 29.9±4.8 mg/dl, 5.7±2.0 mg/dl vs. 6.2±2.4 mg/dl, 4.4±1.1 mg/dl vs. 4.3±1.0 mg/dl and 3.3±4.2 mg/dl vs. 3.1±1.4 mg/dl, respectively) between HCV-G1b and G2 patients. In addition, no significant changes were observed among patients with HCV-G1b showing different outcome of PEG-IFN plus RBV therapy. However, there were substantial differences in apoB and apoCII levels according to the response to PEG-

IFN plus RBV therapy in HCV-G1b patients. In HCV-G1b infection, apoB and apoCII levels were significantly higher in SVR patients than in NVR patients. The levels of apoB and apoCII in HCV-G2 patients were similar to those in HCV-G1b patients who achieved SVR (Figure 2).

These data suggest that the apolipoprotein profile in HCV-G1b patients is basically indistinguishable from that in HCV-G2 patients. Meanwhile, the profile differed in relation to different outcomes of PEG-IFN plus RBV therapy among chronic HCV-G1b infection. From these observations, we presumed that a decrease in LDL and/or VLDL may be a feature of dyslipoproteinemia in HCV-G1b patients who failed to respond to PEG-IFN plus RBV therapy. ApoB-100 (about 95% of apoB is apoB-100 in fasting sera) is an indicator of total VLDL, intermediate-density lipoprotein (IDL) and LDL particles in the blood because each particle of VLDL, IDL or LDL is composed of one molecule of apoB-100. Moreover, the majority of apoCII, which activates the enzyme lipoprotein lipase in capillaries, is associated with VLDL. Therefore, a decrease in apoCII could be related to a decrease in VLDL. Taking into consideration these observations, we concluded that a characteristic of the lipoprotein profile during the pre-treatment period of chronic HCV-G1b patients who subsequently fail antiviral IFN-based therapy is a decline in VLDL levels. Although lower serum apoE has recently been reported to be related to a favorable response to IFN-based therapy [35], we did not find any differences in serum apoE levels.

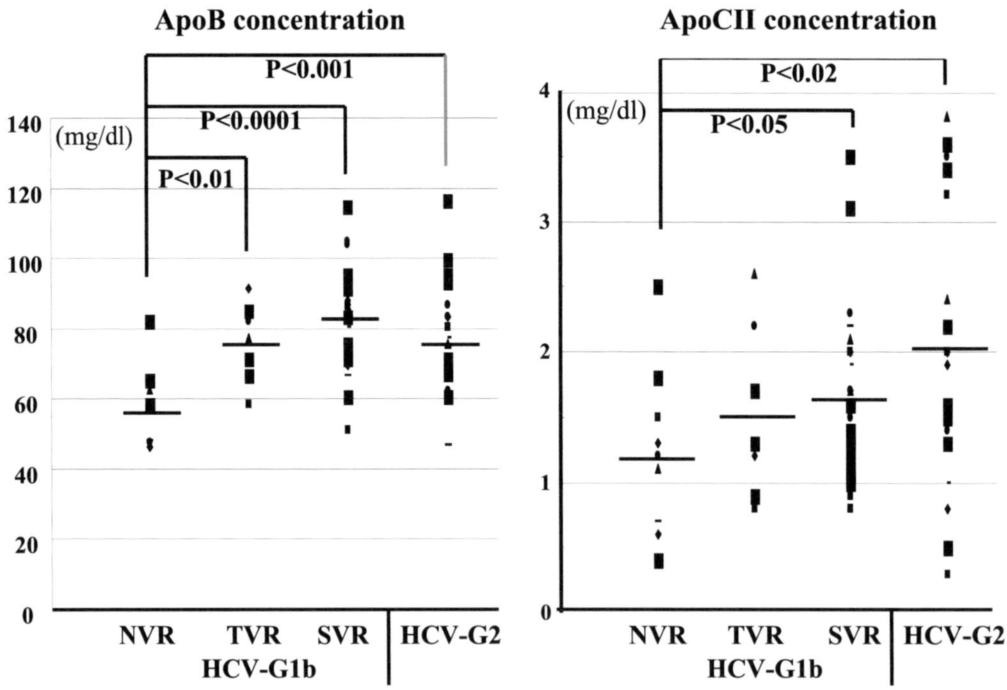

Figure 2. Serum apoB and apoCII levels in patients with chronic HCV-G1b or HCV-G2 infection. All HCV-G1b patients were treated with standard PEG-IFN plus RBV combination therapy and clarified according to the virological response. (NVR; non-viral response, TVR; transient viral response, SVR; sustained viral response)

Next, we examined the profiles of serum lipoproteins in 32 patients who were chronically infected with HCV-G2 and 111 patients with HCV-G1b, along with genotyping of the region near the IL28B gene (SNP ID rs8099917), the strongest predictor of outcome to PEG-IFN plus RBV therapy. Recently, we reported that serum apoB-100 levels are prescribed by genetic polymorphism of rs8099917, and a minor genotype of rs8099917 (TG or GG; also known as non-responder genotype of PEG-IFN plus RBV combination therapy) may be related to a decrease in serum apoB-100 in chronic HCV-G1b infection [36]. In chronic HCV-G1b infections in Japan, the minor genotype was found in 25-35% of cases, whereas it was only 10-20% in chronic HCV-G2 infections. We examined SNP of rs8099917 and the lipoprotein profiles obtained by LipoSEARCH were compared among 29 HCV-G2 patients who had the major genotype of IL28B (rs8099917), 75 HCV-G1b patients who had the major IL28B genotype and 36 HCV-G1b patients who had the minor IL28B genotype.

There were no differences related to gender, age, fibrosis score of liver biopsy, BMI, serum ALT level, viral load or platelet count among these three groups (data not shown). We did not cite the chylomicron fraction because this fraction was too small to analyze precisely.

Lipoprotein fraction	TC	VLDL-C	LDL-C	HDL-C
HCV-G2, IL28B major	167.92±31.20	38.22±17.65	81.18±22.25	44.79±14.49
HCV-G1b, IL28B major	169.65±30.52	45.52±18.64	72.80±17.27*	46.54±12.22
HCV-G1b, IL28B minor	161.69±30.87	36.22±15.18***	68.61±18.40*	52.20±16.64**

(mg/dl)

Lipoprotein fraction	TG	VLDL-TG	LDL-TG	HDL-TG
HCV-G2, IL28B major	104.46±59.41	52.48±36.43	27.22±10.17	16.22±5.91
HCV-G1b, IL28B major	93.30±36.74	43.52±20.99	24.65±7.37	16.82±6.20
HCV-G1b, IL28B minor	92.92±42.21	41.53±20.32	22.10±6.69*	19.14±8.25

(mg/dl)

*P>0.05 vs HCV-G2, **P>0.05 vs HCV-G1b with IL28B major, ***P>0.01 vs HCV-G1b with IL28B major

Table 1. Differences in lipoprotein fractions among patients with chronic HCV-G1b and G2 infection with different IL28 genotype

In patients with HCV-G1b having the IL28B major (responder) genotype, the concentration of cholesterol in the LDL fraction (LDL-C) was significantly lower than that in patients with HCV-G2. In addition, the tendency of a reciprocal increase of VLDL-C (P=0.055) in HCV-G1b patients having the responder genotype was observed. As a result, the level of LDL-C plus VLDL-C was similar between these two groups. In contrast, a reciprocal increase of VLDL-C was not observed in G1b patients who had the non-responder genotype. Decrease of VLDL-C and increase of HDL-C is a feature of the non-responder genotype compared with the responder genotype in chronic HCV-G1b patients.

We also examined serum lipoprotein profiles of five patients who had been infected with HCV-G1b, but achieved SVR by PEG-IFN plus RBV therapy (HCV-free status continuing for longer than 6 months). In these cured patients, low levels of VLDL-C and high levels of LDL-C and HDL-C were observed (Table 2). The lipoprotein profiles of these cured patients were quite normal with a normal composition of cholesterol and TG in each lipoprotein fraction when assessed by C:T ratio. When compared with cured patients, a relative decrease of TG in the VLDL fraction and a relative increase of TG in the LDL and HDL fractions were noted in chronic HCV infection.

In chronic HCV infection, we unexpectedly found that the serum levels of LDL-C measured by an HPLC system (LipoSEARCH) were considerably lower than those measured directly using a conventional method (HCV-G1b: 93.8±26.76 mg/dl; HCV-G2: 101.21±34.19 mg/dl), or measured indirectly by the Friedwald equation. In place of the decreased LDL fraction, the VLDL fraction was increased. This pattern is somewhat reminiscent of hyperlipidemic (high TG) samples [31]. However, in chronic HCV infection, TG levels are not much different from those in cured patients. Therefore, this finding cannot be explained by an increase of TG.

Lipoprotein fraction	TC	VLDL-C	LDL-C	HDL-C
HCV-G1b, achieved SVR	177.49±35.63	26.34±7.24	89.41±31.61	61.28±15.14

(mg/dl)

	TG	VLDL-TG	LDL-TG	HDL-TG
HCV-G1b, achieved SVR	79.15±25.33	44.11±18.77	22.08±4.96	10.47±2.64

(mg/dl)

Table 2. Serum lipid profiles of patients who achieved SVR by PEG-IFN plus RBV therapy for chronic HCV-G1b infection (sera were obtained at least 6 months after HCV was completely eradicated)

An increase of the VLDL fraction in chronic HCV infection could be explained by reduced enzymatic activity of lipoprotein lipase, which may facilitate HCV cell entry [23] while delaying the conversion of VLDL to LDL. Alternatively, discrepancy between chemically determined LDL (conventional measurement method) and levels determined by particle size (HPLC-based method) may be explained by the existence of LDL-associated LVPs in the blood during chronic HCV infection. These particles may have the physicochemical surface nature of LDLs, but particle sizes larger than 55 nm because the diameter of the HCV particle is about 55 nm, and hence must be eluted in the VLDL fraction. Although we must take into consideration that lipoprotein particles could theoretically become fused together during freezing and thawing, thus seriously distorting the lipoprotein fraction pattern determined by the HPLC-based method, freezing and thawing has reportedly been found not to seriously affect lipoprotein profiles (Skylight Biotech, http://www.lipo-search.com).

A recent study by Nishimura et al. [37] suggested that diminished VLDL-TG/non-VLDL-TG is a key feature of chronic HCV infection. They detected VLDL-TG based on the chemical nature of VLDL. Their findings do not conflict with our data. Our results indicate a relative decrease of TG in the VLDL fraction, but a relative increase of TG in the LDL and HDL fractions. Thus, their findings of decreased VLDL-TG/non-VLDL-TG appear to be consistent with our results.

As this kind of lipid abnormality is not easily determined by conventional methodology, the HPLC-based method is extraordinarily useful for the study of lipoprotein profiles in chronic

HCV infection. However, close attention to sample handling is needed because lipoprotein particles are fairly unstable.

The tentative conclusions are as follows: 1. Decrease of apoB (apoB-100) and apoB-100-related cholesterol might be a main feature of dyslipoproteinemia in chronic HCV-G1b infection with unfavorable response to PEG-IFN plus RBV combination therapy; 2. Existence of abnormally large particles (LPVs) eluted in the VLDL fraction by HPLC, in spite of the chemical characteristics of LDL, and/or reduced activity of lipoprotein lipase with delayed VLDL dissociation, may be a feature of dyslipoproteinemia in chronic HCV infection; 3. Relative increase of TG in the LDL and HDL fractions with a relative decrease of TG in the VLDL fraction may be a feature of chronic HCV infection. These conclusions are partially in concordance with those reported by Mawatari et al. [38].

4. Do viral factors participate in the dyslipoproteinemia seen during chronic HCV-G1b infection?

HCV-G1 infection is widely distributed worldwide and the most common genotype in the world, while it is one of the most resistant genotypes to IFN-based therapy. In Japan, almost all G1 subtypes are 1b, contributing to more than 70% of chronic HCV infection cases. Viral factors participating in the response to IFN-based therapies have been extensively studied, especially with regard to HCV-G1b in Japan. Among them, core protein substitution at aa 70/91 [32] and aa substitutions in the ISDR [33], IRRDR [34] and in NS5A are widely accepted as candidates. Among them, substitution at core protein 91, Leu91Met, did not affect serum levels of apoB-100 as reported earlier by us [36]. We further examined the significance of substitution at aa 70, Arg70Gln/His, aa substitutions in the ISDR and IRRDR and aa substitutions at particular positions within the IRRDR.

In that former study, we determined that substitution of Arg70 to Gln/His70 was a distinctive factor participating in the regulation of serum apoB-100 levels in chronic HCV-G1b patients, independent from the IL28B genotype. To clarify the lipoprotein profiles according to substitution at aa 70, we examined the lipoprotein profiles of fasting sera from 113 chronic HCV-G1b patients (68 were Arg70 and 45 were Gln/His70) by LipoSEARCH, as described earlier (Table 3).

Lipoprotein fraction	TC	VLDL-C	LDL-C	HDL-C
HCV-G1b Arg70	169.80±30.75	43.16±18.39	74.23±18.68	48.09±14.03
HCV-G1b Gln/His70	158.54±28.44	40.02±17.40	65.99±15.51*	47.57±14.09

(mg/dl)

Lipoprotein fraction	TG	VLDL-TG	LDL-TG	HDL-TG
HCV-G1b Arg70	89.49±33.68	42.15±20.80	23.53±6.71	16.24±5.31
HCV-G1b Gln/His70	95.95±41.99	42.13±19.50	18.14±7.83	19.42±8.31*

(mg/dl)

(*: $p>0.05$)

Table 3. Differences in lipoprotein profiles related to the substitution at aa 70 in core region (Arg70 to Gln/His70) in patients with chronic HCV-G1b infection

There were no differences related to gender, age, fibrosis score of liver biopsy, BMI, serum ALT level, viral load or platelet count between Arg70 and Gln/His70. However, a significant difference was found in the distribution of the IL28B genotype (SNP of rs8099917). In Arg70, 49 patients were of the major (responder) genotype while 15 were of the minor (non-responder) genotype. In Gln/His70, 22 were major and 20 were minor. A total of 7 patients remained undetermined. The difference in the IL28B genotype distribution between these two groups was significant (*P*=0.0005 by chi-square test with Yate's correction). Therefore, the influence of the IL28B genotype was not excluded in this study. However, the pattern of dyslipoproteinemia seen in Gln/His70 cases is dissimilar to that in the IL28B minor genotype, which was described in detail. In core 70 mutants (Gln/His70), a significant decrease in LDL-C and increase in HDL-TG levels was demonstrated without a decrease in VLDL-C. TC levels tended to be lower than that in the core Arg70 cases (*P*=0.052). These findings may indicate that the core mutation at aa 70 is an important viral feature in relation to the dyslipoproteinemia seen in HCV-G1b, functioning mainly through decreasing LDL-C. As a result of the substitution at aa 70, the nature of the amino acid is substantially changed. Therefore, the configuration and the biological activity of the core protein may be significantly disturbed, which may lead to the disruption of lipid metabolism. However, the precise mechanism of the consequence of aa 70 substitution on lipid metabolism is a matter to be solved in the future.

To exclude the influence of the IL28B genotype, we further compared the lipoprotein profiles in 49 patients with core 70 wild-type, and 22 patients with core 70 mutant phenotypes in whom the IL28B genotype was major (Table 4).

Lipoprotein fraction	TC	VLDL-C	LDL-C	HDL-C
HCV-G1b Arg70	173.37±30.83	45.80±18.29	76.41±18.28	46.38±12.37
HCV-G1b Gln/His70	155.44±24.13*	41.01±17.02	65.71±13.34*	45.72±12.45

(mg/dl)

Lipoprotein fraction	TG	VLDL-TG	LDL-TG	HDL-TG
HCV-G1b Arg70	92.33±35.95	42.57±22.27	25.50±6.77	15.87±5.28
HCV-G1b Gln/His70	93.44±36.33	41.36±17.84	26.12±8.72	18.28±7.38

(mg/dl)

(*: p>0.05)

Table 4. Differences in lipoprotein profiles related to substitution at aa 70 in the core region (Arg70 to Glun/His70) in patients with chronic HCV-G1b having IL28B major (responder) genotype

Although a relatively small-sized study, the features of the dyslipoproteinemia seen in patients with the core aa 70 mutation was clearly elucidated as a decrease of TC due to a decrease of LDL-C. Even after exclusion of the influence of the IL28B genotype, the core 70 aa substitution was found to have a role in dyslipoproteinemia that may be critical.

We also examined aa substitutions in the NS5A region in relation to disturbance of serum lipid/lipoprotein levels, since NS protein may inhibit the secretion of apoB-100 in vitro [12]. Moreover, a polypeptide comprised of aa residues 2135 to 2419 within the NA5A protein co-precipitated with apoB, suggesting a possible interaction between NS5A protein and apoB-

100. Thus, we examined aa substitutions in particular regions of NS5A to elucidate the possibility of a viral factor being the determinant of lipid metabolism. We compared the aa sequence 2209-2248 (ISDR) with the sequence of HCV-J and the number of aa substitutions was classified as wild-type (0 or 1) or non-wild-type (≥2). According to the numbers of aa substitutions in the ISDR [33], 102 of 117 subjects were judged to be wild-type and 15 non-wild-type. No significant differences in serum apoB and lipid concentrations were found between wild-type and non-wild-type ISDR (Table 5).

Lipid profile	T.C.	TG	LDL-C*	apoB
ISDR wild	172.8±32.0	104.3±54.1	92.5±26.6	80.9±19.8
ISDR non-wild	167.1±28.0	82.2±26.2	92..9±22.1	78.2±13.9

(mg/dl)

*measured directly using commercial kit. Note the substantial differences of LDL-C level measured by HPLC system shown in Table 1, Table 2 and Table 3.

Table 5. Lipid profiles of HCV-G1b patients with ISDR wild and ISDR non-wild

Although not described in Table 5, there was no statistical difference between patients with wild-type and non-wild-type ISDR in terms of serum levels of apoAI, apoAII, apoCII, apoCIII, apoE and apoB-48. We also compared lipid profiles between patients with a substitution number of 0 (N=80) and ≥1 (N=37), and found no significant difference (data not shown).

We very recently examined the aa substitution number and the place of substitution in the IRRDR (aa 2334-2376) in 105 patients who were chronically infected with HCV-G1b. By comparison with the HCV-J sequence, the number of aa substitutions was determined. A high degree (≥6 substitutions) of sequence variation in the IRRDR, which is thought to be a useful marker for predicting SVR [34], was found in 34 patients, whereas a less diverse (≤5 substitutions) IRRDR sequence (predictive of non-SVR) was found in 71 patients.

The clinical background is illustrated in Table 6.

Gender (M/F)	42/63	TG	89.8±40.2 (mg/dl)
Age	62.9±12.0 (years)	LDL-C	95.5±29.2 (mg/dl)
ALT	53.5±38.5 (U/L)	apoB-100	73.0±21.1 (mg/dl)
Albumin	4.1±0.4 (g/dl)	rs8099917 (Major/Minor)	65/40
Plt	16.0±6.3x10^4	aa 70 (Wild/Mutant)	64/41
HCV-RNA	6.4±0.6 (logIU/ml)	aa 91 (Wild/Mutant)	63/42
TC	171.3±32.7 (mg/dl)	ISDR (Wild / Mutant)	92/12

Table 6. Clinical characteristics of 105 HCV-G1b patients whose IRRDR sequences were examined

The number of aa substitution detected in this study was illustrated in Figure 3.

As shown in Figure 3, the substitution number in the IRRDR was widely distributed. There was no difference in lipid profile between the two groups (substitution number ≥6 vs. substitution number ≤5) along with other clinical backgrounds except for the distribution of the IL28B genotype (Table 7).

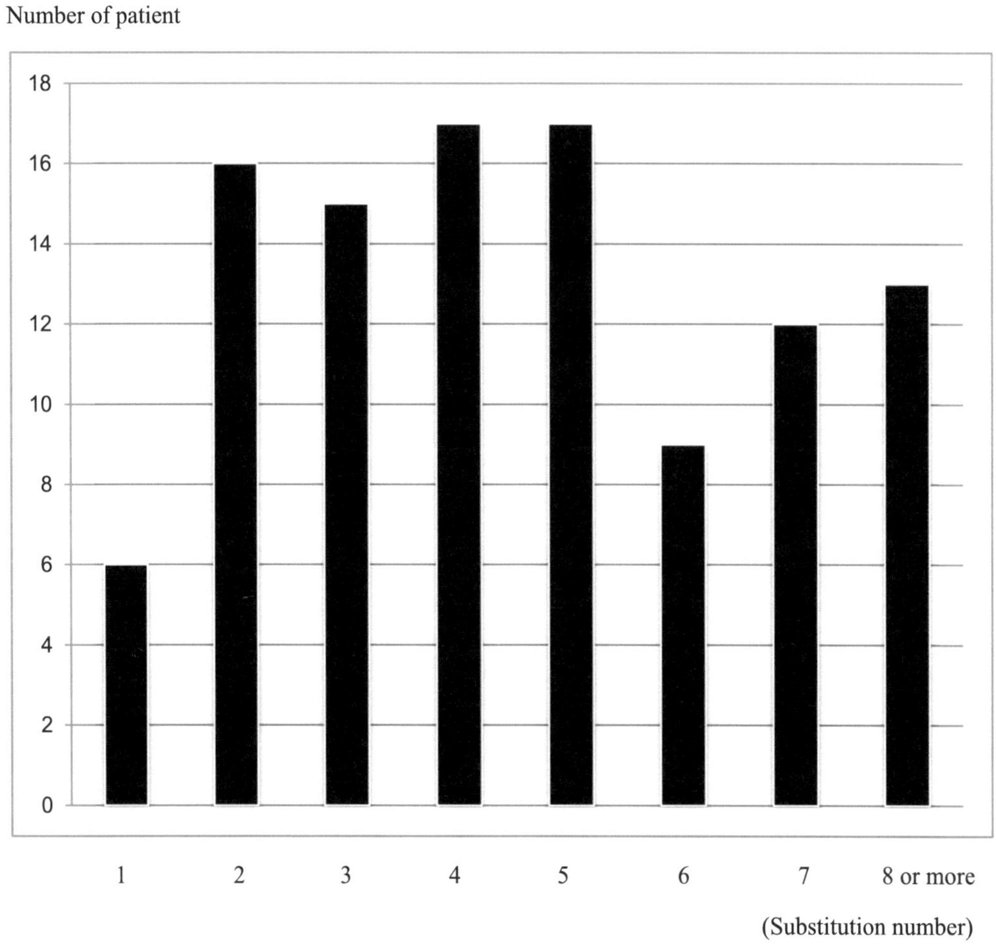

Figure 3. Distribution of aa substitution number in the IRRDR among 105 chronic HCV-G1b patients

	Substitutions (≤5)	Substitutions (≥6)	*P* value
TC	171 (115-253)	160 (96-268)	0.10
TG	81(42-267)	82 (37-207)	0.83
LDL-C	93 (36-193)	86 (50-172)	0.40
apoB-100	75 (39-131)	72 (44-135)	0. 23
IL28-B (TT/nonTT)	39/35	20/5	0.024

Data were expressed as median (range)

Table 7. Differences in lipid profile according to the substitution number in the IRRDR

Next, we examined the relationship between each aa substitution in the IRRDR and dyslipoproteinemia. As illustrated in Figure 4, substitution at aa 2356 may impact serum lipid profiles. There was no difference in lipid profile related to substitutions other than aa 2356.

A substitution at aa 2356 from Gly to Glu, Lys or Ala may be critical for distortion of the serum lipid profile. This substitution was previously shown to be a key substitution determining virological outcome of PEG-IFN plus RBV therapy in HCV-G1b patients [39]. The therapeutic outcome of 63 patients treated with PEG-IFN plus RBV is indicated in the lower right of Figure 4. Although a clear difference was not observed in the outcome of the therapy, NVR tended to be frequent in patients with non-Gly at aa 2356. The substitution at aa 2356 of Gly to Glu or Gly to Lys caused a drastic change in the nature of the amino acid that may influence the nature of the protein and in turn affect the biochemical interaction of apoE or apoB-100 with NA5A protein resulting in the decrease of serum apoB-100 and LDL-C. However, the change in serum apoB-100 level was minor. As aa substitution in the IRRDR is somewhat related to aa substitution in other regions such as the core aa 70, further examination is needed to establish the importance of aa 2356 substitution on lipoprotein metabolism in chronic HCV-G1b infection.

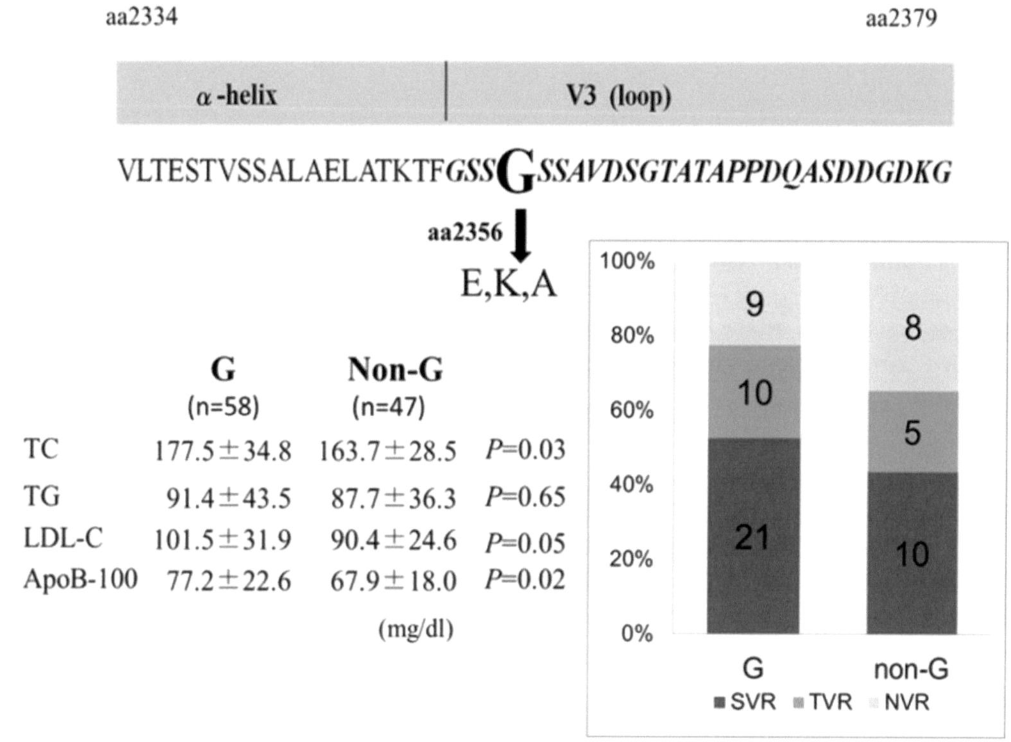

Figure 4. Consequence of substitution at aa 2356 from G to E, K or A in the IRRDR. Serum levels of TC, LDL-C and apoB-100 were significantly higher in patients who had G2356 than in patients who had not. At lower right, the virological response to PEG-IFN plus RBV therapy in 63 patients with HCV-G1b was summarized.

5. Are lipid/lipoprotein profiles independent predictors of therapeutic outcome in chronic HCV- G1b infection?

Chronic HCV infection has been treated with PEG-IFN and RBV combination therapy. HCV-G2 and HCV-G3 responded fairly well to this combination therapy. However, the response was weaker in HCV-G1 patients with only half of the patients achieving a sustained virological response. Before the discovery of the IL28B genotype, a decrease in TC, LDL-cholesterol or apoB-100-related cholesterol had been considered as an independent predictor of the outcome of PEG-IFN plus RBV therapy in HCV-G1. However, after the discovery that the genotype near the IL28B gene was a potent predictor, serum lipid levels have since been considered to be a confounding factor of the IL28B genotype. Actually, as we described earlier in this chapter and also in an earlier manuscript [36], the IL28B genotype may profoundly affect serum lipid levels in HCV-G1b infection.

Our latest knowledge, based on a prospective study examining predictive factors for the outcome of PEG-IFN plus RBV therapy, indicates that an increase in serum apoB-100 levels is an independent factor predicting rapid virological response (decline of serum HCV RNA at 4 weeks) to PEG-IFN plus RBV therapy. However, it does not appear to be an independent predictor of the final outcome of virological response in chronic HCV-G1b infection (data currently under consideration for contribution). These findings may partly indicate that disturbance of the lipid profile may reflect the efficiency of HCV replication due to efficient entry of HCV into hepatic cells and efficient production/secretion of HCV along with VLDL.

Although the evidence described in this chapter is not based on in vitro studies using HCV-secreting cells, our observations based on clinical samples may contribute to furthering the understanding of HCV-lipid metabolism interaction. From this viewpoint, it is noteworthy that lipoprotein lipase may act not only in the conversion of VLDL to LDL, but also in inhibiting HCV entry into liver cells [23]. Interestingly, in in vitro studies using naturally HCV-secreting cells, the enzymatic activity of lipoprotein lipase may be reduced in HCV-infected cells and may act to promote cell entry of HCV. This may contribute to an increase in the number of large-size (suitable for VLDL particle size) LVP having a chemical nature of LDL as suggested in our HPLC-based study.

Finally, hepatic steatosis may be associated with dyslipoproteinemia in chronic HCV infection and has been extensively studied with special attention to its relation to the IL28B genotype [40] and substitution at aa 70 [17] in HCV-G1b. However, in our experience, the relationship between hepatic steatosis and these host and/or viral factors was equivocal, perhaps because there are other environmental and host factors strongly affecting hepatic steatosis other than the factors discussed in this manuscript.

6. Summary

The features of dyslipoproteinemia in chronic HCV infection have been described. Serum lipid/lipoprotein profiles were in part HCV-genotype specific. The HCV core protein of HCV-G1b is closely associated with lipoprotein metabolism, and substitution of amino acid (aa 70) in the core protein may precipitate dyslipoproteinemia, in addition to substitution of aa 2356 in NS5A. In addition, dyslipoproteinemia in chronic HCV-G1b infection is largely affected by the genotype near the human IL28B gene. Therefore, dyslipoproteinemia in chronic HCV infection may involve complicated interplay between viral and host factors that could affect human lipid metabolism. Further study of lipids/lipoproteins in chronic HCV infection will be valuable to clarify the interaction of HCV and host lipid metabolism in detail. Dyslipoproteinemia in chronic HCV-G1b infection may play a role in the rapid decline of HCV during PEG-IFN plus RBV therapy. However, the clinical utility of dyslipoproteinemia as a predictor of final response to PEG-IFN plus RBV treatment remains controversial.

Author details

Yoshio Aizawa*, Hiroshi Abe, Kai Yoshizawa, Haruya Ishiguro and Yuta Aida
Jikei University Katsushika Medical Center, Division of Gastroenterology and Hepatology, Internal Medicine, Aoto, Katsushika-ku, Tokyo, Japan

Noritomo Shimada
Shinmatsudo Chuo General Hospital, Department of Gastroenterology and Hepatology, Shin-Matsudo, Matsudo City, Chiba, Japan

Akihito Tsubota
Jikei University Kashiwa Hospital, Institute of Clinical Medicine and Research (ICMR), Kashiwa City, Chiba, Japan

Acknowledgement

We thank professor Enomoto and doctor Maekawa, University of Yamanashi for kind examination of aa substitution within IRRDR.

7. References

[1] Target-Adams P, Boulant s, Douglas MW, McLauchlan J. (2010) Lipid metabolism and HCV infection. Viruses 2: 1195-1217.
[2] Syed GH, Amako Y, Siddqui A. (2010) Hepatitis C virus hijacks host lipid metabolism. Trends Endocrinol Metab 21:33-40.
[3] Popescu CI, Dubuisson J. (2009) Role of lipid metabolism in hepatitis C virus assembly and entry. Biol Cell 102:63-74
[4] Alvisi G, Madan V, Bartenschlager R. (2011) Hepatitis C virus and host cell lipids: an intimate connection. RNA Biol 8:258-269.
[5] Gastaminza P, Cheng G, Wieland S, Zhong J, Liao W et al. (2008) Cellular determinants of hepatitis C virus assembly, maturation, degradation and secretion. J Virol 82:2120-2129.
[6] Huang H, Sun F, Owen DM, Li W, Chen Y et al. (2007) Hepatitis C virus production by human hepatocytes dependent on assembly and secretion of very low-density lipoproteins. Proc Natl Acad Sci USA 104:5848-5853.
[7] Sabahi A, Marsh KA, Dahari H, Corcoran P, Lamora JM et al. (2010) The rate of hepatitis C virus infection initiation in vitro is directly related to particle density. Virology 407:110-119.

* Corresponding Author

[8] Nielsen SU, Bassendine MF, Burt AD, Martin C, Pumeechockchai W et al. (2006) Association between hepatitis C virus and very-low-density lipoprotein (VLDL)/LDL analyzed in iodixanol density gradients. J Virol 80:2418-2428.
[9] Popescu CI, Dubuisson J. (2010) Role of lipid metabolism in hepatitis C virus assembly and entry. Biol cell 102:63-74.
[10] Chang K.S, Jiang J, Cai Z, Luo G (2007) Human apolipoprotein e is required for infectivity and production of hepatitis C in cell culture. J Virol 81:13783-13793
[11] Zampino R, Ingrosso D, Durante-Mangoni E, Capasso R, Tripodi MF et al. (2008) Microsomal triglyceride transfer protein (MTP) -493G/T gene polymorphism contributes to fat liver accumulation in HCV genotype 3 infected patients. J Viral Hepat 15:740-746
[12] Domitrovich AM, Felmlee DJ, Siddiqui A. (2005) Hepatitis C nonstructural proteins inhibit apolipoprotein B100 secretion. J Biol Chem 280:39802-39808.
[13] Counihan NA, Rawlinson SM, Lindenbach BD. (2011) Trafficking of hepatitis C virus core protein during virus particle assembly. PLoS Pathog 7: e1002302.
[14] Roingeard P, Depla M. (2011) The birth and life of lipid droplets: learning from the hepatitis C virus. Biol Cell 103:223-231.
[15] Koike K, Tsutsumi T, Yotsuyanagi H, Moriya K. (2010) Lipid metabolism and liver disease in hepatitis C viral infection. Oncology 78 Suppl 1:24-30.
[16] Hourioux C, Patient R, Morin A, Blanchard E, Moreau A et al. (2007) The genotype 3-specific hepatitis C virus core protein residue phenylalanine 164 increases steatosis in an in vitro cellular model. Gut 56:1302-1308.
[17] Sumida Y, Kanemasa K, Hara T, Inada Y, Sakai K et al. (2011) Impact of amino acid substitutions in hepatitis C virus genotype 1b core region on liver steatosis and glucose tolerance in non-cirrhotic patients without overt diabetes. J Gastroenterol Hepatol 26:836-842.
[18] André P, Komurian-Pradel F, Deforges S, Perret M, Berland JL et al. (2002) Characterization of low- and very-low-density hepatitis C virus RNA-containing particles. J Virol 76:6919-6928.
[19] Napolitano M, Giuliani A, Alonzi T, Mancone C, D'Offizi G et al. (2007) Very low density lipoprotein and low density lipoprotein isolated from patients with hepatitis C infection induce altered cellular lipid metabolism. J Med Virol 79:254-258.
[20] Meredith LW, Wilson GK, Fletcher NF, McKeating JA. (2012) Hepatitis C virus entry: beyond receptors. Rev Med Virol doi: 10.1002/rmv.723.
[21] Dao Thi VL, Dreux M, Cosset FL. (2011) Scavenger receptor class B type I and the hypervariable region-1 of hepatitis C virus in cell entry and neutralisation. Expert Rev Mol Med 13:e13.
[22] Sainz Jr B, Barretto N, Martin DN, Hiraga N, Imamura M et al. (2012) Identification of the Nieman-Pick C1-like 1 cholesterol absorption receptor as a new hepatitis C virus entry factor. Nature Med doi:10,1038/nm.2581.
[23] Maillard P, Walic M, Meuleman P, Roohvand F, Huby T et al. (2011) Lipoprotein lipase inhibits hepatitis C virus (HCV) infection by blocking virus cell entry. PLoS One 6:e26637.

[24] Sheridan DA, Price DA, Schmid ML, Toms GL, Donaldson P et al. (2009) Apolipoprotein B-associated cholesterol is a determinant of treatment outcome in patients with chronic hepatitis C virus infection receiving anti-viral agents interferon-alpha and ribavirin. Aliment Pharmacol Ther 29:1282-1290.

[25] Moriya K, Shintani Y, Fujie H, Miyoshi H, Tsutsumi T et al. (2003) Serum lipid profile of patients with genotype 1b hepatitis C viral infection in Japan. Hepatol Res 25:371-376.

[26] Akuta N, Suzuki F, Kawamura Y, Yatsuji H, Sezaki H ET AL. (2007) Predictors of viral kinetics to peginterferon plus ribavirin combination therapy in Japanese patients infected with hepatitis C virus genotype 1b. J Med Virol 79:1686-1695.

[27] Ramcharran D, Wahed AS, Conjeevaram HS, Evans RW, Wang T et al. (2010) Association between serum lipids and hepatitis C antiviral treatment efficacy. Hepatology 52:854-863.

[28] Li JH, Lao XQ, Tillmann HL, Rowell J, Patel K et al. (2010) Interferon-lambda genotype and low-density lipoprotein cholesterol levels in patients with chronic hepatitis C infection. Hepatology 51:1904-1911.

[29] Kinoshita M, Kojima M, Matsushima T, Teramoto T (2005) Determination of apolipoprotein B-48 in serum by a sandwich ELISA. Clin Chim Acta 351:115-120.

[30] Usui S, Hara S, Hosaki S, Okazaki M. (2002) A new on-line dual enzymatic method for simultaneous quantification of cholesterol and triglycerides in lipoproteins by HPLC. J Lipid Res 43: 805-814.

[31] Okazaki M, Usui S, Ishigami M, Sasaki N, Nakamura T et al. (2005) Identification of Unique Lipoprotein Subclasses for Visceral Obesity by Component Analysis of Cholesterol Profile in High-Performance Liquid Chromatography. S

[32] Akuta N, Suzuki F, Kawamura Y, Yatsuji H, Sezaki H et al. (2007) Predictive factors of early and sustained responses to peginterferon plus ribavirin combination therapy in Japanese patients infected with hepatitis C virus genotype 1b: amino acid substitutions in the core region and low-density lipoprotein cholesterol levels. J Hepatol 46:403-410.

[33] Enomoto N, Sakuma I, Asahina Y, Kurosaki M, Murakami T et al. Mutations in the nonstructural protein 5A gene and response to interferon in patients with chronic hepatitis C virus 1b infection. (1996) N Engl J Med 1996; 334: 77-81.

[34] El-Shamy A, Nagano-Fujii M, Sasase N, Imoto S, Kim SR et al. (2008) Sequence variation in hepatitis C virus nonstructural protein 5A predicts clinical outcome on pegylated interferon/ribavirin combination therapy. Hepatology 48: 38-47.

[35] Sheridan DA, Bridge SH, Felmlee DJ, Crossey MM, Thomas HC et al. (2012) Apolipoprotein-E and hepatitis C lipoviral particles in genotype 1 infection: Evidence for an association with interferon sensitivity. J Hepatol Mar 10. [Epub ahead of print]

[36] Aizawa Y, Yoshizawa K, Aida Y, Ishiguro H, Abe H et al. (2012) Genotype rs8099917 near the IL28B gene and amino acid substitution at position 70 in the core region of the hepatitis C virus are determinants of serum apolipoprotein B-100 concentration in chronic hepatitis C. Mol Cell Biochem 360:9-14.

[37] Mawatari H, Yoneda M, Fujita K, Nozaki Y, Shinohara Y et al. (2010) Association between lipoprotein subfraction profile and the response to hepatitis C treatment in Japanese patients with genotype 1b. J Virol Hepatitis 17:274-279.

[38] ElHefnawi MM, Zada S, El-Azab IA. (2010) Prediction of prognostic biomarkers for interferon-based therapy to hepatitis C virus patients: a meta-analysis of the NS5A protein in subtypes 1a, 1b, and 3a. Virol J Jun 15;7:130.
[39] Ohnishi M, Tsuge M, Kohno T, Zhang Y, Abe H et al. (2012) IL28B polymorphism is associated with fatty change in the liver of chronic hepatitis C patients. J Gastroenterol. 2012 Feb 18. [Epub ahead of print].

Section 10

Lipoproteins and Hemostasis

Chapter 31

An Apolipoprotein CIII-Derived Peptide, Hatktak, Activates Macromolecular Activators of Phagocytosis from Platelets (MAPPs)

Haruhiko Sakamoto, Masaki Ueno, Wu Bin, Yumiko Nagai, Kouichi Matsumoto, Takao Yamanaka and Sumiko Tanaka

Additional information is available at the end of the chapter

http://dx.doi.org/10.5772/48173

1. Introduction

When thrombi are formed, infiltration of leukocytes including neutrophils and macrophages follows in and around the thrombi. On the other hand, in inflammatory lesions where infiltration of leukocytes is observed, thrombi are often observed. When thrombi are formed, activated platelets release a lot of substances while adhering to subendothelial connective tissues, recruiting other platelets and aggregating together. In these regions with thrombus formation or platelet activation, it is possible that platelets affect leukocytic function via the action of the released products, since the substances released from platelets include factors that activate [1-14] and suppress [15-22] neutrophilic functions.

In experiments using human platelets and neutrophils, we found that platelets release several neutrophilic phagocytosis activators. Among substances released from activated platelets, ATP and ADP have been reported to activate iC3b receptor-mediated phagocytosis [1, 2, 23], and some prostaglandins including PGE2, PGF2α and thromboxane B2 and macromolecular activators of phagocytosis from platelets (MAPPs) activate Fcγ receptor-mediated phagocytosis.

MAPPs have two subsets, l-MAPP ($3x10^5$ Da) and s-MAPP ($1.5x10^5$ Da) [24, 25]. Platelets stored in the form of platelet-rich plasma lose the capacity to release MAPPs but recover it if incubated with the plasma-derived precursors of MAPPs (precursors of l-MAPP and s-MAPP, $3x10^5$ Da and $1.5x10^5$ Da, respectively) and thrombin in the presence of Ca^{++} [26]. It was suggested that the loss of platelet ability to release MAPPs is due to escape of the precursors and thrombin during storage with CPD (citrate-phosphate-dextrose) solution. It is possible to produce MAPPs using stored platelet-derived lysate by stimulation with thrombin or trypsin in the presence of the precursors [27]. It has also been suggested that the precursors of l-MAPP and s-MAPP are polymerized transferrins, probably of tetrameric and dimeric forms, respectively [28].

As for the production of MAPPs in platelets, it was suggested that GP Ibα-bound thrombin reacts with a high-molecular-weight substance (HMW activator) to release a low-molecular-weight substance (LMW activator) [29], which can produce MAPPs from the precursors directly.

In an ultracentrifugation study of the platelet lysate, the HMW activator activity was observed in the HDL fraction. Anti-apolipoprotein A1 antibody abolished the HMW activator function from the HDL fraction of the platelet lysate. These findings suggest that the HMW activator belongs to HDL. In an affinity chromatography study of the protein obtained from the HDL-rich fraction of the platelet lysate using an anti-apolipoprotein CIII (Apo CIII) column, it was suggested that LMW activator is derived from Apo CIII. In fact, it was observed that the commercially available Apo CIII could produce LMW activator by the activity of thrombin [30].

The purpose of this study was to determine the structure of the LMW activator.

2. Materials and methods

This study was approved by the local institutional review board of our university hospital. The procedures followed were in accordance with the Helsinki Declaration of 1975, as revised in 1983.

2.1. Phagocytosis experiment

The phagocytosis experiments were performed according to a method previously described [30]. Briefly, neutrophils were separated from heparinized venous blood from healthy volunteers by centrifugation on MonoPoly resolving medium (ICN Biochemicals Japan, Tokyo, Japan). After washing of neutrophils in phosphate-buffered saline (PBS) (Sigma, St. Louis, MO), one thousand neutrophils in 5 μl of PBS were attached to the surface of a Terasaki microplate well (Nunc, Roskilde, Denmark) by centrifugation at 160 g for 2 minutes. Then, stimulation was performed with 10 μl of the test material in PBS supplemented with 1% bovine serum albumin (Sigma) (PBS-BSA) at 37 °C for 15 minutes. After washing, the neutrophils were incubated with 5,000 sheep red blood cells (SRBCs) in 5 μl of RPMI 1640 medium supplemented with 1% BSA and a one-hundredth volume of anti-SRBC rabbit IgG at 37 °C for 25 minutes under 5% CO_2. After lysing the unphagocytosed SRBCs using a Tris-buffered ammonium chloride solution [31] and fixation with 1% glutaraldehyde (Sigma) in PBS, the phagocytosed SRBCs were counted under a light microscope.

All phagocytosis experiments were performed in triplicate. Mean numbers of ingested SRBCs per neutrophil in a well were calculated. Then, means of three wells were obtained. The results are expressed as phagocytic indices normalized by the phagocytic activity of the PBS-BSA control taken as 100.

2.2. Production of MAPPs in vitro

2.2.1. Preparation of the precursors of MAPPs using holo-transferrin

Precursors of MAPPs were produced artificially using holo-transferrin (Sigma) according to a method described previously [28, 30]. Briefly, the peak fraction of holo-transferrin in Superdex 200 (GE Healthcare UK Ltd., Buckinghamshire, England) gel filtration was incubated on ice for 15 min with 0.2% glutaraldehyde (Sigma) to induce polymerization [32], and subjected to filtration again. The fractions obtained at the same elution volumes as the conventional l-MAPP (tetramer transferrin-rich fraction) and s-MAPP (dimer transferrin-rich fraction) were adjusted to an optical density of 0.04 at 280 nm with PBS and used as precursors of l-MAPP and s-MAPP, respectively.

2.2.2. Preparation of LMW activators

Apo CIII (Chemicon International Inc., Temecula, CA) was adjusted to a concentration of 10 μg/ml in PBS and incubated with 0.1 unit/ml human thrombin (thrombin) (Sigma) or 1 unit/ml bovine trypsin (trypsin) (Sigma) at 37 °C for 30 minutes, and then applied on a PD10 column (GE Healthcare UK Ltd.) to obtain the low-molecular-weight fraction, which was serially diluted x10 and used as the LMW activator.

Another Apo CIII (10 μg/ml) was incubated with 0.1 unit/ml thrombin or 1 unit/ml trypsin and gel-filtered through a Superdex peptide column (GE Healthcare UK Ltd.) (elution buffer, PBS; elution speed, 0.5 ml/minute; fraction volume, 1 ml each) to obtain LMW activator fraction.

Samples of the low-molecular-weight fraction of variously prepared platelet-related materials including platelet release products and platelet lysate (see below for details) were obtained using PD-10 columns.

Some peptides associated with Apo CIII were synthesized artificially, and the activity as an LMW activator was examined.

2.2.3. Preparation of MAPPs using LMW activators and precursors of MAPPs

To form MAPPs, 1 ml of appropriately diluted LMW activator derived from Apo CIII, platelet release products or artificially produced peptides was incubated with 10 μl of precursor of l-MAPP or s-MAPP at 37 °C for 30 minutes.

2.3. Preparation of platelet release products using platelets from platelet-rich plasma

To prepare fresh and stored platelets, heparinized venous blood was mixed with 13 vol% citrate-phosphate-dextrose (CPD) solution and centrifuged at 60 g for 20 minutes at 20 °C. The platelet-rich plasma, which was contaminated by less than one erythrocyte per 1,000 platelets, was transferred into another plastic tube and stored at 20 °C in a water bath with agitation once a second for 0, 72 or 120 hours.

These platelets were washed twice in PBS supplemented with 6.7 mM EDTA and twice in PBS, and adjusted to a concentration of $4x10^5$/μl. The platelet suspension was stimulated with 0.1 unit/ml thrombin at 37 °C for one minute in the presence or absence of 4 mEq/l Ca^{++}, then cooled immediately on an ice-water bath and centrifuged at 960 g for 15 minutes at 4 °C to obtain the platelet release products in the supernatant.

2.4. Preparation of platelet lysate

Platelet suspensions obtained from platelet-rich plasma as described above were frozen at -15 °C, thawed and centrifuged at 1,500 g for 30 minutes. The supernatant was used as the platelet lysate.

2.5. Peptide synthesis

Peptides such as S1-K21, H18-R40, H18-K24, H18-K21 and T22-K24 of Apo CIII were synthesized by Sawady Technology (Tokyo, Japan). Other peptides that contain part or all of H18-K24 (HATKTAK) of Apo CIII were produced by Thermo Fisher Scientific GmbH (Ulm, Germany).

2.6. Inhibition of the LMW activator activity by anti-HATKTAK rabbit IgG antibody

Antibody against HATKTAK was raised in a rabbit by intracutaneous injection of keyhole-limpet-conjugated CHATKTAK (6 injections at 2-week intervals) by Sigma Aldrich Japan (Ishikari city, Japan). The IgG antibody was refined using a protein A column. The control IgG antibody was obtained from the same rabbit at day 0 of immunization.

By indirect enzyme-linked immunosorbent assay (ELISA) [33], specificity of the anti-HATKTAK rabbit antibody was examined. 100 μl of antigens involving Apo CIII and Apo CIII-related peptides (1 μg/ml in PBS) was immobilized to each well of a polystyrene microplate by incubation at 4 °C overnight. 1000-fold-diluted anti-HATKTAK rabbit antibody or l000-fold-diluted anti-Apo CIII gout IgG (Gene Tex, Inc., San Antonio, TX) was used as the primary antibody. Peroxidase-conjugated anti-rabbit IgG (Fab') or anti-goat IgG (Fab') (Histofine simple stain MAX-PO(R) or -PO(G), respectively, Nichirei Bioscience, Tokyo, Japan) was used as the secondary antibody. Substrate solution consisted of 10 ml of 0.5 M citrate buffer, 10 ml of 0.3% hydrogen peroxide and 10 mg of orthophenylene diamine (Wako, Osaka, Japan). 2 M sulfuric acid was used as a stopping solution. Absorbance was determined at 450 nm in a model 550 microplate reader (Bio-Rad, Tokyo, Japan).

To examine the effect of anti-HATKTAK antibody on an LMW activator, a $x10^2$-diluted low-molecular-weight fraction of the platelet release products prepared from fresh platelet- rich plasma and a $x10^6$-diluted one from platelet-rich plasma stored for 120 hours were prepared. Then, they were incubated with a one-hundredth volume of anti-HATKTAK IgG or control IgG at room temperature for 30 min, applied to a protein A column to remove the immune complexes and the residual antibodies, and used as the source of the LMW activator.

2.7. Ion exchange chromatography

Cation exchange chromatography was performed using a MONO S HR5/5 column (GE Healthcare UK Ltd.). Samples were dissolved in 20 mM Tris-HCl, pH 8.0 (buffer A). The adsorbed materials were recovered by elution with a linear increase in buffer B (buffer A + 1 M NaCl).

2.8. Assessment of peptide concentrations using the o-phthalaldehyde (OPA) fluorescent method

Peptide concentrations were assessed according to the OPA fluorescent method with intact protein [34]. Fluorescence was measured with an excitation wavelength of 340 nm and an emission wavelength of 450 nm using a spectrofluorometer (FP-6300, JASCO, Tokyo, Japan).

2.9. Mass spectrometry

Matrix-associated laser desorption/ionization time-of-flight mass spectrometry was performed with a Voyager System 4314 (Applied Biosystems, Foster City, CA) in the reflector time-of-flight configuration at an acceleration voltage of 25 kV with delayed ion extraction. Samples were diluted 1:1 with a freshly prepared matrix solution consisting of 10 mg/ml α-cyano-4-hydroxycinnamic acid (Sigma) in 20% acetonitrile with 0.1% trifluoroacetic acid. Aliquots of 1 μl were deposited on a metallic sample holder and analyzed immediately after drying in a stream of air. Mass scale calibration was performed externally.

2.10. Immunohistochemistry

Coagulation of the fresh peripheral blood taken without anticoagulants was induced in a glass tube for 3 hours. This coagulum was formalin-fixed, and paraffin sections of 3 μm thickness were prepared. After deparaffinization and hydration, antigen retrieval was performed by heating in 10 mM citrate buffer (pH 6.0) for 40 minutes. After rinsing in PBS and incubation with 2% BSA in PBS at room temperature for 60 minutes, incubation with a mixture of rabbit anti-HATKTAK IgG (x1,000 diluted) and murine anti-CD 61 monoclonal IgG (Dako Japan, Tokyo, Japan) (x100 diluted) at 4 °C overnight was performed followed by incubation with a mixture of Alexa Fluor 594-conjugated anti-rabbit IgG (x200 diluted) and Alexa Fluor 488-conjugated anti-murine antibody (x200 diluted) (Molecular Probes, Eugene, OR, USA). The fluorescent signals were viewed under a confocal microscope (Bio-Rad Radiance 2100). As a control experiment, an identical immunohistochemical procedure with omission of the primary antibodies was performed.

2.11. Assay of concentrations of Apo CIII, apolipoprotein AI (Apo AI) and apolipoprotein B100

Concentrations of Apo CIII, Apo AI and Apo B100 in the platelet lysate were assayed by ELISA using commercially available assay kits for Apo CIII (AssayPro, St. Charles, MO), Apo AI (Mabtech AB, Nacka Strand, Sweden) and Apo B100 (Mabtech AB), respectively.

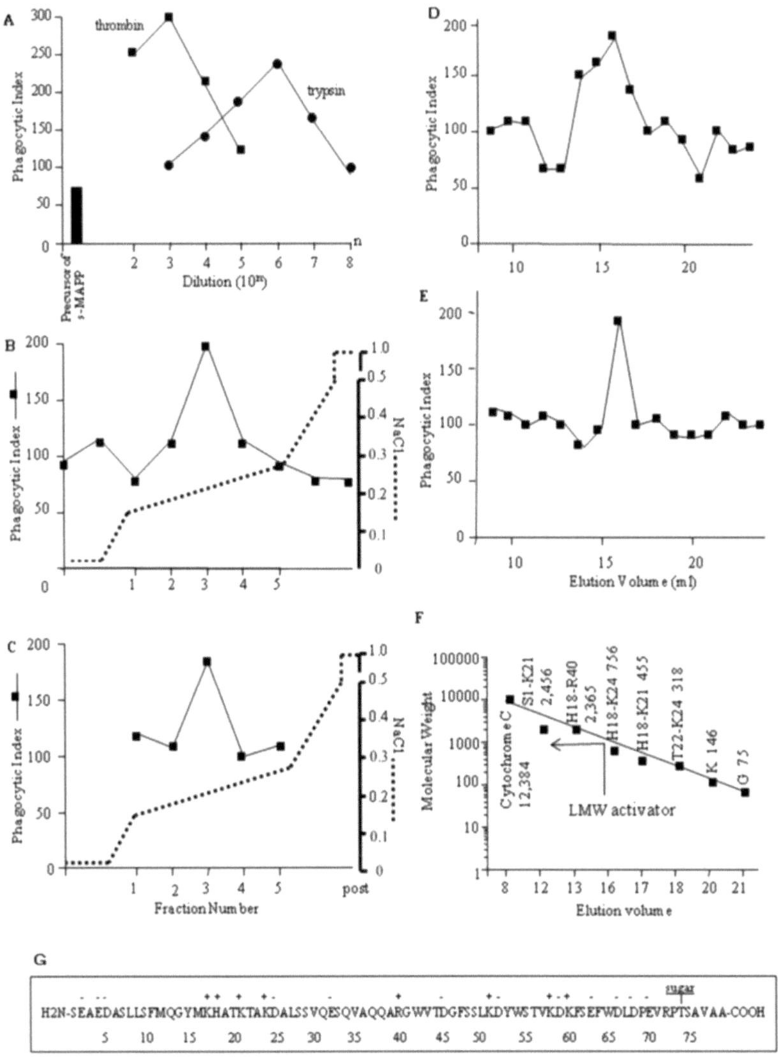

Figure 1. Characterization of the LMW activator produced from Apo CIII. **A.** MAPP formation using Apo CIII. To prepare LMW activator, 10 μg/ml Apo CIII was treated with 0.1 unit/ml thrombin or 1 unit/ml trypsin, and then diluted serially x10. The activity of each diluted sample as LMW activator was examined. **B.** MONO S cation exchange chromatography of thrombin-treated Apo CIII (10 μg). MAPPs were generated using each fraction at a dilution of x10^2 and one of the precursors of MAPPs. **C.** MONO S cation exchange chromatography of trypsin-treated apolipoprotein CIII (10 μg/ml). MAPPs were generated using each fraction at a dilution of x10^4 and one of the precursor of MAPPs. **D.** Superdex peptide gel filtration of thrombin-treated Apo CIII (10 μg/ml). Formation of MAPPs was achieved using each fraction at a dilution of x10^2 and one of the precursors of MAPPs. **E.** Superdex peptide gel filtration of trypsin-treated Apo C-III (10 μg/ml). Formation of MAPPs was achieved using each fraction at a dilution of x10^4 and one of the precursors of MAPPs. **F.** Molecular weight determination of the LMW activator with comparisons with those of Apo CIII-derived peptides. **G.** Amino acid sequence of Apo CIII, distributions of basic (+) and acidic (-) amino acids and sugar binding amino acid (T74) in Apo CIII. The amino acid sequence is cited from the database of GenPex (NCBI Protein DataBase, http://www.ncbi.nlm.nih.gov/protein). **A**, **B**, **C**, **D** and **E** were performed using the precursor of s-MAPP.

2.12. Statistics

In this study, the differences were analyzed using Mann-Whitney test, Kruskal-Wallis test, paired *t* test or Wilcoxon test. A *P* value of less than 0.05 was considered significant.

3. Results

3.1. LMW activator is considered to be HATKTAK

As it was shown in a previous study that both thrombin and trypsin could produce a substance with LMW activator activity from platelet lysate [27] and that incubation of Apo CIII with thrombin resulted in production of LMW activator, Apo CIII was treated with thrombin or trypsin and the LMW activator activities produced were compared to determine whether LMW activators produced by these enzymes are the same substance. Both thrombin and trypsin could produce LMW activator activity, but 1 unit/ml trypsin could produce LMW activator 100 to 1,000 times more effectively than 0.1 unit/ml thrombin (Figure 1A).

The LMW activator activities prepared using thrombin and trypsin appeared at the same NaCl concentration of the MONO S cation exchange chromatography (0.2 M NaCl, pH 8.0) (Figure 1B and Figure 1C, respectively), and at the same elution volume (16 ml) of the Superdex peptide gel filtration (Figure 1D and Figure 1E, respectively), which corresponds to the approximate molecular size of 800 Da (Figure 1F). These findings suggest that the two LMW activators are the same substance.

The fact that LMW activator can be produced by trypsin digestion of Apo CIII suggests that the C terminal of LMW activator consists of K or R [35]. LMW activator is suggested to be one of the basic peptides that would appear after trypsin digestion of Apo CIII (S1-K24, H18-R40, H18-K24, H18-K21, T22-K24) (see Figure 1G for the sequence of Apo CIII, distribution of basic and anionic amino acids and the sugar binding amino acid). These peptides were artificially produced and filtered on the Superdex peptide column to use as molecular size markers. The optical peaks of OPA fluorescence and LMW activator activity by H18-K24 (HATKTAK) appeared at the same elution volume of Superdex peptide gel filtration corresponding to Apo CIII-derived LMW activator (Figure 1D, 1E, 2A and 2B). In MONO S cation exchange chromatography, the action of HATKTAK as LMW activator was recovered at the fraction with the same NaCl concentration (0.2 M NaCl, pH 8.0) (Figure 2C) as the Apo CIII-derived LMW activator (Figure 1B and 1C).

3.2. Only HATKTAK among examined peptides showed MAPP-forming activity

Among the peptides used for calibration on Superdex peptide gel filtration (Figure 1F), H18-R40 and H18-K21 were separated in the fractions near to HATKTAK (LMW activator). The activity of these peptides as LMW activators were examined. Only HATKTAK showed LMW activator activity with a peak at 1 nM (Figure 2D). Then, some Apo CIII-derived peptides that contain part or all of H18-K24 were examined for their LMW activator activity. All of them were used at a concentration of 1 nM. Only HATKTAK showed MAPP-forming activity (Table 1).

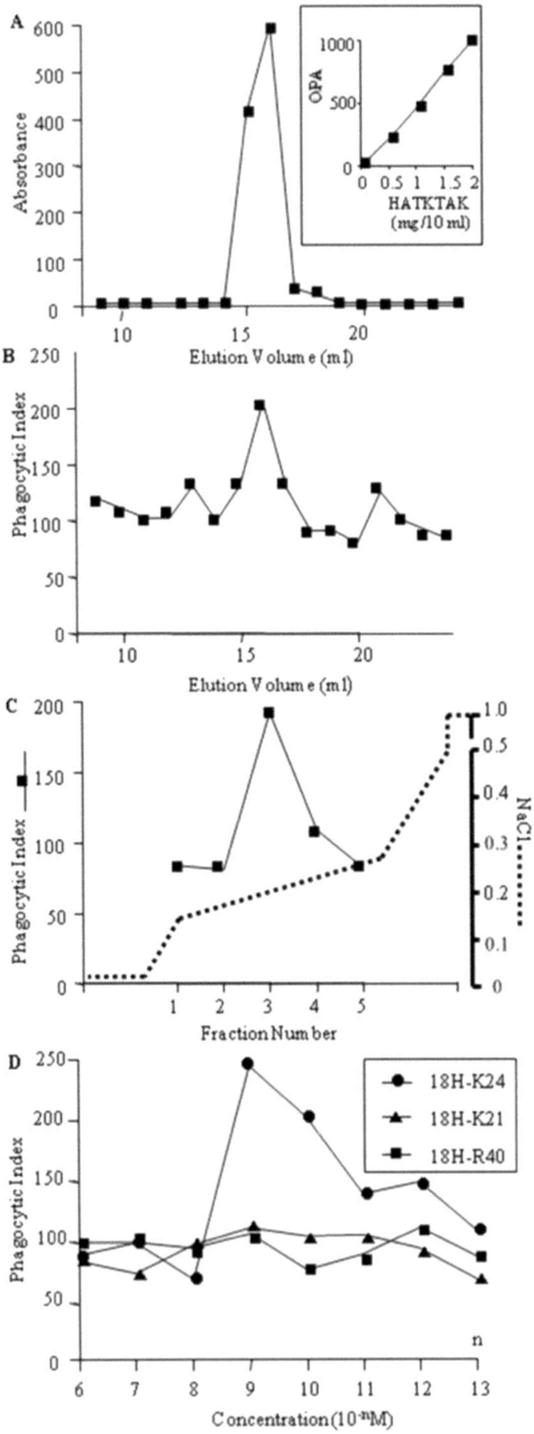

Figure 2. HATKTAK functions as the LMW activator. **A.** Superdex peptide gel filtration of HATKTAK (200 μg, 0.2 ml). The peptide concentration of each fraction was measured by the OPA fluorescent

method. The inset shows the standard graph for the assay of HATKTAK. **B.** MAPP formation using fractions of Superdex peptide gel filtration of HATKTAK (200 μg, 0.2 ml) at a dilution of x10^6 by incubation with one of the precursors of MAPPs. **C.** MONO S cation exchange chromatography of 1 μg/ml HATKTAK. MAPPs were generated by incubation of each fraction from the chromatography at a dilution of x10^4 and one of the precursors of MAPPs. **D.** MAPP formation using peptides containing part of the sequence of apolipoprotein CIII. Each peptide was serially diluted x10 and incubated with one of the precursors of MAPPs. B, **C** and **D** were performed using the precursor of s-MAPP.

Peptides	Precursor of l-MAPP	Precursor of s-MAPP
HATK	-	-
HATKT	-	-
HATKTA	-	-
HATKTAK	+	+
HATKTAKD	-	-
K	-	-
AK	-	-
TAK	-	-
ATKTAK	-	-
KHATKTAK	-	-

Table 1. MAPP formation from precursors of MAPPs with peptides derived from K16 to D25 of Apo CIII (KHATKTAKD)

3.3. LMW activator activity in platelet release products and platelet lysate

In the next part of the study, the LMW activator activities in variously prepared samples including platelet release products and platelet lysate were compared to determine whether platelets release LMW activator. Platelet release products were prepared using platelets from fresh and stored platelet-rich plasma. The reason why we examined platelet release products prepared from stored platelet-rich plasma along with those prepared from the fresh equivalent is that platelet releasate from stored platelet-rich plasma was expected to contain a higher concentration of LMW activator because platelets lose the precursors of MAPPs during storage [26]. From all of these samples, the low-molecular-weight fraction was separated using PD10 columns, and the activity corresponding to LMW activator was compared by the largest dilution from the original sample for which a phagocytic index higher than 150 was recorded (effective dilution).

1 ng/ml HATKTAK and platelet release products from fresh platelets stimulated with 4 mEq/l Ca^{++} and 0.1 unit/ml thrombin showed an effective dilution of 10^2 to 10^3, whereas fresh platelets stimulated only with 4 mEq/l Ca^{++} or 0.1 unit/ml thrombin for 1 minute showed far lower effective dilution, and platelet release products prepared from platelets after storage for 72 hours and 120 hours in the form of platelet-rich plasma showed extremely high effective dilution (10^3 to 10^5, platelets from platelet-rich plasma stored for 72 hours; 10^7 to 10^8, those stored for 120 hours). None of the platelet lysate produced from platelets in fresh and stored platelet-rich plasma showed LMW activator activity (Figure 3).

To confirm whether these LMW activator activities were by the same substance, Superdex peptide gel filtrations of the low-molecular-weight fractions of platelet release products prepared from fresh platelets (A), and those stored for 72 hours (B) and 120 hours (C), with stimulation with 0.1 unit/ml thrombin in the presence of 4 mEq/l Ca^{++} were performed. Fractions of the gel filtrations were diluted $x10^2$ (A), $x10^4$ (B) and $x10^6$ (C). All of these samples showed LMW activator activity at the fraction corresponding to HATKTAK (compare Figure 4A, 4B and 4C with Figure 2B).

3.4. Inhibition of LMW activator activity by anti-HATKTAK antibody

Then, to obtain evidence of the existence of HATKTAK in the platelet release products, actions of an anti-HATKTAK rabbit antibody against LMW activator activity were examined.

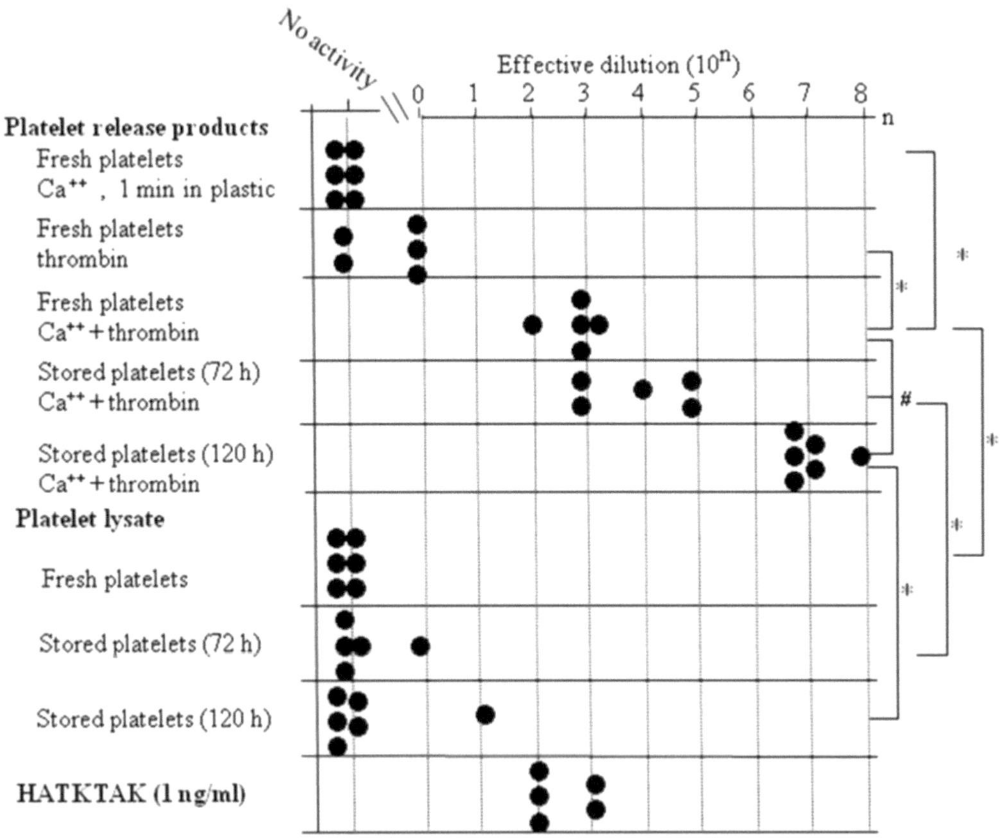

Figure 3. Comparisons of effective dilution of LMW activator in platelet release products and platelet lysate prepared using platelets that were obtained from platelet-rich plasma stored for 0 hours, 72 hours and 120 hours. A black circle means a case with the effective dilution of LMW activator indicated by the horizontal axis. *, *P*<0.01 with the Mann-Whitney test, and #, *P*<0.01 with the Kruskal-Wallis test. All experiments were performed using the precursor of s-MAPP.

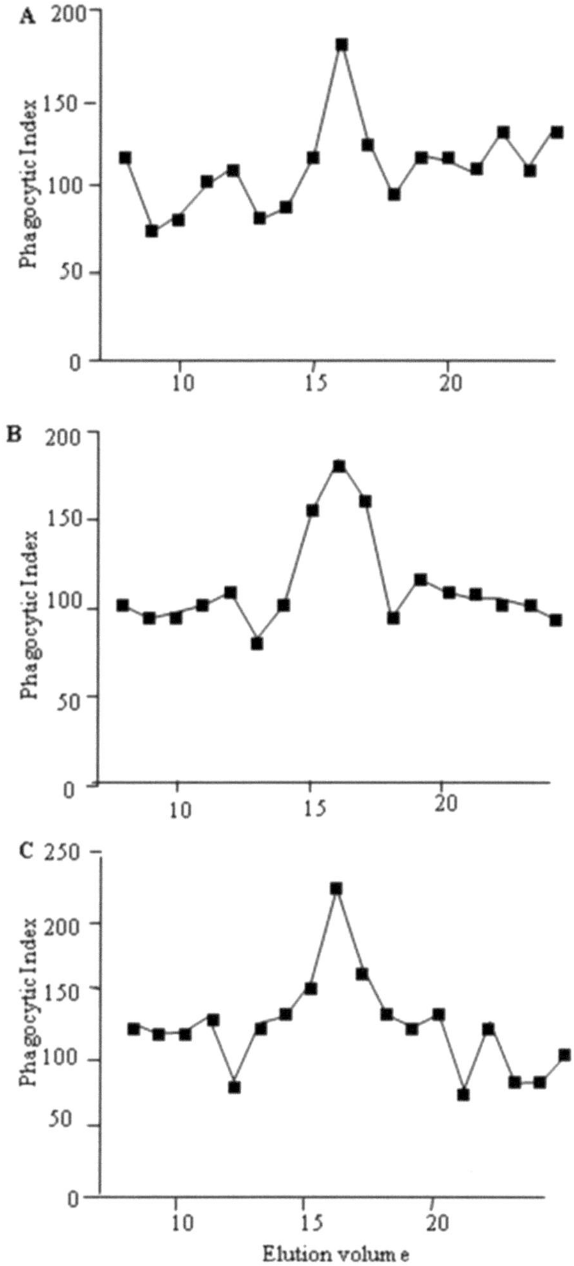

Figure 4. Comparisons of molecular size of LMW activator in platelet release products prepared using (**A**) platelet release products using fresh platelets stimulated with 0.1 unit/ml thrombin in the presence of Ca^{++}; (**B**) platelet release products using platelets stored for 72 hours stimulated with 0.1 unit/ml thrombin in the presence of Ca^{++}; (**C**) platelet release products using platelets stored for 120 hours stimulated with 0.1 unit/ml thrombin in the presence of Ca^{++}. The LMW activator activity of each fraction was examined at dilutions of x10^2 (**A**), x10^3 (**B**) and x10^6 (**C**). All experiments were performed using the precursor of s-MAPP.

ELISA revealed that anti-HATKTAK reacted strongly to H18-K24 (HATKTAK), K17-K24 and S1-K24 of Apo CIII, but very weakly to Apo CIII and H18-D25 of Apo CIII, whereas anti-Apo CIII goat antibody showed a positive reaction only to Apo CIII (Figure 5A and 5B, respectively).

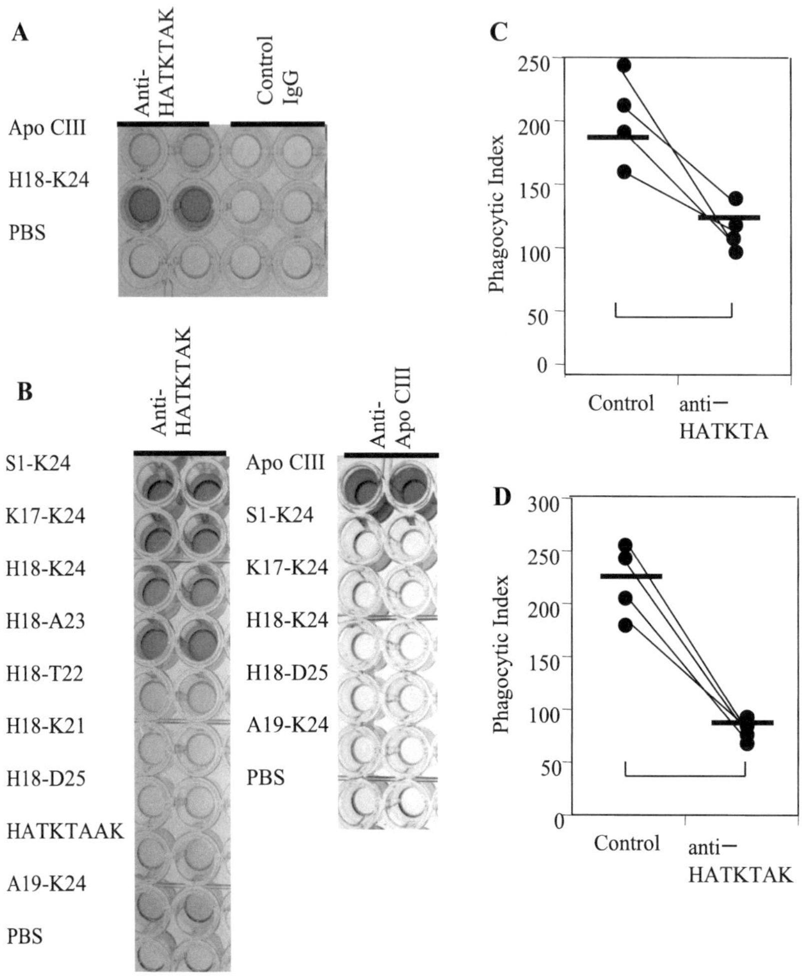

Figure 5. Effects of anti-HATKTAK antibody on LMW activator activity. **A,** ELISA using anti-HATKTAK rabbit IgG and control rabbit IgG against Apo CIII and H18-K24 of Apo CIII. **B,** ELISA using anti-HATKTAK rabbit IgG and anti-Apo CIII gout IgG against peptide associated with Apo CIII, HATKTAAK and Apo CIII. **C,** Effects of control rabbit IgG and anti-HATKTAK rabbit IgG on LMW activator activity in the low-molecular-weight fraction of platelet release products prepared from platelets of fresh platelet-rich plasma. **D,** Effects of control rabbit IgG and anti-HATKTAK rabbit IgG on LMW activator activity in the low-molecular-weight fraction of platelet release products prepared using platelets of platelet-rich plasma stored for 120 hours. In **C** and **D**, figures reveal the result of experiments using the precursor of s-MAPP. The bars in the figure show the average.
*, $P<0.01$; **, $P<0.05$ by paired *t* test.

Both LMW activators derived from fresh platelet and platelets stored for 120 hours were inhibited by the anti-HATKTAK IgG antibody (Figure 5C and 5D, respectively).

3.5. Evidence of existence of HATKTAK by mass spectrometry

In mass spectrometry, HATKTAK showed m/z 756 (Figure 6A). Although it was impossible to show the existence of a substance with m/z 756 in thrombin-digested Apo CIII (10 μg/ml) and in platelet release products released from fresh platelets and those stored for 72 hours (data not shown), a substance with m/z 756 was detected in the trypsin-treated Apo CIII (10 μg/ml) (Figure 6B) and platelet release products prepared from platelets stored for 120 hours (Figure 6C).

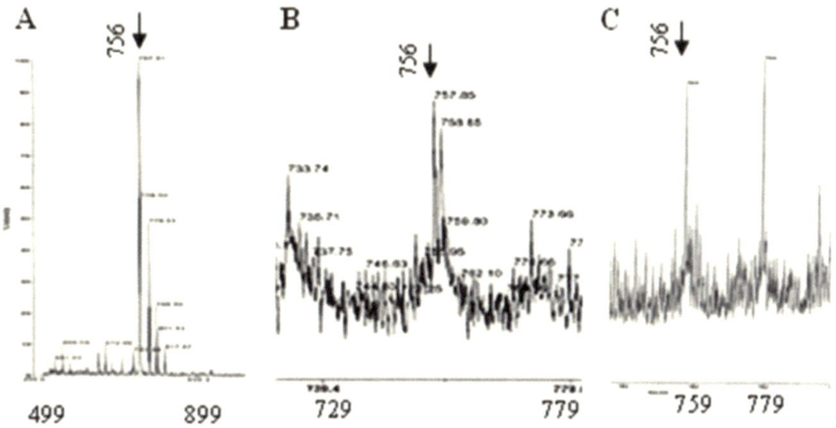

Figure 6. Mass spectrometry of HATKTAK, trypsin-digested Apo CIII and platelet release products from platelets of platelet-rich plasma stored for 120 hours. **A**, 1 mg/ml HATKTAK in PBS; **B**, 10 μg/ml Apo CIII digested with 1 unit/ml trypsin; and **C**, platelet release products using platelets of platelet-rich plasma stored for 120 hours. Platelets were stimulated with 0.1 unit/ml thrombin in the presence of Ca^{++}.

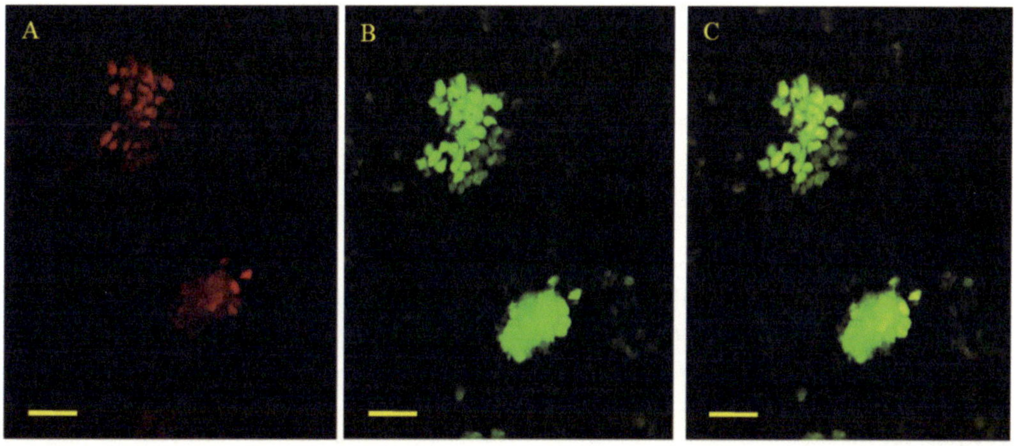

Figure 7. Confocal microscopic images of double staining for the anti-HATKTAK antibody (visualized as red in A), and the anti-CD61 antibody (visualized as green in B). The colocalization of two antibodies is indicated by the conversion of green and red to yellow (C). Scale bar indicate 20 μm.

3.6. Immunohistochemical evidence of existence of HATKTAK in activated platelets

To show the existence of HATKTAK in activated platelets, an immunohistochemical study of the blood coagula was performed. Platelets in the coagula showed positive reactions with both anti-HATKTAK and anti-CD 61 antibodies simultaneously (Figure 7).

3.7. Increases in concentrations of Apo CIII, Apo AI and Apo B100 in the platelet lysate

As the LMW activator activity in the products released from platelets increased markedly if stored platelets were used, concentrations of Apo CIII, Apo AI and Apo B100 in the platelet lysate prepared from platelet-rich plasma stored for 120 hours at 20 °C were compared with those prepared from fresh equivalent plasma to determine whether lipoproteins are internalized by platelets during storage of platelet-rich plasma. All of Apo CIII, Apo AI and Apo B100 in the lysate from 120 hour-stored platelets were high (20 ng/ml, 18 ng/ml and 252 ng/ml in average, respectively), whereas those from platelet-rich plasma before storage were very low (1.4 ng/ml, 0.6 ng/ml and 130 ng/ml on average, respectively) (Figure 8).

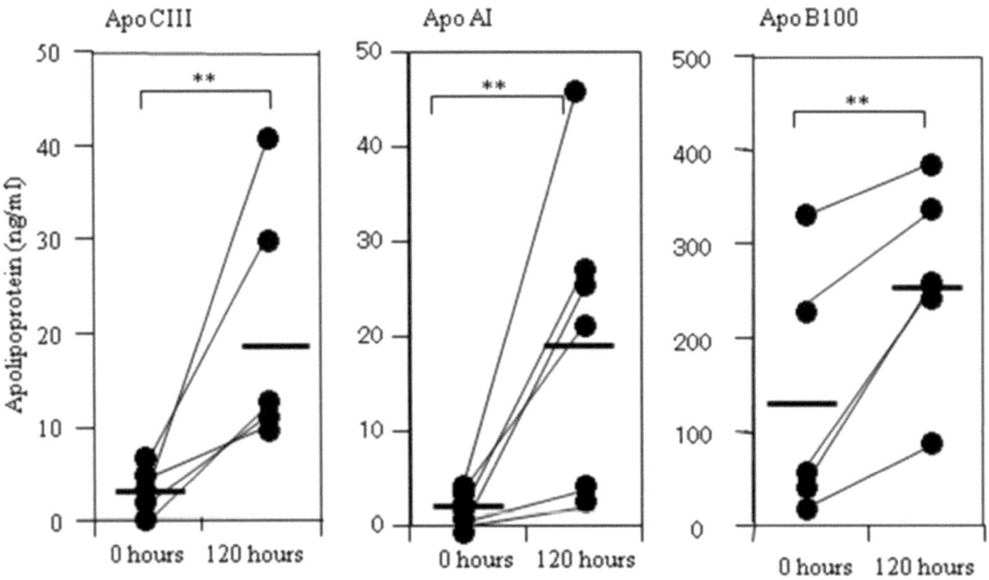

Figure 8. Concentrations of Apo CIII, Apo AI and Apo B100 in the platelet lysate. Concentrations of Apo CIII, Apo AI and Apo B100 in the platelet lysate, which were prepared from platelet-rich plasma before (0 hours) and after storage for 120 hours (120 hours), were determined using commercially available ELISA kits. The bars in the figure show the average. **, P<0.05 by Wilcoxon test.

4. Discussion

Since we found MAPPs in platelet release products, we have investigated their structure and the mechanisms of production and release by platelets. When fresh platelets are stimulated by thrombin, collagen or centrifugation in the presence of Ca^{++}, MAPPs are released. When platelets are stored longer than 72 hours in the form of platelet-rich plasma, they lose the capacity to release MAPPs. It is speculated that this is because platelets lose their contents of thrombin and precursors of MAPPs during storage in the form of platelet-rich plasma. They recover the MAPP-releasing function by washing with PBS and incubation with plasma-derived precursors of MAPPs and thrombin in the presence of Ca^{++} [26]. This suggests that some platelet factor other than thrombin and the precursors of MAPPs is necessary for MAPP production. To analyze this factor, the action of the platelet lysate obtained by freeze-thaw of stored platelets on MAPP formation was investigated. By incubation of the platelet lysate with thrombin and the plasma-derived precursors, it was possible to produce MAPPs. When producing MAPPs using plasma-derived precursors and the platelet lysate, we found that trypsin instead of thrombin can produce MAPPs [27].

After we found that the plasma precursors of MAPPs are dimer and tetramer transferrins [28], we produced precursors of MAPPs by glutaraldehyde treatment of commercially available holo-transferrin and used them to produce MAPPs in vitro.

In the former study [30], we found that Apo CIII associated with the high-density lipoprotein in platelet lysate can be the source of LMW activator. In this study, we compared the LMW activators produced from Apo CIII by the actions of thrombin and trypsin. It was revealed that both LMW activators appeared in the fractions of the same elution volume of Superdex peptide gel filtration and in the fractions with the same NaCl concentration of MONO S cation exchange chromatography. These findings suggest that the LMW activators produced by thrombin and trypsin are the same substance. The fact that trypsin can form LMW activator suggests that LMW activator is a peptide with a C-terminal amino acid of lysine or arginine and that the N-terminal is an amino acid next to lysine or arginine in the amino acid sequence of Apo CIII. It can be asserted that the action of thrombin in LMW activator formation occurs by its trypsin-like activity, although the activity of thrombin to release LMW activator from Apo CIII is far lower than that of trypsin. Cation exchange chromatography in this study revealed that the LMW activator is cationic, suggesting that it is a peptide rich in basic amino acids. In Apo CIII, K17 to K24 is a region rich in basic amino acids, containing four basic amino acids among eight amino acids in total. Several candidate peptides for LMW activator that contain all or part of the K17 to K24 peptide were raised. By comparing the elution volumes of peaks of these peptides with that of LMW activator function on Superdex peptide gel filtrations and by examining MAPP formation using these peptides, it was strongly suggested that HATKTAK is the LMW activator.

As for the reaction of thrombin, the most abundant natural substrate of thrombin is fibrinogen [36], whereas O'Mullan et al. [37] reported that the action of thrombin to various proteins including Apo CIII is more variable and various peptides cleaved from Apo CIII appear by the action of thrombin on Apo CIII. Our study revealed that the trypsin-like thrombin activity that digests Apo CIII to release HATKTAK is weak, but does in fact exist. We have shown that GP-1bα-bound thrombin functions in MAPP production [29]. GP-Ibα is a high-affinity thrombin receptor on platelets [38-39]. Binding of thrombin with GP-1bα might enhance the trypsin-like activity of thrombin in platelets.

To confirm that LMW activator, HATKTAK, is produced by the platelets, we examined the LMW activator activity in the low-molecular-weight fractions of variously prepared platelet release products. We found that, in the platelet release products prepared using $4x10^5/\mu l$ fresh platelets (0.1 unit/ml thrombin in the presence of 4 mEq/l Ca^{++}), the LMW activator involved is as much as 1 ng/ml HATKTAK.

By storage of the platelet-rich plasma, release of LMW activator from platelets induced by thrombin in the presence of Ca^{++} increased prominently. It was suggested that the concentration of LMW activator released from platelets stored for 120 hours is 10,000 times as much as that from fresh platelets.

One reason for the tremendous increase in LMW activator, HATKTAK, in the released products during storage is probably the loss of the precursor of MAPPs during storage, as shown in a previous report [26]. It is speculated that, in activated fresh platelets, LMW activator (HATKTAK) is produced as much as in stored platelets, but it decreases markedly because precursors of MAPPs remove it in fresh platelets. Another possible reason is that lipoproteins, apolipoproteins or fragments of apolipoproteins might be transported at high levels into platelets from the plasma during storage. In fact, it was shown that the concentrations of Apo CIII and Apo A1 in the platelet lysate increased markedly after storage of platelets in the form of platelet-rich plasma, but this was still too small to explain the observed increase in the effective dilution by as much as 1,000 times.

Indirect ELISA of the platelet release products using anti-HATKTAK antibody was undertaken to prove the existence of HATKTAK. The results were satisfactory if synthesized pure peptides were used, and it was shown that the anti-HATKTAK rabbit antibody reacted positively to Apo CIII-derived peptides with C-terminal HATKTAK. However, we have not succeeded in establishment of a method to analyze HATKTAK in platelet release products. It is postulated that some substances derived from the platelet release products interfere with the adherence of HATKTAK on the wall of microtiter plate. Therefore, we examined the effect of the antibody on the LMW activator function. It was confirmed that the anti-HATKTAK antibody cancels the activity of the LMW activator in the platelet release products from fresh platelet-rich plasma and that stored for 120 hours. Mass spectrometry study revealed the presence of a substance corresponding to HATKTAK (m/z 756) in the platelet release products from platelets stored for 120 hours. Immunohistochemistry of the blood coagula revealed the existence of platelets with double-positive reaction to anti-HATKTAK and anti-CD61 antibodies. These findings strongly suggest that the LMW activator is HATKTAK and is produced and released by platelets.

A schematic illustration of the probable mechanism of production and release of MAPPs and HATKTAK by platelets is depicted in Figure 9.

At present, the mechanism of how MAPP contributes to neutrophilic phagocytosis enhancement after binding to neutrophils is not known. Because MAPPs possess transferrin molecules and anti-transferrin receptor antibody inhibits the action of MAPPs [28], it is suggested that the action of MAPPs occurs via the transferrin receptor. As shown previously, treatment of neutrophils with MAPP does not result in an increase in the number of Fc receptors on neutrophils [40]. This suggests that the effect of MAPP is to strengthen the affinity of Fc receptors with the Fc portion of IgG and to internalize more foreign materials inside the cell. The necessity of HATKTAK for MAPP function suggests that the transferrin receptor might have a site for binding to both transferrin and HATKTAK, and the fact that MAPPs consist of multimers (dimer and tetramer) of transferrin suggests that transferrin receptors must be fixed at an appropriate distance when MAPPs induce enhancement of phagocytosis via the Fc receptors. The mechanisms of how the transferrin receptor, which is stimulated by HATKTAK and transferrin, transfers the information to Fc receptors remain to be elucidated.

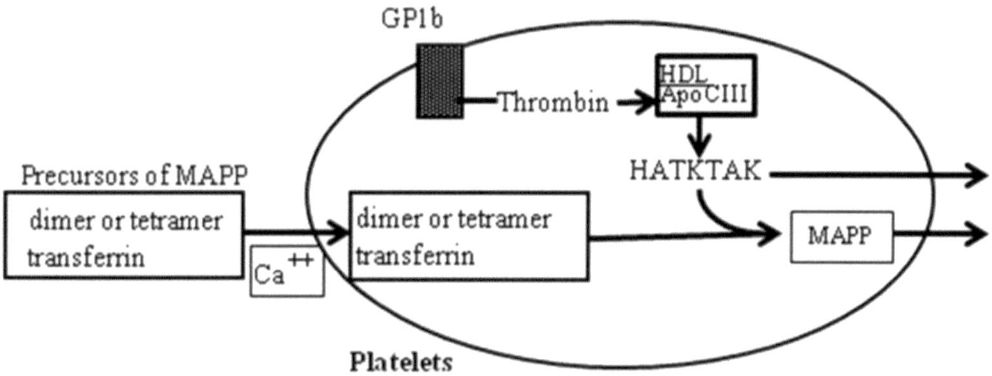

Figure 9. A schematic illustration of the mechanism of production of MAPPs and HATKTAK in platelets and their subsequent release.

In conclusion, we could show that HATKTAK is the LMW activator, which is produced in activated platelets and activates MAPPs, and that the residual HATKTAK is released with other platelet-related substances including MAPPs.

Author details

Haruhiko Sakamoto[1,*], Masaki Ueno[1], Wu Bin[1,2], Yumiko Nagai[3], Kouichi Matsumoto[1], Takao Yamanaka[1,4] and Sumiko Tanaka[1]

[1]*Inflammation Pathology, Department of Pathology and Host Defense, Faculty of Medicine, Kagawa University, Kagawa, Japan*

[2]*Department of Gynecology and Obstetrics, Second Affiliated Hospital, China Medical University, San Hao Road, Shen Yan, China*

[3]*Division of Research Instrument and Equipment, Life Science Research Center, Institute of Research Promotion, Kagawa University, Kagawa, Japan*

[4]*Internal Medicine, Minami-Okayama Medical Center, Incorporated Administrative Agency National Hospital Organization, Okayama, Japan*

[*] Corresponding Author

Acknowledgement

The authors wish to thank staff and students of the Faculty of Medicine, Kagawa University, for their cooperation in providing their blood for this study.

5. References

[1] Sakamoto H, Firkin FC, Chesterman C (1984) Stimulation of leukocyte phagocytic activity by the platelet release reaction. Pathology. 16: 126-130.

[2] Sakamoto H, Firkin F (1984) Characterization of leukocyte phagocytic stimulatory material released by activated human platelets. Br. j. haematol. 57: 47-60.

[3] Ward P, Cunningham TW, McCulloch KK, Phan SH, Powell J, Johnson KJ (1988) Platelet enhancement of $O2^-$ response in stimulated human neutrophils. Identification of platelet factor as adenine nucleotide. Lab. invest. 8: 37-47.

[4] Spisani S, Giuliani AL, Cavalletti T, Zaccarini M, Milani L, Gavioli R, Tranello (1992) Modulation of neutrophil function by activated platelet release factor. Inflammation. 16: 147-158.

[5] Ruf A, Schlenk RF, Maras A, Morgenstern E, Patscheke H (1992) Contact-induced neutrophil activation by platelets in human cell suspensions and whole blood. Blood. 80: 1238-1246.

[6] Tsuji T, Nagata K, Koike J, Todoroki N, Irimura T (1994) Induction of superoxide anion production from monocytes and neutrophils by activated platelets through the P-selectin-sialyl Lewis X interaction. J. leuk. biol. 56: 583-587.

[7] Carulli G, Barsotti G, Cupisiti A (1995) Platelet-neutrophil interactions in uremic patients: effects on neutrophil superoxide anion production and chemiluminescence. Nephron. 69: 248-252.

[8] Zalavary S, Grenegard M, Stendahl O, Bengstsson T (1996) Platelets enhance Fc(gamma) receptor-mediated phagocytosis and respiratory burst in neutrophils: the role of purinergic modulation and actin polymerization. J. leuk. biol. 60: 58-68.

[9] Piccardoni P, Evangelis Vt, A. Piccoli, de Gaetano G, Walz A, Celetti C (1996) Thrombin-activated human platelets release two NAP-2 variants that stimulate polymorphonuclear leukocytes. Thromb. res. 76: 780-785.

[10] Neumann F, Max JN, Gawaz M, Brand K, Ott I, Rokitta C, Sticherling C, Meinl C, May A, Schomig A (1997) Induction of cytokine expression in leukocytes by binding of thrombin-stimulated platelets. Circulation. 95: 2387-2394.

[11] Peters MJ, Dixon G, Kotowicz KT, Hatch DJ, Heyderman RS, Klein NJ (1999) Circulating platelet-neutrophil complexes represent a subpopulation of activated neutrophils primed for adhesion, phagocytosis and intracellular killing. Br. j. haematol. 106: 391-399.

[12] Petersen F, Bock L, Flad HD, Brandt E (1999) Platelet factor 4-incduced neutrophil-endothelial cell interaction: involvement of mechanisms and functional cosequences different from those elicited by interleukin-8. Blood. 94: 4020-4028.

[13] Kirton CM, Nash GB (2000) Activated platelets adherent to an intact endothelial cell monolayer bind flowing neutrophils and enable them to transfer to the endothelial surface. J. lab. clin. med. 136: 303-313.

[14] Brandt E, Petersen F Ludwig A, Ehlert JE, Bock L, Flad HD (2000) The beta-thromboglobulins and platelet factor 4: blood platelet-derived CXC chemokines with divergent roles in early neutrophil regulation. J. leuk. biol. 67: 471-478.

[15] MacGarrity ST, Heyers TM, Webster RO (1988) Inhibition of neutrophil functions by platelets and platelet-derived product: Description of multiple inhibitory properties. J. leuk. biol. 44: 93-100.

[16] Dallegri F, Ballestero, Ottonello AL, Patrone F (1989) Platelets as scavengers of neutrophil derived oxidants: A possible defence mechanism at sites of vascular injury. Thromb.haemost. 61: 415-418.

[17] Moon DG, Der Zee HV, Weston LK, Gudewicz PW, Fenton JW, Kaplan JE (1990) Platelet modulation of neutrophil superoxide anion production. Thrombosis. haemost. 63: 91-96.

[18] Naum C, Kaplan SS, Basford RE (1991) Platelet and ATP prime O_2^- generation at high concentration. J. leuk. biol. 49: 83-89.

[19] Bengtsson T, Zalavary S, Stendahl O, Genegard M (1996) Release of oxygen metabolites from chemoattractant-stimulated neutrophils is inhibited by resting platelets: role of extracellular adenosine and actin polymerization. Blood. 1996: 87: 4411-4423.

[20] Jancinova V, Drabikova K, Nosal R, Danihelova E (2000) Platelet-dependent modulation of polymorphonuclear leukocyte chemiluminescence. Platelets. 11: 278-285.

[21] Losche W, Temmler U, Redlich H, Vickers J, Krause S, Spangenberg P (2001) Inhibition of leukocyte chemiluminescence by platelets: role of platelet-bound fibrinogen. Platelets. 12: 15-19.

[22] Reinisch CM, Dunzendorfer S, Pechlaner C, Ricevuti G, Wiedermann CJ (2001) The inhibition of oxygen radical release from human neutrophils by resting platelets is reversed by administration of acetylsalicylic acid or clopidogrel. Free. radical. res. 34: 461-466.

[23] Miyabe K, Sakamoto N, Wu YH, Mori N, Sakamoto H (2004) Effect of platelet release products on neutrophilic phagocytosis and complement receptors. Thromb. res. 114: 29-36.

[24] Sakamoto H, Ooshima A. (1985) Activation of neutrophil phagocytosis of complement coated and IgG coated sheep erythrocytes by platelet release products. Br. j. haematol. 16: 173-181.

[25] Sakamoto H, Yokoya Y, Ooshima A. (1987) In vitro control of neutrophilic phagocytosis of IgG coated SRBC by macromolecules involved in released products from platelets. J. leuk. biol. 41: 55-62.

[26] Sakamoto H, Ogawa Y, Sakamoto N, Oryu M, Shinnou M, Hirao T (1996) Recovery of macromolecular activators of phagocytosis from platelets (MAPP) producing and releasing function in stored human platelets. Int. j. hematol. 63: 145-148.

[27] Ogawa Y, Sakamoto H, Oryu, M Shinnou M, SakamotoN, Wu Y, Khatun R, Nishioka M (2000) Production of macromolecular activators of phagocytosis by lysed platelets. Thromb. res. 97: 297-306.

[28] Sakamoto H, Sakamoto N, Oryu M, Kobayashi K, Ogawa Y, Ueno M, Shinnou M (1997) A novel function of transferrin as a constituent of macromolecular activators of phagocytosis from paltelets and their precursors. Biochem. biophys. res. com. 230: 270-274.

[29] Sakamoto H, Ueno M, Wu Y, Khatun R, Tanaka S, Miyabe K, Ogawa Y, Onodera M (2000) Glycoprotein Ibα-bound thrombin functions as a serine protease to produce macromolecular activators of phagocytosis from platelets. Biochem. biophys. res. com. 270: 377-382.

[30] Sakamoto H, Wu B, Nagai Y, Tanaka S, Onodera M, Ogawa T, Ueno M (2011) Platelet high-density lipoprotein activates transferrin-derived phagoctosis activators, MAPPs, following thrombin digestion. Platelets. 22: 371-379.

[31] Boyle W (1968) An extension of the 51Cr-release assay for the estimation of mouse cytotoxins. Transplant. 62: 761-764.

[32] Avrameas S, Ternyck T (1971) Peroxidase labeled antibody and Fab conjugates with enhanced intracellular penetration. Immunochem. 8: 1175-1179.

[33] Nakamura R, Voller M, Bidwell DE (1986) Chapter 27 Enzyme immune assay: heterogenous and homogeneous system. In: Weir DM, Herzenberg LA, Blackwell C, Herzenberg LA, editors. Handbook of Experimental Immunology. 1 Immunochemistry. Oxford: Blackwell Scientific Publications. pp. 27.1-27.20

[34] Peterson GL (1983) O-phthalaldehyde fluorescent methods for protein quantitation. In: Hirs CHW, Timasheff SN editors. Methods in enzymology 91. New York: Academic Press. pp. 95-98.

[35] Hafon ST, Baird TT and Craik CS. (2004) 452. Trypsin In: Barret AJ, Rawlings ND, Woessner JF, editors. Handbook of Proteolytic Enzymes, Vol 2. Second edition. London: Elsevier Academic Press. pp. 1483-1488.

[36] Brown MA, Sternberg LM, Stenflo J (2004) 510. Thrombin. In: Barret AJ. Rawlings ND, Woessner JF, editors. Handbook of Proteolytic Enzymes, Vol 2. Second edition. London: Elsevier Academic Press. pp. 1667-1672.

[37] O' Mullan P, Draft D, Yi J, Gelfand CA (2009) Thrombin induces broad spectrum proteolysis in human serum samples. Clin. chem. lab. med. 47: 685-693.

[38] Harmon JT, Jamieson GA. (1985) Thrombin binds to a high-affinity approximately 900,000-dalton site on human platelets. Biochem. 24: 58-64.

[39] Jamieson GA (1997) Pathophysiology of platelet thrombin receptors. Thromb. haemost. 78: 242-246.

[40] Sakamoto N, Sakamoto H, Tanaka S, Oryu M and Ogawa Y. (1998) Effects of platelet release products on neutrophilic Fcγ receptors. J. leuk. biol. 64: 631-635.

Chapter 32

Lipoprotein (a) – An Overview

Anna Gries

Additional information is available at the end of the chapter

http://dx.doi.org/10.5772/45986

1. Introduction

Lipoprotein (a) [Lp(a)], first described in 1963 is an inherited cholesterol-rich particle found in a density range of 1.055-1.120 g/ml. The suggestion that Lp(a) might be a risk factor for cardiovascular diseases was first made by Dahlen et al. [1] who found out that individuals with angina pectoris exhibit an "extra pre-β-band" in lipid electrophoresis. In whites the concentration of Lp(a) in plasma varies from undetectable up to 200 mg/dl in different individuals but seems to be rather constant in the same person [2]. Chemical and physicochemical properties of Lp(a) in comparison with LDL are summarized in table 1.

Plasma Lp(a) concentrations above 30 mg/dl, as measured in about 20 percent of white people, are associated with an approx. two-fold relative risk of coronary atherosclerosis [3] rising to the range of five-fold when LDL and Lp(a) are both elevated [4]. Interestingly, blacks with high levels of Lp(a) do not experience greatly increased atherosclerotic progression and mortality. In those cases it is assumed that the atherogenicity of Lp(a) must be decreased or counterbalanced by other factors [5].

Till now the site and mechanism of Lp(a) synthesis are quite unclear. Measurements of serum Lp(a) levels of patients suffering from liver disease or from cholestasis who showed significantly lower concentrations than healthy controls gave indications that Lp(a) might be synthesized by the liver [6]. On the other hand there are studies which suggest that apo-a is associated with the postprandial d < 1.006 lipoproteins induced by fat feeding [7] but it is not yet clear however whether apo-a determined in this fraction is really of intestinal origin or whether it originates from free apo-a in serum which might bind to freshly secreted chylomicrons [8]. Because of the chemical similarities between Lp(a) and LDL it is possible that Lp(a) is formed during the metabolic catabolism of chylomicrons, VLDL or LDL. As Lp(a) levels stay nearly constant within one individual and as lipid-rich diet as well as fasting have no influence on Lp(a) concentrations it is assumed that Lp(a) exhibits a metabolic behaviour completely different from other apo-B containing lipoproteins. Turnover studies in vivo performed with labelled VLDL confirmed these assumptions. Nearly all the activity of labelled VLDL could be detected in LDL whereas only trace amounts could be found in Lp(a) [9] confirming the hypothesis that unlike LDL, Lp(a) probably has no triglyceride-rich lipoproteins as precursors but seems to be secreted directly by the liver [10]. On the other hand the site of catabolism of Lp(a) in humans is unknown so far although the kidney is favourized to be implicated [11].

Despite extensive work on Lp(a) its possible physiological function remains unclear till now.

	LDL	Lp(a)
Hydr. Density [g/ml]	1.034	1.085
Mol. Wt. [x 10^6]	2.4	5.5
Diameter [Å]	210	250
E. Mobility	β	pre-β1
Chem. Composition [%]		
Free cholesterol	11	10
Cholesterolester	40	30
Triglycerides	4	4
Phospholipids	21	20
Protein	22	28
Carbohydrates	2	8

Table 1. Chemical and physicochemical properties of LDL and Lp(a)

2. Structural arrangement and catabolism of Lp(a)

The major protein component of this LDL-like particle is apolipoprotein B (apo-B-100) which carries an additional protein called apolipoprotein-a [apo-a] linked to apo-B-100 via disulphide bridges (Fig.1) the lipid moiety however being almost indistinguishable from that of LDL [12]. Human apo-a itself consists of multiple so-called kringle repeats, sequences consisting of 80-90 amino acids arranged in a tripleloop tertiary structure and tandemly arrayed resembling kringles IV and V of plasminogen and a protease domain [13]. Copy number variants in the LPA gene on chromosome 6 coding for apo-a are responsible for a variation of plasma Lp(a) levels of up to 1000-fold among individuals. The most influential is the kringle IV-2 size polymorphism [14] while kringle IV types 1 and 3-10 as well as kringle V occur only once in Lp(a) [15]. The number of kringle IV type 2 structure repeats results in a large number of different sized isoforms of apo-a and correlates inversely with the plasma concentration of Lp(a) [16]. Although the exact mechanism responsible for this inverse correlation has not been elucidated so far an isoform dependent retention and degradation in the endoplasmatic reticulum has been implicated [17].

Contradictory results have been reported about the clearance of Lp(a) and till now it remains unclear whether Lp(a) binds to the B/E receptor via apo-B like LDL or whether it is catabolised independently of the LDL-receptor mediated pathway. Whereas in one study using fibroblasts from normal subjects and from subjects with autosomal dominant hypercholesterolemia the conclusion was reached that Lp(a) enters fibroblasts independently of the LDL-receptor [18] others concluded that Lp(a) is also bound to the LDL-receptor, internalized and degraded but with a degradation capacity of only 25% of that of LDL [19]. Binding studies of native and reduced Lp(a) with different monoclonal antibodies against apolipoprotein B revealed that there was no antibody that failed to react with native Lp(a) but some of the antibodies recognized apoB of Lp(a) to a lesser degree than that of LDL. This favoured the idea that certain regions on apo-B of Lp(a) could be different from those on LDL and led to the assumption that certain domains close to the binding domain of Lp(a) to the B/E-receptor could be covered by apo-a or that apo-a causes conformational changes in the binding region of apo-B thereby constricting the binding of Lp(a) to the LDL-receptor [20] being in agreement with the fact that normal unreduced Lp(a) seemed to be taken up by fibroblasts through B/E-receptor-mediated endocytosis but showed poorer specificity for the receptor than LDL [21].

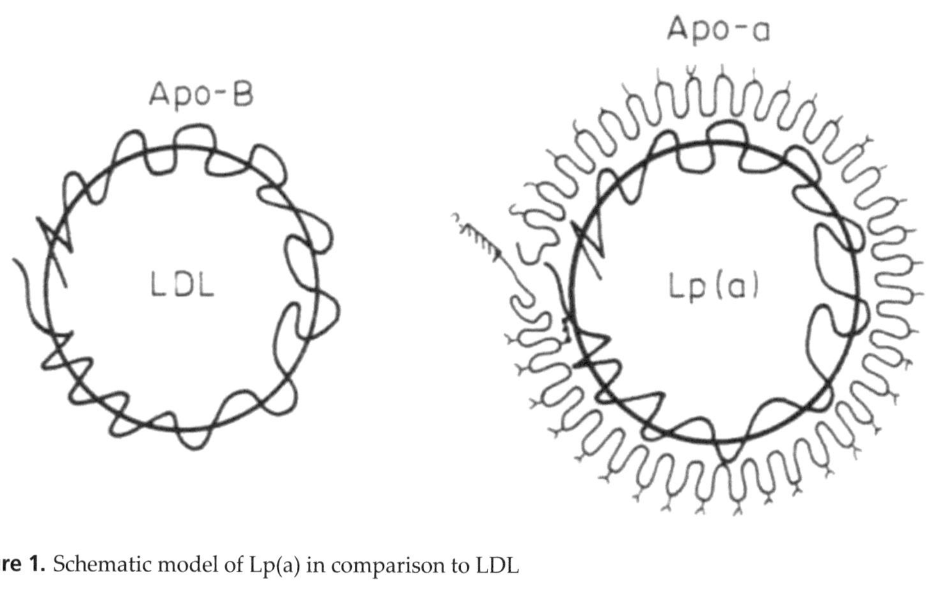

Figure 1. Schematic model of Lp(a) in comparison to LDL

3. Free apolipoprotein-a in human serum

In the beginning of "Lp(a)-research" this lipoprotein was believed to represent a genetically polymorphic form of LDL [22]. According to this assumption apo-a should distribute uniformly between all apo-B-containing lipoproteins. Investigation of this problem in more detail revealed that Lp(a) forms a particular lipoprotein class found primarily in the HDL-2 density region [23] but can also be detected in LDL class (d = 1.019-1.063 g/ml) [24] and even in chylomicrons induced by fat feeding [25]. The fact however that a portion of the Lp(a)-specific antigen can be found in the d > 1.21 g/ml lipoprotein free bottom fraction after ultracentrifugation of plasma [26] led to further investigation of this phenomenon. Apo-a is virtually absent in the VLDL fraction (d < 1.006 g/ml) of freshly drawn fasting human sera while 95% of total Lp(a) can be obtained in the d > 1.006 g/ml bottom fraction. Approximately 5% of the total serum Lp(a) are found in the d > 1.125 g/ml bottom fraction after ultracentrifugation as well as with polyanionic precipitation agents irrespective of the Lp(a) concentration in serum [8]. Due to the lack of Sudan Black staining this bottom Lp(a) is considered as a lipid free "apo-a protein" raising the question whether or not free apo-a can reassociate with LDL to form "native Lp(a)".

4. Lp(a) and platelet aggregation

One of the physiological roles of platelets involves binding to subendothelial tissue after vascular injury [27]. The adherence of platelets to the exposed connective tissue, preferably collagen, leads to aggregation followed by the release of ADP, 5-hydroxytryptamine and Ca^{2+} from their dense granules, causing passing platelets to adhere to the primary clot [28].

There is little doubt that lipoproteins interfere with platelets in vivo being reflected by the fact that platelets from hyperlipoproteinemic patients are hyperreactive [29]. This is confirmed by the fact that incubation of platelets with physiological concentrations of atherogenic apoB-containing lipoproteins such as LDL or VLDL results in enhanced platelet aggregability [30] while antiatherogenic lipoproteins such as HDL exert the opposite effect [31]. Concerning Lp(a) it is generally accepted that elevated plasma concentrations of this lipoprotein are connected with premature atherosclerosis [32] but much uncertainty remains about the influence of Lp(a) on platelet activation, a phenomenon that is believed to be involved not only in long-term processes of plaque formation but also in acute events such as stroke and myocardial infarction [33]. Moreover a two-fold increase in the risk of coronary heart disease (CHD) and ischaemic stroke could be demonstrated especially in subjects with small apolipoprotein(a) phenotypes [34] and prospective findings in the Bruneck study have revealed a significant association specifically between small apolipoprotein(a) phenotypes and advanced atherosclerotic disease involving a component of plaque thrombosis [35]. Indeed, Lp(a) is a "sticky" lipoprotein that self-aggregates, attaches to all sorts of surfaces [36], and precipitates not only in vitro but possibly in vivo. Moreover, Lp(a) binds to proteoglycans and glycosaminoglycans [37] and it has high affinity for fibronectin [38], tetranectin [39], collagen [40], and other connective-tissue structures [41]. Therefore the influence of Lp(a) on platelet aggregation induced with various triggers was investigated measuring serotonin release and thromboxane A_2 formation during collagen-triggered aggregation as well as adhesion of platelets to collagen in flowing blood under the influence of Lp(a). As Lp(a) represents an LDL-like particle an elevated platelet reactivity was expected under the influence of this lipoprotein similar to that described for LDL [42].

Unlike LDL, Lp(a) revealed no proaggregatory effects on platelets, contrary collagen-induced platelet aggregation was inhibited by up to 54% and the aggregation rate was attenuated by 47% compared with platelets incubated with Tyrode's solution (Fig. 2), being accompanied by a significant reduction of serotonin release and TXA_2 formation. Furthermore Lp(a) significantly reduced platelet adhesion to collagen by about 20% and the size of platelet aggregates up to 63% especially at high shear rates (Fig. 3) suggesting that Lp(a) exerts antiaggregatory effects at least under well-defined in vitro conditions [43]. If these observations are relevant for the in vivo situation, a variety of potential platelet-collagen binding sites such as GPIa/IIa or GPIV could be covered by Lp(a) the more that binding of Lp(a) to platelets could be demonstrated [44]. As there is conflicting evidence on the role of Lp(a) in thrombosis in vivo and in vitro work has been done to elucidate the mechanisms whereby Lp(a) is influencing platelet aggregation and a variety of mechanisms is suggested how Lp(a) interferes with platelet aggregation and hence fibrin bound clot formation. Lp(a) binds to resting, non-stimulated platelets on the IIb subunit of the fibrinogen (IIb/IIIa) receptor via binding sites distinct from the arginyl-glycyl-aspartyl (RGD) epitope of apo-a [45]. By this way the RGD binding site of Lp(a) could be exposed via conformational change induced by platelet agonist stimulation leading to binding of the RGD epitope of apo-a to the RGD binding site on the IIb protein of the fibrinogen (IIb/IIIa) receptor of the platelet [46] thereby reducing fibrinogen binding to the platelet [47]. Low concentrations of Lp(a) (1-25 mg/100 ml washed platelets) increase intracellular levels of c-AMP of in vitro resting platelets leading to an antiaggregatory condition [48] while at higher in vitro levels of Lp(a) (50-100 mg/100 ml washed platelets) resting platelet intracellular c-AMP levels return to normal [49] which cannot explain the reported progressive Lp(a)-mediated decrease in collagen-induced aggregation [43, 50]. Further investigations strongly support an apo-a mediated, Lp(a) induced reduction of collagen and ADP-stimulated platelet aggregation via diminished production of thromboxane A_2 [43, 51]. Concerning the in vivo situation only one study has been published to date looking at adult human type 2 diabetics all of whom where obese (BMI >30). In this in vivo study of human type 2 diabetics there is a positive correlation between fasting serum concentrations of Lp(a) and bleeding time, a strong correlate of in vivo platelet aggregation [52] favouring the inhibitory effect of Lp(a) on platelet aggregation. On the other hand there are studies reporting an apparent proaggregatory action of Lp(a) possibly mediated by the apo-a subunit. While no effect of recombinant apo-a [r-apo-a] derivatives on primary ADP-induced platelet aggregation was observed weak platelet responses stimulated by the thrombin receptor-activating peptide SFLLRN were significantly enhanced by the r-apo-a derivatives accompanied by a significant enhancement of [^{14}C]serotonin release of the dense granules [53]. Further investigations showed that r-apo-a isoforms and Lp(a) do not cause platelet aggregation by themselves but preincubation of platelets with r-apo-a derivatives promotes an aggregation response to otherwise subaggregant doses of thrombin receptor activation peptide (TRAP) and arachidonic acid while inversely platelet stimulation with arachidonic acid enhanced platelet binding of apo-a [54]. In both studies it turned out that the size of r-apo-a determined by the number of KIV type 2 modules seems not to play a crucial role in its proaggregant effect.

Summarizing, in vitro studies indicate that Lp(a) induced decreases, increases or no change at all in platelet aggregation [43, 45, 50, 51, 53, 54]. In all cases the mechanisms involved are quite unclear and only speculative. A recent work strongly supports the evidence to suggest that Lp(a) binds to platelets via its arginyl-glycyl-aspartyl (RGD) epitope of the apo(a) but not via apo(a)'s lysine binding region in both strong and weak agonist-stimulated platelets and inhibits the binding of fibrinogen thus reducing aggregation [55]. On the other side there are in vivo studies published quite recently suggesting that Lp(a) concentrations greater than 30 mg/dl are a frequent and independent risk factor for venous thrombosis [56] and that high levels of Lp(a) could be a more frequently thrombophilic risk factor in young women [57]. To date disagreement exists about increased arterial thrombosis due to elevated blood levels of Lp(a). The fact that this procedure is a result of collagen-exposed platelets in case of plaque rupture followed by clot formation argues against the proaggregatory nature of Lp(a) and maybe procedures others than platelet activation account for the atherogenic nature of Lp(a).

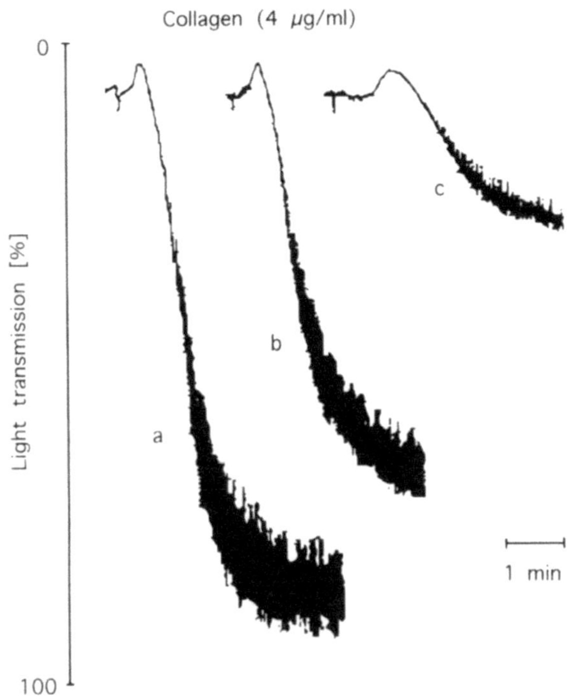

Figure 2. Aggregation curves showing the influence of lipoproteins on collagen-induced platelet aggregation. Gel filtered platelets (200 µl; 2x108/ml) were incubated for 30 min at 37°C with a) LDL 5 mg/ml, b) Tyrodes's buffer or c) Lp(a) 0.5 mg/ml. Aggregation was triggered with 10 µl collagen (final concentration 4 µg/ml).

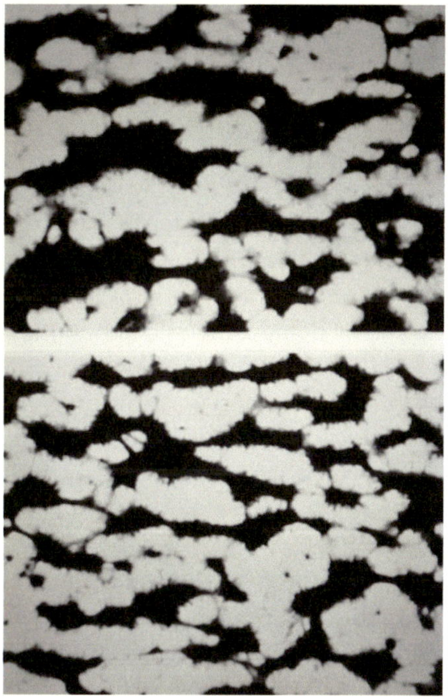

Figure 3. Aggregate formation of fibrillar collagen at a shear rate of 1600/s for a control (top) and under addition of 1 mg/ml Lp(a) (bottom). Aggregates are shown in black.

5. Lp(a) and plasminogen

The mechanism by which Lp(a) accelerates atherosclerosis could not yet been clarified. One possible explanation leads via the connection of Lp(a) to the fibrinolytic system as in 1987 it was found out that Lp(a) and plasminogen are immunochemically related [58] leading to speculations whether Lp(a) might interfere with fibrinolysis. Through partial amino acid sequencing it could be shown for the first time that apo-a has a striking homology of about 70% to plasminogen, the precursor of the proteolytic enzyme plasmin which dissolves fibrin clots [58]. This could be confirmed in our own studies demonstrating that polyclonal antisera from rabbit, sheep and horse as well as three monoclonal antibodies from mouse raised against apo-a reacted with plasminogen on immunoblots and similar to plasminogen, Lp(a) bound selectively but with somewhat lower affinity to lysine-Sepharose [59]. Plasminogen, a protein of 791 amino acids and a molecular weight of about 92 000 D is a plasma serine protease zymogen that consists of five cysteine-rich sequences of 80-114 amino acids each, called kringles, followed by a trypsin like protease domain [60]. The highly glycosylated apo-a exists in various polymorphic forms with molecular weights higher, lower or equal to apoprotein B ($M_r \approx 550\,000$ D) [61] which are covalently linked to apoprotein B via disulfide bridges [62]. It contains a hydrophobic signal sequence for secretion followed by up to >50 copies of kringle IV of plasminogen predicting the risk for coronary heart disease in the way that apo-a alleles with a low kringle IV copy number (<22) and high Lp(a) plasma concentration are significantly more frequent in the CHD group

($p<0.001$) [63]. Additionally one kringle V as well as protease domains of plasminogen are found in apo-a [58]. Later on cDNA sequencing revealed that human apolipoprotein(a) is homologous to plasminogen but despite the fact that apo-a contains a protease domain it does not act fibrinolytically like plasminogen because the arginine at the cleavage site for tissue plasminogen activator in plasminogen is changed to serine in apo-a [64].

Nevertheless Lp(a) might interfere with the fibrinolytic system in different ways due to its similarity to plasminogen as it may inhibit the binding of plasminogen to its receptor on endothelial cells thereby preventing generation of plasmin and increasing the thrombotic risk [65, 66]. Furthermore it could be demonstrated that Lp(a) accumulates in atherosclerotic lesions maybe via adherence to fibrinogen or fibrin incorporated in atherosclerotic plaques thereby inhibiting fibrinolysis [66]. Another mechanism by which Lp(a) is thought to attenuate fibrinolysis involves direct competition with plasminogen for fibrinogen or fibrin binding sites thus reducing the efficiency of plasminogen activation [67]. Fibrinolysis is initiated by binding of plasminogen to lysine residues on fibrin thereby initiating activation of plasmin and amplifying fibrinolytic processes [68]. Like plasminogen Lp(a) also binds to lysine residues [69] but without catalytical activity leading to interference with or inhibition of fibrinolysis resulting in hypofibrinolysis and accumulation of cholesterol included in the LDL-like component of Lp(a) [66]. The fact that low molecular weight isoforms of apo-a are associated with greater inhibition of fibrinolysis [70, 71] confirms the hypothesis that subjects with small apo-a phenotypes have a two-fold risk of CHD and stroke compared with those with larger isoforms of apo-a [34]. In contrast Knapp et al. [72] observed that the rate of plasmin formation was inversely related to Lp(a) but inhibition of plasmin generation increased with the size of apo-a using a standardized in vitro fibrinolysis model. From the fact that the inhibitory effect of free apo-a was much stronger than that of the complete Lp(a) particle they conclude that the apo-a component is responsible for the observed reduction of plasmin formation maybe due to the availability of additional lysine binding sites in the unbound apo-a which was formerly reported by Scanu et al. [73]. On the other hand there are also data showing that the plasma concentration of Lp(a) is inversely related to plasmin formation but that this relationship is not influenced by the size of apo-a isoforms [74]. Above all there are other reports explaining the inhibitory effect of Lp(a) on fibrinolysis not only by competition of Lp(a) with plasminogen for the binding sites on fibrin, endothelial cells and monocytes but also by reduction of tissue plasminogen activator or streptokinase-induced fibrinolytic activity [75, 76, 77].

A novel contribution to the understanding of Lp(a)/apo-a-mediated inhibition of plasminogen activation comes from results showing the ability of the apo(a) component of Lp(a) to inhibit the key positive feedback step of plasmin-mediated conversion of Glu-plasminogen to Lys-plasminogen an essential step for fibrin clot lysis [78]. Interestingly, with the exception of the smallest naturally-occurring isoform of apo(a), isoform size was found not to contribute to the inhibitory capacity of apo(a).

In summary, the proposed mechanisms modulating the antifibrinolytic effects of elevated Lp(a) levels in vitro are manifold and emphasize the prothrombotic effects of this lipoprotein particle. The in vivo situation however seems to be much more complex the more that there is a strong positive correlation reported between bleeding time and fasting serum concentrations of Lp(a) [52].

6. Lp(a) and lipid lowering drugs

High levels of Lp(a) are strongly associated with atherosclerosis as revealed by numerous studies [4, 79, 80, 81, 82]. As plasma Lp(a) concentrations above 30 mg/dl, as measured in about 20 percent of white people, are associated with an approx. two-fold relative risk of coronary atherosclerosis [3] rising to the range of five-fold when LDL and Lp(a) are both elevated [4] reduction of plasma Lp(a) concentration is recommended. Dietary interventions do not seem to be effective in lowering Lp(a) plasma levels [9, 83] or even lead to an increase of Lp(a) in plasma, alone [84] or at least when combined with exercise 85]. The same phenomenon could be observed in case of exercise where cross-sectional data suggest that a lifestyle of moderate to intense exercise training does not exert a significant impact on the Lp(a) level [86, 87]. Therefore pharmacological reduction of plasma levels of Lp(a) would be desirable.

Innumerable investigations however indicate that the plasma concentration of Lp(a) is resistant to drug therapy in most cases. As Lp(a) resembles LDL especially with regard to the lipid content (Tab.1) medications reducing LDL-cholesterol should be suitable for lowering Lp(a) as well. Bile acid resins such as cholestyramine which actually cause a significant reduction of LDL-cholesterol as well as of apo-B have no effect on Lp(a) levels [88, 89]. Therapies with bezafibrate or clofibrate [90, 91] showed that there is no role for fibrates in the treatment of elevated Lp(a) concentrations and estrogens also do not seem to significantly affect Lp(a) [92, 93].

Stanozolol, an anabolic steroid used in the treatment of postmenopausal osteoporosis, showed a significant reduction of Lp(a) by about 65% after six weeks therapy but five weeks after the drug was discontinued Lp(a) was near pretreatment levels [94]. Although drastic reductions of Lp(a) up to 40-50% are reported in another study [95] these compounds seem to be unsuitable for the routine treatment due to their harmful side effects [96].

Statins, also known as HMG-CoA-reductase inhibitors are another group of lipid lowering drugs which could be interesting with regard to Lp(a). These drugs have proven to be extremely effective in lowering plasma LDL and apo-B levels presumably through inhibition of intracellular cholesterol synthesis concomitant with an increase of the LDL receptors in the liver [97]. Although Lp(a) and LDL are very similar especially concerning the content of cholesterol, inhibitors of HMG-CoA-reductase, the regulating enzyme of cholesterol biosynthesis, show no influence [98, 99], only modest reduction of about 10% [100, 101] or even an increase of serum Lp(a) levels [102]. Altogether the limited magnitude of decrease of Lp(a) by HMG-CoA-reductase inhibitors confirms the assumption that the LDL-receptor does not seem to play a major role in Lp(a) clearance from plasma [103].

Nicotinic acid, also known as niacin has been shown to lower not only plasma total cholesterol, LDL-cholesterol and triglycerides thereby increasing HDL-cholesterol [104] but also Lp(a) in a dose-dependent manner up to 40% [105]. A more pronounced effect could be observed in a combination therapy with niacin and neomycin showing a reduction of LDL-cholesterol by 48% and of Lp(a) by 45% respectively [106]. In a recently published study niacin was applied in combination with omega-3-fatty acids and the Mediterranean diet. The average reduction of Lp(a) after 12 weeks combination therapy was reported to be about 23%. Additionally a significant association with increasing baseline levels of Lp(a) was observed [107].

Diets rich in fish oils containing considerable amounts of omega-3 polyunsaturated fatty acids are recommended to have beneficial effects on plasma lipids thereby lowering the risk of vascular complications [108, 109]. In a study investigating the influence of dietary fish oils on plasma Lp(a) levels a decrease of triglycerides could be observed after six weeks dietary supplementation while total cholesterol, LDL- and HDL-cholesterol as well as Lp(a) remained unchanged [110]. Furthermore collagen- and thrombin-stimulated platelet aggregation and TXB_2-formation in platelets decreased by approx. 45% irrespective of the plasma concentration of Lp(a) [111]. This is in agreement with many other studies showing that fish oils only seem to be able to reduce Lp(a) in combination with other therapies [107] or moderate exercise [112] but not when used alone [113, 114, 115].

Summarizing it can be shown that increased Lp(a) levels are minimally if at all influenced by drug treatment or drugs reducing Lp(a) to a greater extent like nicotinic acid are not widely used due to undesirable side effects. From previous turnover studies it could be demonstrated that plasma Lp(a) levels correlate with its rate of biosynthesis rather than with the fractional catabolic rate [116, 117] and therefore attempts to reduce Lp(a) should focus on an interference with apo-a biosynthesis. This is supported by the fact that adenovirus-mediated apo-a-antisense RNA expression efficiently inhibits apo-a synthesis in vitro in stably transfected liver cells but also in vivo in transduced mice expressing recombinant human apo-a [118]. In a recently published study it was found that patients suffering from biliary obstructions have very low plasma Lp(a) levels that rise substantially after surgical intervention. Consistent with this, common bile duct ligation in mice transgenic for human apo-a lowered plasma concentrations and hepatic expression of apo-a. Treatment of transgenic mice with cholic acid led to farnesoid X receptor (FXR) activation followed by markedly reduced plasma concentrations and hepatic expression of human apo-a [119]. From that it is concluded that transcription of the apo-a gene is under strong control of the farnesoid X receptor which may have important implications in the development of Lp(a)-lowering medications.

7. Conclusion

High levels of Lp(a) are strongly associated with atherosclerosis. About 10-15% of the white population exhibit plasma Lp(a) concentrations above the atherogenic cut-off value of approx. 30 mg/dl. Therefore the European Atherosclerosis Society recommended screening for Lp(a) in a consensus report, in which the desirable cut-off was set at less than 50 mg/dl [82]. On the other hand it is very well known that Lp(a) is an inherited atherogenic plasma component determined to more than 90% by genetic factors a fact that aggravates the influence on plasma levels of this lipoprotein. So far there are only speculations about the mechanism by which Lp(a) accelerates atherosclerosis and the exact mechanism could not yet be clarified. Its prothrombotic effects may be ascribed to impaired fibrinolysis by inhibition of plasminogen activation rather than to amplification of platelet aggregation which is shown to be reduced by Lp(a) in most cases. At present dietary interventions or drug therapies seem to be only minimal if at all successful concerning reduction of plasma Lp(a). Up to now it was assumed that the atherogenicity of high Lp(a) levels in blacks must be decreased by other factors [5]. However data published recently show that associations between Lp(a) levels and cardiovascular disease are at least as strong in blacks compared with whites [120] and emphasize the recommendation that factors such as total cholesterol, LDL-cholesterol, smoking, diabetes mellitus or overweight that can still increase the atherosclerotic risk of Lp(a) should be kept under observation.

Author details

Anna Gries
Institute of Physiology, Medical University of Graz, Austria

8. References

[1] Dahlen G, Ericson C, Furberg C, Lundqvist K, Svärdsudd K (1972) Studies on an extra pre-beta lipoprotein fraction. Acta Med. Scand. Suppl. 531: 1-29.
[2] Albers JJ, Cagana VG, Warnick GR, Hazzard WR (1975) Lp(a) lipoprotein-relationship to sinking pre-beta lipoprotein, Hyperlipoproteinemia and apolipoprotein B. Metabolism 24: 1047-1052.
[3] Armstrong VW, Cremer P, Eberle E (1986) The association between serum Lp(a) concentrations and angiographically assessed coronary atherosclerosis. Atherosclerosis 62: 249-257.
[4] Kostner GM, Avogaro P, Cazzolato G, Marth E, Bittolo-Bon G, Quinci GB (1981) Lipoprotein Lp(a) and the risk for myocardial infarction. Atherosclerosis 38: 51-61.
[5] Guyton JR, Dahlen GH, Patsch W, Kautz JA, Gotto AM (1985) Relationship of plasma lipoprotein Lp(a) levels to race and to apolipoprotein B. Arteriosclerosis 5: 265-272.
[6] Kostner GM (1976) Lp(a) lipoproteins and the genetic polymorphisms of lipoprotein B. From: Low Density Lipoproteins, eds. Day CE, Levy RS. Plenum Press, New York 229-269 p.
[7] Bersot TP, Innerarity TL, Pitas RE, Rall jr. SC, Weisgraber KH, Mahley RW (1986) Fat feeding in humans induces lipoproteins of density less than 1.006 that are enriched in apolipoprotein(a) and that cause lipid accumulation in macrophages. J. Clin. Invest. 77: 622-630.
[8] Gries A, Nimpf J, Nimpf M, Wurm H, Kostner GM (1987) Free and apo-B associated Lp(a)-specific protein in human serum. Clin. Chim. Acta 164: 93-100.
[9] Krempler F, Kostner G, Bolzano K, Sandhofer F (1978) Studies on the metabolism of the lipoprotein Lp(a) in man. Atherosclerosis 30: 57-65.

[10] Krempler F, Kostner G, Bolzano K, Sandhofer F (1979) Lipoprotein(a) is not a metabolic product of other lipoproteins containing apolipoprotein B. Biochim. Biophys. Acta 575: 63-70.

[11] Kronenberg F, Trenkwalder E, Lingenhel A, Friedrich G, Lhotta K, Schober M, Moes N, König P, Utermann G, Dieplinger H (1997) Renovascular arteriovenous differences in Lp[a] plasma concentrations suggest removal of Lp[a] from the renal circulation. J. Lipid Res. 38: 1755-1763

[12] Fless GM, ZumMallen ME, Scanu AM (1985) Isolation of apolipoprotein(a) from lipoprotein(a). J. Lipid Res. 26: 1224-1229.

[13] McLean JW, Tomlinson JE, Kuang WJ, Eaton DL, Chen EY, Fless GM, Scanu AM, Lawn RM (1987) cDNA sequence of human apolipoprotein(a) is homologous to plasminogen. Nature 330: 132-137.

[14] Kraft HG, Kochl S, Menzel H, Sandhofer C, Utermann G (1992) The apolipoprotein(a) gene: a transcribed hypervariable locus controlling plasma lipoprotein(a) concentration. Hum. Genet. 90: 220-230.

[15] Haibach C, Kraft HG, Kochl S, Abe A, Utermann G (1998) The number of kringle IV repeats 3-10 is invariable in the human apo(a) gene. Gene 208: 253-258.

[16] Lackner C, Cohen JC, Hobbs HH (1993) Molecular definition of the extreme size polymorphism in apolipoprotein(a). Hum. Mol.Genet. 2: 933-940.

[17] Brunner C, Lobentanz EM, Pethö-Schramm A, Ernst A, Kang C, Dieplinger H, Müller HJ, Utermann G (1996) The number of identical kringle IV repeats in apolipoprotein(a) affects its processing and secretion by HepG2 cells. J. Biol. Chem. 271: 32403-32410.

[18] Maartman-Moe K, Berg K (1981) Lp(a) lipoprotein enters cultured fibroblasts independently of the plasma membrane low density lipoprotein receptor. Clin. Genet. 20: 352-362.

[19] Armstrong VW, Walli AK, Seidel D (1985) Isolation, characterization and uptake in human fibroblasts of an apo(a)-free lipoprotein obtained on reduction of lipoprotein(a). J. Lipid Res. 26: 1314-1323.

[20] Gries A, Fievet C, Marcovina S, Nimpf J, Wurm H, Mezdour H, Fruchart JC, Kostner GM (1988) Interaction of LDL, Lp[a], and reduced Lp[a] with monoclonal antibodies against apoB. J. Lipid Res. 29: 1-8.

[21] Krempler F, Kostner GM, Roscher A, Haslauer F, Bolzano K, Sandhofer F (1983) Studies on the role of specific cell surface receptors in the removal of lipoprotein(a) in man. J. Clin. Invest. 71: 1431-1441.

[22] Berg K (1963) A new serum type system in man – the Lp system. Acta Pathol. Microbiol. Scand. 59: 369-382.

[23] Harvie NR, Schultz JS (1970) Studies of Lpa lipoprotein as a quantitative genetic trait. Proc. Natl. Acad. Sci. USA 66: 99-103.

[24] Fless GM, Rolih CA, Scanu AM (1984) Heterogeneity of human plasma lipoprotein a. J. Biol. Chem. 259: 11470-11478.

[25] Bersot TB, Innerarity TL, Mahley RW (1984) Fat feeding in humans induces lipoproteins of density less than 1.006 that are enriched in apolipoprotein a and that cause lipid accumulation in macrophages. Arteriosclerosis 4: 536a.

[26] Parra MG (1976) Isolation of human serum lipoproteins by precipitation and column chromatography methods. Thesis at the University of Marburg/Lahn. Marburg; Mauersperger Press 48-55 p.

[27] Rossi EC (1972) The function of platelets in hemostasis. Med. Clin. North. Am. 56: 25-38.

[28] Jaffe R, Dykin D (1974) Evidence for a structural requirement for the aggregation of platelets by collagen. J. Clin. Invest. 53: 875-883.

[29] Bruckdorfer KR (1989) The effect of plasma lipoproteins on platelet responsiveness and on platelet and vascular prostanoid synthesis. Prostaglandins Leukot. Essent. Fatty Acids 38: 247-254.

[30] Aviram M, Brook JG (1987) Platelet activation by plasma lipoproteins. Prog. Cardiovasc. Dis. 30: 61-72.

[31] Aviram M, Brook JG (1983) Platelet interaction with high- and low-density lipoproteins. Atherosclerosis 46: 259-268.

[32] Cushing GL, Gaubatz JW, Nava ML, Burdick BJ, Bocan TMA, Guyton JR, Weilbaecher D, DeBakey ME, Lawrie GM, Morrisett JD (1989) Quantitation and localization of apolipoprotein(a) and B in coronary artery bypass vein grafts resected at reoperation. Arteriosclerosis 9: 593-603.

[33] Ross R (1993) The pathogenesis of atherosclerosis: a perspective for the 1990s. Nature 362: 801-809.

[34] Erqou S, Thompson A, Di AE, Saleheen D, Kaptoge S, Marcovina S, Danesh J (2010) Apolipoprotein(a) isoforms and the risk of vascular disease: a systemic review of 40 studies involving 58,000 participants. J. Am. Coll. Cardiol. 55: 2160-2167.

[35] Kronenberg F, Kronenberg MF, Kiechl S, Trenkwalder E, Santer P, Oberhollenzer F, Egger G, Utermann G, Willeit J (1999) Role of lipoprotein(a) and apolipoprotein(a) phenotype in atherogenesis: prospective results from the Bruneck study. Circulation 100: 1154-1160.

[36] Zioncheck TF, Powell LM, Rice GC, Eaton DL, Lawn RM (1991) Interaction of recombinant apolipoprotein(a) and lipoprotein(a) with macrophages. J. Clin. Invest. 87: 767-771.

[37] Bihari-Varga M, Gruber E, Rotheneder M, Zechner R, Kostner GM (1988) Interaction of lipoprotein(a) and low-density lipoprotein with glycosaminoglycans from human aorta. Arteriosclerosis 8: 851-857.

[38] Salonen E, Jauhiainen M, Zardi L, Vaheri A, Ehnholm C (1991) Lipoprotein(a) binds to fibronectin and has serine protease activity capable of cleaving it. EMBO J. 8: 4035-4040.

[39] Kluft C, Jie AFH, Los P, DeWit E, Havekes L (1989) Functional analogy between lipoprotein(a) and plasminogen in the binding to the kringle 4 binding protein tetranectin. Biochem. Biophys. Res. Comm. 161: 427-433.

[40] McConathy WJ, Trieu VN (1991) Lp(a) interactions. Prog. Lipid Res. 30: 195-203.

[41] Kostner GM, Grillhofer H (1991) Lipoprotein(a) mediates high affinity low density lipoprotein association to receptor negative fibroblasts. J. Biol. Chem. 266: 21287-21292.

[42] Surya II, Akkerman JWN (1993) The influence of lipoproteins on blood platelets. Am. Heart J. 125: 272-274.

[43] Gries A, Gries M, Wurm H, Kenner T, Ijsseldijk M, Sixma JJ, Kostner GM (1996) Lipoprotein(a) inhibits collagen-induced aggregation of thrombocytes. Arterioscler. Thromb. Vasc. Biol. 16: 648-655.

[44] Ezratty A, Simon DI, Loscalzo J (1993) Lipoprotein(a) binds to human platelets and attenuates plasminogen binding and activation. Biochemistry 32: 4628-4633.

[45] Malle E, Ibovnik A, Steinmetz G, Kostner GM, Sattler W (1994) Identification of glycoprotein IIb as the lipoprotein(a)-binding protein on platelets: lipoprotein(a) binding is independent of an arginyl-glycyl-aspartate tripeptide located in apolipoprotein(a). Arterioscler. Thromb. 14: 345-352.

[46] Shattil SJ, Hoxie JA, Cunningham M, Brass LF (1985) Changes in the platelet membrane glycoprotein IIb/IIIa complex during platelet activation. J. Biol. Chem. 260: 11107-11114.

[47] Hu DD, White CA, Panzer-Knodle S, Page JD, Nicholson N, Smith JW (1999) A new model of dual interacting ligand binding sites on integrin alphaIIbbeta3. J. Biol. Chem. 274: 4633-4639.

[48] Farndale RW, Winkler AB, Martin BR, Barnes MJ (1992) Inhibition of human platelet adenylate cyclase by collagen fibres. Effect of collagen is additive with that of adrenaline, but interactive with that of thrombin. Biochem. J. 282: 25-32.

[49] Barre DE (2003) Apolipoprotein(a) mediates the lipoprotein(a)-induced biphasic shift in human platelet cyclic AMP. Thromb. Res. 112: 321-324.

[50] Barre DE (2004) Apoprotein(a) antagonises the GPIIB/IIIA receptor on collagen and ADP-stimulated human platelets. Front. Biosci. 9: 404-410.

[51] Barre DE (1998) Lipoprotein(a) reduces platelet aggregation via apo(a)-mediated decreases in thromboxane A_2 production. Platelets 9: 93-96.

[52] Barre DE, Griscti O, Mizier-Barre KA, Hafez K (2005) Flaxseed oil and lipoprotein(a) significantly increase bleeding time in type 2 diabetes patients in Cape Breton, Nova Scotia, Canada. J. Oleo. Sci. 54: 347-354.

[53] Rand ML, Sangrar W, Hancock MA, Taylor DM, Marcovina SM, Packham MA, Koschinsky ML (1998) Apolipoprotein(a) enhances platelet responses to the thrombin receptor-acitvating peptide SFLLRN. Arterioscler. Thromb. Vasc. Biol. 18: 1393-1399.

[54] Martínez C, Rivera J, Loyau S, Corral J, Gonzalez-Conejero R, Lozano ML, Vicente V, Anglés-Cano E (2001) Binding of recombinant apolipoprotein(a) to human platelets and effect on platelet aggregation. Thromb. Haemost. 85: 686-693.

[55] Barre DE (2007) Arginyl-glycyl-aspartyl (RGD) epitope of human apolipoprotein (a) inhibits platelet aggregation by antagonizing the IIb subunit of the fibrinogen (GPIIb/IIIa) receptor. Thromb. Res. 119: 601-607.

[56] Von Depka M, Nowka-Göttl U, Eisert R, Dieterich C, Barthels M, Scharrer I, Ganser A, Ehrenforth S (2000) Increased lipoprotein (a) levels as an independent risk factor for venous thromboembolism. Blood 96: 3364-3368.

[57] Casals F, Escolar G, Deulofeu R, Casals E (2007) Elevated lipoprotein (a) [Lp(a)] levels: a biological marker of venous thromboembolic risk frequently found in young females. Thromb. Res. 119 (Suppl. 1): S 100.

[58] Eaton DL, Fless GL, Kohr WJ, McLean JW, Xu QT, Miller CG, Lawn RM, Scanu AM (1987) Partial amino acid sequence of apolipoprotein(a) shows that it is homologous to plasminogen. Proc. Natl. Acad. Sci. USA 84: 3224-3228 (1987)

[59] Karàdi I, Kostner GM, Gries A, Nimpf J, Romics L, Malle E (1988) Lipoprotein (a) and plasminogen are immunochemically related. Biochim. Biophys. Acta 960: 91-97.
[60] Brown MS, Goldstein JL (1987) Teaching old dogmas new tricks. Nature 330: 113-114.
[61] Seman LJ, Breckenridge WC (1986) Isolation and partial characterization of apolipoprotein (a) from human lipoprotein (a). Biochem Cell Biol. 64: 999-1009.
[62] Mondola P, Reichl D (1982) Apolipoprotein B of lipoprotein(a) of human plasma. Biochem. J. 208: 393-398.
[63] Kraft HG, Lingenhel A, Köchl S, Hoppichler F, Kronenberg F, Abe A, Mühlberger V, Schönitzer D, Utermann G (1996) Apolipoprotein(a) kringle IV repeat number predicts risk for coronary heart disease. Arterioscler. Thromb. Vasc. Biol. 16: 713-719.
[64] McLean JW, Tomlinson JE, Kuang WJ, Eaton DL, Chen EY, Fless GM, Scanu AM, Lawn RM (1987) cDNA sequence of human apolipoprotein(a) is homologous to plasminogen. Nature 12: 132-137.
[65] Miles LA, Fless GM, Levin EG, Scanu AM, Plow EF (1989) A potential basis for the thrombotic risks associated with lipoprotein(a). Nature 339: 301-303.
[66] Hajjar KA, Gavish D, Breslow JL, Nachman RL (1989) Lipoprotein(a) modulation of endothelial cell surface fibrinolysis and its potential role in atherosclerosis. Nature 339: 303-305.
[67] Rahman MN, Petrounevitch V, Jia Z, Koschinsky ML (2001) Antifibrinolytic effect of single apo(a) kringle domains: relationship to fibrinogen binding. Prot. Eng. 14: 427-438.
[68] Lijnen HR, Bachmann F, Collen D, Ellis V, Pannekoek H, Rijken DC, Thorsen S (1994) Mechanism of plasminogen activation. J. Intern. Med. 236: 415-424.
[69] Harpel PC, Gordon BR, Parker TS (1989) Plasmin catalyzes binding of lipoprotein(a) to immobilized fibrinogen and fibrin. Proc. Natl. Acad. Sci. USA 56: 3847-3851.
[70] Hervio L, Chapman MJ, Thillet J, Loyau S, Angles-Cano E (1993) Does apolipoprotein(a) heterogeneity influence lipoprotein(a) effects on fibrinolysis? Blood 82: 392-397.
[71] Undas A, Stepien E, Tracz W, Szczeklik A (2006) Lipoprotein(a) as a modifier of fibrin clot permeability and susceptibility to lysis. J. Thromb. Haemost. 4: 973-975.
[72] Knapp JP, Herrmann W (2004) In vitro inhibition of fibrinolysis by apolipoprotein (a) and lipoprotein (a) is size- and concentration-dependent. Clin. Chem. Lab. Med. 42: 1013-1019.
[73] Scanu AM, Miles LA, Fless GM, Pfaffinger D, Eisenbart J, Jackson E, Hoover-Plow JL, Brunck T, Plow EF (1993) Rhesus monkey lipoprotein (a) binds to lysine Sepharose and U937 monocytoid cells less efficiently than human lipoprotein(a). Evidence for the dominant role of kringle 4(37). J. Clin. Invest. 91: 283-291.
[74] Testa R, Marcovina SM (1999) The rate of plasmin formation after in vitro clotting is inversely related to lipoprotein(a) plasma levels. Int. J. Lab. Res. 29: 128-132.
[75] Edelberg JM, Gonzalez-Gronow M, Pizzo SV (1989) Lipoprotein(a) inhibits streptokinase-mediated activation of human plasminogen. Biochemistry 28: 2370-2374.
[76] Edelberg JM, Gonzalez-Gronow M, Pizzo SV (1990) Lipoprotein(a) inhibition of plasminogen activation by tissue-type plasminogen activator. Thromb. Res. 57: 155-162.
[77] Donders SHJ, Lustermans FATh, van Wersch JWJ (1993) On lipoprotein(a) and the coagulation/fibrinolysis balance in the acute phase of deep venous thrombosis. Fibrinolysis 7: 83-86.

[78] Feric NT, Boffa MB, Johnston SM, Koschinsky ML (2008) Apolipoprotein(a) inhibits the conversion of Glu-plasminogen to Lys-plasminogen: a novel mechanism for lipoprotein(a)-mediated inhibition of plasminogen activation. J. Thromb. Haemost. 6: 2113-2120.

[79] Költringer P, Jürgens G (1985) A dominant role of lipoprotein(a) in the investigation and evaluation of parameters indicating the development of cervical atherosclerosis. Atherosclerosis 58: 187-198.

[80] Murai AT, Miyahara T, Fujimoto N, Matsuda M, Kameyama M (1986) Lp(a) as a risk factor for coronary heart disease and cerebral infarction. Atherosclerosis 59: 199-204.

[81] Rhoads GG, Dahlen G, Berg K, Morton NE, Dannenberg AL (1986) Lp(a) lipoprotein as a risk factor for myocardial infarction. J. Am. Med. Ass. 356: 2540-2544.

[82] Nordestgaard BG, Chapman MJ, Ray K, Borén J, Andreotti F, Watts GF, Ginsberg H, Amarenco P, Catapano A, Descamps OS, Fisher E, Kovanen PT, Kuivenhoven JA, Lesnik P, Masana L, Reiner Z, Taskinen MR, Tokgözoglu L, Tygjaerg-Hansen A (2010) Lipoprotein(a) as a cardiovascular risk factor: current status. Eur. Heart J. 31: 2844-2853.

[83] Mackinnon LT, Hubinger L, Lepre F (1997) Effects of physical activity and diet on lipoprotein(a). Med. Sci. Sports Exerc. 29: 1429-1436.

[84] Randall OS, Feseha HB, Illoh K, Xu S, Ketete M, Kwagyan J, Tilghman C, Wrenn M (2004) Response of lipoprotein(a) levels to therapeutic life-style change in obese African-Americans. Atherosclerosis 172: 155-160.

[85] Ahmadi N, Eshaghian S, Huizenga R, Sosnin K, Ebrahimi R, Siegel R (2011) Effects of intense exercise and moderate caloric restriction on cardiovascular risk factors and inflammation. Am. J. Med. 124: 978-982.

[86] Hubinger L, Mackinnon LT, Lepre F (1995) Lipoprotein(a) [Lp(a)] levels in middle-aged male runners and sedentary controls. Med. Sci. Sports Exerc. 27: 490-496.

[87] Mackinnon LT, Hubinger LM (1999) Effects of exercise on lipoprotein(a). Sports Med. 28: 11-24.

[88] Vessby B, Kostner G, Lithell H, Thomis J (1982) Diverging effects of cholestyramine on apolipoprotein B and lipoprotein Lp(a). Atherosclerosis 44: 61-71.

[89] Dobs AS, Prasad M, Goldberg A, Guccione M, Hoover DR (1995) Changes in serum lipoprotein(a) in hyperlipidemic subjects undergoing long-term treatment with lipid-lowering drugs. Cardiovasc. Drugs Ther. 9: 677-684.

[90] Kostner G, Klein G, Krempler F (1984) Can serum Lp(a) concentration be lowered by drugs and/or diet? In: Carlson LA and Olsson AG (eds.): Treatment of Hyperlipoproteinemia, Raven Press, New York, 151-156 p.

[91] Neele DM, Kaptain A, Huisman H, de Wit EC, Princen HM (1998) No effect of fibrates on synthesis of apolipoprotein(a) in primary cultures of cynomolgus monkey and human hepatocytes: apolipoprotein A-I synthesis increased. Biochem. Biophys. Res. Commun. 244: 374-378.

[92] Christodoulakos GE, Lambrinoudaki IV, Panoulis CP, Papadias CA, Kouskouni EE, Creatsas GC (2004) Effect of hormone replacement therapy, tibolone and raloxifene on serum lipids, apolipoprotein A1, apolipoprotein B and lipoprotein(a) in Greek postmenopausal women. Gynecol. Endocrinol. 18: 244-257.

[93] Persson L, Henriksson P, Westerlund E, Hovatta O, Angelin B, Rudling M (2012) Endogenous estrogens lower plasma PCSK9 and LDL cholesterol but not Lp(a) or bile acid synthesis in women. Arterioscler. Thromb. Vasc. Biol. 32: 810-814.
[94] Albers JJ, Taggart HM, Applebaum-Bowden D, Haffner S, Chesnut CH, Hazzard WR (1984) Reduction of lecithin-cholesterol acyl-transferase, apolipoprotein D and the Lp(a) lipoprotein with the anabolic steroid stanozolol. Biochim. Biophys. Acta 795: 293-296.
[95] Hartgens F, Rietjens G, Keizer HA, Kuipers H, Wolffenbuttel BH (2004) Effects of androgenic-anabolic steroids on apolipoproteins and lipoprotein(a). Brit. J. Sports Med. 38: 253-259.
[96] Kostner KM, Kostner GM (2005) Therapy of hyper-Lp(a). Handb. Exp. Pharmacol. 170: 519-536.
[97] Ma PTS, Gil G, Sudhof JC, Bilheimer DW, Goldstein JL, Brown MS (1986) Mevinolin, an inhibitor of cholesterol synthesis, induces mRNA for low density lipoprotein receptor in liver of hamsters and rabbits. Proc. Nat. Acad. Sci. USA 83: 8370-8374.
[98] Thiery J, Armstrong VW, Schleef J, Creutzfeld C, Creutzfeld W, Seidel D (1988) Serum lipoprotein Lp(a) concentrations are not influenced by an HMG-CoA reductase inhibitor. Klin. Wochenschr. 66: 462-463.
[99] Kostner GM, Gavish D, Leopold B, Bolzano D, Weintraub MS, Breslow JL (1989) HMG-CoA reductase inhibitors lower LDL cholesterol without reducing Lp(a) levels. Circulation 80: 1313-1319.
[100] Joy MS, Dornbrook-Lavender KA, Chin H, Hogan SL, Denu-Ciocca C (2008) Effects of atorvastatin on Lp(a) and lipoprotein profiles in hemodialysis patients. Ann. Pharmacother. 42: 9-15.
[101] Horimoto M, Hasegawa A, Takenaka T, Fujiwara M, Inoue H, Igarashi K (2003) Long-term administration of pravastatin reduces serum lipoprotein(a) levels. Int. J. Clin. Pharmacol. Ther. 41: 524-530.
[102] Choi SH, Chae A, Miller E, Messig M, Ntanios F, DeMaria AN, Nissen SE, Witztum JL, Tsimikas S (2008) Relationship between biomarkers of oxidized low-density lipoprotein, statin therapy, quantitative coronary angiography, and atheroma: volume observations from the REVERSAL (Reversal of atherosclerosis with aggressive lipid lowering) study. J. Am. Coll. Cardiol. 52: 24-32.
[103] Hobbs HH, White Al (1999) Lipoprotein(a): intrigues and insights. Curr. Opin. Lipidol. 10: 225-236.
[104] Digby JE, Lee JM, Choudhury RP (2009) Nicotinic acid and the prevention of coronary artery disease. Curr. Opin. Lipidol. 20: 321-326.
[105] Linke A, Sonnabend M, Fasshauer M, Höllriegel R, Schuler G, Niebauer J, Stumvoll M, Blüher M (2009) Effects of extended-release niacin on lipid profile and adipocyte biology in patients with impaired glucose tolerance. Atherosclerosis 205: 207-213.
[106] Gurakar A, Hoeg JM, Kostner G, Papadopoulos NM, Brewer jr. HB (1985) Levels of lipoprotein Lp(a) decline with neomycin and niacin treatment. Atherosclerosis 57: 293-301.
[107] Helmbold AF, Slim JN, Morgan J, Castillo-Rojas LM, Shry EA, Slim AM (2010) The effects of extended release niacin in combination with omega 3 fatty acid supplements in the treatment of elevated lipoprotein (a). Cholesterol 2010: 306147.

[108] Harris WS (1989) Fish oils and plasma lipid and lipoprotein metabolism in humans: a critical review. J. Lipid Res. 30: 785-807.

[109] Wei MY, Jacobson TA (2011) Effects of eicosapentaenoic acid versus docosahexaenoic acid on serum lipids: a systematic review and meta-analysis. Curr. Atheroscler. Rep. 13: 474-483.

[110] Gries A, Malle E, Wurm H, Kostner GM (1990) Influence of dietary fish oils on plasma Lp(a) levels. Thromb. Res. 58: 667-668.

[111] Malle E, Sattler W, Prenner E, Leis HJ, Hermetter A, Gries A, Kostner GM (1991) Effects of dietary fish oil supplementation on platelet aggregability and platelet membrane fluidity in normolipemic subjects with and without high plasma Lp(a) concentrations. Atherosclerosis 88: 193-201.

[112] Herrmann W, Biermann J, Kostner GM (1995) Comparison of effects of N-3 to N-6 fatty acids on serum levels of lipoprotein(a) in patients with coronary artery disease. Am. J. Cardiol. 76: 459-462.

[113] Marckmann P, Bladbjerg EM, Jespersen J (1997) Dietary fish oil (4 g daily) and cardiovascular risk markers in healthy men. Arterioscler. Thromb. Vasc. Biol. 17: 3384-3391.

[114] Beavers KM, Beavers DP, Bowden RG, Wilson RL, Gentile M (2009) Effect of over-the-counter fish-oil administration on plasma Lp(a) levels in an end-stage renal disease population. J. Ren. Nutr. 19: 443-449.

[115] Kooshki A, Taleban FA, Tabibi H, Hedayati M (2011) Effects of omega-3 fatty acids on serum lipids, lipoprotein (a), and hematologic factors in hemodialysis patients. Ren. Fail. 33: 892-898.

[116] Krempler F, Kostner GM, Bolzano K, Sandhofer F (1980) Turnover of lipoprotein (a) in man. J. Clin. Invest. 65: 1483-1490.

[117] Rader DJ, Cain W, Ikewaki K, Talley G, Zech LA, Usher D, Brewer HB Jr. (1994) The inverse association of plasma lipoprotein(a) concentrations with apolipoprotein(a) isoform size is not due to differences in Lp(a) catabolism but to differences in production rate. J. Clin. Invest. 93: 2758-2763.

[118] Frank S, Gauster M, Strauss J, Hrzenjak A, Kostner GM (2001) Adenovirus-mediated apo(a)-antisense-RNA expression efficiently inhibits apo(a) synthesis in vitro and in vivo. Gene Therapy 8: 425-430.

[119] Chennamsetty I, Claudel T, Kostner KM, Baghdasaryan A, Kratky D, Levak-Frank S, Frank S, Gonzalez FJ, Trauner M, Kostner GM (2011) Farnesoid X receptor represses hepatic human APOA gene expression. J. Clin. Invest. 121: 3724-3734.

[120] Virani SS, Brautbar A, Davis BC, Nambi V, Hoogeveen RC, Sharrett AR, Coresh J, Mosley TH, Morrisett JD, Catellier DJ, Folsom AR, Boerwinkle E, Ballantyne CM (2012) Associations between lipoprotein(a) levels and cardiovascular outcomes in black and white subjects: the Atherosclerosis Risk in Communities (ARIC) Study. Circulation 125: 241-249.